HANDBOOK OF INTERNATIONAL INSURANCE

Between Global Dynamics and Local Contingencies

Huebner International Series on Risk, Insurance, and Economic Security

J. David Cummins, Editor
The Wharton School
University of Pennsylvania
Philadelphia, Pennsylvania, USA

Series Advisors:
Dr. Phelim P. Boyle
University of Waterloo, Canada
Dr. Jean Lemaire
University of Pennsylvania, USA
Professor Akihiko Tsuboi
Kagawa University, Japan
Dr. Richard Zeckhauser
Harvard University, USA

Other books in the series:
Cummins, J. David and Derrig, Richard A.: *Classical Insurance Solvency Theory*
Borba, Philip S. and Appel, David: *Benefits, Costs, and Cycles in Workers' Compensation*
Cummins, J. David and Derrig, Richard A.: *Financial Models of Insurance Solvency*
Williams, C. Arthur: *An International Comparison of Workers' Compensation*
Cummins, J. David and Derrig, Richard A.: *Managing the Insolvency Risk of Insurance Companies*
Dionne, Georges: *Contributions to Insurance Economics*
Dionne, Georges and Harrington, Scott E.: *Foundations of Insurance Economics*
Klugman, Stuart A.: *Bayesian Statistics in Actuarial Science*
Durbin, David and Borba, Philip: *Workers' Compensation Insurance: Claim Costs, Prices and Regulation*
Cummins, J. David: *Financial Management of Life Insurance Companies*
Gustavson, Sandra G. and Harrington, Scott E.: *Insurance, Risk Management, and Public Policy*
Lemaire, Jean: *Bonus-Malus Systems in Automobile Insurance*
Dionne, Georges and Laberge-Nadeau: *Automobile Insurance: Road Safety, New Drivers, Risks, Insurance Fraud and Regulation*
Taylor, Greg: *Loss Reserving: An Actuarial Perspective*
Dionne, Georges: *Handbook of Insurance*
Sinha, Tapen: *Pension Reform in Latin America and Its Lessons for International Polichmakers*
Lascher, Jr., Edward L. and Powers, Michael R.: *The Economics and Politics of Choice No-Fault Insurance*
Grossi, Patricia and Kunreuther, Howard: *Catastrophe Modeling: A New Approach to Managing Risk*
Cummins, J. David and Venard, Bertrand: *Handbook of International Insurance: Between Global Dynamics and Local Contingencies*

HANDBOOK OF INTERNATIONAL INSURANCE

Between Global Dynamics and Local Contingencies

edited by

J. David Cummins

Temple University, USA

Fox School of Business and Management

and

Bertrand Venard

AUDENCIA Nantes School of Management, France

Springer

Library of Congress Control Number: 2006925828

ISBN-13: 978-0387-34162-0
ISBN-10: 0-387-34162-5

ISBN-13: 978-0387-34163-7 (e-book)
ISBN-10: 0-387-34163-3 (e-book)

Printed on acid-free paper.

9 8 7 6 5 4 3 2

springer.com

Table of Contents

Authors' Biographies

GILLES BENIER

Gilles Bernier is a full professor of finance and insurance in the Faculty of Business Administration at Laval University where he also holds the Industrial-Alliance Insurance Chair. His fields of specialization include market power, efficiency and productivity, personal finance and retirement planning. Currently, he is studying the efficiency and productivity of Canadian insurers using frontier methodologies. Dr. Bernier has published more than twenty scientific and professional articles and three books, including *Risk and Insurance.* He serves regularly as a referee for finance and insurance academic journals and is member of the American Risk and Insurance Association.

Email: Gilles.Bernier@fsa.ulaval.ca

FERNANDA CHAVES PEREIRA

Fernanda Chaves Pereira is the academic coordinator of the Instituto de Gestão de Riscos Financeiros e Atuariais (IAPUC), a new institute created at PUC-Rio for the academic development of insurance and pension in Brazil. Dr. Pereira is an actuary who finished her Ph.D. in 2001 from City University London. She also was awarded the Highly Commended Brian Hey Prize from the Institute of Actuaries, UK in 1999. Her fields of specialization include loss reserving and mortality improvement. She is the coordinator of the education committee at the Brazilian Actuary Institute and is also the organizer of the second (2005) and third (2007) Brazilian Conference on Statistical Modelling in Insurance and Finance.

Email: fernanda@iag.puc-rio.br

J. DAVID CUMMINS

J. David Cummins is the Joseph E. Boettner Professor of Risk Management, Insurance, and Financial Institutions at Temple University's Fox School of Business and Management. In addition, he is the Director of the Advanta Center at Temple University and Professor Emeritus at the Wharton School of the University of Pennsylvania. His fields of specialization include financial institutions, securitization, risk management, and productivity. Dr. Cummins has published more than eighty refereed journal articles. His recent books include *Changes in the Life Insurance Industry: Efficiency, Technology, and Risk Management* and *Deregulating Property-Liability Insurance: Restoring Competition and Increasing Market Efficiency*. He is a co-editor of the *Journal of Banking and Finance* and Associate Editor of nine other journals. Dr. Cummins is a past-president of the American Risk and Insurance Association.

Email: cummins@wharton.upenn.edu

STEPHEN DIACON

Stephen Diacon is a professor of insurance and risk management at the Nottingham University Business School. He has written a number of books and academic articles on insurance and is a member of the editorial board of the *Geneva Papers on Risk and Insurance*.

Email: stephen.diacon@nottingham.ac.uk

PAUL FENN

Paul Fenn is the Norwich Union Professor of Insurance Studies at Nottingham University Business School. His background is in applied microeconomics, particularly in relation to the interaction between law, health, and insurance. He has written or edited four books and numerous articles in peer-reviewed journals (including the *Economic Journal, Journal of Law and Economics*, and the *Journal of Health Economics*) on the general themes of liability insurance, health economics, and the economics of workplace risk. He has recently coordinated research projects on these issues for the UK Department of Health and the Department of Constitutional Affairs.

Email: paul.fenn@nottingham.ac.uk

JOHN F. GARVEY

John F. Garvey is a Junior Lecturer in Risk Management and Insurance at the Kemmy Business School at the University of Limerick. His area of specialization is in modern portfolio theory, volatility modeling, and risk management. He has a B.A. in English and Economics and holds a M.A. in Applied Economics and a M.Sc. in Investment and Treasury. He is currently working on his Ph.D.

Email: John.Garvey@ul.ie

LOFTIN GRAHAM

Loftin Graham is a Huebner Fellow and fourth-year Ph.D. student in the Department of Insurance and Risk Management at the Wharton School of the University of Pennsylvania. His research interests include various life and health insurance topics, actuarial science, and insurance economics.

Email: loftin@wharton.upenn.edu

BRIAN GREENFORD

Brian Greenford is the director of the risk research group at the University of Limerick and the past Assistant Dean of Academic Affairs at the Kemmy Business School at the University of Limerick. His fields of interest are holistic risk management, psychology and sociology of risk and law, and personal injury claims from a comparative approach. He is the author of "Claims Management: Irish Law and Practice" and "Equine Risk Management." He has published numerous refereed articles on comparative claims management and compensation culture as well as various reports for the Irish and UK governments on personal injury claims and competition in the Irish insurance industry. He is a past president of the Limerick Insurance Institute.

Email: Brian.Greenford@ul.ie

MICHEL GUIGUIS

Michel Guirguis is a researcher at Bournemouth University, UK. His research interests currently lie in the area of closed-end funds. Dr. Guirguis's doctoral thesis was entitled "A Multifactor Model of Investment Trust Discounts: A Comparative Study of UK Investment Trusts and US Closed-End Funds."

Email: mguirguis@bournemouth.ac.uk

PHILIP HARDWICK

Philip Hardwick holds the Chair of Financial Services at Bournemouth University, UK and is a Visiting Professor at the Warsaw University of Insurance and Banking and at the Academy of Humanities and Economics in Lodz in Poland. Dr. Hardwick has published more than twenty refereed journal articles and is the author of several books, including a successful economics textbook and two macroeconomics books concerned with unemployment and inflation. His main teaching and research interests currently lie in the areas of efficiency measurement in the financial services sector, international trade in financial services, and quantitative research methods.

Email: phardwic@bournemouth.ac.uk

ROBERT HARTWIG

Robert P. Hartwig is the senior vice president and chief economist for the Insurance Information Institute. Hartwig previously served as director of economic research and senior economist with the Boca Raton, Florida-based National Council on Compensation Insurance (NCCI). He has also worked as senior economist for the Swiss Reinsurance Group in New York and as senior statistician for the U.S. Consumer Product Safety Commission in Washington, DC. Dr. Hartwig is a member of the American Economic Association, the American Risk and Insurance Association, and the National Association of Business Economics. Hartwig received his Ph.D. and M.Sc. degrees in economics from the University of Illinois at Urbana-Champaign. He received a B.A. in economics from the University of Massachusetts at Amherst.

Email: bobh@iii.org

THOMAS HOLZHEU

Thomas Holzheu is a senior economist and deputy head of the New York office of Swiss Re's Economic Research and Consulting Unit. His fields of specialization include North American insurance and global reinsurance markets. Dr. Holzheu researches and publishes on market developments and economic and regulatory factors affecting insurance markets. In addition, he supports Swiss Re's strategic planning and internal consulting on new products. He authored several *Sigma* studies including “Measuring Underwriting Profitability of the Non-Life Insurance Industry” (5/2005), “Commercial Insurance and Reinsurance Brokerage – Love Thy Middleman” (2/2004), “The Economics of Liability Losses” (6/2004), and “The Picture of ART” (1/2003).

Email: Thomas_Holzheu@swissre.com

STEPHANIE HUSSELS

Stephanie Hussels is a full-time lecturer in economics at the University of Bradford School of Management, UK. Her research interests include measuring efficiency and productivity, the European insurance industry, frontier methodologies, and financial services within overall economic development.

Email: S.Hussels@Bradford.ac.uk

JULIAN JAMES

Julian James was appointed Director, Worldwide Markets at Lloyd's in December 2000. He is responsible for all Lloyd's commercial activities outside the UK, including the management of its 72 trading licenses. Additionally, he oversees Lloyd's global branding and market development including communications, media relations, and government affairs. Mr. James is an Associate of the Chartered Insurance Institute and is a Chartered Insurance Practitioner. He is a board member of the Lloyd's Community Programme and until January 2005 was a board member of the General Insurance Standards Council. He is currently Vice President of the Chartered Insurance Institute in the UK.

Email: Julian.James@lloyds.com

PAUL J. M. KLUMPES

Paul J. M. Klumpes is a Professor of Accounting at the Tanaka Business School of the Imperial College, UK. His fields of specialization include financial reporting and performance measurement related to pensions and insurance in various markets. Dr. Klumpes has published more than thirty refereed journal articles on these and related topics. His recent publications on these topics have appeared in the *Journal of Business*, *Geneva Papers on Risk and Insurance*, and *Journal of Banking and Finance*. He is a consultant to Charles River Associates and Gerson Lehrman Group. He is also a qualified Australian CPA and is a member of the American Accounting Association.

Email: p.klumpes@imperial.ac.uk

ROMAN LECHNER

Roman Lechner is Senior Economist on Swiss Re's Economic Research and Consulting team. His many fields of specialization include global reinsurance, the London market, and property and casualty insurance in general. He is the author of several *Sigma* studies including "Measuring Underwriting Profitability of the Non-Life Insurance Industry" (5/2005), "Commercial Insurance and Reinsurance Brokerage – Love Thy Middleman" (2/2004), "The London Market in the Throes of Change" (3/2002), and "The Global Reinsurance Market in the Midst of Consolidation" (9/1998).

Email: Roman_Lechner@swissre.com

ANDREA LEVY

Andrea Levy is an affiliate lecturer at the Instituto de Gestão de Riscos Financeiros e Atuariais (IAPUC). He has a doctorate in automation technology from the Université Paul Sabatier –Toulouse. He has worked in recent years for insurance companies both as a CFO and as a COO. In the 1970s, he worked mainly for property and casualty companies, but more recently focuses on life insurance, pension plans, and capitalization companies. At the present time, he works in mergers and acquisitions in the insurance market providing analysis and evaluation for the embedded value of life portfolios and solutions for reinsurance. Finally, he is an active member of organizations that represent the insurance industry, mainly regarding market regulation.

Email: ALevy@icatu-hartford.com.br

RAIMOND MAURER

Raimond Maurer is Professor of Investment, Portfolio Management, and Pension Finance at the Goethe University of Frankfurt. His fields of specialization include asset management, institutional investors, and pension finance. He is the academic chairman of the International Actuarial Association Financial Risks Section (AFIR) of the German Actuarial Society, member of the International AFIR-Committee, and member of the Pension Research Council Advisory Board at the Wharton School of the University of Pennsylvania. Dr. Maurer has published four books and more than thirty refereed articles.

Email: maurer@finance.uni-frankfurt.de

LOUISE MORRIS

Louise Morris is a Ph.D. student at the Graduate Center of Business at the University of Limerick where she is researching the management of risks related to the use of genetic testing by life insurance companies with Dr. Brian Greenford. Her research is funded by the Irish Research Council for the Humanities and Social Sciences. Her research interests are life insurance and risk management. She is a member of the Risk Research Group.

Email: Louise.Morris@ul.ie

MARTIN MULLINS

Martin Mullins is a Lecturer in Risk Management at the University of Limerick and the Course Director for the B.A. International Insurance and European Studies. He is a member of the Risk Research Group. Dr. Mullins research focuses on political risk, risk management in the public sector, and the construction of public policy in post-conflict societies. He is the author of the book *In the Shadow of the Generals* and has published articles and book chapters on political risk and international bond markets.

Email: Martin.Mullins@ul.ie

ALLI NATHAN

Alli Nathan is a Professor of Finance with teaching and research interests in the areas of international finance, financial institutions and markets, and risk management and regulation of banks and insurance companies. She graduated from Queen's University in Canada and has taught at the University of Calgary, Canada and at Nyenrode University, the Netherlands. Dr. Nathan is also a visiting professor at the Amsterdam Graduate School of Business at the University of Amsterdam, the Netherlands. She is currently on the faculty at Providence College in the US. Her research on the structure, conduct, performance, and efficiency of banks and insurance firms has been published in several academic and professional journals.

Email: a.nathan@nyenrode.nl

CHRIS O'BRIEN

Chris O'Brien is director of the Centre for Risk and Insurance Studies (CRIS) at Nottingham University Business School in England. Prior to joining CRIS in 2000, he had 27 years of experience in the life insurance industry and was the chief actuary of a major life insurer. His interests include the financial management of, accounting in, and regulation of insurance companies as well as the role of actuarial management in life insurers. He is a Fellow at the Institute of Actuaries and was President of the Manchester Actuarial Society from 1999-2000.

Email: christopher.obrien@nottingham.ac.uk

TAISHI OKADA

Taishi Okada is an associate professor of Kwansei Gakuin University, Japan. He has been appointed as a visiting scholar at City University. He graduated from Osaka City University and obtained his MBA from Kobe University. His fields of specialization include insurance regulation, risk management, and corporate governance. He has published papers in insurance and business journals including the *Journal of Insurance Science* and the *Journal of Business Administration.*

Email: tokada@kwansei.ac.jp

LIAM O'MEARA

Liam O'Meara is a post-graduate research student at the Graduate Centre of Business at the University of Limerick. He is currently completing his master's thesis entitled "An Analysis and Review of the Methods of Financial Protection Available against Flooding in Ireland" under the supervision of Dr. Brian Greenford. His main areas of interest include risk management of the occurrence of weather risks and natural disasters and life insurance. He is a member of the Risk Research Group.

Email: Liam.OMeara@ul.ie

ALFRED OOSENBRUG

Alfred Oosenbrug is a former professor with chairs in the economics of the financial services industry and in the accounting for financial services organizations. Dr. Oosenbrug is a business economist, a certified accountant, and an actuary. He graduated at the Faculty of Law of the University of Leiden. Currently, he is a consultant in the financial services industry. Dr. Oosenbrug has published more than thirty books and some four hundred articles with respect to accounting, insurance, pensions, and financial services. He is a past-president of the Dutch Institute of Actuaries and a former member of the executive board of a middle-sized insurance group.

Email: oosenbrug@zonnet.nl

MARÍA RUBIO-MISAS

María Rubio-Misas is Associate Professor of Financial Economics and Accounting at the University of Málaga (Spain). Her research areas include financial institutions and the economic analysis of efficiency and productivity in the insurance industry. Dr. Rubio-Misas has published refereed journal articles in publications such as the *Journal of Banking and Finance* and the *Journal of Money, Credit, and Banking*. She has served as a Visiting Scholar in the Department of Insurance and Risk Management at the Wharton School of the University of Pennsylvania.

Email: maria.rubio.misas@telefonica.net

CLAIRE E. SHERMAN

Claire E. Sherman is a lecturer and Ph.D. student at the University of Adelaide, Australia. Claire teaches within the field of business finance and previously worked in the public sector, banking, and brokerage industries. She completed her Canadian Securities Course in 2002. Claire is also a lecturer in Marketing Communications and provides a unique perspective of both marketing and finance streams, with which she hopes to conduct research relating to the marketing of finance.

Email: claire.sherman@adelaide.edu.au

TAPEN SINHA

Tapen Sinha is the ING Commercial America Chair Professor at the Instituto Technologico Autonomo de Mexico (ITAM) in Mexico City, Mexico. He is also a Special Professor at the School of Business of the University of Nottingham, UK. He has a Ph.D. in Economics from the University of Minnesota. Dr. Sinha is the founder-director of the International Center for Pension Research, ITAM, and an Associate of the Centre for Risk and Insurance Studies at the University of Nottingham. He has published over one hundred papers and five books. Additionally, Dr. Sinha has been a consultant for a number of multinational companies and governments of different continents.

Email: tapen@itam.mx

BARBARA SOMOVA

Barbara Somova is a Ph.D. student at the Faculty of Economics and Business Administration of the Goethe University of Frankfurt, Chair of Investments, Portfolio-Management and Pension Finance. Her fields of specialization include institutional investors, valuation of insurance companies, and merger and acquisition activities in the insurance industry. Ms. Somova studied economics at the University of Hamburg, spent three years as a mergers and acquisitions analyst with Rothschild GmbH in Frankfurt and London and is currently working on her dissertation.

Email: somova@finance.uni-frankfurt.de

QIXIANG SUN

Qixiang Sun holds the C.V. Starr Chair of Risk Management and Insurance and is the Associate Dean of both Academic and International Affairs at School of Economics, Peking University, China. She is also the Chair of the Department of Risk Management and Insurance and the director of the China Centre for Insurance and Social Security Research. Dr. Sun is widely recognized as one of China's leading insurance scholars. Her academic work in insurance has been published in China's top insurance and economics journals as well as international insurance journals. Her textbook on insurance is the top-selling insurance textbook in China.

Email: qusun@pku.edu.cn

LINGYAN SUO

Lingyan Suo is a Ph.D. student in finance at the School of Economics of Peking University, China. Her recent research has focused on the development strategy of China's insurance industry and reform of China's Social Security system. She received her B.A. in Economics from Peking University in 2001.

Email: lingyan_suo@pku.org.cn

BERTRAND VENARD

Bertrand Venard is the Professor of Management at Audencia School of Management in France and Sloan Research Fellow at the Wharton Business School of the University of Pennsylvania. He has received many distinctions for his research such as the Marie Curie Fellowship and the NATO Fellowship. He has published more than 40 academic articles. Before his academic career, he held an executive position in a French insurance company and was a senior consultant for PricewaterhouseCoopers. Furthermore, he held various visiting positions at the London Business School, Cambridge University, and the Budapest University of Economics and Public Administration. His research interests are in the fields of insurance, finance, and international management.

Email: bvenard@audencia.com and venard@wharton.upenn.edu

LUIGI VENTURA

Luigi Ventura is a professor of microeconomics and econometrics at the Faculty of Statistics of the University of Rome "La Sapienza." His research interests include individual choice under risk and uncertainty, measures of risk aversion, prudence and time preference, insurance and allocations with incomplete markets, and the analysis of some testable implications of insurability. In all these fields Dr. Ventura has published various journal articles, both theoretical and applied, in international refereed journals and two books.

Email: luigi.ventura@uniroma1.it

ROBERT W. VIVIAN

Robert W. Vivian is the Professor of Finance and Insurance at the University of the Witwatersrand, South Africa. He spent the first decade of his career as a professional electrical engineer and the second as a director of a risk management company. In 1990, the South African insurance industry funded the IISA Chair of Insurance and Risk Management and he was appointed to this position. Since then, he has authored four books and published several articles in refereed journals. He is the past President of the Economic History Society of Southern Africa.

Email: vivianr@sebs.wits.ac.za

DAMIAN WARD

Damian Ward is a senior lecturer in economics at the Bradford University School of Management. His research interests include the role of financial services within economic development as well as the efficiency, governance, and distribution systems employed by insurance companies.

Email: d.r.ward@Bradford.ac.uk

CLAIRE WILKINSON

Claire Wilkinson, is Vice President, Global Issues for the Insurance Information Institute. She previously served as the US bureau chief of the London, UK-based trade newspaper *Insurance Day* and has more than 10 years of experience in writing on the insurance industry. Prior to her assignment to New York, she was deputy editor of *Insurance Day* in London. She has also worked as a reporter at the *Financial Times* newsletters. Ms. Wilkinson received her M.A. in Anglo-American Literary Relations from University College, UK. She also received her B.A. in

English and American Literature from the University of Warwick and has a post-graduate diploma in journalism from the University of Wales at Cardiff.

Email: clairew@iii.org

XIAOYING XIE

Xiaoying Xie is an assistant professor of finance at California State University at Fullerton. Dr. Xie received her Ph.D. degree in Risk Management and Insurance from the Wharton School of the University of Pennsylvania in 2005. Her current research interests focus on the areas of mergers and acquisitions, financial integration, economies of scale, productivity and efficiency, and insurance economics.

Email: xxie@fullerton.edu

NOBUYOSHI YAMORI

Nobuyoshi Yamori is a professor at Nagoya University, Japan. He has been appointed as a visiting scholar at several institutions including Columbia University and the Federal Reserve Bank of San Francisco. He graduated from Shiga University, received his M.A. from Kobe University, and obtained his Ph.D. from Nagoya University. He has published many papers in international banking and financial journals such as the *Journal of Banking and Finance,* the *Journal of Financial Research*, and the *Journal of Risk and Insurance*. He also serves as an editors of several professional journals including the *International Journal of Business.*

Email: nobuyoshiyamori@hotmail.com

WEI ZHENG

Wei Zheng is an Associate Professor and Vice Chair of the Department of Risk Management and Insurance at the School of Economics, Peking University, China. His academic focus has been on pension, life and health insurance, and insurance law. Dr. Zheng has published more than ten academic journal articles and many newspaper articles. He published a book named *China's Social Pension System: Institutional Change and Economic Effects*. Dr. Zheng serves as Secretary General of the China Center for Insurance and Social Security Research, Peking University. He is a member of the editorial board of the *Asia-Pacific Journal of Risk and Insurance*.

Email: wzheng@pku.org.cn

RALF ZURBRUEGG

Ralf Zurbruegg is a Professor of Finance within the School of Commerce at the University of Adelaide. He has taught economics and finance at several universities around the world, as well as published over forty pieces of work, mainly in the area of international finance. His work has appeared in a number of international finance journals including the *Journal of Empirical Finance*, *Journal of International Money and Finance, Journal of Futures Markets, Journal of Risk and Insurance*, and the *Geneva Papers in Risk and Insurance*. He is also the joint editor of the *International Journal of Managerial Finance* hosted by Emerald Publishing.

Email: ralf.zurbrugg@adelaide.edu.au

1 International Insurance Markets: Between Global Dynamics and Local Contingencies—An Introduction

J. David Cummins
University of Pennsylvania

Bertrand Venard
Audencia Nantes School of Management

1.1 INTRODUCTION

Insurance markets have radically and deeply changed in the last 20 years. Deregulation, globalization of insurance institutions, intensified competition, electronic commerce, bancassurance, and the emergence of new risks are among the challenges faced by insurance markets now. These developing trends pose both global and local challenges for insurance firms.

On one hand, it is clear that some key developments in the insurance and financial services markets influence global insurance markets. Some common issues have had, and will have, determinant influences globally, with limited national or local constraints. For example, the deregulation of financial services markets, advances in computing and communications technologies, and the increasing investment sophistication of financial services customers have affected markets worldwide.

On the other hand, it is necessary to avoid a simplistic view of insurance markets. Indeed, the word "global" carries an implicit hypothesis of homogenous influence of a specific factor everywhere, without any constraints. On the contrary, many insurance market segments remain localized, even in the most developed countries, and various nations and regions have particular strengths and limitations. For example, although global financial services conglomerates have made major inroads into annuity markets outside their home countries, foreign insurer market penetration in other markets such as automobile insurance is much more limited.

Moreover, the volume of electronic commerce is technically constrained by the level of Internet access in some countries. Also, the aging population is not yet a reality everywhere—many developing countries still have a high proportion of youth in their population.

Another example of a national boundary limitation to some key economic trends is the distribution of financial services products. The relative importance of distribution channels varies widely, even among relatively "homogeneous" countries. Direct insurance marketing has become very important in some national markets, such as the U.K., but remains more limited in others, such as France. Bancassurance has achieved significant market penetration in some countries, but not in others.

These developments suggest that remaining cultural, legal, institutional, and demographic considerations can have at least as great an effect as emerging global developments. Finally, while financial services integration has focused on developed markets such as the U.S., Europe, and Japan, large, rapidly developing markets in countries such as China and India will play a much more significant role in the future.

Despite the existing complexity of today's insurance industry, both academics and insurance industry practitioners have recognized for a long time the very limited availability of information about various important national insurance markets. Most of the discussion of financial services integration has focused primarily on developed markets due to the concentration of insurance in a few countries. In fact, the U.S. market represented 34% of the world insurance premiums in 2004 and the G7 market 76% (Swiss Re 2005).[1] However, even in many developed countries, comprehensive analyses of developments and trends in insurance markets are difficult to find; and rapidly developing markets such as China and India have received little attention. Another potential limitation to our understanding is the lack of statistical data on many markets and the inaccuracy and lack of standardization of the data that are available.

It is therefore important to reinforce our knowledge of the various international insurance markets. The purpose of the book is to describe the key trends of the insurance industry in the last ten years in more than 15 important national insurance markets. In addition, the book contains overview chapters on Europe, South and East Asia, reinsurance, Lloyd's of London, and alternative risk transfer. This is the first book to present a comprehensive analysis of global insurance markets.

1.2 CONTENTS AND COUNTRY SELECTION

If the book's purpose seems clear, the completion of the book was not easy. As said previously, there is a lack of information on various markets. This book is significant because it is the first of its type in the insurance industry. To resolve the "data challenge," we have chosen to rely on an important network of academic and insurance industry experts across the globe. More than 34 experts have contributed to

---

[1] The G7 consists of seven of the world's leading industrialized nations – Canada, France, Germany, Italy, Japan, the United Kingdom, and the United States.

this project. The chapter on each national insurance market was written by an expert or experts from that country. The goal was for them to conduct an in-depth analysis of their own national insurance market, utilizing their local contacts to provide data and source materials. For example, two Japanese colleagues, Professor Nobuyoshi Yamori and Professor Taishi Okada were in charge of presenting the Japanese insurance market by giving an "insider" evaluation by insurance specialists. Experts were also chosen to write the overview chapters who have specialized expertise and access to the data needed to provide a comprehensive analysis of these topics.

An important task was to harmonize the gathered data. Indeed, our idea was to enable insurance practitioners and academics to be able to compare the surveyed countries. It was therefore important to require the contributing colleagues to gather and analyze the same types of information. As usual in a book of this scope, the final scientific product does not show the various erratic phases which characterize a research undertaking. The process of discovery involves a sequence of trials and errors before reaching a satisfactory final product.

The idea for the book arose during a conversation between the co-editors, Professors J. David Cummins and Bertrand Venard, after a presentation by the latter about bancassurance in France at the Wharton School, University of Pennsylvania. The immediate conclusion of the conversation was that most insurance markets were quite unknown, even to insurance specialists. At the starting point, we had an idea, but the process to finish the book was another story. We asked colleagues in various countries to analyze their markets and suggested various potential subjects. Step by step, it appeared necessary to organize and systematize as much as possible the data and the analysis of each national market. Despite some differences, the final chapters have therefore some substantial similarities. The chapter on each national insurance market follows more or less the same structure:

- Introduction
- History
- Regulation
- Presentation of the Insurance Companies
- Presentation of the Insurance Products
- Insurance Company Performance (costs, technical and financial results)
- Description of Some Key Insurance Actors, especially some leading firms
- Distribution Channels
- Future Trends
- Conclusion
- Lexicon[2]

A key task was to select the countries to present. Indeed, the selection is based on the importance of the country for the insurance industry, the availability of a minimum amount of data, and the availability of insurance specialists willing to carry out a very demanding project. At the starting point of this project, the question

---

[2] The Lexicon is quite essential since the English insurance technical vocabulary differs around the word. See the last section of this introduction for further discussion about the matter.

was to decide what an important market for the insurance industry was. We chose four criteria to select the relevant countries to study.

First, an evident criterion was the size of the insurance market, measured by annual insurance premiums. We selected the top ten countries in terms of annual insurance premiums. Table 1.1 shows the top twenty-five insurance markets in terms of annual insurance premiums. For example, since the U.S. is the largest insurance market worldwide, it was selected. The U.S. represents 34% of world insurance premiums in 2004, 27% in life and 43% in non-life insurance. Reflecting its importance, the U.S. chapter is the longest in the book. The second largest market in terms of premium volume was Japan, thus it was also selected. Separate chapters also are included on the U.K., France, Germany, Italy, Canada, the Netherlands, and Spain, rounding out nine of the top ten countries in terms of market size. Among the top ten, only South Korea is not included in a separate chapter in this edition due to lack of data about its insurance industry. It is covered briefly in the South and East Asia chapter.

In Table 1.2, we have calculated the concentration of the world insurance market by country in 2004. This concentration is over weighted since the Swiss Re sample included 88 countries rather than all countries in the world. However, the 88 countries comprise nearly all of the world insurance market in terms of annual premiums since the countries not selected by Swiss Re are considered to have very small insurance markets. The countries given chapter-length treatment in the book represent 84% of world insurance premiums and the countries covered in the regional chapters bring the total coverage of the book to 93% of world insurance premiums.

Obviously, selecting countries based on their insurance premium volume implies that we would primarily cover developed countries. In order to broaden the scope of the book, a second criterion was adopted which entailed selecting three important emerging countries with large populations—China, India, and Brazil. The rapid economic growth in these countries is driving a substantial expansion of their insurance markets. China was already the eleventh country in terms of annual insurance premiums in 2004, India the nineteenth, and Brazil the twenty-first. Brazil was also selected as the only country in our sample in South America. Table 1.3 gives some data regarding the population coverage of our sample. As indicated, countries covered in the book represent 62.4% of the total world population and the top five countries in terms of population are covered in either a country or regional chapter.

The third selection criterion was the insurance penetration (measured as annual insurance premiums in U.S. dollars as a percentage of gross domestic product (GDP)). Table 1.4 gives detailed information about insurance penetration in our sample. Among our sample of countries, some insurance markets are characterized by high levels of insurance penetration. In particular, South Africa has the world highest insurance penetration with annual premiums equal to 14.38% of GDP in 2004. This country is also the only one selected in Africa. Other important insurance penetration rates among countries covered in the book are in the U.K. (12.60%), The Netherlands (10.10%), Japan (10.51%), France (9.52%), the U.S. (9.36%), and Ireland (8.97%). Thus, Ireland was also chosen. Despite a relative small size, Ireland is an interesting case for its ability to attract foreign direct investments in the field of financial services.

Table 1.1. Top 25 Insurance Countries by Total Premiums, 2004

Country	Ranking	Surveyed Countries	Total Premiums (US$ Millions)	Life Premiums	Non-Life Premiums
USA	1	C	1,097,836	494,818	603,018
Japan	2	C	492,425	386,839	105,587
UK	3	C	294,831	189,591	105,241
France	4	C	194,624	128,813	65,811
Germany	5	C	190,797	84,535	106,261
Italy	6	C	128,811	82,083	46,728
Canada	7	C	69,741	29,509	40,232
South Korea	8	R	68,623	48,680	19,944
Netherlands	9	C	58,577	31,512	27,064
Spain	10	C	55,903	23,592	32,311
China	11	C	52,171	35,407	16,765
Australia	12		49,404	25,689	23,714
Taiwan	13	R	43,236	33,851	9,385
Switzerland	14	R	42,006	24,067	17,939
Belgium	15	R	38,853	24,112	14,741
South Africa	16	C	30,682	24,381	6,301
Ireland	17	C	27,882	19,068	8,815
Sweden	18	R	24,075	15,790	8,285
India	19	C	21,249	16,919	4,330
Denmark	20	R	19,512	12,453	7,060
Brazil	21	C	18,042	8,199	9,843
Austria	22	R	17,395	7,695	9,701
Russia	23		16,352	3,544	12,809
Finland	24	R	16,330	12,823	3,507
Hong Kong	25	R	15,260	12,969	2,291
World Insurance Premiums			3,243,906	1,848,688	1,395,218

Source: Swiss Re (2005).
Notes: Ranking is determined by premium volume. C denotes that the country is covered in a separate chapter of the book. R denotes that the country is covered in a regional overview chapter.

The fourth criterion used in selecting countries to be covered in the book is the insurance density, defined as annual insurance premiums in U.S. dollars per capita in 2004. Table 1.5 shows the insurance density, life insurance density, and non-life insurance density of the top 25 insurance markets in terms of annual insurance premiums in 2004. All of the top ten countries in terms of insurance density are covered in this book, either with separate chapters (U.K., Ireland, Japan, U.S., the

Netherlands, and France) or through regional chapters (Switzerland, Denmark, Belgium, and Finland). Switzerland, which has the highest insurance density due to its very large reinsurance activities, is also partly covered in the reinsurance chapter.

Table 1.2. Concentration of the World Insurance Market by Country, 2004

	Total Premiums (US$ Billions)	Cumulative Market Share (%)	Life Premiums (US$ Billions)	Life Market Share (%)	Non-Life Premiums (US$ Billions)	Non-Life Market Share (%)
Total Top 1 Country	1,097,836	34	494,818	27	603,018	43
Total Top 5 Countries	2,270,513	70	1,284,596	69	985,918	71
Total Top 10 Countries	2,652,168	82	1,499,972	81	1,152,197	83
Total Top 25 Countries	3,084,617	95	1,776,939	96	1,307,683	94
Total Chapter-Length Sample[1]	2,733,571	84	1,555,266	84	1,178,307	84
Total Chapter and Regional Sample[2]	3,018,861	93	1,747,706	95	1,271,160	91
World Insurance Premiums	3,243,906	100	1,848,688	100	1,395,218	100

Source: Swiss Re (2005).
Notes: [1]Total chapter-length sample includes countries covered in separate chapters.
[2]Total chapter and regional sample includes countries covered in separate chapters and in regional chapters.

In summary, the countries included in the book represent a very comprehensive coverage of the world's most important insurance markets. As mentioned, our sample represents approximately 93% of the world insurance premiums in 2004 (see Table 1.2). The sample is also quite important in terms of world GDP. The surveyed countries generate a total annual GDP in 2004 of US$33,750.5 billion, equaling 83.1% of overall world GDP. Table 1.6 gives some macroeconomic information about our sample such as annual GDP, population, and inflation.

1.3 CHAPTER CONTENTS

In Chapter 2, Loftin Graham and Xiaoying Xie analyze the United States insurance market. With over a trillion dollars in premiums written in 2004 (approximately 9.4% of GDP) insurance operations from the U.S. also generate 33.8% of the worldwide total. Their chapter discusses the characteristics and trends of the U.S. insurance market, including a brief history, a presentation of its regulatory institutions and laws, a description of its life-health and property-liability sectors, and a discussion of current issues of importance to the industry.

Table 1.3. Top 25 Countries by Population

Country	Surveyed Countries	Population in millions	% of the World Population	Population Rank	GDP Rank
PR China	C	1,297.2	20.5	1	7
India	C	1,079.5	17.0	2	11
United States	C	292.4	4.6	3	1
Indonesia	R	217.5	3.4	4	23
Brazil	C	178.5	2.8	5	14
Pakistan		152.1	2.4	6	49
Russia		142.9	2.3	7	16
Bangladesh	R	140.5	2.2	8	55
Nigeria		138.4	2.2	9	54
Japan	C	127.1	2.0	10	2
Mexico		103.8	1.6	11	12
Philippines	R	82.9	1.3	12	47
Germany	C	82.5	1.3	13	3
Vietnam	R	82.1	1.3	14	57
Turkey		71.6	1.1	15	21
Egypt		68.7	1.1	16	50
Iran		67.3	1.1	17	35
Thailand	R	62.4	1.0	18	34
France	C	59.9	0.9	19	5
United Kingdom	C	59.4	0.9	20	4
Italy	C	57.5	0.9	21	6
South Africa	C	44.7	0.7	25	28
Spain	C	41.3	0.7	26	9
Canada	C	31.9	0.5	31	8
Netherlands	C	16.3	0.3	41	15
Ireland	C	4.0	0.1	72	31
Top 5 countries		3,065.0	48.3		
Top 10 countries		3,766.0	59.4		
Countries in the book (C or R)		3,957.5	62.4		
All Swiss Re Sample		6,342.1	100.0		

Source: Swiss Re (2005).

Notes: C denotes that the country is covered in a separate chapter of the book. R denotes that the country is covered in a regional overview chapter.

Table 1.4. Insurance Penetration of Top 25 Insurance Countries by Total Premiums, 2004

Country	Premium Ranking	Insurance Penetration Ranking	Total Penetration (%)	Life Penetration (%)	Non-Life Penetration (%)
USA	1	10	9.36	4.22	5.14
Japan	2	5	10.51	8.26	2.25
UK	3	3	12.60	8.92	3.68
France	4	9	9.52	6.38	3.14
Germany	5	22	6.97	3.11	3.86
Italy	6	19	7.60	4.86	2.74
Canada	7	21	7.02	2.97	4.05
South Korea	8	8	9.52	6.75	2.77
Netherlands	9	6	10.10	5.43	4.67
Spain	10	27	5.63	2.38	3.25
China	11	42	3.26	2.21	1.05
Australia	12	15	8.02	4.17	3.85
Taiwan	13	2	14.13	11.06	3.07
Switzerland	14	4	11.75	6.73	5.02
Belgium	15	7	9.62	6.73	2.89
South Africa	16	1	14.38	11.43	2.95
Ireland	17	12	8.97	5.74	3.23
Swenden	18	23	6.96	4.56	2.39
India	19	44	3.17	2.53	0.65
Denmark	20	14	8.07	5.15	2.92
Brazil	21	49	2.98	1.36	1.63
Austria	22	25	5.95	2.63	3.32
Russia	23	51	2.83	0.61	2.21
Finland	24	13	8.77	6.89	1.88
Hong Kong	25	11	9.27	7.88	1.39
Average		18	8.28	5.32	2.96

Source: Swiss Re (2005).
Note: Insurance penetration is defined as annual insurance premiums as a percentage of GDP.

Table 1.5. Insurance Density of Top 25 Insurance Countries by Total Premiums, 2004

Country	Premium Ranking	Density Ranking	Total Density	Life Density	Non-Life Density
Switzerland	14	1	5,716.4	3,275.1	2,441.2
UK	3	2	4,508.4	3,190.4	1,318.0
Ireland	17	3	4,091.2	2,617.4	1,473.8
Japan	2	4	3,874.8	3,044.0	830.8
USA	1	5	3,755.1	1,692.5	2,062.6
Denmark	20	6	3,620.4	2,310.5	1,309.9
Netherlands	9	7	3,599.6	1,936.5	1,663.1
Belgium	15	8	3,275.6	2,291.2	984.4
France	4	9	3,207.9	2,150.2	1,057.7
Finland	24	10	3,134.1	2,461.0	673.1
Sweden	18	12	2,690.0	1,764.3	925.7
Australia	12	14	2,471.4	1,285.1	1,186.3
Germany	5	15	2,286.6	1,021.3	1,265.3
Italy	6	16	2,217.9	1,417.2	800.7
Hong Kong	25	17	2,217.2	1,884.3	332.9
Canada	7	18	2,188.7	926.1	1,262.6
Austria	22	19	2,159.7	955.3	1,204.4
Taiwan	13	20	1,909.0	1,494.6	414.4
South Korea	8	22	1,419.3	1,006.8	412.5
Spain	10	24	1,355.2	571.9	783.3
South Africa	16	31	686.5	545.5	141.0
Russia	23	52	114.4	24.8	89.6
Brazil	21	55	101.1	45.9	55.2
China	11	72	40.2	27.3	12.9
India	19	78	19.7	15.7	4.0
Average		21	2,426.4	1,518.2	908.2

Source: Swiss Re (2005).
Note: Insurance density is defined as insurance premiums in US$ per capita.

Table 1.6. Macroeconomic Data for Top 25 Insurance Countries by Total Premiums, 2004

Country	2004 Premium Ranking	2004 GDP Ranking	Surveyed Countries	2004 Population (Millions)	2004 GDP (US$ Billions)	Real % change in 2004 GDP	Real % change in 2003 GDP	2004 Inflation (%)	2003 Inflation (%)
USA	1	1	C	292.4	11,735.0	4.4	3.0	2.7	2.3
Japan	2	2	C	127.1	4,683.2	2.6	2.0	0.0	–0.3
UK	3	4	C	59.4	2,125.5	3.1	2.2	1.3	1.4
France	4	5	C	59.9	2,018.4	2.3	0.5	2.2	2.1
Germany	5	3	C	82.5	2,707.1	1.6	–0.1	1.7	1.0
Italy	6	6	C	57.5	1,677.9	1.0	0.4	2.1	2.6
Canada	7	8	C	31.9	993.8	2.8	2.0	1.8	2.7
South Korea	8	10	R	48.4	720.7	4.6	3.1	3.6	3.5
Netherlands	9	15	C	16.3	579.9	1.4	–0.9	1.3	2.1
Spain	10	9	C	41.3	993.1	2.7	2.5	3.0	3.0
China	11	7	C	1,297.2	1,599.0	9.5	9.4	3.9	1.2
Australia	12	13		20.0	616.3	3.2	3.4	2.3	2.8
Taiwan	13	20	R	22.6	306.0	5.7	3.3	1.6	–0.3
Switzerland	14	17	R	7.3	357.4	1.7	–0.4	0.8	0.6
Belgium	15	18	R	10.4	352.5	2.8	1.3	2.1	1.6
South Africa	16	28	C	44.7	213.3	3.7	2.8	1.4	5.9
Ireland	17	31	C	4.0	181.8	4.9	3.6	2.2	3.4
Swenden	18	19	R	8.9	346.1	3.0	1.6	0.4	1.9
India	19	11	C	1,079.5	670.0	7.1	8.5	3.8	3.8
Denmark	20	26	R	5.4	241.9	2.0	0.4	1.1	2.1
Brazil	21	14	C	178.5	604.8	5.2	0.6	6.6	14.7
Austria	22	22	R	8.1	292.3	2.0	0.8	2.1	1.4
Russia	23	16		142.9	578.3	7.1	7.3	11.0	13.6
Finland	24	30	R	5.2	186.2	3.4	2.5	0.2	0.9
Hong Kong	25	33	R	6.9	164.6	8.1	3.2	0.0	–2.1
Total Sample: C				3,372.0	30,782.9				
Total Sample: C + R				3,495.2	33,750.5				
Total World				6,342.1	40,630.0				

Source: Swiss Re (2005).
Notes: C denotes that the country is covered in a separate chapter of the book. R denotes that the country is covered in a regional overview chapter.

In the third chapter, the Japanese insurance market is discussed by Nobuyoshi Yamori and Taishi Okada. It is an interesting market not only due to its size, but also to its important transformation since 1990. In particular, deregulation and liberalization in the insurance business has been accelerated by the Japanese "Big Bang" initiative. Insurers can now provide the public with a wide range of insurance products and services. Furthermore, foreign insurers have increased their market share substantially in the 1990s. The authors explain not only how the Japanese insurance market and insurers have changed, but also why they have changed and, in some respects, have not changed. The authors analyze several key aspects of the Japanese insurance market and insurers that make them unique.

In Chapter 4, Philip Hardwick and Michel Guirguis describe the United Kingdom, the largest insurance market in the European Union (E.U.) and the third largest in the world. Insurance companies are also the largest domestic owners of U.K. shares, holding about a fifth of all U.K. ordinary shares. Yet the insurance industry has never faced greater challenges than it does today. This chapter provides an overview of the structure of the U.K. insurance market and describes the main suppliers of insurance and their financial activities. In addition, the authors focus on the product structures of U.K. insurers and their distribution channels. They outline the major challenges and current trends faced by major insurers and present data on the relative international competitiveness of U.K. insurers. Finally, using cross-sectional and time-series data, they apply data envelopment analysis to estimate and compare the relative performance of U.K. life and general insurance firms.

In the fifth chapter, Bertrand Venard discusses the French insurance market, the fourth-largest in the world in terms of premium volume, in tune with the economic power of the country. Among the chapter's topics are the history of the insurance market, the life insurance market, the property-casualty market, financial results of insurance companies, main economic actors, and distribution channels. The French insurance market is characterized by a long history. For many centuries, an important feature was the important involvement of the French State in the supervision of insurance activities. It is also interesting to note that the country was quite innovative in building its insurance market, leading to new insurance activities, such as employer liability for employee casualties in the workplace. In recent years, the building of the European Union has had an important impact on French insurance markets through new insurance laws, the intensification of competition, and the European directives, all of which created freedom of access to the French insurance market. From 1990 to 2005, both life and property-casualty insurance markets have been constantly growing in accordance with French GDP. In 2003, life insurance represented 63% of the insurance market. A distinctive feature of this market is the importance of bancassurance, France being a leading market in the convergence of banks and insurance companies.

In Chapter 6, Raimond Maurer and Barbara Somova present an overview of the contemporary German insurance market, its structure, players, and trends. After a brief discussion of the history of the insurance industry, the authors analyze the contemporary market in terms of its legal and economic structure, with statistics on the number of companies, insurance density and penetration, the role of insurers in the capital markets, premiums by line of insurance, and the main market players and their market shares. Furthermore, the three biggest insurance lines—life, health, and property-casualty—are considered in more detail, including product range, country specifics, and insurance and investment results. A section on regulation outlines its implementation in the insurance sector, offering information on the underlying legislative basis, supervisory body, technical procedures, expected developments, and sources of more detailed information.

Next, Luigi Ventura describes the Italian insurance market, which has undergone a series of important changes over the last fifteen years. It has become a more mature market, able to compete with those established in the most developed European countries. The author analyzes some key factors which have had an impact on the development of the Italian insurance market in the last decade, both from a descriptive statistics point of view and from a more applied, econometric standpoint. Many crucial events that occurred in the 1990s are also discussed in some depth, such as the evolution of the distributive system, the changes in fiscal treatment of insurance products, and the steps towards the creation of a common European market for insurance.

In Chapter 8, Gilles Bernier and Alli Nathan outline the key features of the structure, conduct, and performance of the Canadian insurance industry—both life and property-casualty insurance—over the period from 1990 to 2003. On the international insurance scene, Canada one of the big players, ranking seventh in 2004 in terms market share of the total insurance premiums sold worldwide. Indeed, Canada is an interesting country to study because its insurance industry is stable, dynamic, and open to foreign ownership and participation. After providing a brief history of the industry and reviewing the economic environment that prevailed during this period, this chapter characterizes many dimensions of Canada's insurance markets, including, among other things, the evolution of premium volumes and concentration in each major line of business, the importance of the industry in the economy, the leading players in each class of insurance, the financial performance of both sectors of the industry, and the past, current, and future trends most influential for the whole sector. Today, the insurance industry and the banks are the two key pillars of the Canadian financial system. As such an important player, the insurance industry contributes to the competitive environment of the system and to the greater efficiency of its financial markets.

In the ninth chapter, Alfred Oosenbrug analyzes the insurance market in the Netherlands. A brief overview of the long and rich history of insurance in the Netherlands is given first. Holland was the homeland of the founding father of modern life insurance mathematics, Johan de Witt, while during the seventeenth and eighteenth century the Netherlands comprised among the most important markets for non-life insurance. The traditional openness of the Dutch insurance market stems from a long history of liberalism and tolerance in Holland in general. As a consequence, more than a thousand insurers are licensed to operate in the Dutch market. Furthermore, the leading Dutch insurance groups are key players in insurance markets all over the world. The last two decades of the twentieth century were characterized by strong growth for the Dutch domestic insurance market. However, recently, booming market trends have ceased and the economy has cooled down, stock markets have been sluggish, and strong fiscal incentives for life insurance business were abolished.

In Chapter 10, María Rubio-Misas presents Spain, another European market. The chapter provides an analysis of the structure, conduct, and performance of the Spanish insurance industry from 1990 to 2003. The industry consists of life insurance specialists, non-life insurance specialists, and diversified firms that offer both life and non-life insurance. Although the majority of firms in the market are organized as stock firms, mutuals have a relatively important market share in automobile insurance. Foreign insurers have an active presence in the Spanish market, basically through Spanish licensed and regulated subsidiaries. The number of firms in the industry has fallen dramatically due to firm retirements, insolvencies, mergers, and acquisitions. However, a large number of firms entered the market during the sample period. Among the key features of the Spanish insurance market, the author notices the concentration increase in non-life insurance, the domination of bancassurance in the life insurance business, and the importance of traditional insurance distribution channels (brokers and agents) in the non-life insurance market.

In the eleventh chapter, the Irish insurance market is analyzed by Brian Greenford, Martin Mullins, John Garvey, Louise Morris, and Liam O'Meara. The property-casualty market is first considered. This section aims to show how the property-casualty market has grown and the problems that have arisen, with special emphasis on the innovative approach to dealing with compensation claims under liability and motor policies. The life insurance market is growing rapidly, having developed dramatically since the inception of the state. The health insurance market is peculiar to Ireland in that private insurers must charge the same premium regardless of risk. Pensions are also of concern in Ireland as, like the rest of Europe, the proportion of elderly has been increasing. Ireland also has become a financial services center with reduced taxation and special concessions for insurers operating within the Irish Financial Services Centre (IFSC) in Dublin. This endeavor has become very successful with many insurance companies entering the IFSC to sell insurance internationally. Thus, financial services have become extremely important to the Irish economy.

The twelfth chapter is the first one about a developing country: China. Qixiang Sun, Lingyan Suo, and Wei Zheng are Chinese colleagues, working in China and thus are able to give a unique insider perspective on a market that is largely unknown outside China, despite its growing importance. During the last two decades, China's insurance industry developed from a virtually nonexistent, minimal, and closed industry into a large, open industry with potential promise. As part of China's reform and opening up, China's insurance industry was inevitably deeply affected. The main purpose of this chapter is to provide an analysis of the development of China's insurance industry in the context of reform and opening up, discuss the major experiences and problems in the process, and examine the future prospects of the industry. The chapter begins with a brief history of development of the insurance industry in three stages: (1) resumption and restoration from 1980 to 1985, following the closure of the domestic insurance industry in 1959, (2) the beginning of market-oriented reform from 1986 to 1991, and (3) the opening up and rapid growth from 1992 to the present. The chapter describes the key features of China's insurance market in each respective stage. Opening to the outside world has contributed significantly to the rapid increase of capital and technologies, a favorable competitive environment, and the steady growth and development of China's insurance industry. In this context, the authors explore China's World Trade Organization (WTO) commitments concerning the insurance sector. Finally, key drivers of and possible obstacles to the future growth of China's insurance industry are discussed.

In Chapter 13, Tapen Sinha analyzes another important emerging country: India. With over one billion people, India is fast becoming a global economic power. With a relatively youthful population, India will become an attractive insurance market in the coming decades. This chapter examines the Indian insurance industry by first examining the details of the regulatory regime that existed before India gained its independence in 1947. This is important because the Insurance Act of 1938 became the backbone of the current legislation in place to regulate insurance. The chapter goes on to highlight the importance of the rural sector, where the majority of Indians reside, and describes how the recent privatization is playing out in the market. Based on recent economic estimates, the chapter provides projections for the principal segments of the insurance market through 2025.

In the fourteenth chapter, Robert Vivian provides an overview of South Africa's insurance market, the only African market covered in this book. In the African context, South Africa is very important. The South African life insurance market accounts for 95% and the property-casualty insurance market accounts for 57% of the total premiums of the African continent. As a percentage of GDP, South Africans purchase more life insurance than any other country in the world. As Vivian explains, life insurance assumes a critically important role because, for all practical purposes, South Africa does not have a meaningful government social security system. As a consequence, financial security must be provided by each individual. Reasons why retirement funding and medical aid have been linked to employment and not the government are examined. Retirement funds are not on a pay-as-you go basis, but rather are funded. The aggregate assets of these funds are approximately equal to the country's GDP. Vivian explains the new political forces which arose when the universal franchise was introduced with the end of the apartheid era in 1994. These forces have reshaped financial markets as various changes have been introduced, including the compulsory transformation in terms of the Black Economic Empowerment (BEE) program.

Another emerging market analyzed in the book is Brazil, the only South American country covered in this book. Andrea Levy and Fernanda Chaves Pereira describe the performance of the Brazilian insurance market in recent years and the current challenges facing the industry. Some comparisons with other Latin American countries are provided. The Brazilian insurance development followed the economic restructuring of the country that took place in recent years. In particular, the new economic order that began with the adoption of the Plano Real in 1994 promoted financial and monetary stability after a long period of very high inflation. Occurring simultaneously with this new economic reality, the Brazilian government became concerned about the formation of long-term savings and began in 1995 to develop tax incentives for investment in pension plans. These factors permitted the explosive development of the market for open enrollment pension plans and insurance beginning in 1995. At the same time, the changes in the age distribution of the Brazilian population and legislative reforms in the National Social Security system are discussed as well as the difficulties that Brazil will encounter as the authorities attempt to deal with the situation of chronic deficits that currently plague the system. Finally, the various types of insurance that are sold in the market today, the distribution of the companies that operate in the market, the premiums collected, and the related technical, commercial, and administrative results are discussed.

In addition to the individual country chapters, we decided to include in the book other chapters which provide a regional perspective. In Chapter 16, Paul Klumpes, Paul Fenn, Steven Diacon, and Christopher O'Brien carried out an overview of the main European insurance markets. Indeed, the European insurance market has recently developed in several different segments. The European Union (E.U.) has highlighted the increasing interdependencies of insurance markets by adopting Insurance Directives with the objective of creating a single European market for insurance. The Third Generation Directives, adopted in 1994, were particularly important because they established a "single passport" system, whereby an insurer licensed in one E.U. member nation can conduct business in all E.U. countries. With banking also becoming less regulated, the traditional separation of insurance as a regulated activity has devolved into increasing integration into broader financial markets that affect risk allocation and transformation. This chapter reviews the recent developments in European countries that are subject to European Union regulations. The authors review the major countries and analyze the current trends and identify areas where European regulation is likely to affect future trends. Furthermore, the authors highlight the role of the financial markets and consumer trends that may limit the development of markets. The analysis suggests that common themes of consolidation, poor performance, reduced capitalization levels, and increased market segmentation are pervasive across the European insurance industry over the last ten years. The analysis of solvency and accounting regulations suggests greater harmonization in the future, with major implications for risk management practices. The authors conclude that the European insurance market is subject to an evolving solvency-based regulation that is more closely aligned with international competitiveness and they provide recommendations to further enhance the degree of harmonization in solvency and accounting rules.

In Chapter 17, South and East Asian insurance markets are analyzed by Stephanie Hussels, Claire Sherman, Damian Ward, and Ralf Zurbruegg. This chapter acts as a guide to both executives and academics who are interested in understanding the determinants of insurance market development and how it may affect general economic development in South and East Asia. By providing a synopsis and evaluation of recent empirical research on the development of insurance markets in general and South and East Asia in particular, this chapter provides a balanced understanding of the factors that promote both the demand and supply sides of South and East Asia's insurance markets. The chapter closes by highlighting certain issues that both insurance companies and executives can utilize in their own markets to design future policies that can be geared to promote insurance market development and growth.

In the eighteenth chapter, Thomas Holzheu and Roman Lechner discuss the global market for reinsurance. Reinsurance is clearly the most globalized segment of the insurance market. For example, international reinsurers have played a major role in paying claims for many natural and man-made catastrophes, including Hurricane Andrew in 1992, the Northridge earthquake in 1994, and the World Trade Center terrorist attacks of September 11, 2001. They also played an important role in insurance payments for Hurricane Katrina, which destroyed part of the southern United States in September 2005 and was the most expensive natural catastrophe in history in terms of insurance losses. The impact of such disasters is global and usually leads to an increase in the world reinsurance premium rates. By providing coverage against adverse fluctuations in claims, reinsurance protects the capital base of the primary insurer and creates a more diversified portfolio, reducing the volatility of underwriting results. In 2003, primary insurers ceded business worth US$170 billion to reinsurers worldwide, of which 83% was property-casualty insurance and 17% life insurance. Around 84% of ceded business was generated in the mature insurance markets of North America and Western Europe, reflecting these regions' importance as sources of primary insurance revenue. U.S. and Continental European reinsurers are the main sources of global reinsurance capacity and Bermuda and the London Market are the other major reinsurance centers. Periodic underwriting and investment losses since 1992 have prompted a wave of structural changes in the industry. The number of domestic market players declined, especially in the United States, while another generation of Bermuda start-ups rapidly gained market share. Price increases following events such as Hurricane Andrew in 1992 and the World Trade Center terrorist attack of 2001 restored the profitability and capitalization of the global reinsurance industry, but the industry is subject to periodic loss shocks such as Katrina.

An interesting market place, especially for reinsurance, is Lloyd's of London and the London company market. In Chapter 19, Julian James provides an insider perspective on these markets from his position as Worldwide Markets Director at Lloyd's. The London market and in particular Lloyd's, one of the world's oldest and best known insurance markets, is a significant contributor to the global insurance industry and plays a leading role in internationally-traded specialist insurance and reinsurance. The chapter looks in detail at the operations of the London market by identifying market participants and the skills offered. Turning to Lloyd's, the chapter describes the market's rich history—from a seventeenth century coffee shop to the sophisticated market place of the present day. The chapter discusses how Lloyd's operates—explaining that it is a market and not a company. Next, the reasons for and extent of Lloyd's global reach are highlighted, focusing in particular on the U.S., which is Lloyd's largest single market. Finally, the chapter discusses what Lloyd's believes to be the current hot topics for the non-life industry around the world: managing the insurance cycle, business processing reform, and climate change. The chapter also considers the London company market, which consists of insurers that also contribute significantly to world insurance capacity, particularly in specialized commercial non-life insurance coverages.

In the twentieth and final chapter, the growing sophistication of markets for traditionally insured risks is analyzed by Robert Hartwig and Claire Wilkinson, who discuss the market for alternative risk transfer (ART). The ART market goes beyond traditional insurance and reinsurance markets to find innovative solutions to risk financing and risk management. An ever-changing risk landscape and the emergence of new and complex risks have driven the development of the ART market. The market often provides risk bearing capacity where little or no traditional insurance and reinsurance coverage is available or where insurance coverage is inadequate. At one extreme, high severity, low frequency property-casualty exposures are well-suited to capital market solutions such as catastrophic risk (CAT) bonds, which can greatly expand available capacity. At the same time, non-insurance commercial enterprises have become increasingly comfortable as bearers of risk, particularly for high frequency, low severity events, leading to increased interest in solutions that accommodate high levels of risk retention such as self-insurance programs and captive insurance companies. This chapter discusses all of the major categories of ART solutions commonly found in the market today, including self-insurance and captives, risk retention groups, finite risk (re)insurance, catastrophe bonds, and supplementary government programs.

1.4 GLOBAL SIMILARITIES AND LOCAL DIFFERENCES

Over the last two decades, globalization has become a controversial and widely-debated concept. The failure to achieve a clear definition of the concept is partly due to the scientific process whereby various researchers have different views on globalization, due in part to their diverse scientific backgrounds ranging from politics to sociology and from management to finance and economics. Globalization can be defined as "a multidimensional set of social processes that create, multiply, stretch and intensify worldwide social interdependencies and exchanges while at the same time fostering in people a growing awareness of deepening connections between the local and the distant" (Steger 2004). As a controversial development, globalization is seen as an economic threat by some and an economic opportunity by others (Bauman 2000; Watson 2002). In part, this also reflects the fact that there are winners and losers in the globalization game. Among the winners are multi-national corporations and consumers, to the extent that globalization improves efficiency, reduces prices, and improves consumer choice. Among the losers are entrenched local political and economic interests and high wage, low-skilled workers in protected industries.

On the one hand, opponents of globalization question the benefits of financial liberalization and the integration of the global capital and currency markets (Jomo and Shyamala 2001). Skeptics denounce the world governance of international agencies, like the World Bank or the World Trade Organization (WTO) for "usurping power" from states and local governments (George and Sabelli 1994; Scholte 2000). Other critics stress the increased power of multinationals, which are said to "rule the world" (Barnet and Cavanagh 1994). For critics, the term globalization will maintain its polemic power, much as "capitalism" or "imperialism" has done (Osterhammel and Petersson 2003).

On the other hand, promoters (or globalists) have praised the economic development that can be fostered by globalization in a borderless world (Ohmae 1990; Naisbitt 1994). The globalists believe that globalization is a real and significant historical development, leading to more economic well-being (Held, et al. 1999; Schirato and Webb 2003). Indeed, continuous improvement in financial technologies, aided by the expansion of low cost communication and widespread deregulation has driven finance to become increasingly international in scope (Litan 2001). It is clear that financial intermediation on a global scale has increased the available capital stock and has enabled investors to allocate capital efficiently at the world level. Empirical evidence confirms that the level of financial development is highly correlated with economic growth (Litan 2001).

The various insurance markets analyzed in the book have some important global similarities, such as deregulation, intensification of competition, generally rapid growth in insurance sales, emergence of new distribution channels, and the convergence of insurance, banking, and other formerly separate segments of the financial services industry. However, as the chapters in this book make clear, there remain many important "local" differences among insurance markets across the world, which makes their analysis both challenging and stimulating.

1.4.1 Global Similarities

Using the word globalization in the insurance sector implies that we assume that there are some key world trends which may be seen at more or less the same level in various countries worldwide. The orthodox view of globalization, whether for globalists or critics, is that the world is becoming more homogeneous. Based on the analysis of various key insurance markets presented in the book as well as our own observations, we can highlight various homogenous trends in the global insurance market, which are discussed in this section.

First, the process of deregulation can be seen in most parts of the world. Significant deregulation at the national and regional levels has occurred in Europe, Japan, and the United States. This deregulation has facilitated the integration of previously fragmented segments of the financial services market, primarily insurance, banking, and securities dealing and underwriting. The European Union's banking and insurance directives, implemented during the 1980s and 1990s, have the objective of creating a single European financial services market. The directives have led to widespread consolidation in the financial services markets in Europe, with large numbers of intra-sector and cross-sector mergers and acquisitions. The E.U. directives also deregulated insurance markets, with the exception of solvency regulation, which is carried out by each insurer's home country. This has meant the introduction of true price competition in European insurance markets, where prices were previously strictly regulated at the national level.

Similarly, in Japan, the "Big Bang" financial reforms of the late-1990s aimed to make the Tokyo financial market comparable in scale and in the variety and sophistication of financial products to markets in London, New York, and continental Europe (Dekle 1998; Hoshi and Patrick 2000). Among the reforms primarily affecting the insurance industry was the elimination of the ban on financial holding companies, which can now own firms dealing in various types of financial services, including insurance. The reforms also eliminated restrictions that prevented financial services firms from competing in each other's markets, such that banks, life insurers, and property-casualty insurers can now offer financial products other than their traditional offerings. The reforms are expected to lead to more competition in financial markets, although research on the effects on insurance markets has so far been limited.

In the United States, significant deregulation of the insurance industry has not taken place during the past twenty years. However, the price, product, and geographical regulations affecting U.S. insurers were much less onerous historically than those affecting banks. Bank deregulation in the U.S. has significantly affected the insurance industry, however. Beginning in the 1980s, banks were permitted to offer specific types of insurance, including life insurance and annuities. Banks had previously been excluded from the insurance markets by the National Banking Act of 1916 and the Glass-Steagall Act of 1933. As a result of this deregulation, banks have captured a significant share of the annuity market (Cummins and Wei 2005). The passage of the Financial Services Modernization Act of 1999 (the Gramm-Leach-Bliley Act) removed most of the remaining barriers that restricted competition across sectors of the financial services industry and allowed the creation of financial holding companies (FHCs), which can engage in bank and non-bank financial activities through subsidiaries. Although there has so far been minimal merger and acquisition activity between the banking and insurance industries, banks and insurers compete intensively in the sale of annuities, life insurance, and mutual funds at the retail level and in pension fund and asset management at the wholesale level. The U.S. has long been a hospitable environment for foreign insurers to enter the market, primarily through acquisitions of U.S. insurance companies; and major global financial services firms have made major inroads in the U.S., particularly in the market for annuities (Cummins, Eckles, and Zi 2005).

A second major trend is that various supranational agencies, including the World Bank, the Inter-American Development Bank, and the World Trade Organization (WTO), have been key promoters of globalization throughout the world, with a particularly significant impact in Asia. For example, with China's entry into the WTO in 2001, the Chinese government made significant commitments with respect to liberalizing the insurance sector and thus led China's insurance industry into a completely new stage of development. The current stage is characterized by an overall opening of the market instead of the previously limited and restricted opening. The changes have meant that it is now much easier for global financial firms to enter the Chinese insurance market. Thus, foreign firms have begun to compete intensively in China. Although the competition currently is mostly in terms of joint ventures, China's WTO commitments will make it even easier for foreign firms to enter the market in the coming years. In addition to the developments in China, various international bodies such as the World Bank have influenced insurance and financial markets in South and East Asia, with significant interventions triggered by the Asian crisis in 1991.

As a third major trend, many countries which previously relied on government insurance companies and programs now have private insurance companies as dominant economic actors. Many such countries have witnessed major waves of privatization. The number of government-owned insurance companies is becoming smaller and some countries have almost entirely eliminated government-run insurance companies. This trend is present not only true for some developed countries such as France but also for former Eastern-block countries and emerging economies. For example, India, a long-time protected insurance market, has authorized the development of the private sector since 2000. China has also seen the gradual dismantling of the former government insurance monopoly.

Fourth, the global insurance markets have seen an increased sophistication in insurance technology, whereby new products are launched much more rapidly than in the past. New technologies have also facilitated the introduction of more sophisticated insurance products such as universal life insurance, variable life insurance and annuities, and guaranteed income and guaranteed benefit features for variable insurance products. Alternative risk transfer products are also examples of the insurance sector's ability to develop and implement worldwide innovation. These changes are naturally facilitated by advances in computer and communication technologies as well as theoretical and empirical modeling advances in financial economics.

Fifth, insurance markets become more international, with formerly national insurers increasing their activities abroad and foreign insurers increasing their market share in other markets. The past twenty years has witnessed the development of truly global financial services firms such as ING, AXA, and Allianz. Globalization enables international insurance companies to diversify their risk in a broader range of countries. For example, in the Netherlands more than 1,050 foreign insurers operated in 2003 with more than 60% of them not supervised by the Dutch authority. Thus, one important common trend is that we can expect to observe more entry by foreign firms into both developed and developing markets worldwide, particularly if regulatory barriers to entry continue to fall.

Sixth, the insurance sector is structurally global through the mechanism of reinsurance. Indeed, risks are spread around the world by various key reinsurers. Each major claim affecting reinsurance markets, such as the September 11, 2001 World Trade Center (WTC) attack or Hurricane Katerina in 2005, has an effect on the global insurance market which is cushioned by reinsurance payments. Although the private insurance market has not recovered its ability to provide coverage against terrorism follow the WTC attack (Cummins 2006), private reinsurance markets have proven to be resilient following natural catastrophes. Although reinsurance prices rise and supply contracts following an event, new capital usually enters the industry and prices eventually decline. However, reinsurers have begun to explore securitization solutions such as CAT bonds to provide additional capacity for the largest natural disasters.

Seventh, many countries already face (or forecast very soon) the need for pension and health insurance reforms. Health systems are in need of huge transformations, whether in developed countries where the systems are too expensive or in developing countries where the health services are insufficient. Furthermore, many countries have to reform their pension systems since the actual pension systems are insufficient to provide future generations the promised benefits. Because many government social security programs have been based on pay-as-you-go funding schemes, the decrease in the ratio of working age to retired persons has been forcing many countries to consider scaling back their public pension systems. This provides an opportunity for financial firms to exploit the growing demand for supplemental retirement programs funded by life insurance and annuities and should provide growth opportunities for insurers in many markets. Of course, local contingencies play an important role here in terms of widely varying tax incentives for private savings through insurance and annuities.

Eighth, the continuing evolution of insurance product distribution systems is another commonality among many national insurance markets, although as discussed below and throughout the book, this does not mean that distribution channels are similar everywhere. Particularly in the consumer lines of insurance, direct marketing and bancassurance are likely to play increasingly important roles. The principal advantage of direct and bank distribution of insurance is that the marketing costs tend to be lower than for traditional insurance distribution channels such as brokers and agents. The result is that direct and bancassurance marketers can offer lower prices and higher rates of investment return on savings-based products, giving them an important competitive advantage. In the commercial property-casualty lines, it is likely that intermediaries such as brokers and agents will continue to play an important role in helping commercial buyers access the increasingly complex and global markets for risk transfer.

1.4.2 Local Differences

Taking into account the previous global similarities, Marshall McLuhan's (1962) prediction in the mid-1960s that world would become a "global village" seems for many countries to be happening as a result of globalization. However, even if globalization seems to make the world a more homogeneous or singular place, there are still many local particularities. Indeed, the convergence of insurance markets worldwide and the convergence between insurance and other segments of the financial market are far from complete and many differences among countries continue to exist.

First, a sign of the major differences between the various insurance markets is the semantic heterogeneity of the technical insurance vocabulary. Considering the current scope of international insurance activities, a rapid assumption could be that the English insurance vocabulary is quite similar worldwide. Indeed, the word "premium" is used everywhere. However, there are some significant originalities. For example, "long-term insurance" is used in South Africa and the U.K. as a synonym for life insurance, but this terminology is largely unknown in many other countries. Likewise, to refer collectively to lines of insurance other than life insurance, the terms "property-casualty insurance" or "property-liability insurance" are commonly used in the U.S., the term "general insurance" is used in the U.K. and some former British Commonwealth countries, and the term "non-life insurance" is used in many other countries. Insurance of motor vehicles is called "automobile insurance" in some countries and "motor insurance" or "car insurance" in others. The term "mutual" is used consistently in most countries to refer to an insurance company owned by its policyholders, but there are some prominent exceptions such as France, where there are various types of mutuals that need to be distinguished. Facing such important variation in words and their meanings, we asked our colleagues contributing to this book to use the most common English words in their country and to add at the end of their chapter a lexicon defining them.

Second, each nation state has its own specific regulatory framework for the insurance industry and financial services in general. Although deregulatory efforts such as those in the E.U. have resulted in a greater degree of homogeneity, regulatory differences among countries remain, even within the E.U. In addition, solvency regulations currently vary widely, even among developed nations and there is currently a movement underway by the Basel Committee ("Solvency II") to harmonize solvency regulation. Generally accepted accounting principles (GAAP) also differ significantly across countries. As a final example of regulatory differences, although automobile liability insurance is compulsory in most developed countries, some countries such as Brazil and Thailand have many uninsured drivers.

Third, the role of electronic commerce (e-commerce) in the insurance industry varies widely among countries. Its importance varies according to the likelihood of access to a computer and the Internet (Venard 1999). For example, in many developing countries, a large share of the population has no access to telephones or bank accounts. Therefore, e-commerce has long way to go. Even in developed nations, insurance still tends to be sold mostly through contacts with insurance agents or bank employees, with the Internet primarily used to provide product information or price quotes. An exception to this rule is the increasing importance of direct insurance marketing in several countries, particularly the U.K. and the U.S.

Fourth, countries have very different dominant insurance distribution channels. Brokers are very important in insurance distribution in the U.K., but have limited influence in France. The distribution of insurance products through bank channels remains marginal in many countries such as Germany but is very developed in others, such as France and Spain.

Fifth, all countries have their own political systems, and this diversity has an impact on the insurance industry. For example, at the end of apartheid in South Africa in 1994, a Black Economic Empowerment program was set up. Its main idea was to transfer equity ownership in firms throughout the economy to the black community. The Financial Sector Charter requires that 25% of equity of financial companies be transferred to blacks, since the former segregation implied that most country assets were in the hands of the white minority. Political instability and government corruption in many developing nations hinders the development of the insurance industry, as well as impeding economic growth in general. Radical changes in government fiscal policies, such as those observed in Brazil over the past twenty-five years, have had profound effects on the development of insurance and financial markets.

Sixth, although many developed nations are confronting problems caused by aging populations, it is important to realize that demographic trends are very heterogeneous around the world. Huge variations in population dynamics exist between developed and developing countries, with most developing nations having age distributions that are generally much younger than in the developed world. The consequences of demographics are especially significant for the development of the pension market (and, therefore, the life insurance branch) and the health insurance market. However, significant demographic differences exist even within the developed countries. In Europe, Ireland and France are able to have significant growth in their population, while Italy and Germany are already witnessing a decline of their population. Without taking into account migration, these intra-European differences are even more significant.

Seventh, the portfolio of insurance products in various countries shows important variations. Some countries have a buoyant life insurance market, reaching sometimes two-thirds of the insurance market share. Other markets still have automobile insurance as their main product. Some countries have a low insurance penetration while others are witnessing more important penetration, especially in developed countries. Tax laws, which differ significantly across countries, are critically important in the development of the life insurance market in particular. Demand for life insurance and annuities as savings, investment, and pension products is heavily dependent upon tax incentives provided by governments. Countries also differ significantly in their mechanisms for financing catastrophic risk, with countries like France and Spain having mandatory coverage backed by government reinsurance, while catastrophe financing is more ad hoc in many other countries (Cummins 2006).

Both global and local trends imply that world insurance markets are facing common worldwide trends and at the same time showing an important diversity. Both elements are explained in the stimulating and challenging work done by our research team. We hope that our readers will share this intellectual stimulation as they follow us in the discovery of the richness and diversity of international insurance markets. We also hope that this book will contribute to a broader and deeper understanding of worldwide insurance markets and the forces that will reshape these markets in the coming decades.

1.5 ACKNOWLEDGEMENTS

The editors are grateful to the authors of the chapters appearing in the book for their insights on insurance markets and their tireless efforts to produce high quality content for the book. We also benefited greatly from administrative assistance from Annick Bertolotti, Heather Calvert, Mélinda Schleder, and Claire Gernigon. Copy editing for a multi-authored book is extensive and was very capably carried out by Leandra DeSilva, Kirsten Ellis, Heather Harris, and Caryl Knutsen.

The authors are grateful for financial support for this project from the Alfred P. Sloan Foundation's Industry Studies Program. We also acknowledge financial support from the S.S. Huebner Foundation for Insurance Education of the University of Pennsylvania. Both financial and infrastructure support were provided by Audencia Nantes School of Management and the Wharton School of the University of Pennsylvania.

1.6 REFERENCES

Barnet, J. Richard and John Cavanagh, 1994, *Global Dreams: Imperial Corporations and the New World Order* (New York: Simon & Schuster).

Bauman, Zyggmunt, 2000, *Globalization* (New York: Columbia University Press).

Cummins, J. David, 2006, "Should the Government Provide Insurance for Catastrophes," *The Federal Reserve Bank of St. Louis Review*, forthcoming.

_____, David Eckles, and Hongmin Zi, 2005, "Exporting Best Practices: Are Foreign-Owned Insurers More Efficient in the U.S. Life Insurance Market?" Working Paper, The Wharton School, University of Pennsylvania.

_____ and Ran Wei, 2005, "Convergence and Contagion in the U.S. Banking and Insurance Industries: Evidence from Operational Risk Events," Working Paper, The Wharton School, University of Pennsylvania.

Dekle, Robert, 1998, "The Japanese 'Big Bang' Financial Reforms and Market Implications," *Asian Economic Review* 9: 237–249.

George, Susan and Fabrizio Sabelli, 1994, *Faith* and Credit: The World Bank's Secular Empire (New York: Westview Press).

Held, David, Anthony McGrew, David Goldblatt, and Jonathan Perraton, 1999, Global Transformations: Politics, Economics and Culture (Stanford, CA: Stanford University Press).

Hoshi, Takeo and Hugh Patrick, 2000, *Crisis and Change in the Japanese Financial System* (Boston: Kluwer Academic Publishers).

Jomo, Kwame Sundaram and Nagaraj Shyamala, 2001, *Globalization versus Development* (New York: Palgrave Macmillan).

Litan, Robert E., 2001, "Economics: Global Finance," in Peter J. Simmons and Chantal de Jonge Oudraat, eds., *Managing Global Issues: Lessons Learned* (Washington, DC: Carnegie Endowment for International Peace).

McLuhan, Marshall, 1962, *The Gutenberg Galaxy: The Making of Typographic Man* (Toronto: University of Toronto Press).

Naisbitt, John, 1994, *Global Paradox* (New York: William Morrow and Company).

Ohmae, Kenichi, 1990, *The Borderless World: Power and Strategy in the Interlinked Economy* (New York: HarperCollins).

Osterhammel, Jürgen, and Niels Petersson, 2003, *Globalization. A Short History* (Princeton, NJ: Princeton University Press).

Schirato, Tony and Jen Webb, 2003, *Understanding Globalization* (London: Sage Publications).

Scholte, Jan Aart, 2000, *Globalization. A Critical Introduction* (New York: Palgrave Macmillan).

Simmons, Peter J. and Chantal de Jonge Oudraat, 2001, *Managing Global Issues: Lessons Learned* (Washington, DC: Carnegie Endowment for International Peace).

Steger, Manfred B., 2004, *Rethinking Globalism* (Lanham, MD: Rowman & Littlefield).

Swiss Re, 2005, "World Insurance in 2004: Growing Premiums and Stronger Balance Sheets," *Sigma*, No. 2 (Zurich, Switzerland).

Venard, Bertrand, 1999, "The Influence of the Information Highway on the Insurance Management," *Geneva Papers of Risks and Insurance, Issues and Practice* 24: 189–202.

Watson, Iain, 2002, *Rethinking the Politics of Globalization* (Aldershot, Hampshire, U.K.: Ashgate).

2 The United States Insurance Market: Characteristics and Trends

Loftin Graham
University of Pennsylvania

Xiaoying Xie
California State University at Fullerton

2.1 INTRODUCTION

The United States (U.S.) is home to the largest insurance market in the world. With over a trillion dollars in premiums written in 2003 (approximately 9.6 percent of gross domestic product (GDP)), insurance operations from the U.S. generated over 35 percent of the worldwide total, a market share in excess of the next four largest countries combined (Insurance Information Institute (III) 2005d, p.1).[3] More than half of the 100 largest publicly traded insurance firms in the world are traded on U.S. exchanges. In 2003, seven of the top ten global insurance brokerage firms were U.S. companies (III 2005b, p. 4). The insurance sector is also a significant source of employment, comprising, on average, 2.1 percent of the entire U.S. workforce in the years 1994 to 2003 (III 2005b, p. 13). In 2003, insurers alone provided employment for over 1.4 million U.S. workers, while another 840,000 workers were employed at insurance agencies, brokerages, and at other firms with insurance-related activities.

Recent changes in the rules for financial entities in the U.S., heightened awareness of terrorist threats, and ever-expanding technological innovations have all contributed to an insurance system that is changing in important ways. In what follows, we discuss the history, regulatory environment, major market segments, and trends of U.S. insurance markets. In particular, the first section discusses the history of the U.S. insurance market from its inception to the current period, including the key developments that have given rise to the insurance markets and institutions that

---

[3] The authors would like to recognize the important use of data and information from various U.S. organizations, especially the National Association of Insurance Commissioners (NAIC), A.M. Best, the Insurance Information Institute (III), and the American Council of Life Insurers (ACLI). The analyses presented in the chapter are not endorsed by any of these organizations and any mistakes are the sole responsibility of the authors.

exist today. In the second section, we provide a description of the relevant institutions and regulations that make up the business environment in which insurers must currently operate. Sections 3 and 4 provide a detailed quantitative overview of the current state of the two primary insurance market segments in the U.S.: life-health (L-H) insurance and property-liability (P-L) insurance. The chapter ends with a brief discussion of important current issues and trends.

2.2 HISTORY OF THE U.S. INSURANCE MARKET

2.2.1 In the Beginning

Like many of its first institutions, the early insurance establishments of the United States reflected the influences of 18th century England. In 1720, the English Crown granted monopoly status among the colonies (and in Great Britain itself) to two British stock insurance companies: The Royal Exchange and The London. Perhaps as a consequence of this impervious barrier to entry and the joint lack of expertise and capital necessary for successful stock company operation, the first insurance organizations in North America championed the mutual principle embodied in the friendly societies of Great Britain rather than the profit motive of stock companies.[4] Mutual insurance companies have the distinguishing characteristic that they are owned and operated by and on behalf of their policyholders, in contrast to stock organizations which operate primarily for the benefit of shareholders.

The first insurance company established in the American colonies was a mutual insurance company called the Friendly Society. Founded in Charleston, South Carolina in 1735 while the country was still under British colonial rule, this organization survived for only five years (Baranoff 2004). Seven years later, Benjamin Franklin and other prominent Philadelphians established a mutual insurance organization called the Philadelphia Contributionship for the Insurance of Houses from Loss by Fire, the oldest continuously operating insurance company in the country. In 1759, the Presbyterian Ministers Fund, an organization financed initially by donor contributions and later by policy premiums—became the first life insurance company in the colonies by providing a form of life insurance for Presbyterian ministers.[5]

Not until 1792, over a decade after the American Revolution, was a stock insurance company established in the country: the Insurance Company of North America. This first and earliest venture was short-lived and largely unsuccessful, selling only six policies in the first five years of operation (Black and Skipper 2000,

---

[4] The term "friendly societies" refers to the non-governmental social institutions that arose in eighteenth and nineteenth century England, mostly among the poorer, disadvantaged classes. These organizations had the purpose of spreading the various financial risks associated with indigence, including sickness, disability, and employment (among others) across large groups of workers. For further information on the history of friendly societies, see Gray (2005).

[5] The Presbyterian Ministers Fund was first organized by the Presbyterian Synod of Philadelphia in 1717 but did not become incorporated until 1759.

p. 52). In 1812, a stock company, the Pennsylvania Company for Insurance on Lives and Granting Annuities was chartered, becoming the first company to sell annuities and life insurance policies to the general public in significant amounts. Prior to 1810, the focus in the nascent American insurance industry had been primarily on marine insurance, though many companies were chartered to deal in other lines like fire and life insurance as well (Baranoff 2004). The first known reinsurance arrangement on record occurred in 1813, when the Union Insurance Company ceded some of its fire risk exposure to the Eagle Fire Insurance Company of New York (III 2005b, p. 137).

2.2.2 Early Insurance Regulation

Early insurance regulation in the United States developed along lines similar to that of colonial era government generally.[6] The several colonies had been established through different dispensations of authority under the English crown: some colonies existed as Royal provinces, while others more proprietary in nature were owned by wealthy individuals or groups. Each colony had its own government which operated on a basis that was largely independent of the governments of the other colonies. Within each individual colony, government institutions mirrored those of their mother country, England. The independence that the American colonies later received through revolution from England was jealously guarded. The new nation, the United States of America, continued to have a great deal of discretion in performing government functions. It is perhaps for this reason that government institutions in the U.S. generally, insurance matters not exempt, developed differently from the integrated way in which they existed and subsequently developed in England.

State versus Federal Jurisdiction

Regulatory authority for insurance companies in the United States developed from the ground up on a state by state basis. This was consistent with the initial balance of power between state and federal governments which, problematic as it proved to be in pursuing independence from Britain, persisted following the Revolutionary War and the establishment of a national Constitution. The states were vigilant in preserving as much power as they possibly could, relegating to the federal government only those powers deemed necessary for the survival of the newly formed republic. While the federal government was granted the authority to regulate interstate trade, insurer operations, like most business organizations at the time, initially existed primarily on a local scale and therefore, fell under the jurisdiction of the states and municipalities.

Left on their own to regulate the commerce within their boundaries, several states began to regulate their domestic insurance operations. Beginning with Massachusetts in 1807, followed by New York in 1827, the states began to implement statutory reporting requirements for insurers (D'Arcy 2002, p. 249). Premium taxation was another form of early insurance regulation pursued by the

---

[6] For further discussion of the history of insurance regulation in the U.S., see D'Arcy (2002).

states. This time it was New York that led the way in 1824 and Massachusetts that followed in 1827.[7]

One consequence of the early influence of local government on the regulation of the insurance industry was the proliferation of protectionist policies among the states. These policies involved adopting measures that would protect the interests of one state's domestic companies from competing insurers domiciled in other states. Massachusetts was the first to adopt protectionist tactics when, in 1827, it imposed differential premium tax rates on in-state versus out-of-state insurers. Other states reciprocated with similar measures. This early protectionism adopted by the states created barriers to geographic diversification of risks that might have otherwise naturally resulted from fair competition between in-state and out-of-state insurers. The devastating effects of a catastrophic fire in New York City in 1835 convinced the industry of the importance of geographic diversification. This fire, the magnitude of which was brought on by an inadequate water supply and insufficient manpower, consumed almost 700 mostly commerce-related buildings and caused damages estimated at between 15 and 26 million in 1835 dollars (Baranoff 2004). As a result of the fire and the local, undiversified nature of insurers in that era, nearly all (23 of 26) of the fire insurers operating in the state went bankrupt. This effectively wiped out the P-L coverage in New York since fire insurance was the primary P-L coverage at that time. One result of this catastrophic event and its impact on the insurance industry was the gradual replacement of insurance companies operating only locally with national companies that relied on networks of local agents.

The Growth of State Authority

The disastrous fire of 1835 in New York City generated concern about insurer solvency and it was not long before the states were adopting regulations requiring insurers to be more financially robust to the hazards of large losses. Beginning with New York in 1849, regulations were adopted by the states requiring minimum initial capital amounts for insurance company formation.[8] Two years later in 1851, New Hampshire established the first state regulatory body, a state insurance department, charged with oversight of that state's insurance companies (III 2005b, p. 137). Other states followed shortly thereafter: Massachusetts in 1855, Rhode Island in 1856, and New York in 1860.[9] This basic arrangement, regulation of insurance via an appendage of the state government, eventually became the model adopted by states across the nation. The development of state insurance departments is thought to have contributed to increased consumer confidence in the industry and to have boosted

[7] Massachusetts had levied the first form of taxation on insurers in 1785 via a stamp tax (D'Arcy 2002, p. 249).

[8] For New York the required deposit amount was $100,000 (Zanjani 2004, p. 11).

[9] Vermont also organized an insurance department in 1852, the same year as New Hampshire. However, it was "administered by the Secretary of State and the Insurance Commissioner was the State Treasurer," whereas the New Hampshire Department of Insurance was administered by a board of three commissioners. Similarly, Indiana and Mississippi both organized some form of an insurance supervisory body prior to 1860, but supervisory authority resided with preexisting state authorities as opposed to residing with a newly formed organization head (Brearley 1916, pp. 267–69).

sales of life insurance policies during the second half of the 19th century (Murphy 2005).

While early in the nation's history the states operated unencumbered, it was only a matter of time before the relative roles of the state and federal governments with respect to regulation of the insurance industry would need to be further clarified. The first big milestone for federal insurance regulation was the 1869 U.S. Supreme Court decision *Paul v. Virginia* (1869). In this landmark case, the court held that, unlike typical economic commodities which are "subject to the trade and barter," insurance policies are "like other personal contracts between parties," namely local in nature and subject, therefore, to the laws of the state or locality in which the contracts are signed. This somewhat controversial decision had a profound impact on the regulatory system for the insurance industry because it effectively established state (as opposed to federal) preeminence in matters regarding the regulation of insurance, including insurance activities that crossed state borders. In contrast, other forms of commerce between the states were clearly subject to federal authority.

By the end of the 19th century, most states had insurance regulations that included minimum capital or deposit requirements, investment restrictions, annual financial reporting, and annual liability valuations. In addition, state insurance regulators typically had the responsibility of conducting company audits and of taking corrective action in the event of insolvency.

2.2.3 Early Insurance Market Developments

Organizational Forms & Fraternal Insurance

Until the late 19th century, insurance companies assumed one of two primary organizational forms: the stock form or the mutual form. While mutual companies may initially have experienced something of an advantage due to the monopoly granted to stock companies loyal to the Crown, following the revolution neither organizational form had a distinct advantage over the other from a regulatory perspective. Notwithstanding this fact, most of the successful early insurance ventures were stock companies. In the 1840s, however, raising capital for stock company formation became difficult due to financial hardships brought on by the panic of 1837. This panic and the resulting period of economic depression were brought on by a speculative fever that enveloped the nation and by President Jackson's decision to stop accepting banknotes instead of specie as payment for public lands as of August 15, 1836 (the so-called Specie Circular). Western demand for the hard currency necessary for continued expansion (e.g., railroad and canal development) coupled with the fact that many banks had insufficient specie to cover their paper commitments resulted in bank runs and subsequent failures. An important consequence of this economic environment was that capital became increasingly scarce. Under such circumstances the mutual company organizational form had a greater relative appeal than the stock form (Murphy 2005). However, this relative advantage did not last long. Beginning with New York in 1849, the states began to impose initial capital requirements on newly forming insurers. Due to the greater difficulty that mutual insurers have in raising capital, this change in the regulatory environment was followed, during the latter half of the 19th century, by a sharp decrease in the number of new mutual insurers formed.

The capital requirements imposed on mutual and stock insurers beginning in the mid-1800s contributed to the increased popularity during the late 19th and early 20th centuries of a third organizational form for insurance companies: the fraternal form.[10] Fraternal insurance came into being early in the nation's history in conjunction with fraternal benefit societies—organizations formed along ethnic, religious, and occupational lines and charged with meeting the social and economic needs of organization members and their familial dependents (Murphy 2005). These fraternal organizations provided some of the earliest group-like insurance coverage for accidents, sickness, death, and disability income losses.

Under the fraternal form, insurers operated on a non-profit basis, had representative forms of government, and frequently covered losses via member assessment: requesting funds of its members as needs (i.e., "insured" events) arose.[11] This last characteristic set fraternal insurance organizations apart from the legal reserve organizational forms (stock and mutual insurers) each of which were (and are) required by regulation to collect premiums in advance and to hold reserves in anticipation of future insured events. In contrast, early fraternal organizations were not required to hold assets and, as a consequence, were technically and perpetually insolvent.

The prevalence of the fraternal organizational form, which was largely free of initial capital restrictions, increased rapidly during the latter decades of the 19th century as large waves of immigrants settled in the new nation and sought ways of managing the risks that they faced within their own communities. By the early 1890s, fraternal insurers had captured more than half of the private life insurance market, a marked contrast to 30 years earlier when fraternal insurance was rare to non-existent (Zanjani 2004). Notwithstanding this early success and rapid growth, it was not long before competition and state regulation effectively reversed this trend. Beginning in the early 1900s, fraternal insurers came increasingly under the scrutiny of state regulators. By 1920, nearly 80 percent of the population lived in states with solvency requirements for fraternal organizations. This increase in regulation, accompanied by additional competitive pressures associated with the introduction of products such as industrial life insurance, likely resulted in a sharp decrease in the market share held by fraternal insurers. Group insurance and workers' compensation laws are thought to have also contributed to this decline such that by 1930, the market share of fraternal insurers for private life insurance had dropped from its earlier high of more than 50 percent to a meager 8 percent. This organizational form has subsequently become almost obsolete.

Competition and Premium Adequacy

Following the New York City fire of 1835, insurers began to show increasing interest in geographic diversification of their exposure to catastrophic losses. With time, they found that they could achieve this by using networks of agents operating across wide geographic areas. By 1860, the industry had largely assimilated this

---

[10] The emergence of this organizational form is typically associated with the 1868 formation of the Ancient Order of United Workmen (AOUW). For more detail on the fraternal form, see Black and Skipper (2000), pp. 254–56 and Zanjani (2004).

[11] With time, some fraternal organizations adopted policies similar to those of legal reserve insurers (Zanjani 2004, p. 7).

principle: whereas local operations had predominated originally, geographically diverse networks had become much more common. An important consequence of this trend towards greater geographic diversification, however, was an increase in the level of competition for insurance business within states and localities. As competition increased, premiums dropped and premium adequacy and its corresponding impact on insurer solvency became a notable concern. Efforts by the industry to cooperate by setting prices at adequate levels largely failed. The Great Chicago Fire of October 1871 and the Boston fire of November 1872 ($95 million and $50 million in insured losses respectively) resulted in the failure of a large number of insurers and seemed to provide the industry with concrete evidence that premium rates were inadequate. Cooperative price setting at the local level became common during the 1870s and 1880s, resulting in a period of relative financial success for insurance firms.

It was not long, of course, before policyholders began to put pressure on their state governments to enact antitrust legislation to limit the ability of insurance companies to set prices. These efforts were largely neutralized by the insurance industry's creation and sponsorship of private rating bureaus to set rates. The persistence of higher premium levels and price stability during the late 19th and early 20th centuries provided the industry with a cushion sufficient to weather the catastrophes of the day. While the $225 million in insured fire losses associated with the San Francisco earthquake of 1906 exceeded the combined losses of the Chicago and Boston fires by 50 percent, the number of insurers that ceased business as a result of the catastrophe amounted to only one-fifth (20 out of a combined total of 100) of those closed due to earlier disasters (Baranoff 2004).

Growth and Industry Reform

Coincident with the drastic decrease in the influence of fraternal organizations near the turn of the century, the life insurance industry as a whole experienced a period of rapid expansion. During the decade from 1910 to 1920 alone, for example, legal reserve companies' (i.e. stock and mutual insurers') life insurance in force increased by 158 percent while among fraternal insurers' there was a mild 9 percent decrease (Zanjani 2003). The latter half of the 19th century also witnessed the expansion of U.S. insurance operations in international markets, with most of the growth occurring during the period from 1885 to 1905. For example, by 1900, the Equitable Life Company had operations in almost 100 different nations and territories (Murphy 2005).

The rapid growth of the industry during the second half of the 19th century was accompanied by an increasing concern over corruption and malfeasance. In response to news reports of corruption among industry executives, in particular, among the executives of the Equitable Life Company, the New York state legislature commissioned the Armstrong investigation. This investigation unearthed a host of abuses and paved the way for regulatory reforms that would spill over into the regulatory activities of states across the nation. In 1910, again in New York, another investigation was commissioned, this time with the purpose of addressing the problem of insolvency in the P-L industry, in particular, the problem of insolvency in the fire insurance industry. This investigation, instigated by state Senator Merritt, against a backdrop of insolvencies precipitated by the catastrophic 1906 fire in San

Francisco, concluded that state-regulated rate setting was justified by a need to ensure that reserve levels remained high enough to avoid insolvency (D'Arcy 2002, p. 251). By the early 1920s, within only a decade and a half of the Merritt Committee report, almost thirty states had implemented some form of rate regulation (Baranoff 2004).

The rate regulation that was implemented in the first half of the 20th century by many state insurance departments followed earlier failed attempts to address the problem of insurer insolvency. These efforts endeavored to keep premiums at high enough levels to cover costs and ensure profits through extra-legal agreements between insurers. The earliest efforts to establish voluntary pricing agreements were again repeated following the 1835 fire in New York, but these voluntary agreements were ineffective due to the low costs of market entry and ultimately, they failed to have the desired effect (D'Arcy 2002, p. 250–51).

McCarran-Ferguson and the NAIC

In the decades that followed the Supreme Court decision in *Paul v. Virginia*, the insurance industry operated under a different set of rules from other industries whose business practices were subject to federal scrutiny. This absence of federal jurisdiction gave state officials unchecked power in insurance matters and left the insurance industry exposed to certain forces that were partially mitigated by the threat of federal intervention in other types of commercial activities. Low entry costs and the ease of incorporation, in addition to the often state-encouraged collusion among insurers to keep prices at levels deemed adequate for the risks involved, created an environment in which corruption flourished. The criticism of rate competition and support for state-regulated rating bureaus contained in the Merritt Committee report of 1910 strengthened the bond between state regulators and the insurance industry.

In 1944, this trend was investigated when, in response to flagrant price fixing by a cartel of 196 fire insurance companies (the South-Eastern Underwriters Association (SEUA)), the Supreme Court reversed the earlier decision of *Paul v. Virginia*. In this case, the Court ruled that the insurance industry should, like other commercial industries, be subject to federal regulation when conducting business across state boundaries. This decision appeared to call into question the appropriateness of the state-based system of insurance regulation that had developed during the previous period of over 70 years. Within eighteen months of this ruling, however, Congress passed the McCarran-Ferguson Act (1945) effectively reestablishing the preeminence of state authority in procedural matters regarding the regulation of insurance companies: federal involvement would be limited to situations in which state regulations and/or procedures were found to violate federal antitrust law (Dorfmann 1998). With some exceptions, it is this general relationship between federal and state regulation of insurance that has persisted to the present time: insurance companies are subject to the laws of the state in which they are domiciled and operate, but this state regulatory authority exists against a backdrop of broad federal regulatory influence that must be respected by and reflected in state laws.[12]

While disputes about state versus federal regulation have continued to the present day in insurance as well as in most other areas of U.S. jurisprudence, the conflicts have not been overwhelming and non-governmental institutions have developed to mediate the occasional frictions. For example, in response to public frustration over the lack of regulatory uniformity across different state insurance regulations, state insurance regulators organized the National Association of Insurance Commissioners (NAIC) in 1871, a private, non-governmental association charged with promoting minimum standards and some level of uniformity in state regulatory requirements. This is accomplished primarily by developing "model" laws, those that embody reasonable principles on which to base statutory requirements, that can be adopted by the states individually. One area of particular importance and significant activity for the NAIC has been the development of uniform standards and procedures for issues relating to insurer solvency. With time, more and more states have adopted the model laws drafted by the NAIC. Thus, the industry now exhibits an increasing level of regulatory uniformity. Further discussion of the NAIC is pursued in the section on the current regulatory environment below.

Distribution Systems, Group Insurance, and Guaranty Funds

Beginning in the late 1860s, the first formal distribution networks for insurance products had begun to develop. In 1869, the earliest organization of life insurance agents began in Chicago. Six years later saw the emergence of the first agency system for sales in industrial life insurance. During this same period, state insurance commissioners began to meet regularly, beginning at a convention held in New York City in 1871. The following decades were characterized by the creation of the nation's first formal employer-provided pension plan arrangements, thus setting the stage for the introduction and later growth in the demand for group annuities. Other forms of insurance-related employee benefits were emerging around the same time: a group life insurance policy for employees was introduced in 1911 and group annuity benefits were introduced a decade later in 1921 by the Metropolitan Life Insurance Company (American Council of Life Insurers (ACLI) 2004, p. 170).

Prior to the 1930s, health insurance primarily consisted of coverage against the medical costs associated with disability or serious illness, while normal health care expenses were typically paid for out of pocket. During the Great Depression, however, workers' savings were depleted and employment levels sank. Hospitals found themselves unable to collect for the services rendered and expenses incurred. As a result, many hospitals instituted a kind of insurance arrangement where individuals would subscribe to a particular hospital and pay monthly premiums in return for coverage of up to some specified maximum number of hospital days per year. The plans grew and in the late 1930s, the Blue Cross organization was formed. It was not long until a similar payment arrangement for physician fees emerged through the organization of the Blue Shield plans.

---

[12] For further details on the history surrounding these regulatory developments, see D'Arcy (2002).

In the early 1940s, the state of New York enacted the first insurance law respecting the establishment of a state guaranty fund, a sort of state-sponsored reinsurance system with the purpose of limiting the damage suffered by policyholders in the event of insurer insolvency. As with many of the common state regulatory initiatives which have been implemented and tested first in large influential states, this development was duplicated in other state insurance departments, though most waited until the 1970s.

During the decades preceding World War I, the insurance industry was not the only thing changing. The nation itself was also undergoing some significant changes. As employment shifted from agricultural to industrial activities, the role of the employer became increasingly important. Increased industrialization and urbanization of the country's workforce spurred the growth of group insurance arrangements, including employer-provided group health insurance (Bluhm 2000). Social interest in mitigating the hazards that often accompany industrial employment eventually gave rise to employer liability law and workers' compensation law. These, in turn, influenced the kinds of benefits employers provided for their workforce, including those provided through group insurance contracts.

2.2.4 Post-War Developments

The Changing Asset Mix

During the period following World War II (WWII), the insurance industry experienced some significant changes.[13] Foremost among these changes for insurers was an expansion in the range of assets available for investment.[14] Following the War, business activity in the nation increased and competition for available capital resulted in an increase in the number and sizes of corporate bond issues. Housing demand also picked up and borrowers sought sources of capital to finance the increased demand. Prior to WWII, life-health insurers were, on average, heavily invested (over 50 percent of financial assets) in government securities. However, shortly after the war, they began to reduce their U.S. Treasury holdings in order to increase their holdings of corporate bonds and mortgage loans. From 1945 to the early 1960s, for example, the respective proportions of insurers' financial assets in U.S. Treasuries, corporate bonds, and mortgages went from 50, 25, and 15 percent respectively to approximately 4, 40, and 35 percent. Since the 1960s, these proportions have remained relatively stable for corporate bonds while Treasuries and mortgage loans have lost ground (approximately 3 percent and 7 percent respectively in 2003) to other asset classes (see also the sections on financial performance of the major industry segments below).

Due to the preeminence of solvency considerations in the regulation of the life-health insurance business, state regulators did not generally allow investment in common stocks until the second half of the 20th century (e.g., 1951 for the state of New York), and, even then, imposed stringent restrictions on the stocks that would be considered acceptable for statutory accounting purposes, including assessments of

[13] The post-WWII section relies heavily on Wright (1991).

[14] For a more complete discussion of life insurer investment practices, see Cummins (1977).

solvency.[15] Corporate bonds and mortgages were also required to meet certain standards of financial strength (e.g., bond quality and value-to-loan ratios) in order to be eligible for insurer portfolios and regulators commonly imposed minimum standards for diversification among risky asset classes (Wright 1991, pp. 76–77).

Variable Annuities

Generally speaking, life insurance during the 1950s and 1960s was a fairly profitable financial endeavor due to two factors: better than assumed mortality and favorable interest rates. Relative to pricing assumptions, policyholder mortality experience was good as a result of 20th century improvements in living conditions and healthcare, primarily medical advances and the proliferation of antibiotics. In addition, interest rates increased following WWII, so insurers were earning significantly more on their assets than pricing assumptions had assumed. However, as these conditions persisted and the financial awareness of the populace increased, potential customers became increasingly interested in sharing the high investment returns earned by insurers. As a consequence, term insurance contracts (insurance contracts providing insurance for a fixed period or "term" for a fixed, level premium with no savings element) became more popular during this period than they had been previously, increasing as a percentage of new insurance purchases from 31 percent in 1955 to 41 percent in 1960, where it remained throughout that decade. New consumers were choosing to purchase term insurance and "invest the difference" in mutual funds and common stocks, assets that provided the high returns of the times.

This change resulted in a response from insurers who were eager to maintain their relevance with customers. They began to sell mutual funds through their extensive distribution networks and also became more innovative in the kinds of insurance products they marketed. During the 1950s and 1960s, insurers began to develop and market products that would merge concerns about income adequacy in retirement with a desire to participate in the high returns that were then available in the equities markets. In 1954, the first variable annuity product for the general public was brought to market. A variable annuity is one for which accumulations (interest credited on past premium considerations) depend on the performance of some pre-specified and specifically targeted portfolio of assets as opposed to depending on a fixed rate or on some other standard return benchmark.[16] While variable annuities mitigated the concerns of investors interested in participating in the market, they also resulted in additional regulatory burdens since they sufficiently resembled other types of investment products and consequently required oversight by the Securities and Exchange Commission (SEC).

Separate Accounts

Another important development in the insurance industry during the years following WWII was a 1949 Supreme Court ruling that made pensions a part of collective bargaining contracts. The immediate impact of this ruling was to increase public interest in pension plans. A direct result was that insurer-administered pension plans

[15] For example, in New York, the stock had to be listed on a major exchange and had to pay dividends in each of the 10 years preceding purchase (Wright 1991, p. 76).

[16] For a more detailed description and discussion on variable annuities, see Black and Skipper (2000), pp. 174–77.

doubled during the 1950s. But, since insurers were restricted, relative to other financial services companies, in the share of their assets that they were permitted to invest in equities, they soon lost market share to groups such as commercial banks that could provide clients with greater expected returns.[17] Efforts on the part of insurers to eliminate their disadvantage in this respect resulted initially in only modest changes to these restrictions. For example, the limits on equity holdings were raised from 3 percent to 10 percent.

Notwithstanding these early improvements, insurers were eager to make better progress on the matter and it was not long before insurers sought and achieved relief from the strict rules that limited their ability to attract pension fund assets. The concept of separate accounts was the solution that emerged. Beginning with Arkansas in 1959, state regulators began to allow life insurers to account separately for contracts in which investment risks would be born by the policyholder. These separate accounts do not have the restrictions on equity holdings typical of other accounts involving traditional life insurance contracts. Assets for which the policyholder chooses the assets to back the contract, bears the subsequent risk associated with his choices, and reaps the return are called separate account assets, while all other assets held by an insurer constitute its general account assets. Because the customer agrees to bear all investment risks, separate account assets are not subject to the same restrictions as general account assets (e.g., they can be invested entirely in common stocks) and are not backed by the insurer's capital and surplus. Both variable annuities and most pension fund assets fall into the separate accounts category. By the mid-1960s, several states had adopted measures allowing for separate accounts and the notion subsequently became thoroughly incorporated into state insurance regulation across the nation (Wright 1991, p. 78).

Disintermediation Wave I

The 1960s also saw the beginning of an insurance phenomenon that would radically alter the types of products that the industry would offer going forward. Insurers had had a preview in the preceding decade of the sensitivity of investors to the difference between interest credited on their insurance policy assets and those offered by financial markets. In late 1966, market interest rates jumped above the fixed maximum rates that insurers could charge policyholders for borrowing from their insurance policy assets, resulting in a wave of policy loan withdrawals as policyholders sought to reap the higher returns that outside financial markets offered (Wright 1991, p. 81). This wave of disintermediation (when policyholders borrow money against their policy assets and then invest at a rate higher than the borrowing rate) resulted in heavy losses for insurers who were forced to liquidate assets at a discount at a time when return prospects were the highest. In 1969, another wave of withdrawals hit as inflation pushed interest rates up around the 7 percent level, well above the typical state statutory policy loan rate caps of 5 percent or 6 percent, for six consecutive quarters.

---

[17] Insurers were typically restricted to investing a maximum of 3 percent of assets in stocks, whereas commercial banks had no limit and could potentially invest a full 100 percent of pension plan contributions in stocks.

Dominance of Auto Insurance
The development of technology and industrial innovation had a great influence on the P-L insurance industry. Before WWII, the industry was dominated by fire insurance. About 30 percent of total P-L insurance premiums came from fire insurance until 1943 (Joskow 1973). After WWII, the use of automobiles became more prevalent and state laws required auto liability insurance. Consequently, automobile insurance became dominant in the P-L industry.[18] Auto insurance accounted for 47 percent of total P-L insurance premiums in 1971 and 43.6 percent in 1989 (Cummins and Weiss 1992). The percentage of fire insurance dropped to 6.3 percent in 1971, less than 5.1 percent in 1981, and less than 3.4 percent in 1989 (Joskow 1973; Cummins and Weiss 1992). Automobile insurance is still the most important line of P-L insurance in the 1990s and early 2000s, accounting for about 50 percent of the total P-L insurance premiums written.

2.2.5 The Seventies and Eighties: A Time for Change

In the 1970s, self insurance became the trend in group insurance—especially on the part of large employers who had sizeable employee pools so as to ensure relative consistency in claims patterns from year to year. The primary advantage of self insurance was lower costs, including gains associated with the elimination or reduction of premium taxes paid. Reaping the benefits of investment earnings on self-insurance reserves was a secondary benefit.

In 1974, Congress enacted the Employee Retirement Income Security Act (ERISA) which fundamentally changed how pension plans were treated. In particular, it increased the competitive position of insurer-provided pension funds by imposing more stringent restrictions on how funds were invested through the establishment of fiduciary responsibility for fund managers. The late 1970s saw insurers begin to provide guaranteed investment contracts (GICs) to employers for their profit sharing and 401(k) plans. These contracts provide a guaranteed return of principal and a fixed or floating rate of return for a specified period of time. This further increased the position of insurers in the pension asset arena.

Disintermediation Wave II
The 1970s and 1980s generally were plagued with episodes of disintermediation associated with illiquidity-related losses and decreased pension plan inflows as GIC business lost market share to the higher returns of government treasuries. Insurers responded to these waves of disintermediation by modifying liquidity standards and investment practices and by introducing new products, the so-called interest-sensitive life insurance products. These products expose contract owners to market interest rates by providing market-linked returns on contract cash values. Increased focus on asset-liability matching and the duration of liabilities had to be rethought given disintermediation risk. The maturity lengths of new bond purchases decreased markedly as insurers sought to improve the liquidity of their asset positions—going from approximately 85 percent with maturities above 19 years in 1980 to less than

---

[18] Many states do not have compulsory liability insurance, but have "financial responsibility laws" which encourage people to purchase auto liability insurance.

40 percent with maturities greater than 10 years in 1985 (Wright 1991, p. 83–84). Insurers also switched to treasuries and regular issue corporate bonds from less liquid private issue bonds. Intense competition among insurers for the best rates resulted in decreased profitability and pressure for insurers to take on more risks. This effectively introduced an opposing force to the trend towards greater liquidity and high asset quality. As a result, expense reduction was increasingly pursued as a means of increasing competitiveness.

Insurance Product Innovations

Insurers innovated in several dimensions during the 1970s and 1980s in response to multiple forces in the changing business environment. Interest sensitive products are products which reflect market returns in the accumulation of policy assets. They responded to consumers' desires to share in the risks associated with market interest rates. Variable life products are products whose assets are invested in an asset mix chosen by the policyholder. These were introduced in 1976 to satisfy consumers' desires to have greater control over the management of their policy assets and to reap the rewards of the additional risks they would consequently bear. Technological advances also paved the way for the more flexible premium and benefit provisions characteristic of universal life products, which were promoted by insurers beginning in 1979. These products allowed policyholders to choose and change from among multiple death benefit patterns and to pay premiums when and in the amounts that they desire, subject to policy minimums and maximums. In 1985, a hybrid of variable and universal life policies was introduced into the market.[19] Initially, these new (interest-sensitive, variable, and flexible-premium) products were quite successful, capturing over 30 percent of the whole life insurance market, but then decreased some since market rates have been lower and more predictable than during the more turbulent decades of the 1970s and 1980s.

The 1980s also witnessed the beginning of the managed care revolution in health insurance and the proliferation of flexible benefit plans for employees. These flexible benefit plans allowed employees some amount of choice regarding how their employer-provided benefit dollars are allocated across available benefit categories (e.g., health insurance, life insurance, and retirement plan benefits). High premium growth, especially in the area of healthcare, spurred the development of significant changes (hence "revolution") in how insurers and healthcare providers interacted with each other and with patients. In particular, this era witnessed the creation of organizations structured to control costs through greater management of the healthcare (hence "managed" care) delivery process as well as through increased cost-sharing on the part of insureds.[20]

Financial Services Regulation

In addition to the significant changes that occurred during the 1970s and 1980s in the insurance product mix, the kinds of organizations allowed to sell insurance products began to change in significant ways during the 1980s. Several decades earlier, in

[19] For more detailed information regarding these and other types of insurance, see Black and Skipper (2000), Chapters 5 and 6.

[20] For more information, see Black and Skipper (2000), Chapters 19 and 20 and Bluhm (2000), Chapters 1 and 2.

1956, motivated by concerns about conflicts of interests that might exist within banks that owned or had significant influence over the operations of non-banking businesses, Congress had passed the Bank Holding Company Act (BHCA). This regulation prohibited national banks from either owning non-bank businesses or from engaging in most kinds of non-banking activities.

While the restrictions imposed by the BHCA naturally included insurance sales and underwriting, these restrictions on insurance gradually eroded during the 1980s. This began in 1985 with a series of rulings at the federal level made through the Office of the Comptroller of the Currency (OCC). The OCC ruled in 1985, for example, that banks could sell fixed-rate annuities without violating the BHCA. Two years later, variable-rate annuities were also exempted from the restriction. A decade later, the Supreme Court confirmed these OCC rulings, firmly establishing the legality of selling fixed- and variable-rate annuities by banks (*Nations Bank of North Carolina, N.A. v. Variable Annuity Life Ins. Co.*, also called *VALIC* 1995).[21]

The Emergence of Liability Insurance

Liability insurance became more important in the United States after WWII. Its development was deeply affected by the U.S. tort system which saw a significant rise in tort costs after WWII. Total tort costs account for 2.2 percent of the U.S. GDP in 2003, while it was only 0.6 percent in 1950 (Tillinghast Towers-Perrin 2004). The increase in tort costs was mainly caused by changes in legal doctrine. Prior to the 1960s, most states held the doctrine of contributory negligence, which stipulated that an injured person would be compensated only if he was not at fault and the defendant was wholly responsible for the injuries. However, as more states adopted the doctrine of comparative negligence which stipulated defendants and plaintiffs to share the damages based on the degree of fault, the number of liability lawsuits increased significantly. The situation worsened by the existence of the doctrine of joint and several liability and the existence of punitive damages which intends to punish a defendant's improper conduct (III 2005n).[22] The rising liability costs in product liability, medical malpractice, environmental liability and other types of liability insurance have made liability insurance more important in the economy than it used to be. Certain services and new products such as high risk surgeries and prescription drugs might not otherwise have become available at all without liability insurance. However, the insurance industry often faces unexpected high liability claims, which sometimes threaten the solvency of the industry. For example, the large unexpected asbestos lawsuits in the 1980s created a solvency crisis in the U.S. P-L insurance industry and aggravated the "liability crisis" in the mid-1980s.

No-Fault Auto Insurance

An important change in the U.S. P-L insurance industry in the 1970s and 1980s was the emergence of no-fault automobile insurance. The result of an effort to reform the U.S. tort liability system, the purpose of no-fault insurance was to control automobile insurance costs and improve the efficiency of auto accident claim adjustments. It has two distinct characteristics. First, in an accident, the policyholders will recover

[21] For a survey of the relevant rulings, see Williams, et al. (1997).

[22] III (2005n).

financial losses from their own insurance company, regardless of fault, and second, the right to sue for severe injuries and pain and suffering is limited. Currently, 12 states and Puerto Rico have no-fault auto insurance laws. A typical opposition to no-fault insurance is that it aggravates the moral hazard problem, leading to less cautious lower care drivers and higher accident rates. There is some empirical evidence that no-fault insurance tends to cause higher fatality rates, but more stringent experience rating plans can be used to reduce the incentives of carelessness (Cummins, Phillips, and Weiss 2001). Another issue with no-fault system is insurance fraud. The problem is especially serious in those no-fault states with higher personal injury protection (PIP). The effectiveness of a no-fault system has yet to be seen. Some states have had successful performance, while others with unsuccessful experiences have chosen to repeal the no-fault laws and reverted back to the tort liability system.[23]

Price Regulation and Deregulation

Price regulation in the U.S insurance industry has also changed since WWII. After the passage of the McCarran-Ferguson Act in 1945, most states adopted the bureau rate prior approval system whereby an insurance company must file the rates and rules of its products with state regulators before it can be approved. The purpose of price regulation is to guarantee the solvency of insurers and the affordability and availability of various insurance products. Two states which did not use the prior approval system were California and Illinois. California adopted a no-filing system, which means insurers need not file rates and rules with state regulator, while Illinois abandoned the bureau rate prior approval system in 1971. Both states believed in competitive markets and thus, allowed the market to regulate the rate. In 1980, the NAIC adopted the first model law for a competitive rating system. By 1984, 25 states implemented various forms of a competitive rating system. The deregulation of commercial insurance rates was more complete than that of personal insurance. In fact, most states have passed laws deregulating commercial insurance policy forms and rates. In 2002, the NAIC enacted a model law to deregulate commercial insurance. Insurers can use new rates 30 days after filing with regulators for commercial lines. In 1988, California reversed its trend of price deregulation because of large price increases in auto, home, and business insurance. With the passage of Proposition 103, California converted its competitive rating system to a prior approval system.[24]

Solvency Regulation and Insurance Crisis

In the U.S. P-L insurance industry, insolvencies usually follow the underwriting cycles. In the 1950s and 1960s, underwriting cycles were characterized regularly by a three year "soft market" followed by a three year "hard market" in almost all lines of insurance (A.M. Best 1993–2003). In the mid 1970s, there was an auto insurance crisis due to escalating auto insurance rates. In response to this insolvency crisis, the NAIC proposed a model guaranty fund program which recommended that states adopt a post-assessment approach to financing the claimants of insolvent companies.

---

[23] III (2005i).

[24] III (2005j).

By 1982, almost all states had created guaranty funds. Meanwhile, the NAIC created the Insurance Regulatory Information System (IRIS) to help monitor and identify firms in trouble. In the mid-1980s, a "hard market" for commercial liability insurance occurred, primarily caused by the plummeting interest rates and rising claims cost for liability insurance, specifically the cost of asbestos claims. The commercial liability insurance cycle thus gave rise to the liability crisis in which when certain types of commercial liability coverage became difficult to obtain and the premiums of some lines, such as medical malpractice insurance, increased dramatically.[25] This crisis led to a more complicated solvency regulation system, the adoption of Financial Analysis and Solvency Tracking System (FAST), and risk-based capital (RBC) standards in the early 1990s.

2.2.6 Recent Developments: The 1990s

The Rapid Growth of Managed Care

During the early 1990s, the expansion of managed healthcare continued at a rapid pace in response to medical care inflation rates in excess of 15 percent annually from 1988 to 1990 (Black and Skipper 2000, p. 485).[26] A range of insurer-provider arrangements emerged to facilitate improved management of healthcare delivery and financing, including health maintenance organizations (HMOs), preferred provider organizations (PPOs), and exclusive provider organizations (EPOs). Each of these arrangements involves a different form of financial risk-sharing between healthcare providers and insurers. Also, hybrid plans have become increasingly common.

HMOs are organizations that operate on a per-capita, prepaid basis to provide healthcare to select groups of individuals by contracting with an organized group of physicians and hospitals. This category of managed care, which covers approximately one-fourth of the population, is the largest in terms of enrollment (Black and Skipper 2000, p. 494). Within this category of managed care organizations, there is a wide variety of different types of insurer-provider arrangements ranging from staff model HMOs, where the care providers (i.e., physicians, nurses, etc.) are directly employed by the HMO, to direct contract model HMOs, where the HMO is primarily a financing/insuring organization that contracts with individual physician providers to provide care for those insured through the organization.

PPOs and EPOs are organizations of healthcare providers that contract with insurers and employers to provide services on a fee-for-service basis. Plan sponsors (employers or insurers) typically provide plan participants (enrollees) with financial incentives for receiving their healthcare from the group of providers and negotiate with the providers as a group for reduced fees. EPOs and PPOs differ in the extent to which plan participants are restricted to receive care from the contracted group of providers. EPO plan participants are restricted to only use participating providers, whereas PPO plan participants often have the option of receiving care outside the plan, but must share a greater proportion of the corresponding costs. PPOs are more prevalent than EPOs and grew significantly in popularity among employers during

---

[25] III (2005h). See also III (2005m).

[26] This subsection draws on Bluhm (2000), Chapters 1 and 2 and Black and Skipper (200), Chapters 19 and 20.

the mid-1990s. In fact, the proportion of employers offering PPOs increased from 36 percent in 1993 to 59 percent in 1996 (Black and Skipper 2000, p. 496).

Financial Services Integration

During the 1990s, annuity products, which had been growing rapidly following the development of separate accounts, continued to increase in sales relative to other types of life insurance products.[27] Individual annuity vehicles (IRAs, Keogh plans, etc.) grew especially quickly, consistently outpacing all other life-health insurance products. Among the most important developments of the 1990s was the entry of non-traditional players into the life-health market. Beginning in the mid-1980s, the regulatory barriers that had previously (following the BHCA in 1956) kept the banking sector from competing with insurers were gradually eroded by OCC rulings that granted banks ever greater access to insurance markets.[28] In addition, new types of competitors entered the scene. Investment banks, pension funds, and mutual fund organizations were all eager to participate in insurance markets and began to compete with insurers and traditional banking institutions for annuity and some L-H insurance business. Insurers themselves broadened the scope of their activities by offering products outside their established lines of business.[29]

These conditions resulted in increased competition in the financial services marketplace and precipitated, during the 1990s, a wave of consolidations among brokerage firms and numerous demutualizations among insurance companies. Several existing studies of the empirical evidence seem to suggest that these demutualizations reflected a desire or need on the part of insurers for greater access to capital, something which is limited for mutual relative to stock companies (Viswanathan and Cummins 2003; Mayers and Smith 2002; Butler, et al. 2000; Erhemjamts 2005). For more detailed exploration of some of the likely economic rationales for the existence of stock and mutual organizational forms in the insurance industry, see Mayers and Smith (1988).

The 1990s also witnessed some significant regulatory developments and trends. With the Supreme Court's 1995 decision in *VALIC* (1995) to allow banks to begin to sell certain insurance products again, barriers for banks to enter the insurance industry began to be removed. In 1996, the trend progressed further when a Supreme Court decision (*Barnett Bank of Marion County, N.A. v. Nelson* 1996) ruled that nationally-chartered banks should be allowed to sell insurance nationwide. The removal of these barriers to bank entry into insurance markets resulted in significant financial losses to insurers without a corresponding increase in the aggregate market

---

[27] For a more detailed and complete discussion of financial services integration in the U.S., see Cummins and Santomero (1999).

[28] As early as 1988, under 12 U.S.C. 24, banks were granted "all such incidental powers as shall be necessary to carry on the business of banking." This enabled national banks to sell any insurance or annuity product with important implications for banking business. This included, for example, credit insurance – a form of insurance which protects the creditor in the event that a debtor experiences adverse financial conditions – as well as involuntary unemployment insurance, vendors single interest insurance, double insurance, mechanical breakdown insurance, title insurance, and municipal bond insurance (Williams, et al. 1997).

[29] One study found that approximately one-half of all U.S. life insurance companies offered products which were manufactured by someone else (Black and Skipper 2000, p. 611).

value of the subset of banks eligible to take advantage of the new opportunities (Carrow 2001).[30] These rulings allowed banks to sell insurance, but they were still not allowed to engage in underwriting activities. This changed in 1999 with the passage of the Gramm-Leach-Bliley Financial Services Modernizations Act (GLB), a culmination of the trend towards financial services integration. This act removed the barriers to affiliation that had existed between insurers and other financial services companies (banks and securities firms), thus enabling greater consolidation and integration within the financial services industry.

GLB removed barriers to affiliation between different kinds of financial services businesses through the creation of financial holding companies—legal entities that can engage in any activity or hold shares in companies engaging in any activity that either is financial in nature or that at least poses no risk to the safety of depository institutions or to the financial system in general. Financial services conglomerates must now be organized as financial holding companies with separate subsidiaries for commercial banking, investment banking, insurance, etc. Whereas after 1985, bank holding companies were able to sell some insurance products, since GLB in 1999 they are able to also underwrite insurance. Under GLB, the various financial activities of financial holding companies (and their subsidiaries) are regulated largely as before according to the nature of the activity and by whichever regulatory body is functionally best situated to do so.

The Importance of Alternative Risk Transfer (ART)

The U.S. P-L insurance industry has observed the emergence of the alternative risk transfer (ART) tools in the past several decades, which is the result of the demand of "integrated risk management," efforts to cope with underwriting cycles, and frequent catastrophic losses. ART has become more popular since the early 1990s. It initially referred to the captives, but more recently has expanded to include all non-traditional risk management methods, such as finite risk (re)insurance, multi-line/multi-year insurance contracts, insurance products with multiple triggers, contingent capital arrangements, and securitization which transfer risk via capital market. The U.S. leads the area of alternative risk financing solutions. Though the growth of traditional commercial insurance premiums has been stagnant in the past decade, ART has steadily gained more importance. Captives became popular in the mid 1990s because of the high price of commercial lines of business and favorable regulatory conditions on captives in some states. The events of September 11, 2001 and the resulting hard market in commercial insurance further spurred the use of ART in the U.S. In 2004, there were 524 captives in Vermont, 147 captives in Hawaii, and 114 captives in South Carolina. U.S. domiciled captives' net premiums written reached $8.9 billion in 2003.[31]

---

[30] Specifically, Carrow (2001) finds significant negative returns to insurance company stock prices for event windows surrounding the important rulings that allowed banks to sell annuity and life insurance products. They conclude that these negative returns suggest a decrease in the value of the industry and imply that "barriers to bank entry protected the insurance industry from competition" (p. 147). While insurers lost value, the banking industry did not appreciably gain in value since the stock prices surrounding the event did not change.

[31] Swiss Re (1999) and III (2005k).

2.3 THE CURRENT REGULATORY ENVIRONMENT

Insurance regulatory authority in the U.S. ultimately rests with the federal government, although most operational control of insurance regulation has been ceded to the states. As set out by the U.S. Constitution, the federal government has ultimate authority by the powers vested in Congress to regulate interstate commerce and to levy taxes. Consistent with this authority, Congress enacts laws and provides the funding necessary for implementation of those laws through government agencies. State governments retain authority to regulate all matters that do not pertain to the jurisdiction of the federal government. As discussed in previous sections, a series of Supreme Court decisions resulted in the granting of broad regulatory discretion to the states, including regulation for insurers operating in multiple states, as long as states comply, both individually and as a group, with the general principles and regulations set forth by the federal government. In addition to the direct roles that federal and state governments have in shaping insurance regulation in the U.S., various non-governmental organizations also contribute to the regulatory environment in which insurers must operate.

2.3.1 The Role of the Federal Government

All three branches of the U.S. federal government participate in one form or another in the regulation of the insurance industry.[32] Congress enacts laws regarding interstate commerce and taxation that directly influence insurance companies individually and as a group and which delineate the basic constraints under which state regulators must operate. The courts interpret these laws and adjudicate specific cases, and the executive branch enforces court judgments and often influences implementation of laws through its oversight of government agencies. Federal regulations directly affect insurers through the following government institutions/agencies: the Internal Revenue Service (IRS) of the Treasury Department, the Department of Labor, the Securities and Exchange Commission (SEC), and the Federal Emergency Management Agency (FEMA). The federal government has a less direct influence on insurers through a number of other agencies.[33] Most of these activities do not relate to regulating insurance per se, but rather involve the application of laws and regulations applicable to businesses in general.

The IRS oversees federal taxation and in this role it exerts influence on the insurance industry in two key dimensions. First, the industry is affected by the income tax regulations to which all corporations, including insurance companies, are subject. Second, the IRS shapes tax legislation and policies that have a direct impact on consumer demand for insurance products. Such things as the deductibility of premiums and taxation of insurance proceeds have important implications for their value to potential customers. The Department of Labor interprets and enforces

---

[32] The authors relied heavily on material from Black and Skipper (2000) for this section.

[33] For example, through the Occupational Safety and Hazards Administration (OSHA), the Department of Health and Human Services (DHHS), the Federal Trade Commission (FTC), and the Office of the Comptroller of the Currency (OCC).

ERISA, the country's primary body of regulations for employee benefit plans, which typically involves such insurance products as group life, annuities, and health insurance. The SEC enforces financial reporting standards based on sound accounting principles, for all publicly-traded corporation including insurers. It also has a regulatory role over the sale of variable life insurance and annuity products. FEMA is charged with responding to events deemed to be federal disasters, hazards, and emergencies usually by using federally-provided funds to minimize or ameliorate the financial consequences that such events have for affected citizens.

2.3.2 The National Association of Insurance Commissioners

In addition to the federal institutions, the National Association of Insurance Commissioners (NAIC) has a significant role in the regulation of insurance activities in the U.S. This non-governmental organization, a voluntary association of the insurance department chief officers of every state or territory of the country, was created in 1871 with the initial purpose of coordinating supervision of insurers operating in multiple states.[34] With time the scope of its purposes has expanded so that its primary objectives are (1) the maintenance and improvement of state regulation, (2) the maintenance of a financially secure and healthy insurance industry, and (3) the promotion of fair treatment of policyholders and claimants.

Following passage of the McCarran-Ferguson Act, the NAIC stepped up its push for uniformity of regulation across the states, primarily through the creation and promotion of model laws. NAIC model laws (of which there are currently well over two hundred) have no binding regulatory authority, but rather serve as models for what state authorities should consider implementing as appropriate minimum regulatory standards. Since 1990, a state's failure to meet certain NAIC requirements, which include among other things, adopting and maintaining a minimum set of NAIC model laws and submitting annual disclosures of financial information, can result in withheld accreditation, with all the subsequent inconveniences for state regulators and insurers that such entails.[35] Accreditation is reviewed on a five-year basis and failure to meet the requirements needed for accreditation is rare.[36] In what follows, the influence of the NAIC on state regulations will be frequently apparent and even explicitly noted.

2.3.3 State Insurance Regulation

In the U.S. most insurance regulation is carried out by the states. Insurance companies are regulated primarily through the Office of the Commissioner of Insurance for the state of domicile, but they must also comply with the relevant laws

---

[34] While we refer generally to the chief state position as that of commissioner, in reality the title varies across states. Director and Superintendent are other titles given for the position with similar responsibilities in various states.

[35] For example, accredited states may refuse to accept a non-accredited state's examination reports for its domiciled insurers and choose rather to send a group of accredited-state examiners to carry out an independent assessment.

[36] As of March 2005, for example, New York was the only state that was not accredited by the NAIC (2005).

for each state in which they write policies. In particular, an inter-state insurer must be licensed in each state in which it does business—there is no "single passport" system in the U.S. State insurance commissioners are charged with protecting the interests of consumers in their states and typically respond with laws that regulate insurer investments, charter licensing, policy contracts, and accounting procedures. Insurers also must comply with applicable federal laws regardless of the state or states in which they operate. Whereas the federal government has left it up to the states to regulate insurance matters, the federal government still holds insurers accountable for the federal laws that apply generally to corporations, including those relating to securities transactions, income taxation, racial and religious discrimination, and employment practices.[37]

While the federal government has ultimate oversight in matters involving insurance, it has ceded to the states most of the responsibility for regulating insurance. The precise nature of the power sharing between these two governmental bodies has developed over time. The currently operative division of influence was outlined by the U.S. Congress when it passed the McCarran-Ferguson Act in 1945. This act had the express purpose of defining the respective roles of federal and state governments in regulating the insurance industry. Broadly speaking, the act calls for cooperation, with the federal government wielding influence in the areas that Congress deems to be of greatest general concern (like fair labor standards, employee relations, etc.) and state governments being responsible for taking care of the remaining dimensions in a way that adequately protects the interests of the public.

Congress reserves the right to enact future legislation regarding insurance and to require the application of existing federal regulations to the insurance industry if and when it deems such measures necessary. The practical result of McCarran-Ferguson has been that the bulk of regulatory activity in the insurance industry is carried out by state governments against a backdrop of federal authority which threatens to step in if the states pursue policies that are ineffectual in serving the needs of both corporations and consumers. One important result of this act was to provide an impetus for the industry to develop and implement more uniform insurance regulations across states and territories. This has been one of the primary objectives of the NAIC.

All three branches of state government are involved in the regulation of the insurance industry. State legislatures enact insurance codes which define the legal framework for insurance regulation and taxation. This generally involves regulating the following: organization and operation of a state insurance department, licensing of insurers and agents, creation and operation of state guaranty funds (most states have more than one), and monitoring of financial strength, marketing practices, and products of insurers operating within the state. The courts adjudicate cases between insurers and contract holders, hand out criminal penalties for violations of insurance laws, and provide recourse to insurers against poorly formulated state insurance regulations. The executive branch of state government exerts its influence primarily through enforcement activities and in the person of a state officer (insurance commissioner, superintendent, or director) who is given administrative authority

[37] For more information on some of these topics, see Black and Skipper (2000), pp. 304, 945, 971–74.

over all insurance matters in the state.[38] In most states, this officer is appointed by the state governor. In 12 states, the insurance commissioner is elected rather than appointed.

Formation and Licensing

Licensing requirements vary slightly for mutual versus stock insurance companies and for foreign and alien companies versus domestic ones.[39] In each case, however, formation of a new insurance company or organization of an existing one to do business in another state requires the presentation of a charter describing the insurer's location, name, and intended lines of business, along with information regarding the company's internal organization and the powers to be enjoyed by its various officers. Incorporation follows successful examination of the company by state insurance officials.

Licensing for a new stock insurer typically also involves satisfying a state's minimum capital and surplus requirements and making a deposit with the state. Initial capital requirements range from several hundred thousand dollars to a couple of million dollars, while initial surplus requirements range from 50 to 200 percent of minimum capital requirement (Black and Skipper 2000, p. 950). Licensing for a new mutual insurance company involves satisfying an initial surplus requirement that is roughly equivalent to the capital and surplus requirement of stock insurers. In addition, mutual insurers must establish a minimum number of policyholders. Both stock and mutual companies must satisfy risk-based capital requirements in addition to the minimums described above. Some states have additional requirements for foreign and alien insurers.

Corporate Affiliation and Holding Companies

In the U.S., corporations are allowed to affiliate in many different ways. The relevant corporate structure which facilitates different arrangements is that of the holding company—an entity which can be incorporated, issue shares, and effectively own other corporations, all without maintaining any core productive operations itself. U.S. corporate law allows for a wide range of different types of corporate affiliations. However, historically, there have been two notable exceptions for the insurance industry. First, until recently, insurers were prohibited from affiliating with either commercial or investment banking institutions. This restriction was eliminated by the passage of the Gramm-Leach-Bliley Act (GLB) in 1999. Second, historically, mutual insurers had to be completely owned by their policyholders. Consequently, they have been able to assume only the highest position in the hierarchy of ownership of any holding company arrangement. The recent interest in demutualization, however, has opened the door for a notable alternative: mutual holding companies (MHCs).

An MHC is a corporate structure that was initially conceived in 1987 by the banking industry to combine elements of mutual savings associations and stock savings and loan companies, but which also admits combining aspects of mutual and

[38] For a small number of states, the authority is vested in an entire board as opposed to in a single individual.

[39] Foreign refers to companies domiciled in a different state of the U.S., while alien refers to companies domiciled in other (foreign) countries.

stock insurance organizational forms.[40] Financial services integration and the related increase in competitive pressure during the 1980s and 1990s motivated insurers to consolidate and expand. The accompanying need for additional capital prompted mutual insurers to consider alternative organizational forms (Viswanathan and Cummins, 2003). Beginning in Iowa in 1996, mutual insurers interested in gaining greater access to capital without entirely giving up insurer ownership by policyholders sought permission from their states of domicile to reorganize as MHCs.

MHC conversions involve the creation of a 100 percent policyholder-owned stock holding company which is the direct parent of the newly formed stock insurer that inherits liabilities and assets from the converting mutual company. The stock holding company is held by the MHC, which may issue additional shares of its holding-company subsidiary to the public in order to raise capital. The MHC is required to maintain a controlling interest in its holding company subsidiary and sometimes also in the newly formed stock insurer itself. Erhemjamts (2005) finds that firms with severe liquidity constraints are less likely to demutualize through an MHC conversion than through a full stock conversion.

Registration/Licensing of Agents and Brokers

Sales agents and brokers of insurance products must register and become licensed in each state in which they intend to do business. The primary purpose of the licensing process is to determine or assess the character, experience, and competence of the applicant. The general process involves several steps including: filing an application with the state, receiving formal recognition of sponsorship by every insurer the agent/broker expects to represent, and passing a state exam on insurance fundamentals.

Individuals applying to become agents and/or brokers must demonstrate their good character by receiving a formal certificate of trustworthiness and competence from the sponsoring insurer and by having a notice of appointment sent to the state on his behalf. Competence in insurance fundamentals and state insurance law is demonstrated by successful completion of the state insurance examination. In addition, some states have formal training requirements (e.g., 40 hours) that must be satisfied prior to taking the licensing exam. Licenses terminate automatically with the termination of appointments by insurers and can be suspended or revoked by the state in the case of illegal, fraudulent, or incompetent actions.

Product Approval and Filing

Prior to marketing and selling a product in any given state, U.S. insurers are required to file a detailed description of the policy provisions (called a policy form) and rates with the state insurance department. Policy forms must be approved by state officials while rates must always be filed, but are not always subject to approval by state

---

[40] This corporate structure was first authorized in 1987 by the Competitive Equality Banking Act (CEBA). Clarifying statements were subsequently made by the Financial Institutions Reform, Recovery, and Enforcement Act (FIRREA) of 1989.

officials.[41] State insurance regulations usually specify, by sector (L-H versus P-L), line of business, and product category, the minimum standards and requirements for policy forms with the ultimate purpose of screening products that are unjust, ambiguous, misleading, or unfair. State regulators frequently either adopt or use as reference the many model laws provided by the NAIC regarding policy provisions that should be included and/or prohibited for various types of products across the different lines of business.

Preparation and filing of policy forms, including lobbying state insurance regulators for approval of changes in contract language, can have a significant effect on the flexibility and speed with which insurers are able to meet the changing business needs of their clients, especially in commercial lines of business. In recent years, less traditional mechanisms for transferring risk, alternative risk transfer (ART) mechanisms, including securitizations, finite-risk reinsurance, captive insurance, and self insurance, have competed successfully with insurers for commercial insurance business.[42] These risk-transfer mechanisms are able to operate outside the sphere of state insurance regulations and as a consequence are, unlike traditional insurance arrangements, unencumbered by product form filing requirements. The regulatory burden associated with the state policy form approval processes can be thought of as an implicit tax on insurer-provided commercial insurance services. Evidence exists which suggests that form deregulation for commercial insurance is associated with a significant increase in the amount of business transacted (Butler 2002).

An important dimension of state regulation of the P-L industry involves the regulation of insurance rates. Currently, seven different types of rate regulation systems coexist in the U.S., with "no filing/record maintenance" the most competitive and bureau rate prior approval system the most stringent. The remaining systems are modified prior approval, flex rating, file and use, use and file, and a state prescribed system. Despite the variations among these systems, they are guided by three common principles: adequate (not too low in order to maintain solvency of insurers), not excessive (not so high as to lead to excessive profits), and fairness (not unfairly discriminatory, price differences must reflect expected loss and expense differences). Among the states, California, North Dakota, Hawaii, and West Virginia require prior approval for all business lines, while Illinois and Idaho use "use and file" for all lines.[43]

Marketing Methods and Trade Practices

State regulations often place restrictions on the activities that insurers, agents, and brokers participate in as they market and sell their products. A sample of activities that are typically proscribed includes: rebating (offering extra-contractual inducements), twisting (encouraging policy replacement for personal gain), misappropriation or misuse of policyholder funds, and unfair discrimination. In the

---

[41] For example, life insurance products do not typically require rate approval since sufficient competition is likely to rule out fair pricing, while minimum reserve requirements and expense limitations effectively guarantee premium adequacy.

[42] During the last two decades, the market share of ART mechanisms has gone from 20 percent to about 33 percent, an average yearly increase of about 2.5 percent (Butler 2002).

[43] III (2005j)

event that such activities are suspected, the insurance commissioner is authorized and endowed with sufficient resources to lead an investigation into the matter. In the event that improper conduct is substantiated, the commissioner serves a notice to the offending party and, after a court hearing, may issue a cease-and-desist order.

Agents and brokers can also be censured and their licenses revoked or suspended if they are found to be untrustworthy or incompetent. NAIC model laws for the regulation of insurance advertisement form and content focus on minimum information disclosures and on enforcement procedures. The NAIC has also developed a system that allows states to share information about individuals that are of concern to regulators, including information about regulatory actions taken in the past.

Disclosure Requirements

On an annual basis, insurance companies are required to file, in each state in which they operate, a statement of financial position and operations accompanied by a statement from management of any events or conditions they are aware of that might compromise the usefulness of the report. In addition, insurers in NAIC-accredited states are required to file similar information with the NAIC. These requirements are in addition to and in some ways materially different from the disclosures provided to management for planning purposes, the information provided to the IRS for income tax purposes, and the financial disclosures that stock insurers are required to file with the SEC as publicly traded companies.

The primary purpose of statutory disclosures is to assess the financial strength (i.e., solvency) of the insurer. As a consequence, calculations are made on the basis of the more conservative Statutory Accounting Principles (SAP) rather than on the basis of the Generally Accepted Accounting Principles (GAAP) used for financial disclosures to the SEC.[44] Because of its focus on solvency, SAP calculations emphasize the requirement that insurers be able to meet all of their contractual obligations. For life-health insurance business, this involves ensuring solvency under a range of different possible future scenarios and taking into account such things as future premium considerations and policy persistence. SAP analysis tends to underweight future operations and overweight current position, treating insurers as if they were facing liquidation rather than as ongoing entities, and ignoring the long-term benefits associated with investments intended to enhance future productivity, efficiency, or growth.[45]

State insurance departments use the information provided in statutory filings to establish priorities so they can make the best use of their time and resources by focusing them on the firms that are likely to require the most oversight and/or regulatory attention. Most states rely on automated programs provided by the NAIC, the Insurance Regulatory Information System (IRIS) and Financial Analysis Solvency Tools (FAST), as aids in this process. IRIS involves two phases: a statistical phase and an analytical phase. During the statistical phase, 12 different financial ratios are calculated using the annual statement information that insurers have submitted to the NAIC. These ratios become publicly available along with the

---

[44] Yet another set of accounting standards is invoked for federal income tax purposes.

[45] For further comparison of SAP and GAAP accounting standards, see Black and Skipper (2000), pp. 905–10.

other detailed financial statement information that insurers are required to provide in their annual filings and are used by regulators to flag potential problems. Insurers with four or more ratios falling outside their predetermined ranges (approximately 20 percent of all insurers in any given year) are identified for further, more detailed investigation by state regulators, a process called the analytical phase (Black and Skipper 2000, p. 267). The FAST system is used by the NAIC to calculate 20 financial ratios and an overall score which is also used to flag insurers for further analysis. FAST is also made available to state insurance departments for the purpose of constructing financial reports on individual insurers. During the analytical phase of IRIS, teams of regulators selected from different state insurance departments more carefully review the annual statement information of flagged insurers and establish priorities for further action and/or investigation. The detailed information produced during the analytical phase is not generally made available to the public. In the event that such an analysis suggests an insurer is experiencing serious problems, accredited states form a panel of on-site examiners (again representing multiple states) to conduct further investigation.

In addition to using the IRIS and FAST systems, regulators also rely on a special solvency measure in order to determine whether intervention is needed and the extent of the need. For a given insurer, this measure is calculated as the ratio of actual capital and surplus levels to an amount of capital deemed adequate for the particular type and mix of the insurer's risks, an amount called risk-based capital (RBC). Risk-based capital requirements are discussed more fully below in the section on capital and surplus requirements.

On-Site Examinations

Most state laws require insurers to submit to routine on-site examinations by insurance officials every three to five years. An examination typically consists of a detailed investigation into the important aspects of insurer operations including verification of assets amounts and investment practices, valuation standards and methods for both assets and liabilities, and verification of income and expense items. The NAIC facilitates triennial examinations by organizing teams of examiners with representatives from different state insurance commissioner's offices. These teams are organized according to geographically-defined zones, and their findings are accepted by all the states in which the examined insurer conducts business. In the event that significant problems are uncovered by an IRIS analysis, a routine state examination, or a CPA audit report, state officials may conduct additional "targeted" examinations that focus in greater detail on the specific aspects of the insurer's operations that are deemed problematic.

Asset/Liability Requirements and Investment Restrictions

State regulatory requirements for assets fall under two basic categories: asset valuation methods and investment restrictions. These requirements vary across different states and the valuation methods required can be especially complicated. In many cases, the commissioner can exercise some discretion in the methods required, though are generally subject to the constraint that they be consistent with the guidelines provided in the NAIC model laws (Black and Skipper 2000, p. 956).

For statutory purposes insurer assets fall into three overlapping categories: invested assets, non-invested admitted assets, and non-admitted assets. Life-health insurers may further distinguish between general and separate account assets. Invested assets are admitted assets that generate income whether through interest, dividends, rent, or capital gains. Non-invested admitted assets do not generate income of any sort, but are still related intimately to insurer operations. Non-admitted assets are those that fail to meet the legal criteria established for recognition in statutory assessment of insurer financial position and operations. These include things like furniture and equipment. Separate account assets are those associated with operations for which the insurer is not exposed to investment risk, including such things as pension plan administration and variable life insurance and annuity products. Invested asset categories include bonds, stocks (common and preferred), mortgage loans, real estate, policy loans, and cash.

Asset valuation standards typically require the use of amortized values for bonds, and market or cost value for stocks (depending on whether they are common or preferred). Mortgage loans are valued at unpaid balances and real estate at the lower of book or market value. Regulatory standards for general account investments of all classes generally involve both quantitative and qualitative standards. For bond investments, regulatory standards include restrictions on the quantities (i.e. the proportion of admitted assets) of certain classes of bonds that can be held (e.g., maximums of 20 percent of aggregate assets held in classes 3 and above, 10 percent held in classes 4 and above, 3 percent held in classes 5 and 6, and 1 percent held in class 6) and on the amount of bond holdings that can come from a single institution (e.g., 5 percent).[46] Standards for stock investments typically involve caps (e.g., 20 percent) on the proportion of general account admitted assets that can be held in equities.[47] Mortgage loan requirements usually involve maximum loan-to-value ratios (e.g., 75 percent) and diversification requirements (e.g., 2 percent of assets per property) in addition to restricting the share of admitted assets that can be held (e.g., 50 percent).

Capital and Surplus Requirements

One of the most important requirements made by state insurance authorities relates to the minimum levels of capital and surplus that insurers are required to maintain in order to demonstrate financial strength. Beyond the initial capital and surplus required for insurance company formation, state insurance regulators seek to ensure that insurers maintain, on an ongoing basis, assets sufficient to cover all of their currently due liabilities and a conservative estimate of all future liabilities associated with current contracts. Since the financial strength or solvency of insurers depends

---

[46] Class definitions are based generally on the ratings provided by the Securities Valuation Office (SVO) of the NAIC and decreasing in order of quality from class 1 to class 6. Class 1 bonds are of the highest grade. Class 3 bonds are considered medium-grade bonds. And class 4, 5, and 6 bonds are low-grade bonds. Class 5 bonds are the lowest grade bonds that remain in good standing, while class 6 bonds are at or near default (Black and Skipper 2000, p. 887).

[47] The restriction on stock investment for life insurers is more rigorous than for property-liability insurers. It is commonly required that investment in stocks cannot exceed 10 percent of assets and cannot exceed the insurers' surplus level for life insurers, but fewer restrictions are placed on the stock investment for property-liability insurers.

on a variety of different elements, state insurance regulators use a range of factors in establishing these "internal" (i.e., not publicly expressed or statutorily defined) minimum capital and surplus standards. Common insurer characteristics used for this purpose may include, but are not limited to, the following: insurer size, mix of business (different lines of business), levels of reinsurance, quality, liquidity, and diversification of assets, reserve adequacy, and surplus trends (Black and Skipper 2000, p. 957). Although these "internal" (i.e., because they're not governed by statute, but rather by the judgment of insurance department regulators) requirements are more stringent than the statutorily defined ones, since 1992 the more concrete requirement (i.e., the one that is legally more binding on insurance companies) is based on insurer RBC ratios.

During the 1980s, episodes of regulatory forbearance led to a number of large insolvencies, which in turn moved the states to give greater power to insurance commissioners and directors, including the option of taking mandatory control of impaired insurers. Beginning in 1992, the NAIC changed the way in which minimum capital and surplus requirements were handled through its adoption of so-called risk-based capital requirements. These requirements involve measuring capital and surplus adequacy relative to target capital and surplus levels that can be considered theoretically adequate given the true size and nature of the risks that insurers face. This target amount is called the risk-based capital or RBC. For the L-H insurance industry, the RBC ratio is determined by a formula and reflects insurer exposure to five primary risk categories: asset default risk, other asset risks, insurance (i.e., mortality/morbidity) risk, interest rate risk, and miscellaneous business risk. RBC regulatory analysis relies on a measurement (called the RBC ratio) that is calculated by the following formula:

$$\text{RBCratio} = \text{TAC}/\text{RBC}$$

where *TAC* is the sum of statutory capital and surplus, any voluntary reserves held by the insurer, one-half of the insurer's policyholder dividend liability, and, in the case of L-H insurance, the asset valuation reserve (AVR). The AVR is an accounting mechanism used in statutory accounting for the L-H industry (i.e., it is not used for the P-L industry) that is designed to smooth non-interest-related security market values by absorbing both realized and unrealized capital gains and losses. In this way, fluctuations associated with capital gains and losses have minimal effects on reported surplus from year to year.

Any actions involving capital and surplus adequacy concerns that are taken by the regulatory authority directly depend on the value of the RBC ratio. RBC ratios greater than two signal that an insurer has adequate provision for risks and no actions are required on the part of regulators or the insurer. RBC ratios between 1.5 and 2.0 require the insurer to submit an RBC plan describing the cause of the problem and outlining the actions it will implement to remedy the situation. RBC ratios between 1.0 and 1.5 trigger regulator involvement in addition to a company RBC plan. In this case, state regulators must investigate the situation via examinations and may order the company to take certain corrective actions. When the RBC ratio falls between 0.7 and 1.0, state regulators are authorized, but not required, to take the insurer under regulatory control. Lower RBC ratios (ratios below 0.7) trigger mandatory seizure of company assets and transfer of control to the state.

The RBC requirement for U.S. P-L insurance industry was passed in 1994. The risks covered by P-L insurance RBC (PRBC) are different from those of life-health insurance. PRBC sets up minimum levels of capital required for a P-L firm to maintain solvency by taking into consideration four major types of risks—asset risk, credit risk, underwriting risk, and growth and off-balance sheet risk. Asset risk (investment risk) mainly refers to the risks related to adverse investment performance. Credit risk mainly refers to the chance that an insurance agent or reinsurer will default on their obligations under contracts. Underwriting risk, also referred to as premium risk, mainly assesses the risk that insurance premiums may inadequately reflect the potential cost of claims, and the risk that the loss reserves set aside by insurance firms for future claims might be inadequate. For details of the PRBC system, see Cummins, et al. (1999).

While RBC ratios are used as an important part of the regulatory process, there is some question as to their true usefulness in addressing the problem of insurer insolvency. Studies suggest that this RBC measure alone is a poor predictor of insurer solvency and that FAST ratios have greater predictive power than RBC ratios (Cummins, et al. 1995; Cummins, et al. 1999). The predictive power of RBC measures appears to improve when other variables (e.g., certain insurer characteristics and some cash flow simulation variables) are taken into consideration.

Rehabilitation and Liquidation

When insurance regulators are confronted with a situation involving a weak or renegade insurance company, they have several possible actions they can pursue, including informal negotiations and agreements, corrective orders, rehabilitation orders, and liquidation. When the situation is serious but not dire, regulators may be able to address the problem by working with insurance company management to improve the situation. If the situation is dire or the insurer does not respond to the informal approach, regulators can issue corrective orders. These actions typically consist of court-approved, written statements enumerating additional requirements or restrictions placed on the insurer, including cessation of certain business practices and/or activities, restrictions on writing new business, or demanding an infusion of capital (Black and Skipper 2000, p. 961).

Under certain circumstances, state regulators can take control of weak or renegade insurance companies by serving a rehabilitation order—a court-approved, written permission to assume control over the company for the purposes of rehabilitation or liquidation. If the courts rule in favor of the regulatory authority, insurance officials take control of the company and create a plan of rehabilitation which ultimately leads either to restoration of the insurer to statutory compliance or liquidation. An important part of the RBC system mentioned above is that regulators are required to seize control of any company for which the RBC ratio falls below the mandatory control level of 70 percent. In such a case, the commissioner assumes control of the insurer's assets and creates a plan to resolve the company's problems. In the event that liquidation becomes necessary, asset distribution is prioritized among the various creditors, with administrative expenses, employees, and policyholders being served prior to other creditor groups. Regulators in the liquidating insurer's state of domicile are responsible for the distribution of assets, including those to creditors (e.g., policyholders) residing in other states.

State Guaranty Funds

In order to protect insurance consumers from the consequences of insurer insolvency, every state has enacted laws to establish and provide for the operation of state guaranty funds—usually one each for the L-H insurance and P-L insurance sectors.[48] The funds represent a guarantee with limits to policyholders residing within the state that their legitimate claims will be honored in the event that their insurer is found unable to meet its financial and contractual obligations. Policyholders are typically guaranteed coverage of their interests (whether claims payments, cash values, etc.) up to some specified amounts which depend on the particular product and line of business involved.[49]

Guaranty funds are non-profit associations made up of all the insurers of a given type writing business in a particular state, and they are governed by a board with members from both the insurance industry and the state insurance department (Hall 2000, p. 417). Insurers are required to be members of the state guarantee associations (whether L-H or P-L, or both) that correspond to the insurance lines they sell in any given state. Assessments are made to association members in order to pay for the general administrative costs of the fund as well as to cover the costs of policyholder claims associated with member insolvency or impairment (Black and Skipper 2000, p. 963).

When insurers fail, liquidators are brought in either by appointment or as a part of the office of the commissioner to sell off the insurers' remaining assets. Liquidators use the proceeds of these asset sales to cover their own expenses and then to pay any outstanding wages or taxes owed by the insurer. What remains goes into the guaranty fund and is used, together with assessments on remaining (i.e., solvent) insurers in the state, to cover the payments necessary to fulfill the failed insurer's contractual obligations to policyholders. However, since liquidators have little incentive to keep their expenses down, only about 33 cents (41 cents in the case of cash and invested assets like stocks and bonds) of every dollar taken over by regulators at the time of failure ends up making it into the guaranty fund. Consequently, the bulk of the subsequent liabilities are covered by assessments to remaining insurers (Hall 2000, p. 416).

In the case of insolvency-related assessments, state laws define the amounts and distributions of payments to be made by surviving insurers to make up for the shortfalls associated with failed insurers. The sizes of these assessments typically depend, in some way, on surviving insurers' respective market shares in the relevant lines of business (i.e., those in which the failed insurer had deficiencies). In order to avoid undue burdens on remaining insurers in any one year, the NAIC model law suggests a cap on the payment made in any one year of 2 percent of the assessed insurers' most recent average three-year premiums written on business covered by

---

[48] Some states also have guaranty funds for workers' compensation plans.

[49] For example, typical coverage limits for life-health industry products are as follows: \$300,000 for death benefits; \$100,000 for cash values, annuities (cash values or payments), and non-medical expense health coverage; \$500,000 for medical expense payments; and \$300,000 for disability income payments (Black and Skipper 2000, p. 963).

the guaranty fund. Remaining costs are carried forward into future years. Insurers are generally also allowed to defer payment for up to 5 years if they can show that it would compromise their ability to meet their own contractual obligations. Payments made to a guaranty fund are deductible for federal tax purposes and, in addition, many states grant insurers a tax credit that can be used to reduce their state premium taxes, franchise taxes, and income taxes by an amount not to exceed 20 percent of the annual assessment paid during every year for a maximum of five years. In such instances, state taxpayers indirectly bear most of the costs associated with insurer insolvencies.[50]

State and Federal Taxes

U.S. insurers face taxes at both the state and federal levels.[51] The most important form of state taxation relates to premium income. State premium taxes range from zero to four percent of the premium tax base for insurers domiciled in the state or from one to four percent for foreign and alien insurers. The premium tax base varies by state, but typically involves all non-reinsurance premium considerations collected less any policyholder dividends paid. Annuity considerations are not taxed in most states in order to avoid putting insurers at a comparative disadvantage relative to competitors providing other savings vehicles. Various additional deductions, credits, and exemptions may apply on a state-by-state basis. Certain non-profit insurers in the health line of business (e.g., Blue Cross/Blue Shield) were not subject to taxes in most states until recently when the trend among state insurance departments has been to increasingly treat them in a manner similar to their for-profit counterparts. For example, as of 1996, almost half of all states imposed some sort of premium taxes on Blue Cross plans. Self-funded insurance arrangements are exempt from premium tax requirements.

In addition to premium taxes, a small number of states require foreign and alien insurers to pay income taxes.[52] In such instances, the state premium tax and the income tax can usually be offset against one another, so that ultimately the tax paid by these insurers is just the greater of the two. Some states also permit counties and municipalities to levy premium taxes. Most state tax regulations provide for retaliatory actions (e.g., higher state taxes on alien insurers from specific states than on domestics) to protect their domiciled insurers from discriminatory taxation by other states.

Insurance companies in the U.S. are subject to the same basic federal income tax requirements as other corporations, with some exceptions that are specific to the insurance industry. Taxable income is defined in a manner similar to that of other financial entities: gross income less the applicable deductions. Of course, in the insurance industry the components of income and the candidates for deductions differ significantly from those of other industries, and insurers are taxed under separate sections of the Internal Revenue Code. Insurer income items include

[50] For a discussion on the effectiveness of state guaranty regulations and proposals for alternative systems, see Han, et al. (1997).

[51] In some states, county and city governments are also permitted to levy taxes on insurance companies.

[52] Nine states subject out-of-state insurers to state income taxes, while 14 states require domestic insurers to pay state income taxes (Black and Skipper 2000, p. 970).

premium income, investment-related income (investment proceeds and capital gains), decreases in required reserves, and other miscellaneous types of income. Insurers are able to take the standard deductions available to all corporations and, in addition, enjoy some deductions that are specific to the insurance industry. These include deductions for insurance business expenses (insurance claims, benefits, and losses), deductions for net annual increases in reserves, and, in the case of participating insurance policies, deductions for policyholder dividends paid or accrued during the year.[53]

Following passage of the Revenue Reconciliation Act of 1990 by the U.S. Congress, L-H insurers became subject to a new tax law that effectively increased government tax revenues paid by the insurance industry. The deferred acquisitions costs (DAC) tax was implemented to simulate the increase in taxes that would be associated with amortizing acquisition expenses over some extended period rather than recognizing them and deducting them from income immediately as was done previously. Rather than requiring complicated calculations, the regulations simply specify a tax rate to be paid per premium collected by line of business (e.g., roughly two percent for life insurance and annuity products and 7.5 percent for health products; Black and Skipper 2000, p. 974). The ultimate effect of such a change was a one-time increase in taxes associated with deferring tax deductions into the future.

The P-L industry was subject to a similar adjustment in taxation. In the P-L insurance industry, liabilities such as unearned premiums and loss reserves generate tax deductions for the insurers. The Tax Reform Act of 1986, however, changed the use of tax shelters by P-L insurers. First, P-L insurers can deduct only the present value not the full nominal value of losses. Second, P-L Insurers can deduct the interest write-up on the present value of loss reserves for long-tail policies. Third, P-L insurers are taxed 15 percent of their otherwise tax exempt interest and dividend income (called the pro-ration provision). Under this provision, the P-L insurers lose part of their benefits from double deduction. This law had a dramatic influence on the P-L insurers' investment behavior. They became more likely to use underwriting losses to shelter taxable investment income, and raised the proportion of dividends in regular income (Cummins and Grace 1994).

2.3.4 Other Non-Governmental Influences

In addition to the direct influence of government institutions and the NAIC, insurers face quasi-regulatory pressures from various other non-governmental organizations including the Financial Accounting Standards Board (FASB), the National Association of Securities Dealers (NASD), and rating agencies (e.g., Standard & Poor's and Moody's). FASB, the organization charged with providing financial accounting standards and practices, has an influence on insurer operations through the annual disclosure requirements established by the SEC. Official FASB statements regarding accounting standards for insurance operations can have

---

[53] For mutual insurers, there is a limit on the size of the deduction that can be taken for policyholder dividends related to the difference in the return on equity between similarly situated stock and mutual insurers. The intent of this provision is to avoid discriminatory treatment of mutual versus stock insurers. For more detail, see Black and Skipper (2000), pp. 973–74.

significant implications for the financial position and profitability that publicly traded insurance companies must report in their financial statements.

The NASD is a self-regulating organization of the securities industry that exerts some amount of influence over insurance representatives selling variable life and annuity products. Rating agencies (e.g., A. M. Best Company) provide insurance buyers with information regarding the financial strength and commercial debt of insurers, something that, due to the complex nature of insurance products and liabilities, would be difficult if not impossible for the average insurance consumer. Since consumers use this information in making insurance purchasing decisions, ratings can potentially affect insurers' financial management and cost of capital.

2.4 THE U.S. LIFE-HEALTH INSURANCE INDUSTRY

In 2003, the U.S. life-health (L-H) industry generated nearly $500 billion in premiums (net of reinsurance), employed almost 800,000 workers, and held nearly $4 trillion in assets, including $1.2 trillion of separate account assets.[54] The life insurance industry alone had approximately $17 trillion of insurance (group and individual) in force (ACLI 2004, p. 92). In keeping with the past trends of consolidation and demutualization, the industry consists of a decreasing number of firms, an increasing proportion of which have the stock, rather than mutual, organizational form. Excluding firms with non-positive surplus and/or net premiums, only about 550 firms currently operate in the country, barely more than half the number of similarly situated companies in operation only a decade ago. Among these, only about 60 (roughly 11 percent) are mutual organizations. This dramatic decrease in the number of L-H insurers has been largely due to mergers and consolidations among the insurance companies in this sector of the industry (ACLI 2004, p. 1).

2.4.1 Key Sectors and Lines of Business

The primary purpose of the U.S. life-health insurance industry is to provide a range of different types of insurance contracts that can protect individuals, families, and, to a much lesser extent, corporations against the adverse financial effects of unanticipated mortality and/or morbidity events. The standard lines of life- and health-contingent insurance business for U.S. insurers are as follows: industrial life, individual life, individual annuities, credit life, group life, group annuities, and accident & health. Group insurance coverage is typically obtained through employer-provided benefit plans, while individual (non-group) coverage is not. Credit life is typically sold to protect creditors from loss in the case of borrowers' deaths. Thus, the face amount is usually decreasing over time for any given policy (e.g., auto loan or mortgage). Industrial life insurance has historically been marketed to low income

---

[54] Unless otherwise noted, the quantitative information provided is based on the authors' tabulations of information from the National Association of Insurance Commissioners (NAIC) annual statement datasets. Tabulations generally exclude insurers with zero or negative capital and surplus or premium income.

families and involved weekly or bi-weekly direct collections by insurance company representatives at the residence of policyholders. This form of life insurance has declined steeply over time as it has been largely replaced by group insurance offered to workers via their employers' benefit plans. Accident and health insurance refer to insurance that pays the additional expenses and lost income associated with impaired health or other loss of capacity to work.

In addition to the insurance products that they provide, a significant proportion of the business conducted by U.S. insurers involves financial contracts that do not involve life and/or health contingencies. These products, which are typically referred to as deposit-type contracts, include guaranteed investment contracts (GICs), annuities certain, premium and other deposit funds, dividend and coupon accumulations, lottery payouts, and structured settlements (ACLI 2004, p. 37). GICs are contracts that specify a rate and a term for a given, usually non-trivial, level of investment. Structured settlements are arrangements made, usually as the result of a law suit, to compensate an injured party for damages caused by another party. In 2003, deposit-type contracts represented inflows of approximately \$125 billion, outflows of about \$110 billion, and reserves of approximately \$405 billion (ACLI 2004, p. 38). As such, this is a major line of business representing approximately 20 percent of premium-type income for the L-H industry.[55]

The various insurance products mentioned above fall into four main groupings: life insurance, annuities, health insurance, and deposit-type contracts. These groups differ significantly in the timing of the cash flows defined by the insurance contracts and consequently, in the liabilities associated with those flows. For example, life insurance often involves premium buildups that occur over long periods of time, but payouts associated with a single event (death). Annuities may or may not involve long buildup periods, but generally do involve long payout periods. Health insurance operates on a year-by-year basis for the most part, in much the same way as property-liability insurance works. Deposit-type contracts do not involve mortality or morbidity risk, but may differ in terms of the duration of the cash flows depending on the type of contract. GICs tend to have short durations, while annuities certain, for example, may have durations that are significantly longer.

While insurance companies in the L-H sector may, and often do, write all four types of business, they tend to differ in the proportion of their businesses associated with each type.[56] As a result of this and due to the fact that differences in the nature of contract liabilities have a large impact on the reserves and surplus requirements required by statute, it is difficult to analyze all L-H insurers similarly. For example, a comparison of insurers operating primarily in the health sector to those writing mostly ordinary life contracts on the basis of asset returns is not meaningful since health contracts allow for only short-term investment and cannot compare in terms of returns to those demanded by those managing life reserves, where access to the funds (on average) tends to be for a much longer period. In the section on insurer profitability below, we deal with this complication in more detail.

---

[55] Beginning in 2001, considerations for this line of business (deposit-type contracts) are technically (i.e., for reporting purposes) excluded from income and are instead recorded directly as changes in reserves.

[56] Indeed they often also write or are affiliated with companies that also write property-liability insurance.

2.4.2 Life Insurance in Force

In 2003, U.S. insurance companies had approximately $16.7 trillion of life insurance in force, an increase of over $5.7 trillion (or 50 percent) from a decade earlier. This in force insurance represents more than 200 million policies for individual and credit life and 163 million certificates for group life.[57] The average face amount was approximately $56,000 per policy for individual life insurance, $44,000 for group life, and under $4,000 per policy for credit life. Table 2.1 presents a summary of this information and related counts by line of business for 1993 to 2003. In 2003, individual life insurance represented the greatest proportion of total life insurance in force at 56 percent, while group and credit life make up 43 percent and 1 percent respectively. From 1993 to 2003, the number of insured individuals for individual life remained pretty constant, while in force amounts steadily increased during the period at an average of 3.8 percent per year. Group life in force increased at an even greater rate of 5.0 percent per year, while credit life decreased both in terms of the number of insureds and face amounts (each decreased at a rate of 2.6 percent per year).

Table 2.1. U.S. Life-Health Insurance in Force by Year and Line of Business, 1993–2003

			Ordinary Life Insurance		Ordinary Credit		Group Life	
Year	Firms	Total In Force ($ Billions)	Policies (Millions)	In Force ($ Billions)	Policies[1] (Millions)	In Force ($ Billions)	Certificates (Millions)	In Force ($ Billions)
1993	1,132	11,105	169	6,449	52	200	142	4,456
1994	1,023	11,081	169	6,449	52	189	145	4,443
1995	1,005	11,696	166	6,890	57	201	147	4,605
1996	946	12,705	166	7,426	50	211	139	5,068
1997	833	13,364	162	7,873	47	212	142	5,279
1998	793	14,471	160	8,523	46	213	152	5,735
1999	708	15,495	162	9,172	46	213	159	6,110
2000	658	15,953	163	9,376	50	201	156	6,376
2001	612	16,290	166	9,346	48	179	163	6,765
2002	577	16,346	169	9,312	42	158	164	6,876
2003	546	16,764	167	9,375	40	153	163	7,236
Average Annual Changes from 1993 to 2003 (%)								
	–7.0	4.2	–0.1	3.8	–2.6	–2.6	1.4	5.0

Sources: NAIC (1993–2003) and ACLI (2004).
Notes: [1]Policies include individual policies and group certificates.

[57] A certificate is provided to policyholders as evidence of their coverage and typically has some description of coverage. The actual contract called the master contract is negotiated on a group level between the contract sponsor, often an employer, and the insurer.

These numbers reflect a trend towards greater provision of life insurance through group contracts than through individual channels. This change has probably been at least in part affected by the proliferation of cafeteria-style employee benefit plans provided by employers during the 1990s. Cafeteria plans provide workers with benefits in the form of a dollar amount and allow the employee to choose (within limits) how the amount will be spent. Often employees can choose between receiving cash on an after-tax basis or using it on a pre-tax basis (where possible) to purchase a variety of benefits through group arrangements made by the employer. Often the menu (hence cafeteria) of options includes limited amounts of life insurance through the employer's group insurance arrangement. Such arrangements effectively provide convenient and cost-efficient alternatives to individually purchased insurance at limited face amounts. Additional amounts can typically be purchased (up to some limit, usually denominated in multiples of earnings) on an after-tax basis through the same group policy.

2.4.3 Life-Health Premiums

The 20th century saw significant increases in the premiums written by L-H insurers for each of the three major segments of the business. Table 2.2 documents the changes in net premium considerations over much of the past century. Life insurance net (of reinsurance) premiums alone increased from \$776 million in 1915 to approximately \$130 billion in 2003. Annuity considerations and health insurance net premiums also exhibited spectacular growth during the last century. Health insurance premiums increased from \$1 billion in 1950 to \$109 billion in 2003, and annuity considerations increased from \$6 million in 1915 to \$265 billion in 2003.

While the general trend for L-H net premiums is clearly increasing for each of the three major categories, the share of the premium income associated with the different lines of business has changed drastically over the years, consistent with demographic changes associated with modernization and population aging. In particular, during the last half-century annuities (group and individual) have overtaken life insurance as the primary source of L-H industry premium income, while health insurance premiums have also grown as a share of total net premiums. Whereas in 1915, annuities represented less than 1 percent of total L-H net (of reinsurance) premium considerations, in 2003 they represent over half of the industry total and twice the amount written for life insurance lines of business.

From 1950 to 2003, health insurance premiums exhibited a similar trend, going from 12.2 percent of the total premiums written to 21.6 percent. Figure 2.1 depicts the trends in the share of premiums written by the three main branches of business: life insurance, annuities, and health insurance. Table 2.3 provides more detailed information for 1993 to 2003 by distinguishing between the seven primary lines in the L-H industry. Individual annuities represent the fastest growing line of business, increasing from 22.4 percent of total premiums to 32.2 percent during this period.

Table 2.2. U.S. Life-Health Net Premiums by Year and Line of Business, 1915–2003

Year	Total ($ Billions)	CPI[1]	Life Insurance ($ Millions)	%	Annuities ($ Millions)	%	Health Insurance ($ Millions)	%
1915	782.0	10.1	776.0	99.2	6.0	0.8	N/A	N/A
1920	1,381.0	19.3	1,374.0	99.5	7.0	0.5	N/A	N/A
1925	2,378.0	17.3	2,340.0	98.4	38.0	1.6	N/A	N/A
1930	3,517.0	17.1	3,416.0	97.1	101.0	2.9	N/A	N/A
1935	3,673.0	13.6	3,182.0	86.6	491.0	13.4	N/A	N/A
1940	3,887.0	13.9	3,501.0	90.1	386.0	9.9	N/A	N/A
1945	5,159.0	17.8	4,589.0	89.0	570.0	11.0	N/A	N/A
1950	8,189.0	23.5	6,249.0	76.3	939.0	11.5	1,001.0	12.2
1955	12,546.0	26.7	8,903.0	71.0	1,288.0	10.3	2,355.0	18.8
1960	14,365.0	29.3	11,998.0	83.5	1,341.0	9.3	1,026.0	7.1
1965	21,574.0	31.2	16,083.0	74.5	2,260.0	10.5	3,231.0	15.0
1970	36,767.0	37.8	21,679.0	59.0	3,721.0	10.1	11,367.0	30.9
1975	58,575.0	52.1	29,336.0	50.1	10,165.0	17.4	19,074.0	32.6
1980	92,624.0	77.8	40,829.0	44.1	22,429.0	24.2	29,366.0	31.7
1985	155,863.0	105.5	60,127.0	38.6	53,899.0	34.6	41,837.0	26.8
1990	264,010.0	127.4	76,692.0	29.0	129,064.0	48.9	58,254.0	22.1
1995	348,438.7	150.3	104,814.0	30.1	160,763.2	46.1	82,861.4	23.8
2000	544,372.4	168.8	132,353.5	24.3	304,334.3	55.9	107,684.6	19.8
2003	503,944.7	181.7	129,872.7	25.8	265,098.3	52.6	108,973.8	21.6
Average Annual Changes from 1915 to 2003 (%)[2]								
	7.6	3.3	6.0	–1.5	12.9	4.9	9.3	1.1

Sources: For years prior to 1995, ACLI (2004), pp. 56–57. For years following 1994, NAIC (1995–2003). CPI data is from the U.S. Bureau of Labor Statistics (www.bls.gov).

Notes: Life-Health Net Premiums include deposit-type contract considerations for all years preceding 2001. For years 2001, 2002, and 2003 deposit-type contract considerations are not included, consistent with NAIC-mandated changes in reporting. N/A is not available. [1]The CPI is for all urban consumers, U.S. city average, all items (CPI-U) consumer price index from the U.S. Bureau of Labor Statistics (www.bls.gov). [2]From 1950 for health insurance.

The amounts of life-health insurance business transacted across the different states and territories that are a part of the U.S. vary significantly. Table 2.4 presents the net premiums and corresponding state revenues (state taxes and fees associated with L-H insurance operations) during the year 2003 by state or territory. The top 5 states for L-H insurers in terms of net premiums are California ($55 billion), New York ($49 billion), Texas and Florida ($37 billion each), and Illinois ($29 billion). Pennsylvania ($24 billion), New Jersey ($24 billion), and Massachusetts ($22 billion) are other important states for L-H insurance.

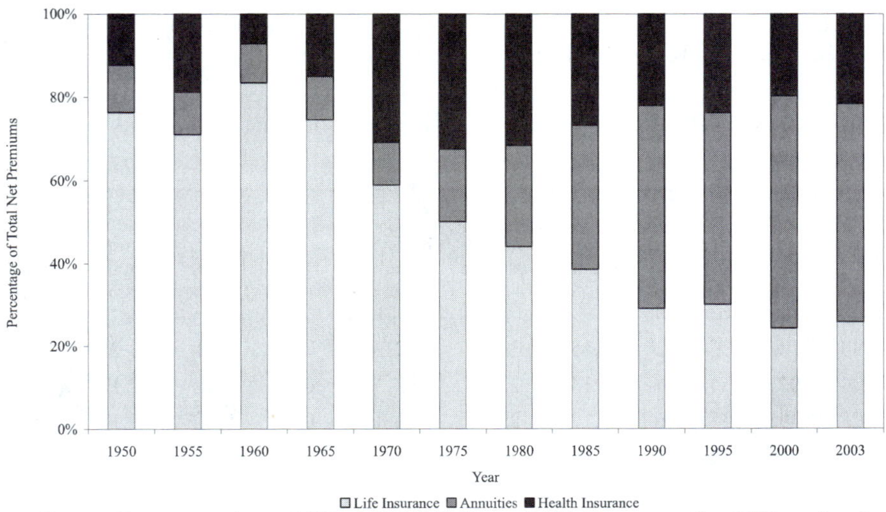

Source: For years prior to 1993, ACLI (2004), pp. 5657. For years after 1992, authors' tabulations using the NAIC dataset.

Figure 2.1. Composition of U.S. Life-Health Premium Income Over Time, 1950–2003

2.4.4 Market Concentration

Concentration in the context of financial markets refers to how a given industry's total revenues (or some other measure of market power) are distributed across its constituent companies. High levels of concentration in an industry are obtained when relatively few companies account for a large proportion of the business transacted (e.g., as reflected by revenues or some other measure). High concentration in an industry can be used to infer the existence of monopoly power (or alternatively the absence of competition) in that industry. From 1993 to 2003, the L-H insurance market has become increasingly concentrated. Tables 2.5 and 2.6 document these changes in market concentration.

Table 2.5 shows the concentration ratios (as measured by net premiums written) for 1993, 1998, and 2003 for the top 1, 5, and 10 firms within each line of business. From 1993 to 2003, all lines of business increased in concentration as measured by the proportions of net premiums written by the largest (i.e. top 5 or 10) firms in each category. The annuities (individual and group) and accident and health lines of business exhibited decreases in concentration from 1993 to 1998 coincident with changing market forces favoring smaller insurers and allowing them to capture greater market share.[58] This brief trend was reversed, however, in the second half of the decade; and by 2003 concentration levels exceeded those of earlier years in all lines of business—including the annuity and accident-health lines. The

[58] During this period (and the period immediately preceding this one), the regulatory barriers to entry by non-insurance financial institutions into the annuities and accident-health insurance markets was gradually eroded, resulting in new opportunities for insurance product distribution. See the section on financial services integration above for more detail.

corresponding concentration for the entire L-H industry increased over the period for both the top 5 and the top 10 firms, but decreased for the largest firm.

Table 2.3. U.S. Life-Health Net Premiums by Year and Line of Business (Detailed), 1993–2003

Year	Firms	Total ($ Millions)[1]	Industrial Life (%)	Individ-ual Life (%)	Individual Annuities (%)	Credit Life (%)[2]	Group Life (%)	Group Annuities (%)	Accident and Health (%)[3]
1993	1,132	324,992	0.15	22.64	22.36	0.58	5.31	23.25	25.50
1994	1,023	341,916	0.14	22.86	23.22	0.61	5.48	22.12	25.23
1995	1,005	357,446	0.12	23.01	21.86	0.59	5.60	23.12	25.17
1996	946	381,561	0.11	22.28	21.89	0.51	5.37	25.40	24.16
1997	833	412,793	0.10	21.15	22.90	0.49	6.21	26.06	22.37
1998	793	466,968	0.09	22.72	20.63	0.43	5.31	29.38	20.28
1999	708	495,877	0.08	18.77	23.47	0.41	5.02	31.18	20.15
2000	658	550,750	0.07	18.69	25.55	0.35	4.93	29.71	19.55
2001	612	476,565	0.07	20.34	29.63	0.35	5.85	23.00	20.76
2002	577	506,167	0.06	20.97	33.08	0.26	5.28	19.93	20.41
2003	546	503,945	0.05	20.40	32.24	0.21	5.10	20.36	21.62

Source: NAIC (1993–2003).
Notes: Life-Health Net Premiums include deposit-type contract considerations for all years preceding 2001. For years 2001, 2002, and 2003 deposit-type contract considerations are not included, consistent with NAIC-mandated changes in reporting. [1]The total includes other miscellaneous lines of business not explicitly listed. [2]Credit life includes group and individual. [3]Accident and health includes group, credit, and other.

Table 2.6 documents changes in the Herfindahl Index over the eleven-year period from 1993 to 2003 both for the industry as a whole and for the different lines of business. A higher index value denotes greater concentration—i.e., less competitive and more subject to the abuses associated with monopoly power. A score of 10,000 indicates a complete monopoly, while highly competitive industries have Herfindahl values nearer to zero. Roughly speaking, the U.S. Department of Justice (DOJ), the federal regulatory body with oversight in antitrust matters, considers markets with Herfindahl values below 1,000 to be unconcentrated, those with Herfindahl values between 1,000 and 1,800 to be moderately concentrated, and those with Herfindahl values greater than 1,800 to be highly concentrated (U.S. Department of Justice and Federal Trade Commission 1997, §1.51).

Table 2.4. U.S. Life-Health and Property-Liability Premiums Written ($ Millions) by State/Territory, 2003

	Premiums Written							Premiums Written					
	Life-Health[1]		Prop-Liab[2]		Total			Life-Health[1]		Prop-Liab[2]		Total	
State	($)	(%)	($)	(%)	($)	(%)	State	($)	(%)	($)	(%)	($)	(%)
Alabama	5,815	0.9	5,677	1.3	11,492	1.1	Nebraska	4,511	0.7	2,906	0.7	7,417	0.7
Alaska	1,026	0.2	1,247	0.3	2,273	0.2	Nevada	3,094	0.5	3,469	0.8	6,563	0.6
Arizona	17,445	2.8	6,982	1.6	24,427	2.3	New Hampshire	2,707	0.4	2,018	0.5	4,725	0.4
Arkansas	3,194	0.5	3,535	0.8	6,729	0.6	New Jersey	24,329	4.0	15,939	3.6	40,268	3.8
California	55,445	9.0	53,555	12.2	109,000	10.4	New Mexico	2,116	0.3	2,150	0.5	4,266	0.4
Colorado	11,702	1.9	7,507	1.7	19,209	1.8	New York	48,912	8.0	32,028	7.3	80,940	7.7
Connecticut	15,778	2.6	6,608	1.5	22,386	2.1	North Carolina	14,269	2.3	10,436	2.4	24,705	2.3
Delaware	12,251	2.0	1,944	0.4	14,195	1.3	North Dakota	902	0.1	1,122	0.3	2,024	0.2
District of Columbia	2,752	0.4	1,396	0.3	4,148	0.4	Ohio	21,785	3.5	16,449	3.8	38,234	3.6
Florida	36,755	6.0	30,196	6.9	66,951	6.4	Oklahoma	4,473	0.7	4,755	1.1	9,228	0.9
Georgia	12,828	2.1	11,949	2.7	24,777	2.4	Oregon	5,083	0.8	4,806	1.1	9,889	0.9
Hawaii	2,235	0.4	1,856	0.4	4,091	0.4	Pennsylvania	24,365	4.0	18,273	4.2	42,638	4.1
Idaho	3,131	0.5	1,560	0.4	4,691	0.4	Rhode Island	1,918	0.3	1,817	0.4	3,735	0.4
Illinois	29,490	4.8	21,707	5.0	51,197	4.9	South Carolina	5,530	0.9	5,379	1.2	10,909	1.0
Indiana	10,517	1.7	9,579	2.2	20,096	1.9	South Dakota	1,526	0.2	1,315	0.3	2,841	0.3
Iowa	8,877	1.4	4,122	0.9	12,999	1.2	Tennessee	9,152	1.5	7,609	1.7	16,761	1.6
Kansas	6,368	1.0	4,144	0.9	10,512	1.0	Texas	37,353	6.1	32,157	7.3	69,510	6.6
Kentucky	4,806	0.8	5,413	1.2	10,219	1.0	Utah	3,454	0.6	2,724	0.6	6,178	0.6
Louisiana	6,301	1.0	7,253	1.7	13,554	1.3	Vermont	1,076	0.2	951	0.2	2,027	0.2
Maine	1,772	0.3	1,836	0.4	3,608	0.3	Virginia	14,394	2.3	9,300	2.1	23,694	2.3
Maryland	12,280	2.0	7,776	1.8	20,056	1.9	Washington	8,424	1.4	7,986	1.8	16,410	1.6
Massachusetts	21,616	3.5	11,267	2.6	32,883	3.1	West Virginia	2,295	0.4	2,151	0.5	4,446	0.4
Michigan	17,333	2.8	15,687	3.6	33,020	3.1	Wisconsin	11,225	1.8	7,340	1.7	18,565	1.8
Minnesota	9,952	1.6	8,406	1.9	18,358	1.7	Wyoming	752	0.1	712	0.2	1,464	0.1
Mississippi	4,268	0.7	3,421	0.8	7,689	0.7	**Total U.S.**	**579,767**	**94.4**	**438,104**	**100.0**	**1,017,871**	**96.7**
Missouri	11,123	1.8	8,357	1.9	19,480	1.9	Other[3]	34,278	5.6	0.0	0.0	34,279	3.3
Montana	1,058	0.2	1,332	0.3	2,390	0.2	**Grand Total**	**614,046**	**100.0**	**438,104**	**100.0**	**1,052,150**	**100.0**

Source: ACLI (2004), pp. 115–116 and III (2005b), p. 29.

Notes: [1]Life-Health includes deposit-type contract considerations. [2]Property-Liability is before reinsurance transactions, excluding state funds, territories, and possessions. [3]Other includes Puerto Rico, American Samoa, Guam, U.S. Virgin Islands, Canada, and other aggregates.

Table 2.5. U.S. Life-Health Market Concentration by Year and Line of Business (% of Total Net Premiums)

Year	Count	Total (%)	Industrial Life (%)	Individual Life (%)	Individual Annuities (%)	Credit Life (%)[1]	Group Life Insurance (%)	Group Annuities (%)	Accident and Health (%)[2]
1993	1	7.5	32.5	9.0	5.9	9.8	18.2	10.3	7.9
	5	23.9	61.3	31.7	22.3	23.7	47.6	40.3	25.4
	10	37.2	73.6	44.9	37.2	35.2	60.1	62.3	39.6
1998	1	5.3	42.6	7.8	5.1	10.9	17.3	8.0	6.7
	5	21.2	75.3	31.9	23.3	37.3	47.0	34.3	22.6
	10	35.4	86.8	49.2	39.9	57.5	61.4	56.2	36.8
2003	1	6.8	44.4	8.5	10.1	21.2	21.4	10.8	10.9
	5	25.3	80.7	35.5	33.3	55.9	49.2	41.5	32.5
	10	39.9	90.4	51.9	51.4	67.8	66.6	66.4	46.9

Table 2.6. U.S. Life-Health Market Concentration by Year and Line of Business (Herfindahl Measures), 1993–2003

Year	Firms	Total	Industrial Life	Individual Life	Individual Annuities	Credit Life[1]	Group Life	Group Annuities	Accident and Health[2]
1993	1,140.0	184.1	1,375.4	292.2	254.1	238.8	637.3	533.6	237.2
1994	1,031.0	168.6	1,501.8	269.7	278.3	262.3	616.0	593.7	220.5
1995	1,012.0	166.0	1,501.8	263.2	329.9	301.6	655.1	570.9	218.1
1996	953.0	173.3	1,684.2	266.4	386.7	339.3	628.1	685.0	217.2
1997	838.0	171.0	2,046.0	267.2	392.7	403.3	606.1	489.6	216.1
1998	794.0	177.9	2,206.3	322.0	406.4	421.9	602.0	1,084.2	220.9
1999	712.0	175.7	2,381.9	297.0	471.2	512.6	630.5	809.1	252.4
2000	659.0	176.7	2,471.5	294.1	469.0	634.4	637.6	936.8	266.3
2001	613.0	243.7	2,757.4	355.2	349.0	766.2	824.7	529.7	280.9
2002	578.0	240.0	2,860.9	375.0	368.3	727.7	762.9	538.3	301.8
2003	548.0	244.2	2,497.2	366.7	376.1	875.5	776.1	541.7	352.3

Source: NAIC (1993–2003).
Notes: Market concentration is here determined by insurance group (i.e., versus by company) and on the basis of premiums net of reinsurance and excluding deposit-type considerations. [1]Credit life includes group and individual. [2]Accident and health includes group, credit, and other.

The competitiveness (or concentration) of an industry has implications for the kinds of merger and acquisition activities that regulatory authorities (e.g., the DOJ) will allow in any given market. Mergers and acquisitions (M&A) resulting in an industry with Herfindahl values below 1,000 typically receive DOJ approval. When M&A activities result in Herfindahl values in the range of 1,000 to 1,800, however, they are subject to DOJ scrutiny. Activities that result in an increase of less than 100 in the Herfindahl Index are typically approved, while those resulting in greater increases in industry concentration are seen as more problematic. Herfindahl values above 1,800 are considered by the DOJ to be evidence of high concentration in an industry and, as a consequence, M&A activity in such industries is subject to greater regulatory scrutiny. Increases of less than 50 points on the scale are generally regarded as benign, while increases exceeding 100 are seen as more troublesome.[59]

As demonstrated by Table 2.6, with the exception of the industrial life line of business, which is practically negligible compared to the other lines of insurance (see Table 2.3), the L-H insurance business lines are unconcentrated by DOJ standards. The credit and group life lines of business are the nearest to being moderately concentrated with Herfindahl values of 875 and 776, respectively, but they are also relatively small lines in terms of market share (i.e. combined they represent roughly 5 percent of total L-H net premiums; see Table 2.3, year 2003). Most lines of business (comprising almost 95 percent of total net premiums) fall well under the Herfindahl-value threshold of 1,000.

The information presented in Table 2.6 largely confirms the concentration ratios presented in Table 2.5. With the exception of the group annuities line of business, the Herfindahl Indices tend to increase over the period of interest.[60] Most notable, perhaps, is the fact that the L-H insurance industry as a whole has become more concentrated over this period as reflected in the change in the number of operating insurers and in the Herfindahl Index over time. From 1993 to 2003, the number of L-H insurers decreased by approximately 50 percent from 1,140 to 548. During this same period, the Herfindahl Index increased from 184.1 to 244.2. The trend towards greater concentration is not surprising given the continued general trend towards consolidation in the insurance industry as a whole since insurers have come under more external competitive pressure from other financial services industries. Table 2.7 presents the top 10 L-H insurance firms in 2003, measured on the basis of total net premiums written across all lines of business, along with key information about assets, surplus, market share, organizational form, distribution system, and state of domicile.

2.4.5 Organizational Forms

In the U.S., insurance companies can take one of several legal forms. While there is no theoretical reason that insurers cannot be owned by an individual or partnership,

---

[59] For more information, see U.S. Department of Justice (2005), §1.51.

[60] The spikes in the Herfindahl Index for group annuities during the years 1998, 1999, and 2000 are due to the merger and acquisition activities of some large insurers during those years. In particular, AIG's life-health branch acquired SunAmerica on January 1, 1999. Later in 1999, Allstate's life-health branch acquired CNA Personal Insurance (now Encompass) and American Heritage Life Investment Corporation.

life insurance operations, in particular, require a certain level of permanence and financial strength. Perhaps, as a result, U.S. insurers must take the legal form of a corporation. Beyond being organized as corporations, most L-H insurers take one of two prominent organizational forms: the stock form and the mutual form.[61] Stock insurers are organized with the primary purpose of generating profits for their shareholders, whether they are individual or institutional investors. In stock insurance companies, policyholders have no legal ownership of the company; they are customers only. Mutual insurers, on the other hand, are actually owned by the policyholders of the company. Unlike stock companies which can be owned by other stock companies or mutual companies, mutual insurers cannot be owned by any entity other than its policyholders. As a consequence, mutual insurers are not beholden to outside owners and can focus exclusively on pursuing policyholders' objectives. This is done primarily by providing insurance and by distributing firm profits to policyholders in the form of dividends to their policies—a phenomenon called participation.

Table 2.8 provides the number and percentages of stock versus mutual insurers from 1993 to 2003. On the basis of numbers alone, the relative proportion of insurers of each category does not appear to have changed significantly during the last decade. Table 2.9 provides the number of overall U.S. L-H insurers and line-of-business market shares by organizational form as of 2003. The table shows that, at about 25 percent of net premiums written, mutual insurers have the greatest presence in the individual life line of business.[62] Group life, individual annuity, and group annuity lines of business tend to be even more dominated by the stock form than the individual life line of business. For these lines, mutual insurers write on average less than 15 percent of total net premiums.

Whereas prior to the 1960s, mutual insurers were larger than stock insurers and tended to dominate the market for life insurance, this has changed in recent decades (Black and Skipper 2000, p. 574). Both the increased importance of separate accounts business and the demutualization of several of the largest mutual insurers have worked to change the status quo. Indeed, the growth and highly competitive nature of business in the separate accounts market may have motivated insurers to switch to organizational forms which provide greater flexibility and managerial discretion (see Viswanathan and Cummins 2003 and Mayers and Smith 1988, 2002). Table 2.10 lists the ten largest demutualizations on the basis of total company assets that have occurred in recent years. Among these are some of the largest stock and mutual firms in the industry (see Table 2.7 for comparison).

---

[61] Other forms that are much less common include Lloyd's and fraternal benefit societies. Fraternal organizations, for example, account for only about 1 percent of the U.S. life-health insurance market. Consequently, they are not dealt with in-depth in this report. For more information, see Black & Skipper (2000), pp. 254–56.

[62] Industrial life is a notable exception, but it represents less than 1 percent of all individual life insurance net premiums written.

Table 2.7. Top Ten Insurers in the U.S. Life-Health Market, 2003

Rank[1]	Company	Assets ($ Billions)	Surplus ($ Billions)	Net Premiums[2] ($ Billions)	Market Share (%)	Organizational Form	Distribution System	State of Domicile
1	Metropolitan Life & Affiliated	287.6	12.8	34.1	6.80	Stock	Agency	Florida
2	AIG Life Group	289.1	29.0	27.5	5.50	Stock	Brokerage	California
3	Hartford Life Group	173.3	9.3	25.5	5.10	Stock	N/A	Connecticut
4	ING Group	147.1	6.9	20.5	4.10	Stock	Agency	Georgia
5	Prudential of America Group	244.7	8.4	19.9	3.90	Stock	Agency	New Jersey
6	AEGON USA Inc.	163.4	10.2	18.3	3.60	Stock	Agency	Iowa
7	New York Life Group	138.7	11.1	15.0	3.00	Mutual	Agency	New York
8	Equitable Group	97.0	4.5	14.9	3.00	Stock	Agency	New York
9	Massachusetts Mutual	97.4	6.9	13.6	2.70	Mutual	Agency	Massachusetts
10	Nationwide	102.1	3.0	12.0	2.40	Stock	Brokerage	Ohio

Source: NAIC (1993–2003).
Notes: [1]Rank is based on net premiums. [2]Net Premiums exclude premiums for deposit-type contracts. N/A is not available.

Table 2.8. U.S. Life-Health Organizational Form Counts and Net Premiums, 1993–2003

Year	Stock Insurers				Mutual Insurers				Other Insurers[1]				Total	
	Count	%	Net Premiums ($ Millions)	%	Count	%	Net Premiums ($ Millions)	%	Count	%	Net Premiums ($ Millions)	%	Count	Net Premiums ($ Millions)
1993	1,019	89.4	196,233	60.4	110	9.6	120,110	37.0	11	1.0	8,649	2.7	1,140	324,992
1994	918	89.0	212,456	62.1	104	10.1	122,201	35.7	9	0.9	7,259	2.1	1,031	341,916
1995	905	89.4	239,660	67.0	98	9.7	109,791	30.7	9	0.9	7,996	2.2	1,012	357,447
1996	836	87.7	241,475	63.3	107	11.2	131,486	34.5	10	1.0	8,600	2.3	953	381,561
1997	733	87.5	287,496	69.6	95	11.3	116,279	28.2	10	1.2	9,019	2.2	838	412,794
1998	693	87.3	334,010	71.5	94	11.8	124,748	26.7	7	0.9	8,209	1.8	794	466,967
1999	614	86.2	356,031	71.8	91	12.8	130,707	26.4	7	1.0	9,140	1.8	712	495,878
2000	567	86.0	444,372	80.7	85	12.9	95,775	17.4	7	1.1	10,603	1.9	659	550,750
2001	530	86.5	395,640	83.0	73	11.9	70,731	14.8	10	1.6	10,194	2.1	613	476,565
2002	501	86.7	426,441	84.2	69	11.9	69,864	13.8	8	1.4	9,862	1.9	578	506,167
2003	477	87.0	416,970	82.7	63	11.5	74,580	14.8	8	1.5	12,395	2.5	548	503,945

Sources: NAIC (1993–2003) and A.M. Best (1993–2003).

Notes: Net premiums include deposit-type considerations for years prior to 2001 and exclude them thereafter. [1]Other Insurers includes several medical indemnity plans and insurers with missing or unidentified organizational forms.

There has been much interest in understanding and explaining in economic terms the simultaneous existence of multiple organizational forms in the insurance marketplace. One major difference between the stock and mutual organizational forms relates to how they manage the incentive conflicts that exist between three primary roles that agents have in insurance company operations, namely managing the company, bearing the risks associated with operations, and acting as customers or policyholders (Mayers and Smith 1988). Different organizational forms are thought to arise, in practice, as optimal responses on the part of insurers to the particular set of circumstances they face, vis-à-vis balancing the potential conflicts between these three forces or roles. Alternative organizational forms have different strengths in dealing with specific types of agency conflicts. For example, stock insurers are better able to deal with owner-manager incentive conflicts because of the superior mechanisms available in the stock ownership form for owners to monitor and control managers, such as proxy fights and incentive compensation through stock options. Mutuals, on the other hand, are better able to control owner-policyholder conflicts because the ownership and policyholder roles are merged in the mutual organizational form. Stocks also have the advantage, unrelated to agency costs, of having superior access to capital through the securities market. It follows that the demutualization trend in recent decades should somehow reflect a change in the circumstances faced by insurers that alters the relative competitiveness of the stock versus mutual organizational forms. We discuss this trend and some of its potential causes further below.

As Table 2.8 indicates, the mix of stock versus mutual insurers has not changed significantly in recent years. This is likely due to the fact that consolidations among stock insurers have accompanied the wave of demutualizations. However, in terms of premiums written the situation is quite different. In 2003, stock insurers wrote over 80 percent of the total industry net premiums, up from about 60 percent only 10 years earlier.[63] As financial markets in the U.S. have developed and grown in size and with the passage of the Gramm-Leach-Bliley Act in 1999, insurers have felt pressure to both strengthen their core insurance business and expand into new financial services markets. In such an environment, access to equity capital is seen as an increasingly important part of success and even survival because competitive forces tend to impose limits on the rate of capital accumulation that is possible through the standard process of retained earnings. As this trend has progressed, the insurance industry as a whole and the L-H industry, in particular, have experienced a clear shift from the mutual to the stock form. Figure 2.2 shows the extent of this shift from 1993 to 2003.

---

[63] Net premiums here (and generally throughout the L-H section) include deposit-type considerations for years prior to 2001 and exclude them otherwise.

Table 2.9. U.S. Life-Health Market Shares by Organizational Form and Business Line (%), 2003

Business Line	Stock	Mutual	Other[1]	Total
Count	477.0	63.0	8.0	548.0
Industrial Life	97.1	2.9	0.0	100.0
Individual Life	73.9	26.1	0.0	100.0
Individual Annuities	86.5	13.5	0.0	100.0
Credit Life	75.6	24.4	0.0	100.0
Group Life	85.5	13.1	1.5	100.0
Group Annuities	87.4	12.6	0.0	100.0
Accident/Health	80.4	8.5	11.0	100.0
Total	82.7	14.8	2.5	100.0

Sources: NAIC (1993–2003) and A.M. Best (1993–2003).
Notes: Market share are computed on the basis of premiums net of reinsurance and including deposit-type contracts before 2001. [1]Other includes insurers with missing or unidentified organizational forms.

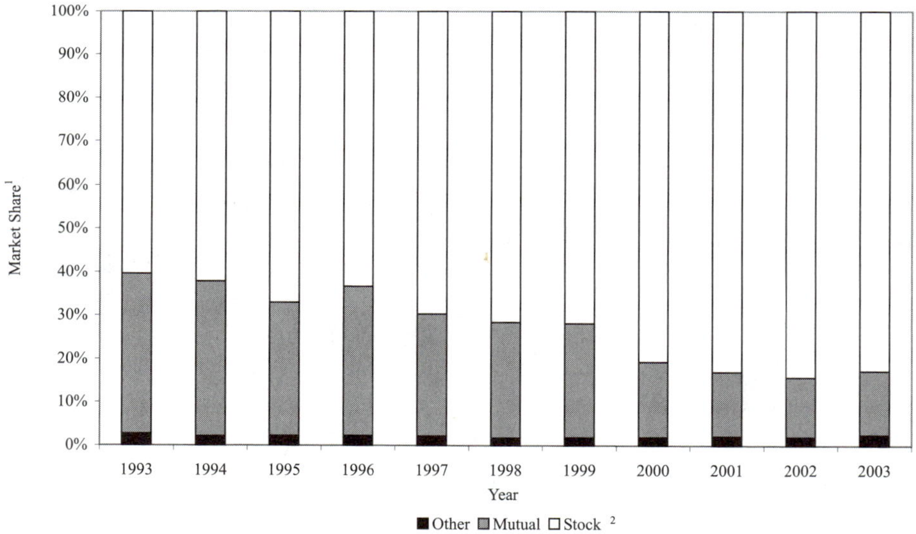

Source: NAIC (1993–2003) and AM Best (1993–2003).
Notes: [1] Determined on the basis of net premiums, including deposit-type considerations for years prior to 2001. [2] Other includes several medical indemnity plans and insurers with missing or unidentified organizational forms.

Figure 2.2. U.S. Life-Health Insurer Market Shares by Organizational Form, 1993–2003

Table 2.10. Largest U.S. Life-Health Insurer Demutualizations, 1993–2003

Rank[1]	Company	Year	Demutualization Type[2]	Assets ($ Billions)	Surplus ($ Billions)	Net Premiums[3] ($ Billions)	Distribution System	State of Domicile
1	Prudential Ins. Co. of America	2001	Full	196.1	8.7	13.9	Agency	New Jersey
2	Metropolitan Life Ins. Co.	2000	Full	183.9	7.6	24.6	Agency	Florida
3	Principal Mutual Life Ins. Co.	1998	MHC	64.0	2.8	12.7	Agency	Iowa
4	John Hancock Life Ins. Co.	2000	Full	60.7	3.5	9.6	Agency	Massachusetts
5	Pacific Mutual Life Ins. Co.	1997	MHC	21.2	0.8	3.8	N/A	California
6	Pruco Life Ins. Co.	2001	Full	19.6	0.8	2.8	Agency	New Jersey
7	Phoenix Life Ins. Co.	2001	Full	17.6	1.3	1.5	Agency	Connecticut
8	Minnesota Mutual Life Ins. Co.	1998	MHC	13.3	0.9	2.2	Direct	Minnesota
9	Mutual Life Ins. Co. of New York	1998	Full	11.8	0.8	1.0	Agency	New York
10	General American Life Ins.	1997	MHC	10.7	0.6	3.0	Agency	Missouri

Sources: Erhemjamts (2005), NAIC (1993–2003), and A.M. Best (1993–2003).
Notes: N/A is Not Available. [1]Rank is based on total assets. [2]MHC is mutual holding company. [3]Net premiums exclude premiums for deposit-type contracts.

Due to the stringent initial capital and surplus requirements typical of state insurance regulations, most new insurance companies are formed as stock insurers. In addition, while it is possible for stock insurers to convert into mutual companies (a process called mutualization), the current trend for existing insurance companies is in the opposite direction: mutual companies are converting into stock companies.[64] Many insurers previously organized as mutuals either have already completed or are actively pursuing the process of demutualization, whereby they either become stock companies or at least fall under the umbrella of a holding company with some (even if somewhat limited) access to equity markets.

While there are many potential motives for insurer demutualization, research suggests that a primary motive is capital access (Erhemjamts 2005). For example, one study identifies a significant increase in the liquidity of assets held by L-H insurance companies after converting from the mutual to the stock form of organization (Viswanathan and Cummins 2003). Since the liquidity of assets is related to L-H insurers' access to the kinds of immediate capital needed to deal with policy surrenders and fluctuating healthcare reimbursements, one might conclude that a primary motivation for demutualization is a desire for greater access to such liquid capital.

Demutualization can proceed in one of several different ways. Full demutualizations involve a complete conversion of the mutual into a stock company by effectively purchasing policyholders' interests in the company and issuing shares in the newly formed company to new shareholders. Sponsored demutualizations occur when the converted mutual company is acquired by another company. A more recent method of conversion involves formation of a mutual holding company (MHC). In this case, a holding company is formed which, through the conversion, will directly own the newly converted stock insurer. Initial shares in the MHC are issued to policyholders and new shares can be issued to other shareholders, generally with the restriction that the policyholders retain a majority of the voting rights in the holding company. Table 2.11 provides breakdown of L-H demutualizations by type from 1985 to 2003.

While access to capital may be the primary motivation for demutualization, there are also other reasons why this might be an attractive alternative for insurers. These include greater flexibility of affiliation (i.e. with other financial entities, such as banks) for stock insurers, tax advantages associated with the stock form, and uses of stock that are unrelated to corporate financing (e.g., as a type of "tradable currency" for use in M&A activities and for use in management compensation). Of these alternative motives, perhaps the most significant is that of greater flexibility in how stock companies can affiliate (i.e., corporate structure arrangements).

---

[64] Indeed a number of the large U.S. mutual companies began as stock companies (Black and Skipper 2000, p. 578). See also Mayers and Smith (1981) for a discussion on mutualizations and their implications for the relative efficiency of the two organizational forms.

Table 2.11. U.S. Life-Health Insurer Demutualizations by Type, 1985–2003

Year	Full	MHC	Sponsored	Total
1985	2	0	0	2
1986	1	0	0	1
1987	0	0	0	0
1988	1	0	0	1
1989	2	0	0	2
1990	0	0	0	0
1991	1	0	0	1
1992	0	0	2	2
1993	0	0	0	0
1994	1	0	0	1
1995	5	0	0	5
1996	0	3	0	3
1997	1	12	0	13
1998	2	13	0	15
1999	1	9	0	10
2000	11	8	0	19
2001	12	6	4	22
2002	0	2	3	5
2003	0	0	0	0
Total	40	53	9	102

Source: Erhemjamts (2005), pp. 31–33.
Notes: MHC is mutual holding company.

In the U.S., the nature of the affiliations of financial entities has important implications for how regulations apply to affiliated corporate entities. Since mutual insurance companies must be owned by policyholders, they cannot be affiliated with other financial services companies under a single, non-policyholder-owned financial holding company: mutuals can own other financial services companies, but other financial services companies cannot own mutuals. The ability of mutuals to affiliate (i.e. own) other financial entities is further limited by the fact that companies engaging in insurance activities, generally speaking, have additional, more restrictive requirements for, among other things, asset quality and capital and surplus levels. Because of these requirements, it is generally unwise for insurers, whether of the stock or mutual form, to hold non-insurance companies as subsidiaries. In an environment in which both the depth and breadth of the financial services that a company can provide has implications for its competitiveness in the marketplace, this lack of flexibility for mutual insurers is seen as a real disadvantage.

The choice of organizational form is likely to influence and/or reflect other aspects of insurers' operations. The efficient structure hypothesis states that the effectiveness with which insurers are able to balance the management, risk-bearing, and customer incentive conflicts, for example, has implications for the optimal organizational form that they should assume in practice (Mayers and Smith 1981, 2002). Stock firms, for example, are characterized by a separation of the roles of management, owners, and policyholders. This is thought to promote specialization within the function that each performs, thus resulting in lower costs relative to mutual companies where the roles of shareholder and policyholder are combined (Mayers and Smith 1988).

On the other hand, the dual role of policyholders as customers and owners of mutual firms is likely to have implications for firm management and company operations. For example, since shareholders are primarily concerned about returns they may want to take greater risks (i.e., by investing assets backing policy reserves in high- rather than low-risk assets) in order to increase the level of expected profits. Policyholders on the other hand, are primarily concerned about solvency and would, as a consequence, generally prefer less risky allocations of firm assets, ceteris paribus. To the extent that mutual insurers are able to more effectively manage the conflict between these two functions (i.e., by merging them into a single owner-policyholder), they may, under some circumstances, be more profitable than stock insurers.

The result is that the choice of organizational form is likely to have implications for the efficiency and profitability of insurance companies. It may even be that stock and mutual organizations can be thought of as representing different technologies, since they differ significantly in terms of their respective production, cost, and revenue frontiers (Cummins, et al. 2004). Table 2.12 presents insurer counts and some standard profitability ratios by organizational form for the years 1993–2003. Unsurprisingly, stock firms are found to have greater capital and surplus to liabilities ratios than mutual firms in every year, though the difference appears to have shrunk in recent years. This is consistent with the greater facility with which stock companies are able to raise capital through equity offerings for example. The table also reports that for most of the early years occurring in the period 1993 to 2003 stock insurers had higher expense ratios than mutual insurers on average, while mutuals had greater benefit ratios. These differences, however, attenuated and even occasionally reversed with the wave of demutualizations (see Table 2.11 above) that began in the late 1990's and early 2000's.

2.4.6 Distribution Systems

In the U.S., life insurance sales can be categorized both by the type of financial services firm associated with the marketing and by the system used to sell or distribute insurance products.[65] In some cases there is a distinct relationship between the two. Increasingly, companies selling L-H insurance products use a range of different distribution channels to sell their products. The various channels available to insurers for distribution of their products can be grouped into three main

[65] This section relies heavily on Black and Skipper (2000), Chapter 24.

categories, including marketing intermediaries, direct response marketing, and other (non-insurer) financial institutions (Black and Skipper 2000, p. 602).

Table 2.12. U.S. Life-Health Various Standard Ratios by Organizational Form, 1993–2003

		C&S Ratio[1]			Expense Ratio[2]			Benefit Ratio[3]		
Year	Total	Stock	Mutual	Other[4]	Stock	Mutual	Other[4]	Stock	Mutual	Other[4]
1993	1,140	0.126	0.094	0.828	0.207	0.167	0.150	0.670	0.846	0.804
1994	1,031	0.125	0.096	0.792	0.203	0.169	0.150	0.692	0.903	0.825
1995	1,012	0.130	0.105	1.025	0.200	0.186	0.152	0.748	0.979	0.866
1996	953	0.142	0.112	0.952	0.200	0.175	0.160	0.769	0.936	0.842
1997	838	0.154	0.118	1.004	0.187	0.188	0.161	0.775	1.012	0.849
1998	794	0.152	0.122	0.791	0.184	0.180	0.178	0.744	0.964	0.836
1999	712	0.156	0.125	0.819	0.178	0.181	0.160	0.848	0.964	0.835
2000	659	0.154	0.123	0.692	0.169	0.179	0.151	0.827	0.895	0.841
2001	613	0.134	0.122	0.800	0.200	0.185	0.134	0.797	0.666	0.845
2002	578	0.120	0.113	0.751	0.185	0.185	0.134	0.733	0.633	0.825
2003	548	0.125	0.123	0.820	0.196	0.183	0.141	0.749	0.612	0.810
		Average Annual Changes from 1993 to 2003 (%)								
		–0.1	2.7	–0.1	–0.6	0.9	–0.6	1.1	–3.2	0.1

Sources: NAIC (1993–2003) and A.M. Best (1993–2003).

Notes: [1]The C&S ratio is equal to the ratio of capital and surplus to liabilities. [2]The expense ratio is equal to the ratio of expenses and commissions to net premiums including deposit-type considerations prior to 2001. Commissions and expenses include payments on both direct and assumed business, general insurance expenses, insurance taxes, licenses and fees, and increases in loading and other miscellaneous expenses. They exclude commissions and expense allowances received on reinsurance ceded. [3]The benefit ratio is equal to benefit payments to net premiums including deposit-type considerations prior to 2001. Benefits include death benefits, matured endowments, annuity benefits, disability/accident/health benefits, surrender benefits and other fund withdrawals, group conversions, coupons and payments on supplementary contracts, and other miscellaneous benefits. [4]Other includes several medical indemnity plans and insurers with missing or unidentified organizational forms.

Marketing intermediaries represent the most traditional and by far the most prevalent of the three marketing channels. Distribution through intermediaries accounts for over 90 percent of all new insurance sold in the U.S. Intermediaries include insurance agents and brokers and salaried insurer employees. They interact with potential purchasers through face-to-face contact in order to execute a sale. Marketing intermediaries generally belong to one of two basic types of distribution organizations: agency-building distribution channels in which the insurer actively hires, trains, and otherwise organizes a sales force of its own and those in which it facilitates, organizes, and authorizes sale of its products through individuals (and groups of individuals) that operate independent of any single insurer.

Most individual life insurance is sold through the agency-building channel by insurance agents that are directly tied to the insurer backing the insurance products sold. In the U.S., this kind of distribution takes several specific forms including the following: career agents, (multiple-line) exclusive agents, and salaried employees of insurers. Career agents are the most common form of an agency-building intermediary for the L-H industry. These intermediaries operate on a commission basis and are organized under either a branch system or a general agency system. Branch systems are the primary distribution system for most large U.S. life insurers. Under this arrangement, sales agents are organized under a manager that is employed directly by the insurer and who bears the responsibility for hiring, training, and otherwise providing the information, clerical services, and other assistance necessary for successful operation within a given geographic area. General agency systems are similar, but are managed by a general agent who typically receives less financial and logistic support from the insurer, but receives a more generous commission arrangement based on sales in the region over which he is responsible. Career agents generally operate from a common or shared office site associated with their sponsoring insurance company.

The other two types of agency-based distribution channels, exclusive agents and salaried employees, represent significantly smaller shares of new individual life insurance sales. Exclusive agents sell a range of insurance products for a single insurer or for a single group of affiliated insurers, including both P-L and L-H insurance products. They operate with less oversight and support from insurers than career agents and typically have their own office as opposed to operating out of a shared office site. A small, but growing number of L-H insurers employ this form of distribution. The recent growth in the use of this distribution system is attributed to lower policyholder attrition rates and to economies of scope associated with producing a range of different insurance products (Black and Skipper 2000, p. 605). Salaried employees of insurance companies are most frequently used to market group life and health insurance products. These salespeople are paid almost exclusively through salary arrangements, though sales bonuses are also used to stimulate and encourage sales efforts.

Among the major non-agency distribution channels for L-H insurance are brokerages and producer groups. Brokers are insurance salespeople who operate on a largely independent basis from any one insurer. Generally, their compensation arrangement is commission-based and they are not required to meet minimum sales levels in order to maintain their association with the insurers whose products they sell. Life-health insurance brokers sell policies for multiple insurers often in the high-end markets (i.e. relatively more expensive coverage usually associated with complexity or difficulty of modeling) where their additional expertise is most valuable. Producer groups are marketing organizations that exist independent of any particular insurance company. These organizations negotiate special arrangements with insurers that provide more generous commissions, but without the extensive support services needed by most other agent categories. Insurer support services are not needed because these groups usually have developed their own sophisticated sales and research support systems which effectively provide them with a competitive edge as they seek to match insurers with customers.

In addition to intermediary channels, many insurers use direct response marketing to reach potential customers. Direct response marketing consists of sales activities involving unmediated direct contact with potential customers. Typical channels used for this kind of marketing include television advertisements, telephone solicitations, direct mailings, and the Internet. While this form of marketing and distribution is relatively insignificant (only 2 percent of 2001 first year life insurance premiums), it may become more important as customers become increasingly sophisticated and are therefore more willing to deal directly with insurers as opposed to going through intermediaries (III 2005b, p. 15). Direct response marketing methods are also likely, however, to continue to be less effective at mitigating information asymmetry problems compared to agency- or broker-based methods, where the intermediary provides additional underwriting services, such as an increased ability to gather information, associated with selling to customers through direct interactions.

Insurers are not the only financial institutions that sell insurance products. Among the other types of financial services firms selling life insurance products are commercial and investment banking institutions, mutual fund organizations, credit unions, and others. Non-insurance financial institutions may offer insurance products to their customers in an effort to increase the range of financial services that they are able to provide to their customers. While these companies have a relatively small share of the total L-H market (less than 5 percent of new sales), they have a large share of the variable insurance and annuities markets (in excess of 20 percent of new annuity sales; Black and Skipper 2000, p. 602). In a 1995 decision, the U.S. Supreme Court decided that annuities are general financial products and should not be categorized strictly as insurance products. This decision accelerated the penetration of non-insurance organizations into the market for annuities.

Table 2.13 provides the allocation of first year premiums by primary distribution system and line of business for 1993, 1998, and 2003. On the basis of first year net premiums, agents and brokers compose nearly all new sales of insurance products in all lines of business. In 1993, for example, they comprised almost 90 percent of new sales of insurance products. In 1998 and 2003, this proportion increased to over 94 percent. While the bulk of total marketing is through the agency and brokerage channels, individual lines of business differ somewhat in the proportion of sales they produce. With the exception of individual and group annuities in 1993 and the accident and health insurance segment, the non-intermediary marketing methods of direct response and mass marketing have produced less than 10 percent of new sales in the other lines of business. On the other hand, accident and health insurance coverage are the least likely to be sold via the two traditional methods of agency and brokers. For this line of business, direct response marketing appears to have increased in popularity significantly in recent years, reaching almost 15 percent in 2003.

Table 2.13. U.S. Life-Health First Year Net Premiums by Primary Distribution System

Year	Distribution System	Total[1] (%)	Individual Life (%)	Individual Annuities (%)	Group Life (%)	Group Annuities (%)	Accident and Health[2] (%)
1993							
	Agency	78.0	89.5	65.1	78.4	79.6	72.9
	Brokerage	10.0	6.2	10.1	11.8	4.7	21.7
	Direct Response	10.8	3.2	24.5	3.5	15.6	2.2
	Mass Marketing	1.3	1.2	0.3	6.3	0.0	3.2
1998							
	Agency	81.0	89.3	76.8	90.5	21.5	69.6
	Brokerage	14.2	9.5	17.5	7.3	70.4	13.8
	Direct Response	4.1	1.1	5.7	0.5	8.1	12.6
	Mass Marketing	0.7	0.2	0.0	1.7	0.0	4.0
2003							
	Independent Agency	60.4	73.7	59.4	78.4	57.2	49.2
	Career Agents	15.0	5.8	16.6	2.2	23.7	4.7
	Exclusive Agents	1.9	2.3	2.4	0.6	0.0	1.6
	Brokerage[4]	18.7	15.7	18.4	17.2	16.5	29.9
	Direct Response	3.9	2.6	3.1	1.6	2.7	14.6
	Other[5]	0.1	0.0	0.1	0.0	0.0	0.0

Source: NAIC (1993–2003) and A.M. Best (1993–2003).
Notes: This table excludes deposit-type contract considerations and insurers for which the distribution system is missing, unavailable, or otherwise unknown. [1]Total includes other miscellaneous lines of business not explicitly listed. [2]Accident and health includes group, credit, and other. [3]A.M. Best changed the basic categories they report for distribution system between 1998 and 2003. [4]Brokerage includes brokers and general agents. [5]Other includes reinsurance, banks, and other miscellaneous categories.

While first year net premiums provide a useful picture of new insurance business written in any given year, they ignore ongoing considerations for policies initiated in prior years. Since many L-H insurance products involve long periods of premium payment prior to contract termination (either by voluntary liquidation or by incidence of insured events), the importance of renewal premiums can be considerable. Figure 2.3 depicts, on a net premium basis, the prevalence of the three categories of distribution for each line of business in 2003. With the exception of credit life insurance, net premiums are primarily generated through the same source that first year premiums are generated, through marketing intermediaries such as insurance agents and brokers. Table 2.14 provides market shares, expense ratios, and benefit ratios by marketing channel for 1993 to 2003 on the basis of total net premiums. With a few notable exceptions,[66] direct marketing insurers have lower expense and benefit ratios than insurers that use marketing intermediaries. These differences, however, appear to have decreased during the period of interest and are less meaningful by 2003.

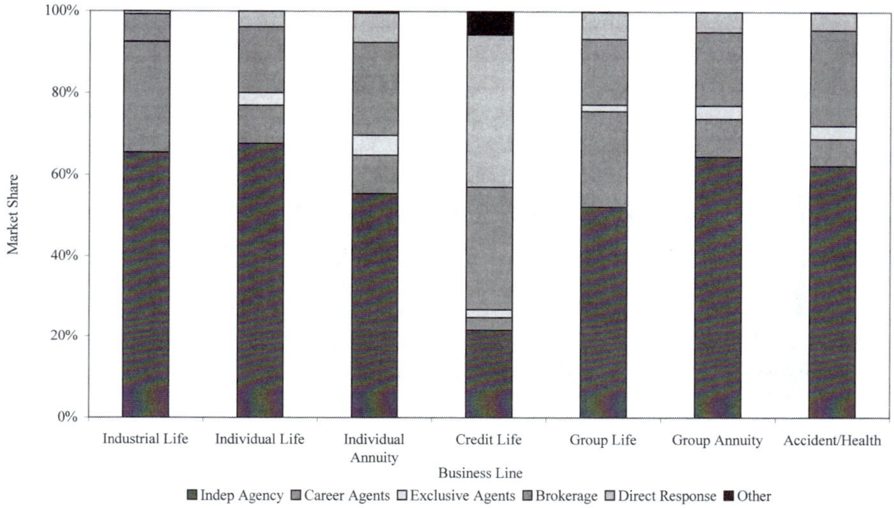

Source: NAIC Annual Statement CD (1993–2003) and A. M. Best Annual Data CD (1993–2003).
Notes: Market share is defined as the percentage of net premiums including deposit-type.

Figure 2.3. Market Shares in the U.S. Life-Health Industry by Distribution System, 2003

2.4.7 L-H Policy Taxation

In the U.S., life-health insurance product purchase is frequently motivated and certainly influenced by the favorable tax benefits that they sometimes confer on their beneficiaries and/or policyowners.[67] The tax benefits associated with life-health insurance products come in three basic forms: deductibility of premiums, tax-free status of death proceeds, and the tax deferral of interest and/or investment income. The deductibility of premiums plays a significant role in health insurance markets, but a lesser role in decisions regarding the purchase of life insurance and annuities. The tax-free status of death benefits and the deferral of interest and investment income, on the other hand, exhibit a converse relationship with the two sectors. They are important considerations for life insurance and annuity purchases, but of no great consequence in decisions regarding health insurance.

[66] These may be driven by reporting errors as the distribution system is reported by insurers and may change over time, depending on what they consider to be their primary form of distribution.

[67] For this section, the authors relied heavily on Black and Skipper (2000), Chapter 13.

Table 2.14. U.S. Life-Health Market Share, Expense Ratio, & Benefit Ratio by Distribution System, 1993–2003

	1993	1994	1995	1996	1997	1998	1999	2000	2001	2002	2003
Market Share[1] (%)											
Agency[2]	80.6	80.0	78.1	75.9	75.1	75.5	78.5	82.7	79.5	70.1	61.2
Career Agents[3]	N/A	N/A	N/A	N/A	N/A	N/A	N/A	N/A	N/A	7.5	9.5
Exclusive Agents[4]	N/A	N/A	N/A	N/A	N/A	N/A	N/A	N/A	N/A	2.1	3.6
Broker[5]	12.2	14.3	15.9	17.1	15.0	20.2	17.6	13.7	16.1	15.7	20.1
Direct	5.1	4.1	3.1	5.0	9.6	4.0	3.9	3.6	4.5	4.4	5.4
Mass Marketing[6]	2.1	1.6	2.9	2.0	0.4	0.4	N/A	N/A	N/A	N/A	N/A
Other[7]	N/A	N/A	N/A	N/A	N/A	N/A	N/A	N/A	N/A	0.2	0.2
Expense Ratio[8]											
Agency[2]	18.9	18.8	19.3	19.3	19.3	19.4	18.2	17.4	19.3	18.1	19.0
Career Agents[3]	N/A	N/A	N/A	N/A	N/A	N/A	N/A	N/A	N/A	19.4	19.9
Exclusive Agents[4]	N/A	N/A	N/A	N/A	N/A	N/A	N/A	N/A	N/A	20.8	24.5
Broker[5]	21.2	20.4	20.9	18.9	16.5	13.9	14.9	14.4	20.0	20.1	21.5
Direct	18.3	16.6	16.5	16.4	13.4	14.8	17.7	15.5	16.9	14.2	20.2
Mass Marketing[6]	22.0	21.8	23.5	24.5	42.3	46.5	N/A	N/A	N/A	N/A	N/A
Other[7]	N/A	N/A	N/A	N/A	N/A	N/A	N/A	N/A	N/A	14.9	16.9
Benefit Ratio[9]											
Agency[2]	75.6	80.3	87.3	88.3	89.0	86.6	90.4	85.5	79.0	69.7	72.6
Career Agents[3]	N/A	N/A	N/A	N/A	N/A	N/A	N/A	N/A	N/A	76.3	71.3
Exclusive Agents[4]	N/A	N/A	N/A	N/A	N/A	N/A	N/A	N/A	N/A	57.0	68.1
Broker[5]	68.5	66.6	70.4	68.8	70.0	62.7	79.5	75.5	76.2	78.2	82.1
Direct	49.7	49.8	61.1	63.6	70.3	64.4	72.2	82.1	73.5	69.0	78.3
Mass Marketing[6]	73.5	80.6	86.4	63.7	41.4	48.1	N/A	N/A	N/A	N/A	N/A
Other[7]	N/A	N/A	N/A	N/A	N/A	N/A	N/A	N/A	N/A	41.9	52.8

Sources: NAIC (1993–2003) and A.M. Best (1993–2003).

Notes: N/A is Not Available. [1]Market share is based on net premiums excluding deposit-type considerations beginning in 2001. [2]Beginning in 2000, Agency excludes career agents (in career agents category), exclusive agents (see exclusive agents category), managing general agents (see broker category), and general agents (see broker category). [3]Career Agents were included in the Agency category prior to 2002. [4]Exclusive Agents were included in the Agency category prior to 2002. [5]Beginning in 2000, Broker included general agents. [6]Beginning in 1999, Mass Marketing was included in the Direct category. [7]Beginning with 2002, Mass Marketing included reinsurance, banks, and other miscellaneous categories. Prior to this, the categorization did not exist. [8]The expense ratio is equal to expenses and commissions divided by net premiums, including deposit-type considerations prior to 2001. Commissions and expenses include payments on both direct and assumed business, general insurance expenses, insurance taxes, licenses and fees, and increases in loading and other miscellaneous expenses. They exclude commissions and expense allowances received on reinsurance ceded. [9]The benefit ratio is equal to benefit payments to net premiums including deposit-type considerations prior to 2001. Benefits include death benefits, matured endowments, annuity benefits, disability/accident/health benefits, surrender benefits and other fund withdrawals, group conversions, coupons and payments on supplementary contracts, and other miscellaneous benefits.

The proliferation of employer-provided medical benefit plans during the second half of the 20th century is directly related to the favorable tax treatment given to those plans. Under the U.S. tax code, premiums paid by employers as a part of a group insurance arrangement for healthcare and related benefits are not considered to be taxable income for their employees. This is tantamount to the full deductibility of health insurance premiums. In addition to any group insurance premiums paid on behalf of their employees, employers are also able to include flexible spending provisions in their plans. These provisions allow employees to deduct additional health-related expenses made during the year. While the premiums paid by policyholders for life insurance and individual annuity products are generally not deductible for income tax purposes, there are several cases in which they are deductible. Premiums paid into annuity products restricted for use as retirement savings vehicles (i.e., tax-qualified arrangements) can be deductible up to a prescribed maximum amount, as can premiums paid for certain healthcare benefits, such as medical expenses and long-term care (LTC) insurance), when they exceed 7.5 percent of income in a year. LTC insurance premiums are limited to maximum levels which vary by and are generally increasing with age.

For life insurance and annuity products, the greater tax incentives stem from the tax-free nature of some contract benefits and the tax deferral of interest income on policy accumulations. An insurance contract is required to meet two basic requirements in order to be considered life insurance for the purpose of receiving tax-favored treatment at the federal level. First, it must meet the requirements of applicable state laws defining life insurance, and second, it must satisfy one of two standards. For any given policy, the standard to be followed is set out in the policy form prior to purchase and must be satisfied throughout the life of the contract. The first standard limits the size of the cash value to be no greater than the net single premium, calculated on a specified actuarial basis. Alternatively, the second standard limits the cumulative amount of premiums paid (relative to the size of the policy net single premium) and the size of policy cash value relative to its death benefits. These requirements are meant to limit the use of life insurance contracts as tax-deferred investment vehicles rather than as protection against mortality risks.

Insurance products that meet the requirements to be considered life insurance for tax purposes fall under one of two possible sets of tax rules. The difference between the two categories revolves around the intended purpose of the insurance product. In particular, according to the IRS, tax treatment for any life insurance or annuity product depends on whether it is deemed to have a large investment and savings element or whether it is primarily intended to provide some kind of insurance protection. Both categories receive favorable tax treatment, but they are treated differently. Contracts with large investment and savings elements are called modified endowment contracts (MECs), while those without the large investment and savings element are non-MECs.

MECs are defined to include all contracts for which premiums are scheduled to be paid on a basis quicker than one-seventh of the net single premium (the seven-pay test). These products are treated differently by tax regulators because due to their short accumulation stage, they are marketed and purchased primarily as tax-deferred investment vehicles, rather than as insurance against mortality risk. Policies that are scheduled to be paid over a longer period naturally fall into the non-MEC category for tax purposes.

For all policies meeting the definition of life insurance (both MECs and non-MECs), death benefits and interest accumulations on policy cash values are tax exempt. For any year in which a contract fails to meet the requirements necessary to be considered life insurance, however, it loses its tax-favored status on a portion of policy interest accumulations for that year. Death benefits cease to be exempt from taxation only under relatively rare circumstances, for example, when a life policy or any interest in such a policy is sold and its benefits are transferred to another party. In this case, unless the policy is transferred as a gift, the new owner is the insured himself, or the transfer is to a corporation, partnership, or business partner of the insured, the purchaser of the policy is must pay taxes on the excess of death proceeds over the sum of the sale price and any net premiums paid on the policy by its new owner. Accelerated death benefits (benefits paid prior to death in the event of a terminal illness) and viatical settlement proceeds (the transfer of one's right to policy death benefits) are treated like death benefits, subject to a medical diagnosis of imminent decease, and thus, are tax-exempt.

For non-MEC life insurance products, dividends are not taxed, but interest earned if left to accumulate is. When such a policy is surrendered, the excess of the proceeds over the accumulated premiums paid net of dividends received is considered income and subject to taxation. Non-MEC endowment benefits are treated in much the same way.

For MEC products, both dividends and their interest accruals, if they are left to accrue, are taxed unless they are used to purchase paid-up additions (additional amounts of insurance). Policy surrender proceeds, including interest accumulations, are typically taxed to the extent that they have not already been taxed. Generally speaking, this means that proceeds in excess of contributions (premiums) and dividends paid (which would have already been taxed as stipulated above) are subject to taxation. Since the tax benefit for these products has the primary purpose of encouraging individuals to provide for themselves in retirement, a 10 percent excise tax is added as a penalty for distributions taken before age 59 ½.

Certain and life annuity proceeds are taxed differently depending on whether they occur during the accumulation or decumulation stages of the contract. During the accumulation stage, the general rule is that accumulations are tax-deferred, while distributions (dividends, policy loans, cash-value withdrawals, etc.) are subject to tax if they have not already been taxed (i.e., excess of distribution amounts over contributions net of dividends received). Similar to MEC product distributions, a 10 percent excise tax applies for the taxable portion of early (before age 59 ½) distributions. The excise tax is waived under certain circumstances.

Taxation of annuity products during the decumulation phase is similar to that of life insurance settlement options where the beneficiary chooses to receive the benefit over a period of time as opposed to all at once in lump sum form. In the case of life insurance settlement options, each payment received is divided into an interest portion and a principal portion based on the particular option chosen and the composition of the funds backing the contract vis-à-vis its division into taxable and tax-exempt portions. The interest portion is considered taxable income, while the principal portion is not. For annuities, the tax-exempt portion of the payment corresponds to the proportion of the accumulated funds backing the contract that consists of annuity contributions less dividends paid.

2.4.8 Financial Condition & Performance

An insurer's separate account consists primarily of the assets backing its variable life insurance, annuity contracts, and its pension contracts. The role that insurers play in managing the assets that comprise this account can range anywhere from active involvement in selecting investments (e.g., selecting individual securities in line with broad customer-defined guidelines) to being primarily administrative in nature (i.e., customer makes the investment decisions and the insurer merely carries them out). In all cases, the policyholder or customer bears the investment risk. This characteristic of separate account products distinguishes them from other insurance products and allows insurers to effectively avoid the usual investment restrictions imposed by state regulations. An insurer's general account, on the other hand, holds assets backing contracts that include guaranteed fixed-dollar benefit obligations, the types of payments made for standard life insurance and annuity arrangements. As Table 2.15 shows, in 2003, the general accounts assets comprised approximately 68.9 percent of total assets, or $2.6 trillion. Of this, approximately 72.0 percent ($1.9 trillion) were bonds—55.4 percent corporate and 16.6 percent government issues. Approximately 22.4 percent of these were privately placed (the terms of sale were negotiated directly between the insurer and the bond issuer), while the remaining 77.6 percent were publicly placed. Separate accounts totaled 31.1 percent (approximately $1.2 trillion) of total assets.

Figure 2.4 depicts the growth of general and separate account assets from 1993 to 2003. The figure suggests a difference in the respective sensitivities of the two asset accounts to stock market performance. During much of the past decade and a half, equities outperformed fixed income investments and this is reflected in the growth of the separate accounts balance relative to that of the general account assets during that period. Beginning in the middle of 1999, the U.S. stock market began a period of stagnation that has persisted to the present. From the middle of 1999 to 2003, in the aggregate, insurers experienced capital losses on their separate accounts assets, whereas general account assets, which primarily consist of non-equity investments, continued to grow.

Table 2.16 provides a fairly detailed description of year-end 2003 asset allocations for the general account. Corporate bonds are the largest category for the general account at 58.3 percent ($1.45 trillion) of total general account assets. Of this amount, the largest two subcategories are unaffiliated corporate issues at 52.0 percent ($1.3 trillion) of total general account assets and public utilities at 5.2 percent ($130 billion) of total general account assets. Most corporate bond holdings (71.3 percent) consist of publicly traded issues, though private placements (the remaining 28.7 percent) also comprise a significant proportion of bonds held. Equities account for just 3.9 percent of general account assets, a smaller class than mortgage and policy loans which constitute 10.0 and 4.1 percent of the total, respectively.

Figure 2.5 presents the distribution of total assets (i.e., combined account = separate account + general account) for the L-H insurance industry during much of the past century (1920 to 2003). The two most noteworthy trends in the combined account asset mix over the last century are (1) the relatively sharp increase in the proportion of these assets held in equities (no doubt associated with the increase in separate accounts business in recent decades, see below) and (2) the significant decrease in the share of assets held in mortgage loans.

Table 2.15. U.S. Life-Health Industry Pro-Forma Balance Sheet, 12/31/2003 ($ Millions)

Assets		**Liabilities**	
Bonds		Policy Reserves	$1,813,254
Government[1]		Deposit-type Contracts	284,165
Private Placement	$ 7,794	Policy Claims	36,609
Publicly Traded	429,116	Dividend Reserve	18,038
Corporate[2]		Asset Valuation Reserve	28,235
Private Placement	415,681	Interest Maint. Reserve	14,110
Publicly Traded	1,035,136	Other	207,938
Total	$1,887,727	**Total Liabilities Excl S/A**	$2,402,349
Preferred Stocks	28,943	Separate Accounts Liabilities	1,176,753
Common Stocks	61,188	**Total Liabilities**	$3,579,102
Mortgage Loans	253,198	**Capital & Surplus**	$ 218,755
Real Estate	19,814		
PolicyLoans[3]	103,849		
Cash & Other Investments	134,212		
Total Cash & Invested Assets	$2,488,931		
OtherAssets[4]	129,248		
Total Assets Excl S/A	$2,618,179		
Separate Accounts Assets	1,179,678		
Total Assets	$3,797,857		

Source: A. M. Best (2004a).
Notes: [1]Government includes U.S. federal, state, and territory governments and their political subdivisions, foreign governments, and special revenue and assessment projects. [2]Corporate includes public utilities, credit tenant loans, industrial, and miscellaneous other. [3]Policy loans include premium notes for 2003. [4]Other Assets include premiums due, accrued investment income, and other miscellaneous assets.

Table 2.17 illustrates the differences between general and separate account asset allocations as of the year-end 2003. Perhaps the most obvious difference between separate and general account allocations is that general account assets are heavily weighted (71.8 percent) towards long-term bond investments, while separate account assets are heavily weighted (78.0 percent) towards equity securities. General account assets are more conservatively invested because they constitute investments that back reserve liabilities and, as such, represent exposure to risk on the part of insurers who must use those assets eventually to meet their contract obligations. Assets held in the general account are also subject to regulatory requirements regarding their quality and regarding their adequacy (i.e., their magnitude) relative to contract liabilities. An important implication associated with the difference in how these two accounts are invested is that the combined return on assets for any given insurer and for the industry in the aggregate will depend in part on the mix of separate account and general account business.

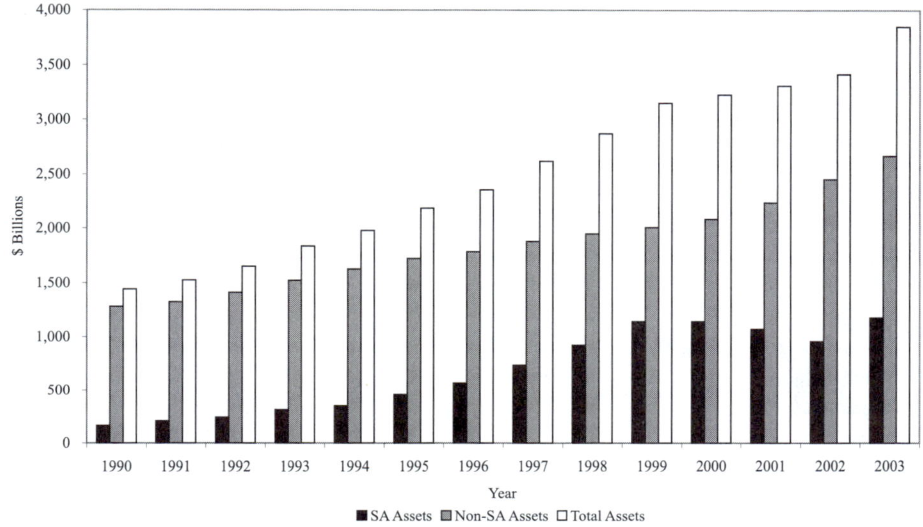

Source: NAIC Annual Statement CD (1993–2003).
Notes: SA is separate account.

Figure 2.4. Growth of U.S. Life-Health Insurer Assets, 1993–2003

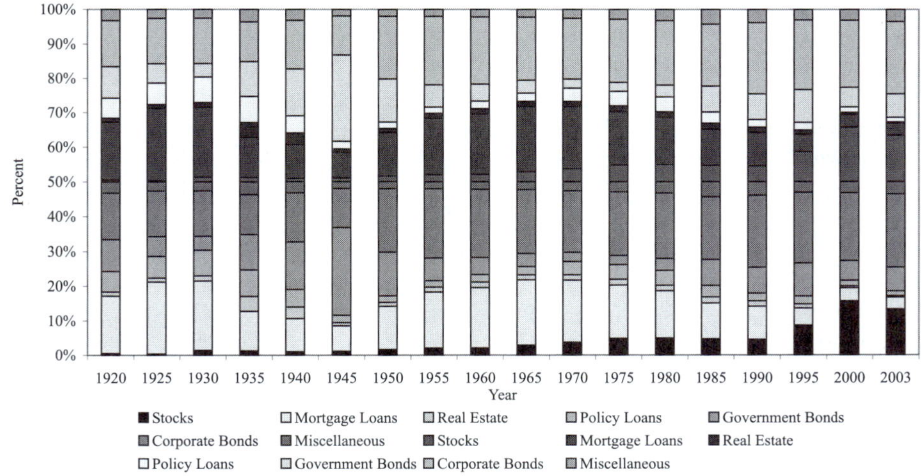

Source: ACLI (2004), p. 31.

Figure 2.5. U.S. Life-Health Industry Total (Combined Account) Asset Allocations by Year, 1920–2003

Table 2.16. U.S. Life-Health Detailed Distribution of General Account Invested Assets, 2003

Asset Category	Publicly Traded ($ Millions)	Privately Placed ($ Millions)	Total ($ Millions)	Total (%)
Government Bonds				
U.S. Federal Government	121,470	519	121,989	4.9
Foreign Governments	32,990	2,731	35,721	1.4
U.S. States and Territories[1]	16,498	1,712	18,210	0.7
Other[2]	258,158	2,832	260,990	10.5
Subtotal Government	429,116	7,794	436,910	17.5
Corporate Bonds				
Unaffiliated Corporations[3]	924,564	369,563	1,294,128	52.0
Public Utilities	100,415	29,661	130,076	5.2
Credit Tenant Loans	680	7,172	7,852	0.3
Other[4]	9,477	9,285	18,762	0.8
Subtotal Corporate	1,035,136	415,681	1,450,817	58.3
Total			1,887,727	75.8
Stocks				
Mutual Funds			6,857	0.3
Preferred			27,917	1.1
Publicly Traded (excl. preferred)			32,081	1.3
Other			28,897	1.2
Total			95,752	3.9
Mortgage Loans				
Commercial			213,544	8.6
Other			35,264	1.4
Total			248,808	10.0
Real Estate			20,326	0.8
Policy Loans[5]			100,895	4.1
Cash & Short-Term Investments			75,426	3.0
Other Invested Assets			69,827	2.4
Total Invested Assets			2,488,931	100.0

Source: A. M. Best (2004a), pp. 4, 89, 97–98.

Notes: [1]U.S. States and Territories include the political subdivisions of states and territories. [2]Other government bonds include special revenue and assessment obligations. [3]Unaffiliated corporations exclude public utilities. [4]Other corporate bonds include parent, subsidiaries, and affiliates. [5]Policy loans include premium notes for 2003.

The main liability items for L-H insurers are policy reserves (for products involving life contingencies as well as for deposit-type contracts), policy claims, dividend reserves, and asset valuation reserves. Policy reserves are by far the largest item at almost $2 trillion. These liabilities consist of approximately $722 billion of life insurance reserves (industrial, ordinary, credit, and group), $974 billion of annuity reserves (individual and group), and approximately $117 billion of reserves for other miscellaneous types of coverage (A. M. Best 2004a). Interest maintenance and asset valuation reserves (IMR and AVR) are of relatively small orders in comparison, but fulfill statutory requirements designed to provide a cushion against the effects on surplus of adverse investment returns and interest rate movements. The AVR absorbs both realized and unrealized capital gains and losses that are not associated with interest rate fluctuations, while the IMR absorbs only realized gains and losses associated with market interest rate changes. Separate accounts represent a little over 1 trillion in liabilities, approximately 85 percent of which are for individual and group annuities (A. M. Best 2005, p. 14).[68] Capital and surplus for the industry reached $218 billion in 2003, up 12 percent from 2002. Figure 2.6 provides the capital-to-asset ratios for the L-H insurance industry from 1983 to 2003. As assets have increased over the thirty-year period, capital-asset ratios have declined slightly from over 7 percent to under 6 percent.

Table 2.18 provides a pro-forma income statement for the L-H industry in 2003. Total income, which was $710 billion in 2003, consists of two major items: premium income and investment income. Premium income represents 70.3 percent of the total, while investment income represents 20.2 percent. The remaining 9.5 percent comes from commissions, expense allowances, and reserve adjustments on reinsurance ceded, income from fees associated with administration of separate accounts, and charges and fees associated with deposit-type contracts. Expenses include benefits paid, changes in policy reserves, commissions, dividends to policyholders, and miscellaneous other operating expenses. Of these, benefit payments ($361 billion) is the largest expense at 53.8 percent of the total. Of this amount surrender benefits and withdrawals constitute the largest category ($173 billion), disability payments the second largest ($82 billion), and annuity and death benefits each at approximately $50 billion. Increases in policy reserves refer to changes in policy reserves from the prior year. In 2003, commissions amounted to roughly 9.2 percent of total premium considerations at $46 billion. Federal income taxes amounted to 19.7 percent of the net gain before taxes, and, in 2003, the industry experienced net capital losses in the amount of $4.6 billion.

Tables 2.19 and 2.20 provide a picture of the operating results and profitability of the L-H insurance industry from 1993 to 2003. Due to the long-term nature of life insurance and annuity contracts and the relatively disparate timing of the benefits versus premium considerations, loss ratio measures are not frequently used to measure efficiency of operation as they are in both the property-liability industry and in accident and health insurance lines of business, where contracts tend to be short-term (often with a single year duration). As a consequence, Table 2.19 only presents expense ratios for the life and annuity lines of business, while both expense and loss

[68] Separate account group annuities are typically associated with pension plans administered by life-health insurers in behalf of corporations, in contrast to the separate account individual annuities which are most likely to be of the variable type.

ratios are provided for the accident and health insurance (combined) lines of business.

Table 2.17. U.S. Life-Health Distribution of Assets by Account Type, 2003

Asset Category	General Account (%)	Separate Account (%)	Combined Account (%)
Bonds			
Corporate	55.2	10.1	41.3
Government	16.6	6.0	13.3
Subtotal: Bonds	71.8	16.1	54.6
Stocks			
Common	2.6	77.9	25.9
Preferred	1.1	0.1	0.8
Subtotal: Stocks	3.7	78.0	26.7
Mortgages	9.5	0.6	6.7
Real Estate	0.8	0.9	0.8
Policy Loans[1]	3.8	0.1	2.7
Cash & Short-Term Investments	2.9	1.2	2.4
Other Invested Assets	2.7	2.5	2.6
Total Invested Assets	95.2	99.4	96.5
Non-Invested Assets[2]	4.8	0.6	3.5
Total Assets ($ Millions)	$2,618,179	$1,179,678	$3,797,857

Sources: A. M. Best (2004a), pp. 97–98 and ACLI (2004), p. 12.
Notes: [1]Includes premium notes. [2]Non-invested assets include premiums due, accrued investment income, and other miscellaneous assets.

The first thing to note is that in the aggregate across all lines of business there has been very little change in the expense ratio during the past decade. While there were some years in the late 1990s and early 2000s in which the industry achieved improved expense ratios, the ratio of 19.3 percent in 2003 is virtually unchanged from that of 1993; this in spite of the fact that in each separate line of business the expense ratios increased. This is mainly due to the shift in the composition of business towards annuity products (which have a lower expense ratio even after the recent decline than the life lines had before the decline). Individual and group life lines increased from 10.6 and 14.6 percent in 1993 to 17.4 and 17.6 percent, respectively, in 2003. At the same time, individual and group annuity lines increased from 10.7 and 4.2 percent to 11.0 and 7.3 percent. The expense ratio for the accident and health lines of business increased from 27.9 percent to 28.4 percent. The loss ratio for accident and health insurance increased during the middle of the period only to decrease somewhat in recent years.

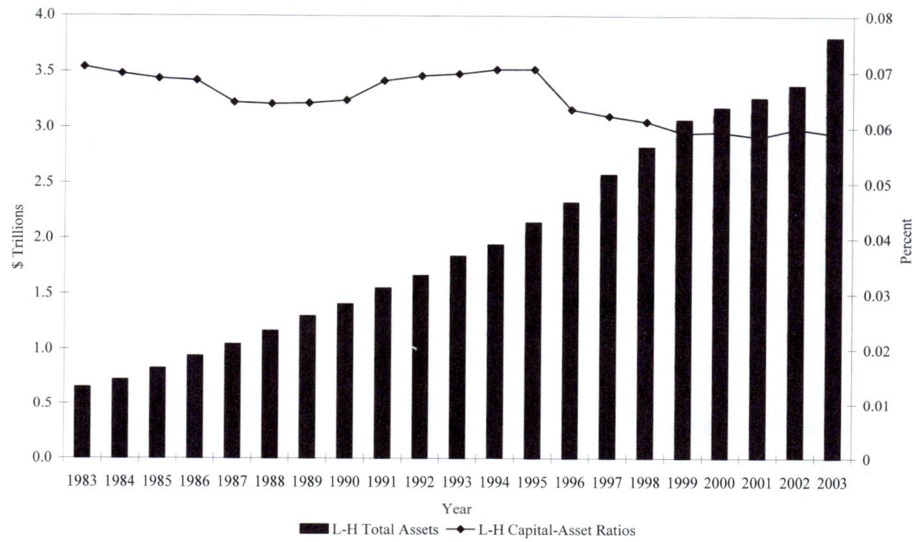

Source: ACLI (2004), p. 31 for assets and pp. 44–45 for capital and surplus data.
Notes: L-H is life-health.

Figure 2.6. U.S. Life-Health Industry Total Assets and Capital-Asset Ratios, 1983–2003

Table 2.20 presents profitability information from 1993 to 2003. These measures fluctuated by relatively small amounts during the period of interest, but appear to have consistently reached their lowest point in 2001 and their highest point in 2003. Return on equity ranged from a low of 8.7 percent in 2001 to a high of 15.5 percent in 2003. During the same period, returns on revenues ranged from 2.2 percent in 2001 to 4.5 in 2003 and returns on assets ranged from 0.5 percent in 2001 to 0.9 percent in 2003.

2.5 THE U.S. PROPERTY-LIABILITY INSURANCE INDUSTRY

2.5.1 Market Structure of the U.S. P-L Insurance Industry

The Unites States has the largest P-L insurance market in the world. U.S. annual P-L insurance premiums represented 43.2 percent of global P-L annual premiums in 2004.[69] More than 1,000 independent firms (insurance groups and unaffiliated single companies) operate in the U.S. P-L insurance market, as shown in Table 2.21.[70]

---

[69] Swiss Re (2005), p. 39.

[70] The information for the property-liability insurance industry is primarily from NAIC (1993–2003). Not all the firms are included in the calculation. We deleted firms that have

Table 2.18. U.S. Life-Health Industry Pro-Forma Income Statement ($ Millions), 2003

Income Statement Item	
Income	
Premium Income[1]	$ 499,551
Net Investment Income	143,655
Other Income	66,811
Total Income	$ 710,017
Expenses	
Benefits Paid	(360,548)
Other Payments[2]	(12,163)
Increase in Policy Reserves	(122,418)
Commissions	(45,806)
Dividends to Policyholders	(18,753)
Other Expenses	(110,607)
Total Expenses Before Taxes	$(670,295)
Net Gain Before Taxes	$ 39,722
Federal Income Taxes	(7,840)
Net Gain After Taxes	$ 31,882
Net Realized Capital Gains (Losses)	(4,586)
Net Income	$ 27,296

Source: A. M. Best (2004a), p. 106.
Notes: [1]Premium Income excludes considerations for deposit-type funds. [2]Other payments include group conversions and interest and adjustments on contract or deposit-type contract funds.

Various property-liability insurance products are sold in this market. Property insurance includes both pure property coverage such as fire insurance, auto physical damage, ocean marine, inland marine, and the property components of multiple peril coverage such as homeowners multiple peril, farm owners multiple peril, commercial multiple peril, and allied lines such as multiple peril crop and federal flood insurance. Liability insurance mainly includes private passenger auto liability, commercial auto liability, medical malpractice, product liability, and other liability. Also an important P-L insurance product in the U.S., workers' compensation differs from property coverage and liability coverage because it is primarily no-fault.

---

negative net worth and highly negative net premiums written or highly negative net premium earned.

Table 2.19. U.S. Life-Health Operating Measures by Line of Business, 1993–2003

	Expense Ratios[3]					Accident and Health[4]		
Year	Total[1]	Ordinary Life[2]	Individual Annuities	Group Life	Group Annuities	Expense Ratio[3]	Loss Ratio[5]	Combined Ratio
1993	19.1	10.6	10.7	14.6	4.2	27.0	69.7	96.7
1994	19.0	9.4	9.2	14.7	4.7	27.9	71.0	98.9
1995	19.4	9.5	10.0	15.9	4.9	28.4	72.9	101.3
1996	19.1	10.2	10.4	15.6	5.4	28.7	73.5	102.2
1997	18.6	10.7	9.9	13.4	5.4	29.2	73.7	102.9
1998	18.3	10.9	12.0	14.8	4.9	30.0	74.7	104.8
1999	17.8	13.9	11.1	16.2	4.8	29.4	75.2	104.6
2000	17.0	13.9	10.2	15.4	4.8	28.3	74.1	102.4
2001	19.7	15.4	10.6	15.3	7.0	29.5	74.2	103.7
2002	18.4	16.3	10.4	16.1	6.9	29.0	72.6	101.6
2003	19.3	17.3	11.0	17.6	7.3	28.4	70.3	98.7

Source: NAIC (1993–2003).

Notes: [1]Total includes life insurance, annuities, accident and health insurance, and other miscellaneous lines of business not explicitly listed. [2]Ordinary life expense ratios include ordinarily life, industrial life, and credit life. [3]The expense ratio is equal to expenses and commissions to net premiums including deposit-type considerations prior to 2001. Commissions and expenses include payments on both direct and assumed business, general insurance expenses, insurance taxes, licenses and fees, increase in loading and other miscellaneous expenses. They exclude commissions and expense allowances received on reinsurance ceded. [4]Accident and health includes group, credit, and other. [5]The loss ratio is equal to benefit payments to net premiums including deposit-type considerations prior to 2001. Benefits include death benefits, matured endowments, annuity benefits, disability/accident/health benefits, surrender benefits and other fund withdrawals, group conversions, coupons and payments on supplementary contracts, and other miscellaneous benefits.

Two important insurance service organizations in the U.S. P-L industry are the Insurance Services Office (ISO) and American Association of Insurance Services (AAIS). The ISO provides statistical, actuarial, and underwriting information to insurers, regulators, agents and other organizations. Additionally, it also provides a standardized policy form for P-L companies. AAIS functions similarly to ISO by providing product development, actuarial analysis, and regulatory filings services for insurance companies and designing insurance policy forms. Both the policy forms from ISO and AAIS are written in standard language and can be compared and packaged with each other.

Table 2.20. U.S. Life-Health Return Measures, 1993–2003

Year	Net Gain After Tax ($ Millions)	Return on Equity[1] (%)	Return on Revenues[2] (%)	Return on Assets[3] (%)
1993	13,963	13.96	3.13	0.80
1994	12,586	11.51	2.66	0.67
1995	15,411	12.88	3.08	0.75
1996	17,207	12.99	3.24	0.77
1997	20,451	13.78	3.61	0.84
1998	15,951	9.92	2.59	0.59
1999	19,598	11.60	2.93	0.66
2000	23,410	13.12	3.17	0.75
2001	16,076	8.72	2.22	0.50
2002	20,040	10.54	2.91	0.60
2003	32,078	15.52	4.49	0.90

Sources: A. M. Best (2004a), pp. 62, 146 and A. M. Best (2002), pp. 2–3, 66.
Notes: [1]Return on Equity (ROE) is calculated as the ratio of post-tax net operating gain to the average of two years' C&S amounts. [2]Return on Revenues (ROR) is calculated as the ratio of post-tax net operating gain to the average of two years' total revenues. [3]Return on Assets (ROA) is calculated as the ratio of post-tax net operating gain to the average of two years' admitted assets.

The number of firms by line of business is shown in Figure 2.7. In 2003, about 375 firms conducted business in private passenger auto liability insurance, 367 firms in private passenger auto physical damage, 367 firms in commercial auto liability insurance, and 325 firms in the commercial auto physical damage. More than 500 firms engaged in fire and allied lines coverage. About 440 firms wrote homeowners insurance businesses and approximately 380 firms underwrote commercial multiple peril insurance. More than 300 firms provided workers' compensation insurance and nearly 600 firms supplied other liability insurance. In contrast, there were a relatively small number (roughly 150 in each line) of firms participating in the medical malpractice, product liability, and reinsurance markets.

In most lines, the number of firms declined from 1993 to 2003. This contradicted the 1980s trend during which the number of firms specializing in the major insurance lines increased (Cummins and Weiss 1992). The phenomenon can be viewed as a signal that the U.S. P-L industry was restructured to improve efficiency of operations since most industry experts believe that the market has become over competitive in the past decades.[71] The emergence of alternative risk transfer (ART) during the period contributed to the competitive nature of the industry. Some firms with bad performance were forced to exit the market and others chose to engage in mergers and acquisitions to compete efficiently with their peers.

Table 2.21 presents the market share by line of business in the U.S. P-L insurance industry. From 1993 to 2003, each year more than 40.0 percent of the

[71] See Bank Director Magazine (2001, 2002).

premiums written in the U.S. P-L industry came from automobile insurance. In 1993, private passenger auto insurance alone composed 37.5 percent of total premiums. Workers' compensation and homeowners multiple peril are the second and third largest categories in the industry. The market share of homeowners insurance increased from 8.6 percent in 1993 to 11.4 percent in 2003 and the market share of workers' compensation insurance declined from 14.3 percent in 1993 to 8.0 percent in 2003. The declining importance of workers' compensation during this period is attributable to workers' compensation reform laws enacted at the state level which reduced claim costs. The average premium growth rate during the period is around 2 to 4 percent, but the growth rate is significantly higher (more than 8 percent) from 2001 to 2003. The higher growth rate coincides with the "hard market" in the U.S. P-L insurance industry that followed the events of September 11, 2001. Specifically, the market experienced a decrease in underwriting capacity and higher premium rates.

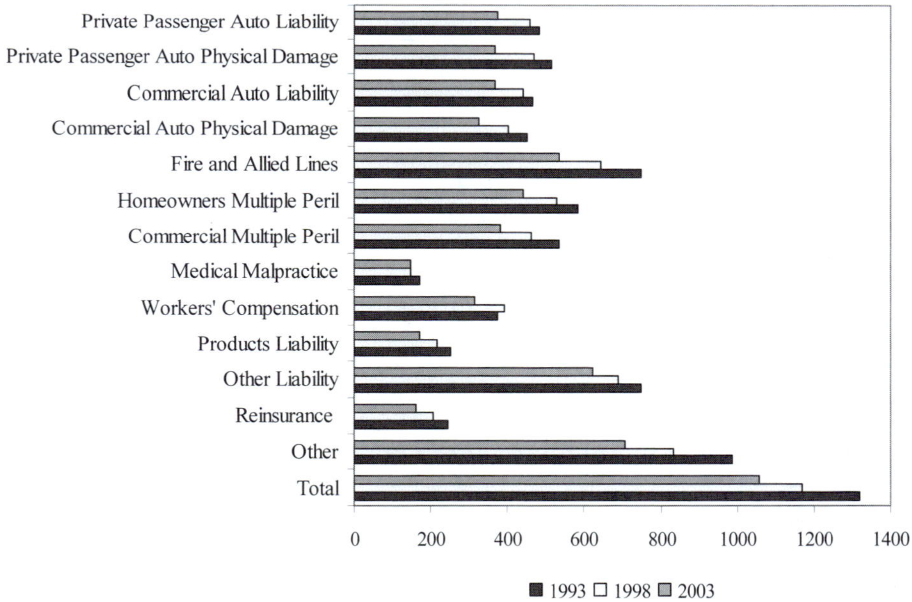

Source: NAIC (1993–2003)
Notes: Includes both group and unaffiliated firms.

Figure 2.7. Number of Firms in the U.S. P-L Industry by Line of Business

In terms of premiums written by state, in 2003, California accounted for the largest proportion of total direct premiums written (12.2 percent) in the nation. Texas and New York accounted for 7.3 percent each, while Florida and Illinois comprised 6.9 percent and 5.0 percent, respectively (see Table 2.4).

Table 2.21. U.S. Property-Liability Insurance Industry Net Premiums Written, 1993–2003 (% of Total)

Line of Insurance	1993	1994	1995	1996	1997	1998	1999	2000	2001	2002	2003
Number of Firms[1]	1,319	1,308	1,289	1,259	1,217	1,169	1,111	1,058	1,063	1,050	1,058
Private Passenger Auto	37.5	37.7	38.7	39.2	40.5	41.2	41.1	39.6	38.9	37.7	37.6
Private Passenger Auto Liability	23.8	24.1	24.8	24.8	25.1	24.8	24.1	22.9	22.6	22.2	22.2
Private Passenger Auto Physical Damage	13.7	13.6	13.9	14.4	15.4	16.4	17.0	16.7	16.3	15.6	15.4
Commercial Auto	6.6	6.5	6.5	6.4	6.4	6.3	6.3	6.5	6.6	6.6	6.1
Commercial Auto Liability	4.9	4.7	4.6	4.6	4.5	4.6	4.5	4.5	4.7	4.6	4.4
Commercial Auto Physical Damage	1.7	1.7	1.9	1.8	1.9	1.8	1.8	2.0	2.0	2.0	1.7
Fire and Allied Lines	2.8	3.1	3.1	3.2	3.0	2.9	2.9	2.8	3.0	3.6	4.0
Homeowners Multiple Peril	8.6	8.8	9.1	9.2	9.6	10.2	10.6	10.7	10.7	10.7	11.4
Commercial Multiple Peril	6.9	6.9	7.1	6.9	6.7	6.7	6.5	6.5	6.8	6.7	6.5
Medical Malpractice	1.7	2.0	1.8	1.8	1.6	1.7	1.8	1.9	1.9	1.9	2.1
Workers' Compensation	14.3	12.7	11.1	10.5	9.3	9.0	7.6	8.4	8.2	8.1	8.0
Products Liability	0.7	0.8	0.7	0.6	0.6	0.6	0.5	0.5	0.6	0.7	0.7
Other Liability	6.4	6.5	6.3	6.4	6.5	6.1	6.0	6.1	6.5	7.8	8.7
Reinsurance	4.0	4.3	4.7	4.3	4.2	3.9	4.2	4.2	3.9	4.0	3.9
Other	10.3	10.7	10.8	11.4	11.6	11.6	12.5	12.8	12.8	12.2	11.1
Total Premiums ($ Millions)	249,233	257,003	263,443	274,875	280,719	285,241	288,707	301,922	327,794	369,655	399,322
% Change (year to year)		3.12	2.51	4.34	2.13	1.61	1.21	4.58	8.57	12.77	8.03

Source: NAIC (1993–2003).
Notes: [1]The firms here include only the decision making units—groups and unaffiliated single companies.

2.5.2 The Top 10 Losses in History

An important issue in the U.S. P-L insurance market is the control of catastrophic (CAT) losses. Table 2.22 lists the top 10 most costly losses in the United States from 1970 to 2004. Except for the World Trade Center and Pentagon terrorist attacks in 2001, all the other major losses were caused by natural disasters such as hurricanes and earthquakes. Interestingly, as shown in the table, the majority of the large losses occurred in the 1990s and early 2000s and four out of the ten events were hurricanes in 2004. The total losses of the four hurricanes were about $28 billion, which exceed the property losses from Hurricane Andrew in 1992 ($21.5 billion) and the property losses of the terrorism attack in 2001 ($20.0 million).

Table 2.22. Top 10 Largest Catastrophes in the United States, 1970–2004 ($ Millions)

Name of Disaster	Date	Incurred Losses (Current Dollar)	Incurred Losses (2004 Dollar)
Hurricane Andrew	Aug. 1992	15,500	21,542
World Trade Center, Pentagon Terrorist Attacks	Sep. 2001	18,800	20,035
Northridge, CA Earthquake	Jan. 1994	12,500	17,843
Hurricane Ivan	Sept. 2004	11,000	11,000
Hurricane Charley	Aug. 2004	8,000	8,000
Hurricane Hugo	Sept. 1989	4,195	6,393
Hurricane Frances	Aug. 2004	5,000	5,000
Hurricane Georges	Sept. 1998	2,900	4,091
Hurricane Jeanne	Sept. 2004	4,000	4,000
Tropical Storm Allison	Jun. 2001	2,500	3,361

Source: III (2005), p. 90 and Swiss Re (2005).
Notes: Catastrophes include property and business interruption losses, excluding life and liability losses.

The U.S. P-L insurance industry is very much influenced by the catastrophes, particularly for homeowners and earthquake insurance. In the U.S., earthquake insurance is not included in standard homeowners insurance policies. Policyholders must either purchase it in the form of an endorsement or as a separate policy. In almost all the states, earthquake coverage is available from the private insurance markets. However, in California, earthquake insurance is also provided by a government institution, the California Earthquake Authority (CEA) which was created after the 1994 Northridge earthquake. This institution is privately-funded but publicly-managed. The incurred losses of the Northridge earthquake were estimated at $12.5 billion and many insurers incurred huge losses. To recover and prevent future losses, some insurance companies either restricted or refused to write new homeowners policies including earthquake coverage or drastically raised their premiums, making policies unaffordable to most homeowners.[72] As a result, in 1996, the California legislature created the CEA to resolve the crisis in earthquake insurance availability. In turn, private insurance companies could choose to participate in the CEA and offer CEA policies or offer their own policies.

The CEA is primarily funded by its participating members which contributed $5 billion to form the agency. Additional funds come from premiums, reinsurance, investment earnings, borrowed funds, and assessments on member insurers and policyholders in the even of a large earthquake. No public money and no funds from the state's general fund are available to cover losses of CEA policyholders. By 2005, the CEA accumulated more than $7.2 billion in reserves. It has five elected officials in the board: California's Governor, Treasurer, Insurance Commissioner, Speaker of the Assembly, and Chairperson of the Senate Rules Committee.[73]

Since late 1997, the CEA has become the world's largest residential earthquake insurer in terms of both premiums and the number of policies. At the end of 2004, about 731,000 policies were in force through the CEA, a decrease of roughly 170,000 from 1998, and 15 insurers participated in the CEA. The cost of CEA earthquake coverage is lower than that of private insurers, but has more restrictions and exclusions. The standard CEA earthquake policy provides $5,000 of personal possessions, $1,500 of extra living expenses, and $10,000 of coverage for upgrades required by new building codes. The deductible of a standard CEA policy is 15 percent of a home's replacement cost. In 2004 and 2005, CEA was reformed to provide broader coverage with lower deductible and a premium cut.

Two large private insurance companies providing residential earthquake insurance in California are Geo Vera and Pacific Select. Private insurers tend to provide broader earthquake coverage than CEA, but at a higher price. CEA accounted for 47.3 percent of total earthquake market share in 2004, down from 52.6 percent in 2000. In the residential market, CEA accounted for 68.2 percent in 1996, increased to 72.3 percent in 2000, and decreased to 69.2 percent in 2003, indicating that some private insurers have returned to the residential earthquake market.[74]

Flood insurance is another line deeply affected by CAT losses. As with earthquake coverage, flood insurance is not included in a homeowners insurance policy. However, flood insurance is not available in the private insurance market because of the severe adverse selection problem, only people who live in the flood zone would buy flood insurance. The National Flood Insurance Act in 1968 created the National Flood Insurance Program (NFIP) which provided flood insurance by the federal government. The NFIP is a self-supporting program, which means that the losses and expenses are paid by the premiums of flood insurance policies, but the program can borrow money from the U.S. Treasury and pay back the loans with interest.[75]

---

[72] In 1986, the California Legislature passed a law that required companies selling residential homeowners insurance to offer earthquake coverage as well. Therefore, to avoid the high risk exposure to earthquake, most insurance companies chose not to write new California homeowners business. See Moore (2001).

[73] California Earthquake Authority (2005). See also Earthquake Engineering Research Institute (1997) and III (2005f).

[74] See California Department of Insurance (1996, 2000a, 2000b, 2003, 2004).

[75] III (2005d) and Guy Carpenter (2003).

The availability and affordability of insurance coverage are always of great concern for people living in CAT loss zones, and regulators intervene in the market whenever there is a crisis. In 1992, Hurricane Andrew forced 11 P-L insurance companies into bankruptcy, 10 of them were located in Florida. Additionally, many remaining insurers re-estimated their exposures in Florida after the event. To stabilize the market, the Florida Hurricane Catastrophe Fund (FHCF) was created in November 1993. The purpose of the FHCF is to maintain the insurance capacity in Florida by proportionally reimbursing the insurers for their catastrophic hurricane losses. Under FHCF, it is mandatory for the private residential property insurers to participate in the program and function as the primary risk bearer, while the FHCF acts as a state-administered reinsurer. The FHCF is financed by three sources: reimbursement premiums charged to participating insurers, investment earnings, and emergency assessments on Florida property and liability insurers. Since the cost of reinsurance coverage through FHCF is significantly cheaper than the cost of private reinsurance, the program has helped stabilize the market. Another program following Hurricane Andrew was the state-sponsored Florida Residential Property and Casualty Joint Underwriting Association (FRPCJUA), which provides insurance to Florida residents unable to obtain coverage from private insurers. In 2002, the Association was combined with the Florida Windstorm Underwriting Association (FWUA), a program that only provided hurricane coverage in coastal areas, to create a new state-sponsored program, Citizens Property Insurance Corporation (Citizens) to provide insurance to homeowners in high-risk areas.[76]

Another important insurance coverage that tends to be affected by natural disasters such as adverse weather condition is crop insurance. Two types of crop insurance: crop-hail insurance and multiple-peril crop insurance are provided to farmers. Private insurers mainly provide crop-hail insurance, while the federal government provides the multiple peril insurance through the Federal Crop Insurance Corporation (FCIC), an organization managed by the federal Risk Management Agency (RMA). The program was created in 1938 and is backed by the resources of the U.S. Treasury Department. The program is highly subsidized by the general revenues of the government because of the low participation rate of farmers, high claims cost, and thus, insufficient loss reserves.[77]

2.5.3 Market Concentration

The U.S. P-L insurance market is a highly competitive market, as shown by the Herfindahl Index of the industry in Table 2.23, Panel A. The industry Herfindahl was 279.0 in 1993 and increased to 348.5 in 2003. This increase in the Herfindahl Index does not necessarily mean that the industry became less competitive, but rather may indicate that more efficient firms gained market share. The market share of the top 1, top 5, and top 10 firms further confirms this tendency. In 1993, the premium share of the top 10 firms was 38.7 percent, while in 2003, this number increased to 50 percent.

---

[76] Guy Carpenter (2003).
[77] III (2005e).

The premium share of the top firms and the Herfindahl Indices for the major lines of insurance are shown in Table 2.23, Panel B. It is important to look at the Herfindahl numbers by line of business because the relevant competition in the insurance market is not in total premiums, but rather by line of business, and, in some areas, by line of business and by state in which the insurance company operates. The market is more concentrated in personal lines, especially the private passenger auto insurance and homeowners insurance. The market observes a slight decrease in Herfindahl Index for these lines of business in 1998, but the Index increased again in 2003. The private passenger auto physical damage insurance ended with a lower Herfindahl Index than that of 1993, but the private passenger auto liability and homeowners insurance ended with a higher index than that of 1993.

The market for commercial lines (such as commercial auto and commercial multiple peril) is less concentrated, but the concentration level increases over time. This difference might reflect the disparity in risks and business characteristics between personal lines and commercial lines. Personal lines of insurance are more standard, less risky, and have a small margin of risk premium. The difference also reflects the fact that there are more established firms in the personal lines market than in the commercial lines market. In contrast, commercial lines are more diversified and have a higher risk premium, and therefore, attract more competitors. However, the increasing Herfindahl Indices could possibly indicate that the market is becoming more efficient over time, by removing poorly-performing firms from the market and awarding efficient firms with more market share.

The most dramatic change is the reinsurance market, which has become more concentrated in the 1990s and early 2000s. The Herfindahl Index of reinsurance was 749.4 in 1993, 979.9 in 1998, and 1,188.3 in 2003. The increase in concentration in this market was due primarily to consolidation in the reinsurance market, which was driven by an increase in the frequency and severity of insured losses caused by natural disasters and the increasing demand for reinsurance for non-catastrophic losses in the 1990s (Cummins and Weiss 2000).

The Herfindahl Indices for fire and allied lines, workers' compensation, product liability, and other liability insurance are all below 1,000 points, indicating competitive markets for these lines. However, all these lines have become more concentrated in 2003 than they were in 1993.

Table 2.24 lists the ten largest firms in the U.S. P-L market by total net premiums written in 2003. All ten firms are insurance groups, with 8 firms of the stock organizational form. However, State Farm, the firm with the largest market share, is a mutual firm. Four firms use independent agents, while four use direct sales or exclusive agency as their distribution system. Only one firm, Progressive Group, uses both direct sales and independent agent distribution system, with direct sales being the primary distribution system. Three of the ten firms are headquartered in Illinois.

2.5.4 Organizational Forms

Mutuals and Stocks

Three major organizational forms, mutuals, stock, and reciprocals, coexist in the U.S. P-L market.[78] A small proportion of firms take organizational forms of Lloyd's of London, risk retention group (captives), or state insurance funds. The earliest insurance companies in North America adopted the mutual form, in which policyholders are also owners of the firms. After the establishment of the first mutual insurance company in the U.S., the Friendly Society, mutuals spread, thus providing insurance for homeowners in cities and towns and farmers nationwide in the U.S.

Today, mutual firms still play a very important role in the U.S. P-L market. As shown in Figure 2.8, in 2003, mutual firms wrote about 33 percent of the total premiums in the market.[79] Mutual firms tend to focus more on personal lines of insurance than commercial ones. Mutuals dominate stock firms in homeowners insurance and represent about half the market in private passenger auto physical damage insurance. Mutuals also have expanded their business from traditional hazards to new risks, such as medical malpractice insurance. Compared with stock firms, the mutuals have much smaller market share in product liability, workers' compensation, other liability, and reinsurance. These lines are more complex, more capital demanding and need more professional services than standard personal insurance and commercial auto insurance. Because of their limited access to capital markets, mutuals have restrictions to raising new capital when large unexpected losses occur. Additionally, mutual firms are more susceptible to owner-manager conflicts than stock firms because policyholders retain minimal control over managers (i.e., they have less access to control mechanisms such as proxy rights, hostile takeovers, and executive stock options compensation). Thus, managers in mutual firms may behave more opportunistically. Liability insurance and workers' compensation require more managerial discretion in pricing and underwriting. Thus, managers must have expert knowledge, ability, or power to make complex decisions wisely. Subsequently, mutual firms are disadvantaged in such lines because the owners have weaker control over managers and their decisions. On the other hand, since policyholders are firm owners, owner-policyholder conflicts are eliminated.

Table 2.25 presents the number of firms, market shares, expense ratios, loss ratios, and the leverage of firms for stock and mutual firms over time. In 2003, about two-thirds of firms in the U.S. P-L market adopted the stock form. Currently, reciprocals only represent a small proportion of firms and operate similarly to mutuals. Therefore, they are combined with mutuals in this analysis. In total, about one-third of firms in the market were mutuals in 2003. As shown in Table 2.25, the market share of mutuals has declined somewhat over time, from 37.9 percent in 1993 to 33.1 percent in 2003. In contrast, the market share of stock firms has increased from 56.1 percent in 1993 to 65.3 percent in 2003.

---

[78] A reciprocal is unincoporated insurance firm that is operated by an attorney. Traditionally, reciprocals maintained separate accounts for each member. Members could be assessed if premiums were insufficient to pay claims. Modern reciprocals do not maintain separate member accounts and are not assessable, making them closer to mutuals.

[79] The mutual category includes both mutuals and reciprocals.

Table 2.23. Concentration of the U.S. Property-Liability Industry

Panel A: Concentration of the U.S. Property-Liability Industry (%), 1993–2003

Year	Top 1	Top 5	Top 10	Industry Herfindahl
1993	12.2	28.3	38.7	279.0
1994	12.5	28.2	38.5	282.3
1995	12.6	29.6	41.4	301.2
1996	12.6	29.7	41.2	298.9
1997	12.4	29.4	41.1	298.3
1998	12.2	31.5	44.5	316.0
1999	11.8	31.9	45.3	324.7
2000	11.0	31.2	45.4	312.9
2001	11.6	31.3	45.1	314.9
2002	11.6	32.4	46.5	322.3
2003	11.7	34.1	50.0	348.5

Table 2.23 *(continued)*. Concentration of the U.S. Property-Liability Industry

Panel B: Concentration of the U.S. Property-Liability Industry by Line of Business (%)

Line of Insurance	Top 1	Top 5	Top 10	Herfindahl Index	Top 1	Top 5	Top 10	Herfindahl Index	Top 1	Top 5	Top 10	Herfindahl Index
Private Passenger Auto	22.2	47.7	56.5	743.1	20.3	46.3	59.2	690.6	20.2	48.6	64.0	710.2
Private Passenger Auto Liability	21.1	46.9	56.2	702.8	20.0	46.2	59.4	677.3	20.0	48.8	65.0	710.4
Private Passenger Auto Physical Damage	24.1	49.0	57.0	821.5	20.8	46.6	59.3	712.9	20.4	48.3	62.8	713.9
Commercial Auto	4.0	18.3	32.1	163.2	5.6	24.4	37.0	210.7	10.1	29.0	43.2	287.5
Commercial Auto Liability	4.2	19.0	33.9	177.3	6.8	26.7	39.6	237.4	11.0	30.3	44.3	312.2
Commercial Auto Physical Damage	5.6	18.4	30.0	159.2	6.1	22.1	34.1	194.0	7.7	29.0	43.3	273.9
Total Auto	19.4	42.1	50.6	585.3	18.1	42.2	55.2	570.6	17.9	44.2	60.4	592.3
Fire and Allied Lines	5.0	17.4	30.7	151.8	6.4	22.2	35.0	193.5	12.7	38.8	50.5	437.1
Homeowners Multiple Peril	24.6	48.2	57.7	847.5	23.8	50.4	62.3	839.9	23.5	53.5	67.4	881.5
Commercial Multiple Peril	5.9	24.2	41.3	244.1	7.4	30.0	46.8	293.8	11.2	34.9	51.6	368.8
Medical Malpractice	12.6	34.1	48.3	375.4	7.6	29.0	47.6	320.3	10.1	38.1	54.8	420.5
Workers' Compensation	7.5	26.7	43.1	255.6	7.4	28.6	46.1	279.1	13.5	39.9	50.7	431.0
Products Liability	9.1	33.7	51.9	369.0	9.1	34.4	53.7	387.8	16.9	49.3	67.2	664.4
Other Liability	20.6	38.2	52.6	582.2	19.6	41.7	54.9	588.5	23.8	49.9	63.2	819.2
Reinsurance	19.6	51.2	66.4	749.4	24.3	56.8	72.0	979.7	30.0	57.0	76.1	1,188.3
Other	7.2	25.1	39.0	213.9	8.3	29.5	42.2	262.0	11.1	36.1	49.2	394.5

Source: NAIC (1993–2003).
Notes: Market share is determined by premiums written.

Table 2.24. Top Ten Property-Liability Insurance Companies in the United States by Premium Income, 2003

Rank	Company Name	Organizational Form	Ownership Type	Distribution System	Net Premiums Written ($ Billions)	Total Assets ($ Billions)	Policyholder Surplus ($ Billions)	Headquarter State
1	State Farm	M	Insurance Group	D	47	105	47	IL
2	American International Group	S	Insurance Group	B	28	80	23	NY
3	Allstate Insurance Group	S	Insurance Group	E	25	46	18	IL
4	St Paul Travelers Group	S	Insurance Group	A	20	75	22	MN
5	Zurich Insurance Group	S	Insurance Group	A	17	42	11	IL
6	Berkshire Hathaway	S	Insurance Group	A	16	101	49	NE
7	Nationwide Corp	M	Insurance Group	E	14	34	11	OH
8	Liberty Mutual Group	S	Insurance Group	D	13	41	11	MA
9	Progressive Group	S	Insurance Group	D,A	12	17	5	OH
10	Chubb & Son Inc.	S	Insurance Group	A	10	28	8	NJ

Source: NAIC (1993–2003).
Notes: M: Mutual. S: Stock. D: Direct Selling. E: Exclusive Agent. B: Broker. A: Independent Agent.

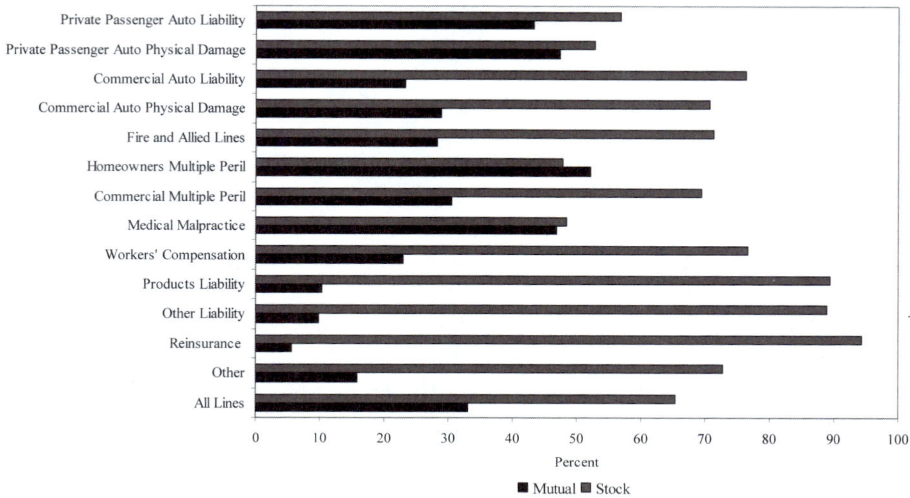

Source: NAIC (1993–2003) and A.M. Best (1993–2003).
Notes: Mutual category includes both mutual firms and reciprocals.

Figure 2.8. The U.S. Property-Liability Insurance Market Share by Organizational Form and Line of Business, 2003

Since owners have weaker control over managers in mutual firms than in stock firms, it is expected that the mutual firm managers are more likely to pursue private benefits and incur a higher cost. This is known as the expense preference hypothesis (Mester 1989). However, the underwriting expense ratio and the total expense ratio of mutuals and stocks provide no evidence on this argument. In contrast, the mutual firms have, on average, 2 percent lower expense ratios than stock firms from 1993 to 2003. A possible explanation for this may be that mutuals face more pressure from competition than stock firms in terms of capital raising and the ability to deal with CAT losses, and, thus, must be more efficient to survive in the market. Furthermore, the loss ratio of mutuals is about 3 percent higher than that of stocks during the sample period. This might be because the owner-policyholder conflict is weak in mutual firms, and, therefore, premium loading is lower for policies written by mutual firms. Mutuals also may have lower expense ratios and higher loss ratios because they write less complex and risky policies, such as personal auto insurance and homeowners insurance, which generally request less underwriting skills, but have a high loss ratio.

A cross-frontier analysis shows that stock firms and mutuals operate on separate production and cost frontiers, thus suggesting that mutuals and stocks do not operate with the same level of efficiency. Stocks specialize in products that require relatively high level of managerial discretion in pricing and underwriting, while mutuals specialize in products with long payout periods (Cummins, et al. 1999). The study also shows that the stock cost frontier dominates the mutual cost frontier, providing some support for the expense preference hypothesis. However, the cross-frontier analysis neglects the fact that mutuals and stocks compete with each other in the same market, and in most of the cases, compete in the same insurance lines. An efficiency analysis pooling together stocks and mutuals shows that stock firms are less cost efficient, but more revenue efficient than mutuals. Big mutual firms are found to be less cost efficient, but have the same revenue efficiency level as big stock firms (Cummins and Xie 2005).

It is commonly acknowledged that it is more difficult and more costly for mutual firms to raise money from external capital markets than stock firms due to the lack of channels and instruments to obtain funding as well as severe information asymmetry between mutual firms and outside investors. Additionally, it also takes longer time for mutual firms to adjust their capital ratios towards their long-run targets, because mutuals' capital ratios are more sensitive to income than those of stock firms. Consequently, mutual firms tend to keep a higher surplus level in the firm rather than paying dividends to policyholders/owners (Harrington and Niehaus 2002). The higher surplus/liability ratio for mutual firms in Table 2.25 corroborates this argument. On average, from 1993 to 2003, the mutual firms in the U.S. P-L market had a surplus/liability ratio of 67.2 percent, while it is only 51.5 percent for stock firms.

Demutualization

Mutuals and stocks have coexisted in the same market for a long period of time. Much research has been done to seek the explanation of the coexistence of the two different organizational forms (Mayers and Smith 2000; Cummins, et al. 1999; Cagle, et al. 1996). A general explanation is that mutuals and stocks manage agency problems in different ways. Mutuals can minimize the owner-policyholder conflicts, while stock firms can minimize the owner-manager conflicts. Moreover, mutual firms and stock firms tend to focus on different segments of market, as discussed above. Mutuals specialize in products that are more standardized, requiring less discretion of managers, and stock firms specialize in businesses that are riskier and require more managerial expertise. Conversion between the two different forms therefore reflects either the special financial needs and strategic changes of the firms or the incentive to maximize value for owners.

An important reason for mutual firms to demutualize is to abolish the financial constraint. It has been found that converting P-L mutuals show significantly lower surplus-to-asset ratios than the non-converting firms, but that this capital constraint eases after demutualization (Viswanathan and Cummins 2003). Some studies compare the pre-conversion performance and post-conversion performance of the demutualized P-L firms and find that these firms operate with similar efficiency before and after the conversion, but incur a capital structure change and a change in business mix (Cagle, et al. 1996).

In the 1990s and early 2000s, while L-H industry experienced a series of notable demutualization cases, the phenomenon was less dramatic in the P-L industry. Unlike the giants in L-H industry, the largest P-L firm, the State Farm group, did not choose to demutualize. One explanation is that the P-L market is less sensitive to product changes than the L-H market in which firms must maintain flexibility to adapt to the development of new products in the market. However, the most significant demutualization in the P-L insurance industry during this period is perhaps Liberty Mutual's conversion to a mutual holding company (MHC) in 2001. Liberty Mutual is headquartered in Massachusetts and is the largest workers'

Table 2.25. U.S. Property-Liability Insurance Market Share and Financial Ratios by Organizational Form, 1993–2003

	Number of Firms		Market Share[1] (%)		Underwriting Expense Ratio (%)[2]		Total Expense Ratio (%)[3]		Loss Ratio (%)[4]		Surplus/Liability Ratio (%)	
Year	Mutual	Stock	Mutual	Stock	Mutual	Stock	Mutual	Stock	Mutual	Stock	Mutual	Stock
1993	423	620	37.9	56.1	22.7	28.3	35.8	41.7	81.1	78.4	50.8	43.0
1994	488	710	38.4	61.0	23.2	27.8	36.8	41.4	84.8	79.0	49.1	41.8
1995	481	710	37.4	58.2	23.5	27.5	37.7	41.2	80.6	77.7	55.3	48.8
1996	468	696	37.3	57.8	23.8	28.1	37.4	41.1	79.7	76.2	62.3	50.7
1997	455	681	37.5	58.2	24.9	28.5	38.5	41.1	73.7	71.9	76.2	59.8
1998	449	650	39.3	58.3	25.9	29.3	41.1	42.0	79.4	73.9	79.4	63.1
1999	441	610	34.5	58.5	26.2	29.2	42.3	42.0	81.4	76.1	85.0	61.6
2000	431	561	36.3	60.3	25.9	28.8	41.5	41.0	85.0	78.3	79.7	55.7
2001	433	564	32.5	60.1	25.6	27.6	40.0	40.3	89.5	89.4	73.2	45.9
2002	429	561	32.9	60.3	25.3	25.6	38.9	37.6	82.3	79.8	60.7	43.6
2003	396	555	33.1	65.3	24.8	24.9	37.6	37.3	74.7	72.8	68.0	52.0
2003	396	555	33.1	65.3	24.8	24.9	37.6	37.3	74.7	72.8	68.0	52.0
Average	445	629	36.1	59.5	24.7	27.8	38.9	40.6	81.1	77.6	67.2	51.5

Sources: NAIC (1993–2003) and A.M. Best (1993–2003).

Notes: [1]Market share is determined by premiums written. [2]The underwriting expense ratio is equal to underwriting expenses incurred divided by net premiums written. [3]The total expense ratio is equal to total expenses divided by net premiums written. [4]The loss ratio is equal to losses and loss expenses incurred divided by net premiums earned.

compensation insurance company in the U.S. Before conversion, it had roughly 2.0 million policyholders with $8.9 billion in net premiums written and assets valued $33.3 billion. This case raised a lot of attention because it was the first P-L firm that chose the mutual holding company form. The firm's major goal after converting to an MHC was to raise funds for acquisitions and to pursue its global expansion strategy.[80] The demutualized Liberty MHC did not launch an IPO, but rather preserved its mutuality. In doing so, the company claimed that:

> Mutuality entails a better alignment of owners' collective interests and customers' individual interests. Further, under a mutual enterprise, management has the benefit of being able to focus on long-term business objectives. The mutual holding company structure will allow us to preserve the mutual culture that the Board of Directors believes has been a key factor in the past and will continue to be important in the future in the generation of significant value for policyholders.[81]

Before Liberty Mutual, there were several less influential demutualization cases, such as the Medical Assurance Inc. in 1991, Farm Family Holdings 1996, Summit Holdings and Old Guard Group in 1997, MEEMIC Holdings, MIIX Group Inc. and NCRIC Group Inc. in 1999, and American Physician's Capital Inc. in 2000. These firms issued demutualization IPOs and earned a positive IPO returns.[82] In the 1980s and 1990s, the demutualization IPOs were, in general, found to generate higher IPO returns than the non-demutualization IPOs in the U.S. insurance market (Viswanathan 2005).

2.5.5 Distribution Channels

The U.S. P-L insurance market provides a natural test for the efficiency of different product distribution systems. There are four major types of distribution systems in the U.S. P-L industry: independent agents, brokers, direct selling (use of the Internet, direct mail, telemarketing and company employees to sell insurance products), and exclusive agents. Usually exclusive agents and direct selling are classified into one category called direct writers because exclusive agents only represent one firm. Direct writers are usually considered to be technically efficient because such firms can fully internalize the benefits from technology investment and, therefore, can quickly adjust to any technology innovations. Independent agents represent more than one insurance firm and play an important role in underwriting procedures such as risk screening, risk classification, and policy renewals. They also provide services to customers in the areas of risk analysis and claims settlement as well as loss prevention and loss reduction. Because independent agents are deemed as providing superior services, they are more costly than other marketing systems. The question is whether the extra costs they charge to insurance firms can be compensated by the extra revenues they bring to the firms.

Although technically brokers represent the policyholder rather than the insurer, brokers and independent agents are similar in many respects and brokers place much

---

[80] Hoover's Online (2005).
[81] Liberty Mutual (2002).
[82] SNL Financial (2005).

of their businesses through appointed agency contracts. The principal difference between independent agents and brokers is that brokers tend to be larger and often deal with more complex national and international clients, whereas independent agents are more likely to be local or regional. Both independent agents and brokers provide a wide range of services to their clients and search for the best price among several insurers. Brokers often provide specialized services to commercial customers who face large and complex exposures to risk, including administration of self-insurance plans and arrangement of captive insurance companies (Cummins and Doherty 2005). In the 1990s, many mergers and acquisitions in the U.S. insurance market involved brokers, especially the mega-mergers of leading brokerage firms, which subsequently made the brokerage market more concentrated. In 2002, the two leading brokers, Marsh and Aon, accounted for 54 percent of the total broker revenues.[83]

Table 2.26 lists the top 10 insurance brokers globally by revenues in 2003. The revenues of brokers mainly come from commercial retail and services. Contingent commissions are also an important revenue resource, especially for large brokers such as Marsh & McLennan.

Figure 2.9 illustrates the market share of different distribution systems by line of business in 2003. The classification of distribution system is based on a firm's primary distribution system, as reported to A.M. Best. Independent agents and direct writers accounted for about 40 percent and 46 percent, respectively, of the total P-L market, while brokers accounted for about 11 percent of the market. The breakdown of distribution channels also differs by line of business. Independent agents have the highest market share in commercial auto insurance, products liability, commercial multiple perils, fire and allied lines insurance, reinsurance, workers' compensation, and other liability insurance. Customers owning these policies are generally more demanding in terms of underwriting services and claim settlements. Direct writers have the highest market share in private passenger auto insurance, homeowners insurance, medical malpractice insurance, and other property insurance. Brokers typically focus on the commercial lines, such as workers' compensation, other liability, reinsurance, medical malpractice, fire and allied lines, and commercial auto liability, which require more complex professional services. The brokers dominate direct writers in the markets of product liability and other liability.

The literature on distribution channels (Berger, et al. 1997) acknowledges that independent agents (including brokers) have higher expenses than direct writers. Table 2.27 shows the premium shares, expense shares, expense ratios, and loss ratios of different distribution systems over time. Compared with direct writers, independent agents have marginally lower premium shares, but higher expense shares. Independent agent firms have a higher underwriting expense ratios and total expense ratios, which is on average 6 percent higher than brokers and direct writers, a statistically significant difference. The higher expense ratios of independent agents are partially compensated by a significantly lower loss ratio, which on average is 3 percent lower than direct writers. Independent agents show no significant difference in loss ratio from brokers.

---

[83] Swiss Re (2004).

Table 2.26. Top 10 Insurance Brokers in the World by Revenues, 2003

Broker	Brokerage Revenues ($ Millions)	Commercial Retail (%)	Wholesale (%)	Reinsurance (%)	Personal Lines (%)	Services (%)	Investments (%)	Other (%)	Contingent Commissions (%)
Marsh & McLennan	9,376	48.0	2.0	7.0	0.0	25.0	1.0	17.0	7.3
Aon	6,734	39.0	4.0	9.0	1.0	16.0	3.0	28.0	2.0
Willis Group Holdings	2,004	65.0	10.0	22.0	1.0	2.0	0.0	0.0	4.0
Arthur J. Gallagher	1,202	54.0	7.0	6.0	1.0	25.0	7.0	0.0	3.0
Wells Fargo	801	68.8	4.2	0.1	9.0	6.7	1.0	10.2	3.4
Jardine Lloyd Thompson	701	33.0	17.0	22.0	1.0	23.0	4.0	0.0	N/A
BB&T Insurance Services	605	60.5	23.2	0.0	12.9	0.0	1.0	2.4	N/A
Hilb Rogal & Hobbs	556	83.5	2.8	0.3	8.3	3.7	0.6	0.8	N/A
Brown & Brown	545	62.0	22.0	0.0	10.0	5.0	1.0	0.0	6.0
Alexander Forbes	537	53.0	11.0	2.0	10.0	9.0	2.0	13.0	N/A

Source: Cummins and Doherty (2005), p. 44.
Notes: N/A is Not Available.

Table 2.27. The U.S. Property-Liability Insurance Market Share and Financial Ratios by Distribution System, 1993–2003

Panel A

	Number of Firms			Premium Share (%)			Expense Share (%)		
Year	Agent	Broker	Direct	Agent	Broker	Direct	Agent	Broker	Direct
1993	685	81	281	43.01	5.60	46.17	48.90	5.25	41.36
1994	660	84	293	43.37	5.83	46.52	49.08	5.45	41.81
1995	659	79	289	42.22	6.06	45.52	46.49	5.76	41.35
1996	683	78	327	41.75	6.77	45.06	46.17	6.66	41.18
1997	664	78	321	41.94	6.84	46.04	45.89	6.50	43.35
1998	632	70	304	40.60	6.76	47.27	43.86	6.05	45.91
1999	547	58	246	44.58	7.10	42.62	48.03	6.47	41.22
2000	512	51	233	41.99	6.55	44.80	44.90	6.13	44.03
2001	495	62	223	44.63	7.48	44.64	47.28	7.18	43.56
2002	499	60	226	42.27	9.67	44.52	45.25	8.81	43.24
2003	539	70	267	40.09	11.40	46.20	43.02	10.76	44.88

Table 2.27 *(continued)*. The U.S. Property-Liability Insurance Market Share and Financial Ratios by Distribution System, 1993–2003

Panel B

	Underwriting Expense Ratio (%)[1]			Total Expense Ratio (%)[2]			Loss Ratio (%)[3]		
Year	Agent	Broker	Direct	Agent	Broker	Direct	Agent	Broker	Direct
1993	30.40	22.76	22.52	44.36	36.58	34.95	77.77	79.86	81.10
1994	30.19	22.65	22.52	44.35	36.68	35.23	76.35	80.01	85.89
1995	29.99	22.95	22.73	43.85	37.83	36.17	75.39	77.81	81.66
1996	30.10	23.76	23.39	43.89	39.04	36.26	76.82	80.10	78.34
1997	30.61	24.71	24.45	43.77	37.99	37.66	71.34	76.14	72.52
1998	31.31	24.76	25.52	44.65	36.99	40.13	74.57	77.30	76.66
1999	31.29	24.87	25.63	44.91	37.94	40.31	76.66	79.04	79.51
2000	30.57	25.38	25.84	43.56	38.11	40.04	79.00	80.67	82.54
2001	29.08	25.65	25.06	42.12	38.13	38.81	91.29	86.45	87.83
2002	28.04	22.56	23.84	40.49	34.49	36.74	77.73	83.84	82.46
2003	27.15	22.28	23.59	39.84	35.02	36.07	72.13	76.48	73.86
Average	29.88	23.85	24.10	43.25	37.16	37.49	77.19	79.79	80.22

Sources: NAIC (1993–2003) and A.M. Best (1993–2003).

Notes: Agent includes all firms whose primary marketing channel is independent agent. Broker includes all firms whose primary marketing channel is brokerage. Direct includes all firms whose primary marketing channel is direct selling and exclusive agent. [1]The underwriting expense ratio is equal to underwriting expenses incurred divided by net premiums written. [2]The total expense ratio is equal to total expenses divided by net premiums written. [3]The loss ratio is equal to losses and loss expenses incurred divided by net premiums earned.

2.5.6 Financial Performance of the Industry

Assets and Liabilities

Since the U.S. P-L industry is mature, its assets and liabilities have been relatively stable over time, though it should be noted that profitability fluctuates with the underwriting cycle. Figure 2.10 shows the assets, policyholders surplus, and liabilities of the industry from 1993 to 2003, adjusted to constant dollars using the 2000 Consumer Price Index (CPI). Liabilities remained stable until the events of September 11, 2001, when total liabilities for the industry increased drastically. The assets and policyholder surplus levels of the industry change with the underwriting cycle. There is a steady increase in assets and policyholder surplus from 1993 to 1996 after Hurricane Andrew and 1994 Northridge Earthquake. The hard market, along with a bull stock market, allowed the industry to accumulate excess capital during this period (Cummins and Nini 2002). This increase peaked in 1998 and subsequently, began to decrease with the rise of the soft market. In a soft market, the industry is characterized by overcapacity of underwriting and thus, insurance companies cut prices to compete with each other. They also compete with the numerous captives that form to deal with the high price of insurance products, especially in commercial lines. As a consequence, the assets and policyholder surplus levels declined. This trend continued until the September 11, 2001 terrorist attacks which pushed the P-L industry into another hard market immediately.

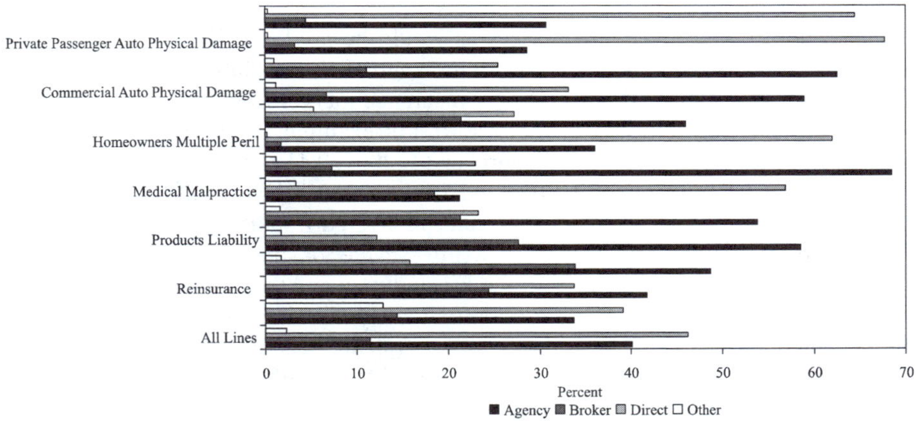

Source: NAIC (1993–2003) and A.M. Best (1993–2003).
Notes: This figure presents the market share by distribution system and line of business in the U.S. property-liability insurance market. "Direct" indicates a firm whose primary distribution system is either direct selling or exclusive agent. "Agent" indicates a firm whose primary distribution system is an independent agent. "Broker" indicates a firm whose primary distribution system is brokerage. "Other" indicates firm whose distribution channel is not direct, agent, or broker or whose distribution system is missing in the databases.

Figure 2.9. The U.S. Property-Liability Insurance Market Share by Distribution System and Line of Business, 2003

Source: NAIC (1993–2003).
Notes: The Consumer Price Index (CPI) is used to deflated assets, liabilities, and policyholder surplus. Base year is 2000.

Figure 2.10. Total Real Assets and Liabilities of U.S. P-L Industry, 1993–2003

The losses due to September 11, 2001 depleted the excess capacity (measured by policyholder surplus) of the industry. This loss, coupled with an increase in demand and decrease in the supply of insurance, caused sharp increases in both direct insurance and reinsurance prices as would be predicted by the capital shock and capacity constraint model (see Harrington and Niehaus 2000). In response to the hard market, new capital flowed into the insurance market from investors who were attracted by the expected high rate of return. For example, more than $28 billion in new capital was raised following September 11, 2001.[84] The overall asset level of the industry increased moderately in 2002 and 2003. In 2003, the overall liabilities of the industry went down and policyholder surplus recovered from its low point in 2002.

Figure 2.11 illustrates the underwriting cycle of the U.S. P-L market by comparing the underwriting profit in the P-L industry and the one-year Treasury Bill yield. From 1959 to 2003, there are approximately six alternations between a hard market and a soft market. The underwriting profits are negatively correlated with the Treasury Bill yield, as suggested by the financial theory of insurance pricing which indicates that the premium income reflects the discounted value of expected future losses.

Table 2.28 provides a pro forma balance sheet for the U.S. P-L industry in 2003. The total assets of the industry amounted to $1.2 trillion. The invested assets, including bonds ($645.3 billion), preferred stocks ($15.4 billion), common stocks ($169.7 billion), mortgage and real estate ($11.7 billion), cash and short-term investments ($89.3 billion) and other invested assets ($32.2 billion), accounted for 82.0 percent of the total assets. Premiums and considerations ($106.1 billion) accounted for 9.0 percent, while reinsurance receivables ($40.8 billion) accounted for 3.5 percent of the total assets.

---

[84] See Kollar (2002), p. 17.

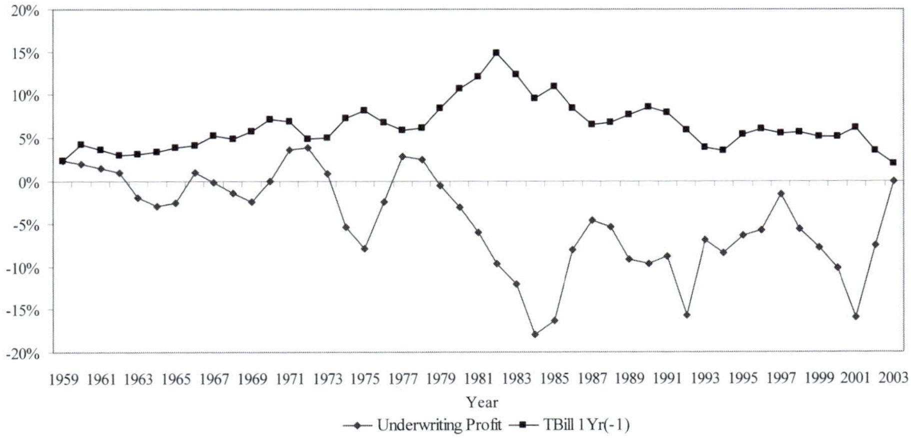

Source: A. M. Best (2004b) and Federal Reserve Bank of St. Louis (1959–2003).
Notes: Underwriting profit = (premium income – expected losses)/premium income. For simplicity, actual losses are used as the proxy for expected losses. Since theoretically, premium income is the present value of expected losses, we have premium income = expected losses/(1 + interest rate). Therefore, underwriting profit = –interest rate. The T-Bill rate is used here to represent the interest rate.

Figure 2.11. U.S. P-L Industry Underwriting Profit versus 1 Year Treasury Bill Yield (Lagged One Year)

On the liability side, the total liabilities of the industry in 2003 were $820.5 billion, giving an overall leverage ratio (liability divided by policyholders surplus) of 231.9 percent. Loss reserves ($445.4 billion) accounted for 54.3 percent of total liabilities, a lower percentage than a decade ago when more than 60 percent of the liabilities were loss reserves. Unearned premiums were the second largest liability, amounting to 21.5 percent ($176.3 billion) of the total liabilities. This is a higher percentage than a decade ago, when unearned premiums were about 17 percent of total liabilities. Reinsurance liabilities (reinsurance premium payable and funds held by company under reinsurance treaties) accounted for 7.9 percent ($64.5 billion) of total liabilities. Policyholders surplus, which included $3.1 billion (0.9 percent) in common stock, $1.7 billion in preferred stock (0.5 percent), $110 billion in gross paid in and contributed surplus (31.1 percent), and $197.4 billion in unassigned funds (surplus) (55.8 percent), amounted to $353.8 billion in 2003.

Income and Expenses

Table 2.29 presents the pro forma income statement of the U.S. P-L industry. In 2004, the net premiums earned of the U.S. P-L industry was $397.1 billion, while the underwriting cost (loss and loss expenses incurred and underwriting expenses) was $399.9 billion, leaving the industry with $2.8 billion in underwriting losses, the best underwriting performance in the industry since 1997 (A.M. Best 2004b). The

Table 2.28. Pro Forma Balance Sheet of the U.S. Property-Liability Insurance Industry ($ Millions), 2003

Assets		Liabilities	
Bonds	645,254	Losses and Loss Adjustment Expenses	445,422
Government Bonds[1]	448,452	Reinsurance Payable on Paid Loss and Loss Adjustment Expenses	9,874
Corporate Bonds[2]	196,802	Other Expenses	18,101
Stocks		Taxes, Licenses, and Fees	5,408
Preferred Stocks	15,367	Current Federal and Foreign Income Taxes	3,746
Common Stocks	169,723	Net Deferred Tax Liability	11,607
Mortgage	2,245	Borrowed Money	2,192
Real Estate	9,465	Unearned Premiums	176,311
Cash and Short-Term Investments	89,298	Ceded Reinsurance Premiums Payable (Net of Ceding Commissions)	25,684
Other Invested Assets	32,170	Funds Held by Company Under Reinsurance Treaties	38,835
Receivable for Securities	2,965	Provision for Reinsurance	5,866
Aggregate Write-Ins for Invested Assets	1,216	Miscellaneous	77,433
Subtotals, Cash, and Invested Assets	967,704	Total Liabilities	820,479
Investment Income Due and Accrued	8,655	Gross Paid In and Contributed Surplus	110,223
Premiums and Considerations	106,079	Unassigned Funds (Surplus)	197,353
Reinsurance	40,771	Common Capital Stock	3,094
		Preferred Capital Stock	1,693
		Miscellaneous	41,486
Miscellaneous	51,119	Policyholder Surplus	353,849
Total Assets	1,174,328	Total Liabilities and Capital	1,174,328

Source: A. M. Best (2004b), pp.138–139.
Notes: [1]Government bonds include public utilities. [2]Corporate bonds include parent, subsidiaries, and affiliates. Government bonds and corporate bonds are imputed from the proportion of bonds in Schedule D.

industry spent $25.1 billion on claim adjustment services and $43.2 billion on commission fees (A.M. Best 2004b).

The net investment income of the industry was $46.6 billion in 2003. The related investment expense of the industry was $2.6 billion (A.M. Best 2004b). After dividends to policyholders and taxes, the industry earned $31.2 billion in net income.

Figure 2.12 illustrates the major expense ratios of the industry from 1993 to 2003. The underwriting expense ratio (underwriting expenses incurred divided by net premiums written) was about 25.7 percent in 1993, increased to 27.9 percent in 1999, and then decreased to 24.6 percent in 2003. The investment expense ratio (total investment expense divided by net premiums written) remained flat at roughly 0.8

percent during the period. The agent expense ratio (expense on agents and brokers divided by net premiums written) indicated that the commissions and allowance paid to agents and brokers accounted for between 10.0 and 11.5 percent of total premiums written. The ratio increased after 1996 and then decreased gradually to its original level after 2001. This is likely due to the introduction of new commission arrangements in the 1990s called placement service agreements (PSAs) and marketing service agreements (MSAs). Under these agreements, insurers paid brokers additional commission income, usually based on premium volume.[85] The average claim adjustment expense ratio (expenses on claim adjustment services divided by net premiums written) was about 6.5 percent before 1995, but declined afterwards to 5.7 percent in 2003.

Table 2.29. Pro Forma Income Statement of the U.S. Property-Liability Insurance Industry ($ Millions), 2003

Premiums Earned	397,069	
Loss Incurred		246,450
Loss Expenses Incurred		51,125
Other Underwriting Expenses		102,364
Aggregate Write-Ins for Underwriting Deductions		126
Total Underwriting Deductions		400,066
Net Income of Protected Cells	1	
Net Underwriting Gain (Loss)	(2,996)	
Net Investment Income Earned	40,337	
Net Realized Capital Gains (Losses)	6,280	
Net Investment Gain (Loss)	46,617	
Total Other Income	(80)	
Net Income before Dividends and Taxes	43,541	
Dividends to Policyholders		1,978
Net Income After Dividends and Before Taxes	41,564	
Federal and Foreign Taxes		10,335
Net Income	31,229	

Source: A. M. Best (2004b), p. 140

Profitability of the Industry

When discussing the profitability of the P-L industry, two important ratios must be considered. The first is the combined ratio which is used to measure underwriting

[85] III (2005a), p. 9.

profitability of the industry. The second is the overall operating ratio which is used to measure the overall profitability of the industry. The combined ratio is calculated as the sum of the loss ratio (loss and loss adjustment expenses incurred divided by net premiums earned), the expense ratio (underwriting expenses divided by net premiums written), and the policyholder dividend ratio (dividends to policyholders divided by net premiums earned). A combined ratio below 100 percent indicates the industry has made an underwriting profit. However, the combined ratio can only partially measure the industry's profitability because it ignores investment income. On the other hand, the overall operating ratio takes into account investment income. It is calculated as the difference between the combined ratio and the investment income ratio (investment income divided by net premiums earned). An operating ratio below 100 percent indicates the industry is profitable from its operations.

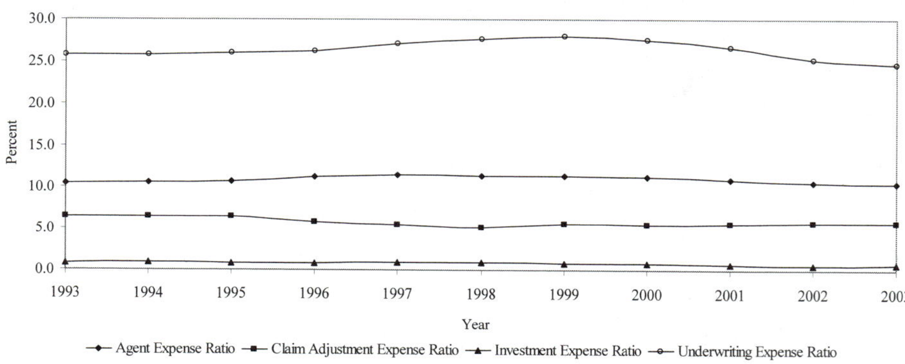

Source: NAIC (1993–2003).
Notes: Investment expense ratio is total investment expense divided by premiums written. The agent expense ratio is expenses on agents and brokers divided by premiums written. Claim adjustment expenses ratio is expenses on claim adjustment services divided by premiums written. Underwriting expense ratio is underwriting expenses divided by premiums written.

Figure 2.12. The U.S. P-L Industry Expense Components, 1993–2003

Figure 2.13 displays the profitability of the P-L industry from 1993 to 2003. The loss ratio of the industry is approximately 80.0 percent for most of the years, except for 1997 when the ratio fell to 72.4 percent and for 2001, when the loss ratio reached 89.3 percent. The expense ratio of the industry ranged from 25.8 percent to 27.9 percent. The policyholder dividend ratio stays flat between 1.2 and 1.8 percent through 1996 and gradually falls to 0.5 percent by 2003. As a result, the combined ratio (sum of the loss, expense, and policyholder dividend ratios) of the industry exceeded 100 percent most of the time in the 1990s and early 2000s, with the highest ratio in 2001 (116.7 percent). This indicated that the industry typically lost money from underwriting.

The investment income ratio fluctuated over time. It gradually increased before 1997, remained relatively constant from 1998 to 2000, and then decreased sharply after 2000. The return was about 16.7 percent in 1997, the highest in the past decade, but only 10.3 percent in 2003. Taking into account investment income, the industry achieved an operating ratio lower than 100 percent in every year except for 2001 in which the ratio reached 103.9 percent as a result of the terrorist attacks of September 11, 2001. Although the investment return in 2003 was low, the industry was still profitable as a result of excellent underwriting performance. Consequently, 1997 and 2003 were the two best years in the industry during the past decade.

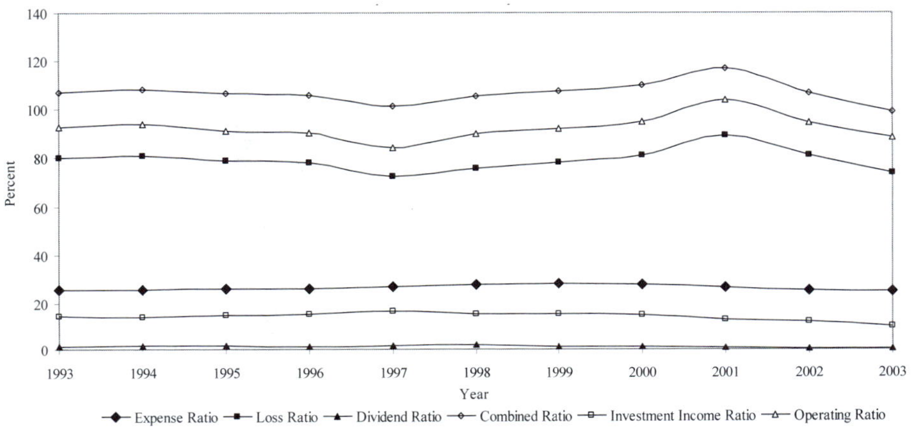

Source: NAIC (1993–2003).

Notes: Loss ratio is the loss and loss adjustment expenses incurred divided by premiums earned. The expense ratio is underwriting expenses divided by premiums written. The policyholder dividend ratio is dividends to policyholders divided by premiums earned. The combined ratio is the loss ratio plus the expense ratio plus the policyholder dividend ratio. The investment income ratio is investment income divided by premiums earned. The operating ratio is the combined ratio minus the investment income ratio.

Figure 2.13. Profitability of U.S. P-L Industry, 1993–2003

The underwriting performance and loss payout periods in the P-L industry differ by the various product lines. Some types of insurance coverage, such as property insurance, incur short loss payment periods, while others, such as liability insurance, pay out over much longer periods of time. The loss payment pattern of insurance lines helps to determine their amount of risk and the investment strategy they should adopt. Table 2.30 shows the loss ratio, expense ratio, combined ratio, investment income ratio, and operating ratio by line of business for the P-L industry in 1993, 1998, and 2003. In 2003, homeowners multiple peril insurance and commercial multiple peril insurance had higher expense ratios than other lines. Medical malpractice and products liability have higher loss ratios and higher combined ratios than other lines in 2003. Additionally, the 2003 investment income ratio is higher for long-tail businesses, such as auto liability, medical malpractice, product liability, other liability, and reinsurance. This phenomenon is expected since the interval between premiums collected and the loss payment for these lines is longer and, thus, insurance companies can make larger investment returns. The operating ratios for auto insurance are stable over time, while fire and allied lines, product liability, and medical malpractice are widely dispersed, implying that these lines are riskier.

Table 2.30. The U.S. Property-Liability Insurance Industry Profitability by Line of Business (%)

	Expense Ratio[1]			Loss Ratio[2]			Combined Ratio[3]			Investment Income Ratio[4]			Operating Ratio[5]		
Line of Business	1993	1998	2003	1993	1998	2003	1993	1998	2003	1993	1998	2003	1993	1998	2003
Private Passenger Auto Liability	22.4	23.5	22.9	85.4	76.4	79.3	108.6	101.9	102.8	9.4	8.5	6.1	99.2	93.5	96.7
Private Passenger Auto Damage	22.0	23.7	23.2	66.9	73.9	68.3	89.8	99.8	92.1	1.8	1.7	1.3	88.1	98.1	90.8
Commercial Auto Liability	30.1	29.7	26.3	77.1	85.6	72.6	107.6	115.6	99.0	14.1	12.4	7.0	93.4	103.2	92.0
Commercial Auto Damage	30.7	30.2	28.4	56.5	78.2	55.9	87.7	108.8	84.4	2.8	1.9	1.1	84.9	106.9	83.4
Fire and Allied Lines	38.9	38.1	24.5	66.6	69.1	53.0	105.9	107.5	77.7	5.0	5.9	3.3	100.9	101.6	74.4
Homeowners Multiple Peril	30.9	30.7	28.3	82.0	77.8	68.6	113.4	109.3	97.3	6.2	4.6	3.3	107.1	104.8	94.0
Commercial Multiple Peril	36.4	36.2	31.5	79.2	83.3	65.3	115.6	119.6	97.7	13.6	11.1	5.4	102.0	108.5	92.3
Medical Malpractice	17.2	21.4	16.4	88.1	97.5	117.5	108.8	122.0	134.2	38.0	26.9	15.4	70.8	95.1	118.8
Workers' Compensation	19.0	24.8	22.2	86.3	70.5	82.7	110.4	101.2	106.4	18.1	15.4	9.7	92.3	85.8	96.7
Products Liability	27.0	30.6	23.2	186.1	107.5	132.0	212.8	138.6	155.2	44.1	49.0	23.9	168.7	89.5	131.4
Other Liability	27.3	29.1	22.5	95.2	82.4	83.4	123.0	111.7	106.1	27.5	27.4	10.6	95.5	84.4	95.5
Reinsurance	25.8	29.3	26.0	78.5	72.4	82.6	104.3	101.8	108.6	20.3	31.3	14.7	84.0	70.5	93.9

Source: NAIC (1993–2003).

Notes: [1]The expense ratio is equal to underwriting expenses divided by premiums written. [2]The loss ratio is equal to losses and loss adjustment expenses incurred divided by premiums earned. [3]The combined ratio is equal to the loss ratio plus the expense ratio plus the policyholder dividend ratio. [4]The investment income ratio is equal to investment income divided by premiums earned. [5]The operating ratio is equal to the combined ratio minus the investment income ratio.

Investment Activity

The above analysis shows investment performance is important in the U.S. P-L industry. Consequently, allocation of these assets is also of great interest. Figure 2.14 shows the allocation of invested assets of the industry from 1993 to 2003. The majority of the investments were in government bonds (including public utilities). In 1993, about 47 percent of the total invested assets are government bonds, but this ratio fell to approximately 42 percent in 2003. Common stocks were the second largest investment category. More than 20 percent of the invested assets were in stocks, with the late 1990s observing the highest level (about 30 percent). Corporate bonds represented the third largest category of investment and their proportion has remained stable over time. In 2003, corporate bonds accounted for roughly 19 percent of the total invested assets. Generally, the industry keeps a considerable percentage of assets in the form of cash and short-term investments to guarantee liquidity. About 4 to 8 percent of invested assets were cash and short-term investments. Only a very small portion of invested assets were invested in real estate and mortgages (less than 1.5 percent). The industry also invested about 1–2 percent of its invested assets in preferred stocks.

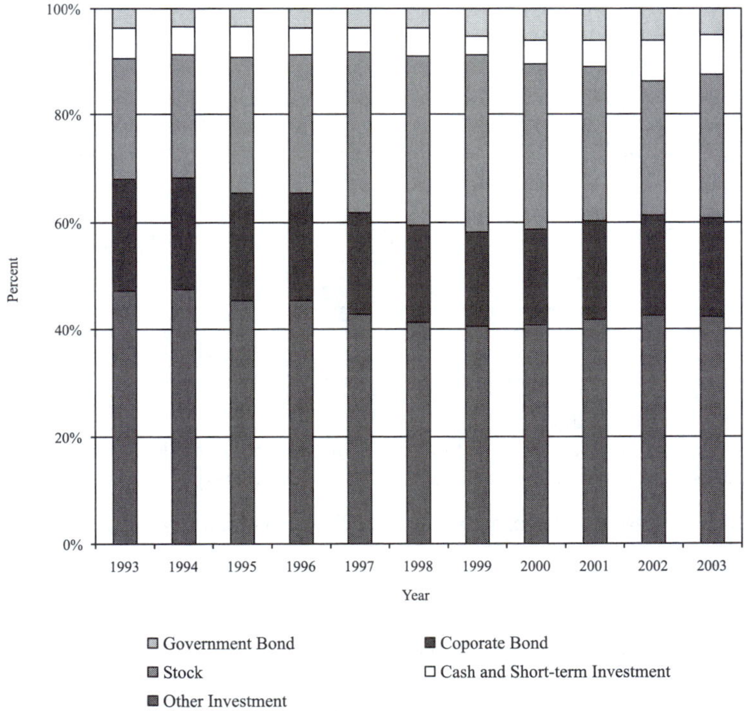

Source: NAIC (1993–2003)

Figure 2.14. Allocation of Invested Assets in the U.S. Property-Liability Industry, 1993–2003

Three important measurements relating to investment performance and profitability of the industry—return on equity (net income divided by policyholder surplus), yield on invested assets (investment income divided by total invested assets), and pretax return on revenues (net income before tax divided by premiums earned)—are displayed in Figure 2.15 for years 1993 through 2003. The yield on invested assets of the industry decreased gradually over time, from 5.5 percent in 1993 to 4.0 percent in 2003. The return on equity fluctuates with the underwriting cycle and with catastrophic events. The average return on equity during this period was 6.8 percent, with 1994 and 2001 witnessing the two major turning points in the return on equity. In 1994, the Northridge earthquake caused insured losses of more than $12.5 billion causing the return on equity for the industry to fall to 5.2 percent. The September 11, 2001 terrorist attacks caused insured losses of $18.8 billion (P-L only) causing the return on equity for the industry to drop to negative 1 percent, coinciding with a recessive stock market. The pattern of pretax return on revenue is similar to that of the return on equity. On average, the pretax return on revenue of the industry is 10.2 percent.

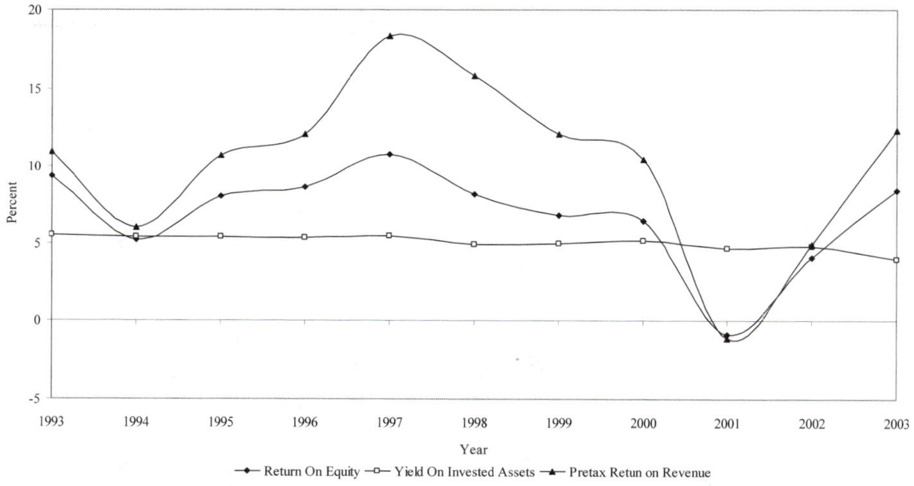

Source: NAIC (1993–2003).
Notes: The return on equity is the net income divided by policyholder's surplus. Pretax return on revenues is net income before tax divided by premiums earned.

Figure 2.15. Investment Performance of U.S. Property-Liability Industry, 1993–2003

2.5.7 The Reinsurance Market

Reinsurance plays a critical role in the U.S. P-L insurance market. The existence of reinsurance helps insurers expand the underwriting capacity of the insurance market, adequately diversify insurable risks, and stabilize and smooth financial performance. There were about 60 professional reinsurance companies in the U.S. in 2003. Their total assets amounted to $177 billion and their net premiums written totaled $29 billion. Unfortunately, these professional reinsurers suffered underwriting losses in late 1990s resulting in a combined ratio higher than 115 percent during the period from 1999 to 2002. The worst year for the professional reinsurers was 2001, in which they had a 141 percent combined ratio and a 120 percent overall operating ratio. Their situation improved somewhat in 2003, with a combined ratio of 100 percent and an overall operating ratio of 80 percent (A. M. Best 2004b).

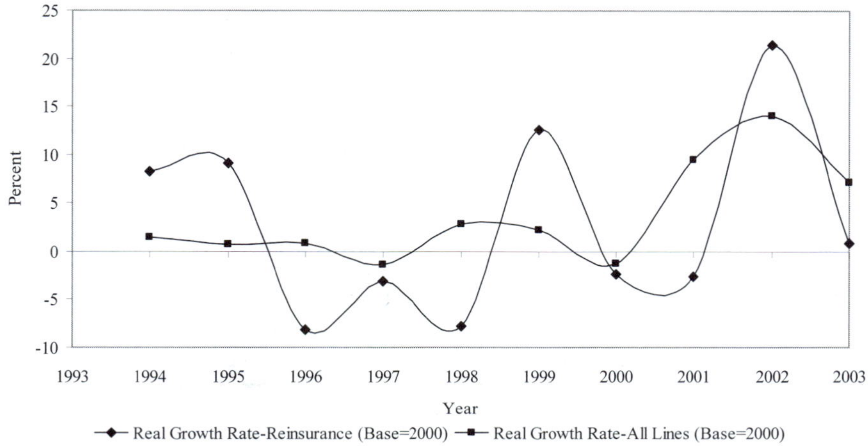

Source: NAIC (1993–2003).
Notes: The Consumer Price Index (CPI) is used to deflate premiums written prior to computing the growth rate.

Figure 2.16. The U.S. Property-Liability Industry Reinsurance Premium Growth Rate, 1993–2003

In addition to the professional reinsurers, some primary insurers also write reinsurance. Figure 2.16 illustrates the total reinsurance premium growth of the U.S. P-L industry. There are three peaks in reinsurance premium growth during the period spanning 1994 to 2003. They occur in 1995, 1999, and 2002 following CAT losses in the preceding year—the 1994 Northridge earthquake, Hurricane Georges in 1998, and the September 11, 2001 terrorist attacks. The average real growth rate of reinsurance premiums written was 2.8 percent during this period and the average real growth rate in all lines premiums was 3.6 percent.

Table 2.31 lists the 20 largest P-L insurance companies writing U.S. reinsurance business in 2003. Three of the top 5 companies are professional reinsurers—Swiss Re, GE Global, and Munich Re.[86] The reinsurance market in the U.S. has become more concentrated in the past decade, as shown by the Herfindahl Index in Table 2.23. The majority of the 20 firms use direct selling or brokerage, while only five of them use independent agents as their marketing channel. Only one of the 20 firms,

[86] GE Global Insurance Holdings writes reinsurance business through its main subsidiary, Employer Reinsurance Corp (ERC). In 1984, GE purchased ERC from Texaco. Since then, GE has been involved in a series of domestic and international acquisitions of insurance firms through ERC. Meanwhile, GE has expanded its insurance businesses into both the life-health and property-liability areas. In 1999, GE ERC became one of the five largest reinsurers in the world. In 2002, it became one of the four largest reinsurers in the world. Today, GE's insurance businesses cover the areas of property-liability, accident and health reinsurance, professional liability, healthcare reinsurance, programs for managing general underwriters, and workers' compensation.

State Farm, is a mutual insurer. This is likely due to the fact that the reinsurance business faces more fluctuations due to unpredictable losses such that the ability to raise new capital is more important for reinsurers. Consequently, stock firms are more popular in the reinsurer market than mutuals due to their flexibility and access to channels for raising new capital from investors and external capital markets.

2.5.8 Major Current Issues

Terrorism and Insurance

Terrorist attacks in U.S. can be traced back to the February 1993 bombing of the World Trade Center, which caused 6 deaths and $753 million in property losses. The second major event is the April 1995 bombing of the Alfred P. Murrah Federal Building in Oklahoma City, Oklahoma, which caused 166 deaths and $150 million in property losses, much of which was uninsured because the loss involved a government building.[87] However, terrorism risk did not become an important issue in insurance industry until 2001.

The events of September 11, 2001 raised the issue of terrorism insurance in the insurance industry. The U.S. insurance industry suffered an estimated loss of $31.7 billion as a result of the attacks. The losses included business interruption claims of $9.8 billion, other property claims of $5.4 billion, liability losses of $4.0 billion, and aviation liability losses of $3.5 billion (Tillinghast Towers-Perrin 2004). After the event, the Congress enacted the Air Transportation Safety and System Stabilization Act on September 22, 2001. The Act provided compensation for the victims of the four airline crashes, limited the liability of the air carriers to the amount of their insured losses, provided assistance to U.S. passenger and cargo airlines through grants and federal credit assistance, extended their deadlines for tax payments, and subsidized them for the increases in insurance premiums. The act also allowed air carriers to purchase certain liability insurance coverage from the federal government and made the federal government responsible for any liability to third parties above $100 million, if the air carrier is a victim of a terrorist attack within 180 days after the enactment of the Act. The Congressional Budget Office (CBO) estimated that the act would cost $13.6 billion over the period from 2001 to 2006.[88]

Right after September 11, 2001, insurance for terrorism risk for most commercial insurance lines became unobtainable because premiums were inestimable due to the sudden increase in expected loss costs and the lack of enough historical data to estimate premiums. Reinsurers also withdrew coverage for terrorism risk. To solve the availability problem, the Terrorism Risk Insurance Act (TRIA) was passed on November 26, 2002. According to this act, the federal government and private insurance companies would share the losses from the risk of terrorism, with the federal government providing reinsurance to private insurers operating in the U.S. The act required insurers to offer terrorism insurance coverage to business customers. The primary goals of the act were to ensure the availability

---

[87] III (2005b), p. 103.

[88] Congressional Budget Office, Pay-as-you-go-estimate, H.R. 2926, Air Transportation Safety and System Stabilization Act, November 2001, p. 1.

Table 2.31. The Top 20 Largest Firms in the U.S. by Reinsurance Premiums Written ($ Millions), 2003

Name	Reinsurance Premiums	Total Premiums	Assets	Best's Rating	Distribution System	Organizational Form
Berkshire Hathaway	4,695	15,762	100,667	A++	A	S
Swiss Re Group	1,252	2,028	11,695	A+	D	S
American International Group	1,066	27,972	80,330	A++	B	S
GE Global Group	1,043	5,204	26,530	A	D	S
Munich Group	851	1,584	16,450	A+	D	S
Everest Reinsurance Holdings Inc.	804	3,124	8,257	A+	D, B	S
Fairfax Financial	742	2,998	12,196	A	B	S
Platinum Underwriters Re	496	668	1,330	A	A	S
W. R. Berkley Corp	493	3,540	9,227	A	B	S
PartnerRe Group	464	1,041	2,516	A+	B	S
State Farm	352	46,581	104,588	A++	D	M
CAN Insurance Group	304	7,365	41,294	A	A	S
Chubb & Son Inc.	237	9,537	28,414	A++	A	S
PENN MFR ASN Insurance	223	1,013	2,889	B++	B	S
Liberty Mutual Group	222	12,514	40,814	A	D	S
GMAC Insurance Holding	171	2,912	8,278	A	D	S
American Agricultural Insurance	170	493	885	A	D	S
X L America	166	727	6,768	A+	B	S
St Paul Travelers Group	159	19,940	74,687	A	A	S
SCORE Reinsurance Co	158	348	2,655	B++	BD	S

Source: NAIC (1993–2003).
Notes: A: Independent Agents. B: Broker. D: Direct Selling. S: Stock Firm. M: Mutual Firm.

and affordability of terrorism insurance by raising insurance capacity through government support and to help businesses and individuals recover from a terrorist attack. The minimum loss incurred before the federal reinsurance payments begin must be at least $5 million and private insurers must pay deductibles defined in the act. The federal government will pay 90 percent of losses above the deductibles up to a maximum event size of $100 billion and insurers have a 10 percent co-payment. Hence, TRIA is basically reinsurance for terrorism risk (Cummins, et al. 2004; Hubbard and Deal 2004).

TRIA will expire at the end of 2005, leaving insurers unprotected. Given the depleted capacity of the insurance industry by the four large hurricanes losses in 2004 (see Table 2.22) and Hurricanes Katrina, Rita and Wilma in 2005, extending TRIA for more years may be necessary before finding possible alternative approaches to TRIA.[89]

Tort Reform

The U.S. has the most expensive tort system in the world. In 2003, estimated tort system costs were $246 billion or $845 per person (Tillinghast Towers-Perrin 2004). This Tillinghast estimate of tort costs included losses (benefits paid or expected to be paid to third parties), defense costs, and administrative expenses. Total tort costs have increased a hundredfold in the past 50 years and, in 2003, the tort costs represented roughly 2 percent of GDP.[90] These costs included insured liability losses of $174 billion, medical malpractice costs of $27 billion, and self-insured or uninsured costs of $45 billion.[91] Of these, the increase in medical malpractice cost is most dramatic. Medical malpractice costs in the U.S. have increased at a rate of 11.8 percent annually since 1975, whereas the growth rate is only 9.2 percent for all other tort costs during the same period (Tillinghast Towers-Perrin 2004).

To understand the need for tort reform, it is important to learn where the tort money really goes. It is estimated that about 22 percent of the total costs are used to cover the economic loss of victims and approximately 24 percent for victim compensation for non-economic losses such as pain and suffering. Of the remaining costs, about 19 percent go to plaintiff's attorneys, 14 percent represent first party defense costs, and almost 21 percent represent administration fees (Tillinghast Towers-Perrin 2003). Obviously, the system is inefficient, because only 46 percent of the money goes to the victims.

The abuse of torts leads to an inefficient tort system, which adversely impacts the U.S. economy. The system has a negative effect on business innovation because of the threat of liability litigation and punitive damages. The skyrocketing cost of medical malpractice has made malpractice insurance unaffordable for some doctors. Physicians practice defensive medicine to protect them from liability lawsuits, cease

---

[89] The estimated insurance loss from Hurricane Katrina is between $40 and $55 billion (excluding losses insured under the National Flood Insurance Program), thus replacing Hurricane Andrew as the costliest hurricane in U.S. history (Tillinghast-Towers Perrin 2005, p. 1). The estimated insurance loss from Hurricane Rita is about $5 billion (Dasgupta 2005). The estimated insurance loss from Hurricane Wilma is between $8 and $12 billion (Risk Management Solutions 2005).

[90] Tillinghast-Towers Perrin (2004), pp. 1, 11.

[91] Ibid, p. 15.

to provide risky practices, and charge higher prices for their services. This causes a "death spiral" in the healthcare system with rapidly growing medical costs and rising health insurance premiums.

Consumers eventually pay for expensive tort costs. Though a portion of the money goes to the victims, the majority of costs are absorbed by the system. With an unpredictable tort system, the insurance industry is exposed to the potential for unexpected tort claim costs and unexpected risks, such as asbestos and environmental claims. The high loss and combined ratios of products liability insurance and the increasing loss and combined ratios for medical malpractice can be partially attributed to the tort system (see Table 2.30). To protect themselves from insolvency, insurers often decline to provide coverage for certain liabilities, such as asbestos, medical malpractice, and pharmaceuticals. To protect against unexpected or unknown liabilities, the insurers have gradually reduced the availability of occurrence-based liability policies and shifted to greater usage of claims-made policies. This is particularly true in the medical malpractice area.[92]

Pressure to reform the tort system is growing. The state and federal governments have initiated legislative efforts to improve the system. The proposed federal reforms mainly focus on three areas. The first one is to reduce "forum" shopping by plaintiffs' attorneys, by permitting the transfer of large interstate class action lawsuits to federal courts, which are considered to be friendly to defendants. The second major reform is to cap punitive and non-economic damages in medical malpractice awards to $250,000 and to speed the resolution process of medical malpractice suits. The third major reform involves asbestos-related diseases and aims at making the losses related to such diseases more predictable. At the state level, some states have revised legislation or enacted new legislation to shield producers from certain liabilities. Efforts are being made to raise the quality of jury service, eliminate the joint and several liability rule,[93] cap pain and suffering compensation, limit punitive damages awards, and limit manufacturers' liability by enforcing the "state-of-the-art defense."[94] It is commonly recognized that the increased use of the tort system today has been driven by the increase in attorneys and the contingency fee system. To discourage suing big companies for huge financial compensation and

---

[92] An occurrence-based policy covers the risks incurred during the term of the policy, regardless whether the claims are made during the policy period or not. The policyholder can put a claim against the insurer years later for any incidents that happened while the policy was in force. A claims-based policy only covers the claims made during the term of the policy. If a claim is made after the policy and the extended reporting period has expired, the policyholder will not get paid even if the incident happened in the term of the policy.

[93] Under the joint and several liability rule, if two or more defendants cause a plaintiff's injury, they are jointly and severally liable. The defendant(s) is (are) therefore liable for the entire amount of the plaintiff's damages regardless of each defendant's relative degree of fault or responsibility. For details, see ATRA (2005).

[94] III (2005d). "State of the art defense" refers to a product liability defense in which a manufacturer should not be liable for harm caused by a product that met the best prevailing standards of design, performance, and safety at the time it was manufactured (i.e., the product represented the state of the art).

dissuade attorneys from pursuing the "lawsuit lottery,"[95] proposals have been made that make the losing party pay all litigation fees and that limit attorney's fees.

Tort reform in the 1980s that aimed to restrain liability costs had a positive effect on the insurance market. The profitability of liability insurance was raised dramatically with these reforms (Born and Viscusi 1994). A similar effect occurred in the medical malpractice insurance market. Most firms providing medical malpractice insurance experienced increasing profitability following the reforms. The effect has been the strongest for firms that would have been most unprofitable without the damage-cap reforms (Viscusi and Born 1995). Whether the proposed new reforms will be enacted or will generate the same positive effects on the insurance market is still unknown. However, at least legislators have made some headway on this issue.[96]

The Issue of Contingent Commission

On October 14, 2004, New York State Attorney General Eliot Spitzer filed a civil complaint against Marsh & McLennan Cos. Inc. (MMC)[97] alleging that MMC, the world's largest insurance brokerage firm, charged an aggressive amount of contingent commissions to insurance companies for the exchange of placing business with them and that the firm used the so called "bid-rigging" arrangement to present false and inflated quotes to steer its clients to a particular insurance company that pays the highest contingent commission. Spitzer also stated that contingent commissions raises the cost of insurance for all customers and ruins the brokers' incentive to seek the best deal for their customers. In 2003, MMC collected $800 million in contingent commissions from insurance companies. The insurance firms involved in the bid-rigging arrangement included ACE Ltd., American International Insurance Group (AIG), Munich American Risk Partners, and the Hartford Financial Services Group. The suit alleged that the MMC's fraudulent practice violated antitrust laws and would be punished with compensatory and punitive damages.[98]

On January 31, 2005, MMC agreed to pay $850 million to settle with Spitzer. The money will go to the clients of MMC, most of which were corporations. Meanwhile, MMC agreed to ban contingent commissions, limit its brokerage compensation to a single fee, and disclose all forms of compensation to its clients.[99]

The impact of Spitzer's case on the insurance practice was dramatic. As a result of the case, many brokers and insurers abandoned the use of contingent commissions. This has the potential to cause vast changes in insurance underwriting. As an incentive compensation agreement between an insurance company and insurance agents or brokers, contingent commissions have a positive effect on

---

[95] A lawsuit lottery is a lawyer-driven lawsuit in which defendants who can pay huge damage awards are sued. Typically, defendants are chosen with little regard to their true responsibility.

[96] In late July, the House of Representatives approved medical malpractice liability reform H.R. 5—The Help Efficient, Accessible, Low-Cost, Timely Healthcare Act of 2005. The reform caps non-economic damages, set standards for punitive damage awards, allow courts to restrict the payment of attorney fees (Hoffman 2005).

[97] Sptizer (2004).

[98] "Sptizer Suit Alleges Bid Rigging" (2004).

[99] "Marsh to Pay $850 Million in Spitzer Settlement" (2005)

insurance industry practices. Brokers provide valuable services to both consumers and insurers by working as consultants, providing risk identification, evaluation, and claim settlement services for customers, and working as a risk screener for insurance firms. The contingent commission gives brokers an incentive to evaluate and provide more accurate underwriting information about their customers. Eliminating this kind of incentive compensation agreement could result in a less efficient insurance market, in which the problems of adverse selection and moral hazard could become more serious.

The Debate on Finite Risk (Re)Insurance

Finite risk contracts have been a popular alternative risk transfer tool since the mid 1990s. It is the product of the so-called integrated risk management or enterprise risk management. Immediately following Spitzer's investigation into contingent commissions in the U.S. insurance industry, Spitzer began to question the use of finite risk reinsurance (FRR) to transfer risks.

Through a finite risk transfer contract, firms set up an insurance premium fund each year and pay a fee equal to a certain percentage of their insurance premium to the insurance company. The balance of the fund each year depends on the investment income earned, the premiums deposited, and losses incurred, which could be pure risk losses or financial losses. At the end of the contract, the firms get back the positive balance of the fund if claims are good and insurance companies bear the negative balances if the claims are bad.

Finite risk contract differs from the traditional insurance in several ways. First, there is a cap for risks that can be transferred and the cap depends on the setup of the contract. Second, a finite risk contract is not renewed annually, but tends to cover a number of years, usually three to five. Therefore, one purpose of such contract is to smooth earnings of a firm.

The central issue on the debate about finite risk reinsurance lies in whether it is insurance. The New York Attorney General is concerned about that the degree of risk transfer included in such product is limited and that the primary purpose of the arrangement is financial statement enhancement. His second concern is that it distorts financial reporting and impedes financial transparency. Some firms purchase FRR to either lower reported losses, inflate reported surplus, or both.

However, insurance companies and brokers defend these claims in two ways by stating that FRR contracts exist legitimately. First, finite risk contracts offer a way to manage the retained risk of firms in a hard market where premiums are high and the availability of insurance is low. Second, FRR contracts work similarly to an excess insurance contract, which technically provide reinsurance above a specific level of losses of firms. It is particularly important for those firms that cannot obtain traditional reinsurance arrangements.[100]

The problem with the investigation into finite risk products is that it is very difficult to draw a line between good contracts and bad contracts. One objective of risk transfer, such as buying insurance, is to smooth earnings of firms. An effective finite risk arrangement will realize the same purpose, but the question is whether it

[100] Felsted (2004).

results in a real smoothing of earnings, hides losses for a short period of time, or creates excessive loss reserves to evade taxes.

Finite risk reinsurance premiums are estimated to account for less than 5 percent of total reinsurance premiums.[101] Currently, the investigation mainly focuses on the purchase of finite risk reinsurance by insurance companies. Through the purchase of FRR, an insurance company transfers its claims, usually, the medical malpractice claims or other long-tail insurance business and sometimes financial losses to the reinsurer. AIG, Berkshire Hathaway, and Ace limited are involved. Recently, MBIA Inc. has also become involved in the investigation.[102]

Asbestos Coverage and Mold Coverage in the U.S.

The U.S. insurance industry experienced a liability crisis in the mid 1980s partially due to the unexpected high costs from asbestos law suits. The first asbestos law suit was filed in 1966 and the number increased to roughly 10,000 cases from 1980 to 1984. About 37,000 cases were filed between 1985 and 1989, but the filings decreased between the mid- and late- 1990s. However, litigation increased in the late 1990s, driven by factors such as expanding law suits to firms using asbestos, nonmalignant filings, and the prevalence of joint and several liability.[103]

By the end of 2002, more than 730,000 people had filed asbestos-related claims. Businesses and insurers paid more than $70 billion to settle these claims. As a result, many firms that were deeply involved in the law suits filed bankruptcy. It is estimated that the cumulative liability for asbestos will reach $200 billion and that insurers will bear 61 percent of the claims, of which U.S. insurers will pay 30 percent and the foreign insurers will pay 31 percent.[104]

Mold contamination has also become another issue of concern in the U.S. insurance industry. Insurers usually do not cover damage from mold, such as rust and mildew, in standard homeowners and commercial property policies. However, mold contamination is covered if the mold results from the covered perils in these policies. Mold claims have increased significantly in the early 2000s. For example, the estimated number of mold claims in Texas rose 1,306 percent from the first quarter of 2000 to the fourth quarter of 2001. The average cost of mold claims rose 152 percent from the first quarter of 2000 to the second quarter of 2001.[105] To avoid any confusion about mold coverage and to finance the higher costs from law suits, many insurers now either insert a clarifying clause in the policy exempting damage from mold or try to cover all mold losses by imposing a higher premium.

---

[101] III (2005g).

[102] McLeod (2005).

[103] Nonmalignant filing is the filing of claims by people with little or no current disability. Because some liability insurance is written on the claims-made basis, the claims must be filed within a specified date to get compensation later if a disability is found.

[104] The information primarily comes from III (2005l).

[105] Hartwig (2003). There were 1,050 mold claims in the first quarter of 2000. The average cost per mold claim in the first quarter of 2000 was $13,719.

2.6 CONCLUSION

The U.S. insurance market is arguably the most important insurance market in the world because of its size, capacity, and impact. It has the world's largest L-H insurance market and the world's largest P-L insurance market. This chapter analyzed the U.S. insurance market from 1993 to 2003. Both the L-H insurance and P-L insurance markets were investigated. Overall, this was a period of fast growth for both markets. The expanding economy, innovations in computing and communication technology, the wave of financial integration, and the booming stock market provided the industry with dramatic growth opportunities.

In the L-H industry, the relative weight continued to shift from traditional life insurance products to the annuity and health insurance lines of business. This trend can be expected to continue in the short-term as baby boomers move into the retirement phase and as life expectancies continue to improve. Significant changes in the nation's healthcare system—e.g., away from an employer-sponsored system—might impact this trend, especially if employers continue to shift more of the healthcare decision-making and payment burdens onto consumers. This notwithstanding, it is difficult to imagine aggregate costs decreasing in the near future. As insurers make progress in understanding the interrelationships between the different risks that individuals face, it is not difficult to envision a future in which multiple risks are increasingly included under a single policy. This would represent a natural continuation of the kinds of changes that have historically accompanied technological and informational advances (e.g., the development of variable insurance products and flexible-premium insurance products during the second half of the 20th century).

In the past decade, a relatively more concentrated P-L insurance market is observed, especially in commercial lines, which may indicate more efficient firms in the industry. However, very frequent and severe natural CATs coupled with the continuously rising cost of liability claims depleted the excess capacity of the market from time to time. Underwriting cycles, the availability of insurance coverage, and the low profitability of the industry were major concerns in this time period. Some insurance companies adapted to the changing market environment through demutualization, mergers and acquisitions, and adjustments in distribution systems.

In the near future, the U.S. insurance industry may continue to face the various challenges it has struggled with for decades. How to manage exceptional and catastrophic risks such as terrorism, hurricanes, and earthquakes, how to deal with the nation's rapidly increasing healthcare expenditures (in both the private and public sectors) and the increasing liability insurance claims will continue to be important issues for the industry. The industry also needs to find better solutions to deal with financial risks and operational risks. Among the most pressing of these are fluctuations in the stock market, constantly changing interest rates, and the assorted challenges associated with markets for alternative risk transfer.

2.7 REFERENCES

A.M. Best Company, Inc., 1993–2003, *Best's Key Rating Guide CD* (Oldwick, New Jersey: A.M. Best Company).

_____, 2002, *Best's Aggregates & Averages, Life-Health United States and Canada, 2002 Edition* (Oldwick, New Jersey: A.M. Best Company).

_____, 2004a, *Best's Aggregates & Averages, Life-Health United States and Canada, 2004 Edition* (Oldwick, New Jersey: A.M. Best Company).

_____, 2004b, *Best's Aggregates & Averages, Property-Liability United States and Canada, 2004 Edition* (Oldwick, New Jersey: A.M. Best Company).

_____, 2005, *Best's Review: The Guide to Understanding the Insurance Industry* (Oldwick, New Jersey: A.M. Best Company).

American Council of Life Insurers (ACLI), 2004, *Life Insurers Fact Book 2004* (Washington DC).

American Tort Reform Association (ATRA), 2005, *Joint and Several Liability Rule Reform*, www.atra.org/show/7345.

Baranoff, Dalit, 2004, "Fire Insurance in the United States," in Robert Whapels, ed., *EH.Net Encyclopedia*, eh.net/encyclopedia/article/Baranoff.Fire.final.

Barnett Bank of Marion County, N.A. v. Nelson, 116 S. Ct. 1103, 1996.

Berger, Allen N., J. David Cummins, and Mary A. Weiss, 1997, "The Coexistence of Multiple Distribution Systems for Financial Services: The Case of Property-Liability Insurance," *Journal of Business* 70: 515–546.

Black, Kenneth and Harold D. Skipper, 2000, *Life & Health Insurance, 13th Edition* (Upper Saddle River, NJ: Prentice Hall).

Bluhm, William F., ed., 2000, *Group Insurance, 3rd Edition* (Winstead, CT: Actex Publications).

Born, Patricia and W. Kip Viscusi, 1994, "Insurance Market Responses to the 1980s Liability Reforms: An Analysis of Firm-Level Data," *Journal of Risk and Insurance* 61: 192–218.

Brearley, Harry Chase, 1916, *Fifty Years of a Civilizing Force: An Historical and Critical Study of the Work of the National Board of Fire Underwriters* (New York: Frederick A. Stokes Company).

Butler, Richard J., 2002, "Form Regulation in Commercial Insurance," in J. David Cummins, ed., *Deregulating Property-Liability Insurance* (Washington DC: American Enterprise Institute-Brookings Institution Joint Center for Regulatory Studies), pp. 321–360.

Butler, Richard J., Yijing Cui, and Andrew Whitman, 2000, "Insurers' Demutualization Decisions," *Risk Management and Insurance Review* 2: 135–154.

Cagle, Julie A. B., Robert L. Lippert, and William T. Moore, 1996, "Demutualization in the Property-Liability Insurance Industry," *Journal of Insurance Regulation* 14: 343–370.

Carrow, Kenneth A., 2001, "The Wealth Effects of Allowing Bank Entry into the Insurance Industry," *Journal of Risk and Insurance* 68: 129–150.

California Earthquake Authority (CEA), 2000, *About the CEA*, www.earthquakeauthority.com/about/about.html#top.

California Department of Insurance, 1996, *Industry: 1996 Earthquake Coverage Experience—Summary of Residential Market Share—1996 Experience Year*, Sacramento, California, www.insurance.ca.gov/0200-industry/0600-data-reports/0100-earthquake-cov-exp/1996-earthquake-coverage-experience.cfm.

_____, 2000a, *2000 California Property and Casualty Market Share Report, Line of Business: Earthquake*, Sacramento, California, www.insurance.ca.gov/0400-news/0200-studies-reports/0100-market-share/upload/MktShr_CmpnyAlpha2000.pdf.

_____, 2000b, *Industry: 2000 Earthquake Coverage Experience—Summary of Residential Market Share—2000 Experience Year,* Sacramento, California, www.insurance.ca.gov/0200-industry/0600-data-reports/0100-earthquake-cov-exp/2000-earthquake-coverage-experience.cfm.

_____, 2003, *Industry: 2003 Earthquake Coverage Experience—Summary of Residential Market Share—2003 Experience Year,* Sacramento, California, www.insurance.ca.gov/0200-industry/0600-data-reports/0100-earthquake-cov-exp/2003-earthquake-coverage-experience.cfm.

_____, 2004, *2004 California Property and Casualty Market Share Report, Line of Business: Earthquake*, Sacramento, California, www.insurance.ca.gov/0400-news/0200-studies-reports/0100-market-share/upload/IndMktShr2004alpha.pdf.

Congressional Budget Office (CBO), 2001, *Congressional Budget Office Pay-As-You-Go Estimate, H.R. 2926, Air Transportation, Safety, and System Stabilization Act: As Cleared by the Congress on September 21, 2001,"* www.cbo.gov/showdoc.cfm?index=3206&sequence=0.

Cummins, J. David, ed., 1977, *Investment Activities of Life Insurance Companies* (Homewood, IL: R. D. Irwin).

_____, ed., 2002, *Deregulating Property-Liability Insurance* (Washington DC: American Enterprise Institute-Brookings Institution Joint Center for Regulatory Studies).

_____ and Neil A. Doherty, 2005, "The Economics of Insurance Intermediaries," Working Paper, The Wharton School, University of Pennsylvania.

_____ and Elizabeth Grace, 1994, "Tax Management and Investment Strategies of Property-Liability Insurers," *Journal of Banking and Finance* 18: 43–72.

_____, Martin F. Grace, and Richard D. Phillips, 1999, "Regulatory Solvency Prediction in Property-Liability Insurance: Risk-Based Capital, Audit Ratios, and Cash Flow Simulation," *Journal of Risk and Insurance* 66: 417–58.

_____, Scott E. Harrington, and Robert Klein, 1995, "Insolvency Experience, Risk-Based Capital, and Prompt Corrective Action in Property-Liability Insurance," *Journal of Banking and Finance* 19: 511–27.

_____, Christopher M. Lewis, and Ran Wei, 2004, "The Market Value Impact of Operational Risk Events for U.S. Banks and Insurers," Working Paper, Social Science Research Network, New York, NY.

_____ and Gregory P. Nini, 2002, "Optimal Capital Utilization by Financial Firms: Evidence from the Property-Liability Insurance Industry," *Journal of Financial Services Research* 21: 15–53.

_____, Richard Phillips, and Mary A. Weiss, 2001, "The Incentive Effects of No-Fault Automobile Insurance," *The Journal of Law and Economics* 44: 427–464.

_____ and Anthony M. Santomero, eds., 1999, *Changes in the Life Insurance Industry: Efficiency, Technology, and Risk Management* (Norwell, MA: Kluwer Academic Publishers).

_____ and Mary A. Weiss, 1992, "The Structure of Conduct and Regulation of the Property-Liability Insurance Industry," in Richard W. Kopcke and Richard E. Randall, eds., *The Financial Condition and Regulation of Insurance Companies* (Boston: Federal Reserve Bank of Boston), pp. 1–18.

_____ and _____, 2000, "The Global Market for Reinsurance: Consolidation, Capacity, and Efficiency," in Robert E. Litan and Anthony M. Santomero, eds., Brookings-Wharton Papers on Financial Services (Washington DC: Brookings Institute Press), pp. 159–210.

_____, Mary A. Weiss, and Hongmin Zi, 1999, "Organizational Form and Efficiency: The Coexistence of Stock and Mutual Property-Liability Insurers," *Management Science* 45: 1254–1269.

_____ and Xiaoying Xie, 2005, "Economies of Scale and Efficiency Analysis in the United States Property-Liability Insurance Industry," Working Paper, The Wharton School, University of Pennsylvania.

D'Arcy, Stephen P., 2002, "Insurance Price Deregulation: The Illinois Experience," in J. David Cummins, ed., *Deregulating Property-Liability Insurance* (Washington DC: American Enterprise Institute-Brookings Institution Joint Center for Regulatory Studies), pp. 248–84.

Dasgupta, Dave, 2005, "Insurers to Pay Hurricane Rita's Victims $4.7 Billion for Insured Property Losses, Says ISO's Property Claim Services," *Insurance Services Office* (November 1), www.iso.com/press_releases/2005/11_01_05.html.

Dorfman, Mark S., 1998, *Introduction to Risk Management and Insurance* (Upper Saddle River, NJ: Prentice Hall).

Earthquake Engineering Research Institute, 1997, *Earthquake Basics Brief, No. 3, Insurance: What Are the Principles of Insuring Natural Disasters?* (April), California Earthquake Authority, Sacramento, California, www.eeri.org/cds_publications/earthquake_basics_series/INS3.pdf.

Erhemjamts, Otgontsetseg, 2005, "What Drives Life Insurer Demutualizations?" Working Paper, J. Mack Robinson College of Business, Georgia State University.

Federal Reserve Bank of St. Louis, 1959–2003, Federal Reserve Economic Data (FRED), St. Louis, Missouri.

Felsted, Andrea, 2004, "Finite Risk Comes Under Scrutiny: Eliot Spitzer Fears the Product Could Be Used To Doctor Earnings," *Financial Times*, December 1, 2004, p. 26.

Florida State Board of Administration (SBA), 2003, *The Investment Funds—Florida Hurricane Catastrophe Fund*, www.sbafla.com/funds.aspx?section=4.

"Forecast for the Top-Tier Institutions: Deals that Increase Value and Competitiveness," 2002, *Bank Director Magazine M & A Supplement*, Brentwood, Tennessee, www.bankdirector.com/supplements/articles.pl?article_id=11153.

Gray, Peter, 2005, "A Brief History of Friendly Societies," The Association of Friendly Societies, www.afs.org.uk/research/researchpgrayhistorypage.htm.

Guy Carpenter, 2003, *The World Catastrophe Reinsurance Market: 2003, The Market Stabilizes in 2003* (September), www.guycarp.com/portal/extranet/pdf/GCPub/worldcat_2003.pdf?vid=1.

Hall, Brian J., 2000, "Regulatory Free Cash Flow and the High Cost of Insurance Company Failures," *Journal of Risk and Insurance*, 67: 415–38.

Han, Li-Ming, Gene C. Lai, and Robert C. Witt, 1997, "A Financial-Economic Evaluation of Insurance Guaranty Fund System: An Agency Cost Perspective," *Journal of Banking and Finance* 21: 1107–1129.

Harrington, Scott E., 2002, "Effects of Prior Approval Rate Regulation of Auto Insurance," in J. David Cummins, ed., *Deregulating Property-Liability Insurance* (Washington DC: American Enterprise Institute-Brookings Institution Joint Center for Regulatory Studies), pp. 285–314.

_____ and Greg Niehaus, 2000, "Volatility and Underwriting Cycles," in Georges Dionne, ed., *Handbook of Insurance* (Boston: Kluwer Academic Publishers), pp. 657–86.

_____ and _____, 2002, "Capital Structure Decisions in the Insurance Industry: Stocks versus Mutuals," *Journal of Financial Services Research* 21: 145–163.

Hartwig, Robert P., 2003, "Mold and Insurance: Is the Worst behind Us?", presentation at the Casualty Actuarial Society Seminar on Ratemaking, San Antonio, Texas (March 27).

Hoffman, Mark A., 2005, "House Passes Med-Mal Reform Bill," *Business Insurance* (July 28), businessinsurance.com/cgi-bin/news.pl?newsId=6045#.

Hoover's Online, 2005, "Liberty Mutual Holding Company, Inc.—Profile View."

Hubbard, R. Glenn and Bruce Deal, 2004, "The Economic Effects of Federal Participation in Terrorism Risk," Analysis Group, New York, NY.

Insurance Information Institute, 2005a, *Background on Insurance Intermediaries*, New York, NY.

_____, 2005b, *Insurance Factbook 2005*, New York, NY.

_____, 2005c, *Tort Excess 2005: The Necessity for Reform from a Policy, Legal, and Risk Management Perspective*, New York, NY.

_____, 2005d, *Hot Topics and Issues Updates. Catastrophes: Insurance Issues* (April), New York, NY.

_____, 2005e, *Hot Topics and Issues Updates. Crop Insurance* (April), New York, NY.

_____, 2005f, *Hot Topics and Issues Updates. Earthquake: Risk and Insurance Issues* (April), New York, NY.

_____, 2005g, *Finite Risk Reinsurance* (May), New York, NY.

_____, 2005h, *Insolvencies/Guaranty Funds* (May), New York, NY.

_____, 2005i, *No-Fault Auto Insurance* (July), New York, NY.

_____, 2005j, *Rates and Regulation* (July), New York, NY.

_____, 2005k, *Captives and Other Risk Financing Options* (September), New York, NY.

_____, 2005l, *Asbestos Liability* (September), New York, NY.

_____, 2005m, *Financial and Market Conditions* (September), New York, NY.

_____, 2005n, *Liability System* (September), New York, NY.

Joskow, Paul L., 1973, "Cartels, Competition, and Regulation in the Property-Liability Insurance Industry," *The Bell Journal of Economics and Management Science* 4: 375–427.

Kollar, John, 2002, *Terrorism Insurance Coverage in the Aftermath of September 11th*, American Academy of Actuaries, Washington, DC.

Liberty Mutual, 2002, *Mutual Holding Company Overview*, www.libertymutual.com/about/mhc.

"Marsh To Pay $850 Million In Spitzer Settlement," 2005, *New York Insurance Law and Litigation Alert* (February 15), p. 1.

Mayers, David and Clifford W. Smith Jr., 1981, "Contractual Provisions, Organizational Structure, and Conflict Control in Insurance Markets," *Journal of Business* 54: 407–34.

_____ and _____, 1986, "Ownership Structure and Control: The Mutualization of Stock Life Insurance Companies," *Journal of Financial Economics* 16: 73–98.

_____ and _____, 1988, "Ownership Structure across Lines of Property-Casualty Insurance," *Journal of Law and Economics* 31: 351–78.

_____ and _____, 2000, "Organizational Forms within the Insurance Industry: Theory and Evidence," in G. Dionne, ed., Handbook of Insurance, (Boston, MA: Kluwer Academic Publishers), pp. 689–707.

_____ and _____, 2002, "Ownership Structure and Control: Property-Casualty Insurer Conversion to Stock Charter," *Journal of Financial Services Research* 21: 117–44.

McLeod, Douglas, 2005, "Regulators Eye Civil Charges against MBIA," *Business Insurance*, August 15: 1, 34.

Mester, Loretta J., 1989, "Testing for Expense Preference Behavior: Mutual versus Stock Savings and Loans," *The RAND Journal of Economics* 20: 483–498.

Moore, Terry D., 2001, "California Earthquake Insurance: What's Shakin'?", Association Services and Insurance Brokers, LLC (March 21), www.asaib.com/CA_Earthquake_Insurance.htm.

Murphy, Sharon Ann, 2005, "Life Insurance in the United States through World War I," www.eh.net/encyclopedia/?article=murphy.life.insurance.us.

National Association of Insurance Commissioners (NAIC), 1993–2003, NAIC Annual Statement CD (Kansas City, MO: NAIC).

_____, 2005, *Financial Regulation Standards and Accreditation Program: Accredited States* (March), www.naic.org/frs/accreditation/map.htm.

NationsBank of North Carolina v. Variable Annuity Life Insurance Co. ("VALIC"), 115 S. Ct. 810, 1995.

Paul v. Virginia. 322 U.S. 533, 1944.

Risk Management Solutions (RMS), 2005, "RMS Estimates U.S. Insured Loss for Hurricane Wilma at $8–$12 Billion," October 27, Newark, CA, www.rms.com/NewsPress/PR_102705_HU_Wilma_Loss_Estimate_2.asp.

SNL Financial, 2005, "Insurance Industry Vital Statistics—Total Return Performance of Former Mutuals after IPO: Ranked by One Year Total Return," www.snl.com/insurance/vitals/total_return.asp.

Spitzer, Eliot, 2004 "Complaint: The People of the State of New York against Marsh & McLennan Companies, Inc. and Marsh Inc.," Office of the Attorney General of the State of New York, Albany, New York.

"Spitzer Suit Alleges Bid Rigging," 2004, *Texas Insurance Law and Litigation Alert* (October 31), p. 2.

Sunders, Anthony and Macia Millon Cornett, 2003, *Financial Institutions Management: A Risk Management Approach, 4th Edition* (New York: McGraw-Hill).

Swiss Re, 1999, "Alternative Risk Transfer (ART) for Corporations: A Passing Fashion or Risk Management for the 21st Century?," *Sigma,* No. 2.

_____, 2004, "Commercial Insurance and Reinsurance Brokerage, Love Thy Middleman," *Sigma,* No. 2.

_____, 2005, "World Insurance in 2004: Growing Premiums and Stronger Balance Sheets," *Sigma*, No. 2.

Bank Director Magazine, "The Impact of Legislation on Acquisitions of Non-Bank Industries," 2001, *Bank Director Magazine M & A Supplement*, Brentwood, Tennessee, www.bankdirector.com/supplements/articles.pl?article_id=10353.

Tillinghast-Towers Perrin, 2003, "U.S. Tort Costs: 2003 Update, Trends, and Findings on the Costs of the U.S. Tort System."

_____, 2004, "U.S. Tort Costs: 2004 Update, Trends, and Findings on the Costs of the U.S. Tort System."

_____, 2005, "Hurricane Katrina: Analysis of the Impact on the Insurance Industry."

U.S. Department of Justice, 2005, "Section 1.51: Concentration and Market Shares," in *Horizontal Merger Guidelines,* Washington, DC, www.usdoj.gov/atr/public/guidelines/horiz_book/15.html.

U.S. Department of Labor Bureau of Labor Statistics, 2005, *Consumer Price Index, All Urban Consumers, U.S. City Average, All Items,* Washington, DC, ftp://ftp.bls.gov/pub/special.requests/cpi/cpiai.txt.

Viscusi, W. Kip and Patricia Born, 1995, "Medical Malpractice Insurance in the Wake of Liability Reform," *The Journal of Legal Studies* 24: 463–490.

Viswanathan, K. S., 2005, "The Pricing of Insurer Demutualization Initial Public Offerings," *Journal of Risk and Insurance*, forthcoming.

_____ and J. David Cummins, 2003, "Ownership Structure Changes in the Insurance Industry: An Analysis of Demutualization," *Journal of Risk and Insurance* 70: 401–437.

Williams, Julie L., Stuart E. Feldstein, and Karen E. McSweeney, 1997, "Current Issues Regarding Bank Sales of Insurance and Annuity Products," *Risk Management and Insurance Review* 1: 65–84.

Wright, Kenneth M., 1991, "The Structure, Conduct, and Regulation of the Life Insurance Industry," in Richard W. Kopcke, and Richard E. Randall, eds., *Conference Series No. 35: The Financial Condition and Regulation of Insurance Companies Conference (June)*, Federal Reserve Bank of Boston (Boston, MA), pp. 73–116.

Zanjani, George, 2004, "The Rise and Fall of the Fraternal Life Insurer: Law and Finance in U.S. Life Insurance, 1870–1920," Working Paper, Federal Reserve Bank of New York (New York, NY).

2.8 LEXICON

AAIS. The American Association of Insurance Services provides product development, actuarial analysis, and regulatory filings services for insurance company. AAIS also designs insurance policy forms.

Accelerated Death Benefit. The accelerated death benefit is a life insurance policy provision which allows the insured to receive payment of all or part of his death benefit in the event of a serious (i.e., life-threatening) adverse health event.

Accident and Health Insurance. Accident and health insurance is insurance that pays when the insured incurs additional expenses or loses income due to incapacity or loss of health (also called accident and sickness insurance or disability insurance).

Accident and Sickness Insurance. See Accident and Health Insurance.

ACLI. American Council of Life Insurers.

AIG. American International Insurance Group.

AOUW. Ancient Order of United Workmen.

ART. Alternative risk transfers are non-traditional risk management methods, such as captives, finite risk (re)insurance, multi-line/multi-year insurance contracts, and insurance products with multiple triggers, contingent capital arrangement, and securitization which transfer risk via the capital market.

AVR. An asset valuation reserve is a statutory accounting mechanism used to prevent large changes in insurers' reported surplus caused by realized capital gains and losses associated with any and all investments whose values do not depend directly on market interest rates.

BHCA. The Bank Holding Company Act was passed in 1956 by U.S. Congress. It prohibited national banks from either owning non-bank businesses or from engaging in most kinds of non-banking activities.

Cafeteria Plan. A cafeteria plan is an employee benefit plan in which employees receive the option of taking cash on an after-tax basis or using it on a payroll deduction basis to purchase benefits from a menu, hence cafeteria, of benefit options. They typically included group life and health insurance plans.

Captive. A captive is an insurance subsidiary owned by one or more non-insurers, to provide owners with coverage.

Captive Agent. See Exclusive Agent.

Career Agents. Career agents are life-health insurance marketing intermediaries who sell exclusively for a single life insurer. Multiple agents typically work out of the same office building and receive training and sales support from the insurer they represent.

CAT. Catastrophe losses.

CBO. Congressional Budget Office.

CEA. The California Earthquake Authority is a government institution that provides earthquake coverage in California. This institution is privately-funded, but publicly-managed.

CEBA. Competitive Equality Banking Act.

Citizens Property Insurance Corporation (Citizens). The Citizens Property Insurance Corporation was created in 2002. It is the combination of the Florida Residential Property and Casualty Joint Underwriting Association (FRPCJUA) and the Florida Windstorm Underwriting Association (FWUA). It serves as a wind-only writer in old FWUA areas and an all-perils writer for FRPCJUA policies.

Claims-Made Policy. A claims-made policy is a type of insurance that only pays claims filed to the insurer during the term of the policy or within a specific period of time after the policy expires. It limits insurers' exposure to unknown future liabilities.

Combined Ratio. The combined ratio is the sum of the benefit ratio and expense ratio for life-health firms. For property-liability firms, it is the sum of the loss ratio, expense ratio, and policyholder dividend ratio.

Common Stock. Common stock represents ownership interest in a corporation that usually carries voting rights and may pay dividends. In the event of bankruptcy, common stock holders generally have claim only on the assets remaining after bondholders, preferred shareholders, and all other debt holders have been paid fully.

Contingent Commission. A contingent commission is a type of incentive payment mechanism for insurance agents and brokers. It is usually based on the underwriting results (such as profitability) of businesses placed or services provided by the agents and broker or the volume of business provided to the insurers.

CPI. Consumer Price Index.

Credit Life Insurance. Credit life insurance is an insurance contract that pays any outstanding debt to the insured individual's creditors upon his death.

DAC Tax. The DAC tax is the deferred acquisition costs tax. It mimics an accelerated recognition of life-health premium income through the capitalization and amortization of acquisition costs.

Demutualization. Demutualization is the process by which a mutual insurance company changes its organizational form to become a stock insurer.

Deposit-Type Contracts. Deposit-type contracts are financial contracts that do not involve life and/or health contingencies. They include guaranteed investment contracts (GICs), annuities certain, premium and other deposit funds, dividend and coupon accumulations, lottery payouts, and structured settlements.

DHHS. U.S. Department of Health and Human Services.

Direct Response Marketing. Direct response marketing includes sales activities involving unmediated direct contact with potential customers, including television advertisements, telephone solicitations, direct mailings, and the Internet. It is also called direct selling.

Direct Selling. See Direct Response Marketing.

Disability Insurance. See Accident and Health Insurance.

Disintermediation. Disintermediation occurs when policyholders borrow money against their policy assets and then lend the money at a rate higher than the borrowing rate.

Distribution System/Channel. The distribution system or distribution channel is the system through which insurance products are sold. It is also referred to as marketing channel.

DOJ. U.S. Department of Justice.

EPO. An EPO is an Exclusive Provider Organization. It is a healthcare organization that contracts with insurers and employers to provide services on a fee-for-service basis. Plan participants are restricted to use only participating providers.

ERISA. ERISA is the Employee Retirement Income Security Act. It was passed in 1974 by U.S. Congress with the purpose of protecting the interests of employee benefit plan participants and their beneficiaries.

Exclusive Agent. An exclusive agent is also known as multi-line exclusive agents or captive agents in the property-liability industry. Exclusive agents are marketing intermediaries who sell policies across different lines of business (e.g., life-health and property-liability) exclusively for a single insurer or affiliated group of insurers.

Expense Ratio. The expense ration is the ratio of expense to commissions in the life-health industry. In the property-liability industry it is the ratio of underwriting expenses to net premiums written.

FASB. Financial Accounting Standards Board.

FAST. Financial Analysis Solvency Tracking System.

FCIC. The FCIC is the Federal Crop Insurance Corporation. It is an organization managed by the federal Risk Management Agency (RMA) that provides the multiple peril crop insurance for U.S. farmers. The program was founded in 1938 and is backed by the resources of the U.S. Treasury Department.

FEMA. FEMA is the Federal Emergency Management Agency. It is the U.S. federal government agency charged with responding to events deemed to be federal disasters, hazards, and emergencies. It usually uses federally-provided funds to minimize or ameliorate the financial consequences of such events to affected citizens.

FHCF. The FHCF is the Florida Hurricane Catastrophe Fund, a state trust fund under the direction and control of the State Board of Administration of Florida (SBA). It was created in 1993 in response to Florida's property insurance crisis resulting from Hurricane Andrew. The FHCF provides reimbursements to insurers for a portion of catastrophic hurricane losses. As the world's largest hurricane reinsurer, the assets of the fund totaled more than $5.5 billion in June 2004 (Florida State Board of Administration 2003).

Financial Holding Company. A financial holding company is a legal entity that can engage in any activity or hold shares in companies engaging in any activity that either is financial in nature or at least poses no risk to the safety of depository institutions or to the financial system in general. This includes banks, securities firms, and insurance companies to name a few.

Finite Risk (Re)Insurance (FRR). Finite risk (re)insurance is a type of ART tool. A firm creates an insurance premium fund each year and pays a fee equal to a percentage of the insurance premiums to the insurance company. The balance of the fund each year depends on the premium deposited, the investment income earned, and the total losses incurred. The amount of risks that can be transferred is capped and the contract is not renewed annually, but tends to cover a number of years, usually three to five.

FIRREA. The Financial Institutions Reform, Recovery, and Enforcement Act of 1989.

Flexible Benefit Plan. A flexible benefit plan is an employee benefit plan that allows an employee some amount of choice regarding how their employer-provided benefit dollars are allocated across available benefit categories (e.g., health insurance, life insurance, and retirement plan benefits).

Franchise Tax. A franchise tax is a tax or fee that is usually assessed annually to corporations, limited liability companies, and similar business entities for the right to exist or do business in a particular state. Failure to pay the franchise tax or similar fees may result in the administration dissolution of the company and forfeiture of the charter.

Fraternal Benefit Societies. Fraternal benefit societies are organizations formed along ethnic, religious, and occupational lines and charged with meeting the social and economic needs of organization members and their familial dependents. Fraternal organizations provided some of the earliest group-like insurance coverage for accidents, sickness, death, and disability income losses in the U.S.

Fraternal Insurance. Fraternal insurance is insurance provided through a fraternal organization.

Friendly Societies. Fraternal societies are non-governmental social institutions that arose in eighteenth and nineteenth century England, mostly among the poorer, disadvantaged classes. These organizations had the purpose of spreading the various financial risks associated with indigence, including sickness, disability, and employment (among others) across large groups of workers.

FRPCJUA. FRPCJUA is the Florida Residential Property and Casualty Joint Underwriting Association. It was created after Hurricane Andrew and primarily provides homeowners insurance to residents in Florida who are unable to obtain coverage from private insurers. All property insurance companies writing coverage in Florida are required to participate in the association (Congressional Budget Office 2001).

FWUA. FWUA is the Florida Windstorm Underwriting Association. It was created in 1970 and provides hurricane and windstorm coverage to undesirable properties in coastal areas designated by the Florida Legislature. Private insurers may therefore exclude hurricane losses in areas covered by the FWUA, which consequently makes offering standard homeowners policies in high-risk areas more attractive to private insurers. This thus reduces the pressure on the FRPCJUA to provide coverage (Congressional Budget Office 2001).

FTC. Federal Trade Commission.

GAAP. Generally Accepted Accounting Standards.

GDP. Gross Domestic Product.

General Account. A general account is an account which holds the assets an insurer uses to back its contracts with fixed and guaranteed benefit payments.

GIC. A GIC is a guaranteed investment contract which offers a guaranteed return of principal and a fixed or floating rate of return for a specified period of time.

GLB. The GLB is the Gramm-Leach-Bliley Financial Services Modernizations Act. Passed by U.S. Congress in 1999, it removed barriers to affiliation between insurers and other financial services companies (banks and securities firms) through the creation of a new class of financial entities: financial holding companies.

Group Insurance. Group insurance is insurance that is marketed and purchased at the group level. Employers often provide this kind of insurance for their workforce.

Hard Market. A hard market is a seller's market in which the price of insurance is high and insurance is in short supply.

Herfindahl Index. The Herfindahl Index is an index used by the U.S. Department of Justice to measure the amount of concentration in any particular industry for purposes of promoting competition.

HMO. An HMO is a Health Maintenance Organization. It is a healthcare organization that operates on a per-capita, prepaid basis to provide healthcare to select groups of individuals by contracting with an organized group of physicians and hospitals.

III. Insurance Information Institute.

IMR. An IMR is an Interest Maintenance Reserve. It is a statutory accounting mechanism that has the purpose of preventing large changes in insurers' reported surplus caused by realized capital gains and losses associated with market interest rate fluctuations.

Individual Annuity. An individual annuity is an annuity marketed to and purchased by individuals as opposed to a group.

Industrial Life Insurance. Industrial life insurance is life insurance marketed directly to and sold to individuals as opposed to a group in small amounts, usually less than $2,000, with premiums collected on a weekly or monthly basis at the policy owner's place of residence.

Interest Sensitive Products. Interest sensitive products are life-health insurance products which reflect contemporaneous market returns in the accumulation of policy assets

IPO. Initial Public Offering.

IRA. An IRA is an individual retirement account that is tax-deferred.

IRIS. Insurance Regulatory Information System.

IRS. The IRS is the Internal Revenue Service. It is the U.S. federal government agency responsible for tax law enforcement and tax collection.

ISO. The ISO is the Insurance Services Office. It is an insurance services organization that provides statistical, actuarial, and underwriting information to insurers, regulators, and other organizations. ISO also provides standardized policy forms for property-liability insurance companies.

Joint and Several Liability. Joint and several liability is a liability rule which specifies that if two or more defendants cause a plaintiff's injury, they are jointly and severally liable. The defendant(s) is (are) therefore liable for the entire amount of the plaintiff's damages regardless of each defendant's relative degree of fault or responsibility.

Keogh Plan. A Keogh plan is a tax-advantaged plan for self-employed individuals.

Legal Reserve Organizational Forms. Legal reserve organizational forms are organizational forms required by law to hold reserves in anticipation of future claims. Mutual and stock organizational forms fall into this category, while fraternal insurers historically do not.

Loss Ratio. In the life-health industry, a loss ratio is the ratio of benefit payments to net premiums. In the property-liability industry, the loss ratio is the ratio of losses and loss adjustment expenses to net premiums earned.

LTC Insurance. LTC insurance is long-term care insurance that pays for care, often up to some fixed amount per month until death, in the event that the insured is unable to adequately perform certain activities deemed necessary for unassisted daily living.

M&As. Mergers and acquisitions.

Managed Care. Managed care is the use of active management of healthcare services and providers, usually through cost-sharing mechanisms and administrative oversight of healthcare processes, in order to reduce healthcare costs.

Market Concentration. Market concentration is a measure of the distribution of market power, measured on some meaningful basis such as sales revenue, across the business units within a given industry or business sector.

Marketing Channel. See Distribution System/Channel.

Marketing Intermediary. A marketing intermediary consists of individuals (agents or brokers) that represent the insurer (or other financial firm) for the purposes of selling insurance products to interested consumers.

Marketing Services Agreements (MSAs). Please see PSA.

MEC. An MEC is a modified endowment contract. It is a life-health contract for which premiums are scheduled to be paid on a basis quicker than one-seventh of the net single premium. This product receives special tax treatment because it is marketed and purchased primarily as a tax-deferred investment vehicle.

MMC. Marsh & McLennan Cos. Inc.

Multi-Line Exclusive Agent. See Exclusive Agent.

Mutual Holding Company (MHC). An MHC provides mutual companies with the organizational and capital raising advantages of stock insurers, while retaining the policyholder ownership of the mutual. In a mutual holding company, policyholders have the right to elect directors to the holding company which holds a majority of the voting stock of its subsidiary stock companies. The balance of the stocks can be sold to outside investors to raise capital.

Mutual Insurer. A mutual insurer is an insurance company that is owned by its policyholders.

Mutualization. Mutualization is the process by which a stock insurer converts into a mutual insurer. This process involves the transfer of control from the shareholders to policyholders.

NAIC. The NAIC is the National Association of Insurance Commissioners. It is a non-governmental organization comprised of the insurance department chief officers of every state or territory of the country which has the purpose of coordinating insurance regulation across the states.

NASD. National Association of Securities Dealers.

NFIP. The NFIP is the National Flood Insurance Program. A self-supporting program, it is a federal program that provides flood insurance. Its losses and expenses are paid by the premiums of flood insurance policies, but the program can borrow money from the U.S. Treasury and payback the loans with interest.

No-Fault Auto Insurance. No-fault auto insurance is auto insurance coverage in which the policyholders will recover financial losses from their own insurance company, regardless of fault. Lawsuits are restricted to severe injuries and pain and suffering.

Nonmalignant filing. Nonmalignant filing is the filing of claims by people with little or no current disability. Because some liability insurance is written on the claims-made basis, the claims must be filed within a specified date to get compensation later if a disability is found.

OCC. The OCC is the Office of the Comptroller of the Currency. It is a bureau established under the Treasury Department of the U.S. government with the authority to charter, regulate, and supervise all national banks.

Occurrence-Based Policy. An occurrence-based policy is an insurance policy that covers the risks incurred during the term of the policy, even if they are filed many years later.

OSHA. Occupational Safety and Hazards Administration.

PIP. Personal Injury Protection.

Policy Loan. A policy loan is a loan taken against the assets accumulated in a life insurance or an annuity policy.

PPO. A PPO is a Preferred Provider Organization. In a PPO, a healthcare organization contracts with insurers and employers to provide services on a fee-for-service basis. Plan participants often have the option of receiving care outside the plan, but must share a greater proportion of the corresponding costs.

Preferred Stock. Preferred stock represents ownership interest in a corporation that has a more senior claim on dividend payments and liquidation assets than common stock. However, owners of preferred stock typically have no voting rights.

Private Placements. Private placements are financial assets for which the terms of sale are negotiated directly between the buyer and the seller.

Producer Groups. Producer groups are organizations of independent insurance salespeople that sell products from multiple insurers or groups of affiliated insurers. They generally require less insurer support services and receive more favorable commission arrangements.

Property-Liability (P-L) Insurance. Property-liability insurance is insurance that provides coverage for policyholders' property loss and their legal liability to a third party for bodily injury or property damage. P-L insurance is one segment of the insurance industry. The other sector is life-health.

Proposition 103. Proposition 103 was enacted in November 1988 in California. It converted California insurance regulation from a competitive pricing system to a prior approval system. The proposition required a statewide auto insurance rate rollback and the pricing of auto insurance based more on driving records and less on geographical location.

PSA. A PSA is a placement service agreement. It is a compensation agreement between a commercial insurance broker and an insurance company. The insurance company pays commissions to a broker based on their overall relationship and the services provided by the broker. PSAs are not linked to placement of a specific policy, but rather linked to the overall business provided by the broker within a certain period of time. The insurance company is required to disclose the PSA arrangements to its commercial customers. Also refers to as marketing services agreements (MSAs).

RBC Ratio. An RBC ratio is a risk-based capital ratio. It is the sum of statutory capital and surplus, any voluntary reserves held by the insurer, one-half the insurer's policyholder dividend liability (and the asset valuation reserve in the L-H industry) divided by the RBC. This ratio is endorsed by the NAIC and used by insurers to evaluate insurer solvency.

Reciprocal. A reciprocal is unincoporated insurance firm that is operated by an attorney. Traditionally, reciprocals maintained separate accounts for each member. Members could be assessed if premiums were insufficient to pay claims. Modern reciprocals do not maintain separate member accounts and are not assessable, making them closer to mutuals.

Regulatory Forbearance. Regulatory forbearance is the reduction or withdrawal of existing regulations (e.g., rate regulations) for an industry or sector in which competition is sufficiently strong to render it unnecessary.

RMA. RMA is the Risk Management Agency. Created in 1996, RMA operates and manages the Federal Crop Insurance Corporation (FCIC) that provides crop insurance to U.S. farmers. Fourteen private-sector insurance companies participate in the program. RMA develops and/or approves the premium rate, administers premium and expense subsidy, approves and supports products, and reinsures the 14 companies.

ROA. ROA is the return on assets. It is the ratio of post-tax net operating gains to admitted assets.

ROE. ROE is the return on equity. It is the ratio of post-tax net operating gains to capital and surplus.

ROR. ROR is the return on revenue. It is the ratio of post-tax net operating gains to total revenues.

SAP. Statutory Accounting Principles.

SEC. The SEC is the Securities and Exchange Commission. It is the U.S. federal government agency responsible for administering federal securities law.

Self-Insurance. Self-insurance is when an employer or other organization representing a large group of people manages the risks of its individual constituents by pooling their risks directly as opposed to pooling their risks with a greater population of policyholders by purchasing insurance from another party.

Separate Account. A separate account is an account separate from the insurer's general account that is used to reflect the experience of a distinct pool of investments.

SEUA. South-Eastern Underwriters Association.

Soft Market. A soft market is a buyer's market in which insurance coverage is easier to obtain at a lower cost.

State Guaranty Fund. A state guaranty fund is a non-profit, involuntary association of a state's solvent insurers formed to pay the policyholder claims of insolvent insurers through a process of assessments to the remaining insurers.

Stock Insurer. A stock insurer is an insurance company that is owned by shareholders (i.e., they need not be and generally are not policyholders).

Structured Settlement. A structured settlement is an arrangement made, usually as the result of a law suit, in order to compensate an injured party for damages caused by another party.

SVO. Securities Valuation Office.

Term Insurance. Term insurance is insurance covering a fixed period or term in exchange for a fixed, level premium with no savings element.

Tort Reform. Tort reform is legislation designed to reduce liability costs through modification of liability rules.

TRIA. TRIA is the Terrorism Risk Insurance Act. Passed on November 26, 2002, the Act provides that the federal government and private insurance companies share the losses from the risk of terrorism, with the Federal government providing reinsurance to private insurers operating in the U.S.

Underwriting Cycle. The underwriting cycle is a special phenomenon in the property-liability insurance market, characterized by the alterations of soft market conditions and hard market conditions.

Universal Life Products. Universal life products are life insurance products that allow policyholders to choose from and change between multiple death benefit patterns and to pay premiums when and in the amounts that they desire, subject to policy minimums and maximums.

U.S. United States.

VALIC. *VALIC* is the abbreviated name for Variable Annuity Life Insurance Co., a series of U.S. Supreme Court decisions that reversed the earlier restrictions on banks selling insurance products.

Value-to-Loan Ratio. The value-to-loan ratio is the ratio consisting of the value of collateral (e.g., a house for the mortgage market) divided by the outstanding loan amount, which can be used to measure the risk of a portfolio of asset-backed loans.

Variable Annuity Products. Variable annuity products are annuity products whose assets are invested in an asset mix chosen by the policyholder.

Variable Life Products. Variable life products are life insurance products whose assets are invested in an asset mix chosen by the policyholder.

Viatical Settlement. A viatical settlement is an arrangement made between a life insurance policyholder and a third party, a viatical settlement company, in which the policyholder receives a payment in exchange for transferring ownership of the death benefits to the third party.

WWII. World War II.

3 The Japanese Insurance Market and Companies: Recent Trends

Nobuyoshi Yamori

Nagoya University

Taishi Okada

Kwansei Gakuin University

3.1 INTRODUCTION

Although the Japanese insurance market in the 1990s was sluggish, it is still the second largest, only behind the U.S. market. According to Swiss Re (2004), U.S. total premium volume (including life and non-life insurance) in 2003 was 1,055 billion U.S. dollars, followed by Japan with $478 billion, the United Kingdom with $247 billion, and Germany with $171 billion.[106]

Although foreign insurers were eager to enter the Japanese insurance market and foreign governments strongly demanded the Japanese government to deregulate the markets, it was hard for foreign insurers to do so because of entry restrictions and the opposition of Japanese business groups, Keiretsu. However, foreign companies now can obtain nationwide sales networks by purchasing failed insurers. Contrary to the Japanese insurance market before the 1990s, foreign insurers have increased their market share substantially and play an important role in the Japanese insurance market. This is because the stagnant economy deteriorates financial conditions of traditional insurers and forces Japanese customers to seek financially sound companies.

Also, the ban on the entry of domestic insurers into the third-sector insurance markets was completely removed in July 2001, and the government reinsurance scheme for compulsory automobile insurance was abandoned in April 2002.[107]

---

[106] The premium volume for the European Union (EU), consisting of 25 countries, was $947 billion. Therefore, if we regard the EU as one market, Japan is the third largest insurance market.

[107] As the life insurance and non-life insurance markets are defined as the first and second markets in Japan, the third-sector insurance market occupies a boundary market between life and non-life insurance markets, such as cancer insurance and medical expense insurance.

Insurers can now provide a wide range of insurance products and services, and competition in the insurance markets is getting more intense.

In sum, the Japanese insurance market has drastically changed since 1990, although there still remain several structural defects. Changes in the economic and financial environments have forced insurers to establish new business models. If a company fails to adjust to the new environment, it loses its market share and, at worst, goes bankrupt. Several big mergers, such as between Yasuda Life and Meiji Life, occurred to strengthen competitiveness.

The main purpose of this chapter is to explain not only *how* the Japanese insurance market and Japanese insurers have changed, but also *why* they have changed and, in some respects, have *not* changed. We also analyze several key aspects of the Japanese insurance market and Japanese insurers that make them unique.

The chapter is organized as follows. The Insurance Markets section describes the key structures of the Japanese insurance market as well as market environments. Because Japanese regulations strictly separated life insurance from non-life insurance businesses for a long time, it is easy to explain them separately. Thus, two sections are devoted to Life Insurers and Non-life Insurers. In the Other Insurance Providers section, we describe important insurance providers other than insurers: postal life insurance provided by the Public Post Agency (Japan Post) and Kyosai (a quasi-insurance provided by informal insurers, such as agricultural cooperatives). The importance of these providers is evidenced by the fact that 50 percent of Japanese households purchase postal life insurance and 15 percent purchase quasi-insurance from agricultural cooperatives. The Current Important Issues section explains two issues that make Japanese insurance markets unique. This section includes a discussion of public earthquake insurance and the resolution of failed insurers. This chapter's structure will enable comparison of the Japanese insurance market with insurance markets in other countries.

3.2 INSURANCE MARKETS

3.2.1 Historical Background

Japan made the transition from feudalism to capitalism after the Meiji Restoration in 1868. Japanese leaders were eager to introduce various institutions and organizations from Western countries to construct a prosperous country. Yukichi Fukuzawa, a founder of Keio University, was the first person to introduce the concept of insurance to Japan. In 1881, Meiji Life was established as the first life insurance company in Japan by some of Fukuzawa's students. Then, Teikoku Life and Nippon Life were established in 1888 and 1889, respectively. These three were stock companies. In 1902, Daiichi Life was established as the first mutual life insurance company. Initially, Japanese people hesitated to purchase death insurance because they felt that purchasers wanted the insured to die. Therefore, endowment insurance was a main product. Life insurance became popular, and the amount insured by life insurance was over 80 percent of gross domestic product (GDP) before World War II.

Regarding non-life insurance business, foreign insurers entered Japan to provide marine and fire insurance to foreigners living in Japan after Japan started international trade in 1853. The first domestic non-life insurance company, Tokyo Marine, was established in 1878. In 1887, Tokyo Fire was established as the first domestic fire insurance company.

Initially, because entry into the insurance business was not regulated, the market was very competitive. To compete, some companies did not accumulate appropriate policy reserves, offered insurance products with very low premium rates, and paid very high dividends. Naturally, such excessive competition forced many companies to fail, resulting in public distrust of the insurance industry. To address the problem, the government studied German insurance laws, and the Insurance Business Law was enacted in 1900. This law required anyone who engaged in the insurance business to be licensed and regulated by the government. The law also strictly separated life insurance business from non-life insurance business.

In 1939, the Insurance Business Law was amended, by which insurance business was placed on a war footing. The government conducted large-scale reorganization and integration during World War II, which reduced the number of companies to less than one half of those operating before. For example, 48 non-life insurers in 1939 were merged into 16 by 1945.

World War II seriously damaged the insurance industry. For example, assets of life insurers totaled 2.5 billion yen in 1936, while they were 18.9 billion yen in 1948. Although the nominal amount increased dramatically, the real value of 18.9 billion yen in 1948 was only 5.9 percent of the prewar 2.5 billion yen due to huge inflation. In 1946, based on the Emergency Act for Financial Institutions Accounts, assets and liabilities of life insurers were separated into "old accounts" and "new accounts." New accounts took over only prime assets and were liable only for small contracts with the amount insured less than 10,000 yen. In essence, contracts with the amount insured over 10,000 yen were written off. Then, 14 new life insurers started to operate as successors to the new accounts. Notably, 13 of the 14 companies were established as mutual companies.

During the rapid growth period, the insurance business developed dramatically. Figures 3.1 and 3.2 show the high growth of insurers' total assets in the postwar period. In addition to assets growth, the amounts insured increased. Assets from life insurance exceeded 1 trillion yen in 1953, 10 trillion yen in 1962, 100 trillion yen in 1972, and 1,000 trillion yen in 1987. Those from non-life insurance exceeded 1 trillion yen in 1947, 10 trillion yen in 1958, 100 trillion yen in 1967, and 1,000 trillion yen in 1977. Regarding the non-life insurance business, it is worth noting the growth of automobile insurance following motorization during the rapid growth period of the 1960s; premiums from automobile insurance have become the largest among the lines of business of non-life insurers in Japan since 1965. Long-term insurance with maturity repayments, which was introduced in 1963, contributed to increases in the total assets of non-life insurers.[108]

[108] The preceding discussion of the historical background of Japanese insurance markets relies on Ninomiya (1997), Mizushima (1999), and Takagi and Nakanishi (1999).

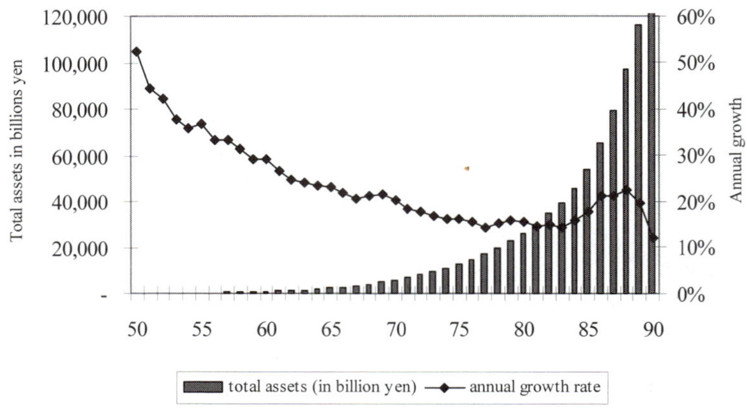

Figure 3.1. Total Assets of Life Insurers, 1950 to 1990

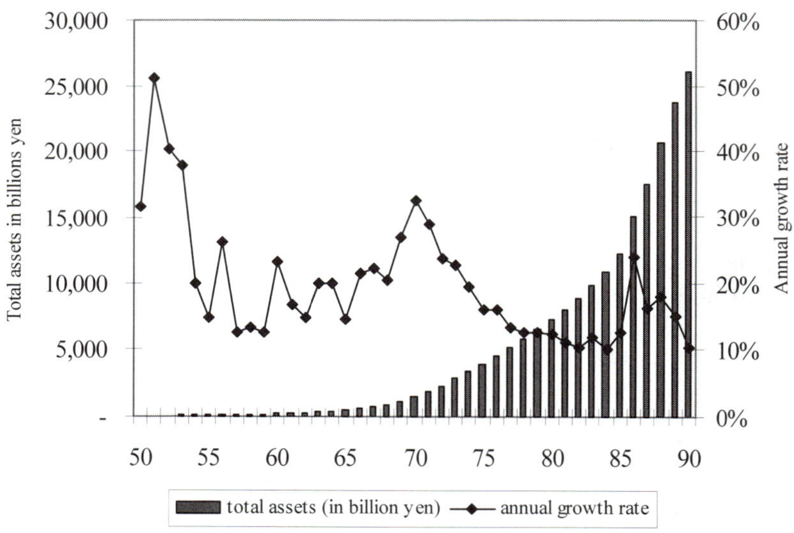

Figure 3.2. Total Assets of Non-Life Insurers, 1950 to 1990

3.2.2 Economic and Social Environments of the 1990s

The Lost Decade

The Japanese economy, whose nominal GDP for 2003 was 498 trillion yen (or about 4.66 trillion U.S. dollars)[109] is the second largest economy in the world. However, Japan has suffered from a stagnant economy since the 1990s—often called the "lost decade" because of low nominal GDP growth, prolonged deflation, and huge

[109] The yen-dollar exchange rate at the end of 2003 was 106.97 yen per U.S. dollar.

declines of asset prices. GDP growth was very low or even negative in the late 1990s. Average annual nominal GDP growth rates were 6.3 percent from 1985 to 1990, 2.5 percent from 1990 to 1995, 0.6 percent from 1995 to 2000, and –0.9 percent from 2000 to 2003. Low economic growth has increased unemployment rates significantly. Although Japan was known as a country with low unemployment rates (e.g., 2.1 percent in 1990), recent unemployment rates are around 5 percent (e.g., 5.5 percent in January 2003 and 4.8 percent in August 2004). Thus, Japanese households cannot afford to purchase insurance policies as they did before.

The second feature of the stagnant economy in the 1990s is prolonged deflation. The GDP deflator has been negative since 1994. As deflation increases the net value of debts, borrowers are more likely to face financial difficulties. As the number of firms that go bankrupt rises to an unprecedented level (e.g., the number of corporate bankruptcies was 19,164 for 2001, three times larger than the number in 1990), Japanese financial institutions suffer from severe bad loan problems. For example, Japanese banks and other depository institutions held 34.6 trillion yen in bad loans at the end of March 2004.[110] Fortunately, the bad loan problem is less serious for insurers than for banks, because insurers have lent mainly to blue-chip firms. However, low interest rates in the 1990s seriously affected insurers. To fight against deflation, the Bank of Japan reduced interest rates to almost zero. Overnight interbank loan interest rates (call rates) have been set at 0.001 percent since 2001. Long-term interest rates are also very low. For example, the yield of newly issued government bonds (10 years maturity) was below 1 percent in 2002. These unprecedented low interest rates help commercial banks but create problems for life insurers, because the companies promised to pay the insurance purchasers at the guaranteed interest rate, which was higher than 5 percent for the insurance contracts during the bubble period (i.e., the late 1980s). The negative spreads between market interest rates and the guaranteed interest rates were so large that they forced some insurers into bankruptcy.

The third characteristic of the stagnant economy of the 1990s is the sharp decline in asset prices, which is often called the burst of the bubble. Land prices have continued to decline since 1992. According to officially published land prices (Chika Koji) issued by the Ministry of Land, Infrastructure, and Transport, average land prices for 2004 were almost half (52 percent) what they were in 1991. Also, stock prices drastically declined in the 1990s. For example, the Nikkei 225 Average, the most popular stock price index, recorded its highest value of 38,915 at the end of 1989, and it had declined to below 7,900 in April 2003. Because insurers invested substantial funds into stocks, these declines in assets prices caused huge losses for them. In sum, insurers have faced serious difficulties in the recent decades in terms of insurance sales and investment activities.

Rapid Aging of the Population

Demographic change also affects the insurance industry. In 1985, the ratio of people age 65 years and over to the total population (old-person ratio) exceeded 10 percent for the first time in Japanese history. As Japanese society is rapidly aging, the old-

---

[110] Although there are several definitions of bad loans in Japan, here we use the definition based on the Financial Revitalization Act of 1998 (Kinyu Saisei Hou) and figures released by the Financial Services Agency (FSA).

person ratio in 2003 reached 19 percent. According to the National Institute on Population and Social Security Research, the old-person ratio is expected to be 26 percent in 2015. This aging society is caused partially by increased life expectancy. In 1960, the mean life expectancy was 63.6 years for men and 67.8 years for women; in 2003, it was 78.4 years for men and 85.3 years for women.

It is quite natural that an aging population affects life insurers. Particularly, as demand for savings-based insurance products and annuities has increased during the aging process, the assets of life insurers have tended to increase. However, once many people reach retirement age and become insurance and annuity beneficiaries, the assets of life insurers will decrease. Lifetime employment practice in Japanese insurers was possible only when constant growth of the company size enabled companies to employ more juniors than seniors. It is apparent that Japanese life insurers will have to seek a new business model.

3.2.3 Historical Development and Current Situations of Insurance Regulations

Amendment of Insurance Business Laws in 1995

The Insurance Business Law of 1940 was substantially revised in 1995 for the first time in 50 years in order to promote competition and enhance efficiency through deregulation, to preserve the soundness of insurance businesses, and to ensure fairness and equity in business operations.

To promote competition, the New Insurance Business Law, effective in April 1996, allowed life insurers to enter the non-life insurance market by establishing non-life insurance subsidiaries and non-life insurers to enter the life insurance market by establishing life insurance subsidiaries. For example, life insurers can underwrite medical insurance by themselves and personal accident insurance through non-life insurance subsidiaries. Non-life insurers are also allowed to found affiliated life insurers to underwrite life and medical insurance. Even though the law was amended, the Ministry of Finance did not allow most Japanese companies to sell the third area (which includes products such as cancer insurance and medical insurance), because this was a main issue for the Japan-U.S. negotiations on insurance deregulation in Japan.

To promote price and product competition, the law introduced the notification system for certain products and premium rates. The old law required non-life insurers to use the premium rates for fire and personal accident insurance that the Property and Casualty Insurance Rating Organization (PCIRO) determined and premium rates for automobile insurance that the Automobile Insurance Rating Organization of Japan (AIRO) determined.[111] The new law gave non-life insurers the power to decide their own premium rates within 10 percent below or above the premium rates determined by the PCRIO and the AIRO. Furthermore, the new law allowed non-life insurers to use their own premium rates for large fire insurance risks.[112]

---

[111] The PCIRO was established in 1948, and the AIRO was split off from the PCIRO in 1964. The PCIRO and the AIRO merged into the Non-Life Insurance Rating Organization of Japan in July 2002.

[112] Although non-life insurers were still forced to use net premium rates decided by the PCIRO, they were allowed to determine their own loading charges for large fire insurance contracts (contracts with

To preserve soundness, solvency margin standards were introduced.[113] The law empowered the Minister of Finance to order the creation of a restructuring plan for a company that had a low solvency margin ratio. Also, to ensure the policyholders' rights, a Policyholders Protection Fund was established. The fund would provide financial support to companies that merged with failed insurers.

To facilitate consumers' protection, a cooling-off period has been introduced for long-term individual insurance contracts. Consumers can cancel insurance contracts within eight days after signing the contract without penalty. To enable customers to purchase the most appropriate insurance contract, an insurance broker system was introduced. While salespersons and insurance agencies sell only insurance products of their affiliated company, brokers should be independent from insurers, and they are expected to advise consumers on what insurance products are best for them. See Preston (1998), which provides more details about the deregulation in the 1990s.

The Japan-U.S. Agreement on Insurance Deregulation in Japan

After the 1995 Insurance Business Law amendment, several more important amendments were implemented to accelerate the liberalization of the Japanese insurance market. The first important turning point was the Insurance Talks between the U.S. and Japanese governments, starting in February 1996 and reaching an agreement in December 1996. The main issues at the negotiations were liberalization of market entry and premium rates deregulation. The Office of the United States Trade Representative argued that entry liberalization in terms of the so-called third-sector insurances (e.g., cancer insurance and medical expense insurance), which were main products of foreign insurance firms operating in Japan, should be permitted only after the Japanese government completed liberalization of other insurance markets that Japanese firms dominated. Regarding insurance premium rates regulation, the U.S. government demanded that the Japanese government introduce a competitive rating system in the property-liability insurance market by depriving the PCIRO and the AIRO of their authority to determine insurance premium rates for fire and personal accident and for automobile insurance, respectively.

Two main articles in the agreement were: (1) Japanese insurance firms would be allowed to enter the third area gradually, and (2) property-liability insurance premium rates would be deregulated. More concretely, property-liability insurance subsidiaries of Japanese life insurers were permitted to sell personal accident insurance from January 1997. Life insurance subsidiaries of property-liability insurers could sell medical insurance from January 2001. After January 2001, Japanese insurers would be entitled to sell any kind of insurance. These gradual

---

insured amounts of more than 20 billion yen) since January 1997, and the minimum amount was reduced to 7 billion yen in April 1998.

[113] At the time, the solvency margin was used as a measure of the early warning system. Because customers would compare insurers by using these simple figures, insurers hesitated to disclose their solvency ratios. Because of strong opposition from insurers, the law did not require insurers to disclose their solvency ratios, but only required them to report the solvency ratios to the Ministry of Finance. The amendment of the Insurance Business Law in 1998 introduced a system of prompt corrective action, by which the solvency margin ratio was designated the most important ratio of the insurance regulation. Also, the amendment required insurers to disclose their solvency margin ratios since March 1998. See Kitamura (2003) for more details.

entries of Japanese insurers to the third area where foreign insurers dominated satisfied the U.S. government. More importantly, both governments agreed that the premium rates for fire, personal accident, and automobile insurances would be liberalized. Based on this agreement, the law concerning Non-Life Insurance Rating Organizations (NLIRO) was revised in July 1998 to deprive the PCIRO and the AIRO of the authority to decide and regulate insurance premiums after July 2000. See Hecker (1996) and Yamori and Kobayashi (2004) for more details about the Japan and U.S. agreement.

The Amendments of Insurance Business Laws as the Financial System Reform Laws

Reflecting the rapidly changing financial environments, the Insurance Business Law has been amended several times since the Japan-U.S. agreement. Particularly, when the Financial System Reform Laws were approved in June 1998, the Insurance Business Law was also substantially amended. The major amendments provided for (1) the above-mentioned reform of the rating organizations; (2) the creation of the Policyholders Protection Corporations to succeed the Policyholders Protection Fund; (3) the acceleration of mutual entry among insurers, securities companies, and banks; (4) the expansion of the business area for insurers (e.g., sale of mutual funds by insurers); and (5) the introduction of a prompt corrective action measure for the insurance business, by which the government can order insurers whose solvency margin ratio is lower than the predetermined level to create a restructuring plan and, in the worst case, to suspend their businesses.

3.2.4 Current Insurance Regulations

Important Provisions of Current Insurance Business Law

Business License

Anyone who does not obtain a license from the prime minister is prohibited from carrying on an insurance business. There are two types of licenses: life insurance business licenses and non-life insurance business licenses. No one is allowed to hold both licenses concurrently. An insurer has to be a stock company or a mutual company with a capital or foundation fund of not less than 1 billion yen.

Permitted Businesses

In addition to core businesses (i.e., underwriting risks and investing insurance premiums), an insurer can conduct the following other business ancillary to its licensed insurance business: (1) agency business connected with the insurance operations of another insurance company,[114] (2) giving of guarantees for debts, (3) dealing in government bonds or handling of their rotation, (4) acquisition or transfer of monetary obligations, and (5) handling of private placement of securities. In addition, an insurer can deal in business relating to specific securities or transactions

---

[114] This agency business includes (1) the documentary arrangement and the delivery of the documents, (2) receiving insurance premiums and paying insurance benefits, (3) investigating the event insured against, (4) educating salespersons for another insurer (such as foreign insurers and their own subsidiary insurers) to underwrite risks.

provided in the Securities and Exchange Law as long as the performance of its licensed insurance business is not adversely affected. In other words, an insurer cannot conduct any business other than those mentioned above and such business as allowed under other laws.

Subsidiaries

The types of subsidiaries in which an insurer can hold more than 50 percent of the stock are: (1) life and general insurers; (2) banks; (3) securities companies; (4) foreign institutions operating insurance, banking, and securities business; (5) companies providing incidental and ancillary business to the parent insurer (e.g., systems development and human resources); (6) companies conducting finance-related business (e.g., investment trusts and investment management); and (7) downstream holding companies, whose subsidiary business activities are the same as those of insurer subsidiaries.[115]

Product Approval and Filing

When an insurer wants to sell new products or change existing products, it must obtain approval from the prime minister. However, in some cases that are defined in the Cabinet Office Ordinance, the insurer needs only to give notice of the amendment to the prime minister. To facilitate innovations in insurance markets, the FSA guideline declares that the approval will not take more than 90 days for new products and not more than 60 days for products that are similar to existing insurance products.

Standard Technical Provision

The Cabinet Office Ordinance requires that an insurer should accumulate various provisions, such as the price fluctuation reserve, premium reserves, and provisions for outstanding claims. Dividends to policyholders are also regulated in order to protect policyholders' rights. Regarding mutual insurers, more than 20 percent of surpluses should be paid to policyholders or reserved as dividends for them.

Limitation on Investment

An insurer is subject to investment regulation. It is allowed to invest funds in financial products that the Cabinet Office Ordinance lists, such as securities, real estate, gold, loans, and some financial derivatives. It cannot invest its funds into new financial products that are not listed. Furthermore, the maximum amount for each product is regulated. For example, domestic stocks should be less than 30 percent of total assets, real estate should be less than 20 percent of total assets, and foreign-currency-denominated assets should be less than 30 percent.

---

[115] Historically, banking, security businesses, life insurance, and non-life insurance were strictly separated from each other. First, the Insurance Business Law of 1995 allowed insurers to hold subsidiary insurers, meaning that life insurers could enter non-life insurance markets by establishing non-life insurance subsidiaries. Furthermore, the 1998 amendment expanded the kinds of subsidiaries allowed to those specified.

Disclosure Requirement
An insurer must draw up an explanatory document describing the condition of its operations and assets, and the documents must be available to the public at the head office and branches.

Registration of Sales Staff
No person, other than (1) officers or employees of a insurer, (2) registered life insurance solicitors or non-life insurance agents, and (3) registered insurance intermediaries (insurance brokers) can engage in insurance soliciting. In April 2001, banks started to sell some insurance products—such as life insurance and long-term fire insurance accompanying housing loans—with restrictive conditions. In October 2002, insurance products that banks can sell were extended to include individual annuities and other savings insurance.

Solvency Regulation
To maintain the sound management of insurers, the solvency margin ratio was introduced as a measure of the early warning system in the Insurance Business Law of 1995. The 1998 amendment requires the ratios to be disclosed to the public since March 1998. Furthermore, since March 2002, not only the solvency margin ratio itself but also its components should be disclosed to the public.

The solvency margin ratio defined in the following equation should be more than 200 percent. A solvency margin ratio of less than 200 percent triggers various prompt corrective actions. For example, a company with a solvency margin ratio between 100 percent and 200 percent is ordered to create an improvement plan and implement it, and a company with a solvency margin ratio less than 0 percent is ordered to suspend its business.

$$\text{solvency margin ratio} = \frac{\text{solvency margin}}{\frac{1}{2} \times \text{risk amount}} \times 100$$

Solvency margin (the numerator of the solvency margin ratio) consists of: capital; capital-like provisions, such as price fluctuation provisions, contingency provisions, and general loan losses provisions; a certain part of unrealized gains/losses on securities and real estate; subordinated debts; and what the director-general of the Financial Services Agency specifies.

Risk amount (the denominator of the solvency margin ratio) consists of: insurance risks (R_1) (i.e., risks that insurers may incur when the probability of insured events is higher than normally expected), which consists of general insurance risks (R_5) and major catastrophe risks (R_6); assumed interest risk (R_2) (i.e., risks that arise when actual investment performance is below the guaranteed interest rates); asset management risks (R_3), such as price fluctuation risks and credit risks; and operational risks (R_4). Although operational risks are defined as unexpected risks that are not classified into the three risk categories listed, it is very hard to quantify them. The current administrative rule assumes the amount of the operational risks as 2 percent (or 3 percent for insurers with undisposed losses) of $R_1+R_2+R_3$. Then, the risk amount (RA) is calculated in the following formula.

For life insurers, $RA=[(R_1)^2+(R_2+R_3)^2]^{1/2}+R_4$,

and for non-life insurers, $RA=[(R_5)^2+(R_2+R_3)^2]^{1/2}+R_4+R_6$.[116]

Other Important Insurance-related Regulations

The Consumer Contract Law has been effective since April 2001. Consumers are allowed to cancel contacts with companies when consumers have misunderstood a contract as a result of a companies' misrepresentation at the time of contract or when consumers have been pressured by companies' persistent sales tactics. This law applies to insurance contracts.

The Law on Sales of Financial Products obliges financial service providers to provide consumers with important information on the financial products that they sell, such as price fluctuation risks and credit risks. Financial service providers are liable for losses incurred by customers resulting from their failure to explain important information. This law also requires financial service providers to draw up and publicize a solicitation policy in order to ensure the protection of customers.

Financial Services Agency

Traditionally, the Ministry of Finance (MOF) oversaw insurance regulation. During the financial crises in the 1990s, the MOF was criticized for its unclear administrative discretion. In 1998, the Financial Supervisory Agency was established to supervise all financial activities, but the MOF still kept some authority regarding financial businesses. However, since the Financial Services Agency (FSA) was established to succeed the Financial Supervisory Agency in 2000, the FSA has independent authority to regulate and supervise insurers as well as banks and security companies. Currently, the FSA is fully responsible for ensuring the stability of the financial system in Japan and protecting depositors, insurance policyholders, and securities investors (information available at the FSA's web site: www.fsa.go.jp/indexe.html).[117]

3.2.5 Life Insurance Markets

Prevalence Ratio

As a testament to the importance of the life insurance industry in Japan, 90 percent of the households in Japan hold one or more life insurance policies. However, the market has undergone tremendous change in recent years. Consumers' needs have rapidly diversified from traditional life insurance products to lifetime beneficial policies such as nursing care and medical insurance. Furthermore, an aging population and a declining birthrate are changing the life insurance business.

The Japan Institute of Life Insurance (JILI) regularly conducts survey research on life insurance. The 2003 survey shows that 76.1 percent of households purchased life insurance (including individual annuities) from insurers, 48.7 percent of

---

[116] The preceding section relies on the presentation material for a Meeting of Solvency and Actuarial Issues Subcommittee of International Association of Insurance Supervisors in September 2003, which is available at FSA's web site (www.fsa.go.jp/indexe.html). See the Notification of Ministry of Finance No. 50 of 1996 (Oukurasho Kokuji) for more details.

[117] See textbooks on Japanese insurance law, such as Yamashita et al. (2004), for more details on current insurance regulations. An English translation of the Insurance Business Law is available at www.sonpo.or.jp/english/english.html or www.mizuho-sc.com/english/ebond/law/insurance.html.

households purchased it from the Postal Life Insurance Agency, and 15 percent of households purchased it from agricultural cooperatives. As some households purchased life insurance policies from two or more suppliers, 89.6 percent of households purchased life insurance policies.

The ratio of households that purchase life insurance policies from these three types of insurance providers (the prevalence ratio) was 88.4 percent in the 1978 survey, and the ratio increased steadily for many years. The highest ratio of 95 percent was recorded in the 1994 survey. Because of the unfavorable economic situation of the 1990s, the prevalence ratio has decreased since 1994.

Evolution of Insurance Premiums

A look at life insurance premiums and their components from 1990 to 2003 reveals several notable trends (see Table 3.1). First, 57.4 percent of insurance premiums in 2003 came from sales of individual life insurance. Group insurance and group annuities tended to decrease their shares in insurance premiums. Second, premium incomes tended to decrease since 1995, while the premium incomes never decreased before 1995. Although the premium incomes in 2003 increased 1.8 percent from 2002, it is not certain whether the trend of decreasing premium incomes changed. Third, premium incomes of individual annuities show remarkable growth since 2002. Although the demand for annuities is expected to grow because of rapid aging of the population, premium incomes of individual annuities decreased by 30 percent from 1995 to 2001. Individual annuities have increased significantly since 2002, because banks started to sell them in October 2002 and because life insurers started to sell various variable annuities. Because premiums for variable annuities are usually paid in a lump sum at the time of a contract, increases in sales of variable annuities lead to sharp increases in premium incomes. Sales of variable annuities amounted to 958 billion yen in 2002, or 530 percent of that in the previous year, and amounted to 1,671 billion yen in 2003.

Finally, despite the unfavorable macroeconomic and financial environments, some companies increased their income premiums (see the **Life Insurers** section). This suggests that only the best performing companies can survive in the wake of market liberalization.

3.2.6 Non-life Insurance Markets

Non-life Insurance Premiums

Growth in the insured amount of non-life insurance policies is influenced by a variety of factors, not only housing investment, automobile holdings, and overseas trade volume, but also by new diversified risks arising from socioeconomic changes and enhanced public awareness about compensation. Direct premiums had shown almost the same steady growth as GDP until fiscal 1996, but thereafter the trend reversed due to competition in premium rates and prolonged economic sluggishness.

The total net premiums written by non-life insurers in fiscal 2003 reached 7,437 billion yen (see Table 3.2). Automobile insurance is the main product for Japanese non-life insurers. Net premiums written from automobile insurance excluding the compulsory automobile liability insurance (CALI) accounted for 47.8 percent of the total net premiums written in 2003 and showed a 1.5 percent decrease from the previous year, although the prevalence ratio of voluntary automobile insurance increased.[118] This decrease of premium incomes was caused by growing price competition.

Premiums written from CALI increased by 76.8 percent in 2002 and by 18.2 percent in 2003 due to the CALI reform.[119] The Automobile Liability Security Law, which was enacted in 1955 and came into effect in February 1956, requires all drivers to purchase CALI. The insurance premium rates are regulated. Although the current maximum limit of liability is 30 million yen for death and 40 million yen for permanent disability, CALI does not cover property damage. Prior to the enforcement of the amended law in April 2002, all CALI premiums were reinsured by the government on a 60 percent quota share basis. The remaining 40 percent was redistributed to each insurer. Since April 2002, the government has not reinsured CALI, and all CALI premiums are redistributed to each insurer. This is why CALI premiums increased greatly in 2002. Because the premium rates of CALI are set at the "no profit and no loss" level, the increase in CALI premiums will not enhance insurers' profits.

Table 3.1. Premium Income for Various Life Insurance Products from 1990 to 2003 (in billions yen)

Fiscal Year	Total of Premium Income	Individual Insurance	Individual Annuities	Group Insurance	Group Annuities
1990	27,321	16,648	1,601	1,368	7,178
1991	28,240	16,999	1,788	1,465	7,580
1992	29,530	17,796	1,990	1,555	7,767
1993	30,393	18,100	2,216	1,630	8,009
1994	30,489	17,455	1,988	1,695	8,896
1995	30,762	18,388	2,938	1,759	7,210
1996	29,354	17,623	2,420	1,729	7,108
1997	30,361	17,740	2,448	1,445	8,260
1998	28,837	17,946	2,276	1,381	6,711
1999	27,607	16,676	2,134	1,375	6,913
2000	26,941	16,165	2,172	1,373	6,839
2001	26,150	15,672	2,135	1,357	6,632
2002	25,512	15,723	3,022	1,319	5,135
2003	25,901	14,867	4,638	1,297	4,871

Source: Life Insurance Fact Book, Japan Institute of Life Insurance. Premiums from other lines of insurance omitted.

[118] The prevalence ratios of bodily injury liability insurance and of property damage insurance were 70.9 percent and 70.4 percent, respectively, at the end of March 2003.

[119] See *Automobile Insurance in Japan 2003,* published by the Non-Life Insurance Rating Organization of Japan for more details (available at www.nliro.or.jp/english/automobile/index.html).

Although net premiums written from fire insurance peaked in 1997 and have generally decreased since then, they increased by 1.4 percent from 2002 to 2003. The prevalence ratio of fire insurance regarding buildings was 5.35 percent in 2002. Net premiums written from personal accident insurance in 2003 decreased by 0.9 percent from the previous year, due to a fall in sales of maturity-refund personal accident insurance policies. The marine and transit lines as a whole showed an increase of 3.4 percent over the previous year in terms of insurance premiums. Finally, miscellaneous liability, including general liability, workers' accident compensation liability, and movables comprehensive insurances, accounts for 10.3 percent of the total net premiums written in 2003. Direct premiums written from general liability, workers' accident compensation liability, and movables comprehensive insurances in 2003 were 3,566 billion yen, 757 billion yen, and 1,132 billion yen, respectively.

In sum, voluntary automobile and personal accident insurances have increased significantly in the past 20 years, while fire, marine and transit, and CALI lines have continued to lose their respective shares. The trend clearly implies that major changes have taken place in the non-life insurance business.

Table 3.2. Premium Income for Various Non-life Insurance Products from 1990 to 2003 (in billions yen)

Fiscal Year	Net Premiums	Automobile	Fire	Marine and Transit	Personal Accident	Compulsory Automobile Liability	Misc. Casualty
1990	5,628.7	2,478.1	973.5	294.0	667.0	614.7	601.4
1991	5,991.7	2,779.7	982.6	290.3	693.6	620.1	625.4
1992	6,229.0	3,030.5	989.9	281.7	700.8	596.9	629.2
1993	6,551.5	3,314.6	1,017.0	265.4	756.7	582.0	616.0
1994	6,765.3	3,466.3	1,076.9	270.1	760.0	570.2	621.7
1995	6,959.3	3,555.3	1,106.0	273.5	782.7	590.5	651.3
1996	7,228.2	3,666.6	1,175.2	293.3	797.1	601.9	694.1
1997	7,215.4	3,697.4	1,186.1	287.6	765.8	576.0	702.5
1998	6,915.1	3,575.9	1,117.2	261.2	718.7	553.9	688.1
1999	6,889.3	3,605.1	1,105.2	232.1	705.8	564.9	676.2
2000	6,874.1	3,650.1	1,053.7	231.5	676.6	569.8	692.3
2001	6,811.6	3,674.5	1,031.9	231.8	645.6	572.2	720.5
2002	7,285.8	3,612.4	1,030.5	233.4	636.7	1,011.7	761.0
2003	7,437.2	3,558.2	1,044.9	241.3	631.0	1,195.6	766.2

Source: General Insurance in Japan Fact Book 2002–2003, General Insurance Association of Japan.

Large Claims

Typhoons have caused huge damages in Japan. Table 3.3 shows the claims paid for the five natural disasters that have caused the largest losses in Japan. Typhoon No. 19 in 1991 caused 567.9 billion yen worth of claims to be paid. As shown in this table, large losses have occurred since the 1990s; three of the five largest disasters have occurred since 1998. Insurers need to manage risks caused by these natural disasters. In addition to typhoons, Japan suffers from destructive earthquakes. Because earthquake risk is hard to diversify, private insurers hesitate to sell earthquake insurance. As the social demand for earthquake insurance has grown, public earthquake insurance has been developed. We discuss earthquake insurance in a later section.

Table 3.3. Five Largest Claims in Japan Caused by Natural Disasters (in billions yen)

Name of Disaster	Date	Claims Paid (Fire and Miscellaneous)	Claims Paid (Automobile)	Claims Paid (Marine)	Total
Typhoon No. 19	Sept. 26–28, 1991	522.5	26.9	18.5	567.9
Typhoon No. 18	Sept. 21–25, 1999	284.7	21.2	8.8	314.7
Typhoon No. 7	Sept. 22, 1998	151.4	6.1	2.4	160.0
Downpour, Sept. 2000	Sept. 10–12, 2000	44.7	54.5	3.9	103.0
Typhoon No. 13	Sept. 3, 1993	93.3	3.5	1.0	97.7

Source: General Insurance in Japan Fact Book 2002–2003, General Insurance Association of Japan.

Recently, the courts have awarded huge damages to victims of traffic accidents. The highest value of the damages awarded was 312 million yen for permanent disability in 2003. All of the highest 20 damages were awarded after 1994. These high awards have a double-edged effect on insurers. It is apparent that people cannot make compensatory payment for these damages without purchasing insurance, which results in higher demand for insurance. However, because high awards increase what insurers have to pay, they have an unfavorable effect on a company's profitability.

Reinsurance Markets (Non-life Insurance)

The outward reinsurance balance had been unfavorable to the Japanese non-life insurance industry as a whole until fiscal 1991, when the balance was reversed due to a sharp increase in claims received for losses caused by a series of typhoons, including Typhoon No.19. In 1992, however, the balance became negative again due to a sharp increase in reinsurance premium rates by overseas reinsurers and has since remained unfavorable. In 2003, premiums paid and claims received for outward reinsurance were 256 billion yen and 123 billion yen, respectively. Therefore, the net balance was minus 132 billion yen.

The inward reinsurance balance has basically been unfavorable to the Japanese non-life insurance industry as a whole for years. This is mainly attributable to payments for reinsurance claims from losses caused by frequent catastrophes abroad. Furthermore, in fiscal 2001, this unfavorable balance was worsened to minus 34 billion yen by the terrorist attack on September 11. In 2003, premiums received and claims paid for inward reinsurance were 225 billion yen and 279 billion yen, respectively. Therefore, the net balance was minus 543 billion yen, while it was 6 billion yen in 2002.

Relative Importance of Insurers in Japanese Financial Sectors
According to the flow-of-funds accounts released by the Bank of Japan, Japanese households held financial assets of 1,405 trillion yen at the end of 2003. Deposits to banks and other depository institutions were the major financial asset for Japanese households, amounting to 745 trillion yen. Insurance and pension assets were 249 trillion yen (or 17.7 percent of household total financial assets) and 146 trillion yen (or 10.4 percent of household total financial assets), respectively. Stocks and other capital investment amounted to 20 trillion yen, and securities other than stocks (e.g., government bond and beneficiary certificate of investment trust) amounted to 68 trillion yen. In sum, insurance and pension assets comprise about 25 percent of households' financial assets.

As shown in Table 3.4, all financial institutions held financial assets of 2,944 trillion yen at the end of 2003. Depository institutions held 1,495 trillion yen, half of which was held by domestic licensed banks (760 trillion yen). Following domestic banks, Japan Post held 277 trillion yen. Insurers and pension institutions held 447 trillion yen, or 15 percent of the total financial assets held by financial institutions. More disaggregated data for insurance and pension are as follows: Private life insurers held 152 trillion yen, and public life insurance institutions, such as postal life insurance (Japan Post, or Kampo), held 121 trillion yen. Private non-life insurers held 32 trillion yen.

3.3 LIFE INSURERS

3.3.1 Industry Organization

Three Groups of Insurers
The Life Insurance Association of Japan represents all life insurers operating in Japan. Forty life insurers belonged to the Association as of March 2004 (see Table 3.5). The number of life insurers remained almost constant until 1995. In fiscal 1996, the number suddenly increased, because the new Insurance Business Law allowed non-life insurers to establish life insurance subsidiaries. The number reached a peak in 2000 and then decreased. The two main reasons for the decreases since 2000 are (1) liquidation of failed insurers (e.g., Daihyaku Life was liquidated, and its all contracts were transferred to Manulife Century Life in 2001), and (2) mergers of subsidiary life insurers following their parent companies' mergers. For example, as Nippon Fire and Koa Fire were merged into Nippon-Koa Insurance in April 2001, their subsidiary life insurers Nippon Fire Partner Life and Koa Fire Magokoro Life were also merged into Nippon-Koa Life.

Insurers are classified into three groups. The first group consists of traditional domestic companies. Nippon Life, Daiichi Life, Meiji-Yasuda Life, Mitsui Life, and Sumitomo Life are major traditional life insurers. The number of traditional domestic companies decreased because of mergers and bankruptcies.

Table 3.4. Financial Assets of Financial Institutions in Japan at the End of 2003 (in billions yen)

Depository Institutions	1,495
Domestic Banks	760
Foreign Banks	43
Agricultural Institutions	194
Small Enterprise Finance Institutions	193
Postal Savings	277
Other	28
Insurance and Pensions	447
Life Insurance	273
Private Companies	152
Non-life Insurance	37
Private Companies	32
Kyosai	43
Pensions	94
Other Financial Intermediaries	843
Securities Investment Trust	50
Finance Companies	105
Public Institutions	572
Securities Companies	72
Other	44
Others	23
Total	2,944

Source: Bank of Japan, Flow of Funds Account.

The second group consists of the subsidiaries of other industrial companies and non-life insurers. ORIX, a leading non-bank financial company, has ORIX Life, and Sony has Sony Life. More important subsidiaries are those of non-life insurers. The amended Insurance Business Act of 1995 allows non-life insurers to establish subsidiary life insurers. In 1996, 11 subsidiaries started operations. These new life insurers have grown steadily because the public wants to purchase long-term insurance products from companies that are more solvent than traditional life insurers. For example, total assets of Tokio-Marine-Nichido-Anshin Life, the subsidiary of Tokio Marine Insurance, increased from 225 billion yen at the end of March 1999 to 1,511 billion yen at the end of March 2004.[120]

[120] The 2,428 percent solvency margin ratio of Tokio-Marine-Nichido-Anshin Life at the end of March 2003 was higher than any traditional life insurer.

Table 3.5. Number of Life Insurers in Japan, End of the Fiscal Year

Fiscal Year	Number
1990	30
1991	30
1992	30
1993	30
1994	30
1995	31
1996	44
1997	44
1998	45
1999	46
2000	48
2001	43
2002	42
2003	40

Source: *Life Insurance Fact Book*, Japan Institute of Life Insurance.

The third group consists of foreign-affiliated companies. The U.S.-based American Life Insurance Company (ALICO) first came to the Japanese market in 1973. In March 2004, 17 out of 40 life insurers were owned by foreign companies. There are two ways that foreign companies' operate in Japan. Some foreign companies establish subsidiary companies in Japan and the subsidiary runs the operation in Japan. Examples are AIG STAR LIFE, Aoba Life, AXA Group Life, AXA Life, Credit Suisse Life, GE Edison Life, Gibraltar Life, Hartford Life, ING Life, Manulife Life, MassMutual Life, PCA Life, Prudential Life, and Skandia Life. Other foreign companies do not establish subsidiaries but open branches in Japan. ALICO Japan, American Family Life, CARDIF Assurance Vie, and Zurich Life are included in the branch operation group.

The traditional group has lost its market share for two reasons. First, because of the burst of the bubble and very low interest rates, most of the traditional companies suffered low performance on their investments and faced financial difficulties. After the Japanese people experienced the bankruptcy of several life insurers, they became overly sensitive to the financial soundness of companies. Newcomers, free from the burden of the bubble's burst, gained market share. Second, most traditional companies have large sales networks. However, many salespeople employed by traditional companies cannot satisfy customers who demand that salespeople have sophisticated financial knowledge, while some newcomers, such as Sony Life, employ well-educated salespeople, who can satisfy customers' consulting demands. Furthermore, because keeping large sales networks is expensive, traditional companies lose cost advantage against newcomers who sell insurance via the telephone and the Internet.

Although its market share has declined, it is notable that the traditional group still dominates the Japanese life insurance market. Table 3.6 ranks life insurers based on the premium incomes for fiscal 2003. Except ALICO and American Family, all of the top ten largest life insurers are traditional ones. Finally, Table 3.7 shows the concentration in the life insurance market. Concentration in the individual insurance market slightly decreased, while that in the group insurance market increased. This variation is because newcomers grew faster than traditional ones in the individual market, while it is still hard for newcomers to enter group insurance because of close relationships between firms and traditional insurers.

Demutualization

The Insurance Business Law of 1995 set up the process of demutualization, and the amendment of 2000 further modified the articles in order to solve practical issues associated with the demutualization.[121] However, all traditional companies except Taiyo Life, Daido Life, and Mitsui Life are still mutual companies.

The first demutualization in Japan was conducted by Daido Life. Although Daido Life announced that it had a plan of demutualization, it had to wait for the amendment of the law in 2000. The board of directors of Daido Life officially decided to start the demutualization process on January 21, 2001. On June 18, 2001, its board of directors decided to ask the general meeting of representatives of members of the mutual to approve the demutualization plan. The general meeting of representatives approved the plan on July 12, 2001. The law gives each insurance policyholder an opportunity to object to the demutualization after the resolution of the general meeting of representatives, and, if over 20 percent of policyholders raise objections, the demutualization plan must be cancelled. Only 18 policyholders (0.002 percent of total policyholders) raised an objection. On December 7, 2001, the Financial Services Agency authorized the demutualization plan. Finally, Daido Life converted from a mutual company to a stock company and issued 1.5 million shares on April 1, 2002. Through this demutualization, 334,530 members received one share or more, 355,810 members received odd lots, and 238,365 members received nothing because their contributions were deemed negative or zero. At the same time, Daido Life started to list its shares on the Tokyo Stock Exchange and on the Osaka Stock Exchange. Daido Life conducted a domestic initial public offering (IPO) of 527,790 shares and an overseas IPO of 105,425 shares. The offered price was 270,000 yen per share, and its initial market price was 320,000 yen.[122] Following Daido Life, Taiyo Life completed demutualization in April 2003. Daido and Taiyo chose the stock company organization because it makes both companies easy to merge with each other. In fact, they established T&D Holdings and became its subsidiaries in April 2004. Since April 2004, Daido and Taiyo Life's shares have been delisted from the exchanges and T&D Holdings' shares are listed.

[121] The Insurance Business Law of 1995 established that policyholders of mutual insurers are given some shares of new stock insurer based on their contributions through the demutualization, but are prohibited from paying cash to policyholders because it would endanger the solvency of the insurer. However, because the Commercial Law requires that the face value of new shares should be worth more than 50,000 yen, most policyholders would not receive unit shares. The amendment allowed insurers to pay cash to policyholders after consolidating their allocated odd lots and selling them.

[122] Its price declined to 212,000 yen in March 2003, then rose to 429,000 yen in March 2004.

Table 3.6. Top Ten Life Insurers in Japan as of March 2004

Rank of Premium Income	Company Name	Organizational Form	Ownership Type	Total Assets (in trillions yen)	Insurance Premiums (in trillions yen)	Principal Distribution System	Number of Sales-people
1	Nippon Life	Mutual	Independent	45.27	5.14	Salesperson	57,565
2	The Daiichi Mutual Life	Mutual	Independent	29.65	3.42	Salesperson	45,809
3	Meiji Yasuda Life	Mutual	Independent	25.33	2.40	Salesperson	39,128
4	Sumitomo Life	Mutual	Independent	21.12	2.70	Salesperson	40,806
5	ALICO Japan	Branch	Foreign (AIG)	2.84	1.40	Agency	4,296
6	Daido Life	Stock	Insurance Holding Company (T&D Holdings)	6.02	0.93	Agency	5,251
7	Mitsui Life	Stock	Independent	7.51	0.92	Salesperson	10,309
8	American Family Life	Branch	Foreign (AFLAC)	4.25	0.89	Agency	n.a.
9	Taiyo Life	Stock	Insurance Holding Company (T&D Holdings)	6.41	0.79	Salesperson	10,282
10	Fukoku Mutual Life	Mutual	Independent	5.00	0.74	Salesperson	11,591

Source: Weekly Toyo Keizai (July 28, 2004).

Table 3.7. Concentration in Life Insurance Markets (Market Share by Premium Volume of the Top 1, 5, and 10) in 1993, 1998, and 2003

		Individual Insurance	Individual Annuities	Group Insurance	Group Annuities
1993	Top 1	20.5%	19.6%	15.5%	16.5%
	Top 5	60.2%	61.2%	59.0%	56.9%
	Top 10	81.7%	87.2%	84.4%	84.6%
	Herfindahl Index	984.7	996.6	880.3	900.1
1998	Top 1	20.4%	19.0%	14.5%	22.4%
	Top 5	59.1%	65.9%	59.5%	67.2%
	Top 10	79.6%	90.3%	85.0%	92.6%
	Herfindahl Index	952.6	1079.1	897.5	1193.8
2003	Top 1	19.5%	17.6%	29.6%	29.9%
	Top 5	59.1%	56.4%	77.2%	86.3%
	Top 10	77.5%	80.1%	92.5%	98.5%
	Herfindahl Index	902.7	856.1	1609.1	1901.4

Source: Insurance, Hoken Kenkyujyo (various issues).

Following Daido and Taiyo, Mitsui Life completed demutualization in April 2004. Mitsui chose the stock company organization in order to strengthen an alliance with Mitsui-Sumitomo Bank and other financial institutions of Mitsui and Sumitomo Groups. Unlike Daido and Taiyo, Mitsui did not list its shares on any stock exchanges at the time of the demutualization. Instead of an IPO, odd lots were consolidated and purchased as treasury stocks by Mitsui Life. Then, Mitsui Life sold a part of the treasury stocks to financial institutions and large firms belonging to Mitsui and Sumitomo Groups in April 2004. Mitsui-Sumitomo Bank held 194,000 shares (14.32 percent of total shares), Chuo Mitsui Trust Bank held 160,000 shares (11.78 percent of total shares), and Mitsui-Sumitomo Insurance held 130,000 shares (9.57 percent) as of April 30, 2004.

Distribution Channels

Only registered life insurance agents can sell life insurance products. As shown in Table 3.8 the number of agents reached a peak of 445,000 in 1990 and decreased to 284,047 at the end of March 2003. An important problem regarding distribution channels is high turnover among sales staff. For example, in 2002, 132,556 persons quit their jobs and companies had to recruit 119,452 new people. This high turnover ratio discourages insurers from investing in educating salespeople, so it is very hard to satisfy customers who want salespeople with sophisticated financial knowledge.

Table 3.8. Number of Agents and Agencies of Life Insurers from 1990 to 2002 (in thousands)

Fiscal Year	Agents	Agency	Number of Agency Employees
1990	445	n.a.	n.a.
1991	443	n.a.	n.a.
1992	428	n.a.	n.a.
1993	421	n.a.	n.a.
1994	400	n.a.	n.a.
1995	395	n.a.	n.a.
1996	390	188	128
1997	354	193	163
1998	342	189	180
1999	330	187	204
2000	313	182	232
2001	298	145	296
2002	284	139	693

Source: Life Insurance Fact Book, Japan Institute of Life Insurance.

The development of information technology such as the Internet and telephone sales has improved upon non-traditional distribution channels. Furthermore, sales of insurance products over banks' counters have increased. Since April 2001, banks have been allowed to sell insurance products related to housing loans, such as life insurance for borrowers of housing loans and fire insurance covering houses that are financed by housing loans. Since October 2002, banks have been allowed to sell individual annuities and property accumulation fund insurance (Zaikei-Hoken in Japanese).[123] Because banks have become agencies of life insurers, the number of employees of agencies dramatically increased in 2002. Further deregulation in the near future will affect the distribution channels of life insurance, which have already significantly diversified.

[123] Bank sales from October 2002 to March 2004 were estimated at about 3 trillion yen in premiums. (Nihon Keizai Shimbun, June 15, 2004). According to survey research conducted by *Toyo Keizai*, a business magazine, Mitsui-Sumitomo Bank sold 391 billion yen in insurance premiums from April 2001 to March 2004, followed by Tokyo-Mitsubishi Bank (261 billion yen), Chuo-Mitsui Trust Bank (207 billion yen), Mizuho Bank (170 billion yen), and Chiba Bank (159 billion yen). ALICO sold their products of 790 billion yen through banks, followed by Hartford (540 billion yen) and Mitsui-Sumitomo Citi-Insurance (466 billion yen). See *Tokyo Keizai* (July 28, 2004).

3.3.2 Financial Performance and Conditions

Assets and Liabilities

Total assets of life insurers reached a peak of 192 trillion yen at the end of March 1999 and have decreased since then (see Table 3.9). At the end of March 2004, total assets of life insurers amounted to 184 trillion yen.

Table 3.9. Total Assets of Life Insurers from 1990 to 2003

	Assets (in trillions yen)
1990	131.6
1991	143.2
1992	156.0
1993	169.1
1994	178.0
1995	187.5
1996	188.7
1997	190.1
1998	191.8
1999	190.0
2000	191.7
2001	184.4
2002	179.8
2003	184.3

Source: Life Insurance Fact Book, Japan Institute of Life Insurance.

Balance sheets of Japanese life insurers on March 31, 2004 are summarized in Table 3.10. Important liabilities of life insurers are policy reserves, reserves for outstanding claims, reserves for policyholders' dividends, and price fluctuation reserves. Policy reserves were 161 trillion yen at the end of March 2004, which accounted for 92 percent of total liabilities. Capital of life insurers increased from 5.5 trillion yen at the end of March 2003 to 9 trillion yen at the end of March 2004. Life insurers' capital consists of capital stocks and foundation funds (2 trillion yen at the end of March 2004), surplus funds (1.3 trillion yen), and appraisal surplus regarding securities in portfolios (4.2 trillion yen). Appraisal surplus has been reported as a capital item since March 2002, when market valuation of assets was required by the accounting reform. The decline of stock prices in 2002 resulted in huge revaluation of marketable securities that reduced capital by 1.4 trillion yen for fiscal 2002, while the recovery of stock prices in 2003 increased capital by 3.1 trillion yen for fiscal 2003. Because it is so variable, the capital position of a company depending heavily on appraisal surplus is often regarded as weak. Many life insurers tried to increase their capital by issuing new stocks and subordinated debts to enhance financial stability. However, some criticize that this is double-gearing because most of the funds come from banks whose stocks are owned by the insurers.

Table 3.10. Balance Sheet of Japanese Life Insurers on March 31, 2004 (in trillions yen)

Assets		Liabilities	
Cash and Deposits	2.1	Policy Reserves	165.9
Call Loans	2.8	Reserve for Outstanding Claims	1.4
Monetary Receivables Purchased	2.8	Policy Reserve	160.7
Money Held in Trust	2.5	Reserve for Dividends to Policyholders	3.8
Investments in Securities	120.5	Accrued Severance Indemnities	1.3
Loans	41.7	Reserve for Price Fluctuations of Investments in Securities	0.8
Real Estate and Movables	7.7	Other Liabilities	6.2
Deferred Tax Assets	0.6	Deferred Tax Liabilities	0.4
Other Assets	3.6	Miscellaneous	0.8
Allowance for Doubtful Accounts	–0.2	Total Liabilities	175.3
Miscellaneous	0.1		
		Capital	
		Funds and Capitals	2.0
		Reserve for Redemption of Funds (Note 9)	0.9
		Surplus	1.3
		Net Unrealized Gains on Securities, Net of Taxes	4.2
		Miscellaneous	0.7
		Total Capital	9.0
Total Assets	**184.3**	**Total Liabilities and Capital**	**184.3**

Source: Life Insurance Association of Japan.

Incomes and Expenditures

As Table 3.11 shows, major incomes of life insurers are insurance premiums (26.4 trillion yen for fiscal 2003) and incomes from invested assets (6.3 trillion yen). Investment incomes have decreased in recent years because of low interest rates and stock price declines. For example, investment incomes were 9.6 trillion yen for fiscal 1997, while they were 5.6 trillion yen for fiscal 2002 and 6.3 trillion yen for fiscal 2003.

The major expenditures of life insurers are "claims and other payments" (24.2 trillion yen for fiscal 2003 or 69 percent of total expenditures) and "brought into policy reserves" (3.9 trillion yen for fiscal 2003). Claims and other payments consist of death benefits (3.2 trillion yen), maturity proceeds (3.9 trillion yen), return premiums for cancellation (7.4 trillion yen), and others. Thanks to cost-cutting efforts by every life insurer, operating expenses were decreased from 4.4 trillion yen for fiscal 1993 to 3.6 trillion yen for fiscal 2003. Other important expenditures are related to asset management activities. For fiscal 2002, life insurers suffered appraisal losses regarding securities of 1.8 trillion yen and losses of securities sales of 0.9 trillion yen. Fortunately, the recovery of stock prices in 2003 decreased asset management costs for fiscal 2003 to 1.7 trillion yen from 4.7 trillion yen for fiscal 2002.

Japanese life insurers have experienced poor profit performance in recent years. Current profits were 1,807 billion yen for fiscal 1999, 1,587 billion yen for fiscal 2000, 632 billion yen for fiscal 2001, 522 billion yen for fiscal 2002, and 1,510 billion yen for fiscal 2003. After adjusting for extraordinary profits and losses (e.g., profits or losses of real estate sales), final surpluses were 808 billion yen for fiscal 1999, 317 billion yen for fiscal 2000, 396 billion yen for fiscal 2001, 227 billion yen for fiscal 2002, and 677 billion yen for fiscal 2003. Although it is hard to define the equity and profits of mutual insurers, we tentatively define the return on equity (ROE) as surplus before tax (in income statements) divided by total capital (in balance sheet). Table 3.12 shows this ROE. The incredibly high value of ROE before 2000 suggests not that life insurers were profitable, but that they could exist without substantial capital positions and their assets in the balance sheet were undervalued.

Solvency margin ratios for 40 life insurers ranged from 7,256.8 percent (Toki Marine & Nichido Financial Life) to 321.9 percent (Aoba Life) at the end of March 2004. Although all insurers satisfied the required level of 200 percent, traditional companies had lower solvency ratios. Even Daido Life, whose solvency ratio was the highest among traditional companies, was 22nd among 40 companies at the end of March 2004.

Negative Spread (Gyakuzaya)

Assumed interest risk is a major risk for life insurers. During the bubble period, private life insurers raised their assumed interest rates to over 6 percent for long-term endowment insurance. These extremely high assumed interest rates were unwise from the present perspective, but they were understandable at that time due to the following reasons. First, life insurers earned approximately 8 percent yields from their portfolio investments because of the sharp appreciation of asset prices. Furthermore, they had massive unrealized capital gains regarding stocks and land. Second, they wanted to grow faster to enjoy the economies of scale. The government started extensive financial deregulations, which would enable banks and securities companies to enter insurance markets. Before other institutions entered the market, life insurers wanted to enlarge their market shares. Also, some insurers planned to enter other financial markets and form a financial conglomerate. Third, postal insurance (Kampo) set high assumed interest rates. To compete with Kampo, private insurers felt that higher assumed interest rates were inevitable.

Table 3.11. Income Statement of Life Insurers for April 2003 to March 2004 (in trillions yen)

Current Revenues	**36.80**	
Income from Insurance and Reinsurance Premiums	26.39	
Investment Income	6.29	
Interest, Dividend and Other Income		3.94
Net Gains from Money Held in Trust		0.11
Gains on Sales of Securities		0.90
Net Gains from Separate Accounts		1.22
Current Expenditures	**35.29**	
Insurance Claims and Other Payments	24.24	
Death and Other Claims		7.30
Annuity Payments		1.58
Health and Other Benefits		4.22
Surrender Benefits		7.38
Other Refunds		3.00
Provision for Policy Reserves	3.88	
Provision for Reserve for Outstanding Claims		0.17
Provision for Policy Reserve		3.65
Expenses for Investment	1.69	
Net Losses from Trading Securities		0.67
Net Losses from Derivative Financial Instruments		0.31
Write-down of Loans		0.24
Depreciation for Rental Real Estate and Others		0.12
Operating Expenses	**3.57**	
Other Expenditures	1.94	
Operating Income	1.51	
Extraordinary Profits	**0.20**	
Gains on Disposal of Assets	0.04	
Reversal of Allowance for Doubtful Accounts	0.08	
Extraordinary Losses	**0.73**	
Losses on Disposal of Assets	0.22	
Provision for Reserve for Price Fluctuations of Investments in Securities	0.30	
Provision for Dividends to Policyholders	**0.10**	
Surplus Before Income Taxes	**0.87**	
Income Taxes (Net of Adjustments)	**0.19**	
Surplus in the Current Year	**0.68**	

Source: Life Insurance Association of Japan. In some cases, subcategories do not add to category totals because of omission of miscellaneous items.

Table 3.12. Surpluses Before Tax and Total Capital of Life Insurers (in billions yen)

	Surpluses Before Tax (A)	Total Capital (B)	A/B
1990	3,989	3,656	1.091
1991	3,071	2,918	1.052
1992	2,166	2,060	1.052
1993	1,678	1,552	1.081
1994	1,334	1,346	0.991
1995	1,604	1,383	1.160
1996	1,762	2,521	0.699
1997	1,471	2,749	0.535
1998	493	4,044	0.122
1999	1,260	4,605	0.274
2000	1,083	6,057	0.179
2001	405	7,071	0.057
2002	320	5,555	0.058
2003	311	5,555	0.056

Source: Insurance, Hoken Kenkyujyo (various issues).

Unfortunately, the investment performance was declining more rapidly than expected. For example, from 1990 to 1994, the investment yields decreased by about 3 percentage points. Although insurers reduced the assumed interest rates for new contracts, they could not reduce rates for existing contracts. Therefore, average assumed interest rates remained higher than actual investment yields. The amount of losses caused by the negative spread has been disclosed since fiscal 1997. For fiscal 2003, the major nine life insurers recorded negative spreads of 1.06 trillion yen; their average investment yield was 2.1 to 2.2 percent and their average assumed interest rate was 3.3 to 3.4 percent.[124] Negative spreads for all life insurers were 1.25 trillion yen for fiscal 2002 and 1.11 trillion yen for fiscal 2003.

These negative spreads caused several insurers to fail. The law allowed insurers to reduce the assumed interest rates of existing contracts only when the companies went to bankruptcy. For example, regarding Toho Life, the assumed interest rates of all existing contracts (approximately 4.75 percent) were reduced to 1.5 percent after the failure. Some argued that the reduction of the assumed interest rates before the failure might be beneficial to policyholders because it would circumvent the destruction of its franchise value. Finally, the Insurance Business Law was amended in 2003 to enable insurers that do not fail to reduce their assumed interest rates. However, because companies must satisfy various conditions (e.g., they have to guarantee at least a 3 percent assumed rate after the reduction) and perform cumbersome procedures, no insurers so far use this program.

[124] Nihon Keizai Shimbun, May 29, 2004.

3.3.3 Investment Activities

Breakdown of Insurance Company Investments

Table 3.13 shows the breakdown of life insurers' total investments. The largest component of assets was securities. Shares of securities have steadily increased from 41.8 percent at the end of fiscal 1993 to 65.3 percent at the end of fiscal 2003. Loans were the second largest component and accounted for a quarter of total assets. Loans used to be the largest component in life insurers' investment portfolios. For example, the share of loans was 67.9 percent at the end of fiscal 1975. Because insurers cannot monitor borrowers as extensively as banks can, insurers usually lend funds for long-term capital investment to blue-chip companies. Life insurers could find borrowers during the rapid economic growth period of the 1950s, 1960s, and early 1970s. As blue-chip companies have accumulated sufficient retained earnings and increased their dependency on securities markets, it became hard for insurers to find appropriate borrowers.

With regard to securities holdings, Japanese government bonds were 35.5 trillion yen at the end of March 2004, which accounted for 19.3 percent of total assets. Recently, insurers have reduced the amount of investment in domestic stocks in order to constrain market risks. The peak amount of domestic stocks was 33.1 trillion yen at the end of fiscal 1994, and their share at that time was 18.6 percent of total assets; they decreased to 17.2 trillion yen and accounted for 9.6 percent of total assets at the end of March 2003. For fiscal 2004, domestic stocks increased to 21.4 trillion yen mainly due to price appreciation. Regional government bonds and corporate bonds were 6.3 trillion yen and 18.9 trillion yen at the end of fiscal 2003, respectively.

Table 3.13. Asset Distribution of Life Insurers as of March 2004 (in trillions yen)

	1998	1999	2000	2001	2002	2003
Cash and Deposits	7.9	7.2	4.5	2.9	2.5	2.1
Money Trust	2.6	2.9	3.6	3.6	3.4	2.8
Call Loan	5.0	4.5	5.7	3.4	2.9	2.8
Loan	59.1	54.8	50.0	47.1	44.5	41.7
Securities	100.1	105.0	110.4	111.0	110.5	120.5
Bonds	47.4	52.1	57.1	57.9	61.4	65.3
Stocks	28.5	28.4	29.5	24.7	17.2	21.4
Foreign Securities	24.2	24.5	23.8	28.4	31.9	33.8
Real Estate	9.7	9.2	8.2	8.0	7.6	7.6
Other	7.4	6.6	9.3	8.4	8.5	6.8
Total Assets	191.8	190.0	191.7	184.4	179.8	184.3

Source: Life Insurance Fact Book, Japan Institute of Life Insurance, prior to 2002, and the Life Insurance Association of Japan for fiscal 2003.

During the bubble period, Japanese life insurers invested large amounts of funds into foreign security markets. As life insurance contracts are denominated in Japanese yen, foreign investments are accompanied by foreign exchange risks. When the yen appreciates against the U.S. dollar, life insurers suffer extensive exchange losses. Life insurers suffered foreign exchange losses of 313 billion yen for fiscal 1994, when the yen substantially appreciated against the U.S. dollar. Consequently, life insurers hesitated to invest substantial funds into foreign countries in the late 1990s. However, as domestic interest rates remained low and domestic stock markets remained stagnant, Japanese life insurers were forced to increase foreign securities investment again. Foreign securities in their portfolios grew and topped the ratio invested in domestic stocks for the first time in 2001. The percentage of foreign securities holdings was a record high of 18.3 percent at the end of March 2004.

Life Insurers' Role in Business Groups

It is well-known that life insurers play an important role in business groups, or Keiretsu, in Japan. Historically, six financial Keiretsu, such as Mitsubishi, Mitsui, Sumitomo, Fuyo, Sanwa, and Daiichi Kangyo were organized around a large city bank. Members of the Keiretsu are linked primarily by cross-shareholdings. Because the financial institutions in the groups hold large shares of stocks of group companies and support the management, group companies do not have to worry about hostile takeovers. In turn, because banks' stocks are owned by group companies, bank managers also do not have to worry about hostile takeovers. However, shareholdings are of great benefit to insurers. Group companies purchase insurance products only from the group insurer. Furthermore, group firms encourage their employees to purchase insurance products from the group insurer (e.g., by allowing its agents to visit the offices and prohibiting agents affiliated with other insurers from doing so).

As mentioned above, insurers have decreased domestic stocks in their portfolios. Life insurers held 12.8 percent of outstanding stocks of all listed companies in 1986, and they held 4.9 percent of them in 2003.[125] The relationship between life insurers and other Keiretsu members has been weaker and will weaken further. However, large insurers are still major shareholders of many companies. For example, at least one life insurer is registered as one of the top ten shareholders for 2,190 companies among 3,621 listed companies in 2003.[126] Nippon Life, the largest life insurer, and Daiichi Life, the second, are registered as among the top ten largest shareholders for 841 companies and 436 companies, respectively. See Hsu (1999) and Nakatani (1984) for more details about life insurers' role in business groups.[127]

---

[125] See the 2003 share ownership survey available at the Tokyo Stock Exchange's web site (www.tse.or.jp/english/data/research/english2003.pdf).

[126] This figure is calculated by using the summer 2003 CD-ROM edition of *Kaisha Shikiho* (Japanese Company Quarterly), published by Toyo Keizai Shinpo-sha.

[127] The preceding section on life insurance in Japan is based on information available at the web site of the Life Insurance Association of Japan (www.seiho.or.jp/english/index.html).

3.4 NON-LIFE INSURERS

3.4.1 Industry Organization

Domestic Non-life Insurers

As of September 2003, there were 30 domestic non-life insurers and 24 foreign non-life insurers in Japan. The number of non-life insurers jumped from 27 to 33 in 1997 (see Table 3.14), because six life insurers entered the market by establishing non-life insurance subsidiaries. Also, Sony, the giant electronics manufacturer, entered the non-life insurance market by establishing Sony Assurance in 1999. Despite new entries, the number of non-life insurers decreased because of bankruptcies and mergers. As will be explained later, two non-life insurers went bankrupt in 2000 and 2001. Many traditional non-life insurers have experienced mergers since 2001. The major mergers are as follows: Dai-Tokyo and Chiyoda were merged into Aioi Insurance in April 2001; Koa and Nippon were merged into Nippon-Koa Insurance in April 2001; Dowa and Nissay were merged into Nissay-Dowa in April 2001; Sumitomo and Mitsui were merged into Mitsui-Sumitomo Insurance in October 2001; Tokio Marine and Nichido established Millea Holdings and became its subsidiaries in April 2004;[128] and Nissann and Yasuda were merged into Sompo Japan in July 2002.

Table 3.14. Number of Non-life Insurers in Japan, from 1992 to 2003

	Domestic	Foreign
1990	24	35
1991	25	35
1992	25	34
1993	25	33
1994	25	30
1995	27	30
1996	27	30
1997	33	31
1998	33	30
1999	36	27
2000	36	27
2001	33	26
2002	31	25
2003	30	24

Source: Insurance, Hoken Kenkyujyo (various issues).

[128] In October 2004, Tokio Marine and Nichido were merged into Tokio-Marine Nichido.

Three domestic non-life insurers confine their activities to reinsurance. We can classify the remaining 27 domestic companies into subgroups based on ownership types: the foreign company group, the life insurance company's subsidiary group, the industrial firm's subsidiary group, and the independent group. Five domestic insurers, such as Allianz Fire & Marine Insurance Japan, AXA Non-Life Insurance, and UNUM Japan Accident Insurance, are owned by foreign companies, but, because they are established as Japanese insurers, we regard these five companies as domestic ones. Four are subsidiaries of life insurers. For example, Sumisei General Insurance and Meiji General Insurance are owned by Sumitomo Life and Meiji Life, respectively. Nippon Life has 33.4 percent of outstanding stocks of Nissay Dowa General Insurance. Some non-life insurers are subsidiaries of or have close alliances with industrial companies. For example, Sony Assurance is a subsidiary of Sony Financial Holdings, which is a 100 percent subsidiary of Sony. Also, Toyota Motors holds 33.3 percent of outstanding shares of Aioi Insurance.

Some non-life insurers maintain independent status. As suggested by the fact that the five largest non-life insurers except Aioi Insurance are classified into this independent category, these independent companies constitute the core of Japanese non-life insurance markets (see Table 3.15). Tokio Marine and Fire Insurance is the largest (premiums for fiscal 2002 were 1.5 trillion yen), followed by Sompo Japan Insurance (1.3 trillion yen), Mitsui Sumitomo Insurance (1.3 trillion yen), Aioi Insurance (0.8 trillion yen) and Nippon Koa Insurance (0.7 trillion yen).

Table 3.15. Top Ten Non-life Insurers in Japan as of March 2003

		Net Premium Incomes (in billions yen)	Total Assets (in billions yen)	Number of Agencies
1	Tokio Marine and Fire Insurance	1,470	6,864	53,655
2	Sompo Japan Isurance	1,318	4,786	76,360
3	Mitsui Sumitomo Insurance	1,235	5,901	80,928
4	Aioi Insurance	837	2,496	46,759
5	Nipponkoa Insurance	721	3,083	52,172
6	Nichido Fire and Marine Insurance	396	1,758	24,962
7	Nissay Dowa General Insurance	308	1,146	20,903
8	Fuji Fire and Marine Insurance	305	1,024	23,164
9	Kyoei Fire and Marine Insurance	173	687	1,454
10	Nisshin Fire and Marine Insurance	149	482	15,403

Note: All listed companies are domestic stock companies.
Source: Insurance, Hoken Kenkyujyo (various issues).

The Insurance Business Law allows non-life insurers to be mutual companies. At one time there were two mutual non-life insurers in Japan: Daiichi Fire and Kyoei Fire. Daiichi went bankrupt in 2000, and Kyoei changed from a mutual company to a stock company in April 2003. All non-life insurers are now stock companies. Table

3.16 shows the market concentration by line of business in the non-life insurance sector. Because of non-life insurers' mega mergers, the concentration rose recently.

Foreign Companies

As explained above, foreign-capitalized insurers that are incorporated and licensed to conduct non-life insurance business under the Insurance Business Law are regarded as *domestic* insurers. Therefore, "foreign" non-life insurers operating in Japan are classified according to the following categories:

(a) foreign non-life insurers licensed on a branch or agent basis to conduct non-life insurance business under the Insurance Business Law,
(b) specific foreign insurers licensed under the Insurance Business Law to conduct non-life insurance business through their general agents in Japan (i.e., the Society of Lloyd's),
(c) insurance or reinsurance brokers acting in Japan as registered insurance agents or insurance brokers,
(d) representative or liaison offices established in order to collect information on the insurance market.

Table 3.16. Concentration in Non-life Insurance Markets (Market Share by Premium Volume of the Top 1, 5, 10) in 1993, 1998, and 2002

		Fire	Automobile	Personal Accident	General Liability	Compulsory Automobile Liability	Total
1993	Top 1	13.6%	17.5%	19.5%	25.0%	17.7%	18.2%
	Top 5	46.9%	53.8%	60.4%	67.4%	55.9%	54.0%
	Top 10	74.6%	83.4%	82.5%	85.0%	83.9%	81.3%
	Herfindahl Index	694.9	878.9	977.7	1223.6	922.3	876.3
1998	Top 1	14.0%	18.2%	18.6%	25.3%	18.8%	18.6%
	Top 5	46.6%	54.5%	58.5%	68.3%	56.4%	54.4%
	Top 10	72.9%	82.6%	80.0%	84.8%	84.2%	80.5%
	Herfindahl Index	685.6	884.3	918.3	1251.9	948.7	874.6
2002	Top 1	16.6%	18.9%	18.8%	26.3%	21.3%	20.1%
	Top 5	67.3%	76.3%	73.1%	82.7%	80.2%	76.2%
	Top 10	91.4%	95.5%	91.7%	96.7%	98.1%	94.3%
	Herfindahl Index	1105.2	1310.1	1271.8	1675.6	1456.9	1314.9

Source: Insurance, Hoken Kenkyujyo (various issues).

Twenty-four foreign companies are classified into (a) or (b) above. Two of them confine their activities to reinsurance, and three offer only protection and indemnity insurance. As of the end of March 2004, foreign non-life insurers' employed 4,066 workers and 20,604 and agents.

The total of direct premiums of foreign non-life insurers grew from 292.7 billion yen in fiscal 1993 to 427 billion yen in fiscal 2003. Because foreign companies have grown faster than domestic firms, the foreign presence in Japan has steadily increased.[129] The main product of foreign non-life insurers is personal accident insurance, maintaining 39.5 percent of total direct premiums for fiscal 2003, followed by voluntary automobile insurance (32.7 percent).

3.4.2 Distribution Channels

The non-life insurance distribution system in Japan is divided into agency, brokerage, and direct distribution by officers or employees of insurers. Agencies and brokers must be registered with the prime minister in accordance with the Insurance Business Law.

In fiscal 2003, agencies collected 92.9 percent of all general insurance premiums including domestic and foreign insurers. The remainder, 6.9 percent and 0.2 percent, was collected through direct distribution and brokerage, respectively. Agencies' sales shares by line of business are shown in Table 3.17. Although the agency is the dominant distribution channel, the number of agencies has declined because many insurers made important organizational changes to their distribution channels (see Figure 3.3). There were 623,741 agencies at the end of March 1997 and 305,836 at the end of March 2004. However, as suggested by the increases in numbers of employees, the average size of agencies is far larger than before. It is very common that an insurance agency has some other main businesses. For example, gas station owners are often registered as insurance agents and sell automobile insurance to their customers. Therefore, 83.2 percent of agencies at the end of March 2004 were part-time agencies. Also, 21.2 percent of agencies dealt with insurance products of two or more companies. The total number of sales staff engaged in agency business amounted to 1,716,006 people at the end of March 2004.

Table 3.17. Percent Shares of Agencies' Sales in Direct Premium by Line of Business, 1998 to 2002

	1998	1999	2000	2001	2002
Fire	85.3	82.9	85.8	87.2	85.3
Personal Accident	92.5	92.0	91.8	89.7	92.5
Auto	93.8	93.6	93.6	93.4	93.8
CALI	98.9	99.0	99.1	99.1	98.9
New	92.0	90.2	91.2	89.7	92.0
Total	91.6	91.0	91.7	91.3	91.6

Note: Because brokers' share is marginal, remaining is classified as direct sale.
Source: Insurance, Hoken Kenkyujyo (various issues).

[129] The market shares of foreign non-life insurers in Japan (including the five foreign-capitalized non-life insurers operating in Japan) were 2.7 percent for fiscal 1993 and 4.5 percent for fiscal 2001, which is the most recent available figure. When the savings portion of maturity-refund type insurance premiums is excluded, their shares increase to 5.2 percent for fiscal 2001 and 3.7 percent for fiscal 1993.

Agencies perform the following functions: consulting for insurance; soliciting (suggesting plans for) and concluding insurance contracts with customers; calculating insurance premiums; accepting application forms from customers and reporting contracts to the insurer; and receiving insurance premiums and issuing receipts to customers. The agency system was reviewed in 2001. The level of agency commissions was regulated to depend on personal qualifications and agency classifications by the FSA's administrative guidelines. Insurers wanted to establish their own commission scheme. The guidelines concerning the level of agency commissions were abolished in April 2001, but companies were required to notify the FSA of their own commission scheme. Since April 2003, companies no longer need to notify the FSA of their own commission scheme. Subsequently, each insurer takes its own measures to promote the further development of agency qualifications.

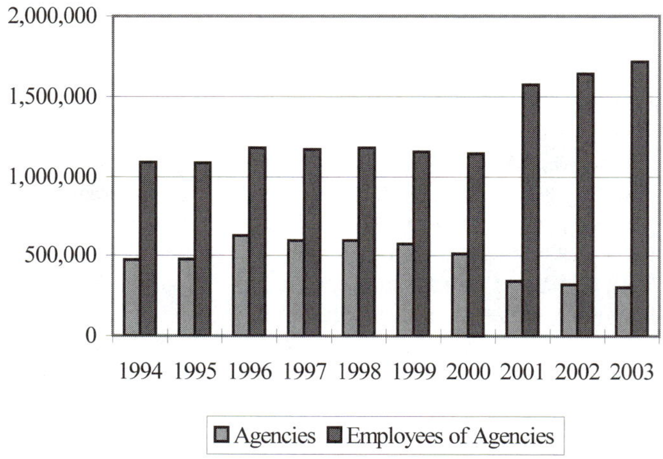

Source: Fact Book, General Insurance Association of Japan.

Figure 3.3. Non-life Insurers' Numbers of Agencies and Numbers of Employees, 1994 to 2003

The insurance broker system was introduced in 1996. Insurance brokers may provide the following services: risk management business; brokerage business; claims handling business; loss prevention services; and other services such as advising on fringe benefit plans. As suggested by the fact that only 43 insurance brokers were registered as of August 1, 2003, they play a marginal role in distributing non-life insurance products. Yamori (1999) investigates what factors affect corporate demand for insurance in Japan.

3.4.3 Financial Performance and Conditions

Assets and Liabilities

As Japanese non-life insurers have been active in selling long-term insurance policy contracts with maturity repayments (i.e., maturity-refund insurance) since the 1980s, their total assets have steadily increased (see Table 3.18). Currently, assets related to maturity-refund type insurance policies account for one-third of total assets. Total assets of non-life insurers amounted to 32.1 trillion yen at the end of March 2004. Due to Japanese accounting rule reform, non-life insurers started to employ market valuation accounting regarding securities in their portfolios. Thus, total assets increased by 4.1 trillion yen to 34.3 trillion yen in fiscal 2000. Recently, due to an increase in cancellations of maturity-refund type insurance policy contracts and a decrease in the unrealized gains on securities caused by the stagnant stock market, total assets of non-life insurers have fallen. Recovery of stock prices in 2003 increased unrealized gains on securities by 1.5 trillion yen, resulting in an increase of total assets from 30.3 trillion yen to 32.1 trillion yen.

Table 3.18. Total Asets of Non-life Insurers, 1990 to 2003

	Assets (in trillions yen)
1990	26,181
1991	26,566
1992	27,445
1993	27,639
1994	28,460
1995	29,453
1996	30,358
1997	31,117
1998	30,824
1999	30,218
2000	34,304
2001	33,121
2002	30,303
2003	32,094

Source: Non-Life Insurance in Japan: Fact Book 2001–2002, Marine and Fire Insurance Association of Japan.

The total liabilities of non-life insurers amounted to 25.7 trillion yen at the end of fiscal 2003 (see Table 3.19). Underwriting funds, consisting of underwriting reserves (20.1 trillion yen) and outstanding loss reserves (2.6 trillion yen), accounted for 88.3 percent of the total liabilities at 22.7 trillion yen. Capital of non-life insurers amounted to 6.3 trillion yen at the end of March 2004. The simple capital ratio (capital divided by total assets) was 19.9 percent at that time. Capital of non-life insurers consists of capital stocks (0.9 trillion yen at the end of March 2003), earned surplus (2 trillion yen), and appraisal surplus regarding securities in portfolios (3.2 trillion yen).

All companies had solvency margin ratios larger than the required ratio (i.e., 200 percent) at the end of March 2004. The solvency margin ratios of major non-life insurers were as follows: Tokio Marine = 1,108.6 percent (1,087.5 percent at the end of March 2003), Sompo Japan = 1,036.3 percent (774.8 percent), Mitsui-Sumitomo =

1,064.3 percent (930.7 percent), Aioi = 910.1 percent (683.4 percent), and Nippon-Koa = 996.4 percent (840.5 percent).

Table 3.19. Balance Sheet of Japanese Domestic Non-Life Insurers on March 31, 2004 (in billions yen)

Assets		Liabilities and Equities		
Deposits	1,516	Underwriting Funds	22,713	
Call Loans	822	Outstanding Loss Reserves		2,596
Receivables under Resale Agreements	44	Underwriting Reserves		20,117
Monetary Receivables Bought	238	Other Liabilities	2,992	
Money Trusts	229	**Total Liabilities**	**25,705**	
Securities	22,370	Capital	853	
National Government Bonds	3,674	Suspense Receipts on Capital Subscriptions	0	
Local Government Bonds	1,020			
Corporate Bonds	4,045	Additional Paid in Capital	423	
Stocks	9,065	Earned Surplus	1,966	
Foreign Securities	3,984	Land Revaluation Excess	–2	
Other Securities	582	Unrealized Gain on Other Securities, Net of Income Taxes	3,238	
Loans	3,080			
Policy Loans	92	Treasury Stock	–89	
Financial Loan	2,988	**Total Equities**	**6,389**	
Real Estate	1,393			
Total Working Assets	**29,692**			
Other Assets	2,402			
Total Assets	**32,094**	**Total Liabilities and Equities**	**32,094**	

Source: General Insurance Association of Japan.

Incomes and Expenditures

Total *net* premiums (direct premiums written + inward reinsurance net premiums – outward reinsurance net premiums – savings portion of maturity-refund type insurance premiums) written by non-life insurers in fiscal 2003 reached 7.4 trillion yen, an increase of 2.1 percent from fiscal 2002 (see Table 3.20). However, as maturity-refund type insurance premiums decreased, total premium incomes decreased to 9.1 trillion yen for fiscal 2003. Automobile insurance premiums (excluding compulsory automobile liability insurance) amounted to 3.6 trillion yen, constituting 47.8 percent of total net premiums.

Net claims paid (direct claims paid + inward reinsurance net claims paid – outward reinsurance claims received) on all classes of insurance business during fiscal 2003 amounted to 3.8 trillion yen. The loss ratio (i.e., the ratio of claims paid plus loss adjustment expenses to net premiums written) was 55.3 percent. Table 3.21 shows direct net premiums and net claims paid by the major line of business.

Table 3.20. Income Statement of Non-life Insurers for April 2003 through March 2004 (in trillions yen)

Underwriting Income	9058.7	
Net Premiums Written		7437.2
Savings Portion of Maturity-Refund Type Insurance Premium		1208.6
Underwriting Expenses	7595.7	
Net Claims Paid		3781.3
Loss Adjustment Expenses		332.8
Agency Commissions and Brokerage		1253.3
Maturity Refunds to Policyholders		1999.7
Investment Income	669.2	
Interest and Dividend Income		470.3
Profits on Sales of Securities		394.1
Investment Expenses	184.0	
Losses on Sales of Securities		110.5
Losses on Devaluation of Securities		30.4
Operating and General Administrative Expenses	1306.3	
Ordinary Profits	**659.6**	
Special Profits and Losses Balance	–151.9	
Taxes and Tax Adjustment	181.9	
Net Profits	**325.8**	

Source: General Insurance Association of Japan. In some instances, subcategories do not add to category totals because of omission of miscellaneous items.

Operating expenses on underwriting (agency commissions and brokerage plus operating and general administrative expenses on underwriting) decreased to 2.5 trillion yen for fiscal 2003. This decrease resulted from efforts to enhance efficiency by each insurer. The operating expense ratio (ratio of agency commissions and brokerage plus operating and general administrative expenses on underwriting to net premiums written) stood at 33.2 percent for fiscal 2003, a 1.3 percent decrease from the previous year (and a 5 percent decrease from fiscal 1993).

Investment incomes and expenses substantially affect the business performance of non-life insurers. Total investment incomes, including profits on sales or redemption of securities, amounted to 669 billion yen in fiscal 2003. On the other hand, investment expenses amounted to 184 billion yen, consisting of losses from devaluation of securities caused by a fall in the price of securities.

Japanese non-life insurers have earned positive profits since fiscal 1947. Ordinary and net profits amounted to 660 billion yen and 326 billion yen for fiscal 2003. However, fiscal 2001 was the exception, because ordinary profits for 2001 were minus 149.2 billion yen. The losses for fiscal 2001 were because of (1) increases in claims payment and the provision related to the terrorist attacks on September 11 and (2) the large losses caused by the devaluation of the book value of securities due to a fall in stock prices. The ROE, defined as current profits (income statement) divided by total capital (balance sheet), is shown in Figure 3.3.

Table 3.21. Direct Net Premiums and Net Claims Paid by the Line of Business, 1990 to 2003 (in 100 million yen)

	Fire		Automobile		Personal Accident	
	Direct Net Premiums	Direct Net Claims Paid	Direct Net Premiums	Direct Net Claims Paid	Direct Net Premiums	Direct Net Claims Paid
1990	17,361	2,605	24,618	13,903	24,950	2,285
1991	16,136	7,693	27,623	16,484	26,778	2,492
1992	16,791	3,237	30,076	18,092	25,422	2,702
1993	18,218	3,972	32,937	19,085	27,937	2,843
1994	18,545	3,638	34,449	18,661	26,197	2,986
1995	18,534	2,955	35,358	19,021	27,306	2,903
1996	19,027	3,367	36,491	19,553	28,930	2,965
1997	19,017	3,279	36,889	19,899	26,668	2,994
1998	17,741	4,476	35,721	19,987	22,954	2,901
1999	17,443	5,444	35,999	20,834	20,498	2,906
2000	15,836	3,638	36,500	21,903	18,313	2,782
2001	15,076	3,250	36,765	21,150	16,253	2,655
2002	14,934	2,916	36,054	20,426	15,781	2,563
2003	14,711	3,105	35,539	20,174	14,855	2,503

Source: General Insurance Association of Japan.

3.4.4 Investment Activities

Table 3.22 shows the breakdown of the investment of non-life insurers. Securities headed the list of investments with 19.5 trillion yen or 64.3 percent of total assets. Stocks have decreased remarkably in recent years—from 10.3 trillion yen at the end of March 2001 to 6.7 trillion yen at the end of March 2003. However, stocks were still the largest category of securities, followed by corporate bonds (4.3 trillion yen), foreign securities (3.9 trillion yen), Japanese government bonds (2.8 trillion yen), and local government bonds (1.2 trillion yen). Loans accounted for 11.3 percent of total assets at 3.1 trillion yen, and real estate accounted for 4.9 percent of total assets at 1.5 trillion yen.[130]

[130] The preceding section on the non-life insurance business in Japan relies on *General Insurance in Japan Fact Book 2002–2003,* published by the General Insurance Association of Japan, which is available at www.sonpo.or.jp/english/english.html. Therefore, the statistics apply to companies affiliated with the association. However, because member companies dominate Japanese insurers, these figures sufficiently

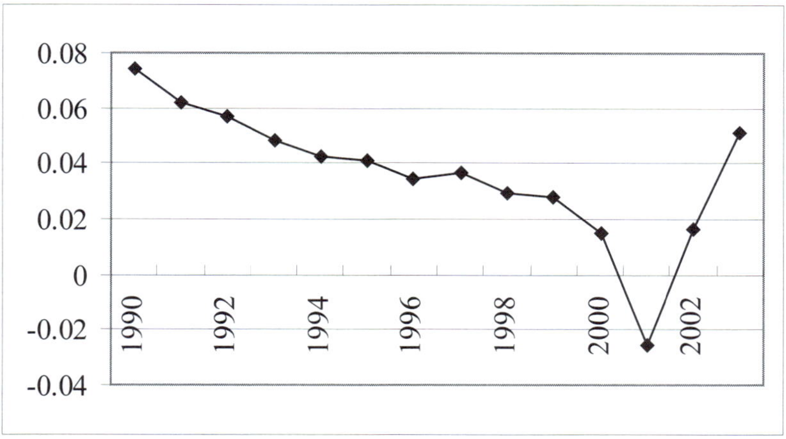

Source: General Insurance Association of Japan.

Figure 3.4. ROE of Non-Life Insurers from 1990 to 2003

3.5 OTHER INSURANCE PROVIDERS

3.5.1 Important Insurers That Are Not Regulated by the Insurance Business Law

The Insurance Business Law prohibits anyone except licensed insurers from conducting insurance business. However, many insurance providers that are *not* licensed under the Insurance Business Law are legally present in Japan. First, the law allows *small* membership organizations (cooperative insurance societies) to collect insurance premiums from their members and pay claims to their members. However, *large* membership organizations are deemed illegal and must be licensed. Second, there are other laws that allow *large* membership organizations to engage in insurance business. For example, the Agricultural Cooperative Society Law allows agricultural cooperatives to engage in insurance business, and the Consumer Livelihood Cooperative Society Law allows consumer cooperatives to do so. These kinds of quasi-insurance authorized by other laws are also called Kyosai in Japanese. Therefore, we refer to Kyosai provided by small membership organizations as "unregulated-Kyosai" in order to differentiate them from the Kyosai that organizations licensed under other laws provide. Third, government institutions engage in insurance businesses. Most of them, such as the Deposit Insurance Corporation and Nippon Export and Investment Insurance, have very special policy purposes, and they are beyond the scope of this chapter. However, Japan Post (Kampo) provides the same insurance policies as private insurers and has substantial shares in the life insurance market.

---

explain non-life insurance businesses and markets in Japan. Note that the Marine and Fire Insurance Association of Japan is the old name of the General Insurance Association of Japan.

Table 3.22. Asset Distribution of Non-Life Insurers (in billions yen)

	1997	1998	1999	2000	2001	2002	2003
Deposits	1,509	1,459	1,460	1,759	1,987	1,619	1,516
Call Loans	1,127	839	909	836	555	619	822
Receivables under Resale Agreements	0	850	0	10	1	32	44
Monetary Receivables Bought	443	0	330	492	276	237	238
Money Trusts	817	673	544	384	342	280	229
Securities	16,152	16,388	16,260	21,867	20,644	19,473	22,370
Government Bond	672	834	1,037	1,401	1,801	2,826	3,677
Municipal Bond	1,239	1,231	1,294	1,387	1,330	1,204	1,020
Corporate Bonds	3,929	4,026	4,180	4,518	4,432	4,248	4,045
Stocks	5,428	5,594	5,501	10,247	8,862	6,770	9,065
Foreign Securities	4,344	4,109	3,665	3,658	3,787	3,944	3,984
Loans	6,664	6,271	5,164	4,452	3,833	3,411	3,080
Real Estate	1,826	1,817	1,764	1,691	1,543	1,497	1,393
Other Assets	2,581	2,527	3,788	3,225	3,940	3,136	2,403
Total Assets	31,117	30,824	30,218	34,716	33,121	30,303	32,094

Source: General Insurance in Japan Fact Book (various issues,) General Insurance Association of Japan. The subcategories under securities do not add to the total due to the omission of miscellaneous securities.

Table 3.23 shows the total assets of major insurance providers. There are many small insurers that are not listed in the table, but their assets are marginal. Therefore, it is fair to conclude that Kampo has one-third and JA Kyosai has one-tenth of total insurers' assets. Their substantial asset sizes strongly suggest their importance in the Japanese insurance market.

3.5.2 Postal Life Insurance

Brief History of Postal Life Insurance

In July 1916, the Postal Life Insurance Law and the Postal Life Insurance Special Account Law were enacted, and the Postal Life Insurance Service (Kampo in Japanese) was launched as a state-run service in October 1916. Since then, post offices have offered Post Services, Postal Saving Services, and Postal Life Insurance Services. The Ministry of Posts and Telecommunications (MPT) operated the postal services before January 2001, when the MPT was reborn as the Ministry of Public Management, Home Affairs, Posts and Telecommunications (MPHPT). At that time, the Postal Services Agency was established to operate the postal services. Furthermore, the government transferred the Postal Services Agency to a public corporation, Japan Post, in April 2003. However, the employees of Japan Post are given the status of civil servants, which means that they will not be laid off even if Japan Post goes bankrupt. Also, Japan Post is given several privileges as a state-run business. It is exempt from paying corporate tax and enterprise tax. More importantly, the government officially guarantees the insurance policies sold by

Japan Post, even though Japan Post does not pay anything to the government for this guarantee. According to the Life Insurance Association of Japan, the value of the privileges that Japan Post enjoyed in the decade from 1991 to 2000 amounts to 3.2 trillion yen. The Koizumi Cabinet is considering the privatization of Japan Post to level the playing field between Japan Post and private insurers.

Table 3.23. Total Assets of Major Insurers in Japan at the end of March 2004 (in billions yen)

	Amount	Shares
Life Insurers	184,330	48.2%
Non-life Insurers	32,094	8.4%
Kampo (Japan Post)	121,912	31.9%
JA Kyosai	42,141	11.0%
Zenrosai	2,010	0.5%
Total	382,487	100.0%

Source: Disclosure materials.

Standout Features

Postal life insurance is distinguished from other life insurance provided by private insurers in the following respects.

(1) Because the postal life insurance is run by the government, profit is not its primary aim. According to the Japan Post Act, the purpose of Japan Post is to contribute to the stable growth of people's daily lives and the steady growth of the national economy. However, because the government does not provide subsidies to Japan Post, it must run on a stand-alone basis. Therefore, it needs to obtain reasonable profits.

(2) Medical examinations are not required for policyholders, and no occupational qualifications are imposed.

(3) Postal life insurance is available all over the country because of the extensive nationwide network of post offices. There were 24,791 post offices at the end of March 2003 and at least one post office is located in all 3,213 administrative regions (i.e., cities, wards, towns, and villages). In contrast, life insurers had only 15,585 offices at the end of March 2003, and most offices are located in large cities. Consequently, the market share of postal life insurance is higher in rural regions.

(4) To restrict the business of postal life insurance within the supplement of private insurers, maximum coverages for postal life insurance contracts are imposed. Maximum coverage for postal life insurance contracts rose from 8 million yen to 10 million yen in 1977, and maximum coverage for policyholders who have insurance policies that have been in force for four years or longer rose from 10 million yen to 13 million yen in 1986. Since 10 million yen is not enough for death security, the savings-based insurance contracts are the main products of Japan Post.

(5) Postal life insurance is fully guaranteed by the government. This government guarantee gives Japan Post a huge competitive advantage over private insurers, whose policyholders suffered from losses due to bankruptcies in the 1990s.

Policies of Postal Life Insurance

Japan Post has suffered from poor sales performance in recent years. New life insurance and annuity policies totaled 6.1 million in fiscal 2000, 5.7 million in fiscal 2001, 5.2 million in fiscal 2002, and 4.4 million in fiscal 2003. New life insurance policies in 2003 numbered 3.9 million, with premiums of 63.8 billion yen and an insured amount of 11 trillion yen. Because of sluggish sales, life insurance policies in force are decreasing in recent years. Those at the end of fiscal 2003 numbered 68.5 million and had a total insured amount of 185 trillion yen, while those for fiscal 1999 numbered 81.3 million. Annuity policies in force at the end of fiscal 2003 came to 7.3 million with an annuity amount of 2.6 trillion yen, resulting in a decrease from the previous year.

Investment

Japan Post held assets of 122 trillion yen as postal life insurance funds (Kampo funds), while private life insurers as a whole held assets of 184 trillion yen at the end of March 2004. Japan Post is required to manage these huge assets based on consideration for the public interest. The Postal Life Insurance Service has played an important role in contributing to the development of local communities through the investment of its funds. Furthermore, to ensure security and profitability of the investment and to avoid the distorted effects of Kampo assets on the market, the investment of Kampo funds has been more strictly regulated than that of private insurers. For example, investments in foreign securities had been prohibited until 1983 and in stocks until 1987.

Kampo funds are invested based on a fiscal year management plan and the public corporation's midterm management plan approved by the Minister of Public Management, Home Affairs, Posts and Telecommunications. The investment of Kampo funds is shown in Table 3.24. The investment rules for Japan Post differ from those for private insurers. Japan Post is still prohibited from providing loans to private firms. Also, Japan Post can only indirectly invest funds in stocks by investing funds into funds entrusted to trust banks. Trust banks use this entrusted fund to buy stocks at their own direction (i.e., Japan Post does not specify what stocks are bought), while Japan Post receives the dividends based on actual investment performance. The reason for using this scheme is to minimize the market misperception that a company whose stock Japan Post holds is supported by the government.

Investment in the form of loans to local governments is carried out based on loan amounts and conditions indicated by the MPHPT. Furthermore, responding to a request from the Minister of Finance, Japan Post underwrites the Fiscal Investment and Loan Program (FILP) bonds to support the FILP reform.[131]

---

[131] The preceding section on postal life insurance is based on *Postal Life Insurance Services in Japan: Annual Report 2003,* available at www.japanpost.jp/top/english/english-index.html.

Table 3.24. Investment of Kampo Funds at the End of Fiscal 2003

Securities	81,671
Japanese Government Bonds	51,403
Local Government Bonds	7,188
Corporate Bonds	21,485
Foreign Bonds	1,595
Funds Entrusted to Trust Banks	11,719
Loans	24,755
Loans to Local Government	19,117
Loans to Government Finance and Public Corporations	3,254
Loans to Policyholders	2,193
Loans to Postal Savings	193
Deposits	2,052
Total Investments	120,197

Source: Japan Post Annual Report 2004.

3.5.3 Cooperative Insurance (Kyosai)

Cooperative Insurance Societies

In addition to Japan Post and private insurers licensed by the Insurance Business Law, other insurance providers in Japan include cooperative insurance programs (Kyosai in Japanese) that are provided by cooperative insurance societies. According to the Japan Cooperative Insurance Association Incorporated (JCIA),[132] with which 60 cooperative insurance societies are affiliated, the insurance contracts undertaken by them were 144.9 million policies in force and 1,112 trillion yen in amounts insured at the end of March 2003. Premiums received and claims paid to members for fiscal 2002 were 6.8 trillion yen and 3.5 trillion yen, respectively. Their total assets were 45.7 trillion yen. In contrast to insurers, the number of insurance policies and premium incomes underwritten by cooperative insurance societies have steadily increased in recent years.

Cooperative insurance societies are different from insurers in the following respects. First, cooperative insurance societies are nonprofit organizations and are allowed to provide insurance services only to their members. However, as membership in several societies is easily obtained, 66 million people (about half of the Japanese population), are members of cooperative insurance societies.[133] Second, cooperative insurance societies provide cooperative insurance programs not based on

[132] JCIA's web site address is www.jcia.or.jp/en/pdf.html. This section on cooperative insurance societies relies heavily on the JCIA's *Fact Book 2003*.

[133] This figure overestimates real membership, because some people have multiple memberships and are counted twice or more..

the Insurance Business Law, but on various other laws.[134] For example, the National Mutual Insurance Federation of Agricultural Cooperatives (JA Kyosai), the largest cooperative insurance supplier, is licensed according to the Agricultural Cooperative Society Law and is regulated by the Ministry of Agriculture, Forestry and Fisheries, not the Financial Services Agency. The National Federation of Workers and Consumers Insurance Cooperatives (Zenrosai) is licensed through the Consumer Livelihood Cooperative Society Law and regulated by the Ministry of Health, Labor and Welfare. Third, in contrast to insurers, most cooperative insurance societies provide both life insurance and non-life insurance products. Fourth, corporative insurance societies are nonprofit organizations, while stock companies are common among insurers. Therefore, insurance premiums of Kyosai are often lower than those of private insurers for similar risks.

JA Kyosai

JA Kyosairen, affiliated with the Japan Agricultural (JA) group, underwrites JA Kyosai. JA Kyosai is sold at the branches of agricultural cooperatives that provide banking services as well as retail businesses. There are 1,182 agricultural cooperatives, which have 23,221 offices across Japan as of the end of March 2002. Their offices are located mainly in rural regions. Due to this nationwide network, JA Kyosairen is the largest cooperative insurance supplier in Japan. JA Kyosairen provides both life mutual insurance and non-life mutual insurance products. Premium incomes for fiscal 2003 were 6.2 trillion yen, and total assets at the end of fiscal 2003 amounted to 42.1 trillion yen.

JA Kyosai's unique and main product is building endowment insurance (*tatemonokousei-kyosai* in Japanese). Policies in force for this type of insurance on March 31, 2004, were 14 million in number and 154 trillion yen in amounts insured. This product insures buildings and movable property owned by the policyholder or the policyholder's relatives against damage caused by fire, earthquake, or other natural disaster, while ordinary non-life insurance products exempt damages caused by earthquake from their responsibility. Additionally, building endowment insurance covers risks for five or more years, while ordinary non-life insurance products insure for one year. If the insured buildings or movable property survive intact beyond the maturity date, building endowment insurance supplies maturity repayments for reconstruction or repurchase.[135]

---

[134] Strictly speaking, there are many Kyosai that are not subject to any special laws and any authority's supervision. Unregulated Kyosai is not considered to conduct insurance business as long as the members receive negligible payment or they target specific groups. Therefore, conducting such business without a license is not regarded as a violation of the Insurance Business Law. However, because many troubles had been reported with Unregulated Kyosai, the Insurance Business Law was amended in April 2005 to regulate the "unregulated Kyosai." First, under the new Law, most "unregulated Kyosai" groups have to register with the Prime Minister as "Small Short-term Insurer(SSI)," and the FSA has a authority to supervise them. Second, the SSIs are allowed to sell only insurance contracts with 2 year or shorter insurance term and with amounts insured of no more than 10 million yen. Third, they are required to invest funds into only government bonds and bank deposits.

[135] The preceding section on other insurance providers relies on *Annual Report 2003*, published by National Mutual Insurance Federation of Agricultural Cooperatives, available at www.ja-kyosai.or.jp/about/annual/e_index.html.

3.6 CURRENT IMPORTANT ISSUES

3.6.1 Earthquake Insurance

Earthquake Risks

Japan has frequently suffered from large-magnitude earthquakes (see Table 3.25). For example, the Great Kanto earthquake, which struck the Tokyo metropolitan area in 1923, resulted in 142,807 deaths.[136] The financial losses were estimated at about 5.5 billion yen, which was three and a half times as large as the annual expenditures of the national government, or 36.8 percent of the GNP at that time.[137] In January 1995, the Hanshin-Awaji earthquake (or Kobe Great earthquake) caused huge losses, including 5,502 deaths and 41,527 injuries. In addition, 100,282 commercial and residential buildings were completely destroyed, 108,402 were half destroyed, and 185,756 were partially destroyed. Insurers paid approximately 77 billion yen in earthquake insurance payments. Other earthquakes of a magnitude higher than 8.0 on the Richter scale are expected in the near future for the Tokai and South Kanto metropolitan areas, which include Tokyo. Because earthquake insurance payments resulting from these earthquakes are estimated to be huge amounts, the earthquakes could seriously damage the financial position of Japanese insurers.

Table 3.25. Claims Paid under Earthquake Insurance

Name of Earthquake	Date	Claims Paid (in millions yen)
Great Hanshin-Awaji (Kobe Great Earthquake)	January 17, 1995	78,347
Geiyo	March 24, 2001	16,688
Tottoriken Seibu	October 6, 2000	2,848
Hokkaido Toho-oki	October 4, 1994	1,333
Mt. Unzen Volcanic Eruption	June 3, 1991	1,288
Sanriku Haruka-oki	December 28, 1994	1,238
Kushiro-oki	January 15, 1993	990
Kagoshimaken Satsuma	March 26, 1997	531
Mt. Usu Volcanic Eruption	March 1, 2000	420

Source: General Insurance in Japan Fact Book 2002–2003, General Insurance Association of Japan.

Public Earthquake Insurance

Although earthquake insurance was badly needed for many years, private insurers hesitated to provide coverage because damage could be too severe and widespread.

[136] This figure includes 43,476 missing.

[137] GNP in 1923 is estimated to have been 14.9 billion yen (Ohkawa, Takayama, and Yamamoto, 1974).

Responding to political pressure after the 1964 Niigata earthquake, the Japanese government proposed the establishment of the Earthquake Insurance Act, which was made effective in 1966. The Earthquake Insurance Act has been modified several times since its establishment. The main features of this public earthquake insurance system are as follows:

(1) Earthquake insurance covers only residential buildings and household belongings. Only people who buy fire insurance are eligible for earthquake insurance coverage. Japanese fire insurance does not cover losses caused by an earthquake. For example, when a house is lost due to a fire caused by an earthquake, the insurer will pay only 5 percent of total losses as earthquake fire expenses insurance. Of course, since the Earthquake Insurance Law does not prohibit insurers from underwriting corporate insurance for earthquake risk at their own risks, some insurers sell corporate earthquake insurance.

(2) Earthquake insurance covers only 30 to 50 percent of losses, and the maximum coverage is 50 million yen for residential buildings and 10 million yen for household belongings.

(3) The Non-Life Insurance Rating Organization of Japan (NLIRO) has the authority to decide the insurance premium rates. The NLIRO sets them at the "no profit and no loss" level, based on losses caused by 375 destructive earthquakes that had happened over a period of 506 years from 1494 to 1999.

(4) The earthquake insurance system is constituted to limit private insurers' risk. When a private insurer assumes an earthquake risk, the insurer reinsures all the risk with the Japan Earthquake Reinsurance Co. Ltd. (JERC), which was established by the Earthquake Insurance Act and is managed by the General Insurance Association of Japan. JERC re-reinsures a small part of these risks with private insurers and a substantial portion of them with the Japanese government (see Figure 3.4).

(5) By law, private insurance firms and the government are protected from any obligation to pay earthquake insurance beyond a predetermined amount of 4.5 trillion yen. Article 4 of the Earthquake Insurance Act allows them to cut the insurance payments on a pro rata base. For example, if total claims are 9 trillion yen (i.e., twice as much as the maximum amount), the insured would receive only 50 percent of their claims.

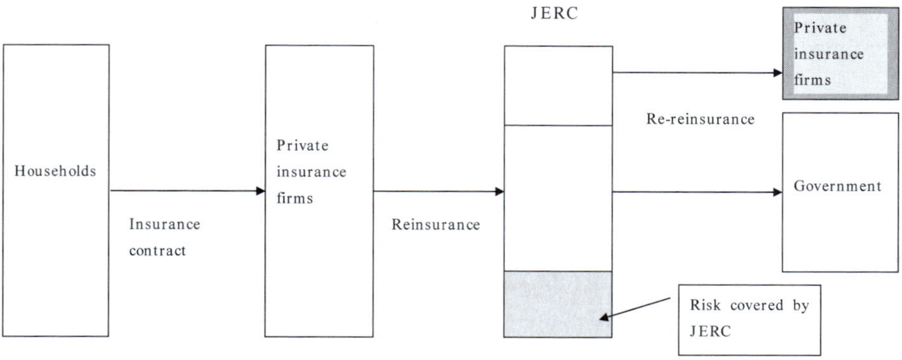

Source: Kinyu Business (Financial Business), Toyo Keizai Shinposha, April 1995.

Figure 3.5. The Structure of Japanese Earthquake Insurance

The government has encouraged households to buy earthquake insurance. However, only 7 percent of households purchased the insurance before the Hanshin-Awaji earthquake and many households faced difficulties in rebuilding their lives. The government has reformed the earthquake insurance by (1) raising the maximum coverage in 1996 from 10 million to 50 million yen for residential buildings and (2) lowering the insurance premiums by approximately 17 percent and introducing a discount scheme based on the building structure in 2001. Owing to these efforts, the prevalence ratio is increasing, but still only 17.2 percent of households purchased earthquake insurance at the end of March 2004.[138]

3.6.2 Policyholder Protection and Resolution of Insurer Bankruptcy

The Failures of Insurers

The convoy system managed by the Ministry of Finance restricted competition in price and products among insurers. The restricted competition had made weak insurers viable before 1997. However, due to the severe economic circumstances and advanced market liberalization, the MOF could not support insurers' survival, and Nissan Life failed in April 1997. As shown in Table 3.26, seven life insurers and two non-life insurers have failed to date. Most failures were attributed to a negative spread (the deficit created by the difference in the guaranteed interest rate to policyholders and the actual investment yields). A remarkable exception is Taisei Fire, which failed because of huge reinsurance payments caused by the September 11 terrorist attacks.

These failures exacerbate public distrust of insurers in Japan, particularly because of their suddenness. Just before the failures, the companies claimed that they were solvent and disclosed incorrect figures. For example, Chiyoda Life disclosed a

[138] The preceding section on earthquake insurance relies on information available at the web sites of the Non-Life Insurance Rating Organization of Japan (www.nliro.or.jp/english/earthquake/index.html) and Japan Earthquake Reinsurance (www.nihonjishin.co.jp/english/index.html). Yamori and Kobayashi (2002) provides a brief view on Japanese earthquake insurance and investigates how the Hanshin Awaji Great Earthquake affected the values of Japanese listed insurers.

solvency ratio of 263.1 percent for March 2000 and Tokyo Life 446.7 percent, but they failed less than one year later.

Table 3.26. Failures of Insurers

April 1997	Nissan Life
June 1999	Toho Life
May 2000	Daiichi Fire
June 2000	Daihyaku Life
August 2000	Taisho Life
October 2000	Chiyoda Life
October 2000	Kyoei Life
March 2001	Tokyo Life
November 2001	Taisei Fire

Policyholder Protection

Although the Japanese government had completely protected all depositors through the blanket deposit insurance until March 2002,[139] it provided only partial protection to insurance policyholders. How to deal with failed insurers and to protect their policyholders is still a hot issue in Japan.

The old Insurance Business Law, effective until March 1996, gave the Minister of Finance the power to order an insurer to merge with failed companies. The new law deprived the Minister of Finance of such power but created the Insurance Policyholders Protection Fund to deal with failed insurers. The fund could assist an insurer who would merge with a failed company. Here, if no company decided to merge with a failed company, the fund could not protect policyholders of the failed company. The Life Insurance and Non-life Insurance Policyholders Protection Corporations were established in December 1998 to shore up the weakness of the funds. All insurers are required to become a member of either corporation and pay contributions, but Kampo and JA Kyosai do not have to join either corporation.

The corporations (1) provide financial aid to the reliever insurer to which the insurance contracts of an insolvent insurer are transferred; (2) undertake the insurance contracts of an insolvent insurer and administer and/or deal with those insurance contracts, when reliever insurers do not appear; and (3) establish a subsidiary (bridge-insurance company) of the corporation to take over the insurance contracts of an insolvent insurer when reliever insurers do not appear. The corporation administers the business operations of the bridge-insurance company.

The scheme for protection of insurance policyholders is as follows: (1) Insurance contracts covered by compensation upon failure are all life insurance contracts (individual insurance, individual annuity, group insurance, and group

[139] Since April 2002, savings and time deposits over 10 million yen were not guaranteed by the Deposit Insurance Corporation, while demand deposits were still fully guaranteed. Also, foreign-currency deposits and negotiable certificates of deposit were not guaranteed at all. Since April 2005, all deposits over 10 million yen except payment deposits, which include checking accounts and ordinary accounts bearing no interest, are not guaranteed. Payment deposits will be guaranteed without limits.

annuity) and some non-life insurance contracts, such as compulsory automobile liability insurance, automobile insurance, fire insurance for individuals or small- and medium-sized enterprises, earthquake insurance, and personal accident insurance.[140] (2) The full value of the policy reserves for compulsory automobile liability insurance and earthquake insurance and 90 percent for other insurance will be compensated for. (3) Contract provisions, such as guaranteed interest rate, may be changed. For example, the Daihyaku Life failure resulted in the decrease of guaranteed interest rates to 1 percent, which caused huge losses for policyholders who had savings-based long-term insurance contracts. In the Daihyaku Life case, the benefits that some policyholders would receive were decreased to less than 30 percent of initially guaranteed benefits.

For non-life insurers, annual contributions to the Non-life Insurance Policyholders Protection Corporation for prior accumulation total 5 billion yen and will be made until the fund reaches 50 billion yen. For life insurers, annual contributions to the Life Insurance Policyholders Protection Corporation for prior accumulation total 40 billion yen and will be made until the fund reaches 400 billion yen. While the amount of each member's annual contribution is decided with due regard to the amount of net premiums written and liability reserves accumulated, payment for the running costs of the corporations are divided equally among members. If compensation exceeds the fund accumulated by the corporations, the corporations are allowed to borrow money from financial institutions. However, the total sum of the funds accumulated and the money borrowed cannot exceed 50 billion yen for the Non-life Insurance Policyholders Protection Corporation (NLIPPC) and 400 billion yen for the Life Insurance Policyholders Protection Corporation (LIPPC). In cases where the costs required for bankruptcy procedures exceed 50 billion yen for NLIPPC and 400 billion yen for the LIPPC, the corporations will consider taking measures, including requesting the authority to review the system in its entirety. Furthermore, the Insurance Business Law allows the government to provide financial support to the LIPPC by March 2006.[141]

---

[140] Therefore, the corporation does not compensate fire insurance (except for individuals and small enterprises), marine and inland transit insurance, credit insurance, aviation insurance, workers' accident compensation liability insurance, general liability insurance, machinery and erection insurance, and movables comprehensive insurance.

[141] The preceding explanation depends on information available at the web site of the General Insurance Association of Japan at www.sonpo.or.jp/english/english.html. The Life Insurance Policyholders Protection Corporation used 366.3 billion yen for Toho Life, 145 billion yen for Daihyaku Life, and 26.7 billion yen for Taisho Life. More recent failures (i.e., Chiyoda, Kyoei and Tokyo) caused no financial costs to the corporation, mainly because the assumed interest rates were substantially reduced and a part of the claims of policyholders was written down. The Insurance Business Law was amended in April 2005, and the policyholder protection scheme was changed. First, while 90 percent value for automobile insurance was compensated under the old law, now 100 percent value will be compensated for in three months after the failure and 80 percent after that. Second, the compensation rates depend on kinds of insurance, guaranteed interest rates, and so on. For example, less than 90 percent value for some insurance contracts will be compensated for. Third, the government emergency support scheme for the LIPPC is extended to March 2009.

3.7 CONCLUDING REMARKS

The Japanese insurance market has drastically changed since 1990 due to extensive liberalization and changes in the economic and social environments. This chapter explains not only how the Japanese insurance market and insurers have changed, but also why they have changed and, in some respects, have not changed. We analyze several key aspects of the Japanese insurance market and Japanese insurers that make them unique.

The Japanese insurance business is facing a major turning point in its history, and several important factors will affect the future of the insurance industry in Japan. First, the rapid aging of the population is particularly important for life insurers. As many people reach retirement age and become insurance and annuity beneficiaries, the assets of life insurers will decrease.

Second, the public pension and health insurance systems face financial difficulties. Some observers argue that current schemes are almost bankrupt, so payments of public pension and medical benefits will be reduced. The government has to encourage the public to purchase pension and insurance from private insurers. Third, global warming may cause huge natural disasters, several catastrophic typhoons and downpours have occurred since the 1990s, and large earthquakes are predicted to occur in the coming years. It is getting harder for Japanese non-life insurers to manage their own risk exposures. At the same time, as consumers and firms need risk management services in more and more uncertain environments, non-life insurers will find more business opportunities. Fourth, the development of information technology is changing the distribution channels. Insurers should utilize these new technologies to decrease operational expenses further and develop more efficient distribution channels.

Fifth, further reform of the policyholder protection scheme is greatly needed. For example, the General Insurance Association of Japan proposes that the insurance contracts of failed companies are fully guaranteed for only three months after the failure and are not guaranteed at all after that period. We have to investigate what protection scheme maximizes policyholders' utility and minimizes the burdens of surviving insurers. Finally, mega-bank mergers have occurred since 2000. Already, some insurers have a close alliance with mega-banks that aim to form financial conglomerates. Insurers will have to choose to keep their independence or join one of the mega-bank groups.

3.8 REFERENCES

Hecker, JayEtta Z. U. S.–Japan Trade: U. S. Company Views on the Implementation of the 1994 Insurance Agreement. Collingdale, PA: Diane Publishing, 1996.

Hoken Kenkyujo (Insurance Research Institute), Insurance: Statistics of Life Insurance Business in Japan (various issues; in Japanese), Tokyo.

Hoken Kenkyujo (Insurance Research Institute), Insurance: Statistics of Non-Life Insurance Business in Japan (various issues; in Japanese), Tokyo.

Hsu, Robert, 1999. "Keiretsu and Business Groups." In The MIT Encyclopedia of the Japanese Economy. Boston: MIT Press.

Kitamura, Masashi. "Commentaries on the Insurance Business Law of Japan (XVIII)." *JILI Journal* No.143 (in Japanese), 2003.

Mizushima, Kazuya. *Gendai Hoken Keizai* (Modern Insurance Economy), 6th ed., Tokyo: Chikura Shobo (in Japanese), 1999.

Nakatani, Iwao. "The Economic Role of Financial Corporate Grouping." In The Economic Analysis of the Japanese Firm, ed. Masahiko Aoki. Amsterdam: North-Holland, 1984.

Ninomiya, Shigeaki, ed. Zusetsu Nihono Seimeihoken (Japanese Life Insurance), Tokyo: Zaikei Shyohosha, 1997.

Ohkawa, K., N. Takayama, and Y. Yamamoto, eds., Estimates of Long-Term Economic Statistics of Japan Since 1868, Vol. 1. Tokyo: Toyo Keizai Shinposha, 1974.

Preston, Claire. Japanese Insurance: The Impact of Deregulation. London: FT Finance/Pearson Professional, 1998.

Swiss Re. "World Insurance in 2003: Insurance Industry on the Road to Recovery," Sigma No. 3, 2004.

Takagi, Hidetaka, and Hiroki Nakanishi, eds. Songai Hoken Dokuhon (Reading on Non-Life Insurance), 4th ed. Tokyo: Toyo Keizai Shinpo-sha (in Japanese), 1999.

Yamashita, Tomonobu, Osamu Takehama, Hiroshi Suzaki, and Tetsuo Yamamoto. Hoken Hou (Insurance Law), 2d ed., Tokyo:Yuhikaku (in Japanese), 2004.

Yamori, Nobuyoshi. "An Empirical Investigation on Japanese Corporate Demand of Insurance." Journal of Risk and Insurance 66(2), 1999, 239–252.

Yamori, Nobuyoshi, and Takeshi Kobayashi. "Do Japanese Insurers Benefit from a Catastrophic Event? Market Reaction to the 1995 Hanshin-Awaji Earthquake." Journal of Japanese and International Economies 16(1), 2002, 92–108.

Yamori, Nobuyoshi, and Takeshi Kobayashi. "Does Regulation Benefit Incumbent Firms? An Investigation of Japanese Insurance Market Deregulation." Journal of Insurance Regulation 22(4), 2004, 35–48.

3.9 USEFUL WEB SITES

- Financial Services Agency
  www.fsa.go.jp/index.html
- Life Insurance Policyholders Protection Corporation of Japan
  www.seihohogo.jp
- Japan Earthquake Reinsurance
  www.nihonjishin.co.jp/english/index.html
- Japan Institute of Life Insurance
  www.jili.or.jp/index.html
- Japan Post
  www.japanpost.jp/top/english/english-index.html
- Non-Life Insurance Policyholders Protection Corporation of Japan
  www.sonpohogo.or.jp

- Foreign Non-Life Insurance Association of Japan
  www.fnlia.gr.jp
- General Insurance Association of Japan
  www.sonpo.or.jp/english/english.html
- Japan Cooperative Insurance Association Incorporated
  www.jcia.or.jp/en/pdf.html
- Life Insurance Association of Japan
  www.seiho.or.jp/english/index.html
- National Federation of Workers and Consumers Insurance Cooperatives (Zenrosai)
  www.zenrosai.or.jp/english
- National Mutual Insurance Federation of Agricultural Cooperatives (JA Kyosai)
  www.ja-kyosai.or.jp/about/annual/e_index.html
- Non-Life Insurance Rating Organization of Japan
  www.nliro.or.jp/english/automobile/index.html
- Tokyo Stock Exchange
  www.tse.or.jp/english

3.10 LEXICON

Basic Profit. Anew index disclosed since March 2003 is a part of disclosure reform for life insurers. Basic profit is defined as current profits less both securities sales profit and loss and temporary profit and loss. Therefore, basic profit reflects the profitability of insurers in terms of their main business.

Financial Services Agency (FSA). The Financial Supervisory Agency was established in 1998 as an administrative organ (more specifically, an external organ of the prime minister's office) responsible for the inspection and supervision of private-sector financial institutions and surveillance of securities transactions. With the establishment of the Financial Reconstruction Commission in December of the same year, the Financial Supervisory Agency became an organization under the jurisdiction of the Commission.

The Financial Services Agency was established in 2000 within the Financial Reconstruction Commission through reorganization of the Financial Supervisory Agency. According to this change, the Financial Services Agency became responsible for planning the financial system that was previously under the control of the Ministry of Finance. By the reorganization of central government ministries, the FSA became an external organ of the Cabinet Office, and after the abolishment of the Financial Reconstruction Commission, the FSA took over operations concerning disposition of failed financial institutions.

Now, the FSA is responsible for ensuring the stability of the financial system in Japan; the protection of depositors, insurance policyholders, and securities investors; and the smooth operation of financial systems through such measures as planning and policymaking concerning the financial system, inspection and supervision of private-sector financial institutions, and surveillance of securities transactions. The FSA performs extremely important roles for the sound development of the national economy. For more information, see www.fsa.go.jp/info/infoe/pamphlet_e.pdf.

General Insurance Association of Japan. The Marine and Fire Insurance Association of Japan was established on April 1, 1946, by all the *domestic* non-life insurance companies. The objective of the Association was to promote the sound development of the non-life insurance industry in Japan. On May 1, 1948, it was reorganized as an incorporated body. On May 20, 2003, it changed its English name to the General Insurance Association of Japan and is composed of 23 member companies (as of October 2004).

The Association engages in the following activities:

- Promotion of a better understanding of non-life insurance and the supply of information about non-life insurance
- promotion of sound development and maintenance of the reliability of the non-life insurance industry
- representing the non-life insurance industry in the presentation of tax reform requests and regulatory reform requests and of opinions to insurance administration
- response to social issues, such as combating automobile theft and taking measures to prevent insurance fraud
- promoting public awareness of fire and disaster prevention and taking measures to prevent and reduce traffic accidents
- international activities, such as promoting dialogue and information exchange with overseas insurance associations, participating in international organizations' activities and international meetings, and providing training programs for and promoting dialogue and information exchange with East Asian regions
- response to environmental issues and promotion of NPO activities
- practicing the earthquake insurance claims settlement seminar and drill.

Foreign non-life insurers organized their own association, the Foreign Non-Life Insurance Association of Japan (FNLIA). The FNLIA consisted of 22 regular members and 3 associate members in February 2005. The General Insurance Association of Japan's Article of the Association is available at www.sonpo.or.jp/e/about_us/pdf/articles.pdf.

Japanese Big Bang. Prime Minister Ryutaro Hashimoto ordered the ministers in November 1996 to start extensive financial reforms in order to allow the Japanese financial market to compete against the New York and London markets. This reform is called the Japanese Big Bang. The reform resulted in the amendment of the Banking Act, Securities and Exchange Law, and the Insurance Business Act as well as many other financial related laws in 1998. Owing to insurance market reforms under the Big Bang, insurance holding companies can be established, insurance premiums for fire and automobile insurance are freely priced by individual insurance companies, and banks can sell insurance products.

The Life Insurance Association of Japan (LIAJ). The Life Insurance Association of Japan was established in 1908. As of February 2005, 40 life insurers belong to the Association. The LIAJ engages in the following activities:

- requesting and representing opinions—for example, the LIAJ submitted its comments to the International Accounting Standards Board (IASB) in 2003.
- social contribution and improving services for policyholders—for example, the Association provides scholarships for foreign students in Japan.
- public relations—the Association publishes several disclosure pamphlets and releases various insurance statistics.
- consultation service—for example, since 1961, the Association has operated the Life Insurance Consultation Center to respond to complaints from policyholders.

- measures taken against moral hazard, such as the Liaison Conference Between Life Insurers and Police, which encourages information sharing between insurance companies and police.
- managing common-to-industry educational system. In 1977, a common-to-industry educational system was founded. Passing the examination under the system is a prerequisite to be registered with the government as solicitors.
- international activities.

For more information on the LIAJ, see www.seiho.or.jp/english/index.html.

Non-life Insurance Policy-holders Protection Corporation of Japan. The Non-life Insurance Policy-holders Protection Corporation of Japan was established to succeed the Policyholders' Protection Fund for Non-life Insurance Companies in December 1998 for the protection of general insurance policyholders. The Insurance Business Law requires all general insurance companies operating in Japan, including foreign insurers and one specific corporation (the Society of Lloyd's), to join the corporation. The members of the corporation must make a contribution to the corporation in order to sustain the policyholders' protection funds and to meet the expenses of the corporation.

The Corporation carries out the following types of business operations:

- provide financial aid to a reliever general insurance company to which the insurance contracts of an insolvent general insurance company are transferred
- provide financial aid to a reliever general insurance company to which the insurance contracts of an insolvent general insurance company are transferred
- undertake the insurance contracts of an insolvent general insurance company and administer the insurance contracts when reliever general insurance companies do not appear
- establish a subsidiary ("bridge-insurance company") of the corporation to take over the insurance contracts of an insolvent general insurer when reliever general insurance companies do not appear. The corporation administers the business operations of the bridge-insurance company
- provide loans to the members of the corporation in the event that they have to stop claims payment to their policyholders due to temporary cash-flow problems
- provide loans to certain policyholders of an insolvent general insurance company within the amount equivalent to claims incurred when the general insurance company has stopped claims payment due to the issuance of an order to suspend its business operations by the supervisory authority
- become an insurance administrator
- purchase policyholders' rights on insurance claims filed with an insolvent general insurance company
- purchase the assets of an insolvent general insurance company.

For more information, see www.sonpohogo.or.jp (available only in Japanese).

Non-Life Insurance Rating Organization of Japan. On July 1, 2002, the merger of the Automobile Insurance Rating Organization of Japan (AIRO) and the Property and Casualty Insurance Rating Organization (PCIRO) resulted in the formation of Non-Life Insurance Rating Organization of Japan (NLIRO) under the Law Concerning Non-Life Insurance Rating Organizations. The Property and Casualty Insurance Rating Organization (PCIRO), whose initial name was the Fire and Marine Insurance Rating Association of Japan, was established as an independent insurance rating organization in 1948. The Automobile Insurance Rating Organization of Japan (AIRO) split off from the Association in January

1964. The members of NLIRO consist of non-life insurance companies licensed under the Insurance Business Law. There are 39 members as of December 2004.

NLIRO has mainly three functions. The first and primary function is to collect large quantities of data from member insurance companies to calculate "reference loss cost rates" (advisory pure risk premium rates) for fire insurance, personal accident insurance, automobile insurance, and nursing care payments insurance as well as "standard full rates" (advisory premium rates) for compulsory automobile liability insurance (CALI) and earthquake insurance and to provide these rates to members.

Its second function is to investigate CALI claims. Under the Automobile Liability Security Law established in July 1955, all drivers are required to have CALI. As independent agencies for conducting fair claims investigation, the CALI Claims Survey Offices have been set up. The third function is a data bank. Insurance data used to calculate Reference Loss Cost Rates and Standard Full Rates as well as other insurance data are collected and used to conduct research and analysis of various risks. The results are provided to members and to the general public. For more information, see www.nliro.or.jp/english/index.html.

Nonperforming Loans. Japanese banks suffered from huge nonperforming loans in the 1990s and have been required to disclose the amounts of nonperforming loans. Insurers are also required to disclose the amounts of nonperforming loans in the same way as banks. The nonperforming loans ("risk management loans") are classified into three categories, depending on the probability of repayments. The lowest category is "loans to borrowers in legal bankruptcy." The second category is "past due loans." The third category is "loans past due three months or more," and the fourth category is "restructured loans," which means that the terms of loans (e.g., interest rates) are relaxed from the original contracts. Usually, the sum of these four categories is regarded as total nonperforming loans. However, it is notable that some of them are covered by collaterals and guarantees. Unfortunately, the insurers and banks do not disclose how much they will lose from such nonperforming loans.

Furthermore, insurance companies as well as banks have disclosed nonperforming assets since 1999 based on the Financial Revitalization Law (FRL) of 1998. Because the FRL-based nonperforming assets include not only loans, but also securities loaned, uncollected interest, suspense payment, and customers' liabilities for acceptance and guarantee, there are some discrepancies between the FRL-based nonperforming assets and the risk-management loans. However, they are usually almost the same. For example, the FRL-based nonperforming assets of Nippon Life at the end of March 2004 were 107,902 million yen, while Nippon Life's risk-management loans at that time were 107,808 million yen.

Public Pension. The Workers' Pension Insurance Act was enacted in 1942 and was entirely revised in 1954. In 1961, the National Pension Act became effective to realize the universal pension system in Japan. Unfortunately, due to rapid aging of the population and bad investment performance in the 1990s, the government has been forced to reform the public pension system several times and there remain serious difficulties.

Current Japanese public pension consists of two tiers. The first tier (i.e., the basic pension) is called National Pension Insurance (NPI, or *Kokumin Nenkin*). All individuals aged 20 to 60 are required to participate in the NPI. There were 70.46 million participants as of the end of March 2003. The monthly premium was 13,300 yen as of April 2004.

The second tier is eligible only to employees of private firms and public institutions. Employees of private firms must join the Employees' Pension Insurance (EPI, or *Kosei Nenkin Hoken*), and employees of public institutions must join the Mutual Aid Pension for

National Public Officials. 36.85 million persons participated in one of these second-tier public pensions as of the end of March 2003. The monthly premium depends on income, and the premium rate for EPI as of April 2004 was 13.58 percent of monthly salary. The premium rates are expected to be raised significantly in order to sustain the pension scheme. So moderate increases of premium rates and reduction of pension benefits are inevitable. Self-employed persons and the unemployed cannot join the EPI. One important exception is spouses of employees (usually full-time homemakers). When these spouses earn less than a certain amount, they can enjoy the benefit of the second-tier public pension at no cost. This was introduced because requiring full-time homemakers to pay pension premiums would force them to fail to pay premiums, resulting in no pension benefits for them. These spouses are called Category 3 Insured, while the self-employed are Category 1 Insured, and employees are Category 2 Insured. There were 11.24 million Category 3 Insured persons as of the end of March 2003.

For more information, see Ministry of Health, Labor and Welfare, "Outline of the Japanese Pension System," April 2004 (www.mhlw.go.jp/english/org/policy/dl/p36–37h.pdf.

Solvency Margin (Capital Requirements for Insurers in Japan) Framework

The common framework, the solvency margin standard, was introduced for both life and general insurers in 1996. It has been improved since then.

$$\text{Solvency Margin Ratio} = \frac{\text{Solvency Margin}}{\frac{1}{2} \times \text{Risk Amount}} \times 100$$

Other capital requirements are the net asset requirement, which is that the liquidation value be greater than or equal to zero, and initial capital of greater than or equal to 1 billion yen.

Use of the solvency margin ratio

- Filtering: The FSA can and does continuously monitor and regularly inspect all insurers, since the number of them is relatively small (i.e., 44 for life and 55 for general). Therefore, unlike other countries, the solvency margin ratio is not used to choose those to which the FSA should pay special attention.
- Early warning system: The FSA monitors the following indicators of insurers in addition to the solvency margin ratio. It requests insurers to report further information and orders them to improve their operation when necessary, even if a solvency margin ratio of the company is well above the regulatory minimum of 200 percent.
  (i) Profitability—profit breakdowns and their projections
  (ii) Credit risk management—concentration of major credit exposures, etc.
  (iii) Market risk management—impact of price fluctuation of securities, etc.
  (iv) Liquidity risk management—asset portfolio and trend of policies, etc.
- Prompt corrective action: If the solvency margin ratio is less than 200 percent, it triggers various prompt corrective actions.
- Disclosure: Not only a solvency margin ratio itself but its components of every insurance company are disclosed to the public.

Solvency margin (numerator of solvency margin ratio)

- Capital
  - (i) Revaluation differentials on securities are deducted to avoid double counting.
- Capital-like provisions
  - (i) Price fluctuation provisions—certain percent of assets is set aside for price fluctuation risks
  - (ii) Contingency provisions—certain percent of premium or net amount at risk is set aside for insurance risks and major catastrophe risks
  - (iii) General loan losses provisions
- Unrealized gains/losses on securities and real estate
  - (i) Some of unrealized gains/losses are on balance sheet
  - (ii) Net unrealized gains/losses on real estate (multiplied by 85% if gains; multiplied by 100% if losses)
- Subordinated debts
- Deduction: Stocks and subordinated debts invested in insurers or any subsidiaries
- Margins for prudence contained in statutory provision

Risk amount (denominator of solvency margin ratio)

- Insurance risks
  Case assumed: insurance claims payment is higher than normal expectations.
  Amount measured: certain level of insurance claims payment minus normally expected level.
- Assumed interest risks
  Case assumed: investment income earned is lower than originally assumed income.
  Amount measured: expected amount of the gap.
- Asset management risks
  - (i) Price fluctuation risks
    Case assumed: capital loss is higher than normal expectations.
    Amount measured: amount at risk with a 90 percent probability.
  - (ii) Credit risks
    Case assumed: counterparty defaults.
    Amount measured: expected amount of loss (including that from credit derivatives).
  - (iii) Other risks for subsidiaries, derivative transaction, reinsurance and reinsurance recoverable
- Major catastrophe risks (general insurance only)
  Case assumed: a natural disaster strikes.
  Amount measured: amount of damage caused by the largest earthquake or typhoon
- Operational risks
  Case assumed: something not in the above categories happens.
  Amount measured: 2 or 3 percent of the total of the other risks.
- Correlation of risks is taken in account to some extent in the following way:

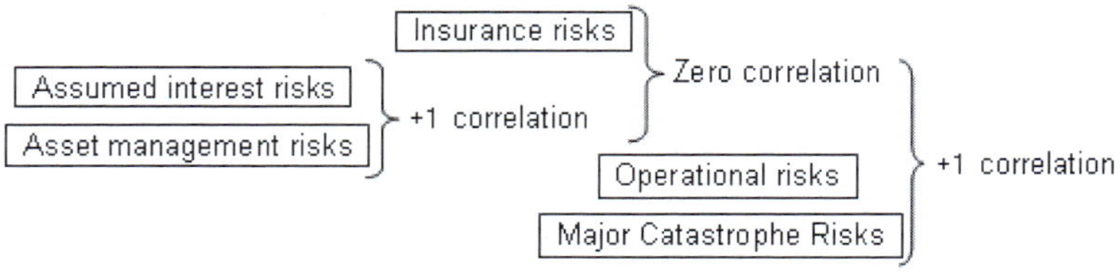

For more information, see www.fsa.go.jp/refer/ins/capital.html.

The Third Area (or Third Sector). In Japan, insurance was traditionally classified into life insurance and non-life insurance. Life insurance is named the first sector, and non-life insurance is named the second sector. The third sector includes such insurances as cancer, medical, and personal accident insurance that are not classified into life or non-life insurance. Due to deregulation, both life insurers and non-life insurers now can treat all insurance products of the third sector.

4 The UK Insurance Industry—Structure and Performance

Professor Philip Hardwick

Bournemouth University

Michel Guirguis

Bournemouth University

4.1 INTRODUCTION

The earliest insurance to be written in the UK was marine insurance, which was introduced into England in the 14th or 15th centuries by the Lombards from northern Italy. Alongside this early development of marine insurance came the earliest life insurance policies, the very first of which is thought to have been written in the late 16th century on the life of a merchant sailing with his goods. Fire insurance in the UK started a little later, with the first policy probably being written by the Phoenix Insurance Company (originally named the Fire Office) in 1680, just fourteen years after the Great Fire of London in September, 1666. The Great Fire destroyed much of medieval London and was an important stepping-stone in the development of fire insurance and organized fire fighting. In June 1861, the great Tooley Street fire occurred in the London docks, destroying warehouses, wharves, shops, offices and even ships. After paying out a total of about £2 million, insurance offices not surprisingly began to impose more restrictive terms on warehouses and wharves and to apply differential rates to encourage owners to think about fire precautions.

It was in the field of marine insurance that Lloyd's of London was founded in the 17th century in the London coffee house owned by Edward Lloyd. In 1760, the Register Society was formed by the customers of the coffee house and printed the first register of ships in 1764 to give both underwriters and merchants an idea of the condition of the vessels they insured and chartered. In 1834, the organization was reconstructed as Lloyd's Register of British and Foreign Shipping. For more than two centuries, safety and reliability have been assured through the Lloyd's Register's surveys of the quality of materials, workmanship and construction.

From these early beginnings, the UK insurance market has grown to become the biggest in Europe and the third largest in the world (after the United States and Japanese markets), accounting for 9 percent of total worldwide premium income.

The UK life insurance market is the largest in Europe and the third largest in the world, while the UK general insurance (i.e. property-liability insurance) market is the second largest in Europe (behind Germany) and the fourth largest in the world (Association of British Insurers, 2004).[142] The UK life insurance market accounts for approximately 33 percent of the total European life market, while the UK general insurance market accounts for approximately 20 percent of the European general insurance market. The UK insurance industry's invested assets account for nearly 34 percent of all European insurance industries' invested assets, equal to that of the German and French industries combined. In addition, UK premium income as a percentage of GDP is the highest in Europe and second only to the United States in the world rankings.

The insurance industry is an important contributor to the UK economy, a major employer and a significant source of overseas earnings. The popularity of insurance as a savings vehicle in the UK, as well as its protective role, gives the UK the highest per capita insurance expenditure in the European Union (EU) and the third highest in the world. According to the Office for National Statistics, average annual *per capita* expenditure in the UK on general insurance increased from £379 in 1991 to £794 in 2002/3. *Per capita* expenditure on life insurance products, excluding personal pensions, also increased steadily from £386 in 1991 to £418 in 2002/3. If personal pensions are included, an even larger increase is recorded, from £600 in 1991 to £780 in 2002/3. As further indicators of its importance to the economy, the industry accounts for more than 20 percent of investment in the stock market, and pays out an average of about £247 million a day in pensions and life insurance benefits and about £50 million a day in general insurance claims.

The total number of people employed in the UK insurance industry was 216,300 in 2003, with an additional 131,700 employees in industries that are auxiliary to the insurance industry. This accounted for more than a third of all jobs in the UK financial services sector. More than a half of all employees in insurance and its auxiliary industries were female in 2003 (a total of 192,400) and about 79,000 employees in insurance and the auxiliary industries in 2003 were part-time (Association of British Insurers, 2004).

During the period 1998–2003, there was an overall increase in the worldwide total premium income written by UK-based insurance companies, from £123 billion in 1998 to £153 billion in 2003 (of which £113 billion was earned from life insurance business and £40 billion was earned from general insurance business). In 1998, the total UK business of UK-based insurance companies amounted to £95.7 billion (of which £72.3 billion was earned from life insurance and £23.4 billion was earned from general insurance). By 2003, the total UK business of insurance companies within the UK had increased to £120 billion (of which £90.3 billion was from life insurance and £29.7 billion was from general insurance). UK life insurers also derive a growing proportion of their income from overseas, with just over 20 percent of the total market for life insurance business coming from overseas in 2003. In terms of premium income written, overseas life insurance business increased from

[142] Life (or long-term) insurance business in the UK consists of life insurance, pensions, annuities and income protection insurance (formerly known as permanent health insurance). General (or non-life) insurance consists primarily of motor, marine and aviation, property, pecuniary loss, accident and health and third-party liability insurance.

£15 billion in 1998 to £22.6 billion in 2003, with the European Union (44 percent) and the United States (27 percent) representing the biggest markets. Overseas business in general insurance is more variable from year to year and premiums actually fell from £13.7 billion in 1998 to £10.2 billion in 2003, with the European Union (44 percent), the United States (18 percent), Canada (15 percent) and Australia (11 percent) representing the biggest markets. (Association of British Insurers, 2004)

The main purpose of this chapter is to discuss the structure and performance of the UK insurance industry. In the next section, we examine the development of the regulatory structure facing insurance companies in the UK, culminating in the establishment of the Financial Services Authority in 2001. Then we analyse the industry structure by calculating and comparing concentration ratios for different sectors of the UK insurance industry. We also describe and examine the principal economic actors in UK insurance, including the largest mutual and proprietary life and general insurance companies, the bancassurers and the London market. The next section focuses on product structure, including a summary of the main lines of life and general insurance offered by UK companies, and describes recent changes in the distribution channels used. Then we concentrate on the international activities of UK-based insurers and attempt to assess the international competitiveness of the industry by calculating a set of net export ratios and 'revealed comparative advantage' indices. Finally, we present the results of a comparative study of the relative cost efficiency of UK life and general insurance companies. In this study, data envelopment analysis is employed to estimate a set of cost efficiency scores for a sample of 50 life and 50 general UK insurance companies. The chapter ends with a brief conclusion.

4.2 REGULATION AND SUPERVISION

Whilst the activities of the insurance industry are undoubtedly of great benefit to the UK economy, the consequences of insurance companies' failing to meet their contractual obligations are potentially extremely serious and far-reaching. In the 19th century, fraud and mismanagement led to frequent failures by UK insurance companies, with serious consequences for policyholders and adverse effects on the reputation of the industry.[143] Even so, it was not until the mid-20th century that a series of acts of parliament began to recognize the need for regulations to protect policyholders. The Insurance Companies Act (1982) (which consolidated the earlier 1974 Act) brought in minimum solvency margins and defined conditions for an insurer to be authorized to transact business in the UK. The overall objective of the legislation was to ensure that only fit and proper persons should transact insurance business

Until 2001, the prudential supervision of UK insurance companies under the Insurance Companies Act (1982) was carried out originally by the Department of Trade and Industry and later by the Treasury. In addition, the promotion and sale of

---

[143] For a survey of recent research on the history of the UK insurance industry, see Pearson (2002).

investment business (which included all life insurance business other than pure protection business) was regulated by the Financial Services Act (1986), which came into effect in April 1988. This act made it a criminal offense to transact investment business without appropriate authorization.

The underlying principles of the Financial Services Act were:

- To maintain confidence in the UK financial system.
- To promote public understanding of the financial markets.
- To reduce the possibilities for authorized business to be used for financial crime.
- To protect consumers.

In the 1980s and 1990s, the system of investor regulation in the UK was one of self-regulation within a statutory framework. The system proved to be unwieldy with several overlapping self-regulatory organizations. For life insurance companies after 1995, the system was simplified somewhat with the establishment of the Personal Investment Authority, whose rules regulated the promotion and sale (conduct of business) of life insurance business. The Personal Investment Authority replaced the Financial Intermediaries, Managers and Brokers Regulatory Association and the Life Assurance and Unit Trust Regulatory Organisation, among other self-regulatory organizations. The Personal Investment Authority's objectives were:

- To reinforce high standards of integrity, fair dealing and competence.
- To make full and proper use of the powers available to protect investors.
- To set rigorous standards of training and professional competence.
- To help investors to protect their own interests by establishing demanding standards of relevant disclosure.
- To provide effective mechanisms for the handling of investor complaints and for securing redress for investors who have been disadvantaged.

In 1997, the UK government announced its intention to set up a single regulator for all financial services (including, for example, banks, investment management and securities traders, as well as insurance companies) to be known as the Financial Services Authority. This was given effect by the passage of the Financial Services and Markets Act (2000), which replaced the Insurance Companies Act (1982) and the Financial Services Act (1986), as well as primary legislation relating to other financial services business. The Financial Services and Markets Act is largely enabling legislation with much of the detail of the previous legislation now covered by the Financial Services Authority, which assumed its full powers from 1 December 2001. In addition, various ombudsman schemes were combined into the Financial Services Ombudsman Scheme and various compensation schemes, including the Policyholders' Protection Board, were combined in the Financial Services Compensation Scheme.

The Financial Services Authority is now the statutory body with responsibility for regulating UK financial services. Its main objectives are to maintain confidence in the UK financial system, promote public understanding of the UK financial system, protect consumers and reduce the incidence of financial crime. Hence it attempts to ensure that UK insurance companies maintain adequate financial strength and are run by 'fit and proper' persons. All UK-based insurance companies have to submit annual returns to the Financial Services Authority, which has formal powers of intervention for firms in financial difficulties.

4.2.1 EU Directives

The fundamental aim of EU legislation under the Treaty of Rome (1957) was to lay the foundations of an ever-closer union among the peoples of Europe. The aim to create a single market applied equally to insurance. As a member of the European Union, the UK is subject to the directives that are issued by the European Parliament. These directives, developed in consultation with the member states, are always looking to develop on the underlying principles of the freedom of establishment and the freedom to provide services.

Three main life and non-life directives have been issued aimed at developing the European single market. The existing solvency margin requirements were established in 1973 under the First Non-Life Directive (73/239/EEC) and in 1979 under the First Life Directive (79/267/EEC). The Third Generation of Life (92/96/EEC) and Non-Life (92/49/EEC) Insurance Directives established the 'single market' for insurance in the mid-1990s. This gave the EU one of the most competitive insurance markets in the world. Insurance companies, on the basis of authorization in any member state, are entitled to sell insurance products throughout the EU without any price control or prior notification of terms and conditions (except for compulsory insurance). Under this 'single passport' system, insurance undertakings authorized by prudential authorities in one member state can sell into another member state, either directly (e.g. by telephone or Internet), or by setting up a branch or a subsidiary there.

4.3 THE STRUCTURE OF THE INDUSTRY

4.3.1 Market Concentration

Table 4.1 shows that in 2003, 772 insurance companies were authorized, either by the UK or by another European Union Member State to carry on insurance business in the UK (Association of British Insurers, 2004). Of these, 568 could carry on general business only (such as motor, household and commercial insurance), 159 were authorized for life insurance business only and 45 were composites (authorized for general and life insurance business). Table 4.1 shows that, during the period 1990–2003, the number of general insurers fluctuated from 570 in 1990, up to 599 in 1997, then down to 568 by 2003. Over the same period, there was a decrease in the number of life insurance companies and composites. Specifically, the number of insurance companies in the life sector decreased from 203 in 1990 to 159 in 2003. Similarly, the number of composite companies decreased steadily from 64 in 1990 to 45 in 2003. The decrease in the number of life and composite insurers has mainly been a result of mergers and acquisitions. These include: the merger of Royal Insurance and Sun Alliance to form the Royal and Sun Alliance in 1996; the merger of Commercial Union, General Accident and Norwich Union in 1998 and 2000 to form CGNU (re-named Aviva from 2002); the acquisition of NPI by AMP in 1998; the acquisition of Guardian Royal Exchange by AXA in 1999; and the acquisition of Scottish Widows by Lloyds TSB in 2000.

Table 4.1. Number of Insurers Authorized in the UK, 1990–2003

	General Only	Life Only	Composite	TOTAL
1990	570	203	64	837
1991	570	202	64	836
1992	565	196	62	823
1993	575	194	59	828
1994	573	191	57	821
1995	594	174	58	826
1996	578	177	59	814
1997	599	177	65	841
1998	594	176	62	832
1999	596	171	62	829
2000	597	165	60	822
2001	594	160	56	810
2002	592	160	54	806
2003	568	159	45	772

Source: Association of British Insurers (2004)

The UK insurance industry has also become more concentrated. In life insurance, the top ten companies now account for more than 50 percent of the market. In each of the individual segments of the general insurance sector, the top ten companies account for more than 70 percent of the market. The growing number of mergers and acquisitions has affected the degree of concentration in recent years. Some indication of the changing structure of the market for life insurance and general insurance in the UK may be gained from an examination of concentration ratios.

For life insurance, Table 4.2 shows the market shares of the largest life insurance company (CR1), the top five life insurance companies (CR5) and the top ten life insurance companies (CR10), together with the Herfindhal index, in 1993, 1998 and 2003. The share of the largest company increased from 10.6 percent to 14.5 percent during the eleven-year period (after a fall to 9.9 percent in 1998), while the share of the top five increased from 35.2 percent to 37.4 percent, and the top ten from 51.3 percent to 54.8 percent. At the same time, the Herfindhal index (defined as the sum of the squared market (percentage) shares of all firms in the industry) increased from 400 to 500. If all the firms in the industry were of equal size, this would be equivalent to a fall in the number of firms from 25 to 20, suggesting a rise in the degree of concentration in the UK life insurance industry.

Table 4.2. Concentration Ratios for the UK Life Insurance Market, Selected Years

Concentration Ratios	1993	1998	2003
CR1	10.6	9.9	14.5
CR5	35.2	36.9	37.4
CR10	51.3	56.2	54.8
Herfindahl index	400	400	500
Equivalent no. of firms	25	25	20

Source: calculated by the authors, using data from Tillinghast-Towers Perrin (2005)

The results summarized in Table 4.3 suggest that the motor insurance market may be the most competitive in the UK general insurance industry. This conclusion is supported by a recent study of the degree of competition in UK general insurance markets by Diacon and Drake (2004). Using the Rosse-Panzar statistic, they report results that suggest that all general insurance markets in the UK "display evidence of monopolistic competition" (p. 11) and that "as might have been expected, the UK motor insurance market appears to be the most competitive overall" (p.10). In general, the UK insurance industry does not exhibit as much competition as might be inferred from a simple examination of the number of firms. This is probably a result of relatively high degrees of asymmetric information and product differentiation.

Table 4.3 summarizes the concentration ratios (CR1, CR5 and CR10) and Herfindhal indices in 1992, 1998 and 2003 for the four main categories of general insurance: accident and health, motor, property and third-party liability. It is clear from these figures that the least concentrated general insurance sector in 2003 was the motor market, where the largest company (Norwich Union) controlled just 11.1 percent of the market, the top ten companies controlled 71.1 percent and the Herfindhal index in 2002 was 600. The most concentrated general insurance sector in 2003 was the accident and health market, where the top company (BUPA) controlled over a third of the market, the top ten companies controlled 78.8 percent and the Herfindhal index was 1,700, equivalent to only six equally sized firms in the industry. It is interesting to see from Table 4.3 that, in contrast to the life insurance market, the general insurance market in the UK became *less* concentrated over the period from 1992 to 2003. In all four sectors highlighted, the Herfindhal index decreased, with the biggest decrease occurring in the accident and health insurance market. In 1992, BUPA was the dominant provider of accident and health insurance in the UK with a market share of 54.5 percent. In 1998, AXA had become the dominant provider with a market share of 26.2 percent, and by 2003, while BUPA and AXA were still the major players, significant increases in market shares had been achieved by other insurers, such as Norwich Union, Lloyds TSB and Royal and Sun Alliance.

Table 4.3: Concentration Ratios for the UK General Insurance Market, Selected Years

	1992	1998	2003
Accident and Health			
CR1	54.5	26.2	34.3
CR5	76.3	72.7	67.1
CR10	91.6	87.7	78.8
Herfindahl index	1,400	1,400	1,700
Equivalent no. of firms	3	7	6
Motor			
CR1	16.9	20.7	11.1
CR5	66.3	63	46
CR10	82.5	81.2	71.1
Herfindahl index	90	100	600
Equivalent no. of firms	11	10	17
Property			
CR1	33.5	35.5	20
CR5	68.7	65.3	54.6
CR10	86.2	81.5	74
Herfindahl index	1,700	1,600	800
Equivalent no. of firms	6	6	13
Third Party Liability			
CR1	17.4	26.7	16
CR5	64.2	54.9	52.3
CR10	78	72	70.2
Herfindahl index	1,000	1,100	700
Equivalent no. of firms	10	9	14

Source: Calculated by the authors, using data from Tillinghast-Towers Perrin (2005)

4.3.2 UK Suppliers of Insurance

The ownership structure of UK insurance companies is complicated by the existence of several large groups, which comprise more than one legal entity. For example, the Royal London Mutual Group includes Refuge Assurance, Scottish Life Assurance and United Friendly Life Assurance, among others. In addition, some companies are proprietary (or stock) companies, which are registered under the Companies Act and owned by shareholders, while others are mutual companies, owned by their policyholder members. Furthermore, some of the proprietary companies are 'bancassurers', which are banking (or building society) groups that have diversified into insurance (mainly life insurance). Both the UK life and general insurance sectors are dominated by proprietary companies, with mutual companies earning only about 15 percent of premiums in life insurance and less than 5 percent of premiums in general insurance in 2003. Bancassurers had larger shares of both markets (about 20 percent and 15 percent respectively). Bancassurers in the UK have been formed mainly through joint ventures and acquisitions, both of which allow the bank or building society to call upon an insurer's expertise in starting up the operation. The major bancassurers in the UK, together with their 2003 premium incomes, are: HBOS (£6.1 billion), Lloyds TSB (£4.8 billion), Abbey National (£1 billion), Barclays Life (£0.5 billion) and HSBC Life (£0.5 billion).

Tables 4.4 and 4.5 present a summary of the UK's ten largest life and ten largest general insurance groups respectively in 2003. The tables also show the country of the ultimate owners of each group, their organizational forms (mutual or proprietary) and the principal distribution channels used.

Table 4.4. Top Ten UK Life Insurance Groups, 2003

Group	Premiums (£m)	Ownership	Distribution
AVIVA	8,184	UK proprietary	I, D, O
Standard Life	7,175	UK mutual	I
HBOS	6,072	UK proprietary bancassurer	B
Prudential	5,420	UK proprietary	I, D, O
Lloyds TSB Group	4,786	UK proprietary bancassurer	B
Legal & General	4,162	UK proprietary	I
AXA	3,664	UK proprietary	I, D
Aegon	2,949	Dutch proprietary	I
Zurich Financial Services	2,750	Swiss proprietary	I
Friends Provident	2,507	UK proprietary	I, D, O

Key:

I: Independent intermediaries
D: Direct sales
O: Own sales force / tied agents
B: Bank/building society

Source: Association of British Insurers (2004) and Standard and Poor's (2004)

It is clear from Table 4.4 that, with the notable exception of Standard Life (a mutual company), the largest life insurers in the UK are the proprietary insurance groups and bancassurers. The top four life insurance groups had premium incomes in excess of £5 billion in 2003, while the top ten all had premium incomes in excess of £2.5 billion. A similar pattern is evident in Table 4.5, where we see that only one company in the top ten general insurers is a mutual (NFU Mutual). Among the top ten, just one group had premium income in excess of £5 billion (Norwich Union) and four had premium incomes in excess of £2.5 billion.

Table 4.5. Top Ten UK General Insurance Groups, 2003

Group	Premiums (£m)	Ownership	Distribution
Norwich Union	5405	UK proprietary	I, D, O
RBS Group	3767	UK proprietary bancassurer	B
Royal & SunAlliance	2726	UK proprietary	I, D
Zurich Financial Services	2547	Swiss proprietary	I
AXA	2392	UK proprietary	I, D
Allianz Cornhill	1352	German proprietary	I, D
BUPA	1240	UK proprietary	I, D, O
NFU Mutual	715	UK mutual	I, D, O
Co-operative	636	UK proprietary	D
Lloyds TSB	588	UK proprietary bancassurer	B

Key:

I: Independent intermediaries
D: Direct sales
O: Own sales force / tied agents
B: Bank/building society

Source: Association of British Insurers (2004) and Standard and Poor's (2004)

Other categories of insurance companies found in the UK include:

- *Foreign Companies* A number of insurance companies with colonial origins have operated in the UK market for many years, such as Sun Life of Canada and Colonial Mutual. In addition, in recent years, many foreign companies have either set up subsidiaries or taken over established UK insurers. Examples include Zurich Financial Services and Allianz.
- *Friendly Societies* In 2003, there were 227 friendly societies registered in the UK (excluding branches). Friendly societies are mutual organizations that were originally unincorporated, but which may now be incorporated under the Financial Services Act of 1992. Whether incorporated or not, friendly societies are permitted to offer a more limited range of products than is the case for mutual and proprietary insurance companies. UK friendly societies vary enormously in size and offer a range of different product types, including with-profits and unit-linked life and pensions products, income protection insurance and general branch accident and sickness plans. The top three in the UK, together with their 2002 premium incomes are: Liverpool Victoria (£553.8 million), Royal Liver (£262.3 million) and Police Mutual (£177.6 million).
- *Captive Insurance Companies* In its simplest form, a captive can be defined as a wholly owned insurance subsidiary of an organization not in the

insurance business whose primary function is to insure some or all of the risks of its parent. There are now many types of captive, including:

- Single-parent captives, underwriting only the risks of related group companies.
- Diversified captives underwriting unrelated risks in addition to group business.
- Association captives which underwrite the risks of members of an industry or trade association. Liability risks such as medical malpractice are frequently insured in this way.
- Agency captives formed by insurance brokers or agents to allow them to participate in the high-quality risks, which they control.
- Rent-a-captives are insurance companies that provide access to captive facilities without the user needing to capitalize its own captive. The user pays a fee for the use of the captive facilities and will be required to provide some form of collateral so that the rent-a-captive is not at risk from any underwriting losses suffered by the user.
- Special purpose vehicles are used in risk securitization. They are reinsurance companies that issue reinsurance contracts to their parent and cede the risk to the capital markets by way of a bond issue.

Evidence of growth in this market is provided, not only by the increase in the number of captives being formed, but also by the increasing number of domiciles available for their incorporation. Long-standing domiciles, such as Bermuda, the Cayman Islands, Guernsey, the Isle of Man and Luxembourg have been joined by the likes of Vermont, the British Virgin Islands, Gibraltar and Dublin.

- *Reinsurance Companies* Reinsurers are the insurers of insurance companies. They help to eliminate the possibility of a large loss from one occurrence and enable an insurer to accept individual risks in excess of the insurer's own limit. The largest reinsurers operating in the UK, together with their 2002 premium incomes were: GE Frankona (£409 million), Swiss Re (£366 million), Munich Re (£273 million), Hannover Life Re (£44 million), Worldwide Re (£30 million), General Cologne Life Re (£12 million) and Gerling Global Life Re (£12 million).

4.3.3 Lloyd's of London

Lloyd's of London is an insurance market in its own right. The 'members' of Lloyd's, who provide capital to underwrite insurance business, consist of individuals, known as Lloyd's 'names', and (since 1994) companies (including Scottish limited partnerships). The members provide capital to the market through 'syndicates', each of which is run by a managing agent. There were 66 Lloyd's syndicates at the end of 2003, specializing mainly in areas such as marine, aviation, catastrophe, professional indemnity and motor insurance.

Clients place risks in the London market via accredited Lloyd's brokers, who use their experience of the market to negotiate appropriate terms and conditions. By the end of 2003, there were 169 firms of brokers at Lloyd's. The regulation of Lloyd's brokers differs from that of other brokers because of the unique nature of the

Lloyd's market. Insurance business may be placed at Lloyd's only by broker's firms specifically accredited by Lloyd's. Other intermediaries that wish to place business at Lloyd's must employ a Lloyd's broker. The system of accreditation of Lloyd's brokers was introduced in 2001. The criteria for accreditation include membership of the General Insurance Standards Council (a self-regulating organization for intermediaries dealing in all types of general insurance), or the meeting of equivalent standards by overseas brokers, and certain financial requirements.

The period 1980–9 was a period of rapid expansion for the UK financial services sector in general, and Lloyd's of London in particular. During this period, there was a rapid rise in the number of Lloyd's names, which peaked at 32,433 in 1988, all non-corporate individuals. However, the recession in the early 1990s was so severe in the UK that the very survival of Lloyd's was put at risk. Increasing underwriting losses, mainly from United States long-term liabilities and a succession of catastrophe losses, proved to be financially disastrous for a significant proportion of the syndicates and membership. At the same time, rising transaction costs undermined the market's competitive position. A 'task force' was set up to study the problems and find ways of dealing with them. Its recommendations led to regulatory changes designed to strengthen central control over agents and syndicates, a streamlining of business processes, the formation of a reinsurance company (Equitas) to assume liability for outstanding claims from the pre-1993 underwriting years and, most importantly, the admission to the market of corporate members (see Lloyd's of London, 1993). As a result of the admittance of corporate funds, which accounted for about 30 percent of total market capacity in 2003, Lloyd's has largely succeeded in replacing the capacity lost due to the large drop in active individual names during the 1990s.

The terrorist attacks of September 11th in the United States created the largest loss that Lloyd's had ever incurred. By December 2001, the market had set aside reserves for a gross loss of over £6 billion, and a net loss of almost £2 billion (Tillinghast-Towers Perrin, 2005). Following an assessment of the effects of September 11th, the Council of Lloyd's took action to improve its solvency, mainly by imposing a premium levy for 2002 and 2003 at an increased rate of 2 percent for most classes, except for life and UK motor where the rate was 1 percent. This rise in premiums contributed to a general hardening of the market in the period 1999–2002, characterized by rising prices. The London market is particularly susceptible to cyclical variations as it deals in lines of business that are particularly affected by the global underwriting cycle.[144]

Product Structure and Distribution Channels

Product structure

The demand for new insurance products in the UK has experienced important changes in recent years that can be attributed to three major factors. First, demographic changes have created different needs for more tailored products. Second, saving patterns have changed (towards capital market investments and away

[144] For a discussion of the present-day structure of the London market and the recent changes to affect Lloyd's of London, see Sigma (2002).

from low-yield bank deposits) due to changing economic conditions. Finally, although the industry is still characterized by asymmetric information, customer awareness has generally increased, forcing life and general insurance companies to offer more competitive and sophisticated products than previously.

The most significant demographic trend in western countries in recent years has been the decline in population growth rates and the associated aging of the population. A major reason for this change is a decrease in birth rates, but the continuing increase in life expectancy in developed economies due to advances in medical care and better living conditions has also contributed to aging (Hoschka, 1994). As a result, the dependency ratio (that is, the proportion of retired people being supported by the working population) has been rising. This demographic effect, together with an upward trend in the sales of 'defined contribution' pensions (in place of 'defined benefit' pensions), is likely to increase the demand for annuities by around 2 percent per year over the next ten years[145] (Stark, 2002). This creates an opportunity for UK insurers to expand sales of investment-linked annuities and to develop price-indexed annuities.

Saving patterns have also changed considerably in recent years, due both to demographic changes and to higher prosperity and higher expectations of standards of living. As a consequence, demand for long-term, high-yield financial savings has been increasing. In addition, a growing emphasis on capital market investments has made sophisticated instruments more attractive than low-yield bank deposits (Lafferty, 1991). Lower inflation rates have also made long-term investments more secure and have tended to boost longer-term contractual savings, such as insurance, to the detriment of short-term deposits. A further incentive to save came in 1999 when the UK government introduced Individual Savings Accounts, in which savers can hold cash deposits, equity or bond market investments and life insurance policies, and benefit from tax exemptions. For example, a saver is allowed to hold a single premium life insurance policy within an Individual Savings Account (up to a maximum investment of £7,000), with the advantage that all income is exempt from income tax and any capital gains are exempt from capital gains tax.

Life insurance business in the UK consists of a range of different products. These include linked and non-linked, with-profit and non-profit life insurance and general annuities, pensions and income protection policies.[146] The total net premiums written in the main lines of UK life insurance business (i.e. life insurance policies, annuities, pensions and income protection policies) from 1990 to 2003 are summarized in Table 4.6. It is evident from the table that premiums from life insurance business, while following a general upward trend during the period, declined each year from 2000 to 2003. Also, the percentage of total premiums for life insurance and annuities business declined from about 42 percent in 1990 to just 32 percent in 2003. Over the same period, premiums from pensions and income

---

[145] A defined contribution pension provides benefits based the amount contributed and is affected by income, expenses, gains and losses. A defined benefit plan guarantees benefits, which are usually based on the participant's salary and length of service.

[146] In 'unit-linked' life insurance or pensions in the UK, the value of the policyholder's fund is linked directly to the value of investment units, such as units in an equity or property portfolio. In 'with-profit' life insurance or pensions, policyholders receive bonuses paid out of the total investment income earned by the insurance company.

protection insurance increased steadily. Overall, net premiums written in life insurance business rose from £32.2 billion in 1990 to £89.8 billion in 2003.

Table 4.6. Main Lines of Life Insurance Business in the UK, 1990–2003

	Net premiums written (£ million)				
	Life Insurance	Annuities	Pensions	Income Protection	TOTAL
1990	12,858	713	18,318	308	32,197
1991	15,391	1,182	21,329	343	38,245
1992	17,886	988	22,628	400	41,902
1993	20,507	938	23,040	438	44,923
1994	19,748	655	20,961	461	41,825
1995	20,784	380	22,072	531	43,767
1996	24,630	851	26,717	597	52,795
1997	26,629	413	32,622	672	60,336
1998	30,019	423	40,266	968	71,676
1999	36,470	128	49,891	1,270	87,759
2000	38,768	199	75,429	1,618	116,014
2001	37,567	181	54,067	1,832	93,647
2002	35,747	209	58,416	1,982	96,354
2003	28,639	244	58,408	2,529	89,820

Source: Association of British Insurers (2004)

Increasing customer awareness and changes in lifestyles have resulted in greater market segmentation. In the non-life sector, consumers tend to opt for simpler and more transparent products, which has led to a further standardization of non-life products. Many common non-life products, such as car insurance and home insurance, require little advice. As a result, it is likely that there will be a growing market for the direct delivery of cost-effective standardized products. The total gross premiums written on the main lines of general insurance business during the period 1990–2003 are shown in Table 4.7. Premiums from all main lines of business increased over the period, with the biggest increases occurring in general liability insurance (up by 156 percent between 1990 and 2003) and pecuniary loss insurance (up by 214 percent between 1990 and 2003). Overall, net premiums written in general insurance rose from £24.3 billion in 1990 to £49.3 billion in 2003.

Table 4.8 summarizes the underwriting and overall trading results for UK general insurers by line of business over the period 1990–2003. Two features of the table are worthy of comment. First, the underwriting result is usually negative for most lines of business, with income from investments making up the shortfall and creating a positive trading result in most years. Secondly, the table provides evidence of an underwriting cycle in UK general insurance: there are regular up and down movements in both the underwriting results of the individual lines of business and in the overall trading result. A casual examination of the figures in this rather short series of data suggests that the cycle has a length of between six and ten years, which is typical of underwriting cycles observed in other countries.[147]

Table 4.7. Main Lines of General Insurance Business in the UK, 1990–2003

	Net premiums written (£ million)							
	Motor	Accident and health	Property	General Liability	Pecuniary Loss	MAT	Other	Total
1990	5,616	2,192	5,462	1,684	1,197	3,591	4589	24,331
1991	6,079	2,405	5,767	1,701	1,394	4,203	5633	27,182
1992	7,033	2,792	6,446	1,767	1,619	5,323	6869	31,849
1993	7,840	2,803	6,975	2,053	1,821	5,067	7097	33,656
1994	7,833	2,836	7,344	2,280	1,901	4,693	5994	32,881
1995	7,179	2,825	6,999	2,264	2,074	4,535	6273	32,149
1996	7,068	2,953	6,667	2,112	2,181	3,640	6371	30,992
1997	7,183	3,239	6,899	2,091	2,253	3,702	6867	32,234
1998	7,761	3,390	7,067	2,034	2,461	3,275	6745	32,733
1999	8,328	3,437	6,573	1,774	2,724	2,930	7290	33,056
2000	9,401	4,112	6,738	1,720	2,955	1,665	9587	36,178
2001	10,652	4,646	7,227	2,531	3,201	5,167	7757	41,181
2002	10,784	4,763	8,127	3,411	3,251	6,949	8956	46,241
2003	10,771	4,793	8,984	4,317	3,762	6,680	9944	49,251

Note: MAT represents marine, aviation and transport insurance
Source: Association of British Insurers (2004)

The UK general insurance industry also faces the risk of major claims from natural disasters and terrorism. The potential threats from terrorism are sadly greater than ever before. The atrocities of September 11th led to the largest claims ever made, running into tens of billions of pounds. The UK has had a long and painful history of terrorist attacks. The Irish Republican Army in the early 1990s threatened to make insurance cover for acts of terrorism unavailable. This led to the establishment in 1993 of Pool Re, following extensive discussions between the insurance industry and the UK government.[148] Pool Re was set up as a mutual reinsurance company to provide cover above a certain threshold for property damage and business interruption caused by terrorist activities. The company is owned by major UK property insurance companies.

[147] For an analysis of the UK cycle in motor insurance, see Fenn and Vencappa (2003). For a study of the cycle in US property-liability insurance, see Fung et al (1998).

[148] For more details on Pool Re, see the Summary Paper published by the UK Government at www.hm_treasury.gov/media/648/AF/ACF1D0D.PDF

Table 4.8: Underwriting Result by Line of Business in General Insurance, 1990–2003

	Underwriting results (£ million)						
	Motor	Accident and health	Property	Other	TOTAL	Investment income	Trading result
1990	–1,439	12	–1,283	–2244	–4,953	3,479	–1,474
1991	–1,532	–78	–1,001	–4350	–6,961	3,691	–3,270
1992	–939	96	112	–4143	–4,874	4,049	–825
1993	–290	200	667	–2783	–2,205	4,051	1,846
1994	4	209	1,201	–1554	–139	4,090	3,951
1995	–369	27	513	–846	–675	4,925	4,250
1996	–987	25	137	–1062	–1,888	4,807	2,953
1997	–1,500	91	–91	–288	–1,789	4,635	2,846
1998	–2,100	–27	–343	–1877	–4,347	5,142	795
1999	–1,804	–74	–412	–1223	–3,513	4,347	834
2000	–1,620	7	–698	–2717	–5,028	5,352	324
2001	–430	62	–276	–2788	–3,432	4,758	1,326
2002	–297	261	–106	–919	–1,062	3,696	2,634
2003	–61	218	367	–1442	–918	3,805	2,887

Note: Underwriting result = net premiums – claims paid – expenses
Source: Association of British Insurers (2004)

The UK life insurance industry has been the subject of much criticism in recent years, particularly following the mis-selling of personal pension policies during the period 1988–94. The mis-selling occurred when people who would have been better off at retirement in their employers' pension schemes, were advised by insurers and independent financial advisers to take out personal pensions instead. During this period, over a million customers were mis-sold personal pensions and the Financial Services Authority has estimated that the scandal has cost insurers and independent financial advisers approximately £12 billion in compensation payments. There have also been more general criticisms of the industry and Chapman (2001) refers to the following factors:

- Life products are too complex.
- As a result, high commissions have to be paid to sales staff.
- High commissions can lead to products being mis-sold, and many products are not kept in force for long.
- High commissions are a factor behind low surrender values.
- Companies have a variety of complex products still in force from past sales. This state, combined with high marketing and sales costs, leads to high charges.

- Competition (which may reduce charges) is limited, due to complex and differentiated products, and to many sales being made by agents tied to one company.

Oliver and Small (1999) and Oliver (2000) suggest an alternative approach that emphasizes simple-structure products that will avoid the above problems. In particular, they refer to some features of unit-linked products, such as capital units, which have complex charging structures and are front-loaded. More recently, the concern about complex products has focused on with-profits policies, where charges are not clear and where the company's discretion in setting bonus rates and surrender values has come under scrutiny (see Blake and Board, 2000).

In 2001, the Chancellor of the Exchequer set up the Sandler Review (2002) to examine the structure of medium- and long-term savings in the UK. Specifically, the Review's remit was "to identify the competitive forces and incentives that drive the industries concerned, in particular in relation to their approaches to investment, and, where necessary to suggest policy responses to ensure that consumers are well served." The Review was seen as necessary because people, particularly those in low- and medium-income groups, were not saving enough for their retirement. The Review identified the need to close this so-called 'savings gap' amongst moderate earners, and identified a number of problems in the UK savings market. First, the charging structure on products was seen as over-complex, and a lack of transparency made it very difficult for savers to compare products. Secondly, because of this complexity, investors had become dependent on financial advisers, whose advice was often compromised by commission payments. Sandler's main recommendations included: a suite of simple saving products sold without the usual burden of highly regulated advice; greater transparency of with-profits products; the removal of unnecessary jargon and clarification of the actual performance of the fund after charges; a re-organization of the independent financial advice industry, with only fee-paid advisers to be called 'independent'.

To date, there has been little progress towards implementing Sandler's recommendations, though in April 2003, the Financial Services Authority proposed the improvement of the governance and transparency of with-profit funds. The Financial Services Authority's proposals include a requirement that firms define and publish the principles and practices of financial management that are applied to with-profit funds. Additionally, the Financial Services Authority intends to require clearer and more directly comparable information about the solvency of with-profit funds and their ability to meet guaranteed benefits. The Association of British Insurers (2003) supported Sandler's proposals, but argued that incremental changes to the current system for regulating the sales of savings products would close the savings gap by only £600–£700 million per year. Only radical reform, they argued, would yield the big improvements of £4–5 billion per year, which would really make a difference.

Distribution channels

Before the Financial Services Act 1986, there were three main types of agents providing investment advice: first, agents of insurance companies, who worked solely for their companies; secondly, insurance brokers; and thirdly, firms or companies paid by commission for introducing business to a small number of providers (mostly banks, estate agents and solicitors). Polarization was introduced as part of the regulatory framework in 1987. The polarization rules apply to advice given on 'packaged' investment products (such as life insurance policies, personal pensions and investment trust savings schemes) and require firms that advise private customers to be either independent intermediaries or tied agents. Independent intermediaries are firms that may not have any tie with product providers, but must act on behalf of their customers by surveying the products available and advising on the most appropriate. Tied agents, on the other hand, are advisers that are tied to a product company or group and only advise customers to buy the packaged products of that company or group. Polarization contributed to the expansion of the independent channel which now consists of different types of intermediaries ranging from independent financial advisers, currently regulated by the Financial Services Authority, to independent intermediaries (such as accountants and solicitors) that are regulated by their professional bodies.

Thus, the main distribution channels currently used by insurance companies in the UK are:

- *Independent intermediaries* These are independent financial advisers and include national brokers, chain brokers (who have a chain of offices), telebrokers (who operate a telephone intermediary service) and other intermediaries. Independent financial advisers are not required to provide advice on all product types and can specialize in certain areas (e.g. annuities and pensions). But where advice is provided, it must be for a product that is as good as any other on the market. Some banks, building societies and big companies have 'independent financial adviser' arms. The top five independent financial advisers in the UK by turnover in 2003 were Bradford and Bingley (£129.6 million), Thomsons (£60 million), Woolwich Independent Financial Adviser Services (£57.9 million), Inter-Alliance (£51.9 million) and Towry Law Financial Services (£49.2 million) (Tillinghast-Towers Perrin, 2004).
- *Company agents* These are tied agents of insurance companies, other than company staff, who can only sell the products of the company or group to which they are tied.
- *Direct sales* These are sales where no intermediary is involved. They occur mostly through telephone or Internet contacts, or through branch offices.
- *Other channels* These include sales through major retailers (such as the supermarkets Tesco, Sainsbury and Asda, and the department stores Marks and Spencer and Debenhams), banks and building societies, and other companies.

Tables 4.9 and 4.10 show the market shares (of new premiums written) by distribution channel for UK life and general insurance respectively over the period 1994 to 2003.

With regard to life insurance (see Table 4.9), there was considerable variation in terms of new premiums generated by distribution channel (independent financial adviser, own sales force or tied agents, direct marketing and others) from 1994 to

2003. Specifically, new premiums generated by independent financial advisers increased steadily from 54 percent in 1994 to 61 percent in 2003. Premiums generated from companies' own sales forces and tied agents decreased from 44 percent in 1994 to 28 percent in 2003. Direct marketing of life insurance products increased from 2 percent in 1994 to 4 percent in 2003, and sales through other channels (including bancassurance) increased from 1 percent to 6 percent. Much of the increased share attributed to independent financial advisers occurred in the sale of pensions. In terms of new premiums written on individual pensions, which include personal pensions, stakeholder pensions and employer-sponsored stakeholder pensions, the contribution of independent financial advisers increased dramatically from 50 percent in 1994 to over 80 percent in 2003.

Table 4.9. Market Shares by Distribution Channels: UK Life Insurance, 1994–2003

	Shares of new premiums written (%)			
	Independent agents	Own sales/tied agents	Direct sales	Other (including bancassurance)
1994	54	44	2	1
1995	57	40	2	1
1996	59	38	1	1
1997	60	37	1	1
1998	58	37	2	2
1999	58	37	2	3
2000	61	34	2	3
2001	63	29	2	6
2002	63	28	3	6
2003	61	28	4	6

Source: Computed by the authors from data in Association of British Insurers (2004)

With regard to general insurance (see Table 4.10), although the independent financial adviser channel is still the largest, it is gradually decreasing in importance (from 69 percent of total new business in 1994 to about 55 percent in 2003). The intermediary's role is to bring together buyers and sellers in the insurance market. But requirements for independent financial advisers to provide best advice have led some of them to become more selective. Many individuals or smaller independent financial advisers and brokers have joined large networks in order to reduce costs. The most notable change over the last ten years has been the growth of direct sales, where the contribution to new business has increased from about 13 percent in 1994 to over 22 percent in 2003. Overall, direct sales and the provision of personal lines by major retailers are on the increase, while sales by intermediaries and company agents are decreasing.

Table 4.10. Market Shares by Distribution Channels: UK General Insurance, 1994–2003

	Shares of total premiums (%)			
	Independent agents	Own sales/tied agents	Direct sales	Other (including bancassurance)
1994	69	15	13	3
1995	68	16	14	2
1996	67	17	14	2
1997	67	17	15	1
1998	67	16	16	1
1999	65	11	17	7
2000	62	11	17	10
2001	57	10	21	12
2002	55	8	21	16
2003	55	7	22	16

Source: Computed by the authors from data in Association of British Insurers (2004)

Numerous insurers and brokers now sell motor and household policies direct to the public. The distinction between independent intermediaries channels and direct selling is set to become increasingly important as investment in technology increases and claims handling becomes more efficient. Specifically, sources of private motor premium income have shifted from using a third party broker to dealing direct with the client, hence the increase in direct and affinity businesses (Association of British Insurers, 2003). The independent financial advisers' proportion of this business has fallen from about 70 percent in 1994 to about 40 percent in 2003, whilst direct selling has increased from 20 percent to 40 percent during the same period.

Other channels generated 16 percent of general insurance business in 2003. These channels include banks and building societies where a significant proportion of the business generated is property insurance offered as collateral for loans. Other sources used, such as direct mail, telesales and e-commerce are becoming important alternative means of distributing insurance products. The market share of utilities, affinity groups and retailers grew from 1 percent in 1999 to 11 percent in 2003 with most growth attributed to personal lines business. There has also been considerable growth in household insurance offered by retailers. In 2001, only 3 percent of household insurance business was sold through this channel, rising to 11 percent in 2003.

4.3.4 The International Competitiveness of UK Insurers

Over the last twenty years or so, the UK insurance industry has faced an increase in global competition, due to decreased costs of communication and deregulation. This increase in competition means that mergers and acquisitions are likely to continue in the insurance market, such as the merger that created the Royal and Sun Alliance Insurance Group in 1996. This merger created a larger, stronger company with wider representation in international markets. Much of the insurance industry's current consolidation reflects insurers' moves to restructure and exploit the most profitable areas, with expansion often accompanied by disinvestment in less profitable sectors.

Overseas expansion through mergers and acquisitions is particularly favoured by insurers whose domestic general markets are highly developed, almost to the point of saturation, such as the UK. Overseas business is becoming increasingly important for UK insurers, enabling them to reduce their reliance on the domestic market. The considerable increase in overseas premium income earned by UK insurers is largely as a result of acquisitions (Association of British Insurers, 2003). Although UK-based insurers are able to sell products in overseas markets from a UK base (that is, sell on a services basis), most UK insurers have preferred to buy existing overseas companies or set up subsidiaries, branches or agencies (that is, sell on an establishment basis). The continuing success of the UK insurance industry in world markets depends on the relative competitiveness of the UK insurance industry and on the relative cost efficiency of UK insurers. We consider competitiveness in this section, and cost efficiency in the next section.

One way of assessing the competitiveness of a country's insurance industry is to calculate indices such as net export ratios and Balassa's 'revealed comparative advantage' index (see Webster and Hardwick, 2004). We consider these indices for the UK, the United States and a selection of EU countries in this section.

The international competitiveness of a country's insurance industry is extremely difficult to measure. We cannot observe differences in competitiveness or comparative advantage directly, but we can use data on a country's exports and imports of insurance to derive conclusions about the competitiveness of the country's insurance industry. For example, if a country were a net exporter of insurance, this would imply that the country had a comparative advantage in insurance markets. This idea has given rise to a number of indicators of *revealed* comparative advantage and we examine two of these indicators here.

The first is the net export ratio (NER_{ij}), defined as:

$$NER_{ij} = (X_{ij} - M_{ij}) \,/\, (X_{ij} + M_{ij})$$

where X_{ij} and M_{ij} represent the exports and imports of insurance services (i) by country j, respectively. The rationale behind the NER is that countries would be revealed as having a comparative advantage in insurance if they exported more insurance than they imported. However, simply to consider net exports (exports less imports) would be misleading if we ignored country size. Dividing net exports by total trade (exports plus imports) controls for this. Net export ratios have a minimum value of −1 (implying that the country only imports insurance) and a maximum value of +1 (implying that the country only exports insurance). Negative values are taken to reveal a competitive disadvantage and positive values a competitive advantage.

The second indicator is Balassa's revealed comparative advantage index (RCA_{ij}), proposed by Balassa (1965). This is given as:

$$RCA_{ij} = s_{ij}/s_{iw}$$

where s_{ij} is the share of insurance in country *j*'s total exports and s_{iw} is the share of insurance in total world exports. Thus, if insurance services are more important in country *j*'s exports than they are in total world exports, then the measure suggests that country *j* has a revealed comparative advantage in insurance. So values of the Balassa *RCA* index greater than 1 are taken to "reveal" a comparative advantage, while values less than 1 are taken to reveal a disadvantage.

Using data extracted from the *OECD Statistics on International Trade in Services* database (Organisation for Economic Co-operation and Development, 2003), we computed net export ratios and Balassa RCA indices for the UK, the United States, Switzerland and ten EU countries for 2001.

From the results shown in Table 4.11, it is clear that, out of the countries selected, Switzerland had the highest net export ratio, with the UK in second place, while the UK had the highest Balassa RCA index, with Switzerland in second place. This suggests that the UK and Switzerland had the most competitive insurance industries in the sample in 2001.

Table 4.11. Net Export Ratios (NER) and Revealed Comparative Advantage (RCA) Indices for Insurance, 2001

Country	NER (rankings in parentheses)	RCA (rankings in parentheses)
Austria	–0.02 (7)	2.86 (3)
Belgium/Lux	0.09 (6)	1.19 (6)
France	0.17 (4)	0.67 (9)
Germany	0.17 (4)	0.66 (10)
Greece	–0.34 (11)	1.08 (7)
Italy	–0.13 (9)	1.00 (8)
Netherlands	–0.40 (12)	0.22 (13)
Portugal	–0.20 (10)	0.44 (12)
Spain	–0.04 (8)	1.44 (4)
Sweden	0.29 (3)	1.33 (5)
Switzerland	0.90 (1)	3.50 (2)
UK	0.67 (2)	3.94 (1)
USA	–0.60 (13)	0.56 (11)

Source: Calculated by the authors

Other countries with a strong revealed advantage according to the net export ratio were Sweden, Germany, France and Belgium-Luxembourg, while Greece, Portugal, the United States and the Netherlands had a revealed competitive disadvantage. Comparing these results to those of the Balassa index does suggest that the different measures occasionally produce some differences in their conclusions. Whilst the UK, Switzerland and Sweden were highly ranked according to both measures, Germany had a low ranking according to the Balassa index. Similarly,

Austria was highly ranked according to Balassa's index, but not by the net export ratio. However, countries such as Portugal, the United States and the Netherlands that received low rankings according to the Balassa measure also received low rankings according to the net export ratio.

Overall, the evidence suggests the UK insurance industry is probably the most competitive in the EU and one of the most competitive in the world. It would seem, therefore, that UK insurers are well placed to take advantage of the globalization of insurance markets.

4.3.5 The Cost Efficiency of UK Life and General Insurers

It is likely that the most cost-efficient UK insurance firms will be the ones to benefit most from the opportunities created by financial liberalization, globalization and further financial market integration in the EU. In this section, therefore, we attempt to assess the overall degree of cost efficiency in the UK life and general insurance sectors. To do this, we apply data envelopment analysis (DEA) to estimate a set of technical, allocative, cost and scale efficiency scores for two samples of 50 life and 50 general insurance companies operating in the UK during the period 1994–2001.

For each sample, we calculate four efficiency measures: pure technical efficiency, allocative (or price) efficiency, cost efficiency and scale efficiency. Pure technical efficiency measures how efficiently technology is employed in the use of inputs to achieve a given level of outputs. Allocative efficiency is a measure of the extent to which a firm produces its chosen outputs with the cost-minimizing mix of inputs, given input prices. The cost efficiency measure for a life insurance firm is then calculated as the product of the pure technical and allocative efficiency scores. A firm is defined as being cost efficient if its costs are equal to the costs of a 'best practice' firm producing the same output bundle under the same conditions. This measure of cost efficiency can be regarded as a (non-volume) measure of economic X-efficiency (Mester, 1997). Scale efficiency measures each firm's costs in relation to the costs incurred by firms with the 'optimum scale'.

Data envelopment analysis

Recent advances in the measurement of efficiency have been made using non-parametric mathematical linear programming approaches (such as DEA) that enable efficiencies for each firm in an industry to be estimated relative to a dominant set of efficient firms with similar characteristics. DEA estimates efficiency in firms with multiple inputs and outputs. Furthermore, it is less demanding than parametric approaches in terms of degrees of freedom; there is no need to specify the form of the production or cost function; and there is no need to make distributional assumptions about a stochastic error term (Cummins and Weiss, 2001). These can be useful attributes in efficiency studies of the insurance industry where the specification of the multi-output production function of life insurers can be difficult to define and measure consistently (Berger and Humphrey, 1997). In addition, DEA is particularly amenable for studies of relatively small samples (like ours), where firms (operating in the same industry) tend to produce reasonably comparable (homogeneous) types of outputs (Cummins et al, 1999). DEA has a further advantage over parametric (stochastic) econometric approaches in that it can easily decompose total firm-specific efficiency into its component parts (Cummins and Weiss, 2001). Berger and Hannan (1998, p. 457) further contend that DEA is an effective estimation technique in efficiency studies as " . . . cost differences owing to inefficiency are relatively stable and should persist over time, while those owing to random error are ephemeral and should average out over time." What is more, Cummins and Zi (1998) compare DEA with parametric approaches using data drawn from the United States life insurance industry and conclude that both approaches derive consistent and similar estimates of the average level of efficiency of life insurance firms. This view is shared by Berger and Mester (1997), though there is evidence (see, for example, Berger and Humphrey, 1997) that the parametric and non-parametric approaches sometimes give very different efficiency rankings of the firms in a sample.

The basic DEA formulation assumes that for a data set of N insurance firms, each employing different amounts of m inputs to produce s outputs:

$$X_i = (x_{1i}, x_{2i}, \ldots x_{mi})^{T} = 0 \text{ and } Y_i = (y_{1i}, y_{2i}, \ldots, y_{si})^{T} = 0 \quad [1]$$

where X_i and Y_i denote the inputs and outputs of the ith insurer. From equation [1], the insurance production set that satisfies the assumptions of convexity, positive monotonicity and the free disposability of inputs and outputs for all insurance firms in our samples can be expressed as:

$$T = \{(x,y): x \geq \sum_{i=1}^{n} \lambda_k x_k,\ y \leq \sum_{i=1}^{n} \lambda_k y_k,\ \sum_{i=1}^{n} \lambda_k = 1; \lambda_k \geq 0,\ i = 1, \ldots, N\} [2]$$

where λ_k are constants. From equation [2], efficiency can be evaluated from the following variable returns to scale specification, proposed by Banker, Charnes and Cooper (1984):

$$\begin{aligned} &\text{Min } \theta \\ &\theta, \lambda \\ &\text{s.t.} \quad x_0\theta - X\lambda \geq 0 \\ &\qquad Y\lambda \geq y_0 \\ &\qquad e^{T}\lambda = 1; \lambda_i \geq 0, i = 1, \ldots N \end{aligned} \quad [3]$$

In equation [3], X and Y represent respectively primal vectors of inputs and outputs with columns x_i and y_i ; e is a vector of ones; $\lambda = (\lambda_1, \lambda_2, \ldots, \lambda_n)^{T}$; and θ is an input radial measure of pure technical efficiency. When $\theta_i = 1$, insurer i is deemed to be on the boundary of total factor efficiency. However, as Schaffnit et al (1997) make clear $\theta_i = 1$ is a necessary but not a sufficient condition for a life insurer to be technically efficient since (x_0, y_0) may contain slack in its allocation of m-inputs and s-outputs. Thus, life insurance firm i is efficient only if $\theta_i = 1$, $X\lambda = x_0$, $Y\lambda = y_0$ and inefficient when $\theta_i < 1$.

Given input price data and assuming cost minimization, pure technical, allocative and overall cost efficiencies can be estimated by running the following cost minimizing DEA:

$$\begin{aligned} &\underset{\lambda,\, x_i^*}{\text{Min}} \; w_i' x_i^* \\ &\text{s.t.} \quad Y\lambda - y_i \geq 0 \\ &\qquad x_i^* - X\lambda \geq 0 \\ &\qquad e^{T}\lambda = 1;\ \lambda \geq 0;\ i = 1, \ldots m; \end{aligned} \qquad [4]$$

where w_i is a vector of input prices and x_i^* is the cost-minimizing vector of input quantities for the ith life insurer, given input prices and the output levels. Hence, for given input prices, pure technical and allocative efficiencies as well as measures of overall cost efficiencies can be derived for our two samples of insurance companies from the use of the DEA procedures described above. The DEA is conducted in this study using DEAP Version 2.1, developed by Coelli (1996).

Variables and data

We use annual data for two separate samples of 50 UK life insurance companies (representing approximately 20 percent of the number of firms in the industry) and 50 UK general insurance companies (representing about 8 percent of the firms in the industry). In order to qualify for inclusion in the sample, a firm must: (a) have operated continuously throughout the eight-year period, 1994–2001; (b) have not been subject to major mergers or acquisitions; and (c) not be a reinsurance company or a trust fund because they do not directly write much life insurance. Our sample of life insurance firms accounts for roughly 30 percent of the annual premiums generated by the UK life insurance industry, and our sample of general insurers accounts for about 40 percent of the annual premiums generated by the UK general insurance industry over the period of analysis. The samples also cover a spread of insurance firms of different sizes and organizational forms and include relatively new entrants (like bancassurers) as well as long-established businesses.

To model the behavior of life and general insurance companies, we adopt a 'production' model in which insurers are assumed to employ labor and capital inputs to produce outputs in the form of policies and investments. Since these outputs are intangible, they are difficult to measure. The pragmatic approach is to identify the services that life and general insurance companies provide and then derive proxies that are likely to be closely correlated with these services. Clearly, all insurance companies engage in *risk-pooling* and *risk-bearing* by selling insurance policies of various kinds. In life insurance the main product lines are life insurance, pensions, annuities and permanent health insurance (or income protection insurance), while in general insurance, the main product lines include accident and health, motor, property, third party liability, marine, aviation and transport insurance. At the same time, all insurance companies engage in *financial intermediation* by investing funds that are not needed to pay claims or to cover expenses.

For the purposes of our study, some degree of aggregation is necessary in measuring insurance outputs. We assume, therefore, that life insurance firms produce just three 'risk-pooling' outputs: (a) life insurance, (b) pensions and annuities, and (c) permanent health and other insurance, and we use annual premium incomes to proxy all three of these outputs. Similarly, we assume that general insurance firms also produce just three 'risk-pooling' outputs: (a) accident and health insurance, (b) motor insurance, and (c) property/liability and other insurance, again all proxied by annual premium incomes. We further assume that life and general insurance companies produce one 'financial intermediation' output: investments, proxied by annual investment income earned[149].

We recognize that there is a problem in using income measures of insurance output. Indeed, there has been much discussion in the literature over the best proxies to use for the risk-pooling and risk-bearing outputs. Many early studies used premium income as the main output indicator, even though this is a form of revenue and so represents price × quantity, rather than just quantity. This problem has led some authors more recently to use the value of benefit payments (i.e. claims) instead (see, for example, Cummins et al, 1999, and Cummins and Santomero, 1999). But there are problems with this too. In particular, *realized* losses differ from *expected* losses, so using current claims as a measure of output may create an 'errors in variables' problem (though econometric methods are available to mitigate this). The problem is to find an output proxy that reflects the number and risk characteristics of the exposure units insured through the risk pool. *Expected* claims (or losses) would seem to be an appropriate proxy, and this being so, we can rationalize the use of premiums as an output proxy since premiums are highly correlated with expected losses.

In addition, as indicated above, we view the life insurance production function as using two key inputs: labor and capital. The price of labor is proxied by taking the average wage rate in 'banking, finance, insurance, business services and leasing' in the UK region where the company's main administrative office is located. The price of capital is proxied by the cost per square meter of office space in the UK region, city or town where the company's main administrative office is located. These definitions enable us to compute wage and rental rates that vary across the firms in the sample as well as over time. The labor input is defined as the number of full-time equivalent employees, while the capital input is computed as total capital expenses divided by the price of capital. The wage rate data were obtained from various issues of *Regional Trends* (published in the UK by the Office for National Statistics) and the rental rate data are taken from various issues of the monthly survey of regional UK office rents published by the property consultants King Sturge (1995–2002). Data on costs and outputs for 1994–2001 were obtained from the 2001 Synthesys Life database, produced by Standard and Poor (2003)[150].

Cost efficiency results

The DEA estimates of pure technical efficiency (PTE), allocative efficiency (AE), cost efficiency (CE) and scale efficiency (SE) are summarized for the UK life and general insurance industries for each of the eight years of the study in Table 4.12.

---

[149] In both samples, investment income is closely correlated with the total value of assets invested in shares and debt securities. The correlation coefficient is 0.94 for life insurance and 0.96 for general insurance. Also, using the value of assets invested as a proxy for financial intermediation output made no significant difference to the efficiency scores.

[150] We are extremely grateful to the Centre for Risk and Insurance Studies at Nottingham University for allowing us access to data from the Synthesys database.

The pure technical efficiency score for firm *i* is measured as the ratio of firm *i*'s actual output to the output that could have been produced by 'best-practice' firms with the same (but not necessarily cost-minimizing) combination of inputs, assuming variable returns to scale. The allocative efficiency score refers to the efficiency (relative to best-practice firms) with which firm *i* uses its inputs to achieve its chosen outputs. Scale efficiency is calculated from the pure technical efficiency score obtained when variable returns to scale are assumed (PTE) and the technical efficiency score that would be obtained if constant returns to scale were assumed.[151] PTE, AE and SE all take values between 0 and 1, where a value of 1 represents 'perfect' efficiency. Cost efficiency is measured as the product of TE and AE and so also takes a value between 0 and 1.

Table 4.12. Summary of the Cost Efficiency Results

	Life Insurance Industry				General Insurance Industry			
	PTE	AE	CE	SE	PTE	AE	CE	SE
1994	0.63	0.94	0.59	0.74	0.76	0.95	0.72	0.85
1995	0.60	0.92	0.55	0.66	0.77	0.94	0.72	0.86
1996	0.62	0.93	0.58	0.73	0.76	0.94	0.71	0.88
1997	0.59	0.92	0.54	0.79	0.66	0.96	0.63	0.75
1998	0.57	0.91	0.52	0.81	0.67	0.92	0.62	0.92
1999	0.58	0.91	0.53	0.79	0.73	0.95	0.69	0.88
2000	0.63	0.91	0.57	0.80	0.65	0.93	0.61	0.70
2001	0.71	0.92	0.65	0.79	0.65	0.93	0.61	0.77
Averages	**0.62**	**0.92**	**0.57**	**0.76**	**0.71**	**0.94**	**0.66**	**0.83**

PTE: Pure technical efficiency
AE: Allocative efficiency
CE: Cost efficiency
SE: Scale efficiency

Source: Calculated by the authors

It is clear from Table 4.12 that the two industries have consistently exhibited high levels of allocative efficiency (both averaging over 0.9), indicating that UK insurance companies are on average reasonably efficient (relative to the best-practice firms in the two industries) in choosing the cost-minimizing input combinations,

[151] To calculate scale efficiency scores, we conducted both a 'constant returns to scale' DEA and a 'variable returns to scale' DEA. If we denote the technical efficiency score for firm *i* obtained from the 'constant returns to scale' DEA as TE_i and the pure technical efficiency score obtained from the 'variable returns to scale' DEA as PTE_i, then the scale efficiency score for firm *i* is given by $SE_i = \frac{TE_i}{PTE_i}$.

given input prices. However, there is evidence of a high degree of pure technical inefficiency, which leads to a high degree of cost inefficiency. The mean PTE scores are 0.62 for life insurance and 0.71 for general insurance. These imply that on average UK life insurers have been producing only 62 percent, and UK general insurers only 71 percent, of the insurance services they could efficiently provide with their chosen levels of resources. The mean CE scores of 0.57 for life insurance and 0.66 for general insurance similarly suggest that, overall, UK life insurers have been producing only 57 percent and UK general insurers have been producing only 66 percent of their attainable output levels. The mean SE scores of 0.76 for life insurers and 0.83 for general insurers indicate that some firms in both industries could reduce their unit costs by changing the scale of their operations. Among the life insurers, over the eight-year period, an average of about 14 percent of the companies exhibited constant returns to scale, about 54 percent exhibited decreasing returns to scale and the remainder (about 32 percent) exhibited increasing returns to scale. Among general insurers, an average of about 14 percent exhibited constant returns to scale, about 48 percent exhibited decreasing returns to scale and the remainder (about 38 percent) exhibited increasing returns to scale.

It should be emphasized that it is not possible to compare efficiencies estimated on different frontiers and using different outputs However, it is interesting to note that our results are 'in the same ballpark' as those reported in Cummins and Santomero (1999) for the United States life insurance industry. They found an average pure technical efficiency index of 0.65, an average allocative efficiency index of 0.71 and an overall cost efficiency index of just 0.46 for a sample of United States life insurance firms in 1995. Hardwick (1997) obtained similar results for the UK life insurance industry for the period 1989–93. Using a parametric approach, Hardwick found that UK life insurance companies' costs were on average 30 percent above the estimated cost frontier[152]. Cummins and Santomero (1999) comment that these low cost efficiency scores are " . . . indicative of an industry that traditionally has not been subjected to much price competition" (p. 89).

Finally, it is interesting to note from our DEA estimates that, relative to the best-practice firms in their own respective industries, UK general insurers had higher average efficiencies (TE, AE, CE and SE) than life insurers over the period 1994–2001. Throughout the period 1994–2000, the average TE and CE scores for general insurers exceeded the TE and CE scores for life insurers in every single year, with the biggest difference (of 0.13) occurring in 1994. However, in 2001 the evidence suggests that the life insurance industry had become more cost efficient, with an overall cost efficiency score of 0.65, compared with only 0.61 for the general insurance industry.

4.4 CONCLUSION

The UK insurance industry will face many challenges in the coming years, not least of which will be the need to address increased competition from insurance

[152] For more efficiency studies concerning the UK insurance industry, see Diacon et al (2002) and Klumpes (2004).

companies around the world. Continuing progress towards globalization means that UK life and general insurers not only have to compete successfully with other UK-based insurance companies, but will face an increasing degree of competition from companies located outside the UK. The survival of smaller UK companies will depend on their ability to secure niche markets and/or to achieve growth and diversification through mergers or acquisitions, thereby potentially benefiting from economies of scale and scope. At the same time, larger companies need to develop product and distribution strategies that enable them to operate successfully in the global market.

In spite of the fact that there are a relatively large number of companies in the UK insurance industry, we find that the industry has a relatively high degree of market concentration. This, together with the existence of asymmetric information and differentiated products, means that the UK insurance industry has a relatively complex market structure. Nevertheless, our study of net export ratios and revealed comparative advantage indices suggests that the UK still has one of the most competitive insurance industries in the world. To maintain this position, however, it would be advisable for the industry to consider ways of improving its productive and cost efficiency.

Our DEA study of the relative cost efficiency of the UK's life and general insurance industries suggests that both industries are operating at relatively high levels of technical inefficiency, and therefore high levels of overall cost inefficiency. The evidence suggests that UK life insurers have been producing their chosen outputs with on average only 57 percent efficiency, while general insurers have been producing their chosen outputs with only 66 percent efficiency. Such high levels of inefficiency, if not addressed, could lead to a loss of international competitiveness in the future.

4.5 REFERENCES

Association of British Insurers (2003), *Insurance Trends: Quarterly Statistics and Research Review,* October 2003, Vol. 39.

Association of British Insurers (2004), Data extracted from the website at: www.abi.org.uk

Balassa, B. (1965), 'Trade Liberalisation and 'Revealed Comparative Advantage'', *The Manchester School of Economic and Social Studies,* Vol. 33, pp 99–123.

Banker, R.D., Charnes, A. and Cooper, W.W. (1984), 'Some Models for Estimating Technical and Scale Inefficiencies in Data Envelopment Analysis', *Management Science*, Vol. 30, pp. 1078–1092.

Berger, A. and Humphrey, D. (1997), 'Efficiency of Financial Institutions: An International Survey and Directions for Future Research', *European Journal of Operations Research*, Vol. 98, pp. 175–212.

Berger, A.N. and Mester, L.J. (1997), 'Inside the Black Box: What Explains Differences in the Efficiencies of Financial Institutions?' *Journal of Banking and Finance*, Vol. 21, pp.895–947.

Berger, A. and Hannan, T.N. (1998), 'The Efficiency Cost of Market Power in the Banking Industry: A Test of the "Quiet Life" and Related Hypotheses', *Review of Economics and Statistics*, Vol. 80, pp. 454–465.

Blake, D. and Board, J. (2000), 'The Next Wave of Success in the Financial Services Industry', *Geneva Papers on Risk and Insurance: Issues and Practice,* Vol 25, pp 539–67.

Chapman, M. (2001), 'Sins of the Industry', *Money Management,* January, pp. 86–90.

Coelli, T.J. (1996), 'A Guide to DEAP Version 2.1: A Data Envelopment Analysis (Computer) Program', CEPA Working Paper No. 8/96, ISBN 1 86389 4969, Department of Econometrics, University of New England.

Cummins, J.D. and Santomero, A. (1999), *Changes in the Life Insurance Industry: Efficiency, Technology and Risk Management*, Boston: Kluwer Academic Publishers.

Cummins, J.D. and Weiss, M.A. (2001), 'Analyzing Firm Performance in the Insurance Industry Using Frontier Efficiency and Productivity Methods', in *Handbook of Insurance Economics* (Dionne, G. ed.), Boston MA, Kluwer Academic Publishers.

Cummins, J.D., Weiss, M.A. and Zi, H. (1999), 'Organizational Form and Efficiency: the Coexistence of Stock and Mutual Property-Liability Insurers', *Management Science*, Vol. 45, pp. 1254–1269.

Cummins. J.D. and Zi, H. (1998), 'Measuring Economic Efficiency of the US Life Insurance Industry: Econometric and Mathematical Programming Techniques', *Journal of Productivity Analysis*, Vol. 10, pp. 131–152.

Diacon, S., Starkey, K. and O'Brien, C. (2002), 'Size and Efficiency in European Long-Term Insurance Companies: An International Comparison', *The Geneva Papers on Risk and Insurance,* Vol. 27, pp. 444–466.

Diacon, S. and Drake, L. (2004), 'Measuring Competition in the UK General Insurance Markets' paper presented at the UK Insurance Economists Conference, Nottingham University, March 2004.

Fenn, P. and Vencappa, D. (2003), 'Cycles in Underwriting Profits from UK Motor Insurance, 1985–2002', paper presented at the UK Insurance Economists' Conference, Nottingham University, March 2003.

Fung, H., Lai, G., Patterson, G. and Witt, R. (1998), 'Underwriting Cycles in Property and Liability Insurance: An Empirical Analysis of Industry and By-Line Data', *Journal of Risk and Insurance,* Vol. 65, pp. 539–62.

Hardwick, P. (1997), 'Measuring Cost Inefficiency in the UK Life Insurance Industry', *Applied Financial Economics,* Vol. 7, pp. 37–44.

Hoschka, T. C. (1994), *Bancassurance in Europe,* London: Macmillan.

Klumpes, P. (2004), 'Performance Benchmarking in Financial Services: Evidence from the UK Life Insurance Industry', *Journal of Business,* Vol. 77, pp. 257–73.

Lafferty (1991), 'The Allfinanz Revolution—Winning Strategies for the 1990s', Lafferty Group Management Research.

Lloyd's of London (1993), *Planning for Profit: A Business Plan for Lloyd's of London*, London.

Mester, L.J. (1997), 'Measuring Efficiency at U.S. Banks: Accounting for Heterogeneity is Important', *European Journal of Operational Research*, Vol. 98, pp. 230–242.

Oliver, M. and Small, M. (1999), 'The Value of Life: Harvesting Lifetime Customer Value in Retail Financial Services', paper presented at the UK Insurance Economists' Conference, Nottingham University, March 1999.

Oliver, M. (2000), 'Goodbye Distribution, My Old Friend . . .', paper presented at the UK Insurance Economists' Conference, Nottingham University, March 2000.

Organisation for Economic Co-operation and Development (2003), *OECD Statistics on International Trade in Services*, Paris.

Pearson, R. (2002), 'Growth, Crisis, and Change in the Insurance Industry: A Retrospect', *Accounting, Business and Financial History*, Vol. 12, pp. 487–504.

Sandler Review (2002), *Medium and Long Term Retail Savings in the UK*, an independent report commissioned by the UK Government.

Schaffnit, C., Rosen, D. and Paradi, J.C. (1997), 'Best Practice Analysis of Bank Branches: An Application of DEA in a Large Canadian Bank', *European Journal of Operational Research*, Vol. 98, pp. 269–289.

Sigma (2002), 'The London Market in the Throes of Change', *Sigma*, 2002, No. 3, Zurich: Swiss Re.

Standard and Poor's (2001), *2001 Synthesis Life,* Database and User Guide, March.

Stark, J. (2002), 'The Future of the Pensions Annuity Market', London: Association of British Insurers.

Tillinghast-Towers Perrin (2005), *Insurance Pocket Book 2005*, NTC Publications.

Webster, A. and Hardwick, P. (2004) 'International Trade in Financial Services', *Service Industries Journal,* forthcoming.

4.6 APPENDIX

The following tables provide more accounting information on the UK insurance industry. Table 4.13 shows the total identified assets for life and general insurance companies during the period 1990–2002. Tables 4.14 and 4.15 show the balance sheets for life and general insurance companies respectively in 2002, while Tables 4.16 and 4.17 show the income statements for life and general insurance companies respectively in 2002.

Table 4.13. Total Identified Assets for Life and General Insurance Companies, 1990–2002

	Life Insurance Companies (£ billion)	General Insurance Companies (£ billion)
1990	234.0	42.7
1991	279.2	45.5
1992	328.0	51.8
1993	434.8	62.5
1994	406.2	62.9
1995	496.7	74.9
1996	549.8	90.7
1997	678.5	96.4
1998	781.2	93.7
1999	941.8	94.6
2000	947.1	90.2
2001	925.0	95.3
2002	852.8	97.9

Source: Association of British Insurers Insurance Trends Quarterly Statistics Research Apr-04 Edition 41

Table 4.14. Balance Sheet for Life Business 2002

Assets	**(£m)**
Long-term business assets	878,979
Net direct investment holdings in:	
Non-insurance subsidiaries and associate companies in the UK	4,577
UK insurance companies and insurance holding companies	4,569
Overseas subsidiaries and associates	5,463
Total assets	**893,588**
Liabilitites	
Borrowing:	
Borowing from UK banks	4,958
Other UK borrowing	7,406
Borrowing from overseas	800
Long-term business:	
Funds	794,177
Claims admitted but not paid	3,234
Provision for taxation net of amounts receivable:	
UK authorities	2,803
Overseas authorities	–20
Provision for recommended dividends	32
Other creditors and liabilities	23,261
Excess of assets over liabilities:	
Excess of assets over liabilities in respect of long -term funds	36,517
Miniority interests in UK subsidiary companies	0
Shareholders capital and reserves in respect of general business	18,629
Other reserves including profit and loss account balances	1,791
Total Liabilities	**893,588**

Table 4.15. Balance Sheet for General Business, 2002

Assets	**(£m)**
General business assets	93,965
Net direct investment holdings in:	
Non-insurance subsidiaries and associate companies in the UK	11,706
UK insurance companies and insurance holding companies	7,190
Overseas subsidiaries and associates	9,014
Total assets	**121,875**
Liabilitites	
Borrowing:	
Borowing from UK banks	1,384
Other UK borrowing	10,472
Borrowing from overseas	2,916
General business technical reserves	62,776
Long-term business:	
Funds	0
Claims admitted but not paid	0
Provision for taxation net of amounts receivable:	
UK authorities	941
Overseas authorities	5
Provision for recommended dividends	958
Other creditors and liabilities	8,025
Excess of assets over liabilities:	
Excess of assets over liabilities in respect of long -term funds	0
Miniority interests in UK subsidiary companies	4
Shareholders capital and reserves in respect of general business	31,983
Other reserves including profit and loss account balances	2,411
Total Liabilities	**121,875**

Table 4.16. Income Statement for Life Business, 2002

Income	**(£m)**
Premiums receivable net of reinsurance ceded (less rebates and refunds)	113,010
Rents, dividends and interest receivable	36,139
Capital gains (losses) and realised profits (losses)	–91,050
Other	944
Total Income	**59,043**
Expenditure	
Claims, net of amounts recoverable from reinsurers	92,573
Administrative expenses, including commissions and taxation	11,959
Dividends and interest payable	1,107
Other	–6,724
Total expenditure	**98,915**

Table 4.17. Income Statement for General Business, 2002

Income	**(£m)**
Premiums, net of refunds, rebates and reinsurance ceded	33,697
Rents, dividends and interest receivable	4,190
Capital gains (losses) and realised profits (losses)	–3,046
Other	2,567
Total Income	**37,408**
Expenditure	
Claims, net of amounts recoverable from reinsurers	21,328
Administrative expenses, including commissions and taxation	9,571
Dividends and interest payable	2,486
Other	1,124
Total expenditure	**34,509**

5 The French Insurance Market: Background and Trends[153]

Bertrand Venard

Audencia. Nantes School of Management

5.1 INTRODUCTION

The French insurance market is in the top four worldwide in terms of its annual insurance premiums. This is not surprising because the country is also the fifth world economic power in terms of gross domestic product. This economic position and the size of the insurance market are sufficient to stress the importance of covering the French insurance market. In addition, the business of insurance in France has a long history.

In the last 20 years, the French insurance market has experienced important changes. For example, the creation of the European Union (EU) brought new regulations affecting the insurance industry. Furthermore, the insurers faced increased competition due to the entrance of various economic actors, most importantly foreign companies and banks. Indeed, bancassurance is a phenomenon in France because the banks have the most important market share of the life insurance market.

This chapter is in six parts: the history of insurance; some key elements of the regulations; a general description of the French insurance market, presenting both life and P&C markets; the main economic actors of the French insurance market; the financial results of insurance companies in France; and bancassurance. Most of the article is based on secondary data, which relies on existing literature on the market. However, the author conducted some interviews to provide the point of view of various market observers.

---

[153] The author would like to thank Prof. David Cummins (Wharton), Pascal Bled-Charreton (FFSA) and Jean-Marc Piéronne (FFSA) for their comments on a previous version of this chapter. The author is solely responsible for eventual omissions or errors.

5.2 A BRIEF HISTORY OF FRENCH INSURANCE

The French insurance market is rooted in the country's vast history. Over centuries, the French insurance industry has grown steadily, as shown in Table 5.1. An important factor of the actual French insurance market history is the prominence of the State. Before the creation of any State, however, the starting point of the insurance sector was the emergence of the need for protection.

Indeed, the need for protection against the hazard in human life is certainly as old as the first human community. For example, the Celtic culture (which was dominant in France 2,500 years ago) emphasized solidarity among its members and with the environment. The Catholic Church collected compulsory payment from the French population (called the dime, in French *dîme*[154]) to fund its activities, such as providing help to the poor. Therefore, the church organized the first "social security" in the country. Starting in the fourth century, various hospitals, *Hôtels-Dieu* (literally hostels of God), were created, including hospitals in Lyon in 542 and Paris in 650. Indeed, the words "hostel" and "hospital" come from "hospitality."

However, the first type of insurance in France was certainly maritime insurance (Richard 1956). In Greek civilization, it was possible to insure merchandise with the "Nauticum Phoenus" (*prêt à la grosse*), under which ship owners covered their goods and their boats (nowadays cargo and hull insurance) against any risk during transport. This was used in most Greek ports such as Marseille (a French city founded by the Greeks circa 600 BCE). In this system, the ship owners received the price of the goods before transport and had to give back the money with significant interest if the transport succeeded. Therefore, this was a loan plus insurance for the boat owner.

The Roman Empire copied this type of risk coverage from the Greeks. The Roman Emperor Claudius (born in Lyon, France, 10 BCE–54 CE) decided to insure the food supply of the Romans. According to historian Caius Suetonius Tranquillus (*Suétone*), Claudius guaranteed certain profits to Roman traders and covered damages in case of destruction during maritime transport (Caius Suetonius Tranquillus 1975). It is during this period that the first table of mortality was designed by the Roman lawyer Domitius Ulpianus (second century CE). Some authors acknowledged the creation of the first term life insurance (life insurance in case of death) with the "cum moriar" contract (Richard 1956:8). Ulpianus was one of three lawyers chosen by the Emperor Severus (Lucius Septimius Severus, 146–211 CE) to codify Roman law.

The Middle Ages retained most of the insurance tradition of the Roman Empire. For example, historians have found a term life insurance contract dated 1228, signed by a *bourgeois* of the town of Tournai, in the north of France (Lambert 1999:28).

However, maritime insurance was the most prevalent insurance in France for many centuries. The French publication of the *Rôles d'Oléron* (regulation decided on the French island of Oléron) in 1150 is one of the earliest French legal documents explaining the different rules governing the sea trade in France (Ripert 1929). The main "insurance" product was still the "Nauticum Phoenus."

---

[154] In this chapter, most French words will be written in italic.

But in 1234, Pope Gregory IX (Grégoire IX, 1145–1241) published a law (*Décrétale*) forbidding any interest on a loan in the Christian world (therefore, also in France) and forced a complete change in the insurance industry (this Pope is also known as the architect of the Inquisition). This new regulation created the opportunity for the first real starting point of modern insurance (Ripert 1929). Indeed, to avoid the restriction of the new Catholic prohibition, the Italian traders from Florence and Genoa decided to invert the system. They decided to pay a premium (amount of money asked before the transport) to a banker (or any rich person ready to "take" the risk). In exchange for the risk coverage, the banker was obligated to pay the insured loss. All elements were explicitly agreed in a "polizzia" (insurance contract). This type of insurance quickly spread into other Italian regions—France, Spain, the Netherlands, and England (Richard 1956:8). Historians have found a maritime insurance contract signed in Genoa in 1347 (Richard 1956) and in 1437 in Marseille (Couibault 2003:8).

Table 5.1. Important Dates in the History of the French Insurance Market

Date	Event
600 b.c.	Introduction of maritime insurance by the Greeks
220	First term life insurance in the Roman Empire
1150	First Maritime Law, *Rôle d'Oléron*
1234 and after	Décrétale of Gregory IX, forbid interest and creation of "modern insurance"
1604	First state pension fund for mine workers
1670	First state pension fund for maritime workers
1686	First non-life insurance company
1717	First fire insurance mutual society (*Bureau des incendies*)
1788	First private life Insurance company
1812	Creation of a state insurance company for mine employees for their retirement, accidents, and health
1861	First company to insure liability of employers regarding potential employee casualty in the workplace
1899	First state controller appointed
1905	First law to protect insured and improve control over insurance companies
1930	Creation of the *Code des assurances terrestres* (insurance law taking into account most of the past insurance laws)
1945	Creation of the Social Security
1946	Insurance nationalization
1973	First EEC Directive
1976	Rewriting of the *Code des assurances* (insurance law taking into account all past insurance laws)
1993	Creation of the European Union
1994	First important insurance privatization
2003	Financial Security Act

Source: Author.

At the end of the Renaissance, in 1604, the French King Henry IV (1553–1610) created a state pension fund for mine workers (Hatzfeld 1971). This fund provided coverage for funeral expenses and other financial help in case of disability (*secours spirituels et matériels aux mineurs*) (Lambert 1999). It is interesting to note that quite early in its history, the French insurance industry was characterized by important involvement of the French state.

With various wars devastating Europe in the sixteenth and seventeenth centuries, wealthy individuals acting as insurers faced important losses. They underwrote mainly hull and cargo insurance policies. But until the seventeenth century, each private investor warranted the payment of the claims based on his or her personal assets. In the case of important losses, the investor could not rely on reinsurance and went bankrupt. Therefore, some people agreed to join their assets to increase their solvency, and the first insurance companies were born. They were limited liability companies. For example, French King Louis the XIV (1638–1715) by legal document (*édit royal*) of May 21, 1686 agreed to the creation of an insurance company: *la Chambre générale d'assurance* (General Insurance Chamber), which started with quite large assets for the period (Ripert 1929).

This decision was part of the economic policy promoted by Jean Baptiste Colbert (1619–1683), minister of Louis the XIV. For example, in 1670 Colbert agreed to create the *Caisses des invalides de la Marine,* which facilitated recruitment of members of the Royal Navy under the navy disability fund. In case of disability, members of the navy received a pension from the state fund. The *Ordonnance du Commerce* (Trade Law of 1673), followed by *Ordonnance de la Marine* (Navy Law of 1681), was passed to promote economic development (Galix 1985). On September 23, 1673, a state pension fund was created for Royal Navy officer retirement (Lambert 1999).

The seventeenth century saw another important insurance innovation: fire insurance. The French Church organized solidarity between Catholics after a fire. A solidarity office (*Bureau de bienfaisance*) managed by the Church already existed in 1630 (Gallix 1985). After a fire, the local clergyman tried to assess the losses and the financial needs and collected afterwards money among the population. The real emergence of fire insurance business occurred after the Great Fire of London in 1666 (Trebilcok 1985:3) and Europeans became more sensitive to fire risk. In France, the first fire insurance company, *Bureau des incendies* (fire office), was created in 1717. It was in fact a mutual society, managed by the Catholic Church to provide financial help in case of fire loss (Gallix 1985). The Church created a fund and asked the French population to give money in advance to provide help for future fires. After a fire, an expert was chosen by the chief of police to evaluate the fire origin and the losses. Premiums were collected among homeowners according to the value of their property. Other insurance companies were created afterwards, such as the *Compagnie d'assurances générales* (General Insurance Company, 1753), or the *Compagnie d'assurance Labarthe* (Labarthe Insurance Company, 1786). One of the first French life insurance companies was the *Compagnie Royale d'assurance Vie* (Royal Life Insurance Company), created in 1788 (Clavière 1990).

During the French Revolution, insurance companies were forbidden (Besson 1977). A member of the *Convention* (revolution parliament), Mr. Cambon declared, "We must kill all corporations, responsible for the destruction of the State System. This will entitle us to establish the Reign of Liberty" (Richard 1956). After the revolution, Napoleon was suspicious about insurance, seeing it as "pure speculation" (Richard 1956). However, his imperial Act of 1812 (*décret impérial du 26 Mai 1812*) created a State insurance company (*caisse locale d'assurance ouvrière*) for miners, which covered their retirement, accidents, and health (Lambert 1999).

In the nineteenth century, French governments (the *Restauration* and the Third Empire) promoted industrialization. With rising industrialization, potential losses increased and the need for coverage rose.

In the first part of the nineteenth century, fire insurance developed in France. Different fire mutual societies were created in various parts of the country, and some still exist: *Mutuelle de la Ville de Paris* (1816), *Mutuelle de la Seine et Oise* (in 1819, now AZUR Insurance Company), the *société d'assurances contre les incendie dans les départements de la Seine inférieure et de l'Eure* (in 1818, later *Mutuelles Unies*, and then part of AXA), Mutuelles du Mans (in 1828, *Mutuelles du Mans Assurances* and now MMA).

For several years, insurance companies covered only buildings. A new law passed March 17, 1839 allowed companies to insure entire properties (building and content) (Richard 1958:40). Mutual societies requested premiums based on the losses of the previous year, similar to an assessment insurance company. The mutual society could charge policy owners additional sums if the mutual's loss experience was worse than had been allowed for in the premium. The client had no idea how much his insurance would cost. Furthermore, losses were paid by mutual societies only at the end of the year. Private insurance companies were created that differed from mutuals in that they charged fixed premiums (at least for the insured year) (Richard 1958:40).[155] Therefore, private companies grew rapidly. Prominent examples of private companies include the *Compagnie d'Assurances Générales* (General Insurance Company), created in 1819 (now AGF), and the *Union*, created in 1834, which became UAP and then part of AXA.

The mortality rate of insurance companies was very high during the nineteenth century. Due to the intense competition of private insurance companies, 173 fire mutual societies disappeared between 1816 and 1903, and out of the 87 fire companies created between 1919 and 1880, only 19 still existed at the end of the twentieth century (Gallix 1985:360). In 1868, a law established requirements for incorporating a new insurance company.

In the second part of the nineteenth century, casualty insurance developed rapidly in France. Nearly 90 percent of the casualty insurance company creation occurred during the second half of the century (Venard 1997). Various innovations came with the transformation of the economy, since new economic activities led to new types of coverage. Pearson highlighted the insurance innovations, which occurred in various key countries, including France (Pearson 1997:239). The prevailing idea is that the United Kingdom has for many years been the most innovative in the insurance industry. But, it is clear that other European countries were at the edge of the insurance innovation, especially Germany, but also France.

Among the first casualty insurance companies we could highlight are the French insurance companies: the *Automédéon* (created in 1825) and the *Seine* insurance

---

[155] We use the word "private insurance company" for stock insurance company, which is not state-owned. Indeed, some stock insurance companies could be owned by the French state.

companies (in 1829), which insured liability of horse and horse-car owners (Richard 1956).

Another innovation was the creation of the mutual society *la Préservatrice* in 1861, which covered the liability of employers regarding potential employee casualty in the workplace (*assurance contre les accidents du travail*) (Richard 1956:70). Following this innovation, the first type of State supervision was institutionalized in France in 1899, with the insurance controller position, the first professionals in charge of controlling work-related insurance activities in the country (*commissaires-contrôleurs des sociétés d'assurances contre les accidents du travail*).

Not only the growth of the insurance in the second half of the nineteenth century in France could be explained by innovation within insurance companies, but this growth is also the result of external factors linked to rapid French industrialization. Indeed, the study of industrialization cannot be limited to a chronological survey of various inventions. The industrialization was also linked to various cultural, sociological, and political phenomena. For example, this was a period of wealth creation for all the French population who could afford insurance policies. During this period insurance switched from a small niche concerning only wealthy businessmen or property owners to a mass market. An important internal migration occurred in the direction of French cities. Breaking family rural links, the new French urban population did not have the benefit of the former solidarity system, which was mainly based on familial relations within French villages (Venard 1997). Relying less on self-insurance, the French population started to find coverage from insurance companies.

The twentieth century continued to witness a rapid growth of the insurance business. One specific invention had an important impact of the insurance industry: the car. Step by step, automobiles became the principal means of transportation. In 1937, car insurance already represented 28 percent of P&C insurance; 41 percent in 1955; 66 percent in 1966; 46 percent in 1994; and 44 percent in 2003 (FFSA, various annual reports). During the twentieth century, the life insurance market witnessed various periods of growth due to the economic expansion of the country, but also periods of relative decline with the two World Wars and the 1929 economic crisis. For example, during the First World War, many insured persons could not pay their premiums (Richard 1956:205).

Most insurance companies took advantage of this economic growth and developed their activities. Their business development was partly due to various laws forcing the underwriting of compulsory insurance products. For example, a law passed in 1898 obliged employers to insure employees in case of work-related casualty. Another law in 1910 directed that employers and employees should contribute to pension funds to provide employee retirement benefits. There are at present 101 compulsory insurance products, among which 87 are linked to general liability (Couilbault 2003:80). Car liability insurance remains the largest compulsory insurance market.

After an important bankruptcy in 1905, the French government passed a law (*Loi du 17 mars 1905*) to increase the State supervision on life insurance companies to protect customers (Lambert 1999). This was expanded to all insurance companies in 1938 (*décret du 14 juin 1938*), which increased State supervision over insurance companies by the creation of various supervision institutions: Insurance Organization Committee (*Comité d'organisation des assurances*) and Private Insurance Bureau

(*Office des assurances privées*) (Richard 1956). In 1930, a law was created to establish contract rules related to most insurance products (*Loi du 30 juillet 1930 concernant le droit des assurances terrestres*).

During World War II, the French insurance market functioned slowly. Portions of the French workforce were in jail in Germany (soldiers or officers) or were working in Germany under compulsory work assignment (*Service de Travail Obligatoire,* STO). The insurers could no longer rely on the British insurance market, especially for reinsurance. At the same time, German insurance companies followed the Reich troops to distribute their products, but never succeeded in growing in occupied France. Thus, Swiss insurers had the prime opportunity to develop their activity in France. Facing various sabotages of the French resistance, the French government created a specific fund to help the P&C companies in fulfilling their insurance obligations. Indeed, P&C companies had to pay claims linked to French resistance sabotages on various French infrastructures (buildings, bridges, railway stations, and such).

After World War II, a government composed partly of communists and socialists (who won the majority at the parliament in 1946) implanted various reforms. In a 1945 law (*Ordonnance du 4 octobre 1945*), social security was created to insure work-related casualty and health, retirement benefits (pensions), and social benefits (money given to households by the State, depending on family size and resources) (Lambert 1999:32). Social state insurance existed to a lesser extent, since 1928. Therefore, an important part of a private insurance company's business was suppressed by the State and private insurance companies could no longer underwrite work-related casualty and health insurance.

The supervision of the State over insurance was centralized to the Finance Ministry (*Ordonnance du 29 septembre 1945*), which controlled contract terms, insurance price regulation, and insurance company management.

In 1946, 34 insurance companies owned by 11 insurance groups were nationalized (Séquanaise, Union, Compagnie d'assurance générales, Caisse fraternelle de capitalisation, Aigle, Compagnie du Soleil, Compagnie générale de réassurances, Nationale, Phénix, Mutuelle Générale Française). Each insurance group had a business of more than one billion French francs (*Ordonnance du 25 avril 1946*). This wave of nationalization represented roughly 50 percent of annual premiums in France (Lambert 1999:35). The government justified this decision by saying, "This project… tries to give back to the state the management of organizations, which are important for the general interest (*intérêt général*) of the population" (Richard 1956:273). In 1950, the market share of state insurance companies was 88 percent in capital redemption bonds, 65 percent of life insurance, and 33 percent of the P&C insurance (Lambert 1996:187).

During the same period, two laws were adopted for insurance agents (*agent général d'assurances*): one for P&C agents with an exclusive sale territory (*exclusivité territoriale*) directed that the agent would be the only one of his/her company in a specified geographical area; and another law for life agents without an exclusive sale territory. Both laws mandated the type of commission, respectively P&C agent law of March 8, 1949 and life agent law of December 28, 1950, and that an agent must be:

- an individual, not a firm:
- of good morality;
- older than 21 (as of 1992, 18); and
- specifically trained.

Both laws were modified by the law of the October 15, 1996, which specified that insurance companies could choose individuals or firms as agents.

The sale exclusive territory means that in case of merger or acquisition, the new company must negotiate with the agents the loss of their exclusivity. Indeed, agents of the former companies may have near-similar sale areas. Such was the case when Generali France recently bought the insurance companies Continent and Zurich France.

In 1968, the nationalized insurance companies were regrouped into three main state insurance companies: UAP (Union des Assurances de Paris); AGF (Assurances Générales de France); and GAN (Groupe des Assurances Nationales). The performance of state-owned insurance companies was poor. Between 1950 and 1968, nationalized insurance companies market shares fell from 50 percent to 38 percent (Lambert 1999). This could be partially explained by the frequent turnover of top civil servants at the head of the nationalized companies, many of whom had no experience in the insurance industry (Baurer, Bertin Mourot 1987; Lambert 1999).

After the election of François Mitterrand as president, the new socialist government decided to nationalize various important parts of the economy (*Loi du 11 février 1982*): 39 banks, two financial institutions (*compagnies financières de Suez et de Paris et des Pays-Bas*) and five industrial firms (Compagnie générale électrique, Compagnie de Saint-Gobain, Pechiney Ugine Kuhlmann, Rhône-Poulenc, Thomson-Brandt).

In 1986, the Socialists lost the parliamentary election to a conservative party. A new government then voted on a law of privatization (*Loi du 2 juillet 1986*). The privatization process started rapidly for the French banks (CECEI, *Comité des établissements de crédit et des entreprises d'investissement, 2001*). It was necessary to vote to create laws to finalize the privatization of the state insurance companies (*Loi du 31 décembre 1987* and *Loi du 15 juillet 1993*) (Dion 1995). UAP was privatized in 1994 (finally bought by AXA in 1996), AGF in 1996 (bought by Allianz in 1997), and GAN in 1998 (bought by Groupama in 1998).

After various waves of nationalization, in the 1990s, the French insurance industry began experiencing a privatization process. Not only were state-owned insurance companies privatized, but also the French state systematically reduced its involvement in the industry. For example, the state chose not to exercise its right of control on insurance tariffs since the middle of the 1980s (Lambert 1999) and abrogated this right in 1991 (Law of June 28, 1991, *Loi du 28 juin 1991*).

5.3 REGULATION OF THE FRENCH INSURANCE MARKET

As we have seen in the preceding section, the French government has always been involved in the insurance market. Therefore, the French regulation concerning insurance is quite important. We will focus in this section of some key characteristics of insurance regulation in France: the influence of the European Union, the privatization process, the state supervision, and the tax system.

For a long time, the evolution of the French regulation was *sui generis*, influenced by France only. A huge shift appeared with the creation of the European Economic Community (EEC) and after the European Union (EU). Various European directives have been adopted since 1964 for reinsurance and 1973 for P&C and life insurance (Boleat 1995).

An important contribution of Europe to the French insurance industry was the regulation simplification. Indeed at the beginning of the 1970s, there were hundreds of texts and laws concerning insurance in France. An EEC directive in 1973 was launched to organize the insurance regulation into one specific act, following specific classification. Three years later, the French Insurance Act was born (*Loi du 16 juillet 1976 concernant le code des assurances*). The Insurance Act is the review and the classification of all insurance laws for insurance companies and mutual insurance companies (*Société d'Assurance Mutuelle, SAM*). In the Insurance Act, most rules governing insurance contracts are of public order (*ordre public*). This means private parties have to contract according to these compulsory rules. For example, law L113–12 of the Insurance Act establishes a clause under which the insurer and the insured can cancel a contract at least two months before its anniversary (*date d'échéance*). This period of two months is compulsory to all contracts regardless of any private agreement between parties (CCA 2002:71). A major exception is life insurance, whose contracts are not cancellable.

Following the first 1964 directive, insurers have seen the number of EU initiatives affecting insurance increase in the past decade to what could be considered a "tidal wave of regulation" (CEA 2004:11). Note that the majority of laws passed at the French parliament are now EU laws. Any law issued by the EU is in principle automatically accepted by the French Parliament (normally in a maximum period of 18 months after the publication of the law in the European Official Journal).

Another main influence of the Europe Union creation was to accelerate the insurance market privatization process (Duncan 1999). There are at least three generations of insurance directives. The directives of the first generation were adopted in 1973 and 1979. They permitted freedom of establishment of European insurers in all European states. All European insurers have the right to set up subsidiaries or a branch in any other European country. This prevents national governments from erecting barriers around their own markets. The first generation directives required all classes of insurance to be supervised by the member state where the head office is located.

The directives of the second generation introduced the freedom to provide service (*Libre Prestation de Services, LPS*) and to sell insurance from any European national market directly to other EU national markets. The French law of December 31, 1989 concerned the adaptation of the Insurance Act to the creation of the single European market. The same law directed the modernization of the insurance supervision with the creation of the Insurance Supervision Committee (*Commission de Contrôle des Assurances, CCA*).

The directives of the third generation provided a single structure for business conducted either on an establishment basis or by cross-border provision services. European insurers were then able to establish their activities everywhere in Europe or to provide insurance services in other European countries from their home country, thus cross-borders. The aim was to develop further the single European market with a complete freedom of establishment. The EU insurance directive came into force July 1, 1994 and began the reshaping of the European insurance landscape (Williams de Broë 1998). Indeed, this directive removed the ability of individual countries to interfere in the establishment of other EU insurers in their territory. It also removed individual states' right to determine either the sort of insurance product offered or its price. The directive allows each EU-domiciled insurer to have a single regulator (the home country regulator) for the entire market. This means that once an insurer is licensed in an EU state, it can transact business and/or establish branches in another EU state, without further hindrance from a regulator. This single licence is known as the "passport." The EU third directive has removed barriers that some countries had put in place, which has encouraged the construction of a single European insurance market and may lead to the creation of a single European financial market. It has already generated more competition in the EU. Thus, all insurance companies in the EU now have the right of establishment and freedom to provide services all over Europe.

An important effort of simplification has been made for life insurance in the EU since the piecemeal development of the European legal regime had made it unnecessarily complex, the directives needed to be reorganized into one single and coherent text. The European Parliament and the Council have adopted a directive concerning life assurance (Directive 2002/83/EC of the European parliament and of the Council of 5 November 2002). The main objective of this directive is to facilitate the comprehension and application of the life assurance directives by recasting them into a clear, coherent, and complete legal text.

P&C insurance is still regulated by many directives. There are for example various directives for the automobile insurance market. The automobile case is interesting because it shows the limits of influence of the EU on member states. In some European countries, there is a bonus/malus system, classifying motorists in function of their past car claims. The fewer claims a driver has, the higher his bonus and the lower his premiums. The European Commission declared that the bonus/malus system was an obstacle to competition since it had a compulsory effect on tariffs. However, the European Court of Justice in its judgment delivered September 7, 2004 declared the bonus/malus systems established in France and Luxembourg for motor vehicle insurance contracts are lawful.

One future evolution will be the creation of a European insurance contract law. With the right of establishment and freedom to provide services, insurance companies could sell their products all over Europe. The risk for the potential European customer is to sign an insurance contract, which is not similar to their "habitual" national contracts. For example, some suggestions have been made to enforce specific disclosures from the insurer to customers before they accept an insurance policy.

An important aspect of insurance regulation is state supervision. France reorganized its financial regulatory authorities in 2003. The Financial Security Act of August 1, 2003 (*Loi de sécurité financière n°2003–706 du 1er août 2003*) reflects the convergence between the bank and insurance industries (CCA 2004; Marini 2004). Table 5.2 shows the various institutions now involved in the regulation, consultation (during decision process), control of conduct of business rules, prudential supervision, and licensing.

In particular, a single entity, the Authority of Financial Market (*Autorité des marchés financiers,* AMF) was set up to handle all the tasks involved in supervising corporate finance, financial information, investment providers, collective investments schemes, financial markets, and post-trade activities (AMF 2004).

The same law directed the merger of various existing supervisory institutions into one *Commission de contrôle des assurances, des mutuelles et des institutions de prévoyanc* (CCAMIP), or Provident Institutions, Mutual Society and Insurance Company Control Commission. Its main duty is to control insurance and reinsurance companies, mutuals and Provident Institutions (*institutions de retraite supplémentaire relevant du Code de la sécurité sociale*). The CCAMIP exercises its supervision through high-ranking civil servants, trained specifically to control insurance companies (*commissaire-contrôleur des assurances).* They have the access to all information and documentations within insurance companies.

Furthermore, a *Comité consultatif du secteur financier* (CCSF), or Financial Sector Consultative Committee, was set up to follow the relationships between firms and consumers and to bring together representatives of the two groups.

The *Comité consultative de la legislation et de la réglementation financière* (CCLRF), or Regulation Consultative Committee, is an advisory committee on financial legislation and regulation. It will issue an opinion on any draft legislation relating to financial matters (i.e., banks, insurance, investment services).

Finally, the *Comité des entreprises d'assurance* (CEA), or Insurance Company Committee, was created to license new insurance companies and mutual societies. The committee has power of decision concerning not only licensing, but also portfolio transfer (*transferts de portefeuilles*), acquisition, or financial participation (*prise de participation*).

The Financial Security Act also places insurers under more stringent obligations to provide customers with better information. For example, when a life policy is purchased, the table of cash surrender values must also indicate the amount of premiums paid. Contracts with unit-linked features must specify the main characteristics of the underlying investment vehicles. Furthermore, the insured should receive annual information about guaranteed returns, dividends, and the average rate of return on assets held by the insurance company (FFSA 2003:16).

Supervisory institutions are also in charge of controlling insurance company solvency. The influence of the EU has been very important here. French law No. 2003–1236 of December 22, 2003 has incorporated into national law the two European directives, Solvency I (2002/12/EC and 2002/13/EC of March 5, 2002 (CCA 2004:36), which improve the existing rules for the calculation of the solvency margin requirement for insurance undertakings. Many features are common to both directives, one of which covers life assurance and the other P&C insurance. Together they represent a package of measures applying to all insurance organizations qualifying under the EU insurance directives for a 'single passport,' which allows them to sell policies anywhere in the EU on the basis of the supervision carried out by their home member state.

Under the directives, the French law requires that the absolute minimum amounts of capital required (the so-called minimum guarantee fund) is set at 3 million euro (2 million euro for some classes of P&C insurance) compared to the previous amounts, which ranged between 200,000 euro and 1.4 million euro. Furthermore, the 2003 Financial Security Act has developed the scope of the existing Car Guaranty Fund (created in 1951) to all compulsory P&C insurances. This fund is the guarantee to protect the right of the insured in case of insurance company defect (*fonds de garantie des assurances obligatoires de dommages)*. This guarantee fund is pre-funded by a compulsory contribution of all insurance companies in function of their P&C premiums (Poiget 2004). However, if the fund is not sufficient, insurance companies could have to post-fund it after assessment. So far, the fund has been sufficient to cope with just a few minor insurance companies' insolvencies.

The thresholds based on levels of premiums and claims below which a higher solvency margin is required have also been increased. The required solvency margin on a premium basis is now 18 percent up to the first 50 million euro of premium, up from 10 million euro. The solvency margin on a premium basis above 50 million euro is now 16 percent. On a claims basis, the margin is 26 percent on the first 35 million euro, up from 7 million euro and also indexed in future, and 23 percent above 35 million euro.

Supervisory bodies have increased powers to intervene early to take remedial action where policyholders' interests are threatened. For example, in a situation where an insurance undertaking currently satisfies the solvency margin requirements, but its financial position is deteriorating rapidly, the supervisor will now be able to take action.

For certain categories of P&C business that have a particularly volatile risk profile (marine, aviation, and general liability), the required solvency margin is increased by 50 percent. Apart from reinforcing solvency requirements, the objective is to better match regulatory capital requirements to the real risk profile of the undertaking.

When the CCAMIP estimates that an insurance company is in peril (which is quite rare), the CCAMIP has the right to increase the solvency margin of the concerned company (*marge de solvabilité renforcée*). The insurance company in jeopardy should within one month propose a financial recovery plan in order to restore the firm's financial equilibrium (CCA 2004:37). No financial recovery plan (*plan de redressement*) was required from French insurance companies in 2002 and 2003 (CCA 2004:41).

Among the regulations, another important topic is the tax system concerning the French insurance market. The level of taxation is quite high and not equal for all the insurance economic actors. On both life and P&C insurance, there are specific taxes. For example, since 1967, a tax has been levied on automobile insurance premiums (mandatory casualty coverage). The rate on car liability insurance is 33 percent (18 percent for car insurance contract tax plus 15 percent for mandatory casualty coverage; FFSA 2003:57). Personal fire insurance policies are taxed at 30 percent of the premiums. Another tax concerns the creation of a fund for victims of terrorism and other offences (*fonds d'indemnisation des victimes d'actes de terrorisme et d'autres infractions*), created in 1986 (*Loi du 9 septembre 1986 and Loi du 23 janvier 1990*), after a wave of terrorism in France. The tax is levied on all property and casualty insurance contracts (3.3 euro per contract in 2004).

Other taxes concern not the insurance products, but insurance company activities. For example, an 8 percent tax is levied on the employer's contribution to group employee retirement benefit plans in companies with more than ten employees (*taxe sur les contributions patronales de prévoyance complémentaire*). Another tax, created in 1983, concerns the surplus of reserves of P&C companies, in an attempt to recover surplus cash where the initially established reserve exceeded the total cost of claims. The rate is set at 0.75 percent per month, or 9 percent per year (FFSA 2003:55). It should be noted that there are some surprising competition distortions in the French insurance market. Indeed, private insurance companies are subject to various taxes, while mutual societies and provident institutions are sometimes exempt. For example, when mutuals and provident institutions are exempt (FFSA 2003:56), private companies are subject to:

- an effective tax of 19.6 percent on long term equities-linked realized capital gain (*taxe sur les plus-values à long terme sur titres de participation*);
- a business tax (*taxe professionelle*); or
- apprenticeship tax (*taxe d'apprentissage*).[156]

Table 5.2. Type of Supervision After the Financial Security Act of August 1, 2003, by Type of Institution

	Mutuals 1945	Insurance Companies, Mutuals of Insurance, Provident Institutions	Credit Institutions	Investment Services	Market Under-takings	Asset Manage-ment
Regulation	Welfare Ministry	Finance Ministry	Finance Ministry	AMF	AMF	AMF
Consultative Powers	CSM	CCLRF / CCSF	CCLRF / CCSF	AMF	AMF	AMF
Conduct of Business Rules	CCAMIP	CCAMIP	CB	AMF	AMF	AMF
Prudential Supervision	CCAMIP	CCAMIP	CB	CB	CB	AMF
Licensing	Welfare Ministry	CEA	CECEI	CECEI	CECEI	AMF

Source: Ministry of the Economy, Finance and Industry (2004), Marini (2004), AMF (2004).
Note: AMF = Autorité des Marchés Financiers; CCAMIP = Commission de Contrôle des Assurances, des Mutuelles et des Institutions de Prévoyance; CCLRF = Comité Consultatif de la Législation et de la Réglementation Financières; CCSF = Comité Consultatif du Secteur Financier; CEA = Comité des Entreprises d'Assurance; CSM = Conseil Supérieur de la Mutualité; CB = Commission Bancaire; and CECEI = Comité des Etablissements de Crédit et des Entreprises d'Iinvestissements.

Furthermore, private companies are subject to income tax at an effective rate of 34.3 percent, while mutual societies and provident institutions have a maximum rate of 24 percent. In 2005, a motion of the FFSA against this competition distortion is under investigation by the European Union.

The tax burden is not expected to decrease quickly because the French government has the tendency to see insurance as a revenue source. For example, the

[156] Tax equivalent to 0.5 percent of all paid salaries to fund technical and professional education.

2004 Finance Law authorized the government to require 300 million euro in the compulsory insurance warranty fund (*fonds de garantie des assurances obligatoires*), to cover accident claims when the responsible party is not known or is not insured, or his insurance company is unable to cover the claim because of financial difficulties. The fund is financed mainly by insurance companies (Weinberg 16 November 2004:34).

5.4 GENERAL DESCRIPTION OF THE FRENCH INSURANCE MARKET[157]

Considering the long history of the insurance industry in France, it is not surprising to see that the insurance sector is significant in this country. In premium income volume, this market is the fourth in the world, with a market share of 6 percent in 2004 (Swiss Re 2005), and second largest in the European market. Four countries in the European Union—the U.K., France, Germany and Italy—represent two-thirds of the European insurance market (with 15 countries) (CEA 2004:9).

If we consider only domestic business (insurance sales in France, not abroad), the French insurance market represented 141.8 billion euro of sales in 2003, including France operation by French and foreign companies licensed to sell insurance in France and branch offices of European companies in France, excluding inward reinsurance. Since 1990, total premiums increased 244 percent (cf. Table 5.3).

The insurance market is dominated by the life business with a market share of 63.3 percent in 2003, as shown in Table 5.4. Automobile insurance is the second largest product and represented 12.1 percent of the French insurance market in 2003. Property insurance accounted for 8.3 percent of the insurance market, and health and bodily injury accounted for 7.9 percent. Note the low market share of liability insurance in France, with only 1.8 percent of the market in 2003.

The French market is characterized by a specialization principle (*principe de specialisation*). The same legal entity cannot underwrite both life and P&C insurance contracts at the same time. French companies that sell both have by law two different legal entities: one for P&C and the other for life insurance. The specialization principle is meant to protect savings of life policyholders. However, both companies could be in fact within the same financial holding company, operating in the same building with the same information technology facilities and using the same distribution channels. Therefore, one single economic actor must have at least two legal entities to underwrite life and P&C risks. However, life insurance companies can sell health complementary insurance contracts. Furthermore, since the third directive, composite insurance companies (*sociétés d'assurance mixtes*) can insure life, health, and bodily injury policies.

---

[157] The part of chapter is based mainly on statistics provided by the FFSA, Fédération Française des Sociétés d'Assurances, French Federation of Insurance Companies. FFSA annual report is (to our knowledge) one of the best professional sector reports in the country. The author thanks FFSA for authorizing the extensive use of data.

Table 5.3. Insurance Premiums in France for Life and Health and Property and Casualty Insurance, 1990 to 2003 (in billions of Euros)(1)

	1990	1991	1992	1993	1994	1995	1996	1997	1998	1999	2000	2001	2002	2003
Life and Health	35.8	40.6	46.7	56.4	66.5	72.2	79.6	86.0	73.9	83.7	99.6	94.2	95.6	103.2
Automobile	10.6	10.9	11.2	11.8	12.6	13.6	14.1	14.0	13.9	14.1	14.6	15.4	16.3	17.1
Individual Property	3.2	3.3	3.5	3.7	3.9	4.2	4.4	4.4	4.5	4.5	4.7	4.9	5.1	5.4
Commercial Property	2.9	2.9	3.0	3.4	3.5	3.8	4.0	3.8	3.8	3.8	3.9	4.2	5.0	5.6
Agricultural Property	0.8	0.8	0.8	0.8	0.8	0.8	0.8	0.8	0.8	0.8	0.8	0.8	0.8	0.8
Subtotal Property Insurance	6.9	7.0	7.3	7.9	8.2	8.8	9.2	9.0	9.1	9.1	9.4	9.9	10.9	11.8
General Liability	1.3	1.3	1.4	1.4	1.5	1.5	1.6	1.6	1.7	1.7	1.8	2.0	2.2	2.6
Construction	0.6	0.6	0.5	0.5	0.5	0.7	0.7	0.9	0.9	1.0	1.1	1.2	1.3	1.5
MAT	1.0	1.1	1.3	1.4	1.5	1.6	1.5	1.4	1.1	1.1	1.2	1.6	1.5	1.2
Natural Disasters	0.6	0.6	0.6	0.7	0.7	0.8	0.8	0.8	0.8	0.8	1.0	1.0	1.1	1.2
Miscellaneous(2)	1.3	1.5	1.8	1.9	2.0	2.1	2.2	2.3	2.3	2.4	2.6	2.8	3.0	3.2
Total Property and Casualty Insurance	22.3	23.0	24.1	25.6	27.1	29.1	30.0	30.0	29.8	30.2	31.7	33.9	36.3	38.6
Total Annual Premiums	58.1	63.6	70.8	82.0	93.6	101.3	109.6	116.0	103.7	113.9	131.3	128.1	131.9	141.8

Source: FFSA (2004), pp. 70–71; FFSA (1999), p. 87.
(1) Premiums prior to the adoption of the Euro have been converted using the exchange rate 1 euro = 6.5597.
(2) Miscellaneous includes credit, legal expenses and assistance.

Table 5.4. Breakdown of Premium Income in France, 1975, 1980, 1985, 1990, and 1999–2003 (Direct Business in the French Market) (Affaires directes)

Insurance Product in % of the total	1975	1980	1985	1990	1995	1999	2000	2001	2002	2003
Life	22.4	23.8	29.7	42.4	59.1	63.2	67.0	64.1	62.9	63.3
Automobile	35.7	32.5	26.6	18.2	13.4	12.4	11.1	12.0	12.4	12.1
Property	17.2	17.0	17.2	11.9	8.7	8.0	7.1	7.8	8.3	8.3
Health and Bodily Injury	11.1	12.4	11.2	9.0	8.2	7.9	7.1	7.4	7.7	7.9
Capital Redemption Bonds	2.6	3.1	4.7	10.3	3.9	2.3	1.9	2.0	1.7	1.6
Liability	6.1	6.4	4.0	2.2	1.5	1.5	1.4	1.5	1.7	1.8
MAT (Transports)	3.5	3.1	2.7	1.8	1.6	1.0	0.8	1.3	1.1	0.8
Natural Disasters	*	*	*	1.0	0.8	0.7	0.8	0.8	0.8	0.8
Construction	**	**	**	1.0	0.6	0.9	0.8	1.0	1.0	1.1
Miscellaneous	1.4	1.7	2.2	2.2	2.2	2.1	2.0	2.1	2.4	2.3
Total	100.0	100.0	100.0	100.0	100.0	100.0	100.0	100.0	100.0	100.0

Source: FFSA (1985), p. 10; FFSA (1990), p. 5; FFSA (1999), p. 6: FFSA (2000), p. 6; FFSA (2001), p. 8; FFSA (2002), p. 8; FFSA (2003), p. 8.
* included in Property.
** included in Miscellaneous.

Considering the specialization principle, we will present the two main lines of business separately: Life and health insurance, *assurances de personnes,* literally "people insurance;" and property and casualty insurance (including liability insurance). In France, they are named *assurance dommages de biens et des responsabilités*, literally "belongings casualty and liability insurance." The acronym IARD is also very often used—*Incendie Accidents Risques Divers*, literally: fire, casualty, and other risks.

5.4.1 Life and Health Insurance Market[158]

The life and health line of business dominates the French insurance market with 103.2 billion euro of premiums in 2003. The life and health branch has grown more than 288.3 percent since 1990, as shown in Table 5.3.

Within life and health insurance, the life activity is based principally on investment-related insurance products, for nearly 85 percent of the life premiums. Sixty-seven percent of the life business corresponds to euro-based contracts, and to French franc-based contracts before 1/1/2002; and 13.6 percent to unit-linked contracts (FFSA 2003), as shown in Table 5.5. Unit-linked is a life insurance policy or capital redemption bond in which the amounts of benefits and premiums are not expressed in euros but in units of an investment vehicle, such as shares of a mutual fund or a real estate partnership.

Capital redemption bond line is a marginal business, with a market share of 2 percent of the life and health insurance market, equivalent to 2.2 billion euro in 2003. Their tax status was modified in 1998, and capital redemption bonds became less attractive.

Term life insurance represented 6.4 percent of the life and health insurance market in 2003, with 6.6 billion euro of premiums. This type of contract is quite often taken in connection to a real estate mortgage loan. If we consider the limited market share of term insurance, it is clear that the life insurance market in France is a

[158] French insurance industry statistics have various sources. Before the creation of the CCAMIP at the end of 2003, insurance entities were controlled by different governmental institutions and each of them provided its own statistics. Furthermore, it was difficult to have up-to-date aggregate data. For example, the 2003 Report of the *Conseil National des Assurances* described the French insurance market until 2001 (CNA 2003). Therefore, to get the most recent data it is necessary to rely on French insurance professional associations.

The FFSA, *Fédération Française des Sociétés d'Assurances* is recognized in France as the best information provider. The FFSA Annual report computes data from insurance companies and Mutual societies of Insurance (SAM), which are regulated by the Insurance Act (*code des assurances*). In quoting the various FFSA reports, we will use the year of reference and not the year of publication. The market share of FFSA members is 95 percent in life insurance and 65 percent in P&C insurance. In this chapter, we will use mainly FFSA statistics.

The GEMA, *Groupement des Entreprises Mutuelles d'Assurance* is a professional association for mutual societies (not Mutuals 1945; see the description of the various types of insurance firms). The statistics provided are quite rare and not very substantial (GEMA 2003, 2004).

The FNMF, *Fédération Nationale de la Mutualité Française* is the association for mutuals 1945. The FNMF does not provide many statistics.

saving market. This saving market has been growing rapidly in the last 30 years and is expected to continue growing, if tax incentives remain similar.

There is concern in France over the funding of population retirement benefits. Mandatory retirement plans paid out benefits totaling an estimated 165 billion euro in 2004 (FFSA 2003). Seventy-five percent of the revenues of people over the age of 65 come from the French state benefits (EUC 2003). In the last 20 years, various reports have been written and different reforms have been introduced. The last reform was enacted August 21, 2003 (Fillion Law, *Loi Fillion*, named for the French minister in charge of reform). Private employees and civil servants should be working for at least 40 years (41 years in 2012) before retiring with a full pension, which is calculated on the basis of average salary of the best 20 years of work (in 2012, the best 25 years) (Bechtel 2004:18). Some financial incentives have been set up to motivate employees to work longer (COR 2004). Retirement benefits will now accrue according to price index, and not salary index.

Table 5.5. Life Insurance Premiums and Market Share (Life and Health) in France by Type of Contract in 2003

	Euros (billions)	Percentage
Pure Endowment Insurance(1)	83.2	80.6
Individual Policies	76.9	
Group Policies	2.2	
Term Life Insurance	6.6	6.4
Individual Policies	2.0	
Group Policies	4.6	
Capital Redemption Bonds	2.2	2.1
Subtotal (Endowment + Term + Bonds)	92.0	
Euro-based Contracts	77.1	
Unit-linked Contracts	14.9	
Health and Bodily Injury Insurance	11.2	10.8
Disability (*incapacité-invalidité*)	5.8	
Health Insurance (*soins de santé*)	5.4	
Total	103.2	100.0

Source: FFSA (2003), p. 13.
(1) Pure endowment insurance could be divided between unit-linked contracts (13.6%) and Euro-based contracts (67% of the total life insurance premiums in 2003).

Two new retirement saving products have been created—the *Plan d'Epargne Retraite Populaire* for individuals, and a employer-sponsored equivalent, the *Plan d'Epargne Retraite Entreprise*—that provide replacement income in the form of a life annuity. Premiums are to a certain extent tax-deductible, and annuities are subject to income tax. The entire population has the right to sign these contracts.

Life insurance is an important part of savings in the French household. Life insurance has represented the majority of household investment flow since at least 1992. The net investment flow doubled between 1990 with 61.9 billion euro and 2003 with 111 billion euro. In 2003, the insurance share of the investment flow was 55.5 percent, as shown in Tables 5.6 and 5.7. In 1990, financial assets invested in life insurance accounted for 13 percent of household financial investments. In 2003, life insurance represented 31.5 percent of the financial investments of French households.

Table 5.6. Net Financial Investment Flow of French Households (Real Currency Unit), 1990 to 2003

	Net Investment Flow		Price Index	Net Financial Investment Flow (%)			
Year	Billions €	% change	% change	Liquid Assets (%)(1)	Securities (%)	Insurance Rerserves (%)	Total (%)
1990	61.9	–18.6	3.4	22.3	35.5	42.2	100.0
1991	79.4	28.3	2.9	13.0	51.0	36.1	100.0
1992	67.8	–14.6	2.0	22.1	27.9	50.0	100.0
1993	82.9	22.2	2.3	39.3	9.9	50.8	100.0
1994	79.1	4.6	1.7	24.5	15.5	60.0	100.0
1995(2)	95.9	21.3	1.7	42.1	–1.3	59.2	100.0
1996	86.2	–10.2	1.5	12.3	–14.0	73.7(3)	100.0
1997	99.7	15.6	1.3	32.1	–8.9	76.1	100.0
1998	91.0	–8.5	0.9	24.8	12.9	62.3	100.0
1999	110.1	20.8	0.5	29.1	10.6	60.2	100.0
2000	73.9	–32.8	0.9	–0.3	9.4	90.9	100.0
2001	106.4	43.8	2.3	26.4	14.8	58.7	100.0
2002	102.8	–3.4	2.1	28.8	15.9	55.3	100.0
2003	111.0	8.2	2.1	36.4	8.1	55.5	100.0

Source: FFSA (1999), p. 93; FFSA (2000), pp. 94–95; FFSA (2003), p. 77. Source Price Index: INSEE, www.insee.fr/fr/indicateur/indic_cons/indic_cons.asp.

(1) Liquid assets (*Actif liquide*) are those that could be negotiable rapidly in cash without any financial gain or cost. Cash on a bank account is a typical liquid asset. This is also the case of the *livrets d'épargne*: French savings accounts with low interest rates.

(2) Since 1995, including private equities.

(3): Read 73.1% of the flow. The flow was multiplied by two between 1991 and 1996.

The second important branch in life insurance is health. Health and accident-related insurance is provided under specific contracts: individual insurance against accidents, supplementary health or hospital care insurance policies, long-term care insurance, non-work-related accidents, etc.; or as supplementary coverage linked to life insurance policies such as long-term or occupational disability. Health insurance contracts provide various coverages:

- reimbursement of healthcare and medical expenses, after social security reimbursement. The French phrase is *complémentaire maladie*, complement for healthcare, describing extra coverage after social security.
- payment of benefits to compensate the loss of job-related income due to illness, unemployment, or long-term disability.

Table 5.7. Total Household Financial Investments in France (Real Currency Unit and %), 1990 to 2003

	Financial Investments		Price Index	Breakdown of Financial Investments			
Year	Billions €	% Change	% Change	Liquid Assets	Securities	Insurance Reserves	Total (%)
1990	1,334.0	–4.1	3.4	41.1	45.8	13.1	100.0
1991	1,513.2	13.4	2.9	36.4	50.1	13.5	100.0
1992	1,611.2	6.5	2.0	35.5	49.6	14.9	100.0
1993	1,865.0	15.8	2.3	32.5	52.2	15.3	100.0
1994	1,640.0	–6.0	1.7	40.6	38.7	20.7	100.0
1995	1,822.0	11.1	1.7	41.0	37.0	21.9	100.0
1996	2,039.9	11.9	1.5	37.5	39.7	22.8	100.0
1997	2,263.0	10.9	1.3	35.3	40.6	24.1	100.0
1998	2,542.0	12.3	0.9	32.1	44.0	23.9	100.0
1999	3,017.5	18.7	0.5	28.4	48.8	22.9	100.0
2000	2,951.8	–2.2	0.9	28.9	45.6	25.5	100.0
2001	2,878.8	–2.5	2.3	31.4	40.8	27.8	100.0
2002	2,685.2	–6.7	2.1	33.6	35.3	31.2	100.0
2003	2,901.9	8.1	2.1	32.3	36.3	31.5	100.0

Source: FFSA (various annual reports); FFSA (1999), p. 93; FFSA (2000), pp. 94–95; FFSA (2003), p. 77.

Bodily injury insurance covers the payment of lump-sum benefits and, in some cases, the reimbursement of medical costs in the event of an accident leading to disability or death. Note that bodily injury related to liability insurance, especially automobile liability, is not in the scope of the life and health branch, but under the property and casualty branch.

The health and bodily injury insurance is a complement to the French state system. In fact, social security is compulsory for nearly all French workers, and health coverage is provided largely within the framework of the state system, representing 78.9 percent of the total health expenses (*Ministère de la Santé* 2004). National health expenditures represented 9.4 percent of the French GDP in 2002 (vs.

14.6 percent in the U.S., 10.9 percent in Germany, and 7.7 percent in the U.K.) (*Ministère de la Santé* 2004:39).

Health and bodily injury had a market share of 7.9 percent in 2003 of the French insurance market, as shown in Table 5.4. This is equivalent 11.2 billion euro, according to data given by FFSA concerning the sales by its members. However, we should add to this number the sales by the mutuals 1945 and the provident institutions. The total sales are in fact 20.5 billion euro in 2003 (Bechtel 2004:55). The mutual societies (any kind) dominate this branch with a market share of 57.1 percent in 2003 (equivalent to 11.8 billion euro), followed by the insurance companies with 22.7 percent (or 4.5 billion euro), and the provident institutions with 20.2 percent (or 4.2 billion euro).

This branch is expected to grow in the next few years because of the decline in social security coverage, which has been facing an important deficit for many years. Various reforms are aimed at reducing the deficit. One simple reform is to reduce the basic coverage of heath expenses, provided by the state system. The percentage of the coverage by Social Security declined over the year for some products. Some medicines are now described as "comfort medicine" and are not covered at all. Also, chemists have the right and are ordered to change prescriptions to cheaper generic medicines for patients if generics are available. In 2005, a reform is the creation of the concept of referee doctor (*médecin traitant*). Each French citizen older than 16 must chose a medical doctor and gives his or her name to Social Security. Since the beginning of July 2005, one must see the referee doctor before visiting any specialist, except for pediatricians, gynecologists and ophthalmologists. If a patient does not see the referee doctor first, Social Security will reduce the health expense coverage.

Investment in information technology offers possibilities. For example, all French citizens have a Social Security card (*carte vitale*) with an electronic chip containing information about the cardholder's past health, medical history and demographics information about family members covered. This information is accessible only to the physician. Another card, *Santé Pharma,* entitles direct payment of certain health expenses by Social Security and other supplementary coverage providers, and allows households to avoid upfront payment for medical care. Originally, this system involved only pharmacist expenses but is now extended to other types of health providers such as medical laboratories. In 2002, a High Council for the Future of Health Insurance was created to prepare future reforms (*Haut Conseil pour l'avenir de l'assurance maladie*). France faces the aging population phenomenon, where the percentage of the population older than 65 increases (16.2 percent of the French population was over 65 in 2003) (INED 2003). Because the health expenses are going to increase "naturally" with the aging population, the health insurance business is growing.

5.4.2 Property and Casualty Market

Despite the domination of the life insurance market, the P&C insurance market is also quite important in France, with a premium volume of 38.6 billion euro in 2003, representing 27 percent of the total insurance market that year. With only 22.3 billion euro in 1990, this business branch grew nearly 170 percent between 1990 and 2003, as shown in Table 5.3.

In the P&C market, automobile insurance and property insurance represented a large majority of the business in 2003, with a market share of 74.9 percent. Car insurance had a market share of 12.1 percent of the total French insurance market, the second largest insurance product, as shown in Table 5.4. Car insurance represented 44 percent of the P&C market in 2003, as shown in Table 5.8. Individual automobile insurance represented 85 percent of the automobile insurance market, while commercial automobile insurance accounted for 15 percent in 2003. The automobile premiums grew from 10.6 billion euro in 1990 to 17.1 billion euro in 2003 (see Table 5.3). For more than 10 years, French state and insurance firms have joined their efforts to decrease the number of annual automobile fatalities, launching various operations: safe-driver programs for young people, national advertisement campaigns, seat-belt laws since 1973, installation of more automated speed radar and so on. Insurance companies are required to spend at least 0.5 percent of the annual car liability premiums on car accident prevention. These measures have been quite successful—the number of fatalities has dropped from 321 per million vehicles in 1994 to 216 in 2002 (FFSA 2003). However, the premiums did not decrease during the same period. Indeed, the average cost of claims increased, especially costs related to body injuries. Furthermore, the market witnessed an increase in reinsurance costs.

Table 5.8. Premiums and Market Share of the Main Property and Casualty Insurance Products in France in 2003

	Premiums (in billions of Euros) in 2003	% Change Between 2002 and 2003	Market Share of Property and Casualty (%)
Automobile	17.1	5.0	44.3
Commercial Property	5.6	12.0	14.5
Individual Property	5.4	6.0	14.0
Agricultural Property	0.8	0.6	2.1
Subtotal Property Insurance	11.8		30.6
General Liability	2.6	18.0	6.7
Construction	1.5	14.7	3.9
Natural Disasters	1.2	9.1	3.1
MAT	1.2	–19.4	3.1
Credit and Surety	0.8	3.8	2.1
Assistance	0.8	6.5	2.1
Legal Expense	0.5	11.8	1.3
Other	1.1	9.0	2.8
Total Primary Business	38.6	6.3	100.0

Source: FFSA (2003), p. 23.

In 2003, the property insurance market (individual, commercial, and agricultural property) accounted to 30.6 percent of the P&C insurance market in France. This type of insurance increased from 6.9 billion euro in 1990 to 11.8 billion euro in 2003, the equivalent of a 170 percent increase (see Table 5.3). Commercial property and individual property insurance are respectively the second and the third insurance product of P&C insurance. Commercial property accounted for 5.6 billion euro of premiums in 2003, with a market share of 14.5 percent of the P&C market (see Table 5.8). Thirty-seven percent of the commercial insurance premiums in 2003 were for fire insurance. Commercial property insurance covers different types of risks: major industrial and commercial; medium-sized firms; self-employed tradesmen, retailers, and service providers; and city councils. Despite many efforts by insurance companies to increase sales of this insurance, business interruption insurance was included in only one-third of commercial property insurance contracts in 2003. Individual property accounted for 5.4 billions of euro of premiums in 2003, with a market share of 14 percent of the P&C market, as shown in Table 5.11.

Liability insurance represented 2.6 billion euro of premiums in 2003, equivalent to a market share of 6.8 percent of the P&C market.[159] The annual premium volume was multiplied by two since 1990. Until 2003, the French Supreme Court of Appeals (decision of December 19, 1990) recognized one criterion for determining the application of coverage over time: the original event must have occurred between the date of effect and the termination date of the policy. Since the Financial Security Act of 2003, the insurers and the insured have the choice between two coverage triggers: "the risk occurrence date" (the date of the event at the origin of the damage) or the "claim date."

Construction insurance corresponds to two types of contracts: "building damage insurance" where the building owners are entitled to receive compensation (*assurance dommages à l'ouvrage*), and decennial liability insurance, which covers builder liability over a period of ten years (*garantie responsabilité civile décennale*). Construction insurance premiums have been multiplied threefold from 1990 to 2003, reaching 1.5 billion euro, as shown in Table 5.3.

Maritime aviation and transport (*MAT*) insurance corresponds to various risks: hull insurance (insurance of vessels), cargo insurance (insurance of goods in transit), aviation insurance, and space insurance. Considering these risk types, it is not surprising that MAT is a prominent international business branch. As a consequence, premiums are usually in U.S. dollars. The recent and important rise of the exchange between euro and U.S. dollars led to a decrease of the MAT book value from 1.6 billion euro in 2001 to 1.2 billion in 2003 for the French market (see Table 5.3). This trend was also accompanied by a premium decline. However, MAT contracts signed in France cover not only France for 1.2 billion euro, but also foreign risks for an extra 0.912 billion euro. The total MAT premium is then 2.112 billion euro in 2003, divided between hull insurance (23 percent of market share, or 482 million euro), cargo insurance (37 percent), aviation insurance (33 percent), and space insurance (6 percent). Space insurance covers satellite launches, satellites already in orbit, and damages to satellites.

---

[159] This figure does not take into account automobile liability insurance (included in car insurance), individual liability insurance (included in many individual property insurance contracts), and construction liability insurance (included in construction insurance).

5.5 INSURANCE ECONOMIC ACTORS IN FRANCE

Various economic actors participate in the French insurance market. We will analyze the market concentration and the various types of companies active in France.

5.5.1 Market Concentration

There are many insurance companies in France: 486 were inventoried in 2003 (see Table 5.11), reflecting more the existence of various legal entities than real economic actors. The market share of the largest P&C insurance company was 10.8 percent in 2003, 9.9 percent in 1998 and 9.8 percent in 1993. The concentration of the top five P&C insurance companies was 38.9 percent in 2003, 35.6 percent in 1998 and 30.9 percent in 1993. The concentration of the top ten P&C insurance companies was 56.7 percent in 2003, 53.1 percent in 1998 and 46.5 percent in 1993 (See Table 5.9). The market share of the largest life insurance company was 11 percent in 2003, 12.4 percent in 1998 and 10.4 percent in 1993. The concentration of the top five life insurance companies was 44.3 percent in 2003, 41.1 percent in 1998 and 41.5 percent in 1993. The concentration of the top ten life insurance companies was 64.2 percent in 2003, 60.3 percent in 1998 and 58.7 percent in 1993 (see Table 5.9). Since at least 1993, there is a movement of insurance concentration in France in both life and P&C insurance. In fact, the concentration of the life insurance market grew with the merger of Prédica and UAF in 2004, making the new group the largest life insurance company in France.

Table 5.9. Insurance Concentration (Market Share by Premium Volume of the top 1, 5, and 10 insurers) in France in 1993, 1998, and 2003

		2003		1998		1993	
		Life	Property and Casualty	Life	Property and Casualty	Life	Property and Casualty
CR1	Top Company	11.0	10.8	12.4	9.9	10.4	9.8
CR5	Top 5	44.3	38.9	41.1	35.6	41.5	30.9
CR10	Top 10	64.2	56.7	60.3	53.1	58.7	46.5

Source: Author.

In 2003, the largest insurance company was CNP, *Caisse Nationale de Prévoyance* (see Table 5.10), created in 1959 and originally state-owned. It became a joint-stock company in 1992. The CNP stocks are owned by the *Caisse des dépôts* (37 percent), the French post office (18 percent), the bank Caisse d'Epargne (18 percent), private investors (24 percent), other partners (2 percent), and the state (1 percent) (Couilbault 2003:37). In 2003, the premium volume in France was 18.6 billion euro, with 92 percent in life business. CNP started foreign operations in 1997. CNP has no direct distribution channel, but sells its products through other institutions. The CNP is then a business-to-business company, using distribution

Table 5.10. Top 10 Insurers in France in 2003 (Premiums in Billions of Euros)

Name of Firm (Year of Creation for Bancassureurs)	World Premiums	Owner	Nation	Premiums Total France	Premiums Life	Premiums Property and Casualty
CNP (1)(1959)	19.5	Bank	France	18.6	17.2	1.4
AXA	71.6	Insurance Company	France	15.5	10.9	4.6
Crédit Agricole(1)*	11.4	Bank	France	11.4	10.7	0.7
Groupama	12.9	Insurance Company	France	10.5	3.6	6.9
AGF	16.5	Insurance Company	Germany	9.6	5.3	4.3
Generali France	9.6	Insurance Company	Italy	9.5	6.1	3.4
BNP Parisbas(1)**(1973)	9.3	Bank	France	6.6	6.6	0
Sogecap(1)***(1973)	5.9	Bank	France	5.7	5.7	0
ACM(1)(1971)	4.9	Bank	France	4.9	3.7	1.2
UAF(1)	4.5	Bank	France	4.1	3.9	0.2

Source: Author and FFSA (2003), p. 49.
(1) This firm is a *bancassureur*. In this table, all firms are stock companies.
* Prédica (1986)/Pacifica (1990).
** Cardif (1973), Natio Life and Natio Insurance (1980).
*** Sogecap (1973 for UMAC Life, then 1984 for Sogecap).

channels of its main shareholders. The distribution of its products is then done mainly over the counter through partnership with the French post office (17,000 branches), the bank Caisses d'Epargne (5,500 branches), and other financial institutions. In French, "over the counter" (*vente aux guichets*) describes the sale of insurance products in bank, financial institutions, or post office branches. A new alliance was signed between CNP and FNMF (*Fédération Nationale de la Mutualité Française*) in order to sell CNP insurance products, mainly retirement funds, through the FNMF members (Weinberg June 4–5, 2004:25).

AXA is the second largest company in the French market. However, it is the biggest French insurance company in consolidated premiums (71.6 billion euro in 2003). The company was created in 1958 by a genial entrepreneur, Claude Bébéar, who transformed a small mutual society, *Mutuelles Unies*, into a private company, which grew rapidly through external, national, and international acquisitions. His company bought Equitable in 1992, Guardian Royal Exchange in 1999, and UAP (*Union des Assurances de Paris*) in 1997. The brand AXA was created in 1985. In 2003, the premium volume in France was 15.5 billion euro, with 70 percent in life business. The main distribution channel of AXA France is through tied agents. It is now one of the largest private insurance companies in the world, listed on the New York, London, and Paris stock exchanges. The CEO, Henri de Castries, pursues a strategy of targeting fast growing markets and international acquisitions (Gay September 14, 2004:41). AXA has recently developed a cost reduction program with off-shoring (in India and Morocco), purchasing centralization, and reorganization (Weinberg November 21–22, 2003:24; Maujean June 3, 2004:29).

The third insurance company, Prédica (life)/Pacifica (P&C) is owned by the bank Crédit Agricole, which also owns UAF, the tenth largest insurance company in the country. Prédica, Pacifica, and UAF combine to create the second largest insurance company, with total premiums in France of 15.5 billion euro in 2003 (*ex aequo* with AXA). The group focuses mainly on life insurance, representing 94 percent of its insurance activities. The distribution is through bank branches.

5.5.2 Types of Companies

The types of insurance companies in France are subject to different laws. Despite the differentiation by French regulators, all companies are governed by the same European insurance directives. There are three sub-groups of French insurance firms.

Insurance Companies

The most important group is composed of insurance companies (*sociétés anonymes d'assurance*), governed by the Insurance Act (*code des assurances*). Insurance companies may sell any insurance products and have no territorial limitation. They may be state-owned or private. After the important privatization process in the 1990s, state-owned insurance companies have nearly disappeared from the insurance industry. There were only one composite and two P&C insurance companies in France in 2003. At present, the state-owned companies are CNP, Caisse Nationale de Prévoyance (in fact two companies), and the CCR, Caisse Centrale de Réassurance. Most of the insurance companies are private-owned stock companies and profit-oriented. There were 247 private companies in 2003, with 65 in life business, 40 composites, and 142 in P&C business (see Table 5.11). Insurance companies usually cover all types of risks. They dominated the life market with a 93.7 percent market share in 2002 (CCA 2003:26). Despite this dominance, their market share is less important in P&C insurance, which holds 65.4 percent of the market (CCA 2003:81).

Mutual Societies

Secondly, the French insurance market is characterized by mutual societies, which in 2003 had 18 million contract holders (*adhérents*) with 38 million insured (CCMIP 2004:7). All of these "mutuals" (*Mutuelles*) are non-profit organizations. Each insured person is both a client and a shareholder (*sociétaire*). Each *sociétaire* has the right to participate at different levels of the mutual structure through internal election. The mutual societies represent a diverse group of insurance firms, governed by various laws. The analysis of mutuals is difficult because some are members of the FNMF (*Fédération Nationale de la Mutualité Française*; French Mutual Societies National Federation), others of the GEMA, or *Groupement des Enterprises Mutuelles d'Assurances* (Insurance Mutual Societies Association); and others of the FFSA, the *Fédération Française des Sociétés d'Assurances* (French Insurance Company Federation). Among French mutual societies, they are no demutualization strategies (Lambert 1999). The trend is more for them to create private entities, completely owned by mutual societies. For example, the mutual insurance company MAAF has created various private legal entities for various products and markets. The mutual societies include:

- The mutual insurance companies (*Sociétés d'Assurance Mutuelle*, SAM), which are civil societies without profit orientation (*sociétés civiles sans but lucratif*). They either use or not use commissioned intermediaries, which are agents or brokers. They have no geographical restriction to sell their products, unless it is stipulated in their charter (status). They are governed by the Insurance Act. An example is *Mutuelles du Mans Assurances* (previously the state-owned MGF, *Mutuelle Générale Française*, privatized in 1987 and now MMA).
- The mutual societies of insurance (*Sociétés Mutuelles d'Assurance*, SMA), which are associations. They always have a charter with geographical or professional specialization. They do not use commissioned intermediaries. They cannot sell life insurance. They are sometimes called pure mutuals (*Mutuelles pures*). They are Assessment Mutual Societies, mutual societies that retain the right to assess policyholders additional amounts if premiums are insufficient for operations.
- The mutuals specialized in agriculture (*sociétés d'assurance mutuelles agricoles*), which are under the supervision of the Agriculture Ministry and the Finance Ministry.
- The "mutuals 1945, " which are governed by a specific act: the Mutuality Act (*code de la Mutualité*). They are called *mutuelles 1945* or *mutuelles or mutualistes* because they were created to provide complementary health coverage to social security created in 1945. They are non-profit associations. Since the law of July 25, 1985, only mutuals 1945 could be called "*mutuelles*" without any precision; the others must add the word "assurance."

> A mutual 1945 is a social insurance regime (*régime social d'assurance sociale*) whose purpose is to cover complementary health expenses. It is a voluntary group of people, bound together to insure their health expenses without any profitability in mind (Abramovici 2001:18).

It is also possible to find in French statistics the segmentation between the mutuals without intermediaries (*Mutuelles Sans Intermediaires, MSI*), such as MAIF or MACIF and mutuals with commissioned intermediaries (*mutuelles rémunérantes*), such as MMA or Azur. Commissioned intermediaries can be agents or brokers.

The introduction of European directives in the Mutual Act (*Code de la Mutualité*) has transformed the French mutual sector. Indeed, by the Law of April 19, 2001, the French regulator has increased the prudential requirements for mutuals 1945. They now must have the same solvency margins as private insurance companies. Because many did not attain the minimum solvency requirements, many mutual 1945 disappeared through mergers or dissolution. There were 8,635 in 1973, 5,780 in 1995, and 2,500 at the end of 2003. CCMIP experts have forecasted that the number of mutuals 1945 should continue to decline to 800 (CCMIP 2005:7). However, their activities are already concentrated. Indeed, 30 mutuals 1945 collected 50 percent of the premiums in 2003 and the top 100 mutuals roughly two-thirds of premiums (CCMIP 2004:8). They collected 16.32 billion euro in premiums in 2002. Mutuals 1945 are mainly focussed on health insurance with 73 percent of their sales figures, providence and retirement plan with 14 percent, and other medical and social activities with 13 percent (*oeuvres sanitaires et sociales*) (CCMIP 2004:7).

Excluding the mutuals 1945, there were 123 mutual societies in 2003 including 14 in life, one composite, and 108 P&C mutual societies. Mutual societies are not fully involved in the life insurance market with a minor market share of only 5 percent (excluding the heath and bodily injury business). They are quite significant on the P&C insurance market with 25.1 percent in 2002 (CCA 2003:87). They have a more important market share in the automobile branch, with 32.5 percent (CCA 2003:116). For example, MACIF is the third car insurer, with a market share of 11 percent. MAIF is the fifth, with 8 percent (Weinberg October 27, 2004:33). The success of mutual societies in the automobile insurance market can be explained by:

- Better risk selection: Mutual societies are created around insured pools that are recognized as better risks, such as teachers, refusing many bad risks, such as young drivers.
- Simpler and cheaper insurance products: French mutual societies were the first in the country to introduce very simple car insurance contracts.
- Important cost reduction strategies: Most mutual societies' headquarters, for example, are outside of Paris, especially in the town of Niort, to avoid the costs of the French capital.

Provident Institutions

The provident institutions (*institutions de prévoyance*) are governed by the Code of Provident Institutions, which is part of the Social Security Act (*Code de la sécurité sociale*). There are non-profit organizations. Among the most important in terms of premiums are Pro BTP, AG2R, Mederic, Prisma Prévoyance, and Malakoff.

Provident institutions provide personal protection, specifically health insurance for extra coverage above the basic social security system. Provident institutions are managed by representatives of employees and employers. They manage mainly corporate contracts (*contrats collectifs d'entreprise*). At the end of 2002, they had 14 million members (*cotisants*) (CCMIP 2004:11). Their activities are divided among health insurance (48 percent of the sales figures in 2002), term life insurance (21 percent), and insurance for payment compensation in the event of disability or accident (18 percent) (CCMIP 2004:12).

Foreign Companies

Among the insurance companies, it is important to analyze the foreign companies in France. A foreign insurance company could invest in France through an unincorporated enterprise, i.e., a branch office (in this case, it could be a division of the foreign insurance company), or an incorporated enterprise, i.e., a subsidiary.

The number of foreign companies with branches in France has been quite stable since 1995 when there were 105 foreign branches, and 113 foreign branches in 2003 (see Table 5.11). Among the 113 branches, 12 are from countries outside the European Economic Area (11 in P&C and one composite) and 101 are from the European Economic Area (84 in P&C, one composite, and 16 life). Ninety percent of the foreign companies with a branch in France are European. The market share of non-European countries is marginal in the French market with a market share of 1.3 percent in the life business and 1 percent in P&C in 2001 (see Table 5.13).

The foreign presence in the French market should also be evaluated when taking into account branches of foreign companies operating in France from their home country and also French licensed subsidiaries of foreign companies.

First, under the free movement of services (FMS) principles, European Economic Area (EEA) companies are authorized to sell insurance policies to French customers directly from the home country. In 2003, 709 FMS foreign companies operated in France, including 559 in P&C (thus, 79 percent of FMS foreign companies are P&C companies), seven composites, and 143 in life business. Over the years, the total number of FMS foreign companies has increased from 382 in 1995 to 709 in 2003, which represents an increase of 185.6 percent since 1995 (see Table 5.12). Hence, the number of foreign companies has increased tremendously. However, the market share of the FMS is very marginal in life (less than 1 percent) and in P&C (around 2 percent) (CCA 2003:16–19).

Second, foreign companies may also have French-licensed subsidiaries. With internal growth and various acquisitions over the years, foreign companies have seen their market shares increase. For example, the German company Allianz bought the French insurance company AGF, Assurances Générales de France, in 1997. Thus, the foreign market share has grown from 10.3 percent of the total premiums in 1990 to 19.7 percent in 2003 as seen in Table 5.13. The augmentation is less notable in the life business—from 6.8 percent in 1990 to 16.3 percent in 2003—than in P&C business, from 13.9 percent in 1990 to 26.3 percent in 2003. The influence of foreign companies, mainly European, is therefore quite important in P&C insurance.

Table 5.11. Number of Insurance Companies in France, 1982 to 2003

	1982	1985	1990	1995	1996	1997	1998	1999	2000	2001	2002	2003
State-owned Companies	10	10	7	5	3	6	3	3	3	3	3	3
Private Companies	127	162	224	258	264	257	256	241	247	250	248	247
Mutual Societies	169	177	208	203	196	165	164	161	159	133	131	123
Total French Companies	306	349	439	466	463	428	423	405	409	386	382	373
Foreign Companies Outside EEE	NA	NA	NA	21	21	19	16	16	17	17	14	12
Foreign Companies from EEE	NA	NA	NA	84	86	96	100	104	101	101	99	101
Total Foreign Companies	168	163	146	105	107	115	116	120	118	118	113	113
Total Number of Companies	474	512	585	571	570	543	539	525	527	504	495	486

Source: FFSA (1982), 5; FFSA (1985), p. 6; FFSA (1990), p. 10; FFSA (1995), p. 66; FFSA (1996), p. 65; FFSA (1997), p. 58; FFSA (1998), p. 60; FFSA (1999), p. 56; FFSA (2000), p. 57; FFSA (2001), p. 51; FFSA (2002), p. 49; FFSA (2003), p. 49.

Note: EEE = Espace Economique Européen; NA = not available. Eight other companies could be included, for a total of 593 companies, in 1990.

Table 5.12. Number of Insurance Companies in France, 1995 to 2003

	1995	1996	1997	1998	1999	2000	2001	2002	2003
Life	143	135	118	107	99	97	95	94	96
Composite	1	13	28	35	36	40	43	44	43
Property and Casualty	427	422	397	397	390	390	366	357	347
Total	571	570	543	539	525	527	504	495	486
Life Under FMS	63	77	89	99	101	115	127	139	143
Composite Under FMS	13	21	24	27	36	33	32	7	7
Property and Casualty Under FMS	306	360	382	435	443	483	496	519	559
Total FMS	382	458	495	561	574	631	655	665	709

Source: FFSA (1982), p. 5; FFSA (1985), p. 6; FFSA (1990), p. 10; FFSA (1995), p. 66; FFSA (1996), p. 65; FFSA (1997), p. 58; FFSA (1998), p. 60; FFSA (1999), p. 56; FFSA (2000), p. 57; FFSA (2001), p. 51; FFSA (2002), p. 49; FFSA (2003), p. 49.
Note: FMS = Free Movement of Services.

Table 5.13. Foreign Company Market Share (%) (Branch Offices and Subsidiaries) of the Total Insurance Premiums in France, 1990 to 2003

	1990	1991	1992	1993	1994	1995	1996	1997	1998	1999	2000	2001	2002	2003
All Companies	10.3	11.7	11.1	10.3	13.0	13.0	12.0	11.6	20.5	19.7	21.2	21.9	21.6	19.7
Life Insurance, Capital Redemption and Composite Companies	6.8	8.1	7.6	7.0	10.4	9.6	9.5	9.8	16.9	16.2	18.2	18.3	18.0	16.3
Property and Casualty Companies	13.9	15.6	15.0	14.8	16.9	18.3	16.7	15.1	26.1	26.3	28.0	28.7	27.8	26.3

Source: FFSA (1999), p. 87; FFSA (2003), p. 70.

5.6 FINANCIAL PERFORMANCE OF FRENCH INSURANCE COMPANIES

In this section, we will analyze company investments, a crucial part of their activities. French financial market results improved in the last years, which enabled French companies to increase earnings. Then, we will analyze insurance company financial performance for life and property and casualty insurance companies.

5.6.1 Insurance Company Investments

The importance of the insurance industry can be assessed by its position as an institutional investor. From 1999 to 2003, the industry's total investments grew from 788.5 billion euro to 1015.9 billion euro (fair market value), as shown in Table 5.14. To give an idea of the importance in value, note that it is roughly the equivalent to the 2003 Annual GDP of India. The growth was thus 29 percent in the last five years. At the same time, the French Stock price index (CAC 40) decreased from 5958.2 to 3556.9, a decrease of 40 percent. French companies are also affected by the appreciation of the euro against the U.S. dollar. In December 2002, the exchange rate was 1.05, and at the end of 2003, 1.26. This implied a reduction of the euro fair market value total investments, while French companies have some investments in the U.S. dollar.

Life companies' total assets are equivalent to 890.3 billion euro, 88 percent of the total assets of insurance companies (fair value). P&C companies' total assets are equivalent to 125.6 billion euro, 12 percent of the total assets of insurance companies (fair value). The total assets grew nearly 30 percent in the last five years. The estimated balance sheet carrying value (acquisition value) of these assets was 953 billion euro in 2003 (FFSA 2003:9).

From 1993 to 2003, fair market value insurance investments increased nearly three times, in fact 285 percent between 1993 and 2003, from 356 billion euro in 1993 to 1015.9 in 2003, as shown in Table 5.14.

If the investment volume increased over a period of ten years, note that the evolution is a little more complex than permanent annual growth. The fair market value investment annual growth was low in 1994 at 2 percent or in 2002 at 3 percent. But other years saw more growth, such as 1996 with an annual growth of 23 percent, or 1997 with 19 percent.

Considering the French stock price index (CAC 40), the financial market grew between 1994 and 2003 from 1881.2 points to 3556.9 points. Using a base 100 for 1994, the French financial market grew from 100 to 189.1, an increase of nearly 90 percent. However, after the collapse of the Internet bubble and the impact of 9/11 on world markets, French financial markets declined in recent years. Using a base of 100 in 1999, the French stock price index decreased to 59.7 in 2003. Considering the decline of French capital markets in the last ten years, it is not surprising to see that the fluctuation of unrealized capital gains has decreased during the same period. Unrealized capital gain declined in 1994 by 83 percent, but also each year between 1999 (–13 percent) and 2002 (–14 percent). On the contrary, unrealized gains were substantial in 1995 (+191 percent), 1996 (+85 percent), 1997 (+27 percent), 1998 (+56 percent), or 2003 (+35 percent).

Table 5.14. Breakdown of Insurance Company Investments in France, for Life and Property and Casualty Companies (Market Value in Euros and %), 1993 to 2003

	1993	1994	1995	1996	1997	1998	1999	2000	2001	2002	2003
Life											
Investments (Euros)	279.9	294.3	354.3	445.7	533.0	615.5	675.2	745.1	782.0	810.9	890.3
Bonds (%)(1)	64.1	68.7	70.6	72.5	71.9	71.4	69.0	65.1	67.3	69.2	69.5
Equities (%)	13.8	14.0	12.8	12.4	14.1	17.2	20.9	25.4	24.6	22.2	22.6
Real Estate (%)	9.4	7.8	6.6	5.5	4.4	4.1	3.8	3.6	3.4	3.4	3.2
Others (%)	12.7	9.5	10.0	9.6	9.6	7.3	6.3	5.9	4.7	5.2	4.7
Total (%)	100.0	100.0	100.0	100.0	100.0	100.0	100.0	100.0	100.0	100.0	100.0
Property and Casualty											
Investments (Euros)	76.2	69.0	73.6	82.3	94.1	102.7	113.3	115.9	11.7	110.3	125.6
Bonds (%)(2)	43.9	45.1	44.5	48.3	48.5	48.8	48.4	49.1	51.3	51.9	53.8
Equities (%)	27.7	29.2	29.9	28.6	27.7	30.6	31.3	31.1	31.0	30.4	29.9
Real Estate (%)	15.9	16.2	15.5	14.7	13.7	12.7	12.1	11.5	9.3	9.2	8.4
Others (%)	12.5	9.5	10.1	8.4	10.1	7.9	8.2	8.3	8.4	8.5	7.9
Total (%)	100.0	100.0	100.0	100.0	100.0	100.0	100.0	100.0	100.0	100.0	100.0
Life and Property and Casualty											
Bonds (%)(3)	60.0	64.5	66.2	68.8	68.6	68.4	66.4	63.3	66.7	68.9	69.2
Equities (%)	16.6	16.7	15.7	14.8	16.0	19.0	22.3	26.1	25.3	23.1	23.4
Real Estate (%)	10.7	9.3	8.1	6.9	5.7	5.3	4.9	4.5	4.1	4.1	3.8
Others (%)	12.7	9.5	10.0	9.5	9.7	7.3	6.4	6.1	3.9	3.9	3.6
Total (%)	100.0	100.0	100.0	100.0	100.0	100.0	100.0	100.0	100.0	100.0	100.0
Total (Euros)	356.1	363.3	427.9	528.0	627.1	718.4	788.5	861.0	893.7	921.2	1015.9

Source: FFSA (2002), p. 85; FFSA (2003), pp. 75–76.
(1) % of the life company total invesments.
(2) % of the property and casualty company total invesments.
(3) % of total insurance company investments.

Note that French insurance companies record all their assets in their balance sheets at their acquisition value after deduction of any amortization and depreciation. But French companies must also disclose in a separate annex to the balance sheet their asset market value: stock or bond trading price at the inventory date, redemption price for mutual funds, and real-estate market value (by an official real-estate expert).

The ability of insurance companies to sustain the depreciation of French stock markets is due to their relatively weak stock market exposure. Indeed, in 2003, French insurance company investment portfolios were mainly composed of bonds for 69.2 percent of the total. This percentage has been stable over the years, with the bond share holding at 64.2 in 1994 (see Table 5.14). In 2003, equity investment corresponded to only 23.4 percent of the total. Thus, a depreciation of 40 percent of the value of 25 percent of the investment portfolio is important, but is also roughly equivalent to a 10 percent decrease of the total. The actual deprecation is even lower if we consider that the equities represented—16.6 percent in 1993 or 14.8 percent in 1996.

Real estate investment was marginal, with 3.8 percent of the total investment. The insurance companies have greatly reduced their exposure to real estate investment risk; it was 17 percent of the total of their investments in 1985 (see Table 5.14). In fact, real estate investment value declined at the end of the 1980s and the beginning of the 1990s.

The investment portfolio composition is regulated by the French Insurance Code, which specifies which types of assets may match insurance liabilities (*engagement d'assurances*). In order to reduce market fluctuation risk and liquidity risk, French insurance companies may have a maximum of 65 percent of their investments (balance sheet carrying value) in equities, 40 percent in real estate, and 10 percent in loans. Furthermore, to avoid over-exposure on a single risk, insurance companies may not invest more than 5 percent of their assets in a single counter party (10 percent in some specific circumstances). For unlisted equity securities, a single counter party could not be more than 1 percent of total investments, and 10 percent for real estate investments.

With different types of engagements, life and P&C insurance companies do not have the same type of investment portfolios. French P&C insurance companies invest less in bonds and more in equities than life insurance companies. In 2003, P&C insurance company investments were 53.8 percent in bonds, 29.9 percent in equities, and 8.4 percent in real estate. But life insurance company investments were 69.5 percent in bonds, 22.6 percent in equities, and 3.2 percent in real estate, as shown in Table 5.14 (FFSA 2003:76).

5.6.2 Insurance Company Financial Performance

Life Insurance Companies

For life insurance companies, net investment income (income plus realized capital gain; *revenues et plus-values réalisées*) was 41.7 billion euro in 2003, as shown in Table 5.15. Underwriting results increased in 2003 to 3.5 billion euro, compared to 1.3 billion euro in 2002.

Actuarial and claims reserves increased to 36.9 billion euro in 2003, or 39.5 percent of earned premiums. Acquisition and administrative expenses were at a low level of 7.1 billion euro in 2003, equivalent to 7.6 percent of premiums. Net earnings improved between 1999 from 2.5 billion euro to 3.6 billion euro in 2003. Net earnings (*résultats comptable de l'exercice*) for a given period of time are equal to the balance of underwriting results (*résultats techniques*) and other financial or non-recurring events (non-underwriting results).

The solvency of French life companies is good. European regulation requires a solvency margin of 4 percent of the total actuarial reserves and 1 percent for unit-linked contracts. Considering the breakdown of contracts in France, the required solvency margin can be estimated at 3.5 percent (FFSA 2003:51). Shareholders' equity was equivalent to 4.1 percent of the total actuarial reserves in 2003. The solvency margin (shareholders equity and unrealized capital gains/actuarial reserves) was 9.7 percent in 2003, 2.8 times the required level. The return on equity has improved over the years from 6 percent in 1994 to 10.7 percent in 2003, but it was lower than in 1990 with 15.2 percent, as shown in Table 5.16.

Table 5.15. Income Statement for Life, Capital Redemption, and Composite Companies (Net of Reinsurance) in France in 2003 (in billions of Euros)

	Billions €	% of Earned Premiums
Underwriting Account		
Earned Premiums	93.3	100.0
Net Investment Income(1)	41.7	44.7
Other Underwriting Income	0.5	0.5
Subtotal A	135.5	
Cost of Claims	57.2	61.3
Actuarial and Claims Reserves	36.9	39.5
Policyholder Dividends	30.2	32.4
Acquisition and Administrative Expenses	7.1	7.6
(+) / (-) Other Underwriting Expenses	0.6	0.6
Subtotal B	132.0	
Underwriting Results = A-B	3.5	3.8
Nonunderwriting Account		
Income from Allocated Investments (Nonunderwriting)	0.5	0.5
Other Nonunderwriting Income (loss)	–0.4	–0.4
Net Earnings(2)	3.6	3.9

Source: FFSA (2003), p. 50.
(1) Includes capital gains on the sale of assets (net of capital losses) and adjustment of ACAV (unit-linked), less investment income to the nonunderwriting account.
(2) Net earnings for a given period of time are equal to the balance of underwriting results (*résultats techniques*) and other financial or nonrecurring events (nonunderwriting results).

Table 5.16. Solvency and Return on Equity of All Insurance Companies (Life and Non-Life) in France (in %)

	1990	1991	1992	1993	1994	1995	1996	1997	1998	1999	2000	2001	2002	2003
Solvency Margin	14.4	14.5	12.4	18.5	6.1	9.6	12.4	12.7	16.9	12.8	11.7	9.5	9.1	9.7
Return on Equity	15.2	14.3	12.0	8.9	6.0	5.7	5.3	7.6	6.7	10.7	11.2	11.1	4.3	10.7

Source: FFSA (1999), p. 90; FFSA (2003), p. 73.
Note: Sovency margin = (shareholders'equity + unrealized gains) / actuarial reserves. Return on equity = net income / shareholders'equity.

Property and Casualty Companies

For property and casualty companies, income from allocated investments and other underwriting income increased between 2002 (2.7 billion euro) and 2003 (3.1 billion euro). The incurred losses increased 6.2 percent between 2002 (32.1 billion euro) and 2003 (34.1 billion euro), as shown in Table 5.17. For the P&C business branch, the ratio of claims to premiums (*Rapport Sinistres à Primes, S/P*) was at 79.1 percent in 2003. Claims reached 34.1 billion euro and premiums reached 43.1 billion euro (FFSA 2004:50). The automobile insurance ratio of claims to premiums was at 80.7 percent in 2003, similar to the rate in 1994 (81 percent), but was much better than the high 1998 ratio of 89.9 percent, as shown in Table 5.18. The personal property ratio of claims to premiums was 79.6 percent in 2003, an important improvement after the 1999 poor results at 137.8 percent, but much more than in 2004 or 1995, just above 64 percent. The commercial and agricultural property ratio of claims to premiums was 62.5 percent in 2003, the best technical results in the last 10 years. The insurance cycle was again at its peak in 1999, when the ratio was 128.3 percent.

Table 5.17. Income Statement for Property and Casualty Companies, Net of Reinsurance, in France in 2003

	Billions €	% of Earned Premiums
Underwriting Account		
Earned Premiums	43.1	100.0
Income from Allocated Investments	3.1	7.2
Other Underwriting Income	0.6	1.4
Subtotal A	46.8	
Incurred Claims(1)	34.1	79.1
Acquisition and Administrative Expenses	9.1	21.1
Other Underwriting Expenses	1.8	4.2
Subtotal B	45.0	
Underwriting Results = A-B	1.8	4.2
Nonunderwriting Account		
Net Investment Income (nonunderwriting)(2)	1.1	2.5
Other Nonunderwriting Income (Loss)	–1	–2.3
Net Earnings	1.9	4.4

Source: FFSA (2003), p. 50.

(1) Includes allowances to claims reserves.

(2) Includes capital gains on the sale of assets (net of capital losses) less investment income transferred to the nonunderwriting account.

Acquisition and administrative expenses reached 9.1 billion euro in 2003, an increase of 5.8 percent in 2002. The average expense ratio (*ratio moyen de chargement, ou frais rapportés aux cotisations*) was at 21 percent in 2003, much higher that the life average expense ratio at 7 percent for the same year. The combined ratio was 101 percent in 2003, and 108 percent in 1994. The combined ratio net of reinsurance is the ratio of paid and incurred claims, acquisition costs and administrative expenses plus other operating expenses to premiums (*ratio combiné net de reassurance = sinistres payés et provisionnés, coûts d'acquisition et d'administration et autres charges techniques/primes*). The peak of the cycle was in 1999 when the combined ratio was 126 percent. Underwriting results improved significantly from 0.9 in 1999, then 0.4 in 2002 to 1.8 billion euro in 2003. The underwriting results/premiums was 4.2 percent in 2003. Net earnings of French P&C insurance companies reached 1.9 billion euro in 2003 (or 4.4 percent of the premiums), an important increase in comparison with 2002 or 1999 with only 0.8 billion euro, but nearly equivalent to 2001, with 2 billion euro.

To understand the French P&C results, we have to stress that the French insurance market was affected by a combination of exceptional events in the last years. In 1999, two important tornadoes, Lothar and Martin, affected the insurance industry. Then, French companies were touched by both the decline of the financial markets and the rise of reinsurance premiums following 9/11 terrorist attacks in the United States and the explosion of an important factory in Toulouse, France.

European regulation requires property and casualty companies to have a minimum solvency margin of 16 percent (shareholders' equity plus unrealized capital gains/premiums), or 23 percent of average claims paid out in the preceding three years if this amount is higher. The solvency margin was 80.4 in 2003, five times higher than the minimum required. The solvency margin has improved along the years, despite being lower during the 1999 peak. The return on equity has improved over the years, from 0 percent in 1994 or 3.2 in 2002 to 7.1 percent in 2003, but much less than in 2000 when it reached 9.7 percent.

5.7 DISTRIBUTION CHANNELS

The French insurance market is characterized by many distribution channels. Tied agents, brokers, salaried sale forces, direct selling from mutual societies, telemarketing, Internet, banks, and financial institutions are among the distribution channels.

5.7.1 Non-bank Distribution Channels

Tied Agents

The tied agents or exclusive agents (*agents généraux d'assurances, AGA*) still hold the largest market share of P&C insurance with 35 percent of the market, but have been losing market shares in the last 60 years. They had 47 percent of the French insurance market in 1990. In the same period, their market share in life insurance declined from 18 percent to 8 percent. There were 13,300 agents with 28,000

Table 5.18. Claims / Premium Ratio for Automobile, Personal Property, and Commercial + Agricultural Property, Combined Ratio, Solvency, and Return on Equity of Property and Casualty Companies in France, 1990 to 2003 (in %)

	1990	1991	1992	1993	1994	1995	1996	1997	1998	1999	2000	2001	2002	2003
Automobile Claims and Allowances to Claims Reserves	8.4	8.9	9.6	10.2	10.2	11.4	11.6	12.0	12.5	12.4	12.6	13.3	13.8	13.8
Automobile Premiums	10.6	10.9	11.2	11.8	12.6	13.6	14.1	14.0	13.9	14.1	14.6	15.4	16.3	17.1
Automobile Claims / Premiums in %	79.2	81.7	85.7	86.4	81.0	83.8	82.3	85.7	89.9	87.9	86.3	86.4	84.7	80.7
Personal Property Claims and Allowances to Claims Reserves	2.4	2.0	2.2	2.4	2.5	2.7	2.7	3.0	2.9	6.2	4.0	3.4	3.7	4.3
Personal Property Premiums	3.2	3.3	3.5	3.7	3.9	4.2	4.4	4.4	4.5	4.5	4.7	4.9	5.1	5.4
Personal Property Claims / Premiums in %	75.0	60.6	62.9	64.9	64.1	64.3	61.4	68.2	64.4	137.8	85.1	69.4	72.5	79.6
Commercial and Agricultural Property Claims and Allowances to Claims Reserves	3.6	3.1	3.6	3.0	3.1	3.1	3.3	2.9	3.0	5.9	5.1	3.9	3.8	4.0
Commercial and Agricultural Property Premiums	3.7	3.7	3.8	4.2	4.3	4.6	4.8	4.6	4.6	4.6	4.7	5.0	5.8	6.4
Commercial and Agricultural Property Claims / Premiums in %	97.3	83.8	94.7	71.4	72.1	67.4	68.8	63.0	65.2	128.3	108.5	78.0	65.5	62.5
Combined Ratio	113.0	112.0	116.0	113.0	108.3	103.8	100.7	102.9	108.2	125.7	111.4	106.3	100.6	101.0
Solvency Margin	80.9	81.3	74.6	80.8	55.9	50.5	63.4	78.9	93.7	97.6	97.3	79.4	68.3	80.4
Return on Equity	12.4	4.2	0.5	0.2	–0.2	5.3	9.4	6.9	1.8	3.8	9.7	8.6	3.2	7.1

Source: FFSA (2004), pp. 71–74; FFSA (1999), pp. 87–88, 91.

Note: Combined ratio for property and casualty insurance = (costs of claims + actuarial and claim reserves + acquisition and administrative expenses)/ premiums. Solvency margin for property and casualty insurance = (shareholders'equity + unrealized capital gains) / premiums (in %). Return on equity for P&C insurance = net income / shareholders' equity (in %).

employees in 2003, compared with 17,400 agents and 35,000 employees in 1995 (FFSA 2003:61). As noted previously, tied-agents are governed by an agent charter (*statuts des agents généraux*) of public order, compulsory for all of the insurance industry. The agent represents only one company and has exclusive rights to the company's products in his geographical territory, except in Paris and for risks not insured by the company.[160] Some agents may have various agencies to manage. They are paid through commissions. As a representative of a single insurance company, the agent is still quite independent in the management of his agency. Among the agent's strengths, we should note: a large presence in France; his or her technical expertise in various insurance products; and his or her local socialization (he or she is often a *notable* in his or her home town). Tied agents compete fiercely with other insurance actors, especially in auto and homeowner insurance. They have not been able to develop their activities in the fast-growing life insurance sector. Their relationship with their companies can be quite conflicted. For example, recently a managing director of MMA had to step down due to the agents' opposition (Weinberg March 24, 2004:34).

Mutual societies

Mutual societies have the second market share of the P&C market in France, as shown in Table 5.19.[161] Their market share is quite marginal in life insurance. Their P&C market share has been quite stable in the last ten years, from 29 percent in 1990 to 33 percent in 2003, as shown in Table 5.19. Part of the P&C success of French mutual societies could be explained by more simple insurance products—at least, until the beginning of the 1990s, they had quite simple insurance products with less warranties than insurance companies—and better risk selection, and thus lower insurance price. At the same time, mutual societies are quite effective in reducing their management costs. For example, many of them have their headquarters outside of Paris, especially in the small town of Niort, where it is obviously cheaper to manage insurance activities. Many mutual societies sell insurance contracts directly to their clients through their employees, without market intermediaries such as agents or brokers.

Brokers

The brokers (*courtiers*) have the third position in the P&C market with a market share of 19 percent in 2003 and also the third position in life market with a market share of 9 percent in 2003. [162] They have been able to increase their position from 17 percent to 19 percent, a small increase but at the same time a success considering the intense competition of this market. Their market share in life insurance was reduced from 11 percent in 1995 to 9 percent in 2003. French brokers often tend to place insurance with various insurance companies. There are 2,800 brokers in France with 17,000 employees (FFSA 2003:61). Eighty percent are small brokers with fewer than

---

[160] If an insurance company does not underwrite a specific risk, its tied agents have the right to be agent of another insurance company for this specific risk.

[161] The data in the table 19 take into account only the mutual societies without intermediaries such as brokers or agents.

[162] Banks are the second leading distribution system and will be discussed in a separate section.

nine employees. Brokers are quite concentrated; 80 percent of the sales figures come from five French regions: Paris (around 50 percent of their premiums), Lyon, Marseille, Strasbourg, and Bordeaux. Legally, brokers represent the insured, which gives them a very good image in the public. Small brokers are considered tradesmen (*commerçant*) and must be entered into the register of commerce (*registre du commerce*). One-third of the brokers are also tied agents: the brokerage is limited to risks refused by their principal company. Large brokers are public limited companies. Brokers are recognized for their technical insurance competence, in particular industrial risks, property, liability, hull, and cargo insurance. Seventy-five percent of their commissions come from P&C insurance. As for the agents, brokers can work long hours and travel to develop their business. Small brokers are in direct competition with banks, mutuals, and agents in auto and homeowner insurance. Small brokers sell insurance products that are quite similar to the products of the competition. Furthermore, brokers have the possibility to switch from one insurance company to another one. But, this weak loyalty may imply a less advantageous situation than agents in terms of commercial premiums and claims.

Direct Sales

The French population is familiar with direct sales (DS), which include commercial mail, coupons in newspapers, television, telephone, and the Internet. For example, Minitel, the ancestor Internet in France, was already an insurance distribution channel in the 1980s (Venard 1999). However, DS is marginal in France, as in many European countries (EIU 2001). Since 1990, the market share in life insurance was around 5 percent of the premiums. In P&C insurance, the market share was smaller at 3 percent. There is no legal obstacle to the development of direct sales in France. The only limitation is due to the exclusive territory of insurance companies. The companies with tied agents cannot develop in principle a direct offer in P&C insurance. Large risks cannot be sold through direct sales. Developing the individual market through DS is difficult due to intense competition—for example, the already low price offers by banks and mutuals.

Salaried Professionals

Salaried sales associates are also marginal in the P&C market, with a 2 percent market share in 2003 (FFSA 2003:61). However, they are quite significant in life insurance, with 16 percent, despite an important decline in the past ten years, from 28 percent of the market in 1990. Salaried sale forces include to an insurance company network such as GPA (Generali Proximité Assurances).

5.7.2 Bank Penetration in the French Insurance Market[163]

The last distribution channel is the banks and financial institutions. In Table 5.19, the bank channel is called over the counter, as in *vente aux guichets*. At present, the penetration of French banks in the insurance sector is very high, with a market share

[163] Due to the important market share, we cover this type of distribution in a specific section. This section is based on secondary data analysis. We also conducted face-to-face interviews with a small sample of bankers and insurers in France.

of 47.3 percent of total insurance in 2003. Sixty-two percent was in the life insurance market and 8 percent in the P& C market.

Combining the two French words *banque* and *assurance* leads to the new word *bancassurance*, the close connection between banks and insurance companies. *Bancassureur* designates the actors (bank or insurance companies) of this global market. In this text, the word designates the bank entering the insurance market. Data concerning various markets show that banks have been more successful in this strategy than insurance companies. Looking at the list of the top ten insurers gives a clear picture of the importance of the bancassurance phenomenon in France. Indeed, five of the top ten insurers in terms of sales are financial institutions: CNP (second), Prédica/Pacifica (fourth), Cardif/Natio (fifth), SOGECAP (seventh), and ACM (tenth), as was shown in Table 5.10.

Bancassurance has been for quite a long time a "life phenomenon," to quote a French CEO. Since 1993, banks have had a dominant position in the French life insurance market, when their market share was over 50 percent. Banks increased their share of the life market from 39 percent in 1990 to 62 percent in 2003. On the contrary, the decrease of the tied-agents is quite significant from 18 percent to 8 percent, as is the decrease for salaried sales associates from 28 percent to 16 percent, as shown in Table 5.19. The director in charge of the insurance activities in a bank said cynically, "The agents were like cows looking at a TGV [high speed train]. They saw us [the banks] entering the life market without understanding what was going on." Not only do the banks have a leading role in terms of volume, but also it is remarkable that their growth rate is often higher than that of the life market. For example, between 1999 and 2000, the market premium increased by 20 percent in the life market, but by 29.5 percent for banks. The same year, some banks have achieved an even larger growth in the life insurance market. For example, Crédit Agricole's life business grew by 37.7 percent and Société Générale's by 45 percent.

Bank activity in the property and personal injury business is quite new (0 percent in 1990) and still marginal. The market share of banks was 8 percent of the total in 2003 (see Table 5.19). But the increase in market share is far from over. "Our goal is to penetrate the P&C market after our success in the life business," said a bank marketing director. Another bank executive recognized, "Our clients are now used to our insurance products. The natural step is to develop the car and homeowner comprehensive markets." For example, between 1999 and 2000, the property and personal injury market grew by 3.4 percent. But at the same time, some banks achieved higher growth. Crédit Mutuel achieved an increase of 12.3 percent; Crédit Agricole, 24.6 percent; Caisse d'épargne, 24.3 percent; CIC, 114 percent; and Société Générale, 124.6 percent. Despite some impressive individual performances, the increased market share in the property and personal injury should be a little slower than in the life market due to strong competition, especially from the mutual societies. On the contrary, a former dominant distribution channel, tied agents, has dropped from 47 percent in 1990 to 35 percent in 2003.

Table 5.19. Premium Income by Distribution Channel for the Life and Property and Casualty Insurance Markets in France, 1990 to 2003 (in %)

	1990	1991	1992	1993	1994	1995	1996	1997	1998	1999	2000	2001	2002	2003
Life														
Over-the-counter	39.0	42.0	46.0	51.0	54.0	56.0	59.0	61.0	59.0	60.0	61.0	60.0	61.0	62.0
Agents	18.0	17.0	17.0	15.0	14.0	12.0	11.0	10.4	10.0	10.0	8.0	8.0	8.0	8.0
Brokers	11.0	8.0	7.0	7.0	7.0	7.0	7.0	6.9	8.0	9.0	9.0	9.0	9.0	9.0
Salaried Sales Associates	28.0	27.0	25.0	22.0	21.0	19.0	17.0	15.9	17.0	17.0	16.0	17.0	16.0	16.0
Direct Marketing and Alternative Channels	4.0	6.0	5.0	5.0	4.0	6.0	6.0	5.8	6.0	5.0	6.0	6.0	6.0	5.0
Property and Casualty														
Over-the-counter	0.0	1.0	3.0	3.0	4.0	5.0	5.0	6.0	7.0	8.0	8.0	8.0	8.0	8.0
Agents	47.0	46.0	45.0	44.0	42.0	40.0	39.0	37.0	36.0	35.0	35.0	35.0	35.0	35.0
Brokers	17.0	18.0	18.0	19.0	19.0	20.0	19.0	19.0	18.0	17.0	17.0	18.0	19.0	19.0
Salaried Sales Associates	4.0	4.0	4.0	4.0	4.0	4.0	4.0	3.0	3.0	3.0	3.0	2.0	2.0	2.0
Mutuals without Intermediairies	29.0	28.0	27.0	27.0	28.0	29.0	31.0	32.0	33.0	34.0	34.0	34.0	33.0	33.0
Direct Marketing and Alternative Channels	3.0	3.0	3.0	3.0	3.0	2.0	2.0	3.0	3.0	3.0	3.0	3.0	3.0	3.0

Source: FFSA (1999), p. 86; FFSA (2003), p. 69.
Note: Over-the-counter distribution channels include financial institutions, the French post office (*la Poste*), and the French government with CNP.

5.7.3 Reasons for Bank Success in the French Insurance Market

The previous data shows the success of French banks in the insurance market, especially in life activity, which is due to several factors. First, we will give the environmental reasons for growth: legal framework, demand, marketing, and competition. Second, we will consider the internal reasons for growth in banks' share of the insurance business: revenues increase and cost reduction. Obviously, each bank has been able to build competitive advantages based on its internal strengths.

Environmental Reasons

The banking and insurance markets are strongly influenced by legal developments. First, the French bank system was, at least until World War II, very liberal (Thiveaud 1997:43). But during two main periods—after the war and after the election of a socialist president in 1981—France experienced two important nationalization processes in its economy. The French state traditionally has had a significant stake in the financial system. An important step toward privatization was the promulgation of various laws in 1966–1967 (Thiveaud 1997:49). One of the results was an increase in the number of bank branches after a period of restriction. This distribution expansion permitted greater contact with clients. In 1984, some banks were gradually privatized. There were 124 state-owned banks in 1984, nine in 1998, five in 2001, and four now, including two created in 2000 by the Caisse des dépôts. (CECEI 2004:176). The process created newly privatized banks, which were more profit-oriented and thus sought new development areas, such as insurance.

Second, French regulations encourage people to save into life insurance-related products (this is also true in other foreign markets). The incentives are very strong. Until 1998, a quarter of the life premium was deductible from income tax, up to 1,500 euro per year. In addition, until 1998, after eight years of holding a life insurance policy, no income tax was charged on life insurance revenues. Now, income tax on a life insurance policy after eight years is charged at 7.5 percent. There is no inheritance tax if the holder bought life insurance contract before the age of 70 and if the savings are less than 152,500 euro for each beneficiary. These incentives have helped the growth of insurance investment flows from the French population, from 42.2 percent of their total investment flows in 1990 to 76.1 percent in 2000, as shown in Table 5.6. Less attractive incentives in 1998, with the taxation of life insurance revenues whatever the duration of the policy, have resulted in a relative decline of the net investment flow to 55 percent in 2003.

It is necessary to explain the reason behind the existence of this "tax haven" in a country that has the reputation for high taxes. The French government is interested in funds invested by insurance companies, equivalent to nearly 1016 billion euro, with 890.3 billion euros from life companies and 125.6 billion euros from property and personal injury companies (see Table 5.14). Of this amount, 69.3 percent is invested in state bonds, around 63.9 billion euro annually, equivalent to the French government's public deficit of 62 billion euro in 2003 (Banque de France 2004:4). At the end of 2003, various European countries had a public deficit higher than the Maastricht limit of 3 percent of the GDP: Greece (4.6 percent), France (4.1 percent), Germany (3.9 percent), U.K. (3.3 percent), the Netherlands (3.2 percent) (Banque de

France 2005:26). Given this large public deficit, it is important for the French government to attract investment flows from French insurance companies. The tax incentives are a means of obtaining funds to cover the public deficit. However, the actual trend is to reduce the tax incentives of life insurance in France.

The legal environment has a strong influence on the insurance market, but this is not the only reason for the market fluctuation. The demand for prospects and clients is also a powerful force. First, clients have more information concerning offers now than they did 30 years ago. This more accessible information helps them choose insurance products. Varying price offers are important information for price-sensitive French consumers, who can compare insurance offers with this in mind. For example, there are some important differences in terms of price positioning of the various distribution channels. With an average price of 100, tied agents sell car insurance at 132, banks at 120, mutual companies at 92, and direct insurers at 62 (Lambert 1999).[164]

Secondly, for a long time the "classic insurance companies" neglected specific niches. The banks decided to exploit those niches. For example, the tied agents were more interested in selling life insurance products to high-income customers, rather than to low- and middle-class clients. A bank sales director described the situation, "The insurance companies didn't cover all the market. We had plenty of room to develop." A life insurance company director explained, "In a fast growing market, the insurance companies didn't understand immediately that the banks were winning in the life market. The competition didn't seem as sharp as it was in fact. Everybody looked like a winner."

Thirdly, in the 1980s, French citizens started to worry about their retirement pensions. An aging population, as is the case in many other industrialized countries, characterizes France. In 2025, one-third of the French population will be over 60 (Ined 2003). Retirement benefits are based on a compulsory system run by the state, in which the active population pays for retirees. Now, because the population is older, there are more pensioners and fewer working people. At the same time, French students are studying longer and arriving later on the job market. It is now necessary for people to look for their own personal retirement solutions. Furthermore, the diminishing role of government in welfare provision is an ongoing question in French society. In terms of preserving a comfortable lifestyle, life-related investment products seem promising.

Fourth, in the 1980s French clients started to ask for complete investment advice. Banks responded to demands concerning savings, investments, loans, and insurance. This response also included specific targets (high earners), who require a more personalized relationship with a wide range of products.

These reasons could be possible advantages for banks and for insurance companies, too. For example, insurance companies have also used anxiety about future pensions to develop their activities in the savings market. In order to explain bank penetration of the insurance market it is necessary to look at some of the competitive advantages of banks compared to insurance companies.

---

[164] It is necessary to be cautious with such comparisons, since average prices do not take into account the car types of each consumer. Mutual societies quite often insure "smaller risks" than tied-agents. However, banks' competitive price positioning create new business opportunities.

Highlighting the demand influence leads, almost naturally, to describing marketing reasons for bank penetration in the insurance market. On one hand, banks have an important resource that is not in the "hands of the insurance companies," to use the words of a bank executive—information about clients. In fact, banks have a lot of information about client spending patterns and purchasing powers. Bankers know their clients' needs and can offer relevant financial products. There is no legal restriction on the banker about using client information for cross-selling, but compulsory cross-selling is forbidden by law. For example, a banker cannot require a client asking for a car loan to also buy a car insurance contract. An insurance company director pointed out, "The information about prospects and clients is a key aspect of any direct marketing operation. It is evident that the banks have more up-to-date information than us."

On the other hand, French banks have a better professional image than insurance companies. "This brand equity is a strong advantage in selling new products for us, such as car insurance," said a bank manager. In many countries, consumers rate their bankers higher than their insurers concerning objectivity of advice and product knowledge (Thomson, Wade 2000; Clinton 2001). The security and trust offered by banks has been a major step in creating consumer confidence in bancassurance. In a 2003 French survey, 80 percent of respondents said they have a good image of their banks (20 percent said they had a bad image); 79 percent said they would recommend their bank to their friends (Observatoire de l'opinion 2004). The CEO of a French bank said, "Our bank was late in the insurance market, but sales become easier with a clearer concept of bancassurance in the mind of the French households, clearer since it started 30 years ago."

Furthermore, banks and their clients have close links. For example, clients receive bank statements each month and use bank means of payment nearly every day.

The links are also strengthened by important bank branch networks. There are more than 26,000 bank branches in France, including Crédit Agricole, Caisse d'épargne, Crédit Mutuel, and Banques populaires (CECEI 2004:137). This is more extensive than in Germany or in the U.K. There are 0.54 branches in France per 1,000 inhabitants but only 0.43 in Germany and 0.23 in the U.K. (CECEI 2001).

Another reason for French bank success is a strong position in the selling process. When people consider buying a car or a house, they start by meeting professionals to agree on financing the transaction. Only after this step will they think about taking out insurance. When the banks thus meet prospects with specific investment needs, they can propose an insurance product adapted to their purchase project, before the client even thinks about it. As a mutual society marketing director said, "Banks have a double proximity with their clients. They are closed physically with large branch networks. They are also closed temporally, being the first to meet the clients in their purchasing cycle."

However, in the future, clients will be less likely to go to their bank branches than ever before. For example, in 1998, 47 percent of French clients said they often go to their bank branches, but only 36 percent reported doing so in 2004 (Observatoire de l'opinion 2004). This means that 64 percent of bank clients rarely go to their bank branches, and this percentage increased in the last few years.

Finally, it is clear that this positioning provides banks with cross-selling opportunities. Despite the fact that tied cross-selling is forbidden under French law, a bank director recognized that temptation is strong to link loans and insurance products: "It is true that some of our employees try to offer good loan rates to their clients in exchange for the purchase of the car or home insurance with us. But it is very rare..." A top executive in a major bancassureur said, "We don't force any client to buy any insurance product, despite the fact that the temptation is important." In such circumstances, it is clear that banks have a very low client acquisition cost. As a consequence, banks have been very successful in selling low-priced insurance products. For example, in 2000 banks started to sell a new contract: *Garantie contre les accidents de la vie*, a P&C policy covering bodily injury and its consequences following an accident, with exceptions such as a car accident or work-related accident. In three years, banks' market share reached 76 percent of the contracts sold in France (Weinberg March 15, 2004:31). French banks were quite efficient in using their competitive advantage (information about clients and strong position in the selling process) and were above all very effective in combining bank and insurance activities. Bank employees may sell both types of products over the counter.

Competition is also another reason to explain banks' success in the French insurance market. Firstly, insurance companies sold life-related savings products, which led some bank clients to withdraw their savings from their bank accounts. Thus, banks faced a haemorrhage of client savings. According to a banker, the "answer of the French banks" was to create insurance subsidiaries to sell their own insurance products.

Second, as inter-bank competition increased, banks had to find new methods of attracting and retaining clients. Selling more products was a possibility. "In our business," said an insurance manager, "clients who hold more than three products are more loyal to their supplier of financial and insurance services."

Third, and most important, for a long time French banks paid no interest on deposit accounts, as prohibited by French law. This was a cheap source of income for banks. Increased client sensitivity concerning profitability of their investments forced banks to offer more attractive products, such as life insurance. Thus, banks faced cannibalisation of bank deposits in favor of their own life insurance products. After a law of March 18, 2005, banks were authorized to paid interest on deposit accounts, and some have begun to do so. Thus, the internal transfer from bank accounts to life insurance contracts is certainly the most important explanation of the French bank success in the life market in the 1980s and 1990s.

Internal Reasons for Growth: A Mixed Set of Motivations

French banks have seen their spreads (the difference between their borrowing and lending rates) fall sharply. They have responded by trying to seek both additional fee-based income using their customer bases and reducing costs in selling and managing insurance activities (CAPA 1999). French banks faced market saturation. In 1982, 90 percent of the population already had a bank account (Thiveaud 1997:57). There are now more bank accounts than French people: there were 65 million non-interest paying deposit accounts in 2000 and 69 million in 2003, and 133 million savings accounts in 2000 and 145 million in 2003 for a population of 62 million (CECEI 2001:44; CECEI 2004:130). Under these circumstances, banks looked for new sources of revenue. The launch of insurance products was an evident strategy to develop revenues, despite the cannibalization of bank deposits.

Banks tried also to increase their return on assets (ROA). One method to increase ROA, assuming a constant asset base, is through fee income. For bankers, insurance activities generate up to 25 percent of operating revenue. Insurance subsidiaries of French banks pay fee income to their bank headquarters due to the distribution of the insurance products.

An important goal for banks was to increase revenue per employee. Starting in the 1970s, French banks invested heavily in information technology, shifting a significant share of administrative from employees to machines and transforming administrative workers into sales staff for both bank and insurance products. In the same time, bank strategists decided to increase the number of products per client, to accentuate client loyalty. The growth in revenue through cross-selling resulted in a decrease in the cost of losing clients.

Table 5.20. Cost Structure of Car Insurance in France in Relation to Different Distribution Channels, in 1996

Distribution Channel	Distribution Cost	Management Cost (%)	Claims / Premiums %)	Combined
Mutual Societies	6.57	5.01	87.16	98.75
Tied Agents	12.35	10.91	74.80	98.05
Brokers	11.29	9.57	80.90	101.76
Salaried Sales Associates	13.93	9.95	80.52	104.40
Bancassureurs	11.71	4.02	86.75	102.49
Direct Marketing and Alternative Channels	83.14	30.90	98.98	213.01
Average for all Distribution Channels	10.55	7.96	82.39	100.91

Source: Lambert (1999).

While they tried to increase their revenue, French banks looked for ways to decrease bank costs. On one hand, they decided to spread distribution costs among more products lines. Insurance development was the evident means to achieve this goal. Lower distribution costs were made possible by their large, loyal customer base. On the other hand, French banks tried to benefit from economies of scale by combining bank and insurance activities. Indeed, it is not clear that they have been successful, because insurance cost is considered "marginal" by banks. Their costs are covered by banking activities and the insurance activities generate "only" marginal costs, which undervalues the real cost. The insurance distribution cost was more a question of decision to undervalue it rather than a reality.

However, French banks did succeed in lowering insurance management costs. For example, some banks decided to externalize non-critical insurance activities. French banks also chose their insurance partners carefully, changing if necessary in view of cost considerations. They also used their client databases intensively, allowing them to target clients based on their buying habits, economic status, spending and asset management practices. Mining their customer databases helped to reduce the cost per sale (Haynor 1998; Freeman, Steckler 2001). Another opportunity to reduce costs was to empower the clients in insurance management, making them responsible for managing their insurance products via venues such as telephone or the Internet. This was (and still is) presented to the clients as a new service, but it is also a transfer of administrative costs from banks to their clients. In 1996, insurance activities generated a management cost of 4.1 percent in premiums, whereas the tied agent costs stood at 10.91 percent, as shown in Table 5.20. In comparison, insurance companies have had difficulties imposing change to decrease their costs. As an insurance executive said: "It was easier for the banks to do better than us from scratch. We have the inertia of tied agents, the inertia of the employee unions, and our own managerial habits."

Finally, banks attempted to lower claim cost. A classic approach in insurance is to apply a better risk selection. In the property and personal injury market, it is not clear if French banks have succeeded. For example, banks have a worse claim/premium ratio for car and home insurance than insurance companies, as shown in Table 5.20 (Lambert 1999). It is possible that poor claim results of banks are the consequence of their development strategy in the property and personal injury market. To enter the market, they were forced to accept bad risks.

Another approach is to manage claims differently. Many banks have decided to centralize claim management, avoiding the costs of local structures and also avoiding the risks of clients asking for non-contractual coverage. To facilitate claim management, banks also decided to design simpler insurance products. A bank director said: "Our insurance products are as simple as possible and they also carry fewer guarantees. They are easier to manage." As a CEO of a bancassureur claimed, "The bancassurance in offering simple products has changed the insurance market, helping to improve the insurance product packaging in the country."

5.7.4 Different Models of Bancassurance: Financial Convergences?

An implicit hypothesis of the preceding description is that there is a single business model for all bancassureurs. The common assumption concerning convergence strategy could be that insurance companies and banks are converging toward the same point, to become financial giants, selling a full range of financial products to their client base in the same way. However, banks entering the insurance markets and insurance companies entering the bank market use obviously different models. French insurance companies have not been very successful in entering the bank market. The French case illustrates that there could be various financial convergences. For several years, bancassureurs have questioned how to fully exploit the potential for growth in the insurance market. The economic actors could choose various business models between alliance and integration to answer this strategic question.

An important decision for banks entering the insurance market is how they will build an alliance in the insurance field. For example, banks could lead the alliance and use independent brokers or agents to distribute through bank branches. For example, Crédit Lyonnais chose the broker status model to develop its insurance activities. The insurance director of Crédit Lyonnais noted in 1999, "We have chosen to keep the broker role, since we believe that it was not necessary to create a new insurance company; the French market already has quite a lot of insurance actors" (Berny 1999). On the contrary, insurers (who have the lead role) could use bank branches to sell their products. In the case of small banks with a few branches, the lead could be taken by the insurer. Those small banks have an insufficient scale to justify a major investment in bancassurance.

Another possibility is to build a joint venture (JV) between a bank and an insurance company. The JV agreement could be only on distribution aspects. However, in some cases, partners have had cross-ownership arrangements or even establish new bancassurance companies through mergers. A main problem of JV could be the difficulty of combining both bank and insurance cultures.

The problem of leadership is certainly an important step in alliance building but is not the only prospective challenge. Another problem is how to disseminate the various insurance tasks. The question is then how bankers and insurers will work together. The answer to this problem is complicated because insurance services are provided by more than one source in insurance companies. Insurance companies have various departments, various services, and different external suppliers. The division of labor and coordination between banks and insurance companies is therefore quite complicated (Jarvis 1998). A habitual division of responsibilities or business model could be putting banks in charge of marketing and sales and the administrative management of the contract (after subscription), with claim management the insurers' responsibility. Even with this simple division of labor, some problems may arise due to a certain overlap. For example, insurance sales management cannot be completely disconnected from administrative management of the contract. The partners will have to find a way to exchange the needed information.

It is clear, however, that all negotiations to split insurance tasks between partners could lead almost naturally to innovative solutions, and therefore often to unique configurations of the insurance value chain. Indeed, the entry of banks into the insurance market could imply the redesign of the insurance value chain. Through outsourcing and intensive use of technology, banks could focus on those areas that enhance their competitive long-term positions. The insurance value chain could be disintegrated from the past strong integration to the new network with an e-business type of management, examples of which are shown in Table 5.21. But, some competitors see out-sourcing as a potential weakness. The marketing director of a mutual society said, "Some banks use various subcontractors to manage insurance, especially claims. But this may reduce the service quality."

The French case is interesting because France was one of the first countries to develop electronic banking and insurance.[165] Minitel, introduced in 1981, allowed

[165] Electronic banking or home banking is an umbrella term for the process by which a customer may perform banking transactions electronically without visiting a brick-and-mortar institution. The following terms all refer to one form or another of electronic banking: personal computer (PC) banking, Internet banking, virtual banking, online banking, home banking, remote electronic banking, and phone banking. PC banking and Internet or online banking are the most frequently used designations. It should be noted, however, that the terms used to describe the various types of electronic banking are often used interchangeably. Electronic insurance refers to the same distribution channel for insurance products.

French consumers to access thousands of services, including bank and insurance products. An example was Cortal, a branchless bank set up to exploit the marketing opportunities of minitel. There were 17 million minitel users at its peak (Strang 1999). However, as Internet technology becomes faster and cheaper, minitel users are decreasing. Since bancassurance development occurred during the minitel and then Internet waves, French banks tried to create e-agencies to sell bancassurance products through the Internet, but with poor results. It is also possible to consider an e-face-to-face marketing, in which firms turn to remote forms of person-to-person communication. It this case, vendors could be in "direct contact" with prospects, face-to face-but using Web cam and video conferencing (McSweeney 2001).

Table 5.21. Type of Configuration of the Value Chain for a Bank Entering the Insurance Field

Product Design and Underwriting	Marketing and Distribution	Risk Financing and Asset Management	Customer Support and Back Office
Data Mining	Price Transparency	Transfer to Risks to Third Parties (Reinsurers, Capital Markets)	Consumer Empowerment
Selection of Clients	Consumer Empowerment	Management of Balance Sheet Risks	
New Type of Quotation			Client Management
Use of New Technology (less labor intensive act)	Faster Marketing		Less Face-to-face Contact with Clients
Outsourcing	Possibility of Simpler Products		Web-based Services
	Decrease of Distribution Cost		

Source: Author.

Therefore, banks have the possibility of "reinventing insurance management," according to a bank CEO. They have the opportunity to design new types of underwriting, policy issue and delivery, premium collection procedures, customer services, and sales approaches. Considering the possibilities, we should recognize that the answers to the previous question concerning the type of alliance are situated between different models. This heterogeneity points to the existence of various bancassurance business models.

Whatever the type of alliance decided, this decision is not definitive. Indeed, banks are able to change their insurance activities over time. Thus, not only are there various bancassurance business models, but they are not stable in the long term. In fact, a certain period of time is necessary to build an insurance management framework. At the beginning of their diversification, financial institutions could choose an alliance. The most common model is for a financial institution to have a partnership that limits resource mobilization at the beginning of the strategy. Then, once the insurance activity is well managed and successful, the bank can gradually integrate the operation. The integration will be limited by the French regulation since insurance underwriting should be done by an insurance company (eventually a separate subsidiary owned by the bank). Despite the fact that the insurance company is a separated legal entity from the bank, the insurance company could be completely integrated at the level of the organization.

Integration can take various forms. A bank could start its own insurance operation and recruit external experts. This internal integration has rarely been done at the beginning; one exception is Crédit Mutuel. Another possibility is full outsourcing at the beginning, to try to learn insurance management techniques as fast as possible and to integrate them fully. Or, banks could choose the option of external growth. Some French banks have decided to buy insurance companies and to distribute their products through bank branches. This is more expensive, but faster.

A crucial integration decision concerns distribution. In an integrative distribution model, the bank recruits and trains bank employees to sell the new insurance products. It is clear that life-related saving products are not new for bankers, unlike P&C insurance. For example, branch bankers sell insurance products to customers by exploiting face-to-face relationships in bank branches.

In an external distribution model, the financial institution requires other agents to deal with the insurance products. Sometimes it is possible to find experts who are employees or representatives of the insurance company within a bank branch. In this case, the branch bankers help identify insurance prospects. Between the two options, the French case shows that the main bancassureurs have chosen complex distribution channels. To sell insurance products, they have mixed various distribution channels such as direct mail, telemarketing, platform bankers (Web portal used by bankers to present and sell various financial products), in-house specialists, and external sales forces, depending on the product complexity.

Another important decision concerns claim management. Again, the final strategy is on a case-by-case basis. Larger banks are able to adopt the strategy of building to the point where they set up their own insurance back offices for application processing, claim management and policyholder service. For small banks, it does not make sense to build their own back offices (Haynor 1998). Life activity was often done internally. On the contrary, P&C claim management activities were externalized to specialists.

Bancassurance case studies

Having given some explanation about bancassurance in France, it is also important to give examples. We have chosen two success stories, in which banks used two different strategies at the beginning. The first case of Crédit Agricole (CA) is an example of an alliance followed by an integration of insurance activities. The case of Crédit Mutuel (CM) is an example of immediate integration. Thus, both banks have now chosen an integration strategy. Indeed, after a long period of development (more than 30 years), French banks seem now to select to integrate insurance and bank activities, especially for life insurance.

Created in 1890 for the agriculture sector, CA has been targeting bank clients outside French agriculture for many years (Thiveaud 1997). It is one the most important banks in the country, with 15 million clients.

In recent years, CA bought Banque Indosuez (1996), Sofinco (1999), and another major French bank, Crédit Lyonnais (CL). The merger between CA and Crédit Lyonnais (CL) was finally accepted by the CECEI on March 26, 2003. As CA, CL also owned an insurance company, the Union des Assurances Fédérales (UAF). CA decided in July 2004 to merge all insurance activities of both CA and CL. Indeed, Prédica (CA Life Insurance Company) and UAF will be merged to create a single-person insurance department (*départment d'assurances de personnes*). The diversification into the insurance business is rather new for this old bank. To start its insurance business, CA signed an agreement with a French insurer, Groupama. In 1986, the new partnership launched a simple interest-bearing annuity product, Sorasaving. Sales through bank branches were very successful. CA then demanded that Groupama immediately share sales profits in addition to the distribution commissions initially agreed upon. When Groupama refused, the bank created its own life insurance company, Prédica, in 1986, which became the second-largest life insurance company in France within two years. The sales of Prédica were 10.7 billion euro in 2003 (see Table 5.10). At the end of 2003, Prédica and UAF combined had 5.5 million clients and 12 million contracts (Crédit Agricole 2004). The products include pure endowment insurance (unit-linked contracts and French franc/euro-based contracts), term life insurance, health insurance, and credit insurance.

After the success of the life business, CA launched a property and personal injury insurance company in 1990, Pacifica, which became one of the largest companies in the sector in less than ten years with 0.7 billion euro in 2003. The company manages 3.2 million P&C contracts. It is now in the third P&C bancassureur. The portfolio is composed of automobile (1 million contracts in 2003), commercial, and personal property insurance, with 1.4 million homeowner comprehensive insurance contracts (Crédit Agricole 2004). Future growth is forecasted on health complementary products (for instance, 162,000 contracts). According to a bank executive, the financial break-even was reached in 1995, five years after the creation of the subsidiary.

Prédica and Pacifica's sales network is the bank branches of CA. Employees, experts in insurance, are devoted to the sale of insurance. In 2003, the group CA had 112,206 employees in France (CECEI 2004:182). CA is structured in various regional federations, with 74 in 1993, but 44 at the end of 2003. Each regional federation has a certain degree of autonomy. Some federations refused to sell insurance at the beginning. Some decided to launch their P&C insurance activity in 1990, others in 1991 or 1996. Once the decision is taken by a federation, the insurance implementation could be quite rapid, mainly due to the hierarchical structure with bank employees accelerating the process. The potential for growth is therefore still significant.

An element of bank differentiation is claim management. The claims are managed only by telephone and much innovation is supposed to be applied at this level of the relationship. For example, clients can manage their contract by Internet—administrative tasks, information about the various products, claims declaration. The claims are in fact handled in ten claim management units in various parts of France.

One of the elements of differentiation of CA was to create simplified product design. The key notion of the design was the industrialization of services, with mass product manufacturing and distribution and costs as low as possible to drive the commercial price down. According to the bank, the products are simple, clear and competitive. As a bank director said, "The contracts are designed to be user-friendly for the clients, as far away as possible from the usual legal jargon."

The second example is Crédit Mutuel, a mutual bank created in 1881 in Alsace-Loraine. This bank expanded its activity after World War II (Thiveaud 1997). In 2004, the bank had 11.3 million clients. Among them, 5.6 million were *sociétaires*, shareholders who are also clients. Twenty-eight thousand participated to the governance of their branch, usually in participating to the branch's annual general assembly (Crédit Mutuel 2004). The 1975 Finance Law (annual law written by the Ministry of Finance) entitled Crédit Mutuel to launch the *livret bleu*, or blue saving account, which was for a long time one of only two bank accounts with interest rates in the French market (Thiveau 1997:47). Most of the French banks do not offer any interest on bank accounts. Money market mutual funds are not developed in France. The bank has 55,770 domestic employees, and is divided into various regional federations. In April 1998, CM bought another French bank, CIC, to become the country's second-largest bank network. In terms of deposits, CM is the fifth-largest French bank.

The strategy of the group can be summarized by three interconnected key elements: bancassurance, proximity, and technology. In 1971, the bank pioneered bancassurance in France and was nearly the only one on the French market for several years. This willingness to develop insurance could be explained by the fact that many CM accounts paid interest. Therefore, transferring savings from a bank account to a life insurance did not affect the cost of financial resources for CM. This was contrary to other banks, which saw the transfer as an increase of their resource cost. With its extensive network, CM is now one of the biggest bancassureurs in the marketplace, now the country's ninth-largest insurance company. Among the bancassureurs, CM is the fifth with 1.2 billion euro in P&C and 3.7 billion euro in the life business in 2003 (see Table 5.10). The insurance activity was firstly a subsidiary, les Assurances du Crédit Mutuel (ACM), of the regional federation of Strasbourg, in the east of France. ACM provided insurance services to the other regional federations that had the choice of using its services. Today, 14 regional federations use ACM services. The remaining four regional federations have their own insurance companies or a specific agreement with an outside insurer. According to a CM top executive, "We were for a long time the only real bancassureur with all our activities being integrated. For example, all new CM employees are trained to sell insurance products." The bancassurance positioning is to create simple insurance products with, most of the time, an innovative and specific warranty, new in the French market. For example, a new car completely destroyed in an accident or crash is reimbursed at the new value during the first five years (*garantie valeur à neuf*) and does not seem to have any moral hazard problems.

The proximity of CM can be illustrated by various elements. As a cooperative and mutual bank, CM is organized on three levels: local, regional, and national. The local and regional levels entitle the bank to be close to its clients. The first level of the structure is the 2,000 local *caisses*, banking institutions subject to banking law. Each local *caisse* is financially autonomous, taking deposits, granting loans, and offering the full range of financial products. Each local caisse has its own board of directors or supervisory board, made up of delegates elected by the members (shareholders and clients). Each *caisse* could have one or more branches. The bank has a large network of 4,760 branches (CECEI 2004:184).

At the regional level, CM is divided into 18 regional federations, which are autonomous banks and organize their development and services according to local markets. A regional federation is the organ of strategy and control, representative of the CM in its region. A regional group provides various banking services: cash management and the provision of financial, technical, and information technology services. An important element of proximity is due to the bank's governance. Indeed, clients are also shareholders of the mutual bank and are involved in various boards of the bank. An important aspect of CM positioning is therefore the large network and the client involvement in its governance.

Technology is another field of differentiation. In the 1970s and 1980s, CM invested in information technology with the goal of leading technological change in France. Once the Internet wave started, the bank devoted much of its energy and budget toward developing its Internet services. As explained by a bank director, "Investing in technology has always been a bank priority. We are constantly looking for every possible innovation which could be used in our bancassurance activities." An Internet specialist within the bank added, "We involve the branch employees in technological development. A lot of them are proud to participate in our tests. They know that it is not only a question of facilitating their work but also improving our service level." The intensive use of IT is also a way to expedite claim management. A top executive said, "In some cases, a bank check is issued at the end of the phone claim declaration."

Potential Difficulties in Bancassurance Development

Our description of French bancassurance could be seen as idealistic, but bancassurance development may face some difficulties. First, the success of bancassurance might be weakened by a lack of strategic commitment. For a bank, the insurance development strategy should be consistent with the bank vision. In fact, it is necessary to fix insurance development as a priority for the bank and within the structure for branch employees. Some banks could experience a lack of strategic involvement by top management or branch managers in the development of insurance potential. It is not always easy to "push" insurance products through bank branches.

In some cases, banks have balanced short-term and long-term issues. They have tried to invest on a long-term basis in a bancassurance strategy with the building of a bancassurance structure. But, some banks could stop their investment because of short-term earnings pressure (Freeman, Knowling 1999).

Second, some organizational problems could arise. Thus, in a case of merger between a bank and an insurance company, some differences of culture could constrain bancassurance development. Employees might have different work habits and practices; for example, different incentive schemes. Furthermore, distribution channels, especially tied agents, could refuse bancassurance, being protected by territorial exclusivity.

Third, bancassurance development could be constrained by marketing problems. Banks may suffer from a lack of experienced insurance professionals who understand the insurance market and may face difficulties in recruiting and training qualified employees. Another difficulty could result from a failure to integrate both bank and insurance marketing plans (Jarvis 1998).

There also might be some danger for the banks. Because of the lack of experience in P&C claim management, some clients might experience what they perceive as bad service. Therefore, in the long term, the bank image could suffer from a repeated "weak" claim-handling image. However, until now, no French bancassureurs seem to have been faced with this type of decline.

Fourth, the emergence of bancassurance might also imply some competitive problems. The main danger would be a too-large concentration among a few huge bancassureurs, which could decrease competition. The French or in general European clients might have less choice in terms of suppliers. Furthermore, the internationalization of the financial and insurance services in Europe occurs at a time when there are still some unequal competitive market structures between the European markets and also in other non-European markets. When financial and insurance institutions do not have the same competition regulations in different markets, it is hard for a European client to understand the ability of a bancassureur to perform its contractual duties. In choosing a supplier, a European client from country A could doubt the solvency of the bancassureur in country B.

The bancassurance phenomenon is already provoking new, intense competition from non-financial institutions. For example, in France, car manufacturers sell both credit and insurance. Food supermarket chains are copying the same strategy, but with difficulties. All new market actors could use information technology to sell financial products at a lower cost.

Fifthly, the emergence of bancassurance implies the possibility of a contagion crisis. A beneficial trend for bancassurance consumers will be the lowering of the price of insurance products. But increased competition could lead to a price war within the financial and insurance markets, which will weaken bancassureur margins. Some observers say the banking industry currently has a much higher return on equity, ROE, than the insurance industry, particularly P&C insurers (Panko 2000). It is not certain that bank shareholders will accept a decline in their ROE.

Any potential price war could weaken financial and insurance institutions. But at the same time, it is difficult to control the new bancassureurs. In fact, the protection of policyholders is essentially provided through financial supervision, first through prevention and subsequently compensation if necessary and possible. The European trend is to have "integrated financial supervisory bodies." In the event of the liquidation of a bank, a guarantee fund was introduced in France in 1999 (Law 99–532 of June 25, 1999 concerning saving and financial security). Such a fund, however, does not have sufficient resources to cope with a systemic crisis. Despite creating integrated supervision and large compensation funds, it is clear that a supervisory body cannot prevent the creation of huge financial conglomerates, a trend that is embedded in a much wider move toward integrated financial services. The possible failure of some huge conglomerates might provoke a crisis within the all-financial and insurance markets.

5.8 CONCLUSION

This study of the French insurance market has enabled us to describe the main features of one of the largest insurance markets worldwide. The size of the market is linked partly to its long history, beginning with maritime insurance, imported by the Greeks more than 2,000 years ago. It is interesting to see that there is also a long tradition of French state involvement in the insurance business. It is also surprising for a non-expert of France to see that the country was among the most innovative countries in insurance, along with the U.K., Germany, Italy and the U.S. In recent years, a trend has been the reduction of the state in the insurance business with various privatizations.

With a notable history, the regulations concerning insurance are obviously quite important and are now influenced by European Union insurance directives. The state supervision changed in 2003 to reflect the convergence between banks and insurance companies.

Some of the important data to consider when analyzing the French insurance industry is the market share of life insurance: 63.3 percent of the insurance market, the life insurance market is mainly a savings market. The financial results of French insurance companies are strong. Despite a decline in stock market value, French insurance companies show a weak exposure to stock market exposure (only 23.4 percent of stocks in the investment portfolio). The volume of insurance investments is significant, with 1,015 billion euro in 2003 (fair market value). Competition is intense with an increasing foreign presence representing 16.3 percent of the life market and 26.3 percent of the P&C market in 2003, the low cost positioning of mutual societies in P&C insurance, and the development of bancassurance.

The last phenomenon of convergence between banks and insurance is quite characteristic of France, because banks dominate the life insurance market with a market share of 62 percent as of 2003 and have also developed their activities in P&C insurance, with a market share of 8 percent in 2003. A significant lesson to be learned from the French experience is that the competition between banks and insurance companies is the main driving force of the future of insurance markets. However, change in other insurance markets will not happen overnight. It took 20 years between the 1970s and 1990s for French banks to dominate the country's insurance market.

Despite the potential difficulties, it is clear that the bancassurance phenomenon is far from being exotic and will likely be commonplace in most of the financial and insurance markets characterized by various heterogeneous business models. Therefore, it is time for traditional insurers to find an efficient answer to bank development in their original markets. France is an excellent example to keep in mind.

5.9 REFERENCES

Abramovici, Gérard, 2001, *Méthodologie générale des comptes de la protection sociale en base 95,* No. 26, September (Paris: Ministère de l'emploi et de la solidarité, Direction de la recherche, des études, de l'évaluation et des statistiques).

Autorité des Marchés Financiers, 2004, *Rapport annuel 2003 (dossier de presse)* (Paris: AMF).

Banque de France, 2004, *Rapport Exercice 2003* (Paris: Banque de France) (www.banque-france.fr).

Banque de France, 2005, *Zone Euro, Principaux indicateurs économiques et financiers* (Paris: Banque de France (www.banque-france.fr).

Bauer, Michel, and Bénédicte Bertin Mourot, 1987, *Les 200. Comment devient-on grands patrons?* (Paris: Seuil).

Bechtel, Julien, Laurent Caussat, and Christian Loisy, 2004, *Les comptes de la protection sociale en 2003,* No. 70, October (Paris: Ministère de l'emploi, du travail et de la cohésion sociale, Direction de la recherche, des études, de l'évaluation et des statistiques).

Berny, Laura, 1999, "Les établissements bancaires font le pari de l'assurance-dommages," *Les Echos,* December 20: 15–16.

Besson, André, 1977, *Les assurances terrestres en droit français* (Paris: LGDJ).

Boleat, Mark, 1995, "The European Single Insurance Market," *Geneva Papers on Risk and Insurance* 20(74) (January): 45–56.

Caius Suetonius Tranquillus, 1975, *La vie des 12 Césars* (Paris: Folio).

CAPA, 1999, *La bancassurance en France: Les raisons du succès et les perspectives d'évolution* (Paris: CAPA).

Clavière, Etienne, 1990, "Prospectus de l'établissement des Assurances sur la Vie," *Risques* 1 (June): 123–136.

Clinton, Stephen, 2001, "How To Apply Global Bancassurance Experience to Developing Markets?" *Lincoln Reinsurance Reporter* 165 (1st quarter).

Comité Européen des Assurances, 2004, *Annual Report* (Paris: CEA).

Comité des Etablissements de Crédits et des Entreprises d'Investissements, 2004, *Rapport Annuel 2003* (Paris: CECEI, www.cecei.org).

Comité des Etablissements de Crédits et des Entreprises d'Investissements, 2001, *Rapport Annuel 2000* (Paris: Banque de France).

Commission de Contrôle des Assurances, 2004, *Rapport d'Activité 2002–2003* (Paris: CCAMIP).

Commission de Contrôle des Assurances, 2002, *Rapport d'Activité 2001–2002* (Paris: CCAMIP).

Commission de Contrôle des Assurances, 2003, *Tableaux de synthèses—exercice 2002* (Paris: CCAMIP).

Commission de Contrôle des Mutuelles et Institutions de Prévoyance, 2004, *Rapport d'Activité 2002–2003* (Paris: CCMIP).

Conseil National des Assurances, 2003, *Rapport au Président de la République et au Parlement* (Paris: CNA).

Conseil d'Orientation des Retraites (Pension Orientation Council), 2004, *Pensions: Reforms in France and Abroad, the Right to Information* (Paris: La Documentation Française).

Couibault, François, Constant Eliashberg, and Michel Latrasse, 2003, *Les grands principes de l'assurance* (Paris: Ed. Argus de l'Assurance).

Crédit Agricole, 2004, *Annual Report 2003* (Paris: Crédit Agricole).

Crédit Mutuel, 2004, *Annual Report 2003* (Paris: Crédit Mutuel).

Daniel, Jean-Pierre, 1992, *La bancassurance en France* (Paris: Verneuil).

Dion, Fabrice, 1995, *Les privatisations en France, en Grande-Bretagne et en Italie,* Notes et études Documentaires, No. 5024 (Paris: La Documentation Française).

Duncan, Matthews, 1998, "Insurance in the Single Market," *European Business Journal* 10(2): 78–84.

Economist [AuQ1]Intelligence Unit, 2001, *E-insurance: Creating a Competitive Advantage* (London: EIU).

European Union Council, 2003, *Joint Report by the Commission and the Council on Adequate and Sustainable Pensions,* No. 7165/03 (Brussels: EUC).

Fédération Française des Sociétés d'Assurances, various years, *Rapport Annuel* (Paris: FFSA).

Flur, Dorlisa, Daren, Huston, and Lisa Lowie, 1997, "Bancassurance. Could Banks Be a New Channel to Sell Insurance?" *McKinsey Quarterly* 3: 126–132.

Freeman, Terrence, and Douglas Knowling, 1999, "So What's the Big Deal (Financially) with Bancassurance?" *LIMRA's Market Facts* 18(5): 37–39.

Freeman, Terrence, and Denis Steckler, 2001, "Technology Drives Bancassurance Gains," *National Underwriter* 23(17): 3–11.

Gallix, Lucien, 1985, *Il était une fois l'assurance* (Paris: l'Argus).

Gay, Pierre-Angel, and Guillaume Maujean, 2004, "Les tarifs d'assurance auto ne peuvent varier brusquement," *Les Echos,* September 14: 41.

Groupement des Entreprises Mutuelles d'Assurances, 2004, *Actualités, La Lettre du GEMA,* No. 141 (October) (Paris: GEMA).

Groupement des Entreprises Mutuelles d'Assurances, 2003, *Actualités, La Lettre du GEMA,* No. 126 (March) (Paris: GEMA).

Hatzfeld, Henri, 1971, *Du paupérisme à la sécurité sociale, Essai sur les origines de la sécurité sociale en France (1850–1940)* (Paris: Armand Colin).

Haynor, William, 1998, "Ways To Bring Insurance Sales Inside a Bank," *Lincoln Reinsurance Reporter* 155 (1st quarter).

Institut National d'Etudes Démographiques, 2003, "La population de la France en 2002," *Populations & Sociétés* 388 (March).

Jarvis, David, 1998, "Bancassurance: Easier To Justify than Execute," *Lincoln Reinsurance Reporter,* 3rd quarter: 14–17.

Jenkins, Patrick, 2000, "World Insurance Industry: Bancassurance," *Financial Times,* April 28, p. 6.

Lambert, Alain, 1999, Assurons l'avenir de l'assurance: Rapport au Sénat No. 45, Commission des Finances, Vol. II (Paris: Sénat).

Lambert, Denis-Clair, 1996, *Economie des assurances* (Paris: Armand Collin).

Marini, Philippe, 2004, *La loi de sécurité financière: Un an après,* Report to the Senate, No. 431, July 27 (Paris: [AuQ2]).

Maujean, Guillaume, 2004, "AXA entend persévérer sur la voie des économies de coûts," *Les Echos*, June 3: 29.

McSweeney, Greg, 2001. "Special Report: Bancassurance: Are We There Yet?" *Insurance and Technology* 26(4): 37–42.

Ministère de la Santé et de la Solidarité Sociale (Social Protection and Health Ministry), 2004, *Health National Account,* No. 65, July (Paris: Direction de la Recherche, des études, de l'évaluation et des statistiques).

Observatoire de l'opinion, 2004, *18ème vague de l'Observatoire de l'opinion AFB/IREQ* (Paris: Observatoire de l'opinion).

Panko, Ron, 2000, "Bancassurance Gets a Boost," *Best's Review,* April: 113–118.

Pearson, Robin, 1997, "Towards an Historical Model of Services Innovation: The Case of the Insurance Industry, 1700–1914," *Economic History Review* 50(2): 235–256.

Poiget, Philippe, 2004, "Défaillances d'entreprises d'assurance. Le nouveau fonds dommages," *La Tribune des Assurances,* September.

Richard, Pierre-Jean, 1956, *Histoire des institutions d'assurance en France* (Paris: Argus).

Ripert, Georges, 1929, *Droit Maritime* (3d ed.), Vol. 2. Rousseau & Cie, Editeurs; Librairie Arthur Rousseau.[AuQ3]

Strang, George, 1999, "Bancassurance and Direct Marketing," *LIMRA's Market Facts,* Sept./Oct.: 32–36.

Swiss Re, 2005, "World Insurance in 2004: Growing Premiums and Stronger Sheets," *Sigma* 2.

Thiveaud, Jean-Marc, 1997, "Les évolutions du système bancaire français de l'entre-deux-guerres à nos jours: Spécialisation, déspécialisation, concentration, concurrence," *Revue d'Economie Financière* 39(1): 27–74.

Thomson, Maria, and Alan Wade, 2000, "The European Model: Much We Can Learn," *National Underwriter*, September 11: 52–53.

Trebilcock, Clive, 1985, *Phoenix Assurance and the Development of British Insurance (1782–1870),* Vol. I (Cambridge, England: Cambridge University Press).

Venard, Bertrand, 1997, "La légitimation historique de l'intermédiation dans le secteur de l'assurance," *Assurances* (Ecole des HEC) 4: 601–620.

Venard, Bertrand, 1999, "The Influence of the Information Highway on the Insurance Management," *Geneva Papers on Risk and Insurance: Issues and Practice* 24(2): 189–202.

Weinberg, Mireille, 2004, "Assurance-dommages: Les bancassureurs veulent confirmer leur percée en 2004," *Les Echos*, March 15: 34.

Weinberg, Mireille, 2004, "Assurance automobile: La baisse des tarifs risque de faire long feu," *Les Echos*, October 27: 33.

Weinberg, Mireille, 2004, "La question de l'indépendance des associations d'assurés reste posée," *Les Echos*, December 9: 30.

Weinberg, Mireille, 2004, "L'alliance entre la CNP et la Mutualité française entre dans le concret," *Les Echos*, June 4–5: 25.

Weinberg, Mireille, 2004, "Le Fonds de garantie des assurances obligatoires ponctionné par Bercy," *Les Echos*, November 16: 34.

Weinberg, Mireille, 2004, "Les agents généraux de MMA contraignent Jacques Lenormand au départ," *Les Echos*, March 24: 34.

Weinberg, Mireille, Pierre-Angel Gay, and Henri Gibier, 2003, "Les assureurs sont moins riches, mais ils sont mieux gérés," *Les Echos*, November 21–22: 24.

Williams de Broe, 1998, *European Insurance Sector Review,* London, December. [AuQ4]

5.10 LEXICON

Source: FFSA: 2003 and author.

Accident and health insurance: Policies that guarantee the reimbursement of health and medical care costs due to sickness or accident, payment compensation in the event of disability, or payment of a lump-sum benefit in the event of accidental death (*assurance en cas de maladie ou d'accident*).

Agent général d'assurances or AGA: See tied-agent.

AMF: Authority of Financial Market, *Autorité des marchés financiers*, was set up to handle all the tasks involved in supervising corporate finance, financial information, investment providers, collective investments schemes, financial markets, and post-trade activities.

Assurances de personnes: Literally "people insurance;" in fact, life and health insurance.

Assessment mutual society: A mutual society that retains the right to assess policyholders' additional amounts if premiums are insufficient for operations.

Average expense ratio: *Ratio moyen de chargement, ou frais rapportés aux cotisations.*

Balance sheet value: *valeur au bilan.*

Bancassurance: French word originally used to describe the penetration of banks into the French insurance market. The word is now used to describe the convergence between both activities. It is possible to see some statistics with the expression over the counter, equivalent to the French expression *Vente aux guichets.*

Brokers (*courtiers*): Legally, brokers represent the insured, which gives them a very good image in the public. Small brokers are considered as tradesmen (*commerçant*) and must be entered into the Register of Commerce (*registre du commerce*). Large brokers are public limited companies.

Capital redemption bonds (*bon de capitalisation* or *capitalisation*): A savings contract that invests and increases built-up savings, guaranteeing a fixed lump-sum benefit at maturity. It is thus a policy of assurance that will mature after a certain period of time with a minimum maturity value being calculated on an actuarial basis.

Car insurance: *assurance automobile.*

Casualty insurance (*assurance accidents*): Insurance coverage for loss or liability arising from a sudden, unexpected event such as an accident. Casualty insurance is a broad category of insurance that includes almost any coverage that is not related to life insurance, health insurance, or property insurance.

CCAMIP, *Commission de contrôle des assurances, des mutuelles et des institutions de prévoyance*: The Provident Institutions, Mutual Society and Insurance Company Control

Comission. Its main duty is to control insurance and reinsurance companies, mutuals and provident institutions.

CCLRF, *Comité consultative de la legislation et de la réglementation financière*: The Regulation Consultative Committee is an advisory committee on financial legislation and regulation. It will issue an opinion on any draft legislation relating to financial matters (i.e., banks, insurance, investment services).

CCSF, *Comité consultatif du secteur financier*: The Financial Sector Consultative Committee must follow the relationships between firms and consumers and to bring together representatives of the two groups.

CEA, *Comité des entreprises d'assurance*: The Insurance Company Committee was created to license new insurance companies and mutual societies. The committee has power of decision concerning not only licensing, but also portfolio transfer (*transferts de portefeuilles*) acquisition or financial participation (*prise de participation*).

Claim experience: *Sinistralité.*

Combined ratio net of reinsurance: The ratio of paid and incurred claims, acquisition costs and administrative expenses plus other operating expenses to premiums (*Ratio combiné net de reassurance = sinistres payés et provisionnés, coûts d'acquisition et d'administration et autres charges techniques/primes*).

Commercial lines (in the property & casualty market): *Risques commerciaux.*

Commissaire-contrôleur des assurances: High-ranking civil servants, trained specifically to control insurance companies.

Complémentaire maladie: Reimbursement of healthcare and medical expenses, after social security reimbursement. Literally, complement for healthcare.

Construction insurance: *Assurance construction.*

Credit insurance: Insurance against financial losses sustained through the failure for commercial reasons of policyholders' clients to pay for goods or services supplied to them. (*assurance de crédit* or *assurance crédit*).

Direct sales (DS): Commercial mail, coupons in newspapers, television, telephone, and the Internet.

FFSA: *Fédération Française des Sociétés d'Assurance,* French Federation of Insurance Companies.

FNMF: *Fédération Nationale de la Mutualité Française,* the French association for Mutuals 1945.

Freedom of establishment: All European insurers have the right to set up subsidiaries or a branch in any other European country.

Funded retirement plan: *Régime de retraite par capitalisation.*

GEMA: *Groupement des Entreprises Mutuelles d'Assurance* is a professional association for mutual societies (not Mutuals 1945).

General Liability insurance: *Assurances responsabilité civile;* covers the liability exposures of manufacturers, building industry professional, trade professionals, service providers, local and regional governments, medical practitioners, and others.

I.A.R.D.: *Incendie Accidents Risques Divers,* fire, casualty and other risks; acronym synonymous for P&C.

Insurance companies: *Sociétés anonymes d'assurance*. They are governed by the Insurance Act (*code des assurances*). Insurance companies could sell any insurance products and they had no territorial limitation. They could be state-owned or private companies. Most of the insurance companies are privately owned stock companies and profit-oriented.

Legal expense: *protection juridique*.

Liquid Asset: *Actif liquide,* an asset that could be negotiated rapidly in cash without any financial gain or cost. Cash in a bank account is a typical liquid asset. This is also the case of the *livrets d'épargne*, French savings accounts with small interest rate.

Long-term care insurance: Insurance policy that guarantees annuities or a lump sum in the event of loss of autonomy (*assurance dépendance*).

Loss ratio: *Ratio sinistres à primes* or S/P.

MAT: Marine, aviation and transport insurance corresponds to various types of policies covering vessels (hull insurance), goods in transit (cargo insurance), aviation (all insurances policies related to aviation), and space.

Market value: *valeur du marché.*

Multi-investment vehicle contract: A contract in which benefits are expressed in terms of one or more unit-linked vehicles and one investment vehicle in euros (*contrat multisupports*).

Mutual societies: Non-profit organizations in which each insured person is both a client and a shareholder (*sociétaire*). The mutual societies include:

The mutual societies of insurance (*Sociétés Mutuelles d'Assurance*, SMA) are associations. They always have a charter with geographical or professional specialization. They do not use commissioned intermediaries. They cannot sell life insurance. They are sometimes named pure mutual (*mutuelles pures*). They are assessment mutual societies.

The "mutuals 1945" are governed by a specific act: the Mutuality Act (*code de la mutualité*). They are named "mutuals 1945" (*mutuelles 1945,* or *mutuelles* or *mutualistes*) because they were created to provide complementary health coverage to social security created in 1945. They are non-profit associations. Since the law of July 25, 1985, only mutuals 1945 could be called "*mutuelles*" without any precision; the others must add the word 'assurance.'

The mutual insurance companies (*Sociétés d'Assurance Mutuelle*, SAM) are civil societies without profit orientation (*sociétés civiles sans but lucratif*). They are governed by the Insurance Act.

The mutuals specialized in agriculture (*sociétés d'assurance mutuelles agricoles*) are under the supervision of the agriculture ministry and the finance ministry.

Mutuelles Sans Intermediaires: It is also possible to find in French statistics the segmentation between the mutuals without intermediaries (*Mutuelles Sans Intermediaires, MSI*), such as MAIF or MACIF and mutuals with commissioned intermediaries (*Mutuelles rémunérantes*), such as MMA or Groupe Azur. Commissioned intermediaries are agents or brokers.

Natural disasters: *Catastrophes naturelles.*

Net earnings: *Résultat comptable de l'exercice;* for a given period of time are equal to the balance of underwriting results (*résultats techniques*) and other financial or non-recurring events (non-underwriting results).

Net investment income: Equals income plus realized capital gain; in French, *revenues et plus-values réalisées.*

Personal protection insurance: Policies offering guaranteed benefits in case of death, occupational disability, or long-time sickness and the reimbursement of covered healthcare costs (*assurances vie en cas de décès, de maladie ou d'accident*).

P&C: Property and casualty insurance, *dommages aux biens et responsabilité.* The main P&C insurance products are: automobile, commercial property, individual property, agricultural property, construction, general liability, natural disasters, MAT, credit and surety, assistance, and legal expenses.

Property insurance:

Individual property insurance (*assurances de dommages aux biens des particuliers*) in particular homeowners' comprehensive insurance (*multirisques habitation*)

Commercial property insurance (*assurance dommages aux biens des professionnels*)

Agricultural property insurance (*assurance dommages aux biens agricoles*)

Provident institutions: *Institutions de prévoyance* or *institutions de retraite supplémentaire relevant du Code de la sécurité sociale.* These are non-profit organizations, governed by the Code of Provident Institutions, which is part of the Social Security Act (*Code de la sécurité sociale*). Provident institutions provide personal protection, specifically health insurance for extra coverage above the basic social security system. Provident institutions are managed by representatives of employees and employers.

Pure endowment insurance: An insurance policy taken out directly by an individual or through an employer or association, which builds up deferred entitlements that are subsequently paid out in the form of a lump-sum benefit or annuity if the insured outlives the term of policy. These policies are usually combined with counter-insurance to cover the event of death (*assurance vie en cas de vie*).

Ratio of claims to premiums: *Rapport Sinistres à Primes, S/P,* loss ratio.

Salaried sales associates: Salaried sale forces correspond to mutual societies' network and also some insurance company network.

Social Security: *Sécurité sociale,* French social security was created in 1945 to insure work-related casualty and disease, retirement benefits (pension), and family benefits (depending on family size and resources). For information in French, see: www.securite-sociale.fr/presentation/presentation.htm#retraites

Solvency margin: *Marge de solvabilité,* the minimum amount of extra capital that an insurance provider must have to fall back on in unforeseen circumstances. Solvency margin equals shareholder's equity and unrealized capital gains/actuarial reserves.

Specialization principle: The French market is characterized by a specialization principle (*principe de specialisation*). The same legal entity cannot underwrite life and P&C insurance contracts at the same time.

Surety insurance: Sureties and guarantees issued to third parties for the fulfilment of contractual liabilities (*assurance caution*).

Term life insurance: Policy guaranteeing the payment of a lump-sum benefit or annuity to a designated beneficiary in the event of death prior to the term of the policy (*assurance vie en cas de décès*).

Tied agents or exclusive agents: *Agents généraux d'assurances, AGA,* are governed by an agent charter (*statuts des agents généraux*) of public order; i.e., compulsory for all of the insurance industry. The agent represents only one insurance company. The agent has exclusive rights to the company's products in his geographical territory (except in Paris and for risks not insured by the company).

Underwriting income: *Résultats techniques.*

Unemployment insurance: Policy offering benefits in the event of unemployment (*assurance chômage*).

Unit linked contract: A life insurance policy or capital redemption bond in which the amounts of benefits and premiums are not expressed in euros but in units of an investment vehicle, such as shares of a mutual fund or a real estate partnership. *Contrats en unités de compte.*

Unrealized capital gain or losses: *Plus-values ou moins-values latentes.*

6 The German Insurance Industry: Market Overview and Trends[166]

Raimond Maurer

Goethe-University Frankfurt

Barbara Somova

Goethe-University Frankfurt

6.1 HISTORY OF PRIVATE INSURANCE MARKETS IN GERMANY

The private insurance business in Germany has its origins in three different lines: mutuals, public, and commercial insurance companies (for details on the German insurance market's history, see Wandel, 1998, pp. 59–65, and Koch, 1988). The first mutuals were organized during the sixteenth century, and they typically provided fire insurance to members of specific groups such as guild members (called *Brandgilden*). In 1821 the *Gothaer Feuerversicherungsbank* and in 1827 the *Gothaer Lebensversicherungsbank* were founded by Ernst Wilhelm Arnoldi, with both companies organized as mutuals. Following these examples, many mutuals were created in all insurance lines during the second half of the nineteenth century. The first public insurer in Germany, the *Hamburger General-Feuercasse*, was formed as a merger of many Brandgilden in 1676. Following this example, other public insurers providing fire insurance to homeowners, who were often required by the authorities to insure their property, were formed in nearly all other German states during the eighteenth and nineteenth centuries. After this period, public insurers also started to offer coverage in other lines of the private insurance market. The first commercial (i.e., profit-seeking) insurers were sea transport insurers, created in 1765 and headquartered in Hamburg and Berlin. It was only in the mid-eighteenth century that the first commercial life insurer was created in Germany (Brockhaus, 1974). One of the major factors in the development of the life insurance industry was

[166] This research was conducted with support from the Förderverein für die Versicherungswissenschaft Frankfurt and DBV-Winterthur Insurance. The authors are grateful for the useful comments provided by Hartmut Nickel-Waninger, Christoph Jurecka, David Cummins, Bertrand Venard and Peter Albrecht. Opinions and errors are only solely those of the authors.

discoveries in mathematics, particularly in probability theory. The German mathematician Carl-Friedrich Gauss (1777–1865), for example, consulted a life insurer in Göttingen on premium calculations for life annuities, where he already used statistical data for death probabilities based on age (Gauss, 1973).

In the first half of the nineteenth century, commercial insurers were mostly active in three market segments: transport insurance, fire insurance, and life insurance. The nineteenth century also marked the beginning of the formation of large insurance stock companies. In 1818 the *Agrippina* insurance company was created (presently the *Zürich-Agrippina*), and in 1890 the *Allianz Versicherungs-Gesellschaft*. A catastrophic fire in Hamburg in 1842 destroyed almost the entire city center. This led to the creation of reinsurance companies, the first being the *Kölnische Rückversicherungsgesellschaft* in 1846, followed by the *Münchener Rückversicherungsgesellschaft* in 1880.

The first law governing state supervision of private insurers, subsequently the Insurance Supervisory Law (*Versicherungsaufsichtsgesetz,* or VAG) of May 12, 1901, created a central supervisory and regulatory body. In addition, the Insurance Contract Law (*Versicherungsvertragsgesetz,* or VVG) of May 31, 1908, regulated the legal rights and duties of the parties involved in insurance contracts. Both laws created the legislative basis for the private insurance industry and increased the influence of the state on the insurance business. General economic growth and the rapid development of industrial manufacturing led to an increased demand for insurance coverage. Before World War I, there were 962 life insurers, 48 property and casualty insurers, and 101 fire and building insurers in Germany. New insurance branches appeared, such as insurance against theft, credit insurance, insurance against air traffic risks, and automobile insurance.

After World War I, the destruction of German industry and the confiscation of assets owned by German insurers in coalition countries combined with high inflation dealt a severe blow to the insurance industry. The currency reform in 1924 left many companies on the brink of bankruptcy, and the number of life insurers was reduced to 691 companies. The calls to intensify state regulation grew louder. The amendment of the insurance company act in 1931 considerably increased the state's influence on private insurers. On the other hand, obligatory insurance coverage in some areas (e.g., liability insurance for master craftspeople and automobile owners) was introduced. During the years 1933 to 1945, the direct influence of the government on insurers further increased. Only the collapse in 1945 prevented the envisaged nationalization of the insurance sector.

The initial years after World War II were challenging. Private insurance was prohibited and substituted by the state-owned monopoly in the Soviet (East) sector, putting an end to the development of the private insurance systems there and forcing existing companies to move to the West.

In the West sector, however, the regulation of insurers differed considerably between the British, French, and U.S. zones of occupation, complicating reconstruction. No functioning state insurance supervision authority was in place until 1951. The liquidation of seven insurance companies was conducted by the Council of the Allied Forces in 1947. These insurers originally were owned by the trade unions and then integrated by force in the national-socialistic state as the companies of the Deutsche Arbeitsfront. Increasing economic growth since the mid-1950s and favorable political developments such as the creation of the European

Economic Community in 1957 facilitated the establishment of strong and reliable structures in the German private insurance market.

The remainder of this article provides an overview of the contemporary German private insurance market and its legal and economic structure; supplies basic information about main product groups and their country specifics; outlines the fundamental regulatory issues; and gives an outlook for some possible avenues of future development.

6.2 GENERAL MARKET OVERVIEW

6.2.1 Legal Structure

German insurers are organized as stock corporations, mutual insurance associations, or insurance companies under public law. Figure 6.1 provides information about the number and the market shares with respect to premium volume of these legal structures from 1997 to 2003. Most insurers are organized as stock corporations and, as such, account for about 80 percent of the premium volume. While mutuals still play an important role with respect to the number of companies in the market, their market share is only about 14 percent of total premium volume. The importance of insurance companies under public law is minor, both with respect to the number of companies and the premium volume written.[167]

A reason that stock corporations are the dominating legal structure in the private insurance market can be at least partly explained by organizational and capital requirements specified in section 7 (organizational requirements), and sections 22 and 53 (capital requirements) of the Insurance Supervisory Law (VAG), which apply to the insurer as long as it is in business. Obviously, stock corporations provide the better basis for compliance with those requirements. Interestingly, only about 20 percent of stock corporations are listed on the stock exchange (Elgeti and Maurer, 2000).[168] One explanation is that German insurance regulation forbids conducting life and health insurance business and property and casualty business under the roof of one legal entity, the *Spartentrennungsprinzip,* codified in section 106(c) of VAG. This requirement aimed to prevent unmanageable cross-subsidies between insurance

[167] An institution under the public law is a corporate body whose existence and activities are regulated not by the set of private law regulations, but by that of public law, usually governing the relationship between the authority and the citizen. An institution under a public law is created by authority's decree and serves a public purpose defined in that decree. This legal form has its roots in the historical economic and state development.

[168] According to the German Companies Act (*Aktiengesetz,* or AktG), the listing on an organized stock exchange is not an obligatory feature of a stock corporation (AktG, §§1, 3). The legal form of a nonquoted company gives all advantages of a public company, such as, for example, simplified procedure of the share transfer, without being burdened by the considerable costs of a stock exchange listing. The requirements regarding the ownership of the nonlisted corporations are identical to the listed one; they are codified in AktG (requirements for all stock corporations) and the Insurarance Supervisory Law (for insurance companies only).

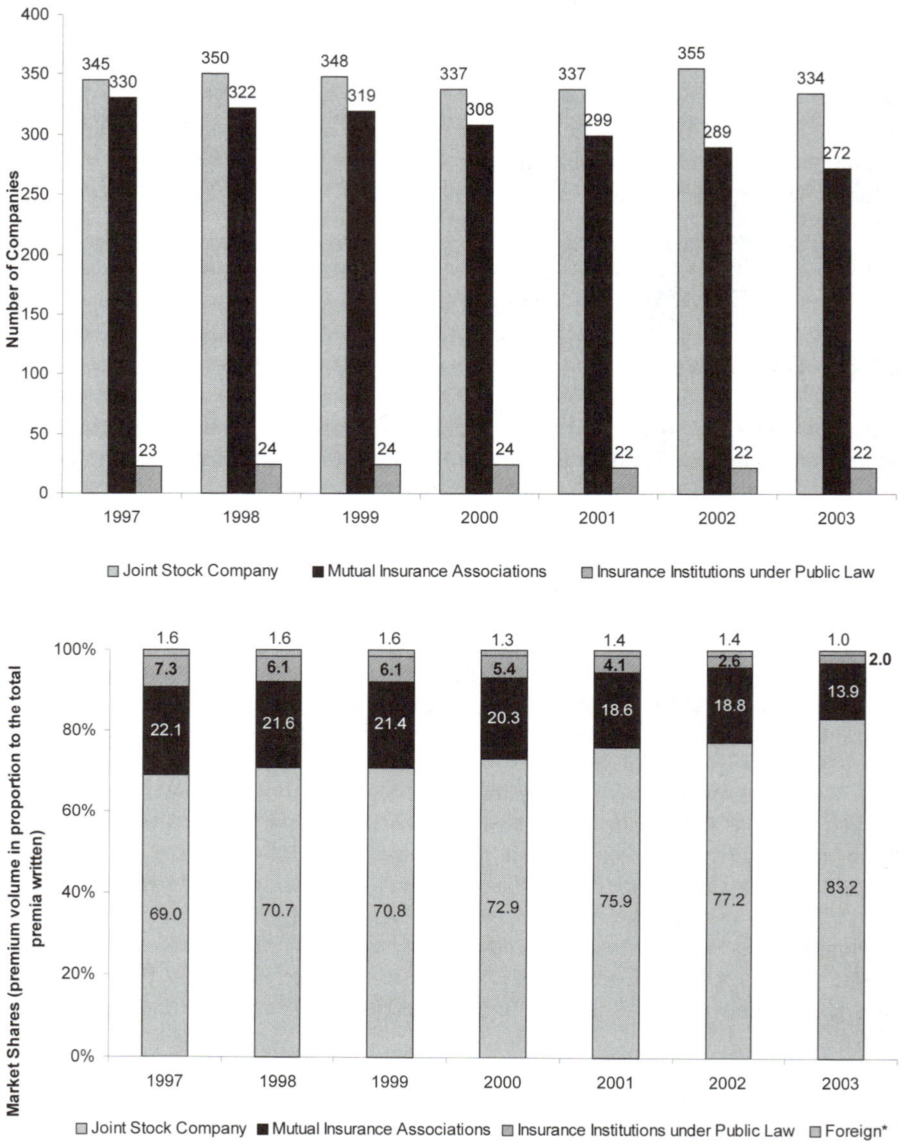

Figure 6.1. Number and Market Shares of German Insurers by Organizational Structure

lines within one insurance company or compensatory pricing that could endanger the financial stability of the company. In order to offer insurance coverage in different lines, holding companies listed on a stock exchange were created whereby the separate legal entities for life, health, and property and casualty insurance are fully owned by the head of the group. Another reason is, in order to combine the advantages of a mutual insurance association with those of a stock company, and to avoid the legally and economically difficult process of outright demutualization, a holding company in the form of a mutual insurance association is often created with wholly owned subsidiaries as stock corporations. (An example of such a structure is the *Gothaer Versicherungen*). On a consolidated basis, the importance of mutuals with respect to premium volume increases.

Foreign insurers are allowed to conduct business in Germany, provided that their relevant subsidiaries are incorporated in Germany and are under the supervision of the Federal Financial Supervisory Authority (*BaFin*). However, insurers from the European Union (EU) are granted the right to conduct business in Germany, not only by establishing a legal entity in the German market, but also by offering services from their original land of incorporation (*Europapass*). Supervision of these companies is incumbent upon the supervisory body in their country of origin.

Currently, the German market is dominated by insurance companies domiciled in Germany, with the share of foreign companies under federal supervision by premiums written comprising less than 2 percent of the market. In fact, the companies with ultimate foreign ownership are very active in the German market: among the market's top ten are, for example, Generali, Zürich, and AXA (see Table 6.1). However, these companies historically conduct business in Germany through their subsidiaries, which are incorporated in Germany and are subject to supervision by German authorities and are thus statistically treated as German companies. Foreign companies are solely those that have only a branch office.

With regard to the participation in insurance companies, there is no general restriction for investors. Particularly, no special legal restrictions with regard to foreign investors, as well as the insurance/bank, bank/insurance or industrial company/insurance ownership combinations exist, and all examples can be found in the market. However, special requirements for owners of significant insurance stakes (i.e., more than 10% of the equity capital) are codified in the Insurance Supervision Act (cf. §104 VAG). For example, if an investor intends to acquire a significant participation of an insurer, the supervisory body must be informed. Under certain quite restrictive circumstances (i.e., if, after the participation, effective supervision of the insurer is not possible) the supervisory body can delay or forbid the transaction.

6.2.2 Economic Structure

As in other countries with developed insurance markets, German insurers offer a wide range of products in various lines. The most important direct lines are life, property and casualty, and private health insurance. Figure 6.2 provides information about the number of companies under the supervision of the German supervisory body from 1990 to 2003 with respect to the most important direct insurances lines. In 2003, there were 679 insurers, with numbers being relatively stable over the last ten years. The life insurance sector has the largest number of companies (324), followed by property and casualty (251). Only 55 companies are active in private health insurance. In addition, 49 companies are active in the reinsurance business, but this sector is not the focus of this chapter.

In terms of the number of insurance contracts, motor vehicle insurance has a clear lead with 98 million contracts in 2002, followed closely by life insurance with 91 million contracts in 2002. Property insurance, with 67 million contracts, is the third-strongest insurance branch based on the number of contracts (Gesamtverband der deutschen Versicherungswirtschaft, 2003, p. 68). The ranks change, however, if the premiums written by insurance sector are considered. As can be seen in Figure 6.3, the premiums written by life insurers constitute approximately 46 percent of the

Table 6.1. Top 10 Insurance Companies by Insurance Line

Ranking Life by Gross Written Premium			Market Share	
2003	(2001)	Group	2003	(2001)
1	(1)	Allianz	15.60%	14.60%
2	(2)	Generali	10.61%	10.74%
3	(3)	ERGO	10.39%	10.05%
4	(4)	Zürich/Agrippina	5.56%	5.71%
5	(5)	R + V	4.61%	4.51%
6	(6)	AXA	3.94%	3.90%
7	(8)	Debeka	3.07%	2.89%
8	(7)	Gerling	2.81%	3.00%
9	(9)	Nürnberger	2.68%	2.71%
10	11	Signal/Iduna	2.21%	2.35%
	Top 3		36.60%	35.40%
	Top 5		46.78%	45.62%
	Top 10		61.49%	60.46%
Ranking Non-Life by Gross Written Premium			**Market Share**	
2003	(2001)	Group	2003	(2001)
1	(1)	Allianz	19.60%	18.64%
2	(2)	ERGO	6.26%	3.93%
3	(5)	HUK Coburg	5.27%	6.10%
4	(6)	R + V	5.01%	4.55%
5	(3)	AXA	4.91%	5.79%
6	(8)	Zürich/Agrippina	4.63%	3.74%
7	(4)	Gerling	4.29%	5.51%
8	(7)	Generali	3.89%	3.93%
9	(9)	Württember-gische/Wüstenrot	2.92%	3.40%
10	(11)	PARION	2.58%	2.66%
	Top 3		31.14%	28.67%
	Top 5		41.06%	39.02%
	Top 10		59.37%	58.24%

Table 6.1 *(continued)*. Top 10 Insurance Companies by Insurance Line

Ranking Health by Gross Written Premium 2003	(2001)	Group	Market Share 2003	(2001)
1	(1)	Ergo	15.41%	15.81%
2	(2)	Debeka	13.33%	12.92%
3	(3)	Allianz	12.07%	12.47%
4	(4)	Signal/Iduna	7.44%	7.59%
5	(5)	Generali	5.58%	5.46%
6	(6)	Continen-tale/Europa	4.72%	4.90%
7	(7)	Barmenia	4.04%	4.27%
8	(8)	DBV Winterthur	3.62%	3.69%
9	(9)	PARION	3.30%	3.30%
10	(11)	AXA	3.20%	2.80%
	Top 3		40.81%	41.21%
	Top 5		53.84%	54.26%
	Top 10		72.72%	73.22%

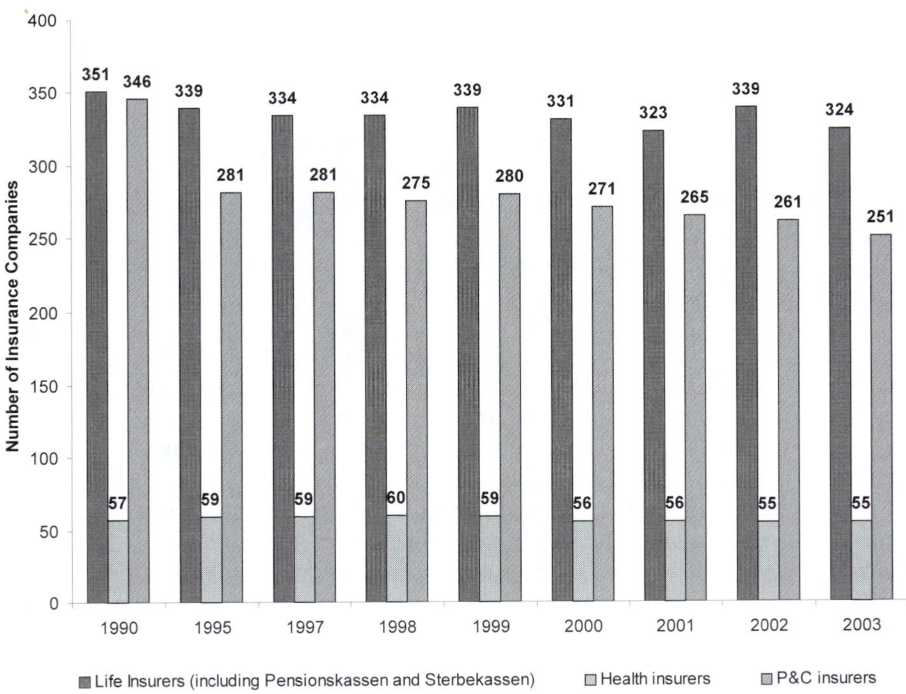

Figure 6.2. Number of Insurance Companies by Segment

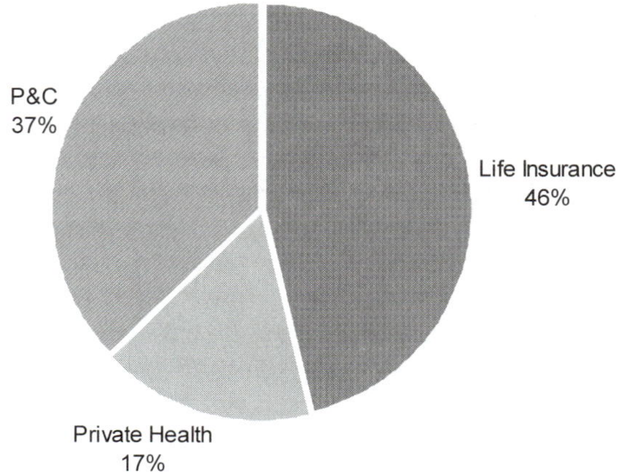

Figure 6.3. Insurance Premium Distribution by Segment, 2003

total premium volume written in 2003, followed by property and casualty, including motor vehicle, liability, casualty, legal claim, property, and transport insurance (37%). Private health insurance is in third place with a market share of 17 percent.

Table 6.1 depicts information about the concentration in the German insurance market as well as in the different lines. Measured by premiums written in all lines, about 35 percent of the market share belongs to the top three companies, whereas the top ten companies cover a market share of more than 60 percent. The picture in the different lines is quite similar: the health sector shows a slightly higher and the property and casualty sector a slightly lower concentration compared to the total market.

With respect to premiums written in 2003, the German insurance industry is the fourth-largest insurance market in the world, after the United States, Japan, and Great Britain. Yet, over the last twenty years, Germany has lost a substantial part of its global market share. Around 6 percent of the world's premium volume was written in the German insurance market in 2003, while in 1980, this ratio was approximately 9 percent.

Figure 6.4 compares important measures for the German insurance market to other countries' markets for the years 1980, 2001, and 2003. The figures for insurance density, defined as premiums received by primary insurers per capita, and insurance penetration, defined as premiums received by primary insurers in percent of the gross domestic product, have also grown more slowly than in the other developed markets. In 1980, Germany was second only to the United States in terms of insurance premiums per capita and third after United States and Great Britain by penetration of insurance, these numbers being above the average of the leading industrial countries. In 2003, the insurance penetration ratio of 7.0 percent was substantially lower than in many developed countries. Moreover, the insurance density of $2,287 also lies below the G7 average. This loss can at least partially be explained by the fact that insurance products have progressively been shifted away from public social security (especially in the health and pension sector) to private contracting in the United States and the United Kingdom, whereas the German public social security system still provides generous benefits. However, the population's awareness of the need to privately provide for retirement and health insurance, together with the impossibility of maintaining public social security benefits at formerly generous levels, is expected to change this situation.

With more than 800 billion euros worth of assets under management in 2003, insurance companies are among the most important institutional investors in the country. As shown in Table 6.2, life insurers account for approximately 75 percent of the amount. For comparison, in 2003, the assets under management of German mutual funds amounted to 436 billion euros and special funds to 520 billion euros (Deutsche Bundesbank, 2004a and 2004b). Special funds are investment vehicles comparable to mutual funds and are regulated like the latter in the Investment Company Act but are available only for institutional investors. With a market share of about 50 percent, measured in terms of assets under management of all special funds, insurers are important users of this vehicle for outsourcing asset management activities within a regulated framework. The assets of the German banking sector, amounting in 2003 to 6.471 billion euros, cannot be used for a proper size comparison: German banks are universal banks and carry a huge amount of loans on their balance sheets, especially to non-banks. More than half of the insurance industry's assets consist of fixed-income instruments. However, the figures vary from 50 percent in property and casualty insurance to over 70 percent in health insurance. Direct investments in stocks and real estate play only a minor role in all insurance lines.

The market standard for the majority of German insurers is the multichannel approach such as agents, brokers, banks, direct insurance, Internet, and other niche distribution channels. Agents (tied—i.e., obliged to exclusively distribute the

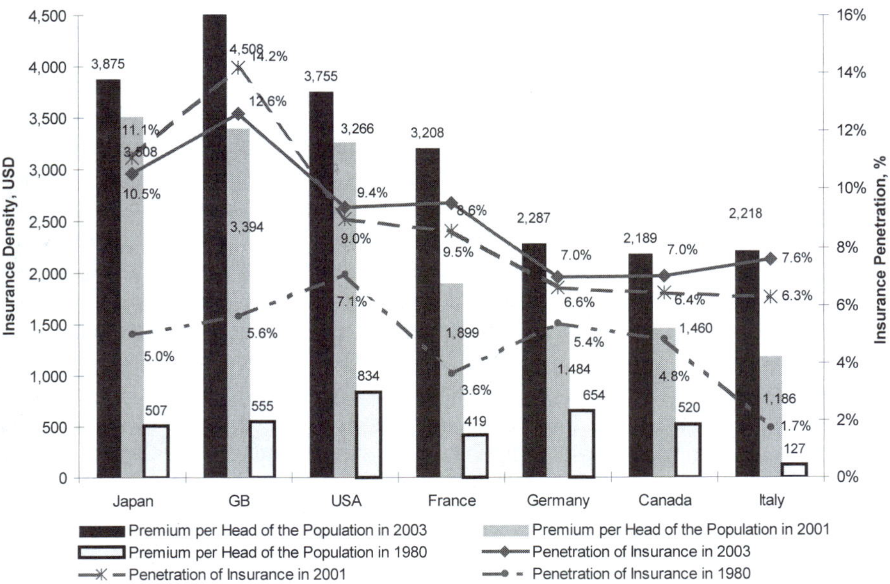

Figure 6.4. Insurance Density and Insurance Penetration in Germany and G7 Countries

Table 6.2. Assets Under Management for German Insurance Companies by Insurance Line

Life Insurance [a]						
In EURm	1998	1999	2000	2001	2002	2003
Real Estate	6,142	15,536	15,478	14,905	15,080	14,134
Growth	*n.a*	*152.9%*	*–0.4%*	*–3.7%*	*1.2%*	*–6.3%*
in % of subtotal	*1.4%*	*2.3%*	*2.9%*	*2.6%*	*2.6%*	*2.3%*
Stocks*	21,825	21,080	25,356	23,209	15,258	11,786
Growth	*n.a*	*–3.4%*	*20.3%*	*–8.5%*	*–34.3%*	*–22.8%*
in % of subtotal	*4.8%*	*4.2%*	*4.7%*	*4.1%*	*2.6%*	*1.9%*
Shares in Pooled Investments	74,559	98,274	116,252	128,400	135,757	141,474
Growth	*n.a*	*31.8%*	*18.3%*	*10.4%*	*5.7%*	*4.2%*
in % of subtotal	*16.4%*	*19.4%*	*21.5%*	*22.5%*	*23.0%*	*23.3%*
Participations**	14,815	15,848	17,942	18,619	20,680	21,621
Growth	*n.a*	*7.0%*	*13.2%*	*3.8%*	*11.1%*	*4.6%*
in % of subtotal	*3.3%*	*3.1%*	*3.3%*	*3.3%*	*3.5%*	*3.6%*
Fixed Income	334,262	352,981	363,906	381,664	399,975	416,733
Growth	*n.a*	*5.6%*	*3.1%*	*4.9%*	*4.8%*	*4.2%*
in % of subtotal	*73.7%*	*69.8%*	*67.2%*	*67.0%*	*67.8%*	*68.6%*
Other Investments	1,795	2,227	2,702	2,770	3,125	2,100
Growth	*n.a*	*24.1%*	*21.3%*	*2.5%*	*12.8%*	*–32.8%*
in % of total	*0.4%*	*0.4%*	*0.5%*	*0.5%*	*0.5%*	*0.3%*
Subtotal Life	453,398	505,946	541,636	569,567	589,875	607,848
Growth	*n.a*	*11.6%*	*7.1%*	*5.2%*	*3.6%*	*3.0%*
in % of total	*75.4%*	*76.0%*	*76.2%*	*75.9%*	*75.3%*	*74.6%*

(continued on the next page)

products of a single insurer—and free) are the main distribution channel for German insurers, contributing more than half of annual new business. However, insurers are not obligated to disclose their distribution statistics. The distribution of insurance products through banks is not widespread. Some cooperations last for many years (for example, R+V Insurance and Volksbanken Raiffeisenbanken), thus giving reason to believe in their value. In contrast, daily media reports reveal some evidence of dissatisfaction of the Allianz with the cross-selling results of its subsidiary, Dresdner Bank (VWD-Vereinigte Wirtschaftsdienste, 2003). Former hopes about selling insurance through the Internet seem to have been dampened by the fact that most insurance contracts have complicated structures and require a considerable amount of consultation, which is difficult to provide via the Internet. This obstacle reduces possibilities of Internet distribution to the provision of simple descriptive information and channeling the request for a call from the insurance company's representative.

Table 6.2 ***(continued).*** Assets Under Management for German Insurance Companies by Insurance Line

Private Health Insurance						
In EURm	1998	1999	2000	2001	2002	2003
Real Estate	1,737	1,795	1,787	1,891	1,770	1,882
Growth	*n.a*	*3.3%*	*–0.4%*	*5.8%*	*–6.4%*	*6.3%*
in % of subtotal	*3.0%*	*2.8%*	*2.5%*	*2.3%*	*2.0%*	*1.9%*
Stocks*	2,234	2,822	2,957	2,804	1,818	1,497
Growth	*n.a*	*26.3%*	*4.8%*	*–5.2%*	*–35.2%*	*–17.7%*
in % of subtotal	*3.9%*	*4.3%*	*4.1%*	*3.5%*	*2.0%*	*1.5%*
Shares in Pooled Investments	9,362	12,579	15,438	17,253	20,129	20,041
Growth	*n.a*	*34.4%*	*22.7%*	*11.8%*	*16.7%*	*–0.4%*
in % of subtotal	*16.4%*	*19.3%*	*21.2%*	*21.3%*	*22.7%*	*20.5%*
Participations**	1,663	1,803	2,120	2,346	3,172	3,017
Growth	*n.a*	*8.4%*	*17.6%*	*10.7%*	*35.2%*	*–4.9%*
in % of subtotal	*2.9%*	*2.8%*	*2.9%*	*2.9%*	*3.6%*	*3.1%*
Fixed Income	42,019	45,961	50,100	56,225	61,422	71,217
Growth	*n.a*	*9.4%*	*9.0%*	*12.2%*	*9.2%*	*15.9%*
in % of subtotal	*73.5%*	*70.5%*	*68.9%*	*69.5%*	*69.2%*	*72.8%*
Other Investments	141	213	283	434	429	218
Growth	*n.a*	*51.1%*	*32.9%*	*53.4%*	*–1.2%*	*–49.2%*
in % of subtotal	*0.2%*	*0.3%*	*0.4%*	*0.5%*	*0.5%*	*0.2%*
Subtotal Health	57,156	65,173	72,685	80,953	88,740	97,872
Growth	*n.a*	*14.0%*	*11.5%*	*11.4%*	*9.6%*	*10.3%*
in % of total	*9.5%*	*9.8%*	*10.2%*	*10.8%*	*11.3%*	*12.0%*

(continued on the next page)

There are, however, some structural changes to be observed, especially in the German life insurance industry. The latest development there shows the declining dominance of the tied agents in the distribution, but strengthening of free agents and banks. In 1999, the tied agents supplied approximately 43 percent of the new annual life insurance business; in 2004, their share was reduced to 33 percent. The share of the free agents increased in the same time from 23 percent to 28 percent and of the banks from 21 percent to 26 percent. The reasons for such trend lie in the increased demand for independent advice and the possibility to easily compare the products of different insurers (Versicherungswirtschaft 2005a).

6.2.3 Merger and Acquisition Activities

The considerable number of insurers in Germany could indicate a strong merger and acquisition capacity in the market. However, this has not been the case. Although

Table 6.2 *(continued)*. Assets Under Management for German Insurance Companies by Insurance Line

P&C Insurance						
In EURm	1998	1999	2000	2001	2002	2003
Real Estate	4,809	4,694	4,465	4,304	4,068	3,971
Growth	*n.a*	*–2.4%*	*–4.9%*	*–3.6%*	*–5.5%*	*–2.4%*
in % of subtotal	*5.3%*	*5.0%*	*4.6%*	*4.3%*	*3.9%*	*3.7%*
Stocks*	5,360	5,898	6,051	5,592	3,491	2,461
Growth	*n.a*	*10.0%*	*2.6%*	*–7.6%*	*–37.6%*	*–29.5%*
in % of subtotal	*5.9%*	*6.2%*	*6.3%*	*5.6%*	*3.3%*	*2.3%*
Shares in Pooled Investments	17,595	20,617	22,790	25,095	28,139	29,494
Growth	*n.a*	*17.2%*	*10.5%*	*10.1%*	*12.1%*	*4.8%*
in % of subtotal	*30.8%*	*31.6%*	*31.4%*	*31.0%*	*31.7%*	*30.1%*
Participations**	11,444	12,146	13,768	15,577	19,255	19,296
Growth	*n.a*	*6.1%*	*13.4%*	*13.1%*	*23.6%*	*0.2%*
in % of subtotal	*12.6%*	*12.9%*	*14.3%*	*15.7%*	*18.5%*	*17.8%*
Fixed Income	51,657	50,810	48,728	48,593	48,949	53,073
Growth	*n.a*	*–1.6%*	*–4.1%*	*–0.3%*	*0.7%*	*8.4%*
in % of subtotal	*56.7%*	*53.8%*	*50.7%*	*48.9%*	*46.9%*	*48.9%*
Other Investments	166	232	324	306	399	325
Growth	*n.a*	*39.8%*	*39.7%*	*–5.6%*	*30.4%*	*–18.5%*
in % of subtotal	*0.2%*	*0.2%*	*0.3%*	*0.3%*	*0.4%*	*0.3%*
Subtotal P&C	91,031	94,397	96,126	99,467	104,301	108,620
Growth	*n.a*	*3.7%*	*1.8%*	*3.5%*	*4.9%*	*4.1%*
in % of total	*15.1%*	*14.2%*	*13.5%*	*13.3%*	*13.3%*	*13.3%*
Total ***	601,585	665,516	710,447	749,987	782,916	814,340
Growth	*n.a*	*10.6%*	*6.8%*	*5.6%*	*4.4%*	*4.0%*

*shares and other variable-yield securities
**shares in affiliates and participating interest
*** without reinsurance
[a] life insurance without Pensionskassen and Sterbekassen

Thomson Financial has counted more than 80 mergers and acquisitions in or with the involvement of the German insurance industry during the last four years, almost all of those transactions have not considerably changed the competitive landscape and were often only a result of restructuring within the insurance conglomerates (for details, see Lier, 2003).

A prominent example of such a merger driven by restructuring is the creation of ERGO by Munich Re in 1997, whereby the insurance subsidiaries of this large reinsurer were gathered together under one name, creating Germany's second-largest insurance group (see Table 6.1). The most important pure merger and acquisition transactions in the industry were Dresdner Bank/Allianz in 2001 and Signal/Iduna in 1999. The Dresdner Bank was taken over by the Allianz AG, one of its biggest shareholders, following failed merger talks with the Deutsche Bank in March 2000 and the Commerzbank in June 2000. The disclosed reasons for the transaction were the building of a global finance conglomerate on the side of the Allianz and pressure on the Dresdner Bank to grow in order to remain a global player. According to the Allianz's financials for 2003 and 2002, its banking business in those years was experiencing considerable operating losses and had undergone considerable restructuring (Allianz Group, 2003, pp. 2, 75–76).

The Signal/Iduna group was created in 1999 by the alliance of two mutual insurance groups: Signal Group, specializing in health and property and casualty insurance, and Iduna, specializing in life insurance and related products. The goal of the alliance was to form an all-around insurance group and secure the position of a countrywide player. Currently, Signal/Iduna belongs to the top ten insurance groups in Germany (see Table 6.1).

In 2005, the takeover of the Karlsruher Versicherungsgruppe by the Wüstenrot & Württembergische AG was completed. By this takeover, the group intends to strengthen its position in the property and casualty business as well in the life, especially old age insurance, business. The synergies are expected in the areas of distribution, information technology, and management. The new group would become one of the top fifteen in the German insurance industry rankings (Versicherungswirtschaft 2005b).

The sluggish merger and acquisition activity may be attributable, at least in part, to the legal structure of the market: about 300 insurance companies have the legal form of mutuals or institutions under public law. The specifics of such legal structures could make mergers with or acquisitions of such companies less attractive for investors. On the other hand, there is merger and acquisition potential in this market segment, especially in view of Solvency II requirements, because mutuals and public insurers have restricted access to the capital markets. Therefore, industry experts anticipate mergers within the same legal forms as well. It is expected that the number of mutual insurers would drop further and that the number of institutions under public law will be substantially reduced by the creation of bigger institutions operating countrywide (Lier, 2003).

The idea of a conglomerate offering both banking and insurance services (also called bankassurance) might be another driver for merger and acquisition activities. However, it has been somewhat damaged by the unfavorable development in the Allianz/Dresdner Bank case. The banking subsidiary is experiencing losses in its core activities (*Dresdner Bank AG,* p. 41) with limited cross-selling advantages for the insurance business. According to the press, less than 40,000 Dresdner Bank customers are so far Allianz-insured (VWD, 2003). Public insurers had a similar but not so publicly known experience as a result of their much longer cooperations with banking institutions under public law. Although the target clients of both insurers and banks under public law are to a large extent the same, only around 15 percent of such a bank's customers have bought insurance from the cooperating insurer (Lier, 2003).

6.3 INSURANCE LINES

6.3.1 Life Insurance

Life insurers provide insurance coverage for dependents against the financial risk of death in exchange for a fixed premium.[169] Additionally, they are important vehicles for long-term savings and draw down accumulated savings for pension payments during retirement. Of the 324 companies that were operating in this sector during 2003, around one-third (119) offered life insurance coverage in the retail market. The other 205 companies were *Pensionskassen and Sterbekassen*, which are a special kind of life insurance only operating in the market for occupational pensions or death benefits. The majority of these companies are organized as mutuals, legally independent from the sponsoring employer (typically a company, public corporation, or industry group) and provide occupational retirement benefits for employees. Since the benefits of this type of occupational pension scheme include substantial insurance features (e.g., mortality and disability coverage) and defined benefit elements, the regulation of *Pensionskassen and Sterbekassen* is similar to that of life insurers. With a market share of about 3 percent of total life insurance premiums written in 2003, less than 1.6 percent of the total insurance premiums written in 2003 (2.25 billion euros vs. 67.66 billion euros and 148.2 billion euros, respectively), *Pensionskassen* play only a minor role in the German insurance market. The market share of *Sterbekassen* is even smaller. The key information for *Pensionskassen and Sterbekassen* is summarized in Table 6.3.

Table 6.3. Key Figures on Pensionskassen

	1980	1990	1995	2001	2002	2003
	In EUR m					
Gross Premium Written	1,496	2,866	1,853	1,919	2,247	3,296
Gross Expenditure on Insurance Claims	690	2,046	1,704	2,600	2,765	2,897
Investment Portfolio (1)	16,500	35,200	49,800	70,700	72,000	75,449
	Pension insurance contracts in force in thousands					
Expectant Pensioners	2,202	2,557	2,634	2,352	3,244	4,182
Disabled and Old-Age Pensioners	387	616	711	320	969	993
Survivors	110	108	104	130	126	129
Number of Pensionskassen	266	162	135	136	154	155

(1) At year-end

[169] This section draws heavily on Maurer (2004).

Individual endowment policies are, with 55 percent of the gross premiums written in 2003, the most important products in the German retail insurance market. The market share of *Sterbekassen* is even smaller. *Life annuities,* which are a traditional and common vehicle to draw down accumulated assets during retirement, enjoy a market share of 25 percent with respect to total premium volume. Collective policies, whereby an individual who is a member of an association becomes eligible to buy individual life policies with the same conditions for all its members subject to a group rebate, show a market share of about 13 percent. Other types of life insurance products (e.g., *supplementary policies*), which provide death benefits within other insurance policies (e.g., disability or private accident policies), are playing a less vital role with a market share of 7 percent (see Figure 6.5).

Endowment policies consist of an insurance component and an investment component. The first provides death benefits (determined by the face value of the policy) if the insured dies within the insurance period. The latter creates a cash value over time, which the insurance company must pay at the end of the insurance period (which is on average about 28 years) or if the contract is terminated. Endowment policies with no investment component—that is, pure term life policies—play only a minor role with respect to premiums written. To back the cash value, the life insurer must generate funding reserves. The registered assets covering these reserves must be kept in a special fund, which must be managed separately from other insurance company assets. Usually, endowment policies are contractual plans—that is, the insured person is obligated to pay a series of fixed premiums (generally on a monthly basis), which are primarily determined by the insured's age upon issue, gender, the face value and the duration of the contract. The life insurance company uses a certain part of the periodic premiums paid by the policyholder to cover the mortality cost for death benefits (based on a mortality table), a second part to cover distribution and administrative expenses, and a third part is invested in specific assets to generate a cash value.

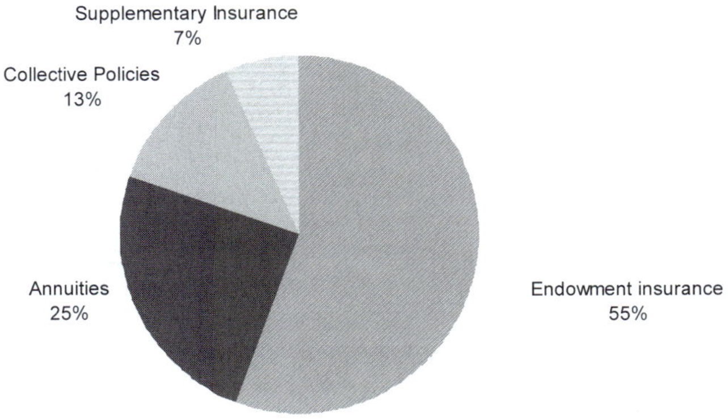

Figure 6.5. Life Insurance Premiums Split by Type of Insurance

Depending on how the cash value is generated, endowment policies can be classified into index-linked, unit-linked, and participating-with-profits policies. In the case of an *index—*or *unit-linked life policy,* the investment component is typically backed by a specific equity or bond portfolio represented by an appropriate index or a mutual fund account as chosen by the policyholder. As a result, the cash value of the policy (and, in part, also the death benefits) depends on the uncertain returns of the specific assets to which the policyholder allocates the investment component. The market for unit-linked or index-linked life insurance products is underdeveloped in Germany. Most insurers consider such policies to be a necessary complement of the product range. Stagnant capital markets further slowed the growth in this segment.

The predominant product in the German life insurance market is the traditional *participating cash-value policy,* whereby the investment component is designed with a guaranteed yearly minimum return and a variable nonguaranteed surplus. The guaranteed return is set when the policy is issued and remains fixed until the contract is terminated. The maximum return an insurer is allowed to guarantee is limited by regulation—that is, it should not exceed 60 percent of the return on long-term government bonds; in 1994, the maximum return guarantee for all life insurers was set at 4 percent per year; in 2000, it was lowered to 3.25 percent per year; and in 2004, to 2.75 percent per year. According to German regulation, life insurers must distribute not only the profits, which were guaranteed as described above, but also at least 90 percent of their annual profits (in the form of an annual/terminal bonus) to policyholders, which are called the variable nonguaranteed surplus. Besides the insurer's experience with mortality and expenses, the nonguaranteed surplus mostly depends on the performance of the investment portfolios.

Despite the fact that the surplus return is not guaranteed, German life insurers traditionally preferred to keep surplus rates stable over time, partly to create an additional impression of stability with their customers. This is realized by using several smoothing possibilities legally permissible under German accounting standards. For example, to calculate investment returns, the assets are evaluated on a book value rather market value basis. This allows the generation of hidden asset reserves arising in "good years"—that is, when (market-based) investment returns are temporary higher than the historical average. In "bad years"—when the insurance company (market-based) investment performance is temporarily less than the historical average—the reserves can be used to keep the surplus stable over the next several years (for details on return smoothing, see Albrecht and Maurer, 2002, and Richter and Ruß, 2002). However, this practice of smoothing return fluctuations in the capital markets is limited. Figure 6.6 depicts the development of net investment returns collected from the annual accounting statements of German life insurers over the years 1994 to 2003. The sample includes 113 life insurers, members of the German Insurer's Association (*Gesamtverband der Deutschen Versicherungswirtschaft,* or GDV), and represents more than 90 percent of the German insurance market.

Looking at the net investment returns over the years 1994 to 2000, the distance between the 25th and 75th percentiles is less than 1.2 percent in each year—that is, the (cross-sectional) investment performance of the various companies is characterized by a high degree of homogeneity. In addition, the investment returns are very stable over that time period, showing a range for the median company of a minimum 7.21 percent in 1994 and a maximum of 7.48 percent in 1998. However, the years 2001 and 2002 were characterized by a sharp decline in the investment results—5.82 percent in 2001 and at 4.54 percent in 2002 for the median company, respectively. This is caused by a substantial downturn in the equity markets (which determines about 10 to 30 percent of the assets held by insurers) together with a sustained decline of interest rates for fixed-income investments since 1995 (which determines about 70 to 90 percent of the average insurance asset allocation). This results in the exhaustion of asset reserves in the books for most life insurers, followed by a substantial reduction in investment returns and surplus participation.

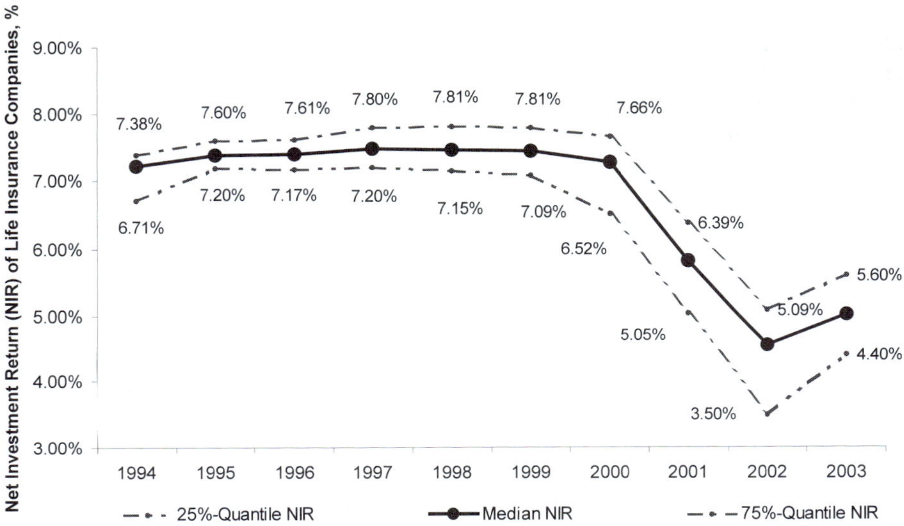

Figure 6.6. Net Investment Return (NIR) of Life Insurance Companies, per annum

Distribution and administrative costs are the major cost positions in the life insurance industry. To offer insight into the development of these positions, both cross-sectional as well as over time, we calculate the administrative expense ratio, the distribution expense, and the total expense ratio for a sample of 113 German life insurers from 1998 to 2003. They are expressed in percent and measured as a ratio of distribution or administrative costs to gross premiums received in the period under consideration. Table 6.4 depicts information about the 25 percent, 50 percent, and 75 percent quantiles of these measures over the course of years. The median total expense ratio, consisting of distribution and administrative expenses, exhibits a level of 14 to 16 percent over a number of years. The administrative expense ratios display a level of approximately 3 to 4 percent and the distribution expense ratios, presented as the difference between total and administrative expenses, of approximately 12to 13 percent over time. However, the (cross-sectional) range of these cost ratios between the various insurance companies is considerable. On the one hand, it illustrates the importance of distribution activity for life insurers; on the other hand, it shows that certain improvement possibilities toward cost structure optimization exist.

The economic significance of distribution expense ratios reported in Table 6.4 should be interpreted with caution. The German cost system in life insurance traditionally uses front-end loads to cover distribution costs. The broker receives a lump-sum payment of about 4 percent of the policy's face value from the insurance company after the contract is signed, both in the case of a policy with a single premium as well as in the case of a contractual plan with a series of regular premiums. Due to the front-end load for the first several years that a policy with regular premiums is in effect, the cash value is usually significantly less than the premiums paid by the policyholder or even zero. The distribution expense ratios reported in Table 6.4 show what part of the premiums received in the year under consideration was spent on the acquisition of new insurance contracts. Without information about the development of the insurance portfolio, this ratio cannot be unambiguously interpreted, especially not as an indicator for the burden with distribution cost per contract. Acquisition costs can be high if the insurer has acquired many new contracts or if a precise risk assessment was practiced. A low distribution expense ratio can be observed if the insurance undertaking has no new business at all. Disproportionately many life insurance contracts with single premiums or with short contract periods, relative to the whole portfolio, could also bias the ratios.

Table 6.4.

Life insurance market	1998	1999	2000	2001	2002	2003
25% Quantile of the market participants by expense ratios	12.59%	15.21%	11.47%	11.97%	12.15%	11.60%
Median expense ratio market	15.24%	18.64%	14.26%	15.93%	15.44%	16.12%
75% Quantile of the market participants by expense ratios	18.84%	23.38%	17.34%	21.36%	18.28%	21.23%
25% Quantile of the market participants by administrative expense ratios	2.98%	2.81%	2.73%	2.72%	2.56%	2.43%
Median administrative expense ratio market	3.65%	3.37%	3.40%	3.48%	3.46%	3.48%
75% Quantile of the market participants by administrative expense ratios	5.34%	4.78%	4.95%	5.02%	4.80%	4.59%

Traditionally, life insurance products have been used as a means to provide for the insured's old age or that of family members and to provide a number of tax advantages for the policyholder and the beneficiary. First, if the insured person dies, the death benefits paid to the beneficiary are free from income taxation. In addition, if an individual pays (unlimited) premiums based on a contractual plan (i.e., a series of at least yearly premiums is necessary) into a cash-value life insurance policy, and the policy runs for at least twelve years, then the periodic increase in the policy's cash value is free from income taxation. In addition, certain professional groups (i.e., civil servants, self-employed) may deduct (within certain limits) the insurance contributions from their taxable income. These tax advantages made cash-value life insurance policies attractive as pension, as well as a general savings vehicle, for all parts of the population. More than 80 million life insurance contracts were in force during 2003; on average, each German citizen owns one life insurance contract. However, these tax privileges for cash-value life insurance policies with premiums paid from taxed income were abolished as of 2005. For all life insurance policies signed after 2004, 50 percent of the proceeds realized at the maturity of the contract will be taxed with respect to the insured's personal income tax rate. Such a move by the government considerably lowers the expected after-tax yield of such life insurance contracts and therefore, its attractiveness as a private pension vehicle.

However, three additional governmental programs offer tax inducements for voluntary funded pension savings during a person's working life and could strengthen the market for products offered by life insurers. All of these programs are designed to facilitate the proper development of the German old-age insurance system, which rests on three pillars: a statutory pension scheme, occupational schemes, and personal retirement provisions. Primarily, within the second-pillar occupational pension system, workers have the option to pay part of their income (up to a certain limit) into a cash-value life insurance policy (so-called direct life insurance policies) on a pretax basis. This type of company pension scheme is frequently used, particularly by small and medium-sized companies and creates the first additional demand for life insurance products. In 2003, about 5.82 million contracts were allocated to direct life insurance policies, covering an amount of 153.73 billion euros.

The second additional demand for life insurance products results from the introduction of a new tax-shielded funded system of supplementary pensions within the Retirement Savings Act (*Altersvermögensgesetz*), which became effective in 2002 (also known as *Riester* pension reform, named after Walter Riester, the German Labour Minister during 1998–2002). Coupled with a substantial cut of public pay-as-you-go pension benefits, individuals will be able to invest a part of their income on a pretax basis into Individual Pension Accounts (IPAs), offered by regulated financial institutions, such as commercial banks, investment management companies, and life insurers. Additional incentives are given in the form of direct subsidies for low-income earners and extra contributions for children. In order to qualify for a tax credit, the IPAs must satisfy a number of criteria, codified in a certification law. For example, certification of IPA requires that, when the age of retirement is reached, a certain fraction of the accumulated assets must be drawn in the form of a lifelong annuity (offered by a life insurer), or a phased withdrawal plan, which must revert into a life annuity not later than the age of 85. Another condition is a "money-back-guarantee" whereby each provider must promise that the contract cash value at retirement is at least equal to the contributions paid into the IPA. Both legal conditions are in line with the features of products offered by life insurers. Until 2003, individuals signed about 3.3 million *Riester* contracts offered by life insurers.

The last program regarding the reform of the German social security system was launched in 2004 and became effective in 2005. This reform was triggered by changes in the Law on Taxation of Retiree Revenues (*Alterseinkünftegesetz*), which aimed to introduce the principle of deferred taxation for all pillars of old-age provisions—the statutory, employer-sponsored, and private. Benefits from the public pension systems before that act are—for most retirees—not subject to income taxation, while, after the act for future retirees, they will be. To compensate for the cut in future state pension benefits due to the introduction of deferred taxation, individuals can invest a part of their income into special registered individual pension accounts (also known as *Rürup* pension plans, named after Bert Rürup, the head of the Commission for the Social Security Reform) on a pretax basis. The criteria to qualify for such *Rürup* pension plans are different than those of the *Riester* reform discussed above. Only those plans with lifelong payments starting not before the age of 60, which cannot be transferred to another person nor bequeathed (even in the accumulation phase) and which can be only partially be borrowed on, can qualify as a *Rürup* plan. It is obvious that deferred life annuities offered by life insurers fit these legal provisions. So far, it can be expected that the life insurance industry will generate new business in such tax-supported funded pensions.

6.3.2 Private Health Insurance

In Germany, the main sources of health benefits are the statutory social security program and individual private health insurance. As a rule, employees and their nonemployed dependents (children, spouse) are compulsory members of the statutory health program, which provides substantial but highly regulated health benefits. Based on the principle of solidarity, premiums in the statutory health security program must be paid (half of this is paid by the employer/employee) as a percentage of the current family working income, while the coverage provided is equal for all members. In addition, students, retirees, the unemployed, and disabled persons pay reduced premiums. Some key data with respect to the statutory health insurance program are provided in Table 6.5.

The demand for policies offered by private health insurers stems from different sources: First, high income workers earning more than the social security ceiling (currently approximately 4,000 euros per month) have the choice between staying in the statutory health program and protecting themselves against medical expense risk via an individual policy offered by commercial health insurers (called substitutive private health insurance). Additional demand for individual private health insurance policies is generated by those who are not compulsory members of the state health program (e.g., self-employed individuals or civil servants). In both cases, individuals usually sign insurance policies providing full coverage for medical expenses and long-term care. Further demand stems from members of the statutory health insurance program, who may purchase different types of complementary insurance policies. Such policies provide coverage for situations in which the statutory health insurance denies or restricts benefits to a minimal extent. A widely used example is a policy giving the statutory-insured person an option to receive an extra level of service in the event of hospitalization: The insurance policy pays the difference between the costs that the statutory insurance is obliged to settle and the actual costs incurred by the treatment, within the limits of the complementary private insurance contract. Another prominent complementary product is the foreign travel health insurance policy, which pays (unexpected) medical costs if the insured individual is temporarily abroad (e.g., for a business trip or on holiday).

The growth of the number of insured individuals over time is shown in Table 6.6. In a country with a population of 83 million, almost 27 million people use commercial insurance companies to purchase foreign travel health insurance, 12 million have optional hospital benefits insurance, and only 8 million have full health insurance coverage with commercial health insurers. However, the full health insurance coverage product line provided 71 percent of the annual premium income in 2003, followed by supplementary insurance (13% in 2003) and long-term care insurance (8% in 2003), as shown in the Table 6.7.

The total insurance expenditures of health insurers consist of more than 60 percent of claims paid out to insureds. More than 30 percent of expenditures are the funds allocated to increasing age provisions. The remaining expenditures (approximately 10%) are bonus and rebate provisions (GDV, 2003, p. 86). They are the expected amount of the premiums to be returned to the insured in an accounting period due to the cost-reduction programs. Such programs were introduced by almost all health insurers and encourage the insured to bear minor health treatment costs themselves, keeping the insurer's administrative costs and claims expenditures low. In turn, the insured person gets back a certain portion of his or her annual premium as a reward. Recent expenditure increases were mainly based on the treatment costs themselves and less on portfolio growth, with the main contributors to the increase being costs for medications and dressings (+9.4% per year), standard hospital

Table 6.5. Key Figures on Statutory Health Insurance

	2001		2002		2003	
	Former FRG Territory	East German States and East Berlin	Former FRG Territory	East German States and East Berlin	Former FRG Territory	East German States and East Berlin
Insured Persons (Millions)	41.3	9.7	41.3	9.6	41.1	9.5
Compulsory (without pensioners)	23.3	5.7	23.2	5.6	23.3	5.5
Share of Compulsory Members in %	*56.42%*	*58.76%*	*56.17%*	*58.33%*	*56.69%*	*57.89%*
Sickness Rate of Compulsory Members in % [1]	4.2	4.3	4.0	4.1	3.6	3.7
Pay-Related Compulsory Membership and Contributions Assessment Ceiling per Month (EUR)	3,336	3,336	3,375	3,375	3,450	3,450
Annual Growth Rate in %	*n.a*	*n.a*	*1.17*	*1.17*	*2.22*	*2.22*
Contribution Rate (as %) of Monthly Income	13.6	13.7	13.5	13.8	14.3	14.3
Annual Growth Rate in %	*n.a*	*n.a*	*–0.74*	*0.73*	*5.93*	*3.62*
	EUR (billions)					
Total Contribution Income	114.6	21.2	118.6	21.7	119.3	21.4
Total Expenditure	115.7	23.1	119.4	24.2	120.0	24.6
Surplus (+) /Deficit (-) Rate in %	*–0.96%*	*–8.96%*	*–0.67%*	*–11.52%*	*–0.59%*	*–14.95%*

[1] Number of compulsory members unable to work due to sickness as % of total number (annual averages)

Table 6.6. Business in Force in Private Health Insurance in Millions of Persons

	1998	1999	2000	2001	2002	2003
Persons with Full Cover	7.21	7.36	7.49	7.71	7.92	8.11
Growth	*n.a.*	*2.1%*	*1.9%*	*2.9%*	*2.8%*	*2.3%*
Insurance of Optional Benefits in Hospital	10.64	10.78	10.91	11.20	11.44	11.80
Growth	*n.a.*	*1.3%*	*1.2%*	*2.7%*	*2.2%*	*3.1%*
Daily Benefits for Disability	2.34	2.54	2.62	2.78	2.94	3.12
Growth	*n.a.*	*8.4%*	*3.3%*	*6.1%*	*5.8%*	*6.0%*
Supplementary Long-Term Care Insurance	0.54	0.57	0.61	0.66	0.69	0.75
Growth	*n.a.*	*5.0%*	*6.1%*	*8.4%*	*5.2%*	*8.7%*
Compulsory Long-Term Care Insurance	8.13	8.23	8.37	8.62	8.83	9.00
Growth	*n.a.*	*1.2%*	*1.7%*	*3.0%*	*2.4%*	*1.9%*
Foreign Travel Health Insurance	23.04	23.35	25.43	27.15	26.82	26.60
Growth	*n.a.*	*1.3%*	*8.9%*	*6.8%*	*–1.2%*	*–0.9%*
Total Privately Health Insured Persons	51.90	52.82	55.42	58.11	58.65	59.37
Growth	*n.a.*	*1.8%*	*4.9%*	*4.9%*	*0.9%*	*1.2%*
Total Population	82.04	82.16	82.26	82.44	82.54	82.53
Growth	*n.a.*	*0.2%*	*0.1%*	*0.2%*	*0.1%*	*–0.01%*

benefits (+9.1% per year), and remedies and aids (+4.2% per year). The costs for hospital accommodation decreased by more than 21 percent per year during the years 2002 and 2003 (GDV, 2003, p. 87), mostly due to patients choosing cheaper accommodation forms such as shared rooms.

Compared to the statutory health insurance program, the pattern of private health insurance policies is much more flexible with the risk coverage. Premiums are calculated by individual risk characteristics, such as defined health benefits, age upon signing the contract, and gender. In contrast to the state health insurance program, premiums must be paid for each family member. Private insurance contracts providing full health coverage are, by regulation, lifelong policies, usually with a fixed monthly premium. If a private health insurer accepts a risk, it may not terminate the contract as long as the premiums are paid on time. The power to terminate or to change the contract is held solely by the insured. The company must—similar to rules for life insurers—calculate a constant premium over the total lifetime of the contract at the time the policy is issued and base the calculation on an actuarial mortality and morbidity table. To take into account the increasing life expectancy of the insured, the mortality tables used for the tariff calculation were modified in 2003.

Table 6.7. Premium Income in Private Health Insurance in EUR Billions

	1998	1999	2000	2001	2002	2003
Full Cover	12.29	13.01	13.72	14.68	15.89	17.53
Share of Total	*63.6%*	*65.3%*	*66.2%*	*67.6%*	*68.8%*	*70.9%*
Daily Benefits Insurance in the Event of Hospitalisation	0.77	0.80	0.78	0.77	0.76	0.76
Share of Total	*4.0%*	*4.0%*	*3.8%*	*3.5%*	*3.3%*	*3.1%*
Supplementary Long-Term Care Insurance, Supplementary Hospitalisation Insurance etc.	2.86	2.84	2.86	2.91	2.99	3.10
Share of Total	*14.8%*	*14.3%*	*13.8%*	*13.4%*	*12.9%*	*12.5%*
Daily Benefits for Disability	0.87	0.88	0.90	0.94	0.98	1.03
Share of Total	*4.5%*	*4.4%*	*4.3%*	*4.3%*	*4.2%*	*4.2%*
Compulsory Long-Term Care Insurance	2.15	1.98	2.01	1.96	1.99	1.85
Share of Total	*11.1%*	*9.9%*	*9.7%*	*9.0%*	*8.6%*	*7.5%*
Special Types of Insurance	0.39	0.41	0.45	0.46	0.49	0.47
Share of Total	*2.0%*	*2.0%*	*2.2%*	*2.1%*	*2.1%*	*1.9%*
Total	19.32	19.91	20.71	21.72	23.08	24.74
Growth	*n.a.*	*3.1%*	*4.0%*	*4.9%*	*6.3%*	*7.2%*

The company can increase (or decrease) premiums if the underlying actuarial assumptions provably change. Yet, it may not raise premiums because of the increasing age of the policyholder. When the policyholder is young, the difference between the required premium and the yearly expected claim payments (plus expense loadings) is positive; it becomes negative as the policyholder ages. Thus, to smooth the premiums over time, the company is required by law to set up an aging reserve. When the policyholder is young, the aging reserve is built up with parts of the premium; it is managed by the insurance company on behalf of the policyholder. When the policyholder becomes old, the health insurance company must use this aging reserve to finance the negative difference between the premiums received and expected claim payments.

Comparable to the mathematical reserves for life insurers, the aging reserve is backed by the registered assets that are specially designated for underlying of this reserve and separated from others; dispositions are only possible with the approval of an appointed trustee. But this smoothing practice has a negative aspect, because the policyholder loses his or her aging reserve if the contract is terminated. By tying the insured to the insurance company, it hinders the policyholder from changing to another private health insurer or, after the insurance contract has been in force for a couple of years, makes it practically impossible for economic reasons. Thus, the insured is deprived of the means to react in the event of unfavorable economic developments of the insurer or the poor quality of its service. Therefore, discussions are currently underway to require health insurers to transfer the insured person's share in the aging reserve to the new insurer of his or her choice, should the insured person desire to change insurers.

Due to increasing deficits in statutory health insurance caused by continuous expenditure increases, unfavorable demographic development, and the absence of individual risk-based premium calculations, the state is forced to gradually reduce the benefits paid in statutory health insurance. Simultaneously, it tries to restrict the outflow of the healthier and wealthier members of the population to private health insurance, which offers a broader range of services in exchange for benefit-related contributions, by annually increasing the salary-related (or wage-related) ceiling of the compulsory membership. A governmental commission was created to investigate the crisis situation in statutory health insurance. The commission's suggestions are still being discussed. Some of them, if implemented, could seriously endanger the further development and even existence of private health insurance.

6.3.3 Property and Casualty Insurance

Property and casualty insurers offer insurance coverage for a wide variety of events: loss, damage, destruction of property, loss or impairment of income-producing ability, claims for damages by third parties from alleged negligence, and loss resulting from injury or death caused by work-related accidents. With a market share of about 41 percent of total premium volume, motor vehicle insurance is by far the largest line in the property and casualty sector, followed by property insurance with 27 percent, general liability insurance with 12 percent, and private accident insurance with 10 percent (data for the year 2003). The rest is unequally shared between legal expense insurance, marine insurance, and tourist assistance insurance (see Figure 6.7).

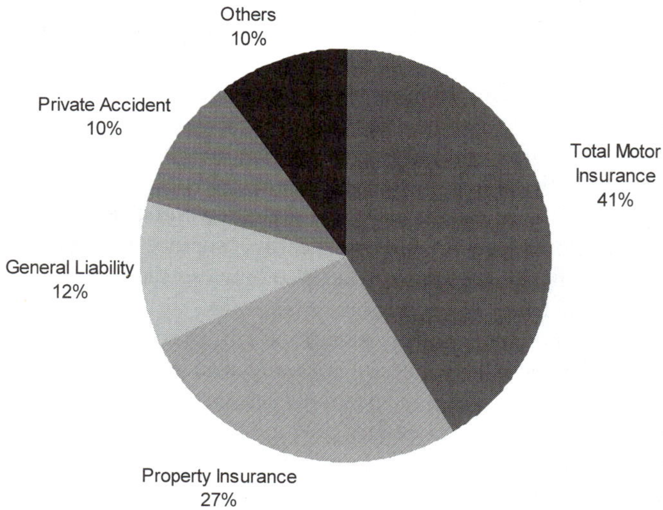

Figure 6.7. Share of Gross Written Premiums of Property and Casualty Lines by Class

The overall loss ratios across the total property and casualty sector have been relatively stable, ranging between 70 and 80 percent, since 1998 (see Table 6.8). However, the situation varies considerably within the insurance lines and thus deserves more attention. Motor vehicle insurance had the highest and most variable loss ratios over the time span considered, with a maximum of 104.70 percent in 1999 and 88 percent in 2003. Within the motor vehicle insurance line, the semi-comprehensive coverage line (covering damage on the insured's motor vehicle that is not caused by the insured's incorrect actions) has loss ratios of under 80 percent since 1998; the third-party liability (TPL) coverage was loss-bringing and the loss ratios only fell below 100 percent once in the same period. This situation forced insurers to increase premiums in TPL by 4.7 percent in 2001 and 3 percent in 2002 (German Insurance Association, 2003, pp. 90, 93), despite considerable competition in the market.

Loss ratios in property insurance increased greatly from 70–80 percent to over 90 percent in 2002, due to extreme storms and floods that year, resulting in the largest claims ever in the history of German insurance. The data available for 2003 indicate positive overall developments due to the absence of major weather-related damages. Loss ratio patterns in general liability insurance declined from 80 percent in 1998 to 73 percent in 2003. However, future claims in many areas of general liability insurance are expected to increase, especially due to pharmaceutical and car recall cost risks. The exposure to those risks has amplified ever since European and German (amendment to the Damage Compensation Law—*Schadensersatzgesetz*) liability laws have become stricter.

Table 6.8. Earned Premiums, Loss and Expense Ratios in Non-Life Insurance

In EUR bn	1998	1999	2000	2001	2002	2003
Motor Business						
Earned Premiums	20.07	19.84	20.40	21.34	21.98	22.27
Loss Ratio	99.5%	104.7%	99.8%	94.4%	92.6%	88.0%
Property Insurance Business						
Earned Premiums	12.47	12.46	12.40	12.41	13.10	n.a
Loss Ratio	79.5%	78.5%	72.2%	71.9%	94.4%	n.a
General Liability Business						
Earned Premiums	5.74	5.88	5.91	5.98	6.17	6.44
Loss Ratio	82.4%	83.1%	78.9%	84.4%	75.3%	73.0%
Private Accident Business						
Earned Premiums	4.31	4.40	4.46	4.51	4.57	4.62
Loss Ratio	57.4%	57.0%	55.2%	54.4%	52.4%	52.0%
Property & Casualty Business Total						
Earned Premiums	45.75	46.97	50.69	52.68	56.91	n.a
Loss Ratio	71.1%	75.0%	74.4%	77.8%	78.3%	n.a
Expense Ratio	25.8%	26.0%	26.7%	27.1%	26.3%	n.a.

The dependence of German property and casualty insurance on investment results (also known as cash flow underwriting) is an important feature of the market. As shown in Table 6.9, the technical results (i.e., the results from the insurance underwriting activity) of the property and casualty insurance segment as a whole were negative from 2000 to 2002. Only the general accident insurance business has exhibited stable positive technical results since 1998 (GDV, 2003a, Table 62). Positive profits in the segment could be achieved mostly due to positive recurrent investment (i.e., not including the results from writeoffs and other exceptional actions) results, which amounted to more than 10 percent of the gross premium received in these periods. This is an important difference to the life and health insurance businesses, where already technical results deliver a positive profit contribution. The reason for such a development is a more intense and less regulated competition in the property and casualty business, as compared to the life and health business. Annual insurance contracts, which are usual for property and casualty insurance, make changing insurers easier. In the largest sector, motor vehicle insurance, insurance benefits are largely predefined by legislation, facilitating comparison between insurers.

Table 6.9. Technical and Investment Results of Property and Casualty Insurers

Total of P&C Insurers, in EUR bn	2000	2001	2002
Gross Premium Received (EUR billions)	55,421	57,444	59,581
Gross Technical Result (EUR billions)	–1,089	–1,944	–2,695
in % of Premium	–1.96%	–3.38%	–4.52%
Recurring Investment Result	6,511	7,977	7,133
in % of Premium	11.75%	13.89%	11.97%

A particularity of German property and casualty insurance accounting is the existence of the legally prescribed equalization fund (*Schwankungsrückstellung*). It is a special balance sheet liability position, permissible only to property and casualty insurers, which aims to smooth reported underwriting results over time. The profits reported in the annual accounting statement of a company is also relevant for taxation—that is, the equalization fund is an important vehicle to immunize the tax burden of a company with respect to the volatility of underlying results. The requirements for the creation of an equalization fund depend on the scale of the insurer's business, the sum of the actual loss and expense ratios, and their volatility in the legally determined time period. As soon as the required minimum level of equalization fund is achieved, the company must reduce it if the loss ratio of the current year is higher than the average loss ratio calculated over the legally prescribed number of years, which depends on the nature of the underwritten business. If the actual loss ratio of a company is lower than the historical average for the period defined by the legislation and differentiated by the type of business, then the equalization fund must be increased. The amounts by which the equalization fund is reduced or increased depend on the difference between the actual and average loss ratios and on the actual premiums written. In 2002, for example, the reserves from the equalization fund were withdrawn in order to reduce the negative technical results.

The property and casualty insurance sector is quite heterogeneous, and various developments might be observed. In motor vehicle insurance, a rise in tariffs could be expected, as well as the introduction of further measures for more precise classification and pricing of risks. The property insurance business would increasingly depend on the climatic influences. On one hand, the many natural catastrophes of 2002 taught the insurance industry to get accustomed to new, bigger damage dimensions. On the other hand, the analysis of the current market situation shows that the insurance density for catastrophe insurance is quite low: According to information supplied by the German Insurer's Association, only one out of ten citizens has insured the contents of his or her household against the risk of a natural catastrophe. This indicates new business opportunities. However, this also indicates the possibility of even bigger claims than currently known as well as of negative risk readouts resulting from the fact that, particularly households in endangered areas such as those close to rivers, would increasingly seek insurance coverage. Currently, the government is discussing a project that would introduce compulsory insurance for the risk of natural catastrophes.

In general liability insurance, the amendment of the Damage Compensation Law, which took effect in 2002, extended claims for pain and suffering to cases where the damage-incurrent incident was not due to the insured's fault. The fears that this would lead to a considerable increase in the claims paid, especially in the event of industrial accidents, have not materialized. However, it is estimated that only in the TPL insurance could the legislation change lead to an additional burden of 50 million euros. Another challenge for property and casualty insurers is the coverage for terrorism risks. In Germany the national solution was the creation of *Extremus Versicherungs-AG* in September 2002. *Extremus* is an insurance company created by sixteen insurance groups engaged in the German market that offers emergency coverage exceeding 25 million euros for property damage and associated business interruption claims resulting from terrorist attacks. One and a half billion euros are provided from *Extremus* founding insurance companies and another 1.5 billion euros by the international reinsurance market. Claims in excess of 3 billion euros are covered by a government guarantee for 10 billion euros.

6.4 REGULATION

6.4.1 Goals and Structure of Current Regulation and Supervision

The insurance industry has always been highly regulated due to the longevity of most insurance contracts, their possibly significant impact on the economic situation of the insured, and the position of the insured as a conditional creditor to the insurance company in case the insured event occurs. This is true for the German market as well. The main goal of state regulation for German insurers is to protect the insured by providing high credibility that the insurance company is able to meet its (contingent) obligations against the insured persons and policyholders. Traditionally, this is done by controlling the financial stability of an insurance company by restricting its insolvency risk to a low level. The Insurance Supervision Law—exercised by the Federal Financial Supervisory Authority (*Bundesanstalt für Finanzdienstleistungsaufsicht,* or BaFin)—subjects German insurers to substantial legal provisions. Here, the supervision body is concerned with every legal entity conducting insurance business, irrespective of its affiliation to a larger group or its holdings. By use of directives and circulars, the BaFin has the opportunity to explicate and/or specify general rules codified in the Insurance Supervision Law.

Like other regulated financial institutions (e.g., banks, mutual funds, pension funds), insurers must obtain official authorization from the supervisory body before they are allowed to offer insurance products in the German market. Additional conditions governing a direct insurer's business after official authorization stem from the permanent supervision of its financial stability, which focuses on four main areas:

- Establishment of sufficient technical provisions
- Solvency requirements
- Investment restrictions
- Stress tests

The establishment of sufficient liability reserves should adequately capture the (uncertain) financial liabilities resulting from the insurer's underwriting activities. The most important positions (i.e., technical provisions) in the various property and casualty lines are the unearned premium and loss reserve; in health insurance, they are the aging reserve (*Alterungsrückstellung*); and in life insurance, they are the funding reserve (*Deckungsrückstellung*). According to current regulation, the volume of such provisions must be based on prudent prospective actuarial valuations, which cover the timing and stochastic nature of future liabilities as determined by the policy conditions. In addition, the appropriateness of the liability reserves of life and health insurers must be checked by an appointed actuary.

The second area is the regulation of the matching assets used to cover the technical provisions. In general, those assets should be adequately diversified and should account for the risk potential, the possible termination dates, and the currency of the underlying insurance business. The third area is the upholding (over and above technical provisions) of a sufficient supplementary reserve, solvency capital, which acts as a buffer against adverse business fluctuations. Finally, with a yearly stress test, the possible impact of substantial losses in the insurer's investment portfolio on its solvency situation must be analyzed and reported to the supervisory body.

Since the deregulation of the Insurance Supervision Law in 1994, the products offered by licensed insurance companies are no longer subject to prior approval by the supervisory authority. There is no direct price regulation for any line in the German insurance market, and the suppliers of insurance coverage are, in general, free to compete on premiums as well. However, the space for such competition is limited by the regulatory requirements regarding the establishment of sufficient liability reserves and solvency capital. In addition, in the life and health lines, an appointed actuary must verify the calculation of premiums and technical provisions. In the property and casualty market, the competition is more rigorous than in the life and health markets, the most severe being in the motor vehicle insurance segment. The short-term contracts with a typical duration of one year—customary for the property and casualty market—make it easier for customers to switch insurers. In contrast, life and health insurance contracts—although the policyholder is not denied the right to terminate them according to the rules defined by legislation—by design contain implicit economic incentives making it more difficult for the policyholder to quit. Examples of such incentives are the existence of the nontransferable old-age reserve in health insurance, which smoothes insurance premiums over time, and the negative surrender values during the first retention years of the life insurance contract.

It is impossible to thoroughly discuss all aspects of regulation areas within the scope of this chapter. Therefore, the remainder outlines the main aspects of regulations on solvency requirements, investment of restricted funds, and stress tests. The details of actuarial valuation techniques to determine the liability reserves and the accounting rule for their representation in the balance sheet are beyond the scope of this chapter.

6.4.2 Current Solvency Requirements

The currently applicable German solvency regulations are based on the European Community Directives from 1973 (for casualty and property/accident insurance) and 1979 (life insurance). Although both directives were modified in 2002 within the framework of the Solvency I project, their basic structure has remained unchanged. Incorporated into the domestic legislation, they control for availability of the risk-adequate level of solvency capital using a two-step procedure: First, the required solvency capital is estimated based on certain measures of the insurer's risk exposure. Second, it is compared with the available solvency capital represented by assets free of any foreseeable liabilities, which is not the same as the insurer's stated equity capital. The solvency situation of an insurance company is sufficient if the available solvency capital exceeds the required solvency capital. If the solvency requirements are not met (i.e., the coverage is less than 100 percent), the supervisory authority intervenes—for example, it has the right to require a plan for the restoration of adequate solvency or to prohibit the underwriting of the new business. In no case should the solvency margin fall below a minimum guarantee fund—an absolute amount, equal for all insurance companies doing the same type of business.

The required and available solvency capital must be reported to the supervisory body on a quarterly basis. However, the insurer is obliged to maintain an adequate level of available solvency capital at all times. The calculation of both the required solvency capital as well as the available solvency capital is different for non-life insurers (property and casualty and health) and life insurers.

6.4.3 Solvency for Non-Life Insurers

The currently enforceable directive governing the solvency margin requirements for non-life insurance undertakings is Directive 2002/13/EC. The basic economic idea of the solvency requirements is to limit exposure due to underwriting risk with respect to a certain level of solvency capital by setting an upper limit with respect to insurance leverage. Technically, this is realized for property and casualty insurers by determining the solvency margin based on the higher premium index or claims index. The premium index is defined as the yearly gross premium proceeds (minus operation costs) across all lines, multiplied by the predefined constant coefficient of 18 percent for premium amounts up to 50 million euros, and of 16 percent for the premium amounts comprising the excess. The claims index is defined as an average of the claims paid during the last three (for some businesses seven) years, multiplied with a constant coefficient of 26 percent for amounts up to 35 million euros and of 23 percent for amounts comprising the excess. Reinsurance coverage can reduce both premiums, as well as the claim index, up to a limit of 50 percent. The ratio of claims paid to the premiums received at which the premium index becomes larger than the claims index is approximately 70 percent. In calculating the solvency margin for health insurers, the same procedure is used as for property and casualty insurers except that applied constant coefficients are reduced to a third. This is done despite the fact that the premiums in German health insurance are calculated according to the principles of life insurance.

It is remarkable that explicit measures of other risk factors that are economically relevant for the overall risk exposure of a non-life insurance company (e.g., investment risk, liability reserve risk, or operational risk) are not taken into consideration.

The available solvency capital for health insurance and property and casualty companies consists not only of the equity capital as stated in the balance sheet, but also participating certificates (special hybrid financing instruments having the features of both equity and debt) and subordinated debt. In exceptional cases and only with the permission of the supervisory body, hidden asset reserves, which result from the differences in accounting and market valuation rules, might also be part of the available solvency capital. For 2002, German property and casualty insurers under the supervision of BaFin had a total required solvency capital of 7.41 billion euros. The surplus coverage for the market amounted on average to 337 percent. Thirty-six percent of the companies had a surplus coverage of up to 100 percent; 21 percent of the companies had 100 to 200 percent; 14 percent had 200 to 300 percent; and 24 percent had over 300 percent. Approximately 5 percent of the companies had insufficient coverage (BaFin, 2004b, p. 138). Only 52 health insurers had to file the solvency report with BaFin. They exhibited the total required solvency capital of 1.298 billion euros. Nine companies showed a surplus coverage of up to 50 percent; ten companies showed 50 to 100 percent; seventeen companies showed 100 to 200 percent; and sixteen showed over 200 percent (BaFin, 2004b, pp. 137–138).

6.4.4 Solvency for Life Insurers

The currently enforceable directive governing the solvency margin requirements for life insurance undertakings is Directive 2002/83/EC. The solvency margin for life insurers should incorporate the underlying investment, management, and mortality risk. In general, the required solvency capital consists of the following components: (1) If the insurance company bears an investment risk (e.g., for endowment policies, annuities, unit-linked policies with minimum return guarantees), a 4 percent fraction of the mathematical provisions is required. (2) For unit-linked life policies whereby the insurer bears no investment risk but the allocation to cover management expenses is fixed for a period exceeding five years, a 1 percent fraction of the technical provisions is required as a solvency margin. (3) The mortality risk for all life policies is estimated as a 0.3 percent fraction of the capital at risk—that is, the insured death benefits minus the provision belonging to the policy. Reinsurance coverage can reduce the required solvency margin up to certain limits.

For life insurers, the available solvency capital consists of three parts: The solvency capital A is to a large extent the same as for property and casualty insurers—that is, the equity capital, participating certificates, and subordinated debt as stated in the balance sheet. In addition to these positions, the solvency capital B includes profit reserves, particularly the noncommitted bonus reserves. With the approval of the supervisory body, hidden reserves and estimated future profits (solvency capital C) can also be part of the considered available solvency capital. A detailed description of the procedures is given in VAG section 53(c) and the Directives 2002/83/EC and 2002/13/EC. For 2002, the supervisory authority reports that German life insurers showed a total required solvency margin of 23.348 billion euros, vis-à-vis an available solvency capital of 39.775 billion euros, resulting in an average coverage ratio of about 170 percent. More specifically, 64 percent of the companies showed a coverage ratio between 100 and 200 percent; 10.3 percent between 200 and 300 percent; 5.6 percent between 200 and 300 percent; and 7.5 percent of the companies possess a coverage ratio of over 300 percent (BaFin, 2004b, pp. 136–137). About 17.3 percent of the total available solvency capital stems from solvency capital A and 81.2 percent from solvency capital B. For about 12 percent of life insurers, the supervisory authority allows the use of solvency capital C (i.e., estimated future profits) to cover the required solvency margin.

6.4.5 Investment of Assets Covering Insurance Provisions

Some insurance company funds economically belong to the insured but are administrated by the insurance company during a considerable period of time. To prevent their abuse, the legislation distinguishes between two types of funds: free and tied. All assets covering the technical provision as given in sections 54 and 66 of VAG (i.e., the loss reserve, aging reserve, funding reserve, and unearned premium reserve) belong to the tied funds. All other assets belong to the free funds. In contrast to the free fund, the investments of the assets belonging to the tied funds are subject to regulation. An investment directive of the supervisory body exactly defines the categories of authorized assets permissible for the tied fund and requires some qualitative investment principles (e.g., safety, yield, diversification, and marketability of investments). The regulation also sets quantitative restrictions regarding the insurer's asset allocation. For certain risky assets, maximum investment weights (with respect to the tied fund) must be taken into consideration—for example, no more than 35 percent in stocks and a maximum of 25 percent in real estate. In addition, short sales in any assets are not allowed; open currency positions that are nonmatched with liabilities are restricted; and financial derivatives such as options, futures, and swaps can only be used for hedging purposes. For life insurance contracts where the benefits are directly linked to the value of units in a mutual fund (i.e., the insurer bears no investment risk), the assets of the corresponding tied fund must be represented by those units as closely as possible.

The currently enforceable modification of the July 9, 2004, Investment Directive (replacing the Directive of December 20, 2001) accommodated the latest changes and trends in the capital markets as well as financial products offered in these markets. It widens the list of authorized assets by new ones, such as asset-backed securities, specific mutual funds, and hedge funds. It also enlarges the list of assets previously authorized by making stocks and interest rate instruments quoted on an organized market outside the European Economic Area qualify (BaFin, 2001, 2004a).

6.4.6 Stress Tests

Since the beginning of 2003, and within the modernization program for insurance supervision methods, the legislation requires a detailed analysis of insurance companies' risks and demands elaborated asset liability management. Within asset liability management, a stress test considering market risk, credit risk, and liquidity risk as well as foreign exchange, operational, and legislative risks has been introduced to estimate the impact of the external changes in the nontechnical factors on the financial stability of the insurance company (BaFin, 2002). The authorities specified the methods of conducting the stress tests, thus making their results comparable. More formally, three (deterministic) scenarios should analyze the possible impact of a loss in the insurer's investment portfolio with respect to its solvency situation. Scenario one considers a 10 percent reduction in the market value of interest rate instruments only; scenario two reflects a 35 percent reduction in the market value of stocks only. Scenario three analyzes the impact of market value change on the combined portfolio of interest rate instruments and stocks assuming a 5 percent drop in the market value of interest rate instruments and a 20 percent drop in the market value of stocks (BaFin, 2004).

The default risks of interest rate instruments must be taken into account through a percentage reduction of their value in accordance with their rating, varying from 0 percent for AAA–BBB grades to minus 30 percent for noninvestment grades (CCC–D). (For a detailed discussion of stress tests, see also Schradin, 2004, pp, 611–664.)

The underwriting fluctuations should be captured by the required solvency capital, which depends on the insurer's underwriting risk and historical claims experience and has to be recalculated and reported to authorities on a quarterly basis.

6.4.7 Solvency II

Current regulation and financial supervision of insurance companies is based on a system of fixed ratios, allowances, and restrictions to quantitatively cope with the investment risks, solvency capital requirements, and their interaction with the technical risks. The low numbers of insolvencies (only one case, that of Mannheimer Lebensversicherung AG, in the last ten years) in the German insurance industry prove that the existing system was—up to now—effective in securing the fulfillment of the insurer's obligations. However, the necessity to optimally use all the available resources in the insurance undertaking, the growing number of financial innovations, the interdependencies between banking and insurance products, the necessity to capture the operational risks, and the interactions between the different risk sources within an insurance company called for the introduction of qualitative, flexible, and integrated control methods.

The Insurance Committee of the European Commission, in close cooperation with supervisory authorities of the member states as well as with auditors' and insurers' associations, works on the development of new European insurance regulation. Due to the fact that Germany is a member of the European Union, the final adopted regulations will have to be incorporated in German regulation and legislation.

The new developments in European insurance regulation, known as Solvency II, follow the example of banks, which introduced the first draft for their new regulation as early as 1999. The modified version of this draft was published in 2001 as The New Basel Capital Accord (Basel II), and aims to create the aggregated financial services legislation in the long term.

For insurance regulations, a three-pillar structure will be introduced as it is in the Basel II: Pillar I will set quantitative requirements for capitalization, distinguishing between the optimum capitalization (target capital) and its absolute minimum level. It is still under discussion whether the calculation of the minimum level requirements would be defined similarly to the existing regulation or whether it will be a fixed percentage of either the technical reserves or the optimum capitalization. For the calculation of the optimum capitalization, a risk-based standard model is being developed under the coordination of the German Insurance Association. Alternatively, companies will be allowed to run their internal models, if approved by the supervisory body. In both models, the following risks should be considered: technical risks (including underwriting and the creation of technical reserves), investment risks (the question whether a distinction between the restricted and unrestricted assets would endure is still open), credit risks (from both investments and reinsurance), and operational risks (including organizational, informational, and personnel risks as well as risks from external influences). Thus, the new regulation seeks to capture the influence of all relevant risk sources on the insurance company and makes the required guarantee capital depend on the total risk of the insurance undertaking. The envisaged dynamic models will be more powerful instruments in measuring the risks as compared to the existing fixed-ratio and discrete models. However, they will also be more complex and more difficult to build, handle, and standardize. The following are just a few of the challenges: modeling the external parameters (for example, interest rates) or modeling risk interdependencies, especially for risks not identically distributed or even belonging to different distribution families (Hartung and Helten, 2004).

Pillar II will impose qualitative requirements for the risk management systems within insurance undertakings, incorporating the harmonized review procedures into the supervisory law. The main goal for these requirements would be the creation of such risk management systems within the insurance undertaking, which take into account all relevant risk sources and their interdependencies. Among others, these qualitative requirements will regulate the investment and asset liability management policies as well as the implementation and functioning of stress tests, which will not only be used to analyze the effects of certain scenarios on investments, but also to model their impact on other balance sheet items.

Pillar III will regulate the disclosure requirements, thus fostering market transparency and market discipline. The requirements for Pillar III will be based on the international accounting standards for insurance companies, which are still pending.

There are no specific regulations for any of the pillars yet. For more detailed descriptions of the Solvency II procedures, a discussion on regulation novelties, drafts of risk models and their historical development, see Perlet (2005) and Schradin (2004). In 2007, new European solvency requirements (Solvency II) are expected to become effective.

6.4.8 Non-Regulatory Means To Ensure the Credibility of Insurance Obligations

The German insurance industry employs several nonregulatory means to enhance the proper functioning of the market: In close cooperation with the supervisory authority, the market players have added a further safety net within their community, which would protect customers in the event that their life insurer slid into financial trouble. In November 2002, following the decline in capital markets and rumors about the deteriorating ability of some insurers to fulfill their obligations, the insurance industry set up Protektor life insurance. It is a registered (nonlisted) stock corporation whose stockholders are members of the German Insurance Association (representing more than 97 percent of German gross insurance premiums) and committed to back the company with up to 1 percent of their asset under management. Thus, Protektor has available funds of about 5 billion euros.

Protektor is not an active participant in the market, but solely a rescue company: If a participating life insurer becomes insolvent, Protektor will take over its existing insurance contracts, leaving their insurance benefits unchanged. The insurance portfolio of a troubled company will thus be rehabilitated and then sold to financially sound insurers. The takeover of the troubled company's existing contracts, whereby that company will cease to exist, represents the last stage of the rehabilitation measures when everything else has proved insufficient. The prior rehabilitation stages, which must be conducted under the supervision of BaFin, include restructuring measures within the troubled company and/or the sale of the company's parts to another insurer. Protektor already took over and sold its first insurance portfolio to another insurer—that of Mannheimer Lebensversicherung AG, which slid into financial trouble due to investment losses in 2003 (cf. BaFin, 2004b, pp. 10, 148). The owners of Protektor invested 240 million euros in the Mannheimer Lebensversicherung case.

This example in the life insurance industry led to the creation of a similar organization for health insurance—the Medicator. It was established in 2003 by eight large German health insurers (BaFin, 2004b, pp. 11, 154). The goal of Medicator is to take over the insurance portfolios of financially endangered health insurers in order to enable the proper functioning of insurance contracts during the transition time, which should end with an integration of those portfolios into financially sound insurance companies. In October 2004, the government put forward a proposal to give legislative backing to the initiatives of the insurance industry, thus trying to make the participation in a rescue fund legally obliging, to fix the contribution criteria, and to eliminate the bottlenecks in the insolvency legislation that occur while creating and using the rescue funds. The suggested changes in the Insurance

Supervisory Law call for the creation of a safety fund, which would be financed by life and health insurers with 0.2 percent of their respective technical provisions, and for the possibility to reduce the insurance payments by 5 percent if the safety fund proves to have insufficient capacity for a single insolvency. The proposed changes in the insolvency law would also regulate the transfer procedure of the insolvent insurer's contract portfolio to a safety fund, which should restructure and sell it.

6.5 CONCLUSION: DEVELOPMENT AND OUTLOOK

The German insurance industry has its origins in medieval mutual insurance companies and, after several setbacks caused by natural and man-made disasters including World Wars I and II, succeeded in creating stable and reliable structures. Currently, the industry is dominated by large stock corporations, mostly organized as holding companies, serving the fourth-largest insurance market in the world (with respect to premiums written). Insurers in Germany are also important investors on the capital market. The biggest line in the German insurance market is life insurance, followed by property and casualty insurance and health insurance. The development of all insurance lines heavily depends on regulation—traditionally rather strong in Germany—and on developments in the tax and social security legislation.

The development of the life insurance sector depends on the taxation of policy payouts and on the reform of the state-owned retirement system. Additional demand for life insurance products is anticipated because of the ongoing reductions of public pay-as-you-go pensions in favor of an extension of tax subsidized private funded pension plans. Due to some legal prerequisites that require annuitization in the decumulation phase (e.g., within *Riester* and *Rürup* plans), especially the demand for annuity products should increase. However, such programmed new business is not without risk. A key factor to covering the financial aspects of longevity risk by organizing risk pools is to develop appropriate mortality tables for annuitants. A further challenge is the provision of the long-term guarantees both with respect to return as well as to mortality tables, which are typical for the contemporary life insurance business. Currently, the further existence of the return guarantees in their traditional form—that is, nominal and fixed at signing for the entire life of the contract (often more than thirty years)—is being questioned. Such factors as the increasing life expectancy and low interest rates, unlikely to change soon, make it more difficult for insurers to securely fulfill their guarantee obligations. Under discussion, initiated by the insurance industry, are adjustment mechanisms that would allow either changing premiums during the lifetime of the insurance contract or altering the benefits. This would, however, weaken the main difference between life insurance and investment fund products in the market for old-age provisions. Another possibility to make the guarantees more bearable is to link them to a consumer price index instead of to a fixed nominal interest rate. For the insured, the advantage of such a solution would be the protection against the risk of inflationary erosion of the contract's value. The insurance company would benefit from the closer link of its liabilities resulting from return guarantees to the development in the capital markets. However, the implementation of this solution would heavily depend on the availability of inflation indexed bonds, which are still not very established financial instruments in Germany.

The main challenge for private health insurers will remain the desolate situation of the state health program and the resulting attempts by the state to improve it by limiting access to private health insurance.

The suggestion of the European Commission about a directive on unisex tariffs and a proposed ban on gender-based tariff differentiation would have considerable impact on the life and health insurance businesses. If implemented, it would cause significant changes in life insurance and the private health insurance business. While life annuities and private health insurance tariffs would become cheaper for women, risk life insurance would become more expensive because of women's longer life expectancy. The proposed changes could also cause an unfavorable development in the portfolio of the insured, leading to the overrepresentation of one gender compared to its numbers in the population. Germany already implemented the idea of unisex tariffs for those private pension insurance contracts that are eligible for state subsidies (e.g., *Riester* plans). Beginning in 2006, those contracts cannot contain a gender-based differentiation; otherwise, the state subsidies would be rejected.

Property and casualty insurance includes several insurance lines that do not always develop similarly. For the main sector, motor vehicle insurance, it will remain important to cost-effectively control the quality of the accepted risks and to strike a balance between the sufficient loads in premium calculations and competition in the market.

Historically, the development of the insurance industry in Germany was closely tied to the overall economic development of the country, satisfying the insurance needs created by economic activities in other sectors. Thus, for further positive development of the insurance industry in Germany, the current economic stagnation must be overcome.

6.6 REFERENCES

Albrecht, Peter, and Raimond Maurer, 2002, "Self-Annuitisation, Consumption Shortfall in Retirement and Asset Allocation: The Annuity Benchmark," *Journal of Pension Economics and Finance* 1: 269–288.

Allianz Group, 2004, *Annual Report of the Allianz-Group,* December 31, 2003.

BaFin/Bundesanstalt für Finanzdienstleistungsaufsicht, 2001, *Investment Directive* from December 20, 2001.

BaFin/Bundesanstalt für Finanzdienstleistungsaufsicht, 2002, *Circulars* 29/2002, 30/2002.

BaFin/Bundesanstalt für Finanzdienstleistungsaufsicht, 2004, *Circular* 1/2004.

BaFin/Bundesanstalt für Finanzdienstleistungsaufsicht, 2004a, *Investment Directive* from July 9, 2004.

BaFin/Bundesanstalt für Finanzdienstleistungsaufsicht, 2004b, *Jahresbericht der Bundesanstalt für Finanzdienstleistungsaufsicht* 2003 (Bonn und Frankfurt am Main).

BaFin/Bundesanstalt für Finanzdienstleistungsaufsicht, 2004c, *Jahresbericht der Bundesanstalt für Finanzdienstleistungsaufsicht* 2002, Teil B (Bonn und Frankfurt am Main).

Brockhaus Lexikonredaktion, ed. Vol., 1974, *Enzyklopädie,* 17th ed. 19 (Wiesbaden), pp. 560–561.

Deutsche Bundesbank, ed., 2004a, *Kapitalmarktstatistik Oktober 2004* (Frankfurt am Main).

Deutsche Bundesbank, ed., 2004b, *Bankenstatistik Oktober 2004* (Frankfurt am Main).

Dresdner Bank AG, ed., 2004, *Finanzbericht des Dresdner-Bank-Konzerns,* December 31, 2003.

Elgeti, Rolf, and Raimond Maurer, 2000, "Zur Quantifizierung der Risikoprämien deutscher Versicherungsaktien im Kontext eines Multifaktorenmodells," *Zeitschrift für die gesamte Versicherungswissenschaft* 89: 577–603.

Gauss, Carl-Friedrich, 1973 (reprint), *Werke*, Vol. 4 (Heidelberg and New York), pp. 119–188.

Gesamtverband der deutschen Versicherungswirtschaft, ed., 2001, *The German Insurance Industry, Yearbook 2001* (Karlsruhe).

Gesamtverband der deutschen Versicherungswirtschaft, ed., 2002, *The German Insurance Industry, Yearbook 2002* (Karlsruhe).

Gesamtverband der deutschen Versicherungswirtschaft, ed., 2003, *The German Insurance Industry, Yearbook 2003* (Karlsruhe).

Gesamtverband der deutschen Versicherungswirtschaft, ed., 2003a, *The German Insurance Industry, Statistical Yearbook 2003* (Karlsruhe).

Gesamtverband der deutschen Versicherungswirtschaft, ed., 2003b, www.gdv.de.

Gesamtverband der deutschen Versicherungswirtschaft, 2003c, *GDV Annual Report Database Online,* www.gdv.org (accessed November 15–19).

Gesamtverband der deutschen Versicherungswirtschaft, ed., 2004, *The German Insurance Industry, Statistical Yearbook 2004* (Karlsruhe).

Hartung, Thomas, and Elmar Helten, 2004, "Modernisierung versicherungswirtschaftlicher Eigenkapitalnormen durch Solvency II," F*inanzbetrieb* 6: 293–303.

HGB/Handelsgesetzbuch *German Commercial Code.*

Koch, Peter, 1988, "Geschichte der Versicherung," in D. Farny et al., eds., *Handwörterbuch der Versicherung,* pp. 223–232.

Lier, Monika, 2003, "Kooperieren geht über fusionieren," *Versicherungswirtschaft* 15/03: 1189.

Maurer, Raimond, 2004, "Institutional Investors in Germany: Insurance Companies and Investment Funds," in Jan. P. Krahnen and Reinhard H. Schmidt, eds., *The German Financial System* (Oxford: Oxford University Press), pp. 106–138.

Perlet, Helmut, and Helmut Gündl, 2005, *Solvency II* (Wiesbaden).

RechVersV/Verordnung über die Rechnungslegung von Versicherungsunternehmen, *Decree About the Accounting of Insurance Companies.*

Richter, Andreas, and Jochen Ruß, 2002, "Tax Arbitrage in the German Insurance Market," *Blätter der Deutschen Gesellschaft für Versicherungsmathematik* 26 (April).

Schradin, Heinrich R., 2004, "Entwicklung der Versicherungsaufsicht," *Zeitschrift für die gesamte Versicherungswirtschaft* 92: 611–664.

VAG/Versicherungsaufsichtsgesetz, *German Insurance Supervision Law.*

VWD-Vereinigte Wirtschaftsdienste—Information Service, 2003, www.vwd.de (accessed November 15).

Versicherungswirtschaft, 2005a, "Ausschließlichkeit büßt ein," *Versicherungswirtschaft* 20/05: 1546–1547.

Versicherungswirtschaft, 2005b, "Allfinanz im Südwesten," *Versicherungswirtschaft* 20/05: 1625.

Wandel, Eckhart, 1998, *Banken und Versicherungen im 19. und 20. Jahrhundert* (München).

6.7 LEXICON

Aging reserve (*Alterungsrückstellung*). An obligatory balance sheet liability position in the health insurance industry containing the capital, which comes from the insureds' contributions as well as the interest on this capital, neither of which are used for the current risk coverage or costs. The position results from the legal obligation to restrict the premium increase for elderly insureds.

BaFin (*Bundesanstalt für Finanzdienstleistungsaufsicht*). Federal Financial Supervisory Authority that was established on May 1, 2002 and brings together under the one roof the three former Federal Supervisory Offices covering banking (the *Bundesaufsichtsamt für das Kreditwesen*—BAKred), the insurance industry (the *Bundesaufsichtsamt für das Versicherungswesen*—BAV) and securities trading (the *Bundesaufsichtsamt für den Wertpapierhandel*—BAWe). Its responsibilities are the supervision of credit institutions, financial services institutions, insurance companies, and securities trading. BaFin supervises just under 2,400 credit institutions, around 800 financial services institutions, and approximately 700 insurance companies (as of July 28, 2003).

Collective policy. In a collective policy an individual who is a member of an association becomes eligible to buy individual life policies with the same conditions for all its members subject to a group rebate.

Comprehensive cover (*Vollkasko*). A component of motor vehicle insurance that covers the damage on the insured's motor vehicle regardless of the origin of the damage, including cases where the originator of the damage cannot be determined.

Equalization reserve (*Schwankungsrückstellung*). A particularity of German property and casualty insurance accounting, this is a special, legally prescribed balance sheet liability position, permissible only to property and casualty insurers, that aims to smooth reported underwriting results over time. The requirements for the creation of an equalization reserve depend on the scale of the insurer's business, the sum of the actual loss and expense ratios, and their volatility in the legally determined time period. As the profits reported in the annual accounting statement of a company are also relevant for taxation, the equalization fund is also an important vehicle to immunize the tax burden of a company with respect to the volatility of underlying results.

Europapass. The right of insurance companies from the European Union to conduct business in Germany or an EU country other than its country of incorporation, not only by establishing a legal entity in the targeted market, but also by offering services from their original land of incorporation.

Expense ratio. Ratio of administrative and distribution costs to annual premium income.

Free agent. A person or a company professionally distributing the insurance products and unrestricted in the choice of insurers for the supply of products.

Free assets (*Freies Vermögen*). The assets belonging to the insurance company, but not to the securing assets, which are not subject to investment regulation.

Funding reserve (*Deckungsrückstellung*). An obligatory balance sheet liability position representing the present value of the existing insurance obligations of a life insurer. The funding reserve is calculated according to the actuarial principles of insurance mathematics; its calculation is supervised by the life insurer's chief actuary. The assets covering the funding reserve must satisfy certain criteria and must be kept separately (see **tied assets,** *Sondervermögen*).

Hidden asset reserves. Differences in accounting valuation rules and market valuation that arise when the market-based valuation is temporarily higher than the historical average. In the opposite case (i.e., when market-based valuation is temporarily lower than the historical average), hidden asset losses are generated.

Institution under public law (*Öffentlich-Rechtliches Unternehmen*). A corporate body whose existence and activities are regulated not by the set of private law regulations, but by that of public law, usually governing the relationship between the authority and the citizen. An institution under public law is created by authority's decree and serves a public purpose defined in that decree. This legal form has its roots in the historical economic and state development. Some insurers are also institutions under public law.

Insurance density. Per-year premium income received by primary insurers per capita.

Insurance penetration. Per-year premium income received by primary insurers as a percentage of the gross domestic product.

Loss ratio. Ratio of claims incurred to premiums earned.

Net investment return. Ratio of net earnings per year to the year's average invested assets.

Non-quoted stock company (*Nicht notierte Aktiengesellschaft*). According to the German Companies Act (*Aktiengesetz,* or AktG) the listing on an organized stock is not an obligatory feature of a stock corporation (AktG, §1,§3): The legal form of a nonquoted company gives all advantages of a public company such as, for example, simplified procedure of the share transfer, without being burdened by the considerable costs of a stock exchange listing. The requirements regarding the ownership of the nonlisted corporations are identical to the listed one; they are codified in AktG (requirements for all stock corporations) and the Insurance Supervision Law (for insurance companies only).

Participating certificate (*Genußschein*). A special financing instrument having the features of both equity and debt, can be traded on a stock exchange or over the counter.

Pensionskasse. A special kind of life insurance operating only in the market for occupational pensions. The majority of these companies are organized as mutuals, legally independent from the sponsoring employer (typically a company, public corporation, or industry group) and provides occupational retirement benefits for employees. Since the benefits of this type of occupational pension scheme include substantial insurance features (e.g., mortality and disability coverage) and defined benefit elements, the regulation of *Pensionskassen* is similar to that of life insurers.

Securing assets (*Sicherungsvermögen*). The assets covering the insurance reserves (also called technical provisions). They must be kept and managed separately from other insurance company assets in a special fund (tied funds). A trustee must authorize any movement in the funds.

Semi-comprehensive cover (Teilkasko). A component of motor vehicle insurance that covers the damage on the insured's motor vehicle that is not caused by the insured's incorrect actions.

Spartentrennungsprinzip. Regulation in the German insurance industry that forbids conducting life and health insurance business and property and casualty business under the roof of one legal entity. It is codified in section 106(c) VAG.

Special funds. Investment vehicles comparable to mutual funds and regulated like the latter in the Investment Company Act, but available only for institutional investors.

Sterbekasse. A special kind of life insurance operating in a niche market, originally created to finance burial services for the insured. Nowadays the insured sum can also be paid out to a designated person(s). The main differences between *Sterbekasse* and life insurance are that *Sterbekasse* has a quicker payout in the case of death, and it has certain age requirements/restrictions at the signing of the contract. The majority of these companies are organized as mutuals. The regulation of *Sterbekassen* is similar to that of life insurers.

Technical results. In the analysis of an insurer's accounts, technical results are from insurance activity, as opposed to results from investing activity (which together comprise the recurring results of an insurer).

Third-party liability (*KfZ-Haftpflichtversicherung*). A component of motor vehicle insurance that covers the damage to third parties that are caused by the insured.

Tied agent. A person or a company professionally distributing insurance products and having a contract with an insurer to offer exclusively the products of that insurer.

Tied assets (*Gebundenes Vermögen*). The assets belonging to the securing assets. The investments of the assets belonging to the tied funds are subject to regulation. An investment directive of the supervisory body exactly defines the categories of authorized assets permissible for the tied fund and requires some qualitative investment principles (e.g., safety, yield, diversification, and marketability of investments). The regulation also sets quantitative restrictions regarding the insurer's asset allocation.

VAG (*Versicherungsaufsichtsgesetz*). German Insurance Supervisory Law.

7 The Italian Insurance Sector between Macro and Micro Facts: Salient Features, Recent Trends, and an Econometric Analysis

Luigi Ventura

University of Rome La Sapienza

7.1 INTRODUCTION

The Italian insurance market has undergone a series of profound modifications over the last decade. This is immediately apparent by looking at the performance of the market in terms of premiums collected, volume of investments, and profits earned, and also by looking at the number and variety of insurance instruments that have been launched on the market and at the distribution of such instruments among households. Moreover, transformation is even more evident if we compare the statistics of the Italian market with those of the markets of other European countries that possess a similar structure.

Insurance policies, especially those of life branches, have become an increasingly important component of households' budgets and finances, along with other financial instruments.[170] At the same time, profound demographic changes in Italian society (of which the most important are the accelerated aging of the Italian population and the increase in life expectancy) gave policymakers a strong incentive to design an updated social security system that can be sustained on a macroeconomic basis, and that also crucially hinges upon some forms of private, complementary social security. Moreover, the complexity of the production processes that constitute an essential part of the economy also calls for the implementation of additional risk-sharing mechanisms, of which damage and liability insurance is a natural component.

This chapter attempts to depict a comprehensive panorama of the dynamics of the Italian insurance market by exploring some descriptive statistics relative to the main aggregates of interest, by illustrating in detail some crucial aspects of the

---

[170] Life and non-life insurance is divided into several lines of business that are denoted here as branches or classes of insurance. There are six such branches for life insurance (branch II, "natality and nuptiality," has always been inactive), and 18 for non-life insurance. For a definition of those classes, refer to Table 3.

development of the market, and by performing some simple econometric analyses based on survey data on Italian households' insurance holdings.

The goal of the chapter is to provide those interested and working in the sector with some useful insights as to the consequences that alternative, internal, and external scenarios might have on the structure of the Italian insurance market. In so doing, we offer some updated and complementary views to the interesting and detailed analysis on the recent developments of the Italian insurance market recently performed by Pontremoli (2002).

7.2 THE MACRO-DYNAMICS OF THE ITALIAN INSURANCE MARKET

La Torre (1995) provides a complete historical perspective of the insurance market in Italy as well as a rich bibliography on the topic. A short outline of this fascinating story is given in the next few paragraphs.

The first bona fide insurance contract in Italy has been recognized in a notary act dated February 20' 1343, and signed in Genoa, which at the time was a lively merchant town (with Venice, Pisa, and Amalfi, it was one of the four so-called Maritime Republics). This should not come as a surprise, as the rhythm and extent of maritime commerce had risen to such levels as to require some form of insurance to be more viable. In fact, the development of insurance contracts further strengthened maritime and, later on, land trades.

Curiously enough, at the outset the insurance policies written in Genoa took the form of a loan contract, for reasons that are still partly unclear. In other towns such as Palermo, Pisa, and Florence, insurance policies featured a more "modern" format (including a precise mention of the subjects and objects of the contract). It is indeed from the insurance contracts signed in Pisa and Florence that most other insurance contracts throughout Europe got their format and inspiration. Needless to say, for a very long time insurance was used only for maritime purposes, and much later for land commerce, and it was not run on a large scale by specialized companies. This came later, probably not until the second half of the eighteenth century.

Moreover, not only were Italian cities important in inventing and spreading the use of insurance instruments across Europe, but they were also key in proposing some specific legal norms concerning these new business tools, and were soon followed by other countries (like Spain, with the so-called Barcelona decrees, in the fifteenth century). The relevance of Italian towns in the development of insurance markets consistently decreased in the sixteenth century, to the benefit of central and northern European countries, who took over the supremacy in trade, with the foundation of the various India Companies (East and West).

To wrap up our brief history of the Italian insurance sector, we must jump to the beginning of the last century, because at that time a very important change occurred in the attitude of Italian lawmakers vis-à-vis the insurance business. Italian lawmakers took a more active stance with respect to the problems of the insurance sector, to which they had been relatively indifferent up to the beginning of the century. This was reflected in several laws, such as Law n. 305 of 1912, which established the Istituto Nazionale delle Assicurazioni (National Institute of

Insurance). This was the first true example of public intervention in the Italian economy, and it paved the way for new public interventions in the insurance sector, whose operations had been, up to that point, strictly confined to the private sector.

A comprehensive reform of the entire insurance business was undertaken in the period 1923 to 1925, whose aim was, above all, ensuring transparency and solvency of insurance firms. In the same years, civil aviation received new and substantial rules, also from an insurance standpoint.[171] The new legislation reflected an awareness of labor problems, which gave impulse to social insurance. This was, in fact, the period in which the newly conceived social security institutions were born.

Subsequent decades witnessed an ever more pervasive role of lawmakers in the regulation of the insurance sector, especially in view of the spectacular progress in technology that occurred during the twentieth century (air and land transportation, in particular motor transportation, nuclear activities, industrial processes, entertainment, etc.) and in light of the ever-increasing importance attributed to health and the progress made by medical sciences. In these respects, from the legislative standpoint, a few measures had a decisive impact on the functioning of the market: the Testo Unico (unified legislative text) n. 449 of 1959 providing rules about the management of private insurance; Laws n. 1862 of 1962 and n. 799 of 1967 about compulsory insurance of nuclear and hunting risks, respectively; Law n. 990 of 1969 about compulsory motor insurance, which has been of key importance in the development of this sector; Law n. 295 of 1978, containing new rules for the management of damage insurance; and Law n. 48 of 1979, which established the register of insurance officers.

Coming now to the more recent history, we sketch an account of the main aspects of the dynamics of the Italian insurance market in the period 1990 to 2003. In doing so we mainly rely on data presented in the yearly accounts by the Associazione Nazionale fra le Imprese Assicuratrici (ANIA; the Italian insurers association) and the Istituto per la Vigilanza sulle Assicurazioni Private e di interesse collettivo (ISVAP, the private insurers' surveillance authority) from 1991 to 2003. As a general remark, and also to explain a partial lack of homogeneity across tables, it should be noted that a major change in insurance firms' reporting was implemented in 1998. In particular, the structure of the balance sheet and income statement was changed to conform to the format decided at the European level. That is why in a few cases we report data only for the subperiod 1998 to 2003, when an adequate (i.e., homogeneous) reconstruction of the entire time series has not been possible.

All in all, the Italian insurance market in last decade of the twentieth century has been marked by very lively dynamics, which have been fostered and accompanied by a number of important changes at the social, demographic, institutional, and normative levels. In particular, on a social and demographic basis, the last decade of the twentieth century witnessed quite a remarkable increase in life expectancy, which rose on average (male and female) from 77.12 to 78.52 years. On the other hand, adult mortality rates have uniformly gone down, which has brought about an important increase in the percentage of people aged 65 or older (from an already substantial 15.3 percent in 1990 to 18.2 percent in 2001). According to demographic

---

[171] It suffices to recall that Title V (about insurance contracts in civil aviation) of the Rules for Air Navigation issued in 1925 extends to civil aviation the insurance procedures adopted for sea navigation, and adapts them to the new context.

forecasts and in view of present fertility rates, this percentage will increase at an even faster pace in the next decade, unless some specific policy measure is taken. The increase in the percentage of total population aged 65 or older has already brought about some important consequences on the insurance market and will bring more. In fact, a higher percentage of elderly people results in a higher dependency ratio and a larger need for social security (both for pension benefits and for health security).[172]

Because Italy no longer experiences high rates of growth (in an economic sense and especially in a demographic sense), and because people of working age, who generate most social contributions, become less and less numerous, the "pay as you go" system (the original, "first pillar" of the Italian social security system) becomes less and less viable. In particular, several studies (see, for example, ANIA 1995) have demonstrated that, at the fertility and mortality rates that prevailed in the 1990s and at a contribution level approximately equal to 30 percent of gross salary, only about 60 percent of social security spending could be financed, which calls for additional funding. The second source of funding (which has been named the "second pillar") should come from collective social security (at the firm or category level), while the "third pillar" should consist in individual security, obtained by investing in private pension funds. A higher life expectancy will push more people to save during their working years for a (possibly) longer retirement period, which might also result in a larger demand for complementary life insurance contracts. The first topic we discuss in some depth is, therefore, complementary social security.

7.2.1 Complementary Social Security: A Challenge for Italian Insurers

The last decade witnessed a comprehensive reform of social security, induced by the lack of macroeconomic sustainability of the welfare system that was in place until 1992. The first step toward the establishment of a new social security system was the Amato Reform, which was essentially contained in the Delegated Law of October 23, 1992, n. 421.[173] This law set the framework for two subsequent legislative decrees, n. 503/92 and n. 124/1993; the former brought about the reorganization of retirement benefits for public and private workers. The reorganization included an increase in the contributions needed for an individual to be entitled to a retirement pension, a gradual increase in the retirement age, an extension of the reference period for the calculation of the retirement benefits, a new indexation of pensions by which benefits were linked exclusively to the cost of living index, and progressive unification of social security systems for all employees. The latter established the most important guidelines for the so-called complementary pension plans and called for additional laws to discipline the sector in greater detail. It was not until 1995 (with the Dini Reform), though, that a comprehensive set of rules was established in a new law, n. 335, which modified both the system of public pension funds and that

---

[172] The dependency ratio is obtained as the ratio of the number of individuals up to age 15 or above age 64 to the number of individuals aged 15 to 64, and gives an immediate idea of the people who need looking after in a given country.

[173] Giuliano Amato was designated head of the Italian government in 1992.

of complementary pension funds.[174] The new law regarding public pension funds governed a transition from a pay-as-you-go system, under which the amount received by the retired worker was largely independent of his or her age (but still dependent on the wage dynamics and the number of years of contribution) to a funded system, where the amount received is closely linked to life expectancy, and therefore also to the age of retirement.

Law n. 335/95 also established a number of rules governing complementary pension funds, in terms of all their relevant aspects: the definition of funds, the definition of agents allowed to run such funds, management principles, information transmission, guaranteed rates of return, and control procedures.

A new regulation was introduced with Law n. 243 of August 23, 2004, which entitled the government to adopt several legislative decrees containing regulations with the following goals: (1) freeing the retirement age, (2) gradually eliminating the ban on the accumulation of pensions with earned income, and (3) supporting the development of complementary social security. The legislative decrees should implement measures designed to increase the volume of cash flows to collective and individual complementary social security schemes, organize the transfer of the accumulated funds to the latter, and remove the obstacle to free circulation of workers in those different pension funds, including individual ones. The "second pillar" and "third pillar," now intertwined, should, therefore, be strongly reinforced by the new measures.

The new law also assigns the power of monitoring the whole system of pension funds to a single supervisory authority, the Commissione di Vigilanza sui Fondi Pensione (COVIP), established with the above-mentioned 1993 decree. COVIP began de facto to operate in its present and full-fledged form in 1996 and has been remarkably strengthened in its powers by the new law.

The fiscal aspects of complementary pension funds were also modified with respect to the original rules, set by the decree n. 124/1993, in the direction of drawing a clearer distinction between pension funds and life insurance. As illustrated in the next section, the fiscal impact of such reforms is far from neutral in determining households' behavior.

7.2.2 Tax Reforms and Insurance: A Short Retrospective

Insurance premiums have been subject to a number of changes in fiscal treatment since the beginning of the 1990s. It is not our purpose to describe in detail the fiscal implications of insurance contracts and their modifications over time. For more detailed information, refer to the interesting work by Persano and Mattei (1998) and all the references cited in their work. We only briefly sketch the problem, because our intent is to show that the past decade has been quite important from this standpoint. Some of the changes concerned the fiscal treatment of premiums. In particular, until 1991 insurance premiums could simply be deducted from taxable income, up to 2.5 million liras (about 1,250 euros). In other words, income used to

---

[174] Lamberto Dini was designated head of the Italian government in 1995. Law August 8, 1995, n. 335, published in GU n. 198, 25–8–1995, suppl. ord. n. 106.

pay for insurance premiums could not be considered as taxable, up to this limit. This was established in 1977 at 2 million and then raised to 2.5 million in 1980.

The subsequent reforms were aimed at curbing a phenomenon that was designated as the spreading of "fiscal policies"—that is, insurance policies deprived of a bona fide insurance content, which people only bought for their fiscal advantages. Starting in 1992 with the Amato Reform, this system was radically changed, and only a given percentage of premiums paid could be subtracted. More importantly, policyholders were allowed to deduct this percentage from gross tax, rather than gross income. This percentage, initially set at 27 percent, was reduced in 1995 to 22 percent, and further reduced in 1998 to 19 percent, its current level. This change was quite radical, because it greatly reduced the incentive to buy insurance policies for fiscal advantages. In fact, marginal tax rates (which determined fiscal savings in the previous tax regime) for relatively wealthy people (who were more likely to buy life insurance products) could be as high as 45.5 percent, yielding a maximal saving of 1,137,500 liras (about 590 euros) versus a maximal tax saving under the new fiscal regime of 475,000 liras (or about 245 euros). In the empirical analyses later in this chapter, we assess the impact of these changes in taxation on the demand for life insurance products.

The fiscal treatment of life insurance policies was also differentiated according to the nature of the insured risk. In case of pure death risk, the treatment was more favorable, in that deduction of premiums from gross income was, and still is, allowed.

As for the taxation of surrender values, or death benefits, received by the insured (or the insured's heirs, in case of death) at the expiration of the contract, the situation has also evolved quite a bit in the last two decades. In the direct tax reform of the 1970s (Decree of the President of the Republic [d.P.R.] 29 September 1973, n. 601), it was clearly stated that the amounts received at the expiration of a life insurance contract could not be taxed, essentially because they had the nature of a compensation. This was not true when the contract paid an annuity, which was assimilated to standard (and taxable) income and was therefore subject to normal taxation rules. A convergence process was initiated in which such payments lost part of their privileges (art. 6 of d.P.R. 26 September 1985, n. 482), in that their "net income component," identified as the difference between the amount to be paid and the sum of premiums paid by the insured (possibly diminished by 2 percent for each year exceeding the first 10 years of the contract), would be subject to taxation at a rate of 12.5 percent. On the other hand, art. 33, c.3, of d.P.R. 4 February 1988, n. 42, established that only a 60 percent fraction of annuities should be taxed (which should correspond to their net income component).

With respect to the tax treatment of pension funds, the maintained principle was that of deferring taxation to the moment of utilization of the funds rather than to the moment of their formation. Moreover, the Amato Reform was even more important in this field, because it set up the necessary guidelines to allow the deduction of premiums paid by workers from their gross income. This was established by Law n. 335 of 1995, which replaced art. 13 and 14 of the fundamental decree n. 124 of 1993. As in the case of life insurance products, a deduction limit at 2.5 million liras (or 2 percent of the overall yearly wage) was introduced for pension funds, but it was also clearly stated that this limit might be increased for policy purposes. In any case, we may safely suggest that the fiscal treatment of premiums made pension funds relatively more convenient with respect to competing saving instruments and certainly contributed to the popularity of this new financial product. More recently, two more decrees by the Ministry of Finance fixed some clearer rules for this sector. Moreover, the new law n. 243/2004 will also, at least partly, modify the fiscal regime affecting pension funds in the direction of diminishing the tax burden on the yield of pension income, but these measures should be contained in the future decrees called for by the law.

Indirect taxation of insurance products, on the other hand, has a simpler story, in that it has been settled by Law 29 September 1961, n. 1261. This law established that the only indirect tax to be applied should be a specific insurance tax, which would have also absorbed, for insurance products, the role of the value added tax and the registry tax.[175] The rate of this tax has been set at 2.5 percent of paid premiums by art. 4, c.2, of Law 11 March 1988, n. 67. This is quite an important indirect tax in the Italian panorama, as it ranks third (in terms of fiscal receipts) among all indirect taxes, after VAT (which is by far the most important source of revenue from indirect taxation) and the registry tax.

It should be self-evident that the fiscal treatment of insurance products is also extremely relevant in terms of the establishment a common European market for insurance and free mobility for workers. Unfortunately, to date many differences can be seen in the various national fiscal systems. This is true both for indirect taxes on insurance products—because these have been abolished almost everywhere but remain in Italy, Austria, Belgium, Greece, and Sweden, at different rates—and for direct taxation (different regimes of deductibility of insurance premiums from income or taxes).

As regards the latter, for example, a blatant obstacle to liberalization and to the creation of a common insurance market, as well as an important hindrance to free mobility of workers, comes from the refusal of various countries to allow (permanent or transitory) residents who have previously contracted life insurance with foreign companies to benefit from fiscal allowances available to home companies. This refusal was declared legitimate by the EU Court of Justice in Luxembourg in the course of the Bachmann decision, but it was subsequently rejected, at least in part, with a new judgment (the Wielockx decision), stating that countries that have previously agreed upon principles of fiscal "equalization" (for example, by the adoption of common rules against double taxation), cannot use any principles of "home country fiscal coherence" (i.e., the need to abide by different national laws) to apply unequal fiscal treatment of insurance policyholders. In this respect, we may signal the establishment of a new European body—the Committee of European Insurance and Occupational Pension Supervisors (CEIOPS). This new organization was created under the terms of the European Commission Decision 2004/6/EC of November 5, 2003, and is composed of high-level representatives from the insurance and occupational pension supervisory authorities from member states of the European Union.

---

[175] This is the tax that must be paid to register public acts.

7.2.3 The Italian Market in a European Common Insurance Market

The second event to be underscored in order to better grasp the Italian insurance market is the liberalization and homogenization of insurance activities in Europe, which led to a huge free exchange area formed by the 18 countries taking part in the European Economic Space (i.e., the 15 countries participating in the European Union during the 1990s plus Iceland, Liechtenstein, and Norway). This free exchange area grew tremendously with the accession of 10 new European Union member states in 2004, and it is going to be enlarged even more in the coming decades. This process was governed, as far as insurance is concerned, by two EEC directives: n. 92/49 (the third generation EC non-life insurance directive) and n. 92/96 (the third generation EC life insurance directive). These directives, which were subsequently introduced in national laws, brought about some important changes:

1. the possibility for a single insurance company to operate in all member states, with a single authorization issued by the country of origin;
2. the introduction of the principle of "home country control," whereby oversight of an insurer's activities is exercised by the relevant institutions of the company's home country. In Italy, surveillance over all insurance activities has been assigned to ISVAP. Although ISVAP was created in 1982, it was not until 1998, with the Legislative Decree n. 373, that it became an independent authority and was assigned almost all powers previously retained by the Ministry of Industry and Productive Activities. For example, ISVAP is responsible for management of professional registers for agents, brokers, and experts, applications of sanctions to insurance companies, and decisions relative to the liquidation of insurance companies;
3. the coordination, at the European level, of all national rules governing investments of technical reserves;
4. the establishment of a solvency control modeled according to the British and North American practice—this means, for a few countries (including Italy) giving up the practice of more intensive and severe controls and, consequently;
5. the elimination of all national rules imposing a standardized or systematic communication of tariffs and of general and specific conditions of insurance policies, as these are thought to be superfluous in a mature market such as the European one.

Particular relevance should be attached to principles 1. and 2., because they aimed to establish much easier procedures for European companies to extend their activities across the European Union by establishing new branches in other countries or simply offering their services abroad. It should be noted, in fact, that, although it had been possible, in principle, to do both things long before (at least since the beginning of the 1970s), bureaucratic difficulties had convinced most companies to operate abroad only by participating in the stockholding of foreign insurance establishments, which seems to be a common practice even in more recent years. The reason for this is the lack of harmonization of national laws defining control rights and of laws concerning the design and functioning of contracts. As for control rights, a clear definition of the boundaries between the freedom of settlement (i.e., the possibility for an insurer to open a branch outside its home country) and the freedom of selling services abroad is still missing, as is a clear definition of "general interest" home rules that foreign companies should respect.

As for contract laws, the lack of common rules has hindered the creation of truly European contracts. Tax harmonization (which is likewise missing) is also extremely relevant, as discussed in the previous sections, for the creation of a European insurance market.

With these elements in mind, we turn to a descriptive analysis of the Italian insurance market, over the period 1990 to 2003. We mainly rely on data obtained by ANIA and ISVAP, the two most important sources of information on this market.

7.2.4 The Dynamics of Italian Insurance: Main Facts

To first get a quantitative idea of the dynamics of the Italian insurance market over the period 1990 to 2003, Tables 7.1 and 7.2 contain, respectively, the number of companies operating in Italy from 1990 to 2003 and a concentration index (market share by premium volume of the top 1, 5, and 10 insurers) for 1993, 1998, and 2003.

Table 7.1. Number of Companies in the Italian Insurance Industry, 1990 to 2003

	Public Companies	Stock	Cooperative	Mutual	Total Registered in Italy	Branches of Foreign Companies	Total
1990	2	185	1	12	200	53	253
1991	2	195	1	12	210	52	262
1992	0	205	1	12	218	50	268
1993	0	211	1	12	224	51	275
1994	0	211	1	12	224	41	265
1995	0	208	1	12	221	40	261
1996	0	213	1	10	224	47	271
1997	0	204	1	10	215	46	261
1998	0	195	1	8	204	49	253
1999	0	192	1	7	200	50	250
2000	0	194	1	6	201	51	252
2001	0	194	1	5	202	54	256
2002	0	190	1	3	194	57	251
2003	0	185	1	3	189	60	249

Source: Author and ANIA annual reports, various years.

The numbers of insurers in the Italian market by organizational form are shown in Table 7.1. Italian insurers can be organized as stocks, mutuals, or cooperatives. A mutual company is a particular form of a cooperative (i.e., a society in which there exists a mutuality purpose), where the mutuality character is prevailing: the insured subjects essentially coincide with the members of the cooperative, whereas a generic cooperative firm may also work in favor of third parties—that is, nonmembers of the cooperative. Moreover, mutual companies must respect stricter rules for the distribution of profits and reserves, with respect to a standard cooperative. Branches of foreign insurers also do business in Italy.

Table 7.1 displays a substantial stability in the number of companies operating in the Italian market. What we can see, however, is a partial substitution of branches of foreign companies for companies registered in Italy. We also can observe that the number of stock firms is overwhelming with respect to other types of organizational forms and that this number increased in the central years of the period under examination and returned to nearly its original values by the end of the period. The number of mutual companies has substantially decreased, and cooperative firms have never played more than a marginal role. In fact, it can be shown that stock companies represent 98.5 percent and 96.6 percent of total premiums, respectively, for life and non-life insurance. Cooperatives account, respectively, for about 0.8 percent and 2.7 percent of total premiums, whereas mutual companies account for about 0.7 percent of total premiums of both branches.

Although the number of companies has been relatively stable, we do observe some interesting dynamics in concentration. Table 7.2 shows a sharp drop in concentration of the life insurance market from 1993 to 1998 and a slight increase in the subsequent years, regardless of the number of companies taken to measure concentration. On the other hand, the reverse is true for the non-life market, where concentration has almost uniformly increased over the years, mainly due to some merging operations (for example, the one between Fondiaria and SAI).

As shown in Table 7.3, representing premium income for non-life and life and capitalization branches in the direct Italian insurance market, we observe a few interesting facts: quite evidently, a sharp difference between the dynamics of life and non-life branches. The non-life branches have been steadily growing at an annual average rate of about 9 percent, while the life branch has featured variable growth, with a first period (from 1990 to 1996) in which premiums increased at an average rate of about 20 percent and a second period (from 1997 to 2003) in which premiums grew at a higher rate of about 27 percent. Within this second period is a subperiod, from 1996 to 1999, with exceptionally high rates of growth, around 40 percent yearly. This is an interesting finding that calls for an explanation. As is well known, and as confirmed in several empirical analyses, an important variable affecting the ratio of premiums to gross domestic product (GDP) is financial development (see, for example, Outreville 1996, and Beck and Webb 2003), which we might proxy by the ratio of total value of stocks traded to GDP. Unlike other structural variables, this ratio can vary quite a bit in the short run and in the medium run, and, indeed, it underwent a spectacular increase in 1997 and 1998; this, we believe, might partly explain the increase in life and capitalization premiums.

A different, though related, reason for the growth of life insurance premiums is linked to the introduction of more precise rules for the management of pension funds. As anticipated in the previous section, the whole matter was first dealt with in a legislative decree issued in 1993, n. 124, which, contrary to initial forecasts, did not immediately ignite a boom of pension funds, as it made reference to a series of other bills that were not promulgated before 1995.

Table 7.2. Concentration (Percent Market Share by Premium Volume) of the Top Italian Insurers

	Top 1			Top 5			Top 10		
	2003	1998	1993	2003	1998	1993	2003	1998	1993
Life Insurance									
I—Human Life[a]	10.4	11.7		45.8	34.4		64.7	51.1	
III—Investment Funds	16.8	14.8		49.1	49.1		67.6	73.1	
IV—Health (long-term)	82.5			97.9			99.5		
V—Capitalization	13.1	19.7		46.8	47.9		67.2	65.2	
VI—Pension Funds	15.2	64.2		62.0	100.0		80.9	100.0	
Total Life Insurance	8.5	8.3	15.6	33.8	32.0	49.5	54.2	51.4	63.9
Non-life Insurance									
Accident	11.6	12.1		44.2	41.1		63.3	61.6	
Sickness	15.5	11.7		46.5	43.0		63.9	62.9	
Land Vehicles	13.5	10.3		40.5	31.2		57.6	51.0	
Railway Craft	58.4	59.2		97.7	99.5		99.4	100.0	
Aircraft	45.9	52.0		92.1	92.9		98.7	98.5	
Ships	30.2	20.8		76.4	70.8		92.6	89.8	
Goods in Transit	15.6	20.7		53.3	51.8		72.8	71.6	
Fire and Natural Forces	14.9	15.3		46.3	43.1		65.4	64.7	
Other Damages	15.5	16.6		45.8	44.2		67.5	67.1	
Motor Liability	12.8	7.4		39.4	31.2		58.7	52.3	
Aircraft Third-party Liability	46.6	67.4		98.6	91.5		99.5	97.4	
Ships Third-party Liability	12.5	14.1		45.5	40.0		63.3	60.2	
General Third-party Liability	14.5	13.2		47.0	43.1		66.0	65.6	
Credit	52.8	62.6		94.3	93.2		99.0	98.9	
Suretyship	14.5	10.1		45.4	41.0		67.5	64.2	
Miscellaneous Financial Loss	12.4	15.0		42.5	39.2		60.0	57.3	
Legal Expenses	13.9	12.4		41.0	43.6		60.3	64.2	
Assistance	11.5	17.2		35.6	42.6		55.8	59.1	
Total Non-life Insurance	11.1	8.8	7.8	40.3	34.1	34.1	58.1	54.1	51.7
Total	5.5	7.8	7.8	24.0	24.6	34.1	40.2	38.4	51.7

Source: Author and ANIA Infobila system and annual reports, various years.
[a] Insurance over the duration of human life only.

Table 7.3a. Premium Volume by Line of Business in the Italian Insurance Industry, 1990 to 2003 (in thousands of euros)

	1990	1991	1992	1993	1994	1995	1996	1997
Aircraft and Ships	80,066	79,626	77,615	103,839	103,692	125,578	132,793	114,058
Automobile Risks	1,584,950	1,840,658	2,125,882	2,185,119	2,128,191	2,174,637	2,288,398	2,451,560
Suretyship	202,913	218,670	224,516	219,434	256,518	304,920	350,131	423,953
Credit	141,266	155,151	170,478	172,502	190,597	181,811	195,869	203,541
Theft	510,757	552,513	599,125	615,206	620,776	648,083	658,276	672,678
Fire and Natural Forces	1,202,712	1,359,163	1,461,859	1,588,366	1,745,131	1,792,708	1,891,971	2,136,334
Accident	1,401,912	1,574,852	1,707,054	1,799,182	1,850,358	1,958,238	2,032,472	2,116,823
Sickness	474,511	571,268	679,077	781,381	862,441	945,744	984,303	1,067,742
Miscellaneous Third-party Liability	854,878	967,968	1,095,502	1,202,964	1,242,091	1,419,749	1,541,015	1,649,729
Mandatory Third-party Liability	5,821,908	6,740,974	7,586,494	8,181,187	8,678,351	9,332,518	9,786,360	10,666,886
Goods in Transit	441,412	525,967	654,735	748,240	777,658	730,958	726,497	732,828
Legal Expenses	19,836	25,017	40,862	51,495	60,866	71,390	80,730	89,209
Other Lines	178,181	217,738	256,675	279,641	338,834	418,707	448,616	462,588
Total Non-life	13,174,450	15,058,770	16,963,144	18,132,898	19,002,345	20,448,665	21,510,275	22,943,460
Life and Capitalization	4,481,341	5,473,191	6,459,979	7,820,729	9,619,232	11,994,914	13,460,215	19,044,224
Total	17,655,791	20,531,961	23,423,123	25,953,626	28,621,577	32,443,580	34,970,490	41,987,684

Table 7.3b. Premium Volume by Line of Business in the Italian Insurance Industry, 1990 to 2003 (in thousands of euros)

	1998	1999	2000	2001	2002	2003
Accident	2,207,659	2,262,820	2,379,862	2,529,859	2,620,870	2,760,428
Sickness	1,127,528	1,164,102	1,255,280	1,343,251	1,426,267	1,508,937
Land Vehicles	2,539,274	2,614,217	2,677,552	2,810,932	2,955,063	3,062,260
Railway Craft	6,975	7,737	8,925	9,519	10,282	12,422
Aircraft	65,962	70,841	117,319	63,939	63,819	64,025
Ships	243,573	219,668	223,860	279,115	318,011	295,855
Goods in Tansit	322,196	288,839	304,670	323,234	320,788	291,989
Fire and Natural Forces	1,659,982	1,657,488	1,700,862	1,771,483	1,978,138	2,038,396
Other Damages	1,588,439	1,652,414	1,740,259	1,860,628	2,079,587	2,158,437
Motor Liability	11,745,031	13,226,148	14,196,143	15,316,673	16,627,769	17,621,608
Aircraft Third-party Liability	12,688	16,766	14,248	29,232	51,250	77,920
Ships Third-party Liability	21,527	22,476	25,029	27,743	25,089	24,179
General Third-party Liability	1,819,673	1,920,359	2,033,135	2,229,297	2,472,297	2,797,786
Credit	203,323	219,485	272,173	314,299	320,862	301,270
Suretyship	524,048	459,272	433,210	483,453	505,451	485,787
Miscellaneous Financial Loss	138,297	150,524	162,574	174,922	234,832	270,012
Legal Expenses	98,398	110,491	126,184	143,549	164,263	183,718
Assistance	176,347	182,756	203,427	214,766	240,275	257,063
Total Non-life	24,500,921	26,246,403	27,874,712	29,925,894	32,414,913	34,212,092
I—Human Life	16,720,500	17,602,529	15,966,622	19,412,878	23,991,484	27,740,300
III—Investment Funds	6,570,858	15,022,263	22,213,516	23,612,624	24,558,959	26,559,997
IV—Health (long-term)		5,784	7,871	11,555	10,359	16,779
V—Capitalization	3,190,613	2,921,319	1,503,073	3,201,301	6,609,514	8,335,121
VI—Pension Funds	590	44,646	93,031	90,638	123,787	127,982
Total Life	26,482,562	35,596,541	39,784,113	46,328,996	55,294,103	62,780,179
Grand Total	50,983,483	61,842,944	67,658,825	76,254,890	87,709,016	96,992,271

Source: Author and ANIA annual reports, various years.
Note: The two subtables account for the change in data collection and distribution over the period.

It was only in 1995, with the legislative decree n. 335 (the Dini Reform) and, more importantly, with the Treasury Decrees n. 673 and 703, issued in 1996, that the organization of private pension funds became more clearly defined, in particular for some crucial operational aspects such as investment of acquired resources, control of investment results, transparency and information to subscribers, and rules about potential conflicts of interest.

Regarding non-life classes of insurance, there is a distinction between motor insurance and other types of damage and liability insurance. Since 1996, motor insurance has featured higher rates of growth and has played the most important role. Its prominence, however, is probably going to decline in the next few years, following the expected decrease in motor liability rates induced by recent regulations introduced by the Italian Parliament and an agreement protocol signed by the government, ANIA, and a large number of consumers' associations on May 5, 2003. Agreement protocols are the outcome of a number of meetings among parties in which some specific issue is debated (in this case, the focus was on the possibility of reducing tariffs) and some solutions are proposed and possibly accepted by all. This "contracting" methodology was successfully employed in 1993–1994 and in 1996, although on those occasions the government did not take part in the deal. In this case, the agreement protocol was assigned a limited duration (12 months), and its main goal was to take the motor liability system back to an equilibrium condition in terms of rates and profitability.

Such a renewed equilibrium condition was to be obtained by means of rate limitations, which should also have been made possible by an improvement in the income statement of the motor liability class. This agreement protocol has been quite successful in achieving its goals, inasmuch as all the insurers engaged to various extents in practices that resulted in cooling down the dynamics of motor liability premiums; in particular, the protocol resulted in a tariff increase of only about 0.62 percent in its 12 months' duration versus a much higher increase of about 5.59 percent in the preceding year (previous years were also characterized by large increases in premium rates).

In any case, the Italian auto insurance market can now be considered a mature one, and it is clear that we cannot expect for the immediate future the same rates of growth in demand that were experienced in the last decade. Among non-life insurance classes, and apart from motor insurance (and collateral risks), land vehicles insurance (i.e., all vehicles moving on land, except for rail cars), health insurance, general third-party liability insurance, credit insurance, and fire insurance all displayed quite lively dynamics in the period 1990 to 2003.

Some detailed comments should be devoted to the peculiarities of health insurance. The Italian health insurance system has two components: the first and most relevant is the public component, which is comprised of some public institutions, in particular the Istituto Nazionale della Previdenza Sociale (National Institute for Social Security) (INPS) and the Istituto Nazionale per l'Assicurazione contro gli Infortuni sul Lavoro (National Institute for Insurance) (INAIL). INPS is responsible, among its other duties, for the payment of sickness allowances (to a maximum of 180 days per year, with some further limits and exceptions), and INAIL is responsible for support in case of work-related accidents. Health benefits in kind (medicine, medical treatment by doctors, hospitalizations, etc.) are provided by the Servizio Sanitario Nazionale (National Health System) via some local agencies and the Azienda Sanitaria Locale (Local Health Agency). The public system is entirely funded by compulsory monthly contributions paid by employers and employees (and, of course, by the self-employed).

The second component of the Italian health insurance system is private and rests on complementary insurance bought from insurers and on private health establishments across the country. Some of these are also connected to the public system by special agreements. Private health insurance has steadily grown, at least as far as premiums are concerned, in the period under analysis. In particular, looking at Table 7.3, we observe that premiums increased from 475.5 million euros in 1990 to 1,508.9 million euros in 2003, or at an average yearly rate of about 8.6 percent. On the other hand, as a fraction of total non-life premiums, we go from a very small percentage of about 3.6 percent to a higher, but not very significantly so, percentage of about 4.4 percent. This means, of course, that the development of health insurance is still limited, both domestically and in comparison to other European countries. To gain an idea of its relative distribution, we may observe that only about 2.8 percent of the overall premiums for health insurance in Europe were collected in Italy, versus an overwhelming percentage of about 51 percent collected in Germany (where private insurance plays a substantial role in providing basic health care to citizens), about 16 percent in France (where the public system is extremely well functioning, and private insurance plays only a subordinate, complementary role), and about 12 percent in the Netherlands (where the whole system is organized very much like the three-pillar system of Italian social security).

Moreover, the management of health insurance in Italy has not, at least until now, provided companies with an adequate profit. If we partly anticipate the subsequent analysis and look at Table 7.7, reporting the results of the technical account for the main non-life insurance branches over the period 1998 to 2003, we immediately observe that health insurance has been run at a loss, varying from a minimum of 0.8 percent of collected premiums in 2001 to a maximum of 5.4 percent in 1998. The last figure, for 2003, was of a loss of about 39 million euros—that is, about 2.6 percent of collected premiums. First indications for 2004 also point at a loss. In other words, health insurance has definitely not been profitable in the period under examination.

What puzzles most observers is the fact that Italian households spend quite a lot of money on private medical items, including medicines, clinical tests, specialist visits, and surgery, and yet buy relatively low amounts of health insurance. The propensity to buy health insurance may also be limited by the "scant awareness of citizens regarding the difficulties of accessing a number of services with reasonable waiting lists" (ANIA 2003, 2004a). Be that as it may, there is much evidence of yet-unexpressed potential insurance demand. For a couple of examples, we look at microdata from the Bank of Italy Survey on Household Income and Wealth (SHIW). In 1989, about 35 percent of the surveyed families declared spending at least 250 euros on private health expenditures (including medicines, hospitalizations, analyses, and specialist visits); yet only around 4 percent of the interviewed families owned any health insurance. According to 1993 SHIW data, the latest wave of the survey in which a specific question about private health expenditures was asked, around 17 percent of surveyed families had private health expenditures of at least 250 euros (not including medicines, which may account for the substantial difference with 1989 data), whereas only about 12 percent of the sample owned health insurance. Of this percentage, less than 8 percent paid a yearly premium equal to or more than 150 euros (which, even at 1993 prices, could not be considered sufficient to obtain adequate coverage).

Coming back to basic facts and figures, technical reserves, which represent commitments undertaken vis à vis the insured, closely mimic the dynamics of total premiums.[176] In fact, total reserves grew at a rate of about 18 percent between 1990

[176] To guarantee an economic equilibrium between cash inflows and outflows, insurers set aside a portion of premiums in special reserve funds called technical reserves. These are mainly made up of actuarial reserves against outstanding risks in the case of life policies

and 1996 and at a rate of about 20 percent between 1997 and 2002, with a large increase (of about 36 percent) in 1998. As one might expect, and as is clear from Figure 7.1, mathematical reserves (computed as the difference between the present value of the insurer's commitments and the present value of the insured's obligations) constitute the largest share of total reserves, and they grew at a much faster pace than reserves relative to non-life branches.

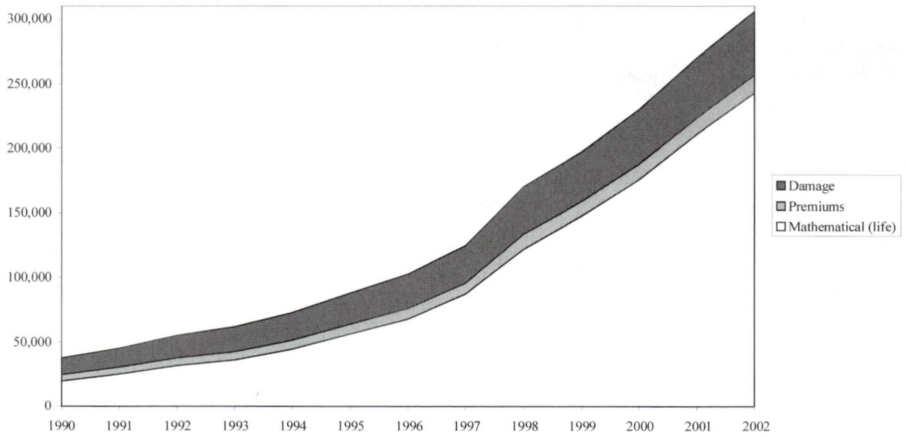

Source: ANIA.
Note: Values are expressed in millions of current euros.

Figure 7.1. Technical Reserves

Coming now to the investments of the insurance sector, we observe in Table 7.4 total investments (invested assets), disaggregated between life and non-life branches. From the very beginning of our reference period (1990), the rate of growth of investments among life classes is somewhat higher than the investment growth within non-life classes. Moreover, while the rate of growth of investment in the non-life branches stays very much the same in the period under examination, this is not the case for life insurance, where an acceleration started in 1997. This is due to the rapid increase in a specific kind of investment linked to life and pension funds. We can check this by looking at the details reported in Table 7.4, where total investments for life branches are divided into their main components: real estate, fixed income assets (bonds), stocks, deposits, and investments for the benefit of the insured (over and above investment of technical reserves), which is the component most rapidly growing since 1997. It is also worth recalling, in this context, the role played by the above-mentioned EEC Directives n. 92/49 (for non-life classes) and 92/96 (for life classes), which established a new and unified discipline of investments by insurance companies and which include a rule whereby "member states do not make prescription as to investing in some specific categories of assets."

(mathematical reserves), reserves that are set aside for risks that will last over several periods and for losses that have already occurred but will give rise to cash outflows in the future for non-life policies.

The dynamics of investments are quite close to those of earnings. Total earnings started to grow at an increasing rate in 1997, which is when pension funds (and, more generally, investment funds linked to insurance products) began to be increasingly appreciated. The share of income pertaining to nontechnical accounts (which include most investment profits) is far larger than 50 percent (for some years it is almost 100 percent).

A clear picture of insurance profitability can be obtained by inspecting Table 7.5, reporting a time series of return on equity indices, for the entire period 1990 to 2003.

Table 7.4. Total Assets in the Italian Life and Non-life Insurance Industries, 1990 to 2003 (in thousands of euros)

	Real Estate	Fixed Income	Stocks	Miscellaneous Financial Investments	Deposits	Total
Non-Life Insurance						
1990	3,850,083	11,410,237	4,744,225	598,470	1,200,452	21,803,467
1991	5,317,853	13,186,487	5,544,062	470,957	1,269,709	25,789,069
1992	5,589,097	15,066,649	6,366,054	981,681	1,200,091	29,203,520
1993	5,579,852	18,255,047	5,992,863	1,142,144	1,184,546	32,154,452
1994	5,996,994	19,504,460	7,309,828	931,068	1,583,922	35,326,272
1995	6,305,784	21,618,834	8,263,517	864,962	1,573,747	38,626,844
1996	6,746,528	22,879,351	9,445,635	887,015	1,572,456	41,530,985
1997	7,080,159	24,526,900	10,052,369	973,005	1,522,205	44,154,638
1998	6,095,886	25,143,988	12,029,650	1,371,813	683,660	45,324,997
1999	6,148,724	25,220,620	14,336,193	1,631,808	766,401	48,103,746
2000	6,107,600	26,976,951	17,381,259	2,008,408	578,886	53,053,103
2001	5,915,200	29,542,538	21,333,590	2,336,616	615,575	59,743,519
2002	4,534,617	32,495,167	22,199,020	3,380,993	1,224,702	63,834,499
2003	3,665,795	35,667,059	23,687,338	3,258,985	1,207,200	67,486,377
Life Insurance						
1990	2,677,313	16,039,499	2,863,650	1,407,810	619,232	23,607,503
1991	4,742,056	21,000,584	3,023,029	1,737,671	850,450	31,353,788
1992	4,858,362	26,815,888	3,344,627	1,453,775	1,058,737	37,531,388
1993	6,620,564	33,444,613	5,448,620	1,008,434	965,206	47,487,437
1994	6,391,619	40,552,712	6,579,041	899,926	1,250,652	55,673,950
1995	6,309,451	49,897,793	7,700,837	1,054,708	1,412,045	66,374,834
1996	5,329,319	62,076,828	8,843,343	1,213,829	1,039,008	78,502,327
1997	5,341,972	78,792,317	12,497,121	1,147,361	1,543,741	99,322,512
1998	2,450,092	79,033,399	13,622,501	22,084,658	6,092,394	123,283,044
1999	2,231,548	88,095,095	18,210,787	41,092,451	7,143,208	156,773,089
2000	2,174,100	95,035,548	22,764,511	113,802,486	9,459,683	243,236,328
2001	2,096,800	109,076,807	18,582,558	78,409,034	10,002,877	218,168,076
2002	903,219	126,686,984	17,485,905	95,239,806	10,435,380	250,751,294
2003	822,517	148,690,593	19,784,851	117,656,471	10,944,578	297,899,010

Table 7.4 *(continued)*. Total Assets in the Italian Life and Non-life Insurance Industries, 1990 to 2003 (in thousands of euros)

	Real Estate	Fixed Income	Stocks	Miscellaneous Financial Investments	Deposits	Total
Total						
1990	6,527,395	27,449,736	7,607,875	2,006,280	1,819,684	45,410,971
1991	10,059,909	34,187,071	8,567,090	2,208,628	2,120,159	57,142,857
1992	10,447,458	41,882,537	9,710,681	2,435,456	2,258,776	66,734,908
1993	12,200,416	51,699,660	11,441,483	2,150,578	2,149,752	79,641,889
1994	12,388,613	60,057,172	13,888,869	1,830,995	2,834,574	91,000,222
1995	12,615,234	71,516,627	15,964,354	1,919,670	2,985,792	105,001,678
1996	12,075,847	84,956,179	18,288,978	2,100,843	2,611,464	120,033,311
1997	12,422,131	103,319,217	22,549,489	2,120,365	3,065,946	143,477,149
1998	8,559,912	104,177,387	25,652,151	23,456,470	6,776,054	168,621,974
1999	8,380,288	113,315,715	32,546,980	42,724,258	7,909,610	204,876,851
2000	8,281,700	122,012,498	40,145,770	115,810,894	10,038,569	296,289,431
2001	8,012,000	138,619,345	39,916,148	80,745,650	10,618,452	277,911,596
2002	5,437,837	159,182,151	39,684,925	98,620,799	11,660,082	314,585,794
2003	4,488,312	184,357,652	43,472,189	120,915,456	12,151,778	365,385,387

Source: Author and ANIA annual reports.

Table 7.5. Return on Equity, 1990 to 2003 (in thousands of euros)

	Non-life Insurance			Life Insurance		
	Profit[a]	Average Capital	Return on Equity (%)	Profit[a]	Average Capital	Return on Equity (%)
1990	–81,652	6,115,805	–1.34	634,622	4,072,466	15.58
1991	–548,064	7,695,256	–7.12	732,542	6,007,951	12.19
1992	–953,534	8,460,576	–11.27	626,617	7,830,052	8.00
1993	–177,713	8,976,069	–1.98	705,893	9,310,895	7.58
1994	–223,368	10,020,642	–2.23	491,305	11,633,268	4.22
1995	422,513	10,826,000	3.90	912,993	12,996,017	7.03
1996	472,558	11,241,073	4.20	947,853	13,597,725	6.97
1997	–129,579	11,568,131	–1.12	996,555	14,456,517	6.89
1998	–115,790	12,943,500	–0.89	1,297,133	14,822,638	8.75
1999	22,441	13,254,005	0.17	1,335,171	15,366,085	8.69
2000	171,057	14,411,617	1.19	1,835,517	17,344,773	10.58
2001	1,306,076	15,223,357	8.58	1,435,484	19,042,652	7.54
2002	1,967,973	15,923,369	12.36	1,483,471	19,744,666	7.51
2003	1,646,193	17,340,297	9.49	1,888,100	21,377,000	8.83
Average		11,714,264	1.00		13,400,193	8.60

Source: Author and ISVAP and ANIA Research and Statistical Offices.

[a] Profit is defined as revenues – costs, both of technical and nontechnical accounts.

Table 7.6. Non-life Insurance Underwriting Results: Combined Ratios and Loss Ratios, 1998 to 2003 (in percent)

	1998	1999	2000	2001	2002	2003
Combined Ratio[a]						
Accident	94.10	93.00	89.00	89.30	87.10	83.50
Sickness	106.90	102.50	102.90	101.80	101.10	101.60
Land Vehicles	72.70	70.20	70.00	66.50	65.70	65.80
Railway Craft	308.20	63.80	112.30	86.50	151.00	32.90
Aircraft	139.60	115.80	111.70	71.40	111.30	76.60
Ships	112.10	150.60	162.20	130.00	84.20	123.80
Goods in Transit	95.90	104.70	100.70	95.80	84.60	75.60
Fire and Natural Forces	88.90	82.50	101.70	84.10	92.80	91.70
Other Damages	111.80	97.10	101.60	100.50	95.00	86.60
Motor Liability	124.10	122.00	116.80	109.40	101.80	98.80
Aircraft Third-party Liability	109.90	101.40	150.80	78.30	31.50	18.10
Ships Third-party Liability	114.20	90.90	115.90	64.80	95.00	141.60
General Third-party Liability	145.10	133.90	131.00	135.50	132.90	121.50
Credit	93.00	94.90	93.30	117.30	106.00	109.40
Suretyship	66.00	61.90	68.20	74.50	80.10	117.70
Miscellaneous Financial Loss	83.10	96.40	173.90	84.30	71.90	81.80
Legal Expenses	76.30	71.50	67.10	66.60	71.80	71.10
Assistance	70.20	70.30	67.50	66.60	70.70	69.20
Loss Ratio						
Accidents	61.70	61.10	57.00	57.70	55.80	51.90
Sickness	79.50	75.70	76.40	75.90	75.70	76.50
Land Vehicles	47.80	45.70	45.40	41.90	41.80	41.70
Railway Craft	295.90	42.40	94.10	70.20	134.60	19.00
Aircraft	123.40	101.90	100.50	55.60	96.00	60.90
Ships	92.30	130.60	142.20	111.10	68.20	106.00
Goods in Transit	64.20	73.30	71.00	66.60	55.60	45.00
Fire and Natural Forces	56.80	50.80	70.30	53.10	62.90	61.10
Other Damages	82.30	67.80	72.70	71.80	66.60	57.50
Motor Liability	105.20	103.70	98.90	91.60	84.20	81.60
Aircraft Third-party Liability	88.60	84.80	129.60	68.10	21.30	10.70
Ships Third-party Liability	83.70	63.20	88.10	43.20	71.80	122.00
General Third-party Liability	113.10	103.20	101.00	105.80	103.60	93.20
Credit	66.90	66.30	67.80	91.00	78.40	80.10
Suretyship	36.80	32.50	38.90	45.80	50.00	87.80
Miscellaneous Financial Loss	46.50	59.10	141.30	52.10	41.50	52.90
Legal Expenses	32.80	29.60	25.70	27.70	33.10	32.40
Assistance	40.30	36.80	36.50	35.20	39.30	37.90

Source: Author and ANIA Infobila system.
[a] Combined ratio = loss ratio + expense ratio (operating expenses/premiums).

Table 7.7. Technical Results by Line of Business, Non-life Insurance, 1998 to 2003 (in thousands of current euros)

	1998	1999	2000	2001	2002	2003
Accident	188,283,142	200,310,391	305,627,831	286,389,811	320,703,000	418,255,000
Sickness	–61,008,537	–23,876,835	–17,987,161	–10,858,506	–22,003,000	–39,165,000
Land Vehicles	717,213,508	754,654,051	800,472,558	917,885,419	943,378,000	971,767,000
Railway Craft	–1,038,078	3,871,361	–2,067,893	–1,015,354	–4,848,000	5,567,000
Aircraft	–10,721,645	–6,369,979	–10,510,414	–3,634,307	–17,140,000	1,178,000
Ships	–28,528,563	–14,347,689	–6,138,607	–11,571,733	7,721,000	2,352,000
Goods in Transit	37,144,097	18,878,566	15,159,559	7,014,518	41,098,000	37,256,000
Fire and Natural Forces	99,399,361	266,511,902	208,682,312	276,844,655	117,057,000	10,044,000
Other Damages	–19,533,949	97,172,915	92,386,392	63,309,353	–42,918,000	165,392,000
Motor Liability	–1,473,025,973	–1,937,508,715	–1,310,552,764	–492,963,791	172,485,000	847,418,000
Aircraft Third-party Liability	–5,342,230	3,418,945	–6,515,620	–7,994,236	9,026,000	18,494,000
Ships Third-party Liability	196,770	1,177,005	2,941,739	8,341,295	1,662,000	–7,336,000
General Third-party Liability	–391,135,017	–319,949,181	–318,438,544	–449,231,254	–455,116,000	–413,031,000
Credit	23,345,918	6,700,512	13,733,105	18,148,295	15,039,000	16,946,000
Suretyship	85,537,658	88,737,108	98,411,379	94,425,881	77,497,000	13,186,000
Miscellaneous Financial Loss	31,339,121	25,944,729	1,990,425	29,683,877	35,888,000	36,511,000
Legal Expenses	23,570,576	28,839,986	38,979,585	44,815,031	42,898,000	46,417,000
Assistance	35,847,790	35,876,195	47,195,897	51,950,400	50,511,000	57,585,000
Total	–748,456,052	–769,958,735	–46,630,222	821,539,351	1,292,938,000	2,188,836,000

Source: Author and ANIA Infobila system.
Note: Technical results are defined as those concerning the insurance operations of companies, strictly speaking.

Until 1998, profitability was negative for the non-life branches of insurance (except for a short interval), and it came back to positive values in 1999, substantially increasing over subsequent years.

To understand this change, we may look at Table 7.6, which features combined ratios and loss ratios for the main non-life branches, and Table 7.7, which reports the technical results of the same lines of insurance. Combined ratios and loss ratios have almost uniformly decreased over time, with a few exceptions, and technical results have improved quite a bit in recent years, particularly in the case of motor insurance. Looking back to Table 7.5, profitability was positive for the life insurance sector throughout the period, given an ever-increasing net income stream. Return on equity, however, was somehow reduced by a more-than-proportional increase in average net capital.

7.2.5 The Distribution Network for Italian Insurance

It is not possible to understand the dynamics of the Italian insurance system without examining the changes that have occurred in the insurance distribution network and without mentioning the increasingly active role played by banks and financial institutions—a role that has been steadily growing in Italy since the 1990s. This is particularly true in the case of life insurance, whose distribution has been characterized by a growing involvement of financial institutions. The percentage of premiums collected through financial and banking intermediaries was about 20 percent in 1992 and rose to about 70 percent (of which around 59 percent was collected by banking intermediaries) in 2003. The main factor in this growth was the spreading of investment funds linked to insurance contracts, which have been placed almost solely by banks and financial promoters. However, the performance of banking and financial intermediaries in selling life insurance products (about 80 percent of the whole set of contracts in 2001) also must be acknowledged. The role of agents and brokers, which is still important in the life insurance branch, dramatically dropped from about 56 percent in 1993 to a much smaller 19.1 percent in 2003. In the damage and liability insurance branch, the situation in terms of distribution channels has remained constant since the 1990s, as the share of agents and brokers went from about 92 percent in 1991 to about 93 percent in 2003.

These dynamics are evident from data presented in Table 7.8, which contains the breakdown of distribution channels (in percent), over the period 1991 to 2003, for life and non-life classes and for the whole market. This breakdown underscores the stability of the various distribution channels for the non-life market and a substantial change in the life market, where agents and brokers have been replaced, over time, by banks and financial advisors. Bank counters alone have accounted for about 60 percent of insurance policy sales in the last few years, almost exactly the percentage of agents at the beginning of the period under examination.

These data underscore the fact that the traditional boundary between banks and insurers was somehow blurred by a rapid process of financial innovation and deregulation that involved virtually all sectors of finance in Italy. As a consequence, a process of progressive integration has taken place, both in terms of participation and in terms of operations. As of 1995, for instance, 20 banks maintained control

Table 7.8. Market Share by Distribution System in the Italian Life and Non-life Insurance Industries, 1991 to 2003 (in percent)

	1991	1992	1993	1994	1995	1996	1997	1998	1999	2000	2001	2002	2003
Life Insurance													
Agents	59.9	55.5	52.7	48.1	44.6	44.0	35.3	30.9	27.2	27.0	17.9	19.6	18.3
Brokers	4.0	3.5	3.7	3.4	2.1	1.9	1.3	1.6	1.2	0.9	0.9	0.9	0.8
Direct Sales	21.0	21.1	21.5	19.5	18.9	17.9	15.5	13.1	9.2	8.6	8.8	8.9	10.9
Banks and Financial Advisors	15.1	19.9	22.1	29.0	34.4	36.2	47.9	54.4	62.4	63.5	72.4	70.6	70.0
Banks							32.7	41.7	49.8	54.1	61.2	56.3	58.9
Financial Advisors							15.2	12.7	12.6	9.4	11.2	14.3	11.1
Non-life Insurance													
Agents	77.0	75.6	75.5	75.4	78.6	80.5	88.3	88.8	88.5	88.2	86.8	86.1	85.2
Brokers	15.0	15.1	15.1	14.9	15.8	15.1	6.8	6.0	7.2	6.4	7.4	7.5	7.5
Direct Sales	6.0	7.2	7.1	7.1	5.1	3.9	4.4	4.7	3.8	4.7	4.8	5.3	6.0
Banks and Financial Advisors	2.0	2.1	2.3	2.6	0.5	0.5	0.5	0.5	0.5	0.7	1.0	1.1	1.3
Banks							0.3	0.3	0.4	0.6	0.9	1.0	1.2
Financial Advisors							0.2	0.2	0.1	0.1	0.1	0.1	0.1
Total													
Agents	72.4	70.1	68.6	66.2	66.0	66.5	64.2	58.9	53.3	52.2	44.9	44.3	41.9
Brokers	12.1	11.9	11.7	11.0	10.7	10.0	4.3	3.7	3.7	3.2	3.4	3.3	3.2
Direct Sales	10.0	11.0	11.4	11.3	10.2	9.3	9.5	9.0	6.9	7.0	7.3	7.6	9.1
Banks and Financial Advisors	5.5	7.0	8.3	11.5	13.0	14.2	22.0	28.4	36.1	37.6	44.4	44.8	45.8
Banks							15.0	21.8	28.8	32.0	37.5	35.8	38.6
Financial Advisors							7.0	6.6	7.3	5.6	6.9	9.0	6.2

Source: Author and ANIA annual reports, various years.

over 35 insurers, and 62 had a minority (but sometimes quite relevant) shareholding position in other insurance firms. Although insurers rarely control banks, it is quite common to find minority shareholding positions of insurance companies in banks. More recently, however, the most important Italian insurance groups have established their own banking branches. Also relevant is operational integration, which ranges from distribution activities to asset investment, assistance in import-export operations, and the design of joint products for both families and firms.

The role that banks (to which we add the Italian Mailing Company, *Poste Italiane*, a corporation that is 65 percent publicly owned) play in distributing insurance products, through their extended network of branches across the country, cannot be underestimated, nor can the importance, for banks, to be able to offer a wider range of products that can better satisfy all customers' needs. The relevance of an appropriate distribution system to create value for an insurance company is now widely recognized and is often discussed in applied research (e.g., Perissinotto 2003).

Commercial integration has been achieved mainly by agreements for insurance policy distribution. As of 1995, 109 banks declared that they had engaged in commercial agreements with 38 insurance companies.[177] Further evidence is given by the fact that 149 banks in a sample of 207 surveyed by the Bank of Italy in the mid-1990s offered life insurance products at their counters, while 92 of them offered non-life insurance products (automobile insurance, fire and theft insurance, etc.). In other words, this integration process seems robust and may possibly lead to more and more cooperation between the two sectors.

7.2.6 Accounting Figures, Solvency, and the Top Italian Insurance Companies

To conclude our presentation of the main features of the Italian insurance market, we present in Tables 7.9, 7.10, 7.11, 7.12, and 7.13, respectively, the balance sheets and income statements for the aggregated life and non-life branches, drawn according to the new accounting rules, and the main features of the top 10 insurers in terms of the group to which each company belongs, the type of control, the main distribution system, and the amount of premiums earned.The five tables illustrate many of the features of the market that have been outlined in the previous paragraphs.

A look at the balance sheet is also useful for understanding the performance of the Italian insurance industry in terms of solvency levels. Table 7.14 shows the number of companies that have undergone bankruptcy procedures in the last 40 years. The number of companies being seriously insolvent has been quite small along the years, except for the periods 1982 to 1987 and 1993 to 1994 (the latter being characterized by an important economic recession). What the table does not reveal is that most failures concerned minor companies, mostly operating in the sector of motor liability insurance. The role of technical reserves in these failures should be stressed, especially in view of the results of a research conducted by ISVAP (2001) for the period 1986 to 1998, showing that for some of those years loss reserves were not sufficient and

[177] Bank of Italy, 1995.

Table 7.9. Balance Sheet for the Italian Life Insurance Industry, 2003

Assets		Liabilities	
Subscribed Share Capital Unpaid	10,560,000	Capital and Reserves	22,605,857,000
		Subscribed Share Capital or Equivalent Funds	5,951,985,000
Intangible Assets	2,120,164,000	Share Premium Reserve	7,431,598,000
		Revaluation Reserves	792,093,000
Investments	190,782,914,000	Legal Reserves	599,241,000
Land and Buildings	822,517,000	Statutory Reserves	145,113,000
Investments in Group Undertakings and Participating Interests:	16,132,925,000	Reserves for Own Shares and Holding Company's Shares	207,915,000
Shares and Participating Interests	14,645,029,000	Other Reserves	5,508,190,000
Debt Securities Issued by Firms	1,231,700,000	Profit (Loss) Brought Forward	94,912,000
Corporate Financing	256,196,000	Profit (Loss) for the Financial Year	1,874,811,000
Other Financial Investments:	162,908,859,000		
Shares and Participating Interests	5,139,822,000	Subordinated Liabilities	1,657,693,000
Units and Shares in Investment Funds	7,714,704,000		
Debt Securities and Other Fixed-income Securities	147,458,893,000	Technical Reserves	
Loans	1,669,794,000	Life Insurance Business	177,597,689,000
Participation in Investment Pools	0	Mathematical Reserves	173,114,290,000
Deposits with Credit Institutions	25,965,000	Ancillary Risks—Reserves for Unearned Premiums	63,716,000
Sundry Financial Investments	899,682,000	Reserve for Outstanding Claims	2,350,349,000
Deposits with Ceding Undertakings	10,918,613,000	Reserve for Bonuses and Rebates	146,599,000
		Other Technical Reserves	1,922,735,000
Investments for the Benefit of Life Insurance Policyholders Who Bear the Risk Thereof and Investments Arising from Pension Funds Management	107,116,095,000	Technical Reserves Where Investment Risk Is Borne by Policyholders and Arising from Pension Funds Management	106,883,613,000

Table 7.9 *(continued)*. Balance Sheet for the Italian Life Insurance Industry, 2003

Assets		Liabilities	
Technical Reserves—Reinsurers' Share	12,385,160,000		
		Reserves Relative to Investments Fund and Index-linked Contracts	106,470,557,000
Credits	6,727,490,000	Reserves Arising from Pension Funds Management	413,056,000
Credits Arising from Direct Insurance Operations	2,303,659,000		
Credits Arising from Reinsurance Operations	954,943,000	Provisions for Other Risks and Charges	968,737,000
Other Credits	3,468,887,000		
		Deposits Received from Reinsurers	11,315,163,000
Other Assets	4,996,129,000		
		Debts and Other Liabilities	5,395,492,000
Prepayments and Accrued Income	2,447,300,000	Debts Arising from Direct Insurance Operations	819,365,000
		Debts Arising from Reinsurance Operations	316,553,000
Total Assets	326,585,813,000	Debenture Loans	750,000,000
		Amounts Owed to Credit and Financial Institutions	111,932,000
		Debts Secured by a Lien on Property	3,000
		Sundry Loans and Other Financial Debts	789,284,000
		Staff Leaving Indemnity	120,311,000
		Other Debts	1,692,512,000
		Other Liabilities	795,534,000
		Accruals and Deferred Income	158,573,000
		Total Liabilities	326,582,818,000

Source: ANIA and ISVAP annual reports for 2003.
Note: Units are millions of euros.

Table 7.10. Profit and Loss Account for the Italian Life Insurance Industry, 2003

Written Premiums, Net of Reinsurance	62,891
Investment Income, Net of Charges	7,632
Investment Income and Unrealized Gains on Investments for Life Insurance Policies Where Investment Risk Is Borne by Policyholders and Arising from Pension Funds Management, Net of Charges and Unrealized Losses	4,313
Other Technical Income	–24
Claims Incurred	–25,349
Change in Mathematical Reserves and Other Reserves, Class C	–20,267
Change in Technical Reserves, Class D	–22,785
Operating Expenses	–3,818
Allocated Investment Return Transferred to the Nontechnical Account	–867
Balance on the Technical Account	1,726
Allocated Investment Return Transferred from the Technical Account	867
Other Income, Net of Charges	–450
Balance on Ordinary Activities	2,143
Extraordinary Income, Net of Charges	632
Tax on Profit	–887
Profit (Loss) for the Financial Year	1,888

Source: ANIA and ISVAP annual reports for 2003.
Note: Units are millions of euros.

Table 7.12. Profit and Loss Account for the Italian Non-life Insurance Industry, 2003

Earned Premiums	30,788
Allocated Investment Return Transferred from the Nontechnical Account	1,735
Other Technical Income	–490
Claims Incurred	–22,645
Changes in Other Technical Reserves	–3
Operating Expenses	–7,163
Change in the Equalization Reserves	–2
Balance on the Technical Account	2,220
Net Investment Income	2,209
Allocated Investment Return Transferred to the Technical Account	–1,735
Other Income, Net of Charges	–485
Balance on Ordinary Activities	2,209
Extraordinary Income, Net of Charges	450
Tax on Profit	–1,010
Profit (Loss) for the Financial Year	1,649

Source: ANIA and ISVAP annual reports for 2003.
Note: Units are millions of euros.

Table 7.11. Balance Sheet for the Italian Non-life Insurance Industry, 2003

Assets		Liabilities	
Subscribed Share Capital Unpaid	33,971,000	Capital and Reserves	18,028,230,000
		Subscribed Share Capital or Equivalent Funds	4,282,313,000
Intangible Assets	1,315,472,000	Share Premium Reserve	4,465,936,000
		Revaluation Reserves	1,047,528,000
Investments	67,486,378,000	Legal Reserves	666,538,000
Land and Buildings	3,665,795,000	Statutory Reserves	187,850,000
Investments in Group Undertakings and Participating Interests	20,090,849,000	Reserves for Own Shares and Holding Company's Shares	106,896,000
Shares and Participating Interests	19,669,402,000	Other Reserves	5,527,448,000
Debt Securities Issued by Firms	252,904,000	Profit (Loss) Brought Forward	97,356,000
Corporate Financing	168,542,000	Profit (Loss) for the Financial Year	1,646,366,000
Other Financial Investments	42,643,364,000		
Shares and Participating Interests	4,017,936,000	Subordinated Liabilities	1,008,223,000
Units and Shares in Investment Funds	1,482,715,000		
Debt Securities and Other Fixed-income Securities	35,414,155,000	Technical Reserves	
Loans	227,127,000	Non-life Insurance Business	61,782,232,000
Participation in Investment Pools	0	Reserve for Unearned Premiums	13,535,022,000
Deposits with Credit Institutions	120,831,000	Reserve for Outstanding Claims	48,052,692,000
Sundry Financial Investments	1,380,601,000	Reserve for Bonuses and Rebates	20,637,000
Deposits with Ceding Undertakings	1,086,369,000	Other Technical Reserves	62,377,000
		Equalization Reserves	111,504,000

Table 7.11 *(continued)*. Balance Sheet for the Italian Non-life Insurance Industry, 2003

Assets		Liabilities	
Technical Reserves—Reinsurers' Share	8,268,277,000		
		Provisions for Other Risks and Charges	1,167,287,000
Credits	12,786,120,000		
Credits Arising from Direct Insurance Operations	8,014,447,000	Deposits Received from Reinsurers	1,490,186,000
Credits Arising from Reinsurance Operations	1,468,448,000		
Other Credits	3,303,224,000	Debts and Other Liabilities	11,648,350,000
		Debts Arising from Direct Insurance Operations	698,439,000
Other Assets	4,885,644,000	Debts Arising from Reinsurance Operations	844,980,000
		Debenture Loans	1,750,000,000
Prepayments and Accrued Income	502,941,000	Amounts Owed to Credit and Financial Institutions	68,460,000
		Debts Secured by a Lien on Property	5,524,000
Total Assets	95,278,802,000	Sundry Loans and Other Financial Debts	2,753,072,000
		Staff Leaving Indemnity	453,039,000
		Other Debts	2,299,139,000
		Other Liabilities	2,775,700,000
		Accruals and Deferred Income	157,297,000
		Total Liabilities	95,281,806,000

Source: ANIA and ISVAP annual reports for 2003.
Note: Units are millions of euros.

Table 7.13. Top 10 Italian Insurers in Life and Non-life Branches, 2003

	Company	Group	Control	Main Distribution System	Premiums (in thousands of euros)
Life Insurance					
1	Creditras Vita	Allianz/Unicredito Italiano	Subsidiaries of joint ventures	Bancassurance	5,320,092
2	Sanpaolo Vita	San Paolo-IMI	Companies owned by banks and financial entities	Bancassurance	4,662,376
3	Poste Vita	Poste Italiane	Companies owned by sate and Italian state holding companies	Bancassurance	4,489,486
4	Intesa Vita S.p.A.	Generali/Banca Intesa	Subsidiaries of joint ventures (bancassurance)	Bancassurance	3,506,168
5	Alleanza Assicurazioni	Generali	Subsidiaries of insurers	Direct sales/agents	3,258,832
6	Ina Vita	Generali	Subsidiaries of insurers	Agents	3,002,576
7	Generali Vita	Generali	Subsidiaries of insurers	Agents	2,946,080
8	Montepaschi Vita	Monte Dei Paschi di Siena	Companies owned by banks and financial entities	Bancassurance	2,389,900
9	Finecovita	Capitalia	Companies owned by banks and financial entities	Bancassurance	2,280,312
10	Fideuram Vita	San Paolo-IMI	Companies owned by banks and financial entities	Financial promoters/SIM	2,148,149
Non-life Insurance					
1	Fondiaria-SAI	Premafin	Subsidiaries of insurers	Agents	3,785,551
2	Riunione Adriatica di Sicurta'	Allianz Aktiengesellschaft	Subsidiaries of EU foreign entities	Agents	2,799,956
3	Assicurazioni Generali	Generali	Subsidiaries of insurers	Agents	2,766,596
4	Compagnia di Assicurazione di Milano	Premafin	Subsidiaries of insurers	Agents	2,508,564
5	Assitalia-LE Assicurazioni d'Italia	Generali	Subsidiaries of insurers	Agents	1,917,262
6	Winterthur (Aurora)	Holmo	Subsidiaries of insurers	Agents	1,333,379
7	Compagnia Assicuratrice UNIPOL	Holmo	Subsidiaries of insurers	Agents	1,323,145
8	Lloyd Adriatico	Allianz Aktiengesellschaft	Subsidiaries of EU foreign entities	Agents	1,296,503
9	AXA Assicurazioni	AXA	Subsidiaries of EU foreign entities	Agents	1,123,049
10	Societa' Reale Mutua di Ass.NI	Reale Mutua	Subsidiaries of insurers	Agents	1,038,929

Source: Author, ISVAP, and ANIA.

Table 7.14. Bankruptcy Procedures in the Italian Insurance Industry as of May 2005

Year	Ended (as of May 2004)	Ongoing	Overall
1964		1	1
1974	1		1
1976		2	2
1977		1	1
1978		3	3
1979		3	3
1980		2	2
1981		4	4
1982	12	4	16
1983	6	2	8
1984	12	6	18
1985	4	3	7
1986	6	3	9
1987	8		8
1988	3	1	4
1989		1	1
1990		1	1
1991	1	1	2
1992	1	3	4
1993		6	6
1994	2	6	8
1995		1	1
1996	1		1
1997		3	3
1998		3	3
Total	57	60	117

Source: Author and ISVAP.

needed to be integrated with those of subsequent years. The most serious case of bankruptcy occurred in 1993 and concerned the life insurer Tirrena, whose portfolio was voluntarily transferred to Praevidentia (later called Nuova Tirrena). No serious insolvency cases have been reported since 1998.

Despite the number of bankruptcy cases, the consequences on policyholders have not been very serious. Almost all failures occurred in the motor liability branch, which is compulsory and for which some specific guarantees are provided. In particular, a specific fund, the road accident victims fund, established by Law n. 990 of 1969 and funded by specific contributions paid by the insurance industry (4 percent of premiums of motor liability insurance) provides a guarantee for the case of bankruptcy of the insurer involved in the accident. Moreover, the bankruptcy procedures adopted in the case of motor liability insurance may oblige other insurers to acquire the portfolio of the insolvent company (a provision that, to date, has never been used).

All in all, therefore, Italian policyholders are in a safe position, and the insurance industry is quite robust. The industry's solvency performance rests on three main pillars: a proper evaluation of technical reserves (which has not always been the case for motor insurance), a correct investment mix to cover those reserves (asset liability management), and an additional solvency margin. The latter is a peculiar feature of the European insurance system, making it quite different from the U.S one (for a thorough and analytical comparison of the two regimes, see an interesting work by ISVAP, Quaderno n. 6).

In fact, the first two EEC insurance directives (73/239/CEE for damages and 79/267/CEE for life, respectively) introduced such a solvency margin, which makes reference to insurers' so-called free capital—that is, that portion of net capital that is not already explicitly covering liabilities. In other words, this solvency margin can be considered a complementary resource with respect to technical reserves, and it constitutes a further element of guarantee for policyholders.

The laws that introduced those EEC directives in the Italian system also have fixed the minimum amount of these complementary guarantees (minimum capital requirements). For life insurance, the measure of the solvency margin is a function of mathematical reserves (4 percent) and capital under risk (0.3 percent), where the latter is defined as the difference between the insured capital and the mathematical reserves. For damage insurance, the solvency margin is a function of either premiums or average losses (of the last three or seven years), and the highest amount is selected.

Also in terms of this additional guarantee, Italian insurers fare quite well; in fact, both in 2003 and in 2004 only two companies fell short of the required margin and were promptly required to bring fresh capital by ISVAP, whereas the average solvency ratio (defined as the ratio between the actual margin and the minimum required) hovered around two in both 2003 and 2004.

This image of stability of the Italian insurance market was reinforced by an initiative taken by ISVAP that asked the top 10 insurers to implement specific stress tests on their accounting statements, under alternative (negative) assumptions about the financial and economic operating conditions and possible catastrophic events (relative to damage insurers). The results were quite encouraging, as the solvency ratio for this group of companies (representing most of the Italian insurance system) would only marginally fall.

The solvency rules will change in the next few years (not before 2008, in any case), as the Solvency II project has been launched by the European Union (and overseen by CEIOPS), with the declared goal of devising new rules concerning the criteria to be adopted by insurers. This has been done also in view of the new accounting criteria to be implemented by insurers in drawing their statements of accounts and that might lead to a reduction of solvency margins computed according to the present methodologies.

7.3 A MICRO PERSPECTIVE OF THE ITALIAN INSURANCE MARKET

Having looked in the previous sections at the main features of the Italian insurance market, we now look at it from a different, micro perspective to assess the role that various social, demographic, and economic factors have played in the formation of insurance demand over the period under scrutiny. The next two sections present a view of the Italian insurance market based on survey observations at two points in time, 1989 and 2000, and a more specific microeconometric analysis of insurance demand, which is offered as a useful complement to the descriptive analysis (both macro and micro).

7.3.1 A Preliminary Look at Micro Data

The following sections illustrate and comment upon some figures representing the distribution of insurance holders across several social, geographic, and economic dimensions; we make a comparison both between insured and noninsured households and with respect to the composition of the insurance mix within the subgroup of insured households. The comments and comparisons are based on survey data on Italian households collected by the Bank of Italy every two years in the course of its Survey on Household Income and Wealth, an investigation run on a very large sample of families. In 2002, when the last survey (whose data have recently been released) was conducted, it concerned 21,000 individuals and more than 7,000 households. The survey contains many interesting variables, of both a demographic and an economic nature: for example, there are variables concerning the different kinds of assets owned by the households, information about savings and insurance, income sources and utilizations, working conditions, and more.[178] In particular, the survey has a section devoted to insurance, where a large amount of data concerning insurance holdings are collected. It should be noted, however, that no account is taken in the survey of automobile liability coverage (which is compulsory in Italy and therefore only depends on the fact of owning a motor vehicle).

To obtain results that can be taken to represent more closely the situation of insurance holdings in Italy, data are weighted, whenever possible, by using the sampling weights provided by the survey itself. In particular, all summary statistics are weighted (weights are contained in the variable PESOFL), as are cross-sectional regressions. (Panel regressions are not weighted due to the fact that the weights change over time in the survey, and this is not allowed by the statistical procedure we use).

In particular, we use data extracted from the 1989 and 2000 waves of the Bank of Italy survey.[179] These two specific waves of the survey were chosen because (1)

---

[178] For a thorough description of the survey, including its features in terms of sampling techniques and reliability, see Brandolini and Cannari (1994) and, more recently, Biancotti, D'Alessio, and Neri (2004), where a Heise reliability index for the main variables in the 1995, 1998, and 2000 waves of the survey is computed.

[179] At the time this essay was written, the 2000 wave was the most recent one publicly available. The 2002 wave was released in 2004.

they are sufficiently far apart in time to describe the evolution of the sector over most of the period under examination and (2) because the samples are close enough to attribute possible differences in insurance patterns to changes in the phenomenon under analysis rather than to changes in the sample itself.[180] For example, the 1989 wave surveyed 8,274 families, randomly extracted from the General Registry lists of 294 towns, for a total of 25,150 individuals, among whom 13,864 were income earners. The 2000 wave, on the other hand, surveyed 8,001 families, drawn from the lists of 333 towns and totaling 22,268 individuals, with 13,814 income earners.

The structure of the two samples in terms of gender, age, importance of towns, geographic area, and education is quite comparable and also is very similar to the actual structure of the Italian population (at least for 1989). For example, we observe the following distribution of frequencies among five classes of education (no education, elementary education, middle school, high school, university): 8.3 percent, 25 percent, 35 percent, 23.2 percent, and 8.5 percent for 1989 and 7.8 percent, 36.4 percent, 25.4 percent, 22.4 percent, and 8 percent for 2000.

Let us now look at the data, considering both the distribution of insured people across the entire population and the distribution of various types of insurance across insured households. First, Table 7.15 reports weighted summary statistics about the distribution of insurance for 1989 and 2000, reflecting the survey sample. The overall percentage of insured households rose from about 25 percent in 1989 to 40 percent in 2000.

Table 7.15. Percentage of Insurance Holders in Italian Households, 1989 and 2000

	1989	2000
Total	25.7	40.0
Life	14.0	19.8
Damage	11.4	20.7
Health	4.4	8.3
Pensions	5.4	12.0

Source: SHIW, Bank of Italy, 1989 and 2000.
Note: Totals are not the sum of partials, because the same household might hold different types of policies.

The distribution of insurance policies across households (both insured and noninsured), as reflected in the SHIW sample for the two selected years, 1989 and 2000, features many important changes, which we will rapidly sketch in the following paragraphs.

[180] With respect to this point, it may be instructive to read the introductory pages of the "Supplementi al Bollettino Statistico—I Bilanci delle famiglie Italiane" for the years 1989 and 2000 and to look at Tables A1 through A4 in Appendix B therein.

In terms of the geographic distribution of insurance policies, there was an increase of the percentage of policyholders in all areas, with a relatively bigger increase in the north and a much smaller increase in the south and isles.[181]

As for marital status, there was a large increase in the percentage of married couples holding at least one insurance policy and a remarkable increase in the insurance holdings among singles and widowers. With respect to the latter, this finding might be linked to the aging of the Italian population, which has progressed at a fairly high rate in the last few years.

Both employees and self-employed workers have increased their recourse to insurance instruments in the last decade, and both categories of workers have overcome those of retired people, who represented the most frequently insured group of people in 1989.

A larger percentage of people living in smaller towns have some sort of insurance than people living in larger cities, and this trend was reinforced in the last few years, as the percentage of insured people living in medium-sized to quite large towns (from 20,000 to 500,000 residents) has gone up by more than 15 percent (but it would be interesting to understand whether this pattern is due to people leaving large urban areas and moving to smaller towns, "exporting" their previous insurance purchasing behavior).

As for age distribution, all age groups of households increased their rate of utilization of insurance instruments, and this is particularly so with people aged 60 and over. This fact, coupled with the aging of the Italian society, should account for a relatively large share of the increase in overall insurance.

Let us now take into consideration only those households possessing at least one insurance policy, and examine the distribution of insurance over several dimensions of interest. This can be regarded as a complementary, but extremely useful, point of view to draw a complete figure of the dynamics of the insurance sector in Italy over the last decade.

First, there was a relatively important change in the household insurance mix that occurred in a relatively short spell of time (see Table 7.15). In particular, we observe a large increase in the percentage of households with damage and liability insurance policies, from 11.4 percent in 1989 to 20.7 percent in 2000. Likewise, there was an impressive increase in the holding of insurance products related to pension funds, from about 5.4 percent in 1989 to 12.0 percent in 2000. The percentage of households with health insurance also nearly doubled (from 4.4 percent to 8.3 percent) over the same period, whereas the percentage with life insurance increased from 14.0 percent to 19.8 percent.

Some descriptive statistics on insurance ownership reveal a remarkable distribution of insurance holdings across towns (see Table 7.16). In 1989, most

---

[181] The three areas are defined according to the geographic divisions of the Italian Statistical Office. Northern Italy includes Piemonte, Valle d'Aosta, Lombardia, Liguria, Trentino Alto Adige, Veneto, Friuli Venezia Giulia, and Emilia Romagna. Central Italy includes Toscana, Umbria, Marche, and Lazio. Southern Italy includes Abruzzo, Molise, Campania, Puglia, Basilicata, Calabria, and the two major islands, Sicilia and Sardegna.

Table 7.16. Distribution of Insured Households in Italy, 1989 and 2000

	1989	2000		1989	2000
Place of Residence (in terms of population)			**Employment Type**		
Less than 20,000	0.2	0.5	Dependent Worker	0.7	0.4
20,000–40,000	0.3	0.1	Independent Worker	0.2	0.2
40,000–500,000	0.1	0.3	Not Working (mostly retired)	0.0	0.4
More than 500,000	0.4	0.1			
Employment Sector			**Home Ownership**		
Agriculture	0.1	0.1	Yes	0.7	0.7
Industry	0.3	0.4	No	0.3	0.2
Commerce	0.2	0.1			
Services	0.1	0.2			
General Government	0.3	0.3			
Other	0.1	0.0			
Geographic Area of Residence			**Number of Household Components**		
North-West	0.3	0.4	1	0.1	0.1
North-East	0.3	0.3	2	0.2	0.2
Center	0.2	0.2	3 or more	0.7	0.7
South	0.1	0.1			
Isles	0.1	0.1			
Number of Income Earners			**Income Classes**		
1	0.4	0.3	0–100,000	1.0	0.9
2	0.5	0.5	100,000–200,000	0.0	0.1
3 or more	0.2	0.2	200,000–300,000	0.0	0.0
Gender			**Education**		
Men	0.9	0.7	None	0.0	0.0
Women	0.1	0.3	Elementary School	0.3	0.2
			Middle School	0.3	0.3
			High School	0.3	0.1
			University Degree	0.1	0.4
Age			**Civil Status**		
< 30	0.1	0.0	Married	0.8	0.8
< 40	0.2	0.2	Single	0.1	0.1
< 50	0.3	0.3	Divorced	0.0	0.0
< 60	0.2	0.2	Widowed	0.1	0.1
60 and older	0.2	0.3			

Source: Author, on SHIW data.

policies (around 45 percent) were held by people living in very large cities; in 2000, most of them were held by households in relatively small towns (with less than 20,000 residents). This is quite important, because it highlights the role that distribution channels have had in the growth of the insurance sector in Italy and the fact that people residing in large cities may feel more protected and less prone to buy insurance instruments.

As for income classes, the largest share of policy holdings (more than 95 percent in 1989 and about 90 percent in 2000) come from households with a net income of less than 100 million lire (about 50,000 euros). This result, which is mostly driven by the breakdown of income (and savings) in the population, should be contrasted with the relative shares of insured versus uninsured households according to income classes.[182] Those statistics led us to conclude that a large majority of relatively wealthy people hold at least one insurance policy (apart from car insurance, whose data are not reported in SHIW), and the proportion is much lower among less affluent households. It is evident that measures aimed at increasing the distribution of insurance instruments should target lower and middle income classes.

Even more striking is the distribution of insurance across sectors and its changes over the decade under scrutiny. There has been a marked increase in the number of policies held by households in the manufacturing and service sectors, with a corresponding decrease for civil servants, merchants, and farmers (see Table 7.16). These changes do not seem to be determined by the relative importance of those sectors, as captured by the ratio of value added, at constant prices, to total value added in the economy. For example, the share of real value added in the manufacturing sector slightly decreased between 1989 and 2000 (from 25.02 percent to 24.8 percent), but the corresponding share of insurance policy holdings rose by about nine percentage points.

Households with three or more members are those in which we find the largest share of insurance policies, whereas holding insurance seems to be inversely related to the number of income earners in a household. The aging of the population is also reflected in the composition by age classes of the insured households, as more than one quarter of insurance holders are aged 60 years or older, against a smaller 16 percent in 1989 (see Table 7.16).

7.4 A MICROECONOMETRIC ANALYSIS OF INSURANCE HOLDINGS

This section presents a microeconometric analysis of insurance holdings in Italy, based on micro data from the Bank of Italy Survey on Household Income and Wealth. The aim of the analysis is to provide a quantitative assessment of the relevance of various economic and sociodemographic variables upon the decision to hold one or more insurance instruments (and consequently the quantity of insurance

---

[182] In terms of income breakdown, about 90 percent of households had an income lower than approximately 31,000 euros in 1989, and about 90 percent of households had an income lower than approximately 54,000 euros (percentages computed out of SHIW data).

held). We will use a very simple model, based on a relationship linking insurance demand to a certain number of economic, social, and demographic variables:

$$I_{ij}^{D} = I_j(\pi_j, p_x, W, H_i), \tag{1}$$

where index i makes reference to a specific household, and index j to a specific insurance product. Expression (1) says that insurance demand, I^D, is a function of insurance prices (unit premiums), π_j; the prices (p_x) of other goods and services (x), possibly including quantities of other real and financial assets; wealth or income W; and a vector of household characteristics, H_i. This relationship could be conceived of as the reduced form of a simple model à la Ehrlich and Becker (1972) or a simple mean-variance model such as the one used by Mayers and Smith (1983) and reformulated in Showers and Shotick (1994).

The dependent variable, I_{ij}, is given by the premiums paid by household i for a given amount of insurance product j, where j in turn indicates total insurance (Total), life insurance (Life), damage and liability insurance (Damage), health insurance (Health), and pension funds (Pensions). Disaggregating over different insurance products may help in ascertaining the effects of the different variables over insurance demand; in fact, if we examine only total insurance demand, the impact of some variables over the dependent one might become much less visible.

Unfortunately, we must consider prices fixed, because the data do not allow us to recover them in a sensible way; however, the second part of the analysis introduces a variable that is related to the cost of insurance, as it includes information on the fiscal treatment of premiums. Moreover, we will be able to account for the joint determination of insurance and other asset holdings, because we have information about whether a given household's portfolio also contains other financial assets. This is indeed relevant, because it has been shown that demand for insurance is determined together with demand for other assets (as is the case, for example, in Mayers and Smith 1983).

Because we are dealing with micro data representing individual decisions, and considering that these decisions are naturally bounded from below at zero, we use the Tobit regression technique (for a thorough discussion of this technique, see Verbeek 2000). This technique is particularly useful, because it allows us to account for the impact of variables on the probability of buying an insurance product and on the amount purchased.

The econometric analysis will proceed as follows: first, separate, cross-sectional, weighted Tobit regressions are performed using data drawn from two waves of the Bank of Italy Survey of Household Income and Wealth, conducted in 1989 and 2000. As was the case with the descriptive statistics, those years were chosen because they are relatively far apart and may provide some insight about whether the structure of insurance demand has changed over a decade that was rich in micro- and macroeconomic events. Several regressions will be run, one for every insurance product for each of the two years.

The estimated coefficients of variables in a Tobit regression have a twofold significance: (1) they express the impact of the corresponding variable on the probability of holding an insurance contract (as if the decision was of a zero-one type), and (2) they express the impact of the variable on the quantity of insurance purchased. In computing elasticities (evaluated at the mean of the relevant variables), therefore, it is the latter component of the coefficients that must be employed. To decompose the coefficients into the two parts, one can use the techniques discussed in McDonald and Moffit (1980). We use this decomposition technique, in some cases, to compute derivatives and elasticities.

The second part of the analysis includes a number of Tobit regressions for a panel data set consisting of the panel components from the 1989, 1991, 1993, and 1995 waves of the survey. The particular selection of waves was made taking into account two needs: First, it was necessary to rely on a sufficiently rich (in terms of number of observations) data set, and it was not possible to encompass more than four consecutive waves without losing too many observations. (Only a percentage of households in any given wave of the survey is, in fact, also interviewed in the following wave). Second, the four years were selected in such a way that they might encompass some remarkable economic events (for instance, a recession in 1993) and some important changes in the fiscal treatment of insurance premiums.

The goal of the panel regressions was to assess the robustness of the cross-sectional estimations and also to have the opportunity to estimate insurance demand in a dynamic way—that is, in a way that is based on first differences of the relevant variables.

7.4.1 Cross-sectional Tobit Estimations

Tables 7.17 and 7.18 contain cross-sectional estimations of demand for total insurance, life insurance, damage and liability insurance, health insurance, and pension funds insurance for 1989 and 2000, respectively. The type of insurance constitutes the dependent variable of the regression, and the coefficients of the independent variables are reported in the corresponding columns (p-values are computed in the last column). As is common in the literature (see, for example, Showers and Shotick 1994), insurance demand is proxied by the volume of premiums paid by households.

In what follows, a general-to-specific approach à la Hendry (see, for example, Hendry and Richard 1982) has been adopted. In other words, starting from the most general specification of an equation, a more parsimonious version of it has been estimated by deleting those variables that turned out to have statistically insignificant coefficients (at usual significance levels). A few exceptions have been made, when it seemed interesting to show that a particular coefficient was not significant or when its sign made it nevertheless noteworthy.

Estimation results for 1989 indicate that demand for total insurance is positively related to income, which also enters nonlinearly (income^2—that is, the square of income), implying that total insurance grows with income, but at decreasing rates. The positive relationship can be rationalized in terms of the increase in potential loss following an increase in income, if that increase is not perceived to be temporary.

Table 7.17. Cross-sectional Tobit Regressions for 1989

Regressors	Total	(P-value)	Life	(P-value)	Damage	(P-value)	Health	(P-value)	Pension	(P-value)
constant	–4,505.88	0.00	–6,216.83	0.00	–2,755.77	0.00	–5,048.57	0.00	–9,761.07	0.00
town size	50.29	0.06	117.53	0.00	76.47	0.00	61.11	0.09	–151.93	0.03
hhold members	46.58	0.13	112.04	0.00			–92.58	0.04		
income earners	–66.28	0.18			–81.69	0.00	136.40	0.02	–353.11	0.00
single	238.22	0.04	–319.58	0.05			497.19	0.00		
married					–227.08	0.00				
widower					–202.07	0.05	–600.87	0.03	–811.86	0.18
d_area1	184.25	0.02	–214.69	0.02	397.23	0.00				
d_area3	–755.03	0.00	–554.40	0.00	–573.75	0.00	–664.93	0.00	–1,257.64	0.00
dep	–546.50	0.00	–602.15	0.00	–169.24	0.00			–812.67	0.00
ind							165.19	0.15		
d_agri	–286.32	0.03	–724.07	0.00	156.60	0.03	–1,583.27	0.00	584.12	0.12
d_ind	–262.93	0.00	–344.44	0.00			–814.57	0.00	397.26	0.04
d_comm							–568.16	0.02		
d_serv	–245.35	0.04	–604.67	0.00			–533.77	0.03	949.34	0.00
d_gg	–311.12	0.00	–347.59	0.00	–144.36	0.00	–764.06	0.00		
education	305.26	0.00	337.51	0.00	69.29	0.00	200.47	0.00	1,336.16	0.02
edu^2									–136.23	0.08
age	113.93	0.00	134.51	0.00	35.03	0.00	107.60	0.00	205.20	0.00
age^2	–1.26	0.00	–1.48	0.00	–0.39	0.00	–1.19	0.00	–2.60	0.00

Table 7.17 *(continued)*. Cross-sectional Tobit Regressions for 1989

Regressors	Total	(P-value)	Life	(P-value)	Damage	(P-value)	Health	(P-value)	Pension	(P-value)
living parents	–198.92	0.00	–327.94	0.00						
banks	545.86	0.00	510.78	0.00	238.79	0.00	362.69	0.00	379.66	0.00
income	0.02	0.00	0.02	0.00	0.01	0.00			0.04	0.00
income^2	0.00	0.00	0.00	0.00	0.00	0.02			0.00	0.00
wealth	0.00	0.00	0.00	0.02	0.00	0.00	0.00	0.00		
wealth^2	0.00	0.00			0.00	0.00	0.00	0.00		
financial assets	–558.16	0.11							–1,562.29	0.03
home			–281.80	0.00	295.52	0.00	–298.41	0.00		
N	6,341.00		6,354.00		6,369.00		6,374.00		6,374.00	
Pseudo R^2	0.04		0.04		0.07		0.06		0.05	

Source: SHIW (1989).

Note: This table contains estimated coefficients (and corresponding P-values) of cross-sectional Tobit regressions, where the role of the dependent variable is played by the various types of insurance mentioned in the column head, and the meaning of regressors is fully specified below and in the text.

Town size is in terms of inhabitants, from 1 to 5. Single, married, and widower are dummy variables indicating the corresponding marital status. D_area1 and d_area3 are dummy variables indicating the area of residence (north and south and isles, respectively). Dep, ind, and un are dummy variables for dependent workers, independent workers, and unemployed, respectively. D_agri, d_ind, d_comm. d_serv, and d_gg are dummy variables for working sectors (agriculture, industry, commerce, services, and general government, respectively). Banks represents the number of banking relationships. Financial assets is a dummy variable indicating the holding of at least one financial instrument other than insurance. Home is a dummy variable for home-ownership.

Table 7.18. Cross-sectional Tobit Regressions for 2000

Regressors	Total	(P-value)	Life	(P-value)	Damage	(P-value)	Health	(P-value)	Pension	(P-value)
constant	–8,491.57	0.00	–14,588.60	0.00	–3,139.18	0.00	–7,090.10	0.00	–10,963.00	0.00
town size	–276.35	0.00	–216.85	0.00	–80.41	0.00	–69.02	0.10	–188.56	0.07
hhold members	221.57	0.00	481.00	0.00	–62.16	0.07				
income earners	–126.12	0.14	–439.37	0.00					39.53	0.82
single	–574.00	0.00	–1,034.33	0.00						
married										
divorced	–755.76	0.00								
widower					–218.35	0.09				
d_area1					878.24	0.00	1,057.82	0.00	3,865.64	0.00
d_area2	752.76	0.00	–338.81	0.14	221.26	0.05	526.87	0.00	3,113.76	0.00
d_area3										
dep			–500.55	0.02	–370.97	0.00	–693.45	0.00	153.78	0.70
ind	880.67	0.00							1,060.26	0.01
un							–465.60	0.00		
d_agri									–570.04	0.36
d_ind			–311.99	0.11						
d_comm					142.80	0.11	325.61	0.02		
d_serv							594.41	0.00		
d_gg					–180.14	0.03			332.05	0.22
education	353.94	0.00	554.47	0.00	31.71	0.22	122.99	0.00		
edu^2										

Table 7.18 *(continued)*. Cross-sectional Tobit Regressions for 2000

Regressors	Total	(P-value)	Life	(P-value)	Damage	(P-value)	Health	(P-value)	Pension	(P-value)
age	103.34	0.00	254.35	0.00	25.18	0.17	112.94	0.00	149.80	0.09
age^2	–1.33	0.00	–3.04	0.00	–0.28	0.09	–1.18	0.00	–2.44	0.00
living parents					164.66	0.06			–563.58	0.07
banks	998.16	0.00	827.72	0.00	186.92	0.00	510.78	0.00	606.65	0.00
income	0.03	0.00	0.02	0.00	0.01	0.00	0.01		0.05	0.00
income^2	0.00	0.00	0.00	0.00	0.00	0.02			0.00	0.00
wealth	0.00	0.00	0.00	0.02	0.00	0.00	0.00	0.00	0.00	0.04
wealth^2	0.00	0.00	0.00	0.00	0.00	0.00	0.00	0.00	0.00	0.05
financial assets	400.74	0.10	1,556.92	0.00						
home					156.93	0.09				
absolute risk av.					2,075.35	0.05			–6,181.44	0.04
N	7,459.00		6,055.00		2,495.00		6,045.00		6,374.00	2,495.00
Pseudo R^2	0.04		0.03		0.05		0.06		0.05	0.07

Source: SHIW (2000).

Note: This table contains estimated coefficients (and corresponding P-values) of cross-sectional Tobit regressions, where the role of the dependent variable is played by the various types of insurance mentioned in the column head, and the meaning of regressors is fully specified below and in the text.

Town size is in terms of inhabitants, from 1 to 5. Single, married, and widower are dummy variables indicating the corresponding marital status. D_area1 and d_area3 are dummy variables indicating the area of residence (north and south and isles, respectively). Dep, ind, and un are dummy variables for dependent workers, independent workers, and unemployed, respectively. D_agri, d_ind, d_comm. d_serv, and d_gg are dummy variables for working sectors (agriculture, industry, commerce, services, and general government, respectively). Banks represents the number of banking relationships. Financial assets is a dummy variable indicating the holding of at least one financial instrument other than insurance. Home is a dummy variable for home-ownership.

This is true, of course, of both life insurance and pension funds, which can be considered a tool to secure a constant stream of consumption for the household. At the same time, it is equally possible that more affluent households have more valuable goods to insure (cars, boats, houses, paintings, etc.). It is also possible that a higher income may trigger the decision to buy some insurance for households not already owning a policy.

The same (positive) relationship holds with respect to total wealth, which again appears with linear and quadratic terms. Quite unpredictably, the coefficients of the number of household members, and the number of income earners are not statistically significant, although their signs (positive and negative, respectively) point to the facts that households with more members need more insurance (because more people depend on the household's receipts) and that a greater number of income earners can substitute for insurance via intrahousehold transfers.[183] As a further, though only hypothetical, confirmation of this, the variable *living parents*, which is a dummy variable indicating whether the head of household has a living mother or father (not necessarily living with the household, though), displays a negative coefficient. These findings indicate that intrahousehold transfers play an important role in the Italian panorama.

That the holding of alternative forms of (financial) assets may substitute for holding insurance (when insurance itself is not compulsory) is indicated by the negative coefficient of the variable *financial assets*, which takes the value one when the household owns at least one financial asset (over and above insurance). In a sense, this may correspond to the fact that the household is also engaging in some form of precautionary saving, which can (indirectly) play the role of insurance.

Demand for total insurance also can be characterized in geographic terms; in fact, dummy variables *d_area1* and *d_area3*, denoting whether the household is in the north or in the south, feature a positive and, respectively, a much larger (in absolute value) negative coefficient. This is in perfect agreement with the qualitative results reported in the previous section (containing descriptive statistics). In other words, people living in southern Italy are less likely than people living in northern Italy to hold an insurance contract, and when they do, the amount of insurance purchased will be smaller.

The variable *age*, indicating the age of the household head, enters with both a positive (and linear) coefficient and with a quadratic (and negative) one, the latter being large enough to offset the positive coefficient of the linear term for all but the youngest heads of households. This is in line with the findings of earlier work (e.g., Duker 1969), indicating that the age of the head of household may be interpreted as a proxy for the maturity of a household. When the household has a young head, it is most likely in a phase in which insurance is badly needed (because more people depend on the household's income and the means to face unforeseen contingencies are scarce). When, on the other hand, the household has an older head, it may be that fewer people depend on its receipts and that it is relatively better off (in terms of both income and, above all, wealth). *Education*, a variable indicating the level of education attained by the head of household, also enters the demand function with a positive sign, indicating that more educated people tend to buy more insurance. This

[183] Multicollinearity does not appear to be a problem for the estimates, as the empirical correlation of all variables involved (including income and wealth) is always lower than 0.6.

finding might be considered as fairly counterintuitive, inasmuch as a higher level of education provides better insurance for future consumption (because it is usually associated with better prospects of future income and wealth) and might be thought to induce people to consume more today than tomorrow (when more and maybe better consumption opportunities will be available). On the other hand, a higher level of education also means less "defective recognition of future utilities" (Becker and Mulligan 1997), which should bring about a greater desire to secure (better) income prospects for the future. In fact, education levels are often used as a proxy for future labor income.

The variable *banks* indicates the number of (different) banking relationships owned by the household. The coefficient is positive, quite large, and extremely significant, which might indicate that the financial sophistication of the household leads to a greater involvement in insurance relationships. On the other hand, a large and positive coefficient might signal the role of intermediation played by commercial banks with respect to insurers. This confirms the importance of banking and insurance integration described earlier in this chapter.

Among other control variables, an interesting role is taken by *dep*, a dummy variable that takes the value one if the head of household is a dependent worker.[184] The coefficient of this dummy variable is highly significant, negative, and large in absolute value. In other words, employees tend to buy less insurance than the self-employed. This could be a consequence of the fact that, in Italy, the degree of protection provided by the publicly funded retirement plans and by the public health care system has been, and in some respects still is, much higher than it is in other countries, such as the United States. For instance, in Italy almost everyone has access to the public health system, whereas in the United States, the covered population is less than 50 percent and private insurance plays a major role (Henriet and Rochet 1998). Moreover, in government-operated medical care facilities, diagnostic and therapeutic treatments are generous and user fees are very limited and mostly proportional to income. Moreover, income or job losses due to illness are less likely to occur in Italy than they are in other countries. In fact, in Italy employees with serious health problems can keep their jobs for a prolonged period of time and the probability of getting fired is extremely low. Earnings losses are also very limited, since the social security system helps employers pay wages to ill workers. This is only partly true for self-employed workers, which might explain the sign and magnitude of this dummy variable's coefficient.

The results for the 2000 wave of the SHIW are similar to those just described, but with a few exceptions that are worth noting. Fewer variables were significant in this estimation. In particular, the coefficient of income in total demand estimation was substantially higher for 2000 than was the case in 1989 (0.029 compared to 0.018). The derivative of total demand for insurance with respect to income also increased (from about 0.015 to 0.024) over the decade.

---

[184] As is done in SHIW, dependent workers are defined as employees (i.e., blue collar workers or similar, office workers, teachers, junior managers, managers, senior officials, headmasters, magistrates) and as independent, or autonomous, workers, the self-employed (i.e., members of the arts and professions, sole proprietors, freelancers, owners or members of family businesses, active shareholders, partners, contingent workers).

The coefficient of the variable related to the number of banking relationships was larger and more significant in 2000 than it was in 1989, and this again stresses the intermediation role played by banks. On the other hand, a couple of important variables had different signs. One is *town size*, a variable representing the importance (in terms of inhabitants) of the place of residence. It takes a value from one to four, indicating towns with a population ranging from 0 to, respectively, 20,000, 40,000, and 500,000 and more. The sign of this variable was relatively small and positive in 1989 and was large and negative in 2000, reflecting the evidence from the descriptive statistics reviewed earlier in this chapter (see also Table 7.2). The variable *financial assets*, indicating whether a household owns at least one financial instrument other than insurance policies in its portfolio, became positive, although it is only significant at the 10 percent level. This might indicate that people believe that (risky) assets can expose the holders to asset-specific shocks that are not perfectly diversified. For instance, the dispersion of consumption flows across households is significantly affected by stockholding, and in some periods between 1989 and 2000 (for example, 1992, 1993, 1994, and 1995), Italian financial markets were occasionally struck by turmoil that markedly increased interest rate volatility. Financial assets, therefore, might not be viewed in more recent years as additional insurance instruments.

The life insurance column of Table 7.17 contains the estimated coefficients of the life insurance demand in 1989. With respect to total demand, there are some interesting differences. First, the number of members of the household, which was barely significant and had a relatively low value in the total demand equation, is extremely significant and large. Life insurance demand increases with the number of people depending on household income. Intergenerational transfers still play a substantive role in determining the demand for life insurance. The coefficient of the *living parents* variable—a dummy variable taking the value one if either parent of the head of household is alive—is negative and quite large in absolute value. Although total wealth is a positive determinant of the demand for life insurance, housing property works exactly in the opposite direction. The coefficient of the dummy variable that takes the value one if the household owns its place of living is very significant, large, and negative. Another interesting difference with respect to total insurance demand concerns the role of marital status. Singles are less likely than married people to acquire life insurance. Again, this points to the effect that has been discussed with reference to the number of household members.

In 2000, the structure of demand for life insurance was quite close to the 1989 estimate. However, marital status was even more important (the dummy variable for singles is more significant and much larger in absolute value), and the variable indicating the number of income earners was extremely significant, with a large negative coefficient. Also in the case of life insurance, the same change of sign, across periods, occurs in the coefficient of the variable *town size,* which indicates the dimension of the place of residence in terms of inhabitants; therefore, although most purchases of life insurance (as was the case with total insurance in 2000) come from households residing in very large cities (because of the absolute number of inhabitants), the very fact of residing in a large city does not enhance, but rather depresses, the probability of buying life insurance and the amount of premiums paid. In both the 1989 and 2000 estimations, multiple banking relationships are positively related to holding a life insurance policy.

Turning now to demand for health insurance (the estimates of which are contained in the health insurance columns of Tables 7.17 and 7.18), the coefficient of income is not significantly different from zero (and thus it has been eliminated from the final version of the estimated equation) for 1989 and is extremely low for 2000. Wealth, though, appears to play a positive, nonlinear role. Demand for health insurance increases with age, although at decreasing rates. Other social and demographic variables play (at least qualitatively) the same role as they do in the demand for total insurance.

The damage insurance column of Table 7.17 reports the estimated coefficients of the equations for damage and liability insurance for 1989. The structure of the demand for this type of insurance is similar to the structure of total demand, except that housing property is a significant and positive determinant, as is working in the agricultural sector. The derivative of demand with respect to income is definitely smaller (0.008 versus 0.015) than the income derivative of total insurance demand. The regressions for the two years are quite similar, except for one variable. The damage insurance column of Table 7.18 includes one additional regressor, *absolute risk av.*, which is a measure of absolute risk aversion computed for each individual in the sample. This has been made possible by a specific question that was asked in the 1995 and 2000 waves of the survey that were related to a hypothetical risky investment offered to an individual.[185] By interpreting the answers given by the respondents as reservation prices beyond which the offer is rejected and using a simple second order Taylor expansion, it is possible to compute a measure of absolute risk aversion (see Guiso and Paiella 2001 and Eisenhauer and Ventura 2003).[186] It is also possible to characterize absolute and relative risk aversion in terms of some social, demographic, and economic variables, as was suggested in Eisenhauer and Ventura (2003).

The question concerning the hypothetical investment, however, was not posed to all heads of households, but only to a limited number of them. Hence, the sample is reduced (from 6,060 to 2,495 households) when we include this variable as an additional regressor. This makes the two equations (i.e., the one with and the one without absolute risk aversion) for the 2000 sample not directly comparable, although some interesting differences emerge. In particular, the coefficient of absolute risk aversion is significant at about 5 percent, positive, and quite large in absolute value. Moreover, it is remarkable that in the equation that does not control for absolute risk aversion (and whose results are not reported), the estimated coefficient of income is much smaller (by about one-half) than what we get by controlling for risk attitudes (in fact, the elasticity declines from about 2.2 to about 1.7). This can be easily explained if we think about it in terms of an omitted variable bias problem. In fact, as was argued in Eisenhauer and Ventura (2003), from SHIW data we obtain some evidence of decreasing absolute risk aversion (i.e., absolute risk

---

[185] The question is: "You are offered the opportunity of acquiring a security permitting you, with the same probabilities, either to gain 10 million lire or to lose all the capital invested. What is the most you are prepared to pay for this security?"

[186] For the entire sample, absolute risk aversion averaged 0.1837, varying from 0.1727 among those with health insurance to 0.1942 among those in poor health, whereas relative risk aversion displays greater variability, ranging from 4.5 to 13.84 (which is mainly due to the wide dispersion in income across sociodemographic groups in the sample).

aversion decreases in income). This means that there is an inverse relationship between income and the variable representing absolute risk aversion. On the other hand, we believe (because it is suggested by theory and appears to be confirmed by the estimation of our model) that the coefficient of absolute risk aversion is positive (i.e., the more risk averse an agent is, the more insurance he or she will buy). This implies that the coefficient of income will be biased downward, if we omit (a proxy for) absolute risk aversion (Johnston and Di Nardo 1997).

The importance of this variable is also confirmed by the fact that the pseudo R-squared increases by about 20 percent (albeit from a very low level) if absolute risk aversion is included among the regressors.[187] Apart from this particular issue, however, estimates for the 2000 sample are quite similar to those obtained for the 1989 sample.

As for pension funds, the estimated equations (whose results are reported in the pension columns of Tables 7.17 and 7.18) feature some peculiarities. In particular, a coefficient for income is more than twice the corresponding coefficient that was obtained for total demand. This is true both for 1989 and 2000. For both years, the variable *financial assets* is significant, negative, and quite large in absolute value. This shows, once more, that financial assets can be regarded as substitutes for pension funds, as some households may decide to build up their portfolios to provide resources for retirement years. The variable *living parents* is also very significant and negative, which, on the one hand, highlights the role of intergenerational transfers and, on the other, might point to the relatively younger age of the head of household interested in this type of insurance. The different nature of pension funds is also underscored by the sign of the variable representing absolute risk aversion, which is strikingly negative and significantly different from zero, at a 5 percent level. This might be taken as an indication of the relative importance that the investment side of the product has over its life insurance counterpart (in some sense, this is also implicit in its being considered as a substitute for other financial products).

7.4.2 A Panel Analysis

The remaining part of this section is devoted to the illustration of panel estimations. A few regressions were run with a smaller number of households (about 1,500) who were interviewed in four waves of the survey (1989, 1991, 1993, and 1995). The aim of this exercise was to confirm the robustness of cross-sectional estimates by enriching the data set with a time dimension (though quite a short one) and controlling for individual specific effects. In fact, this panel data set also presented the possibility of inserting a dummy variable for 1993 to control for the effects of a recession and a variable *dtax*, which represents the fiscal treatment of insurance premiums. Insurance premiums have been subject to a number of changes in fiscal treatment since the beginning of the 1990s. Until 1991, insurance premiums (as well as a number of other items) could simply be subtracted from taxable income, with a maximum of 2.5 million liras (about 1,250 euros). Beginning in 1992, with the Amato Reform, the system radically changed, as only a percentage of premiums

[187] This measure of absolute risk aversion was significant, but only for this particular equation. One possible explanation is that the regressors included in the other equations absorbed the explanatory power of the coefficient of absolute risk aversion.

could be subtracted from gross taxes. This percentage was initially set at 27 percent, was reduced in 1995 to 22 percent, and reduced again in 1998 to 19 percent, its current level. The variable *dtax* holds, therefore, a value of 0.35 for 1989 and 1991 (which captures a sort of average tax rate for the initial years, in which premiums paid did not contribute to the formation of taxable income), a value of 0.27 for 1993, and 0.22 for 1995.

As was done for the cross-sectional estimates, several regressions were run corresponding to various types of insurance. Regressions were estimated both with the dependent variable at levels and at first differences, to capture possibly interesting dynamics in demand.

The results from level regressions fundamentally confirm the results obtained in the cross-sectional estimates. In particular, for the case of total insurance demand, the coefficient of income is relatively close to the one shown in the total columns of Tables 7.17 and 7.18; actually, it lies in between the coefficient estimated for 2000 and the one estimated for 1989, which is what one might expect. The role that other financial assets play as a substitute for insurance is all the more evident in the panel estimation, as the coefficient of the *financial assets* variable is significantly negative and large in absolute value; the importance of banks as intermediaries in the process of buying an insurance contract and in measuring the degree of financial sophistication is also evident in the panel estimation, as the coefficient of the number of banking relationships (the variable *banks*) is positive and quite large. The geographic characterization of insurance demand is also robust to the insertion of a time dimension: residing in the south lowers (relative to the north) the probability of demanding insurance; if a southern household does buy some insurance, moreover, the amount bought will be lower than that demanded by households residing in the northern regions.

Nothing particularly striking, with respect to cross-sectional estimations, emerges in the analysis of panel estimations for specific types of insurance. It is certainly more interesting to look at panel, random effect, Tobit estimates when the dependent variable and some of the explanatory variables are expressed in first differences (indicated by the suffix d) whose results are reported in Table 7.19.

Looking at the change in total insurance demand as the dependent variable, the coefficient attached to the first difference in income is extremely significant, positive, and slightly smaller in absolute value than the coefficient of income in cross-sectional estimates. Quite unexpectedly, the change in the number of household members has a negative coefficient, significant at the 5 percent level. This runs counter to empirical evidence, including our own equation related to total insurance demand for the year 2000 (whereas in 1989 the coefficient, which was positive, was not significantly different from zero), pointing to the relationship between insurance demand and the number of dependents in a household. However, we had already noticed that, concerning specific types of insurance (health insurance and liability and damage insurance), the effect of the number of household members was negative. This is a point that deserves further investigation.

Less ambiguous is the role of the change in the number of banking relationships. The coefficient of the former is significant, positive, and quite large in absolute value, both in the equation modeling total demand and in the equations related to

Table 7.19. Panel Tobit Regressions—First Differences

	Total	(P-value)	Life	(P-value)	Damage	(P-value)	Health	(P-value)	Pension	(P-value)
constant	–1,771	0	–2,704	0	–1,925.83	0	–1,839.9	0	6711.9	0
town size					–96.16	0.03	–118.18	0.09		
dnc	–273.45	0.037								
dnp					–117.99	0.11				
dep										
ind	595.44	0			397.94	0	528.15	0		
edu	177.69	0			157.98	0			507.4	0.025
d93	1,425.71	0	5042	0	–824.87	0	10,628.54	0	34,389.3	0
dbanks	238.97	0.01			117.73	0.08			729	0.02
d1tax	11,832.94	0	5,8605	0	–12,090.5	0	124,166.7	.	41,6962.5	
d(income)	0.015	0	0.02	0	0.0067	0	0.008	0	0.006	0.5
d(wealth)			0.00061	0.15						
N	1,494		2,255		1,803		1,703		1,756	

Source: SHIW, waves 1989, 1991, 1993, and 1995.

Note: This table contains estimated coefficients (and corresponding P-values) of panel Tobit regressions, where the role of the dependent variable is played by the various types of insurance mentioned in the column head, and the meaning of regressors is fully specified below and in the text. Dependent variables are expressed in first differences.

Town size = importance of town in terms of number of inhabitants, from 1 to 5, dnc = first difference in number of household members, dnp = first difference in number of income earners, dep = dummy variable for dependent workers, ind = dummy variable for independent workers, edu = education level, d93 = time dummy variable for 1993, dbanks = first difference in number of banking relationships.

changes in demand for liability and damage insurance and pension funds. (In all other equations, the coefficient is not statistically significant.) As for the number of banking relationships, we might wonder whether the positive and significant relationship is only a consequence of endogeneity, in that people who demand more insurance may seek out banks offering attractive insurance products. (In other words, the two variables, insurance demand and the number of banking relationships, would be jointly determined by other variables, possibly relating to the demand for insurance products.) In this case, the positive relationship could not be attributed a causal interpretation, although it would still signal the relevance of the banking channel in insurance distribution.

Using a panel data set might provide insight into whether such an endogeneity problem exists. To this specific aim, two sets of ordinary least squares panel regressions were run (results not reported) on total insurance and its first difference, with both fixed and random effects. We subsequently ran a Hausman test on both sets of regressions, which indicated (generic) potential endogeneity of regressors in the case of total insurance demand, but not in the case of first differences. In other words, although the demand for insurance and the number of banking relationships appear to be determined by some third common factors, the dynamics of insurance demand and banking relationships do not. Further evidence (perhaps in terms of instrumental variable estimations) could be obtained and analyzed to check the robustness of these findings.

As anticipated in the previous section, we inserted in the short-run equations a variable that captures the changes in fiscal treatment of insurance premiums. This variable, denoted by *d1tax*, is positive and significant in all equations, except in the case of damage and liability insurance, in which it takes on a negative value. The coefficient of this fiscal variable ranges from –12,090.5 (for damage and liability insurance) to 410,844 for pension funds (the coefficient for total demand being equal to 10,565). For example, if we compute the elasticity of changes in total insurance demand with respect to this variable, we obtain 0.13; this means that a 1 percent increase in the percentage of premiums that one can deduct from taxes brings about a 0.13 percent increase in total insurance demand. This finding, which might be contrasted to earlier research on demand for life insurance in Italy by Jappelli and Pistaferri (2003), is of some practical interest, if we consider that taxation of insurance contracts has reached very important levels.

7.5 CONCLUSION

The objective of this chapter is to present an overview of the evolution of the Italian insurance market over the period 1990 to 2003, which was extremely vigorous—so much so as to lead to dimensions that have made the Italian market comparable to those of more developed markets.

All indicators suggest that this dynamic situation will characterize the Italian market for insurance in the near future. The most recent data, about 2004, depict an ongoing vigorous and lively situation, in which premiums have amounted to over 100 billion euros, with a remarkable growth of about 4 percent over the preceding year (with a roughly constant share of about 65 percent going to life insurance and

35 percent to damage and liability insurance). Profitability is also remaining at important levels; overall net income accounted for about 4.8 percent of gross premiums in 2004. The question is, of course, whether this trend will stay for the next few years, and this will very much depend on the capability of the whole system to take profit of new opportunities. These new opportunities might include the market for complementary social security, although one can reasonably forecast a reduction in the corresponding profit margins in the medium run. Other opportunities might appear in the damage and liability sector, where companies should try to devise new instruments to accommodate the increasingly complex (also from a risk perspective) operational contexts in which firms and institutions must work.

Both a macro and a micro analysis may help in analyzing the working of this market, which is the goal of this chapter. The first part contains a synthetic description of the dynamics of the main variables of interest and an overview of the main events that had an important impact on the remarkable development of the market, such as the social security reform process, the homogenization and liberalization of the insurance market at the European level, some important fiscal changes, and the evolution of the distribution system.

The second part of the chapter explores a set of micro data related to insurance holdings in Italy, drawn from the Survey on Household Income and Wealth, run every two years by the research bureau of the Bank of Italy. Once again, the analysis has been carried out at two different levels. The first step presents a series of descriptive statistics, whereby holdings of insurance contracts are characterized in terms of various sociodemographic and economic variables related to households. In the second step, to get a more precise quantitative assessment of the micro determinants of insurance holding, several Tobit regressions were performed, with insurance premiums paid by the households as the dependent variables, both in the aggregate and disaggregated by type of insurance. Some of the regressions were also carried out on a panel data set to check the robustness of the cross-sectional results. The results of the micro analyses pointed out the effects of such variables as age and education, income and wealth, the number of household members, the number of income earners, the possibility of intrafamily and intergenerational transfers, the number of banking relationships, the location of one's residence, and risk aversion, both on the probability of acquiring an insurance contract and on the amount of the insurance purchased. Importantly, the panel Tobit estimation also enabled assessment of the relevance of the fiscal treatment of insurance premiums, as in the period under examination there were many changes also in this respect. This micro analysis, therefore, provides important insights into the possible effects that modifications of the economic and institutional context in Italy might exert on the structure and evolution of the insurance market.

7.6 REFERENCES

ANIA, 1990a–2003a, *The Italian Insurance* (Rome: ANIA).

ANIA, 1990b–2003b, *The Insurance Companies Yearbook*, ED.ASS. (Rome: ANIA).

Banca d'Italia, 1991, *Supplementi al Bollettino Statistico, Note Metodologiche e Informazioni Statistiche, I Bilanci delle Famiglie Italiane nell'anno 1989*, Nuova Serie, Anno I, n. 26.

Banca d'Italia, 2002, *Supplementi al Bollettino Statistico, Note Metodologiche e Informazioni Statistiche, I Bilanci delle Famiglie Italiane nell'anno 1989*, Nuova Serie, Anno XII, n. 6.

Beck, Thorsten, and Ian Webb, 2003, "Economic, Demographic and Institutional Determinants of Life Insurance Consumption Across Countries," *World Bank Economic Review* 17: 51–88.

Becker, Gary S., and Casey B. Mulligan, 1997, "Endogenous Determination of Time Preference," *Quarterly Journal of Economics* 112: 729–758.

Biancotti, Claudia, Giovanni D'Alessio, and Andrea Neri, 2004, "Errori di misura nell'indagine sui bilanci delle famiglie italiane" (Measurement errors in the Survey of Italian Households' Income and Wealth), *Tema di discussione,* n. 520.

Brandolini Andrea, and Luigi Cannari, 1994, "Methodological Appendix: The Bank of Italy's Survey of Household Income and Wealth," in A. Ando, L. Guiso, and I. Visco, eds., *Saving and the Accumulation of Wealth* (Cambridge, England: Cambridge University Press).

Duker, Jacob.M., 1969, "Expenditures for Life Insurance Among Working-wife Families," *Journal of Risk and Insurance* 36: 525–533.

Ehrlich, Isaac, and Gary S. Becker, 1972, "Market Insurance, Self-insurance and Self-protection," *Journal of Political Economy* 80: 623–648.

Eisenhauer, Joseph G., and Luigi Ventura, 2003, "Survey Measures of Risk Aversion and Prudence," *Applied Economics* 35(13): 1477–1484

Guiso, Luigi, and Monica Paiella, 2001. "Risk Aversion, Wealth, and Background Risk," Discussion paper 2728, Center for Economic Policy Research, London.

Hendry, David F., and Jean-F. Richard, 1982, "On the Formulation of Empirical Models in Dynamic Econometrics," *Journal of Econometrics* 20(1): 3–33.

Henriet, Dominique, and Jean-C. Rochet, 1998, "The Political Economy of Public Health Insurance," mimeo.

ISVAP, n.d., "Il margine di solvibilita'delle imprese di assicurazione: confronto tra i sistemi europeo ed americano," *Quaderni dell'Istituto*, n. 6.

ISVAP, 2004, *Relazione sull'attività dell'Istituto nell'anno 2003* (Rome: ISVAP).

Jappelli, Tullio, and Luigi Pistaferri, 2003, "Tax Incentives and the Demand for Life Insurance: Evidence from Italy," *Journal of Public Economics* 87: 1779–1799.

Johnston, Jack, and John Di Nardo, 1997, *Econometrics Methods* (4th ed.). (New York: McGraw Hill).

La Torre, Antonio, 1995, *L'Assicurazione nella Storia delle Idee*, *Monografie di Assicurazioni*.

Mayers, David, and Clifford W. Smith Jr., 1983, "The Interdependence of Individual Portfolio Decisions and the Demand for Insurance," *Journal of Political Economy* 91: 304–311.

McDonald, John F., and Robert A. Moffit, 1980, "The Uses of Tobit Analysis," *Review of Economics and Statistics* 62: 318–321.

Outreville, François J., 1996, "Life Insurance Markets in Developing Countries," *Journal of Risk and Insurance* 63: 263–278.

Perissinotto, Giovanni, 2003, "The Creation of Value Through a Specialized Distribution System," *Geneva Papers on Risk & Insurance—Issues & Practice* 28(3): 495–501.

Persano Adorno, Paola, and Pierfranco Mattei, 1998, "Il trattamento fiscale del risparmio assicurativo sulla vita e delle forme pensionistiche complementari," *Quaderni ISVAP,* n. 2 (Rome : ISVAP).

Pontremoli, Roberto, 2002, "Evolution and Innovation in Insurance Management: The Italian Case," *Geneva Papers on Risk and Insurance* 27(4): 669–687.

Sen, Amartya, 1997, "Human Capital and Human Capability," *World Development* 25: 12.

Servizio Assicurazioni Danni (ISVAP), 2001, *L'Assicurazione R.C. auto in Italia: analisi e proposte* (Rome: ISVAP).

Showers, Vince E., and Joyce A. Shotick, 1994, "The Effect of Household Characteristics on Demand for Insurance: A Tobit Analysis," *Journal of Risk and Insurance* 61(3): 492–502.

Verbeek, Marno, 2000, *A Guide to Modern Econometrics* (New York: John Wiley).

7.7 WEB SITES OF INTEREST

www.ania.it
for documents and data concerning insurance

www.isvap.it
for more documents and data

www.covip.it
for documents and data concerning pension funds

www.bancaditalia.it
for access to SHIW micro data

7.8 LEXICON

ANIA. The Italian Insurers' Association (Associazione Nazionale fra le Imprese Assicuratrici), established in 1944, and representing about 98 percent of the Italian insurance market, in terms of premiums.

ASL. Azienda Sanitaria Locale, local health agency/operating unit of the National Health Service.

Capital under risk. *Capitale sotto rischio,* difference between insured capital and mathematical reserves.

CEIOPS. Committee of European Insurance and Occupational Pension Supervisors (Comitato europeo delle assicurazioni e delle pensioni aziendali o professionali), created in 2003 by the European Commission Decision 2004/6/EC and composed of high-level representatives from the insurance and occupational pensions supervisory authorities from member states of the European Union.

Combined ratio. Algebraic sum of the loss and the expense ratio.

Cooperative. A society with a fundamental mutuality purpose (i.e., its members operate in favor of one another), which also imposes some rules over the formation and distribution of profits.

COVIP. Commissione di Vigilanza sui Fondi Pensione (supervisory commission for pension funds), created in 1993 under the terms of Legislative Decree n.124/1993.

Expense ratio. Ratio of administrative and distribution costs to annual premium income.

INA. Istituto Nazionale delle Assicurazioni (National Institute for Insurance), established in 1912 (and existing to date), to operate a public monopoly on the life insurance sector.

INAIL. Istituto Nazionale per l'Assicurazione contro gli Infortuni sul Lavoro (National Institute for Insurance against Working Accidents), created in 1933.

INPS. Istituto Nazionale della Previdenza Sociale (National Institute for Social Security), created in 1933, charged with the payments of sickness allowances (among other duties).

Insurance penetration. Per annum premium income received by primary insurers in percent of the GDP.

ISVAP. Istituto per la Vigilanza sulle Assicurazioni Private e di interesse collettivo (the private insurers' surveillance authority), established in 1982.

Loss ratio. Ratio of claims incurred to premiums earned.

Mutual. A cooperative firm in which the mutuality purpose is prevailing. The insured subjects essentially coincide, for a mutual company, with its members, whereas a generic cooperative may also insure third parties. A mutual company must also observe stricter rules for the distribution of profits.

Public company. A stock company controlled by the Treasury by more than 50 percent of its stock.

ROE. Return on equity, defined as the ratio of overall profit (of both technical and nontechnical operations) to net capital.

SHIW. Survey of Households' Income and Wealth (Indagine sui bilanci delle famiglie), run on a biannual basis by the Bank of Italy. Since 1989, this survey has a panel component.

Solvency margin. Further guarantee (over and beyond technical reserves) of solvency for policyholders. It is an additional buffer constituted with "free" net capital (i.e., net capital not already employed in covering risks), whose dimensions are a function of mathematical reserves and capital under risk for life insurance and premiums or average losses for damage and liability insurance.

Solvency ratio. Ratio of actual solvency margin to minimum capital requirement (or minimum solvency margin).

SSN. Servizio Sanitario Nazionale (National Health Service), established in 1980.

Technical account. Accounting report containing technical results (see below).

Technical reserves. Mandatory reserve funds whereby insurance firms set aside a portion of premiums to be able to always meet their current and future obligations. Technical reserves are mainly constituted by actuarial reserves against outstanding risks, in case of life insurance, and set as a function of risks extending over several periods, and losses already occurred but not expensed, for non-life insurance.

Technical results. The results from the specific insurance operations, as opposed to the results from the investing activities (nontechnical results).

8 A Descriptive Analysis of Canadian Insurance Markets

Gilles Bernier

Laval University

Alli Nathan

Nyenrode Business University and Providence College[188]

8.1 INTRODUCTION

Canada is the largest country in the western hemisphere and the second largest country in the world in terms of land area, but ranks twenty-fifth in terms of population. It is the smallest of the G7 economies with a population of approximately 31.6 million and a gross domestic product (GDP) of approximately US$867 billion in 2003. It ranks twelfth in the world in terms of GDP per capita. However, in terms of total insurance premium volumes (in U.S. dollars) sold worldwide during 2003, Canada ranked eighth with 2.01 percent of the market. More specifically, Canada ranked twelfth in life insurance with 1.37 percent of the world market, and seventh in non-life insurance with a market share of 2.86 percent (Swiss Re, 2004). These overall world rankings have been quite stable since 2000. With respect to insurance density (premiums per capita in U.S. dollars), Canada ranked seventeenth in 2003, while it ranked twenty-third in terms of insurance penetration (premiums as a percent of GDP).

The purpose of this chapter is to describe and analyze the structure, conduct, and performance of Canada's life and non-life insurance markets over the period 1990–2003. The chapter contains five parts. Part I provides a short historical perspective on the evolution of the Canadian insurance industry. Part II outlines the economic

[188] The authors acknowledge the research and secretarial assistance of Marie-Pascale Koffi and Lise Jacques and the financial support of the Industrial-Alliance Insurance Chair. They would also like to thank the following people for their help and support in gathering the data needed for this research: Jane Voll, Teepu Khawja, and John Tyrrell from the Insurance Bureau of Canada; Yves Millette and Alice Freeburn from the Canadian Life and Health Insurance Association; and, finally, Jacques Carrière and Michel Naud from the Industrial-Alliance Insurance Group.

environment faced by insurance companies operating in Canada during the entire period covered by our analysis. Part III describes the key dimensions of Canada's insurance markets. This includes the development of premium volumes and the degree of concentration for each line of business since 1990. It also highlights the leading players in each class of insurance. Part IV outlines the importance of the insurance industry in the Canadian economy, and analyzes the main indicators of financial performance for both sectors of the industry. Part V recaps the major trends that have been shaping the industry during the period of study and discusses some of the future trends which should be most influential for the whole sector. Finally, we conclude.

8.2 CANADIAN INSURANCE INDUSTRY: A HISTORICAL PERSPECTIVE

The property and casualty insurance industry in Canada predates the banking (the first bank was incorporated in 1817) and the securities industries (the first securities firm was established in 1832). The first general insurance company was the Phoenix Assurance Company, which was licensed in 1804 and opened its operations in Montreal; this was followed by the Halifax Association in 1809, which became the Halifax Insurance Company in 1819. The Quebec Fire Insurance Company was also established in 1819, being the first fire insurance firm to be established in Quebec. This was motivated by a fire in 1815 that destroyed much of the old city of Quebec, similar to the Great Fire of London in 1666 that led to the first fire insurance firms in the United Kingdom (Insurance Bureau of Canada, 2004). The province of Quebec has historically had a special status among the ten Canadian provinces due to its French heritage. As described later, this also extends to some insurance products, services, and regulations that are in force in Quebec. The property and casualty sector also predates the life insurance industry, which expanded more slowly, both in North America and Europe. The first Canadian life insurance company to be chartered was Canada Life in 1847 (Moreau, 1991).

As with the development of the insurance industry in the rest of the world, the first insurance firms fulfilled the need for marine insurance arising out of the trade between the New World and the United Kingdom and Western Europe. This was followed by insurance firms that focused on fire insurance in order to protect against the fire hazards faced by the wooden buildings of the early nineteenth century. In fact, in the early days, many insurers had their own fire brigades, until the responsibility was passed over to local governments in 1866. The property and casualty industry grew in scale and diversified in scope, in parallel with the transition of Canada from an agricultural to an industrial economy. One other key development during the early years was the authorization, by the then Colony of Upper Canada, in 1836, to establish mutual insurance firms, where the policyholders were also the owners of the company. The first mutual general insurance firm was the Gore District Mutual Insurance Company established in 1839, and it is Canada's oldest continuously operating insurance firm. The mutual form was widely embraced by the life firms, with a large percentage of life insurance being provided by mutuals, until the demutualization wave during the last two years of the twentieth century (1999–2000).

About 100 years ago in 1905, there were 50 general insurers operating in Canada, of which 17 were British, 13 were Canadian, and 10 were American. The dominance of foreign firms in the Canadian property and casualty market continues to this day, with the Dutch subsidiary ING Canada being the largest general insurer in Canada today (in terms of net premiums written). However, the Canadian property and casualty market has come a long way since 1905, when the total premium volume was CAN$14.3 million, and claims paid were a mere CAN$6 million, compared to a premium volume of CAN$31.4 billion (of which 51.1 percent were due to foreign firms operating Canada) and claims of CAN$20.4 billion in 2003. The market for life insurance is even larger, with a premium volume of CAN$46.5 billion and claims of CAN$38 billion in 2003. And, unlike the non-life sector, the life sector is dominated by large domestic firms, which also have significant overseas operations. In fact, 54 percent of the life premiums in 2003 were generated from foreign operations.

In keeping with developments in the industrial world, including innovations such as automobiles, the nature of the property and casualty industry has also changed. While 200 years ago, marine and fire insurance were the key products offered by the industry, today they have been replaced by automobile and property insurance as the key products in the general insurance sector. This also has to do with the various risk-management techniques that have vastly reduced marine and fire hazards, often at the initiative of the industry participants. Similarly, in keeping with the changing demographics and sophistication of their clients, life insurers have transformed themselves from being providers of simple term and permanent (whole life) insurance contracts to asset managers, wealth accumulators, and providers of key retirement products.

Further, the industry is active in its efforts to educate and train its clients in preventing automobile accidents and thefts and damages to property (non-life sector), as well as asset management and retirement planning (life sector). Thus, while providing essential risk transfer products to the market, the industry also serves a larger social purpose, as evidenced by the involvement of non-life firms in preventing drunk driving and providing access to various theft prevention mechanisms, and the key role that life firms play in providing private retirement benefits. Also unique is the involvement of some provincial governments in the provision of compulsory auto insurance.

Today the insurance industry (including both life and non-life firms) is a key player in the Canadian financial services market, being the leader in the provision of products that enable its clients to transfer their risks effectively and efficiently. And, among all the financial services sectors in Canada, the property and casualty industry also has the singular, stellar reputation of being one of the most sound and stable sectors that has not experienced any major incidences of failures in its 200-year history.

The two key pillars today of the Canadian financial services industry—banks and insurance firms—have a reputation of being fairly stable, having faced only two serious incidents of corporate failure and bankruptcy in the years after the depression of the 1930s. Two regional western Canadian banks—Northland Bank and Canadian Commercial Banks—failed during the real estate collapse of the mid-1980s. Both banks had invested heavily in real estate without significant due diligence and geographic diversification.

The same reasons also led to the failure of three life insurers in Canada during the early-1990s—Les Coopérants and Sovereign Life in 1992, followed by Confederation Life in 1994. While Les Coopérants and Sovereign were small firms, Confederation Life was the fourth largest insurance company in Canada at the time of its failure and was among the top thirty in North America. With assets of CAN$19 billion, it was the largest life insurer to fail on the North American continent, and the losses of CAN$2.6 billion make it the single largest financial services company to become insolvent globally. In addition to unfettered investments in real estate in Canada, the United States, and the United Kingdom at the peak of the real estate bubble of the late 1980s and the early 1990s, Confederation Life also had made significant forays into the estate and trust business as well as the treasury and derivatives business, but without the skill or the knowledge to manage the risks in these businesses (McQueen, 1996).

8.3 THE ECONOMIC ENVIRONMENT IN CANADA

Insurance markets are an integral part of a nation's economy; therefore, developments in the insurance markets are often analyzed within the context of the overall development of the economy. For example, every year Swiss Re reports on the state of insurance markets around the world by analyzing the links between premium growth and the economic environment by region and by country.

In that context, studies have shown that the business cycle usually has an indirect effect on life insurance premiums via disposable income and the savings rate (Swiss Re, 2001). But the performance of financial markets and the level of interest rates have an impact on the popularity of some life insurance products in any given year. At the same time, growth in non-life insurance is known to be influenced by overall economic development only in the long term (Swiss Re, 2001). In the short term, the non-life insurance price cycle is a dominant factor, affecting both supply and demand of insurance products. Cycles in the Canadian property and casualty insurance industry typically last six to eight years, and they are known to be due in part to falling investment income but also to rising claim costs and heavy regulation (Voll, 2004b).

Therefore, it is interesting to start our analysis by describing the economic environment faced by insurers over the period 1990 to 2003 and then proceed with a more in-depth characterization of Canada's insurance markets for these years. The key features of Canada's economic environment during this period were as follows, some of which are shown in Figure 8.1.

- Canada's GDP per capita (based on current prices and exchange rates) grew from US$20,729 in 1990 to US$27,097 in 2003 (Bank of Canada Web site). This represents a compounded annual growth rate (CAGR) of 2.1 percent per year. During this time, the Canadian economy experienced two recessionary periods—1990–1991 and 2001–2002. Both recessions were followed by slow recovery periods, lasting about two years. But, as with the rest of the world,

Canada also experienced a prolonged period of expansion that lasted until the stock market collapse in 2000.

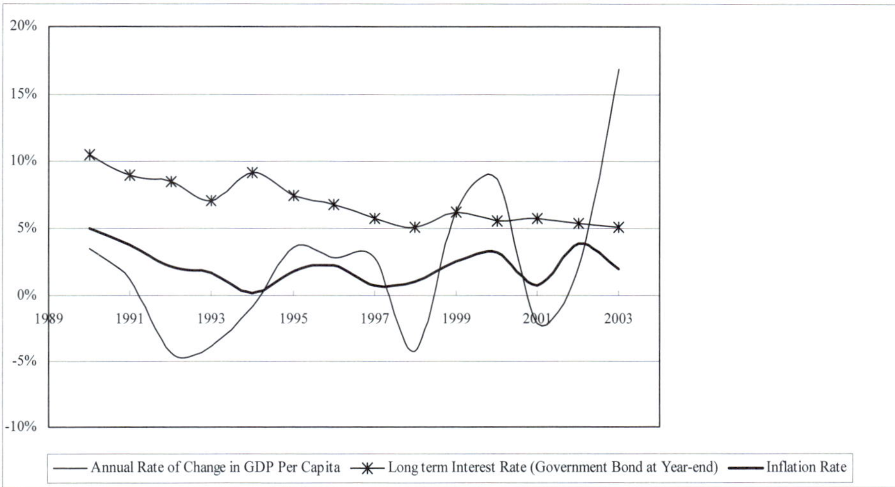

Source: www.bankofcanada.ca/en and unstats.un.org/unsd/snaama.

Figure 8.1. Canadian Economic and Financial Statistics

- Canada saw its household savings rate go down drastically, from 13 percent in 1990 to 7 percent in 1996, 4 percent in 1999, and 1.4 percent in 2003 (OECD Website). This may have been due to the heavy taxation resulting in lower disposable income and, therefore, a lower propensity to save, as well as the phenomenal increases in the value of stocks and real estate. During the same period, the debt level of the households also increased significantly.
- The inflation rate went down significantly during the first half of the 1990s, from 4.37 percent in 1990 to 2.76 percent in 1995. Since then, inflation in Canada has been rather stable and lower than 3 percent, driven by the focus of the central bank (Bank of Canada Web site) on price stability. In 2003, inflation was at 2.61 percent.
- Over the period 1991 to 2001, the Canadian bond market outperformed the stock market six years out of eleven (Bank of Canada Website). Like other G7 countries, Canada's stock markets also experienced a major correction associated with the new economy/dot-com bubble burst. The S&P/TSX composite, Canada's broad stock market index, reached a peak in August 2000 and a trough in October 2002. For the years 2001 and 2002, the Canadian stock market realized returns of –12.6 percent and –12.5 percent, respectively. However, in 2003, the market turned around and realized a highly positive return of 26.73 percent. But, at the end of 2003, market prices in Canada were still some 20 percent below the exceptionally high peak attained in 2000. On the whole, an investor would have achieved a compounded annual rate of return of 7.75 percent by investing in the S&P/TSX composite index from the beginning of 1990 to the end of 2003.

- Long-term federal bonds (more than ten years) were trading at average yields (in December of each year) of 10.51 percent in 1990, compared to 6.77 percent in 1996 and 5.14 percent in 2003. At the same time, short-term T-bills (six-month) were trading at average yields of 11.4 percent in 1990, compared to 3.18 percent in 1996 and 2.5 percent in 2003 (also in December of each year). This shows that Canadian interest rates decreased significantly during the period under consideration in line with low inflation and stable prices (Bank of Canada Website). But such low interest rates combined with poor stock market returns had a severe impact on the financial performance of Canadian insurers, particularly property and casualty insurers, as discussed below.

8.4 THE CANADIAN INSURANCE MARKET

In our discussion, most of the data for the life insurance sector is provided by the Canadian Life and Health Insurance Association (CLHIA), while most of the data for the non-life sector is provided by the Insurance Bureau of Canada (IBC). CLHIA and IBC are privately owned trade associations, representing their respective industry sectors. Both associations provide numerous services to their member companies as well as consumers and act as the voices of the insurance industry in political and regulatory affairs. Hereafter, all data are in Canadian dollars unless otherwise indicated.

8.4.1 Regulation and Corporate Governance

Politically Canada is divided into ten provinces and three territories (the equivalent of states in the United States). In Canada, banks and insurers can be chartered at the federal level by acts of Parliament or by letters of patent issued by the Ministry of Finance authorizing them to conduct business as defined in the respective legislations, such as the Bank Act and the Insurance Companies Act. These legislations are subject to periodic review every five years, allowing the laws to reflect recent changes in the domestic and global financial arenas. Such federally regulated insurance firms form the first tier of financial institutions operating in Canada and include the large domestic banks and insurers, as well as all foreign-owned banks and insurance companies. Foreign institutions were given permission to establish such federally incorporated subsidiaries in Canada as far back as 1965. While foreign subsidiaries are required to hold capital and operate under the Canadian regulatory framework, foreign companies are also allowed to operate branches in Canada, supported by parent capital.

In addition, a second tier of entities operates in the market for financial intermediation. These include provincial deposit-taking financial institutions (such as trust companies and credit unions), as well as provincially incorporated insurance firms, mutual funds, and pension plans, authorized and regulated by provincial governments. These provincial institutions vary in their power and market share in different provinces, with the largest role being played in Quebec (Saunders and Thomas 1997). In general, the federally incorporated firms are larger and tend to dominate the markets in Canada. However, the distribution of insurance is a provincial jurisdiction. This means that all federally incorporated or regulated insurers must nevertheless apply to the provincial authorities for licenses to operate and sell their products and services in each province where they want to do business. In this aspect, the framework is similar to that in the United States.

The Insurance Companies Act (ICA) allows insurers in Canada to be organized as either mutual companies or joint-stock companies. Mutuals are owned and operated for the benefit of policyholders, but there are also nonparticipating policyholders, who do not have an ownership stake in the mutual firm. On the other hand, joint-stock companies (a term reflecting the British heritage) are public companies owned by shareholders and operated by managers for the benefits of the shareholders; policyholders also can be shareholders in a joint-stock company. The ICA requires that mutual life insurers maintain separate accounts for participating and nonparticipating business lines. However, the participating policyholders remain ultimate owners of the firm and are entitled to the full profits and the equity value of both lines. ICA regulations also require that joint-stock companies maintain separate accounts for participating policyholders and allocate income (or losses) and expenses fairly and equitably. Further, joint-stock companies can transfer only a limited amount of profits to their shareholders from their participating accounts. The restrictions range from a 10 percent maximum transfer for participating funds of less than $250 million to 2.5 percent for funds over $1 billion, with a graduated formula being applied for funds in excess of $250 million but less than $1 billion. Finally, the participating policyholders are allowed to retain their voting rights and also are given the right to select at least one-third of the company's directors (CIBC World Markets 1999).

The ICA also allows insurance firms to operate under a holding company structure. This has been designed to allow the companies to expand the scale and scope of their operations with minimal cost and regulatory constraints. Thus, the Canadian insurance market is characterized by large domestic and foreign insurance groups, which own several individual insurance companies under a common holding company name.

8.4.2 Number of Companies

In this section, we focus on the number of individual insurers operating in Canada. Therefore, we do not account directly for the fact that many of these companies are owned by a single parent company, under a holding company structure, thereby impacting on the true number of competing entities. We address this issue later, when we characterize the degree of concentration in the industry and discuss the profile of the big players in the life and non-life sectors.

Life and Health Insurance

As a result of demutualization, merger and acquisition activities, foreign institutions exiting the market, and the three major insolvencies of the early 1990s, the number of life insurance companies operating in Canada has decreased significantly since 1990 (CIBC World Markets 1999). Merger and acquisition activities have been driven by the desire of the key players to prevent future failures (resulting in the stronger players acquiring the weaker ones, with or without regulatory pressures), as well as the benefits from increasing the scale and scope of operations by diversifying across product lines and geographic regions.

On average, the reduction in the number of companies was about 3.1 percent per year between 1990 and 2003. As shown in Table 8.1, 108 life insurers were active in Canada during 2003, compared to 163 in 1990. Among the 108 firms in 2003, 67 were Canadian-registered firms (both Canadian-owned firms with a federal or a provincial charter and Canadian subsidiaries of foreign firms with a federal charter) and 41 were nonresident entities, operating as branches of foreign insurers. The majority of these were branches of firms incorporated in the United States, while others were branches of firms from the United Kingdom and the European Union. Over time, however, foreign branches have become less significant, with their market shares dropping to about 12.2 percent of net premiums written (NPW) in 2003 from 20 percent in 1990.

Table 8.1. Number of Canadian Insurers (Domestic and Foreign), 1990 to 2003

	Life and Health	Property and Casualty
1990	163	not available
1991	155	not available
1992	148	not available
1993	148	not available
1994	146	*221*
1995	140	*225*
1996	131	*225*
1997	132	*220*
1998	129	*225*
1999	127	*205*
2000	120	*198*
2001	116	219
2002	109	207
2003	108	206

Source: Canadian Life and Health Insurance Facts (1991 to 2004) and Insurance Bureau of Canada with data from A.M. Best WinTRAC and MSA Researcher.

Notes: Data in italics represent companies that have net premiums earned in these years. MSA and A.M. Best do not capture all companies. Data reported for 2001, 2002, and 2003 were collected through surveys and are accurate.

On the whole, Canadian-owned life insurers and subsidiaries of foreign institutions accounted for 87.8 percent of life insurance premiums written in 2003. Of course, the bulk of this market share belongs to the larger Canadian life insurance groups that dominate the market. Indeed, when measured in terms of ultimate ownership, the Canadian companies had a market share of 73.3 percent of NPW in 2003. The companies from the United States had a market share of 15.1 percent, while the European firms had a market share of 11.6 percent of NPW in Canada. And, as shown in Figure 8.2, joint-stock firms dominate the life sector with a 94 percent market share.

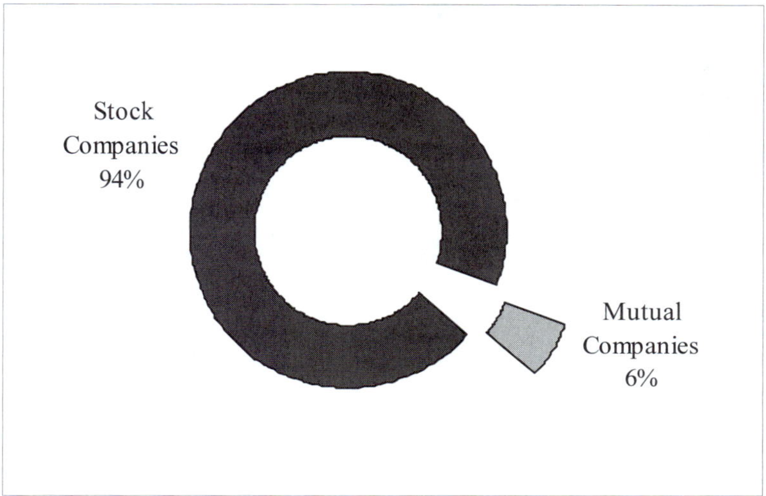

Source: Canadian Life and Health Insurance Facts (2004) available on www.clhia.ca.

Figure 8.2. Market Share by Organizational Form in the Canadian Life and Health Insurance Industry, 2003

Property and Casualty Insurance

As with the life insurance sector, the number of property and casualty (P&C) insurers has also decreased over the years, again due to merger and acquisition activities. On average, however, the reduction in the number of companies in the P&C sector was smaller than that in the life sector, being around 0.8 percent per year since 1994. As shown in Table 8.1, 206 P&C insurers were active in Canada in 2003, compared to 221 in 1994. However, this is twice as many individual firms as in the life sector. Of this total number for 2003, 48.9 percent were Canadian firms, and 51.1 percent were foreign entities. As shown in Figure 8.3, joint-stock companies dominate the P&C market, with a market share of 80 percent of NPW in 2003. The mutuals account for the other 20 percent of the market share.

To summarize, in 2003, a total of 206 P&C firms and 108 life firms operated actively in the Canadian insurance market. In the life sector, joint-stock companies are both numerous and larger than mutual firms, especially since the larger domestic mutuals converted to the stock form in the 1999–2000 demutualization wave. In the non-life sector, joint-stock companies have historically had a more significant market position than mutuals. It is also interesting to note that foreign insurers play a more significant role in the Canadian P&C sector than in the life sector. Finally, the insurance industry as a whole uses the holding-company structure as its preferred organizational form. This means that the actual number of competing entities is a much smaller number of life and non-life groups.

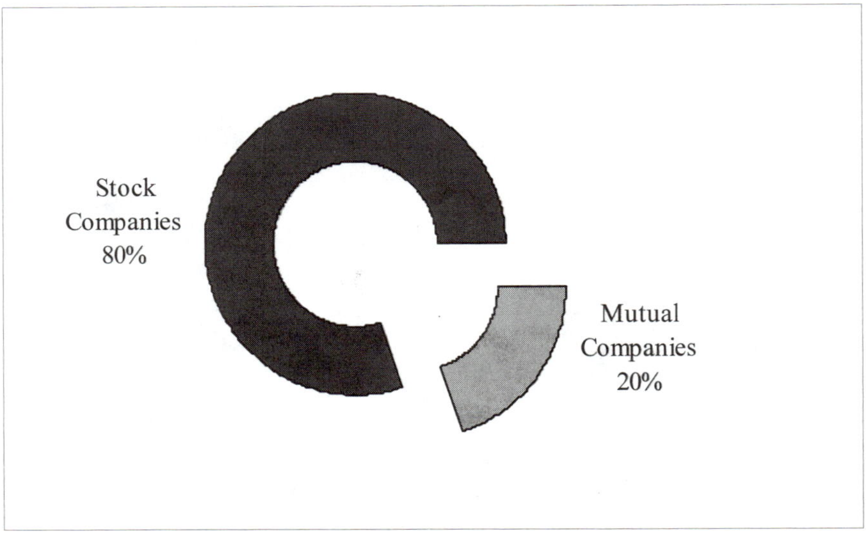

Source: MSA Benchmark Report (2004) provided by the Insurance Bureau of Canada.

Figure 8.3. Market Share by Organizational Form in the Canadian Property and Casualty Insurance Industry, 2003

8.4.3 Employment

Life & Health Insurance

According to CLHIA, there were 118,000 people working in the Canadian life insurance sector in 2003. This includes 45,800 full-time administrative employees (39 percent) and 72,200 career or independent agents (61 percent).

Property and Casualty Insurance

IBC estimates that about 100,372 individuals worked in the Canadian private non-life sector in 2003. This includes 41,147 employees of primary insurers (41 percent), 51,747 brokers and employees (51 percent), 5,978 independent adjusters and appraisers (6 percent), and, 1,500 employees of reinsurers throughout Canada (2 percent).

In summary, although the life insurance industry has a smaller number of operating entities than the P&C industry, both sectors appear to employ more or less similar numbers of personnel.

8.4.4 Premium Volumes

Life and Health Insurance

Premium income is the primary source of revenue for life insurance companies, with investment income and fee income being secondary sources of revenues. Premium income is derived from three distinct sources—life insurance, health insurance, and annuity products (including retirement products)—all of which can be either group or individual contracts.

Group life and health insurance is sold through a competitive bidding process to employers, who in turn distribute it to employees as part of a more comprehensive employer-sponsored benefits plan, which often includes supplemental health insurance products. Individual life insurance is offered in two varieties—term insurance and permanent insurance. Term insurance offers only death benefits, and the policies often terminate at a specific age. Premiums are often lower in the initial years, but increase over time. Permanent insurance contracts, such as whole-life and universal life, in addition to providing death protection, include cash values that accumulate over time, which can be used to meet financial emergencies, pay for specific goals, or provide retirement income. However, during the period under analysis, such life insurance products have lost their market share to the more innovative retirement products offered as annuities (Department of Finance—Canada Website).

As a result of Canada's universal public health care system and strong group insurance penetration, individual health insurance appeals to those not covered by group plans (such as seasonal or contractual workers and retirees) or those supplementing an employer's basic health insurance. Such individual health insurance products vary widely and include such items as dental care, vision care, prescription drug reimbursement plans, disability insurance, extended health care benefits, dread disease coverage, and long-term and critical care coverage (CIBC World Markets 1999).

Annuities have evolved over time to be the preferred retirement products offered by insurers. Group annuities are often in the form of private pension plans offered by employers. These are popularly known as registered pension plans. Individual annuities include deferred contracts that are in the process of being paid out (such as registered retirement income funds—RRIFs), and products that are in the accumulation phase (such as registered retirement savings plans—RRSPs). RRSPs are tax-sheltered, asset-accumulating investment vehicles that permit investors to make tax-deductible contributions and whose investment income accrues tax-free. Limitations exist regarding the level of yearly contributions, based on earned income (similar to 401(K) plans in the United States). RRIFs, on the other hand, are similar to immediate annuities, which provide periodic payments (mostly fixed) over a defined period (term annuities) or over the life of the individual (life annuities). However, RRIFs provide a more flexible payout option than the immediate annuities, where the annuitant is locked in (CIBC World Markets 1999).

Another popular annuity product is the segregated fund (similar to variable annuity in the United States), which is also used to build up RRSPs. Segregated funds are available only from life insurance companies and are similar to mutual funds offered by other financial institutions. Thus, the premiums from these products are invested in a variety of securities, including equities, bonds, and other mutual funds. But the segregated funds operate under the legal requirement that a minimum percentage of the investment (usually 75 percent or more) must be returned to the investor when the product matures, even if the investment performance of the fund has eroded its value to below the minimum level. These are known as segregated funds, because the funds have to be held separate from the other assets of the insurance company (Department of Finance—Canada Website).

By regulation, life annuities (immediate, deferred, variable, and segregated funds) can be sold only by life insurers. But RRIFs and RRSPs also can be offered by other financial institutions, such as banks and mutual funds, creating a very competitive market. However, annuities and other life insurance products do benefit from a favorable tax treatment, albeit less generous than in the United States. For example, when annuities begin periodic income payments, clients may elect to convert the annuity into a prescribed annuity that allows for leveled payments over the annuitant's life expectancy with significant tax deferral. Likewise, many life insurance products are designed to meet many complex income tax requirements, so that they can be categorized as a protection rather than an investment vehicle. This means that the policyholder is not directly taxed on investment gains accruing within the policy; instead, a special uniform corporate income tax rate is applied at the company level. This can be especially beneficial for high-income clients facing a higher tax bracket (CIBC World Markets 1999).

Over the period 1990 to 2003, the yearly average amount of net premiums written by life insurers operating in Canada (in all lines by both domestic and foreign firms) was $33.3 billion. NPW went up from $23.6 billion in 1990 to $46.5 billion in 2003. This corresponds to a compounded annual growth rate of 5.3 percent per year. The amount for 2003 also includes $14.9 billion in segregated fund premiums.

As Table 8.2 indicates, annuity premiums (both individual and group) have been the most important component of premium income for life insurers over the entire time horizon. In 2003, annuity premiums reached $20.9 billion, accounting for 45 percent of the total premium income generated by life companies in that year. This amount includes $9.6 billion of individual annuity premiums and $11.3 billion of group annuity premiums. RRSPs accounted for 37 percent of this total amount. With a persistent low interest rate environment and a buoyant stock market, the demand for fixed annuities in Canada went down through the 1990s, with customers looking to segregated funds that offer returns linked to the stock market (CIBC World Markets 1999). Annuities became a little bit more popular again with the bear market in equities, which lasted until late 2002. Over the period 1990 to 2003, group annuity premiums had a CAGR of 7.3 percent, while individual annuity premiums had a CAGR of only 2.1 percent.

Since 2002, health insurance has become the second most important line of business underwritten by Canadian insurers. Indeed, Table 8.2 shows that health insurance accounted for 28 percent of premium income in 2003 compared to 21 percent in 1990. In 2003, Canadians paid $13.06 billion in health insurance premiums. This amount also includes health insurance premiums from property and casualty insurers who, for example, underwrite accident and sickness insurance just like life insurers. But it does not include the premium equivalents for uninsured group health contracts (known as administrative services only) administered by life insurers on behalf of employers, totaling $6.2 billion in 2003. Individual health insurance premiums had a CAGR of 7 percent over our thirteen-year horizon, compared to 7.7 percent for group health premiums.

Also, looking at Table 8.2, it is clear that life insurance premiums are, since 2002, the least important component of premium income generated by life insurers. They accounted for 27 percent of premium income in 2003 compared to 28.5 percent in 1990. Over this time span, individual life insurance premiums had a CAGR of 5.4 percent, while group life premiums had a CAGR of 3.5 percent. The reduced relative importance of life insurance premiums is an indication that the Canadian life insurance market has matured. In 2003, life insurance premiums (both individual and group) reached a level of $12.5 billion, an increase of 4.87 percent over 2002. Notice that, since the mid-1980s, term (including term 100) and universal life insurance have gained popularity and displaced sales of traditional whole life products.

Table 8.2. Canadian Life and Health Insurance Premium Volume by Line of Business—Net Premiums Written (in millions of Canadian dollars)—1990 to 2003

	Life		Annuity		Health		
	Individual	Group	Individual	Group	Individual	Group	Total Life and Health
1990	4,840	1,879	7,299	4,554	670	4,338	23,580
1991	5,213	1,865	7,591	5,136	684	4,591	25,080
1992	5,627	1,935	7,323	5,612	749	4,742	25,988
1993	5,882	2,056	7,749	5,573	876	5,093	27,229
1994	6,139	2,016	8,412	5,995	940	5,636	29,138
1995	6,385	2,065	7,201	5,723	1,066	6,226	28,666
1996	6,807	2,180	7,054	6,232	1,087	6,388	29,748
1997	7,260	2,273	9,923	7,146	1,109	6,668	34,379
1998	7,887	2,326	9,257	8,185	1,183	7,064	35,902
1999	8,064	2,421	11,111	8,924	1,228	7,638	39,386
2000	8,606	2,541	13,248	10,163	1,301	8,548	44,407
2001	8,707	2,711	9,146	11,906	1,381	9,484	43,335
2002	9,227	2,728	9,212	11,316	1,513	10,637	44,633
2003	9,600	2,937	9,574	11,335	1,621	11,437	46,504

Source: Canadian Life and Health Insurance Facts (2004) available on www.clhia.ca.

The description above focuses only on the life insurance business generated in Canada by both Canadian and foreign firms. It is also important to emphasize that Canadian life insurers generate a significant portion of their premium income outside of Canada. As early as 1960, approximately 35 percent of worldwide total premiums of Canadian life companies were generated outside of Canada, increasing to 46 percent by 1997 (CIBC World Markets 1999) and to 54 percent by 2003 (CLHIA 2004). This shows that Canadian insurers are seeking growth opportunities outside of Canada as the domestic market matures and slows down. Canadian life and health insurers are present in more than twenty countries around the world. This geographical diversification is mostly attributable to the operations of the largest life insurance companies (Great-West Life, Manulife, and Sun Life). According to CLHIA, the majority of Canada's foreign insurance business (data for 2002) comes from the United States (75 percent), with the remainder coming from the United Kingdom and Europe (11 percent), followed by Asia (9 percent) and Latin America (6 percent). Recently, some of the big Canadian players have penetrated emerging markets like China and India. This has often been achieved through joint venture agreements with local life insurers. Such arrangements are made possible and easier now that China has joined the World Trade Organization and India has opened its insurance markets to foreign ownership.

Property and Casualty Insurance

In P&C insurance, premium income is derived from three distinct lines of business—automobile, property and liability—all of which can be either personal or commercial contracts. However, the P&C industry in Canada has an interesting structure due to the fact that automobile insurance is available from either the government or the private sector. British Columbia, Saskatchewan, and Manitoba have government crown corporations that provide compulsory automobile insurance. In these provinces, the government has monopoly over mandatory insurance, while private insurers are allowed to compete in supplemental auto insurance coverages. Quebec has a hybrid of private (for property damages) and government (for bodily injuries) insurance, while the remaining provinces and territories are all served by private insurers. Incomplete coverage of needy users and lower administrative costs are often cited as the reason for such government intervention, including the adoption of no-fault laws in some provinces. Quebec was the first province to adopt no-fault insurance (1978), followed by Ontario (1990), Manitoba (1994), and Saskatchewan (1995). These no-fault laws include specified personal injury protection coverage and restrict tort liability. The other provinces and territories follow a tort system of compensation for automobile accidents (Kleffner and Schmit 1999).

The empirical evidence on the effects of the primary features of each province's automobile insurance scheme on claims cost is rather limited. A recent report released by the Canadian Institute of Actuaries (CIA) focuses on the total claims cost incurred by insurers (both private and public) over the period 1998 to 2002 (except British Columbia). Results show a lot of variation in the average claims cost per car (private passenger only) from province to province. For 2002, costs in the highest-cost province (Ontario) are 80 percent higher than the costs in the lowest-cost province (Québec). Much of this variation is due to differing coverages and benefits in different provinces (CIA 2005). Moreover, there are also studies providing evidence of higher incidence of accidents and higher mortality rates in provinces with government-run insurance schemes (Kovacs and Leadbetter 2003).

The issue of costs and benefits of government intervention in automobile insurance is still under debate. Indeed, issues surrounding the affordability and availability of auto insurance are frequently part of provincial election platforms. During the most recent auto insurance crisis, there were proposals to privatize mandatory insurance in provinces that rely on government insurers and/or no-fault laws. But, at the same time, there were also proposals to nationalize auto insurance in some of the tort provinces. In this context, provinces chose to initiate reforms designed to modify, in the short run at least, some of the parameters of their current system. This means that the different automobile insurance systems in force across Canada continue to reflect the variations in provincial objectives and priorities. In 2003, government insurers in Canada wrote \$4.9 billion in net auto insurance premiums, while private insurers wrote \$15.8 billion.

Also worth mentioning is that, in Canada, the system of workers' compensation is operated at the provincial level by a government monopoly, called a Workers' Compensation Board. This system is one of Canada's largest social programs since it covers work-related injuries and occupational diseases. Harrington et al. (2004) state that: "In 2002, Canadians experienced over a million compensable injuries and received nearly $6.1 billion in workers' compensation benefits" (p. 381).

The discussion below focuses only on the activities of private insurers in the non-life market. Over the period 1990 to 2003, the yearly average amount of net premiums written by general insurers operating in Canada was $18.5 billion. This includes all lines of insurance written within Canada, by both domestic and foreign firms. NPW went up from $13.3 billion in 1990 to $31.4 billion in 2003, realizing a CAGR of 6.8 percent per year. However, domestic private P&C insurers wrote only $283.73 million of net premiums outside Canada in 2003. This is less than 1 percent of the overall premium volume written in Canada during the year, indicating that, contrary to life firms, domestic non-life firms are not very active outside of Canada.

As Table 8.3 indicates, automobile insurance premiums (both personal and commercial, including some third-party liability) have been the most important component of premium income for P&C insurers over the thirteen-year horizon, even if they accounted for only 50.2 percent of the total premium volume in 2003 compared to 53.5 percent in 1990. Over this time period, private automobile insurance premiums had a CAGR of 6.3 percent. It is also worth mentioning that auto insurance premiums grew faster during the subperiod 1998 to 2003, in line with the increase in claims experienced in the provinces dominated by private insurers.

Table 8.3. Canadian Property and Casualty (P&C) Insurance Premium Volume by Line of Business—Net Premiums Written (in millions of Canadian dollars)—1999 to 2003

	Automobile	Personal Property	Commercial Property	Total Property	Commercial Liability	Other P&C Lines	Total P&C
1990	7,119	2,272	1,849	4,121	1,305	759	13,304
1991	7,496	2,492	1,793	4,285	1,302	821	13,904
1992	7,763	2,642	1,866	4,508	1,319	912	14,502
1993	8,158	2,803	2,062	4,865	1,298	918	15,239
1994	8,697	3,042	2,337	5,379	1,430	976	16,482
1995	9,403	3,163	2,553	5,716	1,694	1,258	18,071
1996	9,597	3,246	2,658	5,904	1,867	1,202	18,570
1997	9,553	3,281	2,711	5,992	1,878	1,185	18,608
1998	9,686	3,383	2,469	5,852	1,823	1,198	18,559
1999	9,839	3,293	2,434	5,727	1,846	1,316	18,728
2000	10,705	3,429	2,591	6,020	1,982	1,471	20,178
2001	11,281	3,481	2,768	6,249	2,194	1,518	21,242
2002	13,150	3,971	3,909	7,880	3,145	3,332	27,507
2003	15,781	4,452	4,518	8,970	4,081	2,581	31,413

Source: Insurance Bureau of Canada, based on data from Statistics Canada, A.M. Best Canada's WinTRAC and MSA Researcher.

Property insurance (both personal and commercial) covers homeowners, tenants (lessees who pay rents), and businesses for losses to buildings and contents. Personal property insurance also includes personal liability coverage. Overall, property insurance is the second most important line of P&C insurance in Canada. It accounted for 31 percent of the premium income in 1990 compared to 28.6 percent in 2003. Over the period 1990 to 2003, personal property insurance had a CAGR in premiums of 5.3 percent, whereas commercial property insurance had a CAGR of 7.1 percent. In 2003, Canadians paid almost $9 billion in total premiums for property insurance, split almost evenly between personal and commercial lines.

While the percentage contributions of automobile and property lines of business to the total P&C premiums have decreased over the period, on a relative basis, commercial liability insurance has gained in importance over the period. This includes coverages for product liability and directors' liability. As shown in Table 8.3, commercial liability insurance accounted for 9.8 percent of the overall P&C premium volume in 1990 compared to 13 percent in 2003, with a CAGR of 9.2 percent. In 2003, commercial liability insurance premium income was $4.1 billion. Also seen in Table 8.3 is the fact that there are many other types of (more specialized) P&C insurance such as boiler and machinery insurance, surety and fidelity insurance, and marine and aircraft insurance. A number of P&C insurers also sell some accident and sickness insurance. These other types of P&C insurance lines accounted for about 8 percent ($2.6 billion) of the overall P&C premium business written in Canada in 2003, with a CAGR of 9.9 percent over the period 1990 to 2003.

To summarize, although there were more individual P&C firms (206) compared with life firms (108) operating in Canada in 2003, in terms of NPW, the life sector was a bigger player, and continues to be. The share of the life sector in premium volume for 2003 was 60 percent, compared with a 40 percent share for the non-life sector. This is despite a faster compounded annual growth rate of 6.8 percent per year in the P&C sector, versus a CAGR of 5.4 percent in the life sector.

In the life insurance industry, annuity products (including a variety of tax-sheltered retirement vehicles) are the key premium generators, followed by health insurance and then life insurance. In the P&C sector, automobile insurance dominates the market, followed by property and commercial liability business.

8.4.5 Benefits and Claims in 2003

Life and Health Insurance

During 2003, benefits paid to Canadian policyholders and annuitants totaled $38 billion, compared to $36 billion the year before. The former number does not include payments under uninsured contracts administered by life insurers totaling $5.8 billion in 2003. Of this total amount for 2003, $5.2 billion was paid for life insurance (13.7 percent), $20.1 billion for annuity contracts (52.9 percent), $10.6 billion for health insurance (27.9 percent), and $2.1 billion in policy dividends (5.5 percent). Almost 60 percent of the total amount of benefits paid in 2003 was related to group contracts, and the rest came from individual contracts. Over the period 1990 to 2003, benefits paid by life insurers have grown at a CAGR of 5.6 percent.

It is also interesting to note that 90 percent of those payments went to living policyholders, whereas 10 percent were paid out as death benefits, as was the case in 2002. This is an indication of the changing role of life insurers from being just providers of insurance to managers of wealth and assets on behalf of their clients and providers of retirement products.

Property and Casualty Insurance

Claims incurred by Canadian P&C (private) insurers were $20.4 billion in 2003. Claims incurred are estimates of total outstanding claims and claim expenses at the end of a term, plus all claims paid during the period, minus the total of outstanding claims at the beginning of the term. Over the period 1993 to 2003, claims incurred have grown at a compounded annual growth rate of 5.4 percent.

In 2003, an amount of $12 billion (59 percent) was paid out by insurers in automobile insurance claims, followed by personal property and liability insurance with $2.6 billion in claims paid out for each line (25 percent). Commercial property insurers registered claims of $2.2 billion (11 percent). Lastly, claims incurred for all other lines (boiler and machinery, marine and aircraft, surety and fidelity) were $0.993 billion (5 percent). In 2003, theft accounted for 22 percent of all homeowners' claims. Other causes, such as hail and wind accounted for 39 percent, followed by water damage to homes at 27 percent and fire at 12 percent.

In summary, consistent with the larger share of the premiums written by life insurers, the claims paid out are also higher for these firms (65 percent) in comparison with the P&C firms (35 percent). The dominance of living policyholders as claim recipients in the life industry indicates the changing role of life insurers in line with the demographics of the Canadian population.

8.4.6 Distribution Channels

In Canada, distribution of insurance products is still the domain of the insurance firms, despite repeated efforts by Canadian banks to break this monopoly. This is often attributed to the political power of the large independent insurance brokerage firms, who have steadfastly resisted the pressure to allow banks to compete in their marketplace on an equal footing. As a result, banks have limited access to the distribution of insurance products—both life and non-life. Banks can offer limited insurance products, such as life insurance (often underwritten by traditional insurance firms), if it is offered as part of a package of other banking products such as car loans or mortgage loans, a practice known as tied-selling. Although such tied-selling is permitted, cross-selling insurance products to their banking clients using their extensive branch network is not permitted. Of course, banks are allowed to directly compete in the insurance market through wholly owned subsidiaries, but they cannot use the client list from their banking operations to potentially sell insurance products. This has often been viewed as creating an artificial barrier to the distribution of personal insurance products, making it more costly to provide a holistic package of financial services to customers.

Despite significant lobbying efforts by the major banks in Canada to get entry into the distribution of insurance products, the situation is not expected to change in the near future. Several banks have tried to enter the insurance markets by establishing their own insurance subsidiaries, only to withdraw from the market after finding that such stand-alone insurance operations do not add value in the absence of an extensive agency network for the distribution of products and the lack of actuarial expertise for the design and risk management of the products. Both of these have been and still remain the prerogative of the insurance firms.

Life and Health Insurance

Canadian life insurers distribute their products either through career or independent distributors or a mix of both. Career (full-time) distributors are under contract with one company to sell primarily that company's products. Independent distributors can be part of one of the following channels: brokers/personal producing general agents; managing general agents/independent marketing organizations; stockbrokerage; manufacturers' distribution agreement; and, other less important channels, including direct response (mail, telephone, and Internet) and financial institutions (banks, trust companies, and credit unions).

Life insurance firms, which use independent distributors to sell their products, benefit from not having the costs of setting up and developing their own exclusive/captive career agent networks. But they may lose out on the need to pay higher commissions to the independent distributors or the possibility that their products may not be the preferred choice of the brokers.

Brokers and personal producing general agents (Brokers/PPGAs) are in direct contact with the clients and are free to sell the products of several firms. They own the client list, and the customers often are not aware of the insurance company that underwrites the products they purchase. Managing general agents (MGAs) and independent marketing organizations (IMOs) are firms that appoint independent agents to serve a selected market. They also represent many companies. Stockbrokers are registered representatives employed by a stock exchange member to serve the high net worth niche of the market. The manufacturers' distribution agreement channel is made up of business manufactured by a given company and sold through another (nonaffiliated) insurer's distribution system. Finally, there are other channels with much smaller market shares, including direct home office offerings through mail and media advertising, telemarketing, and Internet efforts and the life insurance sold by banks and other financial institutions.

Despite the relatively generous social security benefits in Canada, more sophisticated and wealthy investors tend to rely on private investments and private pension plans in order to maintain their lifestyle in retirement. In parallel with this trend, life insurers offer products designed to accumulate and protect financial capital as an alternative to the traditional pension products. Although mutual funds tend to dominate these markets, investment-linked insurance products such as universal life and segregated funds are gaining popularity in the marketplace. This, in turn, has led to the increased use of independent channels for distribution of such products, because they are able to market these products more effectively than traditional insurance channels.

During the period under consideration in this study, independent channels have seen their market share of insurance sales rise over time, at the expense of exclusive career agents whose market share fell to 50 percent in 1994 from 59 percent in 1990 (CIBC World Markets 1999). Since 1994, such independent channels have continued to increase their share of new insurance sales at the expense of exclusive career agents. In 2001, approximately one third of the insurance groups operating in the Canadian life insurance market distributed solely through a career channel, one third used only one independent channel, and the remainder used multiple channels (Van Gorder 2004).

In Figure 8.4, we provide a breakdown of the market shares of annualized new individual life premiums by distribution channels in Canada for 2003. The brokers/PPGA channel dominates with a market share of 44 percent, followed by the career channel with 32 percent. The MGA/IMO channel is at 13 percent, an increase of 3 percent over 2002 at the expense of the brokers/PPGA channel. The stockbrokers channel has a market share of 7 percent in 2003. Finally, the manufacturers' distribution agreement channel and the other channels (direct response and financial institutions) each retain 2 percent of the market, as they did in 2002 (Van Gorder 2004).

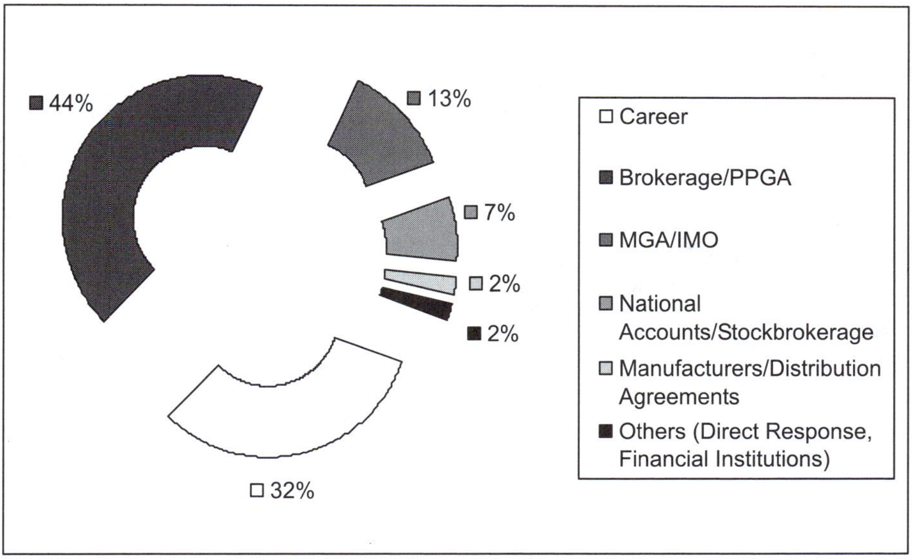

Source: Van Gorder (2004).

Figure 8.4. Market Share of Annualized New Individual Life Premiums by Distribution Channel in the Canadian Life and Health Insurance Industry, 2003

The predominance of independent channels in the distribution of insurance products and the decline of the exclusive agency system can be attributed to the lower cost of distribution, the exclusivity of the client lists, as well as the expertise provided by such independent brokers/agents and advisors. This is consistent with evidence in the United States that exclusive-dealing insurers are significantly less cost efficient than either nonexclusive-dealing or direct-writing insurers (Carr, Cummins, and Regan 1999). However, in Canada the pattern of distribution also has been driven by customer preferences, as many clients still seem to prefer and value the face-to-face contact and the advice provided by independent agents, particularly for the more complex annuity and investment-related life products. This also reflects the changing role of independent insurance agents and brokers into financial planners and asset managers. It is also interesting to note that multichannel distribution is almost nonexistent at the individual firm level. This means that each insurance firm uses either its own exclusive career agents or independent distributors. Although some insurance groups may be characterized as using multichannel distribution for the group as a whole, each company within the group tends to use just one method of distribution.

Property and Casualty Insurance

P&C firms can be classified into three distinct groups in terms of their distribution networks: independent agency (brokers) writers, direct writers (with their exclusive/captive agency networks), and direct response writers (using impersonal electronic channels).

As with the life insurance industry, independent agency writers dominate the P&C market with about 70 percent of the market. This is despite the growth in the market share of exclusive agents to about 25 percent during the 1990s. Despite optimistic predictions about the potential growth of direct response channels, such as the telephone and Internet, such channels (which tend to treat insurance products as commodities) have a market share of only about 5 percent (see Figure 8.5).

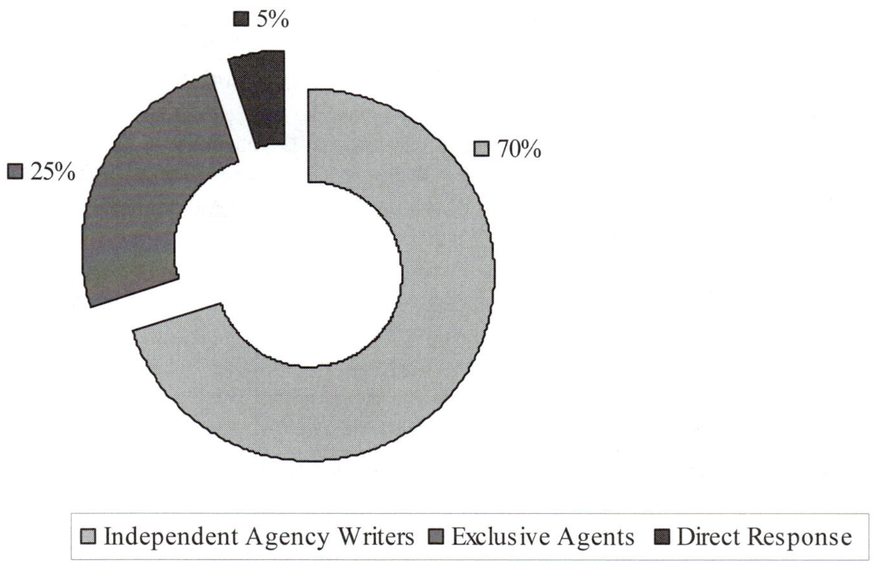

Source: Kelly and Kleffner (2004).

Figure 8.5. Distribution Channels in the Canadian Property and Casualty Insurance Industry, 2003

It is also interesting to note that there are no significant cost differences between independent and exclusive agency writers in the P&C industry in Canada (Kelly and Kleffner 2004). This situation does differentiate Canada from the United States, where such differences have frequently been found in previous studies (Berger,

Cummins, and Weiss 1997). Further, independent agents are viewed as being very important in the commercial business lines, where they tend to specialize in the risk assessment and risk management for their clients. As with the life insurance industry, there is limited use of multichannel distribution by P&C firms.

To summarize, both the life insurance industry and the P&C industry in Canada are dominated by independent agents as the key players in the distribution of insurance products. And, contrary to expectations, nontraditional channels, such as telephone and the Internet have a very small share of the market.

8.4.7 Market Concentration

Concentration is the best-known and most often used indicator of an industry's market structure and its competitiveness. In general, concentration is defined as the number and size distribution of sellers in the marketplace. Many controversies surrounding the construction of concentration indexes exist. Nevertheless, concentration ratios (for example, the market share of top 1, top 5, and top 10 firms) are most commonly used in order to assess the development of the dominance of the largest companies (or groups) and/or special market segments. Another popular indicator is the Herfindahl Index. Easy to calculate, it accounts for the relative size and the distribution of the whole set of companies in the market. The index is calculated by summing the squares of the individual market shares of all competing firms and multiplying the result by 10,000. The Herfindahl index can range from a high of 10,000 (for a single-firm monopolist) to a low of almost zero (for an atomized industry with many, very small firms). For example, an industry with ten companies of equal size would give a Herfindahl index of 1,000 (Swiss Re 1999). Here, we focus on two Herfindahl indexes calculated on the basis of the market shares of the Top 5 and Top 10 companies with respect to the industry as a whole.

Life and Health Insurance

Despite the relatively large number of life insurers operating in Canada (108 in 2003), the life insurance industry is rather concentrated, no matter whether concentration is assessed in terms of assets, number of policies underwritten, or premiums written (McIntosh 1998). Table 8.4 presents key concentration ratios of the life insurance business based on the 1997 and 2003 net premium income written in Canada. The table includes all of the major lines of business and accounts for recent mergers and acquisitions. It also groups companies according to their ultimate affiliation.

Clearly, Table 8.4 indicates that the development of concentration is heterogeneous in the different lines examined. For example, the market shares of the biggest companies have increased more substantially in individual lines than in group lines. This tends to confirm the possibility that the latest restructuring of the industry, through merger and acquisition activities, led to more consolidation of industry structure in individual lines. Thus, individual life business is more concentrated and possibly less competitive than the group life business (all lines included). Based on the evolution of the Herfindahl index for the group annuity line, it appears that this market has become even more competitive between 1997 and 2003.

Table 8.4. Canadian Life and Health Insurance Market Share by Net Premiums Written, 1997 and 2003

	Individual Life	Group Life	Individual Annuity	Group Annuity	Individual Health	Group Health	Total Life and Health
				1997			
Top 1	21.0%	16.8%	4.1%	7.7%	15.4%	19.7%	16.2%
Top 5	56.0%	58.0%	16.9%	25.7%	45.0%	58.0%	53.0%
Top 10	73.0%	81.0%	24.7%	35.1%	60.0%	75.0%	75.0%
5 Herfindahl Index	1,146.57	717.15	58.35	142.71	478.70	755.67	610.70
10 Herfindahl Index	1,211.16	841.77	72.62	167.24	523.83	829.32	720.29
				2003			
Top 1	23.5%	19.2%	9.1%	4.1%	16.4%	19.9%	12.4%
Top 5	69.1%	56.6%	32.4%	17.0%	51.9%	51.5%	39.2%
Top 10	81.2%	62.1%	37.2%	18.9%	57.6%	60.3%	47.4%
5 Herfindahl Index	1,225.92	866.15	238.39	59.42	638.14	718.64	367.39
10 Herfindahl Index	1,256.43	873.71	244.46	61.56	646.20	735.78	386.38

Source: CIBC World Markets (1999) and data from the Office of the Superintendent of Financial Institutions (2004).

Property and Casualty Insurance

In comparison, the Canadian P&C insurance sector is much more fragmented than the life sector. As Table 8.5 indicates, the largest P&C insurance group had a market share (based on total net premiums written) of 10 percent in 2003, while the share of the Top 10 insurance groups was 53 percent. Based on the Herfindahl indexes, it is also quite clear that the largest firms in the non-life sector do not differ much in size, given that the index values are relatively low. But the significant increase in the Herfindahl indexes for the whole P&C market between 1993 and 2003 can easily be attributed to the consolidation of industry structure due to mergers and acquisitions.

On a line-by-line basis for 2003, it appears that liability insurance remains the least concentrated line of business compared to the others. It also seems to be the most stable line in terms of its degree of concentration over the period 1993 to 2003. Indeed, automobile and property insurance experienced quite a significant increase in their respective Herfindahl indexes between 1993 and 2003.

In summary, merger and acquisition activity across Canada increased consistently between 1992 and 1998. As an indication, the total value of 1998 Canadian merger and acquisition deals exceeded $150 billion, with the financial services industry accounting for over 10 percent of this amount (CIBC World Markets 1999). This trend was highly driven by the desire of financial institutions, including insurers, to improve their competitive position. Indeed, it has impacted both the life and non-life sectors of the Canadian insurance industry. As a result, both sectors of the industry are now more concentrated than before. However, there is evidence that the life insurance industry is, on the whole, more concentrated than the P&C industry in Canada. Moreover, within the life sector, group lines of business are more competitive than individual lines. This may be due to the changing demographics of the population and the corresponding increase in the demand for wealth management products. Such products are best provided by firms that tend to be large and specialized in the creation of innovative products, made possible by economies of scale and scope.

Table 8.5. Canadian Property and Casualty (P&C) Insurance Market Share by Net Premiums Written, 1993, 1998, and 2003

	Automobile	Property	Liability	P&C Market
		1993		
Top 1	7.5%	5.9%	7.9%	*6.1%*
Top 5	26.8%	26.3%	30.0%	*26.5%*
Top 10	47.9%	47.0%	48.0%	*43.9%*
5 Herfindahl Index	157.55	139.88	189.77	*142.38*
10 Herfindahl Index	247.49	226.34	254.48	*203.35*
		1998		
Top 1	9.3%	9.8%	8.9%	*8.8%*
Top 5	38.1%	36.3%	30.0%	*34.5%*
Top 10	59.0%	54.6%	47.7%	*52.5%*
5 Herfindahl Index	298.07	281.75	193.41	*247.28*
10 Herfindahl Index	387.69	350.56	257.37	*312.98*
		2003		
Top 1	12.8%	12.3%	8.3%	*10.4%*
Top 5	43.3%	39.1%	33.1%	*34.5%*
Top 10	68.6%	59.6%	55.4%	*52.9%*
5 Herfindahl Index	415.11	345.35	224.02	*266.55*
10 Herfindahl Index	545.19	430.08	324.35	*334.62*

Source: Insurance Bureau of Canada internal data (2004) and *Canadian Insurance* (editions 1994, 1999, and 2004).

8.4.8 The Big Players

Life and Health Insurance

As shown in Table 8.6, in 2003, the Top 10 players in the Canadian life insurance sector were mostly insurance and financial groups of the stock form (eight out of ten), with domestic ownership (seven out of ten) and with a multichannel hybrid distribution system (agency and brokerage) for the group as a whole. But the individual firms within each group focused only on one distribution channel.

Among the largest life insurance groups, four are currently listed on the Toronto Stock Exchange. Great-West Lifeco has always been a public stock company. Manulife, Sun Life, and Industrial-Alliance were listed on the Toronto Stock Exchange following the demutualization wave of 1999–2000. During this period, Canada's five largest mutual life insurers (Canada Life, Industrial-Alliance, Manulife, Clarica, and Sun Life) initiated plans to convert to a more flexible stock company structure in order to expand into new facets of the financial services sector using their stock for acquisitions. In March 1999, the Department of Finance approved the regulations governing the conversion of a mutual life insurance company, irrespective of size, to a stock ownership company. These regulations outline the necessary steps of the demutualization process. In the case of Industrial-Alliance, given that it is regulated by the Province of Quebec, the conversion was handled through a private bill following discussions with the provincial regulator. Whereas the federally incorporated insurance companies were obliged to distribute the firm's value only to participating policyholders, Quebec's legislation considered all policyholders, participating and non-participating, to be the owners of a mutual company like Industrial-Alliance and, therefore, eligible for wealth distribution (CIBC World Markets 1999).

In terms of initial public offering (IPO) performance, there is strong evidence of significant underpricing in the five life and health IPOs that took place during the demutualization wave. The average estimate is 9 percent from the offer price to the first day closing price. As expected, most short-term returns were achieved during the first day of trading. Also, there is evidence of strong returns even if the first day of trading is excluded. Likewise, there is evidence of post-demutualization abnormal performance. Indeed, a secondary market investor would have outperformed the market from day one up to two years after the IPOs (Babin and Bernier 2001).

Once the two-year window imposed by the federal government to restrict post-demutualization merger and acquisition activities was lifted in 2002, Great-West Lifeco acquired Canada Life, and Sun Life merged with Clarica. Both Canada Life and Clarica were listed on the Toronto Stock Exchange prior to being acquired by Great-West Lifeco and Sun Life. Manulife has been the most active in the international arena, expanding its operations in Asia through joint ventures with Chinese and Indian firms, and in the United-States through the recent merger with Boston-based John Hancock Financial Services.

Property and Casualty Insurance

As shown in Table 8.7, in 2003, the Top 10 players in the Canadian property and casualty sector were mostly insurance groups of the stock form (seven out of ten) with domestic ownership (six out of ten) and with a brokerage type distribution system.

The largest provider of property and casualty insurance in Canada during 2003 was ING Canada with $2.97 billion of NPW. ING Canada's principal insurance products are automobile, property, and liability insurance, which are provided primarily to individuals and small to medium-sized businesses through its insurance subsidiaries across Canada. Recently (on December 15, 2004), ING Canada completed an IPO through which it issued 34.88 million common shares at $26 per

Table 8.6. Top 10 Canadian Life and Health Insurers by Net Premiums Written (NPW) in 2003

Rank	Group	NPW (in thousands of Canadian dollars)	Organizational Form	Distribution System	Ownership Type	Country of Ultimate Owner
1	Manulife	5,757,479	Stock	Agency, Broker	Insurance and Financial Group	Canada
2	Sun Life	5,166,649	Stock	Career Agent, Broker	Insurance and Financial Group	Canada
3	Great-West	3,523,648	Stock	Agency, Broker	Insurance and Financial Group	Canada
4	Desjardins Financial	2,000,027	Stock	Agency, Broker	Financial Cooperative Group	Canada
5	Industrial Alliance	1,787,631	Stock	Agency, Broker	Insurance and Financial Group	Canada
6	Standard Life	1,746,035	Mutual	Agency, Broker	Insurance and Financial Group	United Kingdom
7	SSQ	577,851	Mutual	Broker	Insurance and Financial Group	Canada
8	Co-operators	518,893	Stock	Direct	Insurance and Financial Group	Canada
9	UnumProvident	483,577	Stock	Broker	Holding Company	United States
10	AEGON	469,047	Stock	Agency, Broker	Holding Company	United States

Source: Best's Key Rating Guide (2004) and data from the Office of the Superintendent of Financial Institutions (2004).

Table 8.7. Top 10 Canadian Property & Casualty Insurers by Net Premiums Written (NPW) in 2003

Rank	Group	NPW (in thousands of Canadian dollars)	Organizational Form	Distribution System	Ownership Type	Country of Ultimate Owner
1	ING Canada	2,966,609	Stock	Broker	Insurance and Bank Group	Netherlands
2	Aviva Canada	2,766,812	Stock	Broker	Insurance Group	United Kingdom
3	Co-operators	1,705,401	Stock	Direct	Insurance and Financial Group	Canada
4	Economical Ins.	1,449,598	Mutual	Broker	Insurance Group	Canada
5	State Farm Ins.	1,352,505	Mutual	Direct	Insurance and Financial Group	United States
6	Desjardins Financial	1,235,801	Stock	Direct	Financial Cooperative Group	Canada
7	Wawanesa	1,152,836	Mutual	Broker, Direct	Insurance Group	Canada
8	Royal & Sun Alliance	1,131,571	Stock	Broker	Insurance Group	United Kingdom
9	AXA Canada	1,118,893	Stock	Broker	Insurance Group	Canada
10	Dominion of Canada	1,026,658	Stock	Broker	Insurance and Financial Services	Canada

Source: Best's Key Rating Guide (2004) and *Canadian Insurance* (2004).

share. Following the offering, ING Groep N.V. will hold 72.9 percent of the 128,500,000 issued and outstanding common shares. ING Groep N.V., from the Netherlands, is one of the largest bancassurance firms in the world.

8.5 FINANCIAL CONDUCT AND PERFORMANCE

The Canadian insurance industry is a key component of the Canadian financial services industry. However, the life insurance sector dominates in terms of premiums written ($46.5 billion in 2003) and total assets under management ($315 billion in 2003), compared to premiums of $31.4 billion and assets under management of $88 billion in 2003 for the non-life sector. But the insurance industry as a whole manages less than 20 percent of the total assets of the Canadian financial services sector of approximately $2 trillion in 2003.

8.5.1 Assets and Investments Activities

Life and Health Insurance

The first column in Table 8.8 reports the assets held by the life firms, on behalf of Canadian policyholders and annuitants over the 1990 to 2003 period. These numbers exclude mutual fund assets, which have been a major source of expansion for the large Canadian life insurers over the past decade. The compounded annual growth rate of the industry's total assets over those years was 7 percent. This is 1.6 percent more than the CAGR of premium income during the same period. For 2003, the total assets under management was $315 billion, of which $214 billion were assets held on the balance sheet of the firms, while $101 billion were assets held and managed off the balance sheet as segregated funds. As financial intermediaries, life insurers invest most of their premium income and reserves in financial markets. In 2003, more than 80 percent of the assets held in the Canadian life insurance sector were in the form of financial assets, with the rest in real (physical) and other nonfinancial assets. Companies typically formulate their investment policy according to an asset-liability matching program resulting in a higher allocation to fixed-income instruments (cash, bonds, and mortgages) and a smaller proportion to variable-income vehicles (stocks, mutual funds) and hybrid instruments. This is designed to match the long duration of their liabilities.

Since the 1992 adoption of the Insurance Company Act, Canadian life insurers are subject to the "prudent person" rule. This rule aims at ensuring proper diversification and at optimizing the risk-return profile of managed funds. Life insurers also face other regulatory restrictions on their investment strategies, but these are broad guidelines and do not typically hinder a company's asset allocation decisions. For example, life insurers face no restrictions on consumer lending. However, their commercial lending is limited to 5 percent of total assets if regulatory capital is less than $25 million, and any excess over 5 percent requires regulatory approval if the minimum required capital is greater than $25 million. Likewise, real estate or equity investments cannot be greater than the sum of (1) 70 percent of regulatory capital, (2) 15 percent of nonparticipating liabilities, (3) 25 percent of participating liabilities, and (4) 5 percent of liabilities related to prescribed annuities.

At the same time, the sum of equity and real estate investments cannot be greater than the sum of (1) 100 percent of regulatory capital, (2) 20 percent of nonparticipating liabilities, (3) 40 percent of participating liabilities, and (4) 5 percent of liabilities related to prescribed annuities (CIBC World Markets, 1999).

Table 8.8. Total Assets Held in Canadian Insurance Industry (in millions of Canadian dollars), 1990 to 2003

	Life and Health	Property and Casualty
1990	130,862	33,659
1991	145,068	35,569
1992	159,314	36,827
1993	171,195	38,692
1994	176,299	40,011
1995	183,487	43,370
1996	193,094	51,499
1997	208,952	55,871
1998	227,185	57,369
1999	248,938	59,858
2000	267,141	62,478
2001	276,491	66,324
2001	279,522	77,611
2003	314,948	88,254

Source: Canadian Life and Health Insurance Facts (1991 to 2004) and Insurance Bureau of Canada, with data from MSA Researcher, OSFI, CI, and Insurance Information Centre of Canada (2004).

In an effort to avoid another liquidity crunch like the one that led to the failure of Confederation Life in 1994, Canadian life insurers have significantly reduced their domestic mortgage and real estate positions to 15.5 percent of total assets in 2003, dropping from 31 percent in 1997 and 45 percent in 1990. As seen in Panel A of Table 8.9, the remaining allocation of total assets for 2003 was as follows: 4.7 percent in cash, 41.2 percent in bonds (of which more than half are in government bonds), 16.7 percent in mutual funds, 14.4 percent in stocks, and 7.5 percent in other nonfinancial assets.

Companies also are required by regulators to pay close attention to the quality of their investments, in particular the credit quality of their corporate investment portfolio. For this purpose, they monitor quality ratios such as percent of impaired assets, percent of loan loss provisions, percent of corporate bond holdings rated BB and lower, and delinquency rates on mortgages. During the period 1994 to 1997, all major Canadian life insurers saw their percentage of impaired assets and their loan loss provisions fall steadily. However, the situation became critical during the 2001–2002 downturn, when some corporate bonds, especially those in the telecommunications sector, were downgraded. Many life insurers had to write off a portion of their debt portfolio in order to deal with these bond downgrades. For example, Industrial-Alliance took a 100 percent provision for Teleglobe during 2002, with a 2.2 percent impact on the firm's ROE for that year. Since then, the situation is back to normal given the healthier economic environment in Canada.

Table 8.9. Allocation of Total Assets (Life and Health) and of Invested Assets (Property and Casualty) in the Canadian Insurance Industry, 2003 (in millions of Canadian dollars)

	Panel A: Life and Health		Panel B: Property and Casualty	
	Absolute	Relative	Absolute	Relative
Federal Bonds	28,228	9.0%	16,305	30.8%
Provincial Bonds	36,500	11.6%	9,445	17.8%
Municipal Bonds	4,185	1.3%	1,549	2.9%
Total Government Bonds	68,913	21.9%	27,299	51.5%
Corporate Bonds	60,678	19.3%	13,078	24.7%
Total Bonds	129,591	41.1%	40,377	76.2%
Preferred Shares	2,186	0.7%	3,155	6.0%
Common Shares	43,184	13.7%	4,815	9.1%
Total Shares	45,370	14.4%	7,970	15.0%
Term Deposits	n.a.	n.a.	3,237	6.1%
Residential Mortgages	17,449	5.5%	n.a.	n.a.
Non-Residential Mortgages	23,291	7.4%	n.a.	n.a.
Total Mortgages	40,740	12.9%	581	1.1%
Mutual Funds	52,515	16.7%	n.a	n.a.
Real Estate	8,031	2.5%	146	0.3%
Cash	14,943	4.7%	n.a.	n.a.
Other	23,758	7.5%	701	1.3%
Total Assets/Total Invested Assets	314,948	100%	53,012	100%

Source: Canadian Life and Health Insurance Facts (2004) and Insurance Bureau of Canada internal data (2004).

Property and Casualty Insurance

Table 8.8 also reports the assets held by property and casualty insurers over the period 1991 to 2003. The CAGR of the industry's total assets over this period was 7.7 percent, which is about 1 percent higher than the CAGR of premium income during the same period. For 2003, the industry's total assets were $88.3 billion, which represents a significant increase of almost 14 percent over 2002. In 2003, 60 percent of the total assets were invested in long-term financial assets, while the other 40 percent was made up of cash, money market instruments, accrued and due investment income, and receivables from agents/brokers, policyholders, other insurers, subsidiaries, and affiliates.

Panel B of Table 8.9 indicates that the composition of the financial assets of Canadian property and casualty insurers in 2003 was as follows: 51 percent in government bonds (mostly federal and provincial), 25 percent in corporate bonds, 15 percent in shares (preferred and common), 6 percent in term deposits, and, finally, 3 percent in mortgages, real estate, and other vehicles. With more than 80 percent of their financial assets invested in safe and liquid fixed-income instruments, it is fair to say that Canadian property and casualty insurers rely on a prudent investment policy in keeping with the short-term nature of most of their liabilities.

To summarize, it is interesting to note that while 80 percent of the assets of the life firms are financial, the property and casualty firms hold only 60 percent of their assets as financial assets. In addition, safe, liquid, government, and corporate bonds amount to about 75 percent of the P&C portfolios, while the amount is just under 50 percent for life insurers. This may be a reflection of the conservative investment philosophy of the P&C firms, with their need to meet most claims at very short notice. It may also be a reflection of the changing focus of life firms from pure risk management to asset and wealth management for their clients.

8.5.2 Financial Performance

Life and Health Insurance

The early 1990s were marked by poor financial performance in the life insurance sector. This was primarily due to high loan loss provision levels, specifically on commercial real estate, and the slow economic recovery from the recession of 1990–1991. Loan loss provisions consist of specific provisions for loans and debt considered to be impaired and a general provision for other future potential credit losses, as assessed by the insurer. Under Canadian Generally Accepted Accounting Principles (GAAP), when an asset is impaired, interest is no longer accrued and previous interest accruals are reversed. Since the early 1990s, however, declining loan losses due to better credit analysis and improving economy, as well as improving mortality experience, due to better underwriting standards and general improvements in the health of the population, have combined to raise domestic industry earnings from almost break-even in 1992 to much higher levels by the end of the decade (CIBC World Markets 1999).

Thus, Canadian life insurers saw their profit margin (a gauge of profitability) go up from 8.62 percent in 2000 to 10.36 percent in 2003. Meanwhile, foreign life insurers saw their profit margin stay stable at 8.14 percent. This ratio amounts to what is left from each dollar of revenue after deducting policy benefits and other expenses. In terms of asset yield (calculated as total revenues divided by average total assets), Canadian life insurers went from an asset yield of 27.56 percent in 2000 down to 19.40 percent in 2003, while foreign life insurers saw their asset yields increase significantly from 22.84 percent to 30.98 percent during the same two years. This possibly indicates that foreign life insurers were able to recover better than the domestic firms from the recession of 2001–2002.

Through a combination of strong segregated fund sales, additional leverage, and cost reduction efforts, the Canadian life sector has also increased its return on equity (ROE) to record levels, especially since the turn of the century. Here, equity is defined as total surplus for mutually owned firms and shareholder equity for stock-owned companies. As Table 8.10 indicates, over the period 1992 to 2003, the life sector achieved an average ROE of 9.3 percent with a standard deviation of 2.96 percent. The lowest ROE (3.3 percent) was obtained in 1992, while the highest (13.1 percent) was realized in 2003.

Table 8.10. Return on Equity (ROE) and Technical Results in the Canadian Insurance Industry (in millions of Canadian dollars), 1990 to 2003

		Property and Casualty			
	Life and Health ROE	ROE	Underwriting Profit (Loss)	Net Profit	Combined Ratio
1990	n.a.	9.7%	(1,234)	942	110.4%
1991	n.a.	9.6%	(1,421)	996	111.2%
1992	3.3%	8.5%	(1,375)	922	110.6%
1993	4.5%	9.5%	(1,351)	1,077	109.9%
1994	7.6%	6.8%	(1,027)	809	107.0%
1995	9.7%	11.7%	(631)	1,462	104.1%
1996	10.0%	13.6%	(517)	1,876	103.4%
1997	10.4%	13.1%	(421)	2,000	102.6%
1998	8.9%	6.8%	(1,366)	1,101	107.8%
1999	9.2%	6.5%	(1,027)	1,094	105.9%
2000	12.6%	6.3%	(1,614)	1,094	108.7%
2001	11.3%	2.6%	(2,155)	465	111.0%
2002	11.0%	1.7%	(1,390)	340	105.8%
2003	13.1%	11.6%	559	2,531	98.4%

Source: Canadian Insurance (editions 1991 to 2004) and Insurance Bureau of Canada, with data from A.M. Best Canada's WinTRAC and Statistics Canada (2004).
Note: n.a. = data not available.

Major Canadian stock-listed life insurers (Industrial-Alliance, Manulife, and Sun Life) have joined their European counterparts in reporting embedded values (EmV) since the demutualization wave of 1999–2000. EmV is the present value of future earnings on contracts already in force, adjusted for the present value of target surplus, and added to the capital and surplus (Daly and Lombardi 2002). Many actuaries believe that EmV provides more accurate information about the value of life insurance firms and their future prospects. Most life insurance stock analysts would consider a high embedded to book value ratio (EmV/BV) a visible indicator of potential earnings.

Property and Casualty Insurance
In terms of underwriting results in 2003, the Canadian P&C sector made a profit of $559 million for the first time in twenty-five years. According to IBC, this was due to higher premiums, government action to reduce the costs of auto claims, and improvements in insurers' underwriting practices. As shown in Table 8.10, the sector had a combined ratio, all lines included, of 98.4 percent in 2003. It was the first time since 1978 that the combined ratio was below 100 percent. However, this improvement in the average combined ratio was not uniform across all firms (Voll 2004a). Also helping the non-life sector during the year 2003 was its realized investment return of 6.2 percent. This led to a net profit of $2.531 billion for the whole industry. Over the period 1990 to 2003, net profits had a compounded annual growth rate of 7.9 percent, although 2001 and 2002 were the worst years on record for net profits. This poor performance in 2001 and 2002 led to new capital injections from owners for more than one in three insurers in order to restore capacity.

The evolution of the industry's ROE during the entire period is similar to that of the net profits. The two best ROEs were obtained in 1996 and 1997 (13.6 percent and 13.1 percent respectively), while the two worst years were in 2001 with 2.6 percent and in 2002 with only 1.7 percent (see Table 8.10). Over the period 1990 to 2003, the industry's average ROE was 8.42 percent with a standard deviation of 3.58 percent. In fact, for the years from 1998 to 2002, the average ROE was only 4.78 percent, which is 3.64 percent lower than the longer-term average of 8.42 percent. But, with an ROE of 11.6 percent in 2003, the Canadian non-life sector returned to a more normal level of profitability after five straight years of weak performance. This is also in line with the recent reversal of the negative trend in the global general insurance markets.

To summarize, the financial performance of the insurance sector (both life and non-life) tends to follow the business cycle, albeit with a slight lag. In the years after recession (1990–1991 and 2001–2002), the insurance industry has exhibited some of the lowest levels of ROE. It is also interesting to note that the life sector appears to have made significant reforms to counter such cyclical effects, by diversifying their product and service offerings and better risk management. But, the P&C sector appears to be still in the process of doing so. Pro-forma balance sheets and income statements for all life and non-life firms based on financials for the year 2003 are provided in Tables 8.11 and 8.12.

8.5.3 Reinsurance Activities

***Life and Health Insurance* (CIBC World Markets 1999)**
In Canada, most life insurers use reinsurance contracts as a key risk management tool. However, reinsurance is mostly provided by the large life insurance companies. Indeed, in contrast to property and casualty insurance, the majority of the life reinsurance business (approximately 65 percent) is written directly between insurers without the participation of official reinsurance companies.

Table 8.11. Pro-forma Balance Sheet for the Canadian Insurance Industry (in thousands of Canadian dollars), 2003

	Canadian Companies		Foreign Companies	
	Absolute	Relative	Absolute	Relative
Life and Health				
Short-term Assets	23,814,631	8.3%	5,385,289	16.3%
Policy loans	10,000,130	3.5%	1,509,946	4.6%
Bonds, Debentures, and Mortgages	204,352,324	71.2%	23,612,240	71.4%
Preferred and Common Shares	15,567,596	5.4%	471,244	1.4%
Real Estate	8,777,144	3.1%	858,931	2.6%
Other	24,582,912	8.6%	1,219,834	3.7%
Total Assets	287,094,737	100%	33,057,484	100%
Net Actuarial Liabilities	193,790,320	67.5%	22,154,006	67%
Other Liabilities	53,787,435	18.7%	2,441,483	7%
Debts	7,246,559	2.5%	703,339	2%
Policyholders' Equity	2,168,981	0.8%		
Shareholders' Equity	30,101,442	10.5%		
Head Office Account			7,758,656	23%
Total Liabilities and Equity	287,094,737	100%	33,057,484	100%
Property and Casualty				
Short-term Assets	1,446,712	2.7%	1,571,810	6.8%
Total Investments	30,760,361	57.5%	16,243,458	70.5%
Receivables	7,348,320	13.7%	2,043,663	8.9%
Recoverables	9,422,431	17.6%	2,401,028	10.4%
Other Assets	4,563,919	8.5%	768,623	3.3%
Total Assets	53,541,743	100%	23,028,582	100%
Unearned Premiums	12,927,988	24.1%	3,287,194	14.3%
Unpaid Claims and Adjusted Expenses	24,787,927	46.3%	10,608,102	46.1%
Other Liabilities	3,703,852	6.9%	1,574,014	6.8%
Equity	12,121,976	22.6%		
Head Office Account			7,559,270	32.8%
Total Liabilities and Equity	53,541,743	100%	23,028,580	100%

Source: www.osfi-bsif.gc.ca (taken from regulatory returns filed by the insurers).

Table 8.12. Pro-forma Income Statement for the Canadian Insurance Industry (in thousands of Canadian dollars), 2003

	Canadian Companies		Foreign Companies	
	Absolute	Relative	Absolute	Relative
Life and Health				
Total Revenue	55,684,637	100%	10,238,749	100%
Total Benefits and Expenses	49,914,818	89.6%	9,406,031	91.9%
Net Income Before Tax	5,769,819	10.4%	832,718	8.1%
Net Operating Income	4,619,148	8.3%	639,739	6.2%
Net Income	4,403,190	7.9%	625,886	6.1%
Property and Casualty				
Total Underwriting Revenue	17,558,832	100%	7,027,502	100%
Total Claims and Expenses	17,492,835	99.6%	6,673,201	95.0%
Net Investment Income	1,772,555	10.1%	887,799	12.6%
Net Income Before Tax	1,863,298	10.6%	1,232,020	17.5%
Net Income	1,285,871	7.3%	912,973	13.0%

Source: www.osfi-bsif.gc.ca (taken from regulatory returns filed by the insurers).

While traditional life insurance is a low-growth market segment in Canada, life reinsurance experienced a significant jump in premiums during the last few years of the 1990s. For example, an increase of over 49 percent was seen just in 1997. In that year alone, London Insurance Group (acquired by Great West Lifeco in August 1997) wrote net reinsurance premiums of US$1,200 million, Manulife US$531 million, Sun Life US$290 million, and Mutual Life US$42 million. The planning of future transactions—both demutualization and consolidation—may have prompted many insurers to unload blocks of undesirable business to reinsurers.

However, this trend is not expected to continue as the market growth is limited, with life insurers currently retaining approximately 95 percent of their business. This retention rate is expected to increase even further as the Canadian life insurance industry continues to consolidate and the larger size of the new firms raises retention levels.

Property and Casualty Insurance (IBC 2004)

The Canadian P&C insurance industry relies heavily on global reinsurers, who spread their risks by supporting primary insurers in several countries and several regions around the world. Reinsurance is one of many risk management tools used by property and casualty insurers to guarantee that they will meet every obligation arising out of legitimate claims. For example, reinsurers played a major role in meeting the claims from the ice storms in eastern Ontario and southern Quebec in January 1998.

Table 8.13 shows the net premiums written for reinsurance in the P&C Canadian market over the period 1990 to 2003. While NPW for all lines of P&C business in Canada was $31.4 billion in 2003, about 10 percent ($3.1 billion) of NPW was ceded to reinsurance firms. And, over the entire period, P&C reinsurance net premiums written have grown at a compounded annual growth rate of 7.6 percent.

Table 8.13. Canadian Property and Casualty Reinsurance (in thousands of Canadian dollars)

	Net Premiums	Claims
1990	1,117,789	845,286
1991	1,106,925	954,795
1992	1,130,198	842,615
1993	1,207,423	910,989
1994	1,359,431	978,391
1995	1,514,879	1,034,299
1996	1,559,392	1,071,556
1997	1,323,058	923,563
1998	1,297,085	1,131,990
1999	1,395,034	1,046,928
2000	1,486,445	1,164,305
2001	1,979,113	1,504,861
2002	2,460,572	1,896,491
2003	3,108,000	2,183,000

Source: Insurance Bureau of Canada—Perspective (June 2004) and *Canadian Insurance* (editions 1991 to 2004).

In summary, it is indeed fair to conclude that in Canada, while the life sector relies on major domestic Canadian life firms for reinsurance, the P&C sector uses global reinsurance firms for managing part of their insurance risk. It is also fair to say that Canada has not historically faced major natural disasters, such as earthquakes and hurricanes, as seen in other parts of the world. Nevertheless, the Canadian non-life market is still affected by the hard and soft cycles of the global reinsurance markets.

8.5.4 Regulatory Oversight

Life and Health Insurance

The two important regulatory bodies for life insurers are the Office of the Superintendent of Financial Institutions (OSFI) and the Canadian Life and Health Insurance Compensation Corporation (CompCorp) (OSFI 1999). OSFI, the government regulator, performs the same oversight functions for insurance companies as it does for deposit-taking financial institutions. It also requires minimum financial standards, as well as basic risk management standards which are part of a new framework implemented in 1999. CompCorp, on the other hand, is a federally incorporated, private company funded by the life and health insurance industry that provides coverage to policyholders in the event of insurance company failures. Currently, the guarantees stand at $60,000 for life insurance cash values and up to $200,000 for life insurance claims. Originally, CompCorp was not designed to perform a regulatory role in overseeing individual companies. But the failure of life insurers during the early 1990s has forced it to assume enhanced powers to ensure increased coordination of policy protection and regulatory oversight. As of year-end 2003, CompCorp had a liquidity fund of $113.8 million, which is above the minimum of $100 million specified in its by-laws. This fund will provide immediate cash to cover CompCorp obligations in any future insolvency before receiving additional assessments from members (CompCorp Web site).

The licensing of life and health insurers and their agents and brokers is regulated at the provincial level, since the distribution of insurance is a provincial jurisdiction under the Constitution of Canada. But this has resulted in a number of standards and practices that vary from province to province, and today there is a strong need and many calls for greater harmonization of insurance regulations across Canada. Such harmonization of policies is one of the objectives sought by the Canadian Council of Insurance Regulators (CCIR). CCIR is an interjurisdictional association of insurance regulators that was set up in 2000, similar to the National Association of Insurance Commissioners in the United States. Its mandate is to work with all financial service regulators in promoting an efficient and harmonized regulatory framework in Canada.

According to industry advocates, the direct regulatory burden in Canada is estimated to be 14 cents for each $100 of insurance premiums compared to 11 cents in the United States and 5 cents in Australia. For some industry observers, it is not realistic to expect harmonization among all the provinces. Instead, they believe that provincial governments should sign mutual recognition agreements for the regulation of the industry, somewhat like treaties between nations (Praskey 2004). This is, in fact, similar to the framework proposed by the International Association of Insurance Supervisors (IAIS) for supervision of insurance firms that operate in several countries.

Property and Casualty Insurance

P&C insurance underwriting and distribution are also highly regulated. Federally registered companies must meet financial standards set by OSFI, even though they are likely to operate in separate provincial jurisdictions. As with life insurance, registration and licensing of agents, brokers, and adjusters is a provincial responsibility.

The largest line, auto insurance, is subject to provincial laws that require each driver to have minimum third-party insurance coverage. Provincial governments also require insurers to offer insurance to all drivers (IBC 2004). Automobile insurance is the only line of insurance in Canada in which rates are regulated. However, it is fair to say that active rate regulation is a relatively recent phenomenon. Experience is restricted to Ontario, which implemented a "strict prior approval" regime of rate regulation in 1989. In 2000, a change was introduced whereby filing requirements and approval times are streamlined if rate increases fall below a certain threshold determined by the regulator. Among the other regions, Alberta and the four Atlantic provinces maintain variations on the "file and use" system, whereby automobile insurers are required to file rates and, after a period defined in the legislation, acquire "deemed approval" for use. The other provinces maintain a more flexible "use and file" system for competitively delivered optional auto coverages sold by private insurers. Recent research shows that rate regulation has increased premium volatility in the province of Ontario (Leadbetter, Voll, and Wieder 2003). Also worth mentioning is the fact that only three provincial jurisdictions impose risk classification restrictions with respect to the underwriting of automobile insurance.

Finally, the Property and Casualty Insurance Compensation Corporation (PACICC) is an industry-owned fund set up to perform a role similar to CompCorp in the life insurance sector. If a P&C company goes bankrupt, PACICC will pay claims up to $250,000 on most lines of insurance. Policyholders may also claim 70 percent of unused ("unearned") premiums that have been paid in advance up to a maximum of $700. The deductible specified in the policy will be valid. Funding is ensured by an annual levy on all member companies. As of 2003, $30 million was available for such contingencies, and more funding could be raised from members as required (PACICC Website).

In summary, the regulatory framework in Canada, although fragmented, has been fairly effective in maintaining a stable and safe environment for policyholders (Kleffner and Jorgensen 1997). This has been primarily brought about by strict regulatory oversight on the capital adequacy and solvency of the insurance firms. However, now there is a move toward harmonization of the regulatory framework, as well as an initiative to supplement solvency requirements with other measures such as effective corporate governance and risk management.

8.5.5 Capital Adequacy

Life and Health Insurance

The capital adequacy and the financial soundness of life insurers is measured and monitored by the Minimum Continuing Capital and Surplus Requirement (MCCSR) that was designed by CompCorp in 1992 and adopted by OSFI in 1994 (OSFI Website). The MCCSR is a ratio-based system that is similar to the risk-based capital standards that are used by the National Association of Insurance Commissioners to assess the financial soundness of U.S. life insurers. However, MCCSR is only one of several indicators that OSFI uses to assess a company's financial soundness. It is not designed for use in isolation or for ranking companies. In fact, companies must perform dynamic capital adequacy tests annually. This requires the companies to project their future financial situation under a variety of scenarios in order to test whether MCCSR is met at the end of each year in the projection period, for each scenario. Thus, the companies are subject to a series of stress tests, including best, worst, and extreme values of the key variables. In addition to this, the OSFI also requires that individual companies meet up-to-date corporate governance standards, in terms of setting up risk-management committees with the responsibility for identifying, measuring, monitoring, and managing the various risks faced by the firm and assessing the adequacy of capital.

The MCCSR is based on two capital ratios that are measured and monitored for regulatory capital: tier 1 capital/capital required and total capital/capital required. Thus, the definition of capital comprises two tiers—tier 1 (core capital) and tier 2 (supplementary capital)—and includes certain deductions, limits, and restrictions (Saunders and Thomas 1997). The definitions are similar to the capital definitions under Basel I and Basel II for international banks and under Solvency I and Solvency II for European life insurers. Minimum capital requirements for life insurers in Canada are 60 percent for tier 1 capital and 120 percent for total capital (tier 1 and tier 2). OSFI also has established other lower supervisory target levels to serve as early warning signals, requiring the companies to raise additional capital if they are in violation.

The capital required is calculated by focusing on five specific, quantifiable risks, each with its own risk factor. These include (1) risk of asset default, (2) risk arising from mortality/morbidity/lapse assumptions, (3) risk from interest margin pricing, (4) risk due to changes in the interest rate environment and, (5) risk due to the embedded guarantees in segregated funds. OSFI specifies the appropriate risk factors that must be applied to each of five risk dimensions in arriving at the capital required. This then becomes the denominator for the two capital ratios. The minimum amount of available capital required from life insurers is $5 million (Saunders and Thomas 1997).

The above capital requirements apply to all domestic and foreign life insurers registered in Canada. Branches of foreign insurers operating in Canada, in addition to meeting the MCCSR guidelines, must also maintain and vest assets in Canada in accordance with the Test of Adequacy of Assets and Margin Requirement (TAAM). This means that these branches are required to invest a certain amount of the premiums collected in Canada in approved securities and dedicate them as being held for the purpose of meeting the claims arising out of Canadian operations. The margin requirements for such foreign company branches are also based on the five risk components, as in the MCCSR described above, and are calculated by applying the risk factors to the assets and liabilities of their Canadian operations.

As shown in Table 8.14, the average MCCSR ratio of Canadian incorporated life insurers (both domestic and subsidiaries of foreign insurers) fluctuates from year to year. However, during the five years from 1998 to 2003, MCCSR ratios were above 200 percent at the end of each year. Likewise, the TAAM ratios of foreign branches increased steadily over the period to reach 121 percent at the end of 2003.

Between 2001 and 2003, major Canadian life insurers, in their search for new core capital, have issued innovative instruments similar in design to the trust preferred securities (TruPS) found in the United States. These securities are similar to debentures and preferred shares and are generally long term, maturing at face value, with early redemption features and quarterly interest payments. TruPS were issued by companies in the United States for their favorable accounting treatments and flexibility. Specifically, these securities are taxed like debt obligations by the Internal Revenue Service in the United States, while being treated as equities in the company's financial statements, according to GAAP (Investopedia Website).

Table 8.14. Capital Adequacy for Canadian and Foreign Life insurers, 1998 to 2003

	1998	1999	2000	2001	2002	2003
Canadian						
Total Capital Available (in thousands of Canadian dollars)	30,654,752	30,921,727	31,754,200	34,071,452	31,434,271	31,261,590
Total Capital Required (in thousands of Canadian dollars)	13,233,548	13,785,961	13,644,765	16,546,480	13,862,148	14,317,355
Minimum Continuing Capital and Surplus Requirements	232%	224%	233%	206%	227%	218%
Foreign						
Total Capital Available (in thousands of Canadian dollars)	18,844,755	19,580,079	21,001,950	23,402,970	25,603,557	27,704,240
Total Capital Required (in thousands of Canadian dollars)	16,641,599	17,331,948	18,572,546	20,056,584	21,588,787	22,844,601
TAAM (Net Assets Available/Required)	113%	113%	113%	117%	119%	121%

Source: www.osfi-bsif.gc.ca.

In Canada, similar instruments were issued by special purpose vehicles (SPVs), which are 100 percent owned, nonoperating subsidiaries of life insurance firms, established solely for the purpose of raising capital. The SPV's primary asset is a debt instrument, issued by the insurer, often in the form of highly rated debentures that are soft-retractable in the sense that the holder has the option, under specified conditions, to convert the debentures into preferred shares after ten years. Such instruments were first issued by Sun Life in October 2001 and were called SLEECS (Ontario Securities Commission Bulletin 2002).The last issue was done by the Industrial-Alliance Group in June 2003 and was called IATS. Manulife, Clarica, and Great-West Lifeco also issued such innovative instruments during 2001 and 2002 for inclusion as tier 1 capital. In total, Canadian life insurers raised $1.9 billion through such deals.

The security holders have no rights to the trust assets if a loss absorption event ever were to occur. Loss absorption events can include events affecting the solvency of the life insurer or actions taken by the superintendent of financial institutions in respect to the insurer, requiring the liquidation or the use of trust assets for purposes other than repaying the holder. Therefore, the securities could be classified as core capital. For life insurers, this type of financing was a less expensive form of issuing tier 1 capital than preferred stock, because it had a narrower spread on a tax-equivalent basis and was not subject to capital tax. This is somewhat similar to the securitization of catastrophic risks by non-life insurers, particularly in the form of

CAT bonds. It also reflects the trend in the insurance industry for alternative forms of risk transfer and risk financing.

However, in 2004, OSFI announced that it might not allow future issues of such innovative instruments to be considered as tier 1 capital. This is in keeping with the new accounting rule imposed by the Institute of Canadian Chartered Accountants, which calls for considering such soft-retractable debentures as debt and secondary capital, rather than core capital.

Property and Casualty Insurance

In the past, OSFI relied on capital adequacy measures known as the minimum asset test for all Canadian P&C insurers and the deposit adequacy test for branches of foreign P&C insurers.

In 2003, OSFI implemented new risk-based measures for capital adequacy complying with a strong desire from the industry. The new measures came about following extensive discussions with the P&C industry. For Canadian companies there is now the Minimum Capital Test (MCT), while for branches of foreign insurers the new test is called the Branch Adequacy of Assets Test (BAAT). P&C companies must also perform dynamic capital adequacy testing (DCAT) annually, as their counterparts in the life sector.

Both MCT/BAAT are risk-based tests, and they primarily assess the credit risk and liability risk of P&C insurers and specify the general level of capital required to support these risks. This is also a ratio-based measure similar to MCCSR for life insurers based on the ratio of actual to required capital. It is suggested by OSFI that the supervisory target of 150 percent as minimum capital is a benchmark or standard against which individual companies should establish their own higher targets.

At year-end 2003, three-quarters of companies had MCT capital ratios of over 200 percent, and one-quarter had MCT capital ratios between 162 percent and 200 percent. The industry-aggregate MCT ratio rose to 221 percent in 2003, from 211 percent in 2002. These results are obtained for eighty-one active federally regulated Canadian direct writers and reinsurance companies (OSFI Web site).

As experience with the MCT has grown, industry players have gained greater confidence that it appropriately identifies key risk areas. Companies that have relatively higher exposure to the risk areas assessed by the minimum capital test have higher capital requirements. According to OSFI, industry participants have acknowledged that the risk-based MCT is better than the former solvency test, which was based on the minimum levels of assets and deposits. It is OSFI's intention to continue to work with the industry in implementing and improving the process in a collaborative manner.

To summarize, in Canada, both life and non-life insurers are well-capitalized, being the result of strict regulatory oversight. While strong capital regulations are necessary to maintain the stability of the insurance industry, the current fragmented structure in Canada, imposing a different set of rules for domestic and foreign firms, is likely to increase in complexity in the future. Hence, there is a need for harmonization of such capital requirements, both in the domestic and international arena.

8.5.6 Taxation

Life and Health Insurance

The Canadian corporate tax system is based on a worldwide taxation regime under which Canadian residents are taxed on all income, regardless of whether it is earned in Canada or in a foreign country. Generally, Canadian life insurers are subject to this worldwide tax regime. However, earnings of foreign subsidiaries that have an active business outside Canada are generally exempt from tax in Canada. Under this modified territorial regime, if a life insurer carries on an insurance business in Canada and elsewhere, only the income from Canadian insurance operations is subject to tax in Canada. Consequently, insurance business income generated from foreign branches generally is not included in the Canadian tax base.

Canadian life insurers pay taxes on their premiums, income, and capital. Only premiums and related income and expenses from life insurance policies issued to Canadian residents are included in computing the insurer's taxable income from the Canadian insurance business. However, because of the long-term nature of the insurance business and the need for reserves that an insurance firm must hold to ensure that all future claims will be met adequately, the tax rules provide a mechanism whereby an insurer is allowed a deduction for the reserves held, in computing its income for Canadian tax purposes.

The Canadian tax system also contains a complicated set of rules to calculate the investment income of a life insurer that must be included in the Canadian tax base with respect to its Canadian insurance business. The life insurer is not allowed to claim a foreign tax credit with respect to any foreign taxes paid on any investment income. Similarly, there is also a set of rules for capital tax requirements. However, for life insurers, premium taxes (ranging from 2 percent to 4 percent) account for a significant portion of total taxes. In fact, many provinces levy a tax on life insurance premiums instead of the applicable capital taxes that most of the provinces impose on the banks' capital. For the 2003 taxation year, the amount of total taxes (premium, income, and capital) paid by the top three life insurers as a percentage of income is around 26.9 percent. These figures do not include payroll, sales taxes, and property and business taxes (Tang and Vienneau 2004).

Currently, Canadian banks and life insurers with equity greater than $5 billion must be widely held. Widely held Canadian banks and life insurers may be owned through a holding company structure. From a tax perspective, such a holding company structure facilitates the common ownership of a bank and a life insurer, but allows them to exist as separate companies. Canadian tax rules do not address the fiscal implications of the merger of a bank and a life insurer into a single corporation (Tang and Vienneau 2004). Therefore, we expect the holding company structure to be the preferred organizational form if and when banks and insurers are allowed to merge.

Property and Casualty Insurance

According to a study commissioned by the Insurance Bureau of Canada and released in 2003, the Canadian tax structure is most burdensome to P&C insurers. Results of this study show that most other G7 countries have fewer taxes imposed on the P&C

insurance industry, as opposed to the complex cascading transactional taxes and significant capital taxes imposed by the Canadian tax authorities (IBC 2003).

On a comparative basis, the United States and Japan levy premium taxes, solely or partly, in lieu of income tax payable. No other G7 country imposes a capital tax on insurers or sales tax on premiums. Canada's P&C insurance consumers pay all of these taxes when they purchase insurance coverage. No other financial service in Canada is taxed at the retail level.

In 2003, P&C insurers paid $5 billion as taxes to federal, provincial, and municipal governments. Of this total, $3.8 billion was paid in premium-based transaction taxes. Those include the goods and services tax (GST) and the provincial sales tax (PST) on operations and claims, the PST on premiums in some provinces, and premium taxes (at varying rates) in all provinces. Over the last decade, the industry's average tax burden, as a percent of its value added, was more than three times the average tax burden of Canada's other financial services industries (IBC 2004).

In summary, as with capital requirements, the complex and differential tax regime in Canada for the insurance industry needs to be reformed in the interest of increasing the efficiency of the industry and the effectiveness of the tax regime. This is particularly crucial considering the importance of the contribution of the insurance industry to tax revenues.

8.6 CURRENT AND FUTURE TRENDS IN THE CANADIAN INSURANCE INDUSTRY

8.6.1 Trends in the Financial Services Industry Structure

The last two decades of the twentieth century saw significant changes in the structure of the financial services industry in Canada due to forces of deregulation that were also in evidence in the other G7 economies. This has obviously affected the structure and conduct of the insurance industry in Canada.

Until the mid-1980s, the financial services industry in Canada was structured under the "four pillars" framework. *Commercial banks* (and other depository institutions such as credit unions), *trust companies* (defined as organizations engaged as a trustee for individuals or businesses with a fiduciary responsibility for the administration of trust funds, estates, custodial arrangements, including stock registrations and transfers, and other related services), *securities firms* (specializing in investment banking, investment advisory services, and trading activities) and *insurance companies* (focusing on risk transfer) were the four key pillars or sectors of the financial system. Each pillar was restricted to operate only in its own sector, thus limiting cross-sector competition and potential economies of scope through a diversified portfolio of financial products and services.

For example, banks were not allowed to engage in estate and trust business, which was the exclusive domain of the trust companies, in order to prevent conflict of interest between the fiduciary trust fund management activities and the traditional commercial lending activities. For similar reasons of conflicts of interest, commercial banks were also not allowed to engage in investment banking and investment advisory services. This was similar to the restrictions under the Glass-Steagall Act in the United States. Credit unions (known as *caisses populaires* in Quebec) were similar to banks but operated on a smaller, regional scale with a mutual ownership structure. On the other hand, trust companies, securities firms, and insurance companies were not allowed to engage in any type of lending business. Limited deregulation during the 1960s allowed limited entry of trust firms into mortgage loans, securities firms into margin loans for investment purposes, and insurance firms into policy-backed loans.

The first major wave of deregulation occurred in the 1980s and allowed the ownership of securities firms by the other pillars, albeit under a holding company structure. Canadian banks emerged as the key players to take advantage of this, and many of the large securities firms were acquired by banks. However, the banks were required to maintain the separation of commercial banking and investment banking. Hence, investment banking activities of commercial banks are provided by wholly owned subsidiaries. The insurance industry did not become a major player in this sector.

The second wave of deregulation occurred in the mid-1990s, in which the different sectors were allowed free entry into all other sectors, with one key restriction that prevented banks from selling insurance through their extensive branch network.

The structure of the financial services industry in Canada today is the result of the response of both banks and insurance firms to this new opportunity, by entering into the trust business in a major way either by acquiring or setting up trust firms. This was also in keeping with the needed response to the demographic changes in society, which unleashed a demand for financial planning, financial protection, and wealth management from both the middle-income group and high-net-worth individuals.

As a result, the financial sector in Canada is today dominated by two pillars: banks and insurers. While banks are also major players in the securities and the trust businesses, insurers have focused on their traditional risk management business, while adding wealth management to their portfolio of activities.

This has increased the competition in personal financial products, as some banks entered the life and the personal non-life business by acquiring or setting up insurance firms. Life insurance firms, on the other hand, through their acquisition of small banks and trust companies, entered into the domain of deposit-taking and payments intermediation. And, most recently, insurers also have acquired direct access to the clearing payments and settlements network, which was previously the exclusive domain of traditional deposit-taking institutions.

While the idea of bancassurance (banks and insurance activities carried out by a single firm) and the concept of financial supermarket (where customers can get all their financial needs fulfilled in one location) are often discussed in Canada, we expect the current two-pillar framework to continue to exist for the near future.

8.6.2 Trends in Open Architecture

The developments in the structure of the financial services industry have some implications for what is being regarded as the potential “distribution revolution” in

Canada and the rest of the world. Of course, the growth of electronic distribution channels was regarded as inevitable during the Internet boom of the late 1990s. But, however, contrary to expectations, the use of e-channels by banks and insurance firms for the distribution of products has not significantly eroded the proportion of traditional banking products sold through branches and insurance products sold through independent agents, independent marketing firms (financial planners, full-service brokers), and exclusive/captive sales force.

Part of the reason can be attributed to the special nature of insurance products as risk- and wealth-management products. At one end of the spectrum, there exist service-intensive insurance products (such as universal life in the life sector and commercial insurance in the non-life sector), which require direct contact between the agent/broker and the client. And clients who consider and/or require such products will prefer the traditional distribution channels. At the other end, there are the service-light products (such as travel insurance), which are treated as mere commodities, where customers are more price conscious and use nontraditional electronic channels to find the best price.

Insurance firms must therefore choose an optimal point between these two extremes. Current trends indicate that insurers still regard their products as being best sold and serviced through direct contact with clients. This is unlikely to change in the near future, as changing demographics and sophisticated clients demand a package of financial services that are suited to their unique conditions and requirements. Two interesting trends have been observed in Canada in this regard. One is insurers' increasing willingness to enter into joint venture agreements with other insurers, thus providing their agents and brokers the opportunity of presenting their clients with an enhanced package of financial products and services. In addition, many insurance firms are also providing additional support, such as product training, actuarial consultants, and product support analysts to both independent and captive agents, in order to increase the value-added services provided by the agents to the clients (Kelly and Nathan 2000).

Technology also allows for interesting new advances in the area of e-distribution. A good example of this is the CSIO (Centre for Study of Insurance Operations) portal to which more and more P&C companies and brokers are integrated. Through this portal, many P&C insurers are Web-enabling their computer applications and making the processing functions and information available to their brokers/partners. This potentially brings much benefit to those participating in this channel.

However, it is interesting to point out that the success of an insurance firm depends not only on innovations in the distribution of the products, but also on innovations in the products themselves. It is also safe to say that insurers will be able to maintain their supremacy in the risk management business, due to their well-established underwriting and actuarial skills that are lacking in the banking industry. And, even if Canadian banks were allowed to enter the insurance markets in a major way, many of them may choose to only distribute the products that are underwritten by specialized insurance firms. This will apply to both life and non-life products. Similarly, the insurance firms may not enter the traditional banking business of providing credit or the traditional investment banking business of accessing the capital markets for their clients. Instead, they may be willing to underwrite the risks involved through innovative credit insurance products.

Some industry experts and academic researchers are indeed advocates of such an "open architecture" model of production and distribution of financial products and services in general, and insurance products and services in particular. Under such a system, certain firms will seek to specialize in the creation of products and services, while others will choose to specialize in the distribution of such products and services. Such an arrangement may indeed benefit society based simply on the principle of Adam Smith's comparative advantage, which often fosters more innovation both in production and distribution.

8.6.3 Trends in Regulatory and Supervisory Frameworks

Supervision of insurance firms is a difficult process given the dynamic nature of the insurance industry in an uncertain world and the trend toward increased competition and innovations in the financial markets. Further, regulators and supervisors must take into account the fact that the supervisory regime must operate parallel to the market forces, particularly now that a majority of the leading life and non-life firms have moved from a mutual to a stock ownership structure. While early Canadian regulation and supervision primarily focused on issues related to solvency and capital adequacy (as in MCCSR for life firms and MCT for non-life firms), in 1999 OSFI adopted a new supervisory framework that uses corporate governance, risk management, financial conditions, internal audits, and compliance as key components to ensure sustainability of insurance firms. This new framework combines the inherent risks in insurance activities with the quality of risk management of the firm to arrive at a net risk rating, which is then used in conjunction with a forward-looking potential risk assessment to design any required supervisory measures such as intervention and suspension of activities, if deemed necessary.

It is also interesting to note that this is similar to the framework currently under discussion in the European Union under the European Commission's Solvency II initiative. The main focus of the Solvency II initiative is to identify and analyze current and emerging risks that may affect the solvency of insurance firms and to enforce prudential regulation in line with market forces. Organizational framework and corporate governance, strategy and decision-making, and monitoring and information are used as the tools to identify risks in internal controls and processes, risks in investments, credits and asset-liability management, risks in underwriting and technical provisions, and risks in reinsurance and alternate risk transfers. The intention is to identify the diverse nature of risks and to design a supervisory framework that will provide for prioritization and control of risks (Conference of the Insurance Supervisors of the Member States of the European Union 2002).

Finally, it is worth noting that the risk classifications that are used by OSFI for Canadian firms and the proposed risk classifications under Solvency II in the European Union also parallel the risk categories that are in force under Basel I for international banks. Further, there is also talk of adopting the three-pillar framework (internal ratings based approach for risk assessment, along with including stress and scenario testing; supervision by regulators; and market discipline through disclosure) proposed for banks under Basel II, for insurance firms, under the initiative of the International Association of Insurance Supervisors.

8.6.4 Trends in Corporate Governance, Accounting Standards, and Risk Management Practices

The accounting scandals that monopolized the first few years of the twenty-first century have resulted in the tightening of corporate governance through effective board performance, including oversight responsibilities with respect to risk management. Likewise, these scandals have led to tighter accounting standards worldwide. Canada is no exception. While the demand for better governance structures and risk management practices comes from domestic regulators and shareholders, the accounting standards are being set by the International Accounting Standard Board (IASB). International Financial Reporting Standards developed by the IASB, along with the GAAP standards, are likely to emerge as the potential basis for harmonization of insurance accounting standards worldwide. But the impact of such global standards on Canadian firms is hard to assess at this preliminary stage.

In keeping with the new supervisory framework adopted by OSFI in 1999, both banks and insurers are required to disclose in their annual reports the strategies and policies that have been adopted by the board and the management for dealing with key risk issues identified for each sector (OSFI 1999). For insurers this includes a statement of risk measurement and management policies for the key dimensions of risk such as insurance risk (including product design, underwriting, and technical reserves), liquidity risk, investment risks, and solvency risks. While more needs to be done in this area, credit has to be given to the Canadian regulators and the insurers for taking a leading role in this area. Of course, as is the case in most of the other major developed markets, insurers in Canada are also struggling with the concept of aggregating the many risks they face into an integrated risk management framework.

8.7 CONCLUSION

A safe, solvent, sound, and stable financial system is essential for fostering economic growth and development in all economies. This chapter describes and analyzes the key features of the structure, conduct, and performance of the Canadian insurance industry—both life and non-life. It is fair to conclude that as one of the two key pillars of the Canadian financial system (the other pillar being banks), the insurance industry plays a major role not only in providing risk management for its clients, but also in facilitating efficient allocation of resources, wealth management, and asset protection. And, as key players in the financial markets, insurers also contribute to the overall efficiency of the markets by lowering transactions costs, improving liquidity services, and contributing to the competitive environment.

In fact, the Risk and Insurance Management Society has developed a measure called cost of risk to assess the benefits of insurance firms to commercial enterprises within a country (Swiss Re 2003). The measure incorporates insurance premiums and other risk control costs such as retained losses and loss prevention costs to arrive at a dollar cost of risk per $1,000 of revenues. The figures reported for the decade from 1990 to 2000 indicate that in Canada, the cost of risk ranged from $2 to $4 per $1,000 of revenues, compared to a cost that ranged between $4 and $8 for the United States.

Despite the weaker financial performance of P&C firms in Canada over the period 1998 to 2003, it seems that they have indeed contributed significantly to the management of risk of commercial firms in Canada. And, in keeping with the recovery of commercial insurance markets worldwide, it is expected that the Canadian non-life firms will indeed be able to improve their performance over the coming decade.

In contrast with the P&C firms, life insurers in Canada have posted solid financial performance, especially since the turn of the century. A lot of this can be attributed to their willingness to change and adapt to the changing face of the industry. In particular, life insurers have responded to the maturing traditional markets in Canada by seeking overseas markets both in developed and emerging economies and by redesigning their business model to include aspects of wealth and asset management and financial protection, in addition to their risk management functions.

However, it must be pointed out that Canadian insurers are small players in the global market for insurance products and services. It is also interesting to point out that they are slow to embrace some of the more innovative, alternate risk transfer strategies such as self-insurance; finite risk transfers; multiyear, multiline, multitrigger products; derivatives; securitization; and capital markets solutions. This is likely to become important as the global insurance market becomes more competitive and integrated, and as regulators increase their emphasis on better risk management strategies and risk-based supervision.

We conclude by pointing out that 2004 marked the 200th anniversary of the P&C industry in Canada and that Canadian insurers, led by such firms as Manulife, indeed have the potential to be key players in the global markets.

8.8 REFERENCES

Babin, Maxime, and Gilles Bernier, 2001, "Démutualisation des sociétés canadiennes d'assurance de personnes: une caractérisation de leurs premiers appels publics à l'épargne," *Insurance and Risk Management Review* 69(2): 229–258.

Berger, Allen N., J. David Cummins, and Mary A. Weiss, 1997, "The Coexistence of Multiple Distribution Systems for Financial Services: The Case of Property-Liability Insurance," *Journal of Business* 70: 515–546.

Canadian Institute of Actuaries, 2005, *Report of the Task Force on Automobile Insurance Issues,* 1–12.

Carr, Roderick M., J. David Cummins, and Laureen Regan, 1999, "Efficiency and Competitiveness in the U.S. Life Insurance Industry: Corporate, Product, and Distribution Strategies," in J. David Cummins and Anthony M. Santomero, eds., *Changes in the Life Insurance Industry: Efficiency, Technology and Risk Management* (Norwell, Mass.: Kluwer Academic Publishers).

CIBC World Markets (Equity Research), 1999, "Shifting Sands: The Transformation of the Canadian Life Insurance Industry," CEO Seminar/Financial Services, 1–146.

Conference of the Insurance Supervisors of the Member States of the European Union, 2002, *Prudential Supervision of Insurance Undertakings Report,* 1–112.

Daly, Robert, and Mike Lombardi, 2002, "EV Reporting Comes to North America," *Emphasis* (Tillinghast-Towers Perrin), 1–4.

Harrington, Scott E., Greg R. Niehaus, Anne E. Kleffner, and Norma L. Nielson, 2004, *Risk Management and Insurance*, 2d Canadian ed. (New York: McGraw Hill/Irwin/Ryerson).

Insurance Bureau of Canada, 2004, *Facts of the General Insurance Industry—2004,* 1–38.

Insurance Bureau of Canada, 2003, *Taxation of P&C Insurance: A Comparison Between Canada and Other G7 Countries* (Toronto: Insurance Bureau of Canada).

Kelly, Mary, and Alli Nathan, 2000, "Distribution of Personal Banking and Insurance Products in Canada," *International Insurance Society Seminar Proceedings*, 1–26.

Kelly, Mary, and Anne E. Kleffner, 2004, "Same Old, Same Old," *Canadian Insurance*, January: 22–25.

Kleffner, Anne E., and Jerry L. Jorgensen, 1997, "An Analysis of the Difference in Solvency Experience of U.S. and Canadian Life Insurers," *Geneva Papers on Risk and Insurance* 22(84): 536–552.

Kleffner, Anne E., and Joan T. Schmit, 1999, "Automobile Insurance in Canada: A Comparison of Liability Systems," *Journal of Insurance Regulation* 18(1): 35–52.

Kovacs, Paul, and Darrell Leadbetter, 2003, "The Economic Burden of Motor Vehicle Collisions in British Colombia," unpublished research paper, Insurance Bureau of Canada, 1–27.

Leadbetter, Darrell, Jane Voll, and Erica Wieder, 2003, "The Effects of Rate Regulation on the Volatility of Auto Insurance Prices: Evidence from Canada," Unpublished working paper, Insurance Bureau of Canada, 1–17.

McIntosh, James, 1998, "Scale Efficiency in a Dynamic Model of Canadian Insurance Companies," *Journal of Risk and Insurance* 65(2): 303–317.

McQueen, Rod, 1996, *Who Killed Confederation Life? The Inside Story* (Toronto: McClelland & Stewart).

Moreau, Rémi, 1991, "Origine de l'assurance et tableau chronologique," *Assurances* 1: 61–68.

Office of the Superintendent of Financial Institutions, 1999, *Supervisory Framework—1999 and Beyond,* 1–20.

Ontario Securities Commission Bulletin, 2002, Issue 25/14, chap. 2.

Praskey, Salley, 2004, "The Life Line," *Canadian Insurance*, Annual Statistical Issue, 26–29.

Saunders, Anthony, and Hugh Thomas, 1997, *Financial Institutions Management*, 1st Canadian ed. (New York: McGraw Hill/Irwin/Ryerson).

Swiss Re, 1999, "Life Insurance: Will the Urge To Merge Continue? *Sigma* 6: 1–34.

Swiss Re, 2001, "World Insurance in 2000: Another Boom Year for Life Insurance, Return to Normal Growth for Non-life Insurers," *Sigma* 6: 1–38.

Swiss Re, 2003, "The Picture of ART," *Sigma* 1: 1–44.

Swiss Re, 2004, "World Insurance in 2003: Insurance Industry on the Road to Recovery," *Sigma* 3: 1–42.

Tang, Marjorie, and Paul Vienneau, 2004, "Tax, Banks and Life Insurers," *CA Magazine*, October: 1–3.

Van Gorder, Phyllis M., 2004, "Individual Life Insurance Market Share by Distribution Channel—2003 Canada," Research document, *LIMRA International,* 1–11.

Voll, Jane, 2004a, "Turning the Tide," *Perspective* (Insurance Bureau of Canada) 10(1): 1–6.

Voll, Jane, 2004b, "The Hazard of the Average," *Perspective* (Insurance Bureau of Canada) 10(4): 1–4.

8.9 IMPORTANT WEB SITES

www.actuaries.ca (Canadian Institute of actuaries—CIA)

www.assuris.ca (Assuris is the new name for ComCorp)

www.bankofcanada.ca (Bank of Canada)

www.bis.org (Bank for International Settlements)

www.camagazine.com (CA magazine)

www.clhia.ca (Canadian Life and Health Association)

www.fin.qc.ca (Department of Finance Canada)

www.fsa.ulaval.ca/chaire-industriellealliance (Industrial Alliance Insurance Chair, Laval University)

www.haskayne.ucalgary.ca/programs (risk management and insurance programs offered through the insurance chair at the University of Calgary)

www.iais.org (International Association of Insurance Supervisors)

www.iasb.org (International Accounting Standards Board)

www.ibc.ca (Insurance Bureau of Canada)

www.insurance-canada.ca (insurance information and resources)

www.investopedia.com (Investopedia)

www.oecd.org (Organisation for Economic Co-operation and Development)

www.osc.goc.on.ca (Ontario Security Exchange Commission)

www.osfi-bsif.qc.ca (Office of the Superintendent of Financial Institutions)

www.pacicc.org (Property and Casualty Insurance Compensation Corporation)

www.swissre.com (Swiss Re)

8.10 LEXICON

ART. Alternative risk transfer.

BAAT. Branch adequacy of assets test.

CAGR. Compounded annual growth rate.

CAT bond. High-yield insurance-backed bond containing a provision causing interest and/or principal payments to be delayed or lost in the event of loss due to a specified catastrophe—also called CATastrophe bond.

CCIR. Canadian Council of Insurance Regulators.

CLHIA. Canadian Life and Health Insurance Association.

CompCorp. Canadian Life and Health Insurance Compensation Corporation.

CSIO. Centre for Study of Insurance Operations.

DAT. Deposit adequacy test.

DCAT. Dynamic capital adequacy testing.

EmV. Embedded value.

GAAP. Generally Accepted Accounting Principles.

GDP. Gross domestic product.

General insurance. General insurance typically comprises any insurance that is not determined to be life insurance and is called property and casualty insurance in the United States. General insurance is a term used in the United Kingdom and other Commonwealth countries, including Canada.

GST. Goods and services tax.

IAIS. International Association of Insurance Supervisors.

IASB. International Accounting Standard Board.

IBC. Insurance Bureau of Canada.

ICA. Institute of Canadian Actuaries.

ICCA. Institute of Canadian Chartered Accountants.

IFRS. International Financial Reporting Standards.

IMO. Independent marketing organization.

IPO. Initial public offering.

IRB. Internal rating based.

IRS. Internal Revenue Service.

L&H. Life and health.

LIMRA. Life Insurance Marketing Insurance Association.

MAT. Minimum asset test.

M&A. Merger and acquisition.

MCCSR. Minimum continuing capital and surplus requirement.

MCT. Minimum capital test.

MGA. Managing general agency.

NPW. Net premiums written.

OSFI. Office of the Superintendent of Financial Institutions.

PACICC. Property and Casualty Insurance Compensation Corporation.

P&C. Property and casualty.

PPGA. Personal producing general agent.

PST. Provincial sales tax.

RIMS. Risk and Insurance Management Society.

ROE. Return on equity.

RRSP. Registered retirement savings plan.

S&P/TSX Composite Index. Index that measures the total return of the largest companies that trade on the Toronto Stock Exchange (TSX), designed in collaboration with Standard & Poor's (S&P).

SPV. Special purpose vehicle.

TAAM. Test of adequacy of assets and margin requirements.

9 Insurance in the Netherlands: Market Structure and Recent Developments

Alfred Oosenbrug

Economics, Finance & Insurance Centre (EFIC)
AvisO Finance & Insurance Consultancy

9.1 INTRODUCTION

The insurance industry in the Netherlands has a long and rich history. The founding father of modern life insurance mathematics, Johan de Witt, was a famous Dutch statesman. Thanks to the famous Dutch business sense, several insurance groups from a country as small as the Netherlands today are among the top global players in the worldwide insurance market.

On the other hand, the Dutch insurance market is one of the most open markets in the world, mainly due to its liberal regulatory environment and its considerably open distribution structure. As a result, the absolute and relative number of market participants is overwhelming. With "only" 16 million inhabitants in the Netherlands, no less than 786 non-life-insurers and 247 life insurers have licenses to operate in the Dutch market. The fact that the Dutch rank sixth globally in terms of per capita insurance premiums paid is another explanation for the overwhelming number of participants. However, coming from an era of "unlimited exuberance," relatively flexible regulation, and strong fiscal incentives for life insurance business, insurers now have to adjust to a business environment that has changed significantly.

This chapter begins with a short overview of the long and rich history of life and non-life insurance in the Netherlands, followed by a review of recent developments in the domestic market. A detailed overview of the current market is then provided, focusing on the market for life insurance policies; the market for non-life insurance policies; and the existing distribution system for life and non-life policies.

9.2 HISTORICAL BACKGROUND

The insurance industry has a long history in the Netherlands going back to the end of the Middle Ages (Van Barneveld 1984; Bos 1998). The history of non-life insurance even goes back to the eighth century when the first occupational guilds were formed

(Oosenbrug 1999; Van Barneveld 1984). The guild members promised each other mutual assistance in times of adversity and compensated their members for losses caused by the perils of the sea, fire, flood, accident, or cattle thievery. Especially after the Reformation, with its iconoclasm, the financial support of sick guild members and the widows of former guild members became an important part of the guilds' roles.

Before the Reformation, the largest part of the revenues of the occupational guilds was spent for religious purposes such as, maintaining special guild altars or special guild chapels in the parish churches, reading masses for peace of mind of deceased guild members, and worshiping the patron saint (Bos 1998). With the arrival of the Reformation, all those religious activities of the occupational guilds were forbidden; many of the guild altars and chapels in the parish churches were destroyed, along with many of the until-then usual images of (patron) saints. From then on, it was legally no longer possible to spend the available funds of the occupational guilds for religious purposes, so the use of the funds available shifted from financing religious activities to financing "mutual" assistance to poor or disabled guild members and widows of guild members.

The Reformation gave an important impetus to the guilds' social (security) role. In the course of time it became common to put aside the money earmarked for mutual financial support in separate sick, burial, and widows' clubs (Oosenbrug 1999; Gales and Van Gerwen 1988).

9.2.1 Life Insurance in the Netherlands[189]

The Middle Ages

In 1345, Willem V issued life annuities to finance his budget deficits (Stamhuis 1998). In the course of the fifteenth century, issuing annuities to finance municipal budget deficits assumed large proportions. The structure of the financial instrument used to finance the governmental and municipal deficits was the same as the structure of the financial instrument already used in the ancient Greek municipal state Milete in 205 BC (Oosenbrug 1999). The amount of the annuity was independent of the age of the subscriber; nevertheless, the annuity had to be paid out as long as the subscriber was alive (Van der Grinten 1931; Oosenbrug 1999). So those medieval annuities were really true precursors of modern life annuity contracts, although the "premiums" paid still were not founded on any actuarial technique.

Origin of Modern Life Insurance Business

In 1671, the Dutch statesman Johan de Witt published his well-known "*WAERDYE Van LYF-RENTEN Naer proportie van LOS-RENTEN*" ("The value of life annuities in proportion to fixed annuities").[190] Although Domitius Ulpianus had done the same at the end of the second century, De Witt was the first in the modern world to introduce age-dependent survival rates in calculating the value of a life annuity (Van der Grinten 1931).

---

[189] This paragraph is based on Oosenbrug (1999), Chapter 2.

[190] See Schaap and Van den Berg (undated) for a biography of Johan de Witt.

With the ideas in his pamphlet, Johan de Witt became the founding father of modern life insurance mathematics. Nevertheless, it was not until 1762 that the first scientifically based life insurance company was founded, the English Society for Equitable Assurance on Lives and Survivorships (Van Barneveld 1984; Oosenbrug 1999). In the Netherlands itself, only in 1807 was the first scientifically based life insurance company founded, the Amsterdam-based Hollandsche Sociëteit van Levensverzekeringen. While the Equitable eventually went into run-off, the Hollandsche Sociëteit is still going strong. After some mergers, it is now part of Delta Lloyd, the fully owned Dutch subsidiary of the English Aviva Group. Delta Lloyd ranks fourth among Dutch insurers in worldwide premium income earned (AM Jaarboek 2004).

Nevertheless, during the nineteenth century, the Dutch life insurance market was still dominated by traditional funeral funds, selling only funeral insurance policies with a small sum assured (Gales and Van Gerwen 1988; Van Gerwen and Verbeek 1995; Oosenbrug 1999). At the top of their penetration rate, in the 1880s, far more than half of the total Dutch population was insured with one of the approximately 600 funeral funds. Until the end of the nineteenth century, those funds still did not use age-dependent mortality rates to calculate their premiums and did not form sound financial reserves to cover their liabilities. It was not until the second half of the nineteenth century that scientifically based funeral insurance policies came within reach of the general public. Specialized funeral insurance companies and modern funeral funds were founded, and traditional funeral funds were transformed into insurance companies or modern funds.

Rise and Ripening of Modern Life Business

By the end of the nineteenth century, modern life insurance companies began to actively promote their products and the number of players and the diversity of life insurance products increased. In 1880, a long period of strong market growth began, interrupted only by a small dip in the first years of World War I.

The total sum assured grew from dfl. 100 million (€45 million) in 1880 to dfl. 2 billion (€900 million) in 1920 and dfl. 8 billion (€3.6 billion) in 1945 (Oosenbrug 1999). Markedly, more than half of the growth from 1920 to 1945 was during World War II (WWII). Because of the general scarcity of goods, people did not have alternatives to spend their money on, and they tried to protect their money from going into the pockets of the German occupier by buying life insurance policies (Van Gerwen 2000).

The occupation by the Germans also had a more structural and far longer lasting effect on the development of the Dutch life insurance market. In 1941, the German occupier introduced new fiscal rules, considerably enlarging the possibilities to deduct life insurance premiums from taxable income (Barendregt and Van Langenhuijzen 1995). Until a few years ago, those and other fiscal facilities for life insurance business were strong incentives to buy policies.

Growing wealth after WWII resulted in a spectacular growth of the total sum assured, from dfl. 8 billion in 1945 (€3.6 billion) to dfl. 120 billion (€55 billion) in 1970, and to nearly dfl. 2,000 billion (nearly €900 billion) in 2002. But in 2002, the "everlasting" prosperity for life insurance business ended abruptly. With the turn of the century, a new era had started for the Dutch life insurance market.

9.2.2 Non-Life Insurance in the Netherlands

In the sixteenth and seventeenth centuries (the "Golden Age") Holland became the leading trading center in the world. As trade is not possible without insurance—for example, against the perils of the sea—insurance bourses were founded and insurance business flourished. In 1531, skipper Mathias Kuntze insured his ship against the perils of the sea in Antwerpen, a central town in the south of the Netherlands. In the 1560s, during the reign of Charles Quint, insurance broker Juan Henriquez underwrote on average 100 policies a month in Antwerpen.

In the Golden Age from 1580 to 1640, Amsterdam became the most important trading place in the Netherlands, surpassing Antwerpen and Brugge. In 1619, the Amsterdam bourse opened. At this trade bourse, insurance brokers and insurers also met to write marine insurance policies. During the 1700s, Amsterdam became the most important marketplace for transport insurance. Later on, important insurance bourses were also vested in Rotterdam and Middelburg. The occupation of the Netherlands by the French in 1795 and the war between Britain and French in the following years lead to the downturn of the worldwide leading position of the Dutch bourses. The London market especially profited from this downturn, resulting in Lloyd's of London becoming and still being the most important co-insurance market in the world (Oosenbrug, et al. 1996). For that matter, the first chairman of Lloyd's, Marten Kuyck van Mierop, was a Dutchman (Pestman 1983).

In the second half of the sixteenth century, the first fire insurance policies were written, based on mutuality between craftsmen in specific kinds of businesses. Only in 1720, the year of the famous South Sea Bubble, the first non-life insurance company was founded, the Maatschappij van Assurantie, Discontering en Beleening der Stad Rotterdam (Van Barneveld 1984). This company survived as an independent company until the beginning of the new millennium, when it was acquired by Fortis.

After 1795, the interest of insurance brokers and insurers shifted more and more from transport insurance to fire insurance. In spite of the relatively recent and strong rise of auto insurance and disability insurance, fire insurance still remains one of the central branches in the general insurance business in the Netherlands (Oosenbrug 2004).

Although accident insurance and automobile insurance were introduced at the end of the nineteenth century, it was not until the second half of the twentieth century that modern disability insurance and automobile insurance became common. For a long time, only sickness funds and then compulsory health insurance funds—originating partly from the sickness funds of the medieval occupational guilds—dominated health insurance in the Netherlands. Private health insurance became common only in the twentieth century. Compulsory health insurance funds are still dominant players in the Dutch market for health insurance.

9.2.3 Traditional Openness of the Dutch Insurance Market

Holland has long been known by foreigners for its merchants and clergymen. As merchants, the Dutch learned early in their history that an open market policy is the best guarantee to stimulate trade and, as a result, wealth. While their famous business sense drove them all over the world to open trade with local populations, at the same time they opened their home markets for foreign merchants. As a consequence, in its Golden Age the Netherlands became the central trading place for the entire "old" world.

The Dutch borders were open not only for trade but also for immigration. In time, many refugees from less liberal and tolerant countries fled to the Netherlands; skilled craftsmen among them strengthened the Dutch labor force and thus the Dutch economy.

So the Dutch were traditionally well acquainted with the blessings (including economic ones) of liberalism and tolerance. In creating the current insurance industry regulatory system, lessons learned were well remembered, and the resultant system was very liberal. Insurance companies were free to make their own rates and policy terms; and supervisory regulation focused only on solvency rules (Vermaat and Oosenbrug 1994). As long as the solvency rules were adhered to, the market participants were free to do business as they liked. New entrants could come into the market with their own (competitive) premium rates, policy terms, and provision rates by simply having the required solvency margin, which was in principle enough to get a license to operate. With a very well diversified distribution structure and independent agents being one of the most important channels for distribution (Van Voorst Vader 1995), entering the Dutch insurance market was relatively easy from an operational standpoint too.

In the last decades of the twentieth century, supervisory regulation within the European Union was harmonized. On the one hand, the Dutch supervisory rules had to be tightened. On the other hand, and more radically, the supervisory rules in other countries such as Germany, Italy, and France had to be changed fundamentally. "Material" supervision on rates and policy terms had to be abolished and the relatively liberal "normative" Dutch (and British) system of solvency-oriented supervision became the standard for the whole European Union (Pearson 1995; Oosenbrug 1995b; Van den Berghe 1995; Verkerk-Kooiman 1994).

9.3 OVERVIEW OF RECENT DEVELOPMENTS[191]

9.3.1 Introduction

After nearly two decades of strong growth figures for the Dutch domestic market, the booming market trends seem to have ceased since the turn of the century. In line with global developments, the economy is in general cooling down, competition is becoming stronger, and pressures for more market transparency are increasing. In addition, the generous special tax breaks for life insurance policies were changed drastically, partly to enhance the attractiveness of the Netherlands as a domicile for foreign companies by cutting general tax rates.

---

[191] Substantial parts of the rest of this chapter are based on Oosenbrug (2004) and Oosenbrug and Zoon (2002).

9.3.2 Recent Changes in the Tax System

As mentioned, generous tax facilities for life insurance policies were retrenched as a result of the wish to cut general tax rates. Next to those budgetary considerations, the retrenchment of the life insurance facilities also resulted from the long-felt desire to create a more level playing field with straightforward savings and investment products.

In 2001, the maximum for the unconditional tax credit for premiums paid for life annuities was reduced from € 2,804 to not more than € 1,036 per person. As of January 1, 2003, the possibility of an unconditional tax credit for premiums paid for life annuities was totally abolished. Today, only people with deficiencies in their pension rights can deduct premiums paid for life annuities from their taxable income (Oosenbrug loose-leaf).

As of January 1, 2001, the fiscal position of endowment policies also changed. Until 2001, life and death benefits from endowment policies could, up to some limited amount, be received free of income tax, while life insurance policies in general were exempted from personal capital tax. In 2001, the fiscal system for revenues from savings and investments was changed drastically. A general capital gains tax for all kinds of valuable assets was introduced. There is no longer any special exemption for life insurance policies, except one very specific—and limited—exemption for endowment policies related to mortgages.

As a consequence of the changes in the tax system, simple tax-driven production decreased and many life insurers and intermediaries must make a turnaround from a marketing strategy based on "hit-and-run" selling to "advice-based" selling.

9.3.3 Further Recent Developments in the Business Environment

At the same time that many market participants must make a turnaround in their marketing strategy, the regulatory environment for insurers and intermediaries is changing rapidly and accounting rules are being tightened.[192] Thanks to the long history of prosperity, insurers were very well capitalized by international standards and generally could stand the effects of the changes in their business environment. On the other hand, many intermediaries already faced serious problems in the new environment. Some of the "hit and run" intermediaries went bankrupt. In addition, due to the sliding and long-lasting crashes on the stock exchanges during recent years, insurance companies were hit directly by the adverse economic developments.

9.3.4 Impact of Recent Developments on the Financial Condition of Dutch Insurers

As Figure 9.1 clearly shows, the formerly generous solvency positions were hit seriously during the first years of the new millennium. For funeral funds, the average solvency ratio tumbled from almost seven at the end of 2000 to less than three at the

[192] See Verhoog (2003), Oosenbrug (2002), and Oosenbrug (1995a) for detailed information about the generally accepted accounting principles and the accounting rules in the Netherlands, and the changes in the accounting rules since those rules were introduced.

end of 2002. The average solvency ratio of the Dutch life insurers was nearly halved within three years, and the average solvency ratio of the non-life insurers fell from nearly five to about two-and-one-half.[193]

When in 2003 at last the world-wide crash on the stock exchanges stopped, the solvency margins of the Dutch insurers recovered, but only a bit. The average solvency ratio of the funeral funds tumbled again after the consolidation that took place in 2002.

Source: PVK and Central Bureau of Statistics.
Notes: The solvency ratio is the actual solvency margin of a company divided by the required solvency margin of that company. See Oosenbrug (2003) for detailed information about the way the actual and required solvency margins of the different kinds of insurance companies have to be calculated.

Figure 9.1. Average Solvency Ratio of Non-Life Insurers, Life Insurers, and Funeral Funds, 1991–2003

At the same time, relatively low capital market interest rates put the profitability of the insurance business under serious pressure. In addition, for non-life insurers, the business environment was further adversely affected by the huge premium rate increases for reinsurance coverage and the loss of reinsurance capacity in the market following September 11, 2001. For the first time in decades, in 2002, reported net results of life insurers dropped dramatically, from the average percentage of gross premiums written of roughly 10 percent to a very meager 1 percent. At the same time, reported net results of non-life insurers sank from nearly 4 percent to a comparable 1 percent of gross premiums written. Although reported results recovered in 2003, it seems likely, especially in life business, that future results will no longer be as impressive as they used to be.

[193] The solvency ratio is the actual solvency margin of a company divided by the required solvency margin of that company. See Oosenbrug (2003) for detailed information about the way the actual and the required solvency margins of the different kinds of insurance companies have to be calculated.

9.4 MARKET OVERVIEW

9.4.1 The Total Market

From 1985 to 2003, the total life and non-life insurance premiums written grew on average by almost 8 percent a year. This was mainly caused by a nearly double-digit average growth rate for life business, but non-life business also showed a respectable growth rate of more than 6.5 percent a year.

Figure 9.2 shows the development of the total market from 1985 to 2003 in terms of gross premium income written and the division of the market into life and non-life business. In the figures used, some funeral funds are categorized as life insurers because they have a license to operate under the law on the supervision of insurance companies; statistically, those funds are treated and accounted for by the insurance supervisor as insurance companies. The largest funeral funds, in particular, applied for such a license under the stricter supervisory regime for insurance companies.

Total gross premiums written by the funeral funds licensed under the less rigorous special supervisory regime for (specialized) funeral funds amounted only to the relatively marginal sum of €81 million in 2003. Relative to the amounts of premiums written in normal life and non-life business, the premiums written by the specialized funeral funds are so marginal that it is not possible to show the development of that kind of business in Figure 9.2. Total premiums written are not more than approximately one-third of a percent of total premiums written in normal life business. From 1998 until 2003, gross premiums written grew from only €70 million to not more than €81 million, for an average growth rate of approximately 3 percent a year.

In the beginning of the 1980s, non-life business accounted for approximately 60 percent of total insurance business. Although growing at a steady rate most of the

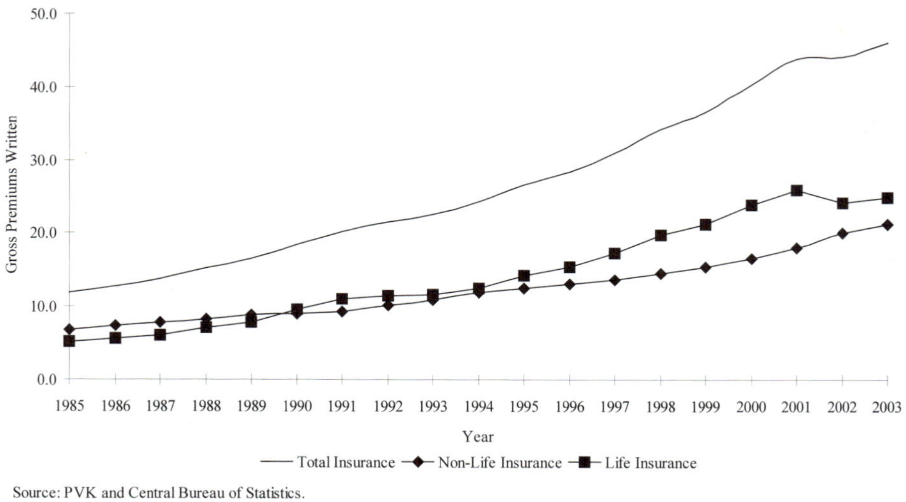

Source: PVK and Central Bureau of Statistics.

Figure 9.2. Gross Premiums Written by Non-Life Insurers and Life Insurers, 1985–2003 (€ Billions)

time, life business was growing still faster. Since 1990 the total life insurance premiums written exceed the total non-life insurance premiums written. In 2001 the situation of the 1980s was reversed, with life business accounting for approximately 60 percent of total business.

As Figure 9.2 clearly shows, life business growth collapsed abruptly in 2002. In one year the relative share of life business shrank from almost 60 percent to less than 55 percent. With a growth history of 5 percent to 12 percent for years, in 2002 the growth rate for total business marginalized to a "pitiful" 0.6 percent.

The turnaround in 2002 was mainly due to adverse changes in the tax system. Because these changes have an adverse effect on the demand for life insurance policies, it seems likely that the years of booming life insurance markets are definitely over, as can be seen from the figures for 2003. Although market growth recovered in 2003, the growth rate recovered to only a meager 3 percent, being far less than the historical growth rates of 5 percent to 12 percent. In addition, the growth figure for 2003 has been boosted by substantially more than two percentage points by a very large one-time single premium paid in the case of a transfer of the assets of a pension fund to an insurance company.

With a growth rate for non-life insurance of almost 6 percent in 2003, the relative share of life business shrank slightly from 55 percent in 2002 to 54 percent in 2003. If this new trend continues for the future, non-life premium income will exceed life premium income within a few years.

With life business in general being far more profitable than non-life business, in the future total profitability for composite insurance groups will no longer be as high as it was in the recent past. Nevertheless, for cost-efficient and administratively well performing insurance companies, the open and still extensive Dutch insurance market still can be a very attractive one.

9.4.2 The Market for Life Business

Until the turn of the century, selling individual life insurance policies was a booming business. Growing prosperity, generous tax facilities, and very high investment returns drove the volume of new business to continuously higher records. At the same time, total premium income earned from group contracts rose at a steady, although less spectacular rate, with relatively high growth rates in periods of accelerating inflation rates and relatively low growth rates in periods with less inflationary pressures. As a result, in 2000, approximately 70 percent of life insurance premium income came from individual contracts and approximately 30 percent from group contracts.

However, in the meantime the fiscal attractiveness of individual life insurance policies decreased, the unshakable belief in sky-high stock market returns faded away, economic growth ceased, and inflation rates accelerated. As a consequence, premium income growth in the individual life market ceased and, in 2001, inflationary pressures on wages and pensions rose.[194] As a result, the market segment

[194] The Dutch pension system is a "three-pillar-system." The first pillar consists of a pay-as-you-go state pension system providing some base-level old age pension for each citizen. The second pillar comprises the privately funded pensions carried out by company pension funds, pension funds for branches of industry and occupational pension funds, and by life insurance

for group contracts could regain some part of the market share lost in the former years. Nevertheless, within a very short time inflation rates slowed down again to turn dangerously close to deflationary pressures. Along with suddenly staggering unemployment rates, the growth rate of premium income from group contracts dropped from approximately plus 20 percent in 2001 to minus 10 percent in 2002.

Meanwhile a general sense of urgency to economize the generous pension plans arose. While the stock markets crashed and capital market interest rates dropped, the supervisory authority, politicians, employers, and employees at last realized that the existing generous system would no longer be sustainable without very substantial premium rate increases. The adverse effects of the "old-age time bomb" became clearly noticeable. Although premium rate increases will also force up premium income from group contracts in the very short run, for the near future widespread retrenchments of pension plans will put the potential growth rates for group life insurance contracts under substantial pressure.

So, all in all, group life business has gone through an era of mutually opposite changes. On the one hand, inflationary pressures have ceased, unemployment rates rose substantially, and pension arrangements retrenched. On the other hand, premium rates are increasing as a result of the transition to a lower discount rate and to updated mortality tables, while at the same time more and more pension funds are liquidated or "outsourced," with the settlement of their pension arrangements being transferred to life insurance companies, other pension funds, or specialized service organizations.[195] So, while group life business is under pressure, higher premium rates automatically increase premium volumes, and the shrinking number of "self serving" pension funds creates new opportunities for life insurance companies offering group contracts or offering service arrangements for pension funds.

At the moment it is still not clear at which level the group life market will find its new equilibrium, and buoyant times are over for the market for individual life insurance policies due to the changed fiscal, economic, and investment environment. As a consequence, the spectacular growth rates for total life insurance business will be over for the years ahead.

Table 9.1 gives the historical figures of the premiums earned each year from 1990 to 2003. The figures are given in total and divided into periodic premiums and single premiums earned.

The Aggregated Balance Sheet of the Life Insurance Industry

The most important items on the balance sheet of an insurer are the investments, the technical provisions, and the guarantee funds (also called net worth or policyholders surplus). Premiums paid by the policyholders create liabilities for the insurer. Because those liabilities are in general longer-term ones, the premiums paid are invested in shares, loans, and fixed-income securities and other assets. The guarantee funds are

---

companies offering group contracts (see also "Competition in the market for life insurance policies" in this chapter). The third pillar consists of the market for individual life insurance policies to create some kind of old-age provision.

[195] This trend is driven by the tightening of solvency rules for pension funds, the introduction of new accounting rules forcing companies to become more transparent about the financial position of their pension arrangements, and growing pressures to better pension fund governance structures.

needed to secure the solvency of the insurer. Table 9.2 shows the aggregated balance sheet of the Dutch life insurance industry as of December 31, 2003.

Table 9.1. Life Insurance Premiums Earned, 1990–2003 (€ Millions)

Year	Periodic Premiums	Single Premiums	Total Premiums
1990	5,176	3,964	9,140
1991	5,871	4,598	10,469
1992	6,426	4,437	10,863
1993	6,458	4,060	10,518
1994	7,013	4,277	11,290
1995	7,508	5,180	12,688
1996	8,262	5,974	14,236
1997	9,274	6,830	16,104
1998	10,383	7,919	18,302
1999	11,469	8,567	20,036
2000	12,040	10,616	22,656
2001	12,319	12,514	24,833
2002	12,493	10,716	23,209
2003	12,579	11,350	23,929

Source: PVK and Central Bureau of Statistics.

To evaluate the level of the guarantee funds, the actual level must be related to the required level, giving the solvency ratio. Developments since 2000 in the average solvency margin for the life insurance industry in the Netherlands were discussed above with reference to Figure 9.1.

The investments of a life insurer can be divided into investments held for the benefit of the insurer itself and investments held for the benefit of the policyholder where the policyholder bears the investment risk. At the end of 2003 over 30 percent of total investments (€238.2 billion) was directly held for the account and the benefit of policyholders (see Table 9.2). Five years ago it was still some 25 percent.

On average 70 percent of the investments for the benefit of the insurer itself are fixed-income investments and approximately 14 percent is in shares (see Table 9.2). The remaining investments consist mostly of investments in land and buildings and in affiliated companies.

Directly related to the division of the investments held is the separation of the technical provisions into provisions held for traditional policies and unit-linked policies. With unit-linked policies, the sum assured is expressed in numbers of units of some specific investment object (Oosenbrug 1999). When the sum assured has to be paid out, the insured number of units is multiplied by the unit value at that moment to obtain the amount to be paid out.

To cover the volatility risk of its liabilities, the insurer invests the money related to the provisions for unit-linked policies in matched assets. By matching assets and liabilities the insurer eliminates its investment and liability risks. Because the total return on assets directly affects the unit value, the investments held for unit-linked policies are actually held for the account of policyholders.[196]

Table 9.2. Aggregated Balance Sheet of the Dutch Life Insurance Industry, 12/31/2003 (€ Millions)

Assets		**Liabilities**	
Investments for the Benefit of the Life Insurer		Guarantee Funds	25,055
Shares	22,441	Technical Provisions	
Loans and Fixed Income	112,998	Traditional Policies	131,430
Other Investments	25,970	Unit-Linked Policies	74,641
Investments for the Benefit and the Account of Policyholders	76,814	Non-Technical Provisions	865
Other Assets	17,374	Other Liabilities	23,606
Total Assets	**255,597**	**Total Liabilities**	**255,597**

Source: PVK and Central Bureau of Statistics.

At the end of 2003, over 35 percent of total technical provisions was related to unit-linked policies (see Table 9.2). Logically this is a little more than the share of the related investments in total investments because some part of the investments for the benefit of the life insurers themselves are held to cover the guarantee funds, the non-technical provisions, and the other liabilities.

On the other side, the 35 percent mentioned is far less than the 45 percent share of premium income from unit-linked policies in total premium income from life insurance policies (see Figure 9.4). The reason for this discrepancy is that unit-linked policies are a relatively new phenomenon in the Dutch market. Because it takes some time to build up technical provisions from premium and investment income received over the course of time, the provision/premium-ratio for the relatively new unit-linked portfolio is lower than for the relatively old traditional portfolio.

The Aggregated Income Statement of the Life Insurance Industry

Since the EU-directive on the accounts and the consolidated accounts of insurance companies was implemented in Dutch civil law, the income statement of insurers consists of a technical account and a non-technical account.[197] Given the technical

---

[196] Oosenbrug (2002) discusses the generally accepted accounting principles for unit-linked policies and related investments.

[197] The new accounting rules meant came into force at January 1, 1995 (see for example Oosenbrug 1995a).

character of the business, the technical account is by far the most important part of the income statement of an insurer. The non-technical account mainly consists of the investment income allocated to the non-technical account, other non-technical revenues and charges, and tax charges. Table 9.3 shows the aggregated income statement of the Dutch life insurance industry for financial year 2003.

Table 9.3. Aggregated Income Statement of the Dutch Life Insurance Industry, 2003 (€ Millions)

Income Statement Item		
Gross Premiums Written	24,838	
Reinsurance Premiums Ceded	(0,787)	
Earned Premiums, Net of Reinsurance		24,051
Investment Income Technical Account		14,087
Other Technical Income		0,732
Total Technical Income		38,870
Claims Incurred, Net of Reinsurance	16,707	
Changes in Technical Provisions, Net of Reinsurance	11,481	
Bonuses and Rebates	2,713	
Total Deductions		30,901
Technical Income minus Deductions		7,969
Operating Expenses and Interest Charges	4,049	
Other Charges Technical Account	1,722	
		5,771
Balance on Technical Account		2,198
Investment Income Non-Technical Account	1,214	
Other Income and Charges and Minority Interests	(0,193)	
		1,021
Profit or Loss for the Financial Year, Before Tax		3,219
Tax Charges		0,426
Profit or Loss for the Financial Year, After Tax		2,793

Source: PVK and Central Bureau of Statistics.

The first part of the technical account shows total revenues accrued from premium income and investment income. As Table 9.3 shows, reinsurance premiums ceded are on average only a small fraction of gross premiums written: that is, approximately 3 percent.

In financial year 2003, investment income accounted for 36 percent of total technical income, but investment income can fluctuate very much from one year to another. For example, investment revenues net of investment income allocated to the non-technical account in 2002 were less than 15 (!) percent of the amount accounted for in 2003, while in 1999 it was just 20 percent more than the 2003 figure.

Due to the enormous difference in total investment income in 2003 and 2002, the aggregate net result for 2003 was substantially more than ten times the aggregate net result for 2002. However, a large part of the difference in investment income is the result of the reversal in 2003 of the write-off in 2002 of large amounts of negative investment revaluation reserves. So although net results for 2003 and the development of net results in 2003 look very good at first sight, in reality net results for 2003 were both in absolute terms, and relative to the non-distorted 2002 figures, much worse. As a consequence the 2003 level of net results and the strong recovery of reported results in this year could not be seen as real and of a structural character.

The second part of the technical account specifies the charges directly related to the insurance liabilities of the insurer to its policyholders. Total technical income is primarily intended to cover the claims to be paid out to policyholders, the building up of technical provisions needed, and the allowance of bonuses and rebates to policyholders.

Claims incurred and changes in technical provisions are normally relatively stable items in the income statement of a life insurer. Nevertheless, in 2002 claims incurred were extraordinarily high and changes in technical provisions were extremely low, even negative. Claims incurred were temporarily boosted because many policies were surrendered as a consequence of the bad performances on the stock exchanges. Also as a result of this bad performance, huge investment losses had to be written off on unit-linked policies. The related write-off on the liabilities to the holders of unit-linked policies combined with the effect of the extremely high surrender frequency made the total change of technical provisions in 2002 negative.

Since the start of the twenty-first century the bonuses and rebates allowed to policyholders are low due to low interest rates on the capital markets and negative or low returns on the stock exchanges.

In the third and last part of the income statement, the operating expenses and other technical charges are specified. In 2003 operating expenses alone amounted to €3,230 million for the whole life insurance industry. So on average 13 percent of gross premium income written was spent on operating expenses. Commissions paid to intermediaries account for a substantial part of these operating expenses.

Changes in the Relative Attractiveness of the Market for Life Insurance Policies

Adverse changes in the business environment for life insurers will show up most directly and clearly in the development of the premium amounts written from single premium policies. Figure 9.3 shows the development of total life insurance premiums earned divided into premiums earned from policies with periodic premium payments and premiums earned from single premium policies.

As can be seen clearly in Figure 9.3, the setback in 2002 was completely due to a substantial decrease of premiums earned from single premium contracts. Although the increase of premiums earned from contracts with periodic premium payment clearly flattened, the premiums earned from these contracts still increased. The same was the case in 1993 when—also caused by adverse ("introductory") changes in the tax system—total premiums earned also showed a minor dip. Nevertheless in 1993 it was a one-time dip with periodic and single premiums recovering growth immediately in the year after. Periodic premium growth remained marginal in 2003 for the third year and single premiums mainly rallied thanks to a very large one-time transfer from a pension fund to an insurer (see "Overview of the total market" in this chapter).

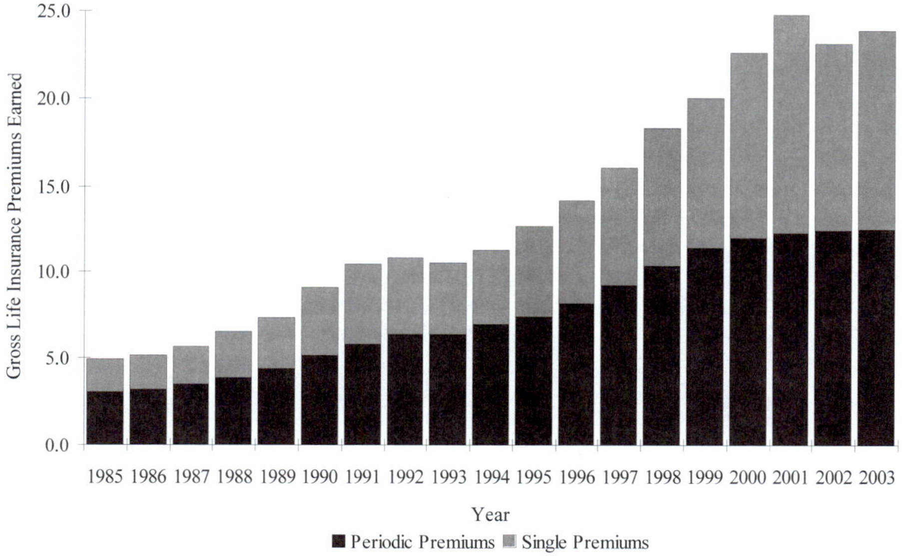

Source: PVK and Central Bureau of Statistics.

Figure 9.3. Gross Life Insurance Premiums Earned, 1985–2003 (€ Billions)

In 2001, premiums earned from single premium contracts accounted for more than 50 percent of total premiums earned. In 1985 it still was not more than some 37 percent. From 1950 to 1973 it was less than 25 percent each year (Nijenhuis, Potjes and Schilder 1994).[198] With an increasing share of single premium contracts, the vulnerability of an insurance company also increases for more or less serious setbacks in the development of total premium income earned.

Due to a setback of almost 15 percent in premium income from single premium policies, total premium income decreased in 2002 by almost 7 percent. The positive point here is that for the first time in years the share of single premium policies decreased. Because single premium contracts do not give any guarantee on the continuity of premium income written, contracts with periodic premium payments

---

[198] From 1947 to 1958, the share of income from single-premium contracts decreased from 26.7% to 14.3%. From 1959 to 1969 it fluctuated between 13.5% and 18.0%. Then it increased rapidly to 27.7% in 1974. In 1975 it peaked at 47.6% to drop to 33.8% in 1976 (Nijenhuis, Potjes, and Schilder 1994). From then on it fluctuated between 30% and 40%, giving the 37% mentioned in 1985.

are in principle more attractive for the insurer. In general, profit margins on single premium contracts are also lower than on contracts with periodic premium payments, so the relative attractiveness (and stability) of the market increases twofold now that the relative share of single premium policies is declining.[199] In 2003 this decline reversed, but only due to the payment of the special one-time single premium mentioned before.

Rise and "Fall" of Unit-Linked Business

It was only recently that the Dutch market became familiar with unit-linked policies. At the beginning of the 1990s unit-linked policies were still considered exotic. With the explosive growth of the general public's interest in stock market investments during the 1990s, the demand for individual and group life insurance policies shifted from traditional policies to unit-linked policies.

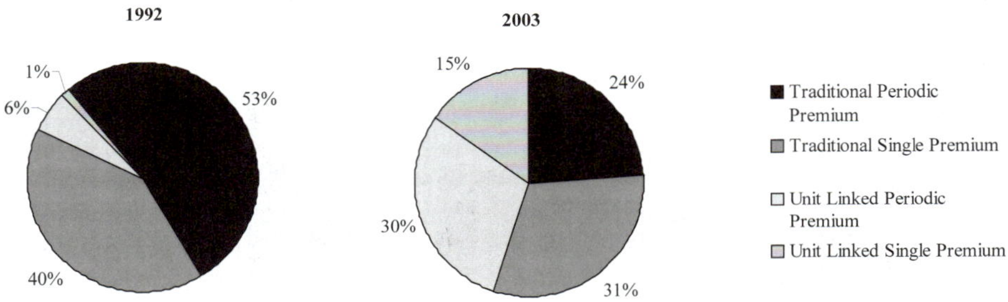

Source: PVK and Central Bureau of Statistics.

Figure 9.4. Composition of Gross Life Insurance Premiums

At the beginning of the twenty-first century unit-linked policies accounted for nearly half of the total life insurance premium income earned. With life insurance contracts generally having durations of decades, it is remarkable that within one decade after the popularization of unit-linked business, the amount of periodic premiums earned from unit-linked policies in 2001 already exceeded the amount of periodic premiums earned from traditional policies. In 2003 it far exceeded the amount of periodic premiums earned from traditional policies (see Figure 9.4).

However, with the end of the euphoric mood at stock exchanges at the beginning of the new millennium, the overwhelming switch to unit-linked policies also came to an end. Currently, reappraisal of traditional life insurance is taking place.

That total premium income from unit-linked policies still does not exceed total premium income from traditional policies is only because the share of single premium unit-linked policies collapsed more than 5.5 percentage points from 21 percent in 2000 to substantially less than 16 percent in 2003. At the same time, the rate at which periodic premium income shifted from traditional policies to unit-linked business slowed down. However, for policies with periodic premium

[199] From 1985 to 2003 the standard deviation of the growth rate for the premium income earned from single-premium policies was more than 12%, with the standard deviation for the growth rate for premium income earned from policies with premium payment being "only" 4.5%.

payments, the share of total premium income written from unit-linked policies can still increase further for years, even if the share of unit-linked policies in new business acquired decreases substantially.

In the segment of individual life insurance policies one can see clearly the effect of the end of the extremely strong and long-lasting rally on the stock markets at the beginning of the new millennium. After the market share of unit-linked policies with premium payment reached its peak of almost 80 percent in this segment in December 2000, the market share fell back to levels common five to six years before. Figure 9.5 shows the recent rise and "fall" of the share of unit-linked business in total new business acquired in the segment of individual life policies with premium payment.

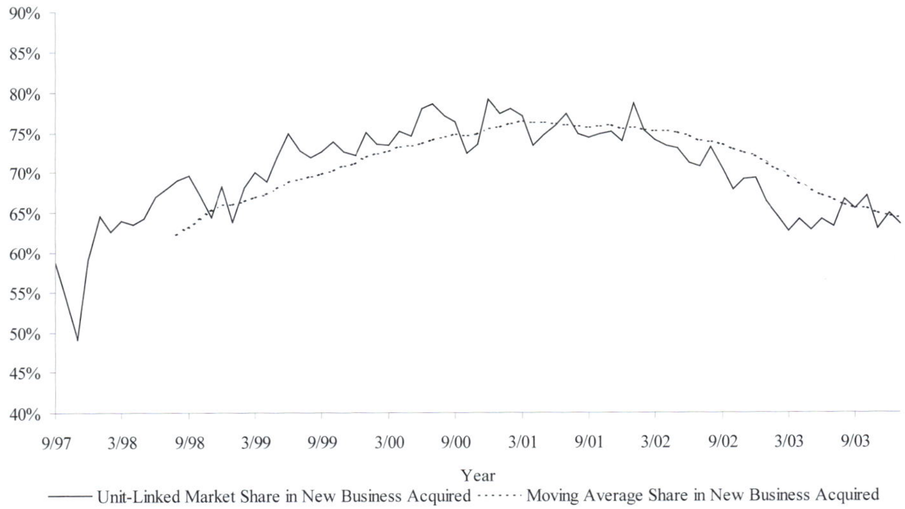

Source: PVK and Central Bureau of Statistics.

Figure 9.5. Unit-Linked Business Market Share of Total New Business Acquired for Individual Life Policies with Premium Payment, 9/1997–1/2004

The trend-break in the first years of the twenty-first century is clear from Figure 9.5. The shift from traditional to unit-linked has reversed in favor of traditional life insurance policies. Nevertheless, the amount of premiums written from unit-linked policies increases still further at a relatively steady rate because the premium income generated by the already existing unit-linked portfolio is still relatively low. Logically, shifts in the composition of new business sold can be seen more directly and clearly in figures concerning premiums written from single premium contracts.

In the segment for single premium contracts the turnaround in December 2000 resulted in a real "landslide." Although the peak reached was less than 60 percent, substantially lower than the peak of nearly 80 percent for contracts with premium payment, the setback was by far more serious (see Figure 9.6). Within three years the 12 months moving average dropped tremendously from more than 43 percent in January 2001 to less than 19 percent in January 2004. So the 12 months moving average was more than halved.

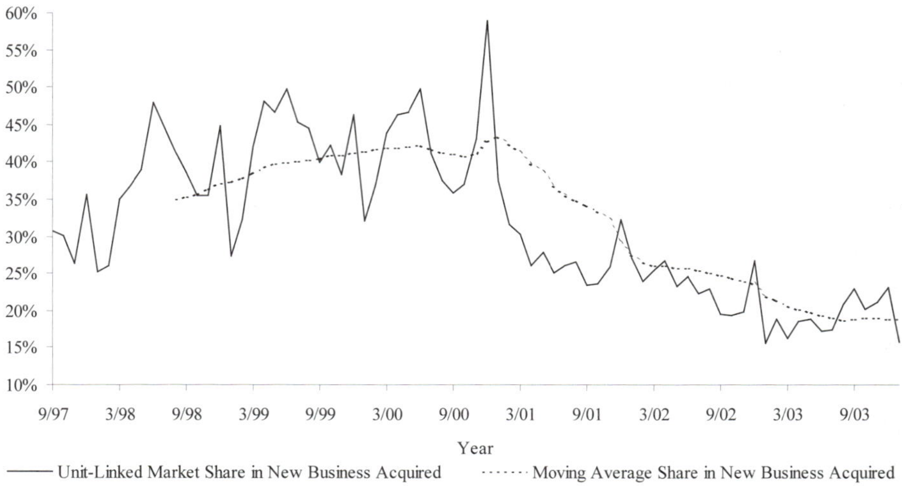

Source: PVK and Central Bureau of Statistics.

Figure 9.6. Unit-Linked Business Market Share of Total New Business Acquired for Individual Single Premium Life Policies, 9/1997–1/2004

Indeed, single premium unit-linked policies never attained the same enormous popularity as periodic premium unit-linked policies because immediate annuities account for a very large part of the segment for single premium policies. Due to the specific character of an immediate annuity, with the client generally in need of a far-reaching income guarantee, this kind of product is less suited for selling as a unit-linked product. Because it is likely that in the years to come the market share of immediate annuities will only increase further, it does not seem likely that in the near future unit-linked business will, in this segment, regain the market share lost.

Growing Importance of Mortgage-Related Life Insurance Policies

During the last few years the dependency of new life business acquired in the market for mortgage loans rose substantially. Figure 9.7 shows the recent development of the monthly share of mortgage-related life business in total new business acquired in the segment of individual life insurance with premium payment.

At the peak of popularity of unit-linked business, almost 85 (!) percent of the new mortgage-related life insurance policies sold were unit-linked policies (in terms of premiums written). The market share of unit-linked business in this segment is still 70 percent. Nevertheless, the growth in this market segment in the most recent years came only from the selling of traditional life insurance policies.

With a 12-months moving average market share of mortgage-related business that has risen from less than 30 percent in 2001 to almost 45 percent at the beginning of 2004, the vulnerability of the sector for the development of the mortgage loans market has risen proportionately. Due to "overheated" housing prices, bottoming interest rates, the economic downturn and the resulting rise in unemployment, and last but surely not least an adverse change in the tax system in effect as of January 1, 2004, the market for mortgage-related life insurance policies stands at a breaking point. In the trade journals some pessimists declared that an adverse effect for new business acquired could be expected of more than €300 million in terms of premium income written.

Source: PVK and Central Bureau of Statistics.

Figure 9.7. Market Share of Mortgage Related Business in Total New Business Acquired for Individual Life Policies with Premium Payment, 9/1997–1/2004 in the Netherlands from September 1997 to January 2004

9.4.3 Overview of the Market for Non-Life Business[200]

The Relative Relevance of the Different Branches

Premium income from non-life business has grown steadily for decades. A substantial part of this steady growth was due to continually rising health care costs and a drastic privatization of the social security system for sickness and disability insurance. As a result, premium income from accident, private health, and disability insurance policies already accounts for nearly half of the total premium income from non-life business. Figure 9.8 compares non-life premiums by line of business for 1981 and 2003, while Table 9.4 gives premiums by line for the period 1990–2003.

In 1981 premium income from accident, private health, and disability insurance still accounted for “only” one-third of total non-life premium income written. Since 1981 the “market share” of accident, health, and disability insurance increased enormously while this rise in the importance of accident, health and disability insurance for the non-life business segment continues. In 2003 approximately 60 percent of the premium income written came from private health insurance policies and approximately 40 percent from sickness and disability insurance policies (AM Jaarboek 2004).

[200] See A. Oosenbrug, et al. (1996).

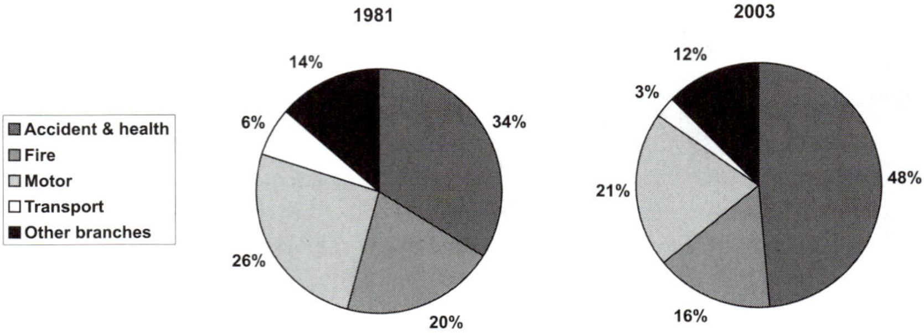

Source: PVK and Central Bureau of Statistics.

Figure 9.8. Composition of Gross Non-Life Premiums

Note that only private health insurance is accounted for in the figures given above. Premium income from compulsory health insurance—almost two times the amount of premium income from private health insurance—is not included. Health insurance is compulsory for each employee with a yearly income of less than approximately €33,000. Of Dutch citizens, roughly two-thirds are insured by one of the compulsory funds.

Also, social premiums imposed by the Exceptional Medical Expenses Act (AWBZ), amounting to almost €13 billion, are not included. Every Dutch citizen is by law insured against exceptional medical expenses, such as the (very high) costs of long-term nursing home care. Approximately 60 percent of the AWBZ-expenses is spent on nursing and care for disabled and aged people. The other 40 percent is more or less evenly spent on mental health care and facilities for disabled people. The licensed health insurers act as administrative offices to execute the AWBZ.

So the relative share of health and disability insurance increased substantially during the past decades. However, due to population growth, growing prosperity, and a continuing progression in the penetration of high-tech and relatively expensive equipment and appliances, premium income from property and automobile insurance policies also showed a steady, although less impressive growth rate. On the other hand, premium income from transport insurance shrank to almost marginal proportions, in spite of the fact that the Dutch ports are among the largest transit ports of the world. Strong competition from the London market was one of the causes.

Approximately 55 percent of total premium income written from automobile insurance policies was generated by automobile liability insurance policies. The other 45 percent was generated by accidental damage insurance policies. In 2003 the growth rate for automobile insurance was below average, with growth mainly resulting from premium rate increases, especially for automobile liability insurance.

In addition to personal and commercial liability insurance, the category "other branches" mainly includes credit insurance, legal expenses insurance, and travel insurance. Especially due to an increasing penetration of liability insurance combined with rising premium rates for it, the growth rate for premium income written from policies within this category has accelerated. In 2001 premium income grew by 9 percent while in 2002 the growth rate accelerated to more than 13 percent. In 2003 the growth rate consolidated on the high level of 13 percent.

Although the U.S. culture of civil litigation has not yet reached the Netherlands, claim awareness and the amounts of claims are nevertheless rising. Against the background of rising claim awareness, it is not surprising that the market penetration of legal expenses insurance grew relatively fast. Nevertheless, legal expenses insurance is still a small branch in terms of premium income written (much less than €400 million).

In the Dutch insurance market employers' liability is also a relatively uncertain risk. With, for example, the DES and asbestosis cases in mind, insurers are reluctant to offer employers' and product liability insurance.[201] As a logical consequence, premium rates are rising. For the near future it seems likely that penetration rates and premium rates for employers' liability insurance will continue to rise.

Table 9.4. Non-Life Insurance Premium Revenues by Line of Business, 1990–2003 (€ Millions)

Year	Accident and Health	Fire	Motor	Transport	Other Branches
1990	3,376	1,833	2,208	406	1,094
1991	3,406	1,881	2,295	433	1,163
1992	3,918	1,999	2,354	502	1,251
1993	4,412	2,159	2,479	529	1,295
1994	4,981	2,236	2,748	498	1,371
1995	5,225	2,362	2,944	506	1,388
1996	5,577	2,417	3,129	520	1,396
1997	6,010	2,459	3,175	523	1,482
1998	6,421	2,527	3,279	520	1,741
1999	7,106	2,512	3,463	490	1,809
2000	7,595	2,674	3,759	515	2,022
2001	8,437	2,814	3,988	543	2,202
2002	9,661	3,129	4,222	579	2,495
2003	10,279	3,314	4,381	598	2,632

Source: PVK and Central Bureau of Statistics.

The Aggregated Balance Sheet of the Non-Life Insurance Industry

The most important items on the balance sheet of a non-life insurer are the investments, technical provisions, and guarantee funds (net worth or equity capital). Premiums paid by the policyholders create liabilities for the insurer. Because non-life insurers have a more or less stable permanent "stock" of liabilities they have a more or less stable amount available to invest in shares, in loans and fixed-income securities, and in other assets. The guarantee funds are needed to secure the solvency

[201] DES (*di-ethyl stilboestrol*) is a synthetic hormone formerly used in some medicine for pregnant women. DES can cause genetic abnormalities in the child.

of the insurer. Table 9.5 shows the aggregated balance sheet of the Dutch non-life insurance industry as of December 31, 2003.

Table 9.5. Aggregated Balance Sheet of the Dutch Non-Life Insurance Industry, 12/31/2003 (€ Millions)

Assets		**Liabilities**	
Investments for the Benefit of the Insurer		Guarantee Funds	10,155
Shares	4,216		
Loans and Fixed Income Securities	21,330	Technical Provisions	22,176
Other Investments	3,772	Non-Technical Provisions	375
Other Assets	7,624	Other Liabilities	4,236
Total Assets	**36,942**	**Total Liabilities**	**36,942**

Source: PVK and Central Bureau of Statistics.

To evaluate the level of the guarantee funds, the actual level has to be related to the required level, giving the solvency ratio. The solvency ratio for Dutch non-life insurers is discussed earlier in this chapter with reference to Figure 9.1.

On average more than 70 percent of the investments of Dutch non-life insurers are fixed-income investments and approximately 14 percent are in shares (see Table 9.5). The other investments consist of investments in land and buildings and in affiliated companies. At the start of the new millennium nearly 28 percent of total investments was invested in shares, almost double the current 14 percent.

The Aggregated Income Statement of the Non-Life Insurance Industry

Since the EU directive on the accounts and the consolidated accounts of insurance companies was implemented in Dutch civil law, the income statement of insurers consists of a technical account and a non-technical account.[202] Given the technical character of the business, the technical account is by far the most important part of the income statement of an insurer. The non-technical account mainly consists of the investment income allocated to or kept back in the non-technical account, other non-technical revenues, and charges and tax charges. Table 9.6 shows the aggregated income statement of the Dutch non-life insurance industry for financial year 2003.

The first part of the technical account shows total revenues accrued from mainly premium income and investment income. As Table 9.6 shows, reinsurance premiums ceded are on average nearly 11 percent of gross premiums earned. In 2002 it was over 12 percent due to the rate increases on the reinsurance market following 9/11. In

[202] The new accounting rules meant came into force as of January 1, 1995 (see for example Oosenbrug 1995a).

2003 reinsurance charges fell back again because the insurance companies on average increased the amounts of their own retentions.

Table 9.6. Aggregated Income Statement of the Dutch Non-Life Insurance Industry, 2003 (€ Millions)

Income Statement Item		
Gross Premiums Written	21,203	
Change in the Gross Provision for Unearned Premiums	(.299)	
Gross Premiums Earned	20,904	
Reinsurance Premiums Ceded	(2,260)	
Earned Premiums, Net of Reinsurance		18,644
Investment Income Technical Account		1,212
Other Technical Income		.258
		20,114
Claims Incurred, Net of Reinsurance	14,064	
Changes in Technical Provisions, Net of Reinsurance	(.009)	
Bonuses and Rebates	.103	
		14,158
		5,956
Operating Expenses and Interest Charges	4,481	
Other Charges Technical Account	.290	
		4,771
Balance on Technical Account		1,185
Investment Income Non-Technical Account	.551	
Other Income and Charges and Minority Interests	(.067)	
		.484
Profit or Loss for the Financial Year, Before Tax		1,669
Tax Charges		(.399)
Profit or Loss for the Financial Year, After Tax		1,270

Source: PVK and Central Bureau of Statistics.

In financial year 2003, investment income accounted only for 6 percent of total technical income. As a consequence, strong fluctuations in the amount of the investment income earned do not impact technical results and net results the same way as in the life insurance industry. Nevertheless, in 2003 net results "exploded" from €197 million in 2002 to €1,270 million in 2003, mainly because total investment income doubled in 2003.

However, a large part of the difference in investment income is the result of the reversal in 2003 of the write-off in 2002 of large amounts of negative investment revaluation reserves. So although net results for 2003 and the development of net results in 2003 look good at first sight, in reality net results for 2003 were much worse both in absolute terms and relative to the non-distorted 2002-figure. As a consequence, the 2003 level of net results and the strong recovery of reported results in 2003 cannot be seen as real and of a structural character.

The second part of the technical account shows the claims incurred during the financial year. Claims incurred plus bonuses and rebates allowed amount to 76 percent of net premium income earned.

In the third and last part of the income statement the operating expenses and other technical charges are specified. In 2003 total operating expenses amounted to 24 percent of net premium income earned. Commissions paid to intermediaries account for a substantial part of the operating expenses.

Together with the loss ratio referred to above, the expense ratio of 24 percent makes a combined ratio of (almost) exactly 100 percent. In 2002 the combined ratio was 103 percent and in 2001 more than 104 percent. The realized improvements were mainly caused by premium rate increases.

Technical Results

The profit or loss at the bottom of the technical account of the income statement as discussed above is the technical result of the insurer. Figure 9.9 shows the development of technical results in total and by branch as a percentage of gross premium income written. As the figure shows, technical results fluctuate from year to year, with negative results for automobile, transport, property, and health insurance in 2002. Health insurance technical results have been near or below zero for several years, but are mostly more than compensated for by positive technical results on sickness and disability insurance policies. Especially in 2001 and 2002, technical results on sickness and disability insurance were again very attractive, resulting in nice overall results for the total branch (see Figure 9.9). In 2003 technical results in all branches were positive.

Due to negative total returns on the stock markets and relatively low capital market interest rates, the dip in relative total net results in 2002 was by far more serious than the dip in relative technical results. Total technical results nearly halved from over 2 percent to a little bit more than 1 percent, while total net results tumbled almost 70 percent, from nearly 4 percent to less than 1 percent.

In 2003 results recovered substantially except for the branch "other branches." Relative total technical results rallied by 400 percent to over 5.5 percent and relative total net results by 200 percent to 6 percent. As discussed above, one has to remember that book results were depressed in 2002 and boosted in 2003 due to the developments on the stock exchanges.

Successive technical results from automobile insurance typically fluctuate around zero. The market for automobile insurance policies is traditionally the most competitive segment of the market for non-life insurers. After some recovery in 2001, automobile insurance technical results slid again in 2002. Nevertheless, in 2003 results recovered strongly. Relative technical results rallied from minus 1.5

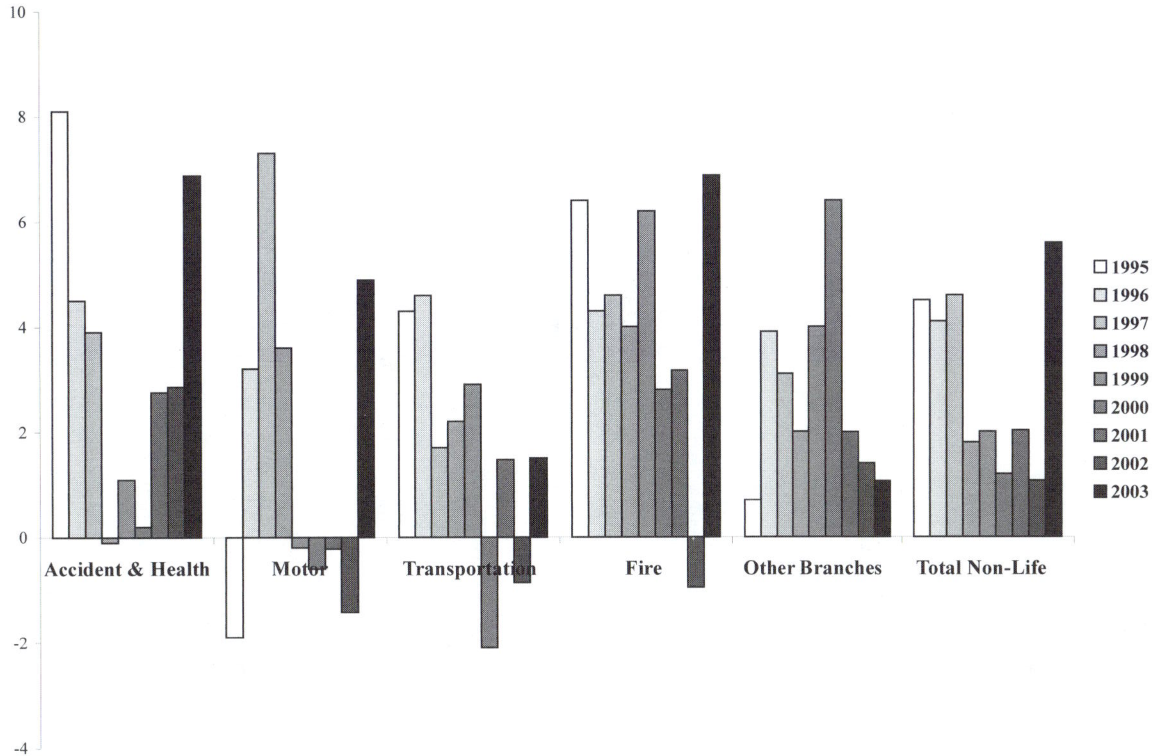

Source: PVK and Central Bureau of Statistics.

Note: The relative technical result of a branch is the profit or loss on the technical account of the aggregated income statement for the branch divided by the aggregated gross premium income written for that branch.

Figure 9.9. Relative Technical Results by Branch, 1995–2003

percent to 5 percent. Besides the turnaround in the book results on the investments made, a substantial drop in the loss ratio of over 3 percent was the cause of this extraordinary improvement. The loss ratio dropped as a result of the lucky combination of good weather conditions during the year and an exceptional decrease in the number of car thefts.

As the second most important branch in terms of premium written, automobile puts a relatively heavy mark upon the development of total non-life technical results. This becomes even clearer by looking at the development of non-life technical results exclusive of technical results for the branch accident, health, and disability. As can be seen clearly from Figure 9.10, the development of technical results for only property and automobile insurance "perfectly" follows the development of technical results for automobile only (compare Figure 9.10 with the automobile section of Figure 9.9).

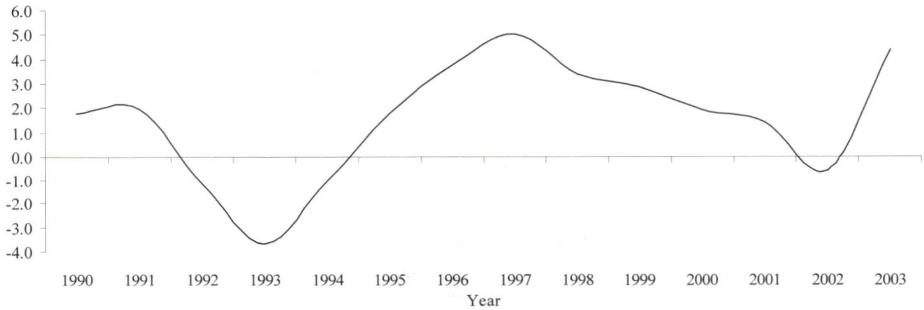

Source: PVK and Central Bureau of Statistics.
Note: The relative technical result of a branch is the profit or loss on the technical account of the aggregated income statement for the branch divided by the aggregated gross premium income written for that branch.

Figure 9.10. Relative Technical Results for Non-Life Excluding Health and Disability Insurance, 1990–2003

Due to restrictive underwriting policies, technical results from transport insurance are positive again. Nevertheless, profit margins remain low due to continuing strong competition from the London market.

Until recently, fire was a relatively profitable and stable branch. In 2000 the results suffered from the great fireworks disaster at Enschede. In the summer of 2000 the storage room of SE Fireworks, in a residential quarter of that town, exploded. Twenty-three people were killed and almost a thousand were injured. The material losses are estimated at half a billion euros. Nevertheless, technical results were still not really bad (see Figure 9.9). In 2002, for the first time in years, technical results were written in red, mainly due to some big losses on industrial fire insurance policies and to higher reinsurance costs resulting from 9/11. However, in 2003 results rallied strongly again to previous levels (see Figure 9.9).

Technical results from the other branches are under pressure. As in other countries, also in the Netherlands experience shows it is difficult to gain effective control over liability claims. Nevertheless, the average profitability of all the other branches together is historically only slightly below average while the volatility of technical results is clearly the lowest of all the non-life branches. The category other branches is also the only category that did not show technical results in red for decades.

Finally, it should be noted that in 2003 a terrorism pool became effective. With support of international re-insurers and the national government, the pool covers losses due to attacks by terrorists. There is a limit for a single policyholder or a single location of €75 million; total coverage is limited to €1 billion a year. As a consequence Dutch non-life insurers jointly limited the coverage provided by their own policies to the coverage provided by the newly created terrorism pool. Of the total coverage provided, the first €600 million is for the account of insurers and re-insurers, while the €400 million coverage above that first layer is provided by the Dutch government.

9.4.4 Distribution Systems

Distribution Channels and Their Relative Importance

In the Netherlands the distribution system for insurance contracts is very diversified. Independent agents, tied agents, banking outlets, direct writers, underwriting agents, and insurance brokers are all common in the Dutch market. Agents and brokers have an estimated market share of approximately 60 percent, direct writers 20 percent, and banks 15 percent.

Remarkably, a bank—the co-operative Rabobank—is by far the insurance brokerage market leader in terms of gross commission fees earned. Most independent agents are relatively small local intermediaries. As a consequence, seven of the top 10 intermediaries in terms of gross commission fees earned are banks (five) or captives owned by insurance groups (two). Two of the remaining three are the Dutch subsidiaries of the worldwide insurance brokers Aon and Marsh.

In the last few years, different kinds of "exotic" intermediaries such as oil companies, supermarkets, football clubs, car dealers, drugstores, and so on have also entered the market. Until now the success of those new initiatives was only limited. For example, although no fewer than 30 automobile brands today have their own brand automobile insurance policy, the total penetration rate of own-brand automobile insurance policies is by rough estimation at most only 5 percent to 7 percent. Moreover, part of this market share was realized by the distribution of some of the auto brand policies by independent agents.

Dynamics in the Business Environment for the Intermediaries

A few years ago the generous special tax facilities for life insurance policies were changed drastically. As a consequence, many intermediaries had to make the turnaround from simple tax-driven "hit-and-run" selling to "advice-based" selling. At the same time stock markets crashed and the economy cooled down. Huge volumes of recently agreed life insurance policies were surrendered, rendered paid up, or cancelled without any value remaining for the policyholder. Due to relatively very high acquisition commissions paid out at the inception of the contract, huge amounts of commissions paid out before were taken back by the insurers in question. The combined effect of a drastically changed business environment (tax changes, crashes on the stock exchanges, a cooling down of the economy, and low interest rates) and huge cash outflows as a result of arisen commission takebacks (high surrender frequencies) resulted in a real shake-out in the "hit-and-run" intermediaries.

In addition, the distribution channel of independent agents is still in search of a new equilibrium since the business environment has changed so drastically. At the same time, independent agents in particular were confronted with important changes in the regulatory environment for the distribution system and have to prepare for the coming supervisory regulation for financial intermediaries.

Dynamics in the Regulatory Environment for the Distribution System

A few years ago, the rules prescribing fixed commission rates for the distribution channel of independent agents were abolished. Nevertheless, the Dutch antitrust authority, the Nederlandse Mededingingsautoriteit (NMa), has reported again that for non-life business the formerly market-wide agreed commission levels are in general still effective. In 2000 the NMa disqualified this situation as being in violation of the antitrust law.

The rules prohibiting offering a gift to potential clients if and only if they buy an insurance policy were abolished. Today, direct writers in particular use this tactic to tempt potential clients. The gifts range from a lottery ticket for a legal expenses insurance policy to a digital camera for a deferred annuity contract.

At the moment there is much talk about (the lack of) transparency with respect to participation of insurance companies in intermediaries or even total ownership of intermediaries by insurance companies. Due to a lack of transparency, clients cannot know if the intermediary is completely free to give the best advice or not. In the near future, the new Act on the Selling of Financial Services will prescribe full disclosure of all financial and/or contractual ties between insurer and intermediary. However, for the time being, the exact contents and wording of the coming act are still under discussion.

Another still longer pending issue is the lack of transparency with respect to the remuneration of financial intermediaries. Especially in life business with its focus on (very) high acquisition commission fees, this lack of transparency is genuinely an issue.

Until now most intermediaries have been against the disclosure of the amount of the commission fee earned for a specific transaction. They compare their lack of transparency about remuneration with that of washing machine and car dealers. The more appropriate comparison with the totally transparent fee-based remuneration system of professional advisors such as accountants, lawyers, and tax advisors is systematically disregarded. Nevertheless, little by little, well-educated people believe a lack of transparency is no longer justified. However, also under the new Act on the Selling of Financial Services, there will be no rules prescribing the disclosure of the commission earned by the intermediary.

In the near future, the Act on the Selling of Financial Services will become law. Under this act intermediaries will have to apply for a license to operate in the market for financial services, with licenses granted by the Authority for the Financial Markets (AFM). This same AFM will be in charge of the supervision of financial intermediaries. To take out and hold a license to operate, an intermediary must satisfy specific rules with respect to:

- the expertise of the agent with respect to the products and the services he offers;
- the integrity of the agent, including his adherence to rules and standards of market conduct;
- the financial position of the intermediary, including holding an adequate commercial liability insurance policy; and
- the management skills of the agent, including a proper administrative system of checks and balances.

As noted, though, the exact contents and wording of the coming act are still under discussion.

E-commerce in the Dutch Insurance Market

Currently, the websites of Dutch insurers are still focused on marketing and counseling; offering insurance policy sign-up electronically is an exception; some "revolutionary" initiatives to start real virtual insurance companies already have failed. Nevertheless, it is possible to take out some standard policies electronically for simple insurance risks, for example, travel insurance. The medium-sized mutual non-life insurer Univé claims that 10 percent of its new business is taken out electronically. In addition, the company expected the share of e-business in total new business acquired would double in 2004 to 20 percent. However, most insurance companies' websites focus on providing information and providing a means to apply for an offer. Insurers working with independent agents also use their websites specifically to generate leads for their intermediaries.

So, real e-business is still not usual business in the Dutch insurance market. Nevertheless, some intermediaries' websites that focus on comparing insurance products seem relatively successful. These websites can be made profitable by selling the leads generated to the intermediaries in question or to outside intermediaries or (direct writing) insurance companies. At the same time, these websites make a welcome contribution to creating more transparency in the insurance market. As products and services will become more transparent, the potential for e-commerce in the insurance market will surely increase enormously. In the Dutch banking sector, e-commerce is already commonplace. It is only a question of time until the same will be the case in the Dutch insurance sector.

9.5 CONCENTRATION, GLOBALIZATION, INTEGRATION, AND THE DUTCH INSURANCE MARKET

9.5.1 Overview of the Total Number of Market Participants

Traditionally, the Dutch insurance market has been one of the most open markets in the world. As a consequence, one of the main characteristics of the market is the large number of licensed companies. Since the European single market was formed in 1993, the number of insurers (excluding the funeral funds under supervision) exploded to 1,072 at the end of 1999 (see Figure 9.11). Due to the single license principle, more than 60 percent of them were not supervised by the Dutch supervisory authority but by the supervisory authority in their home country. Since 1999, the number of market participants has been more or less stable, around 1,050.

In 1996, also the funeral funds with at least 3,000 adult members were brought under supervisory control. Ninety-three funeral funds met the criterion of having at least 3,000 members of age. Due to the planned reconstruction of the sector, this number halved within six years to 46 at the end of 2002 and then stabilized at that level.

At the end of 2003 there were 1,079 licensed market participants, with 247 life insurers, 46 funeral funds, and 786 non-life insurers.

9.5.2 Competition in the Market for Life Insurance Policies

At the end of 2003, 247 life insurance companies were licensed to operate in the Dutch market; in 1985 there had been only 69. Since the European single market was formed in 1993, the number of life insurers exploded from 97 at the end of 1992 to 263 at the end of 2002 (see Figure 9.11). At first, the number of life insurers with a license granted by the Dutch supervisory authority increased steadily to a maximum of 109 at the end of 1999. However, since the start of the new millennium, this number decreased from year to year with 87 remaining at the end of 2003. At the same time, the increase of the number of insurers with a license granted under the "single license principle" continued until the end of 2002. In 2003 this number decreased for the first time, from 171 to 160, for a total of 247 licensed market participants at the end of 2003.

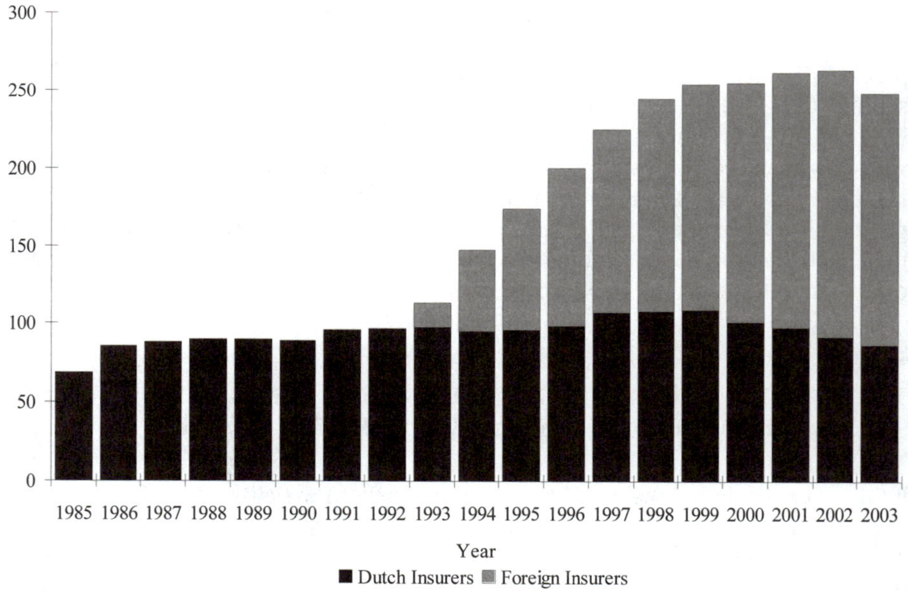

Source: PVK and Central Bureau of Statistics.

Figure 9.11. Number of Licensed Life Insurers in the Dutch Market, 1985–2003

Due to the single-license principle, not less than 65 percent of the licensed market participants are not supervised by the Dutch supervisory authority but by the supervisory authority in their home country. Most of them even do not have their office within the country but do their business through underwriting agents or other kinds of representatives. Unfortunately, it is not known which part of the licensed foreign companies are effectively not operational in the Dutch market. Although it seems likely that by far the largest part of the turnover in a national market for insurance policies will automatically flow to insurers established in the country in question and not to insurers established abroad, it is, however, also likely that a large number of licensed foreign companies will at least lead to some potential competitive pressure in the market.

In addition to the life insurers operating in the Dutch market for individual and group life insurance contracts, no fewer than 867 pension funds were operational in the Netherlands at the end of 2003, the largest part being company pension funds (753), a still substantial part being pension funds for branches of industry (103), and a minor portion being occupational pension funds (11).

Life insurance business both complements and competes with pensions business. At the same time, some pension funds compete directly with insurers by offering their participants supplementary life and/or disability insurance products. However, a few years ago new legislation introduced strict rules for offering insurance products by pension funds or by insurance subsidiaries of pension funds. As a result, some pension funds announced their withdrawal from the insurance market. For example, the second largest pension fund in the Netherlands, the Stichting Pensioenfonds voor de Gezondheid, Geestelijke en Maatschappelijke Belangen PGGM, announced its intention to sell its insurance subsidiaries.

9.5.3 Competition in the Market for Non-Life Insurance Policies

At the end of 2003, 786 non-life insurance companies had licenses to operate in the Dutch market. In 1985 there were still "only" 361 licensed non-life insurance companies. Since the European single market was formed in 1993, the number of non-life insurers exploded from 391 at the end of 1992 to 819 at the end of 1999 (see Figure 9.12).[203]

With the introduction of the single license principle, the number of non-life insurers with a license granted by the Dutch supervisory authority dropped by almost 30 percent, from 391 at the end of 1992 to 280 at the end of 1995. At the same time, no fewer than 402 new licenses were granted to foreign insurers already having licenses to operate in another European Union country.

Since 1998 the number of insurers with licenses granted by the Dutch supervisory authority decreased steadily from 294 at the end of 1998 to 246 at the end of 2003. So in the last five years, more than 15 percent of the domestic non-life insurers disappeared as a consequence of takeovers, mergers, or run-offs. However, the increase in the number of insurers with licenses granted under the single license principle continued until the end of 2002. In 2003, this number decreased for the first

[203] Note that the scale of Figure 12 is not the same as the scale of Figure 11. The total number of licensed non-life insurance companies is more than threefold the total number of licensed life insurance companies.

time, from 555 to 540, for a total of 786 licensed market participants at the end of 2003.

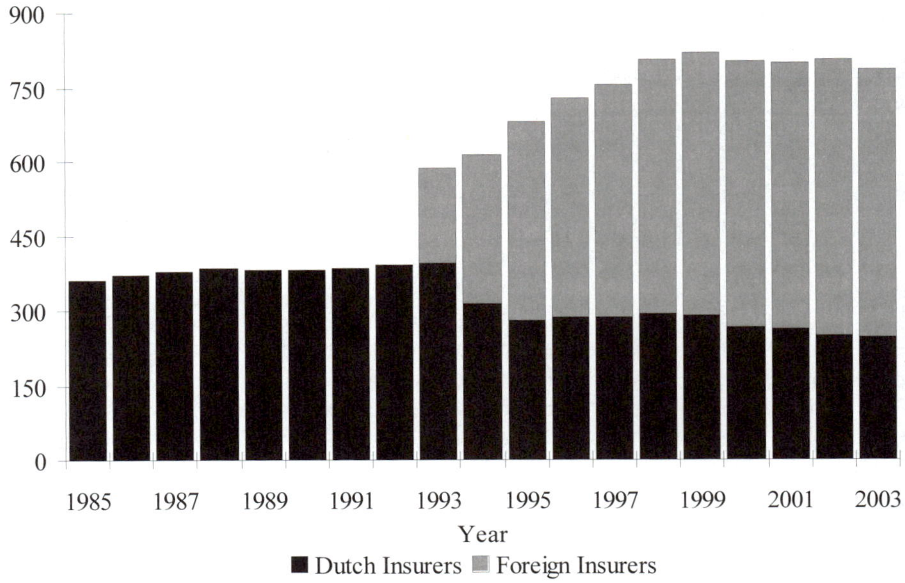

Source: PVK and Central Bureau of Statistics.

Figure 9.12. Number of Licensed Non-Life Insurers in the Dutch Market, 1985–2003

9.5.4 Concentration in the Dutch Insurance Market

Despite the large number of licensed companies in the Dutch market, in 2003 the top 10 insurance groups in the relevant market segments have written almost 90 percent of life business and a respectable 67 percent of non-life business.[204] Table 9.7 specifies the market share of the top 10 insurance groups in the Dutch life and non-life markets.[205]

All top five insurance groups and most top 10 insurance groups are the result of high profile mergers and/or takeovers. In 2002 the Aviva subsidiary Delta Lloyd took over the insurance subsidiaries of ABN Amro. ABN Amro Life Insurance was number nine in the top 10 of life insurers. Due to the take-over by Delta Lloyd, Allianz entered the top 10 of life insurers in 2002. Delta Lloyd itself rose from sixth to fourth in life business. ABN Amro Property & Casualty Insurance was much smaller and had no position in the top 10 of non-life insurers.

---

[204] In 1991 a 71% market share in the non-life market was held by the top 20 in the market. In 1987 the top 20 held a market share of not more than 64% (Buijs 1994).

[205] See Nijenhuis and Potjes (1994) for some comments on the development of the degree of concentration in the Dutch life market from 1956 to 1991.

Table 9.7. Market Share of Ten Largest Insurance Groups in the Dutch Life and Non-Life Insurance Market, 2002–03

Ranks	Life Insurance	Type	2002 (%)	2003 (%)	Non-Life Insurance	Type	2002 (%)	2003 (%)
1	ING	stock	20.6	22.9	Achmea	mutual	13.7	14.6
2	Aegon	stock	15.0	13.2	Fortis	stock	11.5	11.4
3	Fortis	stock	13.1	12.2	ING	stock	9.4	8.7
4	Delta Lloyd's	stock	10.6	8.8	Delta Lloyd's	stock	8.2	8.0
5	Rabobank	mutual	7.4	8.3	Rabobank	mutual	6.0	6.0
6	Achmea	mutual	7.4	7.7	Allianz	stock	5.6	5.5
7	Swiss Life	stock	4.2	6.6	Univé	mutual	3.8	3.9
8	SNS Reaal	mutual	5.9	6.2	Menzis (Amicon)	mutual	2.4	3.4
9	AXA	stock	2.5	2.1	Aegon	stock	3.0	2.9
10	Allianz	stock	1.7	1.6	AXA	stock	2.9	2.5
Total Market Share of Top Ten			88.4	89.6	Total Market Share of Top Ten		66.5	66.9

Source: AM Jaarboek (2003/2004) and (2004).
Note: Market share is defined in terms of premium income written. Delta Lloyd's total market share in the market for non-life insurance as published in AM Jaarboek (2003/2004) is incorrect. In this chapter, the correct figure has been used.

Most recently, Achmea took over Levob in 2003, a smaller all-finance group, while Menzis acquired the smaller health insurance companies NVS and Rijnmond. As a result, Achmea in particular could consolidate its leading position in the non-life market and Menzis could win two positions in the ranking for non-life insurers. Aegon, one of the large international Dutch insurers, sank to ninth. ING and AXA lost market share in the non-life market, viz., 0.7 percent and 0.4 percent.

In the life market AXA lost market share too, but ING could extend its market leadership substantially. The other large international groups lost substantial parts of their market shares, viz., Aegon and Delta Lloyd 1.8 percent and Fortis 0.9 percent. Swiss Life and SNS Reaal changed positions in 2003; however, note that Swiss Life received a very large one-time single premium out of the transfer of the assets and liabilities of a pension fund, so the change does not seem structural.

Based on total market shares in the total Dutch life and non-life markets the top five insurers in decreasing order are ING, Fortis, Achmea, Delta Lloyd, and Aegon. With ING and Achmea winning market share and the others losing market share, the 2003 ranking was still the same as the 2002 ranking.

As can be seen in Table 9.7, mutuality is still an important phenomenon in the Dutch financial services sector. Among the top 10 groups, several still operate more or less on a co-operative basis (for example, Achmea, Rabobank, and Univé).

In April 2005, the merger of Achmea and the insurance subsidiary of Rabobank (Interpolis) was announced. The announced merger of the numbers five and six in the life market will result in a new group ranking second in the life market. However, the number one position is still far out of sight with the number one, ING, having a market share almost half as large as the life market share of the newly created Interpolis/Achmea-group.

On the contrary, with Achmea already being the unchallenged number one in the non-life market, the newly created group will by far be the absolute number one in the non-life market. Its non-life market share will be as large as twice the non-life market share of the number two in the market (Fortis).

9.5.5 Globalization

The openness of the Dutch market typically resulted in many foreign insurance groups playing an important role in the market by having substantial subsidiaries in the Netherlands. Four of the top 10 insurance groups in the life insurance market, as well as three in the non-life market, are full subsidiaries of foreign insurance groups. Delta Lloyd, a full subsidiary of Aviva, ranks fourth for life, as well as fourth for non-life business. Also Allianz, Swiss Life, and AXA are full subsidiaries of insurance groups from abroad.

However, real cross-border trade in insurance policies is until now only a marginal business. The existing major differences in legal rules, tax systems, and languages are still barriers too high for cross-border selling of insurance policies to become common. According to estimates made by the OECD, foreign companies realized only 5 percent of life business and not more than 2 percent of non-life business in the Netherlands. Indeed, it even seems likely that the largest part of the business done by foreign companies was realized through representatives who are still physically present in the local market.

The Dutch-rooted companies in the top 10 are more or less international companies in the real sense, too. The first in non-life and sixth for life business (Achmea) is the leading participant in the European Eureko-group. In addition, the third in life and second in non-life business (Fortis) is really a multinational group based in the Netherlands and in Belgium. Although with pure Dutch roots, ING and Aegon are really international financial groups; ING is a well-known worldwide banking and insurance group with its subsidiaries established in countries from Central Europe to North America and from South America to Asia.

Figure 9.13 shows the geographic spread of total premium income written by ING. Indeed, due to the strong euro, premium income from outside Europe decreased in 2003 from 79 percent to 75 percent. In 2004 ING ranked twelfth on the Forbes 2000 Leadings Companies list. In the category of "Diversified Financials" it ranked third after Fannie Mae and UBS.

As the result of regular and important acquisitions in the past, ING, Aegon and Fortis now belong to a group of key foreign players in the U.S. insurance market. In many other countries around the world the presence of the leading Dutch insurance groups is far from negligible, due to acquisitions and/or greenfields operations started in the past. For example, the Dutch insurer ING was the first foreign insurance group to get a license to operate in China!

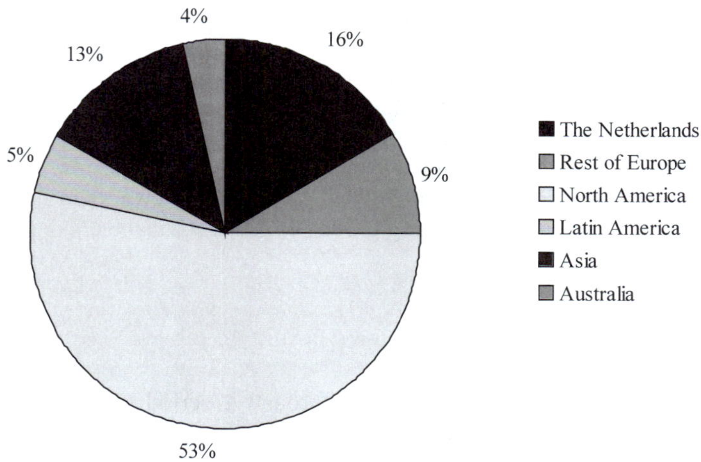

Source: PVK and Central Bureau of Statistics.

Figure 9.13. Geographic Decomposition of Total Premium Income Written by ING, 2003

However, to become a big player abroad a well-developed home market is a prerequisite. In fact with a population of only 16 million, the country ranks tenth globally in terms of premium income earned. Logically, the ranking in terms of per capita premium income is even higher. With per capita premium income in 2002 amounting to €2,626 the Dutch rank sixth in premium income per capita.[206] A long tradition of international business and a well-developed home market explain the existence of various leading Dutch insurance groups.

9.5.6 All-Finance

Holland is the cradle of modern total all-finance (or bancassurance), the origin of combined banking and insurance groups. In 1990, as soon as it became legally possible, insurer Amev acquired VSB-bank which, after the merger with AG from Belgium, resulted in the first multinational all-finance group (Fortis) in the world. Later on NN merged with NMB Postbank to form ING and Rabobank acquired Interpolis.

At the moment, the top five ranking for life as well as for non-life business, as presented in Table 9.7, is dominated by all-finance groups, with ING and Fortis primarily from the insurance sector and Rabobank primarily from the banking sector. The other insurance groups with a position in the top five ranking for life or non-life business (Achmea, Aegon and Delta Lloyd) include more or less substantial banking subsidiaries. As a consequence, the number six life insurance company (Achmea) and the number nine non-life insurance company (Aegon) are companies with more than marginal banking subsidiaries. In addition, the number eight life insurance

[206] Source: Swiss Re. Of total per capita premium income, € 1,376 was spent on life insurance and € 1,250 on non-life insurance.

group is a true financial conglomerate originating from banking groups that merged and took over the Reaal insurance group.

Except for ABN Amro, all the leading financial groups in the Dutch banking sector (ABN Amro, ING, Rabobank and Fortis) are legitimate all-finance groups. ABN Amro only a few years ago split off its insurance activities by selling its insurance subsidiaries to Delta Lloyd. Nevertheless, ABN Amro ranks third in the top 10 of the largest insurance intermediaries. As noted, no fewer than five of the top 10 insurance intermediaries are banking groups or part of a financial conglomerate.

9.6 SUPERVISION

Currently, the supervisory system is going through a period of regulatory change. The supervisory authority for insurers and pension funds, Pensioen- & Verzekeringskamer (PVK) (Pensions and Insurance Chamber), was in 2004 integrated in the Dutch central bank, De Nederlandsche Bank DNB (The Dutch Bank). As part of the regulatory changes over the past few years, the supervisory responsibility for the insurance, pensions, banking, and investments sector was separated into the responsibility for prudential supervision and the responsibility for supervising integrity and market conduct. The PVK was, and today the DNB is, responsible for prudential supervision; the supervisory authority for the financial markets, the Authoriteit Financiele Markten AFM (Authority for Financial Markets), will be responsible for supervising integrity and market conduct. In addition, the AFM will also be responsible for supervising the sector of the financial intermediaries. Finally, the AFM will sometime in the future also be in charge of supervision of the adherence of companies to the accounting rules.

At the moment, a process of re-orientation concerning the rules on solvency for insurance companies and pension funds is going on. Until now, only consultation documents about the way solvency should be monitored were published, the latest being Pensioen- & Verzekeringskamer 2004b.

The most interesting points are the strong focus on "value at risk" measures and the intention to approve the internal model approach for monitoring solvency. In the process of re-orientation, the supervisory rules for insurance companies and pension funds will also be brought more into line with one another.

Of course, all changes in the supervisory system are taking place within the context of the rules set by the European supervisory directives and are partly driven by the (intended) changes in those European rules.

In the Netherlands there are no government guarantees for policyholders of defaulted insurance companies.[207] Instead, but only in life business, the market participants have the duty by law to spread a "safety net" to catch the "falling angel" in case one of them becomes insolvent (Oosenbrug 2001). However, since this duty was introduced a few years ago, no insolvencies have occurred so the safety net has not been activated until now.

---

[207] For the clients of banks the general guarantee based on the European Union Directives of €20,000 per person is applicable (see Garcia and Prast 2004).

Systemic risk is not a large issue for the insurance sector. Nevertheless, the supervisory authorities watch the sector for the risk of systemic problems. Very recently, in 2004, a delegation of the International Monetary Fund made an assessment of the soundness of the Dutch financial system in the context of its Financial Sector Assessment Program. The results were, in general, positive (De Nederlandsche Bank 2004a and 2004b).[208]

9.7 CONCLUSION

The insurance industry in the Netherlands has a long and rich history. Today, several Dutch insurance groups belong to the top global players in the worldwide insurance market. The Dutch insurance market is traditionally one of the most open markets in the world; as a result, the number of market participants is overwhelming, especially since the European single market was formed in 1993, when the number of insurers exploded. At the same time, market conditions were very healthy, especially for life business.

However, with the turn of the century, a new business climate set in. Generous special tax facilities were changed drastically, life markets were depressed by the crashes on stock exchanges and low interest rates on capital markets, the economy cooled down, regulatory and accounting rules were tightened, competition became stronger, and pressures for market transparency increased. Solvency ratios were hit, many intermediaries already faced serious problems, new business and especially unit-linked business came under pressure, and reported results dropped. Although reported results recovered in 2003, it seems likely that especially in life business future results will be no longer as impressive as they used to be. With the still more profitable life business losing market share with regard to the non-life business, total profitability of composite insurance groups is under pressure.

Nevertheless, for cost-efficient and administratively well performing insurers who do not focus on simply selling "tax driven" life insurance products, the open and still extensive Dutch life and non-life market should not lose its attractiveness. For the whole life insurance industry, on average 13 percent of gross premium income written was spent on operating expenses. For non-life business, total operating expenses amounted to 24 percent of net premium income earned. Although commissions paid to intermediaries account for a substantial part of the operating expenses, it is the very well diversified distribution system that guarantees that each serious potential participant who wants to enter the still-attractive Dutch market can have the channels he needs at his disposal.

In the life market, the share of single premium contracts seems to be on the return for the first time in years. At the same time, unit-linked policies are on their return after the tempestuous rise of unit-linked business at the end of the twentieth century. On the other hand, during the last years the share of mortgage related life business grew substantially.

---

[208] See Oosterloo and De Haan (2003) and Houben, Kakes, and Schinasi (2004) for elaborate discussions of the way financial stability can and must be watched.

Group life business went through an era of mutually opposite changes. On the one hand inflationary pressures ceased, unemployment rates rose substantially, and pension arrangements were retrenched. On the other hand, premium rates increased as a result of the transition to a lower discount rate and to updated mortality tables, while at the same time more and more pension funds were liquidated or "outsourced," with the settlement of their pension arrangements transferred to life insurance companies, other pension funds, or specialized service organizations. So, group life business is under pressure while higher premium rates automatically increase premium volumes, and the shrinking number of "self serving" pension funds creates new opportunities for life insurance companies offering group contracts or service arrangements for pension funds. At the moment it is still not clear at which level the group life market will find its new equilibrium.

Premium income from non-life business has grown steadily for decades with premium income from accident, private health, and disability insurance policies accounting for nearly half of the total premium income and still rising. Automobile and fire insurance account for 21 percent and 16 percent of total premium income. Transport has in the course of time decreased to less than 3 percent and the rest—mainly liability insurance—accounts for a relatively stable 12 percent. Technical results are fluctuating, with the results for 2003 looking very nice but partially distorted by the reversal of write-offs on investments accounted for in 2002.

In 2003, the top 10 insurance groups in the relevant market segments wrote almost 90 percent of life business and 67 percent of non-life business, with subsidiaries of foreign insurance groups playing an important role. However, real cross-border trade in insurance policies is so far only a marginal business.

Except for ABN Amro, all the leading financial groups in the Dutch banking sector belong to the leading groups in the insurance sector, too. Also, in a more general sense, all-finance is not an insignificant phenomenon in the Dutch market. Indeed, ABN Amro sold its insurance activities to Delta Lloyd only a couple of years ago.

The distribution channel of independent agents is still in search of a new equilibrium since the business environment has changed drastically and the regulatory system will also be changed drastically. A shake-out in the segment of "hit-and-run" intermediaries was already the result of the changes. E-commerce is still not common practice in the Dutch insurance market, but as products and services become more transparent, the potential for it will surely increase enormously.

Currently, the supervisory system is going through a period of regulatory change. Very recently, in 2004, the PVK was integrated in the Dutch central bank, DNB, with the DNB becoming responsible for the prudential supervision of the total financial sector. Next the AFM will be responsible for supervising integrity and market conduct.

At the same time, re-orientation concerning the rules on solvency for insurance companies and pension funds is still going on. The most interesting points are the strong focus on "value at risk" measures and the intention to approve the internal model approach for monitoring solvency.

In the Netherlands there are no government guarantees for policyholders of defaulted insurance companies. However, since a few years ago, the Dutch life insurers have the duty by law to spread a "safety net" to catch "falling angels."

Because of the still relatively sound financial position of the Dutch life insurers, the safety net has never been activated until now. More generally, the results of a very recent assessment in 2004 of the soundness of the Dutch financial system by a delegation of the International Monetary Fund were predominantly positive. So the Dutch financial sector in general and the insurance sector specifically are still financially sound.

9.8 REFERENCES

AM Jaarboek 2003/2004 (Kluwer, Deventer, The Netherlands).

AM Jaarboek 2004 (Kluwer, Deventer, The Netherlands).

Astin Groep Nederland, 1982, *New motor rating structure in the Netherlands*, Astin Groep Nederland, The Netherlands).

Barendregt, J. and T. Langenhuijzen, 1995, *Ondernemend in risico: bedrijfsgeschiedenis Nationale-Nederlanden 1845–1995* (NEHA, Amsterdam, The Netherlands).

Barneveld, H. van, 1984, *Inleiding tot de algemene assurantiekennis*, 11th ed. (Kluwer, Deventer, The Netherlands).

Berghe, Lutgart A.A. van den, 1995, Marketing en distributie van nationale en Europese verzekeringsprodukten: het economisch kader, *Verzekering en Europa* (W.E.J. Tjeenk Willink, Zwolle, The Netherlands), p. 1–15.

Bos, S., 1998, Beroepsgebonden onderlinges 1500–1800: gilden—en knechtsfondsen, *Studies over zekerheidsarrangementen: risico's, risicobestrijding en verzekeringen in Nederland vanaf de Middeleeuwen*, (Nederlands Economisch Historisch Archief / Verbond van Verzekeraars, Amsterdam / The Hague, The Netherlands), p. 91–140.

Buijs, H.C., 1994, Nederlandse schadeverzekeringsmaatschappijen in een dynamische markt, *Het Verzekerings-Archief*, jaargang 71, 4e kwartaal 1994, p. 121–126.

De Nederlandsche Bank, 2004a, Uitkomsten van het Financial Sector Assessment Program: hoe sterk is het Nederlandse fianciële stelsel?, *DNB Kwartaalbericht*, September 2004, p. 85–90.

_____, 2004b, Stresstesten van de Nederlandse Financiële sector, *DNB Kwartaalbericht*, September 2004, p. 91–99.

Eeghen, Jacob van, Evert K. Greup, Jan A. Nijssen, 1983, *Rate making* (Nationale-Nederlanden NV, Rotterdam, The Netherlands).

Gales, Ben P.A. and Jacques L.J.M. van Gerwen, 1988, *Sporen van leven en schade: een geschiedenis en bronnenoverzicht van het Nederlandse verzekeringswezen* (NEHA, Amsterdam, The Netherlands).

Garcia, Gillian and Henriëtte Prast, 2004, *Depositor and investor protection in the Netherlands: past, present and future* (De Nederlandsche Bank, Amsterdam, The Netherlands).

Gerwen, Jacques L.J.M. van and Nico H.W. Verbeek, 1995, *Voorzorg & de vruchten: het verzekeringsconcern AMEV, zijn wortels en vertakkingen van 1847 tot 1995* (NEHA, Amsterdam, The Netherlands).

_____, 2000a, *Zoeken naar zekerheid III: de ontluikende verzorgingsstaat* (Verbond van Verzekeraars / NEHA, The Hague / Amsterdam, The Netherlands).

_____, 2000b, *Zoeken naar zekerheid IV: de welvaartsstaat* (Verbond van Verzekeraars / NEHA, The Hague / Amsterdam, The Netherlands).

Grinten, W.A.E.A. van der, 1931, *Sporen van verzekering in de oudheid* (De Gebroeders Van Cleef, The Hague, The Netherlands).

Houben, Aerdt, Jan Kakes, Garry Schinasi, 2004, *Towards a framework for financial stability* (De Nederlandsche Bank, Amsterdam, The Netherlands).

Nijenhuis, Odile G.M., Jeroen C.A. Potjes, 1994, The dynamics of the Dutch life insurance market, *Het Verzekerings-Archief*, jaargang 71, 1e kwartaal 1994, p. 2–10.

_____, _____, and G. Schilder, 1994, De dynamische internationale Nederlandse levensverzekeringsmarkt (1947–1992), Lutgart A.A. van den Berghe, Alfred Oosenbrug, Rob Kaas, Henk Wolthuis (eds), *Heterogeniteit in verzekering* (Erasmus Insurance Centre / Instituut voor Actuariaat en Econometrie, Rotterdam / Amsterdam, The Netherlands), p. 287–302.

Oosenbrug, Alfred, 1995a, *Externe verslaggeving door levensverzekeraars* (Delwel Uitgeverij BV, The Hague, The Netherlands).

_____, 1995b, Bedrijfseconomische en actuariële aspecten van een Europese verzekeringsmnarkt zonder grenzen, *Verzekering en Europa* (W.E.J. Tjeenk Willink, Zwolle, The Netherlands), p. 141–164.

_____, 1999, *Levensverzekering in Nederland* (Erasmus Finance & Insurance Centre, Rotterdam, The Netherlands).

_____, 2001, Opvang van wankelende verzekeraars in Nederland, *Het Verzekerings-Archief*, jaargang 78, 4e kwartaal 2001, p. 166–172.

_____, 2002, *Financiële verslaggeving verzekeringsmaatschappijen 1-A* (Erasmus Finance & Insurance Centre, Rotterdam, The Netherlands).

_____, 2003, *Financiële verslaggeving verzekeringsmaatschappijen 1-B* (Erasmus Finance & Insurance Centre, Rotterdam, The Netherlands).

_____, 2004, Trends bij verzekeringsmaatschappijen 2004: wennen aan een nieuw ondernemingsklimaat, Ondernemingsanalyses 2004: trends bij verzekeringsmaatschappijen (LexisNexis Benelux bv, Zwijndrecht, The Netherlands), p. ix-xxxiv.

_____, loose-leaf, *Employee benefits: handboek voor de verzekeringspraktijk* (Kluwer, Deventer, The Netherlands).

_____, Otto Bekouw, F. Karl Gregorius, Frans T.E. Dooren, J. de Vries, J.C.M. Remmerswaal, 1996, *Schadeverzekering in Nederland* (Erasmus Finance & Insurance Centre, Rotterdam, The Netherlands).

_____, J.C.M. van de Griend, Kick (C.) van der Pol, Jaap F. Maassen, Jan A. Nijssen, 1998, *Toekomststrategieën financiële dienstverleners en branchevervaging* (Erasmus Finance & Insurance Centre, Rotterdam, The Netherlands).

_____, Albert H. Zoon, 2002, Country profile: the Dutch insurance market, *Insurance Digest, European Edition,* March 2002 (PricewaterhouseCoopers), p.12–17.

Oosterloo, Sander and Jakob de Haan, 2003, *A survey of institutional frameworks for financial stability* (De Nederlandsche Bank, Amsterdam, The Netherlands).

Pearson, Patrick J., 1995, De interne markt voor verzekeringen: achtergrond, doelstellingen en vooruitzichten, *Verzekering en Europa* (W.E.J. Tjeenk Willink, Zwolle, The Netherlands), p. 17–34.

Pensioen- & Verzekeringskamer, 2004a, *Jaarverslag 2003* (Pensioen- & Verzekeringskamer, Apeldoorn, The Netherlands).

______, 2004b, *Consultatiedocument financieel toetsingskader* (Pensioen- & Verzekeringskamer, Apeldoorn, The Netherlands).

Pestman, P.D., 1983, *Brandverzekeringstechniek* (Stichting Vakontwikkeling Verzekeringsbedrijf, Utrecht, The Netherlands).

Schaap, Dick and Teun van den Berg, undated, *Johan de Witt: een volmaakt Hollander* (Teleboek nv, Bussum, The Netherlands).

Stamhuis, Ida, 1998, Levensverzekeringen 1500–1800, *Studies over zekerheidsarrangementen: risico's, risicobestrijding en verzekeringen in Nederland vanaf de Middeleeuwen*, (Nederlands Economisch Historisch Archief / Verbond van Verzekeraars, Amsterdam / The Hague, The Netherlands), p. 141–156.

Verhoog, Willem, 2003, From profit smoothing to a true and fair presentation of profits at insurance companies and pension funds, Henk Langendijk, Dirk Swagerman, Willem Verhoog (eds), *Is fair value fair?: financial reporting from an international perspective* (John Wiley & Sons Ltd, Chichester, West Sussex, England), p. 237–244.

Verkerk-Kooiman, Annemarie J., Convergentie van toezichtssystemen in de Europese Unie, 1994, Lutgart A.A. van den Berghe, Alfred Oosenbrug, Rob Kaas, Henk Wolthuis (eds), *Heterogeniteit in verzekering* (Erasmus Insurance Centre / Instituut voor Actuariaat en Econometrie, Rotterdam / Amsterdam, The Netherlands), p. 203–222.

Vermaat, Arend J. and Alfred Oosenbrug, 1994, *Bedrijfseconomisch toezicht op verzekeraars en pensioenfondsen* (Erasmus Insurance Centre, Rotterdam, The Netherlands).

Voorst Vader, Alexander A. van, 1995, Het assurantie-intermediair in Nederland: positie en economisch belang mede gezien in Europees perspectief, *Verzekering en Europa* (W.E.J. Tjeenk Willink, Zwolle, The Netherlands), p. 87–107.

9.9 LEXICON

AFM. Authoriteit Financiële Markten (Authority for the Financial Markets).

All-Finance. All-finance is the deliverance of integrated banking and insurance services from within one group.

Annuity. An annuity is a fixed amount of money paid out at regular intervals during some predetermined period.

Asbestosis. Asbestosis is a malignancy of the lungs caused by exposure to asbestos.

AWBZ. Algemene wet bijzondere ziektekosten (exceptional medical expenses act).

Bancassurance. See all-finance.

DNB. De Nederlandsche Bank (The Dutch Bank).

DES. DES is *di-ethyl stilboestrol*, a synthetic hormone formerly used in some medicine for pregnant women, which can cause genetic abnormalities in the child.

Endowment Policy. An endowment policy is a life insurance policy providing some sum assured paid out if the insured person is alive at a pre-specified date.

Group Life Contract. A group life contract is a group insurance contract written by an employer to insure the pension rights awarded to his employees.

Guild. A guild is a society of persons with the same occupation to forward common interests.

Life Annuity. A life annuity is a fixed amount of money paid out at regular intervals as long as the annuitant is alive.

Modern Funeral Fund. A modern funeral fund is a funeral fund calculating its premiums and liabilities using modern life insurance mathematics.

Modern Life Insurance Mathematics. Modern Life insurance mathematics consists of mathematical methods to calculate life insurance premiums and liabilities, taking into account age-dependent mortality rates.

NMa. Nederlandse Mededingingsauthoriteit (Dutch antitrust authority).

OECD. Organization for Economic Co-operation and Development.

"Old Age Time Bomb." The "old age time bomb" is the growing demographic imbalance disturbing the numerical balance between the "young" and the "old" and challenging modern societies.

PVK. Pensioen- & Verzekeringskamer (Pensions & Insurance Chamber).

Relative Technical Result. The relative technical result is the profit or loss on the technical account of the income statement divided by the aggregated gross premium income written.

Required Solvency Margin. The required solvency margin is the minimum solvency margin required by the supervisory rules.

Scientifically-Based Life Insurance Company. A scientifically-based life insurance company is a life insurance company calculating it's premiums and liabilities using modern life insurance mathematics.

Solvency Margin. A solvency margin is a financial margin that can be lost before an organisation becomes insolvent.

Solvency Ratio. A solvency ratio is the actual solvency margin divided by the required solvency margin.

Technical Result. A technical result is the profit or loss on the technical account of the income statement.

Traditional Funeral Fund. A traditional funeral fund is a funeral fund not based on modern mathematical methods to calculate premiums and liabilities.

Unit-Linked Policy. A unit-linked policy is a life insurance policy with the level of the sum assured directly linked to the value of the units of some investment vehicle specifically allocated to the policy.

10 The Structure, Conduct, and Performance of the Spanish Insurance Industry

María Rubio-Misas

Universidad de Málaga

10.1 INTRODUCTION

The Spanish insurance industry experienced a dramatic restructuring during the 1980s and 1990s. In the 1980s, under a 1984 law and a 1985 royal decree, the government began tightening solvency standards and encouraging mergers and acquisitions in the insurance industry to create insurers that would be financially stronger, more efficient, and more competitive both nationally and internationally. The Spanish insurance market also has been affected by the overall deregulation of European insurance markets, particularly through the European Union's (EU) Third Generation Directives, implemented in July 1994. The Third Generation Directives effectively deregulated the EU insurance market, with the exception of solvency regulation, which was carried out by the insurer's home country. As a result of these regulatory changes, the number of insurers operating in the Spanish market declined dramatically during the 1980s and 1990s.[209]

In addition, the Spanish insurance industry has not been detached from important developments in the financial services markets, such as the integration of financial services. Financial conglomerates, both Spanish and foreign, have an active presence in the Spanish insurance market, and the bank distribution channel stands out as the predominant system for selling life insurance products.

As a consequence of deregulation, the integration of financial services, technological changes, and other economic changes, some of the more dynamic developments of the Spanish insurance industry have taken place in the last 20 years.

I am very grateful to the editors of this book, J. David Cummins and Bertrand Venard, for their comments and suggestions that have improved this chapter. I am also very thankful to the people at the DGSFP, ICEA, INESE, and UNESPA for providing information to me.

[209] For further analysis on the effects of deregulation and consolidation in the efficiency of the Spanish insurance industry, see Cummins and Rubio-Misas (2006).

This chapter presents a study of the structure, conduct, and performance of the Spanish insurance industry over the period 1990 to 2003. The objective is to provide an overview and analysis of trends in the market, covering the basics of the industrial organization and providing descriptive statistics and analysis of important developments in the Spanish insurance market.

The chapter begins with a discussion of the history of insurance in Spain in the context of its regulation, including sections on life insurance and taxation. The next section evaluates the industry market structure by focusing on the number and types of firms, the distribution of premiums, life insurance managed savings, total assets in the industry, international activities, market concentration, organizational structure, distribution systems, and leading firms in the industry. The subsequent section analyses technical results, combined ratios, profits, and financial conditions of the Spanish insurance industry.

10.2 REGULATION

10.2.1 History and Regulation

The history of insurance in Spain is joined to the history of its regulation. As a preview to this discussion, the laws on the arrangement and supervision of insurance firms in Spain have been a 1908 law, a 1954 law, a 1984 law, and a 1995 law; the current legislation is contained in the revised text on the private insurance supervision and arrangement law (TRLOSSP) approved by Royal Decree 6 in October 2004, which is comprised of some articles on the 1995 law along with ulterior modifications.

The Spanish insurance contract law originated in the fifteenth century with a set of regulations by Barcelona magistrates (Ordenanzas de Magistrados de Barcelona) published in 1435. This set of regulations and subsequent sets of regulations by Burgos (1537), Sevilla (1555), and Bilbao (1737) deal with forming the insurance contract separately from other traditional contracts (Benitez de Lugo 1930; Tirado-Suárez 1984). These regulations were in force until the Spanish commercial code of 1829 came into force.

With the protection of this legislation, the Spanish pioneer insurance firms were created. These firms were mutual societies (*montepíos*) of disability or old age (precedent of life insurance), and other mutuals covered several risks such as fire or work-related accidents (Martinez-Martinez 1994).

The modern insurance contract was regulated in Spain by the 1885 commerce code, and the law that established the arrangement of private insurance firms in Spain was the Ley de Registro e Inspección de Empresas de Seguros de 1908 (or the registration and inspection of insurance firms law of 1908). This law created the Spanish supervisory authority—Comisaría de Seguros, the predecessor of the current Dirección General de Seguros y Fondos de Pensiones.

Before the 1908 law came into force, insurance firms entered and exited the Spanish market very frequently. Forty percent of the insurance firms that were in the Spanish market in the nineteenth century exited the market during the first five years of their activity. The principal reason for their failure was that they were created with weak technical and financial conditions. This fact revealed the need for legislation that established homogeneous norms for entering the market and operating in it (Frax-Rosales and Matilla-Quiza 1996)

Under a 1927 royal decree, minimum capital requirements to operate in every line of insurance business were established in Spain for the first time. These amounts of minimum capital were modified later by a law of 1944. In 1949 a law temporarily stopped the creation of new insurance firms for a period of three years. This policy was established by the government in an effort to control the extraordinary increase of insurance in Spain—many firms had entered the market but with very low premium volumes. In 1952, the government's policy changed based on a belief that the way to strengthen the Spanish insurance sector would be in a free market that would create new insurance firms and generate new capital. The law of 1952 increased the minimum capital requirements, but only for the new firms (Tirado-Suárez 1984). In 1957 there were 668 insurance firms in the Spanish insurance market (Pons-Pons 2003).

In 1954, a new insurance law came into force. Although the available data on the number of companies and premium volume are not very exact, the period between the 1908 law and the 1954 law was characterized by an increase in the number of firms and premium volume, a decline in foreign company penetration, and an increase in competition. The increased competition in the first three decades of the century triggered the most important insurance firms to diversify and set up control cartels of the market (Pons-Pons 2003).

Although in 1784 a foreign firm called Phoenix Assurance operated in Spain, it was not until the 1870s and 1880s when foreign insurance companies started to penetrate the Spanish market, developing lines of business such as life insurance and accident insurance that were not offered by Spanish insurers. In 1912 there were 106 Spanish insurance firms that accounted for 59 percent of total premium volume and 61 foreign insurance companies that accounted for the other 41 percent. Life insurance was the most important line of business (which accounted in 1910 for almost 40 percent of total premium volume), and it was followed by fire insurance (30 percent). Foreign firms were usually concentrated on life, fire, transport, and accident lines of business. In 1934 there were 247 Spanish insurers that accounted for 65 percent of total premium volume and 155 foreign firms that accounted for the other 35 percent. In 1953 the premium volume of the Spanish insurance market was more than 3,100 million pesetas, and more than 80 percent of it belonged to Spanish insurers (Tirado-Suárez 1984; Frax-Rosales and Matilla-Quiza 1996). These data reveal the importance of foreign firms in the development of the Spanish insurance market at the end of the nineteenth century and at the beginning of the twentieth century, but they also show a decrease in the participation of total premium volume by foreign companies.

This decline of foreign company penetration in the Spanish insurance market is due to several reasons, such as the Spanish insurance legislation that, after 1908, established limits to foreign companies to invest their reserves in foreign countries. The Spanish government encouraged Spanish companies to buy insurance portfolios belonging to foreign companies. Changes in the international economic policy as a consequence of World War I triggered a convolution of international markets that was accelerated after the 1929 crisis. In addition, after the Spanish Civil War (1936–1939), a period of economic autarchy started in 1940 and ended in 1959, and in this period the Spanish economy was closed to the international economy (Frax-Rosales and Matilla-Quiza 1996; Pons-Pons 2003).

At the beginning of the twentieth century, both Spanish and foreign insurers were specialized in one line of business. But in the first decades of the twentieth century, the increase in competition promoted the diversification of the most important insurance firms, both Spanish and foreign, in addition to setting up cartels among them to control the market. This diversification process was reinforced by the 1954 insurance law that allowed insurance firms to extend their activity to other lines of business without costs (Pons-Pons 2003).

The 1954 law created an insurance institution without precedent in Europe named Consorcio de Compensación de Seguros. This institution was the only one allowed to cover extraordinary risks in Spain. This situation of exclusivity to cover extraordinary risks was maintained until the 1990 law, when the European legislation about this issue was incorporated in Spanish legislation.

According to the extraordinary risk insurance regulation approved by Royal Decree 300/2004 of February 20, extraordinary risks are defined as damage to people and property resulting from extraordinary natural disasters (flooding, earthquakes and seaquakes, volcanic eruptions, atypical cyclonic storms, and falling spatial bodies and meteorites); violent acts of terrorism, uprisings, or popular tumult, rebellion, and sedition; and acts or actions of the armed forces or the security forces and corps in peacetime.

During the 1960s, regulation of the Spanish insurance industry was characterized by an interventionist policy. The control authority established not only the tariff system but also uniform models of contract conditions (Tirado-Suárez 1984). The 1954 law did not have the tools to perform effective arrangement and supervision, and it was out of touch with the market situation. A new concept of solvency control, in addition to new rules to improve transparency and competition, were needed.

The end of the 1970s began a new era for the Spanish insurance industry. A royal decree was published in 1978 to improve the Spanish insurance industry; it established solvency margins for firms operating in Spain that were more comparable to the solvency margins for firms operating in other European countries. A 1979 royal decree, started the deregulation process on policy forms and tariffs, and this deregulation was general for all lines of business under the 1984 law. In 1980 the current insurance contract law came into force. In 1982, negotiations on Spain's membership in the European Economic Community, were very advanced, and a royal decree on new conditions for the solvency margin was published. The whole process of regulatory change culminated with the publication of the private insurance arrangement law in 1984 (Ley Ordenadora del Seguro Privado de 1984).

The 1984 law established rules for restructuring the market and strengthening the Spanish supervisory authority. These rules were classified into two groups: the arrangement group of rules and the supervision group of rules. The most important of the arrangement group of rules were:

1. putting all private insurance firms under the control of the Spanish supervisory authority with solvency standards. In this sense, social benefit institutions (*mutualidades de prevision social*) were placed under the

Spanish insurance supervisory authority's control, but with lower solvency conditions than other insurance firms;

2. encouraging mergers and acquisitions in the insurance industry to create insurers that would be financially stronger, more efficient, and more competitive, both nationally and internationally;
3. strengthening the Spanish reinsurance market;
4. requiring firms to specialize according to European Economic Community rules and international trends. In this sense, new firms should be life specialists or non-life specialists. Joint firms that were created before this law are allowed, but they have to separate both activities in the accounting books. Although new firms must be life or non-life specialists, it is allowed for one parent company to have both life and non-life subsidiaries; and
5. classifying the private insurance firms into stocks, mutuals, branches of foreign firms, social benefit institutions, and cooperatives. The 1954 law only allowed the first three types of firms, so it was new to include social benefit institutions as private insurance firms.

As mentioned in the TRLOSSP (2004), social benefit institutions are nonprofit private mutual insurers providing complementary social security coverage. They cover risks of death, widowhood, orphanhood, and retirement and can provide benefits for marriage, maternity, children, and funerals. They can operate on work-related accident; work disability; health; legal defense and assistance; and assure public housing, machinery, and other instruments belonging to mutualists who are small business people and crops that are directly cultivated by farmers.

It was also new to include cooperatives as a type of insurance firm, and this fact was related to the Spanish Constitution of 1978 that encouraged the establishment of cooperative societies (Tirado-Suárez 1984). A cooperative, as mentioned in the 1999 cooperative law, is a society created by people who are joined in a freedom regime of association. The cooperative partners are associated to perform business activities to satisfy their needs or their economic and social ambitions (e.g., work, consumption, credit, insurance). Cooperatives have a democratic structure and a democratic functioning, according to the International Cooperative Alliance principles. It was questioned, when the 1984 law came in force, whether mutuals and cooperatives would coexist in the Spanish insurance market (Tirado-Súarez 1984). So far no cooperative has been created.

In the supervision group of rules of the 1984 law, the most important ones were:

6. increasing the amounts of minimum capital to operate in every line of business;
7. establishing technical and financial rules on insurers' reserves, invested assets, solvency margins, and guarantee funds;
8. establishing rules on compulsory liquidation of insurance firms; and
9. protecting the insured's interests.

Related to social benefit institutions, the 1984 law considered distributing their control and supervision between the central government and the autonomous communities' governments. In this sense, social benefit institutions that operate in only one autonomous community could be supervised by its government. Through different royal decrees, the central government transferred the control of the social benefit institutions that operate in only one autonomous community to the government of the corresponding autonomous community (Maldonado-Molina 2001). An autonomous community is a territorial entity that has legislative autonomy, executive competencies, and can be administrated by its own representatives, under the constitutional umbrella of the Spanish state. The structure of the Spanish state in autonomous communities came from the Spanish Constitution of 1978.[210]

The 1984 law was elaborated, taking into account the European legislation. Few modifications to this law were needed when Spain joined the European Economic Community in 1986. These modifications were established by two royal decrees—RD 1255/1986 and RD 2021/1986—which established the incorporation of the First Generation Directives on both non-life and life insurance. The minimum solvency margin was also regulated to harmonize the Spanish market with other European countries markets.

A commission for insurers (named Comisión Liquidadora de Entidades Aseguradoras) was created by a 1984 royal decree to restructure the market based on the fact that almost 100 firms had had financial difficulties in 1984. Most of these firms were liquidated due to insolvency, and some of them exited the market due to mergers and acquisitions (Esteban-Jodar 1993).

After 1986, the new Spanish insurance regulation incorporated new European legislation. In many cases, transition periods were given to every member nation to adapt to European legislation. What follows is an overview of the regulations that have been enacted in Spain after 1986. Law 21 of 1990 basically incorporated the Second Generation Directives on "freedom of services" for non-life insurance. This law came into force in 1991 and, as mentioned above abolished the monopoly that the Consorcio de Compensación de Seguros had in extraordinary risks.

In 1992, a law about the distribution of insurance products came into force. This law incorporated the European recommendation 92/48/ECC and implied a liberalization in Spanish insurance distribution systems. In 1995, a new law on the arrangement and supervision of the insurance industry incorporated the Second Generation Directives on life insurance, the Third Generation Directives on both life and non-life insurance, and the directive on insurance financial statements. Rules that apply to all the countries belonging to the European Economic Area were also established. With the 1995 law, social benefit institutions were adapted to the regulation of private insurance firms.

The same 1995 law also incorporated Directive 80/987/CEE into the Spanish legislation to protect employers' agreements on employees' private pensions. In the 1970s, public firms and firms of several economic sectors such as the electric utility and banking sectors assumed agreements with their employees to complement public pensions with private pensions. In this sense, these firms formed internal reserves in

---

[210] Spain is formed by 17 autonomous communities: Andalucía, Aragón, Principado de Asturias, Islas Baleares, Canarias, Cantabria, Castilla-La Mancha, Castilla y León, Cataluña, Extremadura, Galicia, Comunidad de Madrid, Región de Murcia, Navarra, País Vasco, La Rioja, and La Comunidad Valenciana. It also has two cities in Africa with a special statute, Ceuta and Melilla.

their financial statements as a guarantee to satisfy these private pensions (Sanz-Arnal 1996). The 1995 law prohibited firms from maintaining internal reserves to satisfy pension agreements. Transition periods were given to materialize these reserves in life insurance policies or external pension funds, and the year 2002 was the last year to perform these outsourcing operations. Nevertheless, anticipating legislation and based on previous recommendations, the special increase in life insurance in 1994 was due to the materialization of internal reserves on single-premium group life insurance (Gonzalez de Frutos 2002).

There were several basic effects of the main European insurance directives. The First Generation Directives introduced "freedom of establishment" with the host country control granting the insurer the right to establish subsidiaries and branches in each member state. The Second Generation Directives established "freedom of services" giving the insurers the ability to conduct business outside their home country without having to establish branches in host countries. And the Third Generation Directives established a single EU license whereby an insurer needs only to obtain one license to operate in the EU rather than having to obtain licenses in each member nation and the principle of home country supervision, through which an insurer is regulated only by the nation that issued its license and not by each host country where it operates. They also abolished "substantive insurance supervision," meaning that regulation is limited to solvency control (Swiss Re 1996).

After 1995, Spanish regulations modified some articles of the 1995 law. First, Law 44 of 2002 on the financial systems reform incorporated the Fourth Generation Directives for auto insurance in addition to Directive 64/2000 on information exchange with third countries. Under this law, the Comisión Liquidadora de Entidades Aseguradoras was dismantled and its functions were assumed by the Consorcio de Compensación de Seguros. Rules to protect financial services clients and infraction characterization and corresponding sanctions were also established. Second, Law 22 of 2003 on creditor's meeting affected the Consorcio de Compensación de Seguros statute. And Law 34 of 2003 incorporated Directive 17/2007 on liquidation of insurance firms in addition to Directives 13/2002 and 83/2002 on new solvency margin requirements for non-life and life insurance, respectively.

The Consorcio de Compensación de Seguros (CCS) is a public business entity attached to the Dirección General de Seguros y Fondos de Pensiones, Ministerio de Economía y Hacienda. It is legally entrusted with both insurance and non-insurance duties. Concerning the first, its subsidiary condition must be emphasized. In general, it behaves as a direct insurer in cases that the private market refuses to deal with (e.g., extraordinary property damage risks, extraordinary personal injury risks, or compulsory and voluntary third-party liability insurance for the use and transit of private motor vehicles not covered by insurance entities). It also functions as a guarantee fund in cases where there is lack of insurance or bankruptcy of the insurer. Concerning an insurer's bankruptcy, the CCS performs the final reckoning that had previously been the responsibility of the Comisión Liquidadora de Entidades Aseguradoras.

The CCS obtains its income by premiums, surcharges, and investment incomes. Its activity is subject to laws that govern private insurers and—as with private insurers—it establishes the relevant reserves and keeps a solvency margin. It is a profitable entity. The 2004 pretax results amounted to 702 million euros, 4 percent more than in 2003. Results were positive in the three groups of activities that the CCS operated, particularly in "general activity," which, at 533 million euros, accounted for 76 percent of the total. In the other two areas of activity, agricultural insurance posted a profit of 54 million euros, all of which was allocated to the stabilization reserve, and winding-up activity generated a surplus of 115 million euros, a figure similar to that of the previous year (CCS 2005).

The articles in force of the 1995 law along with modified articles by ulterior legislation have been combined in a revised text on the private insurance supervision and arrangement law (TRLOSSP). This text was approved by royal decree on October 6, 2004, and it forms the basic legislation on supervision and arrangement of insurers. As mentioned, the law in force concerning insurance contracts is Law 50 of 1980.

10.2.2 Life Insurance and Taxation

In Spain, private pensions that supplement the public pension system are relatively underdeveloped. This is due in part to the coverage offered by the public system, which provides a high level of benefits compared to the social security benefits of other European countries (e.g., the number of contribution years to get a retirement pension of 100 percent of the computation base is 35; the average number of contribution years in other European countries is 41.5. Spanish public pensions are linked to average earnings during the last 15 years of a person's working life, whereas in other countries pensions are linked to a person's entire working life. It is possible in Spain to obtain a pension after 15 contribution years) (Del Valle-Esteve 2002).

The principles of the Spanish social security system are set out in Article 41 of the Constitution of 1978, which states: "the public authorities will maintain a public social security system for all citizens to guarantee sufficient care and social benefits for them in situation of need, especially in case of unemployment. Complementary care and benefits will be optional." The present public social security system consists of two levels: the basic level and the contributory level.

The basic level is universal and includes:

1. heath care benefits that cover people who receive noncontributory benefits and people included in the contributory part of the social security system, as well as all other people registered as living in Spain who are not part of the contributory social security system, as long as their annual income is below the minimum wage;
2. noncontributory pensions when the conditions for receipt of these are that the beneficiaries should prove a minimum period of legal residence in Spain, that they should have financial resources below a minimum level established annually, and that they should be ineligible to receive contributory benefits;
3. family protection; and
4. social services.

In general, people receiving these benefits must prove their lack of sufficient means (e.g., the noncontributory retirement pension is given to people when they reach 65 years of age who have no financial means of support and do not have the right to make use of the contributory system because they have not paid enough contributions to the system).

The resources for expenditure at the public and basic levels come from the central government budget. In addition, a new system of autonomous community finances has been implemented as the result of an agreement between the central government and the autonomous community governments. This establishes the principle of fiscal co-responsibility, under which the autonomous communities pay for the cost of health care and social services in return for a transfer of tax revenues. The benefits are managed by the corresponding social and health services of each autonomous community.

The contributory level of the Spanish social security system is financed by compulsory social security contributions provided by both employers and employees (except in the case of risks of industrial accidents and illnesses, for which the contributions are paid exclusively by the employer). Spanish social security protection consists of income substituting wages in cases of retirement, permanent or temporary disability, maternity risk during pregnancy, and death and survivors' coverage. People legally residing in Spain are covered at the contributory level of the social security system as long as they work in the country. The Spanish social security system is based on intergenerational solidarity because it uses a pay-as-you-go system rather than a funded system.

Nevertheless, despite the high coverage provided by the public system, Spanish legislation has encouraged private systems to complement it. Since 1987 there has been a law (Ley 8/1987 de Planes y Fondos de Pensiones) that encourages private systems in general but especially private pension funds.[211]

Tax regulation has played an important role in the development of private systems to complement public pensions. Tax regulation incentives and changes have been taking place in the life insurance segment since 1978, when the Spanish personal income tax (IRPF) was reformed.

The 1978 law on personal income tax (Ley 44/1978 del IRPF) established that the amounts paid to life insurance premiums were tax deductible. Life insurance yields were treated as investment income, and therefore insurance firms deducted withholding tax on these yields (Gonzalez de Frutos 2002). In general, insurance firms deduct withholding tax on yields if they are treated as investment income or as labor income, but if these yields are treated as increases in the taxable estate, insurers do not deduct withholding tax on them.

In 1985, Law 48 governing personal income tax retained the tax reduction on paid life insurance premiums. Both life and temporary annuities were treated as investment income, but interest earned by the deferred life insurance was taxed as an increase in the taxable estate. This latter type of taxation was more profitable than treating the yields as investment income, and it triggered short-term financial

---

[211] This law established that the management of a private pension fund could be done by stock companies that specialize in the management of pension funds or life insurers (including social benefit institutions). In 2003, 58.6 percent of firms that managed a pension fund were life insurers.

operations considered as insurance contracts, which in reality were capitalization operations without risk or with minimum actuarial risk. Special increases in single-premium insurances in the years after 1985 were due to this fiscal advantage (Gonzalez de Frutos 2002; Sanz-Arnal 2002). In 1989, a royal decree prohibited these short-term single-premium contracts because of previously mentioned special fiscal advantages and because insurers did not deduct withholding tax on these yields, so some clients did not declare these yields (Sanz-Arnal 2002).

In 1991, Law 18 on personal income tax established a tax deduction of 10 percent on paid life insurance premiums in the final amount to be paid as personal income tax. Short-term single-premium contracts were allowed again but with a different taxation structure. Their yields were treated as investment income, whereas yields provided by other savings life insurance contracts with enough length and risk were treated as increases in the taxable estate. Reduction coefficients were established for increases in the taxable estate that allowed tax exemption in insurance contracts with more than 15 years of paid premiums. A system that allowed tax exemption after 21 years of the policy being underwritten was also established for some periodical-premium life insurance contracts. These fiscal advantages triggered increases in savings life insurance contracts that were distributed by bancassurance firms. But the development in life insurance contracts with a savings component was higher after Royal Decree 7 of 1996 established additional fiscal advantages for short-term single-premium life insurance, although reduction coefficients disappeared and fiscal advantages were reduced for retirement insurance contracts. Life insurance experienced spectacular increases in 1997 and 1998 (Sanz-Arnal 2002).

The current law on personal income tax (Ley 40/1998) was approved in December 1998. Since then, several modifications to this law have taken place. Since 1999 there been no tax reductions on paid life insurance premiums in the final amount to be paid as personal income tax, because there is a tax exemption on a minimum taxable estate per person or family, except for qualified retirement plans.

Since 2003, qualified retirement plans have existed in the Spanish insurance market. They have the same illiquidity as pension plans and also have the obligation to provide a positive yield over the product's lifetime. They receive the same personal tax treatment as pension plans: the premiums paid are subtracted from the tax base, subject to specific caps that apply to the sum of premiums paid and contributions to pension plans.

In general, life insurance yields are taxed as investment income, but if they come from group life retirement insurance, they are taxed as labor income. Insurance firms deduct withholding tax on these yields. Reduction coefficients on life insurance yields, depending on the number of years the premium has been paid (e.g., in the case of lump-sum insurance, if it has been more than five years since the payment of the premium, the reduction coefficient in taxable income is 75 percent, so that the taxable percentage is 25 percent of the difference between the surrender value and premiums paid), the recipient's age, and the annuity length have also been established. Under certain conditions, unit-linked products also have these reduction coefficients on their gains, and this is an advantage of unit-linked products compared to direct investment in mutual funds. There is tax exemption for transfer between

Table 10.1. Number of Companies in the Spanish Insurance Industry, 1990 to 2003

	1990	1991	1992	1993	1994	1995	1996	1997	1998	1999	2000	2001	2002	2003
Stocks	391	386	372	354	331	312	306	291	275	264	256	246	244	240
Mutuals	75	72	66	64	63	59	56	56	52	51	49	49	48	46
Branches of Foreign Firms[a]	31	33	30	27	25	27	32	36	39	39	39	37	37	41
Reinsurance Specialists	8	8	7	7	6	6	5	4	4	4	4	3	3	2
Total[b]	505	499	475	452	425	404	399	387	370	358	348	335	332	329
Social Benefit Institutions	328	146	145	140	141	99	95	85	79	78	69	70	65	63
Entering	23	1	3	1	5	12	11	13	6	6	4	7	11	8

Source: Author's compilations from DGSFP(a) *Balances y Cuentas* and DGSFP(b) *Seguros y Fondos de Pensiones Informe.*
[a] The number of branches of foreign firms includes both branches of EU firms and non-EU firms.
[b] This total does not include the number of social benefit institutions.

mutual funds in unit-linked products, and, since January 2003, this tax exemption has been extended to any transfers between mutual funds, not merely those of unit-linked products.

The Spanish government currently is carrying out a reform in direct taxation that could affect life insurance tax treatment.

10.3 THE STRUCTURE OF THE SPANISH INSURANCE MARKET

10.3.1 Number and Types of Companies

Insurance in Spain can be provided by stocks, mutuals, cooperatives, and social benefit institutions. So far no cooperative offering insurance has been created.

Table 10.1 shows, for the period 1990 to 2003, the number of firms that were authorized to operate in insurance, according to the Insurers Administrative Register (Registro Administrativo de Entidades Aseguradoras) from the Spanish supervisory authority (the Dirección General de Seguros y Fondos de Pensiones, Ministerio de Economía y Hacienda). Firms are classified into five categories: stocks, mutuals, branches of foreign firms, reinsurance specialists, and social benefit institutions. The total number of branches of foreign firms—both EU and non-EU—are included in the table. Because branches of EU firms did not have to report to the Spanish regulator after 1994, they were not included in subsequent analysis.

The majority of insurers are organized as stock firms. In 2003, there were 240 stock firms (they were licensed in Spain, although, in some cases, more than 50 percent of their capital belonged to a foreign group), 2 reinsurance specialists organized as stock firms, and 41 branches of foreign firms, most of them organized as stock firms.

The number of firms in the industry fell dramatically over the 1990–2003 period due to firm retirements, insolvencies, mergers, and acquisitions. The number of stock insurers declined from 391 in 1990 to 240 in 2003; the number of mutuals declined from 75 in 1990 to 46 in 2003; and the number of reinsurance specialists declined from 8 in 1990 to 2 in 2003. However, the number of branches of foreign firms increased from 31 in 1990 to 41 in 2003.

During the period 1989 to 1998, firms exiting the market for reasons other than mergers and acquisitions were predominantly small, inefficient, and in poor financial condition; while merger and acquisition targets were larger and roughly equal in efficiency to acquiring firms and those not involved in merger and acquisition transactions. As a result of consolidation, average firm size increased, efficiency trended upward during the sample period, and the market experienced significant total factor productivity gains. Consolidation has been beneficial to insurance buyers—unit prices declined in both life and non-life insurance (Cummins and Rubio-Misas 2006).

Demutualization played an important role in the restructuring of the Spanish insurance market prior to the period of analysis. In the 1980s, many mutuals were transformed into stocks as a way to get significant infusions of new capital. Substantial increases in capital were required to operate in the different lines of insurance business by the insurance law of 1984 (Esteban-Jodar 1993). The wave of demutualization had ended by the time the period of analysis began, but there were seven complete demutualizations, and three mutuals were transformed into social benefit institutions.

The number of social benefit institutions supervised by the Dirección General de Seguros y Fondos de Pensiones declined from 328 in 1990 to 63 in 2003. In addition to firm retirements and insolvencies, this decline is also due to the fact that central government transferred the control of the social benefit institutions that operate in only one autonomous community to the government of this autonomous community during the period of analysis. Therefore, these firms were not supervised by the Dirección General de Seguros y Fondos de Pensiones. From 1992 to 1999, 52 social benefit institutions were transferred to their corresponding autonomous communities (Rubio-Misas 2004).

The restructuring of the Spanish insurance market was also due to entering firms. Table 10.1 shows the number of new entrants each year. During the period 1989 to 1997, 99 new insurers entered the Spanish market—61 were stocks, 31 were branches of foreign firms, 6 were mutuals, and 1 was a reinsurance specialist. New entrants have experienced significant total factor productivity growth, providing evidence of high start-up costs and a relatively steep learning curve, to the extent that new entrants are initially at a disadvantage but improve rapidly thereafter (Cummins and Rubio-Misas 2004).

10.3.2 Distribution of Premiums

Table 10.2 shows the distribution of non-life insurance premiums and trends in annual growth, expressed in nominal monetary units, in the different non-life lines of business. The table shows the importance of auto insurance, multiple peril insurance, and health insurance in the non-life insurance segment of the Spanish insurance market. In 2003, auto insurance accounted for 43.3 percent of non-life premiums, and multiple peril insurance and health insurance accounted for 15.7 percent and 15 percent, respectively.

The annual growth rate in total non-life insurance premiums was 9.2 percent over this time period, but growth varied considerably across lines of business. General liability had the highest annual growth rate, averaging 15.2 percent per year. This increase was due to the greater social awareness of complaints and financial scandals. General liability covers a diversity of risks such as risks related to the performance of members of a board of directors or managing staff, operating risks, work-related accidents, consumer products, or pollution risks. The general liability annual growth rate exceeded the growth rate of assistance (which grew at an annual rate of 14 percent), multiple peril (12.3 percent), other damages (11.9 percent), health (10.4 percent), credit and caution (9.5 percent), and auto (8.5 percent). Fire insurance experienced a negative annual rate growth of –0.3 percent because the risks covered by fire insurance were increasingly insured in the multiple peril lines of business. Multiple peril insurance covers risks corresponding to both individuals and businesses.

Table 10.2. Distribution of Non-life Insurance Premiums in the Spanish Insurance Industry, 1991 to 2003

	Premiums Written (in millions of euros)													Average Annual Growth Rate (%)[a]
Line of Business	1991	1992	1993	1994	1995	1996	1997	1998	1999	2000	2001	2002	2003	
Accident	524	587	631	692	750	777	791	788	612	645	691	730	791	3.5
Transport	257	287	291	311	315	322	302	286	308	356	395	498	512	5.9
Fire	278	236	223	211	201	198	182	169	130	139	152	224	269	–0.3
Other Damages	253	290	301	304	317	368	359	389	426	489	570	841	970	11.9
Auto	3,951	4,490	4,809	4,998	5,163	5,225	5,088	5,403	6,765	7,995	8,840	9,871	10,479	8.5
General Liability	247	307	363	437	488	525	542	576	662	770	842	1,107	1,354	15.2
Credit and Caution	198	227	232	245	250	259	272	300	333	388	459	512	588	9.5
Assistance	87	143	171	201	229	253	278	313	115	361	428	399	420	14.0
Health	1,100	1,315	1,472	1,625	1,789	1,953	2,099	2,270	2,482	2,709	2,994	3,270	3,625	10.4
Mortuary	440	518	576	630	694	762	812	848	842	913	996	1,068	1,141	8.3
Multiple Peril	940	1,137	1,255	1,413	1,551	1,694	1,813	1,974	2,220	2,496	2,771	3,338	3,798	12.3
Income Loss	46	37	44	40	33	48	52	61	67	60	123	82	80	4.7
Legal Defense	82	78	101	102	123	121	170	126	67	99	58	126	137	4.4
Other	1	2	3	2	1	0	0	0	0	0	0	0	0	
Total	8,406	9,654	10,470	11,209	11,904	12,504	12,759	13,502	15,029	17,421	19,319	22,066	24,164	9.2

Source: Author's compilations from DGSFP(b), *Seguros y Fondos de Pensiones Informe.*
[a] Average annual growth rate is calculated as $(\text{prem2003}/\text{prem1991})^{1/12}$.

Accident insurance has the aim of providing a compensation benefit in case of accidents that cause the insured's death or disability as a consequence of activities established in the policy. The cause has to be violent, sudden, external, and nondeliberated by the insured (Ley del Contrato del Seguro 1980).

Caution insurance is a fidelity guarantee insurance through which the insurer is obliged to compensate the insured in case the policyholder does not perform its legal or contractual obligations. Every payment by the insurance firm must be reimbursed by the policyholder. Credit insurance covers the risk when the insured's debtors are insolvent. This insurance requires the insurer to pay indemnity to the insured for losses as a consequence of their debtors' insolvency (Ley del Contrato del Seguro 1980).

Life insurance premium volume per year and the average annual growth rate, expressed in nominal monetary units, are shown in Table 10.3. The annual growth rate in total life insurance premiums was 13.8 percent over the time period 1990 to 2003. Total life insurance premiums experienced special increases in several years of the sample period. The increases in life insurance premiums in 1999 and 2000 were due to the rise of unit-linked products. But the special increases in total life premiums of the years 1994, 2001, and 2002 were triggered by important increases in group single-premium policies as a result of the outsourcing of companies' pension commitments.

Table 10.3. Life Insurance Premiums in the Spanish Insurance Industry, 1990 to 2003

	Direct Premiums Written (in millions of euros)	Distribution (%)	
		Traditional Products	Unit-Linked Products
1990	3,287	—	—
1991	4,610	—	—
1992	4,879	—	—
1993	5,279	—	—
1994	8,664	—	—
1995	8,249	—	—
1996	9,830	—	—
1997	11,467	—	—
1998	12,282	—	—
1999	16,760	62	38
2000	23,855	63	37
2001	22,817	89	11
2002	26,480	95	5
2003	17,737	93	7

Average Annual Growth Rate—$(\text{Prem2003/Prem1990})^{1/13}$—is 13.84%.

Source: Author's compilations from DGSFP(b), *Seguros y Fondos de Pensiones Informe.*

Table 10.3 also shows the distribution of life insurance premiums among traditional products and unit-linked products for the period 1999 to 2003. In the years 1999 and 2000, unit-linked products accounted for 38 percent and 37 percent, respectively, of total life premiums. However, unit-linked products accounted for only 7 percent of total life insurance premiums in 2003. The decline of unit linked products in the life insurance segment can be explained by the evolution of the stock market with negative profitability, falling interest rates, and fiscal regulatory reforms that neutralized the fiscal advantage of unit-linked products on tax exemption for transfer between mutual funds (DGSFP(b) 2004).

10.3.3 Life Insurance Managed Savings

Managed savings that correspond to every type of life insurance at the end of every year of the period 2000 to 2003 as well as the weight of managed savings corresponding to different types of products over the total of managed savings for the year 2003 are presented in Table 10.4. Life insurance managed savings are mathematical reserves. They represent the insurer's best estimate of claims to be paid in the future as a result of future contingencies, so they are savings that are managed by the insurer in favor of its clients. The data are compiled from annual reports on life insurance that the Unión Española de Entidades Aseguradoras y Reaseguradoras (UNESPA) is carrying out.

At the end of 2003, individual life insurance accounted for 59.9 percent of total life insurance managed savings and group life insurance accounted for the other 40.1 percent. The life insurance sector was focused on retirement insurance, which represented more than three quarters of managed savings and the other quarter corresponded to savings insurance on other terms. Qualified retirement plans were new products, and they accounted for only 0.1 percent of life insurance managed savings.

In life retirement insurance, there were differences in the weight of products depending on whether they were for individuals or groups. At the end of 2003, in individual life/retirement insurance, 57.3 percent of accumulated savings were in products designed to be collected as a lump sum, and the remainder were related to annuities. But in group life retirement insurance, lump sum insurance accounted for only 7.5 percent, whereas annuity insurance represented the vast bulk of savings. The main reason for this disparity was the tax treatment of annuities, which was different from the one applied to lump sum insurance and harsher in many age brackets (UNESPA(a) 2004).

In 2002, life insurance managed savings grew 14.4 percent, and in 2003 they grew 6.8 percent. As mentioned above, the special increase in life insurance in 2002 was due to the fact that 2002 was the last year of obligatory outsourcing of companies' pension commitments to employees. At the end of 2002, total group life insurance managed savings accounted for 42,748 million euros, nearly half of which were related to specific outsourcing policies (UNESPA(a), 2003).

Table 10.4. Life Insurance Managed Savings (in millions of euros) in Spain, 2001 to 2003

	2001	2002	2003	Percent of Total Managed Savings in 2003	Growth 2002/2001	Growth 2003/2002
Individual Life Insurance						
Death Indemnity	780	986	1,172	1.1%	26.4%	18.9%
Qualified Retirement Plans	0	0	99	0.1%	0.0%	0.0%
Other Retirement Insurance	27,617	31,488	35,702	32.6%	14.0%	13.4%
Deferred Capital	15,386	18,334	20,444	18.7%	19.2%	11.5%
Inmediate and Deferred Life Annuity	10,139	11,689	13,983	12.8%	15.3%	19.6%
Unit/Index-Linked	2,092	1,465	1,275	1.2%	–30.0%	–13.0%
Nonretirement Insurance	25,955	27,160	28,579	26.1%	4.6%	5.2%
Guaranteed Yield	13,857	17,733	19,523	17.9%	28.0%	10.1%
Unit-/Index-Linked	12,098	9,427	9,056	8.3%	–22.1%	–3.9%
Total Individual Life Insurance	54,352	59,634	65,552	59.9%	9.7%	9.9%
Group Life Insurance						
Death Indemnity	630	881	1,015	0.9%	39.8%	15.2%
Retirement Insurance	34,537	41,867	42,783	39.1%	21.2%	2.2%
Deferred Capital	2,984	2,920	3,227	3.0%	–2.1%	10.5%
Inmediate and Deferred Life Annuity	30,068	38,455	38,870	35.5%	27.9%	1.1%
Unit-/Index-Linked	1,485	492	686	0.6%	–66.9%	39.4%
Total Group Life Insurance	35,167	42,748	43,798	40.1%	21.6%	2.5%
Total Life Insurance	89,519	102,382	109,350	100%	14.4%	6.8%

Source: Author's compilations from UNESPA(a) reports on life insurance in Spain.

Table 10.5. Total Assets in the Spanish Insurance Industry, 1990 to 2003

	Total	Mean	Standard Deviation	n
1990	26,834	53	178	505
1991	29,972	60	181	499
1992	33,500	71	193	475
1993	37,782	83	216	452
1994	45,516	108	276	425
1995	52,024	137	329	381
1996	60,739	164	390	370
1997	70,595	199	451	355
1998	82,761	248	583	334
1999	98,938	307	741	322
2000	114,893	371	910	310
2001	127,308	424	1,103	300
2002	145,015	488	1,270	297
2003	158,196	546	1,392	290

Source: Author's calculations from DGSFP(a) *Balances y Cuentas.*
Note: Monetary variables are expressed in million of euros. Social benefit institutions are not included. Branches of EU foreign firms are not included after 1994.

10.3.4 Total Assets in the Industry

Total assets in the industry as well as average firm size, standard deviations, and number of firms for the period 1990 to 2003 are presented in Table 10.5. The data are drawn from the regulatory accounting statements filed by insurers annually with the Dirección General de Seguros y Fondos de Pensiones, Ministerio de Economía y Hacienda. Accordingly, this analysis includes only those firms for which data reported to the DGSFP were available. Branches of EU foreign firms have not been included after 1994. Social benefit institutions are excluded in this analysis; however, they represented 4 percent of total industry assets in 2003.

Between 1990 and 2003, total industry assets increased by 489 percent and average firm size increased by 920.5 percent. In 2003, total industry assets accounted for 158,196 million euros.

Because the industry consists of three major types of firms—life insurance specialists, non-life insurance specialists, and diversified firms that offer both life and non-life insurance—we conducted a special analysis of total assets classified into these three categories for the period 1998 to 2003.[212]

[212] Since passage of the 1984 insurance law, new firms should be life insurance specialists and non-life insurance specialists. Nevertheless, new life insurance specialists can also operate in health insurance and accident insurance by increasing the amount of capital.

Table 10.6 reports trends in total assets; average firm sizes; standard deviations; numbers of firms for life specialists, non-life specialists, and diversified firms; and the percentages of total numbers of firms and of total assets in the industry.

In 2003, 62 percent of firms were non-life specialists, 22 percent were diversified firms, and 16 percent were life specialists. They represented 11 percent, 68 percent, and 21 percent, respectively, of total industry assets. So non-life specialists are the most representative in the industry in terms of number of firms, but diversified firms are the most representative in terms of total assets.

Between 1998 and 2003, total assets of diversified firms and non-life specialists increased 148 percent and 119 percent, respectively. Their average firm size increased 172 percent and 159 percent, respectively. However, in the life specialist segment, total assets increased 5.3 percent, and the average firm size of life specialists increased 16.9 percent.

Generally, diversified firms are the largest firms, and non-life specialists are the smallest. T-tests for differences between means, reported at the bottom of Table 10.6, show that these differences are statistically significant. The larger size of diversified firms can be justified because they must maintain the capital requirements for the life insurance segment in addition to the capital requirements for the non-life lines of business that they operate.

10.3.5 International Activities

There are three methods for foreign insurers to operate in the Spanish insurance market: (1) through the formation of Spanish-licensed and -regulated subsidiaries of foreign firms that are supervised by the Spanish regulatory authority; (2) through branches of foreign firms; and (3) in a regime of "freedom of services." This method is only allowed to EU insurers and gives them the ability to conduct business outside their home country directly without having to establish branches in host countries. Branches of non-EU foreign firms must obtain a licence in Spain and report to the Spanish regulatory authority, but branches of EU-licensed firms have not had to report to the Spanish regulator since 1994.

As mentioned above, the Third Generation Directives established a single EU licence, but if an EU insurer wants to operate in another EU country in a regime of "freedom of services," it must inform the home country supervisor on risks that it plans to cover in the host country. Within one month after this communication, the home country supervisor must inform the host country supervisor about this firm on the following issues: whether it has enough solvency margin; the lines of business in which it is authorized to operate; and the risks that it plans to cover in the host country. From the date that the home country's supervisory authority relays this information to the host country's authority supervisor, this insurer can operate in the host country in a regime of "freedom of services," which means that this EU insurer is "habilitated" to operate in a regime of freedom to provide services in the host EU country. EU insurers must use similar processes to establish branches in other EU countries, but with more requirements (e.g., it is also necessary to name a general attorney of the branch to represent the firm before third entities and before the host country's judicial authority) (TRLOSSP 2004).

Table 10.6. Total Assets of Diversified Firms, Non-life Specialists, and Life Specialists, 1998 to 2003

	1998	1999	2000	2001	2002	2003
Diversified Firms						
Total	43,512	48,843	60,515	82,319	98,260	108,000
Mean	613	763	903	1,328	1,512	1,665
Standard Deviation	785	978	1,303	2,052	2,322	2,517
n	71	64	67	62	65	65
Percent of Total Number	21	20	22	21	22	22
Percent of Total Assets	53	49	53	65	68	68
Non-life Specialists						
Total	7,591	9,635	12,006	13,627	15,361	16,642
Mean	36	47	62	71	83	92
Standard Deviation	87	130	156	174	196	216
n	213	206	195	191	186	180
Percent of Total Number	64	64	63	64	63	62
Percent of Total Assets	9	10	10	11	11	11
Life Specialists						
Total	31,658	40,460	42,372	31,363	31,395	33,340
Mean	633	778	883	667	683	741
Standard Deviation	925	1,202	1,374	722	741	770
n	50	52	48	47	46	45
Percent of Total Number	15	16	15	16	15	16
Percent of Total Assets	38	41	37	25	22	21
t-tests for Differences Between Means, 1990 to 2003						
Diversified Firms versus Non-life Specialists				11.597***		
Diversified Firms versus Life Specialists				3.605***		
Non-life Specialists versus Life Specialists				−11.427***		

Source: Author's calculations from DGSFP(a) *Balances y Cuentas.*
Note: Monetary variables are expressed in million of euros.
*** Significant at the 1 percent level.

As in other European Union nations, the primary method for foreign insurers to enter the Spanish market has been through the formation of Spanish-licensed and -regulated subsidiaries rather than through branches or agencies. After Spain entered the European Economic Community in 1986, a wave of foreign groups, mostly European insurance groups without a presence in Spain, entered the Spanish market.

In some cases, they entered the market by buying Spanish insurers with financial difficulties (Esteban-Jodar 1993). As a result, the presence of foreign equity capital in the Spanish-licensed insurers' capital has been very important (e.g., in 1996 it accounted for 37 percent of total capital).

Table 10.7 shows the distribution of foreign capital per country for the years 2002 and 2003. In 2003, the percentage of foreign capital over the total capital of Spanish-licensed insurers was 29.5. Most of it belonged to other EU countries (25.6 percent) and involved 29.3 percent of total premium volume, 33.7 percent of life premiums, and 22.9 percent of non- life premiums.

The number of branches of foreign firms per country and their respective activity in the Spanish market is also shown in Table 10.7. In 2003, most of them (39) were branches of EU-licensed firms, 12 of which were U.K.-licensed firms, 9 were French-licensed firms, and 6 were German-licensed firms. There were three branches of EU reinsurance specialists. Branches of EU licensed firms did not have to report to the Dirección General de Seguros y Fondos de Pensiones after 1994, when the Third Generation Directives came out. But the insurance supervisory authority of every country must report a summary of their branch's activity to the insurance supervisory authority of the EU country in which they operate.

In 2002, branches of EU insurance firms accounted for about 489 million euros of total premiums, 445 million euros of which were non-life insurance premiums and 44 million euros were life insurance premiums. French branches had the highest market share (28.2 percent) in the Spanish insurance market, followed by Belgian branches (26.5 percent).

European insurers also operated in the Spanish market in a regime of freedom to provide services. Table 10.7 shows the number of firms habilitated to operate on this regime in the Spanish market with their corresponding activity per country. At the end of 2003, there were 380 EU companies habilitated, although some of them did not operate any insurance business in Spain.

In 2002, habilitated EU companies accounted for 438 million euros of total insurance premiums, 298 million euros of which were life premiums and 140 million euros were non-life premiums. As for premium volume, the most important country was Luxembourg (259 million euros) followed by the United Kingdom (71 million euros).

Table 10.8 shows the Spanish-licensed insurers' activity in the European Union insurance market, the number of branches, and the number of habilitated companies with freedom to provide services along with their respective insurance activity. The international movements of Spanish insurers were within the Portuguese insurance market (Figueiredo-Almaça 1999). In 2003, 43 branches of Spanish insurance firms operated in other European Union countries, 27 of which operated in Portugal. In 2002, branches of Spanish firms accounted for 176 million euros of total premiums, and the most important country for total premiums was Portugal (100 million euros of total premiums). The activity of Spanish habilitated companies with freedom to

Table 10.7. Foreign Activity in the Spanish Insurance Market, 2002 and 2003

	Foreign Capital (percent of Spanish Insurers' Capital)		Number of Branches of Foreign Firms		Number of EU Habilitated Companies		Premium Volume 2002						
							Branches of Foreign firms			EU Habilitated Companies			
	2002	2003	2002	2003	2002	2003	Life Premiums	Non-life Premiums	Total	Life Premiums	Non-life Premiums	Total	Total Activity per Country
EU Countries													
Austria	0.02	0	0	0	7	8	0	0	0	0	168	168	168
Belgium	3.07	2.88	3	3	23	23	75	129,537	129,612	2,096	13,016	15,112	144,725
Denmark	0	0	1	1	5	6	0	862	862	406	5,162	5,567	6,429
Finland	0	0	0	0	2	1	0	0	0	0	162	162	162
France	7.16	7.35	9	9	30	32	16,429	121,433	137,862	223	15,484	15,707	153,569
Germany	1.77	1.62	4	6	39	41	0	55,919	55,919	2,142	5,960	8,102	64,021
Greece	0	0	0	0	1	1	0	0	0	0	0	0	0
Ireland	0	0	2	2	53	63	21,216	0	21,216	33,252	18,039	51,291	72,507
Italy	1.76	1.65	1	1	34	33	0	559	559	1,180	4,784	5,964	6,523
Liechtenstein	0	0	0	0	5	6	0	0	0	112	0	112	112
Luxemburg	0.24	0.22	0	1	36	40	0	0	0	256,513	2,647	259,160	259,160
The Netherlands	4.61	1.74	1	1	9	10	0	0	0	0	2,356	2,356	2,356
Norway	0	0	0	0	5	6	0	0	0	0	1,689	1,689	1,689
Portugal	0	0	3	3	5	6	5,661	82,802	88,463	0	188	188	88,651
Sweden	0	7.78	0	0	13	14	0	1,418	1,418	0	1,486	1,486	2,904
United Kingdom	2.75	2.35	9	12	84	90	0	52,883	52,883	2,155	68,618	70,774	123,657
Total EU	21.4	25.6	33	39	351	380	43,381	445,413	488,794	298,079	139,758	437,838	926,632

Table 10.7 *(continued)*. Foreign Activity in the Spanish Insurance Market, 2002 and 2003

	Foreign Capital (percent of Spanish Insurers' Capital)		Number of Branches of Foreign Firms		Number of EU Habilitated Companies		Premium Volume 2002						
							Branches of Foreign firms			EU Habilitated Companies			Total Activity per Country
	2002	2003	2002	2003	2002	2003	Life Premiums	Non-life Premiums	Total	Life Premiums	Non-life Premiums	Total	
Non-EU countries													
United States	1.75	1.28	2	2									
Switzerland	2.98	2.58											
Total non EU	4.73	3.86	2	2									
Total	26.1	29.5	35	41									

Source: Author's compilations from DGSFP(b) 2003 *Seguros y Fondos de Pensiones Informe.*
Note: Monetary variables are expressed in thousands of euros.

provide services accounted for about 196 million euros of total premiums, and Portugal was again the most important country in terms of total premiums (108 million euros).

10.3.6 Market Concentration

A summary of the market structure in the Spanish insurance industry is presented in Table 10.9. The table shows 1, 5, 10, 20, 50, and 75 firm concentration ratios for non-life insurance and life insurance based on premium revenues for the years 1990, 1993, 1998, and 2003. Number of firms and Herfindahl indices are also shown. The number of companies offering non-life insurance fell from 419 in 1990 to 245 in 2003, and the number of firms offering life insurance declined from 166 in 1990 to 110 in 2003.

Concentration generally increased in the non-life insurance segment of the Spanish insurance industry over the sample period. The 20-firm concentration ratio increased from 46.2 percent in 1990 to 64.4 percent in 2003, and the Herfindahl index rose from 157.5 to 283.7. In contrast, a concentration trend in the life insurance segment of the Spanish insurance industry was not as clear as in the non-life insurance segment. The lack of a clear trend in the life segment was due to large increases in single premiums in some years (e.g., in 2001, when the Herfindahl index was 788.2).

Nevertheless, from 1990 to 2003, concentration decreased in the life insurance segment. The 20-firm concentration ratio fell from 75.7 percent in 1990 to 71.2 percent in 2003, and the Herfindahl index fell from 593.7 in 1990 to 379.5 in 2003. The decline in life insurance concentration is attributable to market share gains by new entrants, including bank-owned firms and foreign firms.

The life insurance segment is also characterized by the coexistence of a high concentration of premium revenues in some firms with the dispersion of a small part of premiums in many companies. Approximately 35 firms account for about 1 percent of total life premiums, and this is due to joint firms that offer both life and non-life insurance and have a residual activity in the life insurance segment.

Further analysis of the concentration in the non-life lines of business for the years 1998 and 2003 is shown in the bottom panel of Table 10.9. In general, the non-life insurance segment, like the life insurance segment, has the structural characteristic of coexistence of a high level of concentration with a high level of dispersion in premium volume.

In the traditional theory of industrial organization, concentration was held to facilitate oligopolistic or collusive practices and thereby to lead to noncompetitive profits. The more modern view is that increasing concentration, at least within limits, may be a natural development in some markets and does not necessarily have adverse consequences. If efficient firms are gaining market shares, prices may fall at the same time as concentration and profits may rise (Cummins and Weiss 1991).

Such a scenario may be applicable to some lines of the Spanish insurance industry in which the level of competition is increasing. The most competitive lines of business in the Spanish insurance industry are life insurance, auto insurance, and multiple peril insurance. In the analysis of the period 1998 to 2003, concentration

Table 10.8. Spanish Insurers' Activity in the European Union, 2002 and 2003

	Number of Spanish Branches in the EU		Number of Spanish Habilitated Companies		Premium Volume 2002						
					Spanish Branches in the EU			Spanish Habilitated Companies			
	2002	2003	2002	2003	Life Premiums	Non-life Premiums	Total	Life Premiums	Non-life Premiums	Total	Total Activity per Country
Austria	0	0	9	9	0	0	0	0	42	42	42
Germany	0	0	16	17	0	0	0	230	2	232	232
Belgium	4	4	9	9	0	21,068	21,069	0	1,021	1,021	22,090
Denmark	0	0	8	8	0	0	0	0	0	0	0
Finland	0	0	7	7	0	0	0	0	0	0	0
France	6	6	22	23	2,094	1,694	3,788	0	524	524	4,312
Greece	1	1	7	7	0	5,261	5,261	0	0	0	5,261
Ireland	2	2	10	10	0	5,717	5,717	0	0	0	5,717
Italy	1	1	13	13	0	0	0	0	7,996	7,996	7,996
Liechtenstein	0	0	1	2	0	0	0	0	0	0	0
Luxemburg	0	0	7	8	0	0	0	0	41	41	41
Norway	0	0	2	3	0	0	0	0	0	0	0
The Netherlands	0	0	9	9	0	0	0	0	1,887	1,887	1,887
Portugal	29	27	26	27	74,519	25,208	99,727	146	7,689	7,835	107,562
United Kingdom	2	2	17	20	0	40,723	40,723	0	0	0	40,723
Sweden	0	0	5	6	0	0	0	0	0	0	0
Iceland	0	0	2	3	0	0	0	0	0	0	0
Total EU	45	43	*	*	76,614	99,671	176,285	377	19,204	19,580	195,865

Source: Author's compilations from DGSFP(b) 2003 *Seguros y Fondos de Pensiones Informe.*
Note: Monetary variables are expressed in thousands of euros.
* The sum of branches is not shown because it does not represent the number of Spanish firms with branches in the EU—a Spanish insurer can have branches in several EU countries.

Table 10.9. Concentration Ratios for the Spanish Insurance Industry, 1990 to 2003

	Concentration Ratios (%)						Herfin-dahl Index	n
	1-Firm	5-Firm	10-Firm	20-Firm	50-Firm	75-Firm		
Life Premiums								
1990	17.8	44.3	60.3	75.7	92.8	97.4	593.7	166
1993	11.9	35.7	53.3	70.1	90.4	96.3	400.1	147
1998	6.8	25.4	43.9	68.0	92.3	98.4	290.1	121
2003	9.0	34.8	52.4	71.2	93.8	98.9	379.5	110
Non-life Premiums								
1990	5.1	18.8	30.1	46.2	72.1	82.2	157.5	419
1993	7.3	20.5	31.1	48.2	73.9	83.9	180.1	380
1998	7.3	24.2	39.3	59.6	82.4	89.6	237.9	284
2003	8.6	27.6	42.9	64.4	85.9	92.2	283.7	245
Non-life Lines of Business								
Accident								
1998	6.8	24.5	37.0	62.0	88.8	95.5	254.4	172
2003	9.9	31.4	47.8	70.6	93.1	98.0	348.9	144
Health								
1998	19.8	57.1	79.5	95.6	99.9	100	946.5	76
2003	18.8	56.9	81.5	95.0	99.8	100	886.8	77
Health Assistance								
1998	22.3	62.9	78.3	87.1	97.0	99.2	1,096.1	103
2003	21.7	69.1	85.3	92.6	98.5	99.7	1,264.4	98
Transport (Hull)								
1998	24.5	67.1	85.9	97.1	100	100	1,167.4	55
2003	19.5	69.9	88.5	99.0	100	100	1,146.7	44
Transport (Goods)								
1998	12.5	45.1	69.1	91.1	99.8	100	622.4	65
2003	12.7	52.3	83.7	96.3	100	100	784.8	51
Fire								
1998	40.9	77.5	86.5	94.4	99.2	99.9	2,220.4	111
2003	46.8	86.8	93.7	97.5	99.8	100	2,896.8	81
Other Damages								
1998	9.2	41.2	64.2	84.1	98.2	99.9	505.6	105
2003	13.9	53.7	78.1	94.0	99.8	100	778.4	78
Auto (Liability)								
1998	15.6	39.9	56.9	76.7	95.2	99.7	509.8	93
2003	17.7	46.4	64.3	82.8	99.0	100	667.6	68
Auto (Other Guaranties)								
1998	16.2	51.2	67.4	83.0	97.7	100	730.7	81
2003	19.3	53.0	70.7	87.7	99.7	100	785.1	60

Table 10.9 *(continued)*. Concentration Ratios for the Spanish Insurance Industry, 1990 to 2003

	Concentration Ratios (%)						Herfin-dahl Index	n
	1-Firm	5-Firm	10-Firm	20-Firm	50-Firm	75-Firm		
General Liability								
1998	10.7	42.5	63.4	83.3	98.2	99.9	520.9	100
2003	16.0	48.5	68.3	88.2	99.7	100	665.0	81
Credit								
1998	55.9	96.7	100	100	100	100	3,735.2	17
2003	64.3	98.5	100	100	100	100	4,611.2	12
Caution								
1998	33.5	79.4	97.4	100	100	100	1,753.7	22
2003	23.6	82.8	98.2	100	100	100	1,661.9	20
Income Loss								
1998	42.6	68.6	80.1	92.5	99.9	100	2,057.5	62
2003	41.5	73.8	85.0	94.9	100	100	2,052.9	48
Legal Defense								
1998	29.1	66.6	83.5	97.5	100	100	1,513.6	34
2003	50.5	92.3	98.3	100	100	100	3,092.0	22
Assistance								
1998	11.7	38.1	59.1	81.6	97.4	99.6	471.6	97
2003	14.0	49.8	80.1	98.6	100	100	757.9	45
Mortuary								
1998	38.8	79.4	89.0	96.5	99.7	100	2,195.4	82
2003	37.4	78.1	89.0	96.9	99.8	100	2,096.6	75
Multiple Peril (Home)								
1998	9.6	34.8	55.6	79.1	97.6	99.8	415.0	108
2003	12.6	37.8	60.8	85.9	98.7	100	510.1	86
Multiple Peril (Commerce)								
1998	11.0	43.2	65.6	84.6	98.9	100	542.4	81
2003	10.8	49.3	74.7	91.4	99.9	100	664.7	67
Multiple Peril (Community)								
1998	16.5	57.0	77.4	92.9	99.9	100	825.5	70
2003	21.7	57.9	77.8	95.3	100	100	963.0	57
Multiple Peril (Industry)								
1998	14.4	52.5	72.0	88.3	99.6	100	692.7	66
2003	20.3	55.9	78.1	92.9	100	100	903.8	57
Other Multiple Peril								
1998	58.0	83.9	94.3	99.5	100	100	3,571.5	41
2003	43.5	89.5	96.3	99.5	100	100	3,120.9	37

Source: Author's calculations from DGSFP(a) *Balances y Cuentas*.

Note: Social benefit institutions are excluded in the analysis of both total life premiums and total non-life premiums, but not in the analysis of the different non-life lines of business.

generally increased in these lines of business. In the life insurance segment, the Herfindahl index rose from 290.1 in 1998 to 379.5 in 2003. In auto (liability), the Herfindahl index rose from 509.8 in 1998 to 667.6 in 2003. In auto (other damages), the Herfindahl index rose from 730.1 in 1998 to 785.1 in 2003. In home multiple peril, the Herfindahl index rose from 415 in 1998 to 510.5 in 2003, and in industrial multiple peril, the Herfindahl index rose from 692.7 in 1998 to 903.8 in 2003. These increases in market concentration provide some evidence of the market structure efficiency that holds that more efficient firms can capture larger market shares and economic rents, leading to increased concentration.

A conclusion about market structure based on the data shown in Table 10.9 is that the Spanish insurance market has experienced dramatic changes, with market declines in the number of firms and changes in industry concentration. This dynamic environment provides opportunities for firms to increase market shares and improve efficiency and profitability. However, this market volatility also presents significant downside risk for firms that are slow to adapt to changing market conditions.

10.3.7 Organizational Structure

Insurance in Spain is primarily provided by stock insurers that are owned by shareholders and mutual insurers that are owned by policyholders. The existence and continued survival of different organizational forms in the insurance market are generally attributed to the fact that they have differential advantages in dealing with particular types of insurance.

According to the modern theory of the firm, agency costs provide an explanation for the structure of organizations, with the organizations that succeed in a given industry being the ones that minimize costs and maximize revenues, where both costs and revenues are potentially affected by agency costs as well as the firm's production process and operating environment (e.g., Jensen and Meckling 1976). Agency costs, including excessive operating costs and lost revenues, are generated because the various stakeholder groups comprising the firm often have conflicting interests. The three principal stakeholder groups in an insurance enterprise are managers, owners, and policyholders. The agency theory hypothesizes that stocks and mutuals coexist because they have comparative advantages in dealing with incentive conflicts among stakeholders.

The stock ownership form is hypothesized to be more effective in controlling conflicts between owners and managers, because it provides superior mechanisms for owners to monitor and control managers. The control mechanism available to mutual owners is much weaker (see Mayers and Smith 1988). Mutuals are hypothesized to succeed in situations in which controlling the policyholder-owner conflict is important, because the policyholder and ownership functions are merged in the mutual ownership form.

These arguments have led to the development of hypotheses such as the managerial discretion hypothesis, which holds that the degree of managerial discretion required to operate in a given line of insurance is an important determinant of the organizational form likely to succeed in that line (Mayers and Smith 1988). The hypothesis predicts that the stock ownership form will be dominant in lines of insurance where managers must be given a relatively large amount of discretion in pricing and underwriting, such as commercial coverages, and in operating over wider geographical areas. Stock insurers are also expected to have an advantage in lines of business where the level of product innovation is relatively high. Mutuals, on the other hand, are expected to be more successful in lines that require less managerial discretion, where the need for individualized pricing and underwriting is relatively low (such as lines with standardized policies and good actuarial tables). Mutuals should also be relatively successful in markets subject to low levels of product innovation.

The stock and mutual organizational forms also differ in another important respect: the access to capital. The stock ownership form provides superior access to capital, giving stock insurers an advantage in risky lines and in lines where the level of product innovation is relatively high. The access to capital argument predicts that stocks will do better in lines characterized by high levels of innovation, whereas mutuals will do better in more stable and predictable lines (Cummins, Rubio-Misas, and Zi 2004).

The above agency-theoretic and non-agency-theoretic components form the efficient structure hypothesis, which holds that stocks and mutuals will be sorted into market segments where they have comparative advantages in minimizing costs and maximizing revenues.[213]

The market shares by organizational form in 2003 are shown in Figure 10.1. The data include all insurers in the Spanish market that are supervised by the Spanish regulatory authority, the Dirección General de Seguros y Fondos de Pensiones, Ministerio de Economía y Hacienda. Three categories of organizational form are presented: stocks, mutuals, and other categories—which include branches of foreign firms, reinsurance specialists, and social benefit institutions. Branches of foreign firms have their highest market share in accidents.[214] Reinsurance specialists accounted for about 2.8 percent of non-life insurance premiums, having a market share of about 70.6 percent in fire insurance. Social benefit institutions accounted for about 2.4 percent of life insurance premiums, a relatively important market share compared to the mutuals' market share in life insurance. In non-life insurance, social benefit institutions accounted for about 0.6 percent and have their highest market share in health insurance (9.4 percent).

As in the analysis of Cummins, Rubio-Misas, and Zi (2004), the data are generally consistent with the efficient structure hypothesis. Mutuals accounted for about 20.5 percent of non-life insurance premiums but only 1.2 percent of life insurance premiums in 2003. Further analysis indicates that, within the non-life category, mutuals have a relatively high market share (more than one-third of the market) in automobile insurance but much lower market shares in commercial lines such as multiple peril insurance (Martínez, Albarrán, and Camino 2001; Cummins, Rubio-Misas, and Zi 2004). The market positioning of the Spanish mutual and stock insurers in auto insurance and life insurance, respectively, provides support for the efficient structure hypothesis.

---

[213] For further analysis on the efficient structure hypothesis and the expense preference hypothesis in the Spanish insurance industry, see Cummins, Rubio-Misas, and Zi (2004).

[214] Only branches of non-EU firms are included, which have to report to the Spanish regulatory authority. Branches of EU-licensed firms did not have to report to the Spanish regulator after 1994.

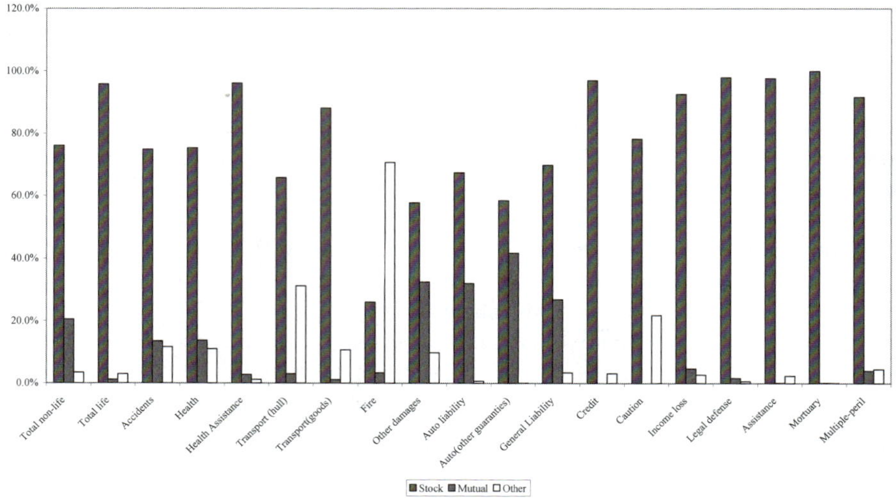

Source: Author's calculations from DGSFP(a) *Balances y Cuentas.*

Figure 10.1. Market Share by Organizational Form and Line, 2003

Automobile insurance is a relatively stable line. It does not require high levels of managerial discretion, and the level of product innovation is very low. The relative lack of success of mutuals in complex commercial non-life lines, where more pricing and underwriting discretion are required, reinforces support for the efficient structure hypothesis in the Spanish insurance industry.

The dominance of stock insurers in life insurance also supports the efficient structure hypothesis. The dynamism of the life insurance market has led to significant product innovation, with the introduction of new products that require high levels of technology investment. Stock insurers have an advantage over mutuals in this market because product innovation requires relatively high managerial discretion and because investment in product research and development has required significant infusions of new capital (Cummins, Rubio-Misas, and Zi 2004). Nevertheless, considering the market share of banks in the Spanish life insurance market, the domination of stock insurers in the life insurance market could also be attributable to the competitive advantage of banks. Banks in Spain can only choose stock organizational structures and never mutual organizational structures to form insurers.

10.3.8 Distribution System

Product distribution channels in the Spanish insurance industry can be classified into six mayor types: (1) agents, (2) brokers, (3) banks, (4) insurers' own offices, (5) direct marketing, and (6) other channels. Agents have a mercantile contractual relationship with a single insurer, but they are not employees of the insurer. Brokers are independent businesses that may sell products of more than one insurer. The banking channel involves the insurance product sale activity distributed by the commercial banks, savings banks, and cooperative credit associations' offices.

Insurance brokers created by banking entities are included in this channel. The insurers' own offices channel involves the distribution within insurer offices by a sales force under direct employment of a single insurer. The direct marketing method involves the insurance product sale activity performed by firms that are subcontracted by the insurer, which are basically call centers. The other channels category refers to distribution via the Internet (DGSFP(b) 2004).

Figure 10.2 shows the market share by distribution system for the year 2002. The data come from the Direccion General de Seguros y Fondos de Pensiones, which requires insurers under its supervision to report data about distribution systems in compliance with the passage of an insurance regulatory order of July 2001.

Banks, with 34.1 percent of premium volume, and agents and brokers, with 45.2 percent of premium volume, are the three most important distribution channels in the Spanish insurance industry. However, because the behavior of the distribution systems is different in the life and non-life insurance segments, Figure 10.2 shows the market share by distribution system for overall life insurance, overall non-life insurance, and selected lines of non-life insurance.

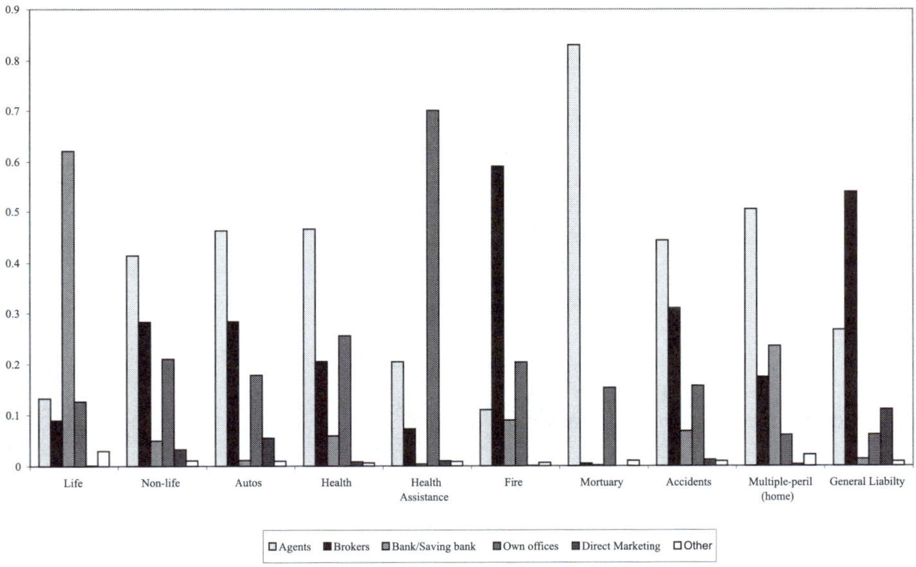

Source: Author's compilations from DGSFP(b) *Seguros y Fondos de Pensiones Informe 2003.*

Figure 10.2. Market Share by Distribution System, 2002

In the life insurance segment, bancassurance is the most important channel, with 62.1 percent of life premium volume and 73.7 percent of premiums for new life insurance contracts in the year of analysis (DGSFP(b) 2004). But in the non-life insurance segment, banks have only 5 percent of premiums, and agents and brokers account for 69.7 percent of non-life premium volume, of which 41.4 percent is distributed by agents and 28.3 percent is distributed by brokers.

Eighty-four commercial banks, savings banks, and cooperative credit associations operated as a distribution channel in 2002. Approximately 40 percent of these financial firms operated as agents and 60 percent as brokers. To better understand the relative presence of banking institutions in the Spanish insurance market, it is important to keep in mind that, at the end of 2002, 275 banking institutions operated in Spain, 143 of which were commercial banks, 47 of which were savings banks, and 85 of which were cooperative credit associations. These 275 banking institutions had a banking network of almost 39,000 branches, and usually banking institutions that operated as an insurance distribution channel were the largest ones with the most bank branches. Bancassurance is the most important distribution system in the two major lines of life insurance, with 69.1 percent of premium volume in individual life insurance and 53.8 percent of premium volume in group life insurance (DGSFP(b) 2004).

In the non-life insurance segment, the predominant distribution channel varies depending on the line of business. Agents are the most important distribution channel in accidents (with a market share of 46.5 percent of its premium volume), health (46.2 percent), automobile (63.1 percent), credit (45.8 percent), assistance (83 percent), mortuary (50.5 percent), home multiple peril (63.2 percent), community multiple peril (63.2 percent), and commercial multiple peril (57.2 percent).

Community multiple peril insurance covers several risks for landlords. Two lines of business comprise health insurance in Spain. The most important one is health assistance that covers medical services in case of illness, and the other covers indemnity to customers for loss of income in case of illness. Assistance insurance provides assistance to people who have difficulties during residence relocation. It also provides assistance to people who have difficulties in other legally established circumstances as long as they are not covered by other types of insurance. Mortuary insurance is a policy for funeral expenses that the Spanish legislation does not include in life insurance policies.

Brokers are the predominant distribution system in transport (with 53.9 percent of its premium volume), fire (50.2 percent), burglary and related coverage (66.7 percent) , nuclear peril general liability (51.1 percent), general liability (53.9 percent), caution (38.2 percent), income loss (44.6 percent), legal defense (63.3 percent), industrial multiple peril (56 percent) and other multiple peril (57.8 percent) lines of business.

Health assistance insurance is typically distributed by insurers' own offices, with a market share of 70.1 percent of premium volume. Bancassurance is the predominant distribution channel for crop insurance, with 41.8 percent of premium volume, although agents are the predominant distribution channel in the premiums for new crop insurance contracts in the year of analysis, with a market share of 41.7 percent of premium volume (DGSFP(b) 2004).

The Internet is used mostly to complement other distribution channels. Although the market share of this distribution channel does not meet the expected results, the Internet distribution system is increasing in the non-life insurance segment. The number of companies with agreements to share their distribution Web sites, according to the 1992 insurance mediation law, in 2002 was 84. Health and health assistance insurance are the most common lines reaching these types of agreements.

Table 10.10. Market Share by Distribution System in the Spanish Insurance Industry, 1994 to 2003

1994	Agents	Brokers	Banks	Insurers' Offices	Other	
Life	31.99	5.73	46.49	8.13	7.65	
Multiple Peril	53.04	18.79	9.65	15.87	2.65	
Auto	47.48	20.74	0.51	28.89	2.37	
Industrial	33.26	48.84	1.69	9.97	6.24	
Health	40.54	16.10	0.17	38.09	5.09	
Other	41.17	25.13	10.42	17.67	5.61	
Total	42.79	17.49	15.03	20.41	4.28	
1995	**Agents**	**Brokers**	**Banks**	**Insurers' Offices**	**Direct Marketing**	**Other**
Life	33.91	5.30	47.39	8.86	0.16	4.38
Multiple Peril	50.23	19.73	14.65	11.41	0.18	3.81
Auto	49.62	19.37	0.64	27.48	1.34	1.56
Industrial	23.16	48.13	3.21	10.41	0.02	15.05
Health	46.82	13.31	1.27	27.14	0.01	11.45
Other	50.66	23.10	8.30	16.26	0.30	1.38
Total	43.92	16.02	17.73	18.27	0.66	3.40
1996						
Life	29.87	8.72	48.69	9.10	0.18	3.45
Multiple Peril	55.20	22.74	13.33	6.06	0.13	2.54
Auto	48.89	19.92	0.11	26.64	3.99	2.45
Industrial	28.88	58.31	1.32	3.41	0.03	8.06
Health	25.06	9.78	2.07	32.80	0.02	30.27
Other	52.80	24.69	13.49	6.98	0.30	1.75
Total	42.33	17.64	19.34	15.39	1.74	3.56
1997						
Life	31.22	7.48	48.82	8.32	0.13	4.03
Multiple Peril	60.32	19.92	11.44	5.70	0.09	2.53
Auto	53.00	18.05	0.17	21.37	4.58	2.82
Industrial	36.88	44.64	1.55	5.60	0	11.32
Health	24.15	5.71	15.69	53.57	0	0.87
Other	63.36	20.52	3.29	7.99	0.12	4.72
Total	45.74	15.00	18.34	15.51	1.91	3.50

Table 10.10 *(continued).* Market Share by Distribution System in the Spanish Insurance Industry, 1994 to 2003

1998	Agents	Brokers	Banks	Insurers' Offices	Direct Marketing	Other
Life	15.42	3.25	71.95	7.36	0	2.02
Health	27.46	5.03	7.86	52.53	0	7.12
Accident	57.01	24.86	7.67	7.66	0	2.81
Auto	51.42	18.29	0.15	24.33	4.54	1.26
General Liability	48.57	42.84	0.64	5.28	0	2.67
Fire	24.36	48.61	6.60	17.63	0	2.80
Multiple Peril	61.81	18.98	12.00	4.93	0.06	2.22
Transport	46.43	40.29	0.67	10.92	0	1.70
Total	31.50	9.47	42.02	13.82	1.01	2.17

1999	Agents	Brokers	Banks	Insurers' Offices	Telephone	Other
Life	17.68	2.70	73.06	4.84	0	1.72
Health	24.63	5.29	7.58	57.08	2.69	2.72
Accident	54.94	26.58	10.61	5.13	0	2.74
Auto	51.32	20.78	0.34	18.61	5.30	3.65
General Liability	39.96	49.48	0.91	6.58	0	3.07
Fire	20.92	40.35	12.62	20.70	0	5.41
Multiple Peril	62.83	17.49	12.66	5.11	0.11	1.80
Transport	37.92	48.76	0.85	11.25	0	1.22
Mortuary	99.03	0.45	0	0.48	0	0.04
Other Lines	36.24	29.93	20.89	10.08	0	2.86
Total	31.22	9.55	44.65	11.03	1.32	2.23

2000	Agents	Brokers	Banks	Insurers' Offices	Telephone	Internet	Other
Life	14.21	2.65	72.22	3.38	0	0	7.55
Health	28.21	7.30	8.22	53.13	2.52	0.29	0.32
Accident	51.43	29.76	7.89	8.74	0	0	2.18
Auto	48.03	22.86	0.86	18.02	6.32	0.06	3.85
General Liability	41.59	43.12	1.45	12.12	0	0	1.72
Fire	18.46	43.43	10.59	23.79	0	0	3.72
Multiple Peril	56.80	21.92	11.96	7.12	0.07	0	2.13
Transport	28.79	58.09	0.70	11.36	0	0	1.06
Mortuary	97.79	1.49	0.03	0.69	0	0	0
Other Lines	32.29	39.22	14.31	7.22	0	0	6.97
Total	27.20	9.76	46.76	9.13	1.36	0.03	5.77

Table 10.10 *(continued)*. Market Share by Distribution System in the Spanish Insurance Industry, 1994 to 2003

	Agents	Brokers	Banks	Insurers' Offices	Telephone	Internet	Other
2001							
Life	14.78	4.93	64.50	7.64	0.02	0	8.13
Health	23.93	5.21	6.39	61.35	1.99	0.91	0.22
Accident	49.82	33.89	7.15	7.65	0	0	1.49
Auto	45.76	24.67	0.48	17.68	7.65	0.12	3.65
General Liability	36.84	52.46	1.72	7.98	0	0	1.00
Fire	14.57	51.22	11.91	19.60	0	0	2.69
Multiple Peril	56.28	23.85	12.91	5.10	0.02	0	1.84
Transport	36.94	52.12	0.33	9.63	0	0	0.98
Mortuary	97.32	1.43	0.18	1.06	0	0	0.01
Other Lines	29.40	41.59	16.35	7.92	0	0	4.74
Total	26.94	11.92	40.53	13.00	1.67	0.08	5.86
2002							
Life	13.40	9.17	66.91	6.68	75.00	0	3.09
Health	22.15	8.50	9.84	57.42	1.41	0.19	0.49
Accident	49.45	31.15	11.37	7.17	0.25	0	0.61
Auto	43.76	25.48	2.30	17.11	8.14	0.44	2.78
General Liability	27.43	61.03	3.21	7.45	0.25	0	0.63
Fire	9.00	57.86	8.40	23.14	1.11	0	0.49
Multiple Peril	50.88	26.70	16.41	4.35	0.32	0	1.35
Transport	14.02	53.67	1.17	30.49	0.12	0	0.52
Mortuary	97.22	1.59	0.25	0.92	0	0	0.02
Other Lines	23.71	48.41	13.74	11.23	0.10	0	2.81
Total	25.56	16.33	40.75	12.36	2.32	0.11	2.56
2003							
Life	16.39	6.49	66.72	7.05	0.76	0	2.60
Health	21.70	8.70	3.22	64.18	1.59	0.18	0.42
Accident	47.23	30.89	13.42	7.62	0.23	0	0.62
Auto	43.20	29.22	2.53	14.53	9.28	0.75	0.49
General Liability	28.82	61.21	3.67	5.80	0.15	0	0.36
Fire	7.61	63.27	7.76	20.14	0.76	0	0.46
Multiple Peril	47.89	28.63	18.08	3.71	0.34	0	1.35
Transport	16.86	59.28	3.02	20.26	0.12	0	0.46
Mortuary	97.08	1.62	0.31	0.96	0	0	0.02
Other Lines	25.31	42.25	25.68	5.98	0.11	0	0.66
Total	29.80	18.64	33.56	13.26	3.02	0.22	1.50

Source: Author's compilations from ICEA reports.

Trends in Distribution Systems

Insurance market shares by distribution system are presented in Table 10.10. The table shows the share of total premiums generated by each distribution system for life insurance and for each major non-life line of business. The data are compiled from annual reports about distribution systems that the Investigación Cooperativa de Entidades Aseguradoras y Fondos de Pensiones (ICEA) has been publishing since 1994. The reports slightly changed in some years with the introduction of new lines of business and distribution systems. The ICEA's definitions of certain distribution systems are slightly different than the definitions used by the DGSFP mentioned above. The definitions of agents, brokers, and banking channels are similar, but the ICEA distinguishes four more channels: the Internet; an insurer's own offices, the telephone, and other channels. The Internet channel includes the insurance sales activity performed via the Internet without using other channels. The insurer's own offices involves the distribution within insurer offices without the use of participating intermediaries, the telephone, or the Internet. The telephone distribution channel includes the sales activity performed by telephone in connection with the entity. Other channels refers to any other channels used by the entity that are not included in the previous categories (ICEA 2004). The period of analysis is 1994 to 2003.

Generally, the predominant distribution channel of every line of business is the same in every year of the period. Bancassurance is the most important distribution system of life insurance products in each year of the 10-year period. Since 1996, health insurance has typically been sold in insurers' offices. The most important distribution channel for multiple peril lines of business were agents followed by brokers and the bank channel. Auto insurance was mostly distributed by agents, but general liability, transport, and fire were principally sold by brokers. Almost all mortuary insurances were distributed by agents (more than 97 percent of total premiums in every year).

Telephone sales represent a small fraction of total premiums, but in the last years of the period of analysis it gained market share. This distribution method is more effective for products such as automobile insurance, a standardized product for which price is seen as an important factor in the buying decision. A conclusion about distribution systems based on the data presented in Table 10.10 is that the different distribution systems are becoming increasingly specialized.

Bancassurance

Spain is a prominent example of the bancassurance phenomenon. The convergence of banking and insurance came about in the late 1980s, when many banks entered the Spanish insurance market and gained significant market share. In 1986, 35 banking institutions had real presence in the Spanish insurance market, and, from 1987 to 1992, 30 of the new insurers that entered the Spanish insurance market were promoted by banking institutions. In 1986, insurers affiliated with banking institutions represented 54 percent of life premium volume and 15 percent of non-life premium volume. In 1991, insurance firms affiliated with banking institutions accounted for 68 percent of life premium volume and 14 percent of non-life premium volume (Esteban-Jodar 1993).

The convergence of banking and insurance has had two important consequences: an increase in the diversity of financial products sold by banks and the use of bank branches as an insurance distribution channel. The bank distribution channel has some significant advantages over the traditional agency channel. Selling insurance using salaried bank employees is less expensive than selling insurance through agents, because the existing bank branches can be utilized and because bank branch employees are less expensive than traditional agents. A second advantage is that banks offer a form of "one-stop shopping" for financial services. Customers who regularly visit bank branches to conduct banking business may find it convenient to shop for insurance at the same time. In addition, because banks have long marketed asset accumulation products such as savings accounts, it is natural for them to sell also life insurance products with a savings component.

However, the lack of experience in insurance underwriting in addition to the diversification discount for the expansion into unrelated products that could increase costs and reduce market value are the most remarkable disadvantages of banks as an insurance distribution channel (Cummins and Rubio-Misas 2005). According to the diversification discount hypothesis, financial conglomerate market values that engage in multiple activities would be lower than if those financial conglomerates were broken into financial intermediaries that specialize in individual activities. This diversification discount is explained by the researchers on the fact that conglomerate firms may destroy value by intensifying agency problems between corporate insiders and small shareholders, making it more difficult to design efficient managerial incentive contracts and inhibiting the efficient allocation of capital across the conglomerate. In addition to these reasons, insiders may expand the range of corporate activities for private gains at the expense of the market's valuation of the firm (Aron 1988; Stulz 1990; Jensen 1986; Jensen and Meckling 1976).

In Spain, bancassurance is the leading distribution channel for life insurance products and has been for the last ten years. In 2003, eight of the ten top life insurers in terms of premium volume were affiliated with bank groups and predominantly used bank branches to sell their products. Currently, 75 of the insurance firms operating in life insurance form the *bancaseguros* group of UNESPA.[215]

Insurers affiliated with banking institutions have experienced the most growth in the Spanish insurance market (Figueiredo-Almaça 1999). The relative efficiency of these firms in the market was investigated in a study by Cummins and Rubio-Misas (2005) about the convergence of financial services markets in Spain for the period 1989 to 1998. The study provides evidence that the most efficient firms are the Spanish savings bank subsidiaries, the Spanish nonfinancial subsidiaries, and the foreign bank subsidiaries. They are followed by the Spanish unaffiliated insurers and the Spanish bank subsidiaries. The less-efficient firms are the Spanish insurer group members and the foreign insurance subsidiaries. The research also revealed that the bank distribution model has been successful in life insurance but not in non-life insurance.

[215] Firms belonging to this group use the bank channel as the main distribution system.

Table 10.11. Ten Top Firms in the Life Insurance Segment of the Spanish Insurance Industry, 2003

Company	Net Premiums	Group	Organizational Form	Principal Distribution System	Ownership Type	Country of the Ultimate Owner
Mapfre Vida	1,571	Mapfre	Stock	Bancassurance	Insurance Group	Spain
Vida Caixa	1,448	CAIFOR	Stock	Bancassurance	Insurance-Bank Group	Spain/The Netherlands
BBVA Seguros	1,279	BBVA	Stock	Bancassurance	Bank Group	Spain
ASEVAL S.A.	1,068	AVIVA	Stock	Bancassurance	Insurance Group	United Kingdom
Bansabadell Vida	737	Banco Sabadell	Stock	Bancassurance	Bank Group	Spain
Ibercaja Vida	716	IBERCAJA	Stock	Bancassurance	Bank Group	Spain
Ascat Vida	698	Caixa Cataluña	Stock	Bancassurance	Bank Group	Spain
Banco Vitalicio de España	627	Generali	Stock	Agents	Insurance Group	Italy
La Estrella	554	Generali	Stock	Agents	Insurance Group	Italy
Mediterraneo Vida	476	CAM	Stock	Bancassurance	Bank Group	Spain

Source: DGSFP(a) *Balances y Cuentas* (2003) and ICEA, INESE, and UNESPA reports.
Note: Bank Group means bank and savings bank group. CAIFOR is a group that 50 percent belongs to La Caixa (a Spanish saving bank) and 50 percent belongs to the FORTIS group. Monetary variables are expressed in millions of euros.

Table 10.12. Ten Top Firms in the Non-Life Insurance Segment of the Spanish Insurance Industry, 2003

Company	Net Premiums (in millions of euros)	Group	Organizational Form	Principal Distribution System	Ownership Type	Country of the Ultimate Owner
Mapfre Mutualidad	1,862	Mapfre	Mutual	Multiple	Insurance Group	Spain
Allianz	1,367	Allianz	Stock	Bancassurance	Insurance Group	Germany
Axa Aurora Iberica	1,086	Axa Aurora	Stock	Multiple	Insurance Group	France
Winterthur	908	Credit Suisse	Stock	Multiple	Bank Group	Switzerland
Mutua Madrileña Automovilística	761		Mutual	Telephone	Unaffiliated Company	Spain
Zurich España	741	Zurich	Stock	Agents	Insurance Group	Switzerland
Adeslas	690	Agbar/Méderic	Stock	Agents	Industrial-Insurance Group	Spain/France
Santa Lucia	619	Santa Lucia	Stock	Agents	Insurance Group	Spain
Asisa[a]	590	Lavinia	Stock	Mail	Cooperative	Spain
Groupama Plus Ultra	580	Groupama	Stock	Bancassurance	Insurance Group	France

Source: DGSFP(a) *Balances y Cuentas* (2003) and ICEA, INESE, UNESPA reports.
[a] Asisa is a stock company belonging to a cooperative of physicians named Lavinia.

10.3.9 The Leading Firms

The ten top firms in the life insurance segment in terms of net premiums in 2003 are shown in Table 10.11. The firms are ranked in order of their premium revenues. All of these firms are stock companies. Eight of the 10 firms used the bank distribution channel, 7 of which were subsidiaries of a bank or savings bank group and 1 of which was a subsidiary of a Spanish insurance group. The other two firms used agents as the predominant distribution system, and they were subsidiaries of a foreign insurance group. Spain was the home country of 7 out of 10 firms, Italy claimed 2 out of 10, and the United Kingdom 1 out of 10.

The leading firms in the non-life insurance segment of the Spanish insurance industry for 2003 are presented in Table 10.12. These firms are ranked in terms of their non-life premium revenues. Eight of these 10 firms were organized as stock firms, and two were organized as mutual firms. One of the mutuals—Mapfre Mutualidad—is the leading firm of the market; it is the parent company of the Mapfre group, which is the most important Spanish insurance group. Mapfre Vida, which was the leading firm in the life insurance market in 2003, also belongs to the Mapfre group. There was less homogeneity in the non-life insurance segment in terms of distribution systems, ownership types, and country of the main owner. Three of the 10 firms used agents as the predominant distribution channel, two firms used the banking distribution channel, one used telephone, another used mail, and the remaining three used several channels and none of them was predominant. Foreign insurers have an important presence in the leading firms of the non-life insurance segment, increasing competition in Spain.

10.4 TECHNICAL RESULTS, PROFITS, AND FINANCIAL CONDITIONS

10.4.1 Pro-forma Balance Sheet and Income Statement

Table 10.13 shows a pro-forma balance sheet for the Spanish insurance industry in 2003, breaking out assets and liabilities into several main categories. In addition to data for the total Spanish insurance industry, the table shows data for non-life specialists, life specialists, and diversified firms that offer both life insurance and non-life insurance.

The Spanish insurance industry invested almost half of its assets in bonds (48.6 percent). Financial investment in affiliated companies accounted for 6.9 percent of total industry assets and was followed by receivables (which accounted for 6.6 percent of total industry assets), unit-linked investment (5.4 percent), real estate (2.8 percent), and corporate stocks (1.8 percent).

The invested assets of Spanish insurers are financed principally by reserves. Life reserves accounted for 60.5 percent of total industry assets, and total reserves amounted to 83.3 percent of industry assets. Capital and earned surplus accounted for 9.1 percent of total industry assets in 2003.

Table 10.13. Pro-forma Balance Sheet of the Spanish Insurance Industry, 2003

	Total Industry		Diversified Firms		Non-life Specialists		Life Specialists	
	Amount	Percent of Total Assets	Amount	Percent of Total Assets	Amount	Percent of Total Assets	Amount	Percent of Total Assets
Assets								
Real Estate	4,574	2.8%	3,176	2.8%	1,009	6.0%	389	1.1%
Financial Investments in Affiliated Companies	11,331	6.9%	7,789	7.0%	1,307	7.8%	2,236	6.1%
Corporate Stocks	3,042	1.8%	1,817	1.6%	362	2.2%	862	2.4%
Bonds	80,143	48.6%	53,868	48.2%	5,465	32.7%	20,811	57.0%
Other Financial Investments	28,005	17.0%	20,895	18.7%	1,777	10.6%	5,333	14.6%
Unit-Linked Investments	8,933	5.4%	5,888	5.3%	0	0%	3,045	8.3%
Reinsurance Reserves	4,736	2.9%	2,523	2.3%	2,111	12.6%	102	0.3%
Receivables	10,891	6.6%	7,387	6.6%	2,309	13.8%	1,196	3.3%
Other	13,251	8.0%	8,369	7.5%	2,353	14.1%	2,527	6.9%
Total Assets	164,906	100%	111,712	100%	16,693	100%	36,501	100.0%
Liabilities								
Capital and Earned Surplus	15,019	9.1%	8,753	7.8%	3,410	20.4%	2,856	7.8%
Unearned Premium Reserves	10,226	6.2%	6,322	5.7%	3,904	23.4%	0	0.0%
Life Reserves[a]	99,782	60.5%	71,409	63.9%	0	0.0%	28,374	77.7%
Loss Reserves	16,428	10.0%	9,639	8.6%	6,281	37.6%	508	1.4%
Other Reserves	2,009	1.2%	1,306	1.2%	564	3.4%	138	0.4%
Unit-Linked Reserves	8,950	5.4%	5,888	5.3%	0	0.0%	3,062	8.4%
Financial Debt	8,703	5.3%	6,013	5.4%	1,637	9.8%	1,053	2.9%
Other	3,789	2.3%	2,382	2.1%	897	5.4%	510	1.4%

Source: Author's calculations from DGSFP(a) *Balances y Cuentas, 2003.*
Note: Monetary variables are expressed in millions of euros.
[a] Life reserves include mathematical reserves and life loss reserves.

As for the fraction that accounts for every category of assets and liabilities with respect to total assets of every type of firm, we can see in Table 10.13 that there are not significant differences between the total industry, life insurance specialists, and diversified firms. However, there are differences between non-life specialists with respect to life specialists, diversified firms, and the total industry.

Non-life specialists invested more in real estate and receivables but less in bonds than the other types of firms and less than the industry average. On the liability side of the balance sheet, non-life specialists have higher equity capital (20.4 percent of total assets) than the other types of firms, but the weight of reserves into total assets is lower than it is for life specialists, diversified firms, and the total industry.

The pro-forma income statement for the Spanish insurance industry in 2003 is shown in Table 10.14. Technical results are also defined in this table where investment incomes and investment expenses are taken into account. Non-life technical results accounted for 73 percent of total technical results.

The intermediation function of Spanish insurers can be analyzed by using the net investment margin: the financial result that is defined as investment incomes minus investment expenses. In the life insurance segment, financial intermediation is a principal function, accomplished through the sale of asset accumulation products such as annuities. In the non-life insurance segment, intermediation is an important but incidental function resulting from the collection of premiums in advance of claim payments (Cummins and Rubio-Misas 2006). As proof of the different importance of the intermediation function in life and non-life insurance, we can see that, in 2003, in the life insurance segment the net investment margin accounted for 769 percent of life technical results, and in the non-life insurance segment the net investment margin accounted for 60.7 percent of non-life technical results.

10.4.2 Financial Condition

Currently, solvency is regulated by Insurance Law 34 of 2003 that incorporates the EU's insurance Directives 2002/13/CE and 2002/83/CE relative to solvency margin requirements for non-life insurance and life insurance, respectively. The main objective of these directives is to strengthen guaranties for policyholders by increasing solvency margin requirements.

However, the legislation did not apply to the 1999 to 2003 period under examination in this chapter. The solvency result presented in Table 10.15 was calculated according to Insurance Law 30 of 1995 that took the EU's insurance directives into account. A minimum solvency margin was calculated based on the insurer's activities, being different for non-life insurance and life insurance. In general, the minimum solvency margin for non-life insurance was the largest resulting amount between the calculation based on annual premium volume and the calculation based on average losses incurred corresponding to the last three years. However, the minimum solvency margin for life insurance was calculated by the sum of two calculations: one was based on mathematical reserves, and the other was based on capital at risk. On the other hand, the insurer's solvency margin included the insurer's net worth free of any expected commitments minus illiquid assets. The solvency margin result was obtained by comparing the firm's solvency margin with the minimum solvency margin (Lozano-Aragües 1999).

Table 10.14. Income Statement of the Spanish Insurance Industry, 2003

		Life	Non-Life
Technical Results			
+	Net Premiums	17,898	20,712
+	Invested Incomes	6,994	1,708
+	Unrealized Invested Appreciation	1,405	0
+	Other Technical Incomes	69	180
-	Net Losses Incurred	15,532	15,312
-	Net Addition to Other Reserves	6,345	136
-	Net Operating Expenses	1,007	4,105
-	Other Technical Expenses	97	389
-	Invested Expenses	1,550	544
-	Unrealized Invested Depreciation	822	0
-	Other items	305	198
=	Technical Results	708	1,916

Net Income

		Total Industry
+	Non-Life Technical Result	1,916
+	Life Technical Result	708
+	Invested Incomes*	792
-	Invested Expenses*	389
+	Other Incomes	384
-	Other Expenses	352
+	Extraordinary Incomes	92
-	Extraordinary Expenses	275
-	Tax	781
=	Net Income	2,095

Source: Author's calculations from DGSFP(a) *Balances y Cuentas.*
Note: Monetary variables are expressed in millions of euros.
* Refers to investment incomes and investment expenses from investments where the insurer's equity capital is materialized.

As mentioned, the Spanish insurance market has been characterized by a large number of firms, many of which are very small. The number of insurance firms in the Spanish market decreased 30 percent between 1980 and 1987 because of insolvency and in minor degree because of mergers and acquisitions of small firms (Esteban-Jodar 1993). The restructuring of the Spanish insurance market in the 1990s was due to mergers and acquisitions followed by firm retirements and insolvencies. In this case, mergers and acquisitions targets were larger and roughly equal in efficiency to acquiring firms, while insolvent firms were predominantly small and inefficient (Cummins and Rubio-Misas 2006).

Table 10.15 shows solvency margin means, medians, and standard deviations, both in the life insurance segment and in the non-life insurance segment, for every year of the sample period and for the whole sample period 1999 to 2003. Numbers of firms with solvency ratios of less than 1 for every year of the sample period are also included. The data used in this table are drawn from the annual regulatory accounting statement files by insurers with the DGSFP.

Table 10.15. Solvency Result in the Spanish Insurance Industry, 1999 to 2003

	Life Insurance Segment				Non-life Insurance Segment			
	Mean	Median	Standard Deviation	Number of firms with ratio<1	Mean	Median	Standard Deviation	Number of firms with ratio <
1999	47.0	3.1	347.0	3	446.6	3.1	5,973.5	5
2000	58.6	2.9	367.0	2	31.8	2.9	193.5	10
2001	48.5	2.7	368.3	0	27.9	2.7	219.8	4
2002	73.5	2.7	460.8	2	32.1	2.7	211.7	9
2003	71.4	2.4	445.1	0	24.8	2.8	181.4	2
1999–2003	59.8	2.7	398.7	7	120.5	2.8	2,707.5	30

Source: Author's calculations from DGSFP(a) *Balances y Cuentas.*
Note: Solvency result is a ratio that compares the insurer's solvency margin with the minimum solvency margin. The actual insurer's solvency margin included net worth free of any expected commitments minus illiquid assets.

Generally, the average solvency margin result is more than 1 for every year of the sample period and for the whole sample period, because the firm's solvency margin is superior to the minimum solvency margin. In the life insurance segment six firms had solvency ratios less than 1 in at least one year from 1999 to 2003, whereas in the non-life insurance segment 22 firms had solvency ratios less then 1 in at least one year from 1999 to 2003.

As mentioned in the TRLOSSP (2004), if a firm's solvency margin is inferior to the minimum solvency margin, this firm can neither distribute dividends nor increase its activity to other lines of business nor increase its operation area inside Spain or in other EU countries both in "freedom of services" or in "freedom of establishment" nor increase its commercial network. In addition to the corresponding sanction, the Spanish authority can establish rules of special control to this firm (e.g., requiring a plan to solve its financial situation). And in case the firm does not execute the requirements in the established time, the supervisory authority could even revoke the license to this firm. As mentioned, the Consorcio de Compensacion de Seguros behaves as a guarantee fund to protect the insured's interests in case of insurer bankruptcy.

10.4.3 Non-life Technical Results and Combined Ratios

Technical results to premium revenues as well as combined ratios of the most important non-life lines of business for the period 1998 to 2003 are shown in Figures 10.3 and 10.4, respectively. Combined ratio is defined as loss ratio plus expense ratio. A combined ratio above 100 indicates that the sum of losses incurred and underwriting expenses are superior to premium revenues.

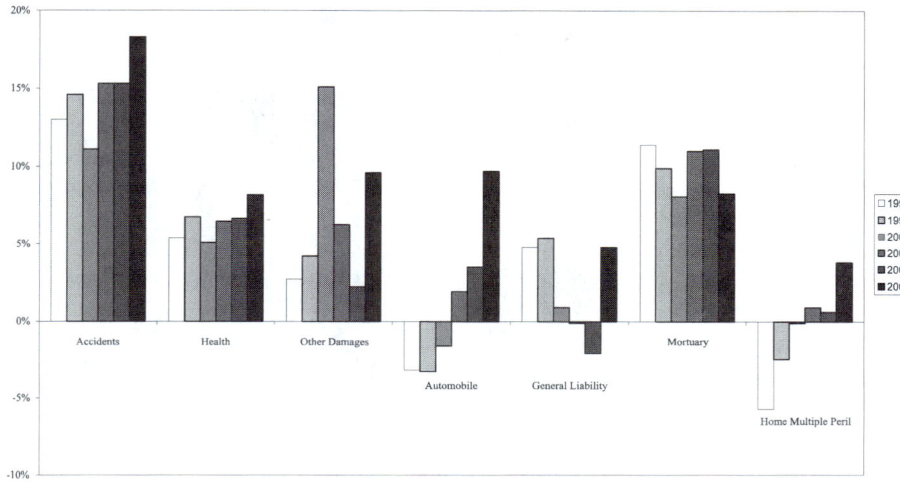

Source: Author's calculations from DGSFP(a) *Balances y Cuentas.*
Note: Firms with technical results to premiums of more than 50 percent and less than –50 percent are excluded in calculating means.

Figure 10. 3. Technical Results to Premium Revenues in Non-life Lines of Business, 1998 to 2003

In 2003, the most important non-life lines of business in terms of non-life premium revenues were auto (general liability), auto (other guarantees), health assistance, health, accidents, other damages, general liability, mortuary, and home multiple peril. The period of analysis started in 1998 because it was the first year that the statutory accounting statements allowed calculation of technical results by the different non-life lines of business, separating also investment incomes and investment expenses by lines of business. As investment incomes and investment expenses are included, the presented technical results are technical-financial results in reality. The way to calculate them is defined in Table 10.14.

Auto (general liability) and auto (other guarantees) are analyzed jointly. Auto insurance (other guarantees) covers several risks such us personal accidents, damage to the vehicle (e.g., broken windshield), burglary, or fire. In the past, the maturity and saturation of the auto insurance market created pressure to reduce prices. There was also an increase in losses, which triggered combined ratios of more than 100 and negative technical results. However, the auto insurance combined ratio decreased during the sample period due to a reduction in both losses and expenses and the proportion of technical results to premiums increased. In 2003, auto insurance technical results accounted for almost 10 percent of premium revenues.

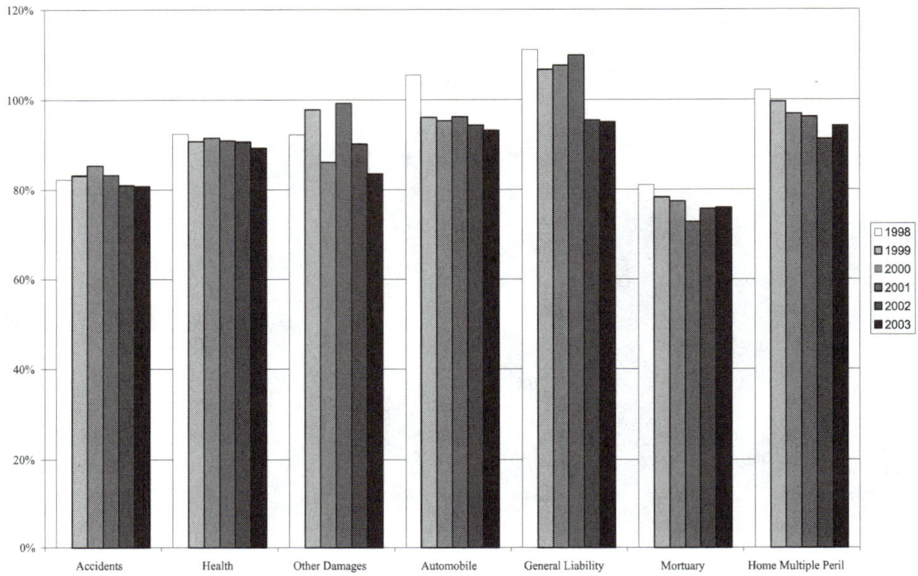

Source: Author's calculations from DGSFP(a) *Balances y Cuentas.*
Note: Firms with technical results to premiums of more than 50 percent and less than –50 percent are excluded in calculating means.

Figure 10.4. Combined Ratios in Non-life Lines of Business, 1998 to 2003

The two lines of business that comprise health insurance in Spain are analyzed together. These two lines of business are private complementary systems to the universal and obligatory public health program of the Spanish social security system. Civil servants can choose between the public health program of the social security and private health programs, but most people have the public health system and can buy complementary private health insurance. Health insurance presented a decreasing trend in combined ratios, being less than 100 in every year of the sample period. It also showed an increasing trend in technical results. In 2003, health technical results accounted for more than 8 percent of premium revenues.

Accidents insurance showed combined ratios of around 80 percent in each of the six years of the sample period and an increasing trend in technical results. Mortuary insurance presented the lowest combined ratios of the analyzed lines of business because it had the lowest loss ratio. Mortuary technical results accounted for more than 8 percent of premium revenues in 2003. Home multiple peril insurance showed a decreasing trend in combined ratios and an increasing trend in the proportion of technical results to premiums. General liability insurance presented combined ratios of more than 100 in the first years of the sample period and also negative technical results, but in 2003 the general liability combined ratio was less than 100 and technical results were positive.

Generally, there was a decreasing trend in the combined ratios of the most important non-life lines of business and an increasing trend in the proportion of technical results to premium revenues that ranged from 18.3 percent in accidents insurance to 3.8 percent in home multiple peril insurance.

10.4.4 Profitability

Return on equity for the overall Spanish insurance industry is presented in Table 10.16. Return on equity is defined as net income for the year divided by the average value of book equity at the beginning and end of the year. Table 10.16 shows return on equity means, medians, and standard deviations for every year of the sample period and for the entire sample period, 1990 to 2003.

Table 10.16. Return on Equity in the Spanish Insurance Industry, 1990 to 2003

	Mean	Median	Standard Deviation
1990	5.96%	4.80%	12.91%
1991	5.54%	5.00%	14.54%
1992	5.65%	5.63%	13.95%
1993	6.70%	6.98%	12.54%
1994	6.53%	6.56%	13.18%
1995	7.21%	7.01%	11.61%
1996	7.76%	7.00%	12.62%
1997	6.86%	7.36%	12.32%
1998	6.19%	6.15%	11.14%
1999	4.88%	5.53%	12.76%
2000	6.21%	5.58%	11.82%
2001	6.69%	6.41%	11.65%
2002	5.99%	5.38%	13.47%
2003	10.01%	9.24%	11.84%
1990 to 2003	6.52%	6.33%	12.77%

Source: Author's calculations from DGSFP(a) *Balances y Cuentas.*
Note: Firms with return on equity of more than 50 percent and less than –50 percent are excluded in calculating means, medians, and standard deviations.

The average per-firm return on equity ranged from 4.9 percent in 1999 to 10 percent in 2003. The average per-firm return on equity for the total sample period 1990 to 2003 was 6.5 percent.

An additional analysis of the return on equity for the period 1998–2003 is shown in Table 10.17. This analysis allows us to draw some inferences about differences in profitability in the Spanish insurance industry. Table 10.17 shows return on equity average per firm, classified into life insurance specialists, non-life insurance specialists and diversified firms, for every year of the sample period 1998–2003 and for the overall sample period.[216]

Generally, life insurance specialists are the most profitable firms. They are more profitable than non-life insurance specialists and more profitable than diversified firms. T-tests for differences between means show that these differences are statistically significant. However, the differences between non-life specialists and diversified firms are not statistically significant, suggesting that diversified firms are not more profitable than non-life insurance specialists. These results provide evidence that, in the Spanish insurance industry, the life insurance segment is more profitable than the non-life insurance segment.

Table 10.17. Return on Equity per Type of Firm, 1998 to 2003

Panel A

	Diversified Firms			Non-life Specialists			Life Specialists		
	Mean	Standard Deviation	n	Mean	Standard Deviation	n	Mean	Standard Deviation	n
1998	5.94%	12.02%	64	5.17%	11.11%	199	10.80%	8.86%	48
1999	3.96%	14.41%	57	4.73%	11.35%	201	6.50%	15.86%	51
2000	6.12%	12.99%	63	5.24%	11.86%	183	10.21%	9.05%	46
2001	6.76%	12.17%	58	6.13%	11.25%	185	8.87%	12.58%	45
2002	7.46%	11.28%	63	4.97%	13.61%	175	7.96%	15.56%	44
2003	10.31%	13.33%	65	9.88%	11.40%	173	10.11%	11.49%	42
1998 to 2003	6.82%	12.78%	370	5.96%	11.87%	1116	9.03%	12.56%	276

Panel B. t-tests for Differences Between Means

	Diversified Firms v. Non-life Specialists	Diversified Firms v. Life Specialists	Non-life Specialists v. Life Specialists
1998 to 2003	1.1445	–2.199**	–3.682***

Source: Author's calculations from the DGSFP(a) *Balances y Cuentas.*
Note: Firms with return on equity of more than 50 percent and less than –50 percent are excluded in calculating means and standard deviations.
** Significant at the 5 percent level.
*** Significant at the 1 percent level.

Results also provide some support for the *strategic focus hypothesis* in contrast to the *conglomeration hypothesis*. The *strategic focus hypothesis* posits that firms can maximize value by focusing on core business and core competencies. According to the *conglomeration hypothesis*, owning and operating a broad range of business is hypothesized to add value due to advantages of sharing inputs in joint productions

[216] A study by Fuentes, Grifell-Tatjé, and Perelman (2001) on Spanish diversified insurance firms finds low rates of productivity growth during the period 1987 to 1994.

(Berger et al. 2000). However, additional research would be needed to provide a more complete analysis of these two hypothesis in the Spanish market.

10.5 CONCLUSION

This chapter presents a study of the structure, conduct, and performance of the Spanish insurance industry. It begins with a discussion of the history of insurance in Spain in the context of its regulation, including information on life insurance and taxation. We analyze the number and types of firms, the distribution of premiums, life insurance managed savings, total assets in the industry, international activities, market concentration, organizational structure, distribution systems, leading firms, technical results, combined ratios, profits, and financial conditions of the Spanish insurance industry. The sample period 1990 to 2003 spans the introduction of the European Union's Third Generation Directives, which effectively deregulated the EU insurance markets with the exception of solvency regulation.

Most firms in the industry are organized as stock firms. In 2003, there were 240 stocks but only 46 mutuals. The industry consists of life insurance specialists, non-life insurance specialists, and diversified firms that offer both life and non-life insurance. Diversified firms are the largest firms in the market in terms of total assets.

In the non-life insurance segment, the most important lines of business in terms of total premiums are auto, multiple peril, and health insurance. The average annual growth rates, expressed in nominal monetary units, in total non-life premiums and in total life premiums were 9.2 percent and 13.8 percent, respectively.

Foreign insurers have an active presence in the Spanish market. In 2003, the percentage of foreign capital over the total capital of Spanish-licensed insurers was 29.4, most of which belonged to other EU countries. There were also 41 branches of foreign firms, and 380 EU insurers were habilitated to operate in the Spanish market in a regime of freedom to provide services. The international movements of Spanish insurers in the European Union were basically addressed to the Portuguese insurance market.

The Spanish insurance industry experienced a dramatic restructuring during the sample period. The number of firms in the industry fell dramatically due to firm retirements, insolvencies, mergers, and acquisitions. However, 111 firms entered the market during the sample period. Between 1990 and 2003, total industry assets increased by 489 percent and average firm size increased by 920.5 percent. Concentration generally increased in the non-life insurance segment of the Spanish insurance industry. In contrast, concentration trends in the life insurance segment were not as clear as they were in the non-life insurance segment, due to important increases in single premiums in some years.

The analysis of market shares by organizational form in 2003 shows that mutuals are most successful in automobile insurance and stocks are most successful in life insurance, providing support for the efficiency structure hypothesis that predicts that the market will sort organizational forms into market segments where they have comparative advantages (Cummins, Rubio-Misas, and Zi 2004).

Bancassurance is the most important distribution channel of life insurance products, but in the non-life insurance segment traditional channels, such as agents and brokers, are predominant. In 2003, the leading firms in the life insurance segment were stocks, which used the bank distribution channel and were mainly subsidiaries of a bank or savings bank group. However, there was not as much homogeneity in the non-life insurance segment as there was in the life insurance segment. Foreign firms had an important presence in the leading firms of the non-life insurance segment.

The most important non-life lines of business experienced a decreasing trend in combined ratios as well as an increasing trend in the proportion of technical results to premiums. The average return on equity per firm for the total sample period was 6.5 percent. Life insurance specialists are the most profitable firms in the Spanish insurance industry.

10.6 REFERENCES

Aron, Debra, 1988, "Ability, Moral Hazard, Firm Size, and Diversification," *Rand Journal of Economics* 19: 72–87.

Benitez de Lugo, Félix, 1930, *Legislación y Técnica de los Seguros Privados* (Madrid: Imprenta Elzeviriana y Librería Camí).

Berger, Allen N., J. David Cummins, Mary A. Weiss, and Hongmin Zi, 2000, "Conglomeration versus Strategic Focus: Evidence from the Insurance Industry," *Journal of Financial Intermediation* 9: 323–362.

Consorcio de Compensación de Seguros, 2005, *Report 2004* (Madrid: Ministerio de Economía y Hacienda).

Cummins, J. David, and María Rubio-Misas, 2005, "Convergence of Financial Services Markets. Evidence from the Spanish Insurance Industry," working paper, Wharton School, University of Pennsylvania, Philadelphia.

Cummins, J. David, and María Rubio-Misas, 2006, "Deregulation, Consolidation and Efficiency. Evidence from the Spanish Insurance Industry," *Journal of Money, Credit and Banking* 38: 232–355.

Cummins, J. David, María Rubio-Misas, and Hongmin Zi, 2004, "The Effect of Organizational Structure on Efficiency. Evidence from the Spanish Insurance Industry," *Journal of Banking and Finance* 28: 3113–3150.

Cummins J. David, and Mary Weiss, 1991, "The Structure, Conduct and Regulation of the Property-Liability Insurance Industry," in Richard W. Kopcke and Richard E. Randall, eds., *The Financial Condition and Regulation of Insurance Companies* (Boston: Federal Reserve Bank of Boston).

Del Valle-Esteve, Silvia, 2002, "La exteriorización de los compromisos por pensiones con los trabajadores y el personal pasivo," *Perspectivas del Sistema Financiero* 74: 37–52.

Dirección General de Seguros y Fondos de Pensiones (a), 1990–2003, *Balances y Cuentas* (Madrid: Ministerio de Economía y Hacienda).

Dirección General de Seguros y Fondos de Pensiones (b), 1992–2004, *Seguros y Fondos de Pensiones. Informe* (Madrid: Ministerio de Economía y Hacienda).

Esteban-Jodar, Lorenzo, 1993, "El Mercado Español de Seguros," *Información Comercial Española* 715: 15–44.

Figueiredo-Almaça, José Antonio, 1999, *El Mercado Ibérico de Seguros* (Madrid: Ed. Mapfre).

Fuentes, Hugo, Emili Grifell-Tatjé, and Sergio Perelman, 2001, "A Parametric Distance Function Approach for Malmquist Productivity Index Estimation," *Journal of Productivity Analysis* 15: 79–94.

Frax-Rosales, Esperanza and Mª Jesús Matilla-Quiza, 1996, "Los Seguros en España: 1830–1934," *Revista de Historia Económica* 1: 183–203.

González de Frutos, Pilar, 2002, "Los seguros de vida. Una pequeña historia," *Perspectivas del Sistema Financiero* 74: 1–12.

Instituto de Estudios Financieros y de Seguros, 1990–2003, "Ranking de primas Totales de los Principales Grupos Aseguradores," *Actualidad Aseguradora.*

Investigación Cooperativa de Entidades Aseguradoras y Fondos de Pensiones, 1994–2003, *Canales de Distribución, Estadística* (Madrid: ICEA).

Jensen, Michael C., 1986, "Agency Costs of Free Cash Flow, Corporate Finance, and Takeovers," *American Economic Review* 76: 323–329.

Jensen, Michael C., and William Meckling, 1976, "Theory of the Firm: Managerial Behavior, Agency Costs and Ownership Structure," *Journal of Financial Economics* 3: 305–360.

Ley del Contrato del Seguro, 1980, Boletín Oficial del Estado, October 17, 1980.

Lozano-Aragües, Ricardo, 1999, *Análisis Práctico de la Normativa Patrimonial de las Entidades Aseguradoras* (Madrid: Centro de Estudios del Seguro).

Maldonado-Molina, Francisco Javier, 2001, *Las Mutualidades de Previsión Social como Entidades Aseguradoras* (Granada: Ed. Comares).

Martínez-Martínez, Miguel A., 1994, *Organización y Estrategia de la Empresa Aseguradora Española* (Madrid: Ed. Mapfre).

Martínez, Miguel A., Irene Albarrán, and David Camino, 2001, "Executive Management in Insurance Entities: A Comparative Study of Mutual Companies and Joint-Stock Companies in Spain with a View to the 21st Century Market," *Geneva Paper on Risk and Insurance* 26: 206–223.

Mayers, David, and Clifford W. Smith, Jr., 1988, "Ownership Structure Across Lines of Property-Casualty Insurance," *Journal of Law and Economics* 31: 351–378.

Pons-Pons, Jeronia, 2003, "Diversificación y cartelización en el seguro español, 1914–1935," *Revista de Historia Económica* 3: 567–592.

Rubio-Misas, Maria, 2004, "Productividad y Eficiencia de las Mutualidades de Previsión Social," working paper, Universidad de Málaga, Málaga, Spain.

Sanz-Arnal, Ernesto, 1996, "Sistemas de Previsión Social Complementaría," *Perspectivas del Sistema Financiero* 56: 53–65.

Sanz-Arnal, Ernesto, 2002, "El seguro de vida como instrumento de captación de ahorro para las entidades financieras," *Perspectivas del Sistema Financiero* 74: 27–35.

Stulz, Rene, 1990, "Managerial Discretion and Optimal Financial Policies," *Journal of Financial Economics* 26: 3–27.

Swiss Re, 1996, "Deregulation and Liberalization of Market Access: The European Insurance Industry on the Threshold of a New Era in Competition," *Sigma* 7.

Texto Refundido de la Ley de Ordenación y Supervisión de los Seguros Privados, 2004, approved by Royal Decree 6/2004 of October 29 (Boletín Oficial del Estado 267, November 5, 2004).

Tirado-Suárez, Francisco Javier, 1984, *Ley ordenadora del seguro privado- exposición y crítica* (Sevilla: Servicio de Publicaciones de la Universidad de Sevilla).

Unión Española de Entidades Aseguradoras y Reaseguradotas (a), 2001–2003, *Life Insurance in Spain,* (Madrid: UNESPA).

Unión Española de Entidades Aseguradoras y Reaseguradotas (b), 2005, *Agrupación Bancaseguros,* (Madrid: UNESPA).

10.7 LEXICON

Accidents insurance: Provides compensation benefits in case of accidents that occurred by a cause that triggered the insured's death or disability as a consequence of activities established in the policy. The cause must be violent, sudden, external, and nondeliberated by the insured (Ley del Contrato del Seguro, 1980, article 100).

Assistance insurance: Provides coverage to people who have difficulties during residence relocation or absence. It also covers assistance to people who have difficulties in other legally established circumstances as long as they are not covered by other types of insurance.

BOE: Boletín Oficial del Estado. Official state bulletin of the Spanish goverment.

Caution insurance: Fidelity guarantee insurance through which the insurer must compensate the insured in case the policyholder does not perform its legal or contractual obligations. Every payment made by the insurance firm must be reimbursed by the policyholder (Ley del Contrato del Seguro, 1980, article 68).

CCS. Consorcio de Compensación de Seguros, www.consorseguros.es.

Commerce multiple peril insurance: Covers several risks corresponding to commercial establishment, offices, bureaus, and medical offices.

Community multiple peril insurance: Covers several risks corresponding to building landlord community.

Cooperative: A society created by an association of people who are joined in a freedom regime of voluntary adhesion and voluntary resignation. The cooperative partners are associated to perform business activities to satisfy their needs or their economic and social ambitions (e.g., work, consumption, credit, insurance). This society has a democratic structure and a democratic functioning, according to the International Cooperative Alliance principles.

Credit insurance: Covers the risk when the insured's debtors are insolvent. Through this insurance, the insurer must pay indemnity to the insurer for losses as a consequence of their debtors' insolvency (Ley del Contrato del Seguro, 1980, article 69).

DGSFP: Dirección General de Seguros y Fondos de Pensiones. Spanish supervisory authority on insurance and pension funds, www.dgsfp.mineco.es.

EU: European Union.

Home multiple peril insurance: Covers several risks corresponding to every type of housing.

ICEA: Investigación Cooperativa de Entidades Aseguradoras y Fondos de Pensiones, www.icea.es.

INESE: Instituto de Estudios Financieros y del Seguro, www.inese.es.

IRPF: Impuesto sobre la Renta de las Personas Físicas. Spanish personal income tax.

Labor accident insurance: Provides compensation benefit in case of bodily injuries suffered by salaried personnel as a consequence of their job. In addition to contributions to the social security, employers must contribute to the obligatory labor accident insurance that is managed by management entities belonging to the Spanish social security or by labor accident mutuals (*mutuas de accidentes de trabajo y enfermedades profesionales*), which are supervised by the Ministerio de Trabajo y Seguridad Social.

Mortuary insurance is a policy for funeral expenses that the Spanish legislation does not include in life insurance policies.

Social benefit institutions: Nonprofit, private mutual insurers that provide complementary coverage to social security schemes. They cover risks of death, widowhood, orphanhood, and retirement and can perform benefits for marriage, maternity, children, and funerals. They can operate in work-related accidents and disability, health, legal defense, and assistance and assure public housing, machinery and other instruments belonging to mutualists who are small business people, and crops that are directly cultivated by farmers. They usually provide insurance to homogeneous groups (e.g., lawyers, doctors)

Solvency margin result: Compares the insurer's solvency margin with the minimum solvency margin, which must be maintained according to the solvency rules.

TRLOSSP: Texto Refundido de la Ley de Ordenación y Supervisión de los Seguros Privados.

UNESPA: Unión Española de Entidades Aseguradoras y Reaseguradoras. Spanish association of insurance entities, www.unespa.es.

11 The Insurance Market in the Republic of Ireland

Brian C. Greenford
Martin M. Mullins
John Garvey
Louise Morris
Liam O'Meara

University of Limerick

11.1 INTRODUCTION

In 1973, Ireland joined the European Economic Community (later known as the European Community), the European Coal and Steel Community, and European Atomic Energy Community. In 1993, these three communities were incorporated into the European Union (E.U.). When Ireland entered these organizations, it was a comparatively poor state with an economy based on agriculture. A small, but growing state, Ireland is now one of the wealthiest states in the E.U., having benefited greatly from its membership. Over the past ten years, the population of Ireland has increased from 3.5 million to over 4 million. This is due largely to Ireland's phenomenal economic growth and net emigration to which, in turn, the continuously changing insurance industry has contributed a great deal.

In this chapter, we first place the insurance industry in context of Ireland's political and economic background by discussing Ireland's history, culture, and relationships with other nations. From its early dominance by Great Britain, through its Independence in 1932, Ireland has become a thriving Republic situated off the coast of Great Britain and Europe. Ireland has, on the whole, been strongly supportive of the E.U. and is a member of the Eurozone, having complied with all economic requirements. Additionally, Ireland has a strong historical and cultural link with the United States (U.S.) and its geographical position provides a bridge between the U.S. and Europe. In the past, Ireland has been used as a means of entering Europe by many businesses in the U.S. including the financial services industry. The growing industrialization of Ireland requires the strong insurance industry to provide financial protection in the event of an unexpected peril. This market is governed by European Directives and regulated by Irish law. Dramatic changes in regulation are occurring and a brief consideration is given to this in the first section of the chapter.

The non-life insurance market consists primarily of the property, liability, and motor insurance markets and in the past five years has been facing several problems. Changing weather patterns have led to severe flooding thus affecting the property market adversely. The central government is still considering this problem. Like many other countries, Ireland also faces a crisis in dealing with personal injury compensation following an accident. Both liability and motor insurance premiums are the costliest in Europe, mainly due to the high cost of compensation (Motor Insurance Advisory Report 2001). To fight this problem, the Irish government implemented an innovative scheme to reduce the cost of personal injury compensation in 2004.

The life insurance market is also thriving and is very innovative. This market is expanding and at least one large Irish company is now an international player. The growth of bancassurance in Ireland has made an impact on the market, but the stated opposition to banks being involved in insurance by the new regulatory authority, the Irish Financial Services Regulatory Authority (IFSRA) may well shape the Irish life market.

Like many other countries, Ireland is facing the aging of its population. However, this has come somewhat later than in other countries because of Ireland's history of emigration. The population of Ireland is on average younger than elsewhere in Europe, but this is changing with the return of émigrés as the country becomes more economically successful. Provisions are now being made for pensions by various means and a very active debate is taking place on how the country should finance its senior citizens.

The Irish health insurance market is unusual that the government insists on community rating. Community rating is the practice of assigning premiums without regards to health risk or health status. Generally, this has been a disincentive to private health insurers to enter the market. However, a new company has just commenced business, thus increasing the number of health insurers from two to three. This market has a great deal of potential for an innovative health insurer and this possibility is considered below.

Ireland, although small, has put into practice a number of pioneering schemes for the insurance market within the European context. It will be extremely interesting to see how the market develops as these changes occur. With the growth of the E.U. and the attempts to integrate the European market, further changes will occur and Ireland is suitably placed as a case study of how the changes in Europe affect insurance.

11.2 POLITICAL AND ECONOMIC BACKGROUND

Few societies have experienced such a radical change as the Irish Republic since the 1960s. From being a conservative, rural, and relatively poor country, Ireland has been transformed into one of the most dynamic economies in Europe. Such impressive economic results gave the country the nickname "Celtic Tiger." In terms of per capita income, Ireland has become one the richest societies in Europe. For much of the twentieth century, Irish per capita income grew at a relatively modest 1.8%. In the early 1990s, economic growth accelerated and as a result Ireland has, in terms of economic growth, outperformed its European neighbors for a decade. This exceptional performance is best explained by a confluence of casual factors. These factors include membership of the E.U. and the accompanying structural funds, inward investment from the U.S., and increased labor inputs.[217] Fiscal policies have also played an important role with Ireland's low corporate tax regime proving to be attractive to investors.

Ireland gained its independence from Great Britain in 1922 after a brutal war, but six counties in the northeast of the island remained part of the United Kingdom (U.K.). Upon independence, the country fell into an extremely bitter civil war. However, by 1926, the 'the crisis of legitimacy' was permanently resolved. In terms of the post-colonial experience, the Irish case is unusual in that stability was quickly established and that continuity rather than change was the defining characteristic of Ireland following independence (Breen, et al. 1990). In its first three decades, the Irish State was marked by deep conservatism and state activity was kept to a minimum. This reflected the philosophy enshrined in the 1931 Papal Encyclical *Quadragesima Anno* and was indicative of the influence the Catholic Church enjoyed over the Irish government at that time. This belief system frowned upon state involvement in the social and economic lives of citizens.

By the late 1950s, persistently low living standards, employment stagnation, and mass emigration had pushed the political elite in Ireland towards a more interventionist approach in the economy. The policy was designed to attract foreign capital to invest in export-oriented enterprises and remained a central element of the state's activities throughout the remainder of the century. Integration into the world economy was also a priority from the 1950's and onwards and accordingly, Ireland finally began to diversify its economic activities. One of the consequences of this process was that Ireland began to disengage from its heavy reliance on Great Britain. From the late 1950s, the state was to take a more active role in the Irish economy. Social expenditures rose during this period, reflecting a more interventionist stance taken by the state. In 1958, social spending stood at 13% of gross national product (GNP), by 1968 this had risen to 18%. The scope of state activity and the share of national income consumed by the state continued to rise during the 1970s and early 1980s and by 1982 it had reached 64% of GNP (Breen, et al. 1990). During the 1980s, the dominant economic paradigm changed and state enterprises ceased to be viewed as the main vehicle to promote economic growth. Ireland would now be dependent on developments in the wider world economy and the state would facilitate the integration of Ireland into this global economy. That said, an unemployment rate of 18% and the rising cost of servicing the national debt represented a substantial commitment on the part of the state.

Throughout the 1970s and 1980s the Irish economy did not perform well. Unemployment remained stubbornly high during this period, prompting a wave of emigration. The traditional locations for these emigrants were the U.K. and the U.S. High unemployment and strong emigration patterns translated into voter dissatisfaction and political instability. The fraught political climate complicated efforts to gain control over public spending and the burgeoning national debt. The ratio of national debt to GNP, which in 2004 stood at just over 31%, peaked at

---

[217] Structural funds are funds paid to developing regions within the E.U. to facilitate their economic growth.

around 130% in the mid-1980s (Bradley 1999, p. 55). It fell to a minority Fianna Fail government under Charles Haughey to tackle the crisis.[218] One characteristic of post-independence Ireland has been the absence of strong ideological differentiation between the main political parties (Foster 1988, p. 575). This has ensured a degree of continuity in terms of economic policy. A combination of social partnership and cuts in public spending stabilized the tenuous economic situation. Certainly the absence of a strong force on the left of the political spectrum has tended to mitigate against abrupt changes in the policies of the state towards the business community.

Since the late 1980s, the policies adopted by successive Irish administrations had also had a positive impact on the macro-economic climate. The inclusion of the rightist Progressive Democrats in the ruling coalition has tended to lock-in neo-liberal economic policies of low taxation and low social spending. During these years the Progressive Democrats retained a high degree of influence within the ruling coalition and remained insistent upon low social spending and a low tax regime. The economic boom of the 1990s was spectacular, with Ireland having one of the highest growth rates in the world at over 8% in real terms per year. During this decade, the real level of GDP almost doubled. At the same time, migratory patterns were reversed, and Ireland had highest net emigration rate in the E.U. (DeBoer-Ashworth 2004).

This success story extends to software development, high-tech engineering, and pharmaceuticals, all of which are substantial exporters. Ireland has a high level of external trade with the combined share exports and imports to GDP of just under 100%. The Organisation for Economic Co-operation and Development (OECD 2004) has found that in terms of GDP per capita, based on current purchasing power parities, Ireland is ranked fourth in the world with an average income per capita of U.S. $32,000. Only the U.S., Norway, and Luxemburg achieved a higher ranking. However, it is instructive to note that World Bank numbers for GNP per capita place Ireland 20th in the world (World Bank Group 2004). The reason for the discrepancy is because GNP adjusts for repatriation of funds made by multi-nationals which would normally be included in GDP. It can be argued that the Irish economy is strongly supported by multi-nationals and this is very evident in the insurance markets. Consequently, there is a large discrepancy between GDP per capita and GNP per capita.

11.2.1 Internationalization and Insurance

As part of its policy of promoting inward investment, successive Irish governments have encouraged the financial sector to establish a strong presence in Ireland. The IFSC, in the city of Dublin, is part of this strategy. In total, there are almost 450 financial institutions operating from Dublin with a further 700 managed entities operating under the IFSC umbrella. In 2002, the financial and insurance sectors amounted to 22.1% of total exports. In terms of the finance/insurance component of total exports, Ireland ranked fourth among the OECD countries (OECD 2004). It is

---

[218] The dominant Irish political party which refers to itself as the Irish Republican party. It is difficult to define in terms of right or left leanings but can be defined as a populist party.

striking that, in percentage terms, exports of financial services from Ireland are of the same order as that of the exports from the U.K.

The international insurance industry has responded positively to the IFSC. Dublin is now host to more than half the world's top 20 insurance companies and there are 186 captive insurance/reinsurance operations trading in the IFSC. Ireland has key advantages over rival offshore centers. First, the E.U.'s third directive in insurance (1992) introduced the 'single passport' scheme to the single market. This allows those insurance companies based in the IFSC to sell into another E.U. member states without having to establish a local presence in that market. Second, the fact that Ireland is the only English speaking, common law jurisdiction in the Eurozone has proved to be attractive. The U.K., although a member of the E.U., has so far resisted becoming a member of the Eurozone. Thirdly, the decision of more than half of the world's top twenty insurance companies to have a presence in Dublin is also based upon the tax regime. Ireland not only has a very low corporate tax rate of 12.5%, but there are also withholding tax exceptions on most interest and dividend payments. Investment income to life insurance policy holders is also free of tax and there are no insurance or reinsurance taxes or levies. Lastly, specific to the establishment of captive insurance companies in Dublin, Ireland is the only E.U. jurisdiction apart from Luxemburg that permits the operation of captives. Consequently, Ireland is a popular destination for international insurance companies, particularly from the U.S., seeking to sell into the Eurozone economy.

Table 11.1 provides evidence of the strength of the IFSC in its endeavors to sell insurance outside of Ireland. While gross risk premium income as a whole for life insurance fell by 1.9% from 2002 to 2003, the Irish risk business decreased by 5.6% and foreign risk business increased by 3.1%. Alternatively, gross risk premium income from 2002 to 2003 for non-life insurance increased by 14.3 %. In this market, the Irish risk business increased by 6.5% and foreign risk business increased by 28.2%.

11.2.2 Regulation

For some years now insurance has been high on the political agenda in Ireland. Insurance premium costs for Irish businesses rose by 53% between 2001 and 2002 and jumped again by 33% between 2002 and 2003 (Enterprise Strategy Group 2004). This has been mainly due to the poor experience in the liability and motor insurance markets, which is discussed later.

The nature of the regulatory regime is the key to understanding the influx of insurance companies into Ireland. IFSRA is the primary public body for insurance regulation. Insurers had long called for a single regulatory authority and in 1998, the government finally decided to centralize the solvency and consumer protection supervision. The Irish Central Bank originally oversaw the supervision of banks, insurers, and investment intermediaries. On May 1, 2003 IFSRA was formed, under the auspices of the Central Bank, and took over this role. In terms of the regulation of solvency and authorization, the rules are largely uniform across the E.U. However, there are variations between consumer protection laws in Ireland and other European countries.

Table 11.1. The Value of Life and Non-Life Premiums by Irish and Foreign Transactions by Insurers Registered in Ireland, 2002–03

	2003		2002	
	Non-Life	Life	Non-Life	Life
Total Gross Premium Income (€ Thousands)	7,354	14,516	6,434	14,790
Irish Risks (€ Thousands)	4,388	7,962	4,121	8,433
Foreign Risks (€ Thousands)	2,966	6,554	2,313	6,357

Source: Irish Financial Services Regulatory Authority (2004), p. 96.
Note: The foreign risk business figure reflects the level of earned premiums, gross of reinsurance, underwritten outside Ireland on a freedom of services basis.

A voluntary insurance ombudsman scheme was established by the insurance industry in Ireland in 1992. Set up by the Irish Insurance Federation (IIF), a body formed by the Irish insurance industry which consists of approximately 90% of the insurance market, the insurance ombudsman scheme only deals with individuals in their private capacity, thus excluding business insurance. There are also limits on the types of complaints that the Ombudsman can oversee. Aside from the provision of information, the principal function of this body is to mediate disputes between policyholders and clients. Insurance companies that are members of the insurance ombudsman scheme agree to be bound by the decisions of the Ombudsman. However, complainants are not bound, nor do they prejudice their legal rights in the Irish court system. IFSRA set up its own insurance ombudsman scheme in 2005, under the auspices of a financial services ombudsman, covering all aspects of insurance. How the existing ombudsman scheme will tie in with the new financial services ombudsman is unclear. More recent insurance legislation is derived from E.U. framework documents and directives. Indeed, the E.U. Commission is now the primary regulator, with IFSRA functioning as the authorized local regulator. The E.U. has had a central role in shaping the current Irish insurance industry.

The aim of the E.U. has been to establish a functioning single market for the provision of financial services, including the insurance industry. In the case of the insurance sector, the goal is to remove obstacles to cross-border trade in order to foster a more competitive market. Skipper and Klein (2000) argue that diverse national regulations and national practices impede new entrants and act as a hindrance to competitive markets. The commission has sought to create a level playing field in the area of insurance across the E.U. As part of this attempt, the authorities have had to tackle the interpretation of the "general good" principle contained within European law. This principle allowed national governments a derogation with regard to the implementation of certain legal instruments in cases where such an implementation would be contrary to the "general good." The Commission has come to realize that uncertainty surrounds the interpretation of European law and, in particular, the relationship between the basic concepts of freedom to provide services and the principle of the "general good" (European Commission 2000). Unfortunately, the application of this principle was not as lawmakers had envisaged, and it has become a legal minefield slowing progress towards a free market. Despite the European Commission's long-held ambition to create a single market in insurance services, a fully-integrated insurance market still does not exist.

The Financial Services Action Plan instituted by the European Union will lend further impetus to the policy of creating a single market in insurance and other financial services (Bolkestien 2004). Moreover, the plan seeks to strengthen confidence in the Europe-wide insurance industry by introducing a new regulatory framework for solvency requirements. A second set of regulations will be established under the auspices of Solvency II. The focus of Solvency II will be to ensure that the reserves of insurance companies more closely reflect the real risks of entities operating in a competitive market. The Appendix contains a list of the main directives, legislation, and regulations, which cover the insurance industry in Ireland. Now that consideration has been given to the context in which the insurance industry is operating, the two major insurance markets, non-life and life are considered.

11.3 THE NON-LIFE INSURANCE MARKET

The major classes of insurance in Ireland in the non-life insurance market are property, liability, and motor. These accounted for 95% of the non-life earned premium income in 2003. The total earned premium income for non-life insurance amounts to over 3 billion euro. Table 11.2 shows how the non-life insurance market has grown between 1990 and 2003. During the 1990s, there was a steady growth in earned premium income of 6% or 7% until 1998 when this increased to 12%. This is consistent with increases in the levels of claims incurred. In 2001 and 2002, there were substantial increases of 20% and 40% respectively as a result of increasing claims costs. This caused a furor in Ireland and the cost of insurance became a prominent political agenda item. The government instructed various bodies to carry out investigations into the causes of these increases. In 2001, the Alliance for Insurance Reform (AIR), a body of concerned business people, was formed to place increasing pressure on the government to reduce the cost of claims. Their main concern was with the high cost of compensation. Once the government had made reforms to increase the efficiency of the market the AIR was dissolved.

The largest market measured in terms of premium income is motor followed by property and liability. As the motor and liability markets are facing similar problems in Ireland these are discussed together and the property market is considered first. There are other small markets, such as credit risk and personal accident classes of insurance in Ireland, but these only make up 5% of the market and therefore are not considered.

Table 11.2. Non-Life Insurance Earned Premium Income and Incurred Claims, 1990–2003

Year	Earned Premium Income (€ Thousands)	Claims Incurred (€ Thousands)	Change in Premiums %	Change in Claims %	Ratio of Claims to Premiums %
1990	955,439	953,588			99.81
1991	1,075,882	1,010,788	12.61	0.06	93.95
1992	1,146,467	991,645	6.56	–1.89	86.50
1993	1,215,649	1,071,903	6.03	8.09	88.18
1994	1,304,703	1,090,257	7.33	1.71	83.56
1995	1,388,147	1,209,641	6.40	10.95	87.14
1996	1,475,219	1,327,499	6.27	9.74	89.99
1997	1,566,436	1,478,774	6.18	11.40	94.40
1998	1,753,320	1,645,453	11.93	11.27	93.85
1999	1,930,161	1,785,696	10.09	8.52	92.52
2000	2,175,513	2,169,897	12.71	21.52	99.74
2001	2,620,302	2,481,178	20.45	14.35	94.69
2002	3,630,009	2,519,042	38.53	1.53	69.39
2003	3,514,758	2,452,173	–3.17	–2.65	69.77

Source: Department of Enterprise, Trade, and Employment (1991–2003); Irish Financial Services Regulatory Authority (2004), p. 64.
Note: Earned premium income is gross of reinsurance.

11.3.1 Property Insurance

Recent History

Property insurance in Ireland is a rapidly growing sector of the non-life insurance industry. The economic boom experienced in Ireland during the mid to late-1990s is a key reason for this expansion. Ireland quickly became one of the wealthiest countries per capita in the E.U. and began building construction on a scale never before seen in the country. In addition, an increasing level of immigration into Ireland has meant that housing requirements have increased significantly. Table 11.3 provides details of the growth of earned premium income in the industry in the period from 1990 to 2003, demonstrating a continuous upward trend, especially in the last three years.

Table 11.3. Property Insurance Earned Premium Income and Incurred Claims, 1990–2003

Year	Earned Premium Income (€ Thousands)	Claims Incurred (€ Thousands)	Change in Premiums %	Change in Claims %	Ratio of Claims to Premiums %
1990	209,871	117,353			55.92
1991	220,078	132,211	4.86	12.66	60.07
1992	229,965	129,460	4.49	–2.08	56.30
1993	245,932	159,622	6.94	23.30	64.90
1994	269,403	159,117	9.54	–0.32	59.06
1995	280,219	179,085	4.01	12.55	63.91
1996	302,237	202,542	7.86	13.10	67.01
1997	320,959	203,003	6.19	0.23	63.25
1998	363,043	263,670	13.11	29.88	72.63
1999	407,915	287,476	12.36	9.03	70.47
2000	455,684	301,956	11.71	5.04	66.26
2001	555,536	445,721	21.91	47.61	80.23
2002	679,064	370,349	22.24	–16.91	54.54
2003	823,751	365,778	21.31	–1.23	44.40

Source: Department of Enterprise, Trade, and Employment (1991–2003); Irish Financial Services Regulatory Authority (2004), p. 74.
Note: Earned premium income is gross of reinsurance.

As Table 11.4 shows, the number of companies, including captives, offering property insurance in Ireland increased significantly in 2002 and 2003, having remained more or less static during the mid to late 1990s. During that period, most changes were brought about as a result of acquisitions by British or other European insurers looking to gain a foothold in the Irish market (e.g. AXA Insurance Ltd. (AXA) acquiring New PMPA Insurance Ltd. (PMPA) and Allianz Insurance Ltd. Acquiring Church & General Insurance Ltd.).

In the past two years, however, 13 companies have either launched new firms or re-established their presence in the Irish market. This amounts to a 30% rise in companies providing property insurance. However, these new entries to the market have had little or no impact on the overall market share. Many are specialist companies interested in niche markets, or captives insuring only their parent company. The dominance of five companies, who control more than 80% of the market, continues.

The greatest threat to the success of the property insurance market is the occurrence of severe weather events in Ireland. In terms of both risk frequency and severity, two types of severe weather events have the largest risk factor attached to them, windstorms and floods (both coastal and fluvial). Indeed, as has been the case in the last five years, these two events often combined, with prolonged and heavy rainfall, lead to an over-saturation of the land and consequential flooding. Table 11.5

outlines the severe weather events that have affected Ireland in recent years and how they have impacted property insurers. It can be seen that 2002, which saw two separate flood events at either end of the year, was a particularly bad year for property insurers, with €87 million in flood-related claims. This figure was exacerbated due to both flood events impacting Dublin, thus affecting a greater concentration of the people than flooding in rural areas. In 2004, further severe flooding affected a number of counties along the south coast of the country causing €36 million of property damage.

Table 11.4. Number of Companies Offering Property Insurance 1990–2003

Year	HQ in Ireland	HQ in E.U.	HQ Outside E.U.	Total
1990	15	16	1	32
1991	14	16	1	31
1992	14	17	1	32
1993	15	15	1	31
1994	22	10	1	33
1995	21	11	1	33
1996	21	11	1	33
1997	21	11	1	33
1998	25	11	1	37
1999	24	10	1	35
2000	22	8	1	31
2001	24	8	1	31
2002	27	10	1	38
2003	32	10	1	43

Source: Department of Enterprise, Trade, and Employment (1991–2003); Irish Financial Services Regulatory Authority (2004), p. 74.

Note: The number of insurance companies offering property insurance includes captives which are not recorded separately from non-captive companies. HQ: Headquarters. E.U.: European Union.

As a direct response to this recent increase in severe weather events, and more specifically those involving flooding, the Irish government set up a National Flood Review Group at the beginning of 2003. This group, comprised of representatives from the Department of the Environment, the Office of Public Works (OPW), the Department of Communications, Marine, and Natural Resources, and the Irish Business and Employers Confederation (IBEC), were tasked with establishing a report investigating the occurrence, management, control, and financing of floods in Ireland (Office of the Public Works 2004). One recommendation called for the creation of a single National Flood Authority, a body that would have overall control of flood management in Ireland. This has been strongly advocated by Irish insurers for many years, as they viewed the prior arrangement, where responsibility was split between the aforementioned government departments, together with local authorities in flood risk areas, as being impractical in the future (Irish Insurance Federation

2003). Currently, there is no government involvement in the insurance of flood risk in Ireland, although limited financial relief, through a humanitarian aid scheme, is usually made available in the immediate aftermath of a major flood event.

Table 11.5. Cost of Previous Weather Events, 1997–2004

Year	Event	Cost (€ Millions)
October 2004	Flood	38
November 2002	Flood	50
February 2002	Flood	37
December 2001	Freeze	30
November 2000	Flood	51
December 1998/January 1999	Storm	56
December 1997	Storm	84

Source: Irish Insurance Federation (2003), p. 14.

Property Insurance Market Share

Concentration of the Irish property insurance market has been high over the past five years, as Table 11.6 indicates, with the dominant companies having 69–80% of the market share determined by earned premium income.

Table 11.6. Market Share of the Top Five Companies in the Irish Market, 1999–2003

Year	Liability %	Property %	Motor %
1999	67.80	69.39	65.35
2000	70.50	77.47	69.62
2001	73.76	80.48	76.96
2002	69.80	80.11	74.97
2003	69.70	77.87	77.93

Source: Department of Enterprise, Trade, and Employment (2000–2003); Irish Financial Services Regulatory Authority (2004), p. 74.
Note: Market share is measured using earned premium income gross of reinsurance.

A further measure of concentration and competitiveness is the Herfindahl Index.[219] Table 11.7 shows that concentration has increased from 1990 to 2001. In

[219] The Herfindahl Index is described as a measure of the size of firms in relationship to the industry as a whole and a measure of competitiveness. It is the sum of the squares of each company's market share so that the final index can range from between 10,000, when there is only one company in the market, to 0, when the market is extremely competitive. As a guideline, it is argued that markets are competitive when a score of 2000 or below is achieved (Krishnamurthy 2000).

2002, concentration within the property insurance market began to decrease due to the entrance of a number of small companies.

Table 11.7. Herfindahl Index for Insurance Market, 1990–2003

Year	Property	Liability	Motor
1990	822	721	1,043
1991	828	829	1,010
1992	850	825	946
1993	881	899	947
1994	909	985	1,003
1995	896	1,012	985
1996	910	982	984
1997	1,113	953	1,318
1998	1,044	969	1,280
1999	1,227	959	1,345
2000	1,525	1,091	1,411
2001	1,638	1,172	1,671
2002	1,539	1,144	1,515
2003	1,441	1,090	1,576

Source: Department of Enterprise, Trade, and Employment (1991–2003); Irish Financial Services Regulatory Authority (2004).
Note: The Herfindahl Index was calculated using earned premium income, gross of reinsurance, for all companies registered in Ireland.

Currently, the property insurance market is split almost equally between household and commercial property insurance. As was seen in Table 11.3, the period from 1999–2003 has seen the upward trend of earned premium income in the market amplify significantly, with each of the last three years showing a 20% increase per year.

Table 11.8 shows the underwriting and technical results of the property market since 1990.[220,221] The underwriting result is based on earned premium income, gross of reinsurance, less cost of claims incurred, management expenses, and commission. The technical result includes income from investments. Data on the technical results has only been available since 1995. During the period from 1999 to 2003, the property insurance market has consistently shown a profit except in 2001.

In general, premiums in the property insurance market increased both from the growth in the housing market and the increasing amount of both commercial and residential construction that the country has experienced in the last few years.

[220] The underwriting result is defined as total earned premium income minus cost of claims incurred, commission paid, and management expenses, plus/minus foreign exchange gains/losses.
[221] The technical result is defined as the underwriting result plus/minus investment income attributable to the underwriting account.

However, the market's profitability, particularly in household insurance, is tenuous and a lot will depend on the level of both new commercial and residential construction, future severe weather events, and on how the government implements the recommendations of the National Flood Review report.

Table 11.8. Underwriting and Technical Results for Property Insurance Market, 1990–2003

Year	Underwriting Result (€ Thousands)	Technical Result (€ Thousands)	Change in Underwriting Results %	Change in Technical Results %
1990	8,929	N/A		
1991	–3,125	N/A	65.00	
1992	12,243	N/A	–491.78	
1993	–12,861	N/A	–5.05	
1994	20,884	N/A	–62.38	
1995	–8,936	19,873	57.21	
1996	–1,298	24,093	114.53	221.23
1997	–2,062	31,359	258.86	230.16
1998	–47,339	–16,902	2395.78	46.10
1999	–18,047	15,368	138.12	9.08
2000	–24,804	7,827	237.44	150.93
2001	–66,526	–22,564	368.21	–188.28
2002	47,658	80,345	28.36	–256.08
2003	185,861	222,914	489.99	377.45

Source: Department of Enterprise, Trade, and Employment (1991–2003); Irish Financial Services Regulatory Authority (2004), p. 74.

Note: The underwriting result is based on earned premium income, gross of reinsurance, less cost of claims incurred, management expenses, and commission. The technical result includes income from investments. Information concerning investment income was not published until 1995. N/A: Not available.

Types of Products

The Irish property insurance market is split almost equally between household and commercial property insurance and the following section details the coverage available under both. Property insurance coverage is divided into two parts, that is, buildings coverage and contents coverage. The insured may choose to have either one or both of these policy types on their property. If coverage includes both policy types, the sum of the contents insured is calculated on value at risk or a percentage of the buildings coverage.

The main players in both the Irish household and commercial insurance markets have their own tailor-made products marketed under specific names (e.g. SafeHome from Royal & Sun Alliance Insurance Co. Ltd. And Homestar from Eagle Star Insurance Co. Ltd.) and each offers the company's own variety of optional extras and extensions to the standard policy. Most also include a series of discounts applicable to policyholders depending on variables such as age of the property, protection measures such as burglar or fire alarms, and voluntary excess increases.

Household Insurance

The basis for coverage provided by household insurance in Ireland is one of material damage to the property and/or its contents. Among the variety of perils covered under this standard material damage policy would be fire, theft, lightning, storm/flood, malicious damage, accidental damage, and water leakage. The storm and flood risk in Ireland necessitates special consideration and generally the two are included together. This eliminates the potential problem of deciding whether flood damage has been instigated by a storm. Personal liability coverage is also usually provided automatically under a household insurance policy.

The buildings coverage provides protection for the main insured buildings as well as other permanent fixtures owned by the insured and covered under the policy. These may be outbuildings, fences or walls, fixtures and fittings within the buildings themselves, or even interior decorations. All of these receive standard coverage from the perils mentioned earlier, with the potential for additional coverage depending on the individual insurer.

The contents coverage includes damage caused to household goods and personal belongings, which can include cash up to an agreed limit. In most cases, the household contents are replaced on a 'new for old' basis and, as a result, the insured person is recommended to insure for the price of full reinstatement. While this means that depreciation of the contents is not taken into account, as is the case if the policy is on an indemnity basis, such coverage is generally more expensive, to eliminate the possibility of moral hazard on the part of the policyholder. Contents coverage can include a number of special features that come either automatically as a part of standard coverage or as part of a series of additional forms of coverage. Among the optional extras available is the ability to have your contents covered on an all risks basis, meaning that they will be insured for loss or damage on the property and when taken away from the property by the insured.

Commercial Property Insurance

As is the case with household insurance policies, material damage from the standard perils (fire, theft etc.) is usually included automatically. However, commercial insurance policies tend to provide more flexible forms of coverage than household policies, given the variety of trade categories such as manufacturing, services/offices, and retail outlets. As a result, selected perils coverage is made available to allow policyholders to choose exactly the level of coverage that most closely suits their line of business. Also, unlike household policies, contents and buildings coverage tend to be automatically covered in the same policy.

A major additional area of concern is also business interruption coverage. This coverage deals with any loss of business resulting from the damage to property and/or contents, and it normally includes any additional expense accrued as a result of trying to avoid or reduce such losses (e.g. if the business is forced to rent out temporary premises, while its own are being repaired). Similarly, employers' liability coverage and public/products liability coverage are also made available to policyholders; although the extent of coverage offered will generally depend on the type of business involved. Both of these forms of coverage are made available to businesses either as additional coverage to a limited, selected perils policy, or as a standard part of a comprehensive, all risks policy. The availability of such comprehensive policies can vary depending on the insurer.

This next section deals with liability insurance which can be purchased, as mentioned above, as part of a packaged policy or bought separately.

11.3.2 Liability Insurance

There are five basic areas of liability insurance in the Republic of Ireland, general liability, employers' liability, product liability, directors' liability, and professional indemnity. None of these are made compulsory by statute although professional bodies do require their members to hold professional indemnity insurance. Liability insurance for motor accidents is covered in the next section relating to motor insurance since this coverage is normally provided together with property coverage. In 1966, benefits for persons injured at work were legislated and consequently, workers' compensation was abolished. The Social Welfare (Occupational Injuries) Act 1966 allowed employees to litigate against their employers following injury at work, as well as claim benefits from the social welfare system. In order to obtain compensation from an employer following an accident, the employee must show negligence on the part of the employer. In the event that the employee was negligent, damages are reduced based on his or her level of contributory negligence. The 1966 act also provides for the deduction of benefits payable from the damages awarded. The main types of liability insurance transacted are general liability and employers' liability, with the other classes representing a very small percentage of the market, 10% at most.

The liability insurance market has been under severe pressure over the past five to ten years with high premiums and large losses. This also occurred in the motor insurance market, which is discussed below. The Irish government tried to reduce the level of premiums by taking action relating to both the severity and frequency of liability losses. Out of this development arose the Safety, Health, and Welfare at Work Act of 1989, which called for safety at work. This was a very general act under which regulations and codes of practice could be developed. This act and its eventual enforcement reduced the frequency of accidents, although the act is in the process of being updated with a new bill before the Dail, the Irish lower house (Central Statistics Office 2004).

Table 11.9 shows the growth of premiums and claims in Ireland from 1990 to 2003. In 2001, the monetary system converted from the Irish punt to the euro. This table shows a steady rise in premiums and claims over the ten-year period with a big increase in 2000 which dropped off in 2003. It was during this three-year period that pressure was placed on the government to take some action over the high cost of insurance in the country. In 2000 and 2001, there was a substantial rise in the Consumer Price Index compared with previous years, which would partially account for the sudden increase in premiums.

The number of companies operating in the liability market has changed over the past ten years with the majority of Irish companies being acquired by the larger

European companies. Until 2001, the number of companies operating in the insurance market remained reasonably steady with 30 to 33 operators. In 2001, this increased to 37 and in 2003 to 43. However, the number of Irish companies decreased during this period from 8 to 4. The Herfindahl Index (see Table 11.7) with respect to the liability market shows a drop off in the competitiveness in the Irish market from 1997 to 2000 and then competition levels began to increase in 2003 as a result of improvements in the market due to the efforts of the government.

Table 11.9. The Irish Liability Insurance Market, 1990–2003

Year	Earned Premium Income (€ Thousands)	Cost of Claims Incurred (€ Thousands)	Change in Earned Premium Income %	Change in Cost of Claims Incurred %	Ratio of Claims to Premiums %	Technical Results as a % of Earned Premium Income
1990	179,918	207,853			115.53	
1991	188,366	218,947	4.70	5.34	116.23	
1992	195,304	214,756	3.68	–1.91	109.96	
1993	213,406	224,775	9.27	4.67	105.33	
1994	241,313	241,027	13.08	7.23	99.88	
1995	268,718	284,526	11.36	18.05	105.88	5.77
1996	277,259	300,278	3.18	5.54	108.30	0.61
1997	282,263	325,947	1.80	8.55	115.48	0.07
1998	301,250	340,011	6.73	4.31	112.87	–3.08
1999	316,114	347,152	4.93	2.10	109.82	–0.08
2000	333,751	394,238	5.58	13.56	118.12	–11.45
2001	419,899	498,285	25.81	26.39	118.67	–17.88
2002	587,108	602,379	39.82	20.89	102.60	–10.98
2003	733,189	622,915	24.88	3.41	84.96	12.23

Source: Department of Enterprise, Trade, and Employment (1991–2003); Irish Financial Services Regulatory Authority (2004), p. 78.

Note: Technical results were not available until 1995. Earned premium income is gross of reinsurance.

The two largest players in the Irish liability insurance market are the German group Allianz, which entered the market in 2000 by purchasing the Church and General and a number of smaller companies and the British group Aviva with its Irish subsidiary Hibernian, a former Irish company partially owned by another British company. In 2003, between them these two companies held 39% of the liability insurance market share, which was similar to the share held in 1999 by the companies acquired. Over the five-year period, 1999 to 2003, the top five companies have held between 67% and 73% of the liability market. Table 11.6 shows the change in market share during this period of the top five companies.

These tables demonstrate that the liability market initially experienced decreasing competition until 2001 when competition increased. This indicates confidence in the steps that the government has been taking in trying to reduce the costs of liability claims. Some of these innovative steps are reviewed below. In general, as Table 11.9 indicates, the liability market has been unprofitable with regards to its technical account (this includes return on investment) since 1998. However, there was a return to profitability in 2003. This could herald the way for further competition as the Irish liability market becomes more attractive to possible new players.

11.3.3 Motor Insurance

The motor and liability insurance markets have experienced similar problems, as their claims are based heavily on the Irish legal system. The difficulties that have arisen and the actions undertaken are discussed below.

There are three basic types of coverage that can be purchased on the Irish motor insurance market. These are comprehensive, third party fire and theft, and third party only. No figures are available as to the breakdown of the types of coverage purchased, but the general consensus is that the main type of insurance purchased is comprehensive coverage. This is an all risks policy on the insured motor vehicle plus third party liability coverage. Certain additional items of coverage are also included, such as windscreen coverage, personal accident insurance, and emergency breakdown, depending on the company. Third party coverage provides protection in the event of the driver of a vehicle incurring a liability while in control of the vehicle. Normally, policies only cover the negligence of named drivers included on the policy. This policy provides basic road traffic act coverage plus liability to passengers. Insurers are reluctant to provide this type of coverage on its own as it is in this section of the policy that the majority of the claims occur. The third type of policy includes third party cover plus losses caused to the insured vehicle due to fire or theft. This policy is only slightly less expensive than the comprehensive policy, and therefore, very few people purchase this type of coverage.

Table 11.10 shows the profits of the motor insurance market since 1990 by relating earned premium income to the cost of claims incurred, as well as profitability on the technical account, which includes investment earnings. Up until 1999, the market made a profit but, in that year, profitability fell and became a loss in the following year. In 2001, the market improved again and went back into profitability in 2002. In 1995, concern was expressed about the high cost of motor insurance premiums in Ireland. Consequently, Deloitte and Touche were commissioned to carry out a survey of insurance premiums throughout Europe. This was published in 1996 and established that motor insurance premiums in Ireland were among the highest in Europe (Deloitte and Touche 1996).

Following this report and subsequent increases in motor and liability insurance premiums, pressure was brought upon the government by various lobby groups to decrease these costs. This includes IBEC, a powerful lobby group for business and AIR, which was formed by various small businesses to lobby the government for changes in the insurance system. This pressure increased when insurance industry was trying to recover from their financial losses between 1999 and 2001. To reduce these losses, the government organized a number of road safety campaigns and changes in safety laws. In particular, the institution of a penalty points system for drivers infringing the law, vehicle testing every two years for vehicles over four years old, and road improvements have all affected the number of road accidents occurring in Ireland, and thus, the number of claims (Central Statistics Office 2004). Despite these actions however, premiums have rise and claims have fallen as Table 11.10 indicates, thus enraging various interest groups. Consequently, it is expected that in 2004 premiums will fall.

Table 11.10. Performance of the Motor Insurance Market, 1990–2003

Year	Earned Premium Income (€ Thousands)	Cost of Claims Incurred (€ Thousands)	Change in Earned Premium Income %	Change in Cost of Claims Incurred %	Ratio of Claims to Premiums %	Technical Results as a % of Earned Premium Income
1990	520,936	596,628			114.53	
1991	617,078	620,984	18.46	4.08	100.63	
1992	662,990	613,569	7.44	–1.19	92.55	
1993	691,737	630,772	4.34	2.80	91.19	
1994	722,230	652,845	4.41	3.50	90.39	
1995	763,331	680,985	5.69	4.31	89.21	16.83
1996	808,808	784,870	5.96	15.26	97.04	4.91
1997	862,721	872,249	6.67	11.13	101.10	4.54
1998	962,122	939,603	11.52	7.72	97.66	2.05
1999	1,072,177	1,074,033	11.44	14.31	100.17	–1.89
2000	1,232,737	1,370,358	14.98	27.59	111.16	–13.48
2001	1,470,534	1,443,279	19.29	5.32	98.15	–0.86
2002	1,600,561	1,390,147	8.84	–3.68	86.85	11.69
2003	1,708,882	1,280,304	6.77	–7.90	74.92	23.11

Source: Department of Enterprise, Trade, and Employment (1991–2003); Irish Financial Services Regulatory Authority (2004), p. 70.
Note: Technical results were not available until 1995. Earned premium income is gross of reinsurance.

Unlike the U.K., there has not been an influx of new direct line companies into Ireland.[222] Quinn Direct is the newest Irish company, which initially was a direct line company, and a number of the larger companies, such as Hibernian commenced a direct line subsidiary. Because of the high rise in premiums and the lack of new entrants to the market, the existence of competition in the market was questioned by the government, who commissioned a report on both the liability and motor insurance markets. This was prepared by Parsons, Dickinson, and Greenford (2004)

[222] A direct line company is an insurance company that provides coverage directly via telephone calls to private individuals without the intervention of an intermediary. The coverage is normally provided through a call center which deals with simple policies such as motor insurance. Generally, clerks with little or no training plug in relevant details to a computer program which subsequently returns a premium. Any client who wishes to insure a non-standard risk will be rejected.

and submitted to the competition authority, together with a report by Europe-Economics (2004). These reports concluded that the Irish liability and motor insurance markets were indeed competitive. The Herfindahl index for the motor market is shown in Table 11.7.

From Table 11.7, it can be seen the motor market is competitive if one applies the assumed level of competitiveness for the Herfindahl index of less than 2000 (Krishnamurthy 2000). Rather than improving over the past five years, the competitive nature of the Irish motor market has worsened, thus reflecting the poor experience of the motor market. Efforts have been made by Irish politicians to encourage new entrants into the motor market, but with little success to date. Hopefully, with increasing profitability and the new procedures discussed below in place, this will change. Profitability is also indicated by the share of the motor market held by the top five companies. Over the last five years, the top two companies have remained the same, holding between 40% of the market in 1999 and 48% in 2003. As Table 11.6 shows, there has been an increasing consolidation of the market shares held by the top five companies holding 65% of the market in 1999 and 78% in 2003. The only Irish new entrant in the market, Quinn Direct, is making quite substantial inroads since entering the market in 1996. Church and General were bought out by Allianz. The only other Irish company is FBD, which originally served the agriculture community. The number of motor insurance companies has remained reasonably static, with 27 companies operating in the Irish market, rising to 29 in 2003.

Both the motor and liability markets have performed relatively poorly, with very high premiums over the last few years, although this is now improving with large profits being reported by the media in 2005. The majority of claims in both markets involve personal injuries and the legal system. In view of this, the reasons for the poor performance in both markets are considered below.

11.3.4 Causes of Poor Performance in the Motor and Liability Markets

The main concern in the Irish liability and motor insurance market relates to what has been dubbed the 'compo culture.' This is a term used by the press to explain why Irish motor and liability insurance premiums are exceptionally high compared to the rest of Europe and is a diminution of the term 'compensation culture.' 'Compo culture' refers to the increased propensity of the Irish to make claims as compared to its European neighbors. It has been estimated by the industry that the Irish are four times more likely to file a claim than the British. It is interesting to note that in 2005, the U.K. began using this term following certain changes in their market that led to increases in the frequency of claims.

Further concerns have been the high cost of compensation awarded in legal liability claims, the slowness of the system, and the high legal costs. For some time now, lawyers have been advertising that they are prepared to take instructions on personal injury claims on the basis of there being no fees to be paid unless damages are obtained, 'no foal, no fee.'[223] In the Irish system, the loser of an action meets the

[223] This saying derives from the fact that Ireland has a large equine industry and it is customary in that industry that if a stallion serves a mare and no foal is produced then no fee is

legal costs for both the defendant and the plaintiff. Thus, insurers on the whole pay both sides costs if compensation is awarded. There is also a perception in the Irish market that the incidence of fraud is high with at least 50% of the claims being fraudulent (Greenford 2003, p. 53).

Another concern is the level of damages awarded by the courts and subsequently paid by insurers. Compensation awarded in Ireland can be up to eight times than that awarded in the U.K. for similar injuries (Greenford 2002, p. 49). These problems are now being addressed by a number of innovative measures. At the end of 2003, the Personal Injury Assessment Board (PIAB) was formed to deal with all personal injury cases, except those where the medical profession is involved. The board consists of representatives from industry, consumers, and insurers, while its staff is composed of personal injury claims handlers whose experience has been with insurers. The legal profession does not support the institution of this board and there have been arguments that it is unconstitutional, although this has not yet been tested in the courts.

The board must review all personal injury claims before they proceed to court. On making a claim against a defendant, the claimant must submit a letter of claim. Once the letter has been received by the defendant, the claimant must prepare a case for the PIAB using written material only. The plaintiff submits a simple form plus a medical report for assessment. Solicitors cannot represent the injured party before the board, although they can assist in preparing the documentation but they cannot charge the defendant or the PIAB a fee. The insurer of the defendant (called the respondent in the act), if they are insured, is then advised by the PIAB that a claim has been received. The insurer may accept or reject their intervention. If the insurers do not respond or accept their intervention, the case is assessed by applying a book of quantum.[224] The claimant and respondent are then informed of the assessment and this can be accepted or rejected. Before any case proceeds to court the PIAB must issue a release certificate giving permission for the injured party to proceed. It is hoped that this procedure will speed up assessments, reduce costs, and also make both the motor and liability insurance markets more attractive to possible new entrants.

11.4 LIFE INSURANCE

11.4.1 Background

The life insurance industry in the Republic of Ireland is very competitive and fast growing, with high levels of coverage in major classes of business such as mortgage protection and pensions (Irish Insurance Federation 2004). The number of operators in the Irish life insurance market has expanded in recent years and this trend appears likely to continue. Gross premiums written in Ireland by life insurance companies in

---

payable to the stallion owner. Thus, if no compensation is obtained the plaintiff does not pay solicitor's fees.

[224] This book is published on the PIAB website (www.PIAB.ie). It is intended that this book will be updated on a regular basis.

2003, fell by 1.9% to €14.52 billion comprising €7.96 billion Irish risk business and €6.55 billion foreign risk business. Given the size of the industry, the life insurance business is a significant contributor to the Irish economy. Life insurance companies are also exempt from Irish tax on the share of their profits allocated to policyholders, although they are liable to corporate tax at the 12.5% on the balance of their profits (KPMG 2003, p. 8).

Critical to the success of the Irish life insurance market was the promotion of life insurance by the Irish government during the last century. After achieving independence, the Irish government pursued a protectionist policy and sought to be self-sufficient economically. The government recognized the need for life insurance to be provided in the Irish market and reacted to this necessity with the creation of the Irish Life Assurance Company. This company was established in 1939, as a result of the amalgamation of nine British and Irish life insurance companies then operating in the market. From the onset of Irish Life Assurance Company, the Ministry for Finance held an 18% share of the company. However, in 1947, the ministry purchased shares held by the British life insurance companies resulting in a government holding of 90%. Between 1965 and 1990, Irish Life Assurance Company commenced its U.K. branch operation, thus being the first Irish insurance company to have an international presence. In order to expand into wider financial services, Irish Life Assurance Company established the Irish Life Building Society.[225] The Irish Life Assurance Company remained a commercial enterprise owned by the government until its privatization by way of flotation on the Dublin and London stock markets in 1991. Irish Life Assurance Company is now part of the Irish Life and Permanent group, which consists of Permanent TSB, which was a building society, Irish Life International, and Irish Life Investment Managers.

Today, Irish Life Assurance Company is Ireland's largest life insurance company and a market leader in the provision of life, pension, and investment products. Table 11.11 shows the top five insurance companies' percentage market share by premium volume.

Irish Life Assurance Company can be considered responsible for most of the major innovations in the life insurance and pensions industry in Ireland, including the launch of serious illness coverage, mortgage protection coverage, tracker bonds, consensus funds, and unit linked investment funds.[226] Irish Life International has

---

[225] A building society was originally a registered company formed mainly for the purpose of lending money for the purchase of private houses. The company provides savings accounts for investors who could only withdraw money after notice was given. These savings were used to fund house loans by means of mortgages registered on the property deeds. These mortgages provided a first call on the outstanding loan if a house was sold. Building societies were not allowed to hold current accounts for investors. These companies have now developed so that loans are available for all purposes and current accounts are provided, thus providing functions very similar to banks.

[226] Tracker bonds are fixed term investment products offering an index-linked return or guarantee on the original capital, whichever is the greater return. Consensus funds allocated their assets and countries of investment based on the average style of investment managers in Ireland. Unit linked funds are a product whereby the value of the policy is linked to units consisting of bundles of stocks and shares quoted on any stock market. These bundles often are constructed based on a theme, such as green companies, blue chip companies, etc. The

grown rapidly since its formation in 1994 and now operates in over 30 countries. New areas of business are constantly being reviewed. Irish Life Assurance Company's products and services are distributed through a wide range of channels, including bank branches, independent intermediaries and brokers, and directly to customers. It can be seen that the main players in the Irish life insurance market have, for the most part, remained the same over the last ten years, as illustrated by the market share for 1993, 1998, and 2003 in Table 11.11.

Table 11.11. Life Insurance Market Share by Earned Income Premium Volume, 1993–2003

	2003	1998	1993
Irish Life	26.33	25.89	30.06
New Ireland	17.53	6.56	6.84
Ark	11.63	16.62	13.41
Hibernian Life and Pensions	11.72	3.45	8.38
Eagle Star Life	8.27	5.37	7.39
Standard Life	7.89	N/A	N/A
Friends' First Life Ass. Co.	5.74	8.78	5.20
Canada Life Ass. (Ireland)	4.40	4.63	1.67
Lifetime Assurance	2.46	11.10	9.66
Scottish Mutual International	0.74	N/A	N/A
Irish Progressive Life	N/A	7.44	N/A
Norwich Union Life	N/A	7.30	N/A
Prudential Life	N/A	N/A	4.63
Rothschild	N/A	N/A	8.99
Total (Top Ten)	96.71	82.40	82.61

Source: Data obtained from Department of Enterprise and Employment, 1994, 1999 and Irish Financial Services Regulatory Authority (2004).
Note: Earned premium income is gross of reinsurance. N/A: Not Available.

One of the most notable recent additions to the Irish life insurance market has been the introduction and development of bancassurance. Bancassurance is the combination of the banking and insurance businesses within the same organization. In particular, backward integrated banks (i.e. banks owning their own life insurance company) emerged as strong competitors in the field during the last twenty years. IFSRA regards banks selling insurance as insurance intermediaries and subject to the same codes of conduct. One of the primary advantages in banks breaking down the traditional border between the deposit taking institution and insurance provider is the

---

values of the units directly reflect the values of the underlying assets of the fund, and fluctuate daily according to the performance of those investments.

banking sector's extensive retail networks, which can be used to sell insurance and investment products.

The two largest national banks, Bank of Ireland and Allied Irish Bank (AIB), have both utilized their wide customer base to create an abundance of life insurance sales opportunities. Bank of Ireland made its initial entry to the Irish life insurance market in 1987, when it formed Lifetime Assurance, while AIB's Ark Life was formed in 1991 (Fagan 2003). In 1997, New Ireland Assurance Company, which held a strong position in the broker market and in the group life and pensions market, became a wholly owned subsidiary of the Bank of Ireland, while Lifetime Assurance was de-registered. New Ireland Assurance Company increased its market share by almost 11% in the space of five years (1998–2003) as shown in Table 11.11. In 2003, AIB Bank and Bank of Ireland together held a third of the life insurance market share as a result of leveraging the strengths of their retail banking outlets and strong relative investment performance. The newly formed IFSRA do not favor banks owning general insurance companies in Ireland and it will be interesting to see how this affects the life insurance market.

11.4.2 Classes of Life Insurance Available

The main life insurance products available on the Irish market include individual insurance policies and annuities, pension schemes, permanent health insurance, critical illness insurance, and long-term care insurance. Table 11.12 breaks down gross written premium by policy type for 2003.

Table 11.12. Life Insurance Gross Written Premium Income by Type of Policy, 2003

	% of Total	Gross Written Premium Income (€ Thousands)
Individual Insurance and Annuities	58.8	4,495.7
Pensions Scheme Business	27.0	2,067.0
Self-employed Pensions and Associated Business	11.4	872.9
Permanent Health Insurance (PHI)	1.8	137.1
Industrial Branch (IB) Business	1.0	71.7

Source: Irish Insurance Federation (2004), p. 34.

Strictly speaking, there are three different classifications of life insurance policy available in the Irish market: term, endowment, and whole life. Term insurance consists of policies that are taken out as a means of insuring against the possibility of death within a specific period. The premiums payable are applied to provide a mix of life coverage, with protection being the main function. Forms of term insurance offered in the Irish market include level, renewable, convertible, increasing, index-linked, family income benefit, credit term, and unit linked term insurances. Unit linked term insurance is a product whereby the value of the policy is linked to units consisting of bundles of stocks and shares quoted on any stock market. These bundles often are constructed based on a theme, such as green companies, blue chip companies etc. The values of the units directly reflect the values of the underlying assets of the fund, and fluctuate daily according to the performance of those investments. These are considered term insurance as the sum insured is only payable on death of the insured during the term of the policy. The amount payable upon death therefore is dependent on the value of the shares at that time. There is also a small guaranteed sum dependent on the company's policy.

A few years ago, it was possible for companies to offer consumers appealing premium rates, however, the low premiums were based partly on the assumption that a proportion of people would lapse their policies each year and that the profits on lapses would fund the enhanced benefits for people who kept their policies in force (Fagan 2003). As it turned out, lapse rates were much lower than expected, and the insurers' finances, or indeed their reinsurers' finances, since the majority of this business was reinsured, suffered severely (Fagan 2003).

A large proportion of the life insurance market consists of investment vehicles, such as universal life and endowments. Universal life is similar to unit funds except written on an endowment basis. The premiums payable are used to purchase a range of investments on which either the surrender or maturity value is based. Endowments have been used as a vehicle for mortgage repayment, whereby only the interest portion of the mortgage is paid by the mortgage-holder and the principal is paid by the endowment policy on maturity. At the beginning of the 1990s, endowment mortgages were so popular in the Republic of Ireland that one in three borrowers were taking them out (Irish Times 2000). However, endowment mortgages have received considerable negative press over the past few years, as many homeowners have been left without enough money to pay off their home loans when the policies matured. This has been attributed to the difficult investment environment.

In the case of whole life policies, a payout occurs on the death of the life insured, whenever this takes place. A range of products is available in the Irish whole life market. Guaranteed whole life coverage provides a guaranteed amount payable upon death for a guaranteed premium. This type of coverage can include compensation for loss of income as a result of the sudden death or serious illness of the primary earner in a household. Advance payment of the death benefit can be made in the case where the insured is diagnosed as having a terminal illness. Products are also available that are designed to mitigate the impact of inheritance tax on the value of an estate. Whole life policies are written in such a way as to provide access to a guaranteed source of funds. These funds can be used to pay any tax arising, thereby maintaining the value of an estate.

Since 2001, many clients have transferred their personal portfolio of stocks, shares, and real estate into a personal portfolio bond, creating an individual personalized managed fund within a life insurance company and, thereby, avoiding tax on both income and capital gains until the policy is surrendered. The tax, standing at 23%, was 21% less than the personal tax rate on dividends on directly held shares and equal to the tax on capital gains. With such 'value' to be had, an unexpected number of investors purchased such products, causing the Minister for Finance to increase the exit tax on gains a further 20%, bringing the total exit tax up to 43%, making such products unattractive and decreasing their popularity. However, property funds involving a life insurance company and a developer creating a package of commercial property and using the rental income to borrow up to 75% or 80% of the property's value have increased in popularity. This portfolio is then provided by insurers to investors as an investment bond whereby income from the bond is generated by rental income and capital growth. Income derived from the investment depends on the nature of the property forming part of the portfolio. This portfolio is then placed in a specialized managed fund within a life insurance company and offered to a number of private investors, typically between 100 and 200 (Fagan 2003). There is a management fee charged by insurers to investors varying, dependent on the bond. The investment is generally held for between 5 and 10 years. Unfortunately, there is no record of the extent of sales in Ireland of this particular product.

11.4.3 Industrial Life Insurance

Life insurance companies also offer special policies and policy combinations, differing from other life insurance products only in that they are aimed at a particular market sector. Industrial life insurance is one such policy, and is sometimes known as 'home service insurance.' As the name implies, it involves an agent or collector calling to people's homes to collect premiums and to service contracts. Industrial insurance was originally designed for low income families, who were unable to afford the amount of protection and premium payments associated with ordinary life and individual health insurance (Black 2000). Industrial insurance today, represents an ever-decreasing share of the life insurance market. New industrial insurance business (gross) effected during the year 1990 stood at 92,082 policies issued grossing €9,924 million by companies with head offices in Ireland and in other E.U. states. This compares with 5,137 policies issued in 2003 and €1,578 million collected in premiums by undertakings with their head offices in other E.U. states. No new industrial life insurance policies were originated by undertakings with their head offices in Ireland in 2003 (Department of Enterprise, Trade, and Employment 1990; Irish Financial Services Regulatory Authority 2004). The decline in industrial insurance may be due to the fact that life insurance policies issuing small amounts of financial remuneration provide little economic security, while being quite expensive to maintain, especially when the life insurance market is becoming increasingly competitive and life insurance products more diverse.

11.5 PENSIONS

In Ireland, the pensions market is under intense scrutiny from a range of government and industry bodies. In common with the rest of Europe and the U.S., the debate is focused on the adequacy of pension provision, with concerns being stimulated by the imminent retirement of the baby boom generation. The high level of emigration throughout the last century has mitigated the demographic problem, partially transferring it to the U.K. and the U.S., where the majority of Ireland's ageing émigrés reside. Ireland has, as a result, one of the youngest populations in Europe.

The Irish pension system consists of a mandatory pillar, which includes a basic old-age pension that is means-tested, and for people who satisfy certain contribution criteria, flat-rate schemes are available. There is also a provision for employees and the self-employed from the age of 65, if contributions commenced before 55. The government indexes payments to a rate usually just above inflation. These provisions make up the first pillar of the Irish pensions system and are supplemented by private and voluntary pension schemes. Private pension schemes cover half of the working population. They are fully-funded and can be either defined contribution or defined benefit.

Concerns over inadequate pension provision remain valid, in particular with regard to the underfunding of pensions by the young and middle-aged Irish. Despite the rise in incomes in this age bracket, spending and lifestyle patterns are absorbing much of the increase, leaving little available for pensions. The culture of property ownership, coinciding with an inflated housing market, has meant that high earners are maintaining above average mortgages, with childcare costs adding to the pressures in many cases. These cost of living issues are increasingly being cited as part of the reason for the apparent neglect in pension provision. Property remains the primary investment asset among the Irish, exposing them to a lack of diversification in their investment assets and its attendant dangers. The savings shortfall is ultimately a public policy issue and the Irish government has introduced a number of initiatives to remedy the situation.

In 1990, the Pensions Board was established, primarily, to provide a regulatory framework for the development of efficient national pension structures. It also provided a policy function and is engaged in the promotion of pension awareness and participation. The most recent stated objectives of the board contained in its 2002–05 strategy, provide a useful context for an analysis of recent developments in pensions. The board is committed to providing ongoing surveys through the Central Statistics Office (CSO) on the level of coverage as part of the National Pensions Policy Initiative (NPPI) instituted by the Pensions (Amendment) Act of 2002.

As part of its regulatory function, the board monitors levels of compliance among occupational pension schemes. Defined benefit schemes are the most common among employers and in its Annual Report of 2003, the Board found that in relation to long-term funding many of these schemes were in a "satisfactory position" and employers had "a genuine commitment to continuing their defined benefit scheme" (Pensions Board 2003, p. 40).

As part of its drive to increase levels of coverage, the NPPI recommended the introduction of Personal Retirement Savings Accounts (PRSAs) in 2002. PRSAs are a possible solution to increasing flexibility and movement in the labor market. This simple and cost effective measure requires an investment account being set up between an individual and an authorized PRSA provider. There are two types of PRSAs, a standard PRSA and a non-standard PRSA. The main difference being that the maximum charges under a standard PRSA cannot exceed 5% of contributions paid and 1% per annum of the PRSA assets. By the end of 2003, the Pensions Board had approved 10 PRSA providers with 55 PRSA products (Pensions Board 2003). Providers are investment firms, insurance companies, and credit institutions. Benefits are normally paid between the ages of 60 and 75 and tax relief exists on member contributions depending on the age of the contributor. For an individual under 30, tax relief is limited to 15% of income. For those between 30 and 39, it is 20% and between the ages of 40 and 49, it is 25%. For individuals who take out PRSAs and are aged 50 and over, tax relief is limited to 30%.

PRSAs are available to the employed, the self-employed, and the unemployed and its features include no charges or penalties when transferring from another PRSA provider or pensions arrangement. Its flexibility allows for termination or suspension and subsequent recommencement of contributions without charges accruing. The lackluster take up of the PRSA scheme, initially, was generally attributed to a muted investment environment. Increasing awareness of the pension deficit has increased the number of PRSA contracts being sold. The period up to the December 31, 2003, saw €41 million being invested in PRSAs, with 58,770 employers having designated a PRSA provider. Every employer in Ireland is required to have a contract with a PRSA provider, who is usually a broker or insurance agent. Employees are then advised who the provider is by their employer and pensions are then arranged between the employee, employer, and provider.

One of the key factors affecting the take up of PRSA, was the introduction of the innovative Special Savings Incentive Scheme on May 1, 2001. The motivation behind this scheme was to reduce spending in the economy and to encourage positive savings habits in the nation. To benefit from the scheme, individuals had to open "special savings incentive accounts" (SSIA) before April 30, 2002. Under the terms of this scheme, for every amount saved in a special account, an additional 25% is contributed by the government to the individual's savings. The cash is credited each month directly to a savings account. Investment can be made through a number of institutions approved for the purpose of the scheme, such as banks, building societies, and life insurance companies.

Individuals saved a stated amount each month and, to benefit fully from the scheme, savings must be left in the special savings account for a period of five years. The government contribution of 25% applies for a period of five years. Income or gains from the savings investment are taxed at 23% and this will be deducted by the participating financial institutions at the end of five years. Penalties were imposed to inhibit people from withdrawing funds from their SSIAs. If there is an early withdrawal from an account (other than on death), the full amount withdrawn (both the savings and investment return) will suffer tax at 23%. Each person is allowed only one SSIA. The minimum amount which must be saved by an individual in any one month, in the first year of the account, is €12.70. The maximum amount that you can lodge in any one month is €253.95 for the remaining 4 years. Strict rules governed who can participate in this scheme.

11.6 HEALTH INSURANCE

Health provision in Ireland is characterized by a two-tier system, public and private. The main feature of the public system is that all residents of Ireland are entitled to a range of public health services that are free of charge or subsidized by the Irish Government. However, the level of coverage depends on the economic status of the individual—those earning a lower income are entitled to a higher level of coverage. In comparative terms, the Irish public health system does not compare well with those of other European countries. Years of under-funding have resulted in a poor health service with long waiting lists being the norm (Wren 2003). Such features have constituted an important driver of the market in private health insurance.

Private health insurance in Ireland differs from other forms of insurance in that it provides payments for sickness or injury, unlike critical illness insurance, permanent health insurance, and hospital cash plans. Unlike private health insurance, these types of coverage do not take into account the medical expenses incurred, and therefore cannot be considered as equivalent products (The Health Insurance Authority 2004). Critical illness insurance provides for payment of a lump sum in the event of the insured being diagnosed with a stipulated disease, rather than indemnifying the medical costs incurred. Permanent health insurance provides for payment of 75% of earnings in the event of an accident or illness, which prevents a person from working after the expiry of a prescribed period of time. This payment can last until retirement age or until the insured is once again employable. Hospital cash plans are designed to supplement private health insurance in that they provide a small lump sum following accident or sickness. Those covered by cash plans can take advantage of a range of direct cash payments for inpatient, outpatient, and nursing home treatments. These are limited payments, not related to the costs incurred, but based on premiums paid and length of time under care. They are not a substitute for private health insurance.

Private health insurance covers all or part of the monetary costs of members' hospital in-patient or day-care treatments and associated hospital accommodation. It plays a significant role in the Irish health system, not only as a means of funding but also in terms of speedy access to health care. In Ireland, policy makers have encouraged health insurance as a means to offer a level of care or choice above that of the public system to those willing to pay (Colombo and Tapay 2004). There are approximately 4 million people in the Irish Republic and 2 million of them are covered by health insurance, even though there are public health entitlements. Individuals with low incomes, about a third of the population, receive medical services free of charge (Hunter 2004). The rest of the population can avail of the public hospital services for a minimal charge, however, other medical expenses, such as doctors' and consultants' fees, must be paid for in full.

Despite billions of euros pumped into the system over the past few years, the healthcare system remains poorly operated. With many accident and emergency wards overcrowded and understaffed, patients suffer long delays for treatment, with waiting lists extending for months or even years (Wren 2003). The healthcare system in Ireland is not based on a patient's rights, that is to say health care is not available as a right to the Irish citizen (Wren 2003). As a result of this poor service, many Irish have opted not to rely solely on the public system.

There has been an increase in demand for health insurance in recent years. Ireland's health insurance market consists of three insurers: Voluntary Health Insurance Board (VHI Healthcare); British United Provident Association, Ireland (BUPA Ireland), and the recently registered Vivas Healthcare, plus a small number of insurers who limit their membership to particular groups such as those for the police force (Garda Síochána) and the national electricity suppliers (Electricity Supply Board).

VHI Healthcare is a semi-state body, whose board is appointed by the minister for health and children. VHI Healthcare was established in 1957, and granted a license by the Minister of Health. No other licenses were granted and VHI occupied a monopoly position until 1994. During this year, the market was opened to competition under the provisions of the Health Insurance Act of 1994, in accordance with the European Union 3rd Directive on Non-Life Insurance (European Commission Interpretative Communication 2000). BUPA, a U.K. registered company entered the Irish market in 1996, and opened for business through its branch, BUPA Ireland, in January 1997. With a continued effort at opening the market to competition, Vivas Healthcare applied to IFSRA for a license to trade as a private health insurer and was registered in October 2004 (Slattery 2004). This company was formed with purely Irish backing and is the only health insurance company registered with IFSRA. All three companies offer a variety of plans that provide, at minimum, semi-private accommodation in a public hospital, while at the higher end of the scale, individuals can choose private accommodation in a selection of private hospitals.

VHI Healthcare has close to 2 million members, representing 40% of the Irish population, and an 80% share of the Irish health insurance market. BUPA Ireland supplies health insurance to approximately 300,000 individuals, representing a significantly smaller market share than VHI, however, this is understandable, given that BUPA Ireland has been operating in Ireland for a considerably shorter length of time. Since BUPA's arrival onto the Irish health insurance market, VHI Healthcare has diversified its product range to include multi-trip travel insurance, coverage for everyday health expenses, and has become the first company in Ireland to offer a separate dental care policy. BUPA Ireland also offers a product that covers everyday health expenses, similar to a cash plan, whereby cash payments are made to the insured upon a claim being processed.

Despite increasing competition, both companies have found it necessary to substantially increase their premiums in recent years. The two companies explain this move as a necessity to cater for rising medical costs, salary inflation, and an aging population in the Republic of Ireland. Despite these cost increases, Irish consumers still have quite affordable private health insurance premiums. In 2004, the average premium per individual was €440 per annum, while in the U.K. and U.S. insured individuals were paying between three and twelve times this amount respectively (VHI Healthcare 2002).

The health insurance regulatory environment in Ireland is unique in Europe and is the area of most change in the Irish health insurance market. The Irish Government, with its commitment to ensure consumer protection, remains the only country in Europe in which insurers are required by national health regulation to offer open enrollment, community-rated premiums, lifetime coverage, and minimum benefits policies (Thomson and Mossialos 2004; Health Insurance Act 1994). Open enrollment provides that insurers must accept all applicants for insurance coverage under the age of 65, regardless of health status, subject to prescribed waiting periods. There is no penalty for those individuals who would like to transfer from one insurer to another. The principle of community rating outlines how insurers must set and charge consumers the same price regardless of age, sex, or health status. Lifetime coverage means that insurers must renew contracts as long as they continue to operate within the market. Health insurance contracts must also provide benefits above a prescribed level, which is referred to as the "minimum benefits level." The minimum accommodation level prescribed is semi-private in a public hospital.

In July 2003, new risk equalization regulations were introduced (Statutory Instrument (SI) No. 261 of 2003). Risk equalization is not unique to Ireland. According to VHI Healthcare risk equalization is operated in Australia, Germany,

Holland, Switzerland, and a number of North American States and is being implemented in South Africa (see VHI position paper on "Lifelong Affordable Health Insurance", 2004 on www.vhi.ie). Risk equalization is a process that aims to neutralize differences in health insurers' costs that arise due to variations in their risk profiles. This results in a transfer of funds from insurers with healthier than average risk profiles to those with less favorable risk profiles (European Commission 2003). According to these regulations, all health insurance insurers must submit bi-annual statutory returns to the Health Insurance Authority (HIA).[227] Using this information, the HIA can evaluate claims by age and gender of the health insurance companies and therefore the risk differentials between insurers. If the level of risk difference between companies is found to be greater than 10%, the HIA shall activate the scheme. If the risk difference level is between 2% and 10%, the HIA has to make a recommendation to the Minister for Health (European Commission 2003, p. 5).

In a statement issued on the May 18, 2005, the HIA announced that it had recommended to the Minister of Health that she should introduce risk equalization. One month later, the Minister announced that risk equalization should not be instituted, and would be considered again in six months upon receipt of additional information from the HIA. In 2002, BUPA Ireland had forwarded a complaint to the European Commission, arguing that risk equalization constituted state aid, since VHI estimated that it could receive up to €20 million from BUPA Ireland if the risk-equalization scheme is introduced (Wall 2004). However, the Commission, upon investigation of this matter, determined that transfers under the equalization scheme do not constitute state aid (European Commission 2003). BUPA Ireland continues to dispute this judgment and has submitted a further challenge to the risk equalization scheme to the European Court of First Instance (Turner 2004).

The HIA is also calling for alterations to be made to the principle of community rating, insisting that a system of lifetime community rating should be introduced. Under lifetime community rating, the premium that a person pays rises with the age at which they enter the private health insurance market, but remains stable in real terms thereafter, subject to medical inflation, etc. Lifetime community rating would also allow first time applicants to get insurance even if over the age of 65, thereby also changing the current system of open enrollment which requires that health insurance be offered for the same premium regardless of health, age or gender up until the age of 65. Turner (2004) believes that it is unlikely that these changes will come about in the short term. Further changes to the health insurance regulatory environment have been presented by the Department of Health and Children, which has indicated its intention to change the legal status of VHI Healthcare, which, until now, has been exempt from certain other insurance legislation, in an effort to counteract any barrier to entry (Turner 2004).

With the above ground rules set for the provision of health insurance in Ireland, and the relatively low cost of purchasing health insurance, it is not surprising that close to half of the Irish population have decided to purchase health insurance. At the end of 2003, in excess of 2 million individuals were covered by health insurance. This figure represents a 3% increase from 2002 (Turner 2004). It is revealing that out of a total population of around 4 million people, all of who should receive free

[227] This is the independent regulatory authority for health insurers in Ireland.

hospital care, so many people opt for private coverage. A recent Economic and Social Research Institute (ESRI) survey of the public found that many choose private health care, for speed of access, choice of doctor and hospital, and the quality of care provided. With a public health service that is heavily criticized, more people are not prepared to take the chance of having to wait on long public hospital lists (Clarke 2004).

11.7 DISTRIBUTION CHANNELS

There are a number of channels through which insurance products are marketed. These are either direct or through various types of intermediaries. Table 11.13 lists the various types of intermediaries to be found in Ireland, together with their shares in the long-term insurance market. The types of intermediaries found in Ireland are discussed below.

Table 11.13. Life Insurance Market Share by Distribution System, 2003

		Brokers %	Agents %	Tied Agents %	Employee and Company Representatives %	Direct %
Annual Premium	Life	16	1	9	5	5
	Pensions	42	1	7	5	9
	Total	58	2	16	10	14
Single Premium	Life	24	1	2	1	10
	Pensions	29	1	2	1	8
	Total	53	3	16	12	18

Source: Irish Insurance Federation (2004), p. 36.

11.7.1 Intermediaries

The Financial Services Regulator maintains a register detailing the status of investment intermediaries in accordance with the Investment Intermediaries Act, 1995. These intermediaries are authorized to provide investment and insurance services from registered product producers. A product producer is usually a bank, building society, or insurance company, which may sell its own products either directly or through intermediaries. Insurance companies and insurance intermediaries were brought under the auspices of the Investment Intermediaries Act, 1995, by the Insurance Act, 2000. Thus, insurance intermediaries are regulated as investment intermediaries, which has led to some confusion as to how the relevant acts are to be applied.

Investment intermediaries are classified under two categories, Multi-Agency Intermediaries (MAI) or Authorised Advisors (AA). An MAI is an investment business firm that may receive and transmit orders in investment instruments and provide advice on investment instruments for product producers from whom it holds an appointment in writing. With the MAI category, some MAIs can receive and transmit orders in insurance policies to Lloyd's brokers or other non-resident intermediaries with whom it has an agreement to refer proposals. They can also act as a product producer. MAIs are required to maintain a minimum level of shareholders funds or a capital account of €10,000; and when acting as a product producer must maintain a minimum level of shareholders funds or a capital account of €50,000. A sub-category of the multi-agency intermediary is the restricted activity investment product intermediary who can only receive orders from specific providers from whom it has an appointment in writing. They must also submit annual audited accounts to the Central Bank within six months of the firm's financial year.

An AA is an investment business firm that can provide investment advice on investment instruments without holding a letter of appointment. In addition, it can receive and transmit orders to product producers from whom it holds a letter of appointment. An AA must recommend the most suitable investment product available in the market, regardless of whether or not it holds an appointment from the relevant product producer. Even though product producers are continually developing their direct sales channels, the volume of business written through intermediaries continues to grow.[228]

Insurance intermediaries are further subdivided into brokers, agents, and tied agents. Brokers are intermediaries that are required to deal with five or more companies in the life and/or non-life markets and are considered to be agents of the insured and experts in the field of insurance. In the collection of premiums, they act as agent of the insurer. Agents must limit their business activities to a maximum of four companies and must include the names of those suppliers on their headed paper. In the same manner as brokers, they are considered as agents of the insured, but agents of the insurer for payment of premiums. However, they are not considered to be experts in insurance. Tied agents only provide a distribution channel for a single company and are agents for that company; they are never agents of the insured.

11.7.2 Duties of the Intermediaries

As intermediaries, the brokers and agents are legally agents of the insured and they are deemed to have the same knowledge about a risk as their client. This is important when it comes to the insured's duty of disclosure, or utmost good faith as it is known in Ireland. The insured, or proposer, is required to reveal all material facts about a risk at the inception of insurance or on renewal. This means the insured must reveal all facts that a reasonable insurer considers would affect the judgment of a reasonable underwriter in underwriting a risk. If the broker or agent is aware of a material fact, even though not informed of it by the insured, they are duty bound to inform the insurer, otherwise the policy could be declared void from inception. If the broker or agent does not inform the insurer of such a material fact and the insured

[228] Further information can be found on the IFSRA web site: www.ifsra.ie.

decides to void the policy, the broker may well be held professionally negligent and have to refund any losses incurred by their client. As the agent is not considered an expert in the realm of insurance, it is possible that they would not be held liable for professional negligence, although they could be in breach of their contract of agency.

11.8 CONCLUSIONS

In the last 30 years, Ireland, the "Celtic Tiger," has experienced an impressive economic growth rate. The insurance sector is part of this success. Over the years, most Irish governments have encouraged competition within the market. The motor insurance market, although competitive, is unattractive to new entrants because of the high cost of liability claims. This is being dealt with through the Personal Injury Assessment Board and the accompanying Courts and Civil Liability Act, 2004. Only time will tell as to how successful these innovative steps will be. Certainly, a lot of interest has been shown in the rest of the E.U. about the PIAB. These steps will hopefully reduce the cost of liability insurance where premiums are also extremely high, especially employer's liability insurance. Again, because of the high potential claims costs, there is no incentive for new companies to enter the market.

The life insurance market is at saturation point, but as the population of Ireland expands, due to its economic success, net immigration, and return of émigrés, there is room for further expansion in this market for an innovative player. Thus, this market is now open to competition.

With health insurance, the market has been considered very unattractive because of the community rating scheme and the threat of risk equalization. This latter threat has been diminished and could thus lead to further new entrants. In the past, the VHI Healthcare had total control on which medical financing schemes could be put in place and this led to a stifling of initiative, but now this market is open to new ideas.

Ireland is the only English-speaking nation in the Eurozone and is an ideal place for non-Europeans, such as Americans and Asians, to enter the European market. In Ireland, all formal documentation is in English, unlike in other nations such as France or Germany. Once registered in Ireland, the insurer can operate anywhere in Europe and be taxed in Ireland. Thus, Ireland could be considered a doorway to Europe, which would be attractive to countries such as the U.S. Furthermore, the IFSC provides a base for all types of insurance organizations to operate internationally at a favorable tax rate, although they cannot operate in Ireland. This has provided opportunities to innovative insurers, brokers, and captives to commence operations selling into the rest of Europe, using Ireland as a base. There has been substantial growth in the number of international brokers, captive insurers, and insurance companies operating out of the IFSC. This rate should continue to increase in the future and contribute even more to the growth of Ireland's economy.

To conclude, Ireland is a favorable place for large insurance organizations to establish themselves and operate into Europe. The country is open to new and innovative ideas and the insurance atmosphere is constantly improving.

11.9 REFERENCES

_____, 2000, "Some Homeowners Face Shortfalls as Endowment Mortgage Matures," *Irish Times*, August 11, 2005, Finance Section.

Bolkestein, Frits, 2004, "Building a Single E.U. Market for Insurance," address at conference organized by the Royal Institute of International Affairs in association with *Reactions* Magazine at Chatham House, London, England, February 2.

Black, Kenneth, 2000, *Life and Health Insurance* (Upper Saddle River, NJ: Prentice-Hall).

Bradley, John, 1999, "The History of Economic Development in Ireland, North and South," in Anthony F. Heath, Richard Breen, and Christopher Whelan, eds., *Ireland North and South: Perspectives from Social Science* (Oxford: Oxford University Press), pp. 35–68.

Breen, Richard, Damien F. Hannan, David B. Rottman, and Christopher T. Whelan, 1990, *Understanding Contemporary Ireland* (Houndmills, Ireland: MacMillan).

Central Statistics Office, 2004, *Statistical Yearbook of Ireland* (Dublin, Ireland).

Clarke, Jim, 2004, "Health Insurance—Where is the Best Deal?" www.irishhealth.com.

Colombo, Francesca and Nicole Tapay, 2004, "Private Health Insurance in OECD Countries: The Benefits and Costs for Individuals and Health Systems," OECD Health Working Papers No.15 Directorate for Employment, Labour and Social Affairs 2004(6), Organisation for Economic Co-operation and Development, Paris, France.

De Boer-Ashworth, Elizabeth, 2004, "The 'Celtic Tiger' in a Global Context," in Neil Collins and Terry Cradden, eds., *Political Issues in Ireland Today* (Manchester: Manchester University Press), pp. 1–14.

Deloitte and Touche, 1996, *Report on the Economic Evaluation of Insurance Costs in Ireland,* Department of Enterprise, Trade, and Employment, Dublin, Ireland.

Department of Enterprise, Trade, and Employment, 1991, *Annual Insurance Report 1990* (Dublin, Ireland).

Department of Enterprise, Trade, and Employment, 1992, *Annual Insurance Report 1991* (Dublin, Ireland).

Department of Enterprise, Trade, and Employment, 1993, *Annual Insurance Report 1992* (Dublin, Ireland).

Department of Enterprise, Trade, and Employment, 1994, *Annual Insurance Report 1993* (Dublin, Ireland).

Department of Enterprise, Trade, and Employment, 1995, *Annual Insurance Report 1994* (Dublin, Ireland).

Department of Enterprise, Trade, and Employment, 1996, *Annual Insurance Report 1995* (Dublin, Ireland).

Department of Enterprise, Trade, and Employment, 1997, *Annual Insurance Report 1996* (Dublin, Ireland).

Department of Enterprise, Trade, and Employment, 1998, *Annual Insurance Report 1997* (Dublin, Ireland).

Department of Enterprise, Trade, and Employment, 1999, *Annual Insurance Report 1998* (Dublin, Ireland).

Department of Enterprise, Trade, and Employment, 2000, *Annual Insurance Report 1999* (Dublin, Ireland).

Department of Enterprise, Trade, and Employment, 2001, *Annual Insurance Report 2000* (Dublin, Ireland).

Department of Enterprise, Trade, and Employment, 2002, *Annual Insurance Report 2001* (Dublin, Ireland).

Department of Enterprise, Trade, and Employment, 2003, *Annual Insurance Report 2002* (Dublin, Ireland).

Enterprise Strategy Group, 2004, *Ahead of the Curve* (Dublin: Enterprise Strategy Group).

European Commission, 2003, "Ireland: Risk Equalization Scheme in the Irish Health Insurance Market," State Aid N46/2003, Brussels, Belgium.

European Commission Interpretative Communication, 2000, "Freedom to Provide Services and the General Good in the Insurance Sector," *Official Journal C043* (February 16): 5–27.

Europe-Economics, 2004, *Report on Competition in the Irish Motor and Liability Insurance Markets,* Irish Competition Authority, Dublin, Ireland.

Fagan, Colm, 2003, "Developments in the Irish Life Assurance Market," presentation to "The Net-Work," Dublin, Ireland (February 7).

Foster, Roy F., 1988, *Modern Ireland.* (London: Penguin Books).

Greenford, Brian C., 2002, "Personal Injury Claims Management: A Comparison of England and Ireland," *Insurance Research and Practice* 17: 49–57.

_____, 2003, *Claims Management, Irish Law, and Practice* (Dublin: Insurance Institute of Ireland).

Health Insurance Authority, 2004, "Private Health Insurance—Your Rights, Your Choices," Health Insurance Authority Consumer Information, Dublin, Ireland.

Hunter, Niall, 2004, "The Medical Card Mine Field," www.irishhealth.com.

Irish Government, 1994, *Health Insurance Act* (Dublin, Ireland).

Irish Government, 2003, *Statutory Instrument No. 261– Risk Equalisation Scheme* (Dublin, Ireland).

Irish Insurance Federation, 2003, "Press Release—February 26, 2003," Dublin, Ireland.

Irish Insurance Federation, 2004, "Factfile—October," Dublin, Ireland.

Irish Financial Services Regulatory Authority, 2004, "Insurance Statistical Review 2003," Dublin, Ireland.

Irish Pensions Board, 2003, *Annual Reports and Accounts,* Dublin, Ireland.

KPMG, 2003, "International Financial Services in Ireland," Dublin, Ireland.

Krishnamurthy, Sandeep, 2000, "Herfindahl Index—Measuring Industry Concentration," *Business Lines-Investment World*, July 30, 2000.

Motor Insurance Advisory Board, 2001, *MIAB Report* (Dublin, Ireland: Brunswick Press).

Office of the Public Works, 2004, "Report of the flood policy review group." Dublin: OPW

Organisation for Economic Co-operation and Development, 2004, "Figures and Statistics on Member Countries," Paris, France.

Parsons, Christopher, Gerald Dickinson, and Brian Greenford, 2004, "Report on Competition in the Irish Motor and Liability Markets," Irish Competition Authority, Dublin, Ireland.

Skipper, Harold D. and Robert W. Klein, 2000, "Insurance Regulation in the Public Interest: The Path towards Solvent, Competitive Markets," *The Geneva Papers on Risk and Insurance* 24: 482–504.

Slattery, Laura, 2004, "AIB to Invest in New Health Insurance Group," *Irish Times* October 21, 2004, p. 3.

Thomson, Sarah and Elias Mossialos, 2004, "Private Health Insurance and Access to Healthcare in the European Union," *Euro Observer* 6: 1–4.

Turner, Brian, 2004, "Recent Developments in the Irish Private Health Insurance Market," *Euro Observer* 6: 5–6.

Voluntary Health Insurance Board, 2002, "Press Release—July 18, 2002—VHI Healthcare Premiums to Increase by 18 Percent," Dublin, Ireland.

Wall, Martin, 2004, "Regulator Rules Out Equalization of Risk in Health Insurance Area," *Irish Times,* September 11, 2004.

VHI Healthcare, 2005, *About VHI Healthcare*, www.vhi.ie/about/re-overview.pdf.

VHI Healthcare, 2004, "Lifelong Affordable Health Insurance," www.vhi.ie.

World Bank, 2004, *World Development Indicators 2004*, www.worldbank.org/data/wdi2004/worldview.htm.

Wren, Maeve-Anne, 2003, *Unhealthy State* (Dublin, Ireland: New Island).

11.10 APPENDIX

Appendix 1. Guide to Main Provisions of Most Significant Irish Insurance Legislation

Legislation	Principal Provisions
European Communities (Non-Life Insurance) Framework Regulations, 1994 (S.I. No. 359 of 1994)	▪ Transposes into Irish Law the Third (E.U.) Non-Life Insurance Directive (92/49/EEC). ▪ Institutes a single authorization and supervision system throughout the E.U. whereby authorization and supervision is done by the Member State in which the insurer's head office is located. ▪ Imposes obligation on insurers to report to the supervising authority, the Minister for Enterprise, Trade & Employment. ▪ Prescribes rules for the valuation of underwriting liabilities and assets and requirements on the localization of assets. ▪ Prescribes information and other requirements for consumers. ▪ Prescribes fees to be paid by an insurer applying for an authorization or an extension.
European Communities (Life	▪ Transposes into Irish Law the Third (E.U.) Life

Legislation	Principal Provisions
Assurance) Framework Regulations, 1994 (S.I. No. 360 of 1994)	Assurance Directive (92/96/EEC). ▪ Contains provisions similar to the Non-Life Regulations described above. ▪ Implements relevant provisions of the Second (E.U.) Life Assurance Directive (90/619/EEC) concerning freedom of services to conduct business on a cross-border basis throughout the E.U. ▪ Prescribes the format for the statutory returns of life insurance undertakings.
European Communities (Insurance Undertakings Accounts) Regulations, 1996 (S.I. No. 23 of 1996)	▪ Transposes into Irish Law the (E.U.) Insurance Accounts Directive (91/674/EEC) on the annual accounts and consolidated accounts of insurance undertakings.
European Communities (Non-Life Insurance Accounts Regulations, 1995 (S.I. No. 202 of 1995)	▪ Prescribes the format for the statutory returns of insurance undertakings.
Insurance Act, 1989 (Number 3 of 1989)	▪ This is the principal framework in domestic legislation for the regulation of the insurance industry. ▪ Article 34 requires life insurance companies to have an appointed actuary whose duty is to carry out any function relating to the business which is required to be undertaken by an actuary. ▪ Article 35 imposes a duty on an insurer's auditor to report independently to the Minister if circumstances exist which could affect the insurer's ability to fulfill its obligations to policyholders. ▪ Part III deals with the payment of commission by insurers. ▪ Part IV deals with the regulation of insurance intermediaries. ▪ Sections 59 and 60 authorize the Minister to appoint 'appointed officers' for the purpose of carrying out investigations.
Insurance (Fees) Order, 1995 (S.I. No. 128 of 1995)	▪ Prescribes statutory fees to be paid annually by authorized insurers to the Minister for Enterprise, Trade, and Employment.
Investment Intermediaries Act 1995	▪ Creates the E.U. investment services directive, which establishes the conditions under which investment firms and fund managers are authorized in E.U. member states.

Legislation	Principal Provisions
Insurance Act, 2000 (S.I. Number 42 of 2000	▪ Introduces a system of statutory authorization and supervision of insurance and reinsurance intermediaries under the aegis of the Central Bank. Strengthens powers and authority in relation to reinsurance entities, including notification procedures.
Directive 2002/87/EC of the European Parliament and of the Council of December 16, 2002 on the supplementary supervision of credit institutions, insurance undertakings, and investment firms in a financial conglomerate.	▪ This directive enhances the prudential soundness and effective supervision of financial conglomerates. It is an important step in the ongoing modernization of E.U. financial legislation. It enhances financial stability and is a significant improvement in the protection of depositors, insurance policy holders, and investors.
Directive 2002/92/EC on insurance mediation	▪ Provides for the harmonization throughout the E.U. of rules applicable to insurance and reinsurance intermediaries and provides a European passport for such entities.
Solvency I Directive 2002/12/EC and 2002/13/EC	▪ Increases the absolute minimum amount of capital required (the so-called minimum guarantee fund) and requires them to be indexed in line with inflation. The new absolute minimum is set at €3 million (€2 million for some classes of non-life insurance) compared to the previous amounts which ranged between €200,000 and €1.4 million. ▪ Increases thresholds, based on levels of premiums and claims below which a higher solvency margin is required. The required solvency margin on a premium basis will now be 18% up to the first €50 million of premium—this figure has been increased from €10 million and will be indexed in future—and 16% above €50 million. On a claims basis, the margin will be 26% on the first €35 million, up from €7 million and also indexed in future, and 23% above €35 million. ▪ Provides supervisory bodies with the added authority to intervene early to take remedial action where policyholders' interests are threatened. For example, in a situation where an insurance undertaking currently satisfies the solvency margin requirements, but its financial position is deteriorating rapidly, the supervisor will now be able to take action. ▪ Increases the required solvency margin by 50% for certain categories of non-life business that have a particularly volatile risk profile (marine, aviation, and general liability).

Legislation	Principal Provisions
Central Bank and Financial Services Authority of Ireland Act 2003	▪ Provides for the establishment of IFSRA. ▪ Restructures the Central Bank. ▪ Amends existing legislation to take account of the functions of IFSRA.

Source: Irish Financial Services Regulatory Authority (2004), p. 12.

11.11 LEXICON

AA. An authorized adviser is an investment business firm that can provide investment advice on investment instruments without holding a letter of appointment. In addition, it can receive and transmit orders to product producers from whom it holds a letter of appointment.

Agent. Agent has two meanings. The first refers to one person acting on behalf of another. The second is a technical meaning where an insurance intermediary is registered with up to four companies on the market and provides their clients with insurance cover from these companies. They are agents of the insured except for premium collection where they become agents of the insurer. They are not considered experts in insurance.

AIR. The Alliance for Insurance Reform is a group of businesses which came together to form a group to lobby government to change the insurance market. This body feels that it succeeded in its task and has now disbanded.

Bancassurance. Bancassurance is the combination of the banking and insurance businesses within the same organization.

Book of Quantum. This is a publication used by the PIAB, insurers and courts to provide guidance on the award of general damages for particular injuries. It was first published in 2004 and compiled from payments made to injured parties during 2003.

Broker. A broker is an insurance intermediary who is registered with at least five life or non-life companies. Their task is to find the best insurance cover for their clients. They are considered agents of the insured except for premium collection when they are agents of the insurer. They are considered experts in insurance.

Building Society. A building society was originally a registered company formed mainly for the purpose of lending money for the purchase of private houses. The company provides savings accounts for investors who could only withdraw money after notice was given. These savings were used to fund house loans by means of mortgages registered on the property deeds. These mortgages provided a first call on the outstanding loan if a house was sold. Building societies were not allowed to hold current accounts for investors. These companies have now developed so that loans are available for all purposes and current accounts are provided, thus providing functions similar to banks.

BUPA. The British United Provident Association is a provider of healthcare insurance registered in the U.K. and operating in the Irish market.

Community Rating. Community rating is the practice of assigning health insurance premiums without regards to health risk or status.

Compo Culture. Compensation culture is the propensity of the public to make insurance claims. In Ireland, this is considered to be very high and the term is used to accentuate the claims making propensity.

Consensus Funds. Consensus funds allocated their assets and countries of investment based on the average style of investment managers in Ireland.

CSO. Central Statistics Office.

Dáil. Dáil Eireann is the lower house of the Oierachtas, houses of parliament.

Direct Line Company. A direct line company is an insurance company that provides coverage directly via telephone calls to private individuals without the intervention of an intermediary. The coverage is normally provided through a call center which deals with simple policies such as motor insurance. Generally, clerks with little or no training plug in relevant details to a computer program which subsequently returns a premium. Any client who wishes to insure a non-standard risk will be rejected.

Earned Premium Income. This is premium income after deduction for reserves for unearned premium income and before deduction of reinsurance premiums.

ESRI. The Economic and Social Research Institute is an Irish government body whose aim is to produce high quality research relevant to Ireland's economic and social development with the aim of informing policy-making and societal understanding.

E.U. European Union.

Eurozone. The Eurozone consists of the countries in Europe using the Euro as their national currency.

Direct Line Insurers. These are insurers that operate telephone call centres for the purposes of providing private insurance to the individual. These deal mainly with motor insurance but also may provide insurance cover for the householder.

Fianna Fail. The dominant Irish political party which refers to itself as the Irish Republican party. It is difficult to define in terms of right or left leanings but can be defined as a populist party.

GDP. Gross domestic product.

Herfindahl Index. The Herfindahl Index is a measure of competitiveness. It measures the size of a firm relative to the industry as a whole. It is the sum of squares of each company's market share. The final index can range from 10,000 (when there is only 1 firm in the market) to 0 (when the market is extremely competitive). Generally, it is argued that markets are competitive if the Herfindahl Index is 2000 or below (Krishnamurthy 2000).

HIA. The Health Insurance Authority ("The Authority") is the independent authority for health insurers in Ireland. It was established on February 1, 2001 in accordance with the terms of the Health Insurance Act of 1994 by Mr. Micheál Martin T.D., Minister for Health and Children. The Authority is a regulatory body for private health insurance in Ireland.

IBEC. The Irish Business and Employers Confederation is the umbrella body for Ireland's leading sectoral groups and associations whose aim is to lobby government and look after member's general interests.

IFSC. The Irish Financial Services Centre in Dublin was set up by the Irish Government with E.U. approval in 1987. It is globally recognized as a leading location for a range of international financial services, including banking, asset financing, fund management, corporate treasury management, investment management, custody and administration, and specialized insurance operations.

IFSRA. The Irish Financial Services Regulatory Authority (Financial Services Regulator) was established on May 1, 2003. The Financial Services Regulator is responsible for the regulation of all financial services firms in Ireland. It also has an important role in the protection of the consumers of products of those firms.

IIF. The Irish Insurance Federation is the representative association for insurance companies in Ireland.

III. The Insurance Institute of Ireland is the professional body for insurance personnel. It provides educational courses in risk management and insurance.

Insurance Ombudsman. The Insurance Ombudsman Scheme is organized by the IIF as a means whereby insureds who feel aggrieved following a decision of an insurer can make an appeal. Only members of IIF may join this scheme and it is entirely voluntary. The members must abide by the rulings of the Ombudsman, but the insured does not. The IFSRA has now appointed there own insurance ombudsman in addition to the one appointed by the IIF.

MAI. A multi-agency intermediary is an investment business firm that may receive and transmit orders in investment instruments and provide advice on investment instruments in respect of product producers from whom it holds an appointment in writing.

'No Foal, No Fee'. Also 'No Win, No Fee.' This phrase implies that solicitors provide services on the basis that if they do not obtain compensation for their clients, then they will not charge any fees to their clients. In Ireland, solicitors are not longer allowed to advertise that they provide this service.

NPPI. The National Pensions Policy Initiative was launched in October 1996 to facilitate debate on how to achieve a fully developed national pension system, to formulate a strategy, and to make recommendations for actions needed to achieve this system.

OPW. The Office of Public Works is an operational arm of the government, which has the expertise, resources, and infrastructure to deliver quality services in property, design, construction, and procurement (on time and in budget) in order to support government policies.

Pensions Board. The Pensions Board is a statutory body set up under the terms of the Pensions Act of 1990. It has two main functions defined under the Pensions Act of 1990, as amended ("Pensions Act"). First, it regulates the occupational pension schemes and personal retirement savings accounts in Ireland as part of its statutory role to monitor and supervise the operation of the Pensions Act. Second, the board advises the Minister for Social and Family Affairs on pension matters generally.

PIAB. The Personal Injuries Assessment Board is a statutory body which provides independent assessments of personal injury compensation for victims of workplace, motor, and public liability accidents.

PRSA. A personal retirement savings account is an account which is set up by providers on behalf of employers so that funds can be provided to set up pensions for employees. All employers must register with a provider so that an account can be established. Self-employed people may also avail themselves of this scheme.

Risk Equalization. Risk equalization is a process that aims to neutralize differences in health insurers' costs that arise due to variations in their risk profiles. This results in a transfer of funds from insurers with healthier than average risk profiles to those with less favorable risk profiles.

Single Passport. The "single passport" scheme was created by the E.U. to enable investment firms to do business anywhere in the E.U. with a minimum of red tape while increasing customer protection.

Solvency II. The Solvency II project is a continuation of the work already undertaken by the E.U. under the auspices of Solvency I. Whereas the Solvency I project aimed at revising and updating the current E.U. solvency regime, Solvency II has a much wider scope. It contains a fundamental and wide-ranging review of the current regime in light of current developments in insurance, risk-management, finance, financial reporting, etc. One key objective of Solvency II is to establish a solvency system that is better matched to the true risks of an insurance company.

SSIA. The special savings investment account was introduced by the Minister of Finance in the Finance Act of 2001 as a new savings scheme to encourage regular savings by individuals.

Structural Funds. Structural funds are funds paid to developing regions within the E.U. to facilitate their economic growth.

Tied Agents. Tied agents are agents for only one insurer.

Tracker Bonds. Tracker bonds are fixed term investment products offering an index-linked return or guarantee on the original capital, whichever is the greater return.

Unit Linked Funds. Unit linked funds are a product whereby the value of the policy is linked to units consisting of bundles of stocks and shares quoted on any stock market. These bundles often are constructed based on a theme, such as green companies, blue chip companies, etc. The values of the units directly reflect the values of the underlying assets of the fund, and fluctuate daily according to the performance of those investments.

U.K. United Kingdom.

U.S. United States of America.

Utmost Good Faith. Utmost good faith is the legal requirement that the insured reveals all material facts when renewing or initializing an insurance policy.

VHI. The Voluntary Health Insurance Board, now known as VHI Healthcare, is an insurance company established by the Irish Government to provide healthcare to the Irish market.

Written Premium Income. This is the premium actually paid by the insured to the insurer before deduction of reinsurance premiums and allowances for reserves.

12 China's Insurance Industry: Developments and Prospects

Qixiang Sun
Lingyan Suo
Wei Zheng[229]
Peking University

12.1 INTRODUCTION

In 1980, the insurance industry in China was revived after twenty years of stagnation when economic system reform and opening up began in China. During the past two decades, the insurance industry has developed from a virtually nonexistent, minimal, and closed industry to a large, open industry with potential promise. The volume of insurance premiums grew rapidly from RMB 0.46 billion in 1980 to RMB 388 billion in 2003,[230] ranking China eleventh in the world with 1.6 percent of the global market share (China Insurance Regulatory Commission (CIRC); Swiss Re 2004, p. 35).[231] This fast growing economy coupled with the largest population in the world, rising personal income, and economic system reform could not only explain the rapid growth of China's insurance industry, but also foretell the increasing importance of China's insurance market in the future.

China's insurance industry was deeply affected by the reform and opening up of China since its development was an integral part of the latter. On one hand, the economic system reforms recreated China's insurance market and thus determined the direction of the industry. China's market-oriented economic system reform was a gradual process of institutional change dominated by the government. As a result, only state capital could be found in China's insurance industry and the market was exceedingly concentrated at the beginning of the industry's restoration. The

---

[229] The authors would like to thank J. David Cummins and Bertrand Venard for their comments and suggestions, which greatly improved the presentation of this chapter. We also thank Ms. Sufang Cui for her editorial assistance on an earlier draft. All remaining errors, of course, are ours. Data in this chapter do not include the insurance industry in Hong Kong, Macao, and Taiwan.

[230] RMB stands for Chinese Currency, of which the full name is Ren Min Bi.

[231] Information was provided to the authors directly by the China Insurance Regulatory Commission (CIRC). This notation is followed through the chapter.

government was and still is heavily involved in the industry, although its role is now changing from an administrator to a regulator. On the other hand, opening up to the outside exposed the infant industry to fierce international competition, bringing great opportunities along with grave challenges to domestic insurers.

This chapter provides an analysis of the development of China's insurance industry in the context of China's economic reform and opening up and its future prospects. The first section briefly covers the development of China's insurance industry before 1949, when the People's Republic of China was founded. The second section is devoted to the story after 1949, in particular, the three stages of China's insurance industry after 1980, when China began its reform and opening up: (1) resumption and restoration, (2) market-oriented reform, and (3) opening up and rapid growth. The third section explores and evaluates the opening up of the industry. The fourth section examines the potential of China's insurance industry, points out the key drivers and possible obstacles of its future growth, and forecasts its future trends. Finally, the last section presents the conclusion.

12.2 CHINA'S INSURANCE INDUSTRY BEFORE 1949[232]

As a country with a history of more than five thousand years, ancient people of China developed various measures by which people with similar conditions could share risks and help each other. These measures can be considered the most primitive forms of insurance. According to the Book of Yizhou, an ancient Chinese literary history book, Chinese people realized the uncertain nature of natural catastrophes and began to make efforts to store grain in abundant years for famine approximately three thousand years ago. Years later, merchants conducting freight on the Yangtze River invented a clever method to diversify risks. Each merchant would place his goods on several different boats, consequently sharing the losses incurred from the sinking of one boat with other merchants.

The modern insurance industry did not emerge until the end of the Qing Dynasty (1644–1911), China's last feudal dynasty. Initially, the Qing Dynasty saw China as the center of the world and pursued a closed door policy resulting in serious trouble both at home and abroad. The western powers, which had already completed the industrial revolution, later compelled the weak and corrupted Qing government to accept trade, thus introducing modern insurance to China in conjunction with western commodities to China.

Guangzhou used to be the only trade port in China before the Opium War (1839–1842) between China and England. Usually, foreign insurance companies entrusted their Chinese business to foreign trade firms established in China, the so-called Yang Hang.[233] In 1805, British merchants established the first foreign insurance institution in Guangzhou, China, the Canton Insurance Society. This kind

[232] Ye, et al. (1998), pp. 1–235.

[233] China's foreign trade then consisted of overland trade with Russia and sea trade with other countries. The foreign trade firms actually participated in almost all of the business related to trade, e.g., buying and selling on principals' behalf, banking, insurance, and shipping.

of effort did not gain popularity until the Qing government was defeated in the Opium War and was forced to open more trade ports. Then, the sharp increase of foreign trade resulted in a tremendous demand for insurance. Many foreign firms divested their insurance business and established separate insurance companies in China, which monopolized China's insurance market for a period of time.

In May 1865, the Shanghai Yihe Insurance Society, a freight insurer, was established. It was the first domestic insurer in China and introduced insurance policies written in both English and Chinese for the first time. Previously, policies were written only in English in China. In December 1875, the Insurance Promotion Bureau was founded and it ended the monopoly of foreign insurers for hull insurance. The emergence of domestic insurers was welcomed and supported by the domestic business community. Subsequently, the Renhe Marine Insurance Company, Jihe Fire and Marine Insurance Company, and other domestic insurance companies were set up. Until 1911, there were approximately forty-five domestic insurers operating in China which were highly concentrated in trade ports in Shanghai. In fact, thirty-seven out of forty-five insurers were in Shanghai (Ye, et al. 1998, p. 54).

Although domestic insurers began to develop their market share, they were at a disadvantage in competing with the foreign insurance companies. China's foreign trade was controlled by foreign firms and consequently, it was very difficult for domestic insurers to expand their business. Furthermore, they had to rely on their foreign rivals for reinsurance because of their weak financial strength. As a result, foreign companies continued to dominate China's insurance market and earn substantive profits. For example, the return on equity of the Hong Kong Fire Insurance Company, established in Hong Kong in 1868 by British businessmen, was about fifty percent (Ye, et al. 1998, p. 27).

The Revolution of 1911 not only ended the Qing Dynasty, but also stimulated the development of industry and commerce in China. Consequently, insurers began to expand their business to other trade ports and inland trade cities. Again, foreign insurers maintained their dominant position due to their predominance in capital, technology, experience, and ability to accept ceding business.

It should be noted that American businessmen became heavily involved in China's insurance industry after World War I (WWI). Cornelius Vander Starr, an American entrepreneur, founded an insurance agency called American Asiatic Underwriters in Shanghai in 1919 and then Asia Life Insurance Company in 1921.[234] In the 1930s, American insurers gained an equal position to their British associates and foreign insurance companies thus had about seventy-five percent of China's insurance market (Ye, et al. 1998, p. 120).

Following the War against Japanese Aggression (1937–1945), the Chinese People's Revolutionary War (1945–1949), and the founding of the People's Republic of China in 1949, China's insurance industry was quite volatile as China's economy fluctuated greatly in time of the war. In turn, foreign insurers began to exit the

---

[234] These companies were the start of the famous international insurance conglomerate known as the American International Group. Mr. Starr moved his company's headquarters to New York in 1939. His business in China was compelled to pause in 1941 by the Japanese army and was resumed after China's Victory in War against Japanese Aggression. In 1949, the center of these companies' business was transferred from Shanghai to Hong Kong and operation in Shanghai ended in June 1951.

market. Furthermore, people, unable to anticipate the unstable economy and in need of economic security, did not purchase life insurance coverage. Therefore, insurance premiums were mainly collected from trade-related businesses and operations subsequently became concentrated in the trade centers.

To sum up, modern insurance was introduced to China by the West, which controlled China's insurance market profitably for a long period of time. This may have contributed to the concern that China's insurance industry might again be controlled by foreign capital and the prudence in granting foreign insurers licenses to operate in China, which was shared by many people even at the end of the 20th century.

12.3 CHINA'S INSURANCE INDUSTRY IN THE ECONOMIC REFORM: PAST AND PRESENT

On October 20, 1949, the People's Insurance Company of China (PICC), the first state-owned insurance company in China's history, was established (Wu 2004, p. 4). PICC developed a mission to "protect state assets, ensure industry safety, promote materials flow, provide stability," and, more importantly, "collect surplus funds and enriching state assets." However, the waning necessity for commercial insurance in a centralized planning economy finally led to the closure of the domestic insurance industry in 1959 (Wu 2004, p. 4). Only a small amount of foreign currency and export related insurance was left. It was not until the 3rd Plenary Session of the 11th Chinese Communist Party (CPC) Central Committee led by Deng Xiaoping at the end of 1978 that economic system reform and economic construction became the main theme of China's development. Soon, nation's financial system, governed by the People's Bank of China (PBOC), came into being. Concurrently, the insurance industry was restored and developed rapidly in the ensuing years. To explore the development of the insurance industry in China, this section is divided into three stages consistent with the gradual development of China's economic system reform: (1) resumption and restoration, (2) market-oriented reform, and (3) opening up and rapid growth.

12.3.1 Period I (1980–1985): Resumption and Restoration

At the initial stage of economic system reform, the control of firms' operations was gradually transferred from the state to firms. The resulting gradual changes in the risk sharing system led firms to seek economic compensation in the market and thus the demand for insurance was created. To meet this new demand, and more importantly, for better management of the financial surplus accumulated in private sector, the government initiated reforms in 1980 aimed at developing the commercial insurance system and encouraging the domestic insurance business. These reforms came twenty-one years after the closure of the insurance industry.

At this stage, all insurance business was monopolized by PICC, which was directly controlled by the central government. Insurance regulation was carried out by the PBOC, China's central bank. The terms and rates of policies were set and/or approved by the "regulator." As a result, the insurance product portfolio lacked variety, there was narrow use of complex actuarial science in pricing, and the price structure was simplistic. Additionally, the distribution channels were limited. Products were distributed primarily through sales staff of branch offices located throughout China (except Tibet and Taiwan), which were established by PICC and specialized in insurance distribution. Also, insurance assets were simply deposited in banks without consideration to an investment policy of diversification between bonds, equities, or real estate.

In 1980, the total gross premium volume of RMB 460 million was solely attributable to non-life insurance. In 1982, gross life insurance premiums totaled only RMB 1.6 million (CIRC). From this very small base, both the non-life and life insurance businesses grew significantly with annual growth rates exceeding GDP growth rates. The nominal growth rate of life insurance once exceeded 500 percent. In 1985, total premiums reached RMB 3.31 billion (CIRC). As Table 12.1 indicates, during this period, the life insurance business was insignificant relative to non-life insurance market. At this stage, enterprise property insurance and transportation equipment insurance (insurance coverage on motor vehicles, keel boats, and fishing boats) dominated the market. In 1985, these two types of non-life insurance had approximately thirty percent and twenty-seven percent of China's insurance market respectively (National Bureau of Statistics of China 1996, §17.14).

Table 12.1. Economic Growth and Insurance Premiums in China, 1980–1985

	GDP		Total Premiums		Non-Life Insurance Premiums		Life Insurance Premiums	
Year	Volume (RMB 100M)	% Change*	Volume (RMB 100M)	% Change**	Volume (RMB 100M)	% Change**	Volume (RMB 100M)	% Change**
1980	4,517.8	7.8	4.6	N/A	4.6	N/A	N/A	N/A
1981	4,860.3	5.2	7.8	69.6	7.8	69.6	N/A	N/A
1982	5,301.8	9.1	10.3	32.1	10.3	31.9	0.0	N/A
1983	5,957.4	10.9	13.2	28.2	13.1	27.3	0.1	550.0
1984	7,206.7	15.2	20.0	51.5	19.3	47.2	0.7	597.1
1985	8,989.1	13.5	33.1	65.5	28.7	48.9	4.4	508.3

Source: National Bureau of Statistics of China (2004), §3.1, §3.3, China Insurance Regulatory Commission (CIRC).

Notes: * Real growth rate. ** Nominal growth rate. As a reference, the annual growth rate of CPI from 1981–1984 was about 2% and in 1985 it was 9.3% (China Economic Information Center 2005; National Bureau of Statistics of China 2004, §9.1).

12.3.2 Period II (1986–1991): Beginning of Market-Oriented Reform

In March 1985, the State Council of China enacted *Interim Regulations on the Administration of Insurance Enterprises*, setting out the requirements for the establishment of insurance companies. The Farming Insurance Company of Xinjiang Production and Construction Group (FICX) was then established on July 15, 1986.

Although this company specialized in farming and animal husbandry insurance and was a quasi captive insurer within the Group, it effectively ended the monopoly of PICC. The number of suppliers in the market started to climb initially with Ping An and Pacific insurance companies established as shareholding multi-line insurers in 1988 and 1991 respectively.

At this stage, PBOC's direct administration of the insurance industry began to give way to supervision as the market-oriented reforms progressed. In 1995, PBOC established the Division of Insurance Supervision under its Department of Non-Banking Financial Institutions Supervision, which took over the responsibility of insurance supervision.

During this period, the limit on insurance investment channels and risk control mechanisms were relaxed. This resulted in extensive investment in real estate, securities, and trusts, which led to substantial investment in non-performing assets. However, there were no visible changes in the distribution system and insurance product portfolio. As Table 12.2 shows, while the volume of life insurance premiums was lower than that of non-life insurance, the fast growth of insurance premiums remained higher than GDP growth.

Table 12.2. Economic Growth and Insurance Premiums in China, 1986–1991

	GDP		Total Premiums		Non-Life Insurance Premiums		Life Insurance Premiums	
Year	Volume (RMB 100M)	% Change*	Volume (RMB 100M)	% Change**	Volume (RMB 100M)	% Change**	Volume (RMB 100M)	% Change**
1986	10,201.4	8.8	45.8	38.4	34.5	20.1	11.3	157.1
1987	11,954.5	11.6	71.0	55.1	46.1	33.6	25.0	120.5
1988	14,922.3	11.3	110.4	55.4	72.9	58.3	37.5	50.0
1989	16,917.8	4.1	97.6	(11.6)	78.1	7.1	19.6	(47.8)
1990	18,598.4	3.8	135.2	38.5	106.8	36.8	28.4	45.1
1991	21,662.5	9.2	178.2	31.9	136.8	28.2	41.4	45.8

Source: National Bureau of Statistics of China (2004), §3.1, §3.3; China Insurance Regulatory Commission (CIRC).

Notes: * Real growth rate. ** Nominal growth rate. As a reference, the annual growth rate of CPI from 1986–1991 was 7%, 8.8%, 18.8%, 18%, 3.1%, and 3.4%, respectively (China Econonomic Information Center 2005; National Bureau of Statistics of China 2004, §9.1).

12.3.3 Period III (1992–Present): Opening Up and Rapid Growth

The 14th National Congress of CPC in 1992 and the 3rd Plenary Session of the 14th CPC Central Committee in 1993 confirmed China's irreversible commitment to market-oriented reforms. Individuals and enterprises lost their traditional economic safeguard and were responsible for assuming financial risks. As a result, many factors contributed to the rapid growth of the insurance industry, including the fast growing economy, rising personal income, accelerating industrialization and urbanization, and social security system reform. In the remainder of this section, the current state of various aspects of China's insurance market is examined.

Business Volume

From 1992 to 2003, the average annual growth rate of gross premiums remained at over 30 percent. Table 12.3 shows the volume of insurance, both life and non-life and GDP. It can be seen that the volume of life insurance premiums surpassed that of non-life insurance for the first time in 1997 and remained higher in the subsequent years. The rapid growth of premium income and an expanded capital base explain the substantive increase in the total assets of the industry. Figure 12.1 shows that the total assets in the insurance industry increased 2.5 times from 1999 to 2003 and reached RMB 912.3 billion by the end of 2003. Total assets were RMB 1,111.65 billion by the end of August 2004 (CIRC). In contrast, aggregate total assets were less than RMB 400 million when the insurance industry had just resumed (CIRC). Despite the rapid growth and large volume of China's insurance business, it is generally agreed that the operation is far from efficient since the overall return on assets was below 1 percent in 2003 (Editorial Committee of China's Insurance Yearbook 2004).[235]

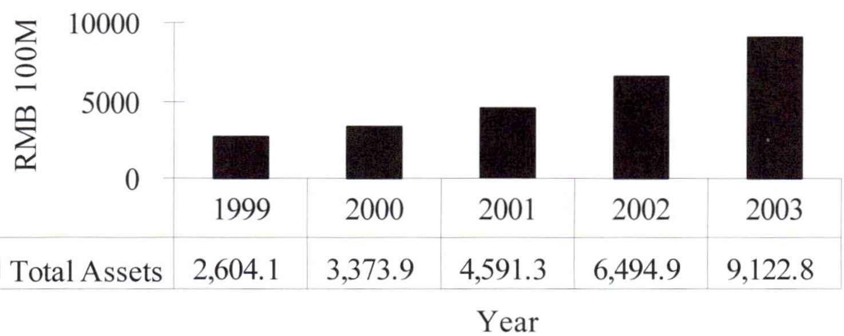

Source: China Insurance Regulatory Commission (CIRC).

Figure 12.1. Total Assets of the Insurance Industry in China, 1999–2003

Operating Structure

The first *Insurance Law of China* was enacted in 1995. It developed a segregated insurance market structure between the life and non-life insurance businesses by prohibiting the same insurer from engaging in both life and non-life insurance. It was possible, however, for an insurance group to have both life and non-life subsidiaries. As a result, the former four multi-line insurers, including PICC, FICX, Pacific, and Ping An, underwent organizational reform. FICX cast off its life insurance business and turned itself into Property Insurance Company of Xinjiang Production and Construction Group. The other three companies followed the pattern of setting up a holding group with life and non-life insurance subsidiaries. In 1996, PICC was restructured into the PICC Group, which controlled the newly established PICC

---

[235] The after-tax-profit for the whole insurance industry was RMB 4.61 billion of which 3.15 billion was from the non-life insurance business and 1.46 billion was from life insurance sector (Editorial Committee of China's Insurance Yearbook 2004).

Property and Casualty Insurance Company, PICC Life Insurance Company, and PICC Reinsurance Company. In October 1998, the three subsidiaries were divested and renamed the People's Insurance Company of China (it inherited the original abbreviation, i.e., PICC), the China Life Insurance Company, and the China Reinsurance Company.

Table 12.3. Economic Growth and Insurance Premiums in China, 1992–2003

	GDP		Total Premiums		Non-Life Insurance Premiums		Life Insurance Premiums	
Year	Volume (RMB 100M)	% Change*	Volume (RMB 100M)	% Change**	Volume (RMB 100M)	% Change**	Volume (RMB 100M)	% Change**
1992	26,651.9	14.2	211.7	18.8	147.4	7.7	64.3	55.3
1993	34,560.5	13.5	395.5	86.8	251.4	70.6	144.1	124.1
1994	46,670.0	12.6	500.4	26.5	336.9	34.0	163.5	13.5
1995	57,494.9	10.5	594.9	18.9	390.7	16.0	204.2	24.9
1996	66,850.5	9.6	777.1	30.6	452.5	15.8	324.6	59.0
1997	73,142.7	8.8	1087.9	40.0	480.7	6.2	607.2	87.0
1998	76,967.2	7.8	1247.6	14.7	499.6	3.9	748.0	23.2
1999	80,579.4	7.1	1393.2	11.7	521.1	4.3	872.1	16.6
2000	88,254.0	8.0	1595.9	14.6	598.4	14.8	997.5	14.4
2001	95,727.9	7.5	2109.4	32.2	685.4	14.5	1424.0	42.8
2002	103,935.3	8.3	3053.2	44.7	778.3	13.6	2274.9	59.8
2003	116,603.2	9.3	3880.4	27.1	869.4	11.7	3011.0	32.4

Source: National Bureau of Statistics of China (2004), §3.1, §3.3, China Insurance Regulatory Commission (CIRC).
Notes: * Real growth rate. ** Nominal growth rate. As a reference, the annual growth rate of CPI from 1992–2003 was 6.4%, 14.7%, 24.1%, 17.1%, 8.3%, 2.8%, –0.8%, –1.4%, 0.4%, 0.7%, –0.8%, and 1.2%, respectively (China Economic Information Center 2005; National Bureau of Statistics of China 2004, §9.1).

The financial sectors, including the banking, securities, and insurance businesses, are still operating independently in China. According to the *Banking Law of China* effective on February 1, 2004, China's commercial banks can not invest in non-bank financial institutions except that such investment is allowed by other regulations or state provisions. Additionally, insurance companies are prohibited from establishing securities firms and other enterprises except insurance institutions according to the *Insurance Law*. In a word, banking, securities, and insurance services can only be provided by separate firms; banks cannot own insurance companies and vice versa.[236] Industrial groups, however, are not prohibited from

[236] At the present time, there is no legislation on financial holding companies, although many researchers deem the pattern as a good approach to achieve financial convergence. The only "financial holding company," CITIC International Financial Holdings Limited (CIFH),

owning both bank subsidiaries and insurance companies. There are several historical exceptions to this law. For example, China Ping An Insurance (Group) Company, Ltd. does offer insurance, banking, and securities through its different subsidiaries.

Suppliers of Insurance

On the supply side, China's insurance market is characterized by two things: (1) an increase in the number of insurers and (2) organizational restructuring of the state-owned insurers. In 1992, China began to open its insurance market by using Shanghai as an experimental city. According to *the Temporary Measures for Administration of Foreign-Funded Insurance Institutions in Shanghai* enacted by the PBOC in September 1992, foreign insurers applying to establish branches or joint ventures shall satisfy the prudential requirements stipulated by the insurance regulator to get licenses including the "533" requirements meaning that the applicant's total assets should be more than $5 billion at the end of the year before application, the applicant should have established its representative office(s) three years before application, and have had over three decades of experience in the insurance business. In September of the same year, American International Assurance Company (AIA) and American International Underwriters, Ltd. (AIU), subsidiaries of the American International Group (AIG), were the first to be approved to establish affiliates in Shanghai and thus opened the gate for foreign capital to flow into China's insurance industry.

Since then the number of insurers, both domestic and foreign, has kept rising. After China's entry into World Trade Organization (WTO) in 2001, the State Council enacted *The Regulations on Administration of Foreign-Funded Insurance Companies*, which became effective on February 1, 2002. Many restrictions to foreign insurers to obtain licenses in China were relaxed in a sense, such as the "533" requirements were changed to the "523," allowing foreign insurers who have just established representative offices in China two years before to submit an application.[237] Actually, the requirements to get licenses could easily be satisfied by many foreign insurers and as Figure 12.2 shows, a number of famous foreign insurers obtained licenses to establish branches or joint ventures in China. Most significantly, Munich Re, Swiss Re, and General Cologne Re were approved to set up branches in China in 2003, and this effectively ended the state-owned China Reinsurance Company (China Re)'s role as the only reinsurer in China. At of the end of 2003, there were sixty-one insurance companies, six insurance holding groups, and two insurance asset management companies in China. Of the sixty-one insurers there are twenty-six non-life insurers, thirty life insurers, and five reinsurers (CIRC). However, many applications were turned down because of deficiencies in their application materials or business plans since the requirements and the review process for approval in China are quite prudential, although there are no limits on the number

---

was approved to establish as a "holding company" in 2002 by the State Council and it operates outside of mainland China.

[237] The qualified foreign applicant can apply to set up a company in China and will be given one year to finish preparation work once the application is approved by CIRC. Once the preparation work has been completed, CIRC will conduct a review and make a decision as whether to grant a license or not in 30 days after they receive the entire set of formal application materials.

of licenses issued. For example, fourteen foreign insurers, including AXA and Cigna, received notices from CIRC that they had failed to submit all the materials requested and should begin another application round on April 29, 2002. Many of them, however, including AXA and Cigna, finally obtained their licenses upon submitting all the required materials.

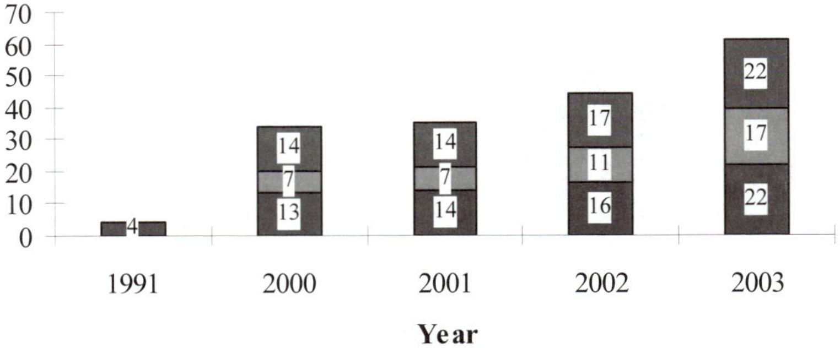

Source: China Insurance Regulatory Commission (CIRC); National Bureau of Statistics of China (2003), p. 709 and (2002), §19.16.

Note: The numbers on the pillars indicate "domestic," "joint ventures," and "foreign branches" from the bottom up. For example, in 2003 there were 22 domestic insurers, 17 joint ventures, and 22 foreign branches in China.

Figure 12.2. The Number of Insurers in China in Selected Years

Corporate governance system reform in China's financial sector started first in the insurance industry. In 2003, the People's Insurance Company of China (PICC), China Life Insurance Company, and China Reinsurance Company, the three purely state-owned insurance companies, successfully underwent organizational restructuring.

- On July 19, 2003, PICC incorporated PICC Property and Casualty Company Limited (PICC P&C). PICC was then renamed and re-registered as PICC Holding Company, which injected all of its commercial insurance operations, together with related assets, liabilities, and subsidiaries into PICC P&C. The initial public offering (IPO) of PICC P&C on the Hong Kong Stock Exchange in November 2003, which was proved to be a great success, raised $693 million (CIRC). For more information regarding PICC P&C, please refer to Appendix 1.
- China Life Insurance Company incorporated China Life Insurance Company Limited (China Life) on June 30, 2003 and transferred its well-performing policies and related assets and liabilities to the latter. China Life Insurance Company was then restructured to China Life Insurance (Group) Company in July 2003. The IPO of China Life on New York Stock Exchange (NYSE) and the Hong Kong Stock Exchange on December 17 and 18, 2003 respectively was the world's largest that year and raised about $3.5 billion (CIRC). For more information regarding China Life, please refer to Appendix 1.

- China Re was renamed China Reinsurance (Group) Company in August 2003, and began to incorporate China Property and Casualty Reinsurance Company, China Life Reinsurance Company, and China Dadi Property and Casualty Insurance Company. Unlike PICC and China Life, China Re did not begin its IPO efforts right after the restructuring, but committed itself to improve its internal management first.

Competitive Market Structure

China's insurance market has become more competitive as the number of insurers has increased. On the non-life insurance side, the market share of the leading insurer, PICC, was 96 percent in 1992 and dropped to 70 percent in 2003 (CIRC). However, the top 3 non-life insurers, including PICC P&C, Pacific P&C, and Ping An P&C, still accounted for more than 90 percent of China's gross premiums in 2003 (CIRC). As Table 12.4 shows, the non-life insurance market could be characterized as a typical oligopoly market. The situation on the life insurance side is a little bit better, but still highly concentrated as well. Table 12.4 indicates that the market share of the top 3 life insurers in 2003 was as high as 87 percent (Chen 2004, p. 59). Table 12.5 and 6 list the top 5 life insurers and top 5 non-life insurers in 2003.

Table 12.4. Concentration of China's Insurance Market in Selected Years

	1992		1996		2001		2002		2003	
Market Share by Premium Volume	Life	Non-Life	Life	Non-Life	Life	Non-Life	Life	Non-Life	Life	Non-Life
Top 1 Insurer	N/A	96.0%	N/A	77.3%	57.1%	73.9%	56.6%	70.9%	54.0%	69.8%
Top 2 Insurer	N/A	N/A	N/A	89.6%	85.2%	86.3%	80.1%	84.1%	73.9%	82.1%
Top 3 Insurer	N/A	N/A	N/A	96.9%	95.3%	96.1%	91.1%	94.8%	86.6%	91.7%

Source: China Insurance Regulatory Commission; Jiang (2004), p. 22; Chen (2004), p. 59.
Note: N/A is not available.

Geographic Distribution

Table 12.7 shows that geographic distribution of China's insurance industry is highly imbalanced. In 2003, the top 6 provinces, autonomous regions, or centrally controlled municipalities collected 46 percent of China's gross premiums while the last 10 only constituted about 10 percent[238] (Editorial Committee of China's Insurance Yearbook 2004). This situation is principally a result of uneven economic development in different regions. Because of the structural differences in location and resource endowment of different regions, the geographic deployment of China's industrialization was uneven after the reform and opening up began. Benefiting from this strategy, the southeast coastal regions soon surpassed the middle and west regions in economic development and attracted insurers' efforts with their favorable

[238] The top 6 provinces, autonomous regions, or centrally controlled municipalities are as follows: Jiangsu, Guangdong, Shanghai, Shandong, Beijing, and Zhejiang (Editorial Committee of China's Insurance Yearbook 2004).

demand. Most of the newly established insurers are located in the east coastal area and the number of insurers in the middle and the west market has not increased much. As a result, the competition in the eastern market is much more intensive. For example, in 2003, the market share of the top 4 life insurers in Shanghai was about 64 percent, indicating a healthy competition. However, in some west provinces, China Life still holds a monopoly position (CIRC).

Table 12.5. An Overview of the Top 5 Life Insurers in China, 2003

Rank	Name	Premium Volume (RMB 100M)	Organiza-tional Form	Principal Distribution System (Market Share)*	Ownership Type	Country of the Ultimate Owner
1	China Life	1,625.33	Stock	Exclusive Agency (66.83%)	Insurance Holding Company	China
2	Ping An Life	598.58	Stock	Exclusive Agency (63.84%)	Insurance Holding Company	China
3	Pacific Life	382.40	Stock	Bancassurance (39.52%)	Insurance Holding Company	China
4	New China Life	174.33	Stock	Bancassurance (56.43%)	Industrial Firm	China
5	Taikang Life	135.19	Stock	Bancassurance (53.72%)	Industrial Firm	China

Source: Chen (2004), pp. 59, 164.
Note: * Market share is the ratio of premium income collected through the distribution channel to the overall premium income of the company.

Table 12.6. An Overview of the Top 5 Non-Life Insurers in China, 2003

Rank	Name	Premium Volume (RMB 100M)	Organizational Form	Ownership Type	Country of the Ultimate Owner
1	PICC P&C	606.80	Stock	Insurance Holding Company	China
2	Pacific P&C	106.90	Stock	Insurance Holding Company	China
3	Ping An P&C	83.50	Stock	Insurance Holding Company	China
4	Tian An P&C	21.00	Stock	Industrial Firm	China
5	China United P&C*	18.70	State-Owned	State-Owned	China

Source: Editorial Committee of China's Insurance Yearbook (2004).
Note: The principle distribution system is not available for non-life companies. * China United P&C was formerly Property Insurance Company of Xinjiang Production and Construction Group. It was in the process of switching to stock before this chapter was finished.

Table 12.7. The Distribution of China's Economy and Insurance Industry by Region, 2003

			Distribution of the Economy by Region			Distribution of Insurance Premiums			
Region	Distribution of GDP (%)	Distribution of Population (%)	Primary Sectors (%)	Secondary Sectors (%)	Tertiary Sectors (%)	Non-Life (%)	Life (%)	Total (%)	Premium per Capita for Each Region (RMB)
The East Coast	58.5	37.9	42.5	64.7	60.4	53.9	61.8	59.8	473.2
The Southeast Coast*	34.7	19.5	21.6	39.3	35.8	33.1	35.6	35.0	537.0
The Middle	24.6	33.4	31.6	25.5	22.9	16.7	23.3	21.7	194.6
The West	16.9	28.7	25.9	9.8	16.7	29.5	14.9	18.5	192.7
Total	100.0	100.0	100.0	100.0	100.0	100.0	100.0	100.0	304.6

Source: National Bureau of Statistics of China (2004), §4.3, §3.11; Editorial Committee of China's Insurance Yearbook (2004).

Notes: * The Southeast coast is a subset of the East Coast, including Shanghai, Jiangsu, Zhejiang, Guangdong and Fujian. We include the Southeast region separately here because it is a significant region in China's Economy. These four provinces' GDP represented about one-third of the whole country in 2003. 1.) It is generally accepted that the east coast region includes Beijing, Tianjin, Hebei, Liaoning, Shanghai, Jiangsu, Zhejiang, Fujian, Shandong, Guangdong, and Hainan. The middle region includes Shanxi, Jilin, Heilongjiang, Jiangxi, Hubei, and Hunan. The west region includes Chongqing, Sichuan, Guizhou, Yunnan, Tibet, Shaanxi, Gansu,Qinghai, Ningxia, Xinjiang, Guangxi, and Inner Mongolia. 2.) The primary sectors referr to agriculture, forestry, animal husbandry, and fishery. The secondary sectors refer to mining and quarrying, manufacturing, production and supply of electricity, water, and gas, and construction. The tertiary sectors refer to all other economic activities not included in the primary or secondary sectors.

Product Portfolio

Different kinds of insurance products have been continually introduced to the market as the industry has developed. On the life insurance side, the product portfolio before June 1999 consisted of traditional policies, whose premiums, benefits, and assumed interest rate were hard to adjust after selling, such as ordinary life insurance and endowments. PBOC reduced the interest rates seven times in succession from May 1996 to June 1999, and the nominal interest rate of a one-year certificate of deposit decreased from 10.98 percent to 9.19 percent after May 1, 1996 and then to 2.25 percent after June 10, 1999. While the interest rates dropped dramatically, the assumed interest rates of China's life insurance policies were not lowered and thus the policies became very attractive to consumers. Some insurance companies did not realize the seriousness of implied losses and failed to adopt effective measures to prevent agents from selling more of the old polices. The large demand and loss of control on supply side contributed to tremendous losses from interest rate discrepancies to China's insurance companies.[239] CIRC then required the life insurers to lower the assumed interest rates of newly issued policies in June 1999 and encouraged the companies to develop innovative products, which transfer some or all of the investment risk to policyholders.

Ping An Life Insurance Company first introduced variable life insurance policies to the market in October 1999 and collected RMB 1.23 billion of premium income in the following 14 months (CIRC). This kind of product received favorable responses in big cities due to its high potential returns in the then bull market. Therefore, other insurers began to follow suit. In 2001, the premium income from variable life products in Beijing alone accounted for 21 percent of its total life insurance premium income (CIRC). However, the rapid growth of variable life products was impeded by the crash of China's stock market which began in late June 2001. The Composite Index of Shanghai dropped from its highest point of 2,245.44 on June 14, 2001 to its lowest of 1,514.86 on October 22, 2001 and closed at 1,645.97 at the end of the year (Shanghai Stock Exchange 2002). The crash is largely attributable to the withdrawal of capital, the reduction in state-owned stock, and the loss of investors' faith due to disclosure of insider trading (Li 2002).

Participating, universal life, and other innovative products with lower investment risk, increased flexibility, and higher returns relative to bank deposits began to appear and became popular in the market thereafter. In 2003, premium income from innovative products accounted for 58 percent of gross life insurance premiums (China Economic Information Center 2005). Furthermore, as Table 12.8 indicates, there was a notable increasing trend in health insurance premiums as health related risk became a major focus of the public's attention and health insurance became an important supplement to the medical coverage provided by the reformed social security system.

---

[239] Because the loss from investment earning is not realized at once and the return on investment keeps changing, an accurate estimate of the loss is unfeasible. In 2002, CIRC declared the volume of the loss at about RMB 50 billion (Ning and Yu 2003). According Goldman Sachs (2003), the volume was about RMB 32–72 billion.

Table 12.8. Market Share by Lines of Life Insurance Business (%), 1998–2003

	1998	1999	2000	2001	2002	2003
Life Insurance*	89.12	87.62	89.09	90.45	91.33	88.66
Traditional	89.12	N/A	N/A	60.99	37.34	30.19
Variable	0.00	N/A	N/A	7.48	2.99	2.43
Universal	0.00	N/A	N/A	2.83	1.50	0.56
Participating	0.00	0.00	N/A	19.15	49.36	55.48
Health Insurance	3.75	4.72	2.83	4.28	5.33	8.03
Accidental Insurance	7.13	7.66	8.08	5.27	3.48	3.31
Total	100.00	100.00	100.00	100.00	100.00	100.00

Source: China Insurance Regulatory Commission (CIRC); China Economic Information Center (2005); Jiang (2004), p. 16.

Notes: * Including pension insurance. The market share of pension insurance was 16.79% and 14.34% in 2002 and 2003, respectively. Owing to lack of data, we cannot get accurate information on pension insurance in other years.

Table 12.9. Market Share by Lines of Non-Life Insurance Business (%), 1998–2003

	1998	1999	2000	2001	2002	2003
Automobile Insurance	55.56	58.12	61.35	61.61	60.51	62.14
Enterprise Property Insurance	22.27	21.43	19.41	17.66	15.77	14.38
Freight Transport Insurance	7.63	6.69	5.92	5.99	5.38	4.72
Liability Insurance	2.78	3.19	3.45	4.09	4.74	4.03
Household Property Insurance	2.41	2.32	2.14	2.77	3.08	2.19
Hull and Cargo Insurance	2.37	2.10	1.81	1.75	1.67	1.73
Construction and Installation Projects Insurance and Related Liability Insurance	1.19	1.07	0.99	0.88	1.03	1.38
Export Credit Insurance	0.59	0.45	0.49	0.44	0.90	0.92
Agriculture Insurance	1.41	1.20	0.66	0.44	0.64	0.58
Guarantee Insurance	0.22	0.27	0.33	0.58	1.15	0.23
Satellite and Nuclear Energy Insurance	0.43	0.29	0.49	0.73	0.77	0.12
Others	3.14	2.86	2.96	3.07	4.36	7.59
Total	100.00	100.00	100.00	100.00	100.00	100.00

Source: China Economic Information Center (2005).

As for the non-life insurance sector, Table 12.9 shows the market share of the various lines of non-life insurance. The proportion of each line in terms of premium income has been relatively stable these years, and automobile insurance has played the leading role. The market share of enterprise property insurance has decreased due to the fact that many State-Owned Enterprises (SOEs), the major clients of the insurers, performed poorly and had to cut expenses on insurance as the way to reduce the costs. At the same time, the newly established private enterprises often lack insurance awareness.

Distribution Channels

As Figure 12.3 indicates, a diversified distribution system, including direct marketing, insurance intermediaries, and direct response has already taken shape. Direct response refers to the marketing channel, such as the Internet, telephone, and direct mail, through which the insurance companies sell products directly to consumers. Typically, direct response does not involve face-to-face contact during the sales process. In the past, direct marketing, where the salaried staff of insurers deals directly with customers, was the most important distribution channel.

Direct marketing began to lose its dominant position to individual agents in the individual life insurance market in 1992 when AIA introduced the commission-based exclusive agency system in Shanghai. Within a few years, all domestic life insurance companies in China introduced the same system. An exclusive agent is an individual that can represent only one insurance company. They typically do not have an exclusive territory. Insurance agency institutions, as legal entities, can sell products of different insurers. The usual commissions to agents are based on the business volume and vary between 20 percent and 40 percent of premium income collected. Different companies have adopted different commission schedules.

Figure 12.4 shows that the market share of the exclusive agency system in the life insurance market climbed to 73.9 percent in 2000 and direct marketing only accounted for 25.5 percent of the market share instead of 100.0 percent in 1992 (Chen 2004, p. 102). *The Provisional Measures for Insurance Agents Administration* promulgated in 1996 recognized the role of agents by setting standards for the professional practice of agents and requiring agents to pass the Insurance Agents Qualifying Examination organized by CIRC. The number of agents greatly increased and contributed significantly to the high growth of life insurance premium volume. In addition, brokers came into being with foreign enterprises and foreign insurers operating in China. In 1998, PBOC promulgated *The Provisional Measures for Insurance Brokers Administration*. CIRC organized the first Insurance Brokers Qualifying Examination in May 1999 and 162 people out of about 4,800 received the qualification (Wu 2004, p. 172). Individuals who had a criminal record in the past five years, or have ever received a penalty due to unfaithful behaviors in the past three years, were prohibited from being an agent or broker by the financial regulators.

In 2001, CIRC enacted The Measures for Insurance Agencies Administration and The Measures for Insurance Broker Companies Administration and then amended them at the end of 2004. The Measures for Insurance Broker Companies Administration was renewed as The Measures for Insurance Broker Institutions Administration. According to the new regulation, insurance agency institutions and broker institutions in China could only take the form of partnership, limited liability company, or limited stock company and each of them had to deposit 20 percent of its registered capital as a guaranty fund or buy required professional liability insurance.

Institutions other than agency companies and broker companies (referred to as other institutions later), such as commercial banks, auto dealers, and post saving offices, can also act as insurers' agents to distribute insurance products after they get licenses from CIRC and come to an agreement with insurance companies.[240] Bancassurance started in 1996 in China and gained popularity on the life insurance side after Ping An Life Insurance Company developed "Century Dividend" in the year 2000. "Century Dividend" was specially designed for sales via commercial banks. The employees of banks' branch offices recommended the insurance products to their customers while doing banking business with them over the counter or used various direct response techniques to sell insurance policies to their existing customers. This distribution channel proved to be cost-effective and a great success. For example, 75 percent of the life insurance premium income for Taiping Life Insurance Company in 2003 came from bancassurance (Chen 2004, p. 104). As Figure 12.4 shows, direct response, such as marketing through telephone and the Internet, has also arisen lately, but the market shares are trivial by comparison.

As we can see in Figure 12.3, by the end of 2003, there were 506 insurance agency companies operating in China and 207 were preparing to start operation; and there were 83 insurance brokerages in operation while another 82 were preparing to start operation. Other institutions engaged in insurance sales totaled 117,185 with 48,489 for non-life insurance and 68,696 for life insurance. Nationwide there were 26,789 insurance branch offices and 1.28 million exclusive agents. These intermediaries played an important role in insurance sales. Figure 12.4 shows the relative importance of all kinds of distribution channels for life insurance in recent years with Figure 12.5 for non-life insurance side.

Investment Management

At the beginning of this period, investment channels were wide open, but due to the lack of investment expertise and management experience, more than RMB 10 billion of non-performing assets were engendered from 1992 to 1993 (Wu 2004, p. 187). The *Insurance Law* enacted in 1995 thus put strict requirements on insurance investments in order to ensure the solvency of the insurance companies, specifying that insurance funds can only be invested in the form of bank deposits, government bonds, financial bonds, and other investments as approved by the State Council (and these requirements were not changed in the 2002 amendment).[241] Subsequently, China's insurers were gradually allowed to widen their investment channels, as the financial market developed. They were allowed to invest in corporate bonds rated AA+ issued by the central government-controlled SOEs (July 1998), government securities repurchased (August 1998), negotiated bank deposits (October 1999), mutual funds (October 1999), corporate bonds rated AA (June 2003), central bank notes (July 2003), and stocks (October 2004). The proportion of each kind of

[240] In China, only China Post, a government-run postal enterprise, could provide all kinds of postal service. Other enterprises, including foreign companies could conduct only certain express delivery service and logistics service. China Post also conducts savings and remittance businesses. Every post saving office of China Post could sell insurance after it received a license from the CIRC.

[241] Financial bonds refer to securities issued by financial institutions in China's inter-bank bonds market requiring repayment of interest and principal within a given period. Until now, financial bonds were just issued by state policy-related banks.

investment to the total investible fund, however, had to comply with the prescription of the regulator.[242]

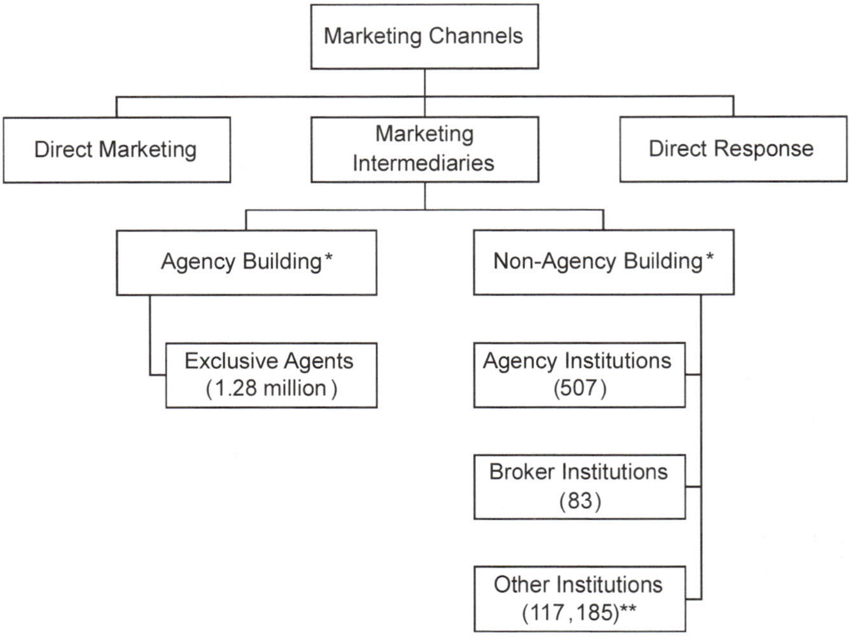

Source: Wu (2004), pp. 176–179.
Notes: * We divide marketing intermediaries into two classes depending on whether the insurer is attempting to build its own agency sales force. Thus, insurers following an agency building distribution strategy recruit, train, finance, house, and supervise their agents. The non-agency building strategy means that the insurers rely only on established agents for their sales (Black and Skipper 2000, p. 603). ** Including 48,489 in non-life insurance business and 68,696 in the life insurance business, which takes insurance marketing as part of their busines, similar to commercial banks.

Figure 12.3. An Overview of China's Insurance Distribution System, 2003

[242] There are no upper limits on bank deposits, government bonds, financial bonds issued by state policy-related banks, and central bank notes. A negotiated bank deposit is a kind of long-term deposit by non-bank financial institutions the terms of which are set by bilateral negotiation. They should exceed RMB 30 million and last more than five years. The proportion of corporate bonds investment should not exceed 20 percent of the insurer's assets at the end of last month and the investment in one corporate bond should not exceed the minimum of 15 percent of the circulation and 2 percent of the insurer's assets at the end of last month. The proportion of mutual funds investment should not exceed 15 percent of the insurer's assets at the end of last month, while the proportion of funds in separate accounts of variable life insurance is 100 percent and for universal life insurance is 80 percent. As to stocks, the stocks of one listed company held by one insurance institutional investor shall not reach 30 percent of the RMB common stocks of the listed company and the specific proportion of stock investment of an insurance institution investor shall be separately specified by CIRC. The limits may be changed when necessary according to the actual situation observed by CIRC.

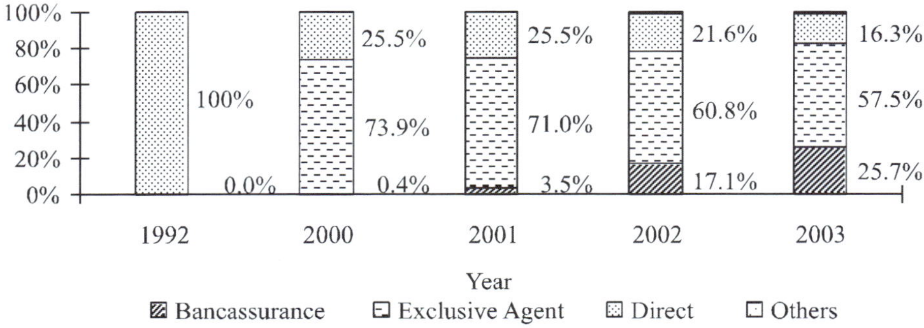

Source: Chen (2004), p. 64, pp. 100–102.

Figure 12.4. Market Share by Distribution Channel for Life Insurance in Selected Years

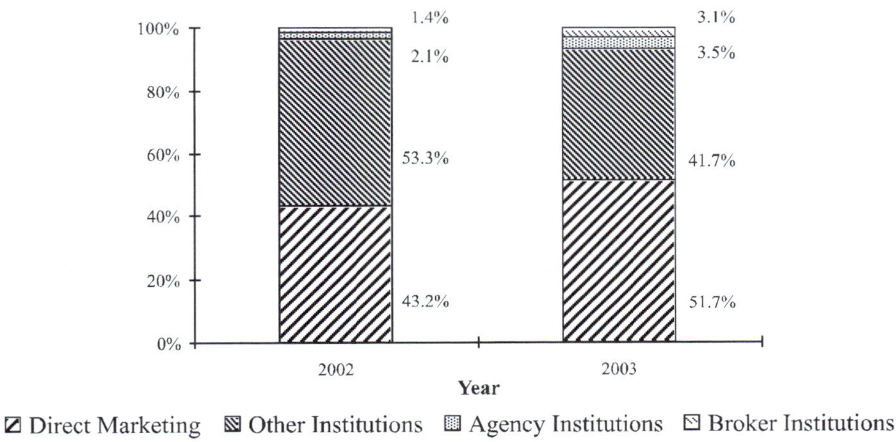

Source: Tang (2004), p. 122.

Notes: The four numbers on the pillars indicate "direct marketing," "other institutions," "agency institutions," and "broker institutions" from the bottom up. For example, in 2003, 51.7% of the premium was collected via direct marketing, 41.7% via other institutions, 3.5% via agency institutions and another 3.1% via broker institutions.

Figure 12.5. Market Share by Distribution Channel for Non-Life Insurance, 2002–03

Table 12.10 provides an overview of insurance investment in China for 2002–03. It is worth noticing that bank deposits, despite the low return, have always been the most important investment channel for insurance funds, and government bonds took second place. Conservative investment strategies resulted in a low rate of return, which was only 2.68 percent in 2003 (Wu 2004, p. 189). Insurance asset management companies, which specialize in fund management for insurance companies, have become the latest trend in insurance investment. PICC Asset Management Company and China Life Asset Management Company, established in 2003 and 2004 respectively, were the first two asset management companies in China.

Table 12.10. An Overview of Insurance Investment in China, 2002–03

	2002 Volume (RMB 100M)	% of Total	2003 Volume (RMB 100M)	% of Total
Bank Deposits	3,026.3	54.7	4,549.7	52.1
Government Bonds	1,107.9	20.0	1,406.9	16.1
Financial Bonds	404.8	7.3	828.7	9.5
Corporate Bonds	84.6	1.5	389.0	4.5
Mutual Funds	307.8	5.6	456.6	5.2
Others	599.2	10.9	1,108.1	12.6
Total	5,530.6	100.0	8,739.0	100.0
Rate of Return	3.1%		2.7%	

Source: Jiang (2004), p. 54; Rate of Return is from Wu (2004), pp. 188–189.

Legislative Environment and Supervision

Progress in market-oriented reform and remolding of the role of government brought forward the need for efficient mechanisms to coordinate the industry and led to great efforts in developing insurance legislation and the regulatory system. *The Insurance Law* enacted in 1995, its amendment in 2002, the *Regulation for the Administration of Insurance Companies* enacted in 2004, the *Regulation for the Administration of Foreign-Funded Insurance Companies* enacted in 2001, and other regulations and measures compose the current framework of China's legislative system for insurance. The *1995 Insurance Law* consisted of insurance contract law and insurance professional law, governing different aspects of the insurance industry such as the insurance contract, insurance business, market access, and insurance regulation. In order to adapt to the new situation of China's opening up, especially its entry into WTO, the amendment of the insurance law began in 2002. The new law abolished mandatory reinsurance, gave prominence to solvency regulation, allowed P&C insurance companies to operate in short-term health insurance and accidental insurance, and deleted the provisions that stipulated that the terms and premium rates of insurance policies should be set by the regulator.

The establishment of CIRC on November 18, 1998 was a milestone in China's history of insurance regulation. It ended PBOC's function of administrating the insurance industry in China, which had continued for quite a long time. At present, CIRC is still the superintendent and regulator of the industry, but its direct involvement in administration of the industry is gradually decreasing as reform progresses. One important aspect of reform is to put more emphasis on solvency regulation, while loosening the former strict restrictions on insurers' market behavior.

In March 2003, CIRC enacted *Regulation on the Solvency of Insurers and Regulation Indicators* based upon the actual situation in China and experiences from the European Union. According to this regulation, insurance companies in China are required to ensure that their actual solvency margin (i.e., the difference between the admitted assets and liabilities) is larger than the minimum, which is calculated based on the methods prescribed in the regulation. CIRC might change the regulation when necessary. The minimum solvency for P&C insurance companies in the 2003 regulation is the maximum of (1) 18 percent of retention of gross premium in last fiscal year (excluding sales taxes and associate charges) below RMB 100 million plus 16 percent of the part above RMB 100 million or (2) 26 percent of average combined loss payment (the sum of loss payment, the difference between reserves in the current year and reserves carried over, and reinsurance loss payment minus reinsurance loss repayment and income of loss payment recovered) below RMB 70 million and 23 percent of the part above RMB 70 million.[243] The minimum for life insurance companies is the sum of minimum solvency of long-term business (carrying an insurance period more than one year) and short-term business. The minimum solvency of long-term business is the sum of (1) 1 percent of mandatory minimum reserve of investment-linked products and 4 percent of other products and (2) 0.1 percent of amount at risk for term life products carrying an insurance period less than three years, 0.15 percent for term life products carrying an insurance period more than three years but less than five years, and 0.3 percent for other life products. The calculation of minimum solvency of short-term business is the same as that of P&C insurance companies.

In addition, CIRC has loosened the former strict limitations on market access, regions of operation, rates, and policy forms. For the non-life insurance side, reform which began in 2003, focuses on automobile insurance. Insurers no longer need to use the rates and forms set by CIRC and can develop their own polices and use them after getting approval from CIRC. For the life insurance side, insurers are permitted to develop their own policy forms and rates and just need to file related materials (including terms and rate schedules, cash value illustration, and legal statements) with CIRC within seven days after sale of the products. However, they must get approval prior to launching certain products according to the *Notice on Several Issues about Administrative Measures on Life Insurance Products* (CIRC 2005n).[244]

All these measures, in return, have inspired many innovations in the industry, e.g., the introduction of automobile premium rate based on insured's personal information such as driving record and sex, the Bonus-Malus system, and the electronic loss assessing system.[245]

Generally speaking, the competitiveness of the market has been on the rise as the number of insurers has increased over time, but China's insurance market is still characterized as a typical oligopoly market. The government remains heavily

---

[243] The minimum for insurance companies, which have been in the business no more than three fiscal years, is what prescribed in (1).

[244] Include mandatory insurance products, products recognized to be related to public interest by CIRC, and life insurance carrying death protection recognized as new type by CIRC. For more information, refer to CIRC (2005o).

[245] The Bonus-Malus system is a kind of no-claims discount strategy which can be chosen by the insurer to eliminate the moral hazard problem. In China, almost all insurers have adopted some form of no-claim bonus schemes in their auto insurance contracts, and the typical maximum premium discount is 10%. A particularly simple Bonus-Malus scheme is termed a "one up/one down" scheme. It rewards individuals for each claim-free period by a reduction in premium to the next lowest category and each claim generates an increase in premiums to the next highest category.

involved in the industry, though not as much as before. In December 2001, China became a member of the World Trade Organization. This further opening up to the outside world has brought significant challenges and great opportunities to China's young insurance industry which are discussed in the next section.

12.4 AFTER OPENING UP: OPPORTUNITIES AND CHALLENGES

12.4.1 Stages of Opening up and China's WTO Commitments

China's insurance industry began to open up to the outside world step by step at its initial stage of restoration. The central government first allowed several foreign insurance companies to set up representative offices in China in the early 1980s which started the first stage of opening up. 1992 was the start of the pilot stage as the State Council chose Shanghai to be the first experimental city for opening up. More cities followed the suit thereafter. Upon China's accession to the WTO in 2001, the government made certain commitments with respect to the insurance sector and thus led China's insurance industry to a completely new stage. The current stage is characterized by an overall opening instead of a limited and restricted opening and it is viewed as more predictable, based on a legal framework, with mutual openings among WTO member countries.

The Chinese government's WTO commitments concerning the insurance sector can be summed up in the following four aspects: corporate forms, regional restrictions, business restrictions, and mandatory insurance.

Corporate Forms
The Chinese government promised that upon WTO accession, foreign non-life insurance companies would be allowed to establish branches or joint ventures with a maximum ownership stake of 51 percent; and 100 percent owned subsidiaries would be allowed two years after WTO accession. Foreign life insurance companies would be allowed to freely choose Chinese partners and establish joint ventures with maximum ownership of 50 percent upon WTO accession. Foreign insurance brokerages may set up joint ventures with maximum ownership of 50 percent upon accession, 51 percent within three years of accession, and 100 percent owned subsidiaries within five years of accession. Foreign reinsurance companies would be allowed to set up branches, joint ventures, or fully-owned subsidiaries to provide life and non-life reinsurance without regional restrictions or numerical restrictions on operating licenses. The new establishments are subject to licenses approved by China's insurance regulator (the approval process has been described in the above paragraphs).

Regional Restrictions
Upon WTO accession, foreign life insurance, and non-life insurance companies were allowed to operate in Shanghai, Guangzhou, Dalian, Shenzhen, and Fuoshan. Within two years of WTO accession, they were further allowed to operate in ten more cities: Beijing, Chengdu, Chongqing, Fuzhou, Suzhou, Xiamen, Ningbo, Shenyang,

Wuhan, and Tianjin. All regional restrictions were removed at the end of 2004, three years after the accession.

Business Restrictions

Upon WTO accession, foreign non-life insurance companies were allowed to engage in the sale of "master policies," which cover the insured's properties and liabilities located in different places without regional restrictions and were also allowed to underwrite large commercial risks. Foreign non-life insurance companies were allowed to provide property coverage, related liability, and credit insurance for subsidiaries of foreign companies operating in China. Within two years of accession, these companies were allowed to provide the whole range of non-life insurance products except mandatory insurance to both Chinese and foreign clients. Foreign life insurance companies were allowed to provide individual (non-group) life insurance to both Chinese and foreign citizens. Within three years of WTO accession, these companies were allowed to provide health, group, and retirement/annuity products to both Chinese and foreign citizens.

Mandatory Insurance

China Re, the former national reinsurance company, was granted a 20 percent mandatory reinsurance from domestic companies before the WTO entry. This percentage was reduced by 5 percent each year after WTO accession and zeroed out four years later. However, foreign companies were still excluded from mandatory insurance for automobile third-party liability, public transportation vehicle liability, and commercial vehicle driver/contract mover liabilities.[246]

Overall, the intense efforts by the developed country members of the WTO on negotiation involving the insurance sector have proven to be effective and China's insurance industry has opened to a greater extent than other emerging insurance markets in the world.

12.4.2 The Status Quo and an Evaluation of Insurance Opening Up

Status Quo of the Opening Up of China's Insurance Industry

By the end of 2004, 17 foreign insurers (including thirteen P&C insurers, one life insurer, and three reinsurers) from 8 countries /regions set up their branches and/or subsidiaries in China and another 20 life insurers from 12 countries/regions accessed Chinese insurance market by setting up joint ventures in China (CIRC).[247] The

[246] Automobile liability insurance is mandatory in China. The compensation to the injured third party in traffic accidents is based on a tort liability system according to the *Road Traffic Safety Law of the People's Republic of China* effective May 1, 2004 (CIRC 2005h). CIRC was preparing *The Regulations on Mandatory Automobile Liability Insurance* when this chapter was being written and details concerning the mandatory insurance, such as the minimum insurance required, were not available when we completed this chapter.

[247] According to China's schedule of commitments in the insurance industry, foreign life companies can only operate in China through the establishment of joint ventures now. As a result, foreign life insurers established after WTO accession are all joint ventures. The existing foreign life branches are affiliated to AIA, which received its licenses in China before China's

foreign insurers had about 1.2 percent of the non-life insurance market and 2.6 percent of the life insurance market in 2004 (CIRC 2005p). With the huge potential of China's insurance market, there are still quite a few foreign insurers waiting for licenses. As of the end of 2004, there were 188 representative offices established by 124 foreign insurance institutions from 18 countries/regions in China (CIRC).

As noted above, China, following its WTO commitments, has allowed foreign insurers to operate in extensive areas of the country and provide many types of insurance. However, the foreign insurers already in business have actually focused on the relatively developed cities, such as Shanghai, Guangzhou, and Shenzhen. These cities offer better infrastructure, high per capita incomes, and the benefits of the cluster effect. Shanghai, the first experimental city to open, was the home of over 45 percent of the foreign insurers (by number) operating in China by the end of 2004 (CIRC).

As the insurance market opens up further, the number and the business volume of foreign firms in China continue to increase. The total assets of foreign insurers reached RMB 29.4 billion at the end of 2004, 2.5 percent of that of China's insurance industry (CIRC). The market share of foreign companies in the overall industry was 1.7 percent in 2003 and increased to 2.3 percent in 2004 (CIRC 2005p). In some regional markets with more foreign players such as Shanghai, Guangzhou, and Shenzhen, the market share of foreign insurers, especially life insurers, far exceeds the national average. For instance, in Shanghai alone, the market share of foreign companies reached 15 percent (Shanghai Insurance Regulatory Agency 2005). It is safe to say that foreign insurance companies tend to grow very fast and show a strong momentum.[248]

In addition, it is notable that foreign insurers tend to choose non-insurance firms, especially large industrial enterprises, as their joint venture partners. Observations on the 18 joint ventures already in business before August 2004 show that only 4 of them took on domestic insurers as their Chinese partners and the other Chinese partners include large industrial firms, such as Haier Group, Beijing Capital International Airport, China National Cereals, Oils and Foodstuffs Corp., and China Minmetals Corp. This fact reveals the market-oriented nature of foreign insurers. Foreign companies, if they invest together with domestic insurers, will not only have to spend a lot of efforts to integrate different corporate cultures, variant mindsets, and modes of operation and management, but also face the possibility of conflicts of interest between the joint venture and the domestic insurer. In contrast, if foreign insurers work with large non-insurance firms who are financially strong, have favorable good will, and enjoy the status of SOEs, they will enhance the likelihood of seeing positive returns sooner rather than later by avoiding the aforementioned risks and benefiting from securing customers, building a good image in the market, and influencing the market direction. Generali China Life Insurance Company

---

WTO entry. Furthermore, since foreign non-life insurance companies are allowed to establish branches, no joint ventures engaging in the non-life insurance business have been established.

[248] Foreign insurance companies collected 19.42 percent of the premium of newly issued individual life policies in Shanghai in 2003, while in 2002 it was 31.55 percent. This decline might prove the relative advantage of domestic insurers in competition, yet it is hard to conclude that the competence of domestic insurance companies has been improved materially without further evidence (Chen 2004, p. 289).

Limited can serve as a convincing example. On March 31, 2005, this company, a joint venture of Assicurazioni Generali S.p.A. (Generali) and China National Petroleum Corporation (CNPC), sold a "Generali China Sunlight Group Pension" policy to CNPC and collected RMB 20 billion premium. This policy raised the rank of Generali China Life from number 14 in 2004 to 4 in 2005 with respect to market concentration in China.[249]

Evaluation of Insurance Opening Up

In general, extensive opening up to the outside world has undoubtedly brought valuable opportunities to China's insurance industry and has proven to be a success in view of its effects on the development and reform of the industry. First, opening up has helped to improve the quality of development, i.e., the industry has seen product and service innovation in addition to sales growth. Foreign insurance companies' access to China's insurance industry through various channels has brought not only substantive needed capital, but also many resources necessary for effective improvement in the quality and productivity of assets. These include experienced talent, advanced technologies and experience, abilities in research and development (R&D), superior management, and the capacity for understanding customers' demand. Actually, China's domestic insurance companies have made great progress profiting from foreign rivals' new ideas and advanced knowledge in actuarial science, product development, management, marketing, and service. What is more, foreign insurers often engage in R&D, with domestic research institutions or organizations, which provide positive spillover benefits to domestic insurers. For example, Swiss Re has completed, together with Beijing Normal University, *The China Natural Catastrophe Maps*, which includes geographic and meteorological data related to major natural catastrophes in the recent 500 years and provides the insurers with a reliable basis to evaluate risks.

Second, opening to the outside has compelled China's domestic insurance industry to reform at a faster pace. Foreign insurance companies have become models for domestic insurers. Thus, fierce competition has erupted in China's insurance market. This will be intensified as foreign insurers further penetrate China. The strengthened competition will also compel domestic insurers to reform their ownership structure and build modern governance, management systems, and operating rules, and fully adopt the market economy mindsets. Furthermore, opening up has accelerated the formation of market-oriented legislation and a regulation system with the characteristics of transparency, equity, and effectiveness.

However, it is, without doubt, apparent that severe challenges to domestic insurers came with these opportunities. In an historical view, China's domestic insurers, who came into being just after the restoration of China's insurance industry, were quite young compared to their foreign rivals, who came from developed

[249] Some may think that the reason that foreign insurance companies search for partners outside China's insurance sector is that domestic companies do not want to partner with foreign firms. However, this is not the case. In fact, China's domestic insurers do desire foreign insurers as partners in order to enhance their skills in management, product development, and marketing. Furthermore, almost every domestic insurer that has a large market share, either has foreign insurers as strategic investors, or is searching for foreign investors, e.g., PICC P&C has AIG as a strategic investor.

markets and were already mature themselves. Early opening has been difficult for domestic insurers, since they are at a disadvantage with regard to many aspects, such as financial strength, administration, governance, management, accumulated knowledge, and expertise.

At the same time, problems resulting from the activities of the domestic insurers themselves, after WTO entry, also magnified the challenges to a great extent. First, domestic insurers mostly adopted a "volume expansion" strategy that focused on short-term sales volume while ignoring product differentiation and core competencies. In the face of intensifying competition, domestic insurers only concentrated on customers and lines which could be easily reached through their available and identical marketing channels, engendering an overcrowded market and identical product portfolios along with considerable unmet demands. Second, domestic insurers generally placed an emphasis on learning and mimicking foreign experience and technologies while neglecting the enhancement of their abilities to innovate. Third, domestic insurers gave most of their attention to learn foreign knowledge in marketing and product development and little to service. The extensive use of exclusive agents became the norm after AIA introduced the exclusive agent channel in 1992. The domestic insurers, especially life insurers, immediately began to heavily employ individual exclusive agents. This action proved to be extremely successful since more than half of the life insurance premiums were collected through exclusive agents over the past several years. However, domestic insurers interpreted "marketing via individual exclusive agents" as simply employing a quantity of salespersons and utilizing them to sell products. This partial understanding led to large numbers of insurance agents trained to compete for market share and premium volume only. The agents often attached little importance to service and made little effort to recommend insurance products that were best suitable for the consumers. While the agents sold new policies to customers in this underdeveloped market, the performance along with the growth and development of the industry were affected negatively in the long run since the creditability of the insurance industry was doubted by the consumers. Generally speaking, the domestic insurers have many things to learn from their foreign competitors to improve their competence.

12.5 THE FUTURE OF CHINA'S INSURANCE INDUSTRY

The accelerated economic globalization is netting more and more national and regional economies into the worldwide market. However, as a process full of conflicts, globalization not only has positive effects, but also inevitably brings a series of problems. China's insurance industry must confront these challenges and make the most of the opportunities offered by globalization.

12.5.1 Great Potential of a Developing Market

With the deepening industrialization in China, various service industries have been established and the framework of the service sector has been completed.[250] However, the development of the tertiary sector does not lend itself to an optimistic evaluation.[251] As Figure 12.6 shows, the ratio of output from the tertiary sector to GDP first decreased and then increased somewhat from 1992 to 2003 with the highest level of 34 percent in 2002 and the lowest of 30 percent in 1996. These figures suggest that the tertiary sector of China falls behind not only that of developed countries, but also behind that of developing countries at a similar stage of economic development. For example, the ratio of output from the tertiary sector to GDP of India was 36 percent in 1980 and kept on rising thereafter (Guo and Wang 1999, p. 7). Moreover, as Figure 12.7 indicates, the share that China's financial service sector contributes to the output of the service sector is quite low in comparison with that of developed countries. Furthermore, the share has been decreasing in recent years despite the increase in the volume of the financial sector output. Along with the progress in China's economic development and economic system reform, many sectors such as information technology, education, and entertainment grew faster than the financial sector grew. As Figure 12.8 shows, the ratio of output from financial sector to that of the whole service sector decreased from 20 percent in 1996 to 17 percent in 2002, while in the U.S., the ratio reached 22 percent in 1970 and kept increasing, although at a decreasing pace. These data suggest that the financial service sector in China is still underdeveloped and has plenty of room to expand. China's insurance industry, as the "developing industry in a developing market," undoubtedly has tremendous potential. It is anticipated that China's insurance industry will soon take a unique role on the world stage and become the key focus of insurers from developed countries.

12.5.2 Key Drivers of Future Growth of China's Insurance Industry

Excellent Economic Prospects
Experts predict that China's economy is expected to continue its robust growth for at least a few years with annual real growth rates of higher than 7 percent (Li, et al. 2003). Increased national income will surely have positive impacts on people's purchasing power. The increase of all types of uncertainties accompanying accelerated industrialization and urbanization will cause individuals to adjust their risk expectation upward and thus raise their willingness to purchase insurance. The combination of strengthening purchasing power and willingness will certainly drive the robust growth of insurance demand. It is in the traditional culture of the Chinese

---

[250] The ratio of output from the primary sector to GDP has decreased from 28.1 percent in 1978 to 14.6 percent in 2003. The employees in the primary sector accounted for 49.1 percent of the entire workforce in 2003 vs. a 70.5 percent in 1978 (National Bureau of Statistics of China 2004, §3.2, §5.2).

[251] The tertiary sector refers to all other economic activities not included in the primary or secondary sectors. The primary sector refers to agriculture, forestry, animal husbandry, and fishery. The secondary sector refers to mining and quarrying, manufacturing, production and supply of electricity, water and gas, and construction.

for individuals to be strongly motivated to prepare against possible economic losses. Thus, one could reasonably assume that increases in income levels will boost insurance demand, other things being equal.

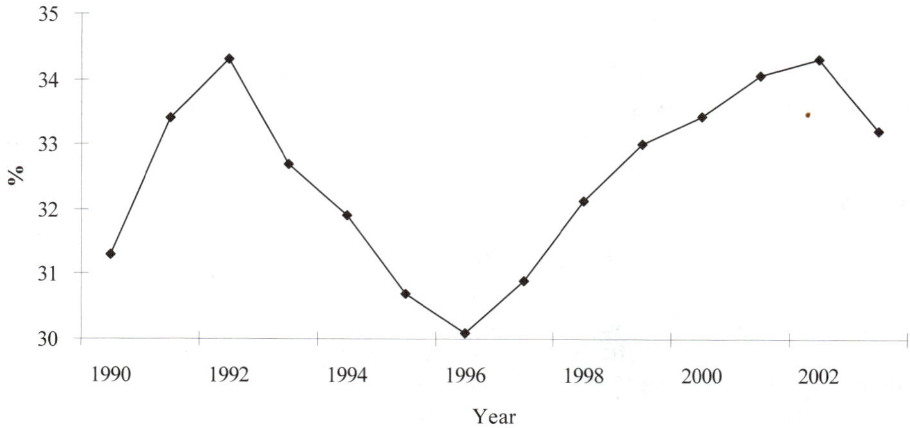

Source: National Bureau of Statistics of China (2004), §3.2;
Notes: There is a crash in the tertiary sectors' output to GDP in 1996. One explanation for this is that although the output from the tertiary sectors of China has kept rising, the secondary sectors grew much faster while the ratio of output from the primary sectors to GDP decreased slowly during the period from 1992 to 1996.

Figure 12.6. The Ratio of Output from the Tertiary Sectors to GDP, 1990–2003

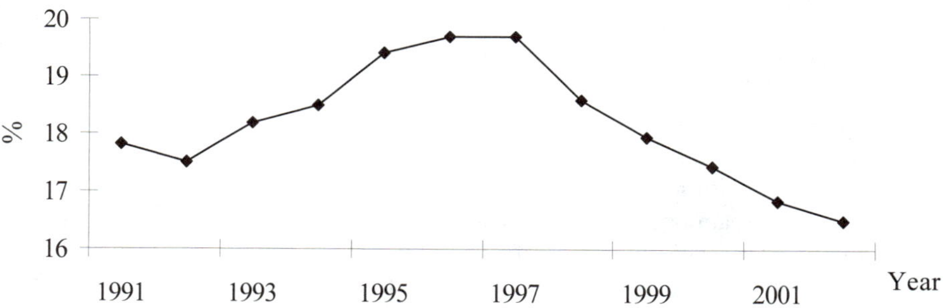

Source: National Bureau of Statistics of China (2004), §3.6 and (1992), §3.6.

Figure 12.7. The Ratio of Output from the Financial Services Sector to the Tertiary Sectors, 1991–2002

Economic System Reform

As mentioned above, market-oriented reform has fueled insurance demand and created the current insurance market in China. We predict that further progress in the reform and opening up will continue to fuel the insurance industry on the supply side. Furthermore, forthcoming reform policies will underlie sustained growth of insurance demand in China. Two of these reform policies are focused upon here. The first is investment system reform. The State Council of China enacted the *Resolution on the Reform of the Investment System* in July 2004, which abolished old administrative measures on the investment activities of firms. The old measures required that every investment must get ex ante approval from local or state government according to its size. Now, most investment decisions are in the hands of firms themselves. The exclusions are those projects that might have negative impacts on public interest. Consequently, the economic power of the firms has increased greatly, while the government's involvement in the economy has decreased as a part of the gradual institutional change in China. We can foresee that the investment enthusiasm will be inspired considerably and the exposure to greater investment risks will surely increase the firms' demand for insurance, especially for non-life insurance, since the government will no longer be responsible for a firm's investment risk.

The second reform is related to the social security system. China's efforts to establish the social security system (previously called the labor insurance system), which began shortly after the People's Republic of China was founded, were held back during the Culture Revolution. The Ministry of Finance promulgated *The Opinions on the Reform of Several State-Owned Enterprises' Financial Systems (draft)* in February 1969, which required that state-owned enterprises stop collecting social insurance funds, and instead treat reimbursements for payments of labor insurance as a non-operating expense. Thus, labor insurance was actually changed into a pay-as-you-go system running within an enterprise. This old system later proved to be a huge obstacle to China's economic system reform. The 3rd Plenary Session of the 14th CPC Central Committee held in 1993 formed *The Decisions on Several Issues about the Establishment of Socialist Market Economic System*. They decided to establish a multi-pillar social security system suitable not only to China's economic development stage, but also appropriate to the financial strength of the government, firms, and individuals. The current social security system in China thus focuses on providing the most basic economic security to laborers. Reform has led to a widening gap between the economic security provided by the social security system and individuals' actual needs. Consequently, there exists a high demand for commercial life and health insurance.[252]

Demographic Changes

China's population was 1.29 billion at the end of 2003, about 1/5 of the world population, and is still increasing. It is expected to reach 1.32 billion by the end of 2005, according to the United Nations (2004). The Chinese aging population will continue to grow due to the family planning policy. It is expected that the percentage of people aged 65 and above will increase from 7.50 percent at the end of 2003 to 13.40 percent by 2025 (National Population and Family Planning Commission of

---

[252] For example, based on the current basic pension system, the target monthly old-age benefit for each retiree is about 58.5 percent of the average wage of the society at the time of their retirement (Ministry of Labor and Social Security of China (MOLSS) 1998, p. 51). If we take the life cycle approach, many retirees' income right before retirement would be higher than the current social average level, for which the target replacement ratio of China's basic pension system is much lower than 58.5 percent.

China (NPFPC) 2003). The average family size (person/household) has decreased from 4.41 in 1982 to 3.38 in 2003 (National Bureau of Statistics of China 2004, §4.4 and §4.7). China's large population and demographic mix will drive the solid growth of China's insurance industry, especially in life insurance, health insurance, and pension insurance.

12.5.3 Factors That Might Have Negative Impacts on China's Insurance Industry

While we are optimistic about the future of China's insurance industry and believe that the growth trend will not be reversed, we have to concede that some factors may impede the fast pace of the development of the industry.

Immature Capital Market

The history of the world insurance industry has made it clear that investment management will be more and more important to insurers' survival and development as the insurance industry expands and matures. Presently, China's insurers are allowed to invest in domestic capital markets and the available investment channels have materially increased in 2004. However, China's capital market has not yet reached a desirable condition and is currently flooded with speculation. The average turnover rate of Shanghai Stock Exchange and Shenzhen Stock Exchange in 2003 was 251 percent and 214 percent respectively (National Bureau of Statistics of China 2004, §20.16), much higher than that of mature capital markets.[253] If insurers invest heavily in these markets, their financial strength may be threatened. In fact, China's insurers are now still reluctant to invest in the stock market when they were authorized to do so in October 2004.

Shortage of Professional Personnel

Over the past approximately twenty years companies were directed by the central government to pursue volume expansion strategies because the government viewed insurance premiums as a form of revenue generation and an approach to collect financial surplus. This strategy impelled companies to focus on short-term growth of sales volume, while profitability, management efficiency, and long-term development were neglected in the long-term. The domestic insurers competed fiercely for the low hanging fruit and did little research and development. To make things worse, the problem was masked by the high growth rate which was attributable only to the high potential of China's insurance market. This strategy not only led to low productivity and high operating costs, but also resulted in a lack of insurance education and career training. Lack of professional personnel, especially the lack of high quality managers, has proven to be a severe bottleneck for the development of China's insurance industry in the long run.

[253] The average turnover rate in year t, TO_t, is defined as: $TO_t = V_t / N_t \times 100\%$ in which V_t represents the volume of transacted shares in year t in the stock exchange, and N_t represents the number of negotiable shares in the exchange during the same time period.

Unfavorable Image of the Industry

The image of China's insurance industry, especially that of domestic insurers, is not desirable and calls for immediate attention. Many consumers believe it is difficult to obtain the promised insurance indemnity and that the procedure of claims filing is inefficient. Some agents and sales staff of insurance companies are perceived as low quality, not acting in good faith and with professional behavior. Some of them concealed important clauses in insurance contracts and even misguided consumers in their insurance purchase decisions. Others refused to meet their contractual obligations. Still, some insurers have acted against regulations and charged exorbitant commissions, used rebates, and/or insufficient premium rates to stimulate sales. The situation mentioned above has been worsened by the fact that the public knows little about insurance. According to a survey conducted in 2002 by the Development and Research Center of the State Council of China (DRC) in 50 big cities, 36 percent of the interviewees said that they were ignorant about insurance on the whole (Market Economy Institute of DRC 2002). Even in Shanghai, the most developed insurance market in China, 73 percent interviewees said they knew just a little about insurance and only 2.4 percent said they were acquainted with insurance (Horizon Group 2004). Thus, it is not surprising that the mistakes made by some individual insurers can be easily spread out quickly and erode the public's trust in some insurers and even in the whole insurance industry.

Lack of Public Good Faith

It is very important to the development of the insurance industry that consumers act on good faith. However, the situation in China at present is unsatisfactory and the development of many product lines, especially credit insurance and guarantee insurance, is restricted. For example, PICC P&C and some other P&C companies were forced to exit the auto loan insurance market temporarily because of the high default risk. Thirty percent of personal auto loans were in default and 10 percent were bad debt resulting in massive losses for the insurers. One can see from the claims practice that the insureds were often concentrated within certain areas or groups and sometimes with the clear intention to cheat (Wu 2003). Additionally, according to a conservative estimate by non-life insurance companies in Beijing, about 20 percent of the automobile insurance payments were frauds, including intentional damage, exaggerated claim amount, and claims of automobiles stolen that are false (Li 2004).

12.5.4 Future Trends for China's Insurance Industry

We forecast that China's insurance industry will maintain a two-digit growth rate in the coming eight to ten years and the growth rate will be descending over time. As to the development of the industry itself, the market system will keep up with international standards gradually and the following trends will be the most prominent features in an environment characterized by globalization, financial integration, aging population, and dynamic information technologies.

Growing Competition in the Insurance Market

China's insurance market will be more competitive as more foreign capital flows in and the number of insurers increases. The Herfindahl index of the life insurance market and non-life insurance market was 2,750 and 3,682 respectively and the respective market share of the top 3 insurers in these two markets has decreased to 75 percent and 80 percent in 2004 (CIRC 2005p).[254] The competitiveness of the insurance industry is growing at an accelerating pace. We anticipate that in next 3 year both of the respective market shares of the top 5 insurers in the life and non-life insurance sectors will fall to about 65 percent and 70 percent respectively and that the largest life insurer will hold about 40 percent of the market while the largest non-life insurer will hold about 45 percent. At the same time, the fierce competition will drive the insurers to pay a great deal of attention to the quality of service and the functions of insurance products. However, on the whole, China's insurance market will still be characterized as a concentrated oligopoly in the medium-term.

Deepening Cooperation with Other Financial Institutions, Especially Banks

Since October 2004, China's insurers are allowed to invest in domestic stock market, after a long time of tradeoff between the possible risks and return, indicating an even closer relationship between the insurance industry and the securities sector. But as China's capital market is still immature, insurers will be very prudent in investing in the stock market due to the potential risk of loss. So the insurers' investment in the stock market will not increase on a large scale in the short-term and the prospective role of insurers in the capital market will depend on the maturing progress of the latter.

The cooperation between insurers and banks will be strengthened greatly. Bancassurance has been a success in China, because banks have advantages over insurers in customer resources and marketing. People in China are acquainted with and have steadfast trust in banks and therefore, bank services can be easily accepted. Bancassurance can be expected to continue to be one of the major distribution channels for life insurance. Besides strengthening strategic cooperation between the two industries, there will be more and more holding companies controlling both banks and insurance companies.

Finally, with a view to the coming financial integration, various innovative measures for the controlling of systematic financial risks, such as insurance securitization, will become popular, but not in the near term. The development of insurance securitization is a complex project and is demanding for both the insurance market and the securities market. Presently, the preconditions for insurance securitization do not exist in China, including theoretical research, the training of professional personnel, and the establishment of required infrastructure (e.g., legislation system). However, alternative risk transfer mechanisms will surely be important means for risk management by China's insurance industry.

---

[254] The Herfindahl index is an index of market concentration calculated by adding the squared value of the individual market shares of all firms in the industry. Market share here is measured by premium volume.

Optimization of the Product Portfolio

With respect to China's life insurance sector, the demand for health and pension insurance will increase rapidly and steadily along with the aging population and the deepening of social security reform. On the health insurance side, while currently the government provides a large portion of health care in urban areas, this portion is declining. Consequently, the role of private health insurance is rising. Note, however, that there is hardly any private health insurance in rural areas of China. However, it is worth indicating that severe adverse selection and moral hazard exist, and domestic insurers lack experience in risk control. Therefore, China's insurers need to improve their health plans and intensify their cooperation with hospitals in order to control the medical costs and reduce adverse selection and moral hazard. Managed care plans will be introduced in China and will likely become the major operation pattern of health insurance in the next five to eight years. As for pension insurance, the insurance industry is faced with great opportunities and foreign insurance companies specializing in pension will play an important role in the market.

With regard to China's non-life insurance sector, property insurance and motor vehicle liability insurance will continue to dominate the non-life insurance market. These types of insurance are expected to account for more than 50 percent of the premium income generated by the non-life industry in the next 5 years, primarily due to the increasing number of motor vehicles (especially private automobiles) and legal requirements for carrying liability insurance. There is likely to be solid growth of liability insurance, construction and installation projects insurance, enterprise property insurance, family property insurance, and credit insurance owing to the gradual clarification of the concept of private property rights, large scale infrastructure construction, fast growth in home ownership, family wealth, and increased public awareness of the value and need for insurance.

Diversification of Distribution Channels with Agents and Brokers Still Playing Important Roles

Innovative distribution channels, including bancassurance, postal insurance, and direct response have already emerged in China and their market share will certainly increase over time. The insurance distribution channels in China will become more diversified. Agents and brokers will remain important and irreplaceable since the applicability of innovative distribution channels will depend on consumers' need for advisory services and the traditional distribution channels are much more effective for complicated policies with long insurance periods and large face amounts. As a result, there will still be a need for agents and brokers to serve consumers face-to-face frequently. In addition, rising wealth levels will drive the demand for comprehensive risk management and financial planning services rather than simple advisory services on insurance purchase decisions. The requirements for well-trained agents and brokers will consequentially be much higher.

12.6 CONCLUSION

China's insurance industry developed rapidly from a virtually nonexistent, minimal, and closed industry into a large, open industry with great potential. This is supported by the growing insurance demand resulting from the government's efforts to create market-oriented reforms in the 1980s. As to the supply side, China's insurance market is characterized by an increase in the number of insurers and organizational restructuring of the state-owned insurers, which gradually lost their former monopoly of insurance market. While the market has become more competitive, it is still characterized as a typical oligopoly and the government remains heavily involved in the industry, though not as much as before. Furthermore, the geographic distribution of China's insurance industry is highly imbalanced.

Opening up to the outside world, especially China's WTO accession, has brought the young domestic insurers into fierce competition with foreign insurance companies. On the one hand, the presence of foreign rivals helps to improve the quality of developments of China's insurance industry and enhance the competitive environment. On the other hand, China's domestic insurers are at a disadvantage in many aspects compared to their foreign rivals, such as financial strength, administration, and governance. Thus, they face severe challenges. However, the domestic insurers themselves magnified the challenges to a great extent. They adopted a so called "volume expansion" strategy while ignoring product differentiation and core competencies, placed an emphasis on learning and mimicking while ignoring the enhancement of the ability to innovate, and paid a lot of attention to learning foreign knowledge in sales and product development while ignoring service.

As a developing industry in a developing economy, China's insurance industry undoubtedly has tremendous potential. Some factors, including excellent economic prospects, accelerated economic system reform, and the large and aging population will drive China's insurance industry to grow continuously and rapidly. However, the development of the industry might be bottlenecked by the immature domestic capital market, shortage of professional personnel, unfavorable image of the industry, and lack of public good faith. Despite these challenges, we forecast that China's insurance industry will maintain a two-digit growth rate in the forthcoming eight to ten years and the growth rate will be decreasing over time. As to the development of the industry itself, the market will catch up with international standards gradually and become more and more competitive. There will be deepening cooperation between insurers and other financial institutions, especially banks. Optimization of product portfolios and diversification of distribution channels are also expected, though agents and brokers will continue to play a crucial role.

12.7 REFERENCES

14th Session of Standing Committee of the 8th NPC, The, 1997, *Insurance Law of the People's Republic of China (1995)* (Beijing, China: China Law Press).

Black, Jr., Kenneth and Harold D. Skipper, Jr., 2000, *Life and Health Insurance* (Upper Saddle River, NJ: Prentice-Hall Inc.).

Chen, Wenhui, 2002, *The Development and Regulation of China's Life Insurance Industry* (Beijing, China: China Financial Publishing House).

_____, 2004, *Development Report on China Life Insurance 2003* (Beijing, China: China Finance and Economics Publishing House).

China Center for Insurance and Social Security Research at Peking University (CCISSR), 2004, *Review and Analysis on Hot Topics in China's Insurance Market (2003–2004), Beida Insurance Review Series* (Beijing, China: Peking University Press).

_____, 2005, *Economic Development and Social Harmony: The Role of Insurance and Social Security* (Beijing, China: Peking University Press).

China Court, 1969, *The Opinions on the Reform of Several State-Owned Enterprises' Financial Systems,* www.chinacourt.org.

China Economic Information Center, 2005, Intranet Business Express Database (Beijing, China: CEInet Data Company), www.cei.gov.cn/.

China Insurance Regulatory Commission (CIRC), 2005a, *Collection of Insurance Laws and Regulations (1998–2005)* (Beijing, China: China Democracy and Law Press).

_____, 2005b, "Provisional Measures for Insurance Brokers Administration (1998)," in *Collection of Insurance Laws and Regulations (1998–2005)* (Beijing, China: China Democracy and Law Press).

_____, 2005c, "Measures for Insurance Broker Companies Administration (2001)," in *Collection of Insurance Laws and Regulations (1998–2005)* (Beijing, China: China Democracy and Law Press).

_____, 2005d, "Regulations on Administration of Foreign-Funded Insurance Companies (2001)," in *Collection of Insurance Laws and Regulations (1998–2005)* (Beijing, China: China Democracy and Law Press).

_____, 2005e, "Measures for Insurance Agencies Administration (2001)," in *Collection of Insurance Laws and Regulations (1998–2005)* (Beijing, China: China Democracy and Law Press).

_____, 2005f, "Insurance Law of the People's Republic of China (2002)," in *Collection of Insurance Laws and Regulations (1998–2005)* (Beijing, China: China Democracy and Law Press).

_____, 2005g, "Regulation on the Solvency of Insurers and Regulation Indicators (2003)," in *Collection of Insurance Laws and Regulations (1998–2005)* (Beijing, China: China Democracy and Law Press).

_____, 2005h, "The Road Traffic Safety Law of the People's Republic of China (2004)," in *Collection of Insurance Laws and Regulations (1998–2005)* (Beijing, China: China Democracy and Law Press).

_____, 2005i, "Measures for Insurance Agencies Administration (2004)," in *Collection of Insurance Laws and Regulations (1998–2005)* (Beijing, China: China Democracy and Law Press).

_____, 2005j, "Measures for Insurance Broker Institutions Administration (2004)," in *Collection of Insurance Laws and Regulations (1998–2005)* (Beijing, China: China Democracy and Law Press).

_____, 2005k, "Provisional Regulations on the Administration of Insurance Asset Management Companies (2004)," in *Collection of Insurance Laws and Regulations (1998–2005)* (Beijing, China: China Democracy and Law Press).

_____, 2005l, "Regulation for the Administration of Insurance Companies (2004)," in *Collection of Insurance Laws and Regulations (1998–2005)* (Beijing, China: China Democracy and Law Press).

_____, 2005m, "Administrative Measures on Examination, Approval, and Filing of Life Insurance Products (2004)," in *Collection of Insurance Laws and Regulations (1998–2005)* (Beijing, China: China Democracy and Law Press).

_____, 2005n, "Notice on Several Issues about Administrative Measures on Examination, Approval, and Filing of Life Insurance Products (2004)," in *Collection of Insurance Laws and Regulations (1998–2005)* (Beijing, China: China Democracy and Law Press).

_____, 2005o, "Notice on Several Issues about Administration Measures on Life Insurance Products Approval and Filing (2004)," in *Collection of Insurance Laws and Regulations (1998–2005)* (Beijing, China: China Democracy and Law Press).

_____, 2005p, *Premium Volumes and Market Shares of China's Insurance Companies in 2004*, www.circ.gov.cn.

_____, 2005q, "Provisional Regulations on Administration of Stock Investment of Insurance Institutional Investors (2004)," in *Collection of Insurance Laws and Regulations (1998–2005)* (Beijing, China: China Democracy and Law Press).

China Life, 2005, *2004 Annual Report of China Life Insurance Company Limited,* www.chinalife.com.cn.

CPC Central Committee, The, 1993, *Decisions on Several Issues about the Establishment of Socialism Market Economy System* (Beijing, China: China People's Press).

Editorial Committee of China's Insurance Yearbook, The, 2004, *2004 Yearbook of China's Insurance*, E-version (Beijing, China: The Editorial Office of China's Insurance Yearbook).

Goldman Sachs, 2003, "The Prospects of China's Insurance Industry: The Key to Create Value is Reform," Beijing, China.

Guo, Kesha and Yanzhong Wang, 1999, *Study on the Change of China's Industrial Structure and Relative Policies* (Beijing, China: Economic Management Press).

Guoyuan Center for Securities Research and Development, 2002, *China's Entry into WTO and the Opening of Financial Sector* (Beijing, China: Intranet Business Express of China).

Herres, Robert T., 2000, "Leading the Commerce Revolution in Financial Services," Seminar Proceedings of the International Insurance Society Conference, Vancouver, Canada (July 8–12).

Horizon Group, 2004, *Insurance is Still Strange to People in Shanghai*, www.horizonkey.com.

Jiang, Shengzhong, 2004, *Development Report on China Insurance Industry 2004* (Beijing, China: China Finance and Economics Publishing House).

Jiang, Xiaojuan, 2002, *Foreign Investment in China* (Beijing, China: China Renmin University Press).

Li, Hai, 2004, "How Low is the Cost of Frauds?" *Beijing Business Today*, November 24, 2004, pp. 1, 9.

Li, Shantong, Yongzhi Hou, and Fan Zhai, 2003, "The Potentials of and Outlook for China's Economic Growth in the Next 50 Years," *References for Economic Studies* 2: 51–60.

Li, Zhilin, 2002, *Explore China's Stock Market* (Shanghai, China: Shanghai Sanlianc Bookstore).

Market Economy Institute of DRC, 2002, *Report on the Insurance Market in 50 Big Cities*, Beijing, China.

Ministry of Labor and Social Security of China (MOLSS), 1998, *Questions and Answers to the Basic Pension System* (Beijing, China: China Labor Press).

National Bureau of Statistics of China, 1996–2004, *China Statistical Yearbook* (Beijing, China: China Statistics Press).

National Population and Family Planning Commission of China (NPFPC), 2003, *Prospects on Population Aging Process in 2001–2050*, Beijing, China.

Ning, Nan and Liqi Yu, 2003, "China's Insurance Industry in Difficulties," *Business Weekly* April 15: 50–52.

People's Bank of China (PBOC) Insurance Division, 1998a, "Interim Regulations on the Administration of Insurance Enterprises (1985)," in *Collection of Insurance Laws and Regulation* (Beijing, China: China Commerce Press).

_____, 1998b, "Temporary Measures for Administration of Foreign-Funded Insurance Institutions in Shanghai (1992)," in *Collection of Insurance Laws and Regulation* (Beijing, China: China Commerce Press).

_____, 1998c, "Provisional Measures for Insurance Agents Administration (1996)," in *Collection of Insurance Laws and Regulation* (Beijing, China: China Commerce Press).

People's Insurance Company of China Property and Casualty Company Limited, 2005, *2004 Annual Report of PICC Property and Casualty Insurance Company Limited*, www.picc.com.cn.

Pritchett, S. Travis, Joan Schmit, Helen Doerpinghaus, and James L. Athearn, 1996, *Risk Management and Insurance* (St. Paul, MN: West Publishing Company).

Shanghai Insurance Regulatory Agency, 2005, "Overview of Shanghai's Insurance Market in 2004," www.circ.sh.gov.cn.

Shanghai Stock Exchange, 2002, *Historical Trading Data of the Composite Index of Shanghai in 2001*, www.sse.com.cn.

State Administration of Foreign Exchange, 2005, *Data and Statistics,* www.safe.gov.cn.

State Council of China, 2004, *Resolution on the Reform of the Investment System* (Beijing, China: XinHua News Agency).

Sun, Qixiang, 2000, *China's Insurance Industry: Conflicts, Challenges, and Countermeasures* (Beijing, China: China Financial Publishing House).

_____, 2003, "The Impact of WTO Accession on the Chinese Insurance Industry," *Journal of Risk Management and Insurance Review* 6: 27–35.

_____, and Wei Zheng, 2005, *Studies on China's Social Security Reform: The Reform of the Social Insurance System and the Development of Commercial Insurance* (Beijing, China: China Financial Publishing House).

Swiss Re, 2004, "World Insurance 2003: Insurance Industry on the Road to Recovery," *Sigma*, No. 3.

Tang, Yunxiang, 2004, *Research on the Development of China's Property and Casualty Insurance Market (2003)* (Beijing, China: China Economic Publishing House).

United Nations, 2004, *World Population Prospects*, Population Division, Department of Economic and Social Affairs of United Nations.

Wei, Hualin, Ziyou Yu, and Guo Yang, 1999, *The Opening-Up and Regulation of China's Insurance Market* (Beijing, China: China Financial Publishing House).

Wu, Dingfu, 2004, *Development and Reform Report on China Insurance Industry (1979–2003)* (Beijing, China: China Economic Publishing House).

Wu, Weiqiang, 2003, "Who on Earth Defaulted?" *China Youth Daily*, October 9, 2003.

Ye, Yide, Yue Wu, and Yuanren Zhu, 1998, *History of China Insurance* (Beijing, China: China Financial Publishing House).

Zhao, Lanliang, 2003, *Studies on Modern Shanghai Insurance Market* (Shanghai, China: Fudan University Press).

12.8 APPENDIX 1: An Overview of the Top Insurance Companies in China's Life and Non-Life Industry

China Life Insurance Company Limited (China Life) and PICC Property and Casualty Insurance Company Limited (PICC P&C) are now the leading insurance companies in China's life and non-life insurance industry. In 2004, China Life accounted for 46.9 percent of China's life insurance market and the PICC P&C accounted for 58.1 percent market share (CIRC).

The following table covers some of the main characteristics of these two companies:

	China Life	PICC P&C
Company Profile	▪ Incorporated on June 30, 2003 by China Life Insurance Company, which transferred its well-performing policies and related assets and liabilities to China Life and was then restructured to China Life Insurance (Group) Company in July 2003.	▪ Incorporated on July 19, 2003 by PICC, which injected all of its commercial insurance operations, together with related assets, liabilities, and subsidiaries into PICC P&C and then renamed and re-registered as PICC Holding Company.
Top Management	▪ The Chairman is appointed by CIRC. The board of directors appoints the president while the candidate is nominated by CIRC, given that the State is still the majority shareholder. ▪ The now Chairman of the Board and President, Mr. Chao Yang, got his appointment in June 2005. He has been the president of China Life Insurance (Group) Company since May 2005. Between March 2000 and May 2005, Mr. Yang was the chairman and the president of both China Insurance (Holdings) Company Limited and China Insurance H. K. (Holding)	▪ The Chairman is appointed by CIRC. The board of directors appoints the president while the candidate is nominated by CIRC, given that the State is still the majority shareholder. ▪ The Chairman of the Board is now Mr. Yunxiang Tang, who joined PICC in 2000 as general manager. He was previously the vice chairman of CIRC and assistant president of the People's Bank of China. Mr. Tang graduated from Shaoyang Normal College of Hunan Province and has 41 years

	China Life	PICC P&C
	Company Limited. He graduated from Shanghai International Studies University and Middlesex University in the United Kingdom. Mr. Yang majored in English language and business administration and has obtained a master's degree in Business Administration. ▪ The former Chairman of the Board and President was Mr. Xianzhang Wang, who served as President of China Life Insurance Company since 2000. Mr. Wang graduated from Liaoning University of Finance and Economics (now Northeast University of Finance and Economics) with a major in foreign economics and trade in 1965.	of experience in business operation and management. ▪ The President and CEO is now Mr. Yi Wang, who joined PICC in 1979 and was previously deputy general manager of PICC. Mr. Wang graduated from the Chinese Communist Party University with a bachelor's degree of economics and management in 2000 and a master's degree of political science in 2004.
Corporate Governance		

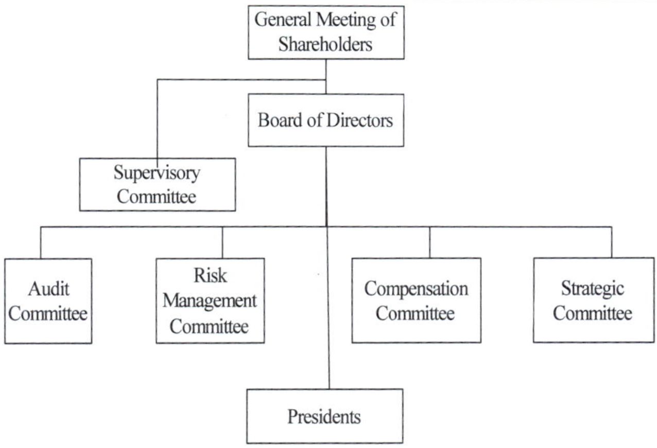

China Life

	China Life	PICC P&C

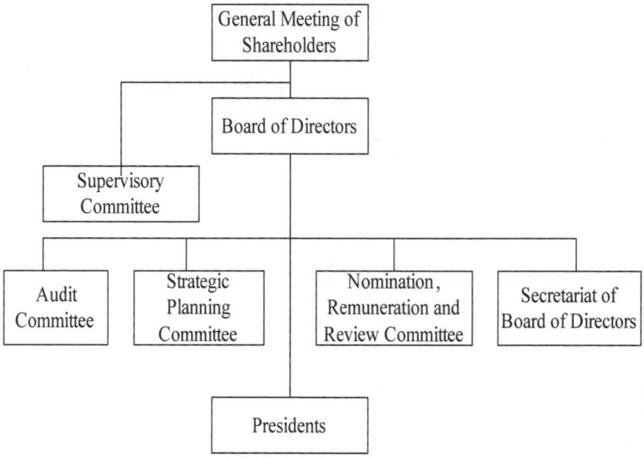

PICC P&C

	China Life	PICC P&C
Financial Results for 2004	▪ Net Premiums Earned and Policy Fees: RMB 65.0 billion ▪ Investment Results: RMB 10.0 billion ▪ Benefits and Claims: RMB 52.8 billion ▪ Net Profit: RMB 7.2 billion ▪ Total Assets: RMB 433.7 billion ▪ Total Shareholders' Equity: RMB 66.5 billion	▪ Net Premiums Earned: RMB 50.6 billion ▪ Investment Results: RMB 304 million ▪ Net Claims Incurred: RMB 38.9 billion ▪ Net Profit: RMB 208 million ▪ Total Assets: RMB 88.3 billion ▪ Total Shareholders' Equity: RMB 16.8 billion (PICC P&C also carries a RMB 2-billion subordinated loan)
Business Segments in 2004	▪ Gross written premiums and policy fees attributable to individual life insurance in 2004 were RMB 54.9 million, 80.6 percent of which were collected from participating products, representing 82.9 percent of its gross written premiums and policy fees for the year. Participating endowment products become one of the most important product series of the Company. ▪ Accidental and short-term health insurance accounted for 7.51 percent and 8.5 percent of its business respectively in 2004.	▪ Automobile insurance products became the most important product line of the company. In 2004, the gross premiums attributable to automobile insurance accounted for 69.2 percent of the gross written premiums. ▪ Enterprise property insurance took second place with a market share of 12.9 percent.

	China Life	PICC P&C
Shareholders	▪ China Life Insurance (Group) Company holds 72.2 percent of China Life's equity. Public investors hold the other 17.8 percent since China Life is a listed company both on New York Stock Exchange (NYSE) and the Hong Kong Stock Exchange.	▪ PICC Holding Company holds 69.0 percent of PICC P&C's equity. Public investors hold the other 31.0 percent since PICC P&C is a listed company on the Hong Kong Stock Exchange. ▪ AIG bought 9.9 percent of PICC P&C's equity at its IPO with a five-year lock-in period and became the latter's strategic foreign investor. AIG has begun to cooperate actively with PICC P&C in the field of accidental and health insurance.

Source: 2004 Annual Report of China Life Insurance Company Limited; 2004 Annual Report of PICC Property and Casualty Insurance Company Limited.

12.9 APPENDIX 2: Average Exchange Rate of RMB to Main Convertible Currencies (Middle Rate), 1980–2003*

Year	100 US$	100€**	Year	100 US$	100€
1980	149.84	N/A	1992	551.46	N/A
1981	170.50	N/A	1993	576.20	N/A
1982	189.25	N/A	1994	861.87	N/A
1983	197.57	N/A	1995	835.10	N/A
1984	232.70	N/A	1996	831.42	N/A
1985	293.66	N/A	1997	828.98	N/A
1986	345.28	N/A	1998	827.91	N/A
1987	372.21	N/A	1999	827.83	N/A
1988	372.21	N/A	2000	827.84	N/A
1989	376.51	N/A	2001	827.70	N/A
1990	478.32	N/A	2002	827.70	800.58
1991	532.33	N/A	2003	827.70	936.13

Source: State Administration of Foreign Exchange (2005); National Bureau of Statistics of China (1996), §16.2 and (2004), §18.2.

Note: N/A is not available. RMB is Ren Min Bi, the Chinese currency.

* Middle rate is the annual average exchange rate is calculated as the weighted mean of the daily exchange rates during the year.

** China began to publicize the exchange rates of the RMB to the Euro on April 2, 2002 after the Euro began circulating.

12.10 LEXICON

AIA. American International Assurance Company.

AIG. American International Group.

AIU. American International Underwriters, Ltd.

Bonus-Malus System. The Bonus-Malus system is a kind of no-claims discount strategy which can be chosen by the insurer to eliminate the moral hazard problem. No claims bonus schemes are an important feature of some insurance contracts, most notably those for auto insurance. A particularly simple Bonus-Malus scheme is termed a "one up/one down" scheme. It rewards individuals for each claim-free period by a reduction in premium to the next lowest category and each claim generates an increase in premiums to the next highest category.

CCISSR. China Center for Insurance and Social Security Research at Peking University.

China Life. China Life Insurance Company Limited.

China Re. China Reinsurance Company.

CIRC. The China Insurance Regulatory Commission is the regulatory body of China's insurance industry directly under the State Council.

CNPC. China National Petroleum Corporation.

CPC. Chinese Communist Party.

Direct Response. Direct response, including Internet, telephone, and direct mail, is the marketing channel through which the insurance companies sell products directly to consumers. This channel does not involve face-to-face contact during the sale process.

DRC. Development and Research Center of the State Council of China.

FICX. Farming Insurance Company of Xinjiang Production and Construction Group.

Financial Bonds. Financial bonds are securities issued by financial institutions in China's inter-bank bonds market. They require the repayment of interest and principal within a given period.

MOLSS. Ministry of Labor and Social Security of China.

Negotiated Bank Deposit. A negotiated bank deposit is a kind of long-term deposit by non-bank financial institutions, the terms of which are set by bilateral negotiation.

NYSE. New York Stock Exchange.

NPC. National People's Congress.

NPFPC. National Population and Family Planning Commission of China.

PBOC. PBOC is the People's Bank of China, the central bank of China.

PICC. PICC is the People's Insurance Company of China. It is the first state-owned insurance company in China's history. It was restructured into the PICC Group in 1996.

PICC P&C. PICC Property and Casualty Company Limited.

Primary Sectors, The. The primary sectors are agriculture, forestry, animal husbandry, and fishery.

R&D. Research and Development.

RMB. RMB is Ren Min Bi, the Chinese currency.

Secondary Sectors, The. The secondary sectors are mining and quarrying, manufacturing, the production and supply of electricity, water, and gas, and construction.

SOE. State owned enterprise.

Tertiary Sectors, The. The tertiary sectors are all other economic activities not included in the primary or secondary sectors.

WTO. World Trade Organization.

WWI. World War I.

13 An Analysis of the Evolution of Insurance in India

Tapen Sinha

Instituto Tecnologico Autonomo de Mexico
and University of Nottingham

13.1 INTRODUCTION

India had the nineteenth largest insurance market in the world in 2003. Strong economic growth in the last decade combined with a population of over one billion makes it one of the potentially largest markets in the future. Insurance in India has gone through two radical transformations. Before 1956, insurance was private with minimal government intervention. In 1956, *life* insurance was nationalized and a monopoly was created. In 1972, *general* insurance was nationalized as well.[255] But, unlike life insurance, a different structure was created for the industry. One holding company was formed with four subsidiaries. As a part of the general opening up of the economy after 1992, a government-appointed committee recommended that private companies should be allowed to operate. It took six years to implement the recommendation. The private sector was allowed into the insurance business in 2000. However, foreign ownership was restricted. No more than 26 percent of any company can be foreign-owned.

This chapter examines the insurance industry in India through different regulatory regimes. A totally regulation-free regime ended in 1912 with the introduction of regulated life insurance. A comprehensive regulatory scheme came into place in 1938. This was disabled through nationalization, but the Insurance Act of 1938 became relevant again in 2000 with deregulation. With a strong hint of sustained growth of the economy in the recent past, the Indian market is likely to grow substantially over the next few decades.

The rest of the chapter is organized as follows. First, we study the evolution of the Indian insurance business before nationalization. This is important because the denationalized structure brought back to play important legal rules from 1938. Second, we analyze the nationalized era separately for life and property-casualty business, because they were not nationalized simultaneously. Much of the post-

---

[255] The term *general insurance* is used in Britain and other Commonwealth countries. Elsewhere, the equivalent term is *property-casualty insurance* or *non-life insurance.*

independence history of insurance in India was the history of nationalized insurance. Third, we examine the new legal structure introduced after the industry was denationalized in 2000. Finally, we examine the current state of play and the projected future of the industry.

13.2 EVOLUTION OF INSURANCE BEFORE NATIONALIZATION

13.2.1 Insurance in the Colonial Era

Life insurance in the modern form was first set up in India through a British company called the Oriental Life Insurance Company in 1818, followed by the Bombay Assurance Company in 1823 and the Madras Equitable Life Insurance Society in 1829. All of these companies operated in India but did not insure the lives of Indians. They insured the lives of Europeans living in India.[256]

Some of the companies that started later did provide insurance for Indians, but they were treated as "substandard." Substandard in insurance parlance refers to life insurance for people with physical disability. In this case, the common adjustment made was a "rating-up" of five to seven years to the normal life expectancy of a British person in India. This meant, treating $q(x)$, the (conditional) probability of dying between x and $x + 1$, for an x-year-old Indian male as if it was $q(x + 5)$ or $q(x + 7)$ of a British male. Therefore, Indians had to pay an ad hoc extra premium of 20 percent or more. This was a common practice of European companies that were operating in Asia or Latin America. The first company to sell policies to Indians with "fair value" was the Bombay Mutual Life Assurance Society starting in 1871.

The first general insurance company, Triton Insurance Company Ltd., was established in 1850. It was owned and operated by the British. The first indigenous general insurance company was the Indian Mercantile Insurance Company Limited, established in Bombay in 1907.

At the time, insurance business was conducted in India without any specific regulation. Insurers were subject to the Indian Companies Act of 1866, but the insurance industry was otherwise unregulated. After the start of the "Be Indian Buy Indian" movement (called the Swadeshi movement) in 1905, indigenous enterprises sprang up in many industries. Not surprisingly, the movement also touched the insurance industry, leading to the formation of dozens of life insurers along with provident fund companies (pension funds). In 1912, two sets of legislation were passed: the Indian Life Assurance Companies Act and the Provident Insurance Societies Act. There are several striking features of these legislations. First, they were the first legislations in India that particularly targeted the insurance sector. Second, they did not apply to general insurance, because the government did not feel the necessity to regulate general insurance. Third, they restricted activities of Indian insurers but not foreign insurers, even though the legislation was modeled after the British Act of 1909.

---

[256] An excellent history of Indian business can be found in Tripathi (2004). It gives a detailed account of the rise of business houses in India during the colonial period.

Comprehensive insurance legislation covering both life and non-life business did not materialize for the next 26 years. During the first phase of these years, Great Britain entered World War I—an event that disrupted all legislative initiatives. Later, Indians demanded freedom from the British. As a concession, India was granted "home rule" through the Government of India Act of 1935, which provided for legislative assemblies for provincial governments as well as for the central government. But supreme authority of promulgated laws remained with the British Crown.

The only significant legislative change before the Insurance Act of 1938 was Act XX of 1928. It enabled the government of India to collect information about (1) Indian insurers operating in India, (2) foreign insurers operating in India, and (3) Indian insurers operating in foreign countries. The last two elements were missing from the 1912 Insurance Act. Information thus collected allows us to compare the average face value of Indian insurers to their foreign counterparts. In 1928, the average policy value of an Indian company was U.S. $619 compared to $1,150 for foreign companies (Indian Insurance Commissioner's Report, 1929, p. 23).

Foreign insurers were doing well during that period. In 1938, the average size of the policy sold by Indian companies had fallen to U.S. $532 (compared to $619 in 1928), and that of foreign companies had risen somewhat to $1,188 (in 1928, the average size was $1,150).

13.2.2 The Insurance Act of 1938

In 1937, the government of India set up a consultative committee. Sushil C. Sen, a well-known solicitor in Calcutta, was appointed the chair of the committee. He consulted a wide range of interested parties, including leaders within the insurance industry. It was debated in the legislative assembly, and, finally, in 1938, the Insurance Act was passed. This was the first comprehensive piece of legislation in India, covering both life and general insurers. It clearly defined the various lines of business in the insurance industry (see Appendix 1). It covered deposits, supervision of insurers, investments, commissions of agents, and directors appointed by the policyholders, among other topics. This piece of legislation lost significance after life insurance was nationalized in 1956, and general insurance was nationalized in 1972. With the privatization in the late twentieth century, the Insurance Act of 1938 returned as the backbone of current legislation of insurers. All legislative changes are enumerated in Table 13.1.

To implement the 1938 Act, an insurance department was established in the Ministry of Commerce by the government of India; later it was transferred to the Ministry of Finance. One curious element of the Act's classification of lines of insurance business was its inclusion of automobile insurance in the "miscellaneous" category. Later in the century, automobiles became the largest single item of general insurance. However, it continued to be included in the miscellaneous category, making it difficult to delineate the effects of losses due to pricing that drove this sector. For example, the Tariff Advisory Committee effectively fixed prices for a number of general insurance lines of business. Most premiums were below what would have been actuarially fair (especially for auto), but reporting auto insurance under the miscellaneous category masked this underpricing.

When the market was opened again to private participation in 1999, the earlier Insurance Act of 1938 was reinstated as the backbone of the current legislation of insurers, as the Insurance Regulatory and Development Authority Act of 1999 was superimposed on the 1938 Insurance Act. This revival of the earlier legislation has created a messy problem in that the Insurance Act of 1938 explicitly forbade insurers to participate in other financial services activities such as banking.

Table 13.1. Milestones of Insurance Regulations in India

Year	Significant Regulatory Event
1912	The Indian Life Insurance Company Act
1928	Indian Insurance Companies Act
1938	The Insurance Act: Comprehensive Act to regulate insurance business in India
1956	Nationalization of life insurance business in India with a monopoly awarded to the Life Insurance Corporation of India
1972	Nationalization of general insurance business in India with the formation of a holding company, General Insurance Corporation
1993	Malhotra Committee established
1994	Recommendations of Malhotra Committee published
1995	Mukherjee Committee established
1996	Setting up of (interim) Insurance Regulatory Authority (IRA) Recommendations of the IRA
1997	Mukherjee Committee Report submitted but not made public
1997	The government gives greater autonomy to Life Insurance Corporation, General Insurance Corporation, and its subsidiaries with regard to the restructuring of boards and flexibility in investment norms aimed at channeling funds to the infrastructure sector
1998	The cabinet decides to allow 40 percent foreign equity in private insurance companies—26 percent to foreign companies and 14 percent to nonresident Indians and Foreign Institutional Investors
1999	The Standing Committee headed by Murali Deora decides that foreign equity in private insurance should be limited to 26 percent. The IRA bill is renamed the Insurance Regulatory and Development Authority Bill
1999	Cabinet clears Insurance Regulatory and Development Authority Bill
2000	President gives assent to the Insurance Regulatory and Development Authority Bill

By 1956, there were 154 Indian life insurers. There were 16 non-Indian insurers, and 75 provident societies were issuing life insurance policies. Most of these policies were centered in the big cities of Bombay, Calcutta, Delhi, and Madras.

13.2.3 Mortality Tables

Before the mortality of Indian lives were used for constructing mortality tables for India, it was common practice to use the British Office Table based on the British experience during 1863 to 1893. As noted earlier, the table was used with a rating up of five to seven years to approximate Indian lives. The first ever Indian table based on assured Indian lives was created based on the experience of the Oriental Government Security Life Assurance Co. Ltd. for the period 1905 to 1925 (Vaidyanathan, 1934). It was noted that the lowest mortality was experienced by the endowment policies, and the highest mortality was experienced by the whole life policies. Subsequent updates were produced by the Life Insurance Corporation in the 1970s (LIC 75–79) and in the 1990s (LIC 94–96).

Given that the Life Insurance Corporation was a monopoly, it had no incentive to update mortality tables frequently. Indeed, the Malhotra Committee (1994) noted this fact: "Quite a few persons including, notably, representatives of consumer groups have told the Committee that Life Insurance Corporation premium rates had remained unrevised for a long period and were unjustifiably high in spite of the fact that trends in mortality rates all over the country are continuously showing improvement" (chap. 5, sect. 5.4, p. 33). The Report recommended that such tables be published every ten years (chap. 5, sect. 5.12, p. 37).

13.3 EVOLUTION OF INSURANCE DURING THE NATIONALIZED ERA: 1956 TO 2000

13.3.1 Rationale for Nationalization

After India became independent in 1947, a National Planning initiative modeled after the Soviet Union's was implemented. Prime Minister Jawaharlal Nehru's vision was to have key industries under direct government control to facilitate the implementation of National Planning. Insurance business (or, for that matter, any financial service) was not seen to be of strategic importance.

Such a planning strategy begs two questions: First, why did the government of India nationalize *life* insurance in 1956? Second, why did it *not* nationalize *general* insurance at the same time?

We deal with the first question first. The genesis of nationalization of life insurance came from a document produced by H. D. Malaviya called "Insurance Business in India" on behalf of the Indian National Congress. Malaviya had written a dozen books, and this was one of the more obscure ones.[257] In that document, he made four important claims to justify nationalization. First, he argued that insurance is a "cooperative enterprise" under a socialist form of government, and, therefore, it is more suited for government to be in the insurance business on behalf of the people. Second, he claimed that Indian companies are excessively expensive. Third, he argued that private competition has not improved services to the public or to the policyholders. Preventative activities such as better public health, medical check-ups, and hazard prevention activities did not improve under privatization. Fourth, lapse ratios of life policies were very high, leading to "national waste."

---

[257] I am indebted to Professor Alan Heston of the University of Pennsylvania for drawing my attention to this document.

Malaviya's argument for the high cost of Indian insurance is the only one that he backed up with data (his other claims were made in vague terms), so we can take a closer look at his evidence on this point. Based on some data, he presents what he called "overall expenses" of insurance business operation in India, the United States, and the United Kingdom. His calculations are shown in Table 13.2. He found that it costs Indian insurers 27 to 28 percent of premium income for insuring lives, whereas in the United States, the corresponding figure is 16 to 17 percent. In the United Kingdom, it is even lower at 13 to 14 percent. On the face of it, this argument seems watertight. Unfortunately, this is not the case. On closer inspection of how the numbers were arrived at, we find that for the calculation, the denominator used for India is not the same as that used for the United States and the United Kingdom. For the Indian data, the denominator uses premium income only, whereas, for the other two countries, the denominator uses *total* income that includes premium income *and* investment income (as is customary the world over).

Table 13.2. Percentages of Premium Income Used to Insure Lives (or Overall Expenses of Life Insurance Business) in India, the United States, and the United Kingdom

Year	India	United States	United Kingdom
1950	28.9	16.8	13.0
1951	27.2	16.5	14.1
1952	27.1	16.7	14.2
1953	27.3	17.0	14.5

Source: Malaviya, n.d.

Finance Minister C. D. Deshmukh announced nationalization of the life insurance business. In his speech, he justified the action as follows.

"With the Second Plan, involving an accelerated rate of investment and development, the widening and deepening of all possible channels of public savings has become more than ever necessary. Of this process, the nationalization of insurance is a vital part." He then went on to declare, "The total [life] insurance in force exceeds Rs. 10,000 millions, that is a little over Rs. 25 per head. Quite recently it was claimed on behalf of a private enterprise that business in force could be increased to Rs. 80,000 millions and per capita insurance to Rs. 200. I am in complete agreement. There can be no doubt as to the possibilities of life insurance in India and I mention these figures only to show how greatly we could increase our savings through insurance." He added, "Thus even in insurance which is a type of business which ought never to fail if it is properly run, we find that during the last decade as many as twenty five life insurance companies went into liquidation and another twenty five had so frittered away their resources that their business had to be transferred to other companies at a loss to the policyholders."

Thus, the nationalization was justified based on three distinct arguments. First, the government wanted to use the resources for its own purpose. Apparently, the government was not willing to pay the market rate of return for the assets (otherwise, it could have raised the capital whether insurers were private or public). Second, it sought to increase market penetration by nationalization. How could nationalization possibly deepen the market that private insurers could not? There are two possibilities. (1) Nationalization would create a monopoly. If there are economies of scale in the market, it would become possible for the government to cut the cost of operation per policy sold below what private companies could. (2) Through nationalization, the government could take life insurance into rural areas, where it was not profitable for private businesses to sell insurance. Third, the government found the number of failures of insurers to be unacceptable and claimed that the failures were the result of mismanagement.

Given that by the end of the century the government did denationalize life insurance, we can examine in some detail whether nationalization did succeed in these three areas. The government did succeed in channeling the resources of life insurance business into infrastructure. The Life Insurance Corporation of India, as of March 2001, had a total sum assured of U.S. $155 billion. The value of its life fund was $40 billion. The book value of Life Insurance Corporation's "socially oriented investments"[258]—mainly comprised of government securities holdings—at March 31, 2001, amounted to $27 billion (73 percent of a total portfolio value of $37 billion). In total, 84 percent of Life Insurance Corporation's portfolio comprises exposure to the public sector (see Bhattacharya and Patel, 2003, and Appendix 2). The Reserve Bank of India Weekly Statistical Supplement of October 11, 2003, shows that 52 percent of the outstanding stock of government securities is held by just two public sector institutions: the State Bank of India and the Life Insurance Corporation of India, approximately in equal proportions.

Did the nationalization and consequent creation of monopoly actually reduce the cost of issuing life insurance policies? If we take a simple view of the world and calculate overall costs, we arrive at the results shown in Table 13.3. If we calculate overall expenses as a percentage of premium income, we arrive at the following. In 1957, the expenses were 27.7 percent. By 1963, the expenses rose to 29.3 percent. They fell back to 27.9 percent by 1982 and to 21.5 percent in 1992. In 2002, expenses rose to 22.9 percent. Could we conclude that in the late 1980s economies of scale kicked in? The answer is negative. The reason is explained in the Malhotra Committee Report (1994, chap. 5, sect. 5.5, p. 34). The expense ratio reported there was 29 percent in 1958 and 25 percent in 1992. The Report excludes group polices from its calculation. Group polices are much cheaper to sell (per policy). These policies did not exist in 1958. But, starting in the 1980s, they became commonplace. Thus, the naïve calculation along this line will lead us to believe that the expense ratio has come down substantially, whereas, in reality, that is an incorrect conclusion. A number of government reports have come to the same wrong conclusion (for example, the Annual Report of the Ministry of Finance 1995–96, p. 3).

The finance minister addressed the question of why general insurance was not being nationalized in 1956 in his speech as follows.

---

[258] Social sector investments mean loans to state electricity boards, housing, municipalities, water and sewerage boards, state road transport corporations, roadways, and railways. These account for about one fifth of the socially oriented investments portfolio, with the balance accounted for by government and government-guaranteed securities.

"I would also like to explain briefly why we have decided not to bring in general insurance into the public sector. The consideration which influenced us most is the basic fact that general insurance is part and parcel of the private sector of trade and industry and functions on a year to year basis. Errors and omission and commission in the conduct of its business do not directly affect the individual citizen. Life insurance business, by contrast, directly concerns the individual citizen whose savings, so vitally needed for economic development, may be affected by any acts of folly or misfeasance on the part of those in control or be retarded by their lack of imaginative policy."

Table 13.3. Financial Performance of Life Insurance Corporation of India 1957 to 1992 (in millions of U.S. dollars)

	1957	1963	1972	1982	1992
Income					
Total premium income	139.02	253.42	516.16	1,284.81	2,836.36
Income from investment	30.28	59.04	180.93	727.22	1,511.72
Total income	169.30	312.45	697.09	2,012.03	4,348.08
Outgo					
Commissions etc. to agents	12.08	23.65	48.74	108.33	274.36
Salaries and other benefits to employees	19.14	37.40	76.95	126.27	284.02
Other expenses of management	7.22	13.25	18.15	41.14	90.91
Taxes etc.			0.26	56.75	150.11
5 percent valuation surplus paid to government		2.85			37.43
Payments to policyholders					
Claims by maturity	32.64	52.49	101.99	369.94	796.73
Claims by death	12.40	21.13	34.57	91.14	180.47
Annuities	0.78	0.67	1.99	8.23	37.00
Surrenders	6.90	8.55	25.43	82.49	257.39
Total outgo	91.16	160.00	308.08	884.28	2,108.42
Excess of income over outgo	78.14	152.45	389.01	1,127.74	2,239.67
Operating cost/premium income	27.70%	29.30%	27.90%	21.50%	22.90%
Operating cost/total income	22.70%	23.80%	20.60%	13.70%	14.90%

Source: Malhotra Committee Report, 1994, Appendix 26, p. 148.
Note: All figures are converted to U.S. dollars using the average exchange rate of that year.

Thus, he did not deem general insurance to be "vitally needed for economic development." The only way this view could be justified is if we view insurance as a vehicle for long-term investment and if we ignore the elimination of uncertainty through insurance as a relatively minor benefit. After all, general insurance reduces uncertainty for the non-life category the same way life insurance reduces uncertainty for life.

13.3.2 Rural Insurance

In his budget speech, Finance Minister Deshmukh had specific hopes for rural insurance. He announced, "It will be possible to spread the message of insurance as far and as wide as possible, reaching out beyond the most advanced urban areas and into hitherto neglected, namely, rural areas."

After nationalization, Life Insurance Corporation has specifically targeted the rural insurance market. To promote rural insurance, it followed a segmented approach to the market. First, it targeted the rural wealthy with regular individual policies. Second, it offered group policies to people who could not afford individual policies. For the very poor, it offered government-subsidized policies.

In India today more than half of the population live in rural areas and contribute a quarter of the gross domestic product (GDP). Thus, the policymakers felt that it was essential to bring life insurance business to the rural population. Did the policymakers succeed in bringing insurance in the rural sector? Exactly to whom in the rural sector did they manage to bring life insurance?

The "success" of the rural expansion can be measured in a number of different dimensions. Here we are not talking about success necessarily in the sense of commercial success of higher profits. The finance minister was talking about a social objective of bringing insurance cover for the "neglected" rural areas. After all, if profits were to be made in rural insurance, the private sector would not have neglected rural areas. Therefore, we will measure success in terms of the penetration of life insurance into rural areas.

First, we examine the penetration of life insurance in terms of individual policies. The earliest figure we have is for 1961. Around 36 percent of all the individual life policies sold were in the rural sector. This proportion fell to 29 percent by 1970. Then, from 1980, the proportion started to climb. It climbed steadily for a decade, and, between 1995 and 1999, it climbed even faster with fully 57 percent of all policies sold being in the rural sector. This increase between 1980 and 1999 is remarkable for two reasons. First, the proportion of people living in rural areas has been falling steadily since 1950. Second, the fall in the proportion had not only halted since 1980 but it had been reversed. Thus, there is no question that the Life Insurance Corporation, through deliberate policy change, managed to penetrate the rural market since 1980 in a way they could not earlier. Of course, the number of individual policies gives us only part of the story.

Another way of examining the rural insurance business is in terms of the value of the insurance policies sold. This is depicted in Figure 13.1. During no period under study did the value of individual life insurance sold in the rural areas as a proportion of the total value of all individual life insurance policies sold in India exceed 40 percent. During the late 1990s, even though the number of rural individual life policies sold kept climbing as a proportion of all individual life policies sold, the *value* of those policies did not. This implies that there were more policies sold with a smaller average value. This would imply that the profitability of each policy sold in rural areas in the late 1990s fell.

Finally, we examine the value of the average individual life policy sold in the rural areas as a percent of the average of all the individual life policies sold nationwide. In 1961, this proportion was 84 percent. It fell almost steadily to 74 percent by 1970 and then climbed back to 84 percent in 1990. It has hovered around that figure since. If we compare the per capita income in the rural areas with the per capita income in the urban areas, for almost all states there is a 50 percent difference. Specifically, if the average rural income is 100, for most states, the average urban income is 150. Therefore, the individual life insurance policies in the rural areas are not being bought by the average rural households but by the relatively wealthy rural households.

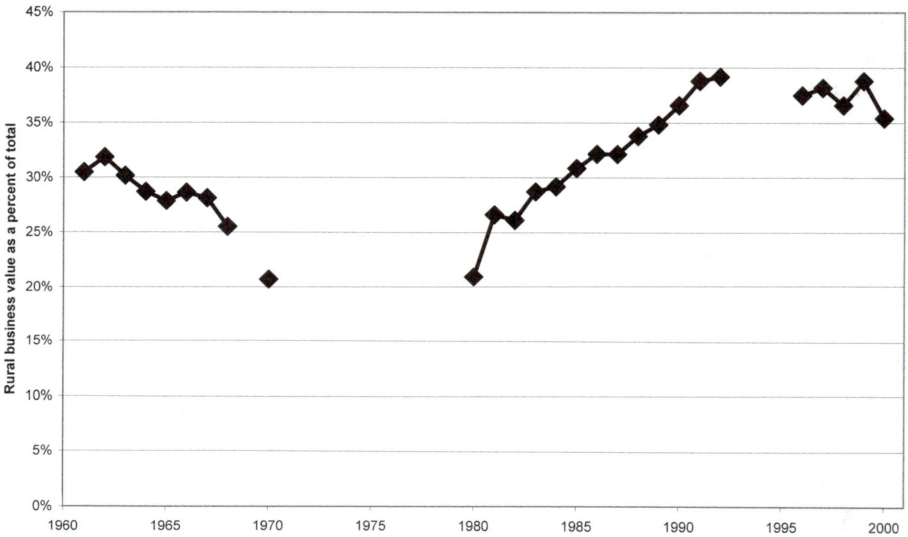

Source: Life Insurance Corporation annual reports, various years.

Figure 13.1. Rural Share of Life Insurance Business by Value, 1960 to 2000

These observations about the value of the policies being sold in rural areas and the issue of who is buying the policies in rural areas raise an uncomfortable question. Has the expansion of the Life Insurance Corporation in the rural areas served the rural rich at the high cost of reduced profitability of the company? Obviously, this was not the intention when the finance minister spoke of serving the neglected rural areas in his speech at the eve of nationalizing life insurance in 1956.

13.3.3 Prelude to Nationalization of General Insurance: Birth of the Tariff Advisory Committee

The first collective measures to regulate rates and terms and conditions go back to 1896, with the formation of the Bombay Association of Fire Insurance agents. By 1950, there was a set of rate regulations accepted by most insurers. The Insurance Act of 1938 was amended in 1950 to set up a Tariff Committee under the control of the General Insurance Council of the Insurance Association of India. Main lines of general insurance came under the Tariff Committee (they included marine, fire, and miscellaneous, which included auto). Over the next 18 years, the Tariff Committee prevailed as the "rate-maker." It was obligatory for all insurers to comply with its terms and conditions.

In 1968, the Insurance Act of 1938 was amended further. A Tariff Advisory Committee replaced the Tariff Committee. The Tariff Advisory Committee became a statutory body. Section 64UC(2) of the Act stated:

> "In fixing, amending or modifying any rates, advantages, terms or conditions relating to any risk, the [Tariff] Advisory Committee shall try to ensure, as far as possible, that there is no unfair discrimination between risks of essentially the same hazard, and also that consideration is given to past and prospective loss experience: provided that the [Tariff] Advisory Committee may, at its discretion, make suitable allowances for the degree of credibility to be assigned to the past experience including allowances for random fluctuations and may also, at its discretion, make suitable allowance for future hazards of conflagration or catastrophe or both."

The introduction of the Tariff Advisory Committee was seen as an independent, impartial, scientifically driven body for ratemaking in general insurance. After the nationalization of general insurance in 1972 (see below), the Tariff Advisory Committee became the servant of the nationalized companies. For example, the chairman of the General Insurance Corporation, the holding company of four nationalized subsidiaries, also became the chairman of the Tariff Advisory Committee. All members of the Tariff Advisory Committee were nominated by the General Insurance Corporation. It came under heavy criticism from the Malhotra Committee. The Report (1994) stated that "the data supplied [by the nationalized companies] was often incomplete and outdated and, over the years, the system has almost broken down" (chap. 5, sect. 5.25, p. 41).

The rates implemented by the Tariff Advisory Committee did not necessarily reflect the market price. For example, after the 1982 and 1988 amendments of the Motor Vehicles Act of 1939, third-party liability became unlimited. It became clear that premium rates had to be revised upward to reflect this change in law. However, political pressure from transporters prevented this rise in premium (Malhotra, 1994, chap. 5, sect. 5.26, p. 41).

Back in 1994, the Malhotra Committee recommended a delinking of the Tariff Advisory Committee from the General Insurance Corporation. It also recommended a gradual phasing out of the Tariff Advisory Committee, with the exception of a few areas. However, it did not set a timetable for this process. A recent report of a special committee set up by the Insurance Regulatory and Development Authority suggests an abolition of the Tariff Advisory Committee by April 1, 2006 (see Report of the Expert Committee, Insurance Regulatory and Development Authority, December 2003).

In India, the Tariff Advisory Committee sets a floor—or minimum—price, and all insurers must charge at least that minimum price. They could charge more (with the approval of the Tariff Advisory Committee). It should be emphasized that tariffication, or setting a floor price, is not unique to India. Neither is the practice followed only in developing countries. The Malhotra Committee Report (1994) noted that Japan and Germany had similar floor prices in place; however, the rates were periodically reviewed and adjusted according to the market experience (chap. 5, sect. 5.17, p. 39). In Japan, tariffs were abolished with the liberalization of insurance in 1998. In Malaysia, motor and fire insurance are still subject to tariff regulations. In Indonesia, a tariff regime was introduced in 1983 and continued until 1996. In India, the removal of tariffs was introduced in 1994 in the marine hull business only—the segment of the market dominated by foreign companies. Due to political pressure, it refrained from removing tariffs in the entire cargo insurance segment (Srinivasan, 2003).

Any move away from a tariffed regime is not always popular. What usually follows is rising premiums in some lines. For example, in India, motor insurance prices will rise with the removal of tariffs, and this will probably be pinned as a folly of privatization. Of course, such premium adjustment has nothing to do with privatization per se. Government-owned insurers could take a loss in the motor insurance business (as they did during the 1990s) and cover the deficit from other lines of business. But privately run insurers would seek to generate profit from every line of business. As a result, motor insurance premiums will rise to cover losses. Similarly, rating systems based on drivers' age and experience will be slowly introduced. It will be a long process because the new systems require individual specific past information that can be gathered cost effectively only with computerization of vehicle accident information and other risks.

13.3.4 Reinsurance

A reinsurer in India was defined clearly for the first time in the Insurance Act of 1938.[259] Indian currency has not been convertible since independence. Thus, to retain the maximum possible premium within India and thereby "preserve" foreign currency became a priority for the reinsurance business. India did not have a floating exchange rate until the 1990s. Therefore, hard currencies (like the U.S. dollar) were considered valuable resources. Thus, the policy was to pay minimum possible premiums in hard currencies. To achieve this goal, the insurance companies in India formed the India Reinsurance Corporation in 1956. This was a voluntary agreement at the time. In 1961, the government created the Indian Guarantee and General Insurance Company. By amendment to Section 101A of the Insurance Act, the government required all companies to give statutory cession of 10 percent each to the India Reinsurance Corporation and to the Indian Guarantee and General Insurance Company. This requirement has been echoed in the Act passed in 2000.[5] The only reinsurance company allowed to operate in India is the General Insurance Corporation.

[259] Section 101A(8)(ii), Insurance Act of 1938 states: "'Indian re-insurer' means an insurer specified in sub-clause (b) of clause (9) of section 2 (any body corporate carrying on the business of insurance, which is a body corporate incorporated under any law for the time being in force in India; or stands to any such body corporate in the relation of a subsidiary company within the meaning of the Indian Companies Act, 1913, as defined by sub section (2) of section 2 of that Act), who carries on exclusively re-insurance business and is approved in this behalf by the Central Government."

13.3.5 Nationalization of General Insurance

General insurance was finally nationalized in 1972 (effective January 1, 1973). There were 107 general insurance companies operating at the time—mainly large city-oriented companies catering to the organized sector (trade and industry). They were of different sizes, operating at different levels of sophistication. They were assigned to four different subsidiaries (roughly of equal size) of the General Insurance Corporation, which was incorporated as a holding company in November 1972 and commenced business on January 1, 1973. Its four subsidiaries were the National Insurance Company, the New India Assurance Company, the Oriental Insurance Company, and the United India Insurance Company. The head offices of the four companies are in Calcutta (now Kolkata), Bombay (now Mumbai), New Delhi, and Madras (now Chennai), respectively.

There were several goals of setting up this structure. First, the subsidiary companies were expected to "set up standards of conduct and sound practices in the general insurance business and rendering efficient customer service" (General Insurance Business (Nationalisation) Act, 1972). Second, the General Insurance Corporation was to help with "controlling their expenses." Third, it was to help with the investment of funds. Fourth, it was to bring general insurance to the rural areas of the country. Fifth, the General Insurance Corporation was also designated the National Reinsurer. By law, all domestic insurers were to cede 20 percent of the gross direct premium in India to the General Insurance Corporation under Section 101A of the Insurance Act of 1938. The idea was to retain as much risk as possible domestically. This was, in turn, motivated by the desire to minimize the expenditure on foreign exchange. Sixth, all four subsidiaries were supposed to compete with one another.

With 30 years of hindsight, we can examine the degree of success of each goal above. It was noted by the Malhotra Committee (1994) that the "behavior of the general insurance employees were not customer friendly" (chap. 2, sect. 2.22, p. 15). Thus, the goal of "efficient customer service" remained elusive. Cost-cutting seemed to have worked between 1973 and 1980. The cost of operation (as a percent of premium income) fell from around 30 percent in 1973 to around 24 percent in 1980 (Malhotra, 1994, Appendix 9, p. 131). Thereafter, there had not been a huge change in cost of operation. For investment funds, it was again noted that funds of the General Insurance Corporation had very low rates of return: "general insurance companies have been far too conservative in managing their equity portfolios" (Malhotra, 1994, chap. 6, sect. 6.9, p. 47). For rural expansion, the subsidiaries introduced a number of schemes such as crop insurance, cattle insurance, and the like. They also tried to stimulate rural business by raising the commissions of rural agents. Unfortunately, the companies did not make much headway in any direction for expanding rural business. With respect to reinsurance, the General Insurance Corporation did retain a large amount of reinsurance domestically in some areas of general insurance (such as motor insurance). Finally, the competition among the subsidiaries of the General Insurance Corporation remained elusive. Effectively they acted like a cartel—carving up the market into their own regional territories and acting like monopolies.

13.3.6 Investment Regimes: Before and After Nationalization

The Insurance Act of 1938 required that the life insurers should hold 55 percent of their assets in government securities or other approved securities (Section 27A). In the 1940s, many insurers were part of financial conglomerates. With a 45 percent balance to play with, some insurers used these funds for their other enterprises or even for speculation. A committee headed by Cowasji Jehangir was set up in 1948 to examine these practices. The committee recommended the following amendments to the Insurance Act: Life insurers should invest 25 percent of their assets in government securities; another 25 percent should go into government securities or other approved securities; another 35 percent should go into approved investment that might include stocks and bonds of publicly traded companies (but they should be blue chip companies); and only 15 percent could be invested in other areas if the board of directors of the insurance company approved them.

In 1958, Section 27A of the Insurance Act was modified to stipulate the following investment regime:

(a) central government market securities of not less than 20 percent
(b) loans to national housing bank including (a) above should be no less than 25 percent
(c) In state government securities, including (b) above, should be no less than 50 percent
(d) In socially oriented sectors—including the public sector, the cooperative sector, housebuilding by policyholders, and own-your-own-home schemes including (c) above—should be no less than 75 percent.

How did the investment regime actually operate on the ground? The figures as presented by the Life Insurance Corporation are reproduced in Table 13.4. Broadly, loans to state and central government and their corporations and boards" has steadily fallen from 42 percent to around 18 percent in 20 years. In their place, in the category of "Central Government, State Government, and Local Government Securities," the proportion of the portfolio has gone up steadily from 57 percent in 1980 to 80 percent in 2000. In fact, the Life Insurance Corporation (along with the State Bank of India) has become one of the two largest owners of government bonds in India.

For General Insurance, Section 27B of the Insurance Act of 1938 was amended in 1976. The guideline for investment was set as follows.

(a) central government securities 25 percent
(b) state government and public sector bonds 10 percent
(c) loans to state governments and various housing schemes 35 percent.

The remaining 30 percent investment could be in the market sector in the form of equity, long-term loans, debentures, and other forms of private sector investment.

The actual level of investment by the General Insurance Corporation is shown in Table 13.5. Investment in central government securities hovered around 20 percent between 1980 and 2000. Investment in state government securities stayed much closer to the target of 10 percent throughout the period. Soft loans to housing rose from 8 percent in 1980 to a high of 29 percent in 1990, only to fall again to a low of 14 percent of the portfolio at the final stage. The Malhotra Committee (1994) recommended that the mandated investment of funds in government securities of the general insurance companies should be reduced to 40 percent (chap. 6, p. 50). In April 1995, the government relaxed the investment policies of the General Insurance Corporation and its subsidiaries. Therefore, in the last half of the 1990s, we see a jump in the other approved investment by the General Insurance Corporation to the 55 to 60 percent range.

Table 13.4. Investment Portfolio of the Life Insurance Corporation, 1980 to 2000 (shown as percentages)

Year	Loans to Government	Government Bonds	Special Central Government	Unapproved	Foreign	Total
1980–1981	41.7	55.0	1.6	1.1	0.6	100
1981–1982	41.1	54.1	3.2	1.0	0.5	100
1982–1983	40.3	54.2	4.0	1.0	0.5	100
1983–1984	39.1	54.5	4.9	1.1	0.5	100
1984–1985	37.7	55.1	5.7	1.1	0.5	100
1985–1986	36.5	55.6	6.3	1.1	0.5	100
1986–1987	35.0	56.8	6.6	1.0	0.6	100
1987–1988	34.1	57.8	6.7	0.8	0.6	100
1988–1989	33.2	58.5	6.7	1.0	0.6	100
1989–1990	33.1	58.8	6.4	1.2	0.5	100
1990–1991	33.6	59.2	5.6	1.1	0.5	100
1991–1992	4.9	85.5	6.9	1.9	0.8	100
1992–1993	34.1	60.1	4.2	1.1	0.5	100
1993–1994	31.4	63.4	3.6	1.1	0.5	100
1994–1995	28.7	66.4	3.3	1.1	0.6	100
1995–1996	26.5	69.0	2.9	1.2	0.5	100
1996–1997	24.8	71.2	2.6	0.9	0.5	100
1997–1998	23.1	73.3	2.4	0.8	0.4	100
1998–1999	21.7	75.4	1.8	0.8	0.3	100
1999–2000	19.8	77.9	1.4	0.6	0.3	100
2000–2001	18.3	79.8	1.1	0.5	0.3	100

Source: Life Insurance Corporation annual reports, various years.

13.3.7 Life Insurance Business during the Nationalized Era

Indian life insurance was nationalized in 1956. All life insurers were merged together to form one single company: the Life Insurance Corporation. By 2000, the Life Insurance Corporation had 100 divisional offices in seven zones with 2,048 branches. There were over 680,000 active agents across India with a total of 117,000 employees in the Life Insurance Corporation employed directly.

Table 13.5. Investment Portfolio of the General Insurance Corporation, 1980 to 2000 (shown as percentages)

Year	Central Government Bonds	State Bonds	Soft Loans	Market Investment	Other Loans	Total
1980	21	9	8	34	27	100
1981	21	10	11	30	28	100
1982	25	10	15	29	20	100
1983	23	10	20	30	17	100
1984	24	11	22	35	8	100
1985	24	10	25	34	7	100
1986	23	10	28	34	6	100
1987	23	10	28	34	5	100
1988–1989[a]	21	11	29	33	7	100
1989–1990	19	11	29	34	8	100
1990–1991	18	11	28	31	12	100
1991–1992	17	10	26	33	14	100
1992–1993	17	10	26	36	11	100
1993–1994	17	10	27	35	11	100
1994–1995	17	10	29	35	9	100
1995–1996	17	5	23	42	13	100
1996–1997	18	6	20	42	15	100
1997–1998	18	7	18	40	17	100
1998–1999	18	8	16	42	16	100
1999–2000	19	9	14	43	14	100
2000–2001	21	11	14	44	10	100

Source: General Insurance Corporation, Annual Reports, various years.
Note: Data up to 1987 are as of December 31, and for the remaining years data are as of March 31.
[a] Denotes figures for 15 months (January 1, 1988–March 31, 1989).

There are two problems with this type of examination of the industry. First, the population of India has grown from 413 million in 1957 to over 1,000 million in

2000. Therefore, we would expect growth in life insurance policies sold by the growth of the population alone. Second, if we measure growth in life insurance in nominal amount, for a country like India, where the annual inflation rate has averaged 7.8 percent per year between 1957 and 2002, we would expected a growth in the sale of life insurance by the sheer force of inflation. Therefore, this discussion will not indulge in such descriptions.

The largest segment of the life insurance market in India has been individual life insurance. The types of policies sold were mainly whole life, endowment, and "money back" policies. Money back policies return a fraction of the nominal value of the premium paid by the policyholder at the termination of the contract. Until recently, term life policies were not available in the Indian market. The number of *new* policies sold each year went from about 950,000 per year in 1957 to around 22.49 million in 2001. The *total* number of policies in force went from 5.42 million in 1957 to 125.79 million in 2001. Thus, on both counts there has been a 25-fold increase in the number of policies sold. Of course, during the same period, the population has more than doubled. On a per capita basis, there were 0.0023 new policies per capita in 1957 compared with 0.0218 new policies in 2001. Total policies per capita went from 0.0131 in 1957 to 0.1218 in 2001. Thus, whether we examine the new policies sold or the total number of policies in force, there has been a 10-fold increase during that period. Therefore, if we consider the number of individual policies to be an indication of penetration, there has been a substantial increase, part of which is directly attributable to a deliberate policy of rural expansion of the Life Insurance Corporation.

Table 13.6 provides the details of different components of life insurance business during the nationalized era. Between 1985 and 2001, total life business has grown from below 18 billion rupees to over 500 billion rupees. During that period, the price index has grown fourfold. Thus, if there had been no change in life insurance bought in real terms, it would have accounted for 78 billon rupees worth of business. Note that, even in 2001, individual life business accounts for 92 percent of the life insurance market.

In recent years, life insurance saving has played a bigger role in national saving. Figure 13.2 plots the nonlinear relationship between total national saving and life insurance premium (both as percentages of the contemporaneous GDP) for 52 years. At relatively lower levels of saving rate (that correspond to lower levels of income), a rise in saving rate does not lead to a rise in life insurance premium expressed as a fraction of GDP. In the case of India, the threshold value of saving rate seems to be around 20 percent beyond that, the life premium as a percent of GDP starts to grow rapidly. India has reached that turning point.

A similar story emerges when we examine the saving through insurers as a component of financial saving (see Table 13.7). Over a period of 10 years between 1991 and 2000, the amount of financial saving as a percent of GDP has varied from around 11 percent to 14.4 percent. Two components of this financial saving—life insurance saving and pension—saving have increased steadily over the years.

Table 13.6. Life Insurance—Direct Premiums in Force for Domestic Risks—in India, 1985 to 2004 (in millions of U.S. dollars)

Year	Total	Individual	Individual Pension	Group	Group Superannuation
1985	1,439.00	1,305.60	0.61	133.40	36.64
1986	1,655.73	1,497.10	1.25	158.63	39.17
1987	2,056.14	1,815.95	10.01	240.19	49.58
1988	2,459.19	2,212.68	99.02	246.52	64.73
1989	2,759.06	2,483.13	131.33	275.92	62.91
1990	3,189.71	2,878.51	147.06	311.19	63.70
1991	3,049.37	2,749.45	131.26	299.92	65.79
1992	2,822.35	2,564.42	26.69	257.93	65.62
1993	3,096.56	2,817.43	18.40	279.12	75.13
1994	3,654.18	3,324.75	16.40	329.43	92.69
1995	4,354.21	3,743.76	13.46	610.45	341.76
1996	4,583.93	4,124.49	41.28	459.43	173.72
1997	5,299.35	4,731.80	39.57	567.54	253.84
1998	5,535.60	4,946.75	54.57	588.85	240.12
1999	6,436.30	5,811.79	121.93	631.62	255.69
2000	7,810.76	6,529.91	64.32	701.47	286.18
2001	10,649.33	9,771.61	611.52	877.71	430.00
2002	12,216.81	10,268.03	593.23	913.71	441.84
2003	14,938.63	12,534.63	789.79	1,079.23	534.98
2004[a]	17,496.40	14,662.19	981.96	1,230.37	621.88

Source: Swiss Reinsurance Company database.
Note: All figures were converted to U.S. dollars by using the average market exchange rate of the year.
[a] 2004 figures are estimates based on the actual results of the first nine months of the year.

13.3.8 General Insurance Business during the Nationalized Era

After the nationalization of general insurance in 1972, the government set up one holding company—the General Insurance Corporation of India. Under the holding company, there were four subsidiaries. Information about general insurance premiums earned is shown in Table 13.8. Between 1985 and 2001, the business of general insurance grew 10-fold to 122 billion rupees. Thus, the general insurance business in India is approximately one fourth the size of the life insurance business. In 1988, fire insurance (in terms of premiums earned) accounted for about a quarter of all business. The percentage remained the same over the next 14 years. Marine insurance has shrunk in relative importance. It went from nearly 20 percent of general insurance to under 10 percent by 2001. The biggest component, called "miscellaneous," has grown to occupy 68 percent of the general insurance market. This unfortunate method of categorizing lines of business can be traced back to the Insurance Act of 1938 (see Appendix 1). It placed whatever could not be classified as life insurance or fire insurance or marine insurance into a "miscellaneous"category . Thus, the biggest component of general insurance—motor insurance—is lumped with a range of other general insurance such as aviation, engineering, and crop insurance. In terms of premiums, motor insurance accounts for around 54 percent of premium income. The Tariff Advisory Committee has been unwilling to revise motor premiums upward (due to political pressure) to be consistent with higher losses experienced. As a result, the losses have mounted in motor insurance. A crude measure (of profitability) of claims as a percent of premiums earned hovered between 120 and 140 percent during 1996 and 2001 for motor insurance business. Therefore, this line of business was not profitable for these state-run enterprises. As a result, overall profitability of four subsidiaries of the General Insurance Corporation has masked the losses incurred in motor insurance.

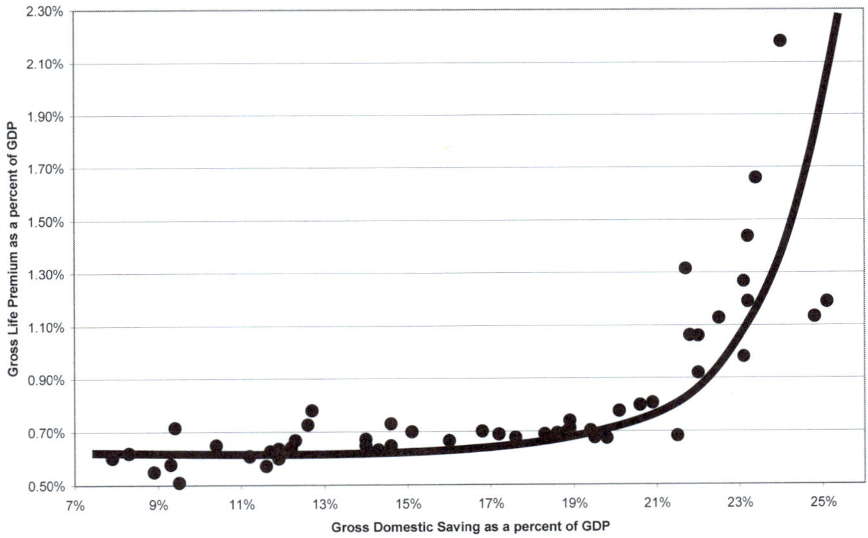

Source: Calculated based on data from the Central Statistical Organization database.
Note: The figure is plotted with gross domestic saving as a percent of gross domestic product (GDP) in the horizontal axis and gross life premium as a percent of GDP on the vertical axis. Thus, there are 52 data points, each pair representing data for a given year.

Figure 13.2. Relationship Between National Saving and Life Insurance Premiums, 1950 to 2001

Table 13.7. Components of Financial Saving as a Percentage of Gross Domestic Product

	1991	1992	1995	1999	2000
Financial Saving	11.0	11.0	14.4	12.5	12.1
Currency	1.2	1.3	1.6	1.2	1.1
Bank deposits	3.7	3.2	6.5	5.2	4.5
Stocks	1.6	2.6	1.7	0.4	0.8
Claims on government	1.5	0.8	1.3	1.6	1.5
Insurance funds	1.0	1.1	1.1	1.3	1.5
Pension funds	2.1	2.0	2.1	2.6	2.8

Source: Central Statistical Organization database.
Note: Each financial year ends March 31.

13.3.9 How Public Insurers Reacted During the Final Countdown

During the final years of the General Insurance Corporation as a holding company, there were a number of suggestions about what to do with the structure of the industry. The Malhotra Committee Report (1994) strongly recommended that the General Insurance Corporation cease to be the holding company and concentrate on reinsurance business only. The four subsidiaries should become independent companies. The report also noted that the subsidiaries were overstaffed (chap. 12, pp. 88–89).

But the Malhotra Report was not the final word. A study conducted by the consulting company PricewaterhouseCoopers, commissioned by the General Insurance Corporation in 2000, recommended just the opposite. It argued that, in the face of impending competition from the private companies, the subsidiaries should be merged to form a single company to better fight the competition. While these discussions were taking place, the four subsidiaries undertook restructuring. The results of this restructuring can be found by looking at the efficiency levels of the companies over the 1997to 2003 period.

To assess the change in efficiency of these four companies, we calculated the technical efficiency of each company. The relative efficiency measure of Farrell (1957) has been formulated in a mathematical programming framework—usually called data envelopment analysis (DEA). We took the number of employees (labor), the number of offices (physical capital) and the commission paid as three inputs. We took two alternative measures of outputs: premiums and claims (see Cummins, Weiss, and Zi, 1997, for a discussion of this issue). The results are similar for both outputs. It shows that, after some initial change, the relative efficiency level converged among the public sector general insurance companies by 2003 (see Figure 13.3). New India Insurance has consistently demonstrated the highest technical efficiency. It is well known in the literature that DEA is biased upward in small samples; thus, the absolute level of efficiency should not be taken at face value. The point here is that there has been a convergence among the public sector companies in terms of efficiency.

Table 13.8. General Insurance in India, 1985 to 2004 (in millions of U.S. dollars)

Year	Total	Fire	Marine	Miscella-neous	Net Claims Incurred	Profit/Loss Before tax	Assets/ Liabilities, Book Value
1979	508.58						
1980	637.52						
1981	718.89						
1982	802.74						
1983	886.14						
1984	911.01						
1985	976.48						
1986	1,129.37						
1987	1,276.66						
1988	1,689.86	418.63	323.09	890.86	1,205.47	243.53	3,394.75
1989	1,405.55	363.23	258.24	732.63	927.02	229.12	3,442.44
1990	1,665.52	390.68	291.25	930.65	1,085.25	275.30	3,819.73
1991	1,542.67	369.44	279.08	818.27	1,014.09	294.45	3,637.74
1992	1,445.42	350.85	273.86	741.76	1,008.24	276.78	3,548.54
1993	1,523.30	371.97	266.03	805.69	974.82	345.67	3,657.56
1994	1,679.26	421.38	263.43	918.89	1,366.52	160.08	4,425.58
1995	1,967.12	486.61	296.51	1,105.98	1,365.52	256.45	5,089.42
1996	2,069.25	506.70	278.91	1,214.22	1,437.51	307.49	5,282.34
1997	2,223.15	549.13	290.02	1,384.05	1,546.49	446.41	5,931.26
1998	2,214.10	527.76	247.78	1,438.56	1,561.97	354.69	5,937.43
1999	2,314.44	558.75	237.49	1,518.20	1,758.78	267.24	6,473.50
2000	2,410.90	492.19	227.22	1,691.50	1,985.70	163.21	6,894.47
2001	2,609.08	618.66	225.60	1,764.80	1,686.05	95.54	7,182.36
2002	2,879.83	682.87	249.01	1,947.94	1,861.02	105.45	7,927.70
2003	3,484.58	826.26	301.30	2,356.99	2,251.83	127.59	9,592.47
2004[a]	4,118.68	976.62	356.12	2,785.90	2,661.60	150.81	11,338.04

Source: Swiss Reinsurance Company database.
Note: All figures were converted to U.S. dollars by using the average market exchange rate of the year.
[a] 2004 figures are estimates based on the actual results of the first nine months of the year.

The management of these companies negotiated a "voluntary retirement scheme" to reduce the level of staffing and implemented the plan in February 2004. Of the 80,000 employees in the four companies, the voluntary retirement scheme option was restricted exclusively to 68,000 employees. Around 12,000 development officers were kept out of the voluntary retirement scheme. A total of 8,500 employees opted for the voluntary retirement scheme from the four companies.

Thirty-four percent of the 13,500 officers in the four companies opted for the voluntary retirement scheme; only 11 percent of the 36,000 clerical staff opted for voluntary retirement. This outcome came as a surprise to the management. They were hoping to eliminate more clerical jobs through the voluntary retirement scheme. The powerful union of clerical workers has strongly opposed a similar plan for the Life Insurance Corporation (Swain, 2004).

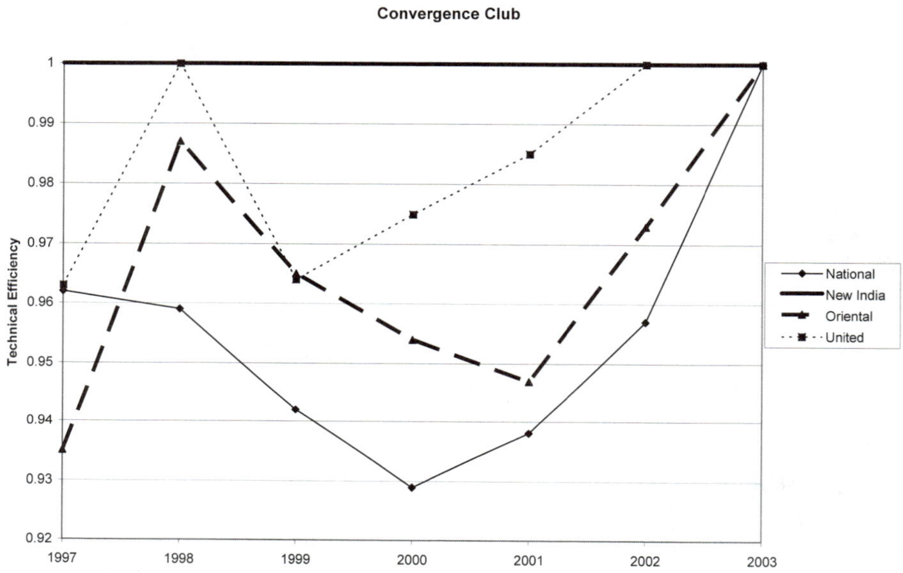

Source: Calculated by the author.

Figure 13.3. Convergence of Efficiency Among the Four Public-Sector General Insurers, 1997 to 2003

13.3.10 Introduction of the New Legal Structure

After the report of the Malhotra Committee was released, changes in the insurance industry appeared imminent. Unfortunately, instability of the central government in power slowed down the process. The dramatic climax came on March 16, 1999, when the Indian cabinet approved an Insurance Regulatory Authority (IRA) Bill designed to liberalize the insurance sector. The government fell in April 1999 just on the eve of the passage of the bill. Deregulation was put on hold once again.

A new government came to power with a late-1999 election, and on December 7, 1999, the new government passed the Insurance Regulatory and Development Authority Act. This Act repealed the monopoly conferred to the Life Insurance Corporation in 1956 and to the General Insurance Corporation in 1972. The authority created by the Act is now called the Insurance Regulatory and Development Authority. Under this bill, new licenses are being given to private companies. The Insurance Regulatory and Development Authority has separated out life, non-life, and reinsurance businesses. Therefore, a company must have separate licenses for each line of business. Each license has its own capital requirements (around U.S. $24 million for life or non-life and $48 million for reinsurance).

13.3.11 Features of the Insurance Regulatory and Development Act

The Insurance Regulatory and Development Act of 1999 set out "To provide for the establishment of an Authority to protect the interests of holders of insurance policies, to regulate, promote and ensure orderly growth of the insurance industry and for matters connected therewith or incidental thereto and further to amend the Insurance Act, 1938, the Life Insurance Corporation Act, 1956 and the General Insurance Business (Nationalisation) Act, 1972."

The Act effectively reinstituted the Insurance Act of 1938 with (marginal) modifications. Whatever was not explicitly mentioned in the 1999 Act referred back to the 1938 Act.

(1) It specified the creation and functioning of an Insurance Advisory Committee that sets rules and regulation.
(2) It stipulates the role of the "appointed actuary." He or she must be a fellow of the Actuarial Society of India. For life insurers, the appointed actuary must be an internal company employee, but he or she may be an external consultant if the company is a non-life insurance company. The appointed actuary is responsible for reporting to the Insurance Regulatory and Development Authority a detailed account of the company.
(3) Under the "Actuarial Report and Abstract," pricing of products must be given in detail, and details of the basic assumptions for valuation are required. There are prescribed forms that have to be filled out by the appointed actuary, including specific formulas for calculating solvency ratios.
(4) It stipulates the requirements for an agent—for example, insurance agents should have at least a high school diploma along with 100 hours of training from a recognized institution.
(5) The Insurance Regulatory and Development Authority has set up strict guidelines on asset and liability management of the insurance companies along with solvency margin requirements. Initial margins are set high (compared with developed countries). The margins vary with the lines of business.

    Life insurers are required to observe the solvency ratio, defined as the ratio of the amount of available solvency margin to the amount of required solvency margin. The required solvency margin is based on mathematical reserves and sum at risk and the assets of the policyholders' fund. The available solvency margin is the excess of the value of assets over the value of life insurance liabilities and other liabilities of policyholders' and shareholders' funds. For general insurers, this is the higher of RSM-1 or RSM-2, where RSM-1 is based on 20 percent of the higher of (i) gross premiums multiplied by a factor A or (ii) net premiums and RSM-2 is based on 30 percent of the higher of (i) gross net incurred claims multiplied by a factor B or (ii) net incurred claims (factors A and B are listed in Appendix 2).

(6) It sets the reinsurance requirement for (general) insurance business. For all general insurance, a *compulsory cession* of 20 percent was stipulated regardless of line of business to the General Insurance Corporation, the designated national reinsurer.

(7) It sets out details of insurer registration and renewal requirements. For renewal, it stipulates a fee of one-fifth of one percent of total gross premiums written direct by an insurer in India during the preceding financial year . It seeks to give detailed background for each of the following key personnel: chief executive, chief marketing officer, appointed actuary, chief investment officer, chief of internal audit, and chief finance officer. Details of the sales force, activities in rural business, and projected values of each line of business are also required.

(8) Details of insurance advertisement in physical and electronic media must be reported to the Insurance Regulatory and Development Authority. The advertisements have to comply with the regulation prescribed in section 41 of the Insurance Act of 1938: "No person shall allow or offer to allow, either directly or indirectly, as an inducement to any person to take out or renew or continue an insurance in respect of any kind of risk relating to lives or property in India, any rebate of the whole or part of the commission payable or any rebate of the premium shown on the policy, nor shall any person taking out or renewing or continuing a policy accept any rebate, except such rebate as may be allowed in accordance with the published prospectus or tables of the insurer."

(9) All insurers are required to provide some coverage for the rural sector.[260]

---

[260] The formal definition of "rural sector" is "any place as per the latest census which has: (i) a population of not more than five thousand; (ii) a density of population of not more than four hundred per square kilometer; and (iii) at least seventy five percent of the male working population is engaged in agriculture. As for obligation to the rural sector, the specific minimum requirements for insurance companies are the following. In respect of a life insurer, (a) five percent in the first financial year; (b) seven percent in the second financial year; (c) ten percent in the third financial year; (d) twelve percent in the fourth financial year; (e) fifteen percent in the fifth year (of total policies written direct in that year). In respect of a general insurer, (a) two percent in the first financial year; (b) three percent in the second financial year; (c) five percent thereafter (of total gross premium income written direct in that year). In addition, each company is obligated to service the 'social sector' as follows. The 'social sector' is a sub-category of the rural sector. It consists of economically underprivileged groups within the rural areas such as the aboriginals. In respect of all insurers, (I) five thousand lives in the first financial year; (II) seven thousand five hundred lives in the second financial year; (III) ten thousand lives in the third financial year; (IV) fifteen thousand lives in the fourth financial year; (V) twenty thousand lives in the fifth year. 'Social sector' includes unorganized sector, informal sector, economically vulnerable or backward classes and other categories of persons, both in rural and urban areas."

13.4 CURRENT STATE OF PLAY AND FUTURE PROSPECTS

Starting in early 2000, the Insurance Regulatory and Development Authority started granting charters to private life and general insurance companies. Appendix 3 lists all companies with charters as of December 31, 2003.

By the end of 2003, 13 life insurers had charters to operate—1 public (the old monopoly) and 12 private companies. *All* of the private companies have foreign partners in life business. Almost all the general insurers also have foreign partners. One such charter was very special. The State Bank of India (SBI) announced a joint venture partnership with Cardif SA of France (the insurance arm of BNP Paribas Bank). Since the SBI is a bank, the Reserve Bank of India (RBI) needed to clear the participation of the SBI because banks are allowed to enter other businesses on a case-by-case basis. Thus, the SBI became the test case.

The latest group to receive an outright charter for operating a life insurance company is the Sahara Group (on March 5, 2004). Sahara's entry is notable for two reasons. First, Sahara would be the only company to enter the Indian *life* insurance market without a foreign partner, thus becoming the only purely domestic company to be granted a license to operate in the insurance sector. Second, it operates the largest non-bank financial company in India, the first to operate in the life insurance sector.

Thirteen general insurance companies were operating in India at the end of 2003. Four are public sector companies—the erstwhile subsidiaries of the General Insurance Corporation that operated as nationalized companies—and the rest are private sector companies.

In India, life insurance business in 2002 amounted to U.S. $12.2 billion, and non-life insurance business amounted to $3.3 billion. Thus, the life business is almost four times as large as the non-life business. The rate of annual growth in the year 2001–2002 was 43 percednt in life insurance and 13.6 percent in non-life insurance. The total premiums underwritten by the Life Insurance Corporation of India was around $11 billion in fiscal year 2001–2002. Income of the Life Insurance Corporation during the same period was $16 billion. Twelve companies in the private sector had foreign company participation up to the permissible limit of 26 percent of equity share capital. Over a period of 10 years (1991 to 2001), the Life Insurance Corporation has averaged a real growth rate of 12 percent.

During the same 2001–2002 period, the gross direct premium income of the four public sector companies, the subsidiaries of the General Insurance Corporation of India, was U.S. $2.9 billion. The real growth rate over a period of 10 years (1991 to 2001) has been 5 percent annually. The general insurance business includes motor vehicle insurance, marine insurance, fire insurance, personal accident insurance, health insurance, aviation insurance, rural insurance, and others. At present, the General Insurance Corporation is the only reinsurer. Foreign capital of around $140 million has been so far invested in the new life insurance companies in India.

13.4.1 DISTRIBUTION CHANNELS

The Life Insurance Corporation has traditionally sold life insurance using tied agents (in-house sales forces are not a traditional feature of the Indian life insurance market). All life insurers have tied agents working on a commission basis only, and the majority of private-sector insurers have followed this approach in distributing life insurance products. Nevertheless, because banks are now able to sell insurance products, bancassurance has made a major impact in life insurance sales. Almost all private-sector insurers have formed alliances with banks, with a few of the insurers using bancassurance as their *major* source of new business. For example, in 2003, SBI Life and Aviva Life sold more than half of their policies through banks. In 2004, private insurers sold more than 30 percent of their policies through the banking channel. In India, banks are used only as a channel of *distribution* because current law prohibits bank employees from accepting commission for selling insurance policies.

13.4.2 Reinsurance

In the reinsurance business, the Insurance Regulatory and Development Authority (General Insurance—Reinsurance) Regulations, 2000, stipulated the following:

> The Reinsurance Program shall continue to be guided by the following objectives to: (a) maximize retention within the country; (b) develop adequate capacity; (c) secure the best possible protection for the reinsurance costs incurred; (d) simplify the administration of business.

For life reinsurance, the government has allowed private reinsurance companies to operate in exactly the same way it has allowed private life insurance companies. A foreign reinsurer is allowed to have up to 26 percent share in a life reinsurance company. However, until the end of 2003, there has been no taker of this offer. The reason that no private reinsurance company has stepped forward is easy to see. For most reinsurers, the main business is (still) general reinsurance. In OECD (Organisation for Economic Co-operation and Development)-member countries, around 95 percent of life insurance risk is retained domestically. For general insurance, however, the average risk retained is around 85 percent (OECD, 2003).

13.4.3 Market Share

We can get an indication of where the life insurance market is heading by examining the new business written during eleven months of the fiscal year of 2003 (April 2003 to February 2004). The distribution of premiums is shown in Table 13.9. The Life Insurance Corporation has slightly over 87 percent of the market share, leaving the rest for the twelve private companies. Among the private companies, ICICI-Prudential has the largest market share at 4.43 percent, followed by Birla Sun Life at 1.90 percent. Tata AIG and HDFC Standard Chartered are the only two other companies with more than 1 percent market share.

Table 13.9. Market Share for Premiums in the Indian Life Insurance Market (March 2003 to February 2004)

Company	Market Share (%)
Public Sector	87.22
Life Insurance Corporation of India	87.22
Private Sector	12.78
Allianz Bajaj Life Insurance Company Limited	0.87
ING Vyasa Life Insurance Company Limited	0.35
AMP Sanmar Assurance Company Limited	0.16
SBI Life Insurance Company Limited	0.89
TATA AIG Life Insurance Company Limited	1.10
HDFC Standard Life Insurance Co. Limited	1.15
ICICI Prudential Life Insurance Co. Limited	4.43
Aviva	0.46
Birla Sun-Life Insurance Company Limited	1.90
OM Kotak Mahindra Life Insurance Co. Ltd.	0.53
Max New York Life Insurance Co. Limited	0.81
MetLife Insurance Company Limited	0.14

Source: IRDA Journal, April 2004, pp. 38–39.

The market share for general insurance companies is shown in Table 13.10. Public-sector companies have close to 86 percent of the premiums, and the rest are with private-sector companies. Once again, ICICI-Lombard has the largest market share among the private companies. Bajaj-Allianz comes next with 2.95 percent of the market share. Tata-AIG has a market share of 2.25 percent, and IFFCO-Tokio has 1.97 percent.

13.4.4 Investment Regimes

Investment regimes in insurance in India have always had quantitative restrictions. Current legal requirements are explained in Table 13.11 for the life insurance business and in Table 13.12 for general insurance. At least half of the investment has to be either directly in government securities (bonds) or for infrastructure investments (which also take the form of government bonds). These investment options are "safe" because they are fully backed by the government. Of course, this also means that they earn the lowest rate of return in the Indian market. The government (both at the federal and state levels) has used insurance business as a way of raising capital. Unfortunately, much of it has been spent on consumption expenditure leading to substantial increase in government debt.

Table 13.10. Gross Premiums Underwritten by General Insurers (April 2003 to February 2004)

Company	Market Share (%)
Public Sector	85.79
National Insurance Company Limited	21.39
New India Assurance Company Limited	24.22
Oriental Insurance Company Limited	18.13
United India Insurance Company Limited	19.38
Private Sector	14.21
Bajaj Allianz General Insurance Co. Limited	2.95
ICICI Lombard General Insurance Co. Ltd.	3.16
IFFCO-Tokio General Insurance Co. Ltd.	1.97
Reliance General Insurance Co. Limited	1.07
Royal Sundaram Alliance Insurance Co. Ltd.	1.58
TATA AIG General Insurance Co. Limited	2.25
Cholamandalam General Insurance Co. Ltd.	0.57
Export Credit Guarantee Corporation	2.66
HDFC Chubb General Insurance Co. Ltd.	0.66

Source: IRDA Journal, April 2004, p. 40.

Table 13.11. Investment Regulation of Life Insurance Business

	Type of Investment	Percentage
I	Government securities	25%
II	Government securities or other approved securities (including (I) above)	Not less than 50%
III	Approved investments as specified in Schedule I	Not less than 15%
	Infrastructure and social sector	
	Explanation: For the purpose of this requirement, infrastructure and social sector shall have the meaning as given in regulation 2(h) of Insurance Regulatory and Development Authority (Registration of Indian Insurance Companies) Regulations, 2000, and as defined in the Insurance Regulatory and Development Authority (Obligations of Insurers to Rural and Social Sector) Regulations, 2000, respectively.	
	Others to be governed by exposure/prudential norms specified in Regulation 5	Not exceeding 20%
IV	Other than in approved investments to be governed by exposure/prudential norms specified in Regulation 5	Not exceeding 15%

Source: Gazette of India Extraordinary Part III Section 4. Insurance Regulatory and Development Authority (Investment) Regulations, 2000.

Table 13.12. Investment Regulation of General Insurance Business

	Type of Investment	Percentage
I	Central government securities	Not less than 20%
II	State government securities and other guaranteed securities, including (i) above	Not less than 30%
III	Housing and loans to state government for housing and fire-fighting equipment	Not less than 5%
IV	Investments in approved investments as specified in Schedule II	
(a)	Infrastructure and Social Sector	Not less than 10%
	Explanation: For the purpose of this requirement, infrastructure and social sector shall have the meaning as given in regulation 2(h) of Insurance Regulatory and Development Authority (Registration of Indian Insurance Companies) Regulations, 2000, and as defined in the Insurance Regulatory and Development Authority (Obligations of Insurers to Rural and Social Sector) Regulations, 2000, respectively.	
(b)	Others to be governed by exposure/prudential norms specified in Regulation 5	Not exceeding 30%
V	Other than in approved investments to be governed by exposure/prudential norms specified in Regulation 5	Not exceeding 25%

Source: Gazette of India Extraordinary Part III Section 4. Insurance Regulatory and Development Authority (Investment) Regulations, 2000.

Recent figures for investment regimes across different companies in the life insurance sector and in the general insurance sector reveal some differences (Rao, 2003).

The Life Insurance Corporation has 64 percent invested in government securities and other approved securities. For the private sector as a whole, the corresponding figure is 60 percent, ranging from 54 percent (ICICI Prudential) to 70 percent (AMP Sanmar). Infrastructure investment for the Life Insurance Corporation was 14 percent and 16 percent for the private companies as a group, ranging from a high of 20 percent for AMP Sanmar and Allianz Bajaj to a low of 0 percent for Aviva. Since Aviva started its operation relatively recently, the figure is somewhat unusual. In the "other approved" investment category, the Life Insurance Corporation has invested 22 percent. The average for the private sector in this category is 19 percent. In the unapproved category, the Life Insurance Corporation has no investment at all, whereas in the private sector 5 percent of the total is in this category. OM Kotak leads this category with 13 percent. Most of the other companies have no investment in this area.

In the general insurance sector, public-sector companies invested 35 percent in government securities, whereas the private sector has 41 percent in that category. For the public-sector companies, it ranges from a low of 29 percent by the Oriental Insurance Company to a high of 42 percent by the National Insurance Company. For the private sector, it varies between 35 percent by Tata AIG and 50 percent by IFFCO Tokio. The other big category is a catchall of "other investments." In that category, the public sector has 49 percent, whereas the private sector only has 39 percent.

13.4.5 Insurance in the Rural Sector in 2003

Insurers have rural sector and social sector obligations (see footnote 6). The requirements are different for life and general business and also differ depending on the number of years in business. Data for private life insurers show that, on average, 13 percent of private life insurance policies have been sold in the rural sector. Among the life insurers, there is great variation. The new entrant—Aviva—virtually did not sell any, whereas ING Vyasa sold almost 35 percent of its policies in the rural areas during the fiscal year 2002–2003. Surprisingly, very few of the agents of ING Vyasa are rural. Similarly, MetLife sold 26% of its policies in the rural areas. Very few of its agents are rural (IRDA, 2002–2003). These companies sold a few large group policies to agribusiness and, thus, they qualified for being "rural" because the locations of these large businesses are in rural areas. For private general insurers, 3.9 percent of all policies sold were rural. The highest proportions are for Bajaj-Allianz and IFFCO-Tokio with 5.87 percent and 5.42 percent, respectively. These figures are surprising, because these companies have very few rural agents. The average for public insurance companies is 6.4 percent (figures are for April 2002 to March 2003).

13.4.6 Price Dispersion of Life Insurance Products

Life insurance products such as whole life or endowment or money back policies have two components: saving and security. Specifically, there is an element that pays even when a policyholder survives the duration the policy is in force. Therefore, the "price" or the premium obscures the protection element offered by such policies. Hence, it is somewhat difficult to compare such products. Many insurance products have additional benefits (called riders). For example, buying bags of fertilizer in villages might include one-year term life benefits. Thus, if a product also has riders, it becomes even more difficult to value it because of the embedded options.

The simplest product to compare across sellers is term life, because it only has one element. It pays the beneficiary in case the buyer dies within a specific time period. When the Life Insurance Corporation was a monopoly, there was nothing else to compare with its term life policy. Now, the buyer has an array of options. How do they compare? This question has been investigated by Rajagopalan (2004), and the results are reported in Table 13.13. The table compares policies of 5, 10, 15, 20, and 25 years for a 30-year-old ordinary man for the Life Insurance Corporation, HDFC, ICICI Prudential, Tata-AIG, Allianz Bajaj, Birla Sunlife, and Max New York with a sum assured of 100,000 rupees. The first striking observation is that ratio of the maximum and the minimum premiums is 2.16 for a 5-year term and 2.60 for a 30-year term. Therefore, the dispersion in price is strikingly high for very similar products. The degree of price dispersion might be expected to decline as the market matures. The second striking feature is that it is not the same company that quotes a consistently low price, but the same company (Tata-AIG) quotes the consistently highest price. There is a problem with comparability here, because Tata-AIG quotes are for a 35-year-old rather than a 30-year-old. The results are quite similar for a sum assured of 500,000 rupees.

Table 13.13. Comparison of Premiums as of October 31, 2002, for Pure Term Insurance for Ordinary Males

	Insurer								
	LIC	HDFC	ICICI Prudential[a]	Tata-AIG[b]	Allianz Bajaj	Birla Sunlife	Max New York		
Age	30	30	30	35	30	30	30		
Sum Assured (in rupees)	500,000	500,000	1,000,000	500,000	500,000	500,000	500,000		
Term (years)	Yearly Premium (in rupees)								
5	n.a.		2,455	2,575	1,655	1,875	1,190		
10	1,140		2,504	2,585	1,805	1,875	1,225		
15	1,285	1,510	2,553	3,010	2,050	1,875	1,265		
20	1,528	1,535	2,680	3,450	2,440	1,905	1,375		
25	n.a.			4,160		1,980	1,600		
30	n.a.	1,790							
Sum Assured (in rupees)	100,000	100,000	100,000	100,000	100,000	100,000	100,000		
Term (years)	Yearly Premium (in rupees/100,000 sum assured)							Best Deal	Max/Min (Ratio)
5			245.5	515	331	375	238	238	2.16
10	228		250.4	517	361	375	245	228	2.27
15	257	302	255.3	602	410	375	253	253	2.38
20	305.6	307	268	690	488	381	275	268	2.57
25				832		396	320	320	2.60
30		358						358	

Source: R. Rajagopalan (2004).

Note: For the year 2002 (the year for which the data was taken), the average value of one U.S. dollar was approximately 49 rupees. SBI Life and ING Vysya do not offer a pure level term plan. Data are not available on the web sites of Om Kotak Mahindra and ANP Sanmar.

[a] The premiums per 100,000 rupees assured is likely to be higher for ICICI Prudential because they are based on the quotes for a sum assured of one million rupees.

[b] TATA-AIG quotes are for 35-year-old male. Its quotes for a 30-year-old male will be lower.

13.4.7 Future Prospects: Market Size

At the end of 2002, the size of the Indian insurance market (in terms of premium volume) was slightly bigger than that of Sweden and 20 percent smaller than that of Ireland (Sigma, 2003). So why would foreign insurance companies be interested in the Indian market? The answer, of course, is that the growth potential of the Indian market is much higher over the coming decades.

Although many commentators have written about India's growth potential, actual estimates of where the Indian GDP will be over the next decades have come only very recently. One is by Wilson and Purushothaman (2003) and the other is by Rodrik and Subramanian (2004). There are two critical ingredients in our approach. First, we need to understand the factors contributing to economic growth in India. Second, we need to understand how the economic growth will influence the saving rate in general and saving in the form of insurance in particular.

Growth Accounting

Wilson and Purushothaman (2003) posit a long-term economic growth rate of 6 percent per annum. However, their model is virtually driven by demographics. Rodrik and Subramanian (2004), on the other hand, show that India has had sustained growth in labor productivity, with a very low variation. In addition, the proportion of people who are economically dependent on others in the labor force (called the dependency ratio) will decline from 0.68 in 2000 to 0.48 in 2025. This alone will increase the saving rate from current 25 percent of GDP to 39 percent of GDP. Just these factors of productivity growth and favorable demographics alone will produce an aggregate growth rate of around 7 percent per year. There is also evidence that the total factor productivity in India is below 60 to 70 percent below where it should be. Institutional reforms (a stable democratic polity, reasonable rule of law, and protection of property rights) needed for the so-called second round of economic growth are already in place. Institutional quality not only helps economic growth, it also makes economic systems reasonably resistant to economic shocks. Moreover, improvement in labor quality due to higher levels of educational attainment will also contribute to economic growth. Conservatively, these factors will contribute 1 percent of additional growth rate to the GDP. Overall, therefore, a conservative estimate shows that the GDP growth rate can be 8 percent per year on a sustained basis.

Insurance Sector and Economic Growth

It is well known that growth in income is positively related to demand for insurance. What is the exact relationship? This question has to be answered empirically for each country. For India, there is a clear nonlinear relationship between total saving and life insurance saving. It can be shown that the relationship is contemporaneous. In other words, there seems to be very few backward and forward linkages: past overall saving does not seem to influence future saving in the form of life insurance and vice versa.

From this, we can project forward the demand for life insurance in real terms using the economic growth estimate described above. This method gives us an estimate of U.S. $140 billion in today's dollars for life insurance in India in 2020. A similar technique also gives us an estimate of general insurance demand in 2020 of $60 billion in today's dollars. Assuming a retention rate of 95 percent in the life insurance market and a retention rate of 85 percent in the general insurance market (OECD, 2003), we arrive at a reinsurance market of $16 billion in today's dollars for 2020.

These projections completely ignore two important elements of the insurance sector: pension and health care. Again, if we add conservative values (of 3 percent of GDP each) in each of these markets, another $240 billion will be added to the above estimates. Excluding pension and health, we have a conservative projection of $180 billion in 2020. If we include health and pension, the estimate will balloon to $440 billion, thus making it as large as the Japanese market in 2002.

13.4.8 Future Prospects: Market Share

How will the life insurance market be divided between the incumbent Life Insurance Corporation and the newcomers in coming years? Models of market share have shown that, in a fast-growing market, the first few years are critical (see Guerrero and Sinha, 2004). There are three important elements to consider when formulating a picture of the future life insurance market in India: (1) The Life Insurance Corporation has a vast distribution network in the rural and semi-urban areas that would be hard to duplicate. One potential way to duplicate this reach would be to sell insurance through banks, and some insurers have already embarked on this road. (2) Since the Life Insurance Corporation started with 100 percent of the market share, it will lose market share simply because of expansion of the market itself and less because of losing existing customers. The Life Insurance Corporation is the *only* financial institution in the top 50 trusted brand names in India by a survey of the *Economic Times* newspaper (economictimes.indiatimes.com/articleshow/362862.cms). (3) As life insurance benefits accrue over time, it becomes more expensive to switch—because switching means a loss of accrued benefits. With the rapid expansion of life insurance, the market share of the Life Insurance Corporation could fall below the 50 percent mark within five years.

For the general insurance business, private companies will have easier access to the market. Unlike life insurance, it is not expensive to switch insurers because most policies are of one year or less. The problem of tariffication makes competitive pricing difficult for the newcomers. In addition, reliable data on hazard rates are not available for many risks.

13.5 CONCLUSIONS

India is among the important emerging insurance markets in the world. Life insurance will grow very rapidly over the next decades in India. The major drivers of this growth include sound economic fundamentals, a rising middle class, an improving regulatory framework, and rising risk awareness. The fundamental regulatory changes that occurred in the insurance sector in 1999 will be critical for future growth. Despite the restriction of 26 percent on foreign ownership, large foreign insurers have entered the Indian market. State-owned insurers still have dominant market positions, but this will probably change over the next decade. In the life sector, new private insurers are bringing new products to the market, and they are using innovative distribution channels to reach a broader range of the population. There is huge market potential in the largely undeveloped private pension market and in the health insurance business. The Indian general insurance segment is still heavily regulated. Three quarters of premiums are generated under the tariff system. Reinsurance in India is mainly provided by the General Insurance Corporation of India, which receives 20 percent compulsory cessions from other general insurers. Finally, the rural sector has potential for both life and general insurance. To realize this potential, designing suitable products is important. Insurers will need to pay special attention to the characteristics of the rural labor force, such as the prevalence of irregular income streams and the preference for simple products.

13.6 REFERENCES

Annual Report of the Ministry of Finance (Section 3, Insurance Division, finmin.nic.in/the_ministry/dept_eco_affairs/budget/annual_report/9596ea3.pdf).

Bhattacharya, Saugata and Urjit R. Patel (2003). "Reform strategies in the Indian financial sector," paper presented to the Conference on India's and China's Experience with Reform and Growth, November.

Central Statistical Organization database.

Cummins, David, Mary Weiss, and Hongmin Zi (1997). "Organizational form and efficiency," Financial Institutions Center, Working Paper No. 97–2, Wharton School, University of Pennsylvania.

Farrell, M. (1957). "The measurement of productive efficiency," *Journal of the Royal Statistical Society,* Series A, 120, 253–281.

Gazette of India Extraordinary Part III Section 4 (2000). Insurance Regulatory and Development Authority (Investment) Regulations.

General Insurance Corporation (various years). Annual reports.

Guerrero, Victor and Tapen Sinha (2004). "Statistical analysis of market penetration in a mandatory privatized pension market using generalized logistic curves," *Journal of Data Science,* 2, 196–211.

Indian Insurance Commissioner's Report (1929). Her Majesty's Stationery Office, London.

Insurance Regulatory and Development Authority Annual Report (IRDA) (2002–2003). Appendix III.

IRDA Journal (2004). "Market share for premiums: Life market (March 2003–February 2004)," April, 38–39.

IRDA Journal (2004). "Gross premium underwritten by general insurance companies (April 2003–February 2004)," April, 40.

Life Insurance Corporation (various years). Annual reports.

Malaviya, H. D. (n.d.). "Insurance in India."

Malhotra Committee Report on Reforms in the Insurance Sector (1994). Government of India, Ministry of Finance, New Delhi.

Office of Economic and Cultural Development (2003). OECD *Handbook on Insurance,* Paris.

Press Trust of India (2004). "India leads the world in road accident deaths," wire report, January 3.

Rajagopalan, R. (2004). "Valuing the term insurance products in the Indian market," paper presented to the Fifth Global Conference of the Actuaries, January 25, New Delhi.

Rao, G. V. (2003). "Playing it safe," *IRDA Journal,* November, 14–16.

Rodrik, Dani and Arvind Subramanian (2004). "Why India can grow at seven percent," *Economic and Political Weekly,* April 17.

Sigma (2003). "World insurance in 2002." Swiss Re.

Srinivasan, K. K. (2003). "Transition from tariff model of pricing non-life insurance in emerging markets: Issues & prospects," paper presented to the Seventh Conference of the Asia Pacific Risk an Insurance Association Conference, Bangkok, July 23.

Swain, Shitanshu (2004). "LIC refusing to see the writing on the wall," *Financial Express,* April 19.

Swiss Reinsurance Company database.

Tripathi, Dwijendra (2004). *The Oxford History of Indian Business* (Oxford University Press, Oxford).

Vaidyanathan, L. S. (1934). "Mortality experience of assured lives in India, 1905–1925," *Journal of the Institute of Actuaries,* 60, 5–66.

Vaidyanathan, L. S. (1939). "Mortality experience of assured lives in India, 1925–1935," *Journal of the Institute of Actuaries,* 70, 42–71.

Wilson, Dominic and Rupa Purushothaman (2003). "Dreaming with BRICs: The Path to 2050," Goldman and Sachs, Global Economics Paper No. 99, October.

13.7 APPENDIX 1: Definitions of Various Lines of Business in the Insurance Act of 1938

Section 2(11), Insurance Act, 1938: "Life Insurance Business" means the business of effecting contracts of insurance upon human life, including any contract whereby the payment of money is assured on death (except death by accident only) and the happening of any contingency dependent on human life, and any contract which is subject to payment of premiums for a term dependent on human life and shall be deemed to include

(a) the granting of disability and double or triple indemnity accident benefits, if so provided in the contract of insurance;
(b) the granting of annuities upon human life; and
(c) the granting of superannuation allowances and annuities payable out of any fund applicable solely to the relief and maintenance of persons engaged or who have been engaged in any particular profession, trade or employment or of the dependents of such persons.

Section 2(6-A), Insurance Act, 1938: "Fire Insurance Business" means the business of effecting, otherwise than incidentally to some other class of insurance business, contracts of insurance against loss by or incidental to fire or other occurrence customarily included among the risks insured in fire insurance policies.

Section 2(13-A), Insurance Act, 1938: "Marine Insurance Business" means the business of effecting contracts of insurance upon vessels of any description, including cargoes, freights and other interests which may be legally insured, in or in relation to such vessels, cargoes and freights, goods, wares, merchandise and property of whatever description insured for any transit by land or water, or both, and whether or not including warehouse risks or similar risks in addition or as incidental to such transit, and includes any other risks customarily included among the risks insured against in marine insurance policies.

Section 2(13-B), Insurance Act, 1938: "Miscellaneous Insurance Business" means the business of effecting contracts of insurance which is not principally or wholly of any kind or kinds included in Section 2 (6-A), 2 (11) and 2 (13-A) of the Insurance Act, 1938.

13.8 APPENDIX 2: Detailed Factors for Each Major Line of Insurance Business

Line of Business	Factor A	Factor B
Fire	0.5	0.5
Marine: marine cargo	0.7	0.7
Marine: marine hull	0.5	0.5
Miscellaneous		
Motor	0.85	0.85
Engineering	0.5	0.5
Aviation	0.9	0.9
Liability	0.85	0.85
Rural insurance	0.5	0.5
Other	0.7	0.07
Health	0.85	0.85

13.9 APPENDIX 3: Insurance/Reinsurance Companies in India (December 31, 2003)

Life Insurers	General Insurers
Public Sector	**Public Sector**
Life Insurance Corporation of India	National Insurance Company Limited
	New India Assurance Company Limited
	Oriental Insurance Company Limited
	United India Insurance Company Limited
Private Sector	**Private Sector**
Allianz Bajaj Life Insurance Company Limited	Bajaj Allianz General Insurance Co. Limited
Birla Sun-Life Insurance Company Limited	ICICI Lombard General Insurance Co. Ltd.
HDFC Standard Life Insurance Co. Limited	IFFCO-Tokio General Insurance Co. Ltd.
ICICI Prudential Life Insurance Co. Limited	Reliance General Insurance Co. Limited
ING Vysya Life Insurance Company Limited	Royal Sundaram Alliance Insurance Co. Ltd.
Max New York Life Insurance Co. Limited	TATA AIG General Insurance Co. Limited
MetLife Insurance Company Limited	Cholamandalam General Insurance Co. Ltd.
Om Kotak Mahindra Life Insurance Co. Ltd.	Export Credit Guarantee Corporation
SBI Life Insurance Company Limited	HDFC Chubb General Insurance Co. Ltd.
TATA AIG Life Insurance Company Limited	
AMP Sanmar Assurance Company Limited	
Dabur CGU Life Insurance Co. Pvt. Limited	

Reinsurer
General Insurance Corporation of India

13.10 LEXICON

Bancassurance. The integration of banking with insurance. In India, so far, it has mostly meant selling insurance products in banks. Given that bank employees are not allowed to accept commissions for selling insurance directly, banks have created other indirect incentive schemes.

General insurance. In India this refers to property-casualty insurance. This term is used in most Commonwealth countries.

General Insurance Corporation (GIC). The company formed after the nationalization of property-casualty insurance business in India in 1972. It formed a monopoly. GIC also became the national reinsurer.

Insurance Act of 1938. The fundamental legal basis for the Indian insurance industry. The Insurance Regulatory and Development Act of 1999 was superimposed on the Insurance Act of 1938.

Insurance Regulatory and Development Act of 1999. This law deregulated insurance business in India. It allowed for private national company participation and limited foreign participation in the Indian insurance market. It created the Insurance Regulatory and Development Authority as the regulatory body of insurance business.

Life Insurance Corporation (LIC). The company formed after the nationalization of the life insurance business in India in 1956. It formed a monopoly.

Malhotra Committee Report of 1994. The basis of recent regulatory changes under the Insurance Regulatory and Development Act of 1999.

Nationalization. In India nationalization of the insurance industry took place in two stages. Life insurance was nationalized in 1956. Property-casualty insurance was nationalized in 1972.

Rural insurance. In India all insurers must have certain proportion of their business conducted in rural India.

Tariff Advisory Committee. The price-setter for property-casualty business in India. The committee is supposed to be phased out in the near future.

Tied agents. Insurance agents who work on a commission basis only. Most life insurance companies in India used tied agents to distribute life insurance products.

14 South African Insurance Markets

Robert W. Vivian

University of the Witwatersrand

14.1 INTRODUCTION

This chapter provides an overview of South Africa's insurance markets and it is important to not only evaluate it within the context of the world, but also the African continent. South Africa's life insurance market accounts for 95% of premiums on the African continent, while its property and casualty market accounts for 57% of premiums (Engels 1999). In the world market, South African insurance markets are important for at least two related reasons. First, South Africans spend more per GDP purchasing life insurance than any other country in the world (Birkmaier and Codoni 2004). It is important to note however, that this figure is misleading since much of South Africa's life insurance market involves savings, not risk products. Second, South Africa does not have a substantial state social security pension system. In turn, the South African government is not facing a pension crisis similar to those faced by most other developed countries. In fact, South Africa is in the position many other countries aspire to reach—a fully-funded privatized pension system. Thus, the South African insurance industry should be of considerable interest to other nations. However, South African's retirement funding system has its flaws and its approach may be more appropriately considered as a warning, rather than as an inspiration.

The current position of South African insurance markets arose for historical reasons. It is thus important to examine the historical path which led to the current state, including the changes which have taken place since the advent of Nelson Mandela's African National Congress (ANC) government. Consequently, the first section provides a historical overview of the development of South African insurance markets including the changes introduced after 1994 when the ANC won South Africa's first universal franchise election, thus ending apartheid.[261] The remaining sections examine the various insurance markets including the public sector markets. Retirement funds, life insurance (called long-term insurance in South Africa), property and casualty (called short-term insurance in South Africa), funding for

---

[261] A large number of books exist on the general history of South Africa including Johnson (2004), Giliomee (2003), Davenport, et al. (2000), Muller (1990), and Lacour-Gayet (1977). Books on the economic history of South Africa include De Kiewiet (1941), Horwitz (1967), and Jones (2002).

medical and health services, and state insurance schemes are discussed. However, equities, bonds, futures markets, banks, and collective investment schemes (mutual funds) are not discussed. Before the insurance markets are discussed, market regulation is examined.

14.2 HISTORY OF SOUTH AFRICA'S INSURANCE MARKETS

14.2.1 White Afrikaners and English-Speaking South Africans

South Africa has a controversial history which has produced complex insurance markets. Originally fashioned by historical forces, South Africa continues to be reshaped, especially after the 1994 election which the ANC won. In South Africa, race is divided into the two broad categories: white and black. Whites can be further divided into two subcategories: Afrikaans and English-speaking. This division is based not only on language, but also on culture. Blacks can be subdivided into three subcategories: Africans, Coloreds, and Indians. The Dutch were the first whites to settle in South Africa. In 1652, they established a victualing station at the Cape, under the authority of the Dutch East India Company (VOC). To incentivize employees to settle permanently in the Cape, the VOC granted them freedom. Today, the descendants of these settlers are considered Afrikaners. In September 1795, Britain sent its Navy to occupy the Cape as a precaution in its war against the French. In 1803, the Cape was restored to Holland under the Treaty of Amiens, but in 1806, the British once again invaded the Cape, this time to remain. Thus, English-speaking whites first arrived in 1795 and continued to do with Britain's second invasion, the largest immigration occurring in 1820 and then again with the discovery of diamonds and gold.[262]

There has always been tension between Afrikaans and English-speaking whites. This reached a breaking point in 1836, when a substantial portion of the Afrikaner population, frustrated with British rule, migrated inland away from the Cape in what is now known as *Die Groot Trek* (The Great Migration). In so doing, they established the independent countries of Natal, Orange Free State, and Transvaal. The British were however not easily deterred, and in 1844, the British annexed Natal to the Cape. Traditionally, Afrikaners have always had strong farming roots; hence they were called Boers (meaning farmers). Transvaal and Orange Free State were Boer Republics. War broke out between the Boers and the British ending with the Boer victory at the Battle of Majuba in 1880. The British army also suffered defeat at the hands of the Zulu army in 1879.[263] The Boers were however defeated in the Second Boer War (1899–1902), a particularly bitter guerrilla war. The British Army resorted to scorched earth policies where they burned the Boers' homes and crops and killed their livestock (Spies 1977). The British also herded women and children

[262] These immigrants are called the 1820 Settlers.

[263] The Zulus, at the time, constituted the independent kingdom of Zululand. Today, after the unification which established the current boundaries of South Africa, they are more correctly regarded as South Africa's dominant (by numbers) black tribe. The best account of the history of Zululand during that period is Morris (1965).

into concentration camps in which tens of thousands died. The result was a lasting sense of bitterness by many Afrikaners towards the British.[264] A distinction is not maintained between the British Army, the British people, and English-speaking South Africans. However, the bitterness was not universal, with a significant section of the Afrikaner population willing to work with the British. In 1910, the Cape, Natal, Orange Free State, and Transvaal were united to form the Union. The Union enjoyed a large measure of self-government and in 1961, it gained its independence, thus becoming the Republic of South Africa.

At the time of the first census in 1911, the white population constituted 21% of the total population. By 2002, it had declined to 10% (South African Institute of Race Relations 2004, p. 2). In 1911, 13% of Africans were urbanized compared to 52% of the whites (English-speaking whites have largely always been urbanized). By 2002, 45% of Africans and 92% of Whites were urbanized (Lipton 1986, p. 401; South African Institute of Race Relations 2004, p. 10). In 1960, Afrikaners represented 60% of the white population. Generally, the Afrikaans population has been greater, with the margin increasing, than the English-speaking population. This has meant that political power has always resided with the Afrikaners. In 1948, the Afrikaner Nationalist Party (NATS) won the general election and subsequently controlled the government until 1994, when the ANC took charge. During this period, consecutive Nat governments organized the political, and to an extent, economic systems based on the forced segregation between whites and blacks. This forced segregation constituted the apartheid system.

A distinction also needs to be made between rural and urban areas. The segregation policy was aimed in part, at encouraging blacks to develop thriving political and economic systems within their traditional, mainly rural homelands, some of which were larger than European countries. In urban areas, separate residential areas were maintained for the various race groups. The apartheid policy was however not successful. There was an ever increasing migration of blacks from the rural areas to the urban areas, a process which has continued and accelerated since 1994. This has resulted in the development of massive squatter camps around major cities and homeless persons occupying buildings in cities and factories in industrial areas.

Until 1994, most of the South African economies, including the financial service institutions were controlled by whites: English-speaking and Afrikaans. With high unemployment levels, the provision of meaningful universal social security benefits has never been an economically viable option. The usual state services and social security benefits, such as free education, old age pension, and first-world medical treatment simply could not be provided to all by the state. Pensions and medical-aid were thus linked to employers, rather than the state. The medical-aid system provided first-world medical treatment for the employed and retirees. However, the state provided adequate medical treatment via public hospitals for those who are not members of a medical-aid. For a long time, primary and secondary education was

---

[264] The Boer Republics of Transvaal and Orange Free State lost their independence and became colonies of Britain and were united with Natal and the Cape to form the Union. The franchise was limited to whites, Afrikaans, and English-speaking, but excluded blacks, thus culminating in apartheid. Initially, coloreds had the right to vote in the Cape Province, but this right also was removed.

provided, free of charge, by the state. About 15 years ago, the system was changed as state schools began to charge fees. These schools are called "Model C" schools. Free schooling has thus become an artifact of the past.

The link between pensions, medical treatment, and employment evolved for historical reasons. In the early 1900s, the country faced a large poor-white population. This population consisted largely of Afrikaners since for various reasons, they migrated from rural areas to the cities.[265] They could not find jobs since they were unskilled and better skilled jobs were filled by English immigrants (who were called *uitlanders*—literally people from outside lands). Consequently, white squatter camps developed around the cities. The government solved this problem by providing public sector jobs mainly for Afrikaners and eventually this problem was overcome as the economy expanded. Soon, Afrikaners became more skilled and educated with many migrating from the public to the private sector. Many whites were employed in lowly-paid government jobs with salaries barely above what was regarded as a "civilized wage" (at the time a major debating point). They would nevertheless be assured of a pension for life and medical treatment after retirement; hence, the link between employment, medical treatment, and pensions. These public sector pensions were and still are funded on a defined benefit basis.

The size of the economy created more opportunities for employment than the number of whites. Consequently, whites, most having more developed skills than blacks, were fully employed. The growth of the economy resulted in an increase in the employment of blacks. By 1995, they constituted the largest percentage of persons employed in the formal sector. Seventy-five percent of all formal jobs were held by blacks (Terreblanche 2002, p. 373; Botha 2003a, 2003b; Dollery 2003; Kenney 2004). On the other hand, the country also had a high level of unemployment, 40%. Furthermore, there has also not been any meaningful growth in the number of formal jobs since the mid-1970s. Indeed, during most of the thirty-year period there has been a loss of jobs and a growing population. The bulk of the unemployed were blacks, with low unemployment rates among whites. Urbanized, employed blacks were increasingly moving into the professional and management positions in corporations, but in 1994, very few held senior positions. Many perceptive economists had correctly predicted the natural demise of apartheid because of this economic growth (Hutt 1964; O'Dowd 1964; Lipton 1986; Williams 1989).

Part of the solution to the poor-white problem was the establishment of Afrikaner private sector financial institutions, such as Sanlam, the long-term (life) insurer, and Santam, the short-term (property and casualty or general) insurer. These were formed by Afrikaners driven by the determination to improve the economic standing of their population and account for the existence today, of what are regarded as Afrikaner institutions. The purpose of these institutions was to encourage Afrikaners to save and to channel those savings into productive assets, thereby giving Afrikaners a stake in the private sector economy (O'Meara 1983). Part of the Afrikaner strategy was thus to create powerful financial companies which would invest in industrial and commercial entities and provide jobs for Afrikaners. Strong, private sector financial institutions were part of the Nationalist government's ideals,

---

[265] The poor-white problem is discussed in most history books but for a specialist discussion see Abedian, et al. (1985).

as was a culture of savings and the conversion of savings into real investments. Persons employed by smaller companies which did not have pension schemes could save for a pension via endowment and retirement annuity policies with these savings also channeled into real investments.

Many British financial institutions, on the other hand, were branches of their overseas parent companies, particularly short-term insurance companies. As such, they maintained stronger ties with overseas institutions and initially operated more in line with international best practices. This was less important to life insurance companies which dealt with a more diversified risk. In the 1970s, the Nationalist government adopted a policy to encourage foreign financial institutions to register local companies (i.e., branch operations were discouraged) and to allow the local population and institutions to acquire shares in these wholly-owned foreign companies (Franzsen Commission 1970). At the time, this was known as the process of domestication and caused most foreign-owned short-term insurance companies to seek local partners (Vivian 1995).

Towards the end of the apartheid period in 1994, overseas companies were increasingly forced to disinvest due to pressure from their own countries. As this happened, foreign companies were increasingly acquired by existing South African companies, in many cases, Afrikaans companies. Strict capital controls on residents were imposed in 1961, to prevent capital flight. The laws are contained in the Exchange Control Regulations. Under these laws, no resident could take any capital out of South Africa and all transactions involving foreign exchange had to be approved by the South African Reserve Bank (SARB). Foreigners, on the other hand, could freely invest and disinvest in South Africa, except during the 1980s, when the government declared a unilateral debt standstill. Consequently, foreign capital could move into the country, but local capital was trapped within the country. This policy, which continues in a modified form to this day, greatly restricted investment decisions of South African financial institutions and individuals, but ensured that funds were available should any foreign company decide to disinvest. Although local funds were available, foreign currency was not always available, resulting in a continual depreciation of South Africa's currency, the Rand, against foreign currencies.

Thus for example, Barclays Bank, an eminent British bank, first operated in South Africa in the early 1800s. Upon disinvestment, it became the First National Bank of Southern Africa (for the early history of this bank see Webb 1992; Jones 1996). It was later acquired by Rand Merchant Bank, which initially was regarded as an Afrikaans bank. Other Afrikaans banks included Volkskas, Bankorp, and Trustbank. They merged with the United and Allied Banks, both English banks, to form ABSA, South Africa's largest retail bank. It was discovered shortly after the merger that some of the Afrikaans banks were in financial difficulties and ABSA was "assisted" or even rescued them by a secret multibillion 'loan' (referred to as a lifeboat) from the SARB (a predominately Afrikaans government institution). Today, ABSA is regarded as a largely Afrikaans managed bank which has a large English-speaking customer base. Recently, Barclays Bank returned to South Africa by acquiring ABSA.

The pre-1994 position can be summarized as follows. Funded, occupational pensions developed in the place of state provided pensions, mainly on a defined benefit basis. A strong culture of contractual savings existed. Financial institutions had developed along cultural lines, English and Afrikaans. However, the earlier division between Afrikaans and English companies began to disappear. Increasingly, the *lingua franca* of the South African commercial world was becoming English, especially in the major commercial centers. Significantly Santam, South Africa's largest short-term insurer, recently appointed an English-speaking CEO.

14.2.2 The Post-1994 Economy

The current events in the financial sector need to be understood within the context of South Africa's historical background. It is a mere decade since a black majority government was elected, too short a time to arrive at any definitive conclusions about the economy. There have been a number of works published to celebrate the first decade of black majority rule—all generally positive (Parsons ed 2004; Bruggermans 2003). The latter half of the past decade has been marked by transformation, Black Economic Empowerment (BEE), charters, and scorecards. These concepts are imposed by legislation, e.g., the Employment Equity Act 55 of 1998, the Broad-Based Black Economic Empowerment Act 53 of 2003, the Preferential Procurement Policy Framework Act 5 of 2000, and the Promotion of Equality and Prevention of Unfair Discrimination Act 52 of 2002. The current state of insurance markets cannot be understood if these concepts are excluded from the discussion.

Transformation

Transformation is a term not clearly defined. Its simplest meaning is to give more importance to the black population in the South African Economy by transforming the economy into a black-controlled economy. In the labor field, transformation factually, is to replace wherever possible whites with blacks, i.e., ostensibly to transform the labor force from white to black. In fact, since whites constituted only 25.0% of the formal labor force, labor transformation is more correctly defined as the transformation of the labor force into a more black one. As pointed out above, in numerical terms, there has not been any meaningful growth of employment in the formal sector since the 1970s. In 1975, 4,669,735 persons were employed in this sector. In 1998, this had increased to 4,926,843 or an increase of 5.6% over the 23-year period. During this period, the private sector employment had decreased by 5.7%, with this decrease being more than compensated by a public sector employment increase of 34.7% (South African Institute for Race Relations 1999, p. 264). It was widely accepted that the public service was bloated. With a stagnant labor force, increasing black employment can only be achieved by causing white unemployment, or white emigration. The net result of labor transformation is thus to restrict the participation of whites in the economy. This has been happening. The percentage of whites in the public service, previously the bastion of the white Afrikaner has declined to 8.9% (Naidoo 2005, p. 119). It is not clear how many whites have emigrated since 1994. The government has embarked on a policy of job creation, but even before it started, the Minister of Finance has admitted that the economy will not create sufficient formal jobs to deal with the employment problem (Hartley 2004).

Black Economic Empowerment

The past few years have been dominated by the Black Economic Empowerment (BEE) program, again not a well-defined term. The closest equivalent concept would be affirmative action, but on a much grander scale. Whereas affirmative action is applied to assist minority groups, BEE is applied by the majority. While the main thrust of affirmative action is employment situations (i.e., transformation above), BEE aims to do more than this. It aims to transfer the equity and the company itself to the preferred group. In addition to equity transfer, it includes increasing the numbers of black employees at all levels. Thus, there is an overlap between transformation and BEE.

Each economic sector has drawn-up a charter indicating how the BEE objectives will be achieved, including specifying measurable objectives within that sector. The Financial Sector Charter (FSC) requires that 25% of the equity of financial services companies be transferred to blacks (§5, paragraphs 10 and 11 of the FSC). Since most major financial companies are quoted on the Johannesburg Securities Exchange (JSE), an equity transfer program is not necessary. The equity is available for anyone to purchase. However, transfer does not imply the sale for value, but rather the giving of equities, even if the word 'sale' is used and the word 'giving' is not. The financing of these transactions have become very complex. The FSC specifically requires financial institutions to deploy funds for BEE deals. In many cases, the transfer of equity is achieved by issuing additional shares which accordingly dilutes the capital stock of existing shareholders. BEE also poses risks to investors. An investor may *inter alia* find its capital diluted, and its ability to select directors, management, staff, suppliers, and products are all restricted in terms of the BEE program. The BEE program has progressed quite far, with virtually every major financial institution having announced its BEE partner. There has been widespread public concern that BEE is benefiting a very small percentage of politically well-connected blacks (BEE Helps Only Elite 2004). The more recent deals have aimed at a broader based spread of the equity.

In order to measure and monitor progress of transformation and BEE, scorecards have been created. In the case of the financial sector, scorecards are made in line with the objectives of the FSC. Included in the charter and scorecard are the number of blacks at each level in the firm (§1 of the FSC), the types of products to be sold (§3 of the FSC), the black status of service providers (§2 of the FSC), the investments financial institutions must make (§4 of the FSC), the transfer of ownership to blacks (§5 of the FSC), the appointment of black directors (§5 of the FSC), and the appointment of black executives to run the company (§5 of the FSC). The inclusion of service providers (§2 of the FSC) poses a particular problem to small and medium enterprises. These are usually closely held family enterprises and if they are owned by whites it is unlikely that they will get any contracts from the government or major corporations. Recently, two white brothers were arrested on the allegation that they lied about their BEE status. Increasingly, the government is also using the licensing process to transform companies into black companies.

14.3 REGULATORS

Three different regulators for the financial markets exist. Banks are regulated by the central bank, the SARB, medical schemes by the Registrar of Medical Schemes, under the auspices of the Department of Health, and most of the remaining markets are regulated by the Financial Services Board (FSB). Each financial market is regulated via its own piece of supervisory legislation. Separate legislation exists covering banks (deposit taking institutions), long-term insurance business, short-term insurance business, medical schemes, pension funds, and collective investment schemes (mutual funds). In addition to specific legislation for each market, a host of other legislation exists to which financial companies must comply (e.g., the Companies Act 61 of 1973, the Income Tax Act 58 of 1962, and the Estate Duty Act 45 of 1955). The legislation has been subjected to a considerable overhaul over the past decade. Although the legislation generally operates at a national level, provincial and municipal legislation, which are of less importance, also exist.[266]

The structures of the various pieces of supervisory legislation all follow a similar pattern. First, the class of financial business is defined, e.g., long-term insurance, short-term insurance, or medical aid. Then, the legislation proclaims it to be a criminal offense to carryout the defined class of business without being registered to do so. In some cases, the legislation goes on to require licensing for each line of business within the class. Thus, to conduct short-term insurance business in South Africa, the insurer must be registered to carry out short-term business and apply for a license for each specific line of short-term insurance business. A registrar is then appointed for each class of financial business. Thus, there is a registrar for medical schemes, long-term insurance, short-term insurance, retirement funding, and so forth. Generally the registrar is the same person, the Chief Executive Officer of the FSB. This peculiar state of affairs exists for historical reasons.

14.3.1 Financial Services Board

The main supervisory body for financial institutions is the Financial Services Board (FSB), a body established especially to supervise financial services.[267] The FSB is funded by way of a dedicated levy on financial institutions (§15A of Financial Services Board Act 97 of 1990) and is empowered in terms of other statutes to impose other levies. The FSB interacts regularly with the various markets in a number of ways, formal and informal. It holds regular meetings with industry bodies and the registrar or one of his senior staff is usually invited to industry conferences and functions. In addition, each Act makes provision for an Advisory Committee, which the FSB can consult on matters pertaining to the industry. The legislation calls for the submission of statutory returns which must be submitted periodically for supervisory purposes. These are usually completed by management and checked by the company auditors. In some cases, the legislation requires directors to sign-off on the statutory returns. The statutory returns are voluminous and it is unlikely that the

---

[266] Copies of the legislation are available at www.acts.co.za.

[267] The Financial Services Board was established after the Melamet Commission's report into the collapse of the AA Mutual Insurance Company was considered.

directors will be in a position to actually verify these returns. In order to maintain a degree of standardization across the markets a Policy Board exists which considers recommendations and proposed legislation from the different registrars and advisory committees (Policy Board for Financial Services and Regulation Act 141 of 1993).

14.3.2 Legislative Review

Over the past decade or so, the FSB embarked on an extensive legislative program, probably the most extensive in the history of South Africa. This includes the passage of the Long-Term Insurance Act 52 of 1998, the Short-Term Insurance Act 53 of 1998, including subsequent amendments to these Acts, the Financial Services Board Amendment Acts 1992, 2000, 2002, various Financial Institutions Amendment Acts, the Financial Advisory and Intermediary Services (FAIS) Act 37 of 2002, the Financial Intelligence Centre Act (FICA) 38 of 2001, the Collective Investment Schemes Control Act 45 of 2002 (i.e., mutual funds), and so on. Secondary legislation has also been promulgated and revised. This includes the Policyholder Protection Rules for long- and short-term insurance.

There is no single reason for the extensive review. Partly the review was to modernize the legislation, some of which have been on the books for decades. For example, the previous insurance Act was passed in 1943. Financial intermediaries have never been regulated, but are now regulated in terms of FAIS. FICA was passed to bring South Africa's money laundering legislation in line with international practices. The legislative review has not been brought about by any serious event in the market place, such as the collapse of an insurance company. This is unlike previously when the FSB was established as a consequence of the collapse of the AA Mutual Insurance Company in 1986.

Nevertheless, the review is not complete. "Clawback of Commissions" regulations are under consideration. In some cases, life insurers involving certain classes of insurance, pay brokers an up-front commission on the acceptance of periodical business. For example, an insured may purchase a retirement annuity in which he intends to make monthly payments for several years. The insurer then pays the broker the full commission over a two-year period and if the insured stops paying, the insurer charges the cost of the commissions to the savings of the insured, resulting in a substantial loss to the insured. This has been a major source of contention as discussed below as the upfront commission saga.

"Decapping of Commissions" has been under discussion for several years. The current regulations specify the maximum commission that an insurer may pay a broker. The purpose of the proposed decapping regulations is to repeal these existing regulations. It has been agreed in principle to de-regulate commissions. The plan was to phase these in by first enforcing disclosure. This was achieved in terms of the policyholder protection rules and now FAIS, but decapping still has not taken place. The regulator seems to have had second thoughts on the issue. The current view is to put the repeal on hold in view of the United States (U.S.) investigation into the American International Group (AIG), which may throw some useful light on the relationship between brokers and insurers (da Silva 2004). The Pension Funds Act 24 of 1956 is also under review and a new Act is expected in 2007.

The implementing of FAIS is currently engaging the industry in considerable effort. In terms of this legislation, all financial advisors and intermediaries must be registered with the FSB. However, because of the broad wording of the legislation, insurers have also decided to register. Consequently, insurers are now registered in terms of multiple pieces of legislation. The general consensus within the industry is that it has had an excess of legislation and time is now needed to consolidate the positions before any further legislation can be digested (Haken 2004).

Table 14.1 shows the number of employees at FSB and in the insurance industry. In line with government policy, the FSB set-about increasing the number of black employees. The number of FSB employees increased by 119% while the number of employees in the industry has decreased. The FSB's CEO announced at a recent conference that further staff increases are being planned. This will result in an increase in the annual levy paid by financial institutions to the FSB.

Table 14.1. Employment Figures for Financial Services Board and the Insurance Industry, 1993–2004

Year	1993	1996	1997	1998	1999	2000	2001	2002	2003	2004
Blacks	10	33	39	53	52	67	78	114	133	143
Male				32	28	36	42	59	64	73
Female				21	24	31	36	55	69	70
White	98	92	89	93	80	81	83	91	93	94
Male				49	42	40	39	38	37	54
Female				44	38	41	44	53	56	40
Unclassified		1								
Total: FSB	108	126	128	146	132	148	161	205	226	237
Number of Employees in the Insurance Industry	69,003	81,051	80,387	78,833	64,373	61,780	58,476	60,134	N/A	N/A

Source: Financial Services Board (1996–2004).
Note: FSB is the Financial Services Board.

In financial markets, disputes are inevitable. Various mechanisms exist to resolve disputes. The matter can be taken to a court of law. South Africa, has, as in the case of Britain, a split bar consisting of attorneys and advocates. Generally lawyers are paid on a fee basis, but a statutorily defined contingency fee system exists, which finds wide-spread application in personal injury claims (Contingency Fees Act 66 of 1997). A range of courts exist, including Small Claims Courts, Magistrates' Courts, High Courts, and the Supreme Court of Appeal (SCA). The more recent addition is the Constitutional Court which is supposed to decide on constitutional matters, but is beginning to act more like a final court of appeal. In recent years, there has been a proliferation of specialized courts.

The financial markets have each introduced less formal mechanisms to resolve disputes, the most common being the ombudsmen, typically a retired judge or senior lawyer (attorney or advocate). There are separate Ombudsmen for the long- and short-term industries. These appointments are in terms of a voluntary agreement between insurance companies which have agreed to abide by the decision made by the ombudsmen. The most recent appointment was the Ombud for the intermediaries, this time appointed, under the terms of the FAIS Act. A Financial Services Ombud Schemes Act 37 of 2004 has recently come into force to regulate the powers of the Ombudsmen who previously had no legislative standing. The retirement fund has a statutory appointed adjudicator, who has powers approximating that of a magistrate. The office of the Pension Funds Adjudicator falls under the FSB. The appointment of all these ombudsmen, have and will give rise to jurisdictional problems. As indicated below (the upfront commissions saga), the long-term insurance market is currently in a state of crisis because of rulings by the Pensions Fund Adjudicator.

14.4 THE RETIREMENT FUNDING MARKET

The key to understanding South Africa's financial markets is to understand South Africa's system of retirement funding. Self-employed persons can save for a pension via endowment and retirement annuity policies, while employed persons save via occupational retirement funds. The aggregate of retirement funds, estimated to be in excess of R 1,000 billion, is the single largest South African financial asset. In the absence of a state social welfare system providing pensions, this is not surprising. Much of the way retirement funds operate are defined by law. The most important laws are the Pension Funds Act and the Income Tax Act. The definitions of the different types of retirement schemes (i.e., pension fund, provident fund, retirement annuities) appear in the Income Tax Act, not the Pension Funds Act. Occupational retirement plans are voluntary, but can be imposed by employers as a condition of employment. Employers may define different classes of employees, but having done so, it is compulsory for all employees of that class to join the fund. Savings (contributions) are, in the case of occupational retirement plans, deducted from employees' salary and paid over to the retirement fund. Most employers establish their own retirement fund. These funds are separate juristic bodies, controlled by a part-time management board (usually incorrectly but universally called Trustees). Trustees are appointed by the employer and elected by members of the fund in equal proportions. Professional administrators are often appointed, as are professional fund managers to manage the assets of the fund.

After collection, the savings are redistributed via the funds themselves (in the case of large employers), insurance companies, or fund managers as investments to equity markets, bond markets, or off-shore investments and other forms of investments. In attempting to get a clear picture of investments in South Africa, the problem of double counting must be borne in mind. Savings are owned by retirement funds, but much of this is paid over to insurance companies and asset managers to invest. Thus, the assets of an insurance company may well primarily be the assets of retirement funds. The market capitalization of the equity market largely reflects the assets of pension funds, insurance companies and other institutions. This chapter does not attempt to de-segregate these assets.

The operation of each fund is defined by the rules of the fund. These have to be approved by the FSB and the South African Revenue Service (SARS). The approved rules are registered with the FSB. Individual funds have a great deal of discretion in defining the operation of the fund and thus wide variations in funds exist. However, as a general statement the following can be noted. The funds belong to the fund itself, not the employer, nor the employee. Since membership in a fund is a condition of employment, employees cannot make any withdrawals from the fund, unless they resign. It is possible for a fund to grant a member a loan for purposes of purchasing property (real estate) to live in. The modern tendency is not to do this, but for the fund to assist the member to get a loan from a bank, secured by a guarantee from the fund. The employer has very limited rights of recourse to the member's funds. The right of recourse is confined to off-setting any outstanding property loans (including securities) and losses suffered by the employer as a result of the admitted dishonesty of the employee or on conviction by a court of law. An employee may withdraw his savings if he resigns or transfer the savings to another fund or a preservation fund. If he or she withdraws the savings the withdrawal is subject to taxation. It is possible for rules to allow for the retention of the employer's contribution, but the Pension Funds Adjudicator has voiced his unhappiness with this and the tendency is for the employee to be entitled to the employer's and employee's contributions together with the investment income. A fund may not invest in the business of the employer without agreement of the members and the permission of the Minister of Finance. Notwithstanding this, there is a case on record where the entire fund consisting of R 80 million was invested in the business of the employer, which subsequently went under and the subsequent criminal prosecution did not succeed.

Individual employees seldom have any direct control over their savings or investment choice, but rather elect trustees and receive regular feedback concerning matters of the fund. Retirement funds constitute 'the major provider of capital for the equity listed on the JSE' (Taylor Report 2002, §9.1.1). Savings, apart from those in retirement funds, in South Africa have diminished substantially in recent years. It has been estimated that savings constitute only 15% of GDP (SC 2004). Thus, the main source of savings available for investments is the savings generated by retirement funds.

As explained above, due to exchange controls, the bulk of South African savings are trapped within South Africa. The government has not fulfilled its earlier assurances to abolish exchange controls. It maintains that its policy is to gradually abolish exchange controls, which it claims it is doing. The SARB is responsible for administering the exchange controls. It, in turn, appointed the major banks as its authorized agents (or authorized dealers). Any South African resident wishing to invest outside of South Africa must apply to a bank, as the SARB's agent for permission to take funds out of South Africa. The SARB communicates the exchange control prohibitions to its authorized dealers in terms of exchange control circulars. Until recently, the contents of the circulars were not available to the public. These circulars span several decades and would not be of much use to the average individual because of their complexity. A detailed summary of the exchange control requirements are now set-out in the exchange control manual (ECM) available on the

SARB's website.[268] The SARB works closely with the South African Revenue Service. A South African resident may not invest more than R 750,000 outside of South Africa (ECM (2005), Rule 6.1.1) and this limit has not changed for several years. South African institutions (retirement funds, insurance companies, fund managers etc.) may not invest, directly or indirectly, more than 15% of their assets outside of South Africa (ECM (2005), Rule 6.1.4.2). In terms of a recent policy ruling, once the 15% is reached, any investment growth must be repatriated back to South Africa. Over the past year, the South African Rand strengthened against the dollar (after consistently weakening for twenty years before that). Consequently, from a South African perspective, non-South African investments are not currently attractive. Therefore, there is currently little incentive for South Africans to invest outside of South Africa.

The present time would have been ideal to abolish exchange controls since it is unlikely that massive outflows would take place. Persons who had emigrated were not allowed to take more than a prescribed maximum out of South Africa. What was left was kept in what are called blocked rand accounts, accounts maintained at a local bank under the strict control of that bank. Once so deposited, funds therein could only be used for purposes specified in the exchange control circulars. The total amount kept in blocked rand accounts was not disclosed. Recently, the government announced that amounts kept in these accounts could be taken out of the country, subject to a forfeit penalty. Many South Africans had taken money out of the country in violation of the exchange control regulations and the government announced an amnesty to those who came forward to regularize their affairs.

In contrast to South African residents, non-residents are free to invest and disinvest at will. Over the past few years there have been substantial inflows of foreign capital into South Africa in the form of the so-called portfolio flows. Increasingly, South African equities are being bought by foreign investors and major South African companies are listing on other stock exchanges. Thus Sasol, South Africa's coal-to-oil company and Sappi, a paper manufacturer, are listed on the New York Stock Exchange. Anglo-American, the mining company, The Old Mutual, South Africa's largest life insurer, and SABMiller are listed on the London Stock Exchange. Presently, South Africa's current account is running at an unprecedented deficit which is being supported by the capital account surplus. The purchase of ABSA by Barclays will, for example, result in a further inflow of R 32 billion on the capital account. The details of the funding have been kept secret.

The aggregate assets of the pension funds are indicated in Table 14.2. South Africa with retirement funds equal to 73% of GDP has one of the largest pre-funded pension-assets in the world. Considering that the unemployment rate is 40%, the level of funding is impressive. In the absence of a meaningful state social security system, this is however not surprising. Most countries have poor-relief schemes, and South Africa is no exception. It can be argued that there is a constitutional duty for the state to provide some income to the destitute elderly in terms of the social assistance obligation set-out in §27 of the Constitution. The poor-relief scheme is an unfunded scheme paid from current taxes and, as of November 2004, pays an amount of R 740 per month (about $120 per month). In order to understand the extent of this

[268] www.reservebank.co.za. The manual can also be downloaded in pdf format or a zip file.

assistance, it should be compared with some items of household expenditure. It is of the same order as a monthly telephone bill, or half the amount of the monthly local authority rates & taxes, including the utility bill (water and electricity).[269] It is hardly sufficient to provide food for a single person. The payment of the amount is also conditional on passing a means test. Accordingly, most formally employed persons will not be entitled to this grant until they have reached specific levels of destitution. It is clear that this amount cannot be regarded as an old age pension sufficient to support a previously employed person at any level commensurate to that during employment. An employed person must on reaching retirement age ensure that he or she has an income from another source.

South Africa does not have a pay-roll tax funded state social security system. The South African government is thus not facing pension crisis most other countries are facing. In the absence of a meaningful social security system, each individual must save for retirement and the norm is to do so via occupational pension schemes. Since occupational pensions are, as the name implies tied to employment, in law, the rights and obligations of the employer and employee are, in theory, defined in the employment contract. However, in reality, the usual employment contract says virtually nothing about pension funds, other than to note that the employer has such a fund and it is compulsory for the employee to be a member of the fund. The precise legal position is thus unclear, which is unsatisfactory (Hunter 2004). Since the rights and obligations are not well-defined, it has been argued that the employer may, at its discretion, change its involvement in pension funds. This view was tested in court, but the court did not agree, maintaining that indeed a valid and enforceable contract exists. The matter has been taken on appeal.

Two types of funds exist: defined benefit pension schemes and defined contribution funds. In a defined benefit pension scheme, as the name implies, the final benefit due to the pensioner is defined. There is a risk that insufficient funds will exist to provide the promised benefit and this risk rests with the employer. This is one of the reasons why there has been a movement away from defined benefit pension schemes. In a defined contribution system, it is the contribution and not the benefit which is defined. This is usually defined as a percentage of the employee's remuneration. The risk that the accumulated funds of the member are insufficient to provide an adequate pension rests with the employee. The older schemes (private and public sector) were almost exclusively defined benefit pension schemes, but there has been a marked movement in South Africa's private sector towards defined contribution schemes over the past two decades. The public sector however continues to have largely defined benefit schemes (RFR 2004, p. 10).

What should a pension fund provide? Generally this question is not answered. As a rule-of-thumb, for every year that a person works, a defined benefit pension

---

[269] Rates & Taxes is the name given to the land tax imposed by local authorities, the main source of income for local authorities.

Table 14.2. Aggregate Net Assets of All Pension Funds, GDP, and Inflation, 1994–2003

	(Rands Millions)									
Fund Type	1994	1995	1996	1997	1998	1999	2000	2001	2002	2003
Self-Administered Funds	170,741	203,706	249,604	266,450	277,750	302,080	329,119	369,996	351,760	367,751
Exempt Funds	84,644	93,954	102,362	117,716	132,431	142,810	115,068	188,581	196,077	213,073
Official Funds	77,266	87,061	97,620	124,131	152,502	161,859	206,733	238,745	277,452	291,644
Transnet Fund	17,694	19,773	28,465	30,057	33,195	30,859	36,965	32,874	34,076	30,209
Telkom Fund	3,366	4,108	261	1,436	479	527	499	185	190	144
Post Office Fund	1,423	269	1,927	2,409	7,801	3,520	4,084	4,580	5,914	4,890
Bargaining Council Fund	885	2,556	2,876	365	680	2,667	1,589	613	1,927	1,388
Foreign Funds						-		-	-	
Total	356,019	411,427	483,115	542,564	604,838	644,322	694,057	835,574	867,396	909,099
GDP	482,120	548,100	617,954	685,730	743,424	813,683	922,148	1,020,007	1,164,945	1,251,468
(At Market Prices)										
Total Funds/GDP	0.74	0.75	0.78	0.79	0.81	0.79	0.75	0.82	0.74	0.73
Inflation (%)	8.90	8.70	7.40	8.60	6.90	5.20	5.40	5.70	9.20	5.80

Source: Registrar of Pension Funds (1994–2003). GDP and inflation figures are from various South African Reserve Bank reports.
Note: Inflation is calculated using the Consumer Price Index (CPI).

scheme should provide a pension at a rate of 2% of his or her final salary. Thus, if someone worked 40 years for the same company, that person would get 80% to 100% of their final salary, adjusted for inflation, for life.[270] Defined benefit pension schemes usually make provision to commute 1/3 of the pension into a lump-sum and in the event of the death of the member, a reduced pension is payable to the surviving partner. It appears that very few persons, other than government employees, obtain these benefits, even where these have been 'promised.'

Occupational pension funds are subject to regulatory supervision. From a regulatory perspective, two different types of schemes exist. First, funds established in terms of specific laws and schemes supervised under the Pension Funds Act 24 of 1956.[271] Schemes established by specific laws are each established in terms of their own parliamentary statutes. Thus, for example the government employees' pension fund is established and operates in terms of its own statute.[272] The rest, mainly private sector pension funds are regulated in terms of the Pension Funds Act 24 of 1956, with the regulatory body being the FSB, falling specifically under the jurisdiction of the Registrar of Pensions (who *ex officio* is the CEO of the FSB). These mainly private sector funds can in turn be divided into two further types, self-administered funds and exempt (or insured) funds. Larger funds are self-administered and must comply with all the statutory requirements. For example, they must appoint a principal officer, administrator, and auditor. They are large enough to justify this infrastructure. This is not justifiable for smaller funds. These funds can then enter into an insurance contract with an insurer which will take over responsibility for the administration of the retirement fund. This usually means the complete administration of the fund, i.e., the administration and investment of the assets of the fund. These funds are then exempt from some of the provisions of the legislation, the insurer having this responsibility. It is for this reason these funds are also called exempt funds.

As Tables 14.5 to 14.7A below indicate, the bulk of long-term insurance business, currently, is actually related to administering pension funds. It is thus a bit of a misnomer to regard South African insurers as insurers, i.e., risk carrying companies. In a defined contribution fund, when the member reaches retirement age, that person may be entitled to a lump sum, a portion of which he or she may retain, tax free. The balance is reinvested to purchase an annuity. Long-term insurers are thus once again involved. Members can use their lump sum to purchase an annuity from an insurer. In a genuine provident fund, the entire capital may be taken by the member. Contributions paid to a provident fund are not tax deductible and hence the final lump sum is not subject to taxation.

It is not compulsory for an employer to establish a pension fund. There is a perception that increasingly employers are questioning their involvement in the provision of pension plans. A recent development in this direction, running ahead of the legislation, is where a number of firms contribute to an umbrella (or multi-employer) fund, with little involvement by the employer. It is conceivable that in the

---

[270] Wassenaar (1989, p. 78) holds that after a life time of employment (a forty year period) the employee should receive 100% of his final salary. The final salary being the average salary during the last three years employment.

[271] This Act is currently under review.

[272] Government Employees Pension Law 1996.

near future, many employers will no longer establish retirement funds for their employees and a new generation of open funds (i.e., funds not linked to employers) will be formed.

Employers are allowed to claim a tax deduction for contributions up to 10.0% of the employees' remuneration and employees can claim an additional 7.5% (see §11(l) and §11(k) of the Income Tax Act 58 of 1962). In terms of the wording of the section, SARS has the discretion to allow a greater percentage and it is accepted that 22.5% of the employee's remuneration will be allowed. Anecdotal evidence, however, indicates that very few companies make the full use of the 22.5% deductions and surveys indicate that, in 2004, the contribution rates are 16.5%, with administration and other costs of 5.7%, giving a net contribution rate of 10.8%, which is at the lower level recommended by the World Bank. The net contribution rate for savings has also been dropping; it was 12.4% in 2002 compared to 10.8% in 2004. Most pension funds also provide risk coverage such as life insurance and these pension contributions may include life coverage (RFR 2004, p. 13).

The single largest pension fund is the Government Employee Pension Fund, estimated at R 400 billion, which, because of its size, has considerable influence on the JSE. It is of course in the interest of government employees and members of parliament, who have considerable influence over Parliament, to pay as little as possible to the fund and to get as large a payout as possible. If the fund is in deficit, this is not a concern for the member of the defined benefit scheme. It is the concern of the government, which has to fund the deficit.

Wassenaar (1989) found that parliamentarians, government, and local government employees had indeed exploited the system for their own benefit. For example, the civil servant's pension is a function of the number of years worked for the government. The government introduced a scheme whereby service could be bought back. So if a person worked for the government for say 15 years he or she could buy additional years of service, to get a full pension of a person who has worked 40 years for the government, i.e., a pension equal to his full final salary. As Wassenaar (1989, p. 80) pointed out "the price at which this could be done was … so ridiculously low that the option to buy back service became a free-gift farce." He pointed out that members of Parliament, ministers, and local authority councilors had created a system whereby they would obtain a 100% pension equivalent to their final salary after a mere fifteen years service. Further, in this case, the final salary used to calculate the pension was not the average salary over the last three years but the salary on the last day of employment. Government officials could then be promoted on the last day of their employment to senior positions, merely to obtain greater pensions. He pointed out that the government dispensed with an actuarial valuation of the government pension fund so that the extent of the deficit could not be determined. He concluded, "All these developments in the public pension schemes and funds and the improvement of pension benefits have converted the public pension funds into one colossal bandwagon" (Wassenaar 1989, p. 81). Similar deficit problems existed in parastatals, which had run up multi-billion rand deficits.[273]

It is not clear if private sector defined benefit schemes were correctly managed or if managers of large companies managed to manipulate these schemes for their

[273] Parastatals are government-owned companies, e.g., Transnet, which operates the state airline and state-owned railway network.

own benefit. This could be done, for example, by ensuring that when employees left the company, they were only refunded their own contributions including interest on their contributions at a nominal rate, but not the employer's contribution. This was the norm until recently when the Pension Funds Adjudicator indicated his displeasure with this arrangement. This, together with the conversion from defined benefit to defined contribution, produced a substantial surplus, taken by the remaining members. In one case, it was discovered that the remaining members (directors) would be entitled to a payout of R 23 million from the surplus left in the fund (Pepkor in a Pickle 2001).

The realization that surpluses existed in pension funds resulted in an analysis of all South African pension funds. It was estimated that an aggregate of R 80 billion existed as a surplus in retirement funds. Very complicated legislation was then passed to deal with this surplus. Attempts are being made to find all former members of funds who would be entitled to portions of this surplus. It is not clear if this elaborate and expensive exercise, which has been on the go for about five years, has been a success and further legislation is now proposed to transfer the surplus, including any unclaimed benefits, past and future, to a fund under the control of the government or its appointees. SARS has also seen an opportunity to obtain a share in the unclaimed benefits. It has issued a directive (GN 35) to the effect that unclaimed benefits are deemed to have passed from the fund to the member and as such are treated as a withdrawal. In terms of the directive, all funds must deduct taxes on the unclaimed portion and forward the deduction to SARS.

All the increased legislation (i.e., fund administration, surplus, taxation) has imposed additional burdens on employers. These additional burdens may contribute to the apparent declining interest employers are showing in providing occupational pension funds.

14.4.1 Pension Funds and BEE

It is clear that since pension funds own a significant portion of equities on the JSE and if 25% of this equity is going to find its way, uncompensated, to blacks in terms of the BEE program, then a great portion of this cost must ultimately be borne by pension funds. This will likely be manifested as lower pensions for future pensioners and it is not clear that future pensioners can afford this generosity.

Providing an Adequate Pension

A question of considerable importance is: Does the South African defined contribution scheme in fact provide an adequate pension? No answer to this question is forthcoming, which is surprising considering that South Africa has a trillion rand industry tasked to provide such a pension. One would expect the answer to this question to be readily available. The author has often made the point (including on national TV) and at the annual national convention of retirement funds that the current system will be, for most, a disaster, a point publicly made by the author on numerous occasions (Vivian 2004a). The vast majority of persons who go on pension will end their lives facing destitution unless they have family members to support them. This assertion has never been debated and the evidence which is available supports this contention.

For example, the Smith Report (1995, p. 18) attempted to determine the level of pensions paid by occupational schemes. It found "the fact that 40% of pensions paid are less than R 410, i.e., the current old age grant, is a matter of concern." Converting this to 2004 values means that 40% of occupational pensioners receive less than the state poverty assistance scheme, currently R 740 per month as discussed above. Although the Smith Report acknowledged that this was a matter of concern, in the past decade nothing has been done to remedy this problem besides introducing a tax on the interest income of pension funds and BEE, both of which will exacerbate the position of future defined contribution pensioners.

Take another example. The author calculated that the pension available to a university professor after in excess of 30 years service would start at one third of his or her final salary, not the 80% of the defined benefit scheme. This is the maximum this university employee would receive without any adjustment thereafter for inflation. In addition, as discussed below, retired members will increasingly have to fund their own medical expenses.

A third example is that virtually every South African pension fund administrator publicly admits that in 96% of the cases where a person retires in South Africa on a defined contribution system, that person will not be financially independent, i.e., the occupational pension will be inadequate. Finally, representatives of old age organizations admit that the vast majority of their members do not earn an income sufficient to warrant filing tax returns.

The anecdotal evidence thus points to serious problems facing members of defined benefit schemes. The evidence is only anecdotal, but one would have thought that this matter would have been investigated after the Smith Report (1995) expressed its concern. The last two government reports which examined retirement funding (Taylor Report 2002; RFR 2004) again never attempted to determine if the defined contribution system can or does work. In both cases, the reports assumed this to be the case. Clearly, there is an urgent need for reliable research to establish the actual position of pensioners in terms of the defined contribution system. On the other hand, Rusconi (2004), in an award winning paper presented at the annual actuarial society's conference, demonstrated that South Africa's fund administration costs, especially where retirement annuities (RA) are concerned, are amongst the highest, if not the highest in the world. The government has recently released a position paper on the reform of the pension fund system in South Africa (RFR 2004). The focus on this paper is to extend pension savings through a National Savings Fund to those who currently do not have pension savings.

There is a growing realization that the final retirement accumulations are well below expectations causing disillusionment with pension funds and insurers. Some of these failed expectations have been referred to the Pension Funds Adjudicator.

14.4.2 Taxation of Retirement Funds

From the above, two points are now emphasized. First, the South African government, unlike other countries, faces no real obligations to provide a meaningful pension. Second, in the place of an obligation, private pension funds are a source of tax income. It is for this reason the taxation principles applied in South Africa to pension funds should be summarized.

Taxation of Investment Income

A controversial decision was to tax the interest income earned by retirement funds, introduced in terms of the Tax on Retirement Funds Act 38 of 1996, on the recommendation of the Katz Commission (1994). This tax is levied at a flat rate of 18% which favors individuals with incomes in the higher marginal tax brackets. The most favored are those with a marginal tax rate of 40%. It initially penalized the poor, low income tax bracket individuals but the government resolved this problem by raising the lowest tax bracket to 18%. This tax produces R 6 billion a year. The Minister of Finance indicated several years ago that this tax will be reconsidered and has appointed several groups over the years to review it, but no report has been published. It is becoming clear that the government is unlikely to abolish this tax. The minister has tried to divert attention away from this tax by repeatedly arguing that the problem is not the tax, but the high administration costs.

Uncapped Contributions to Funds

As indicated, a tax deduction is allowable up to a maximum of 22.5% of the employee's remuneration. The actual average contribution rate for savings is 10.8%. There is no monetary cap on the amount contributed. High net worth individuals can divert millions, tax free to pension funds. For example, an individual who earns R 40 million per annum can divert R 9 million tax fee to pension fund. This can then earn investment income at a marginal tax rate of 18.0% instead of 40.0%.

Raiding of Pension Funds by SARS

Since pension fund assets can be regarded as assets of the member, it is a source to be raided by SARS. Pension fund assets should not be regarded as an asset *per se*, but as the source to provide an income stream, providing a pension. The asset itself should never be subject to attachment. But there are a number of times that SARS can attach pension assets. For example, every time an employee changes jobs he or she must get a tax clearance from SARS. If an employee gets laid-off (i.e., retrenched) and has to withdraw his or her savings, SARS treats the withdrawal as income and levies a tax at the employee's average rate. Many taxpayers decide not to take their savings, but rather to leave their pension assets in the fund, until they find new employment. In terms of a recent circular GN 35 (April 2004), assets left in the fund are deemed to have been withdrawn after a period of six months and the fund must deduct taxes and pay this over to SARS.

Any trustee of a pension fund will testify of the long list of former employees who cannot be paid their savings, sometimes of the order of R 1 million, because of outstanding tax problems. In some cases, it appears as if members simply abandon their accumulated saving rather than address the outstanding tax issues. A current legislative proposal is that any outstanding amounts be transferred to a fund, after taxes have been taken by SARS, to be established by the government, and if not claimed will automatically be declared forfeit to the government after a period of time.

Inequality between Systems

The Katz Commission (1994) found that different types of funds were taxed differently and recommended they be taxed in the same manner. The government has been reconsidering the position of taxation of retirement funds for a number of years. It is expected that it will make its position clear before the end of 2006. Currently, the burden of taxation, especially the taxation of interest income, is borne almost exclusively by the defined contribution fund.

14.4.3 Conclusion

It is clear that the South African experience with a fully-funded privatized, defined contribution retirement system does hold certain general lessons which should be noted:

(a) A method of monitoring the position of pensioners is needed. The success or failure of the system must be known. In South Africa this is unknown.
(b) Pension reform and tax reform should go together. The taxpayer should not be in a worse position than in the case of the payroll social security system. The government may not divest itself of its obligations and profit from pensioners at the same time.
(c) The success of the privatized pension scheme is dependent on investment performance. The investment returns to each pensioner must be known (i.e., the average year-on-year returns on invested amounts) and the performance of the various providers published. Increasingly, individuals are discovering only at the end of their working lives that the performance of their savings fund has been dismal.

The current South African system is unlikely to be an inspiration to other countries, but may well be a warning.

14.5 THE LONG-TERM (LIFE) INSURANCE MARKET

From 1943 onwards, the governing legislation was the Insurance Act 27 of 1943, a single Act covering both the long- and short-term markets, the focus of which was the stability of the industry, not consumer protection.[274] This Act was replaced on the January 1, 1999 by two insurance Acts: the Long-Term Insurance Act 52 of 1998 and the Short-Term Insurance Act 53 of 1998. The division between long- and short-term insurance received further emphasis when, as a consequence of the new legislation, the registration of composite primary insurers (i.e., insurers that underwrote both long- and short-term insurance) was prohibited. It is possible to have a holding company, holding both long- and short-term companies.

The authorities clearly did not want, in the case of insolvency, both the long- and short-term companies to go into liquidation. There has always been reluctance on the part of the authorities to prejudice policyholders who have invested their savings in a life company to lose these if the short-term business got into trouble. Thus, when the AA Mutual got into difficulties in the mid 1980s, it was only the

[274] For a history of South African insurance legislation see Benfield (1997).

short-term business which was liquidated and not the company. The life insurance business was salvaged as the AA Life Insurance Company. Legally, it is awkward and unusual to wind-up a business and not a company. It made more sense to insist that short-term and life insurance businesses operated within two separate companies. Since the promulgation of the two acts, the remaining composite primary insurers separated their two classes of insurance business into two separate insurers. Thus, South Africa no longer has composite primary insurers. Composite re-insurers still exist.

The insurance market, especially the short-term market is based heavily on British insurance practice. A further reason why British insurance practices have been important is that the Roman-Dutch law, the "common-law" of South Africa did not recognize the specific contract of insurance, especially the life contract. It is not surprising that South African courts relied heavily on British law to resolve insurance matters. The Cape Parliament, in 1879 promulgated an Act recognizing the law of Britain as the law applicable to insurance contracts.[275] A similar law was passed, in 1902 in the Orange Free State and by precedent British law was applied to the whole of South Africa in cases involving insurance matters. It was only in 1977 that the statutes applying British law were repealed, but by that time British insurance law was so entrenched in South African case law that the repeal did not make much difference. In isolated cases, the South African courts have deviated from British law.[276]

The South African insurance market started in the early 1800s with foreign companies appointing general agents, mainly in the Cape, who usually accepted both long- and short-term contracts. The first general agents were Houghton and MacDonald who represented the Phoenix in 1806. Lloyd's was also represented in South Africa from a very early date and was operating in South Africa in the early 1800s. The foreign companies came from various countries, not only the United Kingdom. Foreign companies, operating via agents, probably explain why South Africa has a strong intermediary tradition. As the market began to develop, foreign companies sent out full-time representatives to oversee their South African operations which led to these companies opening branch offices in South Africa. The Royal Insurance Company of the U.K. was the first company to send out a full time representative. The Royal maintained a presence in South Africa from the early 1800s until disinvesting in 2004.

The first indigenous company, established in 1831, was the SA Fire & Life Assurance Company. In 1845, the Old Mutual, South Africa's (now listed on the London Stock Exchange) largest life company was established in the Cape by John Fairbairn. Initially, the relationship between the foreign and local companies was acrimonious. The British companies followed the tradition of establishing tariff committees, which the indigenous companies were not keen to join. The tariff committees existed until the late 1970s when, following the U.K. example, they were disbanded. The U.K. monopolies commission had ruled against the existence of tariff committees. Currently, short-term insurers deal with matters of common concern via

---

[275] General Law Amendment Act 8 of 1879.

[276] Probably the most well-known case is that of *Mutual & Federal Insurance Co Ltd v Oudtshoorn* Municipality 1985 1 SA 419 A.

the South African Insurance Association (SAIA) and long-term insurers via the Life Offices Association (LOA).

For reasons set-out above, the traditionally Afrikaans life insurer, Sanlam, South Africa's second largest life company was formed in 1917, also as a mutual.[277] Because of the strong links with the government and the strong saving culture of Afrikaners, Sanlam acquired a substantial volume of business from civil servants.

Table 14.3 indicates that there are currently 71 primary long-term insurers, 3 long-term reinsurers, 89 primary short-term, 4 short-term reinsurers, and 4 composite long- and short-term reinsures operating in South Africa.

Table 14.4 shows the income and expenditure of the life insurance industry for the period 1990–2003. The net annual premiums increased from R 21.81 billion in 1990 to R162.81 billion in 2003 or an average year-on-year annual increase of 17.5% per annum. This is well in excess of both the annual GDP growth and inflation as Table 14.2 indicates. So, by all accounts, the premium growth is more than satisfactory, indicating at least that the long-term industry has access to a stable income stream and/or its marketing campaign was successful. On the other hand, the average year-on-year annum rate of increase in benefits paid was 22.4% or funds were flowing out of the industry at a rate much faster than flowing into the industry. It is not clear what caused this outflow. As indicated below in the discussion of product segmentation, the bulk of the life market's activities are related to long-term savings, not risk products. An explanation could be the retrenchments of whites resulted in them cashing in their savings, especially government employees. In 1994, the benefit withdrawals increased at a rate of 34.0% and 33.0% in 1998. Another explanation is that the public started to lose faith in insurers' ability to produce a meaningful investment return, caused partly by the low investment returns of capital markets over the past few years. The problem with this explanation is that one would expect to see the lack of confidence reflected in diminished flows into the industry, which did not happen until 2001, where the inflows for the first time decreased by 2.0% and continued in 2003 with a decrease of nearly 7.0%. In 2003, for the first time (possibly ever) the benefits decreased compared with the previous year. This may also be indicative of insurers becoming reluctant to pay benefits to policyholders. There is evidence of this in the upfront commissions saga currently unfolding, discussed below.

The high rate of withdrawals in relation to inflows is also reflected in the deteriorating ratio of income to expenditure. This ratio declined from 2.4 in 1990 to 1.1 in 2002 and 1.1 in 2003. Money at the end of the period was flowing out of the industry nearly as fast as it was flowing in. Table 14.4 indicates the average for the market. The position of specific companies could be worse, i.e., with negative growth over the period. Clearly, the industry is in trouble, which has taken it some time to admit.

If inflows deteriorate to mere outflows, a corresponding deterioration in ratio of assets to liabilities can be expected, since institutions are less able to build-up reserves. In an era of defined contribution and linked investment products, insurers promise very little, resulting in low contractual obligations and may remain solvent

---

[277] The history of Sanlam was recorded by Scannell (1968). The author specifically mentions the terrible economic state, the poverty of whites and the desire for Afrikaners also to be employers as reasons for the establishment of Sanlam.

even with high rates of outflows. Table 14.5 shows that a reduction in the ratio of assets to liabilities has occurred. In 1990, the ratio stood at 1.35 but by 2002 it had declined to a mere 1.10 and 1.11 in 2003; the assets covered the liabilities by a mere 10%. The decline should be a matter of concern, but does not appear to have attracted any comment from the regulator other than noting the decline. The regulator attributed the decline to the strengthening of the rand. In 2003, the assets were invested as follows: fixed interest rate stock, deposits and cash (26%), shares

Table 14.3. Number of Registered Insurers by Type, 2001–2004

	2001	2002	2003	2004
Long-Term Insurers				
Primary-Insurers				
Typical	40	42	42	43
Health Insurers	2	2	2	2
Cell Captive Insurers	2	4	6	7
Assistance Insurers	6	6	6	6
Link Investment Only	12	13	13	13
Subtotal	62	67	69	71
Reinsurers				
Long-Term Only	3	3	3	3
Subtotal	3	3	3	3
Total: Long-Term Insurers	65	70	72	74
Short-Term Insurers				
Primary Insurers				
Typical Insurers	29	29	29	29
Niche Insureres	0	0	35	36
Specialists	40	41	0	0
Cell Captive Insurers	0	0	11	10
Captive Insurers	17	19	14	14
Subtotal	86	89	89	89
Reinsurers	0	0	0	0
Short-Term Only	4	4	4	4
Long- and Short-Term	3	3	3	4
Subtotal	7	7	7	8
Total: Short-Term Insurers	93	96	96	97

Source: Financial Services Board (2004).

Table 14.4. Income and Expenditures of Long-Term Insurers, 1990–2003

	(Rands Billions)													
	1990	1991	1992	1993	1994	1995	1996	1997	1998	1999	2000	2001	2002	2003
Primary Insurers														
Income	40.14	43.72	52.23	54.20	64.45	86.97	96.01	114.61	153.40	149.55	191.97	190.94	226.12	206.61
Net Premiums	21.81	24.52	31.22	38.38	46.08	61.77	68.40	82.47	114.52	115.54	147.75	144.47	174.90	162.81
Investment Income	10.58	12.03	13.84	15.45	17.73	23.17	25.84	28.99	36.01	31.96	36.51	34.94	41.09	43.28
Other Income	7.75	7.17	7.17	0.38	0.64	2.02	1.77	3.15	2.87	2.05	7.71	11.53	10.13	0.52
Expenditures	16.63	18.78	26.17	32.93	41.81	53.70	64.10	80.74	108.43	118.91	140.68	139.99	210.50	181.81
Benefits	12.22	13.81	20.11	25.85	34.74	44.61	53.78	69.35	92.47	104.79	118.23	123.19	164.84	156.44
Management Expenses	1.97	2.38	2.95	3.27	3.65	4.49	5.09	5.95	9.50	8.52	9.01	9.36	10.76	11.38
Commissions	1.79	2.03	2.43	2.95	3.42	4.52	4.79	4.90	5.62	5.36	6.06	5.65	7.06	7.17
Other Expenditures	0.66	0.56	0.68	0.87	0.00	0.08	0.44	0.53	0.83	0.24	7.38	1.78	27.84	6.82
Ratio: Income to Expenditures	2.41	2.33	2.00	1.65	1.54	1.62	1.50	1.42	1.41	1.26	1.36	1.36	1.07	1.14
Percentage Year-on-Year Increase of Income		8.93	19.45	3.78	18.90	34.95	10.40	19.38	33.84	–2.51	28.36	–0.54	18.43	–8.63

Table 14.4 *(continued)*. Income and Expenditures of Long-Term Insurers, 1990–2003

	(Rands Billions)													
	1990	1991	1992	1993	1994	1995	1996	1997	1998	1999	2000	2001	2002	2003
Reinsurers														
Income	0.40	0.53	0.66	0.86	0.79	1.04	1.37	1.35	1.57	1.71	2.17	2.72	2.52	2.58
Net Premiums	0.29	0.39	0.50	0.57	0.59	0.72	0.84	1.00	1.20	1.25	1.47	1.54	2.24	1.89
Investment Income	0.10	0.13	0.15	0.27	0.19	0.24	0.31	0.33	0.35	0.38	0.41	0.66	0.41	0.49
Other Income	0.01	0.01	0.01	0.01	0.01	0.08	0.22	0.02	0.01	0.08	0.29	0.53	–0.12	0.20
Expenditures	0.25	0.32	0.43	0.46	0.51	0.59	0.73	1.07	1.24	1.38	1.72	2.28	1.83	1.92
Benefits	0.16	0.21	0.28	0.32	0.37	0.44	0.54	0.71	0.99	1.01	1.06	0.93	1.22	1.33
Management Expenses	0.03	0.04	0.05	0.05	0.06	0.07	0.08	0.10	0.11	0.20	0.16	0.12	0.14	0.13
Commissions	0.06	0.07	0.10	0.09	0.08	0.09	0.12	0.17	0.12	0.17	0.26	0.27	0.42	0.35
Other Expenditures	0.00	0.00	0.00	0.00	0.00	0.00	0.00	0.08	0.02	0.00	0.24	0.96	0.05	0.11
Ratio: Income to Expenditures	1.60	1.67	1.53	1.86	1.55	1.76	1.87	1.26	1.26	1.24	1.26	1.19	1.37	1.35
Percentage Year-on-Year Increase of Income		32.67	23.31	30.49	–8.18	32.44	31.60	–1.61	16.10	9.07	27.18	25.29	–7.35	2.22

Source: Registrar of Insurance (1990–1997) and Registrar of Long-Term Insurance (1998–2003).

Table 14.5. Assets and Liabilities of Long-Term Insurers, 1990–2003

(Rands Billions)

	1990	1991	1992	1993	1994	1995	1996	1997	1998	1999	2000	2001	2002	2003
Direct Insurers														
Assets	141.54	176.96	204.44	243.49	317.34	369.75	443.23	497.98	515.81	645.11	709.38	744.40	771.94	813.31
Liabilities	104.68	150.83	177.82	211.58	259.62	305.36	354.83	402.21	432.70	525.73	601.33	664.03	701.50	733.88
Assets/Liabilities	1.35	1.17	1.15	1.15	1.22	1.21	1.25	1.24	1.19	1.23	1.18	1.12	1.10	1.11
Reinsurers														
Assets	1.12	1.43	1.60	2.03	2.43	2.81	3.18	3.55	3.62	4.59	4.67	5.10	5.26	6.76
Liabilities	0.70	1.88	1.51	1.58	1.67	2.22	2.52	2.92	3.01	3.73	3.71	3.61	4.61	5.58
Assets/Liabilities	1.59	0.76	1.06	1.29	1.45	1.27	1.26	1.22	1.20	1.23	1.26	1.41	1.14	1.21

Source: Registrar of Insurance (1990–1997) and Registrar of Long-Term Insurance (1998–2003).

Table 14.6. Ratio of Surplus Assets to the Capital Adequacy Requirement (CAR), 2000–2003

Covered	2000	2001	2002	2003
0–1 times	1	2	1	3
1–2 times	11	6	11	10
2–5 times	19	24	20	34
5–10 times	7	7	9	6
+10 times	22	24	24	14
Total	60	63	65	67

Source: Financial Services Board (2002, 2004).
Note: The method of calculating the CAR is set out in the schedules to the legislation.

(43%), fixed properties (3%), miscellaneous assets, loans and mortgages (15%), and foreign approved assets (13%) (Financial Services Board 2004, p. 27)

Of course the mere ratio of assets to liabilities is an unsophisticated measure of the market's solvency. The more sophisticated method is the measure of the ratio of surplus assets to the insurer's Capital Adequacy Requirements (CAR). The Long-Term Insurance Act was amended in 2003 to change over to this method of measuring the solvency of life companies. The solvency position of insurers is indicated in Table 14.6. This measure also shows a deteriorating situation. In 2002, twenty-four insurers had a ratio of in excess of 10 times liabilities. In 2003, only fourteen insurers fell into this category. At the lower end, the number of insurers whose free assets exceeded these Capital Adequacy Requirements by a factor of one, increased to three.

The method of reporting market product segmentation changed after 2000 so it is difficult to make comparisons thereafter. Table 14.7 covers the period up to the year 2000 and Table 14.7A indicates the position thereafter. Table 14.7 shows that the single largest line of business was the pension fund and group life business. It represented 35.2% of the market's activities in 1990 increasing to 49.6% in 2000, underlining the importance of this line of business, which usually comes via corporations. As indicated above, most firms have a pension fund which they can either manage themselves or enter into a contract with an insurer to take over this responsibility. Smaller employers usually opt for this route and larger employers could contract with an insurer to manage their pension fund assets. Table 14.2 indicates the assets of pension funds. In 2003, the assets under management of exempt funds (or insured funds) was R 213 billion. It is this line of business in particular which can find its way to insurance companies. By handing over these assets to insurance companies, as explained, pension funds become exempt. This indicates that this line of pension fund business is a substantial line of life insurers' business. Approximately half of the long-term market's net premium income came from managing firms' pension funds and group life benefits, and not from the more

Table 14.7. Product Segmentation in the Long-Term Insurance Market, 1990–2000

	(Rands Billions)										
	1990	1991	1992	1993	1994	1995	1996	1997	1998	1999	2000
Pension Fund and Group-Life Benefits											
Premiums	7.72	8.55	11.48	13.51	15.58	20.98	22.37	30.00	48.72	49.99	66.92
(%)	35.21	34.82	36.47	35.03	33.88	34.20	33.08	37.43	42.97	45.44	49.63
Retirement Annuity Fund Business											
Premiums	2.63	2.43	2.78	3.11	3.59	4.49	4.78	5.64	10.05	7.95	10.39
(%)	12.01	9.89	8.83	8.07	7.81	7.32	7.06	7.04	8.87	7.23	7.70
Immediate Annuity Business											
Premiums	3.83	4.46	5.60	7.16	7.49	10.04	9.93	10.62	12.45	13.38	12.01
(%)	17.49	18.17	17.79	18.56	16.28	16.37	14.69	13.25	10.98	12.17	8.90
Disability and Health Insurance											
Premiums	0.19	0.28	0.57	0.75	1.05	1.41	1.63	2.04	2.47	3.18	2.79
(%)	0.85	1.13	1.82	1.94	2.29	2.30	2.41	2.54	2.18	2.89	2.07
Single Premiums											
Premiums	0.27	0.24	0.56	1.64	3.65	5.42	8.23	8.88	10.90	10.66	17.85
(%)	1.24	0.96	1.78	4.25	7.94	8.84	12.16	11.08	9.61	9.69	13.24
Periodic Premiums											
Premiums	7.28	8.61	10.48	12.40	14.63	18.99	20.69	22.96	28.79	24.83	24.90
(%)	33.21	35.04	33.30	32.15	31.80	30.96	30.60	28.65	25.39	22.57	18.46
Total											
Premiums	21.92	24.57	31.47	38.56	45.99	61.32	67.62	80.13	113.37	110.00	134.84
(%)	100.00	100.00	100.00	100.00	100.00	100.00	100.00	100.00	100.00	100.00	100.00

Source: Registrar of Insurance (1990–1996) and Registrar of Long-Term Insurance (1997–2003).

traditional life insurance sold directly to the public. In addition to pension funds, many firms provide an extensive range of group benefits including group life insurance, group personal accident cover, and so on. In fact, as a general rule anyone employed by a large South African firm has little need for life or personal accident insurance.

Retirement annuity fund business represented 12.0% of market activity in 1990 but declined to 7.7% in 2000. This line of business, also related to pension fund business, is usually bought by persons employed in smaller companies which do not have pension fund facilities as well as by individuals, especially self-employed persons. It is not clear why this line has declined in importance. The relative decline may indicate a decreasing ability of self-employed persons, or persons employed by smaller corporations which do not provide occupational pensions, to provide for retirement or a loss of confidence in insurers' ability to do so.

Immediate annuity business also declined in importance. In 1990, it represented 17.5% of the market, while in 2000 it declined to 8.9%. During the period, there was a swing from defined benefit pension schemes to defined contribution schemes, particularly in the private sector. Usually upon reaching retirement age, it is possible to commute a portion of the pension into a lump sum. In terms of the definition of a pension fund in the Income Tax Act 58 of 1962, not more than one third of the value of the annuity can be commuted into a lump-sum. In the case of a provident fund, a lump-sum becomes available which can be used to purchase an annuity. With the defined contribution pension fund the member may also receive a lump-sum on reaching retirement age and use the balance to purchase an annuity.[278] It could have been expected that the immediate annuity business would have increased in importance with the change to defined contribution schemes. However, this did not happen. An explanation may be that persons did not want an immediate, but rather a delayed annuity. That this could have happened is supported by the increase in the single premium business discussed below.

Disability and health insurance increased in importance during the decade, from 0.9% in 1990 to 2.1% in 2000. The net premium income increased from R 0.187 billion to R 2.787 billion, an increase of 1,390%. This increase was brought about by the expansion of the long-term market into the medical insurance arena. Traditionally, as discussed below medical expenses are provided by the medical-aid industry, but in the early 1990s insurers began to offer medical insurance products. The development resulted in an unseemly demarcation argument between regulators of the respective industries which continued well into the next decade and has only recently been resolved.

Single premium business represented 1.2% of the market's business in 1990. This increased to 13.2% in 2000. This line of business showed the greatest increase in premium flows, from R 0.271 billion in 1990 to R 17.847 billion in 2000, an increase of 6,500%. This increase could possibly be ascribed to two aspects, the increasing withdrawals from life companies and the change from defined benefit to defined contributions pension schemes. When an employee resigns or is laid-off, the employee has a capital sum to invest. This capital can be invested with an insurer as a single premium investment. Second, with the change from defined benefit to

[278] The complex rules governing the taxation of lump sums is set-out in the Second schedule to the Income Tax Act 58 of 1962.

defined contribution, members who leave employment often receive a lump-sum upon leaving or retirement, but may not perceive a need to purchase an immediate annuity. A single premium policy could be purchased with a maturity date in the future. This may be preferable to purchasing an annuity.

Periodic or recurring premium business represents the more historical line of business of life insurers, which is to accept periodic premiums and to pay lump-sums upon defined events. The market-share of this class of business declined from 33.3% in 1990 to 18.5% in 2000 and may be the harbinger of future difficulties for the industry. The FSB's extensive legislative program, especially the FAIS legislation, regulation is likely to add considerable costs to providing insurance. At the lower end of the market, it may no longer be economically viable to provide insurance. The new legislation may thus impact negatively on this class of business. If this happens, it will be unfortunate since it is the lower income group that is in most need of life insurance. It is this group that will be most affected at a time when financial markets are committed to designing products for the lower end of the market.

From the above analysis it can be seen that approximately 80% of the life business is in one way or another related to the retirement or savings business, hence the above observation that the key to understanding South Africa's life market is to understand the country's retirement funding industry.

Table 14.7A indicates the data available since 2001. From the table it is not possible to work out the risk products' contribution to the market. However, it is clear that pensions, or what is now shown as fund business, is still the dominant line of business. The LOA estimates that the premium income from retail reoccurring risk business is R 12.7 billion in 2004 (Hetherington 2005).

Table 14.8 indicates the market-share of the top 22 companies (95% of the market's premium in 2003) for the period 1990 to 2003. The insurers are ranked in the order of largest market-share in 2003. For comparative purposes, the market-shares of the leading companies in previous periods are also indicated. Because of mergers, name changes, and new entrants, the performance of some of the companies are not strictly comparable. In 1990, three insurers dominated the life market: The Old Mutual established in 1845, Sanlam, the Afrikaner insurer established in 1917, and Liberty Life established in 1958.[279] These three insurers accounted for 76% of the market measured by premium income in 1990 which declined to 39% in 2003. The dramatic change which has taken place over the period can be seen from Table 14.8. The Old Mutual lost 42% of its market share, while Sanlam lost 63% of its market share and Liberty Life remained constant at approximately 8%, but had been overtaken by other insurers in some years. The Old Mutual, demutualised and moved its primary listing to London in 1999. Its has been involved in a number contentious investments such as the purchase of Gerrand Private Client in the U.K. for which it paid £525 million only to sell it later for £210 million. It entered the U.S. with the purchase of United Asset Management for which it paid a premium of 22% at a time when the market was at its peak. It also recently paid a fine of $100 million to the U.S. regulators arising out of its purchase of Pilgrim and Baxter. Its investments in

[279] A history of Liberty Life and its founder Donald Gordon is covered in Romain (1989).

Table 14.7A. Product Segmentation in the Long-Term Insurance Market, 2001–2003

	(Rands Billions)		
	2001	2002	2003
Life			
Net premiums	62.55	68.39	72.76
(%)	43.30	39.10	43.10
Disability			
Net premiums	1.88	1.75	2.70
(%)	1.30	1.00	1.60
Health			
Net premiums	2.46	1.40	1.18
(%)	1.70	0.80	0.70
Fund			
Net premiums	64.72	85.18	78.16
(%)	44.80	48.70	46.30
Sinking Fund			
Net premiums	12.14	16.97	12.66
(%)	8.40	9.70	7.50
Assistance			
Net premiums	0.72	1.22	1.35
(%)	0.50	0.70	0.80
Total			
Net premiums	144.47	174.90	168.81
(%)	100.00	100.00	100.00

Source: Registrar of Long-Term Insurance (2001–2003).

South Africa have also been problematic. It had to lend several billion rand to Nedcor, a bank in which it is a major shareholder to get it out of trouble. Sanlam demutalised in 1998 and listed on the JSE. It recently announced a BEE partner. It has suffered the largest loss in market share, which may have prompted the search for a BEE partner. The loss of market share may have been brought about by its loss of business from government employees, which as indicated above had a large Afrikaans complement, committed to saving via Afrikaans institutions, the premier of which was Sanlam. The government stopped the automatic debit order system which prevailed allowing the collection of premiums to insurers directly from the government payroll. Sanlam has been aware of its changed fortunes and has taken a number of steps to improve its prospects. In recent years, the company has changed CEOs three times. The latest CEO is Dr. Johann Van Zyl, formally the Vice-

Chancellor and Principal of the University of Pretoria.[280] Van Zyl oversaw an impressive transformation program of the University of Pretoria and by all accounts, since taking over, has secured the confidence of the investment community. The former CEO, Leon Vermaak is now the CEO of Auto and General, a short-term insurer.

Donald Gordon (now Sir Donald Gordon) the founder of the Liberty Life Group has since relocated to the United Kingdom. Liberty Life was never a mutual insurance company. It is accepted that investors would have done much better to buy shares in Liberty rather than to buy its policies or as Gleason, a leading financial columnist, recently put it, "It was after all common knowledge that if you wanted to follow Donald Gordon, the clever thing to do was to buy the shares, definitely not the policies."

Tables 14.7 and 14.7A clearly indicate that the main function of the life market is to manage the savings of insureds. This begs the question: "Do life insurers provide any meaningful returns to these insureds?" As in the case of pension funds, there is no clear answer to this important question, nor is there any easy way of arriving at that answer. Recently one of South Africa's senior financial investigative reporters, Dion Basson, examined his policies and calculated the returns. The first policy with Liberty Life ran for 13 years and produced a growth of 27.7% over this period (average of 2.1% per annum). The second policy again with Liberty Life ran for 10 years grew by 6.6% (average of 0.7% per annum). The third policy was with Sanlam. This had declined by 36.4% over a sixteen-year period. Old Mutual was unable to provide sufficient information for him to carryout the calculations. He then advised the public to surrender their policies (Basson 2004). Subsequently, other policyholders have done similar calculations and have come to the same conclusion. If what is said above is understood, then it is clear that disappointing returns are a matter of grave concern to persons who will have to survive on their accumulated capital after retirement.

The position is exacerbated by the inability of the public to get a clear picture of what is happening to their savings. Discovering the truth, one by one, as is now happening, on retirement is not the way to go.

14.5.1 The Upfront Commissions Saga

Dismal investment returns will clearly generate complaints. From time to time complaints by policyholders about investment returns have appeared in the press, which have been downplayed by the life industry. Recently, some policyholders, especially persons saving via retirement annuities (RAs), have lodged formal complaints with the Pension Funds Adjudicator, who has ruled against the life industry. He has made eight determinations for different reasons, against the life industry which has indicated its intention to take the determinations on appeal or review to the High Court. It is not clear that he has jurisdiction over the life industry. Some of these rulings concern early termination of RAs. Looking at the facts, it is

[280] In South Africa the Vice-Chancellor and Principal is the highest executive position – the CEO of the university.

Table 14.8. Market Share (%) for Long-Term Primary Insurers by Net Total Premiums, 1990–2003

	Primary Insurer	1990	1991	1992	1993	1994	1995	1996	1997	1998	1999	2000	2002	2003	Cumulative Total
1	Old Mutual	32.96	36.18	33.04	32.35	32.76	32.19	32.86	31.61	38.40	25.98	19.36	19.09	17.80	17.80
2	Sanlam	34.33	33.36	33.60	33.56	29.61	33.44	26.94	26.57	16.97	16.03	15.94	12.86	12.09	29.89
3	Liberty Group (previously Liberty Life)	8.69	9.02	9.35	10.29	11.18	10.05	10.53	10.56	7.47	9.06	7.88	8.12	9.95	39.84
4	Momentum Group (previously Momentum then FirstRand)	4.37	4.08	2.89	2.93	2.86	3.43	4.17	4.66	4.28	8.05	9.83	9.12	9.30	49.14
5	Investment Solutions (previously Forbes Life)									3.19	8.27	9.13	7.01	9.06	58.20
6	Investec								0.13	4.38	3.06	3.58	13.71	7.82	66.02
7	Coronation											1.79	2.19	5.28	71.30
8	Metropolitian	1.92	2.21	2.05	2.08	2.17	2.30	2.60	2.89	2.71	3.57	3.94	3.43	3.75	75.05
9	mCubed Inv											2.58	3.48	3.60	78.65
10	BOE Life Assurance											0.15	1.04	3.59	82.24
11	Capital Alliance											0.42	1.23	2.89	85.13
12	Investec Employee Benefits												2.97	1.99	87.12
13	Charter Life	0.22	0.21	0.26	0.32	0.00	0.63	1.05	1.00	0.85	1.29	1.34	1.37	1.30	88.42
14	Discovery Life (previously Discovery Health)											1.11	1.17	1.07	89.49
15	Allan Grey Life												1.15	1.07	90.56

Table 14.8 *(continued)*. Market Share (%) for Long-Term Primary Insurers by Net Total Premiums, 1990–2003

	Primary Insurer	1990	1991	1992	1993	1994	1995	1996	1997	1998	1999	2000	2002	2003	Cumulative Total
16	BOE Life							0.16	1.31	3.74	4.86	3.42	3.05	1.02	91.58
17	African Life											0.89	0.67	0.99	92.57
18	Sage Life	1.34	0.29	1.12	1.12	1.17	0.92	0.97	1.14	1.06	1.61	1.28	0.92	0.87	93.44
19	Hollard Life											0.58	0.60	0.65	94.09
20	ABSA Life											0.24	0.58	0.61	94.70
21	AIG Life												0.29	0.52	95.22
22	PSG Escher Life (previously PSG Life: PSG Anchor Life)											6.28	3.03	0.32	95.54

Source: Registrar of Insurance (1990–1996) and Registrar of Long-Term Insurance (1997–2003).
Note: Because of name changes and restructuring of individual companies, comparison of some companies may not be accurate. Market share is defined as the percentage of the net total premiums in the market.

not surprising that he ruled against the industry. In the one case, the life office had levied a penalty of R 134,000 against accumulated savings of R 236,000, or the insured forfeited nearly 60% of his savings to the insurance company, in another it was 82% and yet another 85%. Life insurers pay all the commissions over the life span of retirement annuity to brokers during the first two years. Should the insured then cancel, the life office recovers this from the insured. Often the life office recovers from both the broker and insured and retains both amounts. When these figures became known, there was a fair degree of public outrage. Insurers have blamed up-front commissions, that has for some years been the subject of debate in the revision of claw-back regulations. These determinations have caused a crisis in the life industry, already facing a loss of consumer confidence. Stocks of life companies declined and Liberty Life delayed its plan to raise R 2 billion from the bond market until the crisis is resolved. It is estimated that some of these rulings will cost the life industry between R 1 billion and R 4 billion rand. The Old Mutual has set aside R 225 million as a contingency.

Faced with the negative impact of the publicity, the loss of public confidence, and the almost certainty that additional regulations would be promulgated, the LOA revised its position on upfront commissions and presented its proposals to the Treasury on the June 10, 2004. These vastly improved the position of the policyholder. A point which must be made is that policyholders have faced this prejudicial position for decades, but the life market, only moved to correct the situation when it became public knowledge as a consequence of the Pension Funds Adjudicator's rulings.

The point made earlier is again emphasized—there is an urgent need for research to establish the actual position of investors (savers) with pension funds and insurance companies. These returns may well be negative. If one considers that in the past decade there has been no significant Foreign Direct Investment (FDI) in South Africa, BEE is aimed at giving at least 25% of capital away, and directors are paid billions in share options, then poor returns will not be surprising. A method of measuring actual returns to savers, and publishing the outcome of the research is urgently needed.

14.5.2 Conclusions Regarding the Long-Term Market

The last decade has probably been the most traumatic in the 180-year history of South Africa's life market. The nature of the long-term industry seems to be that of a wholesale market and not a retail market. Sales to individuals were of declining importance while sales to employees via corporations increased. It is more efficient to manage the pension savings of 2,000 employees than to sell 2,000 separate Retirement Annuity policies. The well-being of the industry appears to be intimately tied to the country's retirement system. The public is having doubts as to the value of saving via an insurance company.

14.6 THE SHORT-TERM MARKET

Table 14.3 indicates there were 89 primary (direct) short-term insurers registered in South Africa in 2004. These are conventional direct or primary insurers, reinsurers, cell-captive insurers, captive insurers, and bank insurers. Although captive and cell-captive insurers are indicated, technically there are no specific legislative provisions recognizing captive insurers. Starting in the mid 1970s, major industrial companies began to establish their own captive insurers. In the mid 1980s, one of South Africa's largest motor insurers, the AA Mutual failed and the government appointed a judicial commission of inquiry, the Melamet Commission (1988) to investigate the failure. It discovered, unrelated to the failure, the existence of a captive insurer and recommended that these be investigated. The commission, incorrectly, seemed to be of the opinion that captives were set-up to dodge taxation. A second commission, Melamet (1992) was duly appointed (with the same chairman, Mr. Justice David Melamet) as a result of which the operation of captive insurers in South Africa became much better understood. It also gave rise to the cell captive industry, with the leading cell captive insurer being Guardrisk. The position of captive insurers in South Africa is extensively covered by Bawcutt (1997).

The large life insurers have substantial shareholdings in the banks and in short-term insurers. In some cases there are cross-holdings, i.e., banks also holding shares in insurers. Thus, Sanlam had holdings in ABSA (which the directors have agreed to sell to Barclays Bank) and Old Mutual, the life insurer, had holdings in the Permanent Bank and Nedcor. Old Mutual is the dominant owner of the Mutual & Federal Insurance Company, South Africa's second largest short-term company.

Initially, banks placed their client's insurance business with their shareholder's short-term insurer. Thus, the Permanent Bank would place their business with the Mutual & Federal, the Old Mutual's short-term insurer. The Standard Bank has controlling relationship with Liberty which had a relationship with Guardian National, so the Standard Bank placed its business with the Guardian National (taken over by Santam).

A recent development has been the establishment of bank insurers (i.e., each of the banks establishing their own insurers). These insurers are positioned to pick-up insurance from the bank's clients. Thus, if a client takes out a mortgage bond with the bank, the bank can also offer to insure the buildings and credit life. The bank can also offer to insure the household contents and motor vehicles and so on. It is illegal for the bank to make it a condition of any loan that the insurance must be purchased from the bank. The consumer has the free choice of insurer and the bank must advise the client accordingly (§43 of the Short-Term Insurance Act).

Bank insurers generally will insure members of the public, other than its own clients and would like to increase market share. Some bank insurers such as the First National Bank's Outsurance publicizes widely, as a direct insurer, in an attempt to attract insureds directly from the public.

Table 14.9 indicates that the gross short-term premium income for South Africa for the period 1992—2003. In 2003, it was R 35.64 billion. Since a single power station costs R 30 billion to replace, it should be clear that South Africa's short-term insurance market is not large enough to cover all South Africa's risks. The

Table 14.9. Income and Expenditures of Short-Term Insurers, 1992–2003

	(Rands Billions)											
	1992	1993	1994	1995	1996	1997	1998	1999	2000	2001	2002	2003
Primary Insurers												
Income	8.68	9.94	11.20	13.33	15.96	18.25	19.67	20.84	21.20	21.01	25.09	26.74
Gross Premium Income	9.55	11.13	12.58	15.06	18.07	20.74	22.70	21.25	24.67	26.54	31.44	35.64
Net Premiums	7.58	8.71	9.93	11.84	14.08	16.04	17.12	16.21	16.35	17.31	20.96	24.96
Investment Income	1.10	1.23	1.27	1.49	1.88	2.21	2.56	4.63	4.86	3.69	4.13	2.74
Expenditures	6.78	9.39	9.02	10.62	12.36	14.81	16.76	15.00	16.45	16.92	19.81	21.46
Claims Paid	4.96	6.76	6.78	7.87	8.98	10.50	11.98	10.89	11.81	11.59	13.88	15.34
Management Expenses	0.95	1.12	1.15	1.47	1.74	2.27	2.66	2.31	3.00	3.68	4.06	3.75
Commissions	0.87	1.51	1.09	1.29	1.64	2.04	2.13	1.80	1.64	1.64	1.88	2.37
Underwriting Profits	0.54	0.45	–0.21	0.41	0.68	0.50	–0.09	0.00	–0.34	0.02	0.27	1.38
Ratio: Income to Expenditures	1.28	1.06	1.24	1.25	1.29	1.23	1.17	1.39	1.29	1.24	1.27	1.25
Percentage Year-on-Year Increase of Income	1.64	1.68	1.06	1.90	2.56	2.70	2.47	4.63	4.52	3.71	4.40	4.12

Table 14.9 *(continued)*. Income and Expenditures of Short-Term Insurers, 1992–2003

	(Rands Billions)											
	1992	1993	1994	1995	1996	1997	1998	1999	2000	2001	2002	2003
Reinsurers												
Income	0.64	0.83	0.83	1.03	1.33	1.49	2.01	2.43	2.54	2.68	2.79	2.93
Gross Premium Income	1.10	1.39	1.45	1.74	2.34	2.71	3.24	3.18	3.69	4.15	4.96	5.20
Net Premiums	0.54	0.68	0.68	0.86	1.12	1.26	1.75	1.90	2.14	2.03	2.41	2.59
Investment Income	0.11	0.15	0.15	0.17	0.21	0.23	0.26	0.53	0.40	0.65	0.38	0.35
Expenditure	0.49	0.62	0.69	0.75	0.96	1.21	1.33	1.86	2.21	2.24	2.35	2.15
Claims Paid	0.32	0.41	0.46	0.44	0.58	0.76	0.88	1.22	1.53	1.21	1.58	1.31
Management Expenses	0.06	0.06	0.06	0.07	0.09	0.10	0.11	0.15	0.15	0.17	0.14	0.22
Commissions	0.12	0.16	0.17	0.24	0.30	0.35	0.34	0.49	0.53	0.86	0.64	0.62
Underwriting Profits	–0.06	–0.07	–0.12	–0.01	–0.10	–0.09	–0.19	–0.36	–0.14	–0.09	–0.18	0.01
Underwriting Profits plus Investment Income	0.05	0.08	0.02	0.17	0.11	0.14	0.07	0.18	0.26	0.56	0.21	0.35
Ratio: Income to Expenditures	1.31	1.33	1.20	1.38	1.39	1.23	1.51	1.30	1.15	1.20	1.19	1.37

Source: Registrar of Insurance (1990–1996) and Registrar of Long-Term Insurance (1997–2003).

Table 14.10. Assets and Liabilities of Short-Term Insurers, 1990–2003

	(Rands Billions)													
	1990	1991	1992	1993	1994	1995	1996	1997	1998	1999	2000	2001	2002	2003
Primary Insurers														
Assets	10.87	13.33	15.49	19.21	22.45	26.77	31.98	36.63	35.62	39.12	37.73	34.54	38.20	42.01
Liabilities	5.34	5.50	6.04	7.54	8.31	9.64	11.44	13.42	15.24	19.78	17.94	18.34	21.86	24.20
Assets/Liabilities	2.04	2.42	2.57	2.55	2.70	2.78	2.80	2.73	2.34	1.98	2.10	1.88	1.75	1.74
Reinsurers														
Assets	1.07	1.26	1.49	1.98	1.95	2.43	2.95	3.33	3.83	4.98	5.60	5.79	5.50	5.38
Liabilities	0.73	0.83	0.97	1.41	1.62	1.63	2.14	2.41	3.09	3.89	4.52	4.26	4.59	4.49
Assets/Liabilities	1.46	1.52	1.52	1.40	1.21	1.48	1.38	1.38	1.24	1.28	1.24	1.36	1.20	1.20

Source: Registrar of Insurance (1990–1996) and Registrar of Short-Term Insurance (1997–2003).

international market and reinsurance into the international market is essential. It is thus not surprising that until recently, virtually every short-term insurer (an exception being Santam, the Afrikaner short-term company) had a foreign shareholder. The short-term market maintained close links with international insurers especially reinsurers. This situation prevailed for centuries. Both of the large European reinsurers, Munich-Re and Swiss-Re have subsidiary reinsurance companies in South Africa. Swiss-Re made its unhappiness with the idea that it should dilute its ownership through BEE, clear threatening to withdraw from South Africa.

The ratio of income to expenditures has averaged 1.3 over the past decade. The market is generally regarded as well-capitalized and not so long ago Mutual & Federal declared a R 1 billion special dividend, the directors being of the opinion that it was over capitalized. In this case, that may well have been the wrong decision, since shortly thereafter Guardian National came on the market and was acquired by Santam which resulted in Santam taking the first position in the market. An interesting feature of the market is that it returns an underwriting profit. In 2003, the market produced a 6% underwriting profit on net premiums written, possibly the best in its history.

The assets and liabilities of the short-term market are shown in Table 14.10, from which it can be seen that the ratio of assets to liabilities has been fairly stable. In the early part of the decade, the market was probably over-capitalized. In 1996, the assets covered the liabilities by a ratio of 2.8. The larger insurers, in particular were over-capitalized and thus were well-placed to take over foreign insurers (discussed below) or to return excess capital to shareholders. In 2003, the assets were invested as follows; shares (32%), debtors and debentures (21%), stocks (9%), cash and deposits (37%), and fixed assets (1%) (Financial Services Board 2004, p. 31).

The solvencies of the companies which comprise the market are shown in Table 14.11. The solvencies, calculated in terms of legislation, of the short-term companies are currently regarded as more than satisfactory. The short-term insurers are currently (2004) enjoying the best underwriting returns they have achieved in many years. The FSB is working on introducing a system measuring solvency to bring South Africa in line with the best international practice, but this has not yet been promulgated.

Table 14.12 indicates the product segmentation of the short-term market. Traditionally, six insurance classes were reported to the regulator. In terms of the new Short-Term Insurance Act, the classes were redefined and two additional classes were added: liability business and engineering business. Engineering business includes insurance such as machinery breakdown insurance and construction insurance of large projects. The performances of the eight classes are indicated in Table 14.12 from which it will be seen that property (i.e., fire) (31%), motor (36%), and miscellaneous insurance (12%) are the three largest classes of business.

Liability risks are, in many parts of the world, a major cause of concern. There is for example the asbestos crisis which it is estimated to cost American companies in excess of $200 billion and has caused the insolvency of a large number of industrial and insurance companies (Parloff 2002). The separate reporting of this class of business will enable the regulators to monitor the position in South Africa. Until this separate class was reported, it has been difficult to assess the impact of liability

Table 14.11. Solvency of Short-Term Insurers

	Surplus Asset Ratio		
Number of Insurers	2001	2002	2003
Below 15%	0	0	2
Between 15% and 20%	3	4	1
Between 20% and 25%	2	4	3
Between 25% and 30%	2	3	6
Between 30% and 40%	3	4	5
Between 40% and 50%	9	10	8
Between 50% and 100%	22	15	15
Above 100%	34	35	32
Total	75	75	72

Source: Financial Services Board (2002–2004).

claims on the market since liability risks were included as part of the miscellaneous category, which was escalating at a rate in excess of inflation. Not all liability risks are reported separately or covered in the insurance market. Motor passenger liability is covered as a section of the motor policy and will appear in the transportation category. Motor third party bodily injury claims are covered by the government's Road Accident Fund (RAF) as discussed below, as is workers' compensation, also discussed below. Some liability risks are not covered in the market at all, thus recently a R 400 million out-of-court settlement was reached on asbestos claims. This claim does not appear to be covered by insurance, but was paid by shareholders.

The change in the market share is indicated in Table 14.13. Santam and Mutual & Federal are South Africa's largest short-term insurers, accounting for 34% of the market. The past has been characterized by a contraction in the market and the departure of virtually every foreign shareholder of short-term companies in South Africa. SA Eagle is the remaining large, long-established insurer with foreign shareholding. It is part of the Zurich Financial Services Group.

In recent years, the following companies have closed-down or disinvested: IGI, Aegis, Commercial Union (later CGU), General Accident (GA), Guardian National, StanGen, The St. Paul, Allianz (although Allianz has kept a small office going), Gerling General SA (recently returned). The Royal & SunAlliance, a major shareholder of the Mutual & Federal since the early 1800's and for a long time South Africa's largest short-term insurer, sold its shareholding to the Old Mutual in a controversial disinvestment exercise. It had offered the minority shareholders an amount less than the quoted price of the shares. There is no single reason for these departures and most had nothing to do with South Africa. It seems most decisions to depart had more to do with the state of the world market at the time and decisions by the foreign companies to concentrate on their core activities. Though the South

Table 14.12. Product Segmentation in the Short-Term Insurance Market, 1990–2003

	(Rands Billions)													
	1990	1991	1992	1993	1994	1995	1996	1997	1998	1999	2000	2001	2002	2003
Property (Fire) Business														
Premiums written	1.13	1.35	1.47	1.80	2.12	2.40	2.67	3.00	3.13	5.15	5.59	7.63	9.36	10.95
(%)	18.44	18.13	18.05	19.22	19.94	18.89	17.58	17.36	16.56	28.45	30.23	29.00	30.00	31.00
Transportation (Marine) Business														
Premiums written	0.14	0.15	0.16	0.20	0.23	0.33	0.36	0.41	0.47	0.54	0.50	0.79	0.94	1.06
(%)	2.24	2.03	2.02	2.09	2.19	2.56	2.39	2.36	2.47	2.96	2.68	3.00	3.00	3.00
Motor														
Premiums written	2.42	3.06	3.28	3.80	4.22	5.11	6.16	6.96	7.33	6.68	6.94	8.95	11.23	12.71
(%)	39.59	41.00	40.43	40.50	39.79	40.26	40.55	40.22	38.86	36.92	37.52	34.00	36.00	36.00
(Personal) Accident and Health														
Premiums written	0.27	0.21	0.18	0.26	0.40	0.41	0.48	0.49	0.67	0.57	0.88	1.32	1.56	1.41
(%)	4.47	2.82	2.22	2.78	3.74	3.25	3.18	2.82	3.56	3.13	4.73	5.00	5.00	4.00
Guarantee Business														
Premiums written	0.09	0.11	0.13	0.15	0.17	0.19	0.28	0.32	0.38	0.38	0.44	1.32	1.25	1.41
(%)	1.49	1.45	1.65	1.61	1.61	1.48	1.82	1.84	2.04	2.09	2.37	5.00	4.00	4.00
Liability Business														
Premiums written	-	-	-	-	-	-	-	-	-	1.51	0.48	1.58	1.56	2.12
(%)	-	-	-	-	-	-	-	-	-	8.36	2.61	6.00	5.00	6.00
Engineering Business														
Premiums written	-	-	-	-	-	-	-	-	-	0.32	0.39	0.79	1.25	1.41
(%)	-	-	-	-	-	-	-	-	-	1.79	2.08	3.00	4.00	4.00
Miscellaneous Business														
Premiums written	2.07	2.58	2.89	3.17	3.47	4.26	5.24	6.12	6.89	2.95	3.29	3.95	4.06	4.24
(%)	33.77	34.58	35.64	33.80	32.72	33.57	34.48	35.39	36.52	16.31	17.77	15.00	13.00	12.00
Total														
Premiums written	6.12	7.45	8.12	9.38	10.61	12.70	15.20	17.30	18.87	18.10	18.50	26.31	31.19	35.31
(%)	100.00	100.00	100.00	100.00	100.00	100.00	100.00	100.00	100.00	100.00	100.00	100.00	100.00	100.00

Source: Registrar of Insurance (1990–1996) and Registrar of Short-Term Insurance (1997–2003).

Table 14.13. Market Share (%) of Short-Term Insurers by Gross Premiums Written, 1998–2003

	Company	Segment	1998	1999	2000	2002	2003	Cumulative Total
1	Santam	General	13.2	12.1	16.9	20.7	20.0	20.0
2	Mutual & Federal	General	12.0	11.6	11.2	13.3	14.0	34.0
3	Hollard	General	4.6	4.6	4.1	6.5	7.7	41.7
4	SA Eagle	General	6.6	6.4	6.1	7.4	7.4	49.1
5	AIG (SA)	General	1.8	1.8	1.9	3.9	4.0	53.1
6	Guardrisk	Captive	4.1		4.2	3.1	3.7	56.8
7	Lloyd's	General		4.4	5.2	3.5	3.4	60.2
8	Outsurance	General	0.0	0.3	0.4	0.8	2.7	62.9
9	RMB Structured (previously Quantum)	Captive	2.3	4.6	1.8	3.7	2.5	65.4
10	Auto & General	General		2.2	1.8	2.0	2.3	67.7
11	Regent	General	1.4	1.4	1.6	2.1	2.2	69.9
12	ABSA	Bank	2.1	2.1	2.0	2.1	2.2	72.1
13	Guardian National (taken over by Santam)	General	9.9	13.4	8.0	3.7	2.1	74.2
14	Mutual & Federal Risk	Captive		0.6	0.6	1.1	1.8	76.0
15	Santam Risk	Captive					1.8	77.8
16	Nova Risk	Captive		0.2	0.4	1.3	1.6	79.4
17	Stanbic	Bank	0.9	1.0	0.9	1.3	1.4	80.8
18	Credit Guarantee	Specialist	1.6	2.0	2.1	1.7	1.4	82.2
19	Consumer Protection	Specialist	0.7	0.7		1.2	1.2	83.4
20	Constantia	General	0.5	0.6	0.4		1.1	84.5
21	Compass	General	1.5		1.1	1.0	0.9	85.4
22	Lion of Africa	General	-	0.1	0.3	1.0	0.9	86.3
23	Sentrasure	General	1.4	1.4	1.2	1.6	0.8	87.1
24	Ace	Specialist		0.3	0.2	0.1	0.1	87.2
25	CGU (merger of Commercial Union and General Accident)	General		5.3	4.3	0.1	0.1	87.3
26	Outsurance Holdings	General	-	-	-	1.0		87.3
27	Allianz	General	2.5	0.9	2.4	0.2		87.3
28	Commercial Union (merged with General Accident)	General	5.2					87.3
29	General Accident (merged with Commercial Union)	General	2.0	0.9	0.7			87.3
	Total		90.9	95.4	95.3	95.4	87.3	

Source: Registrar of Short-Term Insurance (1998–2003).
Note: Market share is defined as the percentage of the gross premiums written in the market.

African operations may have been regarded as profitable—maybe even the most profitable in the various companies, in absolute terms the contribution to group profits was probably not material.

Santam which had 19% of the market in 1998, acquired the Guardian National which had 14% of the market to become South Africa's largest short-term insurer, over its longtime rival the Mutual & Federal, until then South Africa's largest insurer. The broker community is of the opinion that the market is now too small, and that insufficient competition exists. The presence of, AIG the American insurer, is regarded by the brokers as most welcome and timely. It has more than doubled its market share over the past few years and has overtaken Lloyd's. Lloyd's has had agents in South Africa since the early 1800s and is now also regarded as making a significant contribution to maintaining competition in the market.

The long-term industry makes use of tied agents and independent agents. As is clear from the tables on the long-term industry, a major portion of the premiums comes from corporations and these are usually serviced by the large corporate brokers, e.g., Alexander Forbes or insurers without the intervention of brokers. The South African short-term market is traditionally considered to be a broker dominated market, i.e., most of the business is placed via independent insurance brokers. The broker market includes Alexander Forbes, GlenRand MIB, Aon, and Marsh.

Recently a new short-term insurance distribution channel has developed, the independent underwriting manager (IUM). The underwriting manager accepts business on the basis of an insurer's license, pays claims, and shares the profits with the insurers. In some cases the insurer may have equity in the IUM. The insurer essentially authorizes the underwriting manager to accept risks, settle claims, and possibly even design new products without any reference to itself. The insurer hands over, so to speak, its underwriting pen to a third party. The exact relationship between the insurer and underwriting manager may differ from manager to manager and may not be clearly defined, in fact or law. What is clear however is that recently a large number of underwriting managers have been established. It is estimated that there are about 250 underwriting managers. The reason for their formation is not clear but may be related to the closing down a number of long-established insurers resulting in the laying-off of staff who could not find employment in the remaining companies. A consequence was that a large number of experienced insurance persons were unemployed and it would make sense for them to come up with a scheme to continue to provide a service within the insurance industry with which they are familiar. A further contributing factor is fixed commissions, the abolition of which has been under consideration and agreed to for a number of years, the implementation of which is continually delayed. An independent underwriting manager does not earn a commission. The agency makes a profit which is shared with the insurer and this may well be more attractive than working for a commission.

The forms of policies generally follow British wording, which has had considerable influence on the markets. Each insurer is free to adopt any wording it so chooses. The wording is not subject to approval by the FSB. The standard practice in the personal lines market is to put all polices together in one booklet (as in the case of Santam's Multiplex policies) and for the insured to select which coverage he or she desires (i.e. motor, public liability, house owners, and so on). Most premiums are paid monthly by way of a debit order. However, insureds can select to pay premiums annually in which event the insured will get a discount. Santam claimed credit for introducing this system (Vivian 2002, p. 120). A similar system, the origins of which are not related to the personal lines policies, is used for Small to Medium Enterprises (SMEs). In this case, the MultiMark policies are used. The wording of these policies was agreed to by brokers and insurers, to avoid the necessity of time consuming independent negotiations between brokers and companies. The negotiations on the MultiMark wording took several years to complete and has undergone several revisions. The responsibility for the project currently resides with the SAIA.

14.6.1 Conclusion Regarding the Short-Term Market

The general consensus is that the short-term market is currently in a healthy state, but as is well known, it is subject to cycles. The market goes through periods of negative underwriting profits, during which its overall profit comes from investment income. The negative underwriting profits can become quite severe, threatening the solvency of some companies. The point is reached when the negative profits can no longer be tolerated and premiums rise. Sometimes the rise is brought about by an event, such as the failure of an insurance company or at present, the events of September 11, 2001. Before September 11th, the South African market was in a negative phase. After September 11th, the major reinsurers suffered a loss. The Swiss-Re suffered its first loss in a century. This resulted in a steep increase in reinsurance premiums in South Africa. In many cases, insurance premiums to the public increased by in excess of 100%. At the same time there have been very few natural disasters resulting in lower than normal claims. As a consequence, the South African short-term insurance market is probably enjoying some of the best years in its history. Santam, South Africa's largest short-term insurer, announced a R 1 billion dividend. Once a sudden rise takes place, competition causes the market premiums to decline until the point of unsustainable underwriting losses is reached and the cycle starts all over again. The cycle can be seen by looking at the market's underwriting profit and the solvency. It is probably currently at the top of the cycle and heading downwards—again.

14.7 MEDICAL HEALTH FUNDING

The health system, including the funding thereof, has been undergoing a controversial major overhaul since the advent of the ANC government.[281] Some of the government's proposals which have been implemented were disputed and are currently before the courts.

The funding of medical expenses is another aspect originally tied to employment. Most large employers offer medical-aid to its employees but there is no legal obligation to do so. The employer establishes a medical-aid scheme which, because membership is confined to employees is referred to as a closed or restricted schemes. In terms of these schemes, employees receive medical treatment in the private sector from doctors and suppliers usually of their own choice. Certainly up to 1994, South Africa's private medical sector was regarded as being in line with the

---

[281] For an overview of some of the issues consult Reekie (1997).

best in the world, with world class doctors, hospitals, and treatments. There is virtually no delay in obtaining medical treatment for those on medical-aid schemes. Medical bills are forwarded to the employer's medical-aid department to settle. The cost of medical-aid is funded by contributions from the employer and employee (member). Medical-aid is not regarded as insurance and schemes do not fall under the insurance legislation for supervisory purposes, nor are they supervised by the FSB. Historically, medical-aid continued after retirement, initially at little or no cost to the former employee, now pensioner. This is fast disappearing, leaving pensioners with the additional problem of funding post-retirement medical costs. In some cases, employee organizations have taken employers to court to force them to continue to provide medical-aid after employment. Post retirement funding of medical expenses is a matter of increasing concern, for which no clear answer is forthcoming.

The medical-aid schemes are regulated in terms of the Medical Schemes Acts, the latest being the Medical Schemes Act 138 of 1998. Briefly, every medical-aid scheme must be registered with the Registrar of Medical Schemes and operate in terms of rules adopted by the scheme, but approved by the registrar. The Act lays down a number of provisions which the rules may and may not contain. The registrar reports to the Minister of Health and not the Minister of Finance. The Act requires, for example, that no scheme can be a profit making entity or have shareholders, and minimum benefits to be provided. Medical-aid schemes are thus mutual associations.

A new requirement of the legislation is the so-called community rating system which is applied to determine the levy. The idea of community ratings is that the community stands together, negating the idea of a risk rated premium. Schemes may not differentiate on the basis of age, gender, or health conditions. They may differentiate on the basis of income and number of dependents. There is thus a considerable degree of forced cross-subsidization. Members and their dependents can be covered by the scheme. With the passing of time, specialist medical-aid administration companies have developed. These include very large administrators such as Discovery Health and Medschemes. The system has expanded to include open schemes, i.e., schemes not tied to employment. Currently, more members belong to open schemes, than restricted schemes. Since medical schemes cannot risk rate, this could result in the levy of some funds being much greater than others, where the proportion of high risk individuals is greater. The government has proposed a risk equalization fund to cater for this possibility which has now been approved by the cabinet. The representatives of medical-aid schemes meet with representatives of the medical profession to establish a tariff of fees for services. Individual doctors are not always satisfied with the tariff rates and can contract-out of the tariff rates, in which event the member is liable for the difference.

The Council for Medical Schemes issues a detailed annual report. There are currently about 7 million beneficiaries of medical schemes, 5 million belong to open schemes, and 2 million to closed or restricted schemes. In 2003, the gross contributions to medical schemes was R 48 billion, compared to R 42.7 billion in 2002. The contributions are thus larger than the short-term premiums of R 35.6 billion. Since medical schemes cannot distribute profits to shareholders, profits (if any) are made by the fees charged by medical-aid administrators. A few years ago, the Registrar noticed that some of the schemes had incurred what he considered disproportionate re-insurance charges. He also noted that this happened when the administrator purchased reinsurance from an insurer within the same group of companies as the administrator. To put it succinctly, the administrator was thought to be gaining additional income via incorrectly priced reinsurance premiums—a transfer pricing problem. This led to a very public confrontation between the Registrar and some administrators. Currently reinsurance contracts are very carefully monitored. As a result of the monitoring, the net reinsurance costs (premiums less claims) decreased by 58.5% in 2003 compared with 2002. The reinsurance costs in 2003 constituted a mere 1.9% of gross contributions in 2003 compared to 5.1% in 2002. The Registrar decided that all schemes would have to increase their solvency so as to reduce reliance on reinsurance contracts for solvency purposes, leading to yet another public confrontation. As a consequence, medical schemes have never been as solvent as at present. The intention is that medical schemes must achieve 25.0% free reserves as a percentage of gross contributions by 2004. Once the 2004 results are published, it will be known if this objective has been met.

South Africa does not have a national health insurance scheme but it does have a public health system including hospitals funded out of current taxes. The public health system is regarded as inadequate.[282] Those who are not on medical-aid (the majority) and cannot afford medical treatment have access to public hospitals. It is of course in the interest of the government to shift, as far as possible, the burden of the provision of medical health to the medical-aid system. A number of new proposals are in the public domain. One proposal is that medical-aid members will pay 4.5% of their income to the medical-aid industry and this will be used to expand the medical-aid system to those who are not on the system. This essentially is to increase income taxes by 4.5%. To put it another way, the government would like to transfer some of its health funding obligations to the medical-aid industry. Currently, one of the few deductions allowed against income, for income tax purposes, is a limited deduction for medical expenses including the levies paid for medical-aid (§18 of the Income Tax Act 58 of 1962). It is proposed that these deductions be severely limited.

About fifteen years ago the insurance industry decided to enter the health market and offer medical insurance, which is additional to medical-aid. This caused the then registrar of medical schemes some considerable unhappiness and attempts were made to prevent insurers from entering the field. This resulted in tension between the two industries and regulators. At one time, the Registrar of Medical Schemes threatened to prosecute insurance companies, but this did not happen. It has taken some 15 years to come to an agreement on the demarcation between medical-aid and health insurance. Medical insurance is offered by both long- and short-term insurers and as can be seen from Tables 14.7, 14.7A, and 14.12 but currently does not constitute a major component of the insurance industry's business.

14.7.1 State Insurance Schemes

Five state insurance schemes are of importance, the Compensation of Occupational Injuries and Diseases (i.e., workers' compensation) (COID), the Compensation for Occupational Diseases in Mines Act, the Road Accident Fund (RAF), the

[282] Anecdotal examples may illustrate the point. An elderly lady developed gangrene as a consequence she had a limb amputated. The next day she was sent home and the family told they had were responsible for nursing her. They of course knew nothing about nursing – she died a few days later.

Unemployment Insurance Fund (UIF), and the South African Special Risks Insurance Association (SASRIA), an insurer established to pay for damages caused by politically motivated persons (i.e., terrorism insurance).

Workers' Compensation

South Africa has a compulsory workers' compensation scheme.[283] When mining started to develop, the mine owners decided in 1894, to establish a mutual insurance company to provide compensation for workers injured at work. This company, the Rand Mutual, is still in existence, covering certain mines. The scheme was thereafter extended to cover occupational diseases. Two categories of occupational diseases are considered. First, there are diseases associated to specific work where it is known that a link between the disease and the work (set out in the third schedule to COID). The second category is where it is not clear that the link exists. In the first case, it is presumed that the disease arose from the occupation and in the second, the employee must prove the connection. For example, if an employee works with lead and gets a disease associated with lead, the presumption is that the work caused the disease. If the employee is afflicted with a lead-related disease, but does not work with lead, the employee must prove causation. A second, much smaller, mutual insurance company was founded much later to cover employees in the building industry. By law, only a mutual insurer can offer workers' compensation and to do so requires the approval of the accident commissioner. Following the example of the mines workers, compensation was introduced to cover all wage-earning workers through a central state administered fund.

Workers' compensation initially only applied to employees earning amounts below a prescribed figure, but in 1993 the latest piece of legislation, the Compensation of Occupational Injuries and Diseases (COID) Act 130 of 1993 was passed and virtually all employees are now covered. The wage limitation no longer applies. Exceptions, for example include domestic servants who are excluded because of the difficulties in collecting the levy. The levy is based on the employee's salary, is risk-rated, and is paid exclusively by the employer.

Compensation is on a no-fault basis. It resembles personal accident insurance and not liability insurance. Industries with better than average claims experience, pay less than other industries with bad claims experience. Individual companies qualify for a rebate if their safety record is better than the norm for their industry. The employer is relieved of all common-law liability. However, if an employee is injured as a result of the negligence of a limited number of persons (the so-called vice employers) or a patent defect, then the employee can make an application to the commissioner for additional compensation.

Workers' compensation has operated successfully, until recently, for well in excess of a century, so much so that most employers are hardly aware of its existence. Very few, if any risk managers, include workers' compensation costs as part of their risk costs. Recent reports indicate that the system is in dire administrative straits with lawyers having to sue the fund to get it to process claims (Workers' Compensation 2003; Time to Scrap 2003; The Ills Surround 2004). The

---

[283] For a history of workers' compensation in South Africa consult Budlender (1984) and for a short overview Valsamakis, et al. (2005), p. 127 and the sources quoted therein.

head of the fund was suspended for maladministration. The latest matter to go before the courts concerned a backlog of claims which had arisen in an accumulation of cases where employers had omitted to submit accident reports. The matter was settled with the fund agreeing to clear the back log within the next eight months.

Table 14.14 indicates the income and expenditures of the workers' compensation fund and the number of reportable, compensable occupational accidents and diseases. The average year-on-year income rose at a rate of 12%, while expenditures rose at the rate of 16% and the number of reportable accidents by 2%. These trends are in line with the increased benefits recently introduced and the widening of the scope of compensation to include items such as hearing loss. The static number of claims is in line with the country's stagnant employment figures.

Table 14.14. Income and Expenditures for the Workers' Compensation Fund, 1994–2002

	(Rands Billions)								
	1994	1995	1996	1997	1998	1999	2000	2001	2002
Income	0.89	1.12	1.46	1.75	1.95	1.96	2.17	2.08	2.27
Expenditure	0.52	0.62	0.78	0.85	1.07	1.11	1.08	1.19	1.68
Surplus	0.37	0.50	0.67	0.90	0.88	0.85	1.09	0.90	0.59
Number of Accidents	241,856	268,513	266,159	266,159	289,952	269,259	242,126	223,615	280,631

Source: Valsamakis, et al. (2005), p. 138.

Occupational Diseases in Mines and Works Fund

This is not a well publicized fund, established in terms of the Occupational Diseases in Mines and Works Fund Act 78 of 1973. As the name implies, its purpose is to provide compensation for occupational diseases in mines and the works associated with the mines. Occupational diseases in mines and works have been subject to a large number of commissions of inquiry over the years, and these can be consulted to get a clear picture of occupational diseases in mines in South Africa. It was recently announced that this fund is actuarially insolvent. South Africa thus has two funds which provide compensation for occupational diseases. In the general industry, compensation is provided by COID, but in the mining industry, it is provided by the Occupational Diseases in Mines and Works Fund Act.

Road Accident Fund

South Africa has a non-contributory Road Accident Fund (RAF) which pays common-law damages to persons wrongfully and negligently injured as a consequence of a motor accident caused through the negligence of the driver, owner, or employee of the owner of the vehicle concerned. This is commonly referred to as Third Party motor liability insurance, or in this context simply Third Party insurance. Loss or damage to the vehicle itself can be insured in the short-term market. The RAF does not cover all liability claims, only bodily injury claims wrongfully and negligently caused. It does not for example cover damage to a third party's vehicle. It also does not cover all bodily injury claims; a number of exclusions exist. Liability claims for passengers in the vehicle of the negligent driver, for example, are covered for a very limited amount of R 25 000. Generally the balance can be insured in the short-term motor market.

From 1942, a road accident scheme has been on the books in South Africa (Motor Vehicle Insurance Act 29 of 1942). Because of the war, it only became operational in 1946. Since its inception, the system has had a variety of names, the current being the Road Accident Fund. The operation of the system has also changed over the years. Initially, drivers were simply required to purchase third party liability insurance from an insurer prepared to underwrite the coverage, at its own rates. In 1964, the legislation was amended to allow the Minister to set the premiums. Thereafter, insurers had to approach the Minister to increase the premiums and when, insurers approached the Minister for a 20% increase, he refused and some insurers withdrew from the market. In response, in 1966, the State President restricted involvement in third party insurance to consortium of 16 insurers and set-up a fund called the Motor Vehicle Assurance Fund to reinsure the risk. In 1969, the Act was amended to allow the state to take control of the Motor Vehicle Assurance Fund, but the consortium still consisted of the insurers and dealt with the claims. Insurance was sold, annually (all on the same day) and motorists were issued with a token which they had to display on their cars as proof of insurance. The government decided on the recommendation of a minority report to the Grosskopf Commission (1981) to fund the scheme via a levy on fuel which came in operation in May 1986. Tokens are no longer issued. The consortium nevertheless continued to handle claims as the agent of the fund. Claims were allocated to a consortium member depending on which day of the week the accident occurred.

It has been clear for a very long time that the fund was heading for financial disaster (Vivian 1984). The fund has been insolvent for a long time. Although the operation of the fund has been the subject of a large number of commissions of inquiry, the government appears to be incapable of solving this problem. The latest commission was the Satchwell Commission which took several years to complete its investigation and handed over its comprehensive report in 2002. Nothing has happened thereafter with regard to its recommendations, however a number of contested bills have been promulgated for comment.

The operation of the fund has been riddled with corruption for a long time. Even the chairman of the fund and the entire board was suspended and subsequently reinstated. In the latest development, the chairman was once again dismissed and the rest of the board is also expected to be fired. The corruption was identified several years ago by the Melamet Commission (1992). It was investigated by the Heath Commission, a standing commission specifically set-up to investigate corruption but its investigation into the fund was stopped by the constitutional court. Lawyers have a vested interest in the current system, and get by far the largest portion of the fund's income. The Heath Commission was eventually closed down by the government. The insolvent fund is running out of cash and it announced that it would suspend payments until it finds a resolution. The accumulated deficit was R 23 billion as of March 2003.

An unusual feature is that the fund is financed out of a levy imposed on the fuel. There is thus no separate premium paid by drivers or owners of vehicles. South

Africa appears to be the only country in the world which has adopted this scheme of funding. In essence, the scheme is now a pay-as-you-go scheme and the above deficit is largely the cost of outstanding claims. If the fund pays the claim, the person who caused the accident, i.e., the driver or the owner or the employee of the owner, is absolved from any liability. The wording of the absolving section is interesting. The absolution holds only as long as the fund pays the claim. Technically, after the fund announced it was not going to pay claims (it in fact is paying claims but at a slower rate), the absolution no longer applies. This point does not seem to have been noticed. Although the fund is a government fund, supervisory responsibility was imposed on the FSB, which was never keen to assume this responsibility. The Registrar in his 2003 Annual Report sums up the position as follows:

> "The reality is that the RAF is clearly on the verge of total bankruptcy and financial recovery seems impossible. The recommendations made in the Road Accident Fund Commission Report 2002 [Satchwell Commission] must be discussed at ministerial level and bold decisions must be taken to implement the Road Accident Benefits Scheme as soon as possible."

The state of the fund is not a surprise; its condition has been apparent for well in excess of a decade. The problem is that the government seems incapable of addressing the issues and the frustration of the Registrar, the fund's reluctant supervisor, can be understood.

Unemployment Insurance Fund (UIF)

The UIF is another government administered fund which was in financial difficulties. This has however been solved by extending the operation of the fund to cover high salary earners, knowing that they are unlikely ever to claim and capping the amount that they claim. This levy is simply a payroll tax on high earning persons. As a result, there has been a turnabout in the fortunes of the fund. The turnabout may also have been facilitated by the appointment of Dr. Sandy Sandrock from the private sector as the non-executive chairman of the audit committee. He was previously the CEO of a First National Bank's former insurance company and hence has experience in running insurance companies in the private sector. The UIF, unlike worker's compensation, was extended to domestic workers. The fund has more recently been in the news facing allegations of maladministration. However, these claims do not involve material amounts and do not impact on the solvency of the fund.

SASRIA (South African Special Risks Insurance Association)

SASRIA is the insurer formed to deal with claims caused by persons who are politically motivated. SASRIA was established after the 1976 Soweto Riots, in the face of settling claims from the riots. It was possible before the establishment of SASRIA to get a measure of coverage for damage caused intentionally in terms of the Riot, Strike, and Malicious Damage (RSMD) extension to property policies, but this proved to be inadequate after the Soweto Riots.

On the June 16, 1976, wide-spread rioting broke out in Soweto, the large black township at the South West boundary of Johannesburg (Soweto is derived from (So)uth (We)st (To)wnship). Previously, Soweto had been administered by the

Johannesburg City Council, at the time, largely an English-speaking administration. The government decided to establish separate administration boards and Soweto fell to the administration of the then newly established West Rand Administration Board, staffed, as was often the case as explained above, mainly with Afrikaners. The insurances of Soweto were then transferred to Santam, the Afrikaner short-term insurance company. Days after the transfer, the riots broke out and lasted for some time causing considerable damage to private property, administration property, and other government property including schools, libraries, and post offices. It was not clear if this damage was covered by the existing polices and it was hoped that clarity would be gained once the Cilliers Commission, a commission appointed to investigate the riots, tabled its report. The commission took a long time to complete its investigation and once released did not answer the insurance questions. It was clear that the existing RSMD extension was wholly inadequate when it came to dealing with damage caused by persons who are politically motivated.

The insurance industry decided to establish a section 21 company named SASRIA, which is a non-profit making company, to underwrite risks emanating from damage to property, including consequential losses (business interruption). SASRIA was structured on the basis of a risk pool. Each participating company committed a fixed sum of capital and the liabilities of the companies were limited to their capital commitments. If possible, reinsurance would be arranged and the government was approached first, to grant tax exemption to SASRIA, which was not unreasonable since SASRIA was a non-profit making company and second, to act as the insurer of last resort. The government agreed to both requests. To further limit SASRIA'S liability, it specified a maximum limit on coverage for any one insured (which was a source of contention). It did not insure loss of profits, the government being of the opinion it could not underwrite private sector profits. Loss of profits was insurable in the London market. SASRIA insured standing charges and increased cost of working only.

Since SASRIA was in a sense a collective company of the industry, the industry did not want to establish an expensive administration system for this company. So it was decided that most of the administration would be done by the existing conventional property insurance companies. The main policies, the conventional policies, would be the existing property policies. Each of these would henceforth contain a new exclusion, called the South African Insurance Association (SAIA) exclusion, excluding all claims for damage caused by persons who are politically motivated (including the usual war damage exclusion). SASRIA never issued a separate policy, but a policy subordinate to the conventional policy. This policy was called a coupon policy which covers most of the risks excluded by the SAIA exclusion (not the war risks of course). In a sense then, the SAIA exclusion created a hole in the conventional underlying property policy, much of which was then covered by SASRIA's coupon policy. To ensure that in a case of doubt as to whether the damage was caused by persons who are politically motivated or not, that SASRIA would be liable, the underlying conventional policy contained a reverse onus of proof term. If a loss occurred and it was not clear if it was caused by a person who was politically motivated or not, then the presumption is that the person is politically motivated and SASRIA becomes liable.

SASRIA as an insurer, unlike as is the case with other policies, cannot cancel coverage nor can it refuse coverage. The premium was excessive quickly outpacing the claims resulting in a substantial surplus and was not risk rated. The same premium applied throughout South Africa, irrespective of conditions prevailing at the risk location. The market referred to this as a flat-rated premium. Under these circumstances, the duty to disclose is meaningless; the same premium applies irrespective of the risk. It was possible to get a measure of political unrest coverage in other markets, such as the London market. These policies could however be cancelled. The problem of adverse selection arose. The public would purchase political unrest coverage in these markets when the risk was low. If unrest broke out these insurers could cancel coverage and the insureds would then obtain coverage from SASRIA, which it could not refuse. To overcome this problem, the Reinsurance of Damage and Losses Act 56 of 1989 was passed making it a criminal offence to purchase political damage coverage, except from SASRIA. There is no obligation to purchase political damage coverage, but if it is purchased it must be purchased from SASRIA. South Africa is one of the few countries where the mere purchase of insurance can constitute a criminal offense. The system, however, worked exceptionally well.

The South African government was never called upon as a reinsurer. It never paid a single claim, and after 1994 it became clear that the funds being held in SASRIA would not be used. By this time, it had accumulated several billions as free capital. The government nationalized SASRIA in terms of the Conversion of SASRIA Act 134 of 1998 and once so nationalized, paid itself a multibillion rand dividend. The future of SASRIA is currently unclear. Stripped of its assets, the government tried to sell it back to the industry, but there was not much enthusiasm for the purchase. In 2001, the government decided to hold on to SASRIA for a further period of five years and then reconsider the matter (SASRIA Shelves 2001).

14.8 CONCLUSION

The South African economy is the most successful economy on the African continent. The pensions and life markets played an important role in this achievement by channeling the nation's savings to productive assets. The latest statistics, however, indicate a loss of confidence in the life market as withdrawals increased and premium inflows have declined. The South African economic debate has for the past decade been dominated by the BEE policy. This is an ambitious form of affirmative action. The future of the insurance industry is greatly influenced by this debate.

14.9 REFERENCES

"BEE Helps Only Elite: Tutu," 2004, *Citizen* (December 24).

"Pepkor in a Pickle over Landmark Pension Fund Ruling," 2001, *Sunday Times Business Times* (December 23).

"SASRIA Shelves its Privatization Plans for 5 Years," 2001, *Business Report* (October 3).

"[Workers' Compensation] System of Compensation Mired in Incompetence," 2003, *Business Report* (July 7)

"Time to Scrap Workers' Compensation," 2003, *Business Report* (October 31).

"The Ills Surround Workers' Compensation," 2004, *Business Report* (July 26).

Abedian, Iraj. and Barry. Standish, 1985, "Poor Whites and the Role of the State: the Evidence," *South African Journal of Economics* 53: 141–165.

Basson, Dion, 2004 "Surrender—Why You Should Get Rid of Your Policies," *Finance Week* (November 29): 8–10

Bawcutt, Paul, 1997, *Captive Insurance Companies* (London: Witherby & Co. Ltd.).

Benfield, Brian. C., 1997, "South African Life Assurance Legislation," *South African Journal of Economics* 65: 568–594.

Birkmaier, Ulrike and Camille Codoni, 2004, "World Insurance in 2003: Insurance Industry on the Road to Recovery," *Sigma*, No. 3 (Zurich: Swiss Re).

Botha, Joubert., 2003a, "Review Article—Capitalism, Inequality, Poverty, and Professor Terreblanche—Part I," *South African Journal of Economics* 71: 556–594.

_____, 2003b, "Review Article—Capitalism, Inequality, Poverty, and Professor Terreblanche—Part II," *South African Journal of Economics* 71: 830–886.

Bruggermans, Cees, 2003, *Change of Pace—South Africa's Economic Revival* (Johannesburg, South Africa: Wits University Press).

Budlender, Debbie, 1984, "Workmens' Compensation," *South African Labour Bulletin* 9: 22–40.

Cloughton, David, 1985, *Riley on Business Interruption and Consequential Loss Insurances and Claims* (London: Sweet & Maxwell).

da Silva, Caroline, 2004, "Decapping of Commission, Disclosure, and *Spitzer v Marsh*,' *Cover* (December 2004): 6.

Davenport, Rodney and Christopher Saunders, 2000, *South Africa—A Modern History, 5th edition* (London: MacMillan Press).

De Kiewiet, C. W., 1941, *A History of South Africa—Social and Economic* 1975 reprint (London: Oxford University Press).

Dollery, Brian, 2003, "Review Note: A History of Inequality in South Africa, 1652–2002," *South African Journal of Economics* 71: 595–610.

Engels, Frank, 1999, "Insurance Markets in Africa," *insights* (Zurich: Swiss Re).

Financial Services Board, 1996–2004, *Annual Report of the Financial Services Board* (Pretoria, Republic of South Africa).

Franzsen Commission, 1970, *Third Report of the Commission of Enquiry into Fiscal and Monetary Policy in South Africa*, Paper No. RP87/1970, Pretoria, Republic of South Africa.

Giliomee, Hermann, 2003, *The Afrikaners—Biography of a People* (Cape Town, Republic of South Africa: Tafelberg).

Grosskopf Commission, 1981, *Commission of Inquiry into Certain Aspects of Compulsory Motor Vehicle Insurance*, Paper No. RP23/1985, Pretoria, Republic of South Africa.

Haken, Mark, 2004, "Too Much Legislation?" *Cover* (November 2004): 56.

Hartley, Wyndham, 2004, "Stats SA to Solve 'Second Riddle Economy,'" *Business Day* (December 13).

Hetherington, Bruce, 2005, "LOA Discussion Paper on Costs and Commission Structures for Long-term Insurance Savings Policies," LOA Circular 75–2005, (Johannesburg, Republic of South Africa)

Horwitz, Ralph, 1967, *The Political Economy of South Africa* (London: Weindenfeld & Nicolson).

Hunter, Rosemary, 2004, "Legislative Changes that Should be Made in Order to More Properly Reflect the Pension Promise that an Employer Makes to its Employees and to Better Protect Retirement Funding (A Personal View)," paper presented at the Institute of Retirement Funds conference, Cape Town, South Africa (September 14).

Hutt, William. H., 1964, *The Economics of the Colour Bar* (London: Andre Deutsch).

Johnson, R. W., 2004, *South Africa—The First Man, The Last Nation* (Johannesburg, Republic of South Africa: Jonathan Ball Publishers).

Jones, Stuart, 1996, *The Imperial Banks* (Pretoria, Republic of South Africa: Unisa Press).

_____ , ed., 2002, *The Decline of the South African Economy* (Cheltenham, United Kingdom: Edward Elgar Publishing).

_____, 2005, "Lessons from Affirmative Actions around the World," *South African Economic History Journal* 20: 131–138.

Katz Commission, 1994, *Interim Report of the Commission of Inquiry into Certain Aspects of the Tax Structure of South Africa* (Pretoria, Republic of South Africa).

Kenney, Henry F., 2004, "All Those Radical Capitalists—Sampie Terblanche's A History of Inequality in South Africa, 1652–2002," *South African Journal of Economic History* 19: 42–66.

Lacour-Gayet, Robert, 1977, *A History of South Africa* (London: Cassell London).

Lipton, Merle, 1986, *Capitalism and Apartheid* (Hants, England: Wildwood House).

Melamet Commission, 1988, *Report of the Commission of Inquiry into the Winding-up of the Short-Term Insurance Business of the AA Mutual Insurance Association Ltd.*, Pretoria, Republic of South Africa.

_____, 1990, *Report of the Commission of Inquiry into the Flow of Short-Term Insurance Premiums Out of the Republic and the Operations of Captive Insurers*, Paper No. RP53/1990, Pretoria, Republic of South Africa

_____, 1992, *Report of the Commission of Inquiry into the Affairs of the Multilateral Motor Vehicle Accident Fund*, Paper No. RP87/1992, Pretoria, Republic of South Africa.

Morris, Donald R., 1965, *The Washing of the Spears* (London: Sphere Books).

Muller, C. F. J., ed., 1990, *500 Years—A History of South Africa* (1986, third imprint 1990) (Pretoria, Republic of South Africa: JL van Schaik).

Mutual & Federal Insurance Co. Ltd. v Oudshoorn Municipality, 1 SA 419A (State, Court.1985).

Naidoo, Vino, 2005, "The State of the Public Service," in John Daniel, Roger Southall, and Jessica Lutchman, eds., *State of the Nation South Africa 2004–2005* (Cape Town, Republic of South Africa: HSRC Press), pp. 112–134.

National Treasury of South Africa, 2002, "Study into the Taxation of Retirement Savings," unpublished, Pretoria, Republic of South Africa.

O'Dowd, Michael, 1964, *South Africa in the Light of the Stages of Economic Growth* quoted in Merle Lipton, 1986, *Capitalism and Apartheid* (Hants, England: Wildwood House), p. 3

O'Meara, Dan, 1983, *Volkskapitalisme—Class, Capital, and Ideology in the Development of Afrikaner Nationalism 1934–1946* (Cambridge, United Kingdom: Cambridge University Press).

Organisation for Economic Development and Co-operation, 1998, *Maintaining Prosperity in an Ageing Society*, Paris, France.

Parloff, Roger, 2002, "The $200 Billion Miscarriage of Justice," *Fortune* (March 4): 145 (5): 154–.

Parsons, Raymond, ed., 2004, *Manuel, Markets and Money* (Cape Town: Double Story Books).

Reekie, W. Duncan. 1997, "The Economics of Health and the Health of Economics," *South African Journal of Economics* 65: 283–313.

Registrar of Insurance, 1990–1997, *Report of the Registrar of Long-Term Insurance—Annual Report* (Pretoria, Republic of South Africa: Financial Services Board).

Registrar of Long-Term Insurance, 1998–2003, *Report of the Registrar of Long-Term Insurance—Annual Report* (Pretoria, Republic of South Africa: Financial Services Board).

Registrar of Pensions, 1995–2003, *Reports of the Registrar of Pensions—Annual Report* (Pretoria, Republic of South Africa: Financial Services Board).

Registrar of Short-Term Insurance, 1998–2003, *Report of the Registrar of Short-Term Insurance—Annual Report* (Pretoria, Republic of South Africa: Financial Services Board).

RFR, 2004, "Retirement Fund Reform—A Discussion Paper," National Treasury Republic of South Africa, Pretoria, Republic of South Africa.

Romain, Ken, 1989, *Larger Than Life—Donald Gordon and the Liberty Life Story* (Johannesburg, Republic of South Africa: Jonathan Ball Publishers).

Rusconi, Rob, 2004, "Costs of Saving for Retirement—Options for South Africa," paper presented at the 2004 convention of the Actuarial Society of South Africa, Cape Town, Republic of South Africa, (October 13).

Satchwell Commission, 2002, "Road Accident Fund Commission," Pretoria, Republic of South Africa.

SC, 2004, "Savings Crisis—How Does South Africa Rate?" presentation to the National Treasury, Pretorial, Republic of South Africa (August 17).

Scannell, J. P. 1968, *Uit die volk gebore—Sanlam se eerste vyftig jare* (Cape Town, Republic of South Africa: Nationale Boekhandel Bpk).

Smith Report, 1995, "Report of the Committee on Strategy and Policy Review of Retirement Provision in South Africa," Department of Finance, Pretoria, Republic of South Africa.

South African Institute of Race Relations, 1999, *South African Survey 1999/2000* (Johannesburg, Republic of South Africa).

_____, 2004, *South African Survey 2003/2003* (Johannesburg, Republic of South Africa).

South African Reserve Bank, *Exchange Control Manual*, www.reservebank.co.za.

Sowell, Thomas, 2004, *Affirmative Action around the World—An Empirical Study* (New Haven, CT: Yale University Press).

Spies, S. Burridge, 1977, *Methods of Barbarism?* (Johannesburg, South Africa: Jonathan Ball Publishers).

Taylor Report, 2002 "Transforming the Present, Protecting the Future—Report of the Committee of Inquiry into a Comprehensive System of Social Security for South Africa," RP/53/2002, Department of Social Development, Pretoria, Republic of South Africa.

Terreblanche, Sampie, 2002, *A History of Inequality in South Africa 1652–2002* (South Africa: University of Natal Press).

Vivian, Robert W., 1984, "Identifying the Source of the MVA Rate Crisis," address at the Noschon Conference, Sun City, Republic of South Africa (May 8).

_____, 1995, *The History of the Mutual & Federal* (Johannesburg, Republic of South Africa: Mutual & Federal).

_____, 2002, *Morgan's History of the Insurance Institute Movement in South Africa* (Cape Town, Republic of South Africa: Francolin).

_____, 2004a, "Taxation of Retirement Funding—A Time to Revolt?", paper presented at the Institute of Retirement Funds Conference, Cape Town, South Africa (September 13).

_____, 2004b, "Taxation of Retirement Funding—A Time to Revolt?" (abridged version), *FANews* (October): 20–23.

Valsamakis, Anthony., Robert. W. Vivian, and Gawie. du Toit, 2005, *Risk Management—Managing Enterprise Risks* (Johannesburg, Republic of South Africa: Heinemann).

Wassenaar, Andries. D., 1989, *Squandered Assets* (Cape Town, Republic of South Africa: Tafelberg).

Williams, Walter E., 1989, *South Africa's War Against Capitalism* (New York: Praeger Press).

Webb, Arthur. C. M., 1992, *The Roots of the Tree—A Study in Early South African Banking: The Predecessors of the First National Bank 1838–1926* (Johannesburg, Republic of South Africa: First National Bank of Southern Africa).

14.10 STATUTES

- Broad-Based Black Economic Empowerment Act 53 of 2003
- Collective Investment Schemes Act 45 of 2002
- Companies Act 61 of 1973
- Compensation of Occupational Injuries and Diseases Act 130 of 1993
- Contingency Fees Act 66 of 1997
- Conversion of SASRIA Act 134 of 1998
- Employment Equity Act 55 of 1998
- Estate Duty Act 45 of 1955
- ECM (2005) *Exchange Control Manual* Pretorial, Republic of South Africa, South African Reserve Bank*l*
- Financial Advisory and Intermediary Services Act 37 of 2002
- Financial Intelligence Centre Act 38 of 2001
- Financial Services Board Act 97 of 1990
- Financial Services Ombud Schemes Act 37 of 2004

- General Law Amendment Act 8 of 1898
- Income Tax Act 58 of 1962
- Long-Term Insurance Act 52 of 1998
- Medical Schemes Act 138 of 1998
- Motor Vehicle Insurance Act 29 of 1942
- Occupational Diseases in Mines and Works Fund Act 78 of 1973
- Pension Funds Act 24 of 1956
- Policy Board for Financial Services and Regulation Act 141 of 1993
- Preferential Procurement Policy Framework Act 5 of 2000
- Promotion of Equality and Prevention of Unfair Discrimination Act 52 of 2002
- Reinsurance of Damage and Losses Act 56 of 1989
- Short-Term Insurance Act 53 of 1998

14.11 LEXICON

1820 Settlers. 1820 Settlers comprised the first significant immigration from England into South Africa after the British first arrived in 1795.

AIG. AIG is the American International Group.

ANC. ANC is the African National Congress.

BEE. BEE is the Black Economic Empowerment program, enforced by legislation, in terms of which the economy, in terms of equity, labor, suppliers, and products is to be converted into a black controlled and dominated economy. It is a form of affirmative action but vastly more ambitious, including the compulsory transfer of equity, directorships, procurement, and jobs to the majority group.

Boers. Boers are Afrikaans farmers and in its modern usage refers to Afrikaans-speaking whites.

CAR. Capital Adequacy Requirement.

COID. COID is the Compensation of Occupational Injuries and Diseases Act (i.e., workers' compensation) legislation.

Die Groot Trek. *Die Groot Trek* is The Great Migration of Afrikaners, frustrated with British rule, who migrated inland away from the Cape. They established the independent countries of Natal, Orange Free State, and Transvaal.

FAIS. FAIS is the Financial Advisory and Intermediary Services Act. In recently enacted legislation, FAIS will regulate intermediaries for the first time. However, the wording is so broad that it includes the giving of advice. Since virtually everyone can give advice, it is interpreted to include providers. The odd situation now exists that intermediaries as well as insurers have registered under FAIS. The legislation is administered by the FSB.

FICA. FICA is the Financial Intelligence Centre Act (FICA) 38 of 2001.

FSB. FSB is the Financial Services Board, the statutory body responsible for the supervision of most of the financial services in South Africa (banks and medical-aid schemes are two exceptions). The FSB was established after the collapse of the AA Mutual at the time one of South Africa's largest short-term insurers.

FSC. FSC is the Financial Sector Charter. One of many requirements is that 25% of the equity of financial services companies be transferred to blacks.

Formal Employment. South Africa has a high unemployment rate, in excess of 40%. This 40% would obviously die of starvation if they had no income at all. They therefore must earn an income in anyway which presents itself. They can for example sell flowers at the side of the road, guard cars in car-parks with the expectation that drivers of these cars will give them something for doing so. This type of informal work constitutes the informal sector. The formal sector is thus the opposite of this. It means regular jobs with stable employers. In recent years, the government has redefined unemployment, thus persons in the informal sector can be defined as employed. Persons who have given up all hope whatsoever of ever gaining employment can be excluded from the definition of unemployed and by so doing the unemployment figures look better than one would imagine. The government has also argued that persons employed in the informal sector contribute to the GDP and redefined GDP to include the contribution from the informal sector. The official definition of the formal sector is all businesses that are registered in any way and informal sector consists of those businesses that are not registered in any way. These businesses are run from homes, street pavements, or other informal arrangements. Reports on employment can be accessed from the website of Stats SA /www.statssa.gov.za.

GDP. GDP is gross domestic product.

IUM. IUM is an independent underwriting manager.

JSE. JSE is the JSE Securities Exchange South Africa, previously the Johannesburg Stock Exchange.

LOA. Life Offices Association.

Long-Term Insurance. Long-term Insurance is the name given to life insurance in South Africa. This form of insurance derives its name from the fact that the policies which fall under this class of business are usually of long duration, several years. This class of insurance business is regulated by the Long-Term Insurance Act which defines eight policies or lines of business (1) assistance policy, usually referred to as funeral business, (2) disability policy, (3) fund policy (i.e., retirement or provident funds), (4) health policy, (5) life policy, (6) linked policy, (7) market-related policy, and (8) sinking fund policy. A company wishing to transact long-term insurance business within South Africa must be licensed as an insurer and for each class of business.

Model C Schools. Model C schools are state schools that charge fees.

NATS. NATS is the abbreviation used to refer to the Nationalist Party, which was in political power from 1948 to 1994 when it was replaced by the ANC. It was mainly an Afrikaner political party.

Ombudsmen. When the AA Mutual Insurance company collapsed in 1986, the then State President appointed a judicial commission of inquiry under the chairmanship of Justice David A. Melamet to investigate the collapse. One of the recommendations was that the short-term industry should appoint an Ombudsman to investigate complaints from the public. This industry (short-term, long-term, and banking) accepted this recommendation and each voluntarily appointed an ombudsman, who is usually a qualified attorney, advocate, or retired judge. The agreement of the industry is that it would honor any determination made by the ombudsman. The system has worked very well since its inception. Some companies such as Sanlam have also appointed an internal ombudsman. The ombudsmen issue annual reports on the complaints received from the public and how these have been dealt with. The pension funds industry is different. An adjudicator is appointed in terms of the regulatory legislation. A similar procedure was followed with the promulgation of FAIS. The FSB decided to motivate regulatory legislation for Ombudsmen. It also did not like the gender specific nature of the word Ombudsmen and so the legislation is entitled Financial Schemes Ombud Schemes Act 37 of 2004 which has recently come it force.

Parastatals. Parastatals are government-owned companies, e.g., Transnet, which operates the state airline and state-owned railway network.

Rand. The Rand is South Africa's monetary currency.

RAF. RAF is the Road Accident Fund, the statutory fault based 'insurer' of third party bodily injury claims arising out of wrongful and negligent driving of a motor vehicle. These claims are funded out of a levy on fuel. No separate premium is paid, nor is a policy issued. The fund basically covers all liability claims for bodily injury caused to persons outside of the vehicle. A very limited coverage exists for passengers, the balance of which must be covered by the normal comprehensive motor policy. If the fund can pay the claim, the driver, owner, or employee of the owner is exempt from liability. The fund is currently hopelessly insolvent, but continues to pay claims as it receives funds. The fate of the fund is under consideration by the cabinet.

Rates & Taxes. Rates & Taxes is the name given to the land tax imposed by local authorities, the main source of income for local authorities.

Registrar. Each piece of regulatory legislation creates the office of Registrar. There is thus a Registrar of Short-Term Insurance, a Registrar of Long-Term Insurance, and a Registrar of Pension Funds. When there was a single piece of legislation governing both the long- and short-term insurance there was only one registrar, i.e., the Registrar of Insurance. In fact the registrar is always the same person, the CEO of the FSB. The structure of the legislation exists for historical reasons. Each registrar is obliged in terms of each piece of legislation to prepare an annual report on the industry, which is available to the public.

RSMD. RSMD is the Riot, Strike, and Malicious Damage (RSMD) extension to property policies. It provided a measure of coverage for intentionally caused damage. This coverage was inadequate where damage was caused by politically motivated persons and SASRIA was established to provide the more comprehensive coverage.

SAIA. SAIA is the South African Insurance Association, an association of insurance companies which represents the industry on matters of common interest. It was established in the 1970s after the demise of the Tariff associations which can be traced back to the 1800s.

Sanlam. Sanlam is an Afrikaner private sector life insurer.

Santam. Santam is an Afrikaner private sector property and casualty insurer.

SARS. SARS is the South African Revenue Service, South Africa's tax gathering authority.

SARB. SARB is the South African Reserve Bank, South Africa's central bank.

SASRIA. SASRIA is the South African Special Risks Insurance Association, the specialist insurer established after the 1976 Soweto riots by the short-term market to insure damage caused by politically motivated persons. By 1994, it had a massive surplus of billions of rands and was nationalized to allow the surplus to be taken by the government. Its future is uncertain.

SCA. SCA is the Supreme Court of Appeal.

Short-Term Insurance. Short-term insurance is roughly equivalent to general or property and casualty insurance in other parts of the world. This form of insurance derives its name from the fact that the policies which fall under this class of business are usually renewed annually (even monthly). This class of insurance business is regulated by the Short-Term Insurance Act which defines eight lines of business (1) accident and health policy. This covers the normal personal accident policy but as is well known the personal health policy also provides for medical costs. When the short-term insurance industry decided to enter the medical insurance field they did so via the personal accident policy. The reference to health recognizes this fact. Other lines include (2) engineering (e.g., machinery breakdown), construction policies, (3) guarantee policy, (4) liability policy, (5) motor policy, (6) property policy, (7) transportation policy, and (8) miscellaneous policies.

Soweto. On the June 16, 1976, wide-spread rioting broke out in Soweto, the large black township at the South West boundary of Johannesburg (Soweto is derived from (So)uth (We)st (To)wnship).

Standing Chargers. Standing chargers refer to part of that which is insured in terms of a standard business interruption insurance policy which insures (1) standing charges (also called fixed charges) plus (2) loss of profits before tax plus (3) increased cost of working occasioned by an insured event. SASRIA insures business interruption risks, but because the government was the reinsurer of last resort, it never insured loss of profits. For a more detailed discussion consult Cloughton (1985).

UIF. UIF is the government run Unemployment Insurance Fund. Its operation has recently been extended to cover all employees irrespective of their salary. It is funded by contributions from both employers and employees.

Uitlander. Uitlanders are mainly English-speaking foreigners who flocked into South Africa when gold was discovered. It literally translates to people from outside lands.

VOC. VOC is the Dutch East India Company.

Workers' Compensation. Workers' compensation is provided in terms of the Compensation for Occupational Injuries and Diseases Act and workers' compensation is increasingly being refereed to as COID.

Zulus. In the late 1800s, the Zulus constituted the independent kingdom of Zululand. Today, after the unification which established the current boundaries of South Africa, they are more correctly regarded as South Africa's dominant (by numbers) black tribe.

15 Recent Developments in the Brazilian Insurance Market

Andrea Levy

Icatu-Hartford, Institute of Financial and Actuarial Risks Management at the Pontifical Catholic University (IAPUC)

Fernanda Chaves Pereira

Institute of Financial and Actuarial Risks Management at the Pontifical Catholic University (IAPUC)[284]

15.1 INTRODUCTION

The Brazilian insurance industry has grown rapidly over the last ten years. From a historical level of around 1 percent of the country's gross domestic product (GDP), premiums have doubled to around 2 percent in 1994 and, since then, have been increasing annually, reaching 3.4 percent of GDP in 2004.[285] The jump observed in 1994 is associated with a new economic regime that created a new currency, called the Real, and the growth that followed was driven mainly by the newly created defined contribution pension plans.

With the new economic policy, the Brazilian government became concerned about the formation of long-term savings and began to develop tax incentives, such as deferred income taxes of up to 12 percent, for individual income investment in pension plans that were instituted in 1995. In 1998, the Brazilian legislature also began to regulate programs with defined contributions as well as those with defined benefits. These factors permitted the explosive development of the market for defined contribution pension plans that managed to transfer investments from

---

[284] The authors would like to thank Professor. Luiz Roberto A. Cunha and Ms. Daniela Mendonça for valuable remarks and IAPUC researchers Eduardo Bevilaqua, Felipe Tâmega, Leonardo Czerwinski, and Marcos Guedes, who helped with some data issues. We are also indebted to J. David Cummins and Bertrand Venard for perceptive comments on earlier drafts.

[285] This number does not include pension funds and in the health private insurance system it only includes health insurers as explained later in the text. If pension funds and all private health contributions are included, the percentage would sum 0.8% and 1.3% GDP, respectively, reaching a total of 5.5% GDP (FENASEG 2005).

traditional financial assets to the open pension industry in response to the tax incentives given to pension products.[286] How much of the amounts transferred were really focused on pension benefits remains to be seen, but a major recent change in legislation will clarify this issue: customers now can be penalized if they make withdrawals from the funds before accumulating ten years with the contract. In addition, Brazil has not yet undergone thorough pension reform and public pension plans are still on a pay-as-you-go system and heavily in debt. Thus, it is likely that further growth will occur in the private pension industry.

Reinsurance is still a state monopoly in Brazil, but liberalization is under discussion. Liberalization is expected to increase the participation of foreign investors, continuing the deregulation of the insurance market that began in 1996.

This chapter starts with a general description of insurance markets in South America with a focus on Argentina and Chile. The discussion continues with a description of the Brazilian economic system and its background, followed by an explanation of the public pension system. The chapter concludes by analyzing the insurance market, its players, and their different lines of business.

15.2 THE SOUTH AMERICAN INSURANCE MARKET

South America has managed to overcome high rates of inflation and, today, in most countries inflation is in the single-digit range. The countries in the region have spent the last decade organizing their non-governmental institutions and recovering from the dictatorships previously ruled them during the second half of the last century.

Although many South American countries are now democracies, South America lost much of its appeal to foreign investors with the economic crises in Argentina and Brazil a few years ago. Compared to Asia, South America had lost some of its luster, primarily due to the high level of debt among South American countries, bureaucracies that make it difficult to open and operate local businesses, and low levels of education compared to Asia. Furthermore, the South American economy was very sensitive to the globalization trend and was strongly impacted by the Russian and Asian financial crises. Basic economic data for all South American countries are presented in Table 15.1.

South America has two regional associations that include Brazil: (1) the Mercado Comun del Sur or Mercosur, which includes Argentina, Brazil, Paraguay, and Uruguay and (2) the Free Trade Area of the Americas (FTAA) which includes all countries in the Americas except Cuba. Both regional associations are multi-country pacts to improve the business environment. Although it has a smaller number of countries, Mercosur is broader in the sense that it not only pursues a better environment for business, but also extends its activities to a discussion about a possible single currency for its member countries.

Brazil is by far the largest country in terms of population and GDP in Mercosur while Argentina is the second in terms of GDP. Consequently, the relationship between Brazil and Argentina is of great importance to Mercosur and has, at times,

---

[286] For instance, federal bonds or savings accounts.

Table 15.1. General 2004 Economic Data from South American Countries

	Brazil	Argentina	Venezuela	Colombia	Chile	Peru	Ecuador	Uruguay	Bolivia	Paraguay	Suriname	Guyana
GDP (Current US$ Thousands)	$604,855,100	$151,501,200	$109,321,900	$97,383,930	$94,104,940	$68,394,960	$30,281,500	$13,138,380	$8,773,135	$7,127,176	$1,108,983	$785,733
GDP Growth (Annual %)	5.18	8.98	17.33	3.96	6.06	5.07	6.61	12.27	3.58	2.9	4.61	1.55
Agriculture, Value Added (% of GDP)	5.18	9.64	4.48*	12.58	8.81	10.07	7.46	12.8	14.85 *	27.24*	11.09**	30.82**
Industry, Value Added (% of GDP)	17.15	31.54	41.11*	0	34.3	29.74	29.93	27.3	30.09*	24.23*	19.58**	28.58**
Services, Value Added (% of GDP)	77.67	58.82	54.41*	87.42	56.89	60.19	62.61	59.9	55.06*	48.53*	69.33**	40.59**
Total Population	178,718,400	38,226,050	26,127,000	45,300,000	15,956,000	27,546,700	13,213,080	3,399,400	8,986,396	5,781,569	442,968	772,056
GNI per Capita, Atlas Method (current US$)	$3,090	$3,720	$4,020	$2,000	$4,910	$2,360	$2,180	$3,950	$960	$1,170	$2,250	$990
Population Growth (Annual %)	1.19	0.80*	1.75	1.59	1.15	1.46	1.56	0.57	1.94	2.42	1.10	0.41
Inflation, GDP Deflator (Annual %)	8.09	9.19	30.90	7.01	8.01	5.42	4.42	7.03	8.49	6.66	9.08	6.73
Life Expectancy at Birth (Years)*	68.72	74.46	73.87	71.93	76.44	70.01	71.05	75.38	64.14	70.95	70.37	62.18
Infant Mortality Rate (per 1,000 Live Births)*	33.00	17.00	18.00	18.00	8.00	26.00	24.00	12.00	53.00	25.00	30.00	52.00
Surface Area (Square Kilometers)	8,514,880	2,780,400	912,050	1,138,910	756,630	1,285,220	283,560	176,220	1,098,580	406,750	163,270	214,970

Source: The World Bank Group (2005).
Notes: GDP is gross domestic product. GNI is gross national income. * 2003 data. ** 2002 data.

been the one of the principal obstacles to progress within Mercosur. At various times since 1999, economic crises in Brazil and Argentina have also created crises in Mercosur, thus increasing trade tensions. For example, after Argentina fixed its exchange rate to the U.S. dollar in 1991, its products ceased to be price competitive (i.e., they suddenly became more expensive compared to products from other regions). Furthermore, Argentina was strongly impacted by the financial market crisis of 1997 when capital fled the developing countries. Then, in 1998, when Brazil lost half of its exchange reserves (a total close to US$70 billion) and had to devalue its currency, Argentina's international trading price became even more expensive, starting what has been called the "Mercosur crisis." Argentina then demanded additional compensation from Brazil to offset diminishing exports. Finally, in 2001, Argentina reverted to a floating exchange rate with the U.S. dollar. Only then was the synchronization of economic politics, including the establishment of a single currency, reintroduced into Mercosur discussions. This process of recovery is crucial to the success of Mercosur.

Regarding the South American insurance market, the gross written premiums (in US$ billions) of South American countries are shown in Figure 15.1, with the percentage of GDP shown in parentheses.

Many foreign insurers use Brazil or Argentina for their bases in South America. Some Brazilian insurers are opening subsidiaries in Argentina, but little data is available on the total investment from Brazilian insurers outside of Brazil. There are at least seven Brazilian companies working in the Southern Cone: five in Argentina, three in Uruguay, and one in Paraguay.[287] Some Brazilian companies have been established as reinsurers in Argentina.

Another development in the South American insurance industry was the 1995 implementation of Carta Verde—obligatory auto insurance for travel in the Mercosur countries. It is a third-party liability automobile coverage (bodily injury and material) valid only for accidents that occur outside the insured's original country, but in Mercosur countries. The maximum coverage for bodily injury is US$40,000 per person up to a limit of US$200,000. For material damage, coverage is US$20,000 and is limited to US$40,000. Coverage is usually for one year. In order to proceed to payment, there should be a cross-border agreement among insurance companies so that payments are in local currency. In Brazil, only five insurers offer this coverage. Their total written premiums are R$1.1 million in 2004 and they have a 0.02 percent loss ratio.

Because Chile, Argentina, and Brazil are the largest markets in terms of premiums in South America, we will provide details about the markets in Chile and Argentina as well as for Brazil. Non-life and Life insurance written premiums for these countries are show in Table 15.2 for years 1995–2003.

[287] The Southern Cone or Cone Sul consists of the southernmost areas of South America. Typically, it refers to Argentina, Chile, Uruguay, Paraguay, and southeastern Brazil.

Source: FIDES (2001), Mercosur (2004), and FENASEG (2005).

Notes: * Data as of December 2003, except data for Argentina which are as of June 2003; Ecuador, Colombia, and Peru, which are as of 2001; and Venezuela, which are as of 2002. Brazil also includes capitalization. N/A is not available.

Figure 15.1. South American Premium Revenues (% of GDP), 2003*

15.2.1 Argentina

At the end of 2003, Argentina had US$2.9 billion in written premiums (down from US$6.9 billion in 2001), of which 33 percent was automobile insurance, 31 percent was life insurance and pensions, 9 percent was workers' compensation, and 8 percent was fire insurance. Since 2001, the number of companies dropped by 14 percent, reaching 198 at the end of 2003. Values from 2003 are highly influenced by the currency devaluation and the very high loss ratio, which reached 97 percent (up from 69 percent in 2001). The high loss ratio can be attributed primarily to the life insurance business which experienced a large drop in premiums. The loss ratio for the life insurance business reached 206 percent in 2003 compared to 75 percent in

2001. Alternatively, the loss ratio in the non-life business reached 72 percent in 2003 compared to 66 percent in 2001.[288] The extended combined ratio reached 97.6 percent in 2003, which was an improvement from 2001, when it was 99.7 percent (Mercosur 2004). These values are high compared to the Brazilian values, as described later in this chapter.

Table 15.2. Written Premiums in Argentina, Brazil, & Chile (US$ Billions), 1995–2003

	Argentina		Brazil		Chile	
Year	Non-Life	Life	Non-Life	Life	Non-Life	Life
1995	3.69	1.28	11.68	2.47	0.83	1.30
1996	3.03	1.25	12.34	2.89	0.82	1.58
1997	3.42	1.51	13.90	3.11	0.83	1.84
1998	3.74	2.01	13.78	2.99	0.76	1.71
1999	4.06	2.33	9.27	1.94	0.68	1.87
2000	3.94	2.40	10.38	2.13	0.72	2.14
2001	4.15	2.71	8.50	1.80	0.77	2.04
2002	3.74	2.12	7.85	2.43	0.91	1.78
2003	2.02	0.92	8.23	3.93	0.94	2.01

Source: MercosulMercosur (2004) and Superintendencia de Valores y Seguros (2004).

Given the dimensions of the Argentine economic crisis of 2001, the insurance market reacted fairly well. Before 2001, Argentina had considerable foreign investment and the market's conditions were amenable to further growth. Inflation had been under control since the beginning of the 1990s, and the insurance sector had been gradually deregulated with a liberalization of insurance prices. Additionally, an important pension reform was implemented that encouraged the development of the life insurance business. Furthermore, the national reinsurer Instituto Nacional de Reaseguro (INdeR) lost the reinsurance monopoly and the government implemented requirements to improve the solvency of the market. In fact, when INdeR was finally liquidated in 1992, its unmet liabilities had to be funded by the Tax to Finance INdeR, which charged 8.3 percent on all premiums collected in the insurance market. With so many changes in the market, the number of companies decreased in the beginning of the 1990s. The decline in the number of companies "was an expected and controlled process [due to] many factors that had affected the non-life sector, such as. . .greater competition, new legislation and the insolvency of the INdeR which previously had a role of maintaining many insolvent companies" (Ministerio de Economia y Producción 2004). Although the total number of companies remained unchanged from 1990 to 1996 due to the introduction of new life insurance companies, the number of insolvencies reached 70 in the same period.

[288] Loss ratio data are not available by line of business.

The Argentine market was fully opened to foreign companies in 1998. Presently, more than half the companies have foreign participation. Prior to 1998, only foreign companies that offered credit, property, and workers' insurance were allowed to create a subsidiary. To sell any other insurance products, foreign companies had to obtain licenses from existing firms. After the financial crisis in 2001, many foreign companies left Argentina and subsequently, national companies began to acquire the subsidiaries of international companies.

Today, as in Brazil, the distribution and sale of insurance remains in the hands of brokers, who impose strict conditions on companies. The regulations are still complicated however since different governmental bodies supervise the insurance and pension companies. Superintendencia de Seguros de la Nación (SSN) is the principal regulatory agency responsible for most types of insurance, but there are separate bodies for health, worker's compensation, and pension funds under different government ministries.

Argentina also underwent profound pension reform when workers' compensation was privatized. The reform took place in 1994 (the same year as in Colombia and three years prior to pension reform in Mexico) and included employees and the self-employed. The reform program defined a first and second tier. The first tier is a minimum pension funded by the state, employers, and employees and the minimum pension is at least 28 percent of average wages after at least 30 years of contribution. The second tier is funded by employees and is administered by Administradoras de Fondos de Jubilaciónes y Pensiones (AFJP). Upon retirement, an employee can choose among many types of benefits ranging from a life annuity to a combination of lump sum and term annuities that are usually managed by an insurance company.

15.2.2 Chile

Since early reforms in 1981, Chile has become an interesting example of pension reform. Pension fund assets represent the equivalent of approximately 52 percent of GDP in 2001. These reforms are responsible for the high percentage of the life insurance business in the country. In December 2003, Chile had US$2.9 billion in written premiums, of which 67 percent originated in the life and pension business (45 percent pension and 22 percent life insurance) compared to 21 percent in 1980. This value does not include the amount paid to the Administradoras de Fondos de Pensiones (AFP) which are the main administrators of the funds derived from the contributions to the pension system (Mercosur 2004).

The Chilean pension reform was obligatory for employees. The reform program defines a first and second tier. The first tier is a minimum pension funded by the state, of at least 70 percent of the minimum wage after a minimum 20 years of contribution. The second tier is funded by employees and is administered by the AFP. Upon receiving the pension, the employee can alternatively opt for annuity benefits that are managed by insurance companies.

The Chilean insurance regulator is the Superintendencia de Valores y Seguros, which supervises the entire financial market. Twenty-three percent of the market is non-life insurance, of which the largest line of business is fire insurance (including terrorism and earthquake) and the second largest is auto insurance. The percentage of premiums accounted for by terrorism insurance increased from 2002 to 2003 but decreased in 2004. As of September 2004, the non-life market retention rate was 51 percent. This relatively low rate is mainly due to the extremely low retention rate of 30 percent in fire insurance.[289] The average loss ratio is 53 percent for the non-life market, with the highest loss ratio in auto insurance (74 percent) and the lowest in fire insurance (37 percent).

15.3 THE DEVELOPMENT OF BRAZILIAN INSURANCE

In order to better understand the development of the Brazilian insurance industry, it is important to have some historical background on the financial sector as a whole, the insurance market specifically, and the Brazilian population structure. A brief description of these three elements is presented in this section.

15.3.1 Economic Perspective

By the end of 2004, Brazil had one of the highest real interest rates (around 10 percent per year) in the world,[290] a continuation of an economic program that began in 1994 with a new economic program called the Real Plan. The Real Plan had a positive impact on the country, keeping inflation under control after decades of high and hyper-inflation. The high rates of inflation influenced the insurance sector in many ways: it was hard to anticipate the expected value of pensions for long-term planning and insurance had to create the means to maintain the value of goods. It was with the implementation of the Real Plan that the open pension market experienced a boom.

Beginning in 1980, Brazil's economic performance can be divided into four periods. From 1980 to 1984, at the end of the military regime, the economy was characterized by accelerating inflation, increased use of indexation, and a breakdown of the model of development, all resulting in the worsening of social disparities and increased external dependency. On the one hand, Brazil had a very low rate of savings and, on the other, a very sophisticated financial system in the context of high rates of inflation. From 1985 to 1993, the return to democracy took place in the context of successive economic crises. The failure of various stabilization policies led to strong distortions, legal uncertainties, and fiscal debts ("skeletons") that continued to have an impact on national accounts.[291] A policy of trade liberalization and a privatization process began to be implemented after 1990. From 1994 to 1998, the Real Plan was implemented, named after the new currency that was adopted and is still in place today in Brazil. Although this plan is a landmark and has achieved a substantial reduction in inflation, the costs of stabilization have become more

[289] The retention rate is the proportion of premiums retained by the primary insurer rather than being ceded to the reinsurance market.

[290] Federal bonds had yields of 17.75 percent per year at the end of 2004 (BACEN 2005), while inflation during 2004 was 7.6 percent, as measured by Indice Nacional de Preços ao Consumidor Amplo (IPEA 2004).

[291] Skeletons are non-recognized private or public debts.

evident, because it is no longer possible to finance the fiscal imbalance through high inflation. Furthermore, external economic crises limited internal growth and social inclusion.

From 1999 to 2004, after Brazil abandoned the semi-flexible exchange rate regime, the new economic policies were based on an inflation targeting model, floating rates of exchange, and stricter fiscal measures focused on enforcing government fiscal discipline.[292] The continuation of sound macroeconomic policies by the new left-wing government in 2003 signaled not only a consolidation of democracy in Brazil, but also the possibility of growth as long as the government is able to take effective measures to foster internal savings by taking into account the continuity of structural reforms that were implemented in 1994.

Inflation rates, although high, remained relatively stable during the first half of the 1980s—around 10 percent per month. This underscores the fact that Brazilian inflation was different from inflation experienced abroad, a result of inertia derived from indexed salaries, exchange rates, prices, and contracts—otherwise called *correção-monetária* (freely translated as "monetary correction"). In this context, inflation was supposed to be painless thus resolving the irreconcilable tradeoff between development and stability. Notwithstanding, the near neutralization of the impact of inflation through indexation of the economy was not costless, since this policy generated a high influence of past inflation on current inflation.

After the oil shocks of 1973 and 1979, it was clear that with inflation in continuous acceleration, neutrality could only be maintained at the cost of increasingly higher indexation. Because the economic recession of the first half of the 1980s had not tamed inflation, in the second half of the decade, macroeconomic policy focused on stabilization. Because of the importance of Brazil's inflation process, the so-called heterodox stabilization plans were adopted. Until the advent of the Real Plan, various stabilization plans contained heterodox elements such as price, wage, and, in some cases, exchange rate controls: the Cruzado Plan (1986) was followed by the Bresser Plan (1987), the Summer Plan (1989), the Collor Plan I (1990), and, finally, the Collor Plan II (1991).

The learning process in terms of how to behave in an environment of high inflation caused economic agents to adapt. The consequence was that each unsuccessful stabilization plan was followed by an upsurge in inflation, leading to a level higher than the one that triggered the stabilization plan and consequently, creating legal uncertainty, and fiscal debts or "skeletons." This was a result of the defensive behavior of those agents who desired to protect themselves against new distortions that might be generated by future stabilization plans. Consequently, inflation thus gained a self-fulfilling character. After having stabilized at high monthly rates during the first half of the 1980s, inflation rates embarked on an upward trend that culminated in the hyperinflationary process in 1989, as the first popular vote election of the president since the military took power in 1964 drew near.

Inflation was curbed only with the Real Plan in 1994. Although this new stabilization plan contained some heterodox elements, like the Bresser and Summer Plans, it was a hybrid plan. Instead of freezing prices, it adopted a currency proxy

[292] Discipline in the sense that expenses should not exceed the amount collected.

that had the function of synchronizing price readjustments. Furthermore, the Real Plan also tried to restore the budget balance by implementing several measures in this direction.

In most past stabilization plans, policymakers believed that inflation could be controlled without recession, but Real Plan did not. The main orthodox element of the plan relied heavily on an overvalued fixed exchange rate and high interest rates. The first element had a double function. First, it had a psychological effect, making people believe that the new currency (Real) would finally be reliable. Second, it was pegged to the dollar at an overvalued rate in order to foster imports to meet initial higher demand. Further, the overvalued exchange rate ushered in a competitive shock. In turn, high interest rates would attract foreign capital to fund the current account imbalance caused by the overvalued exchange rate. Finally, in the fiscal area, a so-called Social Emergency Fund was implemented to reduce the portion of funds earmarked for social areas in the federal budget thus giving the federal government greater possibilities budget deficit reduction.

The Real Plan was a tremendous success. Stabilization was rapidly achieved and policymakers were able to control inflation even after the inflation targeting regime was adopted in January 1999, when it was clear that the overvalued exchange rate was unsustainable in the wake of the Asian (1998) and Russian (1998) crises. Table 15.3 shows Brazilian inflation and economic performance over the last 25 years. However, even with inflation relatively under control, economic growth was below its historical average (Note that, in 2004, Brazil had a strong recovery, with a GDP growth rate of 5.3 percent). The Real Plan replaced an economic system based on high inflation with one based on high interest rates. Even after the reversal in the trade balance, which resulted in very high trade surpluses in 2003 and 2004, neither economic growth nor social improvements have been launched on a sustainable basis. The reason that economic growth has remained below expectations seems to reside in the chronic lack of internal savings.

Table 15.3. Brazilian Inflation and Economic Performance (Annual Average Growth), 1980–2003

Period	GDP Growth (%)	Consumer Prices Growth (%)
1980–1984	1.53	131.47
1985–1993	2.51	735.11
1994–1998	3.21	70.91
1999–2003	1.78	8.65

Source: Instituto de Pesquisa Econômica Aplicada (2004).
Note: Consumer price is measured by *Indice Nacional de Preços ao Consumidor Amplo (IPCA)*.

Indeed, the above-mentioned inflationary process is closely connected to the lack of internal savings, which is a *sine qua non* condition for the development of an underdeveloped economy. For decades, the Brazilian inflationary process was viewed by many as a quick and easy way to increase development, even though this solution is not a costless process, as proven by Brazilian history (see below).

Increasing total savings via current account deficits was only considered with the Real Plan. Previous policies were based on the premise that industrial growth could be encouraged via import substitution strategies in the military government. External savings were artificially generated via government-controlled domestic savings, thus creating a fiscal system concentrated on forced savings. The paradigm of this model was the high social insurance cost that provided the government with the funds it needed for public investment plans, since many new taxes were created as social funds, but were used mainly to finance this policy. Today, Brazilian employers still suffer high employment costs that must be paid to the government, which can amount to as much as 100 percent of the real paid salary of an individual.

However, the costs of this strategy were postponed. The social insurance costs have already risen and the costs of inflation can be measured through the poor distribution of income generated by Brazilian development. The domestic savings rate is still insufficient and there is no model to increase private savings to replace the old model of government-controlled forced savings.[293] For instance, in an international comparison, Brazil's domestic savings rate was 17.5 percent of GDP in 2000, while Korea's was 35 percent and Chile's was 25 percent. The Brazilian savings rate is still very low, although it has increased since 2001, reaching 23.2 percent of GDP in 2004 (Instituto de Pesquisa Econômica Aplicada 2004). As shown in Figure 15.2, although the domestic savings rate has recovered from its 1999 level, it is still low from a historical perspective. Similarly, the investment rate is also low from a historical perspective.

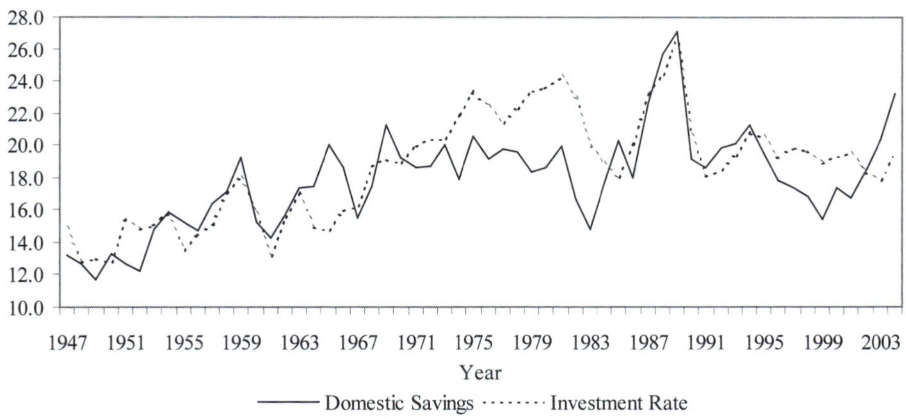

Source: Instituto de Pesquisa Econômica Aplicada (2004).

Figure 15.2. Brazilian Domestic Savings and Investment Rates (% of GDP), 1947–2004

[293] The domestic savings rate is defined as the total national amount in gross income not spent as percentage of GDP.

15.3.2 History of the Insurance Industry in Brazil

Insurance is one of the oldest fully regulated economic activities in Brazil, beginning among the Jesuits in the sixteenth century with Father José de Anchieta who funded "means of assistance" without a proper policy by using mutualism (Povoas 2000). The first published law on insurance was a document from the King of Portugal in 1791 that officially started insurance activity in Brazil. This law submitted not only Brazil, but all Portuguese colonies, to the regulations of the Portuguese regulator.

In 1808, the first Brazilian insurance company, Compania Boa-Fé, was founded in Bahia and was mainly for international marine traders. Following Brazil's 1822 independence from Portugal, an 1850 local law called the Código Comercial Brasileiro provided regulations for Brazil's marine insurance market. By this time, there were three local companies established in Brazil. The regulations for inland marine and life insurance took more than 60 years to be published finally in 1916. With the lack of local regulation, Brazil continued to follow Portuguese laws.

In 1862, international insurance companies began to open local subsidiaries. The pension system started in the nineteenth century. The first pension company, Mongeral, was funded in 1835 by civil servants and still operates today. This was highly significant since it was not until 1923 that the social pension security system was created. At the end of the nineteenth century, the Brazilian insurance market was fully developed, with all the necessary regulations. In 1910, Brazil had 23 local and 6 foreign insurance companies.

In 1937, the Brazilian constitution mandated that all international insurance companies and banks must be transformed in locally-owned enterprises. Under the constitution, no company could be fully-controlled by foreign partners and local majority participation was obligatory. At that time, a high level of solvency was required. Following the nationalization of the insurance industry, the reinsurance company Instituto de Resseguros do Brasil (IRB) was created as a monopoly in 1939. It was half-owned by the government and half-owned by the insurance industry. However, the government held more than half of the voting shares and retains control to this day. Since then, all insurance companies have been obliged to have IRB as their reinsurer, which shares the risk with the entire Brazilian market through retrocession. The IRB also shared the regulation of the insurance market with Superintendência de Seguros Privados (SUSEP) controlling, for instance, coinsurance which is still widely used in the Brazilian market.

Around 1960, when Brazil was under a military dictatorship, the insurance and pension laws underwent another major change that persists to this day. Workers' compensation was transferred from private companies to the government. Foreign investment in insurance was allowed again, to be further discontinued in 1986, when the amount of foreign capital in the insurance market was limited. This regulation limited foreign participation in insurance companies to half of the total capital of the company and one-third of the voting capital, except for those companies that at the time were already owned by foreign institutions. These foreign investment restrictions limited the expansion of foreign companies into Brazil.

As described by the federation of Brazilian insurance companies), "[for] many years...the Brazilian insurance market [was] working under...excessive regulation that would guarantee high returns, [but] would limit competitiveness and efficiency of the sector, [thus], stopping growth [by limiting]...the business basis and jeopardizing the potential for competition by Brazilian companies in the international market" (Federação Nacional das Empresas de Seguros Privados e de Capitalização (FENASEG) 1994). In other words, the Brazilian insurance market was not very competitive. The market stabilized and had approximately 96 companies for many years as well as coverage, premium, reinsurance, and brokerage commissions defined by the government. Most of the insurance lines had a reference cover and tariff which basically governed the entire market and any new insurance product had to have prior authorization from the government before it could be offered to the public.[294]

In 1996, the market was opened again for international investors, and the reinsurance monopoly was partially eliminated from the legislation (although it is still in practice, as discussed further here). From 1996 onward, international investors could have 100 percent ownership in Brazilian insurance companies, creating a new environment for the insurance industry. The market share of foreign companies increased from 6 percent in 1996 to 18 percent in 1997 to a current level of around 30 percent since 1999.

Under the new Brazilian constitution of 1988, the modernization process of the insurance sector began. "The country was [entering] globalization and the fears in [the] near future were justified" (FENASEG 1994). In 1992, during the first democratically elected government, important regulatory changes finally occurred in the market. The most important change was the shift from a standard product and premium schedule to the freedom to offer different products and set premiums, following some minimum rules set in Decreto 605/92 (Brazil 1992).

15.3.3 Population Structure and the Public Pension System

Although Brazil has almost 180 million inhabitants, only 18 million (or 19 percent of the active population) earn more than US$480 a month (dollars at the rate of exchange 2.70:1).[295] Thus, the potential number of customers for insurance is much lower than might be expected for such a populated country. The Brazilian population, similar to other Latin American countries, is young compared to world trends, but it shows signs of aging. In order to understand the changes in the population, the real population triangle is shown in Figure 15.3 for 1980 and 2000 with Instituto Brasileiro de Geografia e Estatística (IBGE) 2005 projections for 2020 and 2040 when Brazil will become a country with a rectangular population structure.

Besides having a youthful population, Brazil has a large debt burden on its public pension system compared to other countries as Figure 15.4 indicates. Among the reasons for this large debt is the usual combination of public social assistance via public pension benefits, a high level of benefits on a pay-as-you-go system, young retirement, and rising life expectancy. For example, the Brazilian public pension program maintains a high level of benefits (the maximum benefit could reach R$2,600 per month) compared to the distribution of population wages. The pension

[294] Usually premiums and covers would have little differentiation among insurers.

[295] Most of the income statistics are the minimum salary, which is also used to calculate the pension on the public system. At the end of 2004, this minimum salary was R$260 per month or US$96 per month if using the same rate of exchange.

system has not shifted to individual capitalization accounts yet, and there is no discussion of decreasing the maximum benefit offered by the state. It is also legally possible for retirees to work after retirement and employees qualify for retirement after working for 30 years (women) or 35 years (men). The minimum retirement age is low—48 years and 25 years of contribution for women and 50 years and 30 years of contribution for men.

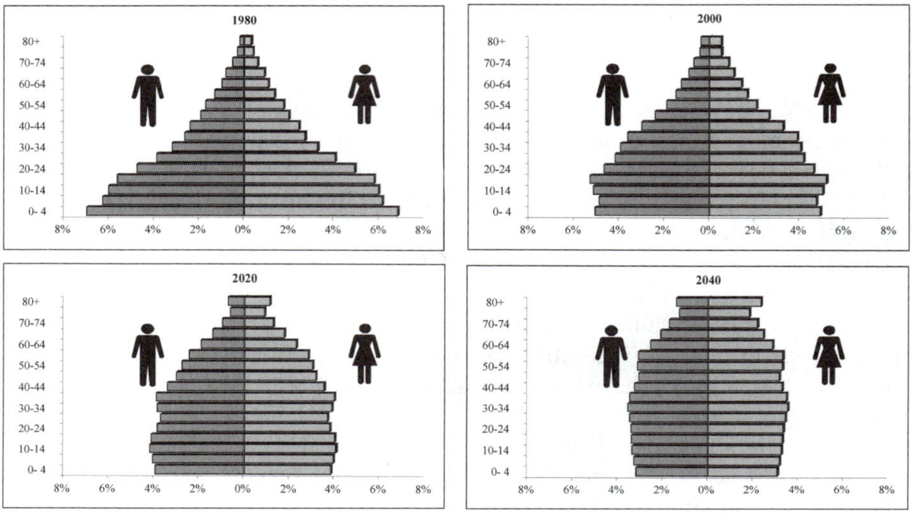

Source: Instituto Brasileiro de Geografia e Estatística (2005).

Figure 15.3. Population Development Projections in Brazil, 1980–2040

Even for those who have made no contribution, the public pension program guarantees a minimum wage for retirement and rural workers can retire without ever contributing to the system. In Brazil, any employee has the right to a pension. If the employee works in the private sector, his or her pension is funded by the obligatory contribution to social security made throughout the employee's working life. With this guaranteed universality, it is not surprising that the system is heavily in debt. Looking at the public pension system, there were 23.1 million beneficiaries at the end of 2004 with an average benefit of R$450. The system had deficits of R$26.4 billion in 2003 and R$32 billion in 2004 (Ministério da Previdência Social 2004).

If the employee is a civil servant, his or her pension used to be defined by rules that would encourage employees to retire early (relying on working years as in the private sector cited above) with 100 percent of their full wages at the time of retirement. One of the first acts of the current government (President Lula was inaugurated in January 2003) was to reform the pension program for civil servants. The reform included a contribution from the pensioners,[296] a top limit on the value guaranteed by the government in line with the private sector, and some rule changes to increase the minimum age for retirement. Although it initiated important changes

[296] The pensioner pays 11 percent of the wage higher than R$1,440 a month for federal civil servants or R$1,200 for other civil servants.

to stop the increase in indebtedness, further reforms are needed, such as the end of retirement by years of contribution and a different source of funding for assistance benefits.[297] These changes are not expected to take place any time soon, but it is already possible to see how some of the changes have impacted the private sector and the growth of the private pension market.

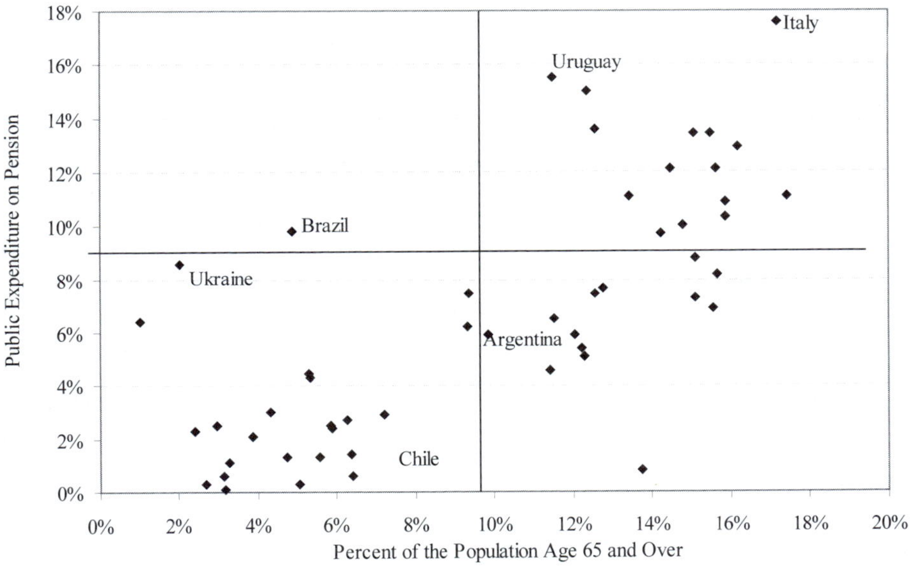

Source: The World Bank Group (2004; 2005).
Notes: Public expenditure data is from the years 1996 or 1997 except for Argentina (1994), Chile (2001), and Mexico (2000). Population data is from 1997.

Figure 15.4. Plot of Public Expenditures on Pensions and Population Age of 65 and Over

The public health system is also universal in Brazil and it has the same funding problems as the public pension system. There is a natural move by individuals toward private health coverage, which is the most widespread type of insurance in the country. A section on the private health insurance system is included later in this chapter.

15.3.4 Players in the Brazilian Insurance Market

Distribution Channels
In Brazil, brokers are required for any insurance transaction. The law does not allow them to be officially linked to any specific insurance company. In practice, however, many brokers are not fully-independent. Brokers are very important to the market and insurers market heavily to attract brokers to sell their products. Brokers can work

[297] Assistance benefits are benefits that are not funded, but are given as a mean of assistance.

as independent brokers or for brokerage companies and they can either sell any type of insurance (called all-products) or be specialized in life. At the end of 2004, there were around 70,000 active brokers in Brazil. Sixty-five percent of the brokers are independent brokers and 35 percent work for brokerage companies. Most brokers and brokerage companies work in all products, but in 1998 the number of new brokers specializing in life products overtook the number of brokers that sell all products—an indication of the increased interest in this segment of the market. In 1999, life insurance brokers represented about 70 percent of the total of new registrations. It is interesting to note that the number of life insurance brokers increased even though the largest life insurance companies are related to banks. Banks circumvent the requirement of using brokers on insurance transactions by having a brokerage company linked to the bank where all transactions are placed.

Bancassurance has been used extensively in Brazil for some time. Before 1994, with high inflation, it is not an exaggeration to say that customers would go to the bank every week, sometimes every day, creating a close relationship with the bank manager. The use of this distribution channel has always been high in Brazil, but precise data are not available and it is difficult to measure how much of the business is sold through bank channels. Bank market share is equivalent to around a quarter of total written premiums. Data are also not available for other distribution channels such as the Internet, but the sense is that it is used primarily as an information source and is not yet used extensively to sell insurance. Brazil is the tenth largest country in the number of Internet connections (International Telecommunication Union (ITU) 2005) with around 22 million as of December 2004. Furthermore, each Brazilian connected to the Internet spends on average more than 14 hours surfing per month (Instituto Brasileiro de Opinião Pública e Estatística (IBOPE) 2005). According to the same research, only Japanese users spend more time on the Internet (over 15 hours per month on average).

Actuaries

Actuaries are changing with the market and over the last decade have moved from their customary bureaucratic activities to become active players in the industry. Before the liberalization of premiums in 1992 and the obligation for calculation of IBNR (incurred but not reported) reserves in 1998, actuaries' activities were mainly related to complying with legislation (e.g., describing new products to the regulator, but they rarely calculated the premium). Nowadays, actuaries can be found controlling companies, something that would have been difficult to imagine in the past. They are also responsible for conducting an annual valuation of pension companies and valuations twice a year for insurance companies and health insurance providers. Actuaries are in high demand for the operation of companies since, it is only with their "agreement" that a company can register its insurance products or make any changes to them and it must be an actuary who is responsible for any changes to company reserve levels.

In Brazil, actuarial qualification occurs at the university level. The Instituto Brasileiro de Atuária (IBA) will accept any new actuary who has an undergraduate diploma from any university up to 2005. Following new directives of the International Actuarial Association, the IBA includes an admission exam for new actuaries in 2005. The number of actuaries in Brazil is still small, with a total of

1,300 at the end of 2004, with 600 active practitioners. This number has been growing steadily over the last ten years, with an increase of 15 percent in the first five years and 35 percent in the last five years. It is expected that this rate will accelerate in the near future, since there are 13 universities offering actuarial undergraduate courses toward a degree, an increase of over 100 percent in the last four years.

Companies

The Brazilian private insurance system has five types of organizations: insurance companies, open private pension entities or Entidade Aberta de Previdência Privada (EAPP), capitalization companies, health providers, and closed pension funds. In December 2004, there were 121 insurance companies in Brazil, of which 14 offered health insurance, 90 offered life insurance, and 17 offered insurance in all other lines of business. In addition, 14 companies specialized in capitalization and 29 in EAPP. Figure 15.5 provides a description of the Brazilian insurance market structure. Also a brief description of each type of entity follows, and more details are provided later in this chapter.

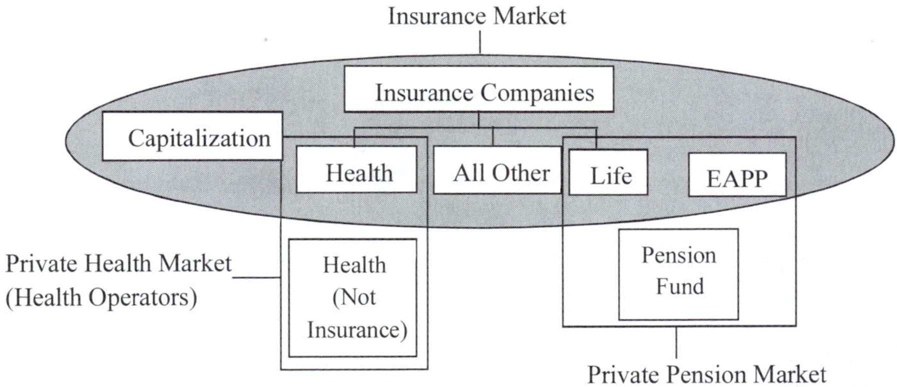

Figure 15.5. Description of Brazilian Insurance Market in Front of Other Types of Companies

In the private health insurance market, insurance companies are not the only players. In fact, there are eight different types of companies, including health insurance providers. They are all called Operadoras de Planos de Saúde and they will be referred to as health operators in this chapter.

Capitalization companies also are part of the insurance market, even though they do not provide an insurance product per se. A capitalization product is a savings account with a lottery attached. Savings products have been on the market for a long time and they reach a broad range of customers, including low-income clients.

The private pension market also has some specific characteristics. There are two types of pension companies: closed and open. The closed market has only one type of economic actor: pension funds which are non-profit companies that are usually linked to a company in order to provide its employees with a complimentary retirement plan. The open market has two types of companies: life insurers and

EAPPs. Since 2002, access to the pension fund system has been extended to members of professional associations and trade unions. In December 2004, there were 366 pension funds in Brazil, covering 2,144 companies and 1.8 million employees, with R$247.9 billion in assets, of which 63 percent originated from pension funds related to the public sector and the remaining 37 percent to the private sector. As a comparison, in December 2004 the open pension market had R$61.4 billion in assets or less than a fifth of the total pension market. One important difference is the level of maturity of the market. Pension funds have 2.4 million participants with 24 percent already taking benefits (613,000 individuals with average monthly benefits of R$2,430), while the open pension market has 8.7 million policies of which 3 percent is taking benefits. It is important to note that this last number could be misleading, since many policyholders hold more than one policy.

Pension funds are not considered part of the insurance market in Brazil. For the sake of comparison, some pension data are presented, but most of the figures in this chapter do not include pension funds. It is important to note that the Brazilian pension system also includes two government-owned institutions: public pensions (described in the previous section) and the Regime Próprio which is the pension system for civil servants related to the direct administration.[298] In 1998, under Law 9,717, each Brazilian city was allowed to create its own retirement funds or Regime Próprio. Approximately 2,000 of Brazil's 5,500 cities had adopted this system by the end of 2004.

As for the open private pension system, both EAPPs and life insurers are able to offer pension products in the market. EAPPs can be non-profit companies and mainly work on the defined benefit type of pension scheme. Since the appearance of the defined contribution schemes in the late 1990s, EAPP has been losing market share. Since 2003, only the non-profit EAPPs still collect premiums. In 2004, life insurance had 98 percent of open private pension system written premiums. This high percentage of market share is also related to the transformation of the EAPPs into life insurance companies.

Foreign participation reached 31 percent of written premiums in 2003 (33 percent in insurance, 36 percent in pensions, and 15 percent in capitalization) with the Netherlands and the United States competing for first place by holding over 50 percent of the total foreign investment. Some new and multinational groups have created partnerships over the last few years—ING with Sul America, AIG with Unibanco, and Mapfre with Veracruz, for example. Brazil also has had long-time partnerships with some multinational companies such as Generali, which was established in Brazil in 1925.

Most of the insurance groups are multiline operations. Consequently, although the legislation requires insurance groups to have a separate company for each line of business (life and EAPP, capitalization, health, and other insurance), the financial group itself participates in many lines. Aggregating the companies into groups, we obtain a distribution of written premiums as of December 2003 which is shown in Figure 15.6.

---

[298] These are civil servants who are hired directly by the national, state, or municipal government.

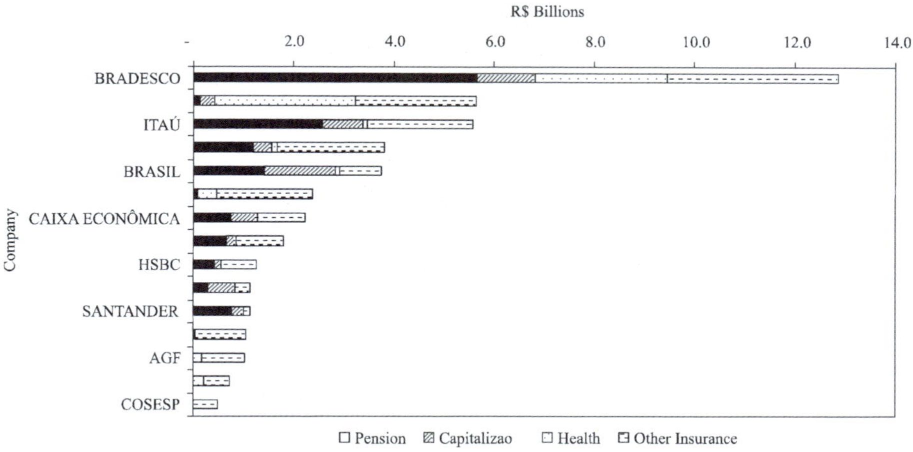

Source: Superintendência de Seguros Privados (2005) and Agência Nacional de Saúde (2005).

Figure 15.6. Written Premiums by Groups in Brazil, December 2003

The aggregation into groups shown here was based on the major asset holder, ignoring partial participation. For example, BrasilVeículos, a joint venture of Banco do Brasil and Sul América, is included in the second group in the graph. The market is highly concentrated. The Bradesco Group holds 25 percent of the total insurance market, followed by Sul América/ING (11 percent), and Itaú (11 percent). Thus, the top three groups account for 47 percent of the market. Because the groups may or may not be related to banks, it is interesting to compare the assets of the banks to the insurers' assets. Table 15.4 lists the largest banks in terms of assets as of December 2004. Comparing the list in Table 15.4 to Figure 15.6, the groups with the lowest rates of participation in pensions and life insurance are also the ones that have no direct relation with banks—for instance, Sul América, Porto Seguro, AGF, Icatu-Hartford, Mapfre, Maritima, and Cosep. This supports the important role of Bancassurance on the distribution of life and pension business.

In terms of total assets, the insurance market is much smaller than the banking sector. In December 2004, the total assets of the insurance market were only 7 percent of those of the banking sector. However, there is an upward trend. Insurance companies had R$11.7 billion in assets in December 1996 and R$84.5 billion in December 2004 with the growth mainly due to the open pension market that increased to R$61.4 billion from December 1996 to December 2004.

The Brazilian stock market, Bolsa de Valores de São Paulo (BOVESPA), is the biggest in Latin America by far. However, there is a very low volume of trading of shares in insurance companies. At the end of 2004, the only insurance company not linked to banks having business on the stock market was Porto Seguro. Among the insurance companies related to banks, there was no direct offer of shares related to insurance, but as the bank shares might be offered, they are also indirectly being traded in the stock market. Among the reasons for such a low level of stock trading, the most important is the lack of maturity of the Brazilian stock market. As of December 2004, insurers invested over 95 percent of their assets in government bonds. Also the characteristics of the insurance market explain insurers' reluctance to offer its shares on the stock exchange. Since the insurance sector used to be extremely regulated and protected, neither the companies nor the investors had an interest in the stock market.

Table 15.4. Total Assets by Bank

Rank	Name	Total Assets (R$ Billions)
1	Banco do Brasil	239.0
2	Bradesco	148.2
3	Caixa Econômica	147.8
4	Itaú	123.4
5	Unibanco/AIG	72.9
6	Santander/Banespa	66.5
7	Real/ABN Amro	59.2
8	Safra	40.8
9	HSBC	34.4
10	Nossa Caixa	31.3
	Total	1,450.6

Source: Sisbacen (2005).

15.3.5 Supervision

Although it is a continent-sized country, Brazil has no regional legislation. Therefore, rules are valid for the country as a whole. Besides health insurance, SUSEP regulates all the traditional insurance product lines since it is the main regulatory body for the Brazilian insurance industry. Insurance activity is supervised by three different ministries as follows. Property-casualty, life insurance, capitalization, and EAPP are regulated by the Conselho Nacional de Seguros Privados (CNSP) and its executive body, SUSEP, which were created in 1966 within the Ministry of Finance. In the insurance market, health insurance is the only line that is not directly supervised by this ministry. Pension funds are regulated by the newly installed supervisory body, Superintendência Nacional de Previdência Complementar (PREVIC), a sub-department of the Ministry of Social Welfare. PREVIC was created in the beginning of 2005,[299] extending the power from the previous Secretaria de Previdência Complementar (SPC). This ministry is also responsible for regulating the public pension system, the pensions of civil servants, the Regime Próprio, and the national worker's compensation program. Health insurance and operators are regulated by Agência Nacional de Saúde (ANS) which

[299] As of mid–2005, PREVIC had not been approved by all legislation levels, thus it could be implemented and subsequently, created a legislation burden. A new body with the same characteristics should be created by the government soon.

was created in 2000 by the Ministry of Health which is also responsible for the public health system. Before that time, health insurance companies were supervised by SUSEP.

The legislation for these ministries is not uniform and there are some discrepancies. For instance, accumulated fund transfers between open and closed pension plans are still under discussion, and the IBNR loss reserving rules for health insurance, health operators, and the other product lines are not homogeneous. It is also notable that the quality of data available from SUSEP is better than from the other supervisory agencies and many statistics are available online from SUSEP. SUSEP is an important driver of the market in many ways. The legislation adopted by SUSEP has led to many important changes in the market in the last few years (e.g., the need of IBNR in 1998 (also for the health sector to some extent), the obligation of annual actuarial reviews in 2001 (reserve review together with comments on assets and reinsurance treaties), and a stimulation to implement an ombudsman program in 2004).

The entire market is in line with the European Union Solvency I project, and discussion of the Solvency II is under way. For example, resolution CNSP 55/01 decrees that the margin of solvency should correspond to the value of the net assets to cover an amount equal to or greater than (a) 0.20 times the total of the net written premiums in last 12 months and/or (b) 0.33 times the annual average of the total of the net losses in the last 36 months (Brazil 2001). [300] These criteria take into consideration the magnitude of operations by making reference to written premiums and the loss experience. Additionally, those criteria are supported by norms that try to limit the exposure to the insurer's risk of investment.

The supervision of risk-based policies including the risks regarding reinsurance agreements and the adequacy of assets was begun in 2004 despite the fact that the legislation is still under discussion. Some other factors, however, have not been considered. For example, underwriting risk, reinsurance bad debt, credit risk, market risks, and operational risks have not yet been incorporated into the supervisory system. In the financial market, Brazil has been following Basel I and II, with some adaptation for local characteristics.[301]

Among the proposals being discussed in the insurance market for the Solvency II project, the most important defines the inclusion of most of the risks discussed above. In addition, there are proposals for an "assets shield" and for the creation of a fund formed by contributions from the whole insurance market in order to guarantee

---

[300] Most of the law related to the insurance sector can easily be found on SUSEP's website (www.susep.gov.br). Whenever defined by CNSP, the law is called *Resolução* (freely translated as "resolution") and by SUSEP, the law is called *Circular*.

[301] The Basel Committee on Banking Supervision, established the the Group of Ten Countries at the end of 1974, sets minimum capital standards for banks. The Committee's objective is to harmonize inter-national bank capital regulation among its member countries, which are from North American and Western Europe, but its regulatory recommendations also are influential in other parts of the world. The Committee's recommendations do not have the f orce of law but are influential in determining bank capital standards at the national level in many countries. The Basel I capitals standards were adopted in 1988, and the revised Basel II standards are being implemented in 2006–2007. For further information, see www.bis.org.

payments to losses under a certain limit (which should be quite low) for policyholders from bankrupted companies.[302]

15.3.6 Reinsurance

Reinsurance reached R$2.9 billion in total premiums in 2004 or 5.5 percent of the market total gross written premiums excluding capitalization. Reinsurance is still a government monopoly in Brazil (as it is in Cuba and Costa Rica), but market liberalization has been under discussion for a while. Although the federal government controls the IRB, the insurance market owns half of its capital. After the founding of IRB in 1939, it was not until the liberal government of Fernando Henrique Cardoso that there was the first important move to end the monopoly. In August 1996, the name "official reinsurance institution" was omitted from the Brazilian constitution. Since then, there have been many debates regarding reinsurance market reforms.

The most recent contribution to the reinsurance debate was a 2004 report from the Finance Ministry that stated, "The reinsurance monopoly in Brazil...creates economic barriers to the continued development of the sector" (Ministério da Fazenda 2004). The report recommended that a gradual opening of the reinsurance market should take place, but did not describe a specific process to be followed. At the same time, some changes in the legislation occurred indicating that the opening of the reinsurance market was again under discussion. It is important to recall that it was the left-wing party of the current president that stopped the privatization of IRB before and that this text goes against any prediction of the market—most reinsurance companies have already left Brazil.

The IRB privatization process started in the midst of a period of privatization in Brazil. In 1997, its name was changed to IRB-Brasil Resseguros S/A and it became a stock company. However, it was maintained as a company-owned and controlled by the federal government. At the end of 1997, the IRB was included in the list of companies to be privatized and after its accounts were reviewed, the IRB was expected to be sold in 1999, but the sale was delayed because of the lack of proper legislation. In order to solve this problem, all legislative power from the IRB (coinsurance, reinsurance, and retrocession) was transferred to SUSEP under Law No. 9,932 (Brazil 1999). It was not until 2002 that it was finally decided that the auction of IRB could not take place and that legislative authorization was required. That decision kept the responsibility of regulating and overseeing the reinsurance market in Brazil under the IRB.

Even with a gap in the legislation, a new date was set for the auction of the IRB in 2000 and, again, the sale was postponed. A minimum bid of R$550 million was set, but even that price was a problem since it was changed twice. In that year, following a lawsuit by the left-wing party of the current president, the auction was considered unconstitutional, because of the gap in legislation regarding IRB as the ruler of the reinsurance market in Brazil. The auction has been canceled and no new

---

[302] Under the "assets shield" proposal, the reserves are owned by the insured and cannot be used to pay any debt of a bankrupt insurance company. The insolvency fund is called Fundo de Proteção dos Mercados de Seguros, Previdência Complementar Aberta e Capitalização (FPM).

date has been set until the legislative problem has been resolved, but it was still under discussion at the end of 2004.

One of the goals of having a reinsurance monopoly is to maintain control by national companies of premiums collected from the reinsurance market. In Brazil, the ceding of risk from IRB was mainly via retrocession to the national insurance market. Unfortunately, no data are available to show historical levels of retrocession. If retrocession to the national market was important when the IRB was created, today national retrocession is minimal (0.2 percent of IRB´s written premiums in 2003) and the total ceding of IRB premiums reached 56 percent in 2003. With over half of its premiums written being ceded to the international market since 2001, the justification of maintaining the IRB as a monopoly—to keep revenues in Brazil—is no longer valid.

IRB also holds some unprofitable businesses in the United States via the United Americas Insurance Company, which has been in run-off since 1984. The default from this company was recorded on IRB´s financial statements (US$14 million), but because it involves asbestos losses, its value is still liable to change. IRB also holds some business and maintains an office in London, and in February 2004, it was still being discussed whether this office would continue to write business.

A proposal is under discussion in the Brazilian Congress to gradually open the reinsurance market, establishing three types of reinsurers: total local, local office-based, and avulses. Total local reinsurers would have a minimum capital requirement of US$50 million that must be deposited in Brazil to start operating. Local office-based reinsurers would need US$5 million and avulses would be able to operate without any local assets and could be based anywhere in the world. Any liability related to the avulses would have to be guaranteed via the reserves of the insurer that wrote the business as if the reinsurance did not exist. It was also proposed that the total local reinsurers retain a minimum of 60 percent of the total reinsurance premium, although that percentage will diminish over time. In this proposal, the IRB would be fully-privatized and become a total local reinsurance company like the others. The establishment of a period of time when the IRB would have preference over the other reinsurers has been discussed, but it does not seem to be the government's proposal. The fact is, the IRB is a profitable company, with an average profit of R$126 million over the last four years.

15.3.7 Various Lines of Brazilian Insurance Business

The Brazilian private insurance market reached R$59.8 billion in gross written premiums during 2004 (US$22.1 billion), distributed by product line as shown in Figure 15.7. Hereafter, unless explicitly stated, this analysis does not include pension funds.

As shown in Figure 15.7, 55 percent of the Brazilian insurance market is in health, life, and pension products, while 34 percent is related to non-life products and 11 percent is related to the financial product capitalization (explained later in this chapter).[303] Individually, pension is the largest product line and captures 31 percent of the market, followed by auto insurance, health insurance, capitalization, and life

[303] Hereafter, pension relates to the open pension market only.

insurance (10 percent), totaling together 86 percent of the overall market. This was not always the case. Pension products historically have been lower than automobile insurance until 2002, when pensions caught up.

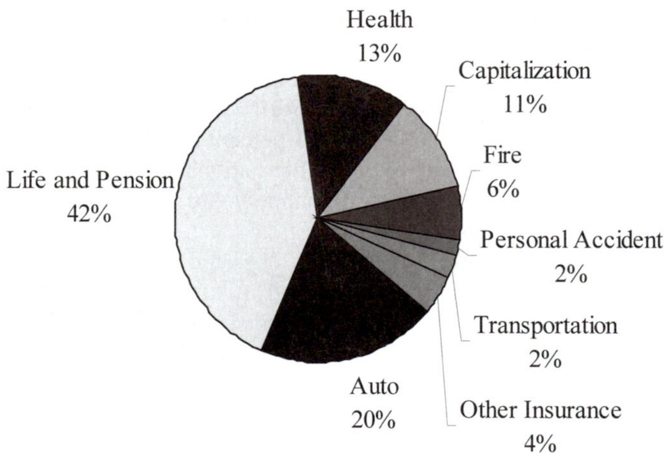

Source: Superintendência de Seguros Privados (2005), FENASEG (2005), Agência Nacional de Saúde (2005), and Associação Nacional da Previdência Privada (2005).
Note: Auto includes DPVAT (Danos Pessoais Causados por Veículos Automotores—mandatory third-party liability automobile insurance). Pension is composed of insurers and EAPP (Entidade Aberta de Previdência Privada—open pension plans). Health only includes health insurance.

Figure 15.7. Written Premiums Market Share in the Brazilian Insurance Industry

The private insurance market has been increasing steadily over the years and the penetration index has increased by approximately 0.1 percent per year in absolute value from 2.9 percent of GDP in 1999 to 3.4 percent of GDP in 2004.[304] If we do not consider capitalization, which is not a traditional line of insurance business in Brazil, the market penetration has decreased by 0.4 percent per year from 1999 to 2004. By product line, the percentage of the GDP evolved from 1995 to 2004 as shown in Figure 15.8.

Pension is the line of business that shows the greatest growth over the years. The main reasons for this growth are implementation of the Real Plan in 1994, the changes in legislation regarding public pensions, and the introduction in 1998 of products based on defined contributions instead of defined benefits. Observe that most of the figures shown here as premiums (and also everywhere in the Brazilian statistics) are, in fact, related to the contributions which will be accumulated by the individuals as opposed to benefits like life annuities.

Further statistics regarding the penetration of insurance among the Brazilian population can be found in a survey developed by the IBGE using data from 2003 (Instituto Brasileiro de Geografia e Estatística 2004). The survey asked about monthly family expenses, including private pensions and health insurance. The most

[304] The penetration index is defined as gross written premiums divided by GDP.

important finding was that health insurance was a product bought by low-income families, with the share of the monthly wages spent on health insurance growing steadily with the income. This is a direct consequence of the low quality of the public health system and how popular the private health products have become. Although the public pension system is also in debt, private pension products are only bought by higher-income families. The low penetration of insurance and private pensions in Brazil can also be seen by the average paid premium per capita of around US$130 in 2004 as compared to averages of over US$1,000 in Europe and the United States.

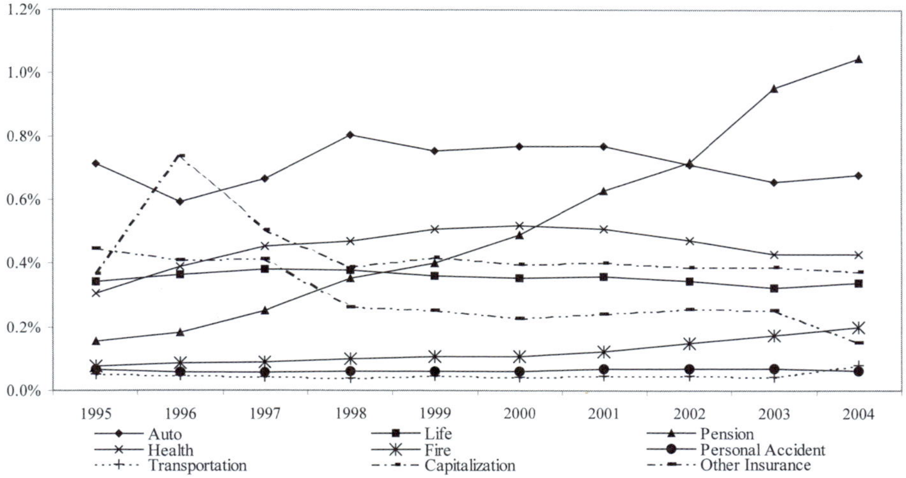

Source: Superintendência de Seguros Privados (2005), FENASEG (2005), Agência Nacional de Saúde (2005), and Associação Nacional da Previdência Privada (2005).
Note: Auto includes DPVAT (Danos Pessoais Causados por Veículos Automotores—mandatory third-party liability automobile insurance). Life insurance is related to death benefits. Pension is composed of insurers and EAPP (Entidade Aberta de Previdência Privada—open pension plans) and includes VGBL (Vida Gerador de Beneficio Livre—a defined contribution pension plan). Health only includes health insurance.

Figure 15.8. Written Premiums by Line of Business (% of GDP), 1995–2004

Excluding pensions and capitalization, the observed average loss ratio in 2004 reached 67 percent, with a combined ratio of 103.6 percent and an extended combined ratio of 89.3 percent. Those figures are in line with the recent past as shown in Table 15.5. While ten years ago, the ratios were 59 percent, 97.9 percent, and 83.6 percent, respectively. The combined ratio reached 100 percent in 1997 and it has not fallen below this value since then, thus showing how dependable the insurance market has become with regards to financial income. The unprofitable performance in the underwriting business has been offset by high investment income from the companies' technical reserves.

The insurance sector managed to adjust to high combined ratios due to strong gains from investment income. Before 1994, inflation worked for insurers by

Table 15.5. Basic Market Information for the Brazilian Insurance Industry, 1995–2004 (R$ Billions)

	1995	1996	1997	1998	1999	2000	2001	2002	2003	2004
Written Premiums							24.3	30.1	37.3	45.1
Net Written Premiums	12.9	14.0	15.1	18.4	20.3	23.0	21.4	26.4	33.1	38.1
Earned Premiums	11.5	12.5	14.0	16.4	17.9	19.8	21.0	23.4	26.5	29.0
Total Losses							16.5	17.4	19.5	20.6
Net Losses	6.7	7.3	8.9	10.9	12.4	13.3	13.8	15.7	18.2	19.4
Commercial Expenses	2.4	2.6	2.7	3.0	3.2	3.2	3.5	3.7	4.0	4.5
Administrative Expenses	2.1	2.3	2.4	2.8	3.7	2.9	3.7	4.2	4.7	5.3
Tax Expenses						0.5	0.5	0.6	0.7	0.8
Financial Returns	2.0	2.1	1.5	1.6	3.2	2.7	2.7	3.7	5.4	4.6
Loss Ratio (%)	58.6	58.5	63.4	66.4	68.9	67.3	65.7	66.9	68.5	67.0
Commercial Expenses/Earned Premiums (%)	20.7	20.7	19.3	18.3	17.6	16.4	16.4	15.7	15.0	15.5
Administrative Expenses/Earned Premiums (%)	18.6	18.5	17.3	17.1	20.9	14.6	17.6	18.1	17.8	18.1
Combined Ratio	0.979	0.977	1.000	1.018	1.075	1.010	1.021	1.033	1.039	1.036
Extended Combined Ratio	0.836	0.835	0.901	0.928	0.913	0.888	0.905	0.891	0.864	0.893

Source: Superintendência de Seguros Privados (2005), FENASEG (2005), Agência Nacional de Saúde (2005), and Associação Nacional da Previdência Privada (2005).

Notes: This table does not include capitalization and all pension products except Vida Gerador de Benefício Livre (VGBL—a defined contribution pension plan). Health only includes health insurance. Previous to 2001, written premiums were recorded net of all coinsurance, reinsurance, and would include retrocession. Tax expenses only started to be recorded in 2000.

devaluing the loss value. Premiums were usually paid in advance and extra premiums would have to be paid in case the insured wanted to maintain in real terms (or increase in nominal terms) the values of its insurance. Nowadays, high interest rates are the main source of profit to the market.[305] Brazil has the world's highest real interest rate as of December 2004—approximately 10 percent per year. The high interest rate is also the reason that insurers mainly choose to invest in government bonds. In 2004, over 95 percent of the assets in the insurance market were allocated in this fashion. A future decline in interest rates will force insurers to refocus on their underwriting profitability, something that has not happened in the recent past.

Net written premiums decreased in 2004, given the large volume of reimbursements due to Free Benefit Generator Life (VGBL).[306] VGBL is a defined contribution pension plan with a high level of deposits, which are mainly short-term savings rather than a real pension accumulation. This is further discussed below, but at this point it is important to highlight this characteristic in order to explain the difference in the written premiums. Because VGBL might have some term life insurance attached to it, it is shown separately in Table 15.6 for data reconciliation. In Brazilian data, it is normally shown as part of life insurance statistics. Looking at individual product lines, 2004 data are shown in Table 15.6, and the main lines of business are detailed below.

Term Life Insurance and Open Pension Plans

In 2004, the term life insurance and open pension market had R$24.7 billion in gross written premiums, 76 percent of which were open pension premiums. Of the pension plan premiums, R$15 billion was related to defined contribution plans and R$3.7 billion to defined benefit plans. At the end of 2004, the pension market had accumulated the highest value in history of R$61.4 billion in reserves related to the pension plans.[307] As previously indicated, life insurers participate in two market segments: term life insurance (payment of a lump sum in case of death) and pension plans (payment of a monthly pension from age of retirement while alive). Term life insurance includes products like death coverage to credit card holders being offered as "free-of-charge," individual credit life insurance for consumer loans, and companies buying collective (group) life coverage for their employees. Term life insurance has a historical loss ratio of around 50%.

There are three types of open private pension plans: defined benefit pension plans, the Free Benefit Generator Plan (PGBL), and the VGBL. Both PGBL and VGBL are the core of defined contribution plans. PGBL was launched in Brazil in 1998 and VGBL four years later. The legislation for defined contribution was set up in 1995. While the defined contribution plans have experienced a boom, the defined benefit plans have seen their total premiums decrease over the last five years.

---

[305] Government bonds pay approximately 10 percent real interest rate.

[306] VGBL, while mainly a survivor's product, is always classified under life insurance in the available aggregate statistics. In the context of this chapter, however, VGBL is included under pension plans whenever possible.

[307] The closed private pensions, or pension funds, are not included in these numbers.

Table 15.6. Basic Market Information for the Brazilian Insurance Industry, 2004 (R$ Billions)

Line of Business	Written Premiums	Earned Premiums	Losses	Loss Ratio (%)	Commercial Expenses	Commercial Expense Ratio (%)
Auto	12.1	10.3	7.6	73.2	1.8	17.8
Life	6.0	5.4	2.9	53.4	1.3	23.8
Health	7.6	7.5	6.6	87.2	0.3	3.8
Fire	3.6	2.0	0.9	43.1	0.5	24.1
Personal Accident	1.1	1.1	0.3	27.9	0.2	19.8
Transportation	1.4	1.3	0.6	48.7	0.3	20.4
VGBL	10.6	0.3	0.1	41.2	0.1	35.2
Other	2.7	1.0	0.5	45.4	0.1	5.0
Total	45.1	29.0	19.4	67.0	4.5	15.6
Capitalization	6.6					
Pension	8.1					
	59.8					

Source: Superintendência de Seguros Privados (2005), FENASEG (2005), Agência Nacional de Saúde (2005), and Associação Nacional da Previdência Privada (2005).

Notes: VGBL is Vida Gerador de Beneficio Livre (a defined contribution pension plan) whose data are shown separately from life and pension figures. Auto includes DPVAT (Danos Pessoais Causados por Veículos Automotores is mandatory third-party liability automobile insurance). Life insurance is related to death benefits. Pension is composed of insurance companies and EAPP (Entidade Aberta de Previdência Privada are open pension plans) and includes VGBL figures. Health only includes health insurance.

PGBL and VGBL are the main drivers of the market, so they deserve specific description. They are defined contribution plans with no return guarantees on investments. All contributions must be directed to mutual funds created for this specific purpose. The main difference between the two types of defined contribution plans is with regard to the payment of taxes. In the case of PGBL, the amount deposited to this plan (up to 12 percent of the individual's total annual wage) can be deducted from income tax payments. Taxes will be charged only when the fund is redeemed, either during the accumulation period or when retirement age is reached. VGBL contributions are not tax deductible, but the future income from accumulated funds is tax-free. It provides tax relief when redeeming the fund with only the investment gains being taxed.

Legislation has changed recently. Now, not only is the tax payment delayed, but also a decrease over time has been established. In the past, policyholders could redeem the fund at any time, with no minimum time of contribution. Now policyholders have the option to maintain the original tax payment of 27.5 percent of the total redeemed fund or opt for a decreasing tax. So, with the new option, the tax would actually increase to 35 percent for those redeeming the fund before it is one year old, but would decrease linearly along the years, reaching 10 percent when redemption occurs ten years after the purchase of the plan. This change in legislation was mainly created to encourage investors to keep their money in the fund. Because it had a high level of refunds in the past, it is believed that most of the investments under PGBL and VGBL are, in fact, short-term investments rather than a provision for a future pension. The change in tax laws have been in force since the beginning of 2005 and will be able to verify this assumption at a later date.

Both PGBL and VGBL are deferred annuities, meaning that the mortality table and the interest rate for annuity conversion are set on the date when the accumulation period is started. This is technically a major problem because there is no room to incorporate improvements in life expectancy from the moment the insured starts to contribute to the fund until he or she reaches the age to retirement, which could take many years. This thus raises the question of how the insurance companies should comply with legislation that imposes the agreement on a mortality table many years before the pension is due. Insurers are using a mortality table like AT 2000 (Annuity Table 2000, a mortality table developed in the USA), which is believed to be conservative for the Brazilian case, but really balances any mortality deficit against reductions in the interest rate that should be guaranteed today for the benefit payment period. Most of the companies guarantee null interest rates.

VGBL had the biggest impact on the written premium increase, starting from R$2.5 billion in 2002 to R$7 billion in 2003 and R$10.6 billion in 2004. Given this performance, the life and pension share of the insurance market has been growing over the years and currently represents about 42 percent of the market. Before the implementation of the economic Real Plan in 1994, life and pension had an average 12 percent market share.

In 2004, the pension market basic amounts were as shown in Figure 15.9. The reserves for each of the plans described have different timing: defined benefit plans have been in effect for more than 50 years—compare that to seven years since the first PGBL has been offered and three years for VGBL. It is also important to note that an estimated 10 percent of the total reserves for defined benefit plans are related to the contribution deficiency reserve, an extra reserve that was implemented in the open private pension market a few years ago to fund any deficit that might exist. The high observed level of refunds for the defined benefit plans is also related to migration policies implemented by insurers to encourage policyholders to migrate from defined benefit plans to defined contribution plans.

Because the defined contribution plans are sold as personal investment products, 77 percent of the total written premiums comes from individuals over 18 years of age, 19 percent comes from companies, and 4 percent comes from individuals under 18 years of age. The defined contribution plans also have high redemption levels, which represent a higher share of reserves than for the defined benefit plans. Most of the plans are sold by companies related to banks. As shown in Figure 15.10, note that only the last two companies are not related directly to banks (Icatu-Hartford and Capemi). We also observe that 65 percent of the written premiums in 2004 went to three companies, with Bradesco as the leader with 37 percent of the market. The distribution channel for this type of business consists of the bank branches, which are a less expensive distribution network than traditional channels, in part because the branches already exist. Selling through bank branches allows companies to target their clients on a more individual basis and has a higher cross-selling potential.

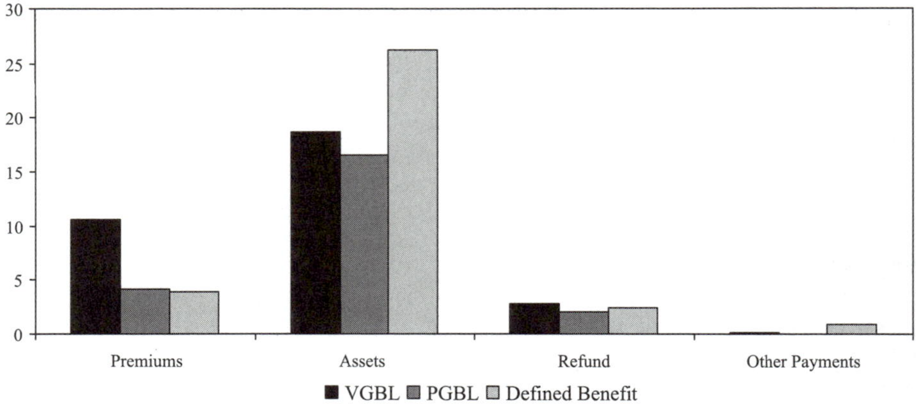

Source: Superintendência de Seguros Privados (2005), FENASEG (2005), and Associação Nacional da Previdência Privada (2005).
Note: VGBL is Vida Gerador de Beneficio Livre, a defined contribution pension plan. PGBL is Plano Gerador de Benefício Livre, another defined contribution pension plan.

Figure 15.9. Basic Values in the Brazilian Pension Market, 2004 (R$ Billions)

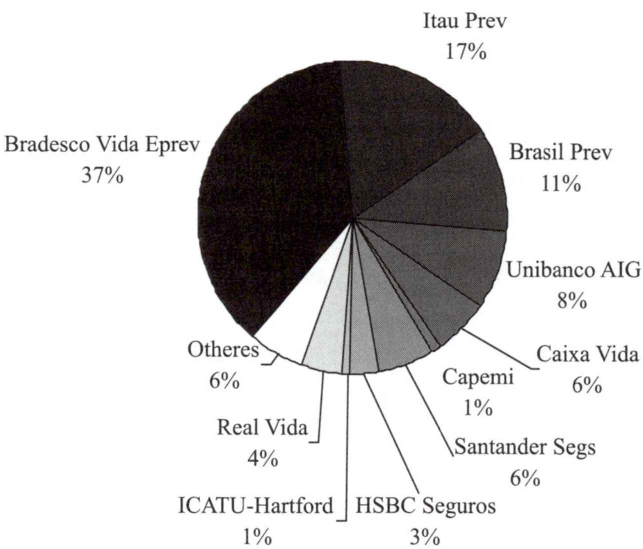

Source: Associação Nacional da Previdência Privada (2005).

Figure 15.10. Open Pension Market Share in Written Premiums, 2004

For term life insurance, the concentration is lower than in other product lines. The first three companies represent only 35.34 percent of the market (Bradesco,

Banco do Brasil, and Itaú). Because VGBL could have part of its premiums related to death benefit coverage, this value could change. Term life insurance also enjoys a very low loss ratio of around 45 percent, while the pension portion is still mainly in a period of fund accumulation.

Health Insurance

Brazil's private health system is complex and widely spread among the population and around the country, with impressive and huge numbers. In 2003, there were more than 86.5 million medical procedures: 19 million doctor's appointments, 47 million exams, and 2.6 million hospital admissions (Agência Nacional de Saúde 2002). Overall, health operators collected R$31.4 billion in written premiums during 2004, but only R$7.6 billion was related to health insurance.

The private companies that offer health coverage against a monthly premium payment or co-participation on the medical expenses are called health operators. There are eight types of companies working in the market with one the so-called "health insurers." The means of payment to providers differs between companies, as does their relationship to their network of services (hospitals, offices and clinics, and doctors). For example, they may use their own network or use accredited services indicated by the health operator, but that would not constitute a third-party relationship. To better understand the differences in the market, the classifications of health operators are shown in Table 15.7. At the end of 2004, there were 2,178 health operators in Brazil with market shares distributed as shown in Figure 15.11.[308]

The health operators are highly specialized, so market share changes depending on the data analyzed. For instance, health insurance is 24.2 percent of total written premiums, but it represents only 11.8 percent of total beneficiaries and 0.6 percent of the health operators in the market. Each health operator can offer corporate and individual plans, but the latter are more controlled by the government. Because of this, insurers are highly concentrated in corporate plans—89.9 percent of their policyholders and 66.5 percent of their written premiums come from corporate plans as of September 2004 (Agência Nacional de Saúde 2005). The coverage also defines which type of medical procedure is available: ambulatory, hospital, or dental. Combinations of these plans are used and the most popular one is a combination of ambulatory and hospital. In practice, the larger companies offer broader coverage, while smaller companies tend to concentrate on specific niches.

Because insurers do not own or work on a cooperative basis with health services providers, they usually charge higher prices. They must have a third-party relationship with the service provider, so they work only via reimbursement and usually having a maximum value for each procedure. Using their choice of doctors, patients pay for their expenses and get a receipt. The health insurance market is highly concentrated, with only 14 players divided into nine groups. Two of them, Bradesco and SulAmerica, own 81.6 percent of the market. These two main groups compete for first position in the market. In comparison with the usual insurance lines, the health insurance loss ratios are the highest in the market, increasing from around 82 percent in 2002 to 87.2 percent in 2004. If the commercial rate were

---

[308] These numbers differ from Table 7 due to a 15 day difference in the source of information[0].

maintained at the same level since 2002 (3.8 percent; it was 7 percent in 1999), the higher loss ratios have been compensated by a lower level of administrative expenses. They declined from 11.3 percent in 2002 to 9.8 percent in 2004. Because of these movements, the combined ratio increased from 98.2 percent in 2002 to 101.7 percent in 2004 and this line of business maintained its profitability via the high interest rates that allowed an extended combined ratio of 94.5 percent and 97.0 percent for 2002 and 2004, respectively

Table 15.7. Description of the Brazilian Health Insurance Market and Differences among Health Operators

Type of Health Operator	Definition	Can Have Own Network of Services?	Number of Companies
Management (Administradora)	Receives a fee to manage the health system for a specific group (mostly corporate). No real risk absorbed.	Yes	12
Self-administrating (Auto-gestão)	A group of people who cover their own costs via a sponsor or through sharing.	Yes	325
Medical Cooperative (Medical Cooperativas)	Doctors in a cooperative to offer health services.	Yes	369
Dental Cooperative (Dental Cooperativas)	Dentists in a cooperative to offer dental services.	Yes	167
Insurance (Seguradoras)	Required to have a third-party relationship with a service provider, working mainly via reimbursement.	No	14
Philanthropic (Filantrópicos)	Non-profit organizations (require recognition as such by the government). Mainly hospitals.	Yes	107
Medical Group (Medicina de Grupo)	A company offering health coverage that cannot be classified as any of the above.	Yes	745
Dental Group (Odontologia de Grupo)	A company offering a dental coverage that cannot be classified as any of the above.	Yes	444

Before June 1998, when Law No. 9,656 and Provisional Measure No. 1,665 (Brazil 1998a, 1998b) went into effect, the health insurance market was close to chaotic. The health operators that were not insurance companies did not have a single set of rules and only insurers had to follow SUSEP's rules. Furthermore, the insurance companies had much more freedom. For instance, insurers could adjust premiums in the event of high loss ratios and could also reduce the number of policies they carried.

At that time, the plans differed substantially from one to the other and there was some discussion about legal disputes about payment for some pre-existing conditions. The changes brought by this new legislation were so important that one can now classify plans as "old" if they were contracted before the new legislation went into effect and "new" if contracted after. Although it has been several years since the legislation was passed, approximately 45 percent of the beneficiaries are under the "old" plans (in the insurance companies, the percentage is lower (40 percent) and they are still a source of technical deficits. These policyholders are becoming older and their coverage has been extended to comply with the new legislation and only small premium increases have been allowed. Finally, the companies cannot simply cancel these policies.

Figure A: Written Premiums by Type of Health Operator (R$ Billions)

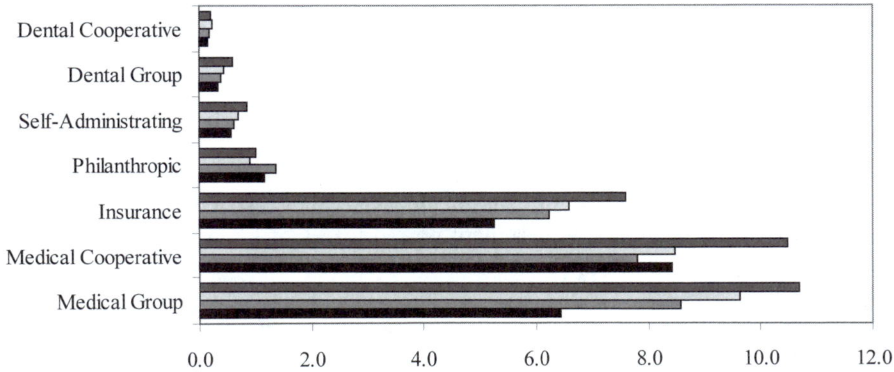

Figure B: Number of Beneficiaries by Type of Health Operator (Millions)

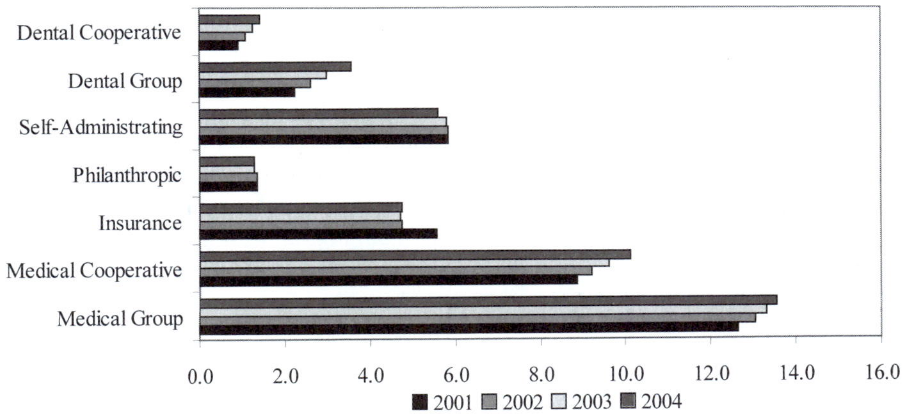

Source: Agência Nacional de Saúde (2005).

Figure 15.11. Development of the Brazilian Health Insurance Private Market, 2001–2004

The Agência Nacional de Saúde was set up in 2000, consolidating the changes in the market. Consequently, many health operators were merged or simply closed. The number of companies decreased by 17.4 percent between 1999 and 2004 and this decrease is expected to continue. The number of beneficiaries and premiums did not follow the same trend, mainly because of the mergers and acquisitions that took place in the market. It is believed that, from now on, small health operators will be forced to merge in order to comply with some rulings by the ANS, such as maintaining databases containing individual information and higher volumes of assets for reserves (e.g., IBNR reserves).

All the changes in legislation from the ANS increased the costs of the industry. For instance, it was not possible to exclude from health coverage some expensive diseases like cancer or AIDS. The unilateral cancellation of individual plans was also prohibited and the use of a new "grace period," risk selection, and exclusion without a good reason for certain users was also prohibited. Also, a strict control of prices and conditions for increasing prices was established. Until 2003, the maximum allowable increase in prices permitted by the ANS was less than the rate of inflation, but in 2004, a higher rate was allowed. One important change that occurred, but which has not yet been fully implemented, was the fee for the use of the public health sector by beneficiaries from the private sector. According to the fee system, any person with a private health plan must have his or her expenses reimbursed by the health operator to the government in the event they use public sector health services. Up to July 2003, the government posted losses of R$217 million, but only R$44 million has been paid and what remains is still being discussed in the Congress.

An important characteristic of health plans is their popularity. In the state of São Paulo, 42 percent of the population has a health plan, while in the state of Rio de Janeiro 32 percent have a health plan. These figures are compared to 21 percent for the total Brazilian population. This concentration of policyholders is likely due to the fact that São Paulo and Rio de Janerio are the two richest states in Brazil. According to the last survey by the IBGE in 2003, among Brazilian families 0.3 percent of the family wage is spent monthly on health plans when the family has an average monthly wage of R$400 or less. This percentage increases with the wage, reaching 2.4 percent for families with average wages between R$4,000 and R$6,000 and 2.1 percent for those with wages over R$6,000. Although considered important to Brazilian households, the percentage of the population with health coverage has stabilized (or slightly decreased as a proportion of the total population) over the last few years. It is understood that this movement might mostly be related to the increase in the premiums.

Automobile Insurance

Auto insurance used to be the most common coverage sold in Brazil, losing first place to life insurance products in 2004 when it reached R$12.1 billion in written premiums. There are two types of auto insurance in Brazil: the mandatory personal damage through the Danos Pessoais Causados por Veículos Automotores (DPVAT) auto insurance and optional car insurance, corresponding to 13 percent and 87 percent of the total auto written premiums, respectively.

Civil liability insurance, which covers the risk of liabilities against third parties in case of a car accident, is not very popular in Brazil. This is mainly due to the small number of lawsuits and the mandatory DPVAT, which is a civil liability product designed to guarantee civil liability coverage without the need to determine guilt or responsibility in automobile accidents. Its operations are highly regulated by the government, which defines and collects payments. It is paid with annual automobile taxes (e.g., road maintenance taxes), but even with this combination, it is estimated that the evasion rate is 25 percent. DPVAT was paid by 29.4 million vehicles in

2004, whereas the total number of registered vehicles was 38.7 million (FENASEG 2005). In case of an accident, medical treatment is to be provided by the public health system and, in the case of a need for private health coverage, the value is limited to R$2,000. If permanent disability or death occurs, the beneficiary has the right to a lump sum, limited to R$10,300.

All the reimbursement, except for the public health benefit, is provided by a pool of insurers (61 companies are listed by the DPVAT website) that receive proportional premiums and share the responsibility. Today, from the total premiums paid to DPVAT, 50 percent goes to the public health system, 35 percent goes to the pool of insurers, 10 percent goes to administrative expenses, and the rest goes to some other entities, including Funenseg.[309] Most analysts agree that the majority of the population is not aware of its rights, but in 2004 the number of death payments were consistent with the number of deaths in automobile accidents—approximately 35,000. So, we can conclude that the death benefit awareness has improved and that most of the unpaid insurance damages are related to disability and private health expenses.

The DPVAT is paid annually and ranges from R$57 for passenger cars to R$456 for buses. Liberalization of this sector has been discussed, but it is not expected to happen soon. In 2005, the government extended the auto coverage to include trucks. Although the private market could provide better coverage (using the private health system or higher amounts of coverage) to policyholders, the government would have to find different sources to pay the public health system expenses for motor vehicle accidents. Furthermore, the private sector would not have the assurance that the new liberalized system would improve the market since it could lead to evasion by old clients and a diminished amount of optional insurance clients (see below).

The optional auto insurance segment comprises extra third-party liability (bodily or material), collision, and theft coverage. It is the second leading product in sales if open private pensions are included and the first leading product if they are not considered. In 2004, it reached R$10.5 billion in written premiums, or approximately 30 percent of the insurance market in terms of premium income (not including pension figures or capitalization). Those who purchase optional auto insurance policies are seeking protection against an auto collision or theft, not protection against civil liability that might be included in the product. Of all 36 million vehicles in Brazil, only 9 million (25 percent) were insured for optional auto insurance in 2003.

Given the importance of auto insurance, the market also has a research center named Centro de Experiência e Segurança Viária-Cesvi Brasil which is responsible for research on auto repairs and road safety. The market is also very well supplied with public data. SUSEP maintains a database called Autoseg that contains statistics related to the auto industry. According to the 2003 data, most of the losses were related to robbery or theft (33 percent), followed by partial or total automobile damage losses (28 percent and 20 percent, respectively).

In 2004, the Brazilian insurance companies with the largest operations in the auto segment were Porto Seguro and Bradesco as indicated by Figure 15.12. In terms

[309] FUNENSEG is the Fundação Escola Nacional de Seguros is the national foundation for technical education. It is responsible for the accreditation of brokers and maintained by the whole insurance market. See www.funenseg.org.br.

of earned premiums, they were tied, but Porto Seguro has a larger share if written premiums are considered. The sector had an average 74 percent loss ratio and 18 percent commercial ratio in 2004.[310] However, Porto Seguro and Bradesco differ substantially from the average. Porto Seguro had 59 percent and 13 percent of loss and commercial ratio, while Bradesco had 90 percent and 16 percent, respectively.

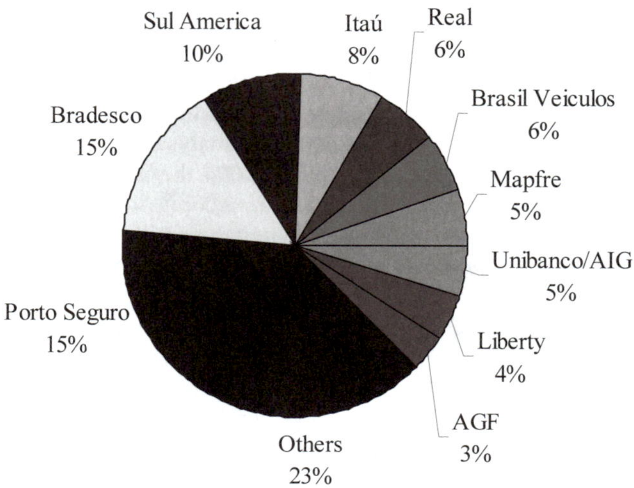

Source: Superintendência de Seguros Privados (2005).

Figure 15.12. Brazilian Automobile Insurance Market Share in Written Premiums, 2004

The five largest auto insurers represent 54 percent of the automobile insurance market.[311] It is one of the least concentrated compared to health insurance and life insurance and open pensions. BrasilVeículos, which is tied for fifth in the ranking, is a joint venture of Sul América and Banco do Brasil, the largest bank in Brazil.

Capitalization

Capitalization is also a part of the insurance market, even thought it is not an insurance product. Basically, it is a savings account that has planned periodic payments into a fund (the premiums) with a lottery attached (rules are very loose on the amount of the lottery). It is offered to the public via a private company; and together with the lottery from the government, it used to be the only private authorized gambling product in Brazil. In 2004 it reached R$ 6.6 billion in written premiums, with 14 companies in the sector.

Since it has an administrative fee, capitalization products are mostly attractive because of the linked lottery. A capitalization product is created as follows: the capitalization product is structured with a redemption period equal to or greater than 12 months and is sold in series. For example, one capitalization product might have a

---

[310] The commercial ratio is the ratio of expenses on brokerage to earned premiums.

[311] This excludes Brasil Veículos which is tied for 5th in the ranking.

series of 100,000 bonds that could be acquired by up to 100,000 different customers for a R$100,000 lottery during one year. Each bond follows the same rules and is in play for the same lottery. The bond foresees policyholder payments that have, in general, three components: the amount to be capitalized, the lottery cost, and the administration fee. The bond can be redeemed at the end of the period.

There are two main types of capitalization bonds: (1) bonds with one payment that is reclaimed one year later and (2) bonds with monthly payments that are usually renewed every year. For the duration of the bond term, the policyholder is eligible for the lottery and the products offer a minimum interest rate.

One famous example of a one-payment product is Telesena, offered by a popular TV channel. It costs around R$5 and is offered in blocks of one million bonds, with a one-year redemption period. During the one-year term of the bond, the lottery draws every month a winner of R$1,000. At the end of the period, it returns half of the face amount of the bond with 6 percent interest (R$2.65). This type of product was responsible for the growth observed in the premiums during 1996 and 1997 (67 percent and 45 percent of total premiums, respectively, compared to a current 26 percent).

One example of the second type of product is one for low-income customers that pays interest on a one-year savings account. This type of account gives a small accumulation, since the contribution was low in the first place. By depositing this amount in a capitalization bond, the customer exchanges this interest for the dream of receiving a lump sum upon winning the lottery attached to the bond. These products can have durations of five years and may include annual or monthly payments.

It is believed that any increase in capitalization products is directly related to the increase in the purchasing power of low- and medium-income consumers. Telesena was an important example of such a relationship, as well as a new capitalization product from Icatu-Hartford that provides low-income customers with a credit card backed by the capitalization bond.

There are 14 companies operating in the capitalization industry, and their market share is presented in Figure 15.13. Similar to the rest of the insurance business in Brazil, capitalization is highly concentrated in the hands of a few players: Brasilcap has the largest market share (26 percent), followed by Bradesco (21 percent), and Itaú (13 percent). All are related to banks, which show that capitalization is of interest to banks for expanding the utilization and profitability of their distribution networks.

15.4 THE FUTURE OF THE BRAZILIAN INSURANCE MARKET

The Brazilian insurance market has experienced a boom in the last few years with defined contribution plans, but no other extreme growth is expected. From now on, development in the insurance market depends mainly on increases in the type of coverage and products offered by the insurance market and on the increasing demand in a growing economy.

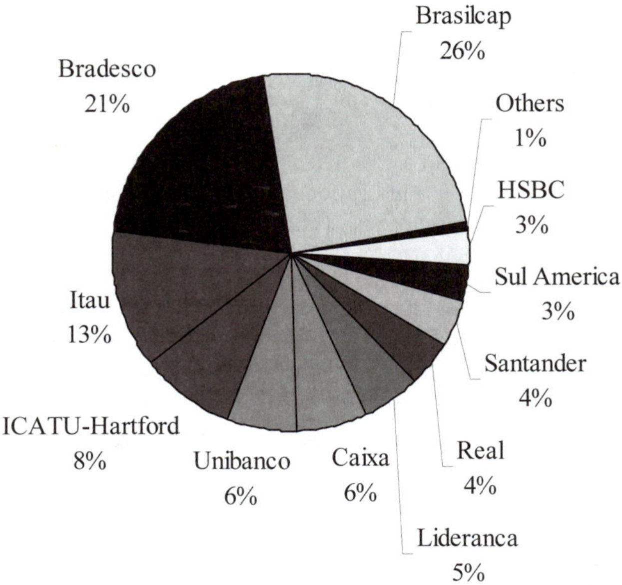

Source: Superintendência de Seguros Privados (2005).

Figure 15.13. Brazilian Capitalization Market Share in Written Premiums, 2004

Since the penetration of insurance products in Brazil is low and many of its inhabitants are not insurance customers due to their low incomes, the insurance market is not standing by and waiting for growth to occur through an improvement in incomes. Rather, it is offering some new products that are being developed for low-income customers, whose demand for products is high. For example, Icatu-Hartford has just launched a capitalization product that offers financial support for a credit card application without any proof of formal employment or income.

In the auto insurance market, following the same idea, SUSEP has launched a proposal for new "popular" products that would have a different basis from those offered today, and some companies have already introduced these types of products. The proposal calls for policies to provide partial coverage, such as only for total loss instead of partial losses. The value of the car would also depend on the amount of use and the possibility of using used parts when fixing the insured car. It remains to be seen whether the proposal will be successful. However, any low-budget bodily injury coverage will not attract new customers since it is already offered on the mandatory DPVAT.

Additionally, the implementation of measures to improve solvency is underway. First, the inclusion of the principles from the Solvency II program has been discussed. Second, loss payments are to be guaranteed up to a certain limit for policyholders from bankrupted companies. These developments are of great importance, because it is feared that the opening of the market to foreign reinsurers could reduce premiums by 25 percent in some product lines as a result of more competition.

15.5 CONCLUSION

This chapter describes the development of the Brazilian insurance industry over the last decade. A comparison with neighboring countries shows that Brazil has a large market as a consequence of its large population. However, insurance penetration in Brazil is smaller than in Chile and Argentina.

The development of the Brazilian economy has involved many plans to control inflation, but the first successful plan was the Real Plan. The financial stability achieved as a result of the Real plan, combined with government concern for the formation of long-term domestic savings, served as an important lever for development in the insurance industry via the private open pension plan market.

Despite the reforms already achieved, several important issues remain on the legislative agenda (e.g., the liberalization of reinsurance and workers' compensation and the reform of the public pension system). Following the example of other Latin American countries, it is necessary that the public pension program be reformed in a fair manner under a regimen of capitalization.

Finally, the Brazilian insurance market is demonstrating a high degree of adaptability by implementing new products targeting low-income customers. This change and the new solvency system are likely to be the major developments affecting the insurance sector in the near future.

15.6 REFERENCES

Agência Nacional de Saúde, 2002, Regulação & Saúde: Estrutura, Evolução e Perspectivas da Assistência Médica Suplementar/Ministério da Saúde, Agência Nacional de Saúde Suplementar (Rio de Janeiro: ANS).

_____, 2005, Tabelas de Operadoras e Dados do Setor (Rio de Janeiro: ANS).

Ardeo, Vagner Laerte, Kaizô Beltrão, Fabio Giambiagi, and João Luis de Oliveira Mendonça, 2004, Texto para Discussão No 1050—Diagnóstico da Previdência Social no Brasil: o que foi feito e o que falta reformar? (Rio de Janeiro: Instituto de Pesquisa Econômica Aplicada).

Associação Nacional da Previdência Privada (ANAPP), 2005, Previdência Complementar Aberta no Brasil Dados Estatísticos Dezembro–2004, monthly research (Rio de Janeiro: Associação Nacional da Previdência Privada).

BACEN, 2005, Histórico das Taxas de Juros Fixadas pelo Copom e Evolução da Taxa Selic, (Brasília: BACEN).

Brazil, 1992, Decreto nº 605, July, 17th, 1992—Changes decreto nº 60.459, March, 13th, 1967–July, 20th, 1992 (Brasília: Diário Oficial da União).

_____, 1998a, Law 9,656, June, 3rd, 1998—Rules the health private coverage—June, 4rd, 1998 (Brasília: Diário Oficial da União).

_____, 1998b, Provisional Measure No. 1,665 , June, 4rd, 1998—Rules the health private coverage—June, 5rd, 1998 (Brasília: Diário Oficial da União).

_____, 1999, Law 9,932, December, 20th, 1999—Rules the transfer of responsibility from IRB-BRasil Resseguros S.A. to SUSEP—December, 21st, 1999 (Brasília: Diário Oficial da União).

_____, 2001, Resolução CNSP nº55, September, 3rd, 2001—Rules solvency margino of insurance companies—September, 13th, 2001 (Brasília: Diário Oficial da União).

Contador, Cláudio and Clarisse B. Ferraz, 2004, "Penetração do Seguro e Preço de Apólices," Revista Brasileira de Risco e Seguro 1: December.

Federação Nacional das Empresas de Seguros Privados e de Capitalização (FENASEG), 1994, Plano Setorial da Indústria do Seguro (Rio de Janeiro: FENASEG).

_____, 2004, 2º Plano Setorial da Indústria do Seguro (Rio de Janeiro: FENASEG).

_____, 2005, Base de Dados do SES—Dados de 1995 a Dezembro 2004 (Rio de Janeiro: FENASEG).

FIDES, 2001, El Seguro Iberoamericano en Cifras (Mexico City: FIDES).

Instituto Brasileiro de Geografia e Estatística (IBGE), 2000, Censo Demográfico 2000 (Rio de Janeiro: IBGE).

_____, 2003, IBGE Pesquisa de Orçamentos Familiares 2002–2003, Primeiros Resultados (Rio de Janeiro: IBGE).

_____, 2004, IBGE Pesquisa de Orçamentos Familiares 2002–2003, Primeiros Resultados (Rio de Janeiro: IBGE).

_____, 2005, Dados Populacionais (Rio de Janeiro: IBGE).

Instituto Brasileiro de Opinião Pública e Estatística (IBOPE), 2005, Resumo da Audiência de Internet Domiciliar no Brasil e Perfil do Internauta Brasileiro (São Paulo: IBOPE).

Instituto de Pesquisa Econômica Aplicada (IPEA), 2005, PIB and Índice Nacional de Preços ao Consumidor Amplo (Brasília: IPEA).

International Telecommunication Union (ITU), 2005, World_Geo3, www.itu.int/ITU-D/ict/statistics/at_glance/Internet04.pdf.

Mercosur, 2004, Boletim Estatístico Anual do Setor de Seguros no Mercosul (Montevideo: Mercosur).

Ministério da Fazenda, 2004, Reformas Microeconômicas e Crescimento de Longo Prazo, Secretaria de Política Econômica (Brasília: Ministério da Fazenda).

Ministério da Previdência Social, 2004, Resultado do Regime Geral de Previdência Social (RGPS)—Dezembro 2004 (Brasília: Ministério da Previdência Social).

Ministerio de Economia y Producción, 2004, Situación del Mercado Asegurador en Argentina 1993–2003, Subsecretaría de Servicios Financieros, Secretaría de Finanzas (Buenos Aires: Ministerio de Economia y Producción).

Povoas, Manuel Soares, 2000, Na Rota das Instituições do Bem-Estar—Seguro e Previdência (São Paulo: Editora Green Forest do Brasil).

Rabelo, Flávio, 2003, "The Structure and Regulation of the Brazilian Private Pension System," Escola de Administração de Empresas de São Paulo (EAESP) of Getúlio Vargas Foundation (FGV) (São Paulo: FGV Escola de Administração de Empresas de São Paulo).

Sisbacen, 2005, Ranking da Instituições Financeiras em 2004 (Brasília: BACEN).

Superintendência de Seguros Privados (2005) Boletim de Seguros Anos de 1995 a 2005 (Rio de Janeiro: Superintendência de Seguros Privados).

Superintendencia de Valores y Seguros (SVS), 2004, Estadística de las Principales Variables del Mercado de Seguros Generales y Vida, Desde 1985 a la Fecha—Evolución de la Prima Directa, Prima Directa, Chile (Santiago: SVS).

Swiss Re, 2002, "Insurance in Latin America: Growth Opportunities and the Challenge to Increase Profitability," Sigma, No. 2.

World Bank Group, The, 2001, Brazil: Critical Issues in Social Security, World Bank Country Study, Report No. 22513, Volume 1, Washington DC.

_____, 2004–2005, World Development Indicators Online, www.worldbank.org, Washington DC.

15.7 LEXICON

Administrative Expenses. Administrative expenses are the general costs of running a company.

AFJP. Administradoras de Fondos de Jubilaciónes y Pensiones.

AFP. Administradoras de Fondos de Pensiones.

ANAPP. Associação Nacional da Previdência Privada is the association of insurance companies working with pension and life products. It was founded in 1974. See www.anapp.com.br.

ANS. Agência Nacional de Saúde is the body that regulates the health operators. See www.ans.gov.br.

Assistance Benefits. Assistance benefits are benefits that are not funded, but are given as a mean of assistance.

Avulse. Avulse is a denomination included in the proposal for reinsurance privatization that is under discussion in the Brazilian congress. This name would be given to reinsurance companies that would operate in Brazil without any local assets and could be based anywhere in the world. The literal definition of avulse is loose.

BACEN. Banco Central do Brasil or the Brazilian Central Bank.

BOVESPA. Bolsa de Valores de São Paulo or the Brazilian Stock Market is based in São Paulo and is the only Brazilian stock market. It is the biggest stock market in Latin America, with around 70% of its trading volume. See www.bovespa.com.br.

Capitalization (free translation to *Capitalização*). Capitalization is a savings-based product that has a lottery attached. It is regulated under the same body as insurance products.

Centro de Experiência e Segurança Viária-Cesvi Brasil. Centro de Experiência e Segurança Viária-Cesvi Brasil is the center of technical studies for automobile repairing costs and technical information for automobiles by model.

Closed Pension Fund. A closed pension fund is a non-profit pension fund that is restricted (or closed) to the employees of the sponsoring organization. It is also called a pension fund.

CNSP. Conselho Nacional de Seguros Privados is the highest regulatory body in the insurance industry. SUSEP is its executive body.

Coinsurance. Coinsurance is insurance held jointly by two or more insurers.

Combined Ratio. The combined ratio is the ratio of net losses, commercial expenses, administrative expenses, and tax expenses to earned premiums.

Commercial Expenses. Commercial expenses are expenses on brokerage.

Commercial Ratio. The commercial ratio is the ratio of expenses on brokerage to earned premiums.

Correção-monetária. Correção-monetária is freely translated as "monetary correction."

Defined Benefit (DB). Defined benefit is a way of funding retirement, where there is a guarantee on the benefit value.

Defined Contribution (DC). Defined contribution is a way of funding retirement, where there is no guarantee on the benefit value.

DPVAT. Danos Pessoais Causados por Veículos Automotores is mandatory third-party liability automobile insurance. There is no need to prove guilt in order to have coverage, so any person involved in a car accident in which there is bodily injury or even death is covered by the insurance. Premiums are collected by the government and the insurers operate a pool to manage it.

EAPP. Entidade Aberta de Previdência Privada are life insurers in the "open pension plans" market. Today, there are only a few non-profit EAPPs that still collect premiums.

Extended Combined Ratio. The extended combined ratio is the ratio of net losses, commercial expenses, administrative expenses, and tax expenses to earned premiums and financial returns.

Fenacor. Federação Nacional dos Corretores de Seguros Privados, de Capitalização, de Previdência Privada e das Empresas Corretoras de Seguros is the national federation of Brazilian brokers founded in 1968. See www.fenacor.org.br.

FENASEG. Federação Nacional das Empresas de Seguros Privados e de Capitalização is the national federation of Brazilian private insurance and capitalization founded in 1951. See www.fenaseg.org.br.

FIDES. Federación Interamericana de Empresas de Seguros is an organization comprised of insurance associations from Latin America, the U.S., and the Caribbean, founded in 1945.

Financial Return. The financial return is the amount earned on asset investments.

FPM. Fundo de Proteção dos Mercados de Seguros, Previdência Complementar Aberta e Capitalização is a fund proposed by SUSEP that would be formed by contributions from the whole insurance market. It would guarantee payments to losses under a certain limit for policyholders from bankrupted insurance companies.

FTAA. Free Trade Area of the Americas is a political and economical regional integration initiative that started in 1994 and includes all North and South American countries, except Cuba.

Funenseg. Fundação Escola Nacional de Seguros is the national foundation for technical education, responsible for the accreditation of brokers and maintained by the whole insurance market. See www.funenseg.org.br.

GDP. Gross domestic product.

Health Operator (*Operadora de Planos de Saúde*). Health Operator (*Operadora de Planos de Saúde*) is a private company that offers health coverage against a monthly premium. There are eight types of health operators which are described in Table 15.7.

IBA. Instituto Brasileiro de Atuária is the professional body of the actuarial profession in Brazil See www.atuarios.org.br.

IBGE. Instituto Brasileiro de Geografia e Estatística is the main supplier of data and information in the country for all levels. For instance, it is responsible for the Brazilian population census. See www.ibge.gov.br.

IBNR. Incurred, but not reported.

INdeR. Instituto Nacional de Reaseguro.

IPEA. Instituto de Pesquisa Econômica Aplicada is a government institute that publishes a substantial quantity of economic data. See www.ipeadata.gov.br.

IRB. Instituto de Resseguros do Brasil is the only company allowed to sell direct reinsurance coverage in Brazil. It is a company half-owned by the government and half-owned by the insurance market.

Local Office-Based Reinsurers. Local office-based reinsurers is a denomination included in the proposal for reinsurance privatization that is under discussion in the Brazilian congress. This name would be given to reinsurance companies that would operate in Brazil with a minimum capital requirement of US$5 million and would need a local office.

Loss Ratio. The loss ratio is net losses divided by earned premiums.

Mercosur. Mercado Comun del Sur is a political and economical regional integration initiative that started in 1991. It includes Argentina, Brazil, Paraguay, and Uruguay as current members.

Net Losses. Net losses are losses paid net of coinsurance, reinsurance, and retrocession including the variation on IBNR.

Net Written Premiums. Net written premiums are total gross premiums written by the insurer plus the retrocession premium, net of coinsurance, reinsurance, and reimbursement.

Open (Private) Pension Plans. Open (private) pension plans are pension plans commercialized either by insurers or EAPP to the general public. It can be sold to individuals or to companies that want to provide their employees with pension coverage but do not want to join a closed pension fund.

Penetration Index. A penetration index is calculated as the gross written premiums divided by GDP.

Pension Fund. See closed pension fund.

PGBL. Plano Gerador de Benefício Livre is the first defined contribution pension plan that was first sold in Brazil in 1998. It is expected that most policyholders who buy PGBL are, in fact, investors who are taking advantage of the tax reduction. PGBL literally translates to Free Benefit Generator Plan.

PREVIC. PREVIC is the Superintendência Nacional de Previdência Complementar. It was established in the beginning of 2005 to substitute SPC in the regulation of Brazilian pension funds. It is still being implemented and its website will be located under the Social Security Ministry. See www.mps.gov.br.

Regime Próprio. Regime Próprio is a type of pension fund that could be created in order to fund benefits to civil servants related to the direct administration (i.e., civil servants who are hired directly by the national, state, or municipal government).

Retention Rate. The retention rate is the proportion of premiums retained by the primary insurer rather than being ceded to the reinsurance market.

Skeletons. Skeletons are non-recognized private or public debts.

Southern Cone (Cone Sul). The Southern Cone or Cone Sul consists of the southernmost areas of South America. Typically, it refers to Argentina, Chile, Uruguay, Paraguay, and southeastern Brazil.

SPC. Secretaria de Previdência Complementar regulates Brazilian pension funds. PREVIC was created to substitute for it.

SSN. Superintendencia de Seguros de la Nación.

SUSEP. Superintendência de Seguros Privados is the executive body of CNSP. It is the supervisor of the EAPP, capitalization, and insurers, but not health operators. It was created in 1901. See www.susep.gov.br.

Tax Expenses. Tax expenses are the amount paid in taxes to the government by the company.

Total Local Reinsurers. Total local reinsurers is a denomination included in the proposal for reinsurance privatization that is under discussion in the Brazilian congress. This name would be given to reinsurance companies that would operate in Brazil with a minimum capital requirement of US$50 million.

VGBL. Vida Gerador de Benefício Livre is a defined contribution pension plan that was first sold in Brazil in 2002 as an option to PGBL. It has experienced a sales boom due to its tax reduction ability. It is thought that most policyholders who buy VGBL are, in fact, investors who are taking advantage of the tax reduction. VGBL literally translates to Free Benefit Generator Life.

Written Premiums. Written premiums are total gross premiums written by the insurance company excluding coinsurance.

16 European Insurance Markets: Recent Trends and Future Regulatory Developments[312]

Paul Klumpes

Imperial College London

Paul Fenn
Steven Diacon
Christopher O'Brien
Canan Yildirim

Nottingham University Business School

16.1 INTRODUCTION

The European insurance market is currently the subject of harmonization of legal, solvency and accounting rules. However, the demand for insurance is still driven by government incentives and by the extent of financial innovation and variations in consumer penetration and wealth across countries. Moreover, the capitalization and taxation incentives are still very focused on a national basis, and thus the existing regulations, industry structure, link between public and private insurance, and consumer demands for particular classes of both life and non-life insurance are idiosyncratic to each country.

The evolution of risk management, consumer sophistication, and market integration with the European Union has also been accompanied by increased regulation and greater transparency and accountability. A key theme of this chapter is to explore whether trends in European markets are becoming more homogeneous

[312] This research was funded by a KPMG-UIUC research grant under the Business Performance Measurement' Program. The authors are grateful to Rashad Abdel-Khalik for his guidance, and appreciate the research assistance provided by Tim Orton. Paul Klumpes dedicates this research in memory of Ab Flikkema. The chapter has benefited very significantly from many helpful comments provided by the editors, Bertrand Venard and David Cummins.

or more heterogeneous. In particular, evidence that solvency regulations or accounting regulations are more or less onerous may impact the credibility and sustainability of the insurance industry both within and without the European Union.[313]

The next section provides an overview of the structure of the European insurance industry, the importance of European Union regulation, and the impact of EU accession countries. It also outlines major developments in international harmonization of solvency and accounting regulation. Then, we provide overviews of each of the European insurance markets and, finally, a conclusion.

16.2 OVERVIEW OF EUROPEAN INSURANCE MARKET

16.2.1 Introduction

This section analyses the major trends and developments affecting the entire European insurance industry. It first identifies the major and minor national European markets to be studied and then discusses various institutional issues—such as regional trends in regulation, the role of European regulation, and the impact of EU accession—and new developments in international harmonization of solvency and accounting rules on the European insurance industry.

Tables 16.1 and 16.2 show the comparative figures for these countries, together with other smaller EU states of Austria, Belgium, Denmark, Finland, Greece, Ireland, Luxembourg, Portugal, and Sweden. Table 16.1 shows the ranking of each country by size (separately for life and non-life insurance) and indicates that European countries have the largest world share of the global insurance market outside the United States and Japan (ranked first and second in each market, respectively). France, Germany, and the United Kingdom are the largest European insurance markets for both the life and non-life insurance markets. Table 16.2 shows the insurance density (per capita) and penetration (as a percentage of GDP) for the life and non-life markets. Switzerland has the highest insurance density of all European markets. The United Kingdom and Switzerland have the highest insurance penetration for the life market and non-life markets, respectively.

Countries selected for inclusion in the major market are defined as those in which total premium income for non-life and life insurance are both at least 1 percent of the world share of the relevant insurance market (Tillinghast-Towers Perrin, 2004). This procedure resulted in the initial selection of the six largest EU markets of the United Kingdom, Germany, France, Italy, Spain, the Netherlands, and the non-EU state of Switzerland.

---

[313] The description of national markets is based on a synthesis of a variety of industry and academic sources. For each country overview, efforts are made to validate the analysis by identifying relevant industry information from multiple industry, government, and consumer organization sources. Implications for discriminating among alternative theoretical interpretations of the regulatory effectiveness of risk management practices by insurance firm managers and regulators in each country are also discussed.

Table 16.1. Ranking of European Insurance Markets in 2003

Country	Life Premium (US$m)	Life World Share, %	Life World Rank	Non-life Premium (US$m)	Non-life World Share, %	Non-life World Rank
UK	154,842	9.26	3	91,891	7.25	4
Germany	76,738	4.59	5	94,073	7.42	3
France	105,436	6.30	4	58,244	4.59	5
Italy	71,694	4.29	6	40,066	3.16	6
Spain	20,042	1.20	16	26,972	2.13	8
Netherlands	25,371	1.52	9	24,895	1.96	9
Switzerland	24,713	1.48	10	16,047	1.27	12
Belgium	21,004	1.26	14	12,810	1.01	14
Ireland	9,037	0.54	22	8,291	0.65	18
Sweden	14,297	0.85	17	6,742	0.53	20
Denmark	10,944	0.65	20	5,793	0.46	22
Finland	11,065	0.66	19	3,058	0.24	31
Austria	6,586	0.39	24	8,410	0.66	17
Portugal	6,122	0.37	26	4,688	0.37	24
Luxembourg	7,130	0.43	23	1,102	0.09	46
Greece	1,628	0.10	36	2,040	0.16	38
Norway	6,031	0.36	27	5,501	0.43	23

Source: Tillinghast (2003), p. 7.

Another nine European countries were identified for analysis. These comprised eight smaller EU states (Austria, Belgium, Denmark, Finland, Greece, Luxembourg, Portugal, and Sweden) and Norway, an Economic Area Association (EAA) member country that is subject to the third-generation EU directives. Iceland, another EAA member state, was excluded from the analysis, because there were insufficient institutional background data available.

16.2.2. Major Regional Trends in European Regulation

The European insurance market is subject to a mix of European Union legislation and regulations and national regulations. In addition, a significant number of new and proposed financial services directives and laws are expected to facilitate the integration of national insurance markets in the EU and EAA areas.

Prior to the recent development of EU regulations, Albert (1987) characterized the European market as having considerable regional differences in regulatory control. Two models of regulatory supervision were identified and briefly discussed by van den Berghe (1990, p. 215).

Table 16.2. European Market Ranking (Density and Penetration)

	Insurance Density Premiums Per Capita (US$)				Insurance Penetration Premiums As % Of GDP			
	Life		Non-life		Life		Non-life	
	$	Rank	$	Rank	%	Rank	%	Rank
Austria	811.0	14	1,035.7	8	2.59	14	3.30	6
Belgium	2,004.8	6	870.9	11	6.81	3	2.96	10
Denmark	2,037.5	5	1,078.5	7	5.18	7	2.74	11
Finland	2,126.8	4	587.7	15	6.81	3	1.88	16
France	1,767.9	7	930.5	10	5.99	6	3.15	9
Germany	930.4	13	1,120.8	6	3.17	12	3.82	4
Greece	152.1	17	190.7	17	0.93	17	1.17	17
Ireland	2,312.5	3	1,356.9	4	6.04	5	3.55	5
Italy	1,238.3	11	674.8	13	4.82	9	2.63	12
Luxembourg	1,161.1	12	1,335.0	5	2.09	16	2.40	13
Netherlands	1,561.7	9	1,532.4	2	4.93	8	4.84	2
Norway	1,322.5	10	998.8	9	2.79	13	2.10	15
Portugal	611.4	15	468.2	16	4.14	11	3.17	8
Spain	488.6	16	657.5	14	2.38	15	3.20	7
Sweden	1,602.3	8	755.6	12	4.74	10	2.23	14
Switzerland	3,431.8	1	2,228.5	1	7.72	2	5.02	1
UK	2,617.1	2	1,441.4	3	8.62	1	4.75	3

Source: Tillinghast (2003), p. 8

Under the normal control system or the "maritime model" (United Kingdom, Ireland, and the Netherlands), regulations are based on a previous license, coupled with financial guarantees, a solvency margin, sufficient technical reserves, congruent investments, and all in a publicity system involving ex post approval.

Under the material control system or the "Alpine model" (Austria, Germany, and Switzerland), the insurance market is highly regulated and covers not only all of the above regulations, but is primarily a prior approval system in which every tariff and every product must be accepted before any activity can begin.

It is unlikely that either model is descriptive of the insurance industry operating in these countries. First, the normal control system model as described implies a system of effective self-regulation. However, the deregulation of financial service firms and the separation of clients from professional advice have resulted in greater government interference (e.g., the U.K.'s Policyholder Protection Act of 1975 guarantees policy default; the Financial Services Authority has imposed new "realistic" reserves). There is a greater emphasis on disclosure and compliance with broader rules of professional conduct in various technical areas. Second, there have been a number of European Union directives on accounting, solvency, and registration of business, as well as integration of economic and social chapters.

On the other hand, the Alpine model can be criticized for effectively imposing a higher rate of tax on small, newer firms. Consequently, questions of feasibility and cost cast doubt on the overall efficiency of rate regulation (Munch and Smallwood, 1980). Moreover, the concept of rate regulation applies to products with capital guarantees; however, the increasing sophistication of consumers in taking on bundled insurance products that involve an aspect of investment risk has facilitated the development of unit-linked products in many European countries. Finally, the traditional demarcation of distribution channels implies the coexistence and separation of independent advisers and direct sales forces. In more recent times, insurance brokerage and intermediation also gradually have been overtaken by Internet-based sales strategies, thus reducing adherence to the strict intermediation practices that are implied by insurance rate regulation.

In addition, since our analysis extends to Scandinavian and Mediterranean countries, one might consider alternative models of regulation that may apply to these geographical areas. In both areas, the provision of insurance by private firms is more peripheral, because it is partly covered by the state or other noninsurance firms (e.g., occupational pensions). Because most countries (except Italy) are also relatively newer members of the EU, the development of homogenous insurance markets, regulation, and laws is relatively more recent than in central and western Europe.

Currently, the regulatory supervision of insurers in many European states is in a state of transition as regulatory authorities are seeking to merge with banking and investment supervisors. During the past ten years, this allocation of duties has become steadily less appropriate to developments in the financial sector. Increasingly, financial institutions are merging into larger organizations with both banking and insurance arms. At the same time, major similarities exist in the nature of the risks that these institutions are exposed to and the type of supervision that the two supervisory bodies perform. However, the lack of independence of supervisors is a concern, especially in light of efforts to harmonize international accounting and solvency regulations (Gugler, 2005). This issue is discussed further in section 4.

16.2.3 Impact of EU Harmonization Rules

The European Commission has promoted a single insurance market in order to facilitate greater economic efficiency and market integration. It has therefore sought to develop a common legal framework to allow insurers to operate throughout the European Union and to establish and provide services freely. This framework is also intended to protect consumers. Although the first directives on life and non-life insurance (79/267/EEC and 73/239/EEC, respectively; updated by the Third Life and Non-Life Directives, 92/96/EEC and 92/49/EEC, respectively) tried to harmonize the regulation of insurance activities in the European insurance industry, they succeeded only in coordinating the rules of financial supervision in relation to the solvency margin. Prior to this time, a financial services firm could face serious problems in offering services across borders, because certain regulations were imposed on all firms supplying services in the host state (Weatherill, 2004, p. 364). The third-generation directives thus facilitated the mutual recognition of insurance regulations by companies that offer insurance in more than one member state.

Nevertheless, EU directives are only binding on the member state as to the result to be achieved. Consequently, national authorities have some discretion about the form and method of implementation. The directives also have been transposed into national law (most countries implemented them by 1996). Member states continue to hold the autonomy to regulate or deregulate all other financial, technical, and commercial aspects of the insurance business. Currently, a single passport system applies whereby insurers authorized by prudential authorities in one member state can sell into another member state. However, the system relies on mutual recognition of the supervision exercised by different national authorities according to rules harmonized to the extent necessary at the EU level. These directives were largely implemented in member states by 1995.

The insurance accounts directive (91/674/EEC) has been more successful in standardizing reporting practices in the EU. This is discussed in more detail in section 2.5 below. Additionally, increasing standardization of solvency positions of insurers are currently being developed, and these are discussed in more detail in section 2.4. The European Union is currently implementing a Financial Services Action Plan, which aims to remove regulatory barriers to the freedom to provide insurance services throughout the European Union.

16.2.4. Impact of EU Accession Countries

This study focuses on the state of the European insurance industry in 2003. Since this chapter was written, the insurance industries of Poland, the Czech Republic, Slovenia, Slovakia, Cyprus, Malta, Hungary, Estonia, Latvia, and Lithuania have become subject to European Union laws. However, unlike established EU countries, prior to this time, the insurance markets in these (mostly) Eastern European countries were not very mature and featured significant tariffs and other forms of state intervention (Eurostat, 2004). Moreover, the historical incidence of extensive government interference in insurance markets in these countries, together with the importance of collaboration between international insurers and former domestic monopolies, significantly influence both regulation and market trends. Further, a relative lack of mature actuarial, insurance, and accounting professional bodies in these countries has inhibited the development of local regulations. Finally, these markets have significantly grown, while most established EU and EAA states have experienced consolidation and rationalization.

Prior to officially joining the EU, all the new countries were required to adjust their national regulations to meet EU requirements. However, given the historical dominance of former state monopolies in their insurance markets, the lack of integration of currency and interest rates with core European Union markets, and the continued reliance on government insurance programs rather than private company insurance policies to underwrite many forms of domestic and commercial activities, it is unlikely that the insurance markets in these countries were comparable to the countries analyzed.

16.2.5 Solvency Regulations

The solvency rules in force in the EU were introduced by the first non-life directive in 1973 and the first life insurance directive in 1979. These directives were subsequently amended, and insurance sector regulations in the European Union are now based essentially on the third-generation directives from 1992. In 1998, these directives were supplemented by a directive on insurance groups, and a directive on insurance intermediaries is now in force. An update of prudential standards also has been planned. The regulation of insurance supervision is therefore largely harmonized throughout the continent of Europe, even if in certain areas the directives leave considerable latitude for application (Organisation for Economic Cooperation and Development [OECD], 2003, p. 9). In 2002, the European Union issued new directives for life and non-life insurers (Directives 2002/83/EC and 2002/13/EC, respectively) to implement standardized solvency margin requirements (called Solvency I). Issued for implementation in 2004, these directives improve the calculation of the solvency margin requirement (i.e., the minimum amount of extra capital that an insurer must fall back on in unforeseen circumstances) and include setting a standardized minimum capital of at least three million euros.

The EU is developing a more wide-ranging review (called Solvency II) in order to further harmonize the determination of solvency using a risk-sensitive approach (Muller, 2003). This project examines the rules governing insurers' assets and liabilities, the matching of assets to liabilities, reinsurance arrangements, and implications of accounting and actuarial policies. The objective of Solvency II is to match the solvency requirements to the risk encountered by insurers and to encourage insurers to improve their measurement and monitoring of the risks they incur. The EU commissioned a study by KPMG (2002) that identified the need to reform solvency margin methodologies and the role of rating agencies, and identified more advanced risk reduction techniques to facilitate greater consistency in measurement. The proposals take into account significant risks and do not impose overly prudent requirements. These proposals are currently the subject of further regulatory discussions.

There is wide variation in existing solvency and accounting regulations governing insurers across the EU. These features are briefly outlined for the most mature insurers listed in Appendix 1. The developments in Solvency II, together with the growing importance of corporate governance, have wide-scale implications for the solvency of the EU insurance market.

Reinsurance firms are subject to various degrees of solvency regulation across Europe. Reinsurers are subject to similar supervisory rules as exist for direct insurers in Denmark, Sweden, and the United Kingdom. Other countries impose more limited supervision, such as financial supervision but no licensing procedure (e.g., France and Germany). By contrast, neither Belgium nor the Netherlands supervise reinsurers on the grounds that professional reinsurers deal only with clients that have the capability of assessing the solvency of firms with which they do business. The European Union is currently planning to introduce a directive on reinsurers.

All European countries use fixed-ratio models to establish solvency requirements as a fixed percentage of the value of a given variable that is assumed to be strongly correlated with a company's degree of risk exposure. Such variables involve a simple function of one or more items on the balance sheet or profit-and-loss account. For example, the Consolidated Life Directive merely requires some harmonization requirements for the valuation of assets and mathematical reserves. It requires minimum solvency capital of the sum of 4 percent of nonlinked reserves and 1 percent of linked reserves, 3/1000 of the sum at risk, and 16 to 18 percent of rider premiums or a claims-incurred basis if higher (Tillinghast-Towers Perrin, 2004). These models are simple to apply but can sometimes lead to illogical requirements. This can be resolved by using and comparing a number of different ratios. A premium-based ratio favorable to companies that undercharge can be offset by a claims-based ratio (OECD, 2003, p. 13).

By contrast, risk-based capital adequacy standards in the United States attempt to address the drawbacks of the fixed-ratio methods by incorporating all of the risks confronting an insurance undertaking—technical risks, investment risks, and other management risks. An effort is made to factor in the correlation between risks; values corresponding to each risk are not simply added together but are plugged into a more complex formula designed to reflect the correlation or independence of the various risks (OECD, 2003, p. 14).

In recent years, the life insurance industry has experienced significant solvency problems due to the combination of volatile equity markets and a historically low interest rate environment. This depleted the capital base of many European insurers between 2000 and 2003 and resulted in many firms either raising new capital or running into financial difficulty. Nine European countries (Austria, Belgium, Greece, Italy, Luxembourg, Norway, Portugal, Spain, and Sweden) are maintaining the minimum requirements without material amendments and are awaiting the formal introduction of Solvency II.

Tillinghast-Towers Perrin (2004) identified another seven countries that are moving, in various degrees, to supplement retrospective models with "prospective" models that calculate solvency requirements using historical data but also factor in assumptions about changes in a company's profile and the market. One approach is to undertake a stress- or risk-based approach, which is analogous to the approach adopted in the United States and is being tested in Finland, France, and Germany. A 1999 law in France requires an analysis of medium- and long-term solvency (OECD, 2003). Denmark also requires a mark-to-market of guaranteed liabilities based on deterministic calculations but does not take into account the value of options or guarantees. Three countries—Netherlands, Switzerland, and the United Kingdom—now require a full mark-to-market of assets and liabilities, with stochastic analysis. Risk capital is then assessed on a number of relevant adverse scenarios, and overall capital requirements are the greater of these "realistic" and current statutory requirements. Section 4 discusses these developments in the context of harmonization of International Financial Reporting Standards (IFRS) and International Accounting Standards (IAS) and discusses how these could impact current risk management practices.

16.2.6. International Accounting Rules

The ongoing developments in insurance accounting are important to both the Solvency II project and to financial reporting needs of shareholders and other users.

Presently, regulatory accounts deviate from publicly issued reporting, while accounting practices for insurance contracts vary widely throughout Europe and the rest of the world; current rules are summarized for the major European countries in the Appendix.

Until recently, there were no specific European Union-wide accounting standards applicable to insurers except for the standardized requirements of the European Union Accounts Directive (91/674/EC), which requires standardized layout of the balance sheet (section 4) and profit-and-loss account (section 5) and valuation rules at current value (section 7). The International Accounting Standards Board (IASB) has been working on a project since 1997 to develop an international standard for insurance accounting. However, it has found it difficult to deal with many practical and conceptual problems in insurance accounting. It therefore decided to develop a standard using a two-phase process. In July 2003, the IASB proposed regulating accounting practices for regulating the accounts of insurance contracts. Finally, in March 2004, the IASB issued IFRS 4.

This standard constitutes part of the IFRS that are applicable to all EU-listed companies from January 2005. It requires limited improvements in accounting practices for insurance contracts. However, it simply codifies existing practices; it does not address contentious issues such as accounting for financial assets and liabilities for insurers, which are presently exempted from the contentious "fair value" recognition and measurement requirements proposed by IAS 32 and IAS 39, respectively (which were amended in December 2003). More fundamental problems such as the valuation of "embedded derivatives" and the development of a fair-value approach to financial reporting have been postponed for the second (unspecified) phase of the project, which is not scheduled for completion until 2007 at the earliest.

The IASB's proposed standard is meeting extensive opposition from elements of the European insurance industry (e.g., Langendijk, Swagerman, and Verhoog, 2004; Dickinson and Liedtke, 2004). Although the European Union is committed to the use of international accounting standards from January 2005, it decided to adopt a "carve out" clause on more contentious aspects of IAS 39; particularly as regards the fair value of insurance liabilities related to embedded derivatives and insurance contracts that do not satisfy the definition of "insurance contracts," and thus are treated as "investment contracts" (e.g., "embedded derivatives"). Consequently, there is a potential conflict between IFRS as applied by most countries beginning in January 2005 and the "modified" IFRS adopted by the European Union for consolidated financial statements of all listed companies.

Although IFRS adoption can facilitate more uniform risk disclosure, it may also be addressed by Solvency II. Some industry observers have suggested that Solvency II should be modeled on Basel II, since both cover the same ground and affect the same organizations (Muller, 2003). However, while a consistent risk-based capital management approach can be adopted toward both standards, they also must take into account material differences between banking and insurance. While Basel II addresses credit and operational risk, it does not in itself provide an integrated risk framework for a bank. By contrast, Solvency II is more applicable to developing a consistent framework for risk analysis, risk capital, and capital allocation. The adoption of "realistic" capital reserve requirements for U.K. life insurers by the Financial Services Authority and relevant risk management rules provides an example for how this might be applied in practice.

These regulatory developments have important implications for the adequacy of existing arrangements for risk management and reinsurance of operations of insurers across the European Union. More importantly, the increasing transparency of earnings and financial reports has drawn attention to the need to align economic and accounting exposures. The remainder of this chapter overviews recent developments in national insurance markets and assesses the extent to which these harmonization efforts have been reflected in the convergence of business trends, capitalization, market concentration, and industry segmentation.

16.3 COUNTRY OVERVIEWS

16.3.1 Introduction

This section provides a brief description and overview of the state of each of the seventeen European insurance markets.[314] Each country overview contains brief descriptions of the market, business trends, and market and regulatory issues. The analysis is supplemented by empirical analysis of major business trends over the last ten years (1994–2003), obtained from Committee Europeene des Assurances (2004). Table 16.3 provides a summary of aggregate premium trends (in local currency), while Table 16.4 reports this data on a per capita basis, inflation adjusted to 1994, and in euros. Table 16.7 summarizes the market concentration of the top five life and non-life insurers operating in each country. Table 16.6 provides a summary of the solvency (free asset ratio of investments to total provisions) position of the industry. Table 16.5 provides some information about the five- and ten-year investment performance of the life and non-life insurance industries over the last decade. Table 16.8 summarizes trends in the total number of companies operating in each European country by ownership status (national, EU branch, or non-EU branch) since 1992. Finally, Table 16.9 shows the type of distribution channel employed by life and non-life insurers in selected countries.

Each of these major markets has experienced recent growth, especially in the newer investment or unit-linked product areas, but the adverse capital market conditions and the deterioration in capitalization and investment conditions have led to significant consolidation. The coming implementation of Solvency II, effective 2009, will force insurers to value first their assets, and then their liabilities, at market value. Significant exposure issues also arise from the implementation of more stringent and consistent risk reporting rules under the proposed IFRS. Consequently,

[314] Below is listed a brief synopsis of each insurance market. It should be noted that this analysis is largely drawn from relevant national market overviews contained within the national industry association and regulatory sources. Consequently, this analysis is limited in scope to the availability, completeness, and accuracy of these secondary sources. Each country overview contains a brief description of relevant growth trends, associations, and regulations, together with relevant discussion and analysis. Relevant industry citations are provided wherever possible.

Table 16.3. Premium Trends 1994–2003 (Local Currency Million)

Austria										
Year	1994	1995	1996	1997	1998	1999	2000	2001	2002	2003
Currency	ATS(m)	ATS(m)	ATS(m)	ATS(m)	EUR(m)	EUR(m)	EUR(m)	EUR(m)	EUR(m)	EUR(m)
Life	40,655	46,530	60,274	50,974	4,103	4,835	5,416	5,831	5,711	5,808
Non-life	80,671	83,368	84,566	84,608	6,120	6,125	6,290	6,639	6,999	7,419
Belgium										
Year	1994	1995	1996	1997	1998	1999	2000	2001	2002	2003
Currency	BEF(m)	BEF(m)	BEF(m)	BEF(m)	EUR(m)	EUR(m)	EUR(m)	EUR(m)	EUR(m)	EUR(m)
Life	160,260	181,786	211,677	251,651	8,289	10,025	12,909	13,235	14,400	18,138
Non-life	249,725	258,240	262,763	263,115	6,709	6,912	7,121	7,119	7,875	8,378
Denmark										
Year	1994	1995	1996	1997	1998	1999	2000	2001	2002	2003
Currency	DKK(m)	DKK(m)	DKK(m)	DKK(m)	DKK(m)	DKK(m)	DKK(m)	DKK(m)	DKK(m)	DKK(m)
Life	29,310	32,420	38,262	42,464	49,836	47,507	52,705	60,601	66,066	72,012
Non-life	23,594	25,000	27,441	29,258	30,119	31,264	32,617	34,240	36,637	38,612
France										
Year	1994	1995	1996	1997	1998	1999	2000	2001	2002	2003
Currency	FRF(m)	FRF(m)	FRF(m)	FRF(m)	EUR(m)	EUR(m)	EUR(m)	EUR(m)	EUR(m)	EUR(m)
Life	383,068	418,860	465,860	506,690	64,913	74,669	90,253	84,635	85,500	93,100
Non-life	230,890	245,588	253,298	254,720	38,749	39,354	41,082	43,424	46,502	50,170

Table 16.3 *(continued)*. Premium Trends 1994–2003 (Local Currency Million)

Germany										
Year	1994	1995	1996	1997	1998	1999	2000	2001	2002	2003
Currency	DEM(m)	DEM(m)	DEM(m)	DEM(m)	EUR(m)	EUR(m)	EUR(m)	EUR(m)	EUR(m)	EUR(m)
Life	82,959	88,406	92,890	98,770	52,505	58,749	61,225	63,287	65,068	68,600
Non-life	130,100	137,510	132,682	134,171	68,686	69,084	70,595	72,837	76,282	79,600
Finland										
Year	1994	1995	1996	1997	1998	1999	2000	2001	2002	2003
Currency	FIM(m)	FIM(m)	FIM(m)	FIM(m)	EUR(m)	EUR(m)	EUR(m)	EUR(m)	EUR(m)	EUR(m)
Life	23,642	29,703	36,664	36,783	7,078	8,150	9,401	9,367	9,680	9,478
Non-life	10,767	10,670	10,885	11,338	2,141	2,248	2,346	2,452	2,567	2,664
Greece										
Year	1994	1995	1996	1997	1998	1999	2000	2001	2002	2003
Currency	GRD(m)	GRD(m)	GRD(m)	GRD(m)	GRD(m)	GRD(m)	EUR(m)	EUR(m)	EUR(m)	EUR(m)
Life	190,078	218,047	248,011	286,095	337,289	456,413	1,321	1,292	1,310	1,435
Non-life	196,370	224,997	265,443	288,082	321,552	364,453	1,252	1,350	1,585	1,800
Italy										
Year	1994	1995	1996	1997	1998	1999	2000	2001	2002	2003
Currency	ITL(m)	ITL(m)	ITL(m)	ITL(m)	EUR(m)	EUR(m)	EUR(m)	EUR(m)	EUR(m)	EUR(m)
Life	18,625,431	23,225,393	26,062,611	36,985,614	26,483	35,597	39,784	46,329	55,294	62,261
Non-life	36,793,670	39,594,136	41,649,700	44,449,851	24,501	26,246	27,875	29,926	32,415	34,373

Table 16.3 *(continued)*. Premium Trends 1994–2003 (Local Currency Million)

Netherlands										
Year	1994	1995	1996	1997	1998	1999	2000	2001	2002	2003
Currency	NLG(m)	NLG(m)	NLG(m)	NLG(m)	EUR(m)	EUR(m)	EUR(m)	EUR(m)	EUR(m)	EUR(m)
Life	27,467	30,129	33,184	38,176	19,749	20,414	23,003	25,814	24,030	24,300
Non-life	26,079	27,380	28,735	30,080	14,490	15,389	16,544	17,655	20,087	21,950
Ireland										
Year	1994	1995	1996	1997	1998	1999	2000	2001	2002	2003
Currency	IEP(m)	IEP(m)	IEP(m)	IEP(m)	EUR(m)	EUR(m)	EUR(m)	EUR(m)	EUR(m)	EUR(m)
Life	1,782	1,689	2,055	2,572	3,564	4,793	5,986	7,168	7,253	7,978
Non-life	1,263	1,308	1,413	1,506	1,672	1,994	2,334	3,350	3,955	4,709
Portugal										
Year	1994	1995	1996	1997	1998	1999	2000	2001	2002	2003
Currency	PTE(m)	PTE(m)	PTE(m)	PTE(m)	EUR(m)	EUR(m)	EUR(m)	EUR(m)	EUR(m)	EUR(m)
Life	216,705	340,564	439,217	457,562	2,929	3,757	3,789	4,487	4,562	5,400
Non-life	411,863	442,345	486,272	508,822	2,699	2,929	3,275	3,502	3,852	4,062
Spain										
Year	1994	1995	1996	1997	1998	1999	2000	2001	2002	2003
Currency	ESP(m)	ESP(m)	ESP(m)	ESP(m)	EUR(m)	EUR(m)	EUR(m)	EUR(m)	EUR(m)	EUR(m)
Life	1,442,982	1,376,740	1,667,893	1,964,698	12,296	17,304	23,800	23,029	26,627	17,675
Non-life	2,033,903	2,143,782	2,122,172	2,216,671	13,375	15,023	17,051	19,034	21,596	23,832

Table 16.3 *(continued)*. Premium Trends 1994–2003 (Local Currency Million)

Sweden										
Year	1994	1995	1996	1997	1998	1999	2000	2001	2002	2003
Currency	SEK(m)	SEK(m)	SEK(m)	SEK(m)	SEK(m)	SEK(m)	SEK(m)	SEK(m)	SEK(m)	SEK(m)
Life	41,561	42,456	58,137	61,430	75,463	96,282	126,017	121,357	107,638	114,083
Non-life	34,401	37,242	36,567	36,634	35,693	37,214	40,110	42,931	47,772	54,989
Switzerland										
Year	1994	1995	1996	1997	1998	1999	2000	2001	2002	2003
Currency	CHF(m)	CHF(m)	CHF(m)	CHF(m)	CHF(m)	CHF(m)	CHF(m)	CHF(m)	CHF(m)	CHF(m)
Life	21,088	24,102	27,158	30,869	34,958	31,341	31,472	33,147	34,639	33,280
Non-life	13,201	13,436	13,288	13,372	13,170	14,337	15,852	17,610	18,394	19,310
United Kingdom										
Year	1994	1995	1996	1997	1998	1999	2000	2001	2002	2003
Currency	GBP(m)	GBP(m)	GBP(m)	GBP(m)	GBP(m)	GBP(m)	GBP(m)	GBP(m)	GBP(m)	GBP(m)
Life	44,185	46,135	55,529	62,755	74,245	91,725	120,727	104,332	103,553	92,000
Non-life	32,039	31,035	29,799	30,339	30,630	30,311	33,765	38,461	43,174	47,286

Source: CEA (2004), Table II-4 and Table II-5, pp. 160–161.

Table 16.4. Life and Non-Life Premiums for Various European Countries (per Capita, Inflation Adjusted 1994, Euros)

		1994	1995	1996	1997	1998	1999	2000	2001	2002	2003
Austria	Life	369.46	436.13	554.59	466.71	527.22	622.41	720.07	793.59	788.89	800.21
	Non-Life	733.11	781.41	778.10	774.66	786.40	788.47	836.27	903.56	966.81	1,022.17
Belgium	Life	398.96	463.41	530.91	628.54	850.59	1,038.18	1,367.19	1,428.76	1,572.65	2,012.41
	Non-Life	621.68	658.31	659.04	657.17	688.46	715.80	754.18	768.52	860.04	929.54
Denmark	Life	737.67	850.53	997.54	1,114.02	1,345.41	1,304.11	1,483.02	1,743.69	1,940.61	2,155.41
	Non-Life	593.81	655.87	715.42	767.57	813.12	858.23	917.78	985.20	1,076.17	1,155.90
Finland	Life	790.91	1,014.07	1,238.07	1,216.30	1,418.62	1,647.02	1,952.42	1,990.60	2,083.98	2,058.69
	Non-Life	360.19	364.28	367.56	374.91	429.11	454.30	487.22	521.08	552.64	578.64
France	Life	987.86	1,117.66	1,240.39	1,349.02	1,147.15	1,321.58	1,627.45	1,542.90	1,577.01	1,753.10
	Non-Life	595.42	655.31	674.43	678.17	684.77	696.53	740.79	791.62	857.71	944.71
Germany	Life	525.30	574.53	591.20	629.03	667.60	750.76	793.20	835.00	867.97	924.23
	Non-Life	823.80	893.64	844.46	854.49	873.34	882.83	914.59	961.00	1,017.56	1,072.43
Greece	Life	56.71	66.78	82.77	99.78	116.35	156.21	158.40	154.25	161.21	182.99
	Non-Life	58.59	68.90	88.59	100.47	110.92	124.74	150.13	161.18	195.05	229.53
Ireland	Life	610.07	574.87	772.44	939.63	1,016.77	1,375.83	1,794.06	2,221.91	2,298.97	2,616.04
	Non-Life	432.39	445.19	531.13	550.19	477.01	572.38	699.52	1,038.42	1,253.61	1,544.11
Italy	Life	155.33	194.68	247.02	351.59	497.74	678.83	776.56	936.31	1,163.38	1,345.25
	Non-Life	306.84	331.88	394.75	422.55	460.49	500.51	544.10	604.80	682.01	742.69
Luxembourg	Life	222.32	637.94	547.77	898.25	1,025.92	1,187.79	1,447.10	1,330.11	1,362.33	1,390.01
	Non-Life	130.00	136.71	144.19	158.11	160.86	167.77	200.75	225.78	239.02	243.88

Table 16.4 *(continued)*. Life and Non-Life Premiums for Various European Countries (per Capita, Inflation Adjusted 1994, Euros)

		1994	1995	1996	1997	1998	1999	2000	2001	2002	2003
Netherlands	Life	821.16	924.06	998.11	1,143.90	1,328.09	1,402.43	1,606.14	1,869.88	1,774.61	1,833.57
	Non-Life	779.66	839.75	864.29	901.31	974.43	1,057.21	1,155.15	1,278.82	1,483.42	1,656.25
Norway	Life	520.57	521.06	567.80	742.89	735.04	813.20	864.61	1,132.61	1,376.55	1,364.15
	Non-Life	663.41	659.97	706.65	722.81	715.61	914.46	1,101.69	901.92	1,047.42	1,065.60
Portugal	Life	107.46	174.27	232.91	240.02	318.17	417.03	111.87	533.97	559.08	660.93
	Non-Life	204.24	226.35	257.87	266.91	293.19	329.56	96.69	416.75	472.07	497.16
Spain	Life	217.72	220.22	268.43	315.32	335.73	482.89	686.27	674.30	795.22	543.94
	Non-Life	306.88	342.91	341.54	355.76	365.20	419.23	491.66	557.32	644.97	733.42
Sweden	Life	504.36	550.07	772.02	809.95	917.47	1,281.86	1,667.33	1,538.82	1,431.69	1,571.07
	Non-Life	417.47	482.52	485.58	483.02	433.95	495.45	530.69	544.37	635.41	757.27
Switzerland	Life	85.50	97.37	113.18	115.48	141.74	129.68	129.68	149.34	158.58	150.68
	Non-Life	53.53	54.28	55.38	50.02	53.40	59.33	65.32	79.34	84.21	87.43
UK	Life	930.52	931.27	1,309.07	1,687.33	1,941.77	2,819.48	3,698.65	3,363.98	3,352.94	2,714.30
	Non-Life	674.73	626.47	702.50	815.74	801.08	931.71	1,034.44	1,240.10	1,397.93	1,395.09

Source: CEA (2003), Tables I-3 and I-4, pp. 155–156; Tables II-4 and II-5, pp. 160–161.

Table 16.5. Investment Performance for Various European Countries

Country	Type	Attribute	1992	1997	2002	2003
Austria	Life	Portfolio (EURm)	14,237	26,714	38,850	40,625
		Return%	10.6	7.8	4.6	
	Non-life	Portfolio (EURm)	9,904	14,535	14,918	15,328
		Return%	4.2	0.5	2.7	
Belgium	Life	Portfolio (EURm)	27,106	48,964	85,084	99,610
		Return%	12.1	11.7	17.1	
	Non-life	Portfolio (EURm)	15,228	25,078	27,804	28,985
		Return%	6.2	2.1	4.2	
Denmark	Life	Portfolio (EURm)	49,579	85,542	127,360	139,097
		Return%	9.9	8.3	9.3	
	Non-life	Portfolio (EURm)	11,209	13,387	14,822	15,939
		Return%	2.8	2.1	8.2	
Finland	Life	Portfolio (EURm)	17,695	36,955	63,722	73,638
		Return%	13.7	11.5	9.9	
	Non-life	Portfolio (EURm)	4,314	9,157	7,960	8,295
		Return%	6.3	–2.8	4.2	
France	Life	Portfolio (EURm)	198,842	526,273	805,751	890,247
		Return%	15.0	8.9	9.8	
	Non-life	Portfolio (EURm)	59,006	92,131	109,673	125,641
		Return%	6.4	3.5	13.9	
Germany	Life	Portfolio (EURm)	273,846	421,857	593,084	609,337
		Return%	8.0	7.1	2.7	
	Non-life	Portfolio (EURm)	152,142	259,558	372,896	450,175
		Return%	10.5	9.7	9.0	
Greece	Life	Portfolio (EURm)		2,299.0	4,040.0	3,900.0
		Return%		11.9	–3.5	
	Non-life	Portfolio (EURm)		917	1,780	NA
		Return%		14.2	NA	
Ireland	Life	Portfolio (EURm)	11,914	25,955	38,047	44,575
		Return%	12.3	7.9	17.2	
	Non-life	Portfolio (EURm)	4,332	8,288	8,597	9,626
		Return%	7.1	0.7	12.0	
Italy	Life	Portfolio (EURm)	41,999	95,917	250,751	297,899
		Return%	19.6	21.2	18.8	
	Non-life	Portfolio (EURm)	32,679	43,827	59,744	67,486
		Return%	6.9	7.8	5.7	

Table 16.5 *(continued).* Investment Performance for Various European Countries

Country	Type	Attribute	1992	1997	2002	2003
Netherlands	Life	Portfolio (EURm)	98,516	172,922	223,767	238,222
		Return%	8.5	5.3	6.5	
	Non-life	Portfolio (EURm)	12,380	22,658	27,073	29,317
		Return%	8.1	3.6	8.3	
Norway	Life	Portfolio (EURm)	23,472	35,082	59,026	56,514
		Return%	9.7	11.0	5.1	
	Non-life	Portfolio (EURm)	6,919	9,608	14,198	13,334
		Return%	7.5	8.1	3.1	
Portugal	Life	Portfolio (EURm)	1,959	9,836	20,777	23,177
		Return%	26.6	16.1	11.5	
	Non-life	Portfolio (EURm)	2,380	4,069	5,205	5,295
		Return%	8.1	5.0	2.0	
Spain	Life	Portfolio (EURm)	15,016	43,164	82,298	91,111
		Return%	18.5	13.8	10.7	
	Non-life	Portfolio (EURm)	9,637	13,601	27,432	30,370
		Return%	11.0	15.1	10.7	
Sweden	Life	Portfolio (EURm)	56,527	118,570	125,550	191,631
		Return%	8.3	1.2	12.0	
	Non-life	Portfolio (EURm)	17,015	35,266	30,808	36,836
		Return%	6.1	–2.7	9.6	
Switzerland	Life	Portfolio (EURm)	83,703	139,830	200,791	195,688
		Return%	9.1	7.5	2.0	
	Non-life	Portfolio (EURm)	29,494	47,290	74,175	73,712
		Return%	9.7	9.4	4.0	
UK	Life	Portfolio (EURm)	462,604	955,262	1,311,280	1,331,914
		Return%	11.0	6.5	14.2	
	Non-life	Portfolio (EURm)	84,685	130,231	112,316	101,286
		Return%	2.9	–2.9	1.4	

Source: CEA (2003), Table A:V2–3, pp. 63–67.

Table 16.6. Free Asset Ratio for Various European Countries

		1994	1995	1996	1997	1998	1999	2000	2001	2002	2003
Austria	Life	1.19	1.04	1.02	0.98	0.99	0.97	1.02	0.96	0.95	1.64
	Non-Life	1.56	1.87	2.02	2.22	2.13	2.08	1.91	1.89	1.88	1.88
Belgium	Life	1.17	1.04	1.19	1.21	1.30	1.20	1.16	1.13	1.08	1.08
	Non-Life	1.35	1.19	1.40	1.44	1.53	1.45	1.37	1.35	1.05	1.05
Denmark	Life	1.12	1.12	1.13	1.14	1.16	1.20	1.19	1.15	1.12	1.18
	Non-Life	1.52	1.67	1.84	1.85	1.83	1.64	2.07	2.02	2.03	2.09
Finland	Life	0.88	0.96	1.01	1.06	1.13	1.16	1.07	1.01	1.01	1.10
	Non-Life	1.24	1.44	1.62	1.74	2.04	2.93	2.47	1.42	1.17	1.22
France	Life	1.01	1.10	1.15	1.15	1.19	1.16	1.16	1.14	1.12	1.24
	Non-Life	1.00	1.73	1.26	1.35	1.41	1.35	1.39	1.28	1.22	1.41
Germany	Life	1.02	1.01	1.01	1.00	1.00	1.01	1.00	1.00	1.01	1.04
	Non-Life	1.00	1.00	1.00	1.00	1.00	1.00	1.00	1.00	1.00	1.00
Greece	Life	0.97	0.97	0.97	0.96	0.95	0.86	1.02	0.97	0.90	0.87
	Non-Life	0.97	0.97	0.97	0.96	0.95	0.86	1.02	0.97	0.90	1.07
Ireland	Life	1.00	1.00	1.00	0.10	1.00	1.00	1.00	1.00	1.00	1.00
	Non-Life	1.73	1.57	1.73	2.14	1.20	1.28	1.31	1.86	2.17	2.39
Italy	Life	1.20	1.15	1.13	1.11	1.13	1.12	1.14	1.11	1.10	1.22
	Non-Life	1.27	1.20	1.19	1.15	1.10	1.07	1.08	1.15	1.14	1.18
Luxembourg	Life	2.12	0.98	40.56	41.62	1.03	1.03	1.02	1.02	1.02	1.02
	Non-Life	1.02	1.02	1.13	1.29	1.15	1.03	1.08	1.06	1.03	1.03

Table 16.6 *(continued)*. Free Asset Ratio for Various European Countries

		1994	1995	1996	1997	1998	1999	2000	2001	2002	2003
Netherlands	Life	1.06	1.20	1.22	1.26	1.39	1.26	1.23	1.21	1.16	1.10
	Non-Life	1.12	1.23	1.29	1.35	1.38	1.35	1.26	1.45	1.36	1.22
Norway	Life	1.30	1.36	1.40	1.45	0.72	0.75	0.71	0.68	0.62	0.62
	Non-Life	1.27	1.45	1.40	1.42	1.37	1.87	1.48	1.30	1.40	1.45
Portugal	Life	1.13	1.19	1.15	1.17	1.13	1.11	1.07	1.01	0.96	1.16
	Non-Life	1.21	1.26	1.23	1.16	1.17	1.26	1.21	1.04	1.07	1.18
Spain	Life	0.93	1.00	1.01	0.95	0.97	0.87	0.80	0.81	0.80	0.85
	Non-Life	0.98	0.95	0.99	1.04	1.18	1.35	1.54	1.78	2.09	2.21
Sweden	Life	1.18	1.14	1.02	1.07	1.06	1.01	1.04	1.05	0.85	1.07
	Non-Life	1.09	1.26	1.85	1.69	1.90	1.68	1.71	1.47	1.13	1.32
Switzerland	Life	1.54	1.49	1.38	1.34	1.34	1.35	1.29	1.30	1.28	1.34
	Non-Life	2.50	2.58	3.11	3.16	3.08	3.06	3.38	3.72	3.60	3.81
United Kingdom	Life	0.98	1.01	0.99	1.00	0.99	1.03	0.99	0.95	0.90	1.02
	Non-Life	0.88	0.88	0.91	0.89	0.93	0.81	0.75	0.72	0.78	0.87

Source: CEA (2003), Tables EII-7 and EII-8, pp. 163–164; Tables EII9 and EII10, pp. 165–166.

Table 16.7. Market Concentration Ratios in European Insurance Markets

	Non-Life Company	%	Life Company	%
Austria	Weiner Stadtische	10.0	Sparkassen-Versicherung	13.5
	Allianz Elementar	11.0	Wiener Stadtische	12.5
	Uniqa sach	10.5	Raiffeisen	10.3
	Generali	9.2	UNIQA Personen	7.3
	Uniqa Versicherungen	7.5	Generali	7.0
		48.2		50.6
Belgium	Fortis	23.0	AXA	17.1
	Ethias	17.7	Fortis AG	14.6
	AXA	12.2	Ethias	11.9
	KBC Assurances/CBC Assurances	11.1	KBC Assurances / CBC Assurances	8.6
	ING	10.1	Winterthur	6.8
		74.1		59.0
Denmark	Tryg	21.7	Danica	19.0
	Topdanmark	18.7	PFA Pension	17.4
	Codan	13.5	Nordea Liv & Pension	8.3
	Alm. Brand	10.3	Kommunernes Pensionsforsikring (KP)	7.6
	If…	5.6	Industriens Pension	5.1
		69.8		57.4
Finland	If P&C Insurance	31.7	Nordea	35.7
	Pohjola Group	21.3	Sampo Life	34.5
	Tapiola Group	17.0	OP-Life	14.8
	Fennia	8.6	Suomi Life	5.2
	Local Insurance Group	3.7	Tapiola Group	4.3
		82.3		94.5
France	AXA	15.0	CNP	15.6
	Groupama	15.0	AXA	11.5
	AGF	11.0	Predica	9.4
	MAAF	9.0	AGF	6.1
	MACIF	6.0	Victoria	5.6
		56.0		48.2
Germany	Allianz	13.2	Allianz	15.1
	AXA Colonia	4.2	Hamburg-Mannheimer	4.9
	Gerling	3.9	Aachener und Munchener	4.6
	R&V	3.6	R&V	4.3
	Zurich	3.0	Deutsche Herold	4.1
		27.9		33.0

Table 16.7 *(continued)*. Market Concentration Ratios in European Insurance Markets

	Non-Life Company	%	Life Company	%
Ireland	Hibernian	21.8	Irish Life	22.6
	Allianz Ireland	14.1	Ark Life	10.3
	AXA	12.7	Hibernian Life	8.7
	Royal & SunAlliance	11.1	Friends First Life	7.7
	Eagle Star Ireland	9.0	Eagle Star Life	7.3
		68.7		56.6
Italy	Fondiaria-SAI	10.5	Creditas Vita	7.9
	Generali	9.8	San Paolo Vita	6.6
	RAS	7.9	Alleanza Assoicurazioni	6.2
	Assitalia	5.2	Fideuram Vita	5.5
	Milano	4.7	Poste Vita	5.1
		38.1		31.3
Netherlands	Achmea	13.8	ING	19.7
	Fortis	11.4	Aegon	15.3
	ING	9.4	Fortis	13.5
	DeltaLloyd	7.2	Aviva	10.9
	Rabobank	6.0	Achmea	7.7
		47.8		67.1
Norway	KLP	28.9	KLP	21.6
	Gjensidge	25.8	Storebrand	18.5
	Vesta	23.4	Vital	14.6
	Spb1	10.9	Gh.Nor.Sp.	12.4
	Gard	3.0	Nordea Liv.	6.5
		92.0		73.6
Portugal	Fidelidade	10.4	Occidental Vida	31.6
	Imperio	9.1	Tranquilidade Vida	18.9
	Tranquilidade	8.9	Fidelidade	16.2
	Axa Portugal Seguros	8.7	Santander Vida	6.5
	Mundial Confianca	8.7	AXA Vida	3.4
		45.8		76.6
Spain	Mapfre Group	7.9	Vida Caixa	11.9
	Allianz Group	6.6	Antares	11.8
	Axa Group	4.8	Mpfre Vida	8.5
	Zurich Group	4.5	BBVA Seguros	5.9
	Winterthur Group	4.0	SCH	5.6
		27.8		43.7

Table 16.7 *(continued)*. Market Concentration Ratios in European Insurance Markets

	Non-Life Company	%	Life Company	%
Sweden	LF Insurance Group	29.2	Skandia	24.4
	If Skadeforsakring	22.9	SEB	14.0
	Trygg-Hansa	18.1	Folksam	14.5
	Folksam	16.4	Handelsbanken	18.2
	Zurich	3.4	Lansforsakringar	9.1
		90.0		80.2
United Kingdom	Royal & Sun Alliance	12.9	Barclays	10.9
	CGU	11.1	Standard life	10.5
	Zurich	7.3	Prudential	8.5
	AXA	7.2	Norwich Union	8.1
	Norwich	5.4	Clerical Medical	4.5
		43.9		42.5

Source: CEA (2003), Table A:II-3 and AII-4, pp. 31–34.

many insurers in the most developed markets have significant exposure to equity. This is especially a problem in the heavily regulated German and Swiss life insurance markets, where smaller insurers are relatively undercapitalized.[315]

16.3.2 Austria

Market Overview

Insurance supervision is under the responsibility of the Ministry of Finance. On April 1, 2002, this duty became part of the newly established Financial Market Supervisory Authority. The insurance undertaking can be formed either as a joint stock company or a mutual association. In early 2004, there were 58 domestic insurers (excluding small mutual associations) listed on the Web site of the Austrian Financial Market Supervisory Authority (www.fma.gv.at). Another 13 nondomestic (EU and non-EU) companies also operate branches in Austria (Committee Europeene des Assurances (CEA) 2004–2005).

Following the significant increase in the number of companies until the mid-1990s, there has been a concentration process taking place in recent years. The number of EU branches has increased by 50 percent, while the number of national firms has decreased by 8 percent.

[315] Extreme caution must be exercised in the interpretation of national industry trends, since they are based on incomplete data provided by either industry association sources or regulatory authorities. Thus, the analysis critically relies on the aggregate market analysis provided by members of the associations and may not be representative of the type of firm against which a firm may wish to benchmark.

Table 16.8. Number of Companies Operating in European Insurance Markets

Country	Type of company	1992	1992%	1997	1997%	2002	2002%	2003	2003%	Growth %
Austria	National	64	86.5	66	85.7	59	81.9	58	81.7	–7.8
	EU Branches	6	8.1	8	10.4	9	12.5	9	12.7	50.0
	Non-EU Branches	4	5.4	3	3.9	4	5.6	4	5.6	0.0
	Total	74	100.0	77	100.0	72	100.0	71	100.0	–2.7
Belgium	National	175	63.4	153	63.2	103	72.0	118	62.4	–41.1
	EU Branches	85	30.8	81	33.5	37	25.9	66	34.9	–56.5
	Non-EU Branches	16	5.8	8	3.3	3	2.1	5	2.6	–81.3
	Total	276	100.0	242	100.0	143	100.0	189	100.0	–48.2
Denmark	National	235	100.0	264	100.0	240	100.0	214	100.0	2.1
	EU Branches	0	0.0	0	0.0	0	0.0	0	0.0	
	Non-EU Branches	0	0.0	0	0.0	0	0.0	0	0.0	
	Total	235	100.0	264	100.0	240	100.0	214	100.0	2.1
Finland	National	54	94.7	51	82.3	47	69.1	46	67.6	–13.0
	EU Branches	1	1.8	9	14.5	19	27.9	21	30.9	
	Non-EU Branches	2	3.5	2	3.2	2	2.9	1	1.5	
	Total	57	100.0	62	100.0	68	100.0	68	100.0	19.3

Table 16.8 *(continued)*. Number of Companies Operating in European Insurance Markets

Country	Type of company	1992	1992%	1997	1997%	2002	2002%	2003	2003%	Growth %
France	National	469	76.4	428	78.8	382	77.2	373	76.7	–18.6
	EU Branches	112	18.2	96	17.7	99	20.0	101	20.8	–11.6
	Non-EU Branches	33	5.4	19	3.5	14	2.8	12	2.5	–57.6
	Total	614	100.0	543	100.0	495	100.0	486	100.0	–19.4
Germany	National	694	89.5	620	86.7	604	85.9	605	86.1	–13.0
	EU Branches	61	7.9	82	11.5	86	12.2	87	12.4	41.0
	Non-EU Branches	20	2.6	13	1.8	13	1.8	11	1.6	–35.0
	Total	775	100.0	715	100.0	703	100.0	703	100.0	–9.3
Ireland	National	51	53.1	106	69.7	139	72.8	178	79.5	172.5
	EU Branches	42	43.8	43	28.3	50	26.2	44	19.6	19.0
	Non-EU Branches	3	3.1	3	2.0	2	1.0	2	0.9	–33.3
	Total	96	100.0	152	100.0	191	100.0	224	100.0	99.0
Italy	National	218	81.3	215	82.4	197	77.6	189	75.9	–9.6
	EU Branches	39	14.6	40	15.3	54	21.3	57	22.9	38.5
	Non-EU Branches	11	4.1	6	2.3	3	1.2	3	1.2	–72.7
	Total	268	100.0	261	100.0	254	100.0	249	100.0	–5.2

Table 16.8 *(continued)*. Number of Companies Operating in European Insurance Markets

Country	Type of company	1992	1992%	1997	1997%	2002	2002%	2003	2003%	Growth %
Luxembourg	National	49	68.1	72	79.1	81	85.3	81	85.3	65.3
	EU Branches	18	25.0	16	17.6	13	13.7	13	13.7	–27.8
	Non-EU Branches	5	6.9	3	3.3	1	1.1	1	1.1	–80.0
	Total	72	100.0	91	100.0	95	100.0	95	100.0	31.9
Netherlands	National	388	72.1	364	72.9	329	72.6	320	72.2	–15.2
	EU Branches	150	27.9	114	22.8	110	24.3	110	24.8	–26.7
	Non-EU Branches	0	0.0	21	4.2	14	3.1	13	2.9	
	Total	538	100.0	499	100.0	453	100.0	443	100.0	–15.8
Norway	National	83	85.6	82	86.3	96	100.0	96	100.0	15.7
	EU Branches	11	11.3	12	12.6	0	0.0	0	0.0	
	Non-EU Branches	3	3.1	1	1.1	0	0.0	0	0.0	
	Total	97	100.0	95	100.0	96	100.0	96	100.0	–1.0
Portugal	National	42	52.5	48	49.5	44	53.0	41	52.6	4.8
	EU Branches	35	43.8	46	47.4	37	44.6	36	46.2	5.7
	Non-EU Branches	3	3.8	3	3.1	2	2.4	1	1.3	–33.3
	Total	80	100.0	97	100.0	83	100.0	78	100.0	3.8

Table 16.8 *(continued)*. Number of Companies Operating in European Insurance Markets

Country	Type of company	1992	1992%	1997	1997%	2002	2002%	2003	2003%	Growth %
Spain	National	404	94.0	346	94.0	297	88.9	287	87.5	–26.5
	EU Branches	23	5.3	14	3.8	35	10.5	39	11.9	52.2
	Non-EU Branches	3	0.7	8	2.2	2	0.6	2	0.6	–33.3
	Total	430	100.0	368	100.0	334	100.0	328	100.0	–22.3
Sweden	National	510	97.1	464	94.7	420	93.8	412	93.6	–17.6
	EU Branches	14	2.7	24	4.9	22	4.9	22	5.0	57.1
	Non-EU Branches	1	0.2	2	0.4	6	1.3	6	1.4	
	Total	525	100.0	490	100.0	448	100.0	440	100.0	–14.7
Switzerland	National	97	78.9	104	76.5	103	72.0	104	73.8	6.2
	EU Branches	21	17.1	29	21.3	37	25.9	33	23.4	76.2
	Non-EU Branches	5	4.1	3	2.2	3	2.1	4	2.8	–40.0
	Total	123	100.0	136	100.0	143	100.0	141	100.0	16.3
United Kingdom	National	682	82.9	681	81.0	637	79.0	594	76.9	–6.6
	EU Branches	60	7.3	76	9.0	97	12.0	102	13.2	61.7
	Non-EU Branches	81	9.8	84	10.0	72	8.9	76	9.8	–11.1
	Total	823	100.0	841	100.0	806	100.0	772	100.0	–2.1

Source: CEA (2003), Table A:II-2, pp. 28–30

Table 16.9. Distribution Channels in European Insurance markets

		Life 1997 %	Life 2001 %	Life 2002 %	Life 2003 %	Non-Life 1997 %	Non-Life 2001 %	Non-Life 2002 %	Non-Life 2003 %
Austria	Employees	31.8	22.4	22.2	26.9	68.8	50.8	48.8	46.0
	Agents	2.4	4.9	3.9	6.0	1.8	7.8	8.8	13.1
	Brokers	14.6	16.0	16.5	15.7	23.1	29.7	31.4	29.1
	Other network	47.8	53.3	55.3	47.9	2.1	4.4	5.2	5.8
	Others	3.4	3.4	2.1	3.5	4.2	7.3	5.8	6.1
	Total	100.0	100.0	100.0	100.0	100.0	100.0	100.0	100.1
Belgium	Employees	3.2	2.0	2.2		2.3	2.5		
	Agents	5.9	4.3	3.4		10.1	8.7		
	Brokers	40.4	23.6	22.5		69.6	62.9		
	Other network	32.4	52.6	50.8		3.6	10.0		
	Others	18.1	17.5	21.1		14.4	15.9		
	Total	100.0	100.0	100.0		100.0	100.0		
France	Employees	16.0	17.0	16.0	16.0	35.0	36.0	35.0	35.0
	Agents	10.0	8.0	8.0	8.0	37.0	35.0	35.0	35.0
	Brokers	7.0	9.0	9.0	9.0	19.0	18.0	19.0	19.0
	Other network	61.0	60.0	61.0	62.0	6.0	8.0	8.0	8.0
	Others	6.0	6.0	6.0	5.0	3.0	3.0	3.0	3.0
	Total	100.0	100.0	100.0	100.0	100.0	100.0	100.0	100.0
Ireland	Employees	15.5	21.5	19.5	15.0				
	Agents	19.5	26.0	24.0	23.5				
	Brokers	62.5	49.5	54.5	55.5				
	Other network	0.0	0.0	0.0	0.0				
	Others	2.5	3.0	2.0	6.0				
	Total	100.0	100.0	100.0	100.0				

Table 16.9 *(continued)*. Distribution Channels in European Insurance markets

		Life 1997 %	Life 2001 %	Life 2002 %	Life 2003 %	Non-Life 1997 %	Non-Life 2001 %	Non-Life 2002 %	Non-Life 2003 %
Italy	Employees	15.5	8.8	8.9	10.9	4.4	4.8	5.3	6.0
	Agents	35.2	17.9	19.6	18.3	88.3	86.8	86.1	85.2
	Brokers	1.4	1.0	0.9	0.8	6.8	7.4	7.5	7.5
	Other network	47.9	72.3	70.6	70.0	0.5	1.0	1.1	1.3
	Others	0.0	0.0	0.0	0.0	0.0	0.0	0.0	0.0
	Total	100.0	100.0	100.0	100.0	100.0	100.0	100.0	100.0
Netherlands	Employees		21.0	26.0	26.0		29.0	31.0	31.0
	Agents		49.0	57.0	57.0		52.0	59.0	59.0
	Brokers						11.0	10.0	10.0
	Other network		17.0	17.0	17.0		8.0		
	Others		13.0	4.0	4.0		0.0	0.0	0.0
	Total		100.0	104.0	104.0		100.0	100.0	100.0
Portugal	Employees	6.0	4.0	4.7	4.7	15.1	12.3	11.1	11.3
	Agents	20.2	12.2	13.0	13.0	57.6	55.6	54.8	57.0
	Brokers	1.9	1.1	1.5	1.5	15.4	16.3	18.7	16.5
	Other network	68.5	81.9	79.5	79.5	4.7	11.0	11.0	11.5
	Others	3.4	0.8	1.3	1.3	7.2	4.8	4.4	3.7
	Total	100.0	100.0	100.0	100.0	100.0	100.0	100.0	100.0
Spain	Employees		0.0	0.0	0.0		0.0	0.0	0.0
	Agents		14.8	13.4	16.4		45.3	41.5	40.6
	Brokers		4.9	9.2	6.5		22.5	25.7	28.4
	Other network		64.5	66.9	66.7		4.3	6.5	6.8
	Others		15.8	10.5	10.4		27.9	26.3	24.2
	Total		100.0	100.0	100.0		100.0	100.0	100.0

Table 16.9 *(continued)*. Distribution Channels in European Insurance markets

		Life 1997 %	Life 2001 %	Life 2002 %	Life 2003 %	Non-Life 1997 %	Non-Life 2001 %	Non-Life 2002 %	Non-Life 2003 %
Sweden	Employees	28.0	28.0	28.0	28.0				
	Agents	0.0	0.0	0.0	0.0				
	Brokers	19.0	19.0	19.0	19.0				
	Other network	45.0	45.0	45.0	45.0				
	Others	8.0	8.0	8.0	8.0				
	Total	100.0	100.0	100.0	100.0				
United Kingdom	Employees	23.1	21.9	29.3	32.1	23.0	25.0	24.0	24.0
	Agents	4.9	3.6	0.0	0.0	18.0	18.0	22.0	15.0
	Brokers	53.3	67.7	64.3	61.8	49.0	38.0	35.0	48.0
	Other network	17.5	0.0	0.0	0.0	9.0	17.0	18.0	12.0
	Others	1.2	6.8	6.4	6.1	1.0	2.0	1.0	1.0
	Total	100.0	100.0	100.0	100.0	100.0	100.0	100.0	100.0

Source: CEA (2003), Tables BIII-2 and BIII-3, pp. 144–145.

In terms of business lines, Austria is different from many European countries in that composite insurers dominate the industry. Of the 58 direct insurers, 35 offer life assurance, 8 offer health insurance, and the remainder offer non-life and accident insurance. Moreover, there existed 5 specialized life assurance companies and 12 specialized non-life and accident insurers.

The life insurance density is relatively low (similar to that of Spain and Portugal), while non-life insurance density is significantly higher (comparable to that of Germany and Denmark). Insurance penetration, on the other hand, is a relatively low 2.59 percent for life business, again similar to Germany, and is 3.30 percent for non-life business in 2003, which is comparable to that of Ireland.

Business Trends

After experiencing rapid growth in the late 1990s, the Austrian insurance industry had a difficult last few years due mainly to the capital market performance and natural disasters resulting in above average claims in 2003.

Total life insurance premium volume approximately doubled in the six years up to 2000, but it increased less than 1 percent from 2001 to 2003. The non-life insurance sector, on the other hand, experienced an increase of 13 percent in premium income over the corresponding period. Over the last five years, claims growth for life insurance business has outstripped premium growth.

Direct sales distribution channels have lost ground to both agents and brokerage over the last five years. Brokerage is particularly important in the non-life insurance business, while branch networks have now become the most important source of sales in Austria for the life insurance business (CEA 2004–2005).

Segmentation and Consolidation

Domestic insurance firms operating in Austria have declined by 8 percent since 1992. The largest five insurers in Austria in both life and non-life business account for around 50 percent of the total life and non-life business. This concentration ratio is comparable to that of other countries, but Wiener Stadische has significant market share in both business lines.

Financial Strength

The Austrian insurance industry is generally well capitalized. However, in common with many other European markets, following years of steady growth, the Austrian life and general insurance sector faced a difficult period in 2001–2002 due to the slump in the stock markets and decrease in interest rates. Heavy losses experienced within the single premium life insurance policies were mainly responsible for the decline in premiums in the life insurance sector. The planned introduction of state-sponsored pension retirement accounts and the associated private equity-lined schemes had a positive impact on the life insurance industry in 2003 (Verband der Versicherungsunternehmen Osterreichs 2005). However, the capital strain is relatively high, and provisions account for nearly 95 percent of total assets. The general insurance industry is also well capitalized, but claims generally amount to over 50 percent of total assets.

Investment Performance
Life insurance business has generated a positive return of 10.6 percent and 7.8 percent over the ten- and five-year horizons to 2002, respectively. However general insurance has generated only 4.2 percent and 0.5 percent over the same period. Latest returns for 2002–3 are 4.6 and 2.7 percent, respectively (CEA 2004–2005). Austrian insurers' investments are relatively diversified and thus are relatively less affected by volatility in share prices (compared to U.K. and German insurers, for example). Total assets under management have grown to €40 billion and €16 billion, respectively, by 2003 (CEA 2004–2005).

16.3.3 Belgium

Market Overview
The Belgian insurance market is generally stable and profitable. The insurance market is also fairly segmented, reflecting the geographical and cultural diversity of the country. Following a law passed in August 2002 on the supervision of the financial sector and on financial services, the Office de Controle des Assurance, or Insurance Supervisory Authority, and the Banking and Finance Commission merged on January 1, 2004, to become the Banking, Finance and Insurance Commission (www.cbfa.be). Life insurance market penetration is the third highest in Europe, and the non-life insurance density and penetration ranks tenth in Europe.

Business Trends
Growth in the market, in which composite insurers are prominent, remains largely limited to the life insurance sector. This sector has experienced a boom in recent years due to the introduction of unit-linked business in 1997. By contrast, non-life business has only gradually increased by approximately 5percent per year after experiencing a decline in the late 1990s. Life insurance now represents 70 percent of total premiums (Assuralia 2004).

Total direct life insurance premium income has gradually grown to reach €18 billion by 2003, which represents a growth of 50 percent in the last five years. The most important shifts in market trends occur in the life insurance sector. In 2003, premium income in unit-linked life insurance policies decreased by €1 billion or 16.6 percent, even though some companies managed to grow in this market through various distribution channels. On the other hand, individual life insurance policies with a guaranteed interest rate increased by 40 percent as a result of the commercial policy of a limited number of big players. However, figures in this line of business had been weak in previous years as a result of lowering maximum interest rates from 4.75 to 3.75 percent (Assuralia 2004).

Overall, individual life insurance policies grew by 2 percent in 2003, but there have been major shifts in the nature of products offered. Premium income became more dependent on single-premium policies and on variable-premium contracts, rather than products calling for a set annual premium. The market has thus grown more dependent on the condition of financial markets and on the financial resources available to Belgium households (Assuralia 2004).

The non-life insurance sector has continued to expand in recent years with greater product demand. Total non-life premiums have grown by approximately 10 percent per year since 2000. However, claims have also increased, resulting in business write-downs.

Brokerage-based distribution channels continue to dominate both the life and non-life segments, although branch networks are now the most important distribution channel for life insurance products (CEA 2004–2005).

Segmentation and Consolidation

Following the European trend, insurers in Belgium are consolidating into larger entities. Between 1992 and 2003, the number of domestic authorized insurers in Belgium declined almost by 50 percent to a total of 118. Nevertheless, the proportion of nondomestic firms operating has increased to 71, at nearly 50 percent, which is nearly the highest in the European countries surveyed in this chapter (after Portugal) (CEA 2004–2005). The top five life insurers accounted for 74 percent of total premiums in 2003 (in 2001 they accounted for 49.8%) while the top five non-life companies accounted for 59 percent of total premiums (in 2001 they accounted for 47.1%).

Financial Strength

The high solvency levels of Belgian insurers, which remains a key feature of the market's stability, has been driven by a cautious regulatory environment and a benign fiscal regime. Solvency (free asset) ratios for both life and non-life firms remain well above the Solvency I threshold and are amongst the highest in Europe. Tax authorities are beginning to recruit insurance specialists, and there is a possibility of a more aggressive tax approach, but both insurers and regulators will defend the reserving pattern in the market.

Investment Performance

Despite the high proportion of unit-linked business and their relatively high exposure to equities, Belgian life insurers have generated persistently high returns over both five- and ten-year horizons (11.7% and 12.1%, respectively). The general insurance industry has been unable to sustain a high investment return since 1997. However latest 2002–3 investment performance impved in both sectors, with a 17.1 percent and 4.2 percent return, respectively. Despite this, assets under management have grown to nearly €100 billion and €29 billion, respectively, by 2003 (CEA 2004–2005).

16.3.4 Denmark

Market Overview

Despite the fact that Denmark is a small economy by international standards, it has a relatively well-developed and mature insurance industry and is extremely competitive. The tariff cooperation that used to exist among the non-life insurers was eliminated in the early 1990s. In the non-life sector there exists no public supervision of insurance conditions or premiums. The life sector, on the other hand, has been subject to extensive public supervision. In recent years, however, there has been a relative deregulation in life and pensions in accordance with EU regulations.

Supervisory provisions are stated in the Financial Business Act and the Insurance Business Act. The Danish Financial Supervisory Authority is responsible for supervision of insurance and pension funds business in Denmark. Insurance business can be undertaken by a public limited company or a mutual insurance association, while pension fund business can be undertaken by a multiemployer pension fund with members typically affiliated to a specific industry or occupation. Life and non-life business cannot be undertaken by the same company. However, a life insurer can write health and accident insurance.

There are 214 insurers listed on the Web site of the Danish Insurance Information Service (www.forsikringsoplysningen.dk, CEA 2004–2005). One hundred forty-five of these were active in non-life insurance, 62 in life insurance and pensions, and 8 in reinsurance. Twenty-eight firms of the 62 life insurance and pensions firms are multiemployer pension funds.

Given the importance of industry-based schemes, it is not surprising that insurance density is relatively high in Denmark; it ranks fifth and seventh in the life and non-life segments, respectively. Insurance penetration is much higher for life than for non-life business. The ratio of life insurance premiums to GDP in 2003 was 5.18 percent, and the ratio of non-life premiums to GDP was 2.74 percent. In 2003, gross premiums and member contributions were 4.5 percent of GDP.

There is little regulation of financial intermediation or distribution channels in Denmark. Consequently, non-life, life, and pension insurers employ various sales channels—company representatives and local insurance offices being the most common channels. In recent years, sales through bank branches and brokers have been increasing. With the backing of the Danish insurance association, a new business model has been developed in which insurance brokers do not receive commissions from the companies but directly charge customers. Many non-life insurers adopted this model and hence terminated existing contracts with the brokers.

Business Trends

Given the high degree of interdependence between the Danish insurance industry and occupational group pension schemes, it is not surprising that premium growth has, except for 1999, exceeded 10 percent over each of the last eight years. By contrast, the Danish non-life sector can be characterized by traditionally profitable, but not significantly growing, insurance business. Accordingly, 2003 was no exception. Although in 2000–2001 exceptional events were responsible for the weak financial performance, there were no such one-off events thereafter. Rising reinsurance premiums were a significant factor for the financial underperformance weakening operating results.

In addition, insurers tended to assume a greater share of the total risks. Total gross premiums in non-life insurance increased by more than 11 percent in 2003, whereas net premiums of reinsurance increased by only 6 to 7 percent, reflecting the steep price increases of reinsurance. Total contributions to life insurers and pension funds, on the other hand, increased by 9 percent, above the average growth rate achieved over the last five years. The strong performance, in turn, was mainly due to strong growth in occupational pension schemes.

In recent years, the pattern of pension savings has changed, with a gradually growing portion of pensions attracted by life insurers and pension funds in the form of occupational group schemes, as opposed to individual savings accounts offered by banks. Unit-linked policies achieved the strongest growth of all pension savings. Strong growth performance of unit-linked policies is expected to continue in the coming years.

Segmentation and Consolidation

In common with other Nordic countries, the Danish insurance industry is relatively highly concentrated. The top five life insurers account for nearly 70 percent of total business, while the top five non-life insurers account for 57.4 percent of total business. The total number of insurers operating in Denmark has declined by only 2 percent since 1992. However, there are few, if any, nondomestic insurers, reflecting the complexity and degree of integration between the insurance industry offerings and government-provided or mandated insurance.

Investment Performance

The Danish life insurance industry has maintained a fairly consistent investment performance over both five- and ten-year horizons. By contrast, the non-life insurance industry has not generated returns above inflation over the corresponding periods. Moreover, the total funds under management for life insurance has increased by 250 percent over the last ten years, while non-life insurance investments have remained virtually static. Latest 2002–3 investment performance improved in both sectors, with a 9.3 percent and 8.2 percent return, respectively. Assets under management have grown to nearly €140 billion and €16 billion, respectively, by 2003 (CEA 2004–2005).

16.3.5 Finland

Market Overview

The Finnish insurance market is small by international standards, notwithstanding Finland's population of 5 million. Its direct contribution to economic activity is also small (around 1% of GDP). The Insurance Contracts Act of 1995 governs the relations between insurers and policyholders, and the Insurance Companies Act regulates the industry. The Ministry of Social Affairs and Health and the Insurance Supervisory Authority issue supplementary regulations. The Finnish life insurance market is the fourth highest in density and the third highest in penetration of all European insurance markets.

According to their line of business, the insurers can be legally divided into life, non-life insurance, and employee pension. Life and non-life business cannot be undertaken in the same company, although the consolidating parent entity may include both life and non-life subsidiary companies. Pension insurers specialize in statutory earnings-related pensions, while life insurers provide life, accident and health, and voluntary pension insurance. They can be formed as limited or mutual companies. Employee pension companies are mostly mutuals. Non-life and personal life insurance, on the other hand, are mostly limiteds. More than half of Finnish insurers are limited companies.

The number of insurance brokers has been increasing in recent years as the life and non-life markets continue to grow. Foreign insurers operate on a moderate scale (under 3% of the Finnish market is assumed to be covered by foreign insurers). They are relatively stronger in life than in non-life business. Finnish insurers, on the other hand, have strengthened their presence in the Baltic States in recent years.

Following the "Nordic welfare state" principle, in Finland, insurance provides wide statutory coverage of the insurance needs of its population. The Social Insurance Institution manages the minimum benefits scheme and statutory health insurance. Privately owned non-life companies provide statutory motor liability and workers' compensation insurance. Privately owned employee pension insurers provide compulsory pension insurance and cover around 80 percent of all private-sector workers, and the rest is covered by pension funds, which are either in the form of company pension schemes or industry-wide schemes (public-sector schemes are insured under specific institutions) (Suomen Vakuutusyhtioiden Keskusliitto 2004). Finnish employee pension insurance is not covered by the EU insurance directives. It is a statutory cover available to all employees and hence covered by the first pillar in the EU classification. The principles of insurance directives cannot be applied, because pension benefits accrue by the same rules regardless of where the employee is insured and because the proportion of benefits funded on a pay-as-you go basis is substantial.

Personal pension plans have gained popularity in recent years. Life insurance can be either unit-linked or traditional with profits policies. Unit-linked policies account for about a third of all new life business written. The volume of non-life insurance (about 2% of GDP in 2003) is low compared to the European average, mainly because personal health insurance is not a significant line and the level of non-life risks are low in Finland due to Finland's special conditions and extensive loss prevention work (Suomen Vakuutusyhtioiden Keskusliitto 2004). The insignificance of health insurance, in turn, is due to the country's extensive public health care services.

Business Trends

Following a period of substantial growth and stability, the Finnish insurance industry is a mature and saturated market, and it experienced stagnation during 2000 to 2003 mainly because of capital market developments and increased reinsurance premiums due to terrorism and natural disasters.

In 2003 the volume of premiums increased by 3.7 percent. Growth was experienced in all the three lines of insurance. Life insurance premiums increased in real terms by 7.3 percent, statutory pensions by 4 percent, and non-life insurance by 4.7 percent. The largest sector in terms of premiums written was statutory pension insurance. That statutory lines generate a major share of premiums written is a typical feature of the Finnish market.

Segmentation and Consolidation

There are 68 insurers, including 21 of foreign insurers' branches as of October 2003; most are members of the Federation of Finnish Insurance Companies (FFIC). FFIC lists 13 life, 7 pension, 21 non-life insurers, 5 reinsurers, and 22 foreign insurers as of January 2004. In addition to insurers, there are more than a hundred insurance associations (local mutual insurers operating in the non-life area). Like most other mature European markets, the Finnish insurance industry has declined by 13 percent since 1992 (CEA 2005).

In common with other Nordic countries, the Finnish insurance sector is very highly concentrated. A period of rationalization occurred following the decision by major insurers to reduce their cross-holdings in 1997. The market has since been traditionally dominated by insurer groups that are made up of companies operating in life, non-life, and statutory pension insurance. The top five insurers dominate 82 percent of the life insurance market and 95 percent of the non-life insurance market. However, in recent years, the companies have been seeking new business models through banking and savings products. In addition, in life insurers established by banks have increased their market shares and hence diminished the degree of concentration.

Financial Strength

The Finnish life insurance market has stabilized since a series of insurer failures in 1995, when Apollo Life, Kansa Life, and Kansa Pensions failed. Despite these failures, the Finnish insurance industry is now well capitalized, particularly in the non-life segment. However, operating profits have reduced as the investment climate has worsened. Non-life insurers' aggregate operating profits fell by 86 percent and solvency capital declined by 26 percent in 2003. Similarly, life insurers' aggregate operating profits almost fell by 50 percent and solvency capital went down by almost 25 percent.

Investment Performance

The proportion of non-Finnish assets in the aggregate portfolio holdings has been increasing, and in 2003 only 43.7 percent of aggregate portfolio holdings were of Finnish origin. The internationalization trend is strongest in the case of bond holdings, with 81.7 percent foreign bonds, mainly issued by Euro zone governments. However, the investment portfolio is more widely exposed to equities than many other European countries. While life insurance investment performance has been over 11 percent in both five- and ten-year horizons, non-life investments have performed poorly over the corresponding periods. Consequently, it should not be surprising that Finnish insurers' investment portfolio reflects an increasingly low-risk one. Debt securities account for 60 percent of total portfolios in all the three insurer types. Latest 2002–3 investment performance improved in both sectors, with a 9.9 percent and 4.2 percent return, respectively. Despite this, assets under management have grown to nearly €74 billion and €8 billion, respectively, by 2003 (CEA 2005).

16.3.6 France

Market Overview

France is the world's fifth largest domestic life market and the fifth largest domestic non-life insurance market. The French life insurance market is in a period of transition, with the life insurance sector experiencing a period of decline and non-life business growing only slowly. The sector is also experiencing segmentation, with insurance firms either specializing or becoming members of larger groups.

Membership of larger consolidated groups has enabled larger firms such as AXA to diversify across markets and sectors and has led to arguments for relaxed EU capital requirements.

Business Trends

Overall life insurance business premiums have grown by over 14 percent in the last decade, but they have somewhat slowed in the last thee years. By contrast, non-life premiums have grown consistently by 10 percent since 2001. Premiums for savings-orientated life insurance products have quadrupled over the last ten years. But during recent years, many of these fiscal advantages have started to disappear. The increase in the CSG (tax) levy on savings from 3.4 percent to 7.5 percent in October 1997 resulted in a reduction in 1998 premium volumes. In 2003, premium volumes rebounded by 14 percent but remain below 1997 levels.

The erosion of the tax advantages historically enjoyed by French life insurance products may lead to a more level playing field for other savings providers—such as banking operations and mutual funds—which have also experienced depressed premium growth. The change in fiscal status of equities has also led to the expansion of "multisupport" linked contracts. These are bundled contracts that provide policyholders with direct access to equity market exposure as well as provide an insurance guarantee element. Consequently, they are also likely to be subject to the investment contract definition that satisfies IAS 39, rather than treated as pure insurance contracts that are subject to IFRS 4. Linked contract premiums represented 30 percent of total premium volumes in 2003, compared with less than 10 percent in 1995.

In 2003, premium volumes rebounded by slightly over 14 percent but remain below 1997 levels. Part of the 2003 growth products arose from the transfer of savings from the bank sector to the insurance sector; savings in the bank sector became less attractive following interest-rate reductions.

The most important growth has been in the area of individual unit-linked products (i.e., policies in which the policy value fluctuates directly in accordance with instantaneous changes in underlying portfolio investments, rather than by a predetermined or guaranteed investment return), while individual guaranteed business has declined, as has group-based business.

Non-life insurance business has also increased gradually over the last ten years. However, growth in the market has been constrained by continuing government involvement, particularly onerous rate regulation, compulsory third-party insurance, and non-life state guarantees of insurance liabilities of many public-sector enterprises. This is despite the EU directives that insurers cannot be required to submit their contracts for approval prior to issue.

Segmentation and Consolidation

Many of the medium-sized players are currently being squeezed. Such insurers may eventually have to choose between slimming down and becoming niche operators or ceding some of their operational independence in order to join or create larger groups. Over 55 percent of the life insurance market (in premium volume) is controlled by five financial groups—distribution is primarily via bank branches. Bancassurers, once content to confine their activities to simple savings contracts, have exploited the proprietary information garnered from their sophisticated customer information systems to diversify into more risk-oriented products—that is, the high net worth segment of the life insurance market and non-life insurance.

Direct sales and agents remain the most common distribution methods for non-life products, and bank branch networks dominate the life insurance sector (CEA 2005). Long-established insurers that rely on either independent financial advisers or brokers to sell their business have suffered from these trends. However, these firms have more recently adapted their business models by focusing on individual client needs. Consequently, they are gradually offsetting the distribution cost advantage of bancassurance firms by realizing economies of scale from increased consolidation and cooperation and improved distribution strategies. Competition is expected to intensify and profit margins are expected to fall unless the long-awaited pension boom finally emerges.

A total of 486 insurance companies (French and non-EU) were operating in France as of February 2004 under French regulation according to the Web site of the Ministère de l'Economie, des Finances et de l'Industrie. Of these, 373 are domestic firms; 84 are operating in the life and 282 in the non-life segments. The remaining are composite firms. Another 113 firms are either EU or non-EU branches (CEA 2005).

The French insurance sector is relatively highly concentrated. The top five life insurers account for 56 percent of all business; while the top five non-life insurers in France account for 48.2 percent of total business.

Bancassurers have increasingly increased their market share of the life insurance sector. This was made possible by exploiting low-cost distribution networks to sell simple tax-advantaged savings products. Bank subsidiaries accounted for almost half of life premium volumes.

Most bancassurers already possess significant low-cost distribution capabilities through their access to bank branches. However, this development must be seen against the European banking sector environment, which is currently undergoing consolidation and regulatory change.

Moreover, competition has evolved over time. More traditional insurers have responded by improving the efficiency of their own distribution systems. Bancassurers have also diversified into non-life insurance, health insurance, and other savings products.

The French insurance market has segmented into specialist niche players or has been consolidated into larger, trans-European insurance or banking groups. The comparative advantage of low-cost distribution enjoyed by bancassurers in the tax-advantaged savings sector of the life insurance market is gradually being reduced. Traditional insurers have more diverse product ranges and distribution media than do the bancassurers. Moreover, the negative publicity associated with the poor experience of some bancassurers' overseas activities (e.g., problems with U.S. authorities over liability for insolvent subsidiaries) has highlighted the dangers of adopting a simplistic, expansionist strategic view toward the insurance market. Medium-sized members of the more traditional segments of the life insurance market that are struggling to compete will be forced to utilize any excess capital to transform themselves into leaner, more cost-effective operations.

Most domestic non-life insurance is sold through a network of tied agents. Brokers mainly sell commercial insurance. Mutuals distributing through noncommissioned sales offices have been very successful in motor and household business. On the other hand, bancassurance has played a significant role in the insurance market. Direct writers' market share remains very limited (Tillinghast-Towers Perrin 2004).

Financial Strength

The French life insurance market is more strongly capitalized (i.e., the ratio of the total market value of insurance assets to the total book value of insurance policy liabilities) than other major European markets that are analyzed in this chapter.

Margins on non-life insurance are generally higher than for life insurance. However, bancassurers have found it harder to penetrate the non-life market due to their over-reliance on banking-related bolt-on products—that is, without undertaking substantial new investment (Benoist 2002, p. 396). The French non-life insurance market has also been subject to exposure to catastrophic losses, such as flooding and fire damage in the south of France during the summer of 2003.

Investment Performance

French insurers have produced relatively high returns on investments over both five- and ten-year investment horizons. The life insurance sector has generated 15 percent and 8.9 percent, respectively. However, the general insurance industry has generated above-inflation returns only over the ten-year investment horizon. The assets under management for both life and non-life sectors has tripled in the last ten years. Latest 2002–3 investment performance improved in both sectors, with a 9.8 percent and 13.9 percent return, respectively. Despite this, assets under management have grown to €890 billion and €126 billion, respectively, by 2003 (CEA 2005).

16.3.7 Germany

Market Overview

Germany is the world's third largest non-life insurance market and the fifth largest life insurance market. Nevertheless, insurance penetration and density have remained fairly static over the past five years and are significantly lower than those achieved in other important insurance markets, reflecting in part a still extensive social security system. In 2003, the share of insurance premiums in the GDP was only 6.5 percent compared with 13 percent in the United Kingdom and Switzerland (Tillinghast-Towers Perrin 2004). This market overview focuses mostly on the life insurance sector, which has been most susceptible to soft capital markets and to poor performance issues.

Business Trends

Weak economic conditions in Germany have translated into a sluggish business environment for the insurance industry. Premium growth in life insurance was exceptionally high in 1999 at 11.8 percent, boosted by the German government's plans to tax life insurance policies, which were abandoned in 2000 according to the German Insurance Association (Gesamtverband der Deutschen Versicherungswirtschaft e.V. [GDV] 2002).

The development of life insurance in 2001–2002 was influenced by the structural pension reform of the federal government. The pension reform, which was due to reduced working capacity and came into force at the start of 2002, increased the demand for private provision. All persons insured in the statutory pension scheme lost their occupational disability protection if they were under 40 years of age as of January 1, 2001. However, assistance for private old-age provision through the Pension Reform Act began in 2002, and many German residents are reorienting their old-age provision.

Pension reform, starting in 2002, which encourages individuals to sign up for supplementary private insurance in order to boost their employer-funded, group-based pensions, has created a boom in the life insurance market, offsetting some of the pressure on earnings due to low investment yields and continuing competition in the property-liability sector. However, only the larger companies—which can afford to fund the regulatory-determined capital constraints that are imposed on the selling of new products and subsequently cross-sell products to new clients—are likely to benefit from these reforms. This trend has been facilitated by greater regulation of financial transparency in the reporting of risks by life insurers (GAS 5–20, DRSC 2004) and auditor opinions on companies' risk-recognition systems (Traas 2003, p. 199).

By contrast, the non-life insurance market remains severely fragmented and soft. Recent flood losses in northern and southeastern regions of Germany in 2002 impacted the loss ratios of some non-life insurers. Many, but not all, German insurers were subject to this exposure, notably in property, motor, and industrial business sectors. In addition, the poor performance of capital markets resulted in many insurers incurring significant losses, especially for the small and mid-size insurers that maintain locally concentrated investment portfolios.

In 2002, premium income increased sharply by 3.7 percent more than in the previous year, especially in the area of supplementary insurance contracts such as accident and occupational disability protection. However, in 2003, new business decreased by 29.1 percent compared to the previous year (GDV 2004).

Life insurance contracts are important for old age and for surviving dependents. The number of existing life insurance contracts declined from 87.82 million in 2002 to 82.49 million in 2003. Nearly half of all main contracts were complemented in 2000 by supplementary insurance contracts. In 2003, private endowment insurance remains the highest share of 61.7 percent of total premiums, compared with 17.2 percent for private pension insurance and 9.3 percent for collective insurance (GDV 2004).

The annual growth in liabilities to policyholders (mathematical provisions for long-term, discounted insurance policy liabilities, after allowing for provisions for premium refunds and profit shares) decreased by 0.4 percent compared to 2002. The growth in mathematical provisions reached €62.24bn, compared to €63.18bn in the previous year. Bonus and rebate provisions reached €8.95bn in 2000, compared to €8.33 billion in 2002; profit shares stayed at €4.06bn in 2004, compared to €4.08bn in 1999 (GDV 2004).

Market Segmentation and Consolidation

According to the GDV, there were 703 foreign and domestic insurers under federal or land supervision at the end of 2003 (CEA 2005). One hundred thirty-two operate in the life and 261 operate in the non-life segments. There were 207 pension funds and providers of funeral expenses insurance, and 55 health insurers.

The top five non-life insurance companies in Germany in 2003 accounted for 28.3 percent of total business. The top five life insurance operating companies accounted for 32.6 percent of gross premiums. Life insurance trends in 2001 included increasing benefits, further growth in premiums, and growth in new business, with periodical premiums increasing at a faster rate than single premiums.

Financial Strength

German companies have historically been better capitalized than those operating in other European markets. The risk-based capitalization for the top fifty German life insurers increased over the period 1995 to 1999, although most of this increase was attributable to favorable investment markets.

The strong capitalization of the market deteriorated after a large proportion of assets that were invested in equities declined in value. Consequently, a number of life insurers are suffering poor financial strength, with Mannheimer Leben being the first German life insurer in fifty years to become insolvent in 2003. A number of other small life insurers are in a poor financial condition. This is occurring during a period of transition, as especially smaller life insurers struggle to adapt to the new challenges offered by the pension reform and the new tax environment. In addition, many insurers are implementing cost efficiency programs. Nevertheless, the growing number of mergers and increased cooperation will lead to further consolidation of the German insurance industry.

Investment Performance

The German life insurance industry has maintained a steady growth rate of 8 percent and 7 percent over ten- and five-year investment horizons, while the non-life sector has maintained above-inflation returns over the corresponding periods. The total funds under management has quadrupled for the life insurance sector and almost tripled in the non-life sector over this period. In contrast to other European markets, latest 2002–3 investment performance declined in both sectors, with a 2.7 percent and 9 percent return, respectively. Despite this, assets under management have grown to nearly €610 billion and €450 billion, respectively, by 2003 (CEA 2005).

16.3.8 Greece

Market Overview

Greece has the smallest insurance market in the EU in terms of both total life insurance premiums and non-life premiums in 2003. This is despite the fact that both total and per capita insurance premiums increased quite dramatically in the 1990s before stabilizing between 1999 and 2001 (life insurance premiums per capita in 2003 were €183 (in 1994 they were €56). Insurance density and penetration is the lowest in Europe; life (general) insurance premiums as a percentage of GDP was only 0.93 percent (1.17 percent).

Business Trends
During the 1990s, the Greek life and non-life insurance sectors grew by over 15 percent per annum to 2003. The life insurance sector, on the other hand, experienced a decrease in premiums in 2000–2002 (–0.83%) before increasing 9.54 percent in 2003. The non-life segment accounted for 51.1 percent of the total direct premiums.

Segmentation and Consolidation
Similar to the developments elsewhere in Europe since the early 1990s, there has been a consolidation process in Greece. The total number of insurers decreased from 161 in 1992 to 107 in 2001. The Greek Association of Insurance Companies on its Web site (www.add.gr) lists 90 firms as its members. A total of 18 companies operate in the life insurance segment, and 60 companies operate in the non-life segments. The remaining 12 companies are composites.

Accordingly, today the insurance market is highly concentrated, especially in the life segment. The top five life insurers control 69.10 percent of the market, and the top five non-life insurers account for 46.95 percent of the market.

Financial Strength
The Greek insurance industry is relatively undercapitalized. Both the life and non-life sectors have the lowest ratios of provisions to assets of all the European insurance markets. This may be due to the relatively high proportion of investment-related contracts sold by Greek insurers over the last few years. The non-life insurance market has had relatively poor underwriting results, and this may have severely strained insurers' capital reserves.

Investment Performance
The five-year investment performance to 2002 shows a relatively high investment return for both the life and non-life sectors, with the latter surprisingly showing a significantly higher return. However, this result may be misleading given that investment assets under management have doubled over the past five years. Latest 2002–3 investment performance information is only available for the life insurance sector, and shows a negative 3.5 percent return. Assets under management are very small by European standards; €4 billion and €2 billion, respectively, in 2003 (CEA 2005).

16.3.9 Ireland

Market Overview
Despite the fact that both the Irish economy and its insurance market are both relatively small in terms of population and market size by European standards, insurance premium income in Ireland is more significant than in many other countries. This is largely due to government incentives, such as the establishment in 1996 of an international financial services center for overseas companies to conduct business, together with a very high rate of growth relative to other European countries. According to per capita premiums in 2003, Ireland is second only to the United Kingdom in life insurance (€1,712) and fourth after Germany in non-life insurance (€991) (Tillinghast-Towers Perrin 2004). Total premium income as a

percentage of GDP was 8.7 percent in 2003 (9 percent in 2001). The life insurance market segment accounted for 64.7 percent of gross premium income in 2003.

The Irish Insurance Federation (IIF) represents the entire domestic life insurance market and over 95 percent of the domestic non-life insurance market. Insurance supervision is under the Irish Financial Services Authority, which took over the supervision of insurers, banks, investment intermediaries, and other financial services firms in May 2003.

Business Trends

The domestic life and non-life premium income of IIF members in 2003 was €15.4 billion. After the strong economic performance and the accompanying premium growth in the late 1990s, the life and pensions sector experienced a downturn after 2001. Total premium income decreased by almost 6 percent, mainly due to a significant drop in pension scheme business. However, new life business sales increased by 7.4 percent.

In 2001 to 2003, the business environment became difficult for the life and pensions market; total premium income increased by only 1.2 percent and there was a significant drop in single premium business. New regular premium sales, however, increased by over 11 percent, resulting in an overall increase of 3.5 percent in new business. Aggregate regular premium business increased by over 21 percent.

In 2003, total life insurance premium income reached €7,970.3 million. Individual assurances and annuities accounted for 60.7 percent of the total income.

Brokers dominate the life insurance business, distributing 55 percent of the total annual premium in 2002, the latest year available. Tied agents and employee and company representatives are the next two major channels (CEA 2005).

In all main classes of non-life insurance, there was strong premium growth in recent years and a much lower growth in the cost of claims. In 2003, the rate of increase in premium volume slowed relatively in motor (+13%) and property insurance (+25%). Gross liability premium income, however, grew by over 40 percent.

Total non-life insurance premium income reached €7,312 million in 2003. Private motor class accounted for 33.2 percent of the total. The net underwriting result of the non-life market in 2003 improved to a loss of €45.4 million. The market made a net operating profit of €216.8 million in 2003 compared to a loss of €67 million in previous years.

Segmentation and consolidation

The IIF lists 56 companies as members, 21 of which operate in the life segment and 22 in the non-life segment. The remaining 12 are designated International Financial Services Centre companies, writing foreign business. In aggregate, there were 178 domestic firms operating in Ireland in 2003, a nearly 350 percent increase over the last decade. In addition, there were 44 EU branches and 2 non-EU branches operating in Ireland in 2003 (CEA 2005).

The insurance industry is highly concentrated. The top five life insurers account for 69 percent of total business, and the top five non-life companies account for 56 percent of total business.

Financial Strength
Information on the capitalization of the Irish life insurance industry is not available. The non-life market is relatively well capitalized by European standards but has been under strain recently. The motor and liability markets have incurred substantial losses in recent years. As a result, a number of insurers withdrew from the market in 2001, especially in the commercial/liability areas. There was significant premium growth in 2003, especially in liability and property insurance. The strong growth was mainly due to increased economic activity and, hence, higher demand for insurance coverage together with significant rate increases. The rates increased mainly because of substantial underwriting losses of previous years, continuing claims inflation on both personal injury and damage claims, and pressures on investment returns in volatile markets.

Investment Performance
The life insurance industry has generally performed well over both five- and ten-year investment horizons, and total investments under management have nearly quadrupled over the last decade. However, the industry is also highly exposed to equities, and this results in greater volatility in performance relative to other European markets. The non-life sector has provided a lower return, showing a net 7 percent return over the last decade, but this fell to just 0.7 percent over the last five years. The non-life sector investment portfolio has doubled over the last ten years. Latest 2002–3 investment performance has dramatically improved in both sectors, with a 17.2 percent and 12 percent return, respectively. Despite this, assets under management have grown to nearly €45 billion and €10 billion, respectively, by 2003 (CEA 2005).

16.3.10 Italy

Market Overview
Italy has the fourth largest life and non-life insurance markets in the EU with respect to total premiums. However, in 2003, insurance premiums accounted for only about 7 percent of the GDP (Association Nationale Italia Annuale 2003). Moreover, insurance density and penetration is relatively low, given that Italy is Europe's sixth largest insurance market.

Business Trends
Premium income in both life and non-life segments continually increase, although growth is much stronger in the former. In the period 2000–2003, premium income in the life insurance segment increased by 18.3 percent (the average growth rate per year between 1996 and 2000 was 26%). Life insurance now represents 63 percent of the total market.

Single premium products have shown strong growth in recent years, while annual and recurrent single premium income has decreased. Traditional products have grown faster than linked and index policies due to stock market developments.

The non-life sector has grown more steadily over the last ten years. Non-life insurance sales are dominated by traditional agency distribution channels, rather than by branch networks.

New regulations concerning individual life products came into force in 2001. Accordingly, new policies sold after January 1, 2001, were no longer subject to premium tax, and policy proceeds on death remains tax free.

Italy is very publicly regulated. Most insurance and pension savings occur via the state (i.e., first pillar). The Italian pension market encompasses elements of the classical second and third pillars with open and closed pension funds and PIPs (Piano Individuale Pensionistico). PIPs can only be sold by life insurers and are subject to the regulatory authority of the insurance regulator rather than the pension funds regulator.

Market Segmentation and Consolidation

There are 189 domestic and 60 non-domestic insurance firms operating in Italy (CEA 2005). The Web site of the Istituto per la Vigilanza sulle Assicurazioni Private e di Interesse Collettivo lists 78 life, 88 non-life, 19 mixed, and 3 reinsurance companies authorized as of January 2004 (www.isvap.it). The top five non-life insurers control 38.1 percent of the total market. The top five life insurers account for the 31.3 percent of the life market. In addition, almost 23 percent of the total Italian market comprises nondomestic branches.

Large regional differences in the Italian market—both economic and cultural—affect the main insurance distribution channels, with local agents and small regional brokers playing a significant role in the non-life area. In the life area, banks dominate the market and have been gaining share to the detriment of all other channels (CEA 2005).

Investment Performance

Investment performance over five- and ten-year investment horizons is one of the highest in Europe. However, since Italian insurers are permitted to smooth earnings over time, it is not clear whether this is sustainable over the longer term. More significantly, the investment portfolio of life insurers has increased five-fold in the last ten years, while that of the non-life sector has grown by less than 40 percent. Latest 2002–3 investment performance has slightly declined in both sectors, with a 18.8 percent and 5.7 percent return, respectively. Despite this, assets under management have grown to nearly €300 billion and €68 billion, respectively, by 2003 (CEA 2005).

16.3.11 Luxembourg

Market Overview

Although it is a relatively small state, Luxembourg operates a significant insurance market, because most operators are captive insurers that cater to high net worth individuals. The Commissariat aux Assurances provides a comprehensive register of the names and addresses of companies with branches in Luxembourg. It lists 32 companies offering life insurance and 25 companies offering non-life insurance. Overall, the number of insurers operating in Luxembourg is 95 firms, of which 81 are domestic entities. The total number of firms has increased by 65 percent since 1992, but only 15 percent are nondomestic branches, a decrease from 25 percent (CEA 2005).

Business Trends
The overall trends of new premium growth show a steady increase in premium revenue, and increasing provisions, except for 2001. Life insurance dominates the premium income, since non-life insurance is generally sold to domestic customers. By contrast, the proportion of life insurance premium growth generated by domestic business (i.e., Luxembourg nationals) is only 5 percent of total business, since the majority of businesses established in Luxembourg are intended for overseas policyholders seeking to maximize their returns in a low-interest and low-tax rate environment. Belgium accounts for a significant and increasing proportion of total business (46%), followed by France (14%), Germany (12%), and the United Kingdom (10%).

Segmentation and Consolidation
Given the small size of the Luxembourg market, the market is segmented between purely domestic firms and those selling tax-friendly life insurance contracts to nondomestic clientele. Moreover, the favorable tax treatment accorded to the insurance industry means that the incidence of international insurers operating in Luxembourg is relatively uncorrelated with the social and economic trends affecting the economy.

Financial Strength
The Luxembourg insurance market is relatively well capitalized, in both the life and non-life sectors.

Investment Performance
Insufficient data are available to provide a meaningful analysis of the investment performance of the diverse and highly segmented Luxembourg insurance market.

16.3.12 Netherlands

Market Overview
The Netherlands insurance market is quite saturated and has been under sustained pressure for the last ten years. For insurers, 2002 culminated a prolonged period of declining production, a steep fall in share prices, and higher reinsurance premiums, although the market recovered somewhat in 2003. These developments placed considerable strain on both the results and the solvency margins of insurers.

The fall in share prices during the first three quarters of 2001 resulted in a loss of $55 billion of insurance company investments (Oosenbrug 2003). Subsequently, many companies had to charge further depreciation in the value of equity portfolios in their results. The coverage of the costs of unit-linked insurance policies also declined, because they are usually related to the value of the underlying investments.

Prior to 2001, large insurers such as Aegon had sought to immunize these volatilities through a book-keeping procedure of applying a 30-year average to the return on a portfolio with a seven-year average, and then gradually "release" this indirect income to the profit and loss account. However, this smoothing procedure resulted in serious misstatement of earnings relative to U.S. GAAP when the first three quarters of 2001 produced substantial unrealized gains and losses (Storm

2003). Moreover, claims arising from September 11, 2001, hit the results of reinsurers, which in turn sharply increased the reinsurance premiums charged to insurers in 2002. In addition, the reinsurance capacity declined in 2002 because of the higher risk profile. Consequently, a number of insurers with a professional reinsurer in the group reviewed their reinsurance activities. In a number of cases, these activities were put up for sale and the companies did not sign any new reinsurance contracts.

Business Trends

Overall business has grown over the last ten years. Total premiums for life insurers tripled on a per capita basis from 1995 to reach €25 billion in 2003. Similarly, total earned premiums for non-life insurers increased from €12,432 million in 1998 to €20,086 million in 2003.

However, total life insurance benefits increased even more substantially, from €4,519 million in 1995 to €28,074 million by 2003. Similarly, claims on non-life insurers increased from €8,578 million in 1995 to 13,713 in 2003.

By contrast, the non-life insurance market has grown more steadily and has doubled on a per capita basis since 1995. The total size of the general insurance industry has also grown faster than the life sector in recent years, contrary to the experience of most other European countries.

The vast proportion of insurance in the Netherlands is sold either via direct sales forces or independent agents. Branch networks accounted for 11 percent of total life insurance sales (CEA 2005).

Segmentation and Consolidation

The number of life domestic insurers operating business declined from 105 in 1998 to only 84 in 2003. Similarly, the number of non-life insurers declined from 302 in 1998 to only 236 in 2003 (Verbond van Verzekeraars 2004). In addition, there are another 123 non-domestic firms operating as of 2003, comprising 110 EU branches and 13 non-EU branches (CEA 2005). Most registered Dutch life insurers that actively sell insurance are also members of the Dutch life insurance association (Verbond van Verzekeraars 2004). The Dutch insurance industry is very international, with nearly 30 percent of the insurers comprising EU and non-EU branches.

The Dutch insurance market is also relatively highly concentrated. The top ten non-life insurers control 47.8 percent of the non-life market. The top five life insurers account for the 67.1 percent of the life market. ING holds a significant market position in both the non-life and life sectors.

Financial Strength

The Dutch insurance market is relatively well capitalized. However, the effect of poor investment returns, increasing longevity, and general economic downturn have dampened insurers' solvency status, where solvency is defined as the percentage of free assets in excess of the present value of insured losses or policy liabilities. The solvency ratio of life insurers declined from 440 percent in 1998 to only 231 percent in 2002. The solvency ratio of non-life insurers declined from 302 percent in 1998 to 249 percent in 2003.

Investment Performance
Return on investments for life insurers declined from €14,065 million in 1998 to €2,185 million in 2002. Similarly, total returns of non-life insurers declined from €1,717 million in 1998 to €883 million in 2002. However latest 2002–3 investment performance improved in both sectors, with a 6.5 percent and 8.3 percent return, respectively. Despite this, assets under management have grown to nearly €238 billion and €29 billion, respectively, by 2003 (CEA 2005).

16.3.13 Norway

Market Overview
The Norwegian insurance market is relatively small by both Nordic and European standards. Although the Norwegian market is an EAA, rather than EU, member, unlike Switzerland the market is dominated by regional and small national companies. Norway was the first country in Europe to adopt a single regulator for integrated financial markets in 1986 (Gugler 2005). The market was deregulated in the 1990s, with financial conglomerates permitted in 1991, and unit-linked insurance products launched in 1997. The insurance regulatory authority (Kredittilsyinet) is also the banking regulator and provides little information on the status and development of the insurance market. Despite the maturity of the market, little comparative data are available with other Nordic countries. Accordingly, the scope of this review is severely limited.

Business Trends
Both life and non-life insurance premiums have grown gradually over the last ten years, but suffered a slight decline in 1998 due to significant industry consolidation. Much of the growth since 2000 has occurred in the life sector, with premiums increasing by nearly 60 percent. By contrast, the non-life sector has remained virtually static. There are 96 firms operating in Norway as of 2003 (CEA 2005).

Segmentation and Consolidation
The number of insurers operating in Norway has stayed about the same, although information of the composition between domestic and non-domestic firms is not fully available. A number of insurers closed business during the late 1990s, as the growth of bancassurance began to enhance competition by placing increased pressure on distribution systems.

The Norwegian insurance industry is the most concentrated in Europe. The top five non-life insurers account for over 92 percent of total business, while the top five life insurers capture nearly 75 percent of total business. This may be the result of the heavy geographical dispersion of the population across a wide area. The largest insurer (KLP) alone captures nearly 29 percent of the non-life and 22 percent of the life business in Norway.

Financial Strength
Solvency regulations in line with the European Union were introduced in 1993. The Norwegian insurance industry is one of the most highly capitalized in the entire European insurance market.

Investment Performance
The Norwegian insurance market has generated consistently high average investment returns over both ten-year and five-year investment horizons. The total funds under management has almost doubled in the life insurance sector and increased by 60 percent in the non-life sector over the last decade. Latest 2002–3 investment performance declined in both sectors, with a 5.1 percent and 3.1 percent return, respectively. Despite this, assets under management have grown to nearly €57 billion and €13 billion, respectively, by 2003 (CEA 2005).

16.3.14 Portugal

3.13.1. Market Overview
Portugal has the second smallest market in the European countries analyzed in this chapter. Direct premiums per capita in 2003 were €660 for life insurance and €497 for non-life insurance, substantially higher than the 1995 figures of €107 for life and €204 for non-life. Similarly, the share of direct premiums in GDP has increased in recent years. In 2003 it was 4.93 percent for life and 4.84 percent for non-life. Nevertheless, insurance density is only above that of Greece.

Similar to the overall European trend in non-life insurance, the Portuguese insurance industry has been consolidating. The number of insurers declined from 97 in 1998 to 78 in 2003 (CEA 2005). Of these, nearly 50 percent are branches of EU or non-EU companies, the highest proportion in Europe. The Associacao Portuguesa de Seguradores (APS) is the association of insurers operating in Portugal. It represents more than 95 percent of the Portuguese insurance market. The APS identifies sixty-four member companies.

Business Trends
In 2003, total insurance direct premium income reached €9,534 million. The life segment accounted for 56.6 percent of the total. Insurance premiums grew by an average of 20 percent per year in the late 1990s. More recently, premium income increased by 20.8 percent in life, but only by 1.8 percent in non-life segments in 2003. This followed a period of relative stagnation in 2000–2001.

Segmentation and Consolidation
The Portuguese life insurance industry is highly concentrated. The top five life insurers account for over 76 percent of total business; the largest company (Occidental Vida) accounts for 32 percent of total business. The non-life insurance market is less concentrated; the top five insurers account for 46 percent of total business.

The dominant distribution network for life insurance is via bank branch networks; by contrast, most general insurance is sold via traditional direct sales, brokerage, or agencies (CEA 2005).

Financial Strength
Both the life and non-life insurance industries are relatively well capitalized. However, capitalization is lower than it was in the late 1990s, particularly for the life insurance industry, which is relatively highly exposed to volatile equity investments.

Investment Performance

The life insurance industry has generated a 26.6 percent return over the last decade, the highest in Europe. However, this declined to 16 percent over the last five years. By contrast, the non-life industry has generated unspectacular but steady returns of 8 percent and 5 percent over the corresponding periods. This result is not surprising given that the life insurance industry's total investment portfolio has increased ten-fold over the last decade, whereas the non-life insurance portfolio has doubled. However latest 2002–3 investment performance declined in both sectors, with a 11.5 percent and 2 percent return, respectively. Despite this, assets under management have grown to nearly €24 billion and €5.3 billion, respectively, by 2003 (CEA 2005).

16.3.15 Spain

Market Overview

Spain has the sixth largest life and fifth largest non-life insurance markets in the EU with respect to total premiums. However, its premium per capita is one fifth and one-half of that of United Kingdom in the life and non-life areas, respectively. In 2001, the share of insurance premiums in the GDP was only 4.8 percent (Tillinghast-Towers Perrin 2004). The density and penetration of insurance is relatively low by European standards, except for the non-life industry as a percentage of GDP.

Latest data available from the Committee Europeen des Assurances (CEA 2005) suggest that there are 287 domestic insurers and an additional 41 EU and non-EU branches as at 2003. This represents a 25 percent decline from 1992.

Business Trends

The rapid growth that the Spanish insurance industry experienced in the 1990s came to an end in 2001 to 2003. Affected by the declines in equity markets, life insurers experienced a difficult few years with total premium income declining by 18.3 percent. After two years of strong growth, sales of unit-linked products collapsed. The non-life market, however, experienced a more steady path of premium growth over this period of +24 percent. Externalization of corporate pension obligations continued, with the numbers of pension funds and pension plans increasing accordingly.

Segmentation and Consolidation

The trend toward market concentration continues through mergers and acquisitions and portfolio cessions. In 2003, the total number of life insurers fell by 2.5 percent and the (top-five insurer) concentration in life insurance market increased from 43.7 percent in 2001 to 44.6 percent in 2003. The concentration in the non-life sector is relatively lower than that in life. The top five non-life insurers have 27.8 percent of total market share.

Distribution channels for insurance products are heavily regulated. Tied agents dominate the non-life business, distributing 42 percent of non-life premium volume in 2003. With 26 percent of the premium volume, direct writing is the next major channel, followed by banc assurance with 6 percent (CEA 2005).

In recent years, traditional life insurers have been losing market share to bancassurers, which accounted for 67 percent of new individual business in 2000.

Tied agents regained their position by 2003 at the expense of banking entities, which had difficulties in gaining new contributions for unit-linked products.

As new products, mortgage indemnity was introduced at the end of 2001 and payment protection insurance reappeared, providing unemployment coverage and being sold with life insurance in the case of mortgage loan or credit card subscription.

Financial Strength

Spanish insurers have historically been less well capitalized than other European life markets, and the rapid expansion of the market—largely due to growth in unit-linked business and pensions—in recent years worsened the situation. A number of firms have expanded more rapidly than is considered sustainable, which resulted in either credit rating downgrades or nonratings.

Investment Performance

The Spanish insurance market has experienced significant declines in investment performance over the past five years and has performed relatively poorly by European standards. Despite this, the total investment portfolio has tripled in both the life and non-life sectors over the last ten years. However latest 2002–3 investment performance declined in both sectors, with a 10.7 percent return, respectively. Despite this, assets under management have grown to nearly €92 billion and €30 billion, respectively, by 2003 (CEA 2005).

16.3.16 Sweden

Market Overview

Two major legislations concern the private insurance activity in Sweden: the Insurance Business Act, establishing the regulations for insurance operations, and the Insurance Contracts Act, regulating the relationship between the insurer and the insured. The Consumer Insurance Act, on the other hand, deals with the most common non-life insurance policies. The Financial Supervisory Authority, under the Ministry of Finance, is responsible for supervision of the insurance industry. The Authority exercises supervision over 512 insurers, 1,250 insurance brokers, and 120 friendly societies. These are institutions established as private mutual cooperatives and tend to be regionally focused. The various European insurance directives were incorporated into Swedish legislation in 1995–1996.

The domestic Swedish private insurance industry is made up of insurers, a number of friendly societies, and 28 branch offices or agents of foreign companies. Out of the 440 Swedish insurers operating in 2003, 412 are domestic, and 28 are non-domestic companies (CEA 2005). Most are member companies of the Swedish Insurance Federation (SIF 2005, www.forsakringsforbundet.com). According to SIF Annual Chapter 2003, the insurance groups and firms that belong to the Federation account for almost 100 percent of the life market and 95 percent of the non-life market. In addition, there are 28 non-domestic companies with EU or non-EU branches. The proportion of nondomestic insurers has grown significantly over the past ten years and is the highest percentage in the entire Nordic region.

Life and non-life insurance cannot be undertaken by the same company. Life insurers are further divided into those offering traditional life insurance and unit-linked insurance. Different regulations apply to each category. Labor market insurance policies are administered by labor market insurers, set up by different parties in the labor market, and provide additional security in relation to retirement, illness, and occupational injury or illness.

Traditionally, insurers employed their own sales forces. Although this is still common, new distribution forms, such as direct sales and telephone marketing, are gaining importance. In addition, brokers have increased their share of the market. In life insurance, brokers have about 20 percent market share. The bank branch, too, has become an increasingly common channel for selling life insurance, although equivalent data for non-life insurance is not yet available (CEA 2005).

Business Trends

Growth in overall business premiums has been mixed, with rapid periods of increase among general stagnation in the late 1990s. In 2003, Swedish life insurers' premium income increased by 8 percent, but overall is still 2 percent lower than in 2001. Non-life insurers' premium income, on the other hand, increased consistently by 12 percent per year in the period 2001 to 2003. The increase was mainly due to the increases in business and homeowner insurance. However, corporate governance failures at the country's largest insurer, Skandia, have led to increased consumer and shareholder activism in the industry.

Segmentation and Consolidation

Deregulation of the Swedish insurance industry in 1992 permitted banks to own insurers and vice versa, which resulted in a period of consolidation. Sweden is the most highly concentrated market in the European Union; most of the insurers are small, local, non-life insurance companies. Over 90 percent of the market for non-life insurance is accounted for by the five largest insurance companies or groups, while 80 percent of the market for life insurance, on the other hand, is controlled by the five largest insurance groups. Some 28 active foreign insurers have a combined share of 5 percent of the non-life market.

Financial Strength

The Swedish insurance industry is very highly capitalized, but capitalization levels have declined significantly since the late 1990s. This is not surprising given the removal of restrictions concerning the equitable treatment of policyholders, and the large exposure of the Swedish market to equity-related investments and unit-linked products.

Investment Performance

Total assets of insurers (excluding AFA) have increased by only 2 percent since 1997 to SEK 156 billion. Stock market developments during 2002 were largely responsible for this development. The percentage of total assets invested in shares and participations went down, while that of bonds increased at the end of 2003 compared to the beginning of the year.

Swedish insurance company performance also has declined significantly from the ten-year horizon to the five-year horizon. The non-life insurance industry performance is negative 2.7 percent over the last five years, which is the lowest in Europe (but is equal to Finland). However latest 2002–3 investment performance improved in both sectors, with a 12 percent and 9.6 percent return, respectively. Despite this, assets under management have grown to nearly €192 billion and €37 billion, respectively, by 2003 (CEA 2005).

16.3.17 Switzerland

Market Overview

Strong competition, a relatively lower growth potential than in other European insurance markets, and continuing pressure on capital have hampered efforts by the Swiss insurance industry to provide adequate returns to shareholders. As a non-EU country, there is a relative lack of comparability of this market with those of other major countries. In addition, the availability of public information about this market is relatively low, and this has severely restricted the scope of this review. The Swiss insurance industry is regulated by the Federal Office of Private Insurance (FOPI). Gugler (2005) provides a detailed description of the broader regulatory framework and arrangements affecting the current regulation of the Swiss insurance industry and discusses ongoing efforts to develop a more unified single regulatory framework for financial markets.

Business Trends

Premium growth has remained fairly static in life insurance and has not managed to exceed the heady levels reached in 1998. By contrast, non-life insurance premiums have grown consistently by 5 percent since 2000, after stagnating in the late 1990s. Claims have nearly doubled for both life insurance and non-life insurance during this period.

Segmentation and Consolidation

The analysis of segmentation was drawn from the Swiss Federal Office of Private Insurance and by reference to the Swiss Insurance Association (www.svv.ch) and CEA (2005). There are a total of 141 companies operating in Switzerland, of which 104 are domestic firms. As of September 2003, there were 26 life insurance, 79 non-life (of which 38 are non-Swiss), and the remainder are reinsurance companies. The Swiss Federal Office of Private Insurance does not regulate non-life insurers.

With their international reach, the market leaders, Swiss Life, Winterthur, and Zurich, far exceed the earnings potential of even their closest rivals—although some mid-size companies have operations in Central and Western Europe. Those smaller companies that have tried to diversify abroad, however, have found it hard to attain critical mass, and the majority have had to retract and rely predominantly on Swiss business.

Financial Strength

The Swiss insurance market is highly capitalized. Despite the fact that Switzerland has one of the richest populations per capita, the population is relatively small by

European standards. Consequently, the market has segmented. Market leaders such as Swiss Life, Winterthur, and Zurich, are able to diversify their operations internationally. The majority of domestic Swiss insurers are competing for limited market share.

Swiss life insurers suffer from legal requirements to provide a minimum return from investments, which increases the competitive pressure to provide returns to shareholders. Mutual insurers have faced capital adequacy concerns in a tough competitive and interest-rate sensitive environment. The new Swiss Solvency Test proposed by FOPI, effective 2007, is likely to further increase the pressure on mutuals to demonstrate that they can achieve an adequate risk margin above a realistically determined level of policy liabilities. However latest 2002–3 investment performance declined in both sectors, with a 2 percent and 4 percent return, respectively. Despite this, assets under management have grown to nearly €196 billion and €74 billion, respectively, by 2003 (CEA 2005).

16.3.18 United Kingdom

Market Overview

U.K. insurers have emerged from the problems of recent years of policy misselling and regulatory overhaul. The U.K. market remains one of Europe's strongest and most competitive insurance industries, although the exposure of a number of companies to the downturn in equity markets has resulted in some significant downgrades. The Lloyd's non-life insurance market, captive insurance markets of the Channel Islands, Isle of Man, and other U.K. dependencies such as Bermuda are not included in this analysis due to space constraints and because of their specialized nature. Special provisions governing accounting, solvency, and conduct regulation apply to Lloyd's in both U.K. and EU law.

Business Trends

Life insurance sales declined by 11 percent in 2003, and overall growth has never exceeded that achieved in the pre–2001 period. There is a marked decline in sales of popular products such as personal pensions, due to scandals of misselling and a lack of trust in life insurers. General insurance premium sales also grew by about 10 percent per year to 2003 after initially declining in the period from 1994 to 1999.

Segmentation and Consolidation

As of 2003, there were a total of 772 insurers are authorized, either by the U.K. or by another European Economic Area member, to carry on insurance business in the United Kingdom. Of these 178 are non-domestic firms. Almost 600 of these insurance companies can conduct non-life business only (such as motor, household, and commercial insurance), 160 are authorized for long-term business only (such as life insurance and pensions), and 54 are composites (able to do both). The U.K. insurance industry is very international by European standards, with 12 percent of firms being EU branches of nondomestic firms, and another 9 percent of the market being non-EU based, which is the highest in Europe. Bank distribution in the U.K. market has evolved since the early 1990s and is now a significant distribution channel for the leading companies.

Historically, most insurance sales were performed by brokerage firms, but, while still dominating the life insurance market (65%), they have declined in importance in the non-life segment. By contrast, direct sales account for 29 percent and 24 percent of the life and non-life sectors, respectively. By contrast, other networks account for a significantly increasing proportion of non-life sales (CEA 2005).

Disclosure of commissions, regulatory scrutiny of persistency rates, and the arrival of a 1 percent cap on stakeholder policies have changed the historical business model for distribution of products. Moreover, the decision by the Financial Services Authority, effective 2004, to drop the former requirement for "polarization" or separation of distribution system between independent financial advisers (IFA) and tied agents has presented new challenges in the marketing of products. High-cost distribution networks for simpler products—that require little or no financial advice—have struggled in the face of increased competition and public and supervisory scrutiny. Distribution has also become more focused on products and consumer segments (Klumpes 2004).

The top five life insurers accounted for 42.5 percent of gross premiums written, while the top five non-life insurers in the United Kingdom account for 44 percent of total premiums written.

Financial Strength

U.K. insurers are relatively well capitalized. Prior to 2004, firms had considerable discretion to alter the assumptions underlying their calculations of reserves. However, the new "realistic" solvency levels established by the Financial Services Authority (FSA) in 2004 under its new prudential standards (FSA 2005) limited the scope for discretion by requiring more market-based valuation of policy liabilities. It also led to a reexamination of the strength of a number of life insurers, some of which have now ceased writing new business. The issue by the Accounting Standards Board of a new financial reporting standard for life insurers (Financial Reporting Standard 27), fully effective 2006, has reduced the information gap by mandating standardized financial statements that are both consistent with U.K. GAAP and international accounting standard IFRS 4.

Investment Performance

U.K. life insurance firms have generated a relatively high performance over both ten- and five-year horizons. By contrast, returns generated by non-life insurers are extremely low over the corresponding periods. The total investment portfolio of the life sector has tripled since 1992, while only increasing by 50 percent for the non-life sector. However latest 2002–3 investment performance improved in both sectors, with a 14.2 percent and 1.4 percent return, respectively. Assets under management have grown to nearly €1332 billion and €101 billion, respectively, by 2003 (CEA 2005).

16.4 CONCLUSION

This chapter has documented the major recent trends in the European insurance markets and examined various market and regulatory-based sources that can explain the apparent diversity in the structure of national insurance markets and in the forms of regulation affecting the European insurance industry. The promulgation of new EU directives (e.g., Solvency II) and greater harmonization in International Accounting Standards (IFRS 4) were identified as regulatory factors impacting the future performance of the European insurance market; the adoption of harmonized accounting and solvency regulations in most EU states has further facilitated convergence in the evolution of national European insurance markets by encouraging competition, mergers and acquisition activity, and cross-country product comparisons by consumers.

Common themes discussed in this chapter across countries include the consolidation of the insurance industry, the integration of supervision between banks and insurers, and the increasing emphasis on retirement savings and policy effects on the development of insurance. There has been a gradual shift from traditional distribution channels toward brokers and banks and Internet channels. The life insurance industry has been subject to the effects of an aging population, as well as increased emphasis on equity-related investments. This has resulted in greater volatility in reported performance over time. The general insurance industry has been subject to greater claims and consequent underperformance of investments. On the regulation side, the reduction of the tariff mechanism in many countries and deregulation of investments have increased the business cycle over time. In most countries, financial strength has gradually weakened over time, possibly due to the adverse capital market conditions combined with increased numbers of claims incurred.

While some of these influences are idiosyncratic and reflect unique sociopolitical environments, the increasing harmonization of insurance regulation across the European Union has also demonstrated the importance of taking a more standardized view of performance. As the new regulations are gradually implemented, more research is needed to examine the implications of convergence in financial reporting and solvency regulations for the incidence of mergers and acquisition activity, cost and profit efficiency, and corporate strategy, particularly as regards risk management and corporate governance effectiveness.

Areas of regional variation include the relatively poor capitalization and growing exposure of Mediterranean insurers to equities. By contrast, the consolidation of financial services in the Scandinavian countries has been accompanied by stagnation in business sales and has resulted in severe market concentration and a general diminution in capital strength. Central Europe is somewhat a mix of these trends, with new business closely linked to public policy efforts to foster retirement income provision. In this respect, Luxembourg is an exception, because most of the insurance is sold to foreign policyholders for tax reasons. Ireland and the United Kingdom share a mix of strong business growth, rapid financial innovation, and regulatory overhaul. In nearly all countries, the regulatory structures have moved away from industry specialization and stability and toward cross-functional financial service firm supervision, which appears to enforce the development of bancassurance as the most important evolution in the distribution and marketing of insurance.

This chapter has not examined some broader issues surrounding these trends, especially the growing segmentation and concentration of domestic markets by a small group of increasingly concentrated international firms. The role of financial innovation in risk management, asset management, and liability management has also not been investigated. The rapidly growing importance of Eastern European insurance markets also requires further analysis. The integration of private insurance markets with government insurance roles, especially in the areas of environmental liabilities, medical coverage, third-party accident claims, litigation risk, and catastrophic events on industry sectors and recent outrages in Spain, also are becoming increasingly seen as national or even European areas. The development of financial innovation in securitization, collateralized debt obligations, catastrophe bonds, longevity bonds, and long-dated debt and derivative instruments is also likely to transform the financial structure of European insurers in the next few years.

Presently only the Netherlands, Switzerland, and the United Kingdom require market-to-market accounting and stochastic modeling of assets and liabilities for solvency purposes. These developments have foreshadowed likely harmonization of European Union solvency regulations away from static accounting-based minimum standards toward risk-based cash flow and scenario models. However, most other European states have not yet moved toward harmonizing their solvency and accounting rules, or still use a combination of book-to-market or other minimum requirements as established by the EU Solvency I directive.

Greater sophistication in risk management will be required to facilitate further financial innovation, consolidation, and development of the pan-European insurance sector. These developments also highlight the likely demand for more sophisticated risk management systems to monitor the increasingly complex financing, investment, and marketing strategies. It is likely that there will be increased segmentation of the market between multinational firms and niche players. Such evolution in turn will likely facilitate greater convergence between the European and international insurance sectors. These and other areas pose the greatest challenges for European insurance firms and warrant further research into the impact of more sophisticated financial innovation, technology, and risk management needs on the mix of market forces and regulatory structures that they are likely to face in the near future.

16.5 REFERENCES

Accounting Standards Board (U.K.), 2004, *Financial Reporting Standard 27: Life Assurance* (London: ASB).

Albert, Michael, 1987, "How To Ensure the Security of Reinsurance in Spite of Economic and Social Insecurity," *Quarterly Letter of NRG* May: 5–13.

Associacao Portuguesa de Seguradores, 2005, *Annual Report* (Lisbon: APS).

Association of Insurance Companies Greece, 2005, *Annual Report* (Athens: AICG).

Associazione Nazionale fra le Imprese Assicuratrici, 2003, *Italian Insurance in Figures* (Rome: ANIA).

Assuralia, 2004, *Assurinfo No. 31, Kerncijfers en voornammste resultaten van de Belgische verzekeringsmarkt in 2003* (Brussels: Assuralia).

Austrian Financial Market Authority, 1996–2003, *Annual Report* (Vienna: AFMA).

Benoist, Gilles, 2002, "Bancassurance: The New Challenges," *Geneva Papers on Risk and Insurance* 27(3): 295–303.

Committee Europeene des Assurances (CEA), 2004–2005, *Facts and Figures: European Insurance Market* (Brussels: CEA).

Danish Insurance Association, 1999–2003, *Summary Annual Report* (Copenhagen: DIA).

Danish Insurance Information Service, 2002–2003, *Insurance and Pensions in Denmark, 2002–2003 Figures and Graphs* (Copenhagen: DIIS).

Deutsche Rechnungslegungs Standards Committee, 2004, *Deutsche Rechnungs-legungs Standards* (GAS 5–20) (Stuttgart, Germany: Schaffer-Poeschel Verlag).

Dickinson, Gerry, and Patrick M. Liedtke, 2004, "Impact of a Fair Value Financial Reporting System on Insurance Companies: A Survey," *Geneva Papers on Risk and Insurance: Issues and Practice* (special issue).

Eurostat, 2004, *Statistics in Focus: Credit Institutions and Insurance in the New Member States and Candidate Countries*, 40/2004.

Federation of Switzerland, Federal Office of Private Insurance, 2004, *Annual Report* (Bern: Federation of Switzerland).

Federation of Finnish Insurance Companies, 1996–2004, *Finnish Insurance Economy* (Helsinki: FFIC).

Financial Services Authority (FSA), United Kingdom 2005, *Annual Report* (London).

German Insurance Association (Gesamtverband der Deutschen Versicherungswirtschaft e.V.–GDV), 2004, *Statistical Yearbook of German Insurance* (Berlin: GDV).

Gugler, Philippe, 2005, "The Integrated Supervision of Financial Markets: The Case of Switzerland," *Geneva Papers on Risk and Insurance: Issues and Practice* 30(1): 128–143.

Hausler, Gerd, 2005, "Risk Transfer and the Insurance Industry," *Geneva Papers on Risk and Insurance: Issues and Practice* 30(1): 121–127.

Irish Insurance Federation, 1996–2004, *IIF Factfile* (Dublin: IIF).

International Accounting Standards Board, 2004, *International Financial Reporting Standard 4: Insurance Contracts* (London: IASB).

International Association of Insurance Supervisors, 2000, *Core Principle* (Brussels: IAIS).

Istituto per la Viglanza sulle Assicurazioi Private e di Interesse Collettivo, 2005, *Annual Report* (Rome: IVAPIC).

Klumpes, Paul J. M., 2004, "Performance Benchmarking in Financial Services: International Evidence," *Journal of Business* 77(2): 257–273.

KPMG, 2002, *Study into the Methodologies To Assess the Overall Financial Position of an Insurance Undertaking from the Perspective of Prudential Supervision* (London: KPMG).

Muller, Bernd, 2003, "The Other Side of the Fence: Allianz's View of the European Commissions' Solvency II Proposals on Risk and Capital Management," *Insurance Risk Management Special* January: 2–5.

Munch, P., and D. E. Smallwood, 1980, "Solvency Regulation in the Property-Liability Insurance Industry: Empirical Evidence," *Bell Journal of Economics*, 261–279.

O'Brien, Christopher, 1994, "Profit, Capital, and Value in a Proprietary Life Assurance Company: A Discussion Paper," *Journal of the Institute of Actuaries* 121: 285–361.

Office de Contrôle des Assurances (www.cdv-oca.be), *Annual Chapter* 2002–2004.

Oosenbrug, Alfred, 2003, "From Profit Smoothing to a True and Fair Presentation of Profits at Insurance Companies and Pension Funds," in Henk Langendijk, Dirk Swagerman, and Willem Verhoog, eds., *Is Fair Value Fair?* (Chichester, United Kingdom: Wiley), pp. 191–202.

Organisation for Economic Cooperation and Development, 2003, *Policy Issues in Insurance: Assessing the Solvency of Insurance Companies* (Paris: OECD).

Pensions en Verzekeringskamer, 1996–2004, *Jaarverslag* (Apeldoorn, Netherlands: PV).

Quintyn, Marc G., and Michael W. Taylor, 2004, "Should Financial Sector Regulators Be Independent?" *IMF Economic Issues No. 32*. (Washington, DC).

Storm, Kees J., 2003, "Introduction of Fair Value Accounting: Little If Any Haste," in Henk Langendijk, Dirk Swagerman, and Willem Verhoog, eds., *Is Fair Value Fair?* (Chichester, United Kingdom: Wiley), pp. 245–254.

Suomen Vakuutusyhtioiden Keskusliitto, 2004, *Finnish Insurance Today* (Helsinki: SVK).

Swedish Insurance Federation, 1996–2004, *Swedish Private Insurance in Figures* (Stockholm: SIF).

Swedish Insurance Federation, 1996–2004, *Fact Sheets* (Stockholm: SIF).

Swiss Insurance Association, 2004, *Facts and Figures (*Zurich: SIA).

Tillinghast-Towers Perrin, 2004, *Insurance Pocket Book* (London: Tillinghast-Towers Perrin).

Traas, Lou, 2003, "Insurers Are Lagging Behind," in Henk Langendijk, Dirk Swagerman, and Willem Verhoog, eds., *Is Fair Value Fair?* (Chichester, United Kingdom: Wiley), pp. 191–202.

Van den Berghe, Lutgart, 1990, "(De)Regulation of Insurance Markets," in Henri Louberge, ed., *Risk, Information and Insurance* (Boston: Kluwer), pp. 199–220.

Verband der Versicherungsunternehmen Osterreichs, 2004, *Annual Report* (Vienna: VVO).

Verbond van Verzekeraars, 2004, *Annual Report* (Amsterdam: VV).

Weatherill, Stephen, 2004,. *Cases and Materials on EC Law* (London: Blackstone Press).

17 South and East Asian Insurance Market Growth and Development

Stephanie Hussels
Bradford University

Claire Sherman
University of Adelaide

Damian Ward
Bradford University

Ralf Zurbruegg
University of Adelaide[316]

17.1 INTRODUCTION

Recent economic research, notably by King and Levine (1993a, 1993b), Levine and Zervos (1998), Levine (1999), Levine, et al. (2000), and Beck, et al. (2000), indicates that financial services and its various components, including insurance and banking, have substantial potential for spreading positive externalities throughout the commercial sector of an economy. Such benefits can stem from improved access to capital by firms, better allocation of capital to investment projects, greater risk management, and enhanced portfolio diversification and liquidity for individual investors. While existing economic research shows the development of financial services is generally important for economic growth, a number of previous studies by Outreville (1990) and Ward and Zurbruegg (2000) provide empirical evidence that insurance market development in its own right can promote economic development. The importance of the insurance industry to the wider economy is seen to stem from the relative size of the insurance industry to GDP in many developed economies, the

[316] The authors wish to thank the editors and an anonymous referee for very instructive and helpful comments in developing this chapter.

transfer of risks, and the scale of insurance companies' financial intermediary functions.

While the link between financial services development and economic growth is well-established, the focus of this chapter is to highlight research that has sought to identify the factors that will promote the demand and supply of the insurance sector. In particular, recent empirical research on insurance markets by Beck and Webb (2003), Browne, et al. (2000), Esho, et al. (2004), and Ward and Zurbruegg (2002) have shown that the level of insurance demand within an economy can be influenced by a number of particular variables, including economic, legal, political, and social factors.

Despite these results, there has been little information presented on which specific factors should be fostered to aid financial development via the insurance market. To address this problem, the objective of this chapter is to present a synopsis of the existing literature and provide some insight on how the development of the South and East Asian insurance market might be achieved. Insurance companies who intend to expand their business activities abroad can also utilize this knowledge to select markets in South and East Asia.

In focusing upon South and East Asia, the chapter is organized as follows: First, the relative size, economic importance, and likely developments of the insurance markets within the region are presented. Second, empirical evidence relating to how demand and supply within the insurance sector can be developed is discussed. Finally, this empirical evidence is then further examined to consider the effects of economic stimulation in the insurance market on insurers operating in South and East Asia.

17.2 DEMAND AND SUPPLY OF SOUTH AND EAST ASIAN INSURANCE

Before delving into the specific demand and supply features of the regional insurance market in South and East Asia, a preliminary statistical review of the countries considered within this chapter is listed in Table 17.1, including 2003 data on population, gross domestic product (GDP), inflation rates, and net written insurance premiums. Net written life and non-life insurance premiums for 1998–2003 are included in Table 17.2. This helps to establish a comparative basis to examine insurance market forces within the region, starting below with an analysis of demand. In many of the emerging markets of Asia, data by underwriting class is unfortunately difficult to obtain. Therefore, in Table 17.3 we only provide a breakdown of non-life underwriting classes for the developed economies of Hong Kong, Japan, and Singapore.

Table 17.1. Population, GDP, Inflation, and Aggregate Net Written Insurance Premiums for Countries in South and East Asia, 2003

Country	Population (Millions)	GDP (US$ Billions)	Inflation (%)	Life Premiums (US$ Millions)	Non-Life Premiums (US$ Millions)
Bangladesh	138.1	52.0	5.9	194.0	102.0
Hong Kong	6.8	159.0	–2.1	10,117.0	2,377.0
India	1,056.3	601.0	3.9	13,590.0	3,712.0
Indonesia	214.5	208.0	6.6	1,373.0	1,733.0
Japan	127.0	4,429.0	–0.3	381,335.0	97,530.0
Malaysia	24.7	105.0	1.1	3,455.0	2,154.0
Philippines	81.4	80.0	3.0	702.0	489.0
PR China	1,290.8	1,410.0	1.2	32,442.0	144,468.0
Singapore	4.3	91.0	0.5	5,561.0	3,337.0
South Korea	48.1	621.0	3.5	41,998.0	17,760.0
Sri Lanka	19.2	18.0	7.4	102.0	137.0
Taiwan	22.6	287.0	–0.3	23,739.0	8,662.0
Thailand	62.0	143.0	1.8	3,222.0	1,711.0
Vietnam	81.5	38.0	3.2	331.0	218.0

Source: Various National Annual Statistical Reports and Swiss Re (2004a), pp. 37, 39.
Note: GDP is gross domestic product.

17.2.1 Demand for Insurance in South and East Asia

With the recent growth of the South and East Asian Insurance market, it is no wonder that it is attracting much attention from the global insurance market. Life insurance premiums for South and East Asia reached a phenomenal growth rate of 18.3 percent in 2002 and non-life insurance premiums grew by 13.8 percent in the same period (Swiss Re 2003b). As Figure 17.1 indicates, patterns of growth within the region's insurance industry suggest that changes in premiums are correlated with GDP growth, and with GDP predicted to grow at a real rate of 5.7 percent per annum between 2003 and 2006, this growth is likely to continue (Swiss Re 2003b). This link between GDP and demand for insurance can be explained through the S-curve relationship (Enz 2000), which indicates that as GDP grows, people begin to have sufficient income and assets to warrant insurance protection. This leads to a steep increase in insurance demand greater than the growth in GDP. However, eventually, as more and more people reach a level where they have already attained insurance, growth in premiums stabilizes. This particular level tends to be different depending on the social security system in place within the particular country.

Table 17.2. Life and Non-Life Net Premiums, 1998–2003

Life Insurance Net Premium Volumes and Insurance Penetration Rates

	1998		1999		2000		2001		2002		2003	
	Premiums (US$ Millions)	% of GDP	Premiums (US$ Millions)	% of GDP	Premiums (US$ Millions)	% of GDP	Premiums (US$ Millions)	% of GDP	Premiums (US$ Millions)	% of GDP	Premiums (US$ Millions)	% of GDP
Bangladesh	75.00	N/A	76.00	0.22%	130.00	0.28%	129.00	0.28%	175.00	0.28%	194.00	0.37%
Hong Kong	4,680.00	3.26%	5,771.00	3.63%	5,969.00	3.73%	8,404.00	5.13%	8,400.00	5.20%	10,117.00	6.36%
India	5,463.00	1.90%	6,066.00	1.40%	7,615.00	1.77%	9,418.00	2.15%	11,515.00	2.59%	13,590.00	2.26%
Indonesia	487.00	0.62%	927.00	0.66%	816.00	0.54%	773.00	0.53%	1,135.00	0.66%	1,373.00	0.66%
Japan	363,393.00	9.35%	392,908.00	8.87%	400,993.00	8.69%	356,731.00	8.85%	354,109.00	8.63%	381,335.00	8.61%
Malaysia	1,610.00	1.90%	1,774.00	2.16%	2,671.00	2.13%	3,082.00	3.38%	2,921.00	2.94%	3,455.00	3.29%
Philippines	473.00	0.72%	585.00	0.76%	590.00	0.79%	493.00	0.69%	681.00	0.87%	702.00	0.88%
PR China	9,031.00	0.86%	10,535.00	1.02%	10,476.00	1.12%	15,556.00	1.34%	25,202.00	2.02%	32,442.00	2.30%
Singapore	2,670.00	3.53%	2,762.00	3.19%	2,942.00	3.12%	2,914.00	3.33%	6,111.00	2.55%	5,561.00	6.11%
South Korea	35,703.00	10.32%	35,638.00	8.39%	44,237.00	9.89%	36,392.00	8.69%	40,182.00	8.23%	41,998.00	6.76%
Sri Lanka	71.00	0.45%	69.00	0.43%	87.00	0.53%	83.00	0.53%	91.00	0.56%	102.00	0.57%
Taiwan	11,928.00	4.56%	13,968.00	4.83%	15,771.00	5.10%	16,953.00	5.66%	20,767.00	7.35%	23,739.00	8.27%
Thailand	1,362.00	1.21%	1,619.00	1.28%	1,883.00	1.53%	2,127.00	1.86%	2,219.00	2.09%	3,222.00	2.25%
Vietnam	15.00	0.06%	N/A	N/A	91.00	0.12%	169.00	0.52%	302.00	0.87%	331.00	0.87%

Table 17.2 *(continued)*. Life and Non-Life Net Premiums, 1998–2003

Non-Life Insurance Net Premium Volumes and Insurance Penetration Rates

	1998		1999		2000		2001		2002		2003	
	Premiums (US$ Millions)	% of GDP	Premiums (US$ Millions)	% of GDP	Premiums (US$ Millions)	% of GDP	Premiums (US$ Millions)	% of GDP	Premiums (US$ Millions)	% of GDP	Premiums (US$ Millions)	% of GDP
Bangladesh	80.00	N/A	78.00	0.22%	75.00	0.21%	81.00	0.18%	92.00	0.18%	102.00	0.20%
Hong Kong	1,834.00	1.13%	1,727.00	1.09%	1,829.00	1.13%	1,987.00	1.21%	2,342.00	1.45%	2,377.00	1.49%
India	2,177.00	0.71%	2,325.00	0.53%	2,338.00	0.55%	2,459.00	0.56%	3,149.00	0.67%	3,712.00	0.62%
Indonesia	666.00	0.65%	1,054.00	0.76%	977.00	0.64%	868.00	0.60%	1,415.00	0.83%	1,733.00	0.83%
Japan	101,977.00	2.38%	91,999.00	2.30%	102,521.00	2.23%	89,114.00	2.21%	92,032.00	2.24%	97,530.00	2.20%
Malaysia	1,399.00	2.12%	1,414.00	1.72%	1,486.00	1.59%	1,636.00	1.80%	1,960.00	1.97%	2,154.00	2.05%
Philippines	446.00	0.78%	466.00	0.61%	471.00	0.63%	414.00	0.58%	474.00	0.61%	489.00	0.61%
PR China	6,035.00	0.63%	6,295.00	0.61%	7,228.00	0.67%	9,928.00	0.86%	11,760.00	0.96%	144,468.00	10.25%
Singapore	1,027.00	1.08%	944.00	1.09%	1,024.00	1.09%	1,091.00	1.25%	2,799.00	2.36%	3,337.00	3.67%
South Korea	10,970.00	3.55%	12,291.00	2.89%	14,112.00	3.16%	14,145.00	3.38%	16,526.00	3.38%	17,760.00	2.86%
Sri Lanka	117.00	0.74%	112.00	0.71%	110.00	0.68%	106.00	0.68%	121.00	0.74%	137.00	0.76%
Taiwan	4,973.00	1.90%	6,009.00	2.08%	7,075.00	2.29%	7,299.00	2.49%	7,956.00	2.81%	8,662.00	3.02%
Thailand	1,198.00	1.08%	1,208.00	0.95%	1,211.00	1.00%	1,239.00	1.08%	1,928.00	1.15%	1,711.00	1.20%
Vietnam	126.00	0.45%	N/A	N/A	127.00	0.41%	146.00	0.44%	205.00	0.58%	218.00	0.57%

Source: Swiss Re (1999), pp. 25, 27, 29; Swiss Re (2000), pp. 29, 31, 33; Swiss Re (2001a), pp. 33, 35, 37; Swiss Re (2002a), pp. 31, 33, 35; Swiss Re (2003b), pp. 33, 35, 37; Swiss Re (2004a), pp. 37, 39, 41.

Note: Insurance penetration rate is defined at net written insurance premium divided by GDP. N/A is not available.

Table 17.3. Distribution of Net Written Premiums by Line, 2003 for Developed Countries in South and East Asia (% of Total)

	Aviation-Cargo	Marine-Hull & Liability	Fire	Motor	Workers' Compen-sation	Miscellan-eous	TOTAL
Hong Kong	4.90%	3.10%	16.90%	15.70%	21.50%	37.90%	100%
Japan	2.20%	0.70%	16.50%	54.60%	17.30%	8.70%	100%
Singapore	6.30%	3.20%	10.80%	37.70%	7.90%	34.10%	100%

Source: Singapore MAS (2005), General Insurance Association of Japan (2004), and the Hong Kong, Office of the Commissioner of Insurance (2004).

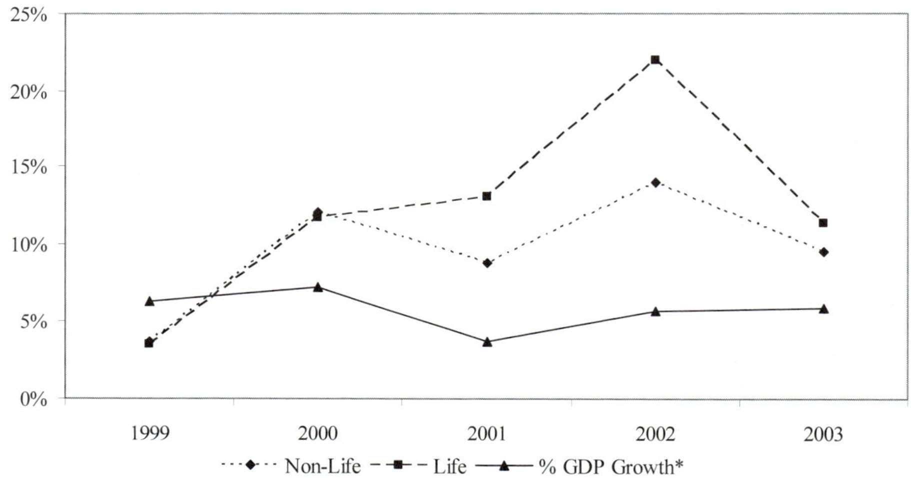

Source: Datastream International (year) for GDP. Swiss Re (1999), p. 25, 27, 29; Swiss Re (2000), pp. 29, 31, 33; Swiss Re (2001a), pp. 33, 35, 37; Swiss Re (2002a), pp. 31, 33, 35; Swiss Re (2003b), p. 33, 35, 37; Swiss Re (2004a), pp. 37, 39, 41 for insurance premiums.
Note: Data for 1999 excludes Sri Lanka and Vietnam.

Figure 17.1. % Change in Net Written Insurance Premiums and % Change GDP, 1999–2003 (Adjusted for Inflation)

Table 17.4 provides evidence of the increase in premiums in most of the countries in South and East Asia. In particular, China has seen outstanding growth in premiums for both life and non-life insurance, at 62.2 percent and 20.1 percent in 2002, respectively (Swiss Re 2003b). This is not surprising when considering the changing socio-economic structure of the Chinese population. A rising middle class with a greater income to protect has opened a new market for insurers. Assisted by a decline in restrictive regulations, insurers have been able to offer more insurance products to this developing middle class resulting in a higher growth rate in premiums. The threshold where GDP growth will begin to outpace premium growth has yet to be reached, and hopeful insurers in this market will continue to capitalize on this opportunity until it is attained.

Table 17.4. The Percentage Growth in Net Written Insurance Premiums in South and East Asia, 2001–2003

	2001		2002		2003	
	Life	Non-Life	Life	Non-Life	Life	Non-Life
Bangladesh	–0.4%*	–0.4%*	4.1%*	4.1%*	10.8%*	10.8%*
Hong Kong	43.2%	10.4%	18.8%	22.3%	22.8%	3.5%
India	22.5%	3.3%	14.1%	21.4%	8.4%	8.2%
Indonesia	3.9%	3.5%	11.7%	9.1%	20.5%*	20.5%*
Malaysia	13.8%	2.2%	–9.1%	13.4%	16.9%	8.7%
Philippines	–9.4%	4.3%	17.0%	13.2%	3.2%*	3.2%*
PR China	47.4%	11.4%	62.2%	20.1%	27.2%	21.6%
Singapore	1.9%	9.6%	4.2%	26.6%	–11.9%	6.5%
South Korea	–11.9%	7.4%	0.7%	6.5%	–2.0%	0.7%
Sri Lanka	–4.4%*	–4.4%*	5.8%	13.0%	12.8%*	12.8%*
Taiwan	16.4%	11.3%	25.2%	11.4%	14.2%	8.8%
Thailand	23.1%	11.3%	19.8%	12.6%	15.4%	11.6%
Vietnam	95.3%	21.5%	61.1%	36.4%	8.9%*	8.9%*

Source: Swiss Re (2002a), pp. 31, 33; Swiss Re (2003b), p. 33, 35; Swiss Re (2004a), pp. 37, 39.
Note: * Not adjusted for inflation.

In general, South and East Asian countries have relatively high savings rates. The financial volatility experienced within the region's previous decade has meant that people are looking for safer options to invest their savings, and insurance has recently been seen as one way to reduce risk. By capturing a wide customer base, insurance companies have been able to diversify this risk on behalf of the investors within the region. A culture of self-reliance within many of these countries has also fueled the demand for insurance. As faith in the ability of some governments to provide for people in later life fades, people are procuring insurance policies to ensure their well-being. Also, as with much of the world, an ageing population in South and East Asia is driving demand for pension and investment-linked insurance products, either through direct personal investment or through governments searching for ways to fund their pension schemes.

The 1997–98 financial crisis in the region has had numerous effects on the insurance industry, one of which is the increased emphasis on risk awareness and risk minimization. Initially, the real estate market experienced a downturn that significantly changed the security holdings of banks. Additionally, bad corporate debt instigated the collapse of many banks within the region (Roubini 2004). As a result, the currencies within Asia started to plummet and central banks unsuccessfully tried to support their weakened position. As international funds were being funneled out of the region, and assets were declining in value, companies, governments, and the public within the region became more aware of the risks of investing and relying on established financial bodies to provide good credit assessments. For the insurance industry, this has fueled an increase in demand as a currently limited, but growing public awareness of insurance develops.

Another factor that has enhanced awareness was the Severe Acute Respiratory Syndrome (SARS) epidemic, which first appeared in Southern China in November 2002 and spread to Canada, China, Taiwan (China), Germany, Hong Kong, Singapore, Slovenia, Thailand, Vietnam, and the United Kingdom by March 2003 (World Health Organization 2003). In particular, SARS has increased the necessity of health-related insurance, and although initially having a marginally negative impact on demand where travel is a major component, it should stimulate further demand for insurance. Furthermore, due to other events such as the Indonesian riots in 1998, the 1999 earthquake in Taiwan, and the tsunami that struck several South and East Asian countries in late 2004, a need to insure against the risks of both natural and man-made disasters has become apparent within the region.

There are differences in the demand for both commercial and personal non-life insurance that are often dependent on the country's stage of development. More rapidly developing countries such as China, India, and Indonesia found a greater proportion of their demand came from commercial insurance in 2001 when compared to their counterparts (Swiss Re 2003a). This is partly due to the lack of awareness for personal insurance and the low income of members of the general population. However, as stated previously, this is changing. Additionally, the rise in total premiums may be due to a relative increase in the price of commercial premiums compared to personal insurance.

In summary, the growth of GDP, changes in socio-economic structures within the region, the increased awareness of the need for insurance, and the need for risk minimization highlighted by recent damaging events have lead to an increased demand for both life and non-life insurance within the South and East Asian region. Nevertheless, coupled with this ongoing increase in demand, is an unlocking of the supply of insurance for this region, which is explored in following section.

17.2.2 Supply of Insurance in South and East Asia

Both Singapore and Hong Kong have well-established insurance markets and, due to the deregulation of financial markets, insurance markets in other countries within this region have also begun to grow over the past five to ten years. Changes to the structure of companies and the regulatory environment have been pivotal in opening new avenues for the supply of insurance within the region.

Structure of Insurance Companies

The structure of insurance companies can be defined in several ways, namely, whether they are domestic or foreign, government-owned or private, listed or non-listed, or by their size. Data on the market share held by government and foreign owned insurance companies are listed in Table 17.5. Also, broadly speaking, within

South and East Asia's non-life insurance market in 2002, 39 percent of companies were foreign (including joint ventures and branches) versus 61 percent domestic companies (Swiss Re 2003a). Interestingly, this 39 percent of companies represents only 10 percent of the dollar share of net written premiums within the region. These divisions vary widely depending on the country of focus. For example, in Singapore 52 percent of the premiums are held with foreign companies, whereas, only 0.6 percent are held by foreign companies in South Korea (Swiss Re 2003a). The opening of insurance markets to foreign companies has allowed investment of much needed capital into the region, where the effects of the Asian financial crisis of 1997 and sluggish world investment market continue to linger.

Table 17.5. Market Shares of State and Foreign Owned Insurers (Net Premiums Written), 2003

	> 50% state owned		> 50% foreign owned		≥ 50% foreign owned	
	Life	Non-Life	Life	Non-Life	Life	Non-Life
Bangladesh	N/A	N/A	N/A	N/A	61%	29%
Hong Kong	0%	0%	87%	74%	N/A	N/A
India	92%	86%	0%	0%	33%	12%
Indonesia**	10%	10%	48%	25%	48%	25%
Japan	N/A	N/A	N/A	N/A	58%	53%
Malaysia*	0%	0%	71%	25%	N/A	N/A
Philippines*	0%	0%	61%	29%	N/A	N/A
PR China*	57%	72%	2%	1%	41%	7%
Singapore*	0%	0%	58%	53%	87%	74%
South Korea	23%	4%	10%	1%	2%	1%
Sri Lanka	N/A	N/A	N/A	N/A	56%	6%
Taiwan	11%	0%	33%	12%	10%	1%
Thailand*	0%	N/A	N/A	N/A	N/A	N/A
Vietnam	44%	94%	56%	6%	81%	25%

Source: Swiss Re (2004b), p. 40.
Note: Market share is calculated using the total premium volume of those firms with ≥50% ownership. > (≥) highlight the % held by firms with exactly 50% foreign ownership or equal ownership (greater than 50% ownership) for some countries. * Ownership data is from 2002. ** Ownership data is from 2001. N/A is not applicable.

Between 1997 and 2001, the performance of foreign companies was more impressive than domestic companies, with growth in premiums for foreign firms amounting to 5.3 percent, whilst domestic companies experienced a slight decline (Swiss Re 2003a). Despite this, domestic companies have experienced greater profit margins, albeit, they have not had to incur the start up costs that foreign insurers have. Foreign companies have also shown a better ability to assess risk within the region with lower loss ratios in most countries. However, this is only within the short-term as foreign firm admittance is relatively recent and long-term losses may not yet be apparent. In general, there are no discernable differences between the types of businesses that foreign and domestic insurers underwrite, with the exception of motor insurance, which is more prevalent in the domestic realm.

Within environments characterized by deregulation, consolidation, and a decline in government ownership, the withdrawal of underwriting restrictions and premium controls have provided an increasing opportunity for insurance companies to become progressively more commercially driven and strategically focused. Large insurance companies have generally taken the initiative, now representing 8 percent of the insurance companies in the region, but holding 70 percent of the dollar share of premiums.[317] As with many markets, the large consolidated firms have been able to utilize their economies of scale and breadth of expertise to become more competitive. This push for consolidation has meant that a considerable number of smaller companies have become prime merger and acquisition targets for the large insurance companies. Large companies may also be able to diversify to a greater extent when compared to the smaller companies. However, they tend to follow regional market trends, which would not make them as resilient as global companies when considering shocks such as the Asian economic crisis. An indication of the degree of market concentration in the non-life sector is provided in Table 17.6.

Table 17.6. Non-Life Insurance Market Concentration, 2001

	Total Number of Insurers	Market Share of Top Five
India	5	100%
China	17	98%
Vietnam	10	90%
South Korea	17	78%
Japan	57	54%
Taiwan	27	48%
Thailand	71	39%
Indonesia	104	39%
Singapore	45	38%
The Philippines	112	32%
Malaysia	50	32%
Hong Kong	125	28%

Source: Swiss Re (2003), p. 27.
Note: Market concentration is measured in terms of 2001 net written premiums.

Public ownership is on the decline, particularly within the previously protected markets such as China and India, where 90 percent of non-life insurance business is undertaken by government-owned entities (Swiss Re 2003a). Like any other public entity, non-profit objectives are often sought and subsidizations of rural services take

[317] Large companies refer to the top 1–5 firms within the country.

place within government-owned insurance companies. However, now that the markets are opening to private companies, it is likely that these matters will receive lesser consideration. Interestingly, performance of government-owned insurers is surprisingly consistent with their private counterparts (Swiss Re 2003b). Nonetheless, this may be attributed to their greater ability to secure government business (Swiss Re 2003a) and use of established distribution networks and databases. It is also worth pointing out that in Islamic nations there is a general preference for mutual insurance, which is a form of risk sharing. This would be in contrast to private, profit-seeking insurance companies that profiteer from speculation on life expectancies through the issue of price-related premiums, which does not conform well to Sharia Islamic law.

The Regulatory Environment

The World Trade Organization (WTO) has made a significant impact in the South and East Asian region by encouraging insurance market liberalization. In particular, market deregulation is breaking down barriers for foreign insurers and creating a more market-based system in countries such as China, India, Taiwan, Malaysia, and Thailand.

In China, amendments to the 1995 Insurance Law, stipulate a market-based insurance system where companies have more control over their policies and premium rates (Allens Arthur Robinson 2003). The introduction of new regulation on the administration of foreign-invested insurance companies (China Insurance Regulatory Commission 2004) also provides more detail for foreign companies who wish to enter the market. Although restrictions still remain, there have been a number of foreign licenses granted and this step is seen as an indication of further liberalization to come.

Another country that has undergone substantial deregulation is Thailand. It has gone from a market where a limited amount of foreign investment has been allowed to a market that encourages foreign capital and involvement. There is considerable foreign involvement in the Thai insurance market with much of the foreign investment in complex share agreements and joint venture structures. The reason for this complexity is that foreign insurance companies are limited to a 25 percent holding in any Thai insurance company; and there are still restrictions on direct foreign investment, where companies with sole licenses can only operate from one location unless they are in a joint venture with a Thai insurance company (Allens 2003). Also, until 2004, motor vehicle insurance was not compulsory in Thailand. The introduction of compulsory motor insurance will obviously increase the market and, if further commitments to WTO agreements are met, this should allow a greater opening of this large sector of the Thai insurance industry to foreign investment.

De-tariffication of the non-life sector is particularly relevant in China, Taiwan, and India. Tariffs are placed on most lines of insurance within India, with fire (22 percent) and motor insurance (38 percent) being the most prevalent insurance provided (The Tariff Advisory Committee 2004). Currently, information regarding claims is not complete and this means that the tariffs set by the Tariff Advisory Committee are based on assumptions rather than calculated risk estimates. Reforms are in place to change the tariff-based system to a risk-based premium setting, which will assist in a better match of assets coming in to the risk being taken.

In general, both the life and non-life insurance markets in South and East Asia are becoming more liberalized, although some countries are far more regulated than others. Given the above discussion of the overall structure and developments in the insurance markets in South and East Asia, there is a particular trend in the regional market that deserves particular attention, the recent rise of bancassurance.

17.2.3 The Rise of Bancassurance

Career agents are agents that are usually licensed to write insurance exclusively for one company and are employed as the most common form of distribution for insurance in South and East Asia for both life and non-life insurance (see Tables 17.7 and 17.8, respectively) (Swiss Re 2002b). Although their prevalence as the primary means of distribution was a result of their ability to capture a dispersed customer base with a physical presence, there are other burgeoning alternative avenues for distribution such as direct marketing, insurance brokers, financial planners, and the topic in question, bancassurance, which is when a bank distributes insurance products. Although only a small percentage of both the life and non-life insurance markets, bancassurance is now a fast-growing distribution channel in South and East Asian markets. The bancassurance market may seem to contribute to only a small percentage of the insurance market as a whole. However, as Figure 17.2 indicates, it varies greatly among individual country markets. For example, the life bancassurance market is as high as 26 percent in Singapore and the non-life bancassurance market as high as 16.7 percent in Hong Kong (Swiss Re 2002b).

Life insurance is the most common type of insurance sold through banks, which can be attributed to its similarity with other financial products sold by banks, such as mutual funds and time deposits. The main products offered in the life bancassurance market are whole-life, term-life, health/disability, retirement/pension insurance, endowments, annuities, and policies linked to loan agreements. The benefit of tying the bank's existing financial products to insurance also influences the types of insurance products offered. The majority of bancassurance products are linked in some manner to deposits, mortgages, or other investment products. In some cases, entirely new insurance products have been created, such as depositor's insurance, which provides a payout in the case of an event (i.e. death) that is a multiple of the depositor's cash balance and over-draft insurance, which provides payment of an existing overdraft in case of a stipulated event (i.e. death) (Munich Re Group 2001). However, the similarity of some of these products to existing bank products should cause some concern, as there is a risk that substitution may take place.

Although a smaller section of the bancassurance market, non-life insurance offered by banks also includes a wide variety of products. This includes fire insurance, home insurance, and in some cases travel and accident insurance. Once again these products are linked to the banks' existing products. For example, home insurance is typically linked with mortgage loans. The motor insurance market is one form of non-life insurance that has not been developed heavily by the South and East Asian bancassurance merchants, as it is less related to the banks' existing products.

Table 17.7. Percentage of Direct Life Insurance Premiums by Distribution Channel, 2002

Country	Year	Career Agency & Direct Sales Force	Brokers	Direct Marketing	Bancassurance	Others
China	N/A	Main Channel	Insignificant	Insignificant	> 20% in Major Cities	N/A
Hong Kong	N/A	Main Channel	Insignificant	<1%	15.10%	N/A
India	N/A	Main Channel	Effectively Prohibited	Insignificant	<5%	N/A
Indonesia	2000	Main Channel	About 1%	Insignificant	Negligible	N/A
Japan	N/A	Main Channel	Insignificant	<1%	<1%	N/A
Malaysia	1998	86%	0.60%	2%	6%	4%
Philippines	1998	75%	1.80%	1%	Not Allowed	3%
Singapore	1998	77%	Insignificant	<0.5%	26%	1%
South Korea	N/A	Main Channel	<1%	Insignificant	Not Allowed	N/A
Taiwan	1998	93%	4.40%	1%	1%	2%
Thailand	1998	97%	<0.5%	Insignificant	2%	N/A
Vietnam	N/A	Main Channel	Small	0%	<1%	N/A

Source: Swiss Re (2002), p.35.
Note: N/A is not available.

Varying degrees of regulation in different countries also dictate the use of this form of insurance distribution. Countries such as Singapore and Hong Kong have been using bancassurance for some time and have little or no restrictions as opposed to countries, such as South Korea, where bancassurance is gradually being introduced through the staged relaxation of rules (Asia Insurance Review 2005). Where restrictions do apply, they are quite varied between countries. For example, in Thailand, the restrictions are focused on particular products and licenses. For instance, a bank cannot issue non-life insurance products unless it is under a brokerage license and until recently, an insurance company was not permitted to pay commissions to banks (SCOR 2003). Alternatively, in China, restrictions focus on the parties of the arrangements, where it is often stipulated that insurers have to deal with each bank branch individually. In South Korea, the introduction of bancassurance has been gradual and restrictions have focused on the types of insurance products permitted to be sold by banking institutions (Asia Insurance Review 2005). Overall, restrictions governing affiliations between banks and insurance companies as well as restrictions on the distribution of insurance products by non-insurers have been relaxed after the introduction of market liberalization, thus contributing to the rise of bancassurance.

Despite these current barriers, there has been a push for financial deregulation throughout South and East Asia. As a result of this deregulation, boundaries between providers of financial products are blurring and countries that currently disallow bancassurance, such as the Philippines, are expected to accommodate this

Table 17.8. Percentage of Direct Non-Life Insurance Premiums by Distribution Channel, 2002

Country	Year	Career Agency & Direct Sales Force	Brokers	Direct Marketing	Bancassurance	Others
China	N/A	Main Channel	<1%	Insignificant	<1%	N/A
Hong Kong	2000	Main Channel for Personal Lines	Main Channel for Commercial Lines	<1%	16.70%	N/A
India	N/A	Main Channel	Effectively Prohibited	Insignificant	Negligible, but Increasing	N/A
Indonesia	2000	Main Channel for Personal Lines	65.60%	Insignificant	Insignificant	N/A
Japan	2001	91.70%	0.10%	8.10%	0.40%	N/A
Malaysia	2001	Main Channel for Personal Lines	19.40%	Insignificant	<5%	N/A
Philippines	2000	Main Channel for Personal Lines	42%	Insignificant	Not Allowed	N/A
Singapore	N/A	Main Channel for Personal Lines	Approximately 40% of Corporate Accounts	Insignificant	5–10%	N/A
South Korea	2000	99%	0.10%	Insignificant	Not Allowed	1% (pools)
Taiwan	N/A	73.30%	12.30%	Insignificant	<5%	N/A
Thailand	N/A	Main Channel	<30%	Insignificant	<5%	N/A
Vietnam	N/A	Main Channel	N.A.	0%	Insignificant	N/A

Source: Swiss Re (2002), p.35.
Note: N/A is not available.

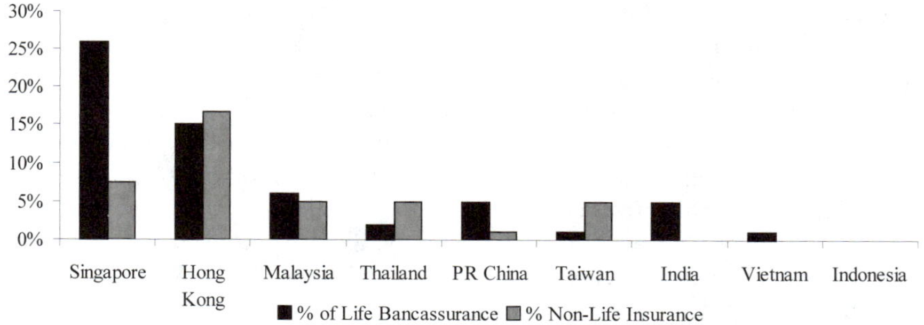

Source: Swiss Re (2002b).
Note: The Philippines and South Korea do not permit bancassurance. All other countries not listed have negligible or nor bancassurance distribution as yet.

Figure 17.2. Market Penetration of the Bancassurance System (%), 1998–2001

distribution channel through banking subsidiaries within the Philippines or through banks that are affiliated with insurance companies. South Korea is also hoping to introduce bancassurance; however those with concerns about the polarization of a market that is already dominated by large banks and insurers may inhibit this.

Some restrictions shape the operations of the bancassurance markets in South and East Asia. In particular, restrictions on foreign insurer entry have meant that the majority of bancassurance agreements are between domestic banks and foreign insurers. This form of alliance allows the foreign insurer to tap into the pertinent market without having to form a partnership with direct competitors and to gain access to the market by using existing networks. As a consequence, the level with which these partners are collaborating is mainly through distribution agreements (69 percent) with the remaining operations based on the more integrated joint ventures, subsidiaries, holding companies, and financial conglomerates (Swiss Re 2002b). These foreign insurers, which are involved in approximately 77 percent of bancassurance operations in South and East Asia (Swiss Re 2002b), also have more experience with this type of distribution elsewhere in the world.

The rise of bancassurance in South and East Asia is also due to synergies between insurance products and banks' particular channel attributes. For example, this would include a bank's staff already being knowledgeable about such products, an established customer base, and customer information being used to target key segments with great accuracy. Along with the increase of experienced foreign insurers in the market, synergies have fueled a major push for bancassurance within the region.

The banks' incentive to decrease costs and recoup diminishing interest margins has been heightened due to the Asian financial crisis. It has been recognized that the efficiency of using existing bank channels and various other distribution channels is greater than using the current career agents system (LIMRA International 2000). With this acknowledged, bancassurance has been difficult for governments, banks, and insurers to ignore. Introducing these new higher-yield products also helps to delay the outflow of a bank's funds, which has been significant recently due to low rates on deposits. Another issue highlighted by the financial crisis was the lack of stringency in the credit culture within many of the South and East Asian economies. By tying in insurance products to bank loans, for example, fire insurance for a housing loan, the risk of the asset is reduced somewhat.

As a rising middle class is fueled by greater economic growth, a new market has emerged. Therefore, the traditional distribution systems are no longer targeted towards the entire market. As many of the financial products that are consumed by the new middle class are accessed through existing bank systems, banks provide the perfect avenue to provide their insurance needs as well. This new market is also not necessarily looking for the more traditional insurance products, but is looking for tailored products as the market becomes more financially savvy. Consequently, banks and insurers are now collaborating to provide comprehensive coverage and more integrated products. With the lack of pension schemes throughout the region, likely through deficient government and employment pensions, consumers are also becoming increasingly motivated to provide for their own future.

In summary, the main issues for bancassurance in this region are the restructuring of the industry with an emphasis on consolidation and integration, a shift in previously restrictive regimes, creating synergies, the need to tap into new markets through more efficient channels, product innovation, and the influx of global competition and partnerships. Another fundamental change in the region's insurance industry is the deregulation and liberalization of the market. This also leads on to two final areas that need further discussion as they underlie the movement of the insurance sector in the region that has arisen from deregulation: the type of business that is underwritten in the region as well as the importance of financial intermediation.

17.2.4 Types of Businesses Underwritten

In general, commercial lines of insurance have seen greater growth than personal, with commercial growth of 3.5 percent versus 1.5 percent for personal lines of non-life insurance in 2001 (Swiss Re 2003b). As would be expected, growth in commercial net premiums written increases as the amount of business undertaken within the country increases. This explains the greater proportion of commercial lines within the more developing nations such as India and China compared to the more developed markets such as South Korea and Taiwan.

Throughout the region much of the non-life insurance growth can be attributed to motor vehicle insurance, in particular, over half the premiums in both Taiwan and Thailand relate to this line. Other countries within the region also show a large representation with most attributing approximately a third of all premiums to motor vehicle insurance. Insurers will come under further pressure to compete within the particular motor vehicle insurance markets in the region as motor vehicle premium rates in India, China, and Taiwan are liberalized.

The SARS epidemic is also thought to have generated a rise in both personal and commercial policies. Medical insurance may be affected directly through an increased awareness of health issues such as SARS. To add to this, individuals are now conscious of the advantages of having coverage for the use of private medical services because public resources were put under strain during the epidemic. Commercial lines of insurance should receive increases in premiums due to SARS as disruptions to business and cancellation of events make businesses wary of further health epidemics, even those unrelated to humans, such as the avian flu crisis.

Although declining, government providers often play a big role within the insurance markets of South and East Asia. In particular, governments often provide medical and unemployment insurance as well as pensions. This varies from country to country quite substantially. For instance, in India and Hong Kong, governments do not usually provide for any of these three lines of insurance. Alternatively, Singapore, South Korea, and Taiwan provide all three types of insurance as well as other lines such as crop insurance, workers' compensation, and motor vehicle insurance (Swiss Re 2003b). Some governments have even started to provide insurance against terrorism, particularly for the aviation industry.

With greater competition and a more client-oriented focus, South and East Asian insurance is becoming more diverse and the scope of contracts available to consumers and businesses are tailored to more specific risks. Large insurers, who are now segmenting their considerable customer base to a greater extent, are also developing several innovative products. Specialist firms are also continuing to service niche markets. Overall, these changes show that the traditional role of insurance companies when providing policies is changing. However, so too is their role as a financial intermediary.

17.2.5 The Role of Insurers as Financial Intermediaries

Despite the world downturn that began in late 2000, optimism within the global financial markets in the beginning of 2005 has strengthened the financial sector. Also, growth in bank lending within the region has not been as affected from 2002 to present. However, increases in both consumer and commercial lending have been met with more prudence, to avoid any future investment bubble. The continued confidence of financial intermediaries is an important facet of the recovery from the previous decade's slump. In particular, insurance companies' role in allowing risks to be spread throughout the region (and the world) and their injection of funds back into the region is pivotal. Large and foreign insurers are more conservative in their investment activities focusing on more passive investments such as cash and deposits. Wide ranging regulation of insurance investment is still prevalent, although changes are expected to increase flexibility and allow for higher yield investments (Organization for Economic Co-operation and Development (OECD) 1999).

In general, the majority of insurance companies' investments are kept in cash and cash-equivalents (including other relatively risk-free assets, such as government bonds), and more so for non-life insurers than life insurers due to the different liability requirements. Over the 1997 Asian financial crisis, equity investments declined and bond investments increased (Swiss Re 2003a). This is not surprising as the volatility of the South and East Asian stock markets increased over the last decade. Thus, insurers are adjusting portfolios because regulators have consequently maintained a stringent policy on investing in equity markets and because many insurers were negatively affected by their higher weighting of equity holdings during the financial crisis. Moreover, despite the fact that the proportion of real estate investments remained relatively stable, there was an overall decrease in the value of insurers' investments. Consequently, solvency concerns have come to light as regulators are now looking at insurers' balance sheets with more scrutiny. Losses incurred on local investment assets have meant that reshuffling of investments has had to take place to secure appropriate reserves, particularly within Thailand, Indonesia, and South Korea.

Finally, reinsurance has started to be used to alleviate the mismatch between the assets and liabilities of insurance companies, although restrictions to foreign company entry have hindered the entry of some of the large global reinsurers into the region. Due to the long-term nature of life insurance, companies within this sector experienced a greater mismatch between assets and liabilities. In Indonesia, problems arose because too few investments were denominated in U.S. dollars to offset the policies that were based in the same currency. This currency risk, which was a by-product of insurers trying to gain from higher returns on domestic investment, proved to be quite detrimental once the financial crisis emerged. There was a mismatch in the maturities of assets and liabilities, where longer-term policies were matched with what proved to be shorter-term investments with diminishing value. Perhaps a closer look at what is determining demand and supply within the region will help to alleviate problems such as that experienced in Indonesia. This, therefore, leads us into the next section that focuses on academic research of these determinants.

17.3 UNDERSTANDING THE DRIVERS OF INSURANCE MARKET GROWTH

The insurance market growth discussed previously has been enormous across Asia. But growth in economies such as Hong Kong, Singapore, India, and China has been more pronounced. For key decision makers, such as managers and policy makers, it is essential to understand those factors that promote higher growth rates. The purpose of this section is to provide a basic insight into the drivers of insurance market growth by drawing upon current empirical studies. The discussion highlights the factors that promote the demand and supply of insurance and relate these findings to the insurance market of Asia.

17.3.1 Drivers of Demand

Economies flourish under economic, legal, and political stability. It is therefore of little surprise that economic sectors, such as insurance, should also do well under similar conditions. Economic growth and economic stability, coupled with effective political and legal institutions, have been found to promote the development of insurance. However, for a better understanding of the differing rates in insurance market growth experienced in the markets across Asia, it is necessary to explore these factors more fully.

There is broad agreement that growth in GDP and/or GDP per capita leads to higher demand for insurance (Beenstock, et al. 1986; Truett and Truett 1990; Browne and Kim 1993; Outreville 1996; Ward and Zurbruegg 2002; Beck and Webb 2003). As economic prosperity increases, the need and ability to purchase insurance also increases, although not at a constant rate. As consumers become increasingly wealthy, they can afford to retain risks within their current financial portfolios. Therefore, the strongest link between income and insurance consumption occurs in those countries with moderate levels of GDP per capita, which Ward and Zurbruegg (2002) found to be the Asian 'Tiger' economies.[318] However, as these economies continue to grow, the demand for insurance will wane, which must be an important consideration for insurance companies seeking to exploit growth in Asia on a long-term basis.

Economic stability is equally important for insurance consumption. High rates of inflation devalue the net present value of insurance; and therefore reduce its attractiveness to consumers. Moreover, excessive inflation can be linked to macroeconomic instability, with consequences for consumption, investment, and exports. Ward and Zurbruegg (2002) provide evidence that inflation is around 2.5 times more important for insurance demand in Asian economies than it is for developed OECD countries. This emphasizes the role that broad macroeconomic

[318] In this discussion, the 'Asian tiger economies' are categorized as consisting of Hong Kong, South Korea, Malaysia, Singapore, Taiwan, and Thailand.

stability plays in the long-term development of the demand for insurance, a role that is mirrored by the accelerated growth of insurance within the relatively stable economies of Singapore, Hong Kong, India, and China.

Recent debate on economic development and economic stability has been dominated by discussions of the function played by legal and political institutions. Prominence, however, has usually been given to the effectiveness of legal systems in promoting commercial transactions within an economy. As a contractual exchange of risks, insurance market development is arguably very sensitive to legal and political effectiveness. Ward and Zurbruegg (2002) and Esho, et al. (2004) find clear support for this hypothesis in both the life and non-life sectors. Importantly, a direct positive correlation seems to exist between an improvement in the legal systems within a country and life insurance demand. It is, therefore, highly evident that the effectiveness of the legal environment within an economy is very important for the development of the insurance sector.

The significance of legal and political variables is usually measured from constructed indices. A good example is from Knack and Keefer (1995, 2000), who use an index of governance constructed from five International Country Risk Guide (ICRG) variables that reflect the security of private property and the enforceability of contracts: 'Corruption in Government,' the 'Rule of Law,' 'Expropriation Risk,' 'Repudiation of Contracts by Government,' and 'Quality of the Bureaucracy.' The higher the index the more effective the legal and political system is. In the case of Indonesia where the index was low, weak functioning legal systems are associated with low GDP growth and low insurance market growth. The message for international insurance companies is, therefore, very clear. Along side considerations of economic stability, legal effectiveness is also a necessary condition for economic and insurance market development, especially within the Asian economies.

Social factors can also play a role in the development of insurance market demand. Hofstede (1995) argues that the level of insurance within an economy depends on the national culture and the willingness of individuals to use insurance as a means of dealing with risk. While an appealing argument, Esho, et al. (2004) and Park, et al. (2002) fail to identify national culture as a significant driver of insurance demand. However, of particular interest to Asia is the role of Muslim dominated countries, where religious beliefs inhibit those forms of insurance that facilitate speculation of future events, thereby discouraging growth of the insurance sector. Since Asia has a number of important Muslim based societies, including Indonesia and Malaysia, these maybe expected to have a lower rate of insurance market development.

17.3.2 Drivers of Supply

The supply of insurance, more so within many Asian economies is composed of domestic and international supply. The amount of supply, the number of firms, and the openness of competition, are all important for an understanding of market development, pricing, and ultimately profitability.

The Domestic Supply of Insurance Services

Unfortunately, little is empirically known about the determinants of domestic supply in the insurance industry. However, there has been an enormous research agenda in measuring the efficiency of various insurance industries around the world, see Cummins and Weiss (2000) for a review. However, little has been done to develop this work into an understanding of growth rates in domestic supply. Admittedly, the use of Malmquist indices to measure productivity developments over time is useful, but these indices have not been linked to the dynamics of the insurance market, including such topics as growth rates in premiums, number of insurance companies, or the types of risks being traded. All of which are important supply related topics.

Competition and Financial Stability

A tradeoff is often seen to exist between increased competition and financial stability. However, this view, arguably, places too great an emphasis on regulation being supply constraining, rather than supply enhancing. Regulation can restrict flawed products and facilitate the development of competition, innovation, and new beneficial products. Regulation can also be an ally of the sector, rather than simply being an enforcer of rules. Swiss Re (2003a) provides a brief review of regulatory characteristics and expected changes in the South and East Asian region. It is notable that the fastest growing insurance markets, Hong Kong, Singapore, and South Korea have the most liberal regulatory environments.

Admittedly, prudent regulation is still important and the need to regulate foreign insurance companies in Asia will only grow as globalization continues, but even globalization brings risk reduction. For example, globalization enables international insurance companies to diversify a broad range of risks, from the obvious examples of underwriting to the investment opportunities presented by different capital markets. In addition to these operational risk reductions, globalization also offers world players in the insurance market the opportunity to strategically reduce risk by broadening activities across many different international markets. As a result, financial stability can be strengthened by accessing a broad array of strategic positions in a variety of developing, emerging, and developed economies.

Access to Capital

A key development in the leading insurance markets of the world has been the emergence of alternative risk financing, characterized by CAT bonds, insurance derivatives, and the securitization of insurance lines (Swiss Re 2001b and Cowley and Cummins 2005). The benefits of such developments are the ability to draw on additional capital and to diversify risk beyond the insurance companies' balance sheet, or that of its reinsurer. These capital market developments highlight to the emerging economies of the world that the development of insurance supply is intrinsically linked to the development of the entire financial system and ultimately the depth and diversity of the accessible financial markets. Economies, such as Singapore and Hong Kong with well-developed capital markets and access to the capital markets of North America and Europe, should be able to bolster underwriting capacity and supply by broadening capital market access and financial innovation.

17.3.3 The International Supply of Insurance and Services

The international supply of insurance services can take two broad forms, the export and import of insurance services and the provision of insurance services through foreign direct investment.

In assessing the attractiveness of international markets to insurance companies Ma and Pope (2003) reveal that the number of domestic competitors and high levels of existing demand are most important. In essence, international insurance companies appear to seek out high growth markets, with an abundance of existing suppliers; and therefore, no dominant incumbent. This may ease entry into the market by facilitating accommodation and switching between suppliers. In Asian countries, where state provision of insurance has led to the creation of a dominant incumbent, international entry into these markets maybe limited. This may then constrain future price competition and long-term growth of the market.

In addition to examining the factors that attract the import of insurance services, it is necessary to understand why insurance may be supplied internationally across national borders. A key argument is that the continued and growing supply of financial services across international boundaries is suggestive of competitive advantages gained from access to superior and cheaper factors of production, particularly labor and capital. Recent research conducted at the World Bank also sheds light on recent liberalization commitments in financial services trade, and how there are domestic and international forces that drive this liberalization process (Schuknecht, et al. 2003).

International insurance companies can use this analysis to benchmark their own domestic resource base and level of competitive advantage against the international markets into which they are contemplating entrance.

17.4 IMPLICATIONS FOR DECISION MAKERS

Promoting economic development stimulates both life and non-life consumption. In particular, national income seems to have a stronger impact on life insurance consumption in Asia, than on the OECD markets. Decision makers should, however, not overvalue these results. Recent research indicates that the effect of national income on insurance demand decreases enormously after controlling for legal and political factors. The empirical results generally suggest that economic stability, conditioned on legal and political stability, appears to be more crucial for long-run success than economic development alone.

These findings are also reflected in the impact of income per capita on the demand for life insurance in the Asian 'Tiger' economies. After controlling for legal and political factors, income elasticities tend to be smaller. While economic development may occur at faster rates in Asia, the link to insurance market development does not appear to be as strong.

Insurance companies intending to expand their business activities abroad should consider these factors when choosing which markets to enter. In the non-life sector, empirical findings highlight the fact that the link between insurance demand and the legal environment is focused on the single issue of property rights enforcement. This indicates that the effectiveness of the legal system in enforcing contracts is paramount to the development of the insurance market. However, in contrast to the life insurance industry, existing research highlights that the non-life insurance business is largely unaffected by cultural and institutional factors. In fact, Esho, et al. (2004) state that the development of non-life insurance is 'technically rather than culturally located' indicating the importance of the legal and regulatory environment.

On the supply side, decision makers need to be aware of how a number of key findings from the literature, in particular, the factors that promote a comparative advantage in insurance, drive international insurance supply. An important determinant is the capital to labor ratio. Applying this finding to the insurance sector is likely best achieved by arguing that the level and quality of technical infrastructure is important for supply side development in the insurance sector. To price risks and manage losses, decision makers should include the quality of office space, the clustering of insurance companies, services for shared experience and technical advice, and the reliability and breadth of telecommunication and computer systems.

The quality of capital and labor is also important since a comparative advantage is found to be related to the amount of schooling and the level of research and development (R&D) expenditures. Again, relating these directly to the insurance sector, R&D becomes important in the development and implementation of risk management and control systems. Fire systems, construction methods, motor vehicle security, and safety are further examples, along with the development of underwriting models and the ability to model financial and capital market risks. When coupled with an educated workforce and the development of professional training, the insurance sector can exploit and leverage knowledge available within its broader environment.

In terms of attracting an international supply of insurance, Ma and Pope (2003) show that international involvement in domestic markets is promoted by increased domestic competition and increased liberalization. Therefore, in order to promote supply, the emerging economies of Asia need to enable greater competition among existing and potential entrants and present flexibility and growth through increased liberalization. Deregulation, increased licensing of insurance companies, reduction in rate setting, and the broadening of underwriting lines and potential investment options are all suitable policies to promote.

17.5 CONCLUSION

The South and East Asian economies offer a variety of risks and opportunities for global and domestic insurance companies. In the relatively developed and economically stable economies of Singapore, India, South Korea, Hong Kong, and Japan growing economic prosperity is strongly associated with insurance market growth. With increasing deregulation, freedom of market access to foreign companies, and a willingness to integrate financial services through bancassurance, these markets are competitive growth spots for the world's leading insurers.

In the more emergent economies of the region, where economic, legal, and political effectiveness are less-assured, commercial confidence and insurance market development is more open to macroeconomic shocks and potential financial crises.

Governments need to foster a commercial environment, which creates greater confidence for economic exchange and insurance risk transfers. Legal stability and the enforcement of property rights are essential for an economy to grow and to make insurance sectors attractive and profitable for both domestic and international players.

In the future, it will be interesting to see how domestic competitors rise to the challenge of international rivals; and the fight may not be one sided. Singapore, in the areas of telecommunications, aviation, and shipping, has shown itself to be a dominant regional competitor; and the same is possible in financial services, and in particular insurance. International rivals should not assume that the markets of South and East Asia offer effortless opportunities.

Perhaps most interestingly, the insurance sector stands in stark contrast to many other industries reflecting the trends of globalization. While many manufacturing industries view South and East Asia as a source of low-cost inputs, global insurance companies view South and East Asia as a potential source of premium growth. The empirical evidence suggests that intra-industry trade in insurance is promoted by a convergence in economic activity across economies. Therefore, an increasing international supply of insurance in South and East Asia will benefit from increased economic growth, and more importantly, salaried employment and rising personal disposable income. All factors which are likely, however, to impede further development of low cost manufacturing and subsequent export growth. It is without a doubt that the South and East Asian economies are now arriving at an important economic crossroad, with pressures to expand into the financial services industry while also trying not to lose their cost-advantages in their already established manufacturing sector. How successfully these economies migrate from manufacturing to service-based economies may have important implications for the medium term economic development of their respective markets, and the consequential desire to purchase insurance.

17.6 REFERENCES

Allens Arthur Robinson, 2003, *Focus: Insurance and Reinsurance Asia*, www.aar.com.au.

Asia Insurance Review, 2005, *Interview with the Regulator—Bancassurance in Korea: Future Supervisory Directions*, www.asiainsurancereview.com.

Beck, Thorsten, Ross Levine, and Norman Loayza, 2000, "Finance and the Sources of Growth," *Journal of Financial Economics* 58: 261–300.

Beck, Thorsten and Ian Webb, 2003, "Economic, Demographic, and Institutional Determinants of Life Insurance Consumption across Countries," *World Bank Economic Review* 17: 51–88.

Beenstock, Michael, Gerry Dickinson, and Sajay Khajuria, 1986, "The Determination of Life Premiums: An International Cross Section Analysis 1970–1981," *Insurance: Mathematics and Economics* 5: 261–270.

Browne, Mark J., Jae Wook Chung, and Edward W. Frees, 2000, "International Property-Liability Insurance Consumption," *Journal of Risk and Insurance* 67: 73–90.

Browne, Mark J. and Kihong Kim, 1993, "An International Analysis of Life Insurance Demand," *Journal of Risk and Insurance* 60: 616–634.

China Insurance Regulatory Commission, 2004, *Regulations of the People's Republic of China on Administration of Foreign-Invested Insurance Companies*, Beijing, China.

Cowley, Alex, and J. David Cummins, 2005, "Securitization of Life Insurance Assets and Liabilities," *Journal of Risk and Insurance*, 72: 193–226.

Cummins, J. David and Mary A. Weiss, 2000, "Analyzing Firm Performance in the Insurance Industry Using Frontier Efficiency and Productivity Methods," in Georges Dionne, ed., *Handbook of Insurance,* (Boston: Kluwer Academic Publishers), pp. 767–829.

Dickinson, Gerry, 1999, "The Changing Focus in the Supervision of Insurance Company Investment," in *OECD Proceedings Insurance Regulation and Supervision in Asia* (Paris: OECD Publications Service) pp. 307–314.

Enz, Rudolf, 2000, "The S-curve Relation between Per-Capita Income and Insurance Penetration," *Geneva Papers on Risk and Insurance: Issues and Practice* 25: 396–406.

Esho, Neil, Anatoly Kirievsky, Damian Ward, and Ralf Zurbruegg, 2004, "Law and the Determinants of Property-Casualty Insurance," *Journal of Risk and Insurance* 71: 265–284.

General Insurance Association of Japan, 2004, *General Insurance in Japan Fact Book 2003–2004*.

Hofstede, Geert, 1995, "Insurance as a Product of National Values," *Geneva Papers on Risk and Insurance: Issues and Practice* 20: 423–429.

Hong Kong, Office of the Commissioner of Insurance, 2004, *OCI Annual Report: 2004*

King, Robert G. and Ross Levine, 1993a, Finance and Growth: Schumpeter Might Be Right, *Quarterly Journal of Economics* 108: 717–37.

_____ and _____, 1993b, "Finance, Entrepreneurship, and Growth: Theory and Evidence," *Journal of Monetary Economics* 32: 513–542.

Knack, Steven and Philip Keefer, 1995, "Institutions and Economic Performance: Cross-Country Tests Using Alternative Institutional Measures," *Economic and Politics* 7: 207–227.

_____ and _____, 2000, "Polarization, Politics, and Property Rights: Links between Inequality and Growth," Policy Research Working Paper No. 2418, World Bank, Washington DC.

Levine, Ross, 1999, "Law, Finance, and Economic Growth," *Journal of Financial Intermediation* 8: 8–35.

Levine, Ross, Norman Loayza, and Thorsten Beck, 2000, "Financial Intermediation and Growth: Causality and Causes," *Journal of Monetary Economics* 46: 31–77.

Levine, Ross and Sara Zervos, 1998, "Stock Markets, Banks, and Economic Growth," *American Economic Review* 88: 537–558.

LIMRA International, 2000, *Survey of Alternate Distribution Channels: Asia,* www.limra.com.

Ma, Yu-Luen and Nat Pope, 2003, "Determinants of International Insurers' Participation in Foreign Non-Life Markets," *Journal of Risk and Insurance* 70: 235–248.

Munich Re Group, 2001, *Bancassurance in Practice*, www.marclife.com.

Outreville, J. Francois, 1990, "The Economic Significance of Insurance Markets in Developing Countries," *Journal of Risk and Insurance* 18: 487–498.

Outreville, J. Francois, 1996, "Life Insurance Markets in Developing Countries," *Journal of Risk and Insurance* 63: 263–278.

Park, Hoon, Stephen. F. Borde, and Yoon. Choi, 2002, "Determinants of Insurance Pervasiveness: A Cross-National Analysis," *International Business Review* 11: 79–96.

Roubini, Nouriel, 2004, *Basic Readings and References on the Causes of Crisis*, www.stern.nyu.edu/globalmacro/asian_crisis/basic_readings.html.

Schuknecht, Ludger, Philipp Harms, and Aaditya Mattoo, 2003, "Explaining Liberalization Commitments in Financial Services Trade," World Bank Working Paper Series No. WP2999, World Bank, Washington DC.

SCOR, 2003, *Bancassurance across the Globe: Meets with a Very Mixed Response*, www.scor.com.

Singapore, Monetary Authority of Singapore (MAS), 2005, *Annual Statistical Report 2004/2005.*

Swiss Re, 1999, "World Insurances in 1998: Deregulation, Overcapacity, and Financial Crisis Curb Premium Growth", *Sigma*, No. 7.

_____, 2000, "World Insurance in 1999: Soaring Life Insurance Business," *Sigma*, No. 9.

_____, 2001a, "World insurance in 2000: Another Boom Year for Life Insurance: Return to Normal Growth for Non-Life Insurance," *Sigma*, No. 6.

_____, 2001b, "Capital Market Innovation in the Insurance Industry," *Sigma*, No. 3.

_____, 2002a, "World Insurance in 2001: Turbulent Financial Markets and High Claims Burden Impact Premium Growth," *Sigma*, No. 6.

_____, 2002b, "Bancassurance Developments in Asia—Shifting into a Higher Gear," *Sigma*, No. 7.

_____, 2003a, "Asia's Non-Life Insurance Markets: Recent Developments and the Evolving Corporate Landscape," *Sigma*, No. 6.

_____, 2003b, "World Insurance in 2002: High Premium Growth in Non-Life Insurance," *Sigma*, No. 8.

_____, 2004a, "World Insurance in 2003: Insurance Industry on the Road of Recovery," *Sigma*, No. 3.

_____, 2004b, "Exploiting the Growth Potential of Emerging Insurance Markets: China and India in the Spotlight," *Sigma*, No. 5.

The Tariff Advisory Committee, 2004, *India: Transition from Uniform Insurance Tariff System to Risk-Based Approach*, www.firstinitiative.org.

Truett, Dale B. and Lila J. Truett, 1990, "The Demand for Life Insurance in Mexico and the United States: A Comparative Study," *Journal of Risk and Insurance* 57: 321–328.

Ward, Damian and Ralf Zurbruegg, 2000, "Does Insurance Promote Economic Growth? Evidence from OECD Countries," *Journal of Risk and Insurance* 67: 489–506.

_____ and _____, 2002, "Law, Politics, and Life Insurance Consumption in Asia," *Geneva Papers on Risk and Insurance: Issues and Practice* 27: 395–412.

World Health Organization, 2003, *Update 3—Severe Acute Respiratory Syndrome (SARS)*, www.who.int.

17.7 LEXICON

Annuity. An annuity is a life insurance policy in reverse, whereby the purchaser gives the life insurance company a lump sum of money and the life insurance company pays the purchaser a regular income, usually monthly.

Bancassurance. Bancassurance is the amalgamation of assurance and banking business within a financial environment, whereby banks are used as distribution channels to sell insurance products.

Broker. A broker is an individual who arranges and services insurance policies on behalf of the insurance buyer. The broker is the representative of the insured, although the broker receives compensation in the form of a commission from the company.

Career Agents. Career agents work on behalf of the insurance company, selling products to potential policyholders. They are also referred to as a tied agent.

CAT bonds. CAT bonds are risk-based securities that allow (re)insurance companies to transfer natural catastrophe insurance risk to institutional investors in the form of bonds. CAT bonds help to spread peak exposures. They are also called Catastrophe bonds.

Claim. A claim is a demand by an insured for indemnity for loss incurred from an uninsured peril.

Commercial Insurance. Commercial insurance is sold by privately formed insurance companies with the objective of making a profit.

Depositor's Insurance. Depositor's insurance provides a payout that is a multiple of the depositor's cash balance in the occurrence of a certain event (i.e. death).

Endowment Insurance. Endowment insurance provides the insured with the face value of a policy if the insured or the beneficiary is alive on the maturity date stated in the policy.

FDI. Foreign direct investment reflects the objective of obtaining a lasting interest by a resident entity in one economy ('direct investor') in an entity resident in an economy other than that of the investor ('direct investment enterprise'). The lasting interest implies the existence of a long-term relationship between the direct investor and the enterprise and a significant degree of influence on the management of the enterprise. Direct investment involves both the initial transaction between the two entities and all subsequent capital transactions between them and among affiliated enterprises, both incorporated and unincorporated.

GDP. Gross Domestic Product represents the total value of final goods and services produced within a country's borders during a specific time period, usually a year.

GNP. Gross National Product is the total value of final goods and services produced by domestically owned factors of production.

ICRG. International Country Risk Guide.

Income Elasticity. Income elasticity of demand measures the responsiveness of the quantity demanded of a good to the income of the people demanding the good. It is measured as the percentage change in demand that occurs in response to a percentage change in income.

Insurance Derivatives. Insurance derivatives are investment instruments which are determined directly and solely by the loss patterns of natural catastrophes.

Liability Insurance. Liability insurance is insurance for damages that a policyholder is obliged to pay because of bodily injury or property damage caused to another person or entity based on negligence, strict liability, or contractual liability

Life Insurance. Life insurance provides for a payment of a sum of money upon the death of the insured. More specifically, in exchange for a series of premium payments or a single premium payment, upon the death of the insured, the face value minus outstanding policy loans and interest is paid to the beneficiary.

Malmquist Index. The Malmquist index is a measure of the degree of concentration in a market. It is the sum of the squares of the percentage market shares of all companies in the market.

Mutual Insurance Company. A mutual insurance company is an insurance carrier, without capital stock, that is owned by the policyholders. It may be incorporated or unincorporated.

OECD. Organisation for Economic Co-operation and Development is an international organization of those developed countries that accept the principles of representative democracy and a free market economy. It originated as the Organisation for European Economy Co-operation (OEEC) to help administer the Marshall Plan for the re-construction of Europe after World War II. Later, its membership was extended to non-European states, and in 1961 it was reformed into the OECD.

Over-Draft Insurance. Over-draft insurance covers the repayments of an overdraft facility in the case of certain events (i.e. death).

Premium Density. Premium density is premiums per capita.

Premium. A premium is the payment, or one of the periodical payments, a policyholder agrees to make for an insurance policy reflecting his/her expectation of loss or risk.

Premiums Earned. Premiums earned are premiums an insurance company has recorded as revenues during a specific accounting period.

Premiums Written. Premiums written are premiums for all policies sold during a specific accounting period.

Product Liability Insurance. Product liability insurance is insurance for the manufacturer or supplier for of goods for damage caused by their products.

R&D. Research and Development can be defined as any project to resolve scientific or technological uncertainty aimed at achieving an advance in science or technology. Advances include new or improved products, processes, and services.

SARS. Severe Acute Respiratory Syndrome is a viral respiratory illness caused by a coronavirus, called SARS-associated coronavirus (SARS-CoV). SARS was first reported in Asia in February 2003. Over the next few months, the illness spread to more than two dozen countries in North America, South America, Europe, and Asia before the SARS global outbreak of 2003 was contained.

Term Life Insurance. Term life insurance is a life insurance policy that has a set duration limit on the coverage period. Once the policy is expired, it is up to the policy owner to decide whether to renew the term life insurance policy or to let the coverage terminate.

Permanent Life Insurance. Permanent life insurance, also referred to as ordinary life insurance or whole life insurance, is a life insurance policy that covers an insured for their entire lifetime, assuming premiums are paid as specified in the policy. Whole life insurance provides a guaranteed fixed sum (sum assured) upon death of the life or lives assured.

Time Deposits. Time deposits are savings accounts which require notice of withdrawal.

WTO. World Trade Organization.

18 The Global Reinsurance Market

Thomas Holzheu

Swiss Re

Roman Lechner

Swiss Re

18.1 INTRODUCTION

Insurers buy reinsurance for risks they cannot or do not wish to retain and to benefit from the capital relief reinsurance provides. Reinsurance and primary insurance represent two separate elements in the insurance value chain. By providing coverage against adverse fluctuations in claims, reinsurance protects the capital base of the primary insurer and creates a more diversified portfolio, thus reducing the volatility of the underwriting result. Another important benefit of reinsurance is that it allows insurers to accept more business with a given amount of capital, e.g., an insurer can afford to accept larger individual risks or risks exposed to larger loss accumulations from a single event.[319]

Reinsurance is a global business, covering risks from insurers in many geographic locations. Although some primary insurers are international, most operate through local subsidiaries in different territories. Large reinsurers write business in all major markets in order to maximize the geographic diversification of their portfolios. The diversification of risks is enhanced by retrocession, the reinsurance of reinsurers. By increasing the efficiency of the insurance industry's use of its capital base, reinsurance facilitates the insurability of risks which are characterized by major uncertainty or large, potential losses.[320] This chapter gives an introduction to the reinsurance market, which is driven by international trends and cycles. The worldwide reinsurance market serves as a force which synchronizes primary market

---

[319] For a comprehensive discussion of the history, main product types, key lines of business, and other fundamental aspects of the reinsurance market refer to Gastel (2004), Gerathewohl (1983), Schwepcke (2004), or Swiss Re (2004c, 2005a).

[320] For a discussion of the efficiency of the insurance industry's use of capital see Cummins and Weiss (2000).

developments across the world along with interest rates and worldwide stock markets.[321]

In 2003, primary insurers ceded business worth $170 billion to reinsurers worldwide, of which 83% was non-life insurance and 17% was life insurance.[322] The average cession rates, or ceded premiums as a percentage of direct premium volume, were 13.1% in non-life insurance and 1.9% in life insurance. Around 84% of ceded business is generated in the mature insurance markets of North America and Western Europe, reflecting those regions' importance as sources of primary insurance revenue. Only 2% of ceded business comes from Japan, a fact that can be explained by the high concentration and large average company size in the Japanese insurance industry, which accounts for 11% of global direct premiums. By contrast, the emerging markets tend to show strong demand for reinsurance due to the relatively lower capitalization of many local insurers and high proportion of industrial risks in their portfolios. Their share of worldwide cessions is 13%, compared with an 11% share of direct insurance premiums.

From 1990 to 2003, cessions in non-life insurance rose from $60 billion to almost $141 billion. This rise was fueled mainly by increased demand from North American insurers. Thanks to a rapid growth in direct premiums, the relative importance of emerging markets grew as well. Global non-life cessions posted an annual average real growth of 5%, exceeding the 3% real growth rate for primary insurance. The ceded premiums of the life reinsurance market rose from $7 billion to $29 billion, an annual average real growth rate of 9%, well in excess of non-life cession growth as well as the average 4% real growth rate of direct life premiums.

The crisis in U.S. liability insurance in the mid-1980s and the accumulation of natural catastrophe losses at the start of the 1990s prompted significant changes in the reinsurance market. The subsequent capacity shortage and withdrawal of many providers from the market provided a market entry opportunity, of which Bermuda firms took advantage. Since 2001, severe underwriting and investment losses have prompted another wave of structural changes in the industry. The number of domestic market players declined, especially in the United States and Europe, while another generation of Bermuda startups rapidly gained market share. The entry of billions of dollars of fresh capital and the ability of the new market entrants to gain significant market share both indicate that the barriers to entry in many reinsurance markets are low.

---

[321] For a discussion of international underwriting cycles see Cummins, et al. (1991), Cummins and Outreville (1987), or Swiss Re (2001b).

[322] The data used in this report have mostly been taken from the relevant market statistics published by insurance associations, regulatory authorities, and rating agencies. They are based on the business of those insurance companies domiciled in the countries in question. The subsidiaries of foreign companies are usually included as well. The overall market figures are generated by summation of consolidated company data. There are several issues with regards to double counting. The first is when the business activity of the subsidiary is recorded in the country where a subsidiary is based and once again as part of the consolidated annual accounts of the parent company. Second, to avoid double-counting of ceded premiums and retrocessions, retrocession figures are excluded from the calculation for each country. Finally, the study attempts to deal only with "free" market transactions, excluding intragroup reinsurance. For countries where no explicit distinction for intragroup reinsurance is available in the market statistics, estimates are used.

The large specialized reinsurers, also labeled as professional reinsurers, were able to expand their market shares through organic growth and a wave of mergers and acquisitions (M&A) during the mid- and late 1990s. From 1990 to 2003, the four largest non-life reinsurance providers increased their market share from 20% to 36%. The concentration was higher in life reinsurance in 2003 with a market share of 57% for the top four players. The winners of consolidation include the Bermuda based reinsurers, who raised their market share from 0% to 14% between 1990 and 2003. Lloyd's of London saw its market share erode from 7% in 1990 to 4% in the late 1990s, recovering to 6% by 2003.

The first section of this chapter describes regional structures and growth trends in the demand for reinsurance. The next section outlines the structure and trends in the supply of reinsurance with regards to consolidation and distribution channels. The third section introduces the characteristics of major reinsurance centers including primary regulatory developments. Finally, recent profitability trends and concluding remarks are discussed.

18.2 THE DEMAND FOR REINSURANCE

In 2003, primary insurers ceded business worth $170 billion to reinsurers worldwide, of which 83% or $141 billion was non-life insurance and 17% or $29 billion was for life insurance. The average cession rates, or ceded premiums as a percentage of direct premiums, were 13.1% in non-life insurance and 1.9% in life insurance. Within the non-life insurance segment, commercial lines of insurance have usually higher cession rates than personal lines of insurance such as motor insurance. Life insurance has a much lower cession rate since it is largely restricted to mortality risk. The volume-intensive savings component of primary life insurance is rarely reinsured.

18.2.1 The Main Sources of Non-Life Reinsurance Cessions: North America and Western Europe

About 82% of ceded non-life reinsurance business is generated in the markets of North America and Western Europe, reflecting these regions' importance as sources of primary insurance revenue. Table 18.1 provides a breakdown of direct and ceded non-life premiums in 2003.

The importance of North America as the largest reinsurance market in the world (51% share) is mainly due to the size of the primary insurance market and the relatively high exposure to natural catastrophes and liability losses. Hence, despite the substantial cessions of U.S. insurers to pools and low cession rates of Canadian insurers, North America's share of the world's ceded volumes is marginally higher than its share of direct business.[323]

---

[323] Most of the U.S. pools are state-run residual markets covering hard to insure risks in homeowners insurance, non-standard auto insurance, workers' compensation, and health insurance. One reason for the low cession rates of Canadian insurers is the existence of large state-owned motor insurers, which barely use reinsurance.

Western Europe's share of worldwide ceded business is 31%, slightly below its percentage of direct insurance business (34%). Western Europe's cession rate of 12% is slightly lower than North America's 14% cession rate. Western Europe's reinsurance markets are characterized by:

- A heavy fragmentation with 18 markets and about 5,000 non-life insurance providers.
- Greater prevalence of proportional reinsurance, where premium volume is higher in relation to the transferred risk exposure. In Germany, for example, proportional reinsurance accounts for some 80% of business.
- The London market has a high cession rate (approximately 30%), thus reflecting a strong demand for reinsurance. This arises from the fact that more risk-bearing commercial risks like marine, aviation, and industrial property and liability risks are insured with a large number of relatively small providers.

The Asian-Pacific reinsurance demand falls marginally short of its share of global non-life insurance. Within the region, cession rates range from below average demand in Japan, to average demand in Australia and New Zealand, and above average demand in emerging Asian economies. Despite its high exposure to natural catastrophe risks, the Japanese insurance industry accounts for just 2% of worldwide non-life reinsurance demand. This low cession rate is due to the high concentration and capitalization of the Japanese primary insurance market, and the existence of a state-organized pool offering earthquake insurance coverage.

With 14% of ceded non-life premiums compared to a 10% share of direct insurance business worldwide, emerging markets account for a disproportionately high share of reinsurance demand.[324] This above average reinsurance demand is partly due to the smaller capital bases of many local primary insurers. Another reason for high overall cession rates is the fact that the business mix is heavily weighted toward industrial and commercial risks, which usually have high cession rates.

18.2.2 Life Reinsurance Cessions

From 1990 to 2003, the life reinsurance market grew from $7 billion to $29 billion. Reinsurance market growth significantly exceeded non-life cession growth and direct premium growth in the life insurance industry.

The market volume of life reinsurance was about $29 billion in 2003. The cession rates of life reinsurance are generally very low, between 0.1% in Japan and 3.8% in North America. The life reinsurance market is therefore much smaller than the non-life reinsurance market. The main reason is that a predominant fraction of life primary insurance premiums cover savings products, which are usually not reinsured. Life reinsurance mostly covers mortality risk.

---

[324] Emerging markets comprise the countries of Latin America, Middle East, Asia excluding Japan, and the formerly communist countries of Central and Eastern Europe.

Table 18.1. Regional Breakdown of Ceded Non-Life Premiums, 2003

	Direct Premiums		Ceded Premiums		Cession Rates
	US$ Billions	% of Total	US$ Billions	% of Total	% of DP
North America	515.9	48.0	71.8	50.9	13.9
Western Europe	370.1	34.5	44.1	31.3	11.9
Japan	65.6	6.1	2.8	2.0	4.2
Oceania	17.8	1.7	2.5	1.8	14.2
Asia-Pacific	53.0	4.9	11.0	7.8	20.7
Latin America	21.0	2.0	4.7	3.3	22.4
Eastern Europe	21.6	2.0	1.9	1.3	8.7
Rest of the World	8.8	0.8	2.3	1.6	26.5
Total World	1,073.7	100.0	141.2	100.0	13.1
Industrialized Nations	969.4	90.3	121.3	85.9	12.5
Emerging Nations	104.3	9.7	19.9	14.1	19.1

Source: Swiss Re, Economic Research and Consulting.
Note: DP is Direct Premiums.

Table 18.2 provides a regional breakdown of 2003 direct and ceded life premiums worldwide. Demand for life reinsurance is more heavily concentrated in North America and Western Europe than is non-life reinsurance, with 92% of ceded premiums coming from these two regions. Japan, accounting for just 1% of the worldwide life reinsurance market, is substantially underrepresented given the size of its primary insurance market (15% market share) due to the large average size of Japanese life insurers and the high savings content in the Japanese life primary insurance business. Despite their relatively high cession rates, emerging markets still only account for 5 % of worldwide reinsurance market volume.

18.2.3 Cyclical Growth in Non-Life Reinsurance

Growth of the non-life reinsurance market is cyclical. Market volume rose from $60 billion in 1990 to $90 billion in 1995, and then fell back to $83 billion in 1998. Figure 18.1 shows the cyclical growth of the worldwide non-life insurance and reinsurance premium volume in U.S. dollars, adjusted for inflation.[325] The decline in the non-life market volume between 1995 and 1998 can be attributed to an appreciation of the U.S. dollar against the European currencies and to falling prices

[325] Regional growth rates are calculated using a weighted average of the real growth rates of the individual countries. The weighting is based on the relevant premium of the previous year in U.S. dollars.

due to a soft market.[326] Since the U.S. dollar appreciated significantly against European currencies and the Japanese yen from 1995 to 1997, the U.S. dollar value of premium volumes in these regions fell. Market growth resumed in 1999 and accelerated after 2001 to reach $141 billion globally in 2003.

Table 18.2. Regional Breakdown of Ceded Life Premiums, 2003

	Direct Premiums		Ceded Premiums		Cession Rates
	US$ Billions	% of Total	US$ Billions	% of Total	% of DP
North America	531.7	34.8	19.9	68.0	3.8
Western Europe	564.8	37.0	7.0	23.7	1.2
Japan	223.9	14.7	0.2	0.8	0.1
Oceania	21.7	1.4	0.7	2.4	3.2
Asia-Pacific	142.9	9.3	0.7	2.3	0.5
Latin America	13.6	0.9	0.4	1.5	3.1
Eastern Europe	6.8	0.4	0.0	0.0	0.0
Rest of the World	22.7	1.5	0.4	1.4	1.8
Total World	1,528.1	100.0	29.3	100.0	1.9
Industrialized Nations	1,342.1	87.8	27.8	94.9	2.1
Emerging Nations	186.0	12.2	1.5	5.1	0.8

Source: Swiss Re, Economic Research and Consulting.
Note: DP is Direct Premiums.

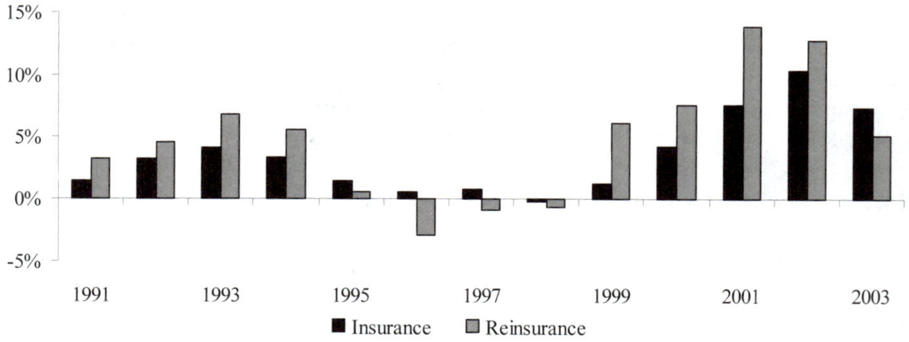

Source: Swiss Re, Economic Research & Consulting.

Figure 18.1. Real Growth in Global Non-Life Insurance and Reinsurance Premiums, 1991–2003

[326] During a soft market, rising insurance capacity and fierce price and market share competition lead to widely available coverage and low premium rates. Profitability of business written during a soft market is below average. Declining capacity and competition lead to hard markets with rising premium rates and improving profitability of new business.

From 1990 to 2003, worldwide non-life reinsurance market volume rose by a 4.6% annual average in local currency terms, after adjusting for inflation. This was substantially higher than the 3.4% annual real growth recorded in direct premiums, resulting in an increase in cession rates, which measure ceded premiums as a percentage of direct premiums. This strong growth in reinsurance volume is due to several factors.

Figure 18.2 examines the regional patterns in non-life cession rates from 1990 to 2003. In North America, reinsurance demand continued to rise despite stagnation in the primary insurance market throughout the 1990s. Reasons for the increased demand include continuous growth in exposure to natural catastrophes and surging uncertainty regarding liability claims. In addition, a large volume of reinsurance that was previously ceded to pools shifted back to the free market when market conditions softened. Consequently, North American cessions grew rapidly in excess of primary market growth from 1999 through 2002. In contrast, in Western Europe, direct premiums and reinsurance market volumes grew roughly at a similar pace, although consolidation and deregulation encouraged slightly lower cession rates in general. The recent hard market also resulted in stronger cession growth from 2000 through 2002. In emerging markets, the reinsurance business enjoyed higher real growth than the direct insurance market, due mainly to privatization and deregulation. Primary and reinsurance premium growth benefited from the emerging market boom years during the first half of the 1990s. Reinsurance market volumes rose substantially faster in emerging markets than in the mature markets, due to more dynamic growth in the direct business and, in some cases, rising cession rates. Emerging markets increased their share of the non-life reinsurance market from 10.2% in 1990 to 14.1% in 2003.

18.2.4 Strong Growth in Life Reinsurance

From 1990 to 2003, ceded premiums of the life insurance market rose from $7.4 billion to $29.4 billion. Figure 18.3 shows the inflation adjusted growth of the global life insurance and reinsurance volume in U.S dollars. The annual real growth rate of 8.9% significantly exceeded cession growth in the non-life reinsurance business and the growth of life direct premiums, which was 4.2%. As Figure 18.4 indicates, the global cession rate rose from 1.1% in 1990 to 1.9% in 2003. Unlike the non-life reinsurance sector, there was no clear cyclical pattern in life reinsurance growth rates.

Figure 18.4 shows cession rates for life reinsurance, defined as ceded life premiums as a percentage of life direct premiums. In North America, the cession rates for the life insurance business increased strongly from 1.8% in 1990 to 3.8% in 2003. In Western Europe, the cession rate remained rather flat (1.2% in 1990, 1.2% in 2003, with a temporary decline to 0.9% in 1999). In Japan, life cessions also grew marginally faster than direct premiums, which stagnated in real terms. However, this trend did not make much of an impact globally, because of the extremely low cession rate. Latin America and Eastern Europe recorded high growth rates in both the direct business and cessions.

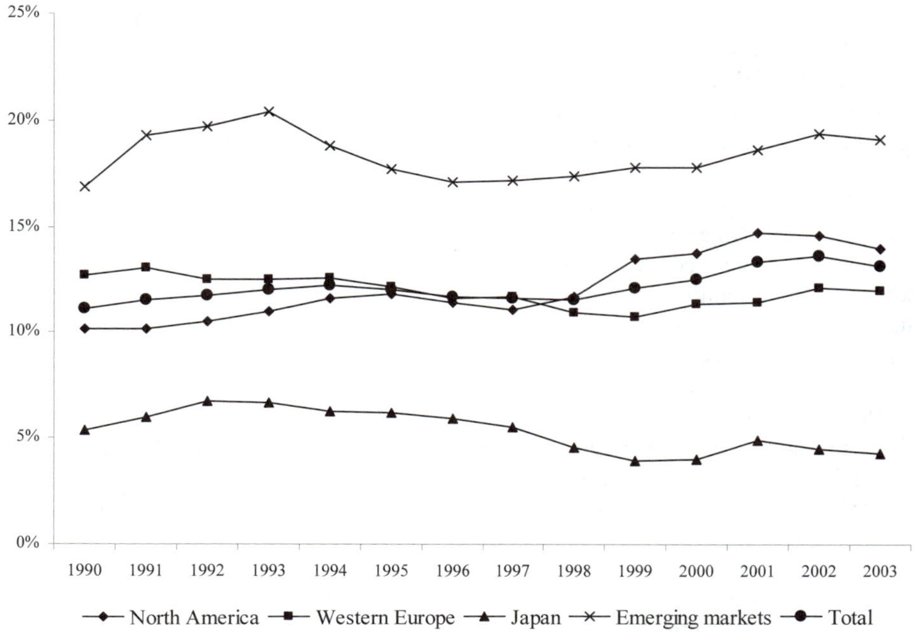

Source: Swiss Re, Economic Research & Consulting.

Figure 18.2. Regional Patterns in Non-Life Cession Rates, 1990–2003

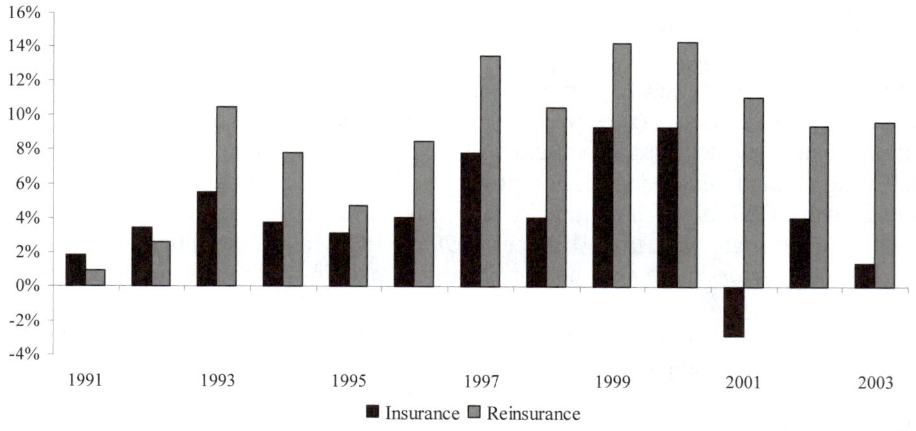

Source: Swiss Re, Economic Research & Consulting.

Figure 18.3. Real Growth in Global Life Insurance and Reinsurance, 1990–2003

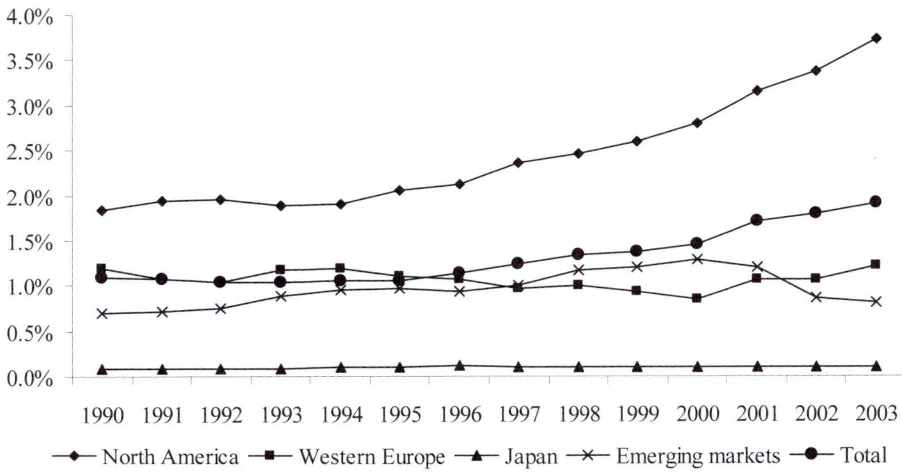

Source: Swiss Re, Economic Research & Consulting.

Figure 18.4. Regional Patterns in Life Cession Rates, 1990–2003

18.3 STRUCTURE OF THE GLOBAL REINSURANCE MARKET

18.3.1 Consolidation in the Reinsurance Industry

The development of the reinsurance industry during the last decade relates closely to the underwriting cycle. The heavy financial losses caused by the liability insurance crisis of the 1980s and the accumulation of heavy natural catastrophe losses at the start of the 1990s created serious solvency problems for Lloyd's and several other primary insurers and reinsurers. After Hurricane Andrew in 1992, the number of active reinsurers fell as many went bankrupt or withdrew from the market.

The capacity shortage and the problem of uncollectible debts from insolvent reinsurers led to much greater importance being attached to the security and financial rating of reinsurers. This resulted in a shift of ceding insurers' preferences away from reinsurers with smaller capital resources and towards the large professional reinsurers. A wave of mergers and acquisitions followed between 1994 and 1996 with non-life reinsurance companies representing roughly 13% of global market share changing hands. The buyers spent a total of more than $40 billion on acquisitions.[327] Consequently, as Figure 18.5 indicates, the non-life reinsurance market share of the four largest professional reinsurers rose from 22% in 1990 to 29% in 1996.[328]

---

[327] For a more detailed description of the merger and acquisition wave in the mid-1990s refer to Swiss Re (1998).

[328] Occasionally, higher market shares are quoted for the four largest reinsurers. This is due to a narrower definition of market volume (e.g., only those professional reinsurers in the Standard & Poor's Top 100 list). Here the market volume is defined as the total available cession volume.

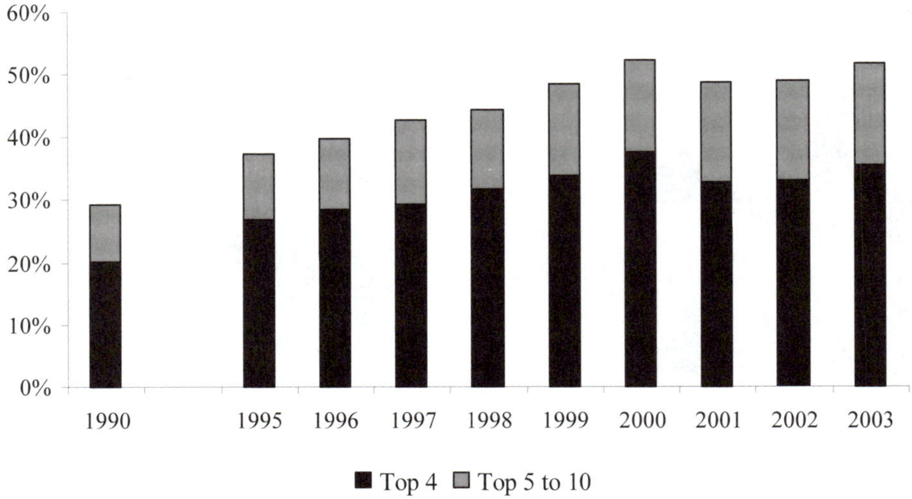

Source: Swiss Re, Economic Research & Consulting.

Figure 18.5. Shift in Global Reinsurers Market Shares, 1990–2003

Another major change in the market structure has been caused by the growing importance of Bermudan reinsurers, which were originally set up in the aftermath of Hurricane Andrew in 1992. Due to its favorable tax and supervisory regime and its geographical proximity to the United States, Bermuda became the most important offshore reinsurance center in less than two years. After a period of consolidation, only two independent companies, Partner Re and Renaissance Re, remained out of the original eight specialist catastrophe reinsurers of the "Bermuda Class of 1992."

The next wave of structural change in the international reinsurance market came after 2001. The performance of the industry declined between 2001 and 2003 due to a combination of losses from the events of September 11, 2001, adverse development of business written during the soft market years, and the equity markets decline. The Insurance Information Institute currently estimates the total losses from September 11, 2001 at $32.5 billion. This comprises $9.6 billion in property claims, $11 billion in business interruption claims, $3.5 billion in aviation liability claims, $4 billion in other liability claims, and $3.4 billion in claims from other lines of business (Hartwig, et al. 2004). The resulting capacity shortage and rate increases were widespread over all geographic regions and lines of business.[329]

These shocks were managed without any major defaults or general unavailability of reinsurance coverage. However, there were a series of market exits via runoffs and a wave of ratings downgrades. Between 2001 and 2003, over $30 billion of new capital was raised globally, of which $9 billion funded another wave of Bermuda-based startups.[330] Strategic sponsors of the startups included large

[329] For an analysis of these effects on the entire property and casualty industry see Swiss Re (2002b).

[330] The largest startup companies were Awac, Arch, Aspen, Axis, Endurance, Montpelier, Olympus, and Quanta.

insurers and insurance brokers. The recent entry of billions of dollars of fresh capital and the ability of the new market entrants to gain significant market shares suggests that barriers to entry in many segments of the reinsurance business are low. This is particularly true for the short-tail property lines of business.[331] The last flight to quality benefited the top four reinsures, increasing their market share to 36% in 2003.[332] Bermuda reinsurers also benefited, as their market share rose to 13%. Table 18.3 shows the market shares of the top 10 non-life reinsurers in 2003 as measured by net premiums earned.

Table 18.3. Market Shares in the Non-Life Reinsurance Market, 2003

Rank	Company	Domicile	Rating*	NPE (US$ Billions)	Market Share
1	Munich Re**	Germany	A+	18.3	13.0%
2	Swiss Re	Switzerland	AA	15.2	10.0%
3	Berkshire Hathaway	USA	AAA	10.8	7.0%
5	GE Insurance Solutions	USA	A+	7.0	5.0%
6	Converium	Switzerland	BBB+	3.3	2.0%
7	PartnerRe	Bermuda	AA-	3.2	2.0%
8	Transatlantic Re	Bermuda	AA	3.2	2.0%
9	Everest Re***	Bermuda	AA-	2.9	2.0%
10	Scor	France	BBB+	2.6	2.0%
	Top 10			73.5	50.0%

Source: Swiss Re Economic Research & Consulting, Annual Reports.
Notes: NPE is net premiums earned. Market share is defined as a firm's percentage of the industry's total net premiums earned. S&P Rating as of November 2004. ** Reinsurance segment of Munich Re, including intra-group transactions. *** Excludes the primary insurance segment.

The latest trend towards cycle management and more effective risk diversification is reflected in the growing interest in life reinsurance. Unlike the very volatile catastrophe reinsurance business, life reinsurance provides a much steadier cash flow and more stable results for the reinsurers. Table 18.4 lists the top 10 life reinsurers by their 2003 markets shares, measured by net premiums earned. After a recent wave of consolidation between 1997 and 2003, cedents have seen a decline in the number of reinsurance providers and capacity (Standard & Poor's 2004).

[331] Short-tail businesses are all classes of business in which claims are reported and settled in a relatively short period of time (e.g. most classes of property business).

[332] The "first flight to quality" with respect to reinsurance buying behavior occurred in the early 1990s when an accumulation of large losses highlighted the relevance of reinsurers' security. Ceding insurers shifted their demand to larger reinsurers with strong balance sheets.

Table 18.4. Market Shares in the Life Reinsurance Market, 2003

Rank	Company	Domicile	Rating*	NPE (US$ Billions)	Market Share**
1	Swiss Re	Switzerland	AA	7.6	22.0%
2	Munich Re***	Germany	A+	7.2	21.0%
3	GE ERC	USA	A+	3.0	9.0%
4	RGA	USA	A-	2.6	8.0%
5	Hannover Re	Germany	AA-	2.2	6.0%
6	Gen Re	USA	AAA	1.8	5.0%
7	Scor	France	BBB+	1.5	4.0%
8	ING Re	Netherlands	AA	1.5	4.0%
9	Transamerica	USA	AA-	1.2	3.0%
10	Revios	Germany	A-	1.1	3.0%
	Top 10			29.8	87.0%

Source: Swiss Re Economic Research & Consulting, Annual Reports.
Notes: NPE is net premiums earned. Market share is defined as a firm's percentage of the industry's total net premiums earned. * S&P Rating as of November 2004. ** Market volume used for market share calculation includes the health reinsurance business. *** Reinsurance segment of Munich Re, including intra-group transactions.

18.3.2 The Role of Intermediaries

Unlike the life reinsurance market, in which transactions are rarely intermediated, brokers play a significant role in non-life reinsurance. In the U.S., brokers intermediated about 66% of property and casualty reinsurance premiums in 2004. In continental Europe, direct customer contacts between cedents and reinsurers still dominate, although the share of brokered business is rising. Bermuda and London are both almost exclusively brokered markets.

Cedents often use both direct and broker distribution channels for placing reinsurance. The key considerations are the financial condition, willingness to partner, past relationships, expertise, reputation, and pricing of the reinsurer, rather than a certain distribution channel. In 2000, over 80% of the U.S. cedents used a combination of broker and direct channels, 14% used brokers exclusively and 5% used direct writers only (Conning 2000).

Reinsurance brokers assess and analyze the insurer's risks, structure, and market and then negotiate an insurance program. This includes, for example, aggregate policy limits, deductible or retention levels, and coverage terms. During the last two decades, the role of the broker has evolved from middleman to service provider. Brokers' services include claims and risk management that go beyond the structuring and placement of reinsurance cover. Reinsurance brokers also advise on retrocession—the reinsurance of reinsurance (Swiss Re 2004a).

The non-life reinsurance brokerage and primary brokerage markets share similar characteristics. Both are highly concentrated, with U.S. and London-based players dominating and Bermuda holding the number three position and gaining rapidly. Figure 18.6 lists the most important markets for the placement of brokered reinsurance, measured by their share of global net premiums written in 2003 (Allen 2004). The London Market accounted for 30% of the placement of brokered reinsurance, United States 26%, and Bermuda 14% in 2002. Other important markets in terms of placement of reinsurance were Germany, France, and Switzerland.

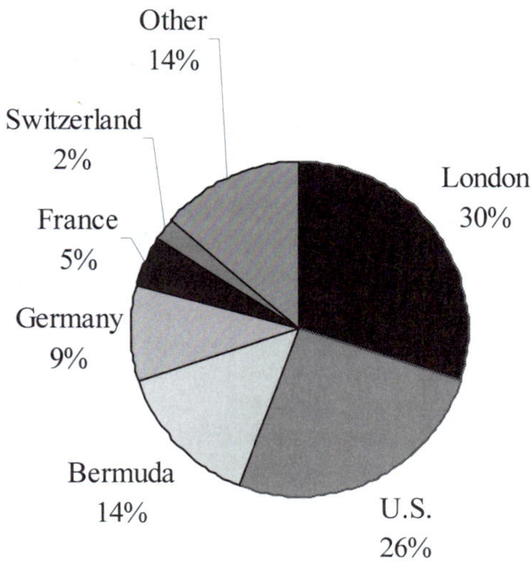

Source: Allen (2004).
Note: Market size is measured by a nation's share of global net premiums written in 2003.

Figure 18.6. Markets for the Placement of Brokered Reinsurance, 2003

18.4 THE WORLD'S MAIN REINSURANCE CENTERS

For 2003, Standard & Poor's publication Global Reinsurance Highlights listed 246 companies from 51 countries accounting for net premiums written of $164 billion (Standard & Poor's 2004). Taking into account their affiliations, there were 119 independent groups and single companies that provided reinsurance coverage to primary insurers. This tally does not include some smaller niche players, reinsurance pools, and primary insurers active in providing reinsurance coverage. Although Lloyd's consisted of 71 syndicates in 2003, many of which wrote reinsurance business, it is viewed as one consolidated provider. Thus, it can be concluded, that currently there are approximately 150 active specialized reinsurers.

The reinsurance business is dominated by specialized reinsurance companies and is concentrated in a small number of centers. Figure 18.7 shows the regional distribution of reinsurance companies measured by net premiums written in 2003. Some 92% of the global reinsurance market is domiciled in eight countries: Germany, the United States, Switzerland, the United Kingdom (the London market),

Bermuda, France, Japan, and Ireland.[333] In other countries, particularly the emerging markets, there are often just one or two reinsurers, usually focusing on their home markets.

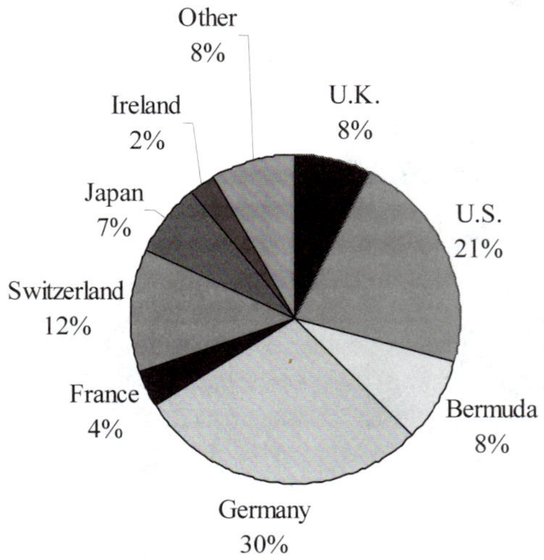

Source: Standard & Poor's (2004).
Note: The regional distribution of reinsurance firms is measured by net premiums written in 2003.

Figure 18.7. Domiciles of Reinsurance Companies, 2003

Of the above mentioned eight major reinsurance centers, Germany, Switzerland, Bermuda, London, France, and Ireland are "net exporters" of reinsurance services, assuming more business from foreign companies than they cede to foreign reinsurers.[334] The U.S. is traditionally the most important net importer of reinsurance. The Japanese insurance industry has usually been a net importer as well. However, most of the Japanese reinsurance business is domestic, with more than 70% stemming from pooling arrangements for compulsory automobile liability insurance.

The international reinsurers from Germany, Switzerland, and France have traditionally accounted for a large share of global reinsurance volume. In 2003, their share was 46%, a figure that increases to 50% if foreign subsidiaries are included.

Despite some difficulties at the start of the 1990s, the London market is still a very important trading centre for reinsurance coverage. Lloyd's and the London-based (re)insurers account for 8% of the worldwide market. However, this figure probably understates the importance of the London market, since much of the business transacted via London brokers is placed with insurers and reinsurers in

[333] The market shares are based on the domicile of individual insurance companies rather than the domicile of insurance groups. Data source is Standard & Poor's (2004) and own calculations.

[334] See International Association of Insurance Supervisors (2004) for an example.

Europe and elsewhere. During the last decade, Bermuda, and to a smaller degree, Ireland emerged as important offshore centers for reinsurance.

18.4.1 The United States and the Increasing Importance of Foreign Players

The United States is the largest reinsurance market in terms of demand (ceded premiums) and supply (assumed premiums). Furthermore, there is a strong international involvement in the U.S. reinsurance market due to the size of its risks. American insurers ceded about $77 billion in premiums to reinsurance companies, $15 billion more than premiums assumed by U.S. (re)insurers, which amounts to a net importer of reinsurance. Continental European reinsurers are the most important foreign providers of reinsurance for U.S. insurers. Bermuda, supported by its geographical proximity and its favorable tax and supervisory regime, has developed into a very important offshore reinsurance market for U.S. (re)insurance business.

Two of the unusual features of the U.S. insurance market are the $18 billion of non-life reinsurance premiums written by primary insurers and $8 billion assumed by various pools (A. M. Best 2004a). This compares with $36 billion accepted by professional reinsurers. Domestic professional reinsurers therefore account for 58% of the U.S. capacity (measured by premium volume). The capacity offered by pools and primary insurers has declined over the course of the last decade. Professional reinsurers, by contrast, expanded both their premium volume and capital funds.

Reinsurers in the U.S. were heavily involved in the recent wave of worldwide consolidation. There were several withdrawals from the U.S. reinsurance market via spinoffs and runoffs.[335] The market share of foreign-owned U.S. domiciled reinsurers has increased from 45% in the year 2000 to 49% in 2004. Bermuda (re)insurers in particular have increased their U.S. market presence. As shown in Figure 18.8, the number of U.S. reinsurers has declined by more than half since 1990 as a result of acquisitions, mergers, and business closures, illustrating the extent of consolidation in the United States.[336]

In the life reinsurance segment, there was a wave of consolidation among North American life reinsurers with nine significant acquisitions since 1997 (Standard & Poor's 2004, p. 46). As a result, the market is heavily concentrated. Eight companies wrote 90% of new life reinsurance business in 2004.[337]

---

[335] Examples of runoffs are AXA Corporate Solutions, Gerling Global Re, Hartford Re, PMA Re, and Trenwick America Re. Zurich Re was divested into Converium in 2001 which put its U.S. reinsurance operation in runoff in 2004. St. Paul Re was spun off into Platinum Underwriters Holdings in 2002, which is domiciled in Bermuda.

[336] Sources are the quarterly statistics of the Reinsurance Association of America (2004) and A.M. Best (2004a).

[337] See Society of Actuaries and Munich American Reassurance Company (2005).

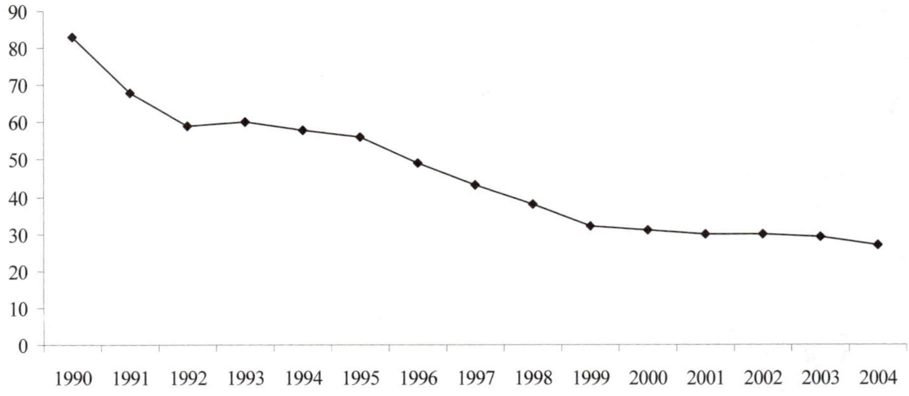

Source: Reinsurance Association of America (1990–2004).

Figure 18.8. Number of Professional Non-Life Reinsurers in the U.S., 1990–2004

Reinsurers licensed in at least one U.S. state are subject to the same spectrum of laws and regulations as primary insurers. These include minimum capital and surplus requirements, risk-based capital requirements, investment restrictions, and numerous disclosure and licensing requirements.[338] The exceptions to this general rule are rates and forms, which are free of regulation. State regulation requires foreign reinsurers without a carrier or branch in the U.S. to provide collateral for all obligations (claims reserves) to their insurance company clients. Collateral can be provided with letters of credit, trust, or funds withheld. This money is held in the U.S. to protect U.S. insurance companies and their policyholders, since foreign reinsurers are not subject to U.S. supervision. There is increasing lobbying from European insurance associations, notably the International Underwriting Association (IUA) and Comité Européen des Assurances (CEA), to reduce or drop these requirements given the quality of regulation and supervision in the E.U.

18.4.2 Continental European Reinsurers

Continental European reinsurers are the world's second largest source of reinsurance capacity. Many of the world's largest internationally active reinsurers are domiciled in Germany, Switzerland, and France. These companies write large amounts of international business through cross border transactions or subsidiaries. In 2003, their share was 46%, a figure that increases to 50% if foreign subsidiaries are included.

Europe's reinsurers have seen their financial strength stabilize in 2003 after a two-year period of downward pressure due to heavy underwriting losses and severe capital losses from the global equity market meltdown between 2000 and 2002. European reinsurers were significantly more invested in equities than U.S. (re)insurers, partially due to large strategic participations. Many of the major players raised fresh capital after 2001, in order to participate in the hard market. A rebound

[338] For a discussion of U.S. reinsurers' capital requirements see Harrington (2005) or Swiss Re (2000).

of equity markets and strong operating earnings helped to restore the solvency of the top tier players. However, there were market exits of medium-sized local reinsurers. Recent consolidation also resulted in the withdrawal of several European reinsurers form the U.S. market due to poor performance.

Many European reinsurers have suffered adverse developments in the claims reserves for their U.S. business, although they had lower relative exposure compared to their U.S. competitors in most cases. However, some European liability markets have also shown substantial adverse developments in recent years (e.g., employment practices and professional liability in the U.K. and medical malpractice in France). While liability claims are still lower than the U.S., there appears to be a trend towards a more litigious environment in Europe (Swiss Re 2004b).

With the adoption of the E.U. Reinsurance and Retrocession Directive in 1964, conduct of cross-border reinsurance business is wholly liberalized. So far, the solvency regulation of reinsurance companies is not harmonized, since the various solvency directives do not apply to reinsurers. Therefore, the intensity of reinsurance supervision varies among the member states of the E.U.

The E.U. is preparing a reinsurance directive called "Fast Track" which introduces a harmonized system for reinsurance supervision, based on the principles of the existing primary insurance directives: single passport, home country control, and solvency rules. According to the current timetable, the new provisions will go into effect by mid-2007. The applicable solvency requirement is based on the current solvency margin for primary property & casualty business. It will apply to the entire reinsurance business, including life reinsurance. The introduction of a uniform standard will level the playing field for reinsurers operating in different jurisdictions. Several jurisdictions will face capital requirement rules for reinsurers for the first time.

18.4.3 Bermuda's Emancipation from the Niche

Bermuda, no longer the newcomer as a reinsurance center, has developed in several waves into the fourth-largest reinsurance market. For about thirty years, it attracted primarily captive companies. In the mid-1980s, a number of specialized liability insurers set up business there. The breakthrough as a reinsurance center did not come until the period of 1992 to 1993, when specialized catastrophe reinsurers were formed to benefit from the hard market following Hurricane Andrew. After 2001, Bermuda has been the domicile for almost all startup companies, which offer exclusively or partly reinsurance coverage. Over $8 billion in fresh capital was raised within a few months after September 11, 2001 and invested in these new ventures (Benfield 2002). By 2003, Bermuda-based insurance groups wrote 12% of the world's professional reinsurers' premiums.

The Bermuda offshore market should be analyzed closely alongside the U.S. insurance and reinsurance markets. Of all the foreign reinsurance centers, Bermuda is currently the most important source of reinsurance capacity for U.S. companies. Seven billion dollars in premiums were ceded directly from U.S. insurers to unaffiliated Bermuda reinsurers in 2003 (Reinsurance Association of America 2004). Another $7 billion in reinsurance business was written by Bermuda-owned firms in the U.S. In addition, Bermuda is the world's premier captive location, with about

60% of its sponsors domiciled in North America. Apart from geographical proximity, an advantageous tax and supervisory climate makes Bermuda an attractive location for captives and reinsurers.

Bermuda reinsurers have diversified from their initial niches in excess liability or property-catastrophe reinsurance and now provide cover in all major lines of business. There is also more geographic diversification than ten years ago, although there remains a bias towards U.S. business. Mostly unburdened by long-tail claims, the new companies benefited from the last hard market and generated excellent returns.[339]

Bermuda is a pure business-to-business market where buyers and sellers of insurance are both sophisticated. Therefore, there are no consumer protection issues at stake and supervision and regulation are flexible and pragmatic. There are no regulations regarding the calculation of premium or loss reserves. There are four classes of insurers' licenses, ranging from single parent captives to companies conducting international insurance business including excess liability business and property-catastrophe reinsurance. Capital and solvency requirements differ for insurers according to their type of license.

18.4.4 Lloyd's Rebirth and Return to Profitability, London

London has a long tradition as an important center for international primary insurance and reinsurance business.[340] In 2003, London market participants wrote $42 billion in premiums, about half of which was reinsurance business.[341] Apart from the brokers who act as intermediaries between the involved parties for almost 100% of the London market, the main operators in London are insurance companies and Lloyd's. Although Lloyd's often has a high profile in reports on the London market, the insurance companies' own market is bigger in terms of volume. The IUA forms the hub of this market. There are currently 40 active ordinary members of IUA and a similar number of affiliated and associated members.[342] Other participants in the London market include the Marine Protection & Indemnity Clubs, which are largely restricted to the insurance of maritime risks.

Lloyd's, an institution with a more than 300 years of history, is itself a market, not an insurance company. Currently, it consists of 62 syndicates offering insurance and reinsurance coverage. Traditionally, the only risk bearers of Lloyd's were

---

[339] Long-tail business is a class of business in which claims take a long period of time to occur or be settled (e.g., products liability and many classes of third party liability).

[340] For more general information on the London market see Carter and Diacon (1990), Carter and Falush (1998), and Swiss Re (2002a). For more specific information on Lloyd's see their web site www.lloyds.com, where their annual report and other publications can be found.

[341] See Association of British Insurers (2004). Unfortunately, the available statistics concerning the London market do not allow for a deep analysis for the following reasons: (a) Some London market participants such as those with branch offices whose headquarters are located in other E.U. countries are not covered; (b) While the treaty reinsurance is separately shown, the facultative reinsurance business, which is of particular importance for the London market, is combined with direct business insurance.

[342] See the web site www.iua.co.uk.

wealthy individuals who provided insurance capital on an unlimited basis, the so-called "names."

Lloyd's went through a severe crisis during the last decade, which also affected the London Market as a whole (Swiss Re 1995). Between 1988 and 1992, Lloyd's incurred total losses of more than £8 billion. Some of the "names," completely lost their fortune in the crisis. These losses were caused by the accumulation of the U.S. liability crisis, the dramatic increase in catastrophic losses, and the syndicates' practice of mutually providing retrocession. Robust earnings in the early 1980s created a lot of additional capacity that was offered for retrocession at increasingly ruinous terms. After each step of retrocession, less and less premium net of the brokers' commission was left to actually cover the risk. Also, transparency regarding the true exposure to the original risk was significantly reduced. The losses that created the crisis were concentrated in three lines of business: (1) Marine insurance, with the explosion on the Piper Alpha oil rig and various oil tanker disasters; (2) Property insurance, with losses caused by hurricanes and winter storms; and (3) Liability insurance, mainly as a result of U.S. liability claims.

The crisis was resolved and since 1993 profits have resumed. All Lloyd's liabilities that arose before the treaty year 1993 were transferred to the rescue company Equitas in 1996, set up by Lloyd's with an initial capital base of £11.2 billion. This freed the syndicates of the uncertainties regarding adverse developments and facilitated the inflow of fresh capital (Swiss Re 2002a).

A number of far-reaching reforms were also introduced to achieve greater transparency and efficiency. The number of syndicates has shrunk from 401 in 1990 to 66 by 2004 as a result of mergers and retirement from business. The initial threat of a capacity shortage following the withdrawal of Lloyd's "names" was successfully averted. Since 1994, additionally to names also corporations, which operate on limited liability basis, were allowed as capital providers. In 2004, corporates accounted for 88% of the total capacity of £15 billion (Lloyd's 2005b). Figure 18.9 illustrates the development of Lloyd's capacity, measured in £ billions.

The figures show that after restructuring and restoring confidence, Lloyd's managed to recapitalize and to take full advantage of the latest hard market. Net premium volumes, which had suffered a sharp drop in the early 1990s from £8.3 billion in 1992 to £6.3 billion in 1996, almost doubled to £11.7 billion in 2003. The share of reinsurance business sank from just under 50% to 39% and rebounded to 48% in 2003. Between 1992 and 1996, the market share of Lloyd's syndicates in the global reinsurance industry fell from 7% to 4% and rebounded to 5% in 2003.

The present regulatory environment in the United Kingdom allows all overseas reinsurers (E.U. or other) to freely provide reinsurance and retrocession to U.K. insurers and reinsurers, for all types of risks. Reinsurers that are domiciled in the U.K. are supervised by the Financial Services Authority (FSA), the independent regulatory body with statutory authority to regulate the U.K. financial services industry. All regulations regarding taxation, reserving, capital, and solvency apply equally for non-life primary insurers and reinsurers. In anticipation of the E.U.'s Solvency II initiative the FSA has started work on a new capital adequacy framework designed to improve the financial soundness of regulated companies. In 2004, the FSA introduced variable, risk-based capital requirements. This is a combination of asset-related and insurance-related capital requirements minus equalization reserves. All companies are required to model their own capital needs, which are used by the FSA to provide individual capital guidance. These rules will also affect Lloyd's.

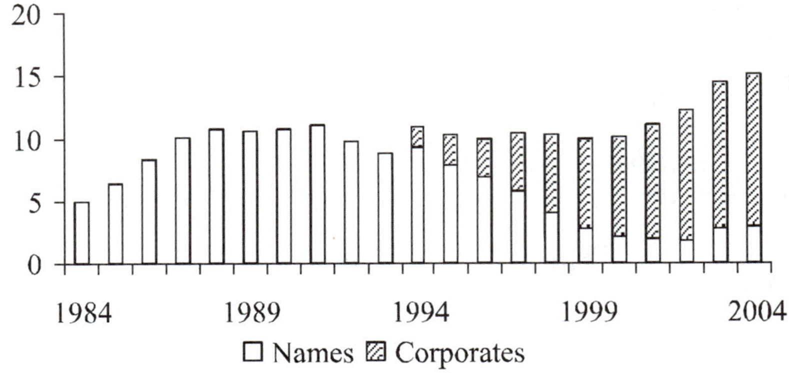

Source: Lloyd'\'s (2005a).

Figure 18.9. Lloyd's Capacity, 1984–2004 (£ Billions)

18.5 MARKET IMPROVES RESULTS AND CAPITALIZATION

Overall, performance of the reinsurance industry declined between 2000 and 2002 due to a combination of adverse developments on both the asset and liability side of the balance sheet, especially the events of September 11, 2001 and the equity markets decline. While these shocks were managed without any major defaults or retraction in the provision of reinsurance coverage, the impact was severe and fueled ratings downgrades of most of the major reinsurers.

As the industry has moved from excess capital to tight capitalization, behavior of the companies has changed. The focus has shifted from expansion and diversification to concentration on the core business and cost management; from an offensive strategy in which capital is not a constraint, to a defensive strategy of funding and ensuring liquidity, concentrating on freeing up capital and maintaining flexibility on assets and liabilities. After 2002, profitability and capitalization of the industry improved and many reinsurers regained financial flexibility. In the current financial environment, where investment returns have reached very low levels and where no fundamental improvements can be expected, sound technical underwriting remains a prerequisite for profitability.

Reinsurance prices have been increasing since 2001, fueling a recovery of profitability. As Table 18.5 indicates, non-life underwriting results have improved significantly, supporting double digit returns on equity (ROEs) in 2003 and 2004. In 2004, third-quarter catastrophic losses overshadowed reinsurers' improving operating profitability, pushing the combined ratio to 100%, from 96% a year ago. The ROE decreased to 10%, from 13% in 2003. Nevertheless, the industry's shareholders' equity grew by 12% in 2004, due primarily to strong operating earnings.

Operating earnings will continue to increase reinsurers' surplus in the next few years. Profitability for the first half of 2005 was strong. However, the losses from Hurricane Katrina will severely lower the ROE for 2005. The solvency ratio (surplus in % of net premiums) has recovered from its low in 2002 and is expected to continue increasing from 112% at end of 2004.[343]

Table 18.5. Performance Indicators of Major Non-life Reinsurers, 1999–2004

	1999	2000	2001	2002	2003	2004
Underwriting Results as a % of NPE	–16.4	–12.6	–34.3	–12.7	3.7	0.2
Investment Results as a % of NPE	25.1	23.5	19.2	13.6	12.6	13.2
ROE	7.4	9.7	–6.7	3.4	12.9	10.2
Shareholders' equity as a % of NPE	143.4	136.6	113.8	94.8	100.7	109.5

Sources: Annual Reports, Swiss Re Economic Research & Consulting
Note: ROE is return on equity. NPE is net premiums earned.

18.6 CONCLUSION

The current market situation is characterized by strong operating earnings and an ongoing recovery of the capital base of global reinsurers. Cycle management is one key issue for the industry going forward. This entails a strict adherence to actuarial pricing and underwriting standards, portfolio steering, and active capital management. There is a long-term trend towards a more flexible deployment of risk capital. Increased scrutiny from investors, analysts, regulators, and rating agencies, paired with improved transparency regarding price and profitability, should help dampen and shorten future cyclicality.

The risk landscape is dynamic. Reinsurers need to adapt to the changing degree of insurability of the underlying risks. Traditionally-insured risks can turn uninsurable due changes in the legal system, the economy, or the socio-political environment. Examples are asbestos, terrorism, or directors and officers liability insurance for certain industry groups. At the same time, new emerging risks, such as nanotechnology, electro-magnetic fields, and obesity, need to be monitored by reinsurers in order to avoid another asbestos-like situation. Product adaptations and innovations can be used to overcome obstacles to covering hard to insure risks (Cummins 2005; Swiss Re 2005b).

[343] For information on the current market conditions see A.M. Best (2004b), Benfield (2005), Guy Carpenter (2004), Moody's (2005), Standard & Poor's (2004), or Willis (2004b).

One of the most promising tools for enhancing the flexibility of risk capital and facilitating insurability is the securitization of insurance risks.[344] Insurance risk securitization involves transforming insurance risks into bonds that are sold to investors. Securitization provides capacity to primary or reinsurers, making coverage more available and affordable. The insurance-linked securities (ILS) market developed in reaction to a series of major natural catastrophes in the early 1990s.[345] Since its inception, the ILS market has witnessed worldwide issuance of around $19 billion (Swiss Re 2005b). The two major segments of the market are catastrophe (cat) bonds and life insurance bonds. Since 1996, more than $11 billion of life bonds and $8 billion of cat bonds have been issued. At year end 2004, there was over $10.3 billion of ILS outstanding, up 60% from a year earlier. The investor base for ILS is broadening to hedge funds, dedicated catastrophe funds, and money managers, suggesting a potential to expand to other hard to insure risks.

18.7 REFERENCES

Allen, Nigel, 2004, "Multi-Purpose Broker," *Global Reinsurance* December 2004: 28–37.

A. M. Best, 2004a, *Best's Aggregates & Averages: Property-Casualty* (Oldwick, NJ: A.M. Best Co.).

_____, 2004b, *Global Reinsurance* (Oldwick, NJ: A.M. Best Co.).

Association of British Insurers, 2004, *General Insurance Overview Statistics*, www.abi.org.uk.

Benfield, 2002, *Old Money, New Money: A Review of the Bermuda Reinsurance Market,* Benfield Industry Analysis and Research, London, England.

_____, 2005, *Outrageous Fortune: Reinsurance Market and Renewals Review*, Benfield Industry Analysis and Research, London, England.

Carter, Robert L. and Stephen R. Diacon, 1990, "The Microstructure of the London Insurance Market," in Henri Louberge, ed., *Risk Information and Insurance* (Boston: Kluwer Academic Publishers), pp. 231–253.

Carter, Robert L. and Peter Falush, 1998, "The London Insurance Market," report prepared for the London Insurance Market Strategy Committee, London, England.

Conning and Company, 2000, "The Cedant Pipeline Remains Split in Two," *The Conning Commentary* 9: 1–7.

Cowley, Alex and J. David Cummins, 2005, "Securitization of Life Insurance Assets and Liabilities," *Journal of Risk and Insurance* 72: 193–226.

Cummins, J. David, 2005 "Convergence in Wholesale Financial Services: Reinsurance and Investment Banking," *The Geneva Papers* 30: 187–222.

---

[344] For more detailed discussion on the use of insurance-linked securitization see Froot (2001), Cummins, et al. (2002), and Swiss Re (2001a, 2005b). For more specific information on life securitization see Cowley and Cummins (2005) and Millette, et al. (2002).

[345] The first non-exchange-traded capital market product that insured against catastrophe losses was a US$85 million Hanover Re cat bond issued in 1994.

_____, Scott E. Harrington, and Robert W. Klein, 1991, "Cycles and Crises in Property-Casualty Insurance: Causes and Implications for Public Policy," *Journal of Insurance Regulation* 10: 50–93.

_____, David Lalonde, and Richard D. Philips, 2002, "Managing Risk Using Index-Linked Catastrophic Loss Securities," in Morton Lane, ed., *Alternative Risk Strategies* (London: Risk Books), pp. 10–46.

_____, and J. François Outreville, 1987, "An International Analysis of Underwriting Cycles in Property-Liability Insurance," *Journal of Risk and Insurance* 54: 246–262.

_____, and Mary A. Weiss, 2000 "The Global Market for Reinsurance: Consolidation, Capacity, and Efficiency," in Robert E. Litan and Anthony M. Santomero, eds., *Brookings-Wharton Papers on Financial Services* (Washington, DC: Brookings Institution Press), pp. 159–209.

Froot, Kenneth A., 2001, "The Market for Catastrophe Risk: A Clinical Examination," *Journal of Financial Economics* 60: 529–571.

Gastel, Ruth, ed., 2004, *Reinsurance: Fundamentals and New Challenges* (New York: Insurance Information Institute).

Gerathewohl, Klaus, 1983, *Reinsurance: Principles and Practice, Volumes I & II* (Karlsruhe, Germany: Verlag Versicherungswirtschaft).

Guy Carpenter & Company, 2004, *The World Catastrophe Reinsurance Market: 2004* (New York: Guy Carpenter & Company Inc.).

Harrington, Scott E., 2005, "Capital Adequacy in Insurance and Reinsurance," in Hal S. Scott, ed., *Capital Adequacy Beyond Basel* (New York: Oxford University Press), pp. 87–122.

Hartwig, Robert P., Gordon Stewart, and Claire Wilkinson, 2004, "Terrorism, Insurance, and the United States Government," Insurance Information Institute, New York, NY.

International Association of Insurance Supervisors, 2004, *Global Reinsurance Market Report 2003* (Basel, Switzerland).

Lloyd's, 2005a, *Annual Report 2004* (London, England).

_____, 2005b, "Close Up," *Lloyd's Review 2005* (London, England).

Millette, Michael J., Shiv Kumar, Omar J. Chaudhary, Jacqueline M. Keating, and Steven I. Schreiber, 2002, "Securitization of Life Insurance Business," in Morton Lane, ed., *Alternative Risk Strategies* (London: Risk Books), pp. 395–426.

Moody's, 2005, *Global Reinsurance Industry Outlook—September 2005* (New York: Moody's Investor Service).

Reinsurance Association of America, 2004, *Alien Reinsurance in the U.S. Market—2003 Edition* (Washington, DC: Reinsurance Association of America).

_____, 1990–2004, *Reinsurance Underwriting Report (Quarterly)* (Washington, DC), community.reinsurance.org/ScriptContent/index_datacenter.cfm.

Schwepcke, Andreas, ed., 2004, *Reinsurance: Principles and State of the Art* (Karlsruhe, Germany: Verlag Versicherungswirtschaft).

Society of Actuaries and Munich American Reassurance Company, 2005, *Life Reinsurance Survey 2005* (Atlanta, GA: Munich American Reassurance Company).

Standard & Poor's, 2004, *Global Reinsurance Highlights—2004 Edition* (London: Reactions Publishing Group).

Swiss Re, 1995, "The London Market," *Sigma*, No. 2.

_____, 1998, "The Global Reinsurance Market in the Midst of Consolidation," *Sigma*, No. 9.

_____, 2000, "Solvency of Non-Life Insurers—Balancing Security and Profitability Expectations," *Sigma*, No. 1.

_____, 2001a, "Capital Market Innovation in the Insurance Industry," *Sigma*, No. 3.

_____, 2001b, "Profitability of the Non-Life Insurance Industry: It's Back to Basics Time," *Sigma*, No. 5.

_____, 2002a, "The London Market in the Throes of Change," *Sigma*, No. 3.

_____, 2002b, "Global Non-Life Insurance in a Time of Capacity Shortage," *Sigma*, No. 4.

_____, 2004a, "Commercial Insurance Brokerage—Love Thy Middleman," *Sigma*, No. 2.

_____, 2004b, "The Economics of Liability Losses—Insuring a Moving Target, *Sigma*, No. 6.

_____, 2004c, *Understanding Reinsurance: How Reinsurers Create Value and Manage Risk* (Zurich, Switzerland: Swiss Reinsurance Company).

_____, 2005a, *Reinsurance Matters: A Manual of the Non-Life Branches* (Zurich, Switzerland: Swiss Reinsurance Company).

_____, 2005b, "Innovating to Insure the Uninsurable," *Sigma*, No. 4.

Willis Ltd., 2004a, *Lloyd's Review* (London: Willis Ltd.).

_____, 2004b, *Marketplace Realities and Risk Management Solutions 2004* (London: Willis Ltd.).

18.8 LEXICON

CEA. The Comité Européen des Assurances is a federation of European national insurance associations. See their web site www.cea.assur.org/cea.

Ceding Company or Cedent. An insurer who reinsures its risks is referred to as the ceding company or cedent.

Cession. Cession is insurance that is reinsured.

E.U. European Union.

Facultative Reinsurance. Facultative reinsurance is reinsurance of an insurer's risks on an individual basis. The reinsurance company looks at each individual risk and determines whether to accept or decline coverage.

Financial Reinsurance. Financial reinsurance is a form of reinsurance that provides risk financing which may be combined with elements of traditional risk transfer.

Flight to Quality. The first "flight to quality" with respect to reinsurance buying behavior occurred in the early 1990s when an accumulation of large losses highlighted the relevance of reinsurers' security. Ceding insurers shifted their demand to larger reinsurers with strong balance sheets.

FSA. The Financial Services Authority is the regulatory body with statutory authority to regulate the U.K. financial services industry.

Hard Market. A hard (re)insurance market is characterized by rising or high insurance rates. During a hard market, insurance capacity is scarce and competition is low.

Intragroup Reinsurance. Intragroup reinsurance categorizes reinsurance relations between affiliated insurance companies belonging to the same group. Large insurance conglomerates tend to first diversify their insurance portfolios internally with intragroup reinsurance before approaching the reinsurance market.

ILS. Insurance-linked securities.

IUA. International Underwriting Association.

Long-Tail Business. Long-tail business is a class of business in which claims take a long period of time to occur or be settled (e.g., products liability and many classes of third party liability).

M&A. Mergers and acquisitions.

Non-Proportional Reinsurance (also Excess Loss of Reinsurance). Non-proportional reinsurance is a form of reinsurance in which the reinsurer assumes part of the insurer's claims which exceed a certain amount.

NPE. Net premiums earned.

Premiums Earned. Premiums earned are premiums an insurance company has recorded as revenues during a specific accounting period.

Premiums Written. Premiums written are premiums for all policies sold during a specific accounting period.

Proportional Reinsurance. Proportional reinsurance is a form of reinsurance in which the premiums and claims of the insurer are shared proportionally by the insurer and reinsurer.

Quota-Share Reinsurance. Quota-share reinsurance is a form of proportional reinsurance in which a defined percentage of all risks held by the insurer in a specific line is reinsured.

Reinsurance. Reinsurance is insurance for insurance companies which spreads the risk of the direct insurer, making the risk of its portfolio more homogenous. It includes various forms such as facultative, financial, non-proportional, proportional, quota-share, surplus, and treaty reinsurance. Also, reinsurance is the passing of an insurer's risks to the reinsurer against payment of a premium.

Retrocession. Retrocession is the amount of risk accepted by the reinsurer which is then passed on to other reinsurance companies.

ROE. Return on equity.

Securitization. Securitization is the process of aggregating similar instruments, such as credit portfolios or catastrophe risks, and marketing them to investors as a negotiable security.

Short-Tail Business. A short-tail business is one in which claims are reported and settled in a relatively short period of time (e.g., most classes of property business).

Soft Market. During a soft market, rising insurance capacity and fierce price and market share competition lead to widely available coverage and low premium rates. Profitability of business written during a soft market is below average. Declining capacity and competition lead to hard markets with rising premium rates and improving profitability of new business.

Surplus Reinsurance. Surplus reinsurance is a form of proportional reinsurance in which risks are reinsured above a specified amount.

Treaty Reinsurance. Treaty reinsurance is participation of the reinsurer in certain sections of the insurer's business as agreed to by treaty, as opposed to single risks.

U.K. United Kingdom.

U.S. United States.

19 Lloyd's and the London Insurance Market: An Overview

Julian James

Lloyd's of London

19.1 U.K. INSURANCE MARKET CONTEXT

19.1.1 Market Size and Structure

Nobel Prize winning economist Kenneth Arrow stated that the world would be a better place if we could insure against future possibility. This would enable more risk taking, creating a source of economic progression. As one of the largest insurance markets in the world, the U.K. insurance market has a pivotal role in the growth of this economic progression that Arrow speaks of.

It is in within the overall context of a highly mature and developed U.K. insurance industry that we must approach the particular features of the London market. The U.K. market is currently the largest in Europe and the third largest in the world after the United States and Japan (Association of British Insurers (ABI 2005, AXCO 2005). U.K. premiums per capita are the second highest of any country, after Switzerland, and insurance penetration (the ratio of insurance premiums to gross domestic product), at 14.2 percent, is second only to South Africa (International Financial Services London, 2004).

In studying the London market, it is also important to bear in mind the size of the United Kingdom as a non-life insurance market. According to the ABI, U.K. non-life market premiums were almost $90 billion in 2003.[346] Further, over the four years 1999 to 2003, the non-life market grew much faster than the long-term (life insurance) market, yielding an overall growth rate of 15 percent over the period.

The international flavor of the U.K. insurance industry is also a clue to understanding the London market better. Its international nature is confirmed by the fact that the United Kingdom accounts for 8.4 percent of worldwide premium income, and some $12 billion of British export sales (ABI 2005). Overseas insurers also have a notably high presence in the U.K. market, due in part to its relative size and its regulatory openness, among other factors.

---

[346] Exchange rate used is US$1.9 2= £1 at year end 2004 unless otherwise stated.

19.1.2 Market Distribution

Because the London insurance market is, by and large, an intermediated market, it is important to appreciate the overall distribution structure of the U.K. industry. The U.K. insurance market is, in distribution terms, one of the most sophisticated in the world, with a wide and increasing variety of channels available. Traditionally, brokers have played a key role in the U.K. insurance market, often a greater role than in many other European markets, acting as intermediaries in the placing of insurance business for their clients, negotiating price, and designing risk programs for their clients.

It is estimated that 80 to 85 percent of U.K. commercial lines business is distributed by brokers; however, brokers' share of the personal lines market is considerably lower and has fallen significantly in recent years due to the rise of direct selling, Internet distribution, bancassurance, and sales through affinity groups including "high street" supermarkets.[347] The structure of the U.K. broking sector and the role of the intermediary have changed over the last decade. In particular, brokers have had to adapt and extend their range of services offered to include consultancy services such as risk management advice and control services, the development of new financial products for the management of risk, claims consultancy, and employee benefits consulting. As has happened with general insurance companies, the broking sector has become highly concentrated in recent years.

19.1.3 Regulation

The London market's character has also been significantly influenced by the approach taken by the U.K. supervisory authorities over the years. In the past, regulation of the insurance market in the United Kingdom was fragmented. However, the Financial Services Authority (FSA) was formed in 2001. The FSA is an independent nongovernmental body, given statutory powers by the Financial Services and Markets Act 2000, and is a company limited by guarantee and financed by the financial services industry. The U.K. Treasury appoints the FSA Board, which consists of a chairman, a chief executive officer, three managing directors, and 10 nonexecutive directors, and this board sets overall FSA policy. The FSA describes its mission as follows:

> In January 2000 we set out our proposed approach to regulation in A New Regulator for the New Millennium. We stated that our overriding aim was "to maintain efficient, orderly and clean financial markets and help retail consumers achieve a fair deal". Our approach is based on a clear statement of the realistic aims and limits of regulation; it recognises the proper responsibilities of consumers and of firms' own management, as well as the impossibility and undesirability of removing all risk and failure from the financial system (FSA 2005).

Prior to the establishment of the FSA, the United Kingdom had experienced some insurance market failures, most notably that of Independent Insurance. This company was a leading underwriter of commercial business, and in particular

[347] Estimates vary. This is based on ABI data (2005) and AXCO data (2005).

employers' liability business, before going into liquidation in 2001. Many have commented since that the new regime and risk-based capital requirements put in place by the FSA have significantly raised regulatory standards in the United Kingdom. At the same time, the regulator's approach means that the United Kingdom retains its reputation for a "liberal attitude" and "stable environment," which have been identified as critical factors for an international market (Herve-Bazin 1994). In contrast to the United States, rates and forms are not regulated in the United Kingdom, and as part of the European Union (EU), the United Kingdom benefits from what is essentially free trade in commercial lines insurance with other EU member states. Taken together, these conditions create an excellent insurance infrastructure and a mature insurance market for both domestic and international risks, and overall close to a quarter of U.K. insurance companies' net premiums come from overseas business including from subsidiaries overseas (International Financial Services London 2004).

19.2 THE LONDON MARKET

19.2.1 Introduction

The London market plays a unique role within the United Kingdom's insurance industry, well summarized as follows almost a decade ago:

> The London Insurance Market is a distinct, separate part of the UK insurance and reinsurance industry centred on the City of London. It comprises insurance and reinsurance companies, Lloyd's syndicates, and brokers who handle most of the business. While there is no watertight definition of the market, there is general agreement that the core of its activity is the conduct of internationally traded insurance and reinsurance and reinsurance business, which would include the very large risks of UK companies and multinationals. The business is almost exclusively non-life...with an...emphasis on high exposure risks (Jim Bannister Limited 1999).

Taken together, the London market employs around 40,000 insurance professionals within the so-called Square Mile of offices in the City of London, accounting for 1.2 percent of national gross domestic product and ranking as the world leader for large, complex, or unusual specialist commercial risks, with an estimated 10 to 15 percent share for large industrial risks overall, and a considerably higher share of specific specialist sectors, as highlighted later (International Financial Services London 2004). However, the London market does not, in general, underwrite much life business. Total London market premiums grew steadily throughout the late 1990s, reaching an estimated $45 to 50 billion by 2004.[348]

Insurance and reinsurance coverage within the London market is provided by a number of key groups, each distinct but often working together:

Lloyd's insurance market. Discussed in detail below, which is estimated to account for over half of London market business.

---

[348] This is an estimate based on a total London market premium figure of £25 billion for 2002 (IFSL 2004, p. 17) and two years of subsequent premium growth.

London company market. London is the only market in the world where the world's 20 largest insurance and reinsurance companies all have offices, whether headquarters, branch, or representative offices. In total, over 100 companies are estimated to have a London operation, the purpose often being to access and underwrite the international business that the market sees.

The London company market is represented by the International Underwriting Association of London (IUA), and their members are headquartered in 40 countries around the world, according to the Corporation of London. The IUA is the world's largest representative organization for international and wholesale insurance and reinsurance companies. It is both a trade association, representing its members' interests internationally, and a market association supporting the business environment in London.

IUA membership is available at ordinary, affiliate, and associate levels. Ordinary members enjoy full membership status, while the affiliate category comprises companies in run-off and service providers to full membership companies. Associate membership is the most limited form of membership for non-London based companies. A list of the 42 ordinary members, at September 2005, is provided in Table 19.1.

It is difficult to give definitive premium statistics for the London company market, due in part to the different processing and accounting treatments adopted by different member companies. However, it is generally reckoned that the London company market accounts for less than half of the total London market premiums, and the business mix is similar to that experienced by Lloyd's. The IUA also indicated in April 2004 that premium figures had been increasing for the past three years (International Underwriting Association 2004).

The London company market shares a common processing bureau with Lloyd's, and the two sectors cooperate on a number of processes and other related initiatives. The general thrust of these initiatives is to improve efficiency and drive down the cost of processing within the London market. Further details and current updates on these initiatives can be found on their respective web sites.

19.2.2 Protection and Indemnity (P&I) Clubs

P&I clubs are insurance associations of shipowners and charterers who operate on a mutual basis. They originated in order to underwrite liability risks not covered by marine policies usually available in the insurance market. Of the world's marine insurance markets, London is the largest for P&I clubs, with about 70 percent of such global business directed through London (International Financial Services London 2004).

Table 19.1. IUA Membership, September 2005: List of Ordinary Members

Ace European Group (was ACE Insurance SA-NV)
Alea London Limited
Allianz Cornhill Insurance Plc (was Cornhill Insurance plc)
Allianz Marine & Aviation Versicherung AG *(was AGF Marine Aviation Transport)*
Arch Insurance Company (Europe) Ltd
Aspen Insurance UK Limited
Assicurazioni Generali S.p.A. (UK Branch)
AXA Corporate Solutions Assurances SA
AXA Corporate Solutions SA
Chubb Insurance Company of Europe
CNA Insurance (Europe) Company Limited *(now incorporating Maritime Insurance Company Limited)*
Ecclesiastical Insurance Office plc
Endurance Worldwide Insurance Limited
Everest Reinsurance Company (UK Branch)
FARADAY Reinsurance Company Limited
Federal Insurance Company
GE Frankona Reinsurance Limited
GE Frankona Ruckversicherungs Aktiengesellschaft
GE Specialty Insurance (UK) Limited
Global Aerospace Underwriting Managers (GAUM)
Great Lakes Reinsurance (UK) Plc
Groupama Transport
Hannover Re
Houston Casualty Company London
International Insurance Company of Hannover Limited
Liberty Mutual Insurance Europe Limited
Markel International Insurance Company Limited
Munich Reinsurance Company
New Hampshire Insurance Company
New India Assurance Company Limited
QBE International Insurance Limited
QBE Reinsurance (Europe) Ltd
Royal & Sun Alliance Insurance plc
SCOR UK Company Limited
Sirius International Insurance Corporation (publ)
SR International Business Insurance Company Limited
Swiss Reinsurance Company UK Limited
Tokio Marine Europe Insurance Limited (TMEI)
Tokio Marine Global
XL Re Ltd
Zurich Insurance Company
Zurich Specialties London Limited

Source: International Underwriting Association (2005).

19.2.3 London Market Brokers

Almost all risks underwritten within the London market are intermediated, and brokers therefore also make up a key part of the London insurance market. The majority of the world's largest insurance broking firms have a significant presence in London. One commentator (Jim Bannister Limited 1999) has attempted to group them as follows:

- Major global brokers who have a significant share of the London market and global business
- Multiniche brokers based in London but with sizeable operations outside the London market
- London-based wholesale brokers who specialize in placing business from other non-Lloyd's brokers
- Many smaller and specialist niche brokers

In addition to these participants are many firms of specialist insurance lawyers, loss adjusters, actuaries, consultants, and other insurance experts who complete the hub of insurance expertise.

19.2.4 London as an International Insurance Market

The available literature on international markets is quite limited; however, one of the best surveys was prepared in 1994 by Eric Herve-Bazin of AXA for the French insurance association. It was well researched, involving several months visiting insurance centers in Europe, North America, and Asia. The survey produced a long list of requirements for a successful international center as follows:

- Political stability
- Geographical location
- Quality transportation
- Developed communications systems
- Highly qualified personnel
- Office space at competitive prices
- Multilingualism
- Stable legal and regulatory environment
- Liberal attitude by authority
- Quality family life
- Appropriate time zone
- Strong foreign presence

He also added a number of secondary advantages, ranging from a developed financial center to a strong currency, and emphasized the need for strong international "network" links to potential sources of business.

Herve-Bazin concluded that only London qualified as an international insurance center according to all of the above criteria. He also specifically mentioned its strong ties with the United States as a positive factor. A decade later, while various other financial and insurance locations have continued to emerge and develop, London has nevertheless retained its preeminence as the world's international insurance center. The two lines of business where the London market is most prominent are discussed further below.

Marine, Aviation, and Transport Market

The London market, and in particular Lloyd's, is regarded as the leading global center for marine, aviation, and transport insurance and reinsurance. The London market's share of the global marine market is estimated at up to 19 percent. Having backed and secured the fortunes of mariners and merchants on the high seas for over 300 years, Lloyd's in particular continues to lead a significant proportion of the world's marine business, underwriting $1.8 billion dollars in marine premiums in 2003 alone (Asia Insurance Review 2004; Tolle 2005). Lloyd's heritage and expertise in the sector is further demonstrated by the existence of around 350 Lloyd's agents spread across every major global port. Their presence provides valuable specialist surveying and adjusting services throughout the world.

In recent years, the global marine market has been intensively competitive, particularly the hull sector. Profits in the early 1990s ebbed in the face of high loss events until 2001, when things finally began to improve from an extremely low starting point. Too many losses in 2002 crippled the sector further, with the hull sector on its sixth consecutive annual loss, leading to a general call for a return to sanity based on profitable underwriting.[349] In 2003, Lloyd's did see marginally better hull results, but the pricing and coverage conditions were still far from levels capable of supporting consistent profits.

The cargo and marine liability sectors, where Lloyd's is also a major player, have returned a more consistent performance over time, although once again there continue to be concerns about pricing adequacy, and cargo profits for 2005 have been adversely impacted by hurricane damage in the American Gulf. In the future, exchange of technical data and greater transparency of results should help the marine market move toward the need to underwrite at the right price to allow underwriters to make sustainable profits.

The London market's share of aviation business is even higher than that of marine, standing at around 39 percent (IFSL 2004). However, the aviation sector is not immune to the cyclical pressures and increased competition in recent years. Notwithstanding, the global aviation insurance market continues to grow in spite of increased potential problems with security, from $51.3 billion in 2003 to $56.7 billion in 2004, an increase of 12 percent. Passenger numbers have also increased, rising from 197 million to 229 million in 2004, an increase of 17 percent (*Insurance Day* 2005).

London Market Reinsurance Sector

The London market's share of nonmarine treaty reinsurance has remained relatively stable in recent years, at around 5 to 10 percent, and London is estimated to account for a full quarter of the world's internationally available reinsurance (IFSL 2004;

---

[349] Publicly expressed at the IUMI Conference, 2003.

Swiss Re 2004). Lloyd's alone is ranked as the world's fifth largest reinsurer (Standard & Poor's 2004a).

After over three centuries of experience of dealing with risk, the expertise, underwriting skill, and knowledge residing within the London market for specialist reinsurance needs is certainly extensive. Like their direct counterparts, reinsurance risks, whether facultative or treaty business, local or international, reach the London market exclusively through intermediaries. The leader may choose to accept the entire risk on behalf of his syndicate or company, but in most cases, especially with large reinsurance treaties or facultative placements, the risk will be greater than the syndicate or company wishes to bear alone. When that happens, other syndicates and companies are invited to take a share of the risk, spreading the risk throughout the wider London market, and beyond. This unique subscription system allows individual syndicates and company market insurers to combine their capital and compete with the largest of the world's reinsurers, yet retain the responsiveness and short decision-making hierarchy which ensure that underwriters are able to provide clients with individual attention.

Current conditions suggest a stabilizing global reinsurance market after years of poor results. Positive factors often cited include high current accident year profitability and a trend in superior quality underwriting control and analysis. Some offsetting factors include concerns about continued undesirable development in U.S. casualty lines, capital depletion following two record windstorm seasons, and ongoing concern about the ability of reinsurers to maintain underwriting discipline throughout the insurance cycle.

19.3 LLOYD'S BACKGROUND AND HISTORY

19.3.1 Early History

Founded in 1688, Lloyd's is the world's oldest and best known insurance market. Lloyd's first existed as a coffee house owned by a man called Edward Lloyd who encouraged ships' captains, merchants, ship owners, and others with an interest in overseas trade to visit his establishment. Within a short period of time, Lloyd's had gained a strong reputation for trustworthy shipping news, ensuring that Lloyd's became the popular place for obtaining information on all things related to marine trade and the risks that accompanied it.

In the seventeenth century, the growing importance of such trade led to demand for ship and cargo insurance, a natural extension of this sharing of information. Business in those days was conducted very informally, and a merchant with a ship to insure would ask a "broker" to take the risk from one wealthy merchant to another seeking insurance cover. Each merchant would take a portion of the risk, and the broker would visit several merchants until the risk was completely covered. The broker's skill lay chiefly in ensuring that policies were underwritten only by people who could meet their share of a claim, if need be, to the full extent of their personal fortunes. This was the process that became known as underwriting.

In 1769, a number of the more serious players at Lloyd's broke away and set up a rival, more professionally organized establishment. This "New Lloyd's Coffee

House" eventually became too small, so a committee was elected and 79 merchants, underwriters, and brokers each paid £100 into the Bank of England for new premises. In 1774, rooms were leased in the Royal Exchange and, no longer a coffee house, the modern Lloyd's was born.

19.3.2 A Period of Rapid Growth

Over the next century, the society of underwriters at Lloyd's evolved further. Membership was regulated, and the elected committee was given increased authority. A process of formal investment into the Lloyd's market began, and those who invested in the market were called Names, sole capital providers for the market. Lloyd's then took another step forward in 1871 with the incorporation of Lloyd's by Act of Parliament. The Lloyd's Act 1871 gave Lloyd's a formal legal basis, allowing it to acquire property and make by-laws backed with the full authority of Parliament, establishing Lloyd's as a unique business institution, a status it maintains today. The Lloyd's market grew steadily thereafter, both in terms of financial size and product range. Lloyd's first began to diversify from its marine history in 1887, underwriting property risks for the first time, which was considered a controversial innovation at the time.

The trend toward growth and innovation continued throughout the twentieth century. As a result, Lloyd's is home to some of the most skilled and experienced insurance specialists in the world and has led the industry in creating new areas of coverage over the last 50 years, such as kidnap and ransom, space and aviation, cyber-liability, specialist policies for satellites, contingency coverage for large international and national sporting events, and terrorist threats.

19.3.3 Recent History

However, in the mid-1980s and the early 1990s, the Lloyd's insurance market suffered an unprecedented wave of losses due to events including the Piper Alpha oil spill, Hurricane Andrew, and mounting asbestos claims. With the market in a state of financial uncertainty, Lloyd's implemented a Reconstruction and Renewal Plan in 1996 aimed at rebuilding the market. The Plan involved the establishment of a reinsurance company called Equitas to allow members of Lloyd's to run off their 1992 and prior year liabilities, while allowing current and future members of Lloyd's to trade forward. The Plan succeeded, securing support from the vast majority of Lloyd's members. Equitas is a completely separate entity from Lloyd's.[350]

19.4 LLOYD'S: A MARKET, NOT A COMPANY

Today, Lloyd's is probably the world's best known insurance brand, but it is not always well understood. This is because Lloyd's is an insurance market, and not an insurance company. Lloyd's is very different to the coffee shop of its origins, but it

[350] A.M. Best comments briefly upon Equitas in its rating rationale of the Lloyd's market in A.M. Best (2005b).

still operates using the same principles. It is best represented diagrammatically as shown in Figure 19.1.

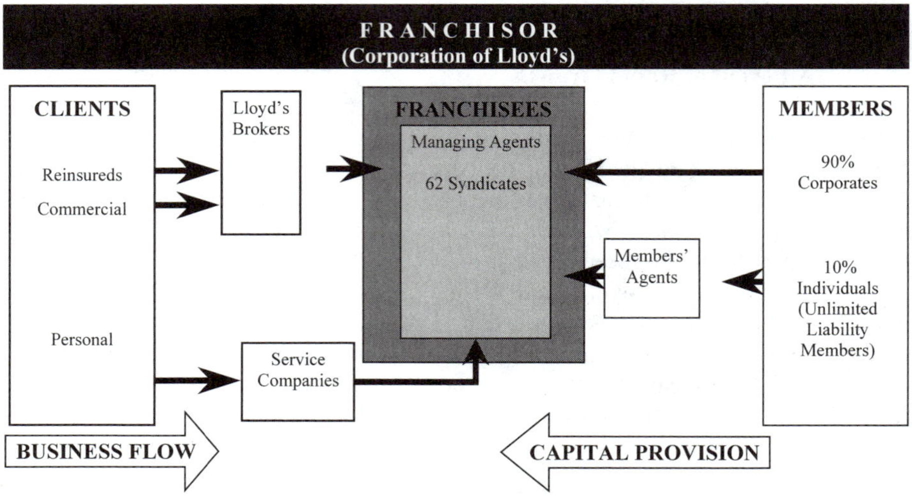

Source: Members Services Unit, Admissions, Nov. 2003; Broker Services, Nov. 2003; Lloyd's Global Results 2002, May 2003

Figure 19.1. How the Lloyd's Market Works

19.4.1 Underwriting

The market comprises a number of underwriting businesses called syndicates. Working for the individual syndicates, underwriters accept risks on their behalf from the community of London market brokers. The number of syndicates operating within the Lloyd's market has varied widely over time, reaching a peak of 437 in 1980 (Statistics Relating to Lloyd's, Syndicate section, 2001). Since then, consolidation has taken place within Lloyd's, as elsewhere in the insurance markets, and the number of syndicates has decreased while syndicate size and strength has conversely increased, and market structure has become more efficient. In 2005, there are 62 syndicates operating within the Lloyd's market, of different shapes and sizes, and each with its own specialties and expertise (Lloyd's 2005b). Each syndicate is in turn managed by a managing agent, also known as a Franchisee under the Lloyd's structure. A number of managing agents manage more than one syndicate.

Given that Lloyd's is a marketplace, different underwriters, working for different syndicates, will sometimes compete on price and terms and conditions to win business and the lead role in underwriting the risk, a practice that has traditionally encouraged a flexible and innovative approach and ensures that a competitive spirit remains alive within Lloyd's. The broker may need to visit a number of syndicates in order to cover the risk: frequently, larger risks are divided between more than one syndicate in the market, with each syndicate taking a percentage of the risk, usually at the same price and terms as the lead underwriter who assumes the first portion of the risk. This is known as the subscription market. In any circumstance, the underwriter works with the broker and calculates a premium based on the likelihood of the risk occurring, using a mixture of scientific and actuarial modeling tools and a degree of intuition based on specialist and detailed knowledge of what are often highly specialized areas.

Lloyd's has long been viewed as being at the leading edge of the specialist insurance market. Lloyd's is known as a place for "firsts": the first reinsurance, aviation, motor, and satellite policies were all underwritten at Lloyd's during the nineteenth and twentieth centuries, and it continues to lead the development of innovative insurance products today, such as insurance policies to protect businesses against computer hacking attacks, and coverage for intangible assets, including intellectual property risks. See Table 19.2 for a breakdown of the lines of business within Lloyd's.

Table 19.2. Lloyd's Business Split by Category, 2004

Line of Business	Percent
Reinsurance	30
Casualty	26
Property	22
Motor	7
Marine	7
Energy	5
Aviation	3

Source: Lloyd's (2005b), p. 12.

19.4.2 Capital Provision

Capital providers to Lloyd's are traditionally known as members. They accept insurance business through syndicates on a several basis for their own profit or loss (in other words, members of Lloyd's are not jointly responsible for each other's losses). The membership of Lloyd's is currently made up of companies that operate with limited liability and individuals who continue to operate with unlimited liability. Individual members tend to support a number of syndicates, whereas some corporate members only underwrite through a single syndicate. Figure 19.2 shows Lloyd's capital provision for 2005.

In recent years, corporate membership has grown rapidly; currently, around 90 percent of Lloyd's market capacity is provided by corporate, limited liability capital (Lloyd's 2005b). American, British, and Bermudian insurance companies provide the largest tranches of capital, with the remainder coming from other overseas companies and individual Names that have converted to limited liability by forming limited companies (Lloyd's Member Services Unit 2005). It is no longer possible for new members to enter Lloyd's on an unlimited liability basis; thus, all new members of Lloyd's do so on a limited liability basis.

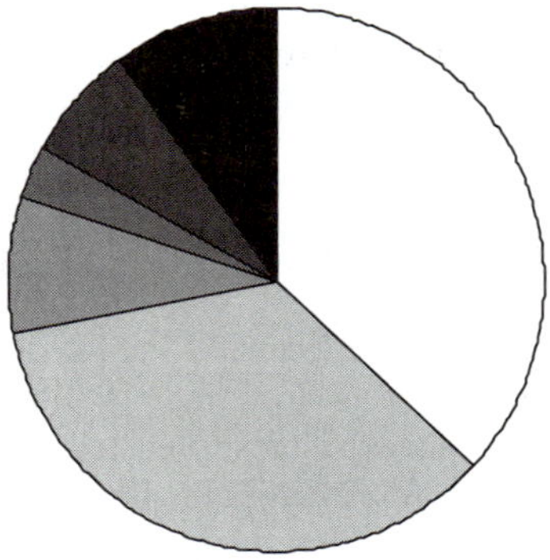

Source: Lloyd's (2005b), p. 9.

Figure 19.2. Lloyd's Capital Provision 2005 (Percent of total)

Each member of Lloyd's is required to demonstrate its ability to support its total Lloyd's underwriting business by holding sufficient capital. In particular, members are required to hold funds in trust at Lloyd's as security for the policyholders and to support a member's overall underwriting activities. The amount is related to a member's premium income limit, in accordance with the ratios laid down by the Council of Lloyd's. The administration of these funds is subject to the various trust deeds and regulatory requirements. The funds must be in a form approved by the Council of Lloyd's and be maintained in value. The amount of funds required from members will vary, depending on the perceived level of risk in the business that they underwrite. Thus Lloyd's operates what is known as a "risk-based capital," or RBC, system. In the past, Lloyd's RBC system was a rule-driven system based on market average performance. Implementation of the new Individual Capital Assessments regime at Lloyd's will enable Lloyd's to align capital with risk in a more efficient manner.

19.4.3 Intermediaries

The Lloyd's market is also made up of Lloyd's brokers. A Lloyd's broker is a partnership or corporate body permitted by the Council of Lloyd's to bring insurance business to Lloyd's on behalf of its clients. In order to become a Lloyd's broker, the intermediary must go through an application and approval process, with criteria based upon levels of professionalism, integrity, and customer service. Lloyd's positively welcomes applications from brokers, in the United Kingdom or overseas, who want direct access to the Lloyd's market.

The number of Lloyd's brokers has remained relatively constant over the past 50 years, and there are currently over 160 firms of Lloyd's brokers working with the Lloyd's market (Lloyd's 2005a). The brokers act on behalf of clients who are seeking to cover their risks by arranging insurance with underwriters. The brokers have a thorough understanding of the Lloyd's market, and, like the underwriters, many specialize in finding insurance for particular risks. Brokers use their specialist knowledge of the insurance available in the market to choose the best syndicate and negotiate competitive prices for their clients. Brokers add value to the underwriting process for their clients by representing their interests, achieving the best value for money, ensuring the quality of the coverage offered meets the client's need, and doing the best to find the most competitive price.

Sometimes Lloyd's managing agents develop particular working arrangements with intermediaries whereby brokers are given delegated authority to underwrite locally on their behalf. Such arrangements usually apply only for particular categories of risk up to a certain size, and in practice tend to involve personal lines and smaller commercial lines policies. Managing agents may also establish service companies in particular locations in order to improve the sourcing and servicing of local business.

19.4.4 Franchisor

The Corporation of Lloyd's, also known as the Franchisor, provides the franchise or platform from which Lloyd's businesses can operate. An administrative body of the Lloyd's market, the Corporation itself does not provide insurance. Rather, it provides the infrastructure for underwriters and brokers to do business and to support the market in its commercial aims. Where authority is delegated by the FSA, the Corporation also acts in a supervisory capacity to ensure that the market operates efficiently and effectively and that those involved meet certain standards.

19.4.5 Infrastructure

In recent years, the Lloyd's market has begun to invest in modern infrastructure systems for processing the millions of risk its deals with every year. Certain classes of insurance, including fine art and cargo, can now be bought on-line from Lloyd's syndicates. Lloyd's own web site, www.lloyds.com, allows insurance professionals from all over the world to find information on the market, underwriters, and brokers. More recently, Lloyd's has invested heavily in a new technology platform, Kinnect, which is designed to transform the traditional risk placement process and make processing of business more efficient. In addition, Lloyd's has been leading the call for change in the London market in respect of improving contract certainty and standards of business processing generally, with several important initiatives currently underway. Lloyd's and the IUA cooperate on such business processing reform initiatives, and the latest developments are available from their respective web sites.

19.4.6 Financial Information

Lloyd's derives considerable strength from being a market, but the manner in which it structures its finances and conducts its businesses can be difficult to understand if compared to a standard insurance company. In the past, financial analysts have commented that Lloyd's tends to outperform the rest of the insurance industry over the course of the insurance cycle (HSBC 2000).

All Lloyd's syndicates benefit from Lloyd's central resources, including the Lloyd's brand, its network of global licences, and the Central Fund. One of the unique competitive strengths of Lloyd's, the funds that make up the Central Fund are derived from a small percentage of each syndicate's annual income. The Central Fund is designed to operate as a fund of last resort at the discretion of the Council of Lloyd's, to meet any valid claim that cannot be met by the resources of any member syndicate. All Lloyd's policies are ultimately backed by this common security, and it was worth $1.1 billion dollars at the end of 2004 (Lloyd's 2004b). It is impossible to say that the central resources will never be exposed because of poor underwriting, but the operating environment is designed specifically to avoid such problems. Since Reconstruction & Renewal, systems have been enhanced dramatically, enabling Lloyd's to monitor risk at both the syndicate and market levels. According to Standard and Poor's (2004b), "management is very strong and independently minded and the lessons of the recent past have been well learned."

For 2004, Lloyd's recorded $2.6 billion in reported annual profit. This is more than 2002's reported profit of $1.6 billion but less than 2003's annual reported profit of $3.6 billion—largely as a result of Florida hurricane losses (Lloyd's 2004b). Lloyd's combined ratio of 96.9 percent for 2004 compares favorably with an estimated average of 98.7 percent for U.S. property and casualty insurers, 106.1 percent for U.S. reinsurers, and 98.2 percent for European insurers and reinsurers. Financial ratings agencies Standard & Poor's and A.M. Best currently rate Lloyd's A (Strong) and A (Excellent), respectively.

19.4.7 The September 11, 2001 Terrorist Attacks

The terrorist attacks of September 11, 2001, created the largest-ever loss for Lloyd's to date. Lloyd's loss proved to be larger than that of any other insurer or reinsurer, highlighting its role as an integral part of the U.S. market. The majority of the loss was caused by property and aviation claims, and over 60 percent related to coverage granted by Lloyd's syndicates under reinsurance treaties. In fact, 10 U.S. insurance groups account for over 50 percent of Lloyd's reinsurance exposure to the event (Lloyd's 2004a). Lloyd's has paid out around $5 billion in claims for the tragedy of 9/11 to date (Lloyd's 2004a). Despite the magnitude of its exposure, the Lloyd's market has not only survived, but has emerged strongly. Its ability to withstand the event is testament to the diversity of its capital base and to its financial security. At the same time, Lloyd's franchise system aims to improve the market's financial performance over the longer term, by generating more consistent results and long-term profitability.

19.5 LLOYD'S GLOBAL IMPACT

Although it is synonymous with the City of London, Lloyd's is a truly global organization, attracting businesses and capital from around the world (see Table 19.3). More than two-thirds of its business comes from overseas and the Lloyd's market conducts business in more than 200 countries and territories and has offices and staff in over 25 countries (Lloyd's Worldwide Markets 2005).

Table 19.3. Lloyd's Business Geographic Split, 2004

Region	Percent of Total
United States	34
United Kingdom	24
Europe	18
Americas	11
Asia/Africa	7
Rest	6

Source: Lloyd's (2005b), p. 12.

Of course, since Lloyd's began, entire national economies and insurance industries have grown up in the developing world, and new types of insurance have been pioneered, many by Lloyd's itself. Today, Lloyd's underwrites a huge range of international business risks and projects, including oil rigs, underground transport networks, airlines, and the world's top four manufacturers of personal computers. Over 94 percent of the companies listed on the main U.K. stock exchange, the FTSE 100, have policies at Lloyd's, and insurance capital from all over the world invest in, and operate businesses at, Lloyd's (Lloyd's 2004c). These include, among others, well-known insurers and reinsurers such as Munich Re (Germany), Mitsui (Japan), AIG (United States), and ACE (Bermuda) (Lloyd's Commercial Directorate and Member Services Unit 2005).

19.5.1 Lloyd's in Europe

Today, Lloyd's provides coverage for a wide range of European industrial risks outside of the United Kingdom. In reinsurance, too, Lloyd's plays an important role, supporting the balance sheets of many of Europe's largest insurance companies. The continental Western European market represents 23 percent of global non-life premiums, making it the second largest in the world; it accounted for 18 percent of Lloyd's total business in 2004 (AXCO and Lloyd's Market Reporting 2004). Overall, Europe's risk environment is changing rapidly, and the Lloyd's market is serious about playing a leading role in its future.

19.5.2 Lloyd's in Asia

Asia has also experienced tremendous economic growth, which is impacting the insurance industry lately and is currently especially strong in Northeast Asia—although the Asian economy is becoming increasingly integrated, and the benefits of economic success are therefore being felt throughout the whole region.

This reflects the gradual change in the global economy with a definitive shift in the economic center of gravity toward the east. Many of the 62 syndicates in the Lloyd's market have particular expertise and knowledge in the Asian market. The market's first purely Asian syndicate was established in 2005, writing marine and associated risks. The Lloyd's market provides insurance companies in Asia with access to a unique pool of underwriting expertise and talent. Lloyd's is able to support the development of insurance markets in the region by providing technical knowledge, specialist expertise, and insurance education to the market. Lloyd's is also a firm supporter of liberalizing markets in the region, and as China and India open up in particular, Lloyd's believes that it will have considerable expertise to share with these markets.

19.5.3 Lloyd's in North America

However, Lloyd's modern character is undoubtedly founded upon a long history of innovation in North America. Lloyd's underwriters have been writing business in the United States for more than 150 years. By 1840, the United States was already an important source of marine business for Lloyd's underwriters. By 1900, with the aid of an array of attractive new products, Lloyd's had become a major force in non-marine insurance as well. In the early years of the twentieth century, Lloyd's supported the rapid development of the U.S. economy by underwriting a wide range of risks, both directly and as a provider of reinsurance in support of domestic insurance companies. In these years, the most popular line of business offered by Lloyd's was fire insurance. Many of the companies that purchased fire coverage were also attracted to an additional form of protection recently pioneered at Lloyd's: loss of profits, or consequential loss insurance. But it was the 1906 San Francisco earthquake and its aftermath that did the most to establish Lloyd's reputation in the United States. The earthquake and fire that followed it destroyed 28,000 homes and cost insurers $180 million (Lloyd's 1998). Lloyd's, unlike some domestic insurance companies, paid claims swiftly and without dispute. But just as important in the long run was the way Lloyd's underwriters responded to the opportunities the disaster created. Other Lloyd's innovations also proved attractive to North American businesses. Jewelers protected their stocks with jewelers' block insurance. Bankers covered their securities and liabilities with bankers' blanket bond policies. A new generation of art collectors also sought coverage for their precious acquisitions at Lloyd's. The bond between U.S. insurance coverage and Lloyd's was strengthening at a rapid pace.

Today, Lloyd's role in the world's largest insurance market is greater than ever with more than a third of Lloyd's business generated in the United States. Approximately 93 percent of the companies that make up the Dow Jones Industrial Average have policies with Lloyd's. Lloyd's business underwritten in the U.S. insurance market totaled approximately $8.3 billion in 2003, followed by the United Kingdom, Canada, and Bermuda. Hence, Lloyd's continues to shape the agenda for the North American markets (Lloyd's 2004c).

19.6 CURRENT CHALLENGES FOR THE U.K. INSURANCE INDUSTRY

19.6.1 Managing the Insurance Cycle

It is well known that underwriting results in the non-life insurance market have tended to follow a cyclical pattern over time, with a wide variation in profitability from year to year (see Table 19.4). This pattern, known as the "insurance cycle" or "underwriting cycle," creates uncertainty for both shareholders and capital providers in the insurance industry (who expect a stable and adequate return on investment) and for clients (who see both availability and pricing of insurance change dramatically over the cycle).

Table 19.4. Lloyd's Market Underwriting Capacity, 1993 to 2005

Year	US$ bn
1995	19.6
1996	19.2
1997	19.8
1998	19.6
1999	19.0
2000	19.4
2001	21.7
2002	23.4
2003	27.6
2004	28.8
2005	26.3

Source: Lloyd's (2005b), p. 8.

Taking the U.S. property-casualty industry as a measure, the insurance industry has recorded an underwriting profit in only one of the past 25 years, which was 2004 (Insurance Information Institute 2005). Traditionally, insurers have tried to compensate for poor underwriting performance with investment income derived from investing their assets on the stock markets. However, given current global economic conditions that are generating sustained lower investment returns, this has become much more difficult, and Lloyd's believes that it is imperative for insurers to focus on achieving a profit from their underwriting activities alone. Given a simultaneous apparent upward trend in insurance losses in recent years, Lloyd's is convinced that the need for underwriters to manage the insurance cycle is now greater than ever.

As A.M. Best stated in its recent report on the 2005 European renewal season, "Forthcoming renewals throughout 2005 and the first quarter of 2006 will be decisive" (A.M. Best 2005a). Performance management therefore remains one of Lloyd's key themes for the future, because it is from here on that the pressure is likely to build and test underwriters' and management's resolve. In 2005, Lloyd's paid particular attention to direct and facultative property, aviation, energy, non-U.S. professional indemnity lines where it has seen the strongest signs of softening in terms and conditions.

Of course, it is important not to generalize. At this point in the cycle, the reinsurance market has exhibited more underwriting discipline than the direct market, which is encouraging, and any softening is often from a high level. However, barring any major catastrophes, underwriters must expect further pressure on pricing and conditions in the short to medium term. Reinsurers still believe there is a significant margin in the business. Competition will continue to be high. The only way to counter this pressure is for the market—and the reinsurers in particular—to remain disciplined and restrained. At Lloyd's, the 2005 business planning cycle proved its determination to take action where it is needed. Few of the plans submitted by the businesses in the Lloyd's market were approved without some degree of challenge or further consultation with the Franchise Performance team, and the fact that Lloyd's market's underwriting capacity is down 9 percent in 2005 reflects the market's disciplined approach to current market conditions (Lloyd's 2005c). This approach will continue in coming years; Lloyd's focus is on working with the businesses in the market, to ensure that they have the tools and robust processes in place which they need in order to manage the cycle, and to expose and reverse underperformance.

19.6.2 Business Processing

Lloyd's also believes that, within the commercial insurance market, the processing of insurance business needs radical improvement. Recent events in the global market have proved that improved contract certainty, including final policy wordings fully agreed between all the participants, is critical to the future of the insurance industry. For this reason, Lloyd's is working (often in conjunction with the rest of the London market) to deliver a program of business processing reform.

Lloyd's has mandated a standard form of slip for recording deals done in the London market, for the first time in over three centuries. Used properly, it helps ensure that important features of the contract are documented in a more uniform way and are clear from the start. The use of this slip by market businesses is improving month by month.

Lloyd's also recognizes that, in order to improve standards, brokers and underwriters must exchange reliable data and agree on contract details in a secure and transparent way. Lloyd's is therefore sponsoring the Kinnect electronic platform to drive this change using efficient modern technology. The insurance industry owes it to its customers as well as to itself to ensure that coverage is fully agreed and clearly documented right from the start.

19.6.3 Rising Catastrophe Trends

Finally, recent catastrophe events will intensify the debate on global warming, with the rate of natural disasters expected to grow further as temperatures rise. In recent years, a number of important and credible studies have raised a question about the link between global warming and the rise in natural disasters around the world. Research shows that the 10 hottest years on record have all occurred since 1990, and while scientists say it is not yet possible to link recent record storm seasons directly to global warming, the debate is increasing. From an insurer's perspective, the facts are certainly dramatic enough: 2004 was the most expensive year for natural catastrophes ever, with around $50 billion of insured losses, driven by the unprecedented string of hurricanes that pounded the United States and Caribbean, and the worst typhoon season ever recorded in Japan.[351]

Looking forward, the signs are that things are going to get even worse. The Intergovernmental Panel on Climate Change, a group including hundreds of leading scientists from around the world, says that extreme events will increase further. As a result of climate change, insurance markets may well be forced to look again at how they model, plan for, and manage weather related losses. Global warming is undoubtedly a debate in which the insurance industry must become more involved as its contribution has the potential to make a significant impact to future business performance.[352]

19.7 CONCLUSION

As one of the world's largest and most sophisticated insurance markets, the United Kingdom makes a significant contribution to the global financial services sector. In particular, the international focus of the London market means that the U.K. insurance industry plays a pivotal role in helping the global economy manage its business risk. All business involves risk, and as the economy becomes ever more complex and interlinked, business owners will increasingly need access to appropriate, internationally available insurance solutions to help them manage it. At the heart of the London market, with a 300-year history of paying valid claims and an innovative approach to risk, is Lloyd's. Unlike any other insurer or reinsurer, Lloyd's is a market and not a company. Its underwriters interact daily with intermediaries in London and around the world on some of the largest, most complex, and most difficult international risks. With a strong recent financial performance and commercially focused structure, Lloyd's plays a strong and leading role in the development of the London and global markets. The current challenges facing the U.K. industry are managing the insurance cycle, reforming the industry's business processing, and engaging in the debate on global warming.

---

[351] Swiss Re (2004), plus estimate of $10 billion for tsunami insured losses.

[352] A useful insurance industry study on the subject is "Opportunities and Risks of Climate Change," published by Swiss Re and available on its web site.

19.8 REFERENCES

A.M. Best Company, 2005a, Report on the 2005 Renewal Season, February.

A.M. Best Company, 2005b, *A.M. Best Special Report: Lloyd's*, September.

Asia Insurance Review, 2004, roundtable with Lloyd's Marine Underwriters, July.

Association of British Insurers, 2005.

AXCO Insurance Information Services, 2005, *AXCO Global Statistics, UK Non-Life Insurance Market Report 2004*.

Corporation of London, 2003, presentation to Lloyd's.

Financial Services Authority, 2005.

Herve-Bazin, Eric, 1994, Survey of AXA for FFSA (French Insurance Association).

HSBC, 2000, Destroying the Myths.

Insurance Day, 2005.

Insurance Information Institute, 2005, www.iii.org.

International Financial Services London, 2004, *City Business Series 2004 Insurance*.

International Underwriting Association, 2004, "IUA Statistics Show Market Fundamentally in Good Health," news release, April.

Jim Bannister Limited, 1999, The London Insurance Market and Its Future.

Lloyd's, 2005a, Admissions.

Lloyd's, 2005b, *Close-Up Lloyd's Review 2005*.

Lloyd's, 2005c, Franchise Performance Directorate.

Lloyd's, 2005d, *Lloyd's 2004 Global Results*, April.

Lloyd's Worldwide Markets, 2005.

Lloyd's, 2004a, Claims, Reinsurance & Open Years.

Lloyd's, 2004b, Global Report & Accounts.

Lloyd's, 2004c, *A Look at Lloyd's*, September.

Lloyd's America Inc., 1998, *New Risks, New Frontiers—Lloyd's in North America.*

Lloyd's Member Services Unit, 2005.

Standard & Poor's, 2004a, Global Reinsurance Highlights.

Standard & Poor's, 2004b, Rating of the Lloyd's Market.

Lloyd's, 2001, *Statistics Relating to Lloyd's 2001*.

Swiss Re, 2004, Sigma, report on London market, 2002.

Tolle, Rolf, 2005, presentation to IUMI conference, September.

19.9 USEFUL WEB SITES

- **www.abi.org.uk,** Association of British Insurers
- **www.airmic.com,** British Risk Managers Association
- **www.biba.org.uk,** British Insurance Brokers Association

- **www.cbi.org.uk,** Confederation of British Industry—the voice of business
- **www.cii.co.uk,** Chartered Institute of Insurers
- **www.ifsl.org,** International Financial Services London—a trade association
- **www.iii.org,** Insurance Information Institute—U.S. information resource
- **www.insuranceday.com,** *Insurance Day*—daily insurance newspaper
- **www.iua.co.uk,** Association of International Underwriters of London—London market
- **www.lloyds.com,** Lloyd's insurance market
- **www.swissre.com,** Swiss Re—reinsurance company

19.10 LEXICON

ABI: Association of British Insurers.

Capacity: A measure of company or market size. The ability to underwrite insurance premiums based on capital provided.

Central Fund: A fund of last resort and the final part of Lloyd's chain of security for policyholders.

Combined ratio: A measure of an insurer's underwriting profitability based on the ratio of the net incurred claims plus net operating expenses to net earned premiums. A combined ratio of 100 percent represents the break-even point. A ratio of over 100 percent represents an underwriting loss, while a ratio of under 100 percent represents an underwriting profit.

Commercial property insurance: Property insurance for businesses rather than homeowners.

Council of Lloyd's: The governing body for the Lloyd's market.

Cyber-liability: Liability specifically created by the advent of e-commerce and the Internet.

Domestic property insurance: Insurance for homeowners rather than businesses.

Equitas: The corporate entity into which the general business insurance liabilities of Lloyd's Syndicates allocated to the 1992 and prior years of account have been reinsured.

EU: European Union.

Franchisee (of Lloyd's): Managing agents that manage syndicates.

Franchisor (of Lloyd's): The Corporation of Lloyd's, which manages the Lloyd's Franchise on behalf of managing agents, who are the Franchisees.

FSA: Financial Services Authority, the U.K. regulator of the insurance industry.

Hardening: The increase in insurance pricing and tightening of rates and terms that takes place as part of the insurance cycle.

ICAs: Individual Capital Assessments. Method of calculating risk-based capital requirements.

Insurance cycle: Underwriting results in the non-life insurance market are notoriously inconsistent and have tended to follow a cyclical pattern over time, with a wide variation in profitability from year to year. This pattern is known as the "insurance cycle" or "underwriting cycle."

Insurance penetration: The ratio of premiums collected to gross domestic product expressed as a percentage.

IUA: International Underwriting Association of London, also known as the London company market.

Long-term market: Includes life and annuities insurance; individual, group, and occupational pensions; and income protection insurance.

MAT: Marine, aviation, and transportation business.

Members (of Lloyd's): Capital providers.

Motor: Known as automobile in U.S. English.

Non-life market: Also known as general insurance. Excludes all long-term insurance.

P&I club: Protection and Indemnity Club; a mutual insurer usually focused on marine liability business.

R&R: Lloyd's Reconstruction & Renewal, which took place in 1996 to secure the market's future following unprecedented losses.

Softening: The decrease in insurance pricing and relaxation of rates and terms that takes place as part of the insurance cycle.

Subscription market: Large risks may be covered by more than one insurer within the London market. The process by which the risk is shared among syndicates and/or company market insurers is called subscription.

Syndicate: A group of underwriting members at Lloyd's.

20 An Overview of the Alternative Risk Transfer Market

Robert P. Hartwig
Claire Wilkinson

Insurance Information Institute

20.1 WHAT IS ALTERNATIVE RISK TRANSFER?

The concept of alternative risk transfer (ART) defies a precise definition. One reason for this is that the range of risk products that can reasonably be defined as ART has expanded over time as product innovation continues. ART is not one product, but rather a way of doing business. It is generally accepted that there are two segments in the ART market—risk transfer through alternative carriers and risk transfer through alternative products (Swiss Re 2003). The market for alternative carriers (i.e., risk-bearers) consists of self-insurance, captives, risk retention groups, and pools. Alternative products include finite risk reinsurance, runoff solutions, committed capital, multiline, multiyear products, multitrigger programs, structured finance and new asset solutions, and capital market solutions for weather risk.

Despite the fact that ART is difficult to define, virtually all alternative risk transfer products have at least one of the following features:

- custom-tailored to the unique needs of the client;
- coverage provided on a multiyear basis;
- coverage applicable to multiple lines; or
- payoff can be triggered by multiple factors, rather than a single event.

Most alternative risk "solutions" require the skills of an interdisciplinary team of insurance and financial professionals. The more complex the solution, the broader the set of skills required. For example, structuring a transaction designed to transfer a unique catastrophe risk would require input not only from experienced insurance and reinsurance professionals, but also from catastrophe modelers, capital-market experts, and accountants, as well as tax and legal experts. On the other hand, formation of captives, a form of self-insurance, is easy even for relatively small businesses, and can be accomplished quickly with nominal expenses by soliciting the services of an experienced captive manager.

One characteristic of alternative risk solutions for larger-scale exposures is their reliance on nontraditional sources of capital. In other words, while traditional insurance contracts involve the transfer of risk from the policyholder to the insurer (or insurer to reinsurer), ART often seeks to lay off risk to the capital markets, rather than relying solely on the claims-paying capital of the insurer (known as policyholder surplus in the United States). The capacity of global capital markets to absorb risk dwarfs that of the world's insurers and reinsurers, a point that makes ART an attractive solution for large, unique problems.

Alternative risk financing and risk transfer is also proving attractive to the life insurance industry. The first U.S. insurance securitizations took place in 1988 and involved sales of rights to emerging profits from blocks of life insurance policies and annuities (Cowley and Cummins 2005). Life insurers, like property-casualty insurers, are increasingly looking to these opportunities as a way to raise additional capital and transfer risk. To date, most insurance-linked securities issues have been either catastrophe bonds or life bonds. Life bonds are typically based on the income flow of premiums from life and savings products and tend to provide capital to the life insurer (Swiss Re 2003). In this way, life bonds have more of a focus on financing than catastrophe bonds and are more akin to the securitization of income flows and asset-backed securities, such as mortgage-backed securities. Captives are also increasingly being used to fund life insurance exposures, such as employee benefit risks. In the United States, the Labor Department must approve any move to transfer employee benefit risks to a captive insurer.

The complexity of some categories of alternative risk solutions means that they are both time-consuming and expensive. These two issues have proven to be formidable obstacles to the growth in some segments of the ART market.

20.2 MARKET OVERVIEW

The beginnings of the alternative risk transfer market can be traced back to transactions that enabled businesses to insure their own risks. Traditionally, businesses have managed risk by transferring it to an insurance company through the purchase of an insurance policy or, alternatively, by retaining the risk and setting aside funds to meet expected losses through an arrangement known as "self-insurance." Most businesses, in fact, use a combination of traditional and self-insurance. The distinguishing feature of many corporate risk management programs is the degree to which potential losses are retained. For instance, a company may opt for a modest retention (deductible) if it is particularly risk averse. If the business has sufficient capital, it may decide to establish a special self-insurance fund to cover a substantial first layer, or amount, of any loss, with insurance coverage to pay for losses in excess of that layer.

Over the years, a number of alternatives to traditional commercial insurance have sprung up to respond to fluctuations in the marketplace. These entities include *captives*—a special type of insurance company set up by a parent company, trade association, or group of companies to insure the risks of its owner or owners. The oldest captive insurance domicile is Bermuda (see later section on captives). Other options include *risk-retention groups* (RRGs)—in which entities in a common industry join together to provide members with liability insurance coverage—and *risk-purchasing groups* (RPGs)—which were introduced in the 1980s. Like RRGs, risk-purchasing groups must be made up of persons or entities with like exposures and in a common business. However, whereas RRGs are essentially liability insurance companies owned by their members, purchasing groups buy liability coverage for their members from admitted insurers, surplus lines carriers, or RRGs.

During the 1980s, a combination of rising liability claims and falling premiums inevitably resulted in dwindling profits and ultimately losses for liability insurers. The situation reached a crisis level as the severe losses on liability insurance led to sharply increased prices and reduced availability of insurance in 1985 and 1986. Many U.S. businesses began to have difficulty in obtaining some types of commercial insurance coverage. It was at this time that new mechanisms for transferring risk developed, facilitated by passage of the federal Product Liability Risk Retention Act of 1981 (P.L. No.97–45 1981). The 1981 law, which provided for the formation of risk-retention groups and purchasing groups in the areas of products and completed operations liability, was expanded by 1986 amendments to include all areas of commercial liability, except workers' compensation. The final act was known as the Liability Risk Retention Act of 1986. To facilitate their operation, the Act exempts both risk-retention and risk-purchasing groups from many of the state laws that normally apply to insurance organizations (see later section on RRGs and RPGs). The hardening of the commercial property insurance market since the September 11 terrorist attack has led to a push by some groups, including schools and real estate firms, to expand the scope of the Risk Retention Act to commercial property, surety, and commercial automobile lines.

Following Hurricane Andrew in 1992 and the Northridge Earthquake in 1994, insurers increasingly looked to the capital markets to diversify their risks and expand capacity. Insurance-linked securities, such as catastrophe bonds, developed at this time. Another new ART approach was finite reinsurance, which can be traced back to Bermuda and the formation of Centre Re (later known as Centre Solutions) by Steven Gluckstern and the late Michael Palm in 1988. These developments were a reflection of the growing convergence between the insurance and financial markets. The late 1990s also saw the introduction of various multiline, multiyear integrated insurance products.

20.3 WHY ALTERNATIVE RISK TRANSFER DEVELOPED

Alternative risk transfer solutions help to expand the set of possible insurable risks. They can also pick up where traditional insurance and reinsurance leave off or provide coverage where no traditional coverage is available. By tapping the vast resources of the capital markets, many times greater than those of insurers, ART solutions can also greatly increase capacity.

ART not only provides solutions for unique problems, it can also improve the situation of the risk bearer. The fact that most insurance risks (natural catastrophes, for example) are uncorrelated with traditional types of investment risk (such as changes in interest rates or fluctuations in stock prices) produces a favorable diversification effect on investor portfolios.

The demand for alternative risk transfer products should theoretically be quite high. The frequency and severity of catastrophic events appears to be on the rise, both in the United States and elsewhere in the world. In terms of insured losses, seven of the ten most expensive catastrophes in U.S. history have occurred within the past four years (see Table 20.1). This phenomenon is not isolated to the United States. Table 20.2 shows a distinct rise in the severity of catastrophes worldwide post–2000 and during the 1990s as compared with the 1980s.

Table 20.1. Top Ten Insured Property Losses in the United States (in billions of 2004 dollars)

Event (Year)	Insured Loss
Hurricane Jeanne (2004)	$3.7
Hurricane Rita (2005)*	$4.1
Hurricane Frances (2004)	$4.6
Hurricane Hugo (1989)	$6.4
Hurricane Ivan (2004)	$7.1
Hurricane Charley (2004)	$7.5
Northridge Earthquake (1994)	$15.9
September 11 Terrorist Attack (2001)**	$20.1
Hurricane Andrew (1992)	$20.8
Hurricane Katrina (2004)*	$35.0

Source: ISO/PCS; Insurance Information Institute.
* Estimate, stated in 2005 dollars, as of September 26, 2005.
** 9/11 loss figure is for property claims only.

Mega-catastrophes, such as Hurricane Andrew in 1992, the Northridge earthquake in 1994, and Hurricane Katrina in 2005, created severe financial strain as well as a number of insurer insolvencies. Alternative risk transfer solutions have played an important role since then by improving insurers' capacity to withstand catastrophic-risk losses, but not to the extent that most industry observers had anticipated. Hurricane Katrina, the Category 4 storm that struck the states of Louisiana, Mississippi, and Alabama in late August 2005, is now the most costly U.S. catastrophe ever and is expected to stimulate greater use of alternative risk transfer solutions.

Interest in alternative markets is in part a function of the cost of traditional insurance and reinsurance. Property-casualty insurance pricing is highly cyclical. Typically, several years of declining prices and relaxed underwriting is followed by a few years of rising rates and more restrictive underwriting. When prices are high and the traditional market hardens, alternative market products become relatively less expensive and therefore attractive to buyers as substitutes for traditional insurance products.

Table 20.2. Global Insured Catastrophe Losses 1970–2004 (in billions of 2004 dollars)

Year	Loss	Year	Loss
1970	$4.83	1988	$11.11
1971	$2.09	1989	$22.23
1972	$4.49	1990	$24.68
1973	$4.82	1991	$22.84
1974	$7.74	1992	$38.58
1975	$4.07	1993	$15.62
1976	$4.93	1994	$29.11
1977	$4.69	1995	$21.50
1978	$4.68	1996	$15.58
1979	$9.29	1997	$9.69
1980	$6.32	1998	$22.09
1981	$2.81	1999	$36.61
1982	$7.48	2000	$14.17
1983	$8.49	2001	$37.65
1984	$5.22	2002	$15.20
1985	$9.66	2003	$19.04
1986	$4.99	2004	$48.63
1987	$14.18		

Source: Swiss Re.

It was projected that by the end of 2003, 50 percent of the U.S. commercial market would migrate to the global alternative risk transfer market, up from about 40 percent in 2000 and up from about 30 percent in 1996 (A.M. Best Company 2003a). Other sources indicate that alternative risk transfer accounts for a somewhat smaller share of the commercial marketplace. Some 40 percent of the respondents to a 2003 survey of corporate risk management said they placed business in captives or risk-retention groups (Risk and Insurance Management Society 2004). In certain industries, such as telecommunications and information services, the percentage was higher, with some 60 percent of respondents saying they used captives.

Rising rates and a shortage of sufficient capacity in some commercial insurance lines, a trend that began in 2000 and intensified following the September 11 terrorist attacks, incented businesses to consider a number of alternative risk transfer vehicles, including captives and risk-retention groups. By 2003 and 2004, however, the hard market had passed its peak and this easing of conditions appears to have reduced interest in alternative markets. In 2004, 83 percent of the commercial brokers responding to a trade press survey said that some of their clients had formed a captive, down from 93 percent the previous year (Friedman 2004).

The alternative market in the United States also has been affected by the Terrorism Risk Insurance Act (TRIA), enacted in November 2002 to provide a federal backstop for terrorism insurance. The U.S. Treasury department has taken the position that domestic captive and risk-retention groups chartered in the United States or its territories are to be considered "insurers" under the Act, which requires commercial insurers to offer terrorism insurance coverage. This means that all domestic RRGs and captives, except those writing medical malpractice and other lines excluded by the Act, are required to offer terrorism coverage to their insureds and are subject to the law's 3 percent surcharge provision.[353] Some observers believe TRIA will prompt companies without significant terrorism exposure to locate their captives offshore to avoid the surcharge, on the one hand, and spur captive formations on the part of companies seeking to take advantage of the federal backstop, on the other.

20.4 GLOBAL DEVELOPMENTS

The global market for alternative carriers, consisting of self-insurance, captives, risk-retention groups, and pools, accounted for about $88 billion in premiums in 2001, with estimated growth rates of approximately 10 percent per year through 2005 (Swiss Re 2003) (see Table 20.3). Self-insurance (with $44 billion in gross premiums written) accounted for the largest ART segment in 2001, followed by captives ($38 billion), U.S. state pools ($5 billion), and risk-retention groups ($1 billion).

Table 20.3. Size of the Alternative Risk Transfer Market (Based on 2001 Direct Written Premiums)

Type	Size (in billions of dollars)	Market Share (%)
Traditional Carriers	$370	80.9%
Self-insurance and Other Alternative Carriers (U.S.)	$49	10.7%
Captives	$38	8.3%

Source: Swiss Re.

It should be noted that self-insurance, state pools, and risk-retention groups are largely U.S. concepts, whereas captives are a global phenomenon that are

[353] Under TRIA, if the federal government pays for insured losses during the course of a year, the Treasury Secretary is required to recoup the difference between total industry costs (individual insurers' losses up to their deductibles, plus the industry's 10 percent cost share above the deductibles) and $15 billion for 2005. The recoupment is accomplished through a surcharge on all policyholders that cannot be more than 3 percent of the premium charged for a policy in a given year.

increasingly used by non-U.S. corporations.[354] While more than 50 percent of the 4,000-plus captives worldwide are registered in Bermuda, the Cayman Islands, and onshore United States, other popular captive domiciles include Guernsey, Dublin, Luxembourg, and the Isle of Man, which tend to service the European markets. At the end of 2004, Guernsey, Dublin, Luxembourg, and the Isle of Man had an estimated 1,018 captives between them ("Spotlight" 2005).

Capital market solutions, such as catastrophe bonds, appear to be driven by the United States in terms of peril and location. While market participants continue to explore the possible securitization of additional perils and geographies, the dominant risks continue to be U.S. earthquakes and U.S. hurricanes, followed by Japanese earthquakes and European windstorms (Guy Carpenter 2005). However, some of the bonds cover multiple perils, which may contain a blend of U.S. and non-U.S. risks.

20.5 TYPES OF ALTERNATIVE RISK TRANSFER

20.5.1 Captives

A captive is an insurer or reinsurer owned by a corporation or an association of businesses for which the primary business is not insurance. The principal distinguishing feature of a captive is that the insured (generally the captive's owner) exercises direct control over the insurer (captive). Under traditional arrangements, the insured (a policyholder of the company transferring risk) has no influence over the operation of the insurer. In this sense, captives may be viewed as a form of self-insurance.

Captives may be owned by one entity or several and they may insure the risks of organizations other than their major owners. Wholly owned captives are companies established by large corporations to finance or administer their risk financing needs. If such a captive insures only the risks of its parent or subsidiaries, it is called a "pure" captive.

Captives are the oldest form of an alternative risk transfer vehicle, dating back to the 1950s. In 2004, there were more than 4,000 captives spread across the globe, accounting for approximately 8.3 percent ($38 billion) of the world's commercial insurance market, based on 2001 direct premiums written. Large corporations account for 80 percent of the captive market.

The number of captives in the top 10 captive domiciles increased from 3,992 in 2003 to 4,157 in 2004 ("Spotlight" 2005) (see Table 20.4). Bermuda, with 1,150 captives, tops the list, followed by the Cayman Islands and Vermont, with 694 and 524 captives, respectively. Guernsey, Luxembourg, and Barbados are also significant centers for captives. In May 2003, Anguilla, a British territory, introduced an insurance act that would establish the Caribbean island as a captive insurance domicile. In August 2004, Bahrain licensed its first captive insurer. A list of the five domiciles with the largest net increase in captives in 2004 is headed by the Cayman Islands (50 new captives), followed by South Carolina (47), the British Virgin Islands (43), Guernsey (25), and Hawaii (25) ("Spotlight" 2005).

[354] In the United States a wide range of pools, funds, and plans insure a variety of risks from workers' compensation, to windstorm, to high-risk auto.

Table 20.4. Top Ten Captive Domiciles: 2003 and 2004

	Number of Captives	
Domicile	2003	2004
Bermuda*	1,150	1,150
Caymans	644	694
Vermont	507	524
Guernsey	383	410
British Virgin Islands	307	350
Barbados	248	257
Luxembourg*	216	219
Dublin	205	214
Isle of Man	173	175
Turks and Caicos	159	164

Source: "Spotlight" 2005; Insurance Information Institute.
* "Spotlight" 2005 estimate.

U.S.-domiciled captives' net premiums written increased by 4.6 percent from 2002 to $8.9 billion in 2003 (A.M. Best 2004). Medical malpractice was the top product line in 2003, followed by auto physical damage and private passenger auto liability. While declining prices for commercial insurance could trigger a loss in captives' market share, many captives are expected to retain the commitment of their parent organizations. U.S.-domiciled captives' net premiums written and admitted assets rose by 45 percent and 29 percent, respectively, in the five-year period from 1999 to 2003 (A.M. Best 2004).

Bermuda

Bermuda remains the leading global captive domicile with 1,150 captives. Still, the growth of captive formations in Bermuda has slowed in recent years. New insurance incorporations in Bermuda, most of them captives, dropped from 89 in 2003 to 77 in 2004, with the decline attributed to increasing competition from U.S. domiciles. It is not known how many captives were liquidated in 2004, but the total number of Bermuda captives is holding steady, between 1,100 and 1,200 ("Spotlight" 2005). Offshore captives covering U.S. risks are predominantly located in Bermuda, where they enjoy tax advantages and less onerous regulation.

Fred Reiss, who is credited with coining the term "captive," established a management company in Bermuda in 1962 called International Risk Management Ltd. (Higginbottom 2002). The captive insurance market on the island developed slowly at first, but then gained momentum. Until the 1980s, the Bermuda market was almost entirely focused on captives, and many new concepts were developed to expand their use.

Cayman Islands

The Cayman Islands is the second leading global captive domicile and continues to see substantial growth, with 694 active captives at year-end 2004, writing about $6 billion in premiums and reporting assets of $22 billion. This was a net increase of 7 percent in 2004, taking into account the closure of 26 captives during the year. Most of the new captives in 2004 were health care-related. Business is spread among a diverse group of companies, and types of coverage range from workers' compensation and product liability to life and annuity business. Segregated portfolio companies, known elsewhere as segregated cell captives (described below), are gaining popularity, accounting for 13 of the new licenses in 2004, bringing the island's total to 93 at year-end. Among the benefits of establishing a captive in the Cayman Islands are the highly developed captive support infrastructure and professional yet flexible regulation.

Vermont

Vermont, with 524 captives in 2004, is the leading U.S. domicile for captives, followed by Hawaii and South Carolina, with 147 and 114 captives, respectively (see Table 20.5). While the state licensed 43 captives in 2004, it also saw 26 captives dissolve, giving it a net gain of 17 captives. It is estimated that gross written premiums amounted to more than $10 billion in 2004, up from $9.4 billion in 2003. Vermont's success as a captive domicile has been attributed to its flexible regulatory approach as well as its well-established infrastructure of management firms, specialist accounting and legal advisors, and knowledgeable bankers to assist captive owners in setting up and operating a captive insurance company (Moriarty and Davis 2003). Vermont also has its own captive insurance trade association, the Vermont Captive Insurance Association.

Table 20.5. Top Ten Captive Domiciles in the United States: 2003 and 2004

	Number of Captives	
Domicile	2003	2004
Vermont	507	524
Hawaii	122	147
South Carolina	67	114
District of Columbia	20	40
Arizona	18	39
Nevada	19	38
New York	18	28
Georgia	16	14
Montana	9	10
Colorado	11	10

Source: "Spotlight" 2005; Insurance Information Institute.

Captives may be established to provide insurance to more than one entity. An association or group of companies may band together to form a captive to provide insurance coverage. Professionals—physicians, lawyers, accountants—have formed many captives over the years. Captives may, in turn, use a variety of reinsurance mechanisms to provide the coverage. In particular, many offshore captives use a "fronting" insurer to provide the basic insurance policy. Fronting typically means that underwriting, claims, and administrative functions are handled in the United States by an experienced commercial insurance company, since a captive generally will not want to get involved directly in running the insurance operation. Also, fronting allows a company to show it has an insurance policy with a U.S.-licensed insurance company, which may be required for legal and business reasons.

Although a record 462 new captive insurance companies were formed in 2002, there were also a record number of captive liquidations, causing the overall number of captives to remain relatively flat, with 4,526 active captives at year-end 2002 compared with 4,521 at the end of 2001 (A.M. Best 2003b). The large number of liquidations (311 in 2002, up from 202 in 2001) has been attributed to corporate collapses during the difficult economic period of the early 2000s, as well as events such as the Enron and Tyco debacles in which captives failed along with their parents. There were also consolidations within the captive industry itself (see Table 20.6). In addition, a number of medical malpractice captives have fallen victim to the escalating loss costs and claims severity plaguing the traditional medical malpractice market. Growth of the captive market has also been impeded by a shortage of traditional insurers willing to provide underwriting, claims administration, and other so-called fronting services to captives.

Table 20.6. New Captives vs. Liquidated Captives: 1993 to 2002

Year	Number of New Captive Formations	Number of Liquidated Captives
1993	238	150
1994	289	113
1995	243	103
1996	290	102
1997	294	135
1998	305	154
1999	250	156
2000	245	170
2001	316	202
2002	462	311

Source: A.M. Best Company 2003b.

Rent-a-Captives

The "rent-a-captive" concept was introduced in Bermuda 20 years ago and remains a popular alternative market mechanism. Rent-a-captives serve businesses that are unable to capitalize a captive but are willing to assume a portion of their own risk and share in the underwriting profits and investment income. Generally sponsored by insurers or reinsurers, which essentially "rent out" their capital for a fee, the mechanism allows users to obtain some of the advantages of a captive without having the expense of setting up a single parent captive and meeting minimum capital and surplus requirements.

Segregated/Protected Cell Captives

A variation on rent-a-captives, the segregated or protected cell captive (PCC), was introduced in Guernsey in 1997. A PCC offers participants many of the benefits of a group captive but with lower startup costs. A PCC offers more security to policyholders by isolating each participant's assets and liabilities as if they were a separate company, called a cell, doing business with the core company. The mechanism has helped fuel the growth of the captive market. Actual numbers of segregated cells may be underreported because some captive domiciles don't report the number of cells within them. Such mechanisms accounted for some 10 percent of new captive formations in 2002 (A.M. Best 2003b). Overall, the number of segregated cell companies increased 20 percent in 2002, with Guernsey and the Cayman Islands recording a combined growth of 45.8 percent.

Tax Status and Captives

Since 2001 a number of regulatory rulings have favored captives. In June 2001, the Internal Revenue Service (IRS) ruled that it would allow premiums paid for captive insurance to be tax deductible, signaling it would no longer invoke its long-held "economic family" theory. This theory was first espoused by the IRS in 1977 in Revenue Ruling 77, which held that a policyholder, its noninsurance subsidiaries, and its captive form one "economic family," and that the premiums paid within the family are not deductible because the risk was not shifted from the policyholder.

In 1981, the IRS had used the economic family theory to deny the deductibility of insurance premiums paid by Carnation Company. The U.S. Tax Court upheld the IRS ruling that premiums paid to Carnation's Bermuda captive for underwriting Carnation's risk did not constitute insurance and therefore that Carnation could not claim a tax deduction.

However, the economic family line of reasoning was later rejected by courts. In 1992, in *Sears, Roebuck and Company v. Commissioner*, the tax court ruled that risk shifting existed because the insurance subsidiary of Sears wrote virtually all unrelated risks so premiums were therefore deductible.

In September 2001, captive owners won another victory when a federal appeals court overturned a controversial 1999 U.S. Tax Court decision that had labeled Overseas Partners Ltd., a Bermuda-based United Parcel Service of America Inc. captive reinsurance program, a fraud and a tax dodge. The ruling held that the program served a legitimate business purpose.

20.5.2 Self-insurance

Estimates of the self-insurance market vary widely, possibly as a result of definitional differences and an absence of uniform reporting requirements. The segment is estimated to be around $44 billion in gross premiums (Swiss Re 2003). The amount of self-insurance in force also fluctuates with the insurance cycle and in response to the operating environment for specific coverages—for example, directors and officers coverage. With the improvement in the workers' compensation environment in the mid-1990s, some of the many businesses that had been self-insuring their workers' compensation programs returned to the traditional marketplace.

Pricing and availability problems, fueled by the September 11 terrorist attacks, are once again making self-insurance groups (SIGs) an attractive option. Workers' compensation SIGs generally consist of firms in similar industries that pool resources to insure each other's exposure. As such, SIGs are functionally similar to mutual insurance companies. While SIGs are separate, legal, risk-bearing entities, they are formed by a trade association, common industry, or other related group for the purpose of self-insuring the risks of its members. SIGs provide the benefits of self-insurance to small to medium sized businesses that would not be able to self-insure individually. Members of the SIGs are joint and severally liable, that is, they are liable for the losses of other group members. SIGs can only operate in a single state, and 37 states have enacted legislation authorizing SIGs, sometimes referred to as trusts. There were 178 government pools and 240 self-insurance funds in 2002, up slightly from 177 government pools and 237 self-insurance funds in 2001 (A.M. Best 2003b). A government pool is a pooling arrangement formed by one or more governmental agencies and/or subdivisions for the purpose of self-insuring its risks.

The use of higher retentions/deductibles is increasing in most lines of insurance, especially for commercial casualty and liability coverages (see Table 20.7). In workers' compensation, many companies are opting to retain a larger portion of their exposure through policies with large deductible amounts of $100,000 or higher (see Table 20.8). Large deductible programs, which were first introduced in 1989, now account for a sizable portion of the market. In workers' compensation, large deductible programs accounted for 43 percent of the total workers' compensation market in 2003 and 33 percent in 2000, up from less than 1 percent in 1990 (National Council on Compensation Insurance 2005).

Table 20.7. Average Underlying Limits (in millions of dollars), 2000 to 2004

Year	Attachment Point
2000	$1.8
2001	$1.9
2002	$1.9
2003	$2.0
2004	$2.2

Source: Marsh 2004.

Self-insurance pools and trusts have financial results that compare favorably to captives and traditional commercial insurers. Overall, the five-year average combined ratio after policyholder dividends was 95 for the self-insurance pools in the study, compared with 116.9 for captives and 109 for a representative group of commercial insurers (A.M. Best 2004). The pools' success is attributed to their strict admission requirements and emphasis on loss control and antifraud activities. All the firms in the study met A.M. Best's rating requirements.

Table 20.8. How the Risk Dollar Is Spent (2004) (Firms with Revenues >$1 billion)

Budget Item	Percentage
Liability Premiums	27%
Property Premiums	22%
Workers' Compensation Premiums	16%
Administrative Costs	9%
Total Management Liability	8%
Retained Workers' Compensation	6%
Retained Property	4%
Total Professional Liability	4%
Retained Liability	3%
Other	1%

Source: Risk and Insurance Management Society 2004; Insurance Information Institute.

20.5.3 Risk-Retention Groups and Risk-Purchasing Groups

Risk-retention groups are a U.S. phenomenon, established in 1981 when Congress passed the Product Risk Retention Liability Act, now known as the Liability Risk Retention Act. A risk-retention group (RRG) is a corporation owned and operated by its members. It must be chartered and licensed as a liability insurance company under the laws of at least one state. The group can then write insurance in all other states. It need not obtain a license in a state other than its chartering states. In 2005, there were calls for state regulators to enact uniform regulatory standards for RRGs and for Congress to consider enacting corporate governance standards (Government Accountability Office 2005).

As of July 2005, there were 197 RRGs in the United States, up from 186 at year-end 2004. As of April 2005, Vermont and South Carolina were the most favored domiciles for RRGs, with 167 and 40 groups, respectively. The District of Columbia, which added six new groups during the six-month period from November 2004 to April 2005, moved into third place with 21 RRGs, overtaking Hawaii (18 RRGs).

Premiums written by risk-retention groups increased to $2.2 billion in 2004, up 26 percent from $1.7 billion in 2003 (Risk Retention Reporter 2004) (see Table 20.9). There were 54 new RRGs formed in 2004. The majority of the new groups were in five domiciles: South Carolina (14), Vermont (11), the District of Columbia (9), Arizona (8), and Nevada (7). The risk-retention group premium marketplace is dominated by three business segments: health care, professional services, and government and institutions (see Table 20.10). Thirty-six of the new groups were in the health care industry, a sector that includes doctors, nursing homes, and hospitals.

Table 20.9. Risk Retention Group Premiums, 1988 to 2004 (in millions of dollars)

Year	Premiums
1988	$250.2
1989	$358.4
1990	$419.3
1991	$493.6
1992	$493.7
1993	$527.2
1994	$585.8
1995	$575.5
1996	$707.6
1997	$751.9
1998	$790.5
1999	$875.3
2000	$775.5
2001	$944.0
2002	$1,265.1
2003	$1,737.7
2004	$2,156.5

Source: Risk Retention Reporter.

Table 20.10. Risk Retention Group Premium by Business Area

Business Area	Percentage
Health Care	49%
Professional Services	25%
Government and Institutions	10%
Property Development	7%
Manufacturing and Commerce	4%
Transportation	3%
Environmental	2%

Source: Risk Retention Reporter 2004.

RRGs are helping to fill gaps in the medical malpractice marketplace. Fourteen of the 36 health care RRGs provide liability for doctors. Premiums for RRGs in nursing homes are projected to increase 400 percent from $10.1 million in 2003 gross written premiums to $40.9 million in 2004.

Like risk-retention groups, risk-purchasing groups (RPGs) must be made up of persons or entities with like exposures and in a common business. However, as discussed previously, RRGs are liability insurance companies owned by their members, whereas risk-purchasing groups buy liability coverage for their members from admitted insurers, surplus lines carriers, or RRGs. Risk-purchasing groups are regulated by each state in which they operate. The insurance that these groups purchase on behalf of their members must meet the laws and regulations of the state designated as the domicile of that group. However, like RRGs, RPGs are exempted from certain elements of state law (Webel 2003). For example, laws in some states prohibit insurers from giving groups formed to purchase insurance advantages over individuals. Many states have laws known as "fictitious group laws" that specifically prohibit or limit groups from purchasing insurance for the members of the group, particularly if the group exists solely for the purchase of insurance. Purchasing groups are not subject to these fictitious group laws, however. RPGs are also exempted from countersignature laws, which are laws requiring a local broker's or agent's signature on an insurance contract. The Risk Retention Act of 1986 specifically provided for purchasing groups to be created to purchase liability insurance for members of the sponsoring groups.

The number of new risk-purchasing groups being formed in the United States has begun to rise, as commercial insurance coverage became more available and affordable following the hard market of the early 2000s. Eleven purchasing group formations and 14 retirements were recorded in the first quarter of 2005, compared with seven formations and 25 retirements during the same period in 2004 (Risk Retention Reporter 2005). There had been a dramatic drop in the number of formations after 2000, accompanied by a corresponding rise in retirements.

Four states with hospitable regulatory climates accounted for the majority of the risk-purchasing groups formed during the 18-year period from 1987 to September 2004: Texas (289), California (227), Illinois (183), and Delaware (102). When retirements are taken into account, the five states with the greatest number of operational purchasing groups, as of September 2004, were Illinois (110), Texas (80), Delaware (75), California (57), and New York (55). Purchasing groups are now domiciled in 44 states.

20.5.4 Catastrophe Bonds

Historically the capacity to finance catastrophic risk was limited by the claims-paying ability of insurers and reinsurers or by the insured's ability to retain (or pool) that risk. Today, the ability to securitize catastrophe risk unlocks the vast potential of the global capital markets (Lane 2002).

Investors in catastrophe bonds benefit not only from the relatively high yields offered on these securities, but also from the additional diversification that results from the fact that catastrophic risk is essentially uncorrelated with financial risk (e.g., fluctuations in interest rates or the stock market). Investors must balance the advantages of high yield and greater diversification with the risk assumed (Litzenberger, Beaglehole, and Reynolds 1996).

Insurers, reinsurers, and corporations interested in securitizing catastrophic risk usually enlist the aid of one or more intermediaries, including investment bankers, reinsurers, and brokers. The typical structure of a catastrophe-bond offering is displayed in Figure 20.1.

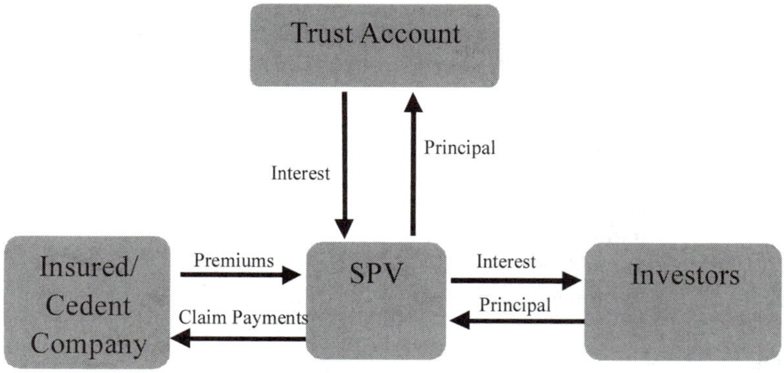

Figure 20.1. Structure of Typical Catastrophe Bond Issue

Funds raised from investors are used to establish a "special purpose vehicle" (SPV), which is similar to a captive. The SPV then issues a reinsurance policy to the insurer or corporation transferring (ceding) the risk, which pays a premium to the SPV. This formal reinsurance structure is necessary so that the transaction is formally recognized for tax and regulatory purposes.

Investors receive notes (bonds) from the SPV with an agreed-on coupon (interest payment) (Froot 1999). In 2004 to 2005, the fixed coupon rates paid to investors in catastrophe (cat) bond transactions ranged from a low of 1.9 percent to a high of 13.5 percent (Lane Financial 2005[AuQ5]). For example, the Hartford Financial Services Group's first cat bond issue, Foundation Re Ltd. (described in more detail later) is a two-tranche issue. The Class A tranche was issued with a spread of 4.10 percent over the London Inter-Bank Offer Rate (LIBOR), while under the Class B tranche, the investor is paid 1.95 percent over LIBOR regardless of whether there has been a triggering event. Depending on how the deal is structured, investors face the prospect of losing some or all of the investment income produced by the bond—and even some of the principal—in the event of a catastrophic loss. During the policy period, the capital raised from investors is held in trust in a conservative, highly liquid portfolio.

Tapping into the capital markets allowed insurers to diversify their risk and expand the amount of insurance available in catastrophe-prone areas. Catastrophe bonds developed in the wake of Hurricanes Andrew and Iniki in 1992 and the Northridge earthquake in 1994—mega-catastrophes that resulted in a global shortage of reinsurance (insurance for insurers) for such disasters (Swiss Re 2003).

Catastrophe bonds have been used to cover a wide variety of exposures. Earthquakes (both in the United States and Japan) and East Coast windstorms have accounted for the majority of bond issues to date. Bonds with European windstorm exposure also have been established. The issuance of bonds is becoming an increasingly popular option for transferring noncatastrophe risks as well. Examples include risks arising from the issuance of life insurance policies, acquisition costs, auto residual (i.e., post-lease) value, and mortgage loans. Power failures and sport events are among the risks covered by catastrophe bonds. The world football (soccer) federation, Fédération Internationale de Football Association (FIFA), worked with Credit Suisse First Boston and other banks to develop and issue $260 million in cancellation bonds in August 2003 covering the 2006 World Cup in Germany. The arrangement covers FIFA's losses in the event the World Cup competition cannot be completed (or rescheduled) due to a variety of events, including terrorism. Some bonds include a blend of uncorrelated risks, such as U.S. and European windstorm risk.

The market for natural catastrophe bonds continued its steady growth in 2004, with total announced outstanding risk capital reaching $4.04 billion at year-end 2004, representing a 17 percent increase over the year-end 2003 total of $3.45 billion and a 41 percent increase over the $2.86 billion outstanding at year-end 2002 (Guy Carpenter 2005).[355] These figures do not include other insurance securitization transactions such as weather derivatives and instruments that secure life mortality risks. For example, in April 2005 Swiss Re sponsored its second insurance-linked security related to life insurance to transfer mortality risk coverage to the capital markets.

A distinct class of investors who manage funds dedicated to investing capital in risk-linked securities, including cat bonds, has clearly emerged (Guy Carpenter 2005). In recent transactions, reinsurers account for a smaller share of the investor base, while dedicated cat bond investors, including hedge funds and large institutional investors, play an increasing role in pricing and structure. Investor demand for catastrophe bonds now outstrips the supply of bonds brought to market on an annual basis. Dedicated cat bond funds are estimated to now have capital under management exceeding $3 billion.

The catastrophe bond market saw total issuance of $1.14 billion in 2004, a decline from the record issuance of $1.73 billion in 2003 and a slight drop from issuance of $1.22 billion in 2002 (see Table 20.11). During 2004, a total of six transactions were completed by five separate sponsors, a slight decline from the seven issues in each of the three previous years. However, an increase in the number and volume of transactions issued on a private nondisclosed basis may be skewing the 2004 results downward. Since 1997, the first year that multiple transactions occurred, 59 cat bonds have been issued with total risk limits of $8.66 billion.

In terms of perils and geographies securitized, the catastrophe bond market held steady in 2004. While potential sponsors and other market participants continue to explore the possible securitization of additional perils and geographies, the dominant risks continued to be U.S. earthquakes, U.S. hurricanes, Japanese earthquakes, and European windstorms (see Tables 20.12 and 20.13). Notably in 2004, the Hartford Financial Services Group participated in the market for the first time (Guy Carpenter 2005). Its first cat bond issue, Foundation Re Ltd., is a two-tranche issue providing

---

[355] Total risk capital outstanding measures the total bond principal currently at risk in the market as of the relevant year-end, regardless of issuance year, and is distinct from total risk capital issued (which measures the incremental risk capital issued in a given year).

coverage for qualifying first-event Atlantic and Gulf Coast U.S. hurricanes as well as subsequent year coverage for qualifying U.S. hurricane or earthquake events.

Table 20.11. Catastrophe Bonds, Issuance (in millions of dollars) and Number of Deals, 1997 to 2004

Year	Issuance	Number of Deals
1997	$633.0	5
1998	$846.4	8
1999	$985.0	10
2000	$1,139.4	9
2001	$966.7	7
2002	$1,219.4	7
2003	$1,729.7	7
2004	$1,143.0	6

Source: Guy Carpenter 2005.

Table 20.12. Catastrophe Bond Market by Peril, 1997 to 2004

Peril	Percentage
U.S. Earthquake	31.1%
U.S. Hurricane	31.6%
European Windstorm	15.7%
Japanese Earthquake	15.4%
Japanese Typhoon	2.9%
Other	3.3%

Source: Guy Carpenter 2005.

Meanwhile, the occurrence of four consecutive hurricanes in Florida in 2004 forced insurers and reinsurers to refocus on event-specific catastrophe protection. Hurricanes Charley, Frances, Ivan, and Jeanne between them caused estimated insured losses of $22.8 billion. It is reported that potential cat bond sponsors have shown increased interest in structures that would provide coverage for cumulative event losses over a specified period (Guy Carpenter 2005). However, it remains to be seen whether this will lead to any significant structural innovations in 2005 and beyond. There were no reports of outstanding cat bonds being triggered by any of the Florida or Japanese windstorms that occurred in 2004.

Despite gains, the dollar value and number of catastrophe securitization transactions is still very modest, accounting for a small percentage of global reinsurance capacity. Between 1997 and 2004, 59 catastrophe bonds were issued with total issuance value of $13.4 billion, relative to $300 billion in reinsurance capacity (see Table 20.14).

Table 20.13. Securitization by Peril, 2000 and 2004

	Risk Capital (in millions of dollars)	
Peril	2000	2004
U.S. Earthquake	$486.5	$803.3
U.S. Hurricane	$506.5	$660.8
European Windstorm	$482.5	$220.3
Japanese Earthquake	$217.0	$310.8
Japanese Typhoon	$17.0	$0.0
Other	$129.0	$0.0

Source: Guy Carpenter 2005.

Table 20.14. Total Catastrophe Bonds Issued Worldwide, 1997 to 2004

Peril	Risk Capital (in billions of dollars) 1997 to 2004
U.S. Earthquake	$4.17
U.S. Hurricane	$4.24
European Windstorm	$2.11
Japanese Earthquake	$2.06
Japanese Typhoon	$0.39
Other	$0.44
Total	$13.41

Source: Guy Carpenter 2005.

Although transactions securitizing catastrophe risk involve risk transfer, they do not qualify for accounting treatment similar to that for traditional reinsurance transactions. This difference in accounting treatment is cited as one reason for the measured growth in the securitization of catastrophe risks to date. As noted previously, insurers and investment bankers have established special purpose vehicles offshore that then issue a reinsurance policy. This structure allows insurers to take advantage of the accounting treatment for reinsurance transactions. The principal value of reinsurance to a ceding company (the purchaser of reinsurance) for regulatory purposes is the recognition on the ceding company's financial statement of a reduction in its liabilities in terms of two accounts: its unearned premium reserve and its loss reserve. The unearned premium reserve is the amount of premiums equal to the unexpired portion of insurance policies. The loss reserve is made up of funds set aside to pay future claims. The transfer of part of the insurance company's business to the reinsurer reduces its liability for future claims and for return of the unexpired portion of the policy. However, this process adds cost and complexity to the catastrophe bond process. The location of these structures and captive insurers in offshore jurisdictions also allows them to take advantage of a less onerous regulatory environment and more favorable tax treatment. Debate continues over how regulation, accounting practices, and tax laws might be changed to make the securitization of risk easier and less costly.

20.5.5 Finite Risk

Finite-risk solutions represent a type of ART where individual risks are spread over time (Cummins 2005, Swiss Re 1999). This is distinct from traditional insurance or reinsurance where the transfer of risk takes place through pooling with a large number of similar risks.

Finite-risk solutions have the following characteristics:

- The transfer of risk from the policyholder to the insurer is limited (finite). In other words, finite-risk solutions contain a significant element of risk-sharing between the client and the insurer.
- Because finite-risk solutions rely on the smoothing and diversification that occurs with time, policies are usually written on a multiyear basis.
- Costs to the policyholder are primarily a function of individual experience. Much of the premiums not used to fund claims are repaid to the policyholder at the end of the policy period. In this sense, finite-risk solutions are similar to retrospectively rated policies.
- Investment income earned during the policy period is factored directly into the premium calculation. Because of the multiyear nature of finite-risk policies, the time value of money can have a considerable influence on premium.

The global market size for finite (re)insurance is estimated to be around $27 billion for both corporate and insurance clients (Swiss Re 2003). A small number of deals are executed each year, and the market size fluctuates annually.

Finite-risk solutions also can be blended with traditional forms of risk transfer. Such blended solutions are gaining in popularity with clients. The blending of finite and traditional approaches permits coverage that smoothes out annual fluctuations in claims costs, while also eliminating the financial risk associated with catastrophic perils.

Finite risk or nontraditional insurance products have come under increasing regulatory scrutiny since late 2004. Critics of these products have expressed concerns that they are actually financial transactions more akin to loans rather than true risk transfer products. Widening investigations into various finite reinsurance transactions have prompted calls for new regulations and accounting standards and disclosure requirements at both the state and federal level.

Finite-risk contracts must meet requirements as to the amount of risk transfer to qualify as reinsurance for accounting purposes. To be considered reinsurance for accounting purposes, a reinsurance contract must involve some transfer of risk to the reinsurer. If there is insufficient risk transfer, the transaction is considered a financing mechanism and is booked as a loan or liability instead of an asset. A well-established rule of thumb for assessing whether true risk transfer has occurred is the so-called 10–10 rule: there is a 10 percent probability of a 10 percent loss of premium. For a long time this rule has been considered sufficient, although in some instances it has been argued that the standard is 15/15 (Fitch Ratings 2004).

20.5.6 Multiline/Multiyear Products

The concept of blending experience across multiple lines over a period of time is intuitively appealing (Cummins 2005). In the same way that diversification benefits stock market investors by reducing volatility, a portfolio consisting of several distinct categories of risk (e.g., liability, commercial auto, property, and workers' compensation) would generally also be less volatile. The multiyear nature of the policies also produces an additional smoothing effect (over time). The volatility of results over a period of several years will generally be less than over shorter intervals.

Policyholders benefit from multiline/multiyear products because loss costs will generally be more stable over a period of several years (Harrington, Niehaus, and Risko 2002; Swiss Re 1999). Bad years will likely be balanced out by good ones. Premium stability is another advantage.

Integrated multiyear/multiline products (MMPs) provide several key advantages for clients (Swiss Re 2003). First, they allow clients to take advantage of the risk consolidation within their own portfolio of risks. They also combine uncorrelated risks into an insured portfolio, allowing for efficient risk transfer and avoiding over-insurance. Finally, MMPs provide an alternative to traditional insurance during a hard market, although the availability of multiyear products is limited during hard markets.

20.5.7 Multitrigger Products

Coverage under a traditional insurance policy is "triggered" when the policyholder suffers a loss as the result of an event caused by an insured peril. As the name suggests, multitrigger coverages require more than one triggering event. The first event is insurance-related (e.g., an earthquake or fire), while the second is often a noninsurance event (e.g., a specified increase in interest rates or decline in the stock market index). Payments for losses from the insurance risk are only paid if the second event or risk is triggered.

Multiple-trigger products (MTPs) are attractive to corporations whose earning power is heavily affected by fluctuations in commodity prices, exchange rates, or interest rates (Swiss Re 2003). Large, well-capitalized corporations with a relatively high risk appetite are especially well suited. Insurance risks, which usually would be retained by the company, could become a severe financial problem if coupled with another adverse economic event. A hedge for the combined risk can be provided by defining a trigger that is highly correlated with the company's profits.

The remote likelihood of the simultaneous occurrence of two uncorrelated events means that coverage can be provided relatively cheaply. Examples of dual trigger policies include the following (Swiss Re 2003):

- Electric utility company: A dual-trigger policy pays for the actual losses caused by the following two events occurring simultaneously: (1) a power outage resulting from equipment failure or storm-related damage and (2) the spot market price for power exceeding a preset threshold during the storm or equipment-related failure.
- Hospital: A dual-trigger policy pays (1) actual medical malpractice claims above a specified level only if (2) the value of the hospital's equity portfolio falls below a specified level during the same period.
- Iron ore mining company: A dual-trigger policy pays (1) a specified level of workers' compensation claims (not to exceed actual claims) if the claims exceed a specified level at the same time (2) iron ore prices decrease below a specified level.

20.5.8 Contingent Capital

Risk is generally transferred or financed before the occurrence of a loss event. Contingent capital represents one way of financing a loss after the event has occurred. Contingent capital may be particularly useful in financing low-frequency/high-severity exposures.

Contingent capital is similar to a line of credit except that access to the capital is conditional (contingent) upon the occurrence of (i) an insured event and (ii) an impact of a predetermined size on some measure of company financial performance (such as certain financial statement items) (Culp 2005). If both (i) and (ii) occur, then the company is assured of a cash infusion in the form of a loan at its time of greatest need. Put options (which give the owner the right to sell at a predetermined price) on a company's own stock also can be used in the case where item (ii), the financial trigger, is the company's stock price.

While contingent capital solutions can provide liquidity to a company when it needs it the most—possibly even sparing it from insolvency—the financial impact can still be severe. Risk is not actually transferred but is instead merely spread over time, nor is risk diversified across other lines or pooled with other policyholders.

The market for contingent capital is still small, but shows great potential. It is estimated that in 2000 there were several committed capital deals, totaling nearly $1.5 billion (Swiss Re 2003). The number of deals subsequently declined in 2001, but then increased in 2002 due to the hardening insurance market.

20.5.9 Governmental Participation in ART

A variety of mechanisms are in place around the world that take a public/private risk-sharing approach to natural and manmade disasters. These mechanisms, in which the government typically acts as a backstop for private industry, are an essential part of the global risk transfer market.

One of the key advantages of such mechanisms is that they cap the potentially unlimited economic consequences private insurers may face from extreme events, such as hurricanes, earthquakes, and terrorist attacks. Insurers' pool of capital is limited and supports not only potential losses arising from future mega-catastrophes, but also dozens of other lines of insurance written across hundreds of millions of commercial and personal lines policies. Government participation in these mega-risks ensures that insurance remains affordable and available for the vast majority of insurable exposures.

Such mechanisms therefore provide an important additional source of capital and capacity to finance catastrophic risk. Although it is debatable whether government assumption of risk can be described in its pure form as alternative risk transfer, there can be no argument that it is increasingly becoming an alternative to traditional forms of risk transfer.

In the United States, the Terrorism Risk Insurance Act is one such example. Enacted in November 2002—14 months after the September 11 terrorist attacks—TRIA is a public/private risk-sharing partnership between the federal government and the insurance industry that provides a backstop (effectively reinsurance) for incurred losses resulting from certain acts of foreign terrorism. The program, which caps the federal government's liability at $100 billion, is designed to ensure that adequate resources are available for businesses to recover and rebuild if they become the victims of a terrorist attack. Government involvement is triggered only after losses from a certified event exceed the commercial insurance industry's aggregate retention (approximately $30 billion in 2005). By sharing potential losses from terrorist attacks between private insurers and the government, TRIA was responsible for bringing significant additional capacity and stability to the terrorism insurance market and established a cost-efficient way to manage terrorism risk.

Terrorism risk presents particular challenges for private insurers because the frequency of attacks is unknown and the potential severity of attacks is virtually unbounded, making it very difficult for insurers to appropriately price the risk. Initially established as a three-year program, TRIA was scheduled to expire at the end of 2005; a possible renewal has been the subject of intense debate in Congress and the business community. A public/private risk-sharing approach to terrorism risk is expected to survive in some form in the United States for at least the next several years. In fact, this is the preferred approach to terrorism risk adopted by many countries around the globe, including the United Kingdom, Spain, Austria, France, and Germany.

For example, in Great Britain, a mutual reinsurance pool named Pool Re was set up in 1993 to provide terrorism coverage following acts of terrorism by the Irish Republican Army. The U.K. Treasury is the reinsurer of last resort for Pool Re, protecting it in the event that it exhausts its financial resources following claims payments. Until 2002, Pool Re covered only commercial property damage and business interruption costs arising from acts of terrorism resulting in fire or explosion. However, after the September 11, 2001, terrorist attacks, Pool Re's coverage was expanded to cover all risks, including nuclear and biological contamination, aircraft impact, and flooding if caused by terrorist attacks.

20.6 THE BERMUDA INSURANCE MARKET

Bermuda holds a key position in the global insurance and reinsurance industry. At the end of 2003, the island was home to 1,247 insurers with total assets of $235.9 billion and capital and surplus of $87 billion.

Up until the 1980s, the Bermuda market was almost entirely focused on captives, and many new mechanisms were developed to expand their use (Fox 2004, Higginbottom 2002). Group captives were created to enable smaller companies, and those with similar interests, to benefit from greater control over their insurance programs by pooling their risks. The rent-a-captive concept was created in the 1970s to enable smaller firms to benefit from managing their own risks.

In the mid-1980s, large U.S. corporations were finding it difficult to buy excess liability insurance. In response to the market crisis, Bermuda-based ACE Ltd. was formed in 1985, led by Marsh & McLennan with the backing of over $200 million in initial capital from 34 U.S. companies. In 1986, Marsh again led investors to form XL Capital Ltd (formerly EXEL Ltd). Both companies initially only offered excess liability coverage but later expanded to become large multiline global players.

In 1988, Centre Re (later Centre Solutions) was formed with $250 million of capital provided by the Zurich Insurance Group and a number of other investors. Centre Re's success made Bermuda the focus for much of the finite reinsurance and finite insurance business that later developed. A number of companies on the island now offer these structured risk solutions.

Bermuda continued its role, reacting quickly to market needs. Mid Ocean Reinsurance Ltd. was formed in 1992 in response to the severe lack of capacity in the property catastrophe reinsurance market. In the same year Hurricane Andrew hit the insurance industry with a $15.5 billion loss. Later, Bermuda received an influx of over $4 billion in capital to form an additional seven property catastrophe reinsurers.

Bermuda continued to foster innovation and to expand the number of products and solutions being offered. The period from 1996 to 2001 was one of expansion and diversification, with ACE and XL both making significant acquisitions. During the same period, Bermuda saw the development of some new products with the formation of several life reinsurers and financial guaranty companies.

The year 2001 was another turning point for the Bermuda insurance market, with over 108 new companies formed, both captives and commercial insurance companies. In the wake of the terrorist attack of September 11, 2001, the year saw the raising of significant new capital to replace the billions that left the market. Bermuda was the location for more than half of the new capital that flowed into the reinsurance market.

Bermuda's insurance regulation is designed to facilitate the creation of companies and insurance products while ensuring the companies operate responsibly within specific margins of solvency. The Insurance Act 1978 requires registration of all insurers, reinsurers, insurance managers, brokers, agents, and sales people doing business in Bermuda. The Act also provided for the establishment of the Insurance Advisory Committee, which provides a formal way for the government to obtain advice from the industry on all insurance matters.

In 1995, the Act was amended in response to the changing marketplace, and new regulations were implemented. The Act created four classes, ranging from single parent captives (Class 1) to highly capitalized, publicly traded companies (Class 4). The Supervisor of Insurance, a nonpolitical appointee under the independent Bermuda Monetary Authority, incorporates insurance and reinsurance companies under a tiered system of regulation.

20.7 CONCLUSION

An ever-changing risk landscape has led to the need to develop a range of alternative methods of risk transfer that can provide coverage where no traditional insurance and reinsurance coverage is available and greatly increase capacity. Alternative risk transfer is a concept that defies exact definition but covers a multitude of carriers and products, from self-insurance, to captives, to risk-retention groups, to finite risk reinsurance, catastrophe bonds, and even government participation in the global risk transfer market (see Table 20.15).

The development of the ART market is both an opportunity for and a challenge to insurers. We have shown how pricing and availability problems can make ART an attractive solution, especially for unique and large risks, but also how the very complexity of some ART solutions can be a barrier to their development. In a world of increasing technological advancement, a time-consuming and expensive option is unlikely to meet with approval. In addition, the impact of increasing regulatory scrutiny and compliance standards may cause greater hesitance among risk bearers to pursue certain categories of ART in the future.

Nevertheless, the growing numbers of captives and increasing levels of self-insurance among corporations around the globe, especially in the United States, indeed, are an indication of the critical need met by alternative methods of risk transfer. Catastrophe securitizations also continue their steady growth, reflecting investors' increasing appetite for risk. Another alternative to traditional insurance and reinsurance markets in the shape of government participation in risk is an unfolding area, particularly in response to risks such as terrorism, that are widely accepted to be uninsurable beyond certain limits and that so far have attracted little support from the global capital markets.

20.8 REFERENCES

A.M. Best Company, 2004, "Domestic Captives: Can Solid Results Be Sustained?" *BestWeek,* August 9.

A.M. Best Company, 2003a, "Quick Cover," *Best's Review,* February.

A.M. Best Company, 2003b, "Special Report, Sizing Up the Captive Market: Growth in the Number of Active Captives Remained Flat in 2002," April 14.

A.M. Best Company, 2001, "Adding to the Mix," *Best's Review,* December.

Cowley, Alex, and J. David Cummins, 2005, "Securitization of Life Insurance Assets and Liabilities," *Journal of Risk and Insurance* 72(2): 193–226.

Culp, L. Christopher, 2005, *Structured Finance and Insurance: The Art of Managing Capital and Risk* (New York: John Wiley).

Cummins, J. David, 2005, "Convergence in Wholesale Financial Services: Reinsurance and Investment Banking," *Geneva Papers on Risk and Insurance: Issues and Practice* 30: 187–222.

Fitch Ratings, 2004, *Special Report: Finite Risk Reinsurance,* November 17.

Fox, David, 2004, "The Bermuda Market," in Ruth Gastel, ed., *Reinsurance: Fundamentals and New Challenges* (New York: Insurance Information Institute).

Friedman, Sam, 2004, "State of the Market Survey," *National Underwriter*, May 24 (sponsored by Zurich North America and conducted by The Response Center).

Froot, Kenneth A., 1999, "The Evolving Market for Catastrophic Event Risk," *Risk Management and Insurance Review* 2(3): 1–28.

Government Accountability Office, 2005, *Risk Retention Groups: Common Regulatory Standards and Greater Member Protections Are Needed* (Washington, DC: GAO).

Guy Carpenter, 2005, *The Growing Appetite for Catastrophe Risk: The Catastrophe Bond Market at Year-End 2004,* March.

Harrington, S.E., G. Niehaus, and K.J. Risko, 2002, "Enterprise Risk Management: The Case of United Grain Growers," *Journal of Applied Corporate Finance*, Winter: 71–81.

Hartwig, P. Robert, 2000, "ART 101: A Primer on Alternative Risk Transfer," *John Liner Review* 14(1) (Spring).

Higginbottom, Dennis, 2002, "The Development of the Bermuda Reinsurance Market," *Journal of Reinsurance* 9(2) (Spring): 1.

Insurance Information Institute, 2005, *The Insurance Fact Book* (New York: III).

ISO/PCS, 2005, "Preliminary Estimate Puts Insured Losses from Hurricane Katrina at $34.4 Billion: ISO Property Claim Services," October 4.

Lane Financial LLC, 2005, "The 2005 Review of the Insurance Securitization Market," April 30.

Lane, Morton, 2002, *Alternative Risk Strategies* (London: Risk Waters Group).

Litzenberger R.H., D.R. Beaglehole, and C.E. Reynolds, 1996, "Assessing Catastrophe Reinsurance-linked Securities as a New Asset Class," *Journal of Portfolio Management*, Winter: 76–86.

Marsh Inc., 2004, Limits of Liability Report.

Moriarty, Kevin, and H. Kathleen Davis, 2003, "Vermont's Thriving Captive Insurance Industry: Alternative Risk Financing in the Green Mountain State," *John Liner Review* 17(1) (Spring).

National Council on Compensation Insurance, 2005.

P.L. No.97–45 Stat. 949 (1981) (codified as amended at 15 U.S.C. §§ 3901–3906).

Risk and Insurance Management Society and Advisen Ltd., 2004, *2003 RIMS Benchmark Survey* (New York: RIMS).

Risk Management Solutions, 2004, *Catastrophe Securitizations in 2003 Reveal Increased Innovation, Modest Growth in Volume,* February.

Risk Retention Reporter, 2004, Survey, October.

Risk Retention Reporter, 2005, 19(3) March.

"Spotlight: Self Insurance & Captive Management," 2005, *Business Insurance,* March 7.

Swiss Re, 2003, "The Picture of ART," *Sigma,* No. 1.

Swiss Re, 1999, "Alternative Risk Transfer (ART) for Corporations: A Passing Fashion or Risk Management for the 21st Century? *Sigma,* No. 2.

Webel, Baird, 2003, *The Risk Retention Acts: Background and Issues*, Congressional Research Service Report for Congress (Washington, DC: CRS).

20.9 LEXICON

Alternative Markets—Mechanisms used to fund self-insurance. This includes captives, which are insurers owned by one or more non-insurers to provide owners with coverage. Risk-retention groups, formed by members of similar professions or businesses to obtain liability insurance, are also a form of self-insurance.

Captives—Insurers that are created and wholly owned by one or more non-insurers to provide owners with coverage. A form of self-insurance.

Catastrophe Bonds—Risk-based securities that pay high interest rates and provide insurance companies with a form of reinsurance to pay losses from a catastrophe such as those caused by a major hurricane. They allow insurance risk to be sold to institutional investors in the form of bonds, thus spreading the risk.

Commercial Lines—Products designed for and bought by businesses. Major coverages include business interruption, commercial auto, comprehensive general liability, directors and officers liability, product liability, professional liability, and workers' compensation.

Contingent Capital—A way of financing a loss after an event. An agreement, entered into before any losses occur, that enables an organization to raise cash by selling stock or issuing debt at prearranged terms following a loss that exceeds a certain threshold.

Federal Liability Risk Retention Act—The 1986 law that expanded the Federal Product Liability Risk Retention Act of 1981, to allow for the formation of risk-retention groups and risk-purchasing groups in all areas of commercial liability, except workers' compensation.

Federal Product Liability Risk Retention Act—The 1981 law that provided for the formation of risk-retention groups and risk-purchasing groups in the areas of products and completed operations liability.

Finite Risk Reinsurance—Contract under which the ultimate liability of the reinsurer is capped and on which anticipated investment income is expressly acknowledged as an underwriting component. Also known as financial reinsurance because this type of coverage is often bought to improve the balance sheet effects of statutory accounting principles.

Fronting—A procedure in which a primary insurer acts as the insurer of record by issuing a policy, but then passes the entire risk to a reinsurer in exchange for a commission. Often, the fronting insurer is licensed to do business in a state or country where the risk is located, but the reinsurer is not. The reinsurer in this scenario is often a captive or an independent insurance company that cannot sell insurance directly in a particular country.

Insurance Pool—A group of insurance companies that pools assets, enabling them to provide an amount of insurance substantially more than can be provided by individual companies to ensure large risks such as nuclear power stations.

Protected Cell Captive (PCC)—A form of captive that is a variation on rent-a-captives. The PCC isolates each participant's assets and liabilities as if they were a separate company, called a cell, doing business with the core company. Also known as segregated cell captive.

Personal Lines—Property-casualty insurance products that are designed for and bought by individuals, including homeowners and automobile policies.

Rent-a-Captives—A form of captive serving businesses that are unable to capitalize a captive but are willing to assume a portion of their own risk and share in the underwriting profits and investment income. Generally sponsored by insurers or reinsurers, which essentially "rent out" their capital for a fee.

Risk-Retention Groups—Insurance companies that band together as self-insurers and form an organization that is chartered and licensed as an insurer in at least one state to handle liability insurance.

Securitization of Insurance Risk—Use of the capital markets to expand and diversify the assumption of insurance risk. The issuance of bonds or notes to third-party investors directly or indirectly by an insurance or reinsurance company or a pooling entity as a means of raising money to cover risks.

Self-insurance—The concept of assuming a financial risk oneself, instead of paying an insurer to take it on. Every policyholder is a self-insurer in terms of paying a deductible and co-payments. Large firms often self-insure frequent, small losses such as damage to their fleet of vehicles.

Special Purpose Vehicle—A one-off reinsurance company whose sole purpose is to transform a traditional reinsurance transaction into an insurance-linked securitization.

Index

Printed in the United States of America.